LearnSmart™ Study Modules

Powered by Connect Biology, **McGraw-Hill LearnSmart™** provides students with ~~ ccess).~~ It is an adaptive diagnostic tool based on artificial intelligence that constantly ~~assesses a student's~~ knowledge of the course material. Sophisticated diagnostics adapt to each student's individual knowledge base, and vary the questions to determine what the student doesn't know and how best to improve his or her knowledge level. Students actively learn the required concepts, and instructors can get specific LearnSmart reports to monitor overall progress.

Learn Fast. Learn Easy. Learn Smart.

Learn more at www.mhlearnsmart.com.

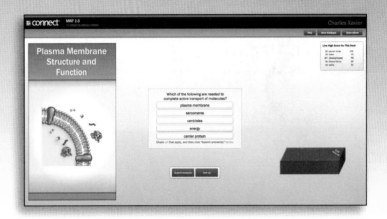

My Lectures—Tegrity

McGraw-Hill Tegrity Campus™ records and distributes lectures with just a click of a button. Students can view anytime/anywhere via computer, iPod, or mobile device. It indexes as it records PowerPoint® presentations and anything shown on the computer, so students can use keywords to find exactly what they want to study.

Instructor Resources

Connect Biology provides easy access to the following resources:

- Enhanced image PowerPoints with editable art
- Lecture PowerPoints with animations
- Animation PowerPoints
- Labeled and unlabeled Jpeg files of art, photos, and tables from the textbook

Chapter	Enhanced Image PPTs (includes photos, and editable art)	Lecture PPTs with Animations	Animation PowerPoints	Labeled Jpeg Images	Base Art Image Files (.jpgs, no labels or leader lines)
All Chapters	Enhanced Image PPTs (707,634 KB)	Lecture Animation PPTs (649,609 KB)	Animation PPTs (1,64,060 KB)	Labeled Images (859,337 KB)	Base Images (599,793 KB)
Ch01	Ch. 1 Enhanced Image PPTs (23,977 KB)	Ch. 1 Lecture Animation PPTs (14,860 KB)	There are no Animation PPTs correlated to this chapter.	Ch. 1 Labeled Images (28,669 KB)	Ch. 1 Base Images (21,333 KB)
Ch02	Ch. 2 Enhanced Image PPTs (9,907 KB)	Ch. 2 Lecture Animation PPTs (6,012 KB)	Ch. 2 Animation PPTs (6,794 KB)	Ch. 2 Labeled Images (15,054 KB)	Ch. 2 Base Images (8,805 KB)
Ch03	Ch. 3 Enhanced Image PPTs (16,605 KB)	Ch. 3 Lecture Animation PPTs (14,220 KB)	Ch. 3 Animation PPTs (1,833 KB)	Ch. 3 Labeled Images (18,346 KB)	Ch. 3 Base Images (14,607 KB)

Powerful Reporting Solutions

Connect Biology offers detailed reporting so instructors can quickly assess how students are doing in regards to overall class performance, individual assignments, and each question.

All practice and test bank questions are tagged to the textbook by chapter, section, topic, and Bloom's level of difficulty, and aligned with the learning outcomes in the textbook to aid these reporting features.

Second Edition

BIOLOGY

Concepts and Investigations

Mariëlle Hoefnagels

The University of Oklahoma

Connect
Learn
Succeed™

BIOLOGY: CONCEPTS AND INVESTIGATIONS, SECOND EDITION
Published by McGraw-Hill, a business unit of The McGraw-Hill Companies, Inc., 1221 Avenue of the Americas,
New York, NY 10020. Copyright © 2012 by The McGraw-Hill Companies, Inc. All rights reserved. Previous
edition © 2009. No part of this publication may be reproduced or distributed in any form or by any means,
or stored in a database or retrieval system, without the prior written consent of The McGraw-Hill
Companies, Inc., including, but not limited to, in any network or other electronic storage or
transmission, or broadcast for distance learning.

Some ancillaries, including electronic and print components, may not be available to customers
outside the United States.

This book is printed on acid-free paper.

3 4 5 6 7 8 9 0 DOW/DOW 1 0 9 8 7 6 5 4 3 2

ISBN 978-0-07-340347-2
MHID 0-07-340347-4

Vice President, Editor-in-Chief: *Marty Lange*
Vice President, EDP: *Kimberly Meriwether David*
Vice-President New Product Launches: *Michael Lange*
Publisher: *Janice Roerig-Blong*
Executive Editor: *Michael S. Hackett*
Senior Developmental Editor: *Anne L. Winch*
Senior Marketing Manager: *Tamara Maury*
Lead Project Manager: *Sheila M. Frank*
Senior Buyer: *Kara Kudronowicz*
Lead Media Project Manager: *Judi David*
Manager, Creative Services: *Michelle D. Whitaker*
Cover/Interior Designer: *Elise Lansdon*
Cover Image: *Chrysochus auratus © 2004 Bev Wigney*
Senior Photo Research Coordinator: *John C. Leland*
Photo Research: *Emily Tietz/Editorial Image, LLC*
Art Studio and Compositor: *Electronic Publishing Services Inc., NYC*
Typeface: *10/12 Times Roman*
Printer: *R. R. Donnelley*

All credits appearing on page or at the end of the book are considered to be an extension of the copyright page.

Library of Congress Cataloging-in-Publication Data

Hoefnagels, Mariëlle
   Biology : concepts and investigations / Mariëlle Hoefnagels. — 2nd ed.
      p. cm.
   Includes index.
   ISBN 978-0-07-340347-2 — ISBN 0-07-340347-4 (hard copy : alk. paper) 1. Biology — Textbooks.
I. Title.
   QH307.2.H64 2012
   570 — dc22
                              2010020431

www.mhhe.com

About the Author

Mariëlle Hoefnagels is an associate professor in the departments of Botany/Microbiology and Zoology at the University of Oklahoma, where she teaches both traditional and online courses in introductory biology. She has received the University of Oklahoma General Education Teaching Award and the Longmire Prize (the Teaching Scholars Award from the College of Arts and Sciences). She has also been awarded honorary memberships in several student honor societies.

Dr. Hoefnagels received her B.S. in environmental science from the University of California at Riverside, her M.S. in soil science from North Carolina State University and her Ph.D. in plant pathology from Oregon State University. Her dissertation work focused on the use of bacterial biological control agents to reduce the spread of fungal pathogens on seeds. In addition to this textbook, her recent publications have focused on the creation of investigative teaching laboratories and methods for teaching experimental design in beginning and advanced biology classes. She frequently gives presentations on study skills and related topics to student groups across campus.

Dedication

To my students

Mariëlle Hoefnagels

Brief Contents

v

Preface | Thinking for Life

Biology—the science of life—is central to our lives and our planet. Nutrition, cancer, HIV/AIDS, global climate change, water quality, endangered species, stem cells, the spread of drug-resistant bacteria, and countless other matters have their foundation in biology. I hope that after reading this book you will be better able to understand and evaluate items in the news, become a more thoughtful voter, and, most importantly, develop a greater appreciation for the amazing, ever-changing world around you. I designed this book to convey the general concepts of biology and to connect them to your life.

Focus on Scientific Inquiry

Every biology textbook explores the process of science as a way of learning about the natural world, but this book is unique in that each chapter reinforces the importance of scientific inquiry with a section titled "Investigating Life." These capstone concepts each explain one study that sheds light on an evolutionary topic related to the chapter's content. In each case, the focus is on how scientists developed and tested a specific hypothesis.

Biologists study every imaginable species, but a few have made truly extraordinary contributions to our understanding of life. These species are called "model organisms" because so much of what we learn from them is applicable to life in general. The "Focus on Model Organisms" boxes in unit 4, The Diversity of Life, shine the spotlight on these laboratory workhorses. For example, Chapter 16 has a box on the bacterium *Escherichia coli*, Chapter 19 profiles *Arabidopsis thaliana*, and Chapter 20 has boxes for a nematode (*Caenorhabditis elegans*), a fruit fly (*Drosophila melanogaster*), and the zebra fish (*Danio rerio*).

Focus on Evolution

All around us, we can see that life seems perfectly suited to its habitat. Centuries of scientific research tell the compelling story of how this came to be. When Charles Darwin published *On the Origin of Species* in 1859, he set into motion the science of evolutionary biology. But it didn't stop there. Generations of scientists have built on that foundation, and we now have a richly detailed understanding of the evolutionary processes that have brought life to this point.

A famous journal article is titled, "Nothing in biology makes sense except in light of evolution," a profound statement that is not to be taken lightly. Like other textbooks, this one includes units dedicated to evolution and the diversity of life. But if evolution permeates our understanding of biology, it should be included in every chapter. This book does just that. The "Investigating Life" section at the end of each chapter provides a tangible focus on the evolutionary forces behind biology at every scale, from chemistry to ecology.

Focus on Learning

This book is full of features that will help you learn to think scientifically. Each chapter begins with an attention-grabbing essay, a learning outline that previews the main concepts, and a study tip. Each main section finishes with a set of questions designed to help you assess your understanding before moving on. Many sections also feature a Figure It Out question, usually focusing on quantitative skills. Moreover, the Investigating Life section that concludes each chapter includes data and one or more critical thinking questions. The end-of-chapter Multiple Choice, Pull It Together, and Write It Out questions reinforce basic content, conceptual understanding, and integration. Illustrated tables and strategically placed mini-glossaries will help you organize the information and understand the connections between details.

Scattered throughout the book are "Apply It Now" boxes, brief readings that explain the biology behind phenomena that you may have noticed for yourself. For example, you can learn why some (but not all) artificial sweeteners are calorie-free, why leaves change color in the fall, and why purebred dogs often suffer from health problems.

Asking questions is an integral part of the scientific process. "Burning Question" boxes are based on questions that my own students have asked me over many years of teaching biology at the University of Oklahoma. On the first day of class, I always ask my students to submit a question about biology that they would like to have answered during the semester. The Burning Questions selected for this book represent a tangible connection between my own students and anyone else who has wondered the same things.

Focus on Biology as a Visual Science

Biology is difficult to explain with words alone. This book features an art program developed by a talented team of professionals. The illustrations are bright and colorful, often combining art and photos or micrographs into appealing and informative combinations. Repetition aids in learning, so the illustrators were careful to use consistent colors for membranes, DNA, proteins, cell organelles, molecules, atoms, and other structures that occur throughout the book. Numbered steps help you work through complex processes, and figure legends add additional explanation.

A Commitment to Educators and Students

Many of us who choose teaching as a profession are passionate about our subject, and we want our students to share our enthusiasm. I hope this book will help bring biology to life for faculty and students alike. I have tried to make sure that all of the information is accurate, complete, up-to-date, and explained at a level that a beginning student can understand, but I continue to welcome your suggestions on how to serve you better. Please write me at marielle_hoefnagels@mcgraw-hill.com with ideas on how to improve this book and what you'd like to see in future editions.

Key Changes for the Second Edition

In the second edition, the table of contents has changed in two major ways. First, the genetics unit has been rearranged. The chapter on gene function has now been combined with the material on DNA structure and moved to the beginning of the genetics unit (chapter 7). The material on DNA replication has moved to the chapter on cell division (chapter 8). Chapter 9, Sexual Reproduction and Meiosis, remains in its original location. The two chapters on inheritance patterns have been rolled into one to improve continuity and reduce redundancy (chapter 10).

Several arguments justify these changes. Describing the function of DNA at the start of the genetics unit is a logical followup to the material in unit 1, which describes the chemical constinuents of cells. In addition, this order allows students to better understand why high-fidelity DNA replication is an essential precursor to cell division. A third justification is that a background in DNA function helps students understand important topics such as the origin and inheritance of new alleles, the relationship between dominant and recessive alleles, and inheritance patterns. Mendel may not have known that DNA existed when he did his famous experiments, but now we do. I have found in my own classes that students make a smooth transition between DNA function, DNA replication, cell division, and inheritance when the material is presented in this order.

The second change to the table of contents is in the diversity unit, where two chapters on animal diversity have been combined into one. The new combined chapter enhances the focus on animals as a unified group rather than emphasizing the artificial distinction between invertebrates and vertebrates.

Many students need instruction on how to study, so the pedagogical features of this second edition have been enhanced in several ways. Each chapter now begins not only with a learning outline but also with a study tip designed to help students take a concrete step toward improved learning. Many chapters now contain one or more "Figure It Out" questions that allow students to apply new ideas right away. At the end of each chapter, a new section called "Pull It Together" depicts a simplified concept map that integrates chapter content; followup questions ask students about the relationships illustrated in the map. In addition, a new end-of-chapter section called "Write It Out" combines the best of the open-ended questions from the first edition and adds many new ones. Asking students to write what they know helps them to see that recognizing material from books or notes is very different from being able to recall and integrate the material for themselves.

Other, smaller changes should also enhance student learning. In response to reviewer suggestions, many passages and illustrations have been clarified, streamlined, or expanded. Throughout the diversity unit, new summary figures should help students see evolutionary patterns. The addition of scale bars and microscopy information to each micrograph in the text should help students begin to understand the size of biological structures. In addition, to reinforce the importance of statistics in science, error bars have been added to graphs where appropriate, and a new appendix entitled "A Brief Guide to Statistical Significance" introduces students to statistical analysis and P values. Finally, the overall paging and visual appeal of the book have improved, helping students stay focused on the most important ideas of biology.

A Student's Guide
To Using this Textbook

Ask yourself, "What's the Point?"

Go to the website to listen to the author briefly describe the key points of each chapter. While you're there, try the other learning aids available on the site.

Learn how to be a better learner.

Learning biology takes daily practice. Each chapter begins with a study tip you can employ to help you learn, rather than memorize, the material in this and other courses you take.

Check out the chapter road map.

The numbered concepts and detailed outline lay out a clear course through the chapter. To get you started, the opening essay offers insights into why this chapter matters.

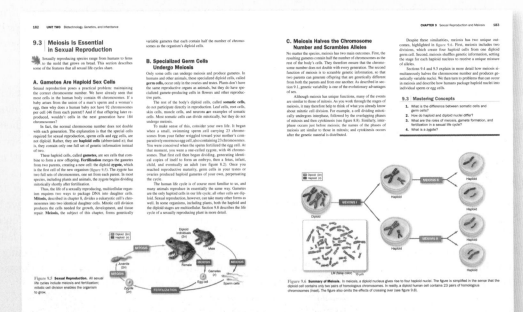

Build your understanding one concept at a time.

Pay attention to headings. Each one is carefully written to summarize the main idea of the section to come.

Note key terms that appear in bold-faced, definitional sentences.

Take time to test your understanding after each numbered concept.

Become an expert—use your visual learning skills.

Combination Figures use art and photos to connect idealized drawings with what is seen in the natural world or laboratory.

Macro-to-Micro Figures take you from a familiar starting point down to what is happening at a microscopic level.

Step-by-Step Figures present concepts in easy-to-follow steps.

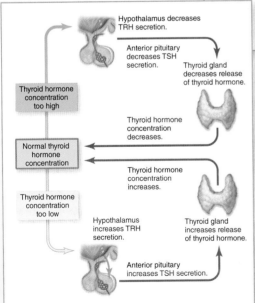

Schematic Figures consistently demonstrate homeostasis throughout the animal unit.

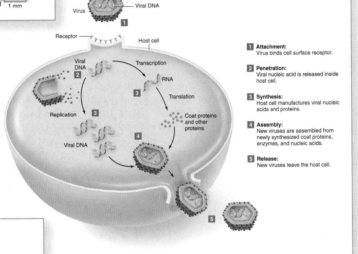

Illustrated Tables help organize information and connect details.

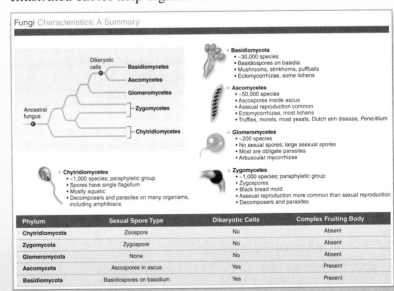

Fungi Characteristics: A Summary

Phylum	Sexual Spore Type	Dikaryotic Cells	Complex Fruiting Body
Chytridiomycota	Zoospore	No	Absent
Zygomycota	Zygospore	No	Absent
Glomeromycota	None	No	Absent
Ascomycota	Ascospores in ascus	Yes	Present
Basidiomycota	Basidiospores on basidium	Yes	Present

Connect this material to your own life.

Burning Questions came from the author's own students. Submit your own *Burning Question* to the author for possible inclusion in future editions.

Burning Question

What does the ozone hole have to do with global climate change?

Many people confuse global climate change and the ozone hole. These two problems are largely separate, but they do share two common threads. First, the chlorofluorocarbon gases that deplete the ozone layer are also greenhouse gases, contributing to a warmer atmosphere. Second, the greenhouse effect may cause the hole in the ozone layer to grow. A thick, heat-trapping "blanket" of greenhouse gases in the troposphere (the lowest part of the atmosphere) means less heat reaches the stratosphere, where the ozone layer is. A cooler stratosphere, in turn, extends the time that stratospheric clouds blanket the polar regions in winter. These clouds of ice and nitric acid speed the chemical reactions that deplete stratospheric ozone.

Submit your burning question to:
marielle_hoefnagels@mcgraw-hill.com

Burning Question

Are there areas on Earth where no life exists?

Sand, bare rock, and polar ice may seem devoid of life, but they are not. Scientists using microscopes and molecular tools have discovered microbes living in the hottest, coldest, wettest, driest, saltiest, highest, most radioactive, and most pressurized places on the planet. Bacteria and archaea colonize every imaginable habitat, including places where no other organism can survive.

There are a few places, however, that humans keep artificially microbe-free for the sake of our own health. For example, people in many professions use autoclaves, radiation, and filters to sterilize everything from surgical tools, to medicines and bandages, to processed foods. Artificial sterilization eliminates microbes that could otherwise cause infections, food poisoning, or other illnesses.

Our own bodies are home to many, many microbes, both inside and out. Yet we manage to keep many of our internal fluids and tissues germ-free, including the sinuses, muscles, brain and spinal cord, ovaries and testes, blood, cerebrospinal fluid, urine in kidneys and the bladder, and semen before it enters the urethra. These areas are among the few places where microbes do not ordinarily live; if a bacterial infection does occur, the resulting illness can be deadly.

Submit your burning question to:
marielle_hoefnagels@mcgraw-hill.com

Second Edition *Burning Questions*

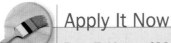

Apply It Now

Bony Evidence of Murder, Illness, and Evolution

Skeletons sometimes provide useful clues to past events. Hard, mineral-rich bones and teeth remain intact long after a corpse's soft body parts decay. These durable remains can help solve crimes, lend insight into human history, and shed light on evolution.

Detectives can use bones to identify the sex of a decomposed murder victim. This technique relies on the differences between male and female skeletons. Most obviously, the average male is larger than the average female. In addition, the front of the female pelvis is broader and larger than the male's, and it has a wider bottom opening that accommodates the birth of a baby. These same features allow anthropologists to determine the sex of ancient human fossils.

Bones can also reveal events and illnesses unique to each person's life. Healed breaks may

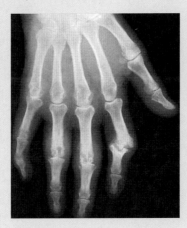

indicate accidents or abuse. Egypt's King Tut, for example, suffered a severe leg break shortly before he died. Crooked joints, such as those in the hand shown in the photo, may be evidence of arthritis, and patterns of bone thickenings tell whether a person spent his or her life in hard physical labor.

The shapes and sizes of fossilized bones also reveal some of the details of human evolution. Section 14.4 explains how the skeletons and teeth of primate fossils provide clues to brain size, diet, and posture in our ancestors. Animal skeletons also tell the larger story of vertebrate evolution. For example, paleontologists can examine skeletal features to determine whether an extinct animal was terrestrial or aquatic. Air is much less supportive than water, so land-dwellers tend to have sturdier skeletons than their aquatic relatives.

Biology is central to our lives. *Apply It Now* boxes explain the biology behind phenomena you may have noticed yourself.

Second Edition *Apply It Now*

Become an investigator!

Figure It Out like an investigator. Exercise your problem-solving skills by answering these quantitative or predictive questions that appear throughout the text.

Figure It Out

Consider the following nutritional facts. Bacon cheeseburger: 23 g fat, 25 g protein, 2 g carbohydrates; large fries: 25 g fat, 6 g protein, 63 g carbohydrates; large soda: 86 g carbohydrates. How many kcal are in this meal?

Answer: 1160 kcal.

Focus on Model Organisms showcases the remarkable contributions that model organisms have made to our understanding of life. Each box highlights what makes that species so beneficial for scientific study and summarizes a few key discoveries.

1.4 | Investigating Life: The Orchid and the Moth

Each chapter of this book ends with a section that examines how biologists use systematic, scientific observations to solve a different evolutionary puzzle from life's long history. This first installment of "Investigating Life" revisits the story of the orchid plant pictured in figure 1.11.

In a book on orchids published in 1862, Charles Darwin speculated about which type of insect might pollinate the unusual flowers of the *Angraecum sesquipedale* orchid, a species that lives on Madagascar (an island off the coast of Africa). As described in section 1.3C, the flowers have unusually long nectar tubes (also called nectaries). Darwin observed nectaries "eleven and a half inches long, with only the lower inch and a half filled with very sweet nectar." Darwin found it "surprising that any insect should be able to reach the nectar; our English sphinxes [moths] have probosces as long as their bodies; but in Madagascar there must be moths with probosces capable of extension to a length of between ten and eleven inches!"

Alfred Russel Wallace picked up the story in a book published in 1895. According to Wallace, "There is a Madagascar orchid—the *Angraecum sesquipedale*—with an immensely long and deep nectary. How did such an extraordinary organ come to be developed?" He went on to summarize how natural selection could explain this unusual flower. He wrote: "The pollen of this flower can only be removed by the base of the proboscis of some very large moths, when trying to get at the nectar at the bottom of the vessel. The moths with the longest probosces would do this most effectually; they would be rewarded for their long tongues by getting the most nectar; whilst on the other hand, the flowers with the deepest nectaries would be the best fertilized by

Charles Darwin

Alfred Russel Wallace

Figure 1.12 **Found at Last.** More than 40 years after Darwin predicted its existence, scientists finally discovered the sphinx moth *Xanthopan morgani*.

should search for it with as much confidence as astronomers searched for the planet Neptune—and I venture to predict they will be equally successful!"

A taxonomic publication from 1903 finally validated Darwin's and Wallace's predictions. The authors described a moth species, *Xanthopan morgani*, with a 225-millimeter (8-inch) tongue (figure 1.12). Given the correspondence between lengths of the orchid's nectary and the moth's tongue, the authors concluded that "*Xanthopan morgani* can do for *Angraecum* what is necessary [for pollination]; we do not believe that there exists in Madagascar a moth with a longer tongue. . . ."

This story not only illustrates how theories lead to testable predictions but also reflects the collaborative nature of science. Darwin and Wallace asked a simple question: Why are these nectar tubes so long? Other biologists cataloging the world's insect species finally solved the puzzle, decades after Darwin first raised the question of the mysterious Madagascar orchid.

Darwin, C. R. 1862. *On the Various Contrivances by Which British and Foreign Orchids are Fertilised by Insects, and on the Good Effects of Intercrossing.* London: John Murray, pages 197–198.

Each chapter closes with *Investigating Life*, a section that follows a real experiment focusing on an evolutionary topic related to the chapter content.

Read and analyze the data in each case and answer the critical thinking questions that follow.

Practice what you have learned.

Use the chapter study tools to become more comfortable with the material.

Chapter Summary highlights the key points and terminology from each section.

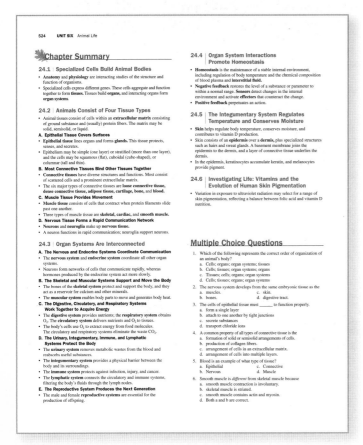

Pull It Together is a concept map with associated questions. After you've answered the questions, try creating your own concept maps for the chapter.

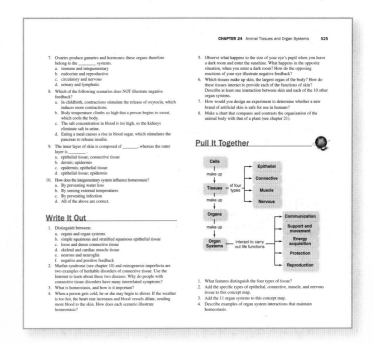

Use the *Multiple Choice* and *Write It Out* questions as a practice test. The questions test not only your ability to recall what you've read, but also to integrate those details and apply it in new situations. These skills are key to learning the material and doing well on exams.

Additional practice tests and study aids are available on the website www.mhhe.com/hoefnagels.

In addition, your instructor may choose to use some of the resources described on the following pages.

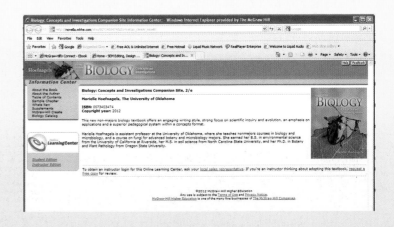

Connect offers...

Everything You and Your Students Need in One Place!

 **connect™**

McGraw-Hill Connect™ interactive learning platform provides auto-graded assessments, an adaptive diagnostic tool, and powerful reporting against learning outcomes and level of difficulty—all in an **easy-to-use interface**.

A Customizable, Interactive ebook

ConnectPlus

The **connect™** plus+ version of Connect includes the full textbook as an integrated, dynamic **ebook**. ConnectPlus provides all of the Connect features plus the following:

- An integrated, printable **ebook**, allowing for anytime, anywhere access to the textbook.
- **Dynamic links** between the problems or questions you assign to your students and the location in the ebook where those problems or questions are covered.
- You can assign fully integrated, self-study questions.
- Pagination that **exactly matches** the printed text, allowing students to rely on ConnectPlus as the complete resource for your course.
- Embedded media, including **animations and videos**.
- Customize the text for your students by **adding and sharing your own notes and highlights**.

www.mcgrawhillconnect.com

Save Time...

with Auto-Graded **Assessments** and Impressive **Reporting** Solutions

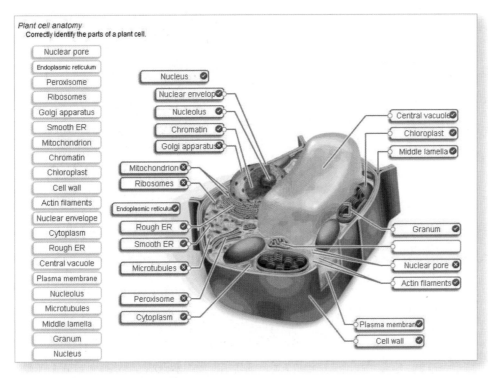

Fully editable, customizable, **auto-graded assignments** using high-quality art from the textbook take you way beyond multiple choice questions.

With Connect's detailed reporting, instructors can **quickly assess** how students are doing in regards to overall class performance, specific objectives, individual assignments, and more.

Graded reports can be **easily integrated** with a Learning Management System, such as Blackboard.

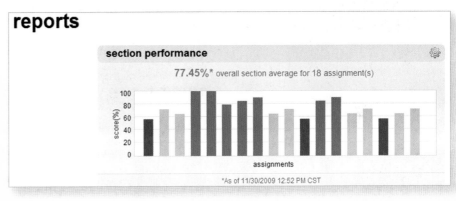

Quickly identify content in your assignments that is causing your students trouble.

Instructors can easily modify a lecture if students aren't ready to move on.

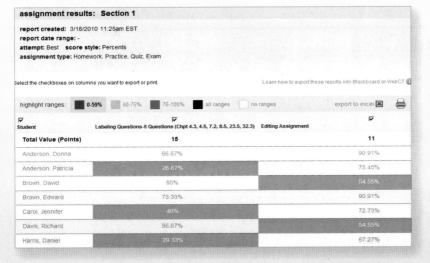

www.mcgrawhillconnect.com

Assess...

Student Knowledge Using a **Diagnostic, Adaptive** Learning System

McGraw-Hill LearnSmart™

ADAPT

McGraw-Hill LearnSmart™ is an adaptive diagnostic tool based on artificial intelligence. LearnSmart constantly assesses a student's knowledge of the course material and adapts based on his or her performance.

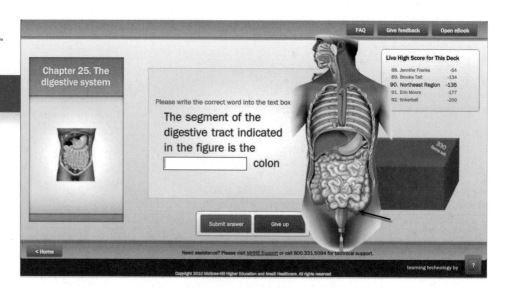

LEARN

LearnSmart pinpoints concepts the student doesn't understand and provides a personalized study plan for success. Students who require remediation are provided with appropriate learning material based upon their knowledge level. A direct link to the eBook is provided for ConnectPlus users.

MEASURE

Assigned or used as a self-study tool, both instructors and students can measure student knowledge, view completion level and time on task, discover areas of proficiency and weakness, and more.

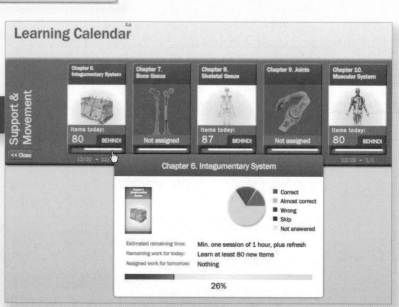

Create...
Teaching Resources to Match the Way *You* Teach

With McGraw-Hill Create™, **www.mcgrawhillcreate.com**, you can easily rearrange chapters, combine material from other content sources, and quickly upload content you have written like your course syllabus or teaching notes. Find the content you need in Create by searching through thousands of leading McGraw-Hill textbooks. Arrange your book to fit your teaching style. Create even allows you to personalize your book's appearance by selecting the cover and adding your name, school, and course information. Order a Create book and you'll receive a complimentary print review copy in 3–5 business days or a complimentary electronic review copy (eComp) via email in minutes. Go to **www.mcgrawhillcreate.com** today and register to experience how McGraw-Hill Create™ empowers you to teach your students *your* way.

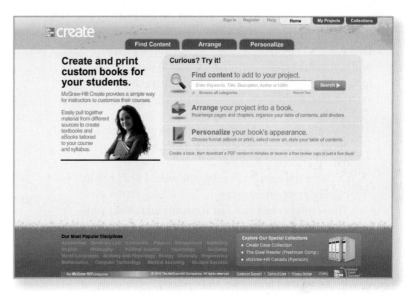

McGraw-Hill Higher Education and Blackboard have teamed up.

Blackboard, the Web-based course-management system, has partnered with McGraw-Hill to better allow students and faculty to use online materials and activities to complement face-to-face teaching. Blackboard features exciting social learning and teaching tools that foster more logical, visually impactful and active learning opportunities for students. You'll transform your closed-door classrooms into communities where students remain connected to their educational experience 24 hours a day.

This partnership allows you and your students access to McGraw-Hill's Connect™ and Create™ right from within your Blackboard course—all with one single sign-on.

Not only do you get single sign-on with Connect™ and Create™, you also get deep integration of McGraw-Hill content and content engines right in Blackboard. Whether you're choosing a book for your course or building Connect™ assignments, all the tools you need are right where you want them—inside of Blackboard.

Gradebooks are now seamless. When a student completes an integrated Connect™ assignment, the grade for that assignment automatically (and instantly) feeds your Blackboard grade center.

McGraw-Hill and Blackboard can now offer you easy access to industry leading technology and content, whether your campus hosts it, or we do. Be sure to ask your local McGraw-Hill representative for details.

Engage...
with **Dynamic, Customizable** Content from McGraw-Hill

Animations for a new generation

Dynamic, 3D animations of key biological processes bring an unprecedented level of control to the classroom. Innovative features keep the emphasis on teaching rather than entertaining.

- An options menu lets you control the animation's level of detail, speed, length, and appearance, so you can create the experience you want.
- Draw on the animation using the white board pen to highlight important areas.
- The scroll bar lets you fast forward and rewind while seeing what happens in the animation, so you can start at the exact moment you want.
- A scene menu lets you instantly jump to a specific point in the animation.
- Pop ups add detail at important points and help students relate the animation back to concepts from lecture and the textbook.
- A complete visual summary at the end of the animation reminds students of the big picture.

Animation topics include: Cellular Respiration, Photosynthesis, Molecular Biology of the Gene, DNA Replication, Cell Cycle and Mitosis, Membrane Transport, and Plant Transport.

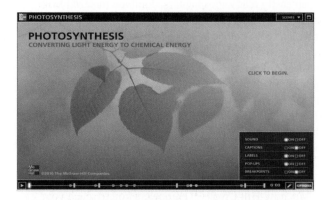

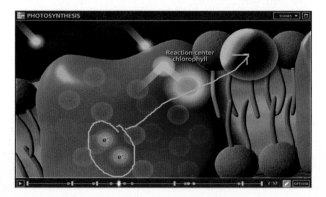

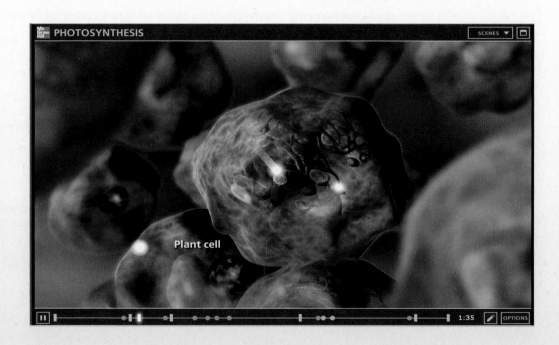

Find...

Everything You Need for Outstanding Presentations

FlexArt

- **FlexArt Image PowerPoints:** Every piece of art is sized and cropped specifically for superior presentations, and labels can be edited. Key figures can be deconstructed and reshaped to fit your needs. Tables, photographs, and unlabeled art pieces are also included.
- **Lecture PowerPoints with Animations:** Animations illustrating important processes are embedded in the lecture material.
- **Animation PowerPoints:** Animations are embedded in PowerPoint slides to easily paste into your existing lecture.
- **Labeled JPEG Images:** Download folders with full-color digital files of all illustrations and most photos from the book that can be readily incorporated into presentations, exams, or custom-made classroom materials.
- **Base Art Image Files:** Download unlabeled digital files of all illustrations.

Make your lectures available anytime/anywhere

Tegrity Campus™ records and distributes your class lecture, with just a click of a button. Students can view anytime and anywhere via computer, mp3 player, or smartphone. It indexes as it records your PowerPoint presentations and anything shown on your computer so students can use keywords to find exactly what they need to review. Tegrity is available as an integrated feature of McGraw-Hill Connect™ or standalone.

Presentation Center

Access the depth of McGraw-Hill content

Available to adopters, this online digital library contains photos, artwork, animations and other media from McGraw-Hill textbooks covering a broad array of subject areas. Create customized lectures and assessment or enhance your online course materials with our peer-reviewed content.

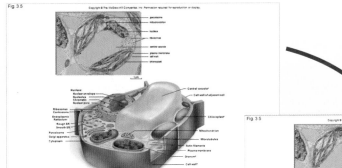

Move figure components.

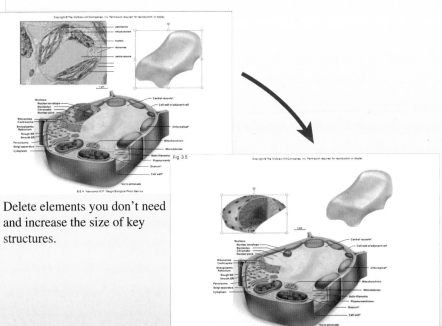

Delete elements you don't need and increase the size of key structures.

Acknowledgments

I owe a huge debt of gratitude to Ricki Lewis, Bruce Parker, and Doug Gaffin for providing the foundation of the first edition of this book. In addition, an army of people at McGraw-Hill works behind the scenes to carry a textbook project from start to finish. Publisher Janice Roerig-Blong leads the way for the entire biology program. I am particularly grateful to Michael Lange, who has been with this project since the beginning; I always welcome his insights about everything from higher education to page layouts. Michael Hackett came on board as Executive Editor shortly before the first edition was complete. Our conversations are always enlightening and entertaining, and I especially value his thoughts on biology from the perspective of the non-major. Both Michaels have supported me in countless ways, and each has put his own stamp on this project.

Developmental editor Lisa Bruflodt ably kept the first edition on track and coordinated reviews when the book came out. She then passed the torch to the wonderful Anne Winch, who is always quick to offer excellent advice and thoughtful suggestions with patience and good humor.

I also appreciate the efforts of the rest of my book team. Project manager Sheila Frank keeps all the wheels turning, along with buyer Kara Kudronowicz, designer Michelle Whitaker, and photo research coordinator John Leland. Kari Voss rescued me from computer glitches many times. Digital product manager Eric Weber's dedication to electronic media, combined with marketing manager Tamara Maury's ability to connect with people, continue to amaze me. Thanks also to Jane Peden and JoAnn Mohr for administrative assistance. Finally, I thank McGraw-Hill's Patrick Reidy and Suzanne Guinn for their continued friendship.

The folks at EPS are incredibly skilled at transforming my terrible sketches and handwritten notes into beautiful, informative art. And I very much enjoy working with photo researcher Emily Tietz, who is kind enough to chase down requests and offer advice when I am stuck on a photo selection.

Several faculty colleagues have helped me as well. Steve Vessey provided the initial draft of the animal behavior chapter for the second edition. Lena Ballard contributed most of the new How to Solve a Genetics Problem feature in chapter 10. In addition, Steven Black helped me sort out questions about embryology, and Andy Baldwin offered suggestions on how to depict the evolutionary history of invertebrates.

Many faculty and staff members at the University of Oklahoma have answered technical questions and helped generate ideas. They include Gordon Uno, Anne Dunn, Brad Stevenson, Tyrrell Conway, Ingo Schlupp, Rich Broughton, Scott Russell, Wayne Elisens, Ben Holt, Phil Gibson, Ken Hobson, Mark Walvoord, Rosemary Knapp, Laurie Vitt, Randy Hewes, Bing Zhang, and Greg Strout. Several OU students have also been extremely helpful to me. In particular, Matt Taylor and Caleb Cosper have offered valuable, thoughtful feedback and assisted me in many other ways; Elise Knowlton has brainstormed with me and shared her perspective on textbooks and teaching; and Erin Dwinnell has helped me expand my repertoire of study tips. All of these amazing students, and many others at OU, are an inspiration to me.

My family and friends—Cees, Clarke, Karen, Kelly, Lucelle, Marika, Nicole, Robin, Scoops and Sidecar—continue to support me in this project. I appreciate their love, pride, and companionship. Finally I thank my husband, Doug Gaffin, for being my cheerleader, confidante, and mentor. I could not do this work without him by my side.

360° Development

McGraw-Hill's 360° Development Process is an ongoing, never ending, market-oriented approach to building accurate and innovative print and digital products. The process includes market research, content reviews, faculty and student focus groups, course and product specific symposia, accuracy checks, art reviews and boards of advisors. This is initiated during the early planning stages of our new products, intensifies during the development and production stages, and then begins again upon publication, in anticipation of the next edition.

Focus Group Participants

Caroline Ballard, *Rock Valley College*
Peggy Brickman, *University of Georgia*
Lori Buckley, *Oxnard College*
Jane Caldwell, *Washington and Jefferson College*
James Claiborne, *Georgia Southern University*
Scott Cooper, *University of Wisconsin– La Crosse*
David Cox, *Lincoln Land Community College*
Bruce Fink, *Kaskaskia College*
Brandon Foster, *Wake Tech Community College*
Douglas Gaffin, *University of Oklahoma*
Carlos Garcia, *Texas A&M University– Kingsville*
Phil Gibson, *University of Oklahoma*
Kelly Hogan, *University of North Carolina at Chapel Hill*
Jessica Hopkins, *University of Akron*
Tim Hoving, *Grand Rapids Community College*
Dianne Jennings, *Virginia Commonwealth University*
Leslie Jones, *Valdosta State University*
Hinrich Kaiser, *Victor Valley Community College*
David Lemke, *Southwest Texas State University*
Mark Lyford, *University of Wyoming*
Richard Musser, *Western Illinois University*
Ikemefuna Nwosu, *Lake Land College*
Murad Odeh, *South Texas College*
Karen Plucinski, *Missouri Southern State University*
Maretha Roberts, *Olive-Harvey College*
Frank Romano, *Jacksonville State University*
Felicia Scott, *Macomb Community College–Clinton Twp*
Cara Shillington, *Eastern Michigan University*
Brian Shmaefsky, *Lone Star College– Kingwood*
Wendy Stankovich, *University of Wisconsin–Platteville*
Bridget Stuckey, *Olive-Harvey College*
Sharon Thoma, *University of Wisconsin– Madison*
Kip Thompson, *Ozarks Technical Community College*
Sue Trammell, *John A. Logan School*
Valerie Vander Vliet, *Lewis University*
Mark E. Walvoord, *University of Oklahoma*
Daniel Ward, *Waubonsee Community College*
Scott Wells, *Missouri Southern State University*
Stephen White, *Ozarks Technical Community College*
Sonia Williams, *Oklahoma City Community College*
Jo Wu, *Fullerton College*

Ancillary Contributors

Companion Site:

Caroline Ballard, *Rock Valley College*
Cara Shillington, *Eastern Michigan University*
Brian Shmaefsky, *Lone Star College– Kingwood*

Connect:

Jane Caldwell, *Washington and Jefferson College*
John P. Gibson, *University of Oklahoma*
Matthew Taylor, *University of Oklahoma*
Mark E. Walvoord, *University of Oklahoma*

Presentation Tools:

Brenda Leady, *University of Toledo*
Sharon Thoma, *University of Wisconsin– Madison*

Test Bank:

Scott Cooper, *University of Wisconsin– La Crosse*

Reviewers

Dennis Anderson, *Oklahoma City Community College*
Kenneth D. Andrews, *East Central University*
Nina L. Baghai-Riding, *Delta State University*
Lee C.D. Baines, *South Texas College*
Neil R. Baker, *The Ohio State University*
Caroline Ballard, *Rock Valley College*
Keith Bancroft, *Southeastern Louisiana University*
Stephen Barnett, *Bryan College*
Donald R. Baud, *University of Memphis*
Lisa L. Behm, *Tidewater Community College*
Michael C. Bell, *Richland College*
Daniel S. Bickerton, *Ogeechee Technical College*
Benjie Blair, *Jacksonville State University*
William D. Blaker, *Furman University*
Dennis Bogyo, *Valdosta State University*
Mark Bolyard, *Union University*
Laurie J. Bonneau, *Trinity College*
Janice M. Bonner, *College of Notre Dame of Maryland*
Alicia Bosela, *Indiana University– Purdue University Fort Wayne*
Jacqueline K. Bowman, *Arkansas Tech University*
Peggy Brickman, *University of Georgia*
Katherine D Buhrer, *Tidewater Community College*
Jamie Burchill, *Troy University*
Matthew Burnham, *Jones County Junior College*
Nancy Buschhaus, *University of Tennessee at Martin*
Brian P. Butterfield, *Freed-Hardeman University*
Jane Caldwell, *Washington & Jefferson College*
Emily B. Carlisle, *Pearl River Community College*
Nicholas J. Cheper, *East Central University*
Roger Choate, *Oklahoma City Community College*
Genevieve C. Chung, *Broward College*
George R. Cline, *Jacksonville State University*
Clifton Cooper, *Linn-Benton Community College*
Scott Cooper, *University of Wisconsin– La Crosse*
Douglas Creer, *Concord University*
Gregory A. Dahlem, *Northern Kentucky University*
Marc DalPonte, *Lake Land College*
Deborah Dardis, *Southeastern Louisiana University*
Juville Dario-Becker, *Central Virginia Community College*
Leigh Delaney-Tucker, *University of South Alabama*

Jean DeSaix, *University of North Carolina at Chapel Hill*

Kimberly Dudzik, *Cuyamaca College*

P. K. Duggal, *Maple Woods Community College*

Kathryn A. Durkee, *Central Virginia Community College*

John A. Ewing, III, *Itawamba Community College*

Jeff D. Fennell, *Linn-Benton Community College*

Susan W. Fisher, *The Ohio State University*

Edward R. Fliss, *St. Louis Community College at Florissant Valley*

Jason Flores, *The University of North Carolina at Charlotte*

Patricia Flower, *Miramar College*

Diane Wilkening Fritz, *Northern Kentucky University*

Bernard L. Frye, *The University of Texas at Arlington*

Dennis W. Fulbright, *Michigan State University*

Ann D. Gathers, *The University of Tennessee at Martin*

John P. Gibson, *University of Oklahoma*

Paul J. Gier, *Huntingdon College*

Jennifer L Greenwood, *University of Tennessee at Martin*

Kristi L. Haik, *Northern Kentucky University*

Katina L. Harris-Carter, *Tidewater Community College*

Juliana G. Hinton, *McNeese State University*

Eva Horne, *Kansas State University*

Melba A. Horton, *McNeese State University*

Martin J. Huss, *Arkansas State University*

Dianne Jennings, *Virginia Commonwealth University*

Michael Kempf, *University of Tennessee at Martin*

Amine Kidane, *Columbus State Community College*

Scott S. Kinnes, *Azusa Pacific University*

Dennis J. Kitz, *Southern Illinois University Edwardsville*

Amy L. Kovach, *The Ohio State University*

Carolyn J. Lebsack, *Linn-Benton Community College*

Stephen G. Lebsack, *Linn-Benton Community College*

Eric C. Lovely, *Arkansas Tech University*

Ann S. Lumsden, *Florida State University*

Paul D. Luyster, *Tarrant County College District, South Campus*

Craig E. Martin, *University of Kansas*

Mike Maxwell, *National University*

Kamau W. Mbuthia, *Bowling Green State University*

Dax R. McDonald, *Gadsden State Community College*

Tiffany B. McFalls, *Southeastern Louisiana University*

Ashley Rall McGee, *Valdosta State University*

Susan L. Meacham, *University of Nevada Las Vegas*

Mark E. Meade, *Jacksonville State University*

Mark A. Melton, *Saint Augustine's College*

Jon Milhon, *Azusa Pacific University*

Beth A. Miller, *Pulaski Technical College*

J. Jean Mitchell, *Northwest Florida State College*

Jeanne Mitchell, *Truman State University*

Ronald S. Mollick, *Christopher Newport University*

Daniel Moore, *Southern Maine Community College*

Juan M. Morata, *Miami Dade College*

Thabiso M'Timkulu, *Laney College*

Scott Murdoch, *Moraine Valley Community College*

Rajkumar Nathaniel, *Nicholls State University*

Kathryn M. B. Nette, *Cuyamaca College*

Howard Neufeld, *Appalachian State University*

Meredith Somerville Norris, *University of North Carolina, Charlotte*

Richard J. Nuckels, *Austin Community College*

Ikemefuna Theodore Nwosu, *Lake Land College*

Karen E. Plucinski, *Missouri Southern State University*

Subbarayan R. Pochi, *University of Miami*

Kathryn Stanley Podwall, *Nassau Community College*

Steve Pollock, *Louisiana State University*

Meenakshi Rajan, *Mesa College*

Raul Ramirez, *Oklahoma City Community College*

Marceau Ratard, *Delgado Community College*

Darrell Ray, *The University of Tennessee at Martin*

James Rayburn, *Jacksonville State University*

Wenda Ribeiro, *Thomas Nelson Community College*

David A. Rintoul, *Kansas State University*

Darryl Ritter, *Northwest Florida State College*

Frank A. Romano, III, *Jacksonville State University*

Amanda Rosenzweig, *Delgado Community Colleges*

Albert S. Rubenstein, *Ivy Tech Community College*

Michael L. Rutledge, *Middle Tennessee State University*

Sanghamitra Saha, *Harold Washington College*

Deemah N. Schirf, *The University of Texas at San Antonio*

Michael Scott, *Lincoln University*

Anju H. Sharma, *Stevens Institute of Technology*

Erin M. Sheehan, *Sterling College*

Brian Shmaefsky, *Lone Star College– Kingwood*

Eric M. Sikorski, *University of Tampa*

John B. Skillman, *California State University at San Bernardino*

Marc A. Smith, *Sinclair Community College*

Anna Bess Sorin, *University of Memphis*

Fitzgerald Spencer, *Southern University, Baton Rouge*

L. Brooke Stabler, *University of Central Oklahoma*

Anthony J. Stancampiano, *Oklahoma City Community College, University of Central Oklahoma*

Peter Stiling, *University of South Florida*

C. Michael Stinson, *Southside Virginia Community College*

Kirby C. Swenson, *Middle Georgia College*

Robin Taylor, *The Ohio State University*

Chad Thompson, *SUNY/Westchester Community College*

Willetta Toole-Simms, *Azusa Pacific University*

Richard Trout, *Oklahoma City Community College, Oklahoma Christian University*

Lin Twining, *Truman State University*

Fred Vogt, *Elgin Community College*

Stephen Wagner, *Stephen F. Austin State University*

Mark E. Walvoord, *University of Oklahoma*

Hong Li Wang, *University of Arkansas at Little Rock*

Van Wheat, *South Texas College*

Jeff White, *Lake Land College*

K.L. Wilsen, *University of Northern Colorado*

Heather Wilson-Ashworth, *Utah Valley University*

Lynette Winters, *Piedmont Virginia Community College*

David E. Wolfe, *American River College*

Geraldine W. Wright, *Tidewater Community College*

Matthew Wund, *The College of New Jersey*

Todd C. Yetter, *University of the Cumberlands*

Martin Zahn, *Thomas Nelson Community College*

Ted Zerucha, *Appalachian State University*

Michelle Zurawski, *Moraine Valley Community College*

Changes by Chapter

Unit 1

Chapter 1 (The Scientific Study of Life): reorganized the "Life Is Organized" section to better reflect the corresponding art; moved much of the taxonomy material to the evolution unit; reorganized the description of scientific method to better recognize discovery science as a way to generate data; added a real-life example of an experimental test of a rotavirus vaccine; created a new Investigating Life on the Madagascar orchid and the discovery of its pollinator.

Chapter 2 (The Chemistry of Life): incorporated additional photos to improve the connection between chemistry and students' lives; redrew the figures on hydrogen bonds and protein structure to improve clarity; incorporated recent findings into the Investigating Life section.

Chapter 3 (Cells): improved the section on microscopes to better highlight the strengths and weaknesses of each type; improved and expanded the explanation of surface area to volume ratio; placed the material on the three domains between the sections on microscopes and membranes; improved the explanation of membranes to explicitly address selective permeability; defined the endomembrane system and clarified the relationship between its organelles; improved the summary table and moved it to the end of the chapter.

Chapter 4 (Energy of Life): improved the organization and labeling of the art for clarity and simplicity; added new art to illustrate the electron transport chain, negative feedback, and the effect of temperature on enzyme activity; shortened the sections on enzymes and membrane transport to improve the focus on core issues.

Chapter 5 (Photosynthesis): rearranged material on chloroplast anatomy and photorespiration for clarity and to improve paging; improved the illustrations of chloroplast structure and the C_3, C_4, and CAM pathways; created a new Investigating Life piece on photosynthetic sea slugs.

Chapter 6 (How Cells Release Energy): rearranged content and improved art for greater consistency with chapter 5; better distinguished between the theoretical and actual yield of ATP in respiration.

Unit 2

Chapter 7 (DNA Structure and Gene Function): brought this chapter to the front of the genetics unit; improved and expanded the presentation of transcription factors; added a figure that simultaneously summarizes protein synthesis and the regulation of protein synthesis; improved the figure on transgenic bacteria; clarified many other figures to improve their correspondence with the narrative and their consistency with other figures.

Chapter 8 (DNA Replication, Mitosis, and the Cell Cycle): moved content on DNA replication into this chapter; better explained the connection of mitosis and meiosis to the human life cycle; expanded the coverage of cancer staging and risks, stem cells, and cloning; modernized the coverage of DNA profiling.

Chapter 9 (Sexual Reproduction and Meiosis): improved the overall organization and reduced redundancy with subsequent chapters on inheritance; improved the presentation of crossing over and independent orientation.

Chapter 10 (Patterns of Inheritance): combined all material on inheritance patterns from former chapters 10 and 11 into one chapter; explicitly connected gene function (chapter 7), meiosis (chapter 9) and inheritance (chapter 10); added guidelines for solving genetics problems involving one gene, two genes, and X-linked genes.

Unit 3

Chapter 11 (Forces of Evolutionary Change): added illustration to summarize how evolutionary thought has expanded since Darwin's time; tightened the focus of the Apply It Now box to better show the consequences of artificial selection in dogs; expanded the discussion of sexual selection; moved the discussion of Hardy-Weinberg ahead in the chapter to more clearly show how mechanisms of evolution change allele frequencies; improved Hardy-Weinberg art to make it easier for students to calculate allele frequencies on their own; added new art to better explain genetic drift and gene flow.

Chapter 12 (Evidence of Evolution): added Wallace's line and fossil evidence of continental drift to the description of biogeography; expanded the coverage of evolutionary developmental biology, including new figures showing the effect of genes on development; improved coverage of molecular evidence for evolution, including a new figure of molecular clocks; created a new Apply It Now box describing how homologous structures explain hiccups.

Chapter 13 (Speciation and Extinction): improved the narrative and art on reproductive barriers; improved the examples and art pertaining to modes of speciation; added material on the taxonomic hierarchy; combined all material on cladistics and systematics into one section; improved, modernized, and expanded presentation of cladistics, including explanation of how to read and construct a cladogram using both morphological and molecular data.

Chapter 14 (Origin and History of Life): added mini-figures that improve the connection of this material with the geologic time scale and continental drift; improved the presentation of endosymbiosis, including new art to illustrate secondary endosymbiosis; created a new Apply It Now box on coal; added illustrations of Mesozoic and Cenozoic life; streamlined and expanded the material on human evolution.

Unit 4

Chapter 15 (Viruses): clarified the explanation of lytic/lysogenic cycles; reorganized and improved the coverage of animal viruses and how symptoms arise; improved the coverage of plant viruses; tightened and clarified the Investigating Life section; added a Burning Question connecting sex, human papilloma viruses, and cervical cancer.

Chapter 16 (Bacteria and Archaea): combined content on characteristics that are useful in identification; improved the coverage of the diversity of bacteria and archaea; improved the coverage of beneficial and disease-causing bacteria.

Chapter 17 (Protists): added a new, more comprehensive introduction to protists, including new figure on chloroplast origin; rearranged the chapter's sections to better reflect evolutionary relationships; improved the coverage of malaria; added an Apply It Now box on *Cryptosporidium* outbreaks in public water sources and recreational water.

Chapter 18 (Plants): rewrote the introductory section to improve consistency with chapters 19 and 20 and to better emphasize evolutionary trends in plants; added evolutionary and utilitarian information to each section; added mini-evolutionary trees to each diversity section; added a table with phylum names, examples, and number of existing species; added a new Burning Question on biofuels; updated the information on angiosperm diversity, including a new phylogenetic tree; improved the angiosperm life cycle figure; added a new summary figure and table.

Chapter 19 (Fungi): updated the content to reflect the polyphyletic nature of chytrids and zygomycetes; added a table with phylum names, examples, and number of existing species; added a section for Glomeromycota; added mini-evolutionary trees to each diversity section; added a discussion of endophytes to the section describing fungal interactions with other species; added a new summary figure and table.

Chapter 20 (Animals): combined the two animal diversity chapters from the first edition into one; combined birds with nonavian reptiles to better reflect evolutionary history; added information about the diversity of mammals; added mini-evolutionary trees to each diversity section; added summary figures and tables for both invertebrates and vertebrates; added a Burning Question on "minor" animal phyla (placozoa, rotifers, tardigrades); added an Apply It Now box on worm farming; added anatomical diagrams for fishes,

amphibians, non-avian reptiles, plus additional photos for most invertebrate and vertebrate groups.

Unit 5

Chapter 21 (Plant Form and Function): added an illustration depicting the locations of meristems; added illustrated summary tables for plant cell types; added a Burning Question on difference between fruits and vegetables; created a new Investigating Life piece exploring the stem anatomy of ant-harboring trees.

Chapter 22 (Plant Nutrition and Transport): expanded the section on soils; added a section on parasitic plants; created a new Investigating Life section on nutritional tradeoffs in carnivorous pitcher plants.

Chapter 23 (Plant Reproduction and Development): improved the angiosperm life cycle figure; added an explanation of single-sex flowers; added content on the advantage of double fertilization and evolutionary tradeoffs involving seed size; improved connections with other chapters in the plant physiology unit and in the plant diversity chapter; explained how hormones move within a plant; added an Apply It Now box on keeping flowers and foods fresh by blocking ethylene receptors; combined all material on phytochrome responses to improve the section on plant responses to light; created a new Investigating Life piece on hot chili peppers.

Unit 6

Chapter 24 (Animal Tissues and Organ Systems): added an illustration explaining anatomical terms; improved the presentation of the four tissue types; added an Apply It Now box on plastic surgery; created a new Investigating Life section on the relationship between skin pigmentation, UV radiation, and vitamin D nutrition.

Chapter 25 (Nervous System): added content on nervous system diversity in invertebrates; expanded and improved the explanation of synapses and synaptic integration; improved and clarified material on brain anatomy; added material on memory; created a new Investigating Life piece on toxins in paralytic shellfish poisoning.

Chapter 26 (The Senses): added material on invertebrate eyes; added information about the path of visual information in the brain; improved explanations of all senses; added photos to improve relevance to student perceptions.

Chapter 27 (Endocrine System): created a new chapter opening essay on plastics as endocrine disruptors; improved the explanation of the role of the second messenger in peptide hormone action; created uniform summary figures for the glands and hormones of the endocrine system; expanded information about diabetes trends and the connection with obesity; simplified the Investigating Life section.

Chapter 28 (Skeletal and Muscular Systems): improved discussion of the evolutionary tradeoffs of endo- and exoskeletons; improved the description of muscle anatomy and function; improved the figure illustrating the sliding filament model; added a figure showing how calcium ions and ATP interact in muscle contraction.

Chapter 29 (Circulatory System): updated the chapter opening essay; added material on the advantages of each type of circulatory system; moved information on the components of blood to a new position earlier in the chapter; clarified the material on blood vessels; created a new Investigating Life section on the colorless blood of Antarctic icefishes.

Chapter 30 (Respiratory System): expanded information on principles and diversity of respiratory surfaces; improved the explanation and illustration of countercurrent exchange in gills; improved the illustration of alveoli; expanded the explanation of ventilation and promoted this topic to its own section; added definitions of tidal volume and vital capacity.

Chapter 31 (Digestion and Nutrition): clarified the dual role of nutrients in food (as energy and building blocks); expanded material on the role of metabolic rate in determining an animal's diet; improved the consistency of the treatment of vitamins and minerals; added coverage of leptin and ghrelin; created a new Investigating Life piece on sexual cannibalism in redback spiders.

Chapter 32 (Regulation of Temperature and Body Fluids): added new information on heterotherms and temperature regulation in ectotherm; added new material on osmoregulation in freshwater and marine fishes; added information on nitrogenous wastes, including a new figure on Malpighian tubules; clarified and improved the coverage of nephron function.

Chapter 33 (Immune System): created a new chapter opening essay on myths about vaccination; improved the coverage of inflammation; improved coverage of the adaptive immune response, including streamlined figures and more extensive explanation of antibodies; promoted coverage of vaccines to an A-level head; added a Burning Question on why we need repeated vaccinations; created a new Investigating Life piece on the connection between worm infection and suppression of allergies.

Chapter 34 (Animal Reproduction and Development): created a new chapter opening essay on the role of males in reproduction (whiptail lizards and seahorses as two extremes); expanded the explanation of animal development in general; updated the table of birth control methods; added an Apply It Now box on assisted reproductive technology; added a new section on sexually transmitted diseases; expanded the section on birth defects.

Unit 7

Chapter 35 (Animal Behavior): totally rewrote the chapter, with the exception of the chapter opening essay, Apply It Now box, and Investigating Life section. The new material, which includes nearly all new illustrations and a new Burning Question on whether lemmings commit mass suicide, has a much greater emphasis on the fitness value of behaviors.

Chapter 36 (Population Ecology): improved the connection between populations and evolution; modified the Apply It Now box on counting populations to apply to both plants and animals; improved the explanation of age structures; expanded the explanation of life tables and survivorship curves, including a new figure; added examples, explanations, and figures to the sections on exponential growth and logistic growth; added an example of a fluctuating population (collared lemmings); added graphs to better explain the guppy experiments that reveal the effects of natural selection on life history traits; reorganized and improved the explanation of human growth trends to better connect to earlier concepts and to better reflect current trends in birthrates and mortality; supplemented the discussion of water scarcity to include the overall ecological footprint.

Chapter 37 (Communities and Ecosystems): reorganized the chapter opening essay to better focus on community interactions; expanded the explanations of species interactions; clarified the role of the climax community in succession; added material about pharmaceuticals and household chemicals in sewage treatment; improved and expanded the coverage of biogeochemical cycles.

Chapter 38 (Biomes): improved the explanation of the connections among biomes, nutrients, and primary productivity; added information about ocean currents; added a description of polar ice cap biomes; improved descriptions and illustrations of lake and ocean zones.

Chapter 39 (Preserving Biodiversity): expanded the introductory section to explain more about biodiversity and to define extinct, endangered, vulnerable, and threatened species; added material about the Gulf of Mexico's dead zone; added a description of plastics in the ocean; expanded the examples of conservation biology tools; expanded the Burning Questions on what an ordinary person can do to help the environment.

Contents

10 | Patterns of Inheritance 198

UNIT 3 | The Evolution of Life

11 | The Forces of Evolutionary Change 228

UNIT 5 | **Plant Life**

21 | Plant Form and Function 454

UNIT 6 | Animal Life

26 | The Senses 550

27 | The Endocrine System 566

28 | The Skeletal and Muscular Systems 582

33 | The Immune System 672

34 | Animal Reproduction and Development 692

UNIT 7 | **Behavior and Ecology**

35 | Animal Behavior 720

36 | Population Ecology 740

37 | Communities and Ecosystems 760

38 | Biomes 782

39 | Preserving Biodiversity 802

The Scientific Study of Life

DNA technology has revolutionized biology. This computer screen is showing a DNA microarray display, which allows researchers to learn exactly which genes are turned "on" and "off" in a cell.

Biology Is Everywhere

WELCOME TO BIOLOGY, THE SCIENTIFIC STUDY OF LIFE. Biology is everywhere: you are alive, and so are your friends, your pets, and the plants in your home and yard. Countless other organisms thrive on and in your body. The food you have eaten today was (until recently, anyway) alive. And the news is full of biology-related stories about newly discovered fossils, weight loss, cancer prevention, genetics, global climate change, and the environment.

Stories such as these enjoy frequent media coverage because this is an exciting time to study biology. Not only is the field changing rapidly, but its new discoveries and applications might change your life. DNA technology has brought us genetically engineered bacteria that can manufacture life-saving drugs—and genetically engineered plants that produce their own pesticides. The same technology may one day enable physicians to routinely cure hemophilia, cystic fibrosis, and other genetic diseases by replacing faulty DNA with a functional "patch."

The ability to sequence DNA has led to a wealth of new information. In a field of biology called genomics, scientists compare the DNA of many species, from bacteria to pine trees to humans. Genomics is yielding unprecedented insight into everything from the history of life to the function of individual disease-causing genes.

The biotechnology revolution doesn't stop with genomics. Each of your cells has the same DNA sequence, but cells don't all use that genetic information in the same way. DNA is organized into genes, each of which acts as a recipe for a specific protein. Thanks to unique combinations of genes that are turned "on" and "off," each cell type has a specialized function. The study of the proteins that a cell produces has blossomed into yet another new field of biology: proteomics. Proteomics may someday save your life. Once scientists learn which proteins are produced by breast cancer cells but not by their healthy neighbors, better treatments or perhaps even a cure may follow. Likewise, improved vaccines or antibiotics may come from studies of the proteins that disease-causing microbes produce.

DNA technology, genomics, and proteomics are but a small part of modern biology. This book will bring you a taste of what we know about life and help you make sense of the science-related news you see every day. Chapter 1 begins your journey by introducing the scope of biology and explaining how science teaches us what we know about life.

Learning Outline

Learn How to Learn
Real Learning Takes Time

You got good at basketball, running, dancing, art, music, or video games by putting in lots of practice. Likewise, you will need to commit time to your biology course if you hope to do well. To get started, look for the "Learn How to Learn" tip in each chapter of this textbook. Each hint is designed to help you use your study time productively.

3

1.1 | What Is Life?

Biology is the scientific study of life. The second half of this chapter explores the meaning of the term *scientific,* but first we will consider the question, "What is life?" We all have an intuitive sense of what life is. If we see a rabbit on a rock, we know that the rabbit is alive and the rock is not. But it is difficult to state just what makes the rabbit alive. Likewise, in the instant after an individual dies, we may wonder what invisible essence has transformed the living into the dead.

One way to define life is to list its basic components. The **cell** is the basic unit of life; every **organism,** or living individual, consists of one or more cells. Every cell has an outer membrane that separates it from its surroundings. This membrane encloses the water and other chemicals that carry out the cell's functions. One of those biochemicals, deoxyribonucleic acid (DNA), is the informational molecule of life (**figure 1.1**). Cells use genetic instructions—as encoded in DNA—to produce proteins, which enable cells to carry out specialized functions in tissues, organs, and organ systems.

A list of life's biochemicals, however, provides an unsatisfying definition of life. After all, placing DNA, water, proteins, and a membrane in a test tube does not create artificial life. And a crushed insect still contains all of the biochemicals that it had immediately before it died.

In the absence of a concise definition, scientists have settled on five qualities that, in combination, constitute life (**table 1.1**). An organism is a collection of structures that function together and exhibit all of these qualities. Note, however, that each of the traits listed in table 1.1 may also occur in nonliving objects. A rock crystal is highly organized, but it is not alive. A fork placed in a pot of boiling water absorbs heat energy and passes it to the hand that grabs it, but this does not make the fork alive. A fire can "reproduce" and grow very rapidly, but it lacks most of the other characteristics of life.

A. Life Is Organized

Just as the city where you live belongs to a county, state, and nation, living matter also consists of parts organized in a hierarchical pattern (**figure 1.2**). At the smallest scale, all living structures are composed of particles called **atoms,** which bond together to form **molecules.** These molecules form **organelles,** which are compartments that carry out specialized functions in cells (note

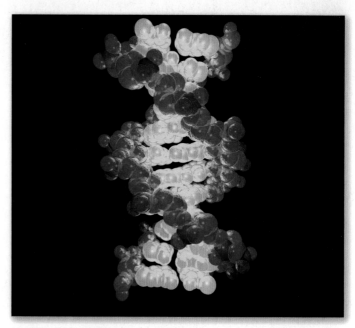

Figure 1.1 Informational Molecule of Life. All cells contain DNA, a series of "recipes" for proteins that each cell can make.

that not all cells contain organelles). Many organisms consist of single cells. In multicellular organisms such as the tree illustrated in figure 1.2, however, the cells are organized into specialized **tissues** that make up **organs** such as leaves. Multiple organs are linked into an individual's **organ systems.**

Organization in the living world extends beyond the level of the individual organism. A **population** includes members of the same species living in the same place at the same time. A **community** includes the populations of different species in a region, and an **ecosystem** includes both the living and nonliving components of an area. Finally, the **biosphere** refers to all parts of the planet that can support life.

Biological organization is apparent in all life. Humans, eels, and evergreens, although outwardly very different, are all organized into specialized cells, tissues, organs, and organ systems. Single-celled bacteria, although less complex than animals or plants, still contain DNA, proteins, and other molecules that interact in highly organized ways.

An organism, however, is more than a collection of successively smaller parts. When those components interact, they create

Table 1.1   Characteristics of Life: A Summary

Characteristic	Example
Organization	Atoms make up molecules, which make up cells, which make up tissues, and so on.
Energy use	A kitten uses the energy from its mother's milk to fuel its own growth.
Maintenance of internal constancy	Your kidneys regulate your body's water balance by adjusting the concentration of your urine.
Reproduction, growth, and development	An acorn germinates, develops into an oak seedling, and, at maturity, reproduces sexually to produce its own acorns.
Evolution	Increasing numbers of bacteria survive treatment with antibiotic drugs.

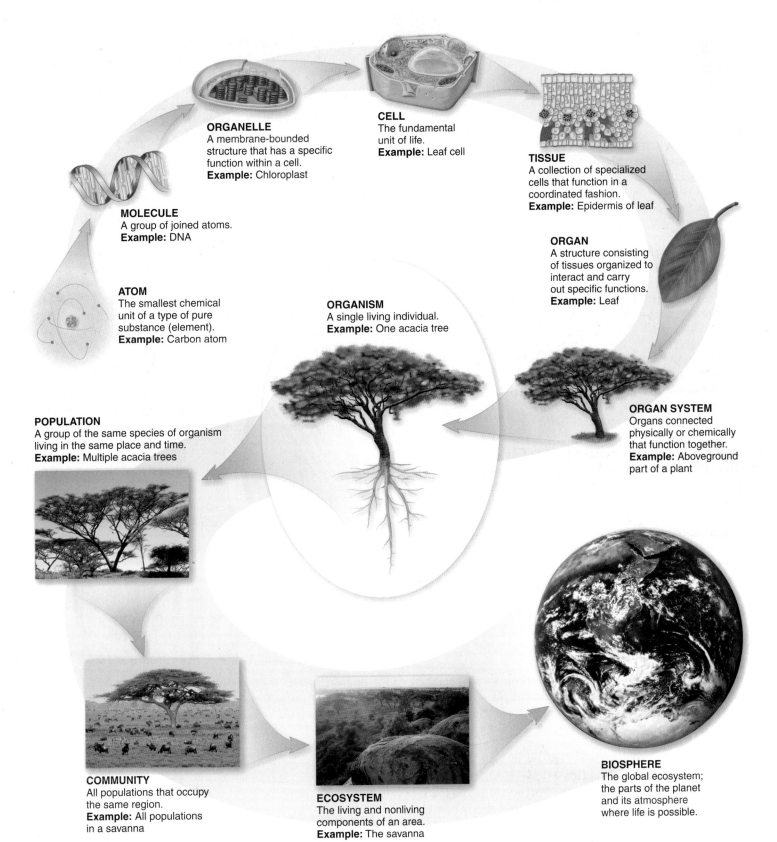

ORGANELLE
A membrane-bounded structure that has a specific function within a cell.
Example: Chloroplast

CELL
The fundamental unit of life.
Example: Leaf cell

TISSUE
A collection of specialized cells that function in a coordinated fashion.
Example: Epidermis of leaf

MOLECULE
A group of joined atoms.
Example: DNA

ORGAN
A structure consisting of tissues organized to interact and carry out specific functions.
Example: Leaf

ATOM
The smallest chemical unit of a type of pure substance (element).
Example: Carbon atom

ORGANISM
A single living individual.
Example: One acacia tree

POPULATION
A group of the same species of organism living in the same place and time.
Example: Multiple acacia trees

ORGAN SYSTEM
Organs connected physically or chemically that function together.
Example: Aboveground part of a plant

COMMUNITY
All populations that occupy the same region.
Example: All populations in a savanna

ECOSYSTEM
The living and nonliving components of an area.
Example: The savanna

BIOSPHERE
The global ecosystem; the parts of the planet and its atmosphere where life is possible.

Figure 1.2 Levels of Biological Organization. Atoms arranged into molecules make up the parts of a cell. Multiple cells are organized into tissues, which make up organs and, in turn, organ systems. An individual organism may consist of one or many cells. A population consists of individuals of the same species, and communities are multiple populations sharing the same space. Communities interact with the nonliving environment to form ecosystems, and the biosphere consists of all places on Earth where life occurs.

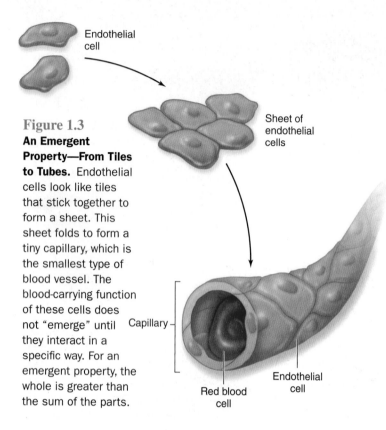

Figure 1.3

An Emergent Property—From Tiles to Tubes. Endothelial cells look like tiles that stick together to form a sheet. This sheet folds to form a tiny capillary, which is the smallest type of blood vessel. The blood-carrying function of these cells does not "emerge" until they interact in a specific way. For an emergent property, the whole is greater than the sum of the parts.

new, complex functions called **emergent properties** (figure 1.3). These characteristics arise from physical and chemical interactions among a system's components, much like flour, sugar, butter, and chocolate can become brownies—something not evident from the parts themselves.

Emergent properties explain why structural organization is closely tied to function. Disrupt a structure, and its function ceases. Shaking a fertilized hen's egg, for instance, disturbs critical interactions and stops the embryo from developing. Likewise, if a function is interrupted, the corresponding structure eventually breaks down, much as unused muscles begin to waste away. Biological function and form are interdependent.

B. Life Requires Energy

Inside each living cell, countless chemical reactions sustain life. These reactions, collectively called metabolism, allow organisms to acquire and use energy and nutrients to build new structures, repair old ones, and reproduce.

Biologists divide organisms into broad categories, based on their source of energy and raw materials (figure 1.4). **Producers,** also called autotrophs, make their own food by extracting energy and nutrients from nonliving sources. The most familiar producers are the plants and microbes that capture light energy from the sun, but some bacteria can derive chemical energy from rocks. **Consumers,** in contrast, obtain energy and nutrients by eating other organisms, living or dead; consumers are also called heterotrophs. You are a consumer, using energy and atoms from food to build your body, move your muscles, send nerve signals, and maintain your temperature. **Decomposers** are heterotrophs

that obtain energy and nutrients from wastes or dead organisms. Fungi and many bacteria are decomposers.

Within an ecosystem, organisms are linked into elaborate food webs, beginning with producers and continuing through several levels of consumers (including decomposers). But energy transfers are never 100% efficient; some energy is always lost in the form of heat (see figure 1.4). Because no organism can use heat as an energy source, it represents a permanent loss from the cycle of life. All ecosystems therefore depend on a continuous stream of energy from an outside source, usually the sun.

C. Life Maintains Internal Constancy

An important characteristic of life is the ability to sense and react to stimuli. The conditions inside cells must remain within a constant range, even if the surrounding environment changes. For example, a living cell must maintain a certain temperature—not too high and not too low. The cell must also take in nutrients, excrete wastes, and regulate its many chemical reactions to prevent a shortage or surplus of essential substances. **Homeostasis** is the process by which a cell or organism maintains this state of internal constancy, or equilibrium.

Your body, for example, has several mechanisms that maintain your internal temperature at about 37°C. When you go outside on a cold day, you may begin to shiver; heat from these involuntary muscle movements warms the body. In severe cold,

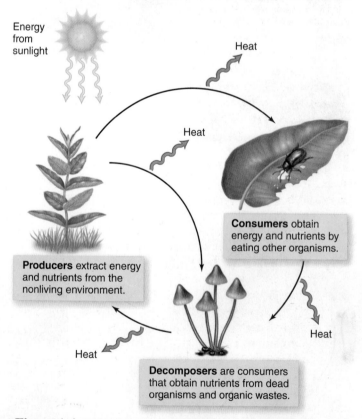

Energy from sunlight

Heat

Heat

Consumers obtain energy and nutrients by eating other organisms.

Producers extract energy and nutrients from the nonliving environment.

Heat

Decomposers are consumers that obtain nutrients from dead organisms and organic wastes.

Heat

Figure 1.4 Life Is Connected. All organisms extract energy and nutrients from the nonliving environment or from other organisms. Decomposers recycle nutrients back to the nonliving environment. At every stage along the way, heat is lost to the system.

a. SEM (false color) ⊢5 µm⊣ b. c.

Figure 1.5 **Asexual and Sexual Reproduction.** (a) This fungus, *Pencillium,* asexually produces identical cells called spores on brushlike structures. (b) A coconut tree seedling and (c) a newborn deer are products of sexual reproduction.

your lips and fingertips may turn blue as your circulatory system diverts blood away from your body's surface. Conversely, on a hot day, sweat evaporating from your skin helps cool your body.

D. Life Reproduces Itself, Grows, and Develops

Organisms reproduce, making other individuals similar to themselves (**figure 1.5**). Reproduction transmits DNA from generation to generation; this genetic information defines the inherited characteristics of the offspring.

Reproduction occurs in two basic ways: asexually and sexually. In **asexual reproduction,** genetic information comes from only one parent, and all offspring are virtually identical. One-celled organisms such as bacteria reproduce asexually by doubling and then dividing the contents of the cell. Many multicellular organisms also reproduce asexually. For example, a strawberry plant's "runners" can sprout leaves and roots, forming a new plant identical to the parent. The green, white, or black powder on moldy bread or cheese is made of the countless asexual spores of fungi (figure 1.5a). Some animals, including sponges, reproduce asexually when a fragment of the parent animal detaches and develops into a new individual.

In **sexual reproduction,** genetic material from two parent individuals unites to form an offspring, which has a new combination of inherited traits. By mixing genes at each generation, sexual reproduction results in tremendous diversity in a population. Genetic diversity, in turn, enhances the chance that some individuals will survive even if conditions change. Sexual reproduction is therefore a very successful strategy, especially in an environment where conditions change frequently; it is extremely common among plants and animals (figure 1.5b,c).

If each offspring is to reproduce, it must grow and develop to adulthood. The fawn in figure 1.5c, for example, started as a single fertilized egg inside its mother. That cell divided over and over, developing into an embryo. Continued cell division and specialization yielded the newborn fawn, which will eventually mature into an adult that can also reproduce—just like its parents.

E. Life Evolves

One of the most intriguing questions in biology is how organisms become so well-suited to their environments. A beaver's enormous front teeth, which never stop growing, are ideal for gnawing wood. Tubular flowers have exactly the right shapes for the beaks of their hummingbird pollinators. Some organisms have color patterns that enable them to fade into the background (**figure 1.6**).

Figure 1.6 **Blending In.** (a) The superb camouflage of the adder snake, *Bitis peringueyi,* makes it virtually undetectable buried in the sand in the Namib Desert, Namibia. (b) It is little wonder that the sand lizard, *Aporosaura anchietae,* soon became the meal of the snake.

a.

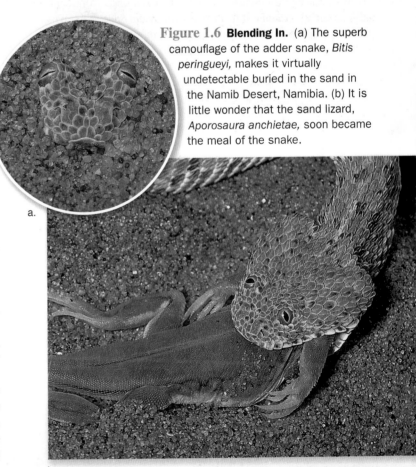

b.

These examples, and countless others, illustrate adaptations. An **adaptation** is an inherited characteristic or behavior that enables an organism to survive and reproduce successfully in its environment.

Where do these adaptive traits come from? The answer lies in natural selection. The simplest way to think of natural selection is to consider two facts. First, resources such as food and habitat are limited, so populations produce many more offspring than will survive to reproduce. A single mature oak tree may produce thousands of acorns in one season, but only a few are likely to germinate, develop, and reproduce. The rest die. Second, no organism is exactly the same as any other. Genetic mutations—changes in an organism's DNA sequence—generate variability in all organisms, even those that reproduce asexually.

Of all the offspring in a population, which will survive long enough to reproduce? The answer is those with the best adaptations to the current environment; poorly adapted organisms are most likely to die before reproducing. A good definition of **natural selection,** then, is the enhanced reproductive success of certain individuals from a population based on inherited characteristics (**figure 1.7**). Over time, individuals with the best combinations of genes survive and reproduce, while those with less suitable characteristics fail to do so. Over many generations, individuals with adaptive traits make up most or all of the population.

But the environment is constantly changing. Continents shift, sea levels rise and fall, climates warm and cool. What happens to a population when the selective forces that drive natural selection change? Only some organisms survive: those with the "best" traits in the *new* environment. Features that may once have been rare become more common as the reproductive success of individuals with those traits improves. Notice, however, that this outcome depends on variability within the population. If no individual can reproduce in the new environment, the species may go extinct.

Natural selection is one mechanism of **evolution,** which is a change in the genetic makeup of a population over multiple generations. Although evolution can also occur in other ways, natural selection is the mechanism that selects for adaptations. Charles Darwin became famous in the 1860s after the publication of his book *On the Origin of Species by Means of Natural Selection,* which introduced the theory of evolution by natural selection; another naturalist, Alfred Russel Wallace, independently developed the same idea at around the same time.

Evolution is the single most powerful idea in biology. As unit 3 describes in detail, evolution has been operating since life began, and it explains the current diversity of life. In fact, the similarities among existing organisms strongly suggest that all species descend from a common ancestor. Evolution has molded the life that has populated the planet since the first cells formed almost 4 billion years ago, and it continues to act today.

1.1 | Mastering Concepts

1. What characteristics distinguish the living from the nonliving?
2. List the levels of life's organizational hierarchy from smallest to largest, starting with atoms and ending with the biosphere.
3. What are the roles of natural selection and mutations in evolution?

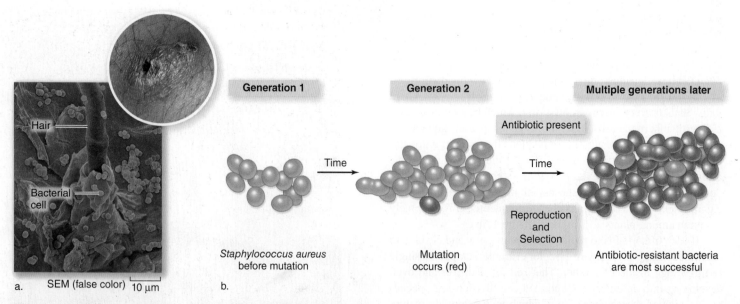

Generation 1   Generation 2   Multiple generations later

Antibiotic present

Time   Time

Hair

Bacterial cell

Reproduction and Selection

a.   SEM (false color) 10 μm   b.

Staphylococcus aureus before mutation   Mutation occurs (red)   Antibiotic-resistant bacteria are most successful

Figure 1.7 Natural Selection. (a) *Staphylococcus aureus* is a bacterium that causes skin infections. (b) By chance, some *S. aureus* bacteria are resistant to the antibiotic methicillin (the commonly used abbreviation *MRSA* refers to methicillin-resistant *S. aureus*). The presence of the antibiotic increases the reproductive success of the resistant cells, which pass this trait to the next generation.

1.2 | The Tree of Life Includes Three Main Branches

Biologists have been studying life for centuries, documenting the existence of everything from bacteria to blue whales. An enduring problem has been how to organize the ever-growing list of known organisms into meaningful categories. **Taxonomy** is the biological science of naming and classifying organisms.

The basic unit of classification is the **species,** which designates a distinctive "type" of organisms. Closely related species, in turn, are grouped into the same **genus.** Together, the genus and species denote the unique scientific name of each type of organism. A human, for example, is *Homo sapiens* (note that scientific names are always italicized). By assigning each type of organism a unique scientific name, taxonomists help other biologists communicate with one another.

But taxonomy involves more than simply naming species. Taxonomists also strive to classify organisms according to what we know about evolutionary relationships; that is, how recently one type of organism shared an ancestor with another type of organism. The more recently they diverged from a shared ancestor, the more closely related we presume the two types of organisms to be. Researchers infer these relationships by comparing anatomical, behavioral, cellular, genetic, and biochemical characteristics.

Section 13.6 describes the taxonomic hierarchy in more detail. For now, it is enough to know that genetic evidence suggests that all species fall into one of three **domains**, the broadest (most inclusive) taxonomic category. **Figure 1.8** depicts the three domains: Bacteria, Archaea, and Eukarya. Species in domains Bacteria and Archaea are superficially similar to one another; all are single-celled prokaryotes, meaning that their DNA is free in the cell and not confined to an organelle called a nucleus. Major differences in DNA sequences separate these two domains from each other. Domain Eukarya, on the other hand, contains all species of eukaryotes, which are unicellular or multicellular organisms whose cells contain a nucleus.

The species in each domain are further subdivided into **kingdoms;** figure 1.8 shows the kingdoms within domain Eukarya. Three of these kingdoms—Animalia, Fungi, and Plantae—are familiar to most people. Within each one, organisms share the same general strategy for acquiring energy. For example, plants are autotrophs. Fungi and animals are consumers, although they differ in the details of how they obtain food. But the fourth group of eukaryotes, the Protista, contains a huge collection of unrelated species. Protista is a convenient but artificial "none of the above" category for the many species of eukaryotes that are not plants, fungi, or animals.

1.2 | Mastering Concepts

1. What are the goals of taxonomy?
2. How are domains related to kingdoms?
3. List and describe the four main groups of eukaryotes.

Figure 1.8 Life's Diversity. The three domains of life (Bacteria, Archaea, and Eukarya) arose from a hypothetical common ancestor.

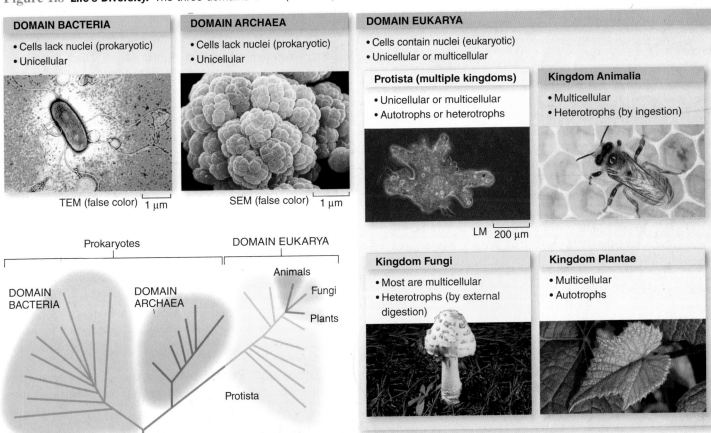

DOMAIN BACTERIA
- Cells lack nuclei (prokaryotic)
- Unicellular

TEM (false color) 1 μm

DOMAIN ARCHAEA
- Cells lack nuclei (prokaryotic)
- Unicellular

SEM (false color) 1 μm

DOMAIN EUKARYA
- Cells contain nuclei (eukaryotic)
- Unicellular or multicellular

Protista (multiple kingdoms)
- Unicellular or multicellular
- Autotrophs or heterotrophs

LM 200 μm

Kingdom Animalia
- Multicellular
- Heterotrophs (by ingestion)

Kingdom Fungi
- Most are multicellular
- Heterotrophs (by external digestion)

Kingdom Plantae
- Multicellular
- Autotrophs

Prokaryotes

DOMAIN EUKARYA

Animals
Fungi
Plants

DOMAIN BACTERIA

DOMAIN ARCHAEA

Protista

1.3 | Scientists Study the Natural World

The idea of biology as a "rapidly changing field" may seem strange if you think of science as a collection of facts. After all, the parts of a frog are the same now as they were 50 or 100 years ago. But memorizing frog anatomy is not the same as thinking scientifically. Scientists use evidence to answer questions about the natural world. For example, if you compare a frog to, say, a snake, can you determine how those animals are related? How can the frog live both in water and on land, and how does the snake survive in the desert? Understanding anatomy simply gives you the vocabulary you need to ask these and other interesting questions about life.

Biology is changing rapidly because technology has expanded our ability to make observations. New microscopes allow us to spy on the inner workings of living cells, DNA sequencing machines are faster than ever, and powerful computers allow us to process huge amounts of data. Scientists can now answer questions about the natural world that previous generations could never have imagined.

A. The Scientific Method Has Multiple Interrelated Parts

Scientific knowledge arises from application of the **scientific method,** which is a general way of using evidence to answer questions and test ideas. Scientific inquiry consists of everyday activities: observing, questioning, reasoning, predicting, testing, interpreting, and concluding (**figure 1.9**). It includes thinking, detective work, communicating with other scientists, and noticing connections between seemingly unrelated events.

Observations and Questions   The scientific method begins with observations and questions about the natural world. The observations may rely on the senses of sight, hearing, touch, taste, or smell, or they may be based on existing knowledge and experimental results. Often, a great leap in science happens when one person makes mental connections among previously unrelated observations. Charles Darwin, for example, developed the idea of natural selection by combining the study of geology with his detailed observations of organisms. His understanding of Earth's long history and the variation he saw in life led him to the insight that organisms change over long periods. Another great advance occurred decades later, when biologists realized that mutations in DNA generate the variation that Darwin saw but could not explain.

Hypothesis   A **hypothesis** is a tentative explanation for one or more observations. The hypothesis is the essential "unit" of scientific inquiry. To be useful, the hypothesis must be testable—there must be a way to collect data that can support or reject the hypothesis. Interestingly, no hypothesis can be proven true, be-

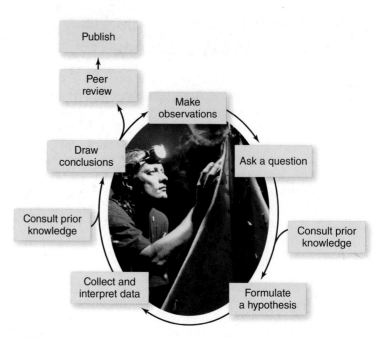

Figure 1.9 Scientific Inquiry. This researcher is collecting insects; the resulting observations could lead to questions and testable hypotheses. Additional data, combined with prior findings, can help support or reject each hypothesis. Note that the step-by-step layout of this figure is simplified; in reality, teams of scientists work on multiple "steps" simultaneously.

cause scientific thinking is open to future discoveries that may contradict today's results. A scientist would therefore avoid the phrase "scientific proof," although advertisers commonly use it.

Data Collection   Investigators draw conclusions based on data, which they can collect in many ways. Often, a scientist devises an experiment to test a hypothesis under controlled conditions; section 1.3B explores experimental design in more detail. But not all data are collected in the context of an experiment; instead, many scientific investigations are based on discovery. For example, British anthropologist Jane Goodall used careful observations, not experimentation, to learn about the dynamics within chimpanzee social groups. Likewise, taxonomists use words and pictures to describe newly discovered species of microbes, plants, and animals.

Experimentation and discovery work hand in hand. For example, determining the sequence of DNA building blocks in a human cell is discovery-based science. But learning the functions of individual genes requires experiments. Likewise, we now understand the connection between cigarettes and cancer because scientists noticed that smokers are far more likely than nonsmokers to develop lung cancer. Laboratory experiments with cells growing in culture help fill in the details of how cancer develops.

Analysis and Peer Review   After collecting and interpreting data, investigators decide whether the evidence supports or falsifies the hypotheses. Often, the most interesting results are those that are unexpected, because they force scientists to rethink their hypotheses. Figure 1.9 shows this feedback loop. Science advances as new information arises and explanations continue to improve.

Once a scientist has enough evidence to support or reject a hypothesis, he or she may write a paper and submit it for publication in a scientific journal. The journal's editors then send the paper to anonymous reviewers knowledgeable about the research. In a process called **peer review,** these scientists independently evaluate the validity of the methods, data, and conclusions. Peer review is not perfect. Some published papers are recalled or amended as unnoticed mistakes are later discovered. Overall, however, peer review ensures that published studies are of high quality.

B. An Experimental Design Is a Careful Plan

Scientists can test many hypotheses with the help of experiments. An **experiment** is an investigation carried out in controlled conditions. This section considers a real study that tested the hypothesis that a new vaccine protects against rotavirus. This virus causes severe diarrhea that takes the lives of hundreds of thousands of young children each year. An effective, inexpensive vaccine would prevent many childhood deaths.

Sample Size   One of the most important decisions that an investigator makes in designing an experiment is **sample size,** which is the number of individuals that he or she will study. For example, several hundred infants participated in the rotavirus vaccine study. In general, the larger the sample size, the more credible the results. But obtaining a large sample is not always practical. When medical researchers study very rare disorders, only a few patients may be available. So they conduct small-scale experiments instead, which are valuable because they may indicate whether continuing research is likely to yield valid, meaningful results.

Variables   A systematic consideration of variables is also important in experimental design (table 1.2). A **variable** is a changeable element of an experiment, and there are several types. The investigator manipulates the levels of the **independent**

variable to determine whether it influences some other phenomenon. For example, in the rotavirus study, the independent variable would be the dose of the vaccine. The **dependent variable** is the response that the investigator measures, such as the number of children with rotavirus-related illness during the study period.

A **standardized variable** is anything that the investigator holds constant for all subjects in the experiment, ensuring the best chance of detecting the effect of the independent variable. For example, rotavirus infection is most common among very young children. The test of the new vaccine therefore included only infants younger than 12 weeks. Furthermore, vaccines work best in people with healthy immune systems, so the study excluded infants who were already ill or who were known to have weak immunity.

Controls   Well-designed experiments compare a group of "normal" individuals to a group undergoing treatment. Ideally, the only difference between the normal group and the experimental group is the one factor being tested. The normal group is called an experimental **control** and provides a basis for comparison.

Experimental controls may take several forms. Sometimes, the control group simply receives a "zero" value for the independent variable. If a gardener wants to test a new fertilizer in her garden, she may give some plants a lot of fertilizer, others only a little, and still others—the control plants—would get none. In other types of experiments, a control group might receive a **placebo,** an inert substance that resembles the treatment given to the experimental group. In medical research, a placebo is often a stand-in for a drug being tested: a sugar pill or a treatment already known to be effective. In the rotavirus study, the control infants received a placebo that contained all components of the vaccine except the active ingredient.

The investigators in the rotavirus study used a **double-blind** design, in which neither the researchers nor the participants knew who received the vaccine and who received the placebo. Double-blind studies are common safeguards to avoid bias in medical research. The investigators break the "code" of who received which treatment only after the data are tabulated or if one group does so well that it would be unethical to withhold treatment from the placebo group.

Table 1.2   Types of Variables in an Experiment: A Summary

Type of Variable	Definition	Example
Independent variable	What the investigator manipulates to determine whether it influences the phenomenon of interest	Dose of experimental vaccine
Dependent variable	What the investigator measures to determine whether the independent variable influenced the phenomenon of interest	Number of children with illness caused by rotavirus
Standardized variable	Any variable intentionally held constant for all subjects in an experiment, including the control group	Age of children in study

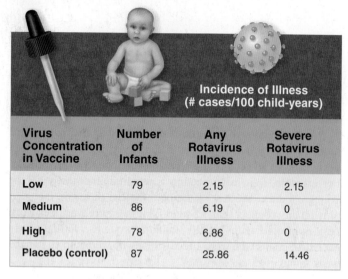

Virus Concentration in Vaccine	Number of Infants	Incidence of Illness (# cases/100 child-years)	
		Any Rotavirus Illness	Severe Rotavirus Illness
Low	79	2.15	2.15
Medium	86	6.19	0
High	78	6.86	0
Placebo (control)	87	25.86	14.46

Figure 1.10 **Vaccine Test.** In this experimental test of a new vaccine against rotavirus, the independent variable was the dose of the vaccine. Control infants received a placebo. The data suggest that the vaccine was probably effective.

Statistical Analysis   Once an experiment is complete, the investigator compiles the data and decides whether the independent variable affected the dependent variable. Look at the experimental results in figure 1.10. Did the rotavirus vaccine prevent illness? Apparently so, but the only way to know for sure is to apply a statistical analysis, using mathematical tools that help the researchers interpret the data.

Researchers use many different statistical tests, all of which measure variation in the data. The less variation within each treatment level, the more likely that the independent variable is responsible for the apparent difference among the treatments. The analysis considers both variation and sample size to yield a measure of **statistical significance,** which is the probability that the results arose purely by chance. Appendix B shows how scientists illustrate statistical significance in graphs.

Creativity and Logic in Experimental Design   Good experimental design requires both creativity and logic. Sometimes, the best experiments are very simple, prompting colleagues to wonder how they could have overlooked what seems obvious in retrospect. The creative inspiration may come at any stage of design: selecting the subjects, deciding on treatment levels and controls, or applying the treatments. Throughout the design process, sound logic ensures that the experiment is a convincing test of the hypothesis.

C. Theories Are Comprehensive Explanations

Former U.S. President Ronald Reagan famously dismissed the idea of biological evolution, saying "It is a scientific theory only." Outside of science, the word *theory* is often used to describe an opinion or a hunch. For instance, immediately after a plane crash, experts offer their "theories" about the cause of the disaster. These tentative explanations are really untested hypotheses.

In science, the word *theory* has a distinct meaning. Like a hypothesis, a **theory** is an explanation for a natural phenomenon, but a theory is typically broader in scope than a hypothesis. For example, the germ theory—the idea that some microorganisms cause human disease—is the foundation for medical microbiology. Individual hypotheses relating to the germ theory are much narrower, such as the suggestion that rotavirus causes illness. Not all theories are as "large" as the germ theory, but they generally encompass multiple hypotheses. Note also that the germ theory does not imply that *all* microbes cause disease or that all diseases have microbial causes. But it does explain many types of human diseases.

A second difference between a hypothesis and a theory is acceptance. A hypothesis is tentative, whereas theories reflect broader agreement. This is not to imply that theories are not testable; in fact, the opposite is true. Every scientific theory is falsifiable, meaning that there must be a way to prove it wrong. The germ theory remains widely accepted because many observations support it and no reliable tests have disproved it. The same is true for the theory of evolution and other scientific theories.

Another quality of a scientific theory is its predictive power. A good theory not only ties together many existing observations, it also suggests predictions about phenomena that have yet to be observed. Both Charles Darwin and naturalist Alfred Russel Wallace, for example, used the theory of evolution by natural selection to predict the existence of a moth that could pollinate orchid flowers with unusually long nectar tubes (figure 1.11). Decades later, scientists discovered the long-tongued insect (see section 1.4). A theory weakens if subsequent observations do not support its predictions.

At some point, a theory is so widely accepted that people regard it as a fact. The line between theory and fact is fuzzy,

Nectar tubes

Figure 1.11 **Prediction Confirmed.** When Charles Darwin saw this orchid, he predicted that its pollinator would have long, thin mouthparts that could reach the bottom of the elongated nectar tube. He was right; the unknown pollinator turned out to be a moth with an extraordinarily long tongue.

but the late paleontologist Stephen Jay Gould recognized a useful difference: "In science 'fact' can only mean 'confirmed to such a degree that it would be perverse to withhold provisional assent.'" Although a theory can never be proven 100% true, some theories are so well-supported that no educated person questions its validity. Gravity, for example, is a fact.

Biologists also consider biological evolution to be a fact. Yet the phrase "theory of evolution" persists, because evolution is both a fact *and* a theory. Both terms apply equally well. The evidence for genetic change over time is so persuasive and comes from so many different fields of study that to deny its existence is unrealistic. Nevertheless, biologists do not understand everything about how evolution works. Many questions about life's history remain, but the debates swirl around *how,* not *whether,* evolution occurred.

Science is just one of many ways to investigate the world, but its strength is its openness to new information. Theories change to accommodate new knowledge. The history of science is full of long-established ideas changing as we learned more about nature, often thanks to new technology. For example, people thought that Earth was flat and at the center of the universe before inventions and data analysis revealed otherwise. Similarly, biologists thought all organisms were plants or animals until microscopes unveiled a world of life invisible to our eyes.

D. Scientific Inquiry Has Limitations

Scientific inquiry is neither foolproof nor always easy to implement. One problem is that experimental evidence may lead to multiple interpretations, and even the most carefully designed experiment can fail to provide a definitive answer. Consider the observation that animals fed large doses of vitamin E live longer than similar animals that do not ingest the vitamin. So, does vitamin E slow aging? Possibly, but excess vitamin E also causes weight loss, and other research has connected weight loss with longevity. Does vitamin E extend life, or does weight loss? The experiment alone does not distinguish between these possibilities. Can you think of further studies to clarify whether vitamin E or weight loss extends life?

Another limitation is that researchers may misinterpret observations or experimental results. For example, centuries ago, scientists sterilized a bottle of broth, corked the bottle shut, and observed

bacteria in the broth a few days later. They concluded that life arose directly from the broth. The correct explanation, however, was that the cork did not keep airborne bacteria out. Science is self-correcting, in the sense that scientific thought is open to new data and new interpretations. But it is also fallible, especially in the short term.

A related problem is that the scientific community may be slow to accept new evidence that suggests unexpected conclusions. Every investigator should try to keep an open mind about observations, not allowing biases or expectations to cloud interpretation of the results. But it is human nature to be cautious in accepting an observation that does not fit what we think we know. The careful demonstration that life does not arise from broth surprised many people who believed that mice sprang from moldy grain and that flies came from rotted beef. More recently, it took many years to set aside the common belief that stress causes ulcers. Today, we know that a bacterium (*Helicobacter pylori*) causes most ulcers.

Although science is a powerful tool for answering questions about the natural world, it cannot answer questions of beauty, morality, ethics, or religion. Nor can we directly study some phenomena that occurred long ago and left little physical evidence. For example, many experiments have attempted to re-create the chemical reactions that might have produced life on early Earth. Although the experiments produce interesting results and reveal ways that these early events may have occurred, we cannot know if they accurately re-create conditions at the beginning of life.

Scientific research seeks to understand nature. Because humans are part of nature, we sometimes view scientific research, and particularly biological research, as aimed at improving the human condition. But knowledge without any immediate application or payoff is valuable in and of itself—because we can never know when information will be useful.

1.3 | Mastering Concepts

1. What are the components of scientific inquiry?
2. Identify the elements of the experiment summarized in the Apply It Now box on the next page.
3. What is the difference between a hypothesis and a theory?
4. What are some limitations of scientific inquiry?

Burning Question

Why am I here?

The Burning Questions featured in each chapter of this book came from students. On the first day of class, I always ask students to turn in a "Burning Question"—anything they have always wondered about biology. I answer most of the questions as the relevant topics come up during the semester.

Why not answer *all* of the questions? It is because at least one student often asks something like "Why am I here?" or "What is the meaning of life?" Such puzzles have fascinated humans throughout the ages, but they are among the many questions that we cannot approach scientifically. Biology can explain how you developed after a sperm from your father fertilized

an egg cell from your mother. But no one can develop a testable hypothesis about life's meaning or the purpose of human existence. Science must remain silent on such questions.

Instead, other ways of knowing must satisfy our curiosity about "why." Philosophers, for example, can help us see how others have considered these questions. Religion may also provide the meaning that many people seek. Part of the value of higher education is to help you acquire the tools you need to find your own life's purpose.

Submit your burning question to:
marielle_hoefnagels@mcgraw-hill.com

Apply It Now

The Saccharin Scare

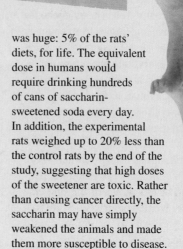

You probably sometimes hear reports that a food previously considered healthy is actually bad for you, or vice versa. Eggs are good and caffeine is bad! No, they're both bad! No, they're both good! If scientists can't make up their minds, does it mean that scientific studies are invalid? Or are scientists just out for publicity?

The reality is a bit more complex. Take, for example, the artificial sweetener saccharin (see the Apply It Now box on sugar substitutes in chapter 2). Saccharin was discovered in 1879, and its popularity as a low-calorie sugar substitute peaked in the 1960s. But in 1977, the U.S. Food and Drug Administration (FDA) proposed a ban on saccharin, based on a handful of studies suggesting that the sweetener caused bladder cancer in rats. Because few alternative artificial sweeteners were available at the time, however, Congress immediately placed a moratorium on the ban and instead required warning labels on products containing saccharin. In 1991, the FDA withdrew its proposed ban, and in 1998, the International Agency for Research on Cancer rated saccharin as "not classifiable as to its carcinogenicity to humans." Two years later, legislation removed the warning label requirement.

This tangled legislative history raises an important issue: Why can't science reply "yes" or "no" to the seemingly simple question of whether saccharin is bad for you? To understand the answer, consider one of the studies that prompted the FDA to propose the ban on saccharin in the first place. Researchers divided 200 rats into two groups. The control animals ate standard rodent chow, whereas the experimental group got the same food supplemented with saccharin. At reproductive maturity the animals were bred, and the researchers fed the offspring the same dose of saccharin throughout their lives as well. They measured the incidence of cancer in both generations of rats for 24 months or until the rats died, whichever came first. The results are summarized in the table 1.A.

At first glance, the conclusion seems inescapable: saccharin causes cancer in male lab rats. But closer study reveals several hidden complexities that make the data hard to interpret. First, the dose of saccharin was huge: 5% of the rats' diets, for life. The equivalent dose in humans would require drinking hundreds of cans of saccharin-sweetened soda every day. In addition, the experimental rats weighed up to 20% less than the control rats by the end of the study, suggesting that high doses of the sweetener are toxic. Rather than causing cancer directly, the saccharin may have simply weakened the animals and made them more susceptible to disease.

The researchers could have tested for that possibility by adding additional treatments with lower, less toxic saccharin concentrations. Then, if the sweetener really did cause cancer, they could have looked for a predicted "dose-response" relationship: low doses should yield just a few cases, and high doses should produce more. But the experiment was not designed to test for such a relationship. Furthermore, studies using mice, hamsters, and monkeys were inconclusive. The researchers who used these other animals created different designs for each experiment, so it is difficult to compare the results. The rat study, for example, followed two generations of animals; those with other animals used just one generation.

The animal studies did not yield uniform results, so maybe the scientists should have studied the saccharin–cancer connection in humans instead. Such research, however, is extremely difficult. It is obviously unethical to keep humans in captivity, control every facet of their environment and breeding, intentionally expose them to potentially harmful chemicals, and then kill and dissect them to check for tumors. The only way to approach the question in humans, therefore, would be to measure the incidence of cancer in saccharin users versus nonusers. But with so many other possible causes of cancer—smoking, poor diet, exposure to job-related chemicals, genetic predisposition—it is difficult to separate out just the effects of saccharin.

So what are we to make of the mixed news reports on eggs, caffeine, chocolate, wine, and soy? It is hard to say, but one thing is certain: No matter what the headlines say, one study, especially a small one, cannot reveal the whole story. Good, bad, or neutral? The complexities of real-world science mean that in most cases, the jury is still out.

Table 1.A   Tumors in Two Generations of Rats

	Rats with Tumors/Rats Examined (% with tumors)	
	Parents	**Offspring**
Male rats		
Saccharin-fed	7/38 (19%)	12/45 (27%)
Controls	1/36 (3%)	0/42 (0%)
Female rats		
Saccharin-fed	0/40 (0%)	2/49 (4%)
Controls	0/38 (0%)	0/47 (0%)

1.4 | Investigating Life: The Orchid and the Moth

Each chapter of this book ends with a section that examines how biologists use systematic, scientific observations to solve a different evolutionary puzzle from life's long history. This first installment of "Investigating Life" revisits the story of the orchid plant pictured in figure 1.11.

Charles Darwin

In a book on orchids published in 1862, Charles Darwin speculated about which type of insect might pollinate the unusual flowers of the *Angraecum sesquipedale* orchid, a species that lives on Madagascar (an island off the coast of Africa). As described in section 1.3C, the flowers have unusually long nectar tubes (also called nectaries). Darwin observed nectaries "eleven and a half inches long, with only the lower inch and a half filled with very sweet nectar." Darwin found it "surprising that any insect should be able to reach the nectar; our English sphinxes [moths] have prosbosces as long as their bodies; but in Madagascar there must be moths with prosbosces capable of extension to a length of between ten and eleven inches!"

Alfred Russel Wallace

Alfred Russel Wallace picked up the story in a book published in 1895. According to Wallace, "There is a Madagascar orchid—the *Angraecum sesquipedale*—with an immensely long and deep nectary. How did such an extraordinary organ come to be developed?" He went on to summarize how natural selection could explain this unusual flower. He wrote: "The pollen of this flower can only be removed by the base of the proboscis of some very large moths, when trying to get at the nectar at the bottom of the vessel. The moths with the longest prosbosces would do this most effectually; they would be rewarded for their long tongues by getting the most nectar; whilst on the other hand, the flowers with the deepest nectaries would be the best fertilized by the largest moths preferring them. Consequently, the deepest nectaried orchids and the longest tongued moths would each confer on the other an advantage in the battle of life."

At that time, the pollinator had not yet been discovered. However, as Wallace wrote, moths with very long tongues were known to exist: "I have carefully measured the proboscis of a specimen . . . from South America . . . and find it to be nine inches and a quarter long! One from tropical Africa . . . is seven inches and a half. A species having a proboscis two or three inches longer could reach the nectar in the largest flowers of *Angraecum sesquipedale* . . . That such a moth exists in Madagascar may be safely predicted; and naturalists who visit that island

Figure 1.12 Found at Last. More than 40 years after Darwin predicted its existence, scientists finally discovered the sphinx moth *Xanthopan morgani*.

should search for it with as much confidence as astronomers searched for the planet Neptune—and I venture to predict they will be equally successful!"

A taxonomic publication from 1903 finally validated Darwin's and Wallace's predictions. The authors described a moth species, *Xanthopan morgani*, with a 225-millimeter (8-inch) tongue (figure 1.12). Given the correspondence between lengths of the orchid's nectary and the moth's tongue, the authors concluded that "*Xanthopan morgani* can do for *Angraecum* what is necessary [for pollination]; we do not believe that there exists in Madagascar a moth with a longer tongue. . . ."

This story not only illustrates how theories lead to testable predictions but also reflects the collaborative nature of science. Darwin and Wallace asked a simple question: Why are these nectar tubes so long? Other biologists cataloging the world's insect species finally solved the puzzle, decades after Darwin first raised the question of the mysterious Madagascan orchid.

Darwin, C. R. 1862. *On the Various Contrivances by Which British and Foreign Orchids are Fertilised by Insects, and on the Good Effects of Intercrossing.* London: John Murray, pages 197–198.

Rothschild, W., and K. Jordan. 1903. A revision of the lepidopterous family Sphingidae. *Novitates Zoologicae* 9, supplement part 1, page 32.

Wallace, Alfred Russel. 1895. *Natural Selection and Tropical Nature: Essays on Descriptive and Theoretical Biology.* London: MacMillan and Co., pages 146–148.

1.4 | Mastering Concepts

1. What observations led Darwin and Wallace to predict the existence of a long-tongued moth in Madagascar?
2. How does this story illustrate discovery science?

Chapter Summary

1.1 | What Is Life?

- A combination of characteristics distinguishes life: organization, energy use, internal constancy, reproduction and development, and evolution.

A. Life Is Organized

- An **organism** consists of **atoms**, which form **molecules.** These molecules form the **organelles** inside **cells.** In most multicellular organisms, cells form **tissues,** then **organs** and **organ systems.**
- Multiple individuals of the same species make up **populations;** multiple populations form **communities. Ecosystems** incorporate the nonliving environment; the **biosphere** includes all of the world's ecosystems.
- **Emergent properties** arise from interactions among the parts that make up an organism.

B. Life Requires Energy

- Life requires energy to maintain its organization and functions. **Producers** make their own food, using energy and nutrients extracted from the nonliving environment. **Consumers** eat other organisms, living or dead. **Decomposers** are consumers that recycle nutrients to the nonliving environment.
- Because of heat losses, all ecosystems require constant energy input from an outside source, usually the sun.

C. Life Maintains Internal Constancy

- Organisms must maintain **homeostasis,** an internal state of constancy in changing environmental conditions.

D. Life Reproduces Itself, Grows, and Develops

- Organisms reproduce asexually, sexually, or both. **Asexual reproduction** yields virtually identical copies, whereas **sexual reproduction** generates tremendous genetic diversity.

E. Life Evolves

- In **natural selection,** environmental conditions select for organisms with inherited traits that increase the chance of survival and reproduction. The result of natural selection is **adaptations,** features that enhance reproductive success.
- **Evolution** through natural selection explains how common ancestry unites all species, producing diverse organisms with many similarities.

1.2 | The Tree of Life Includes Three Main Branches

- **Taxonomy** is the science of classification. Biologists classify types of organisms, or **species,** according to probable evolutionary relationships. A **genus,** for example, consists of closely related species.
- The two broadest taxonomic levels are **domain** and **kingdom.**
- The three domains of life are Archaea, Bacteria, and Eukarya. Within each domain, mode of nutrition and other features distinguish the kingdoms.

1.3 | Scientists Study the Natural World

A. The Scientific Method Has Multiple Interrelated Parts

- Scientific inquiry, which uses the **scientific method,** is a way of using evidence to evaluate ideas. Science involves observing, questioning, reasoning, predicting, testing, interpreting, concluding, and posing further questions.
- Scientific inquiry begins when a scientist makes an observation, raises questions about it, and uses reason to construct a testable explanation, or **hypothesis.**
- After collecting data and making conclusions based on the evidence, an investigator may seek to publish scientific results. **Peer review** ensures that published studies meet high quality standards.

B. An Experimental Design Is a Careful Plan

- An **experiment** is a test of a hypothesis carried out in controlled conditions.
- The larger the **sample size,** the more credible the results of an experiment.
- Experimental **controls** are the basis for comparison. The **independent variable** in an experiment is the factor that the investigator manipulates. The **dependent variable** is a measurement that the investigator makes to determine the outcome of an experiment. **Standardized variables** are held constant for all subjects in an experiment.
- **Placebo**-controlled, **double-blind** experiments minimize bias.
- Experimental results are **statistically significant** if they are unlikely to be due to chance.

C. Theories Are Comprehensive Explanations

- A **theory** is more widely accepted and broader in scope than a hypothesis.
- The acceptance of scientific ideas may change as new evidence accumulates.

D. Scientific Inquiry Has Limitations

- The scientific method does not always yield a complete explanation, or it may produce ambiguous results. Science cannot answer all possible questions, only those for which it is possible to develop testable hypotheses.

1.4 | Investigating Life: The Orchid and the Moth

- Charles Darwin and Alfred Russel Wallace knew of an orchid in Madagascar with an extremely long nectar tube. They predicted that the orchid's pollinator would be a moth with an equally long tongue.
- Years later, other scientists discovered the moth, illustrating the predictive power of evolutionary theory.

Multiple Choice Questions

1. All of the following are characteristics of life EXCEPT
    a. evolution.
    c. homeostasis.
    b. reproduction.
    d. multicellularity.

2. Which property of life can a scientist directly observe in a single plant fossil?
    a. Homeostasis
    c. Energy use
    b. Organization
    d. Growth

3. Which of the following lists is ordered from smallest (least inclusive) to largest (most inclusive)?
    a. Cell < Tissue < Organelle < Individual < Community
    b. Community < Population < Ecosystem < Biosphere
    c. Organelle < Cell < Organ < Individual < Population
    d. Individual < Ecosystem < Community < Biosphere

4. Because plants extract nutrients from soil and use sunlight as an energy source, they are considered to be
    a. autotrophs.
    c. heterotrophs.
    b. consumers.
    d. decomposers.

5. Evolution through natural selection will occur most rapidly for populations of plants that
    a. are already well adapted to the environment.
    b. live in an unchanging environment.
    c. are in the same genus.
    d. reproduce sexually and live in an unstable environment.

6. Which of the following statements is true?
    a. Two of the three domains contain eukaryotes.
    b. The three main branches of life are animals, plants, and fungi.
    c. Humans and plants share the same domain.
    d. Two species in the same genus can be in different domains.

7. In an experiment to test the effect of temperature on the rate of bacterial reproduction, temperature would be the
    a. standardized variable.
    b. independent variable.
    c. dependent variable.
    d. control variable.

8. What is the role of a placebo in medical research?
   a. It ensures that all patients in a study have the same illness.
   b. It is the highest possible value of the dependent variable.
   c. It is a standardized treatment given to all patients.
   d. It is given to some patients as a control.

9. Can a theory be proven wrong?
   a. No, theories are the same as facts.
   b. No, because there is no good way to test a theory.
   c. Yes, a new observation or interpretation of data could disprove a theory.
   d. Yes, theories are the same as hypotheses.

10. Which of the following questions cannot be answered using the scientific method?
    a. What was the first living organism on Earth?
    b. Does a particular gene influence aging in mice?
    c. How does migration affect the reproductive success of monarch butterflies?
    d. How does coastal development affect wetland biodiversity?

Write It Out

1. Describe each of the five characteristics of life, and list several nonliving things that possess at least two of these characteristics.

2. Draw and explain the relationship between producers and consumers (including decomposers).

3. What is homeostasis? Give an example other than those mentioned in the book.

4. Describe the main differences between asexual and sexual reproduction. Why are both types of reproduction common?

5. Describe a specific adaptation in an organism familiar to you, and explain how the environment could have selected for that adaptation.

6. How are the members of the three domains similar? How are they different?

7. Find an example of a news story that describes an experiment. Which components of the scientific method can you identify in the article?

8. Give two examples of questions that you cannot answer using the scientific method. Explain your reason for choosing each example.

9. If you dissect and label the parts of an earthworm, are you "doing science"? Why or why not? Give an example of a testable hypothesis that could result from a dissection.

10. Studies show that drug company-funded research is more favorable to new drugs than is publicly funded research. How can scientists avoid such systematic biases?

11. For each of the following examples, state whether each of the following faults occurred: (I) experimental evidence does not support conclusions; (II) inadequate controls; (III) biased sampling; (IV) inappropriate extrapolation from the experimental group to the general population; (V) sample size too small.
    a. "I ran 4 miles every morning when I was pregnant with my first child," the woman told her physician, "and Jamie weighed only half as much as a normal baby. This time, I didn't exercise at all, and Jamie's sister had normal birth weight. Therefore, running during pregnancy must cause low birth weight."
    b. Eating foods high in cholesterol was found to be dangerous for a large sample of individuals with hypercholesterolemia, a disorder of the heart and blood vessels. It was concluded from this study that all persons should limit dietary cholesterol intake.
    c. Osteogenesis imperfecta (OI) is an inherited condition that causes easily fractured bones. In a clinical study, 30 children with OI were given a new drug for 3 years. Afterward, the children all had less fatigue, improved bone density, and a lower incidence of fractures than they had before treatment started. The researchers concluded that the drug is effective in treating OI.

d. Researchers studied HIV in blood and semen from 11 HIV-infected men. In eight of the men, the virus was resistant to several medications. In two men, viruses from the blood were resistant to the class of drugs called protease inhibitors, but viruses from semen were not resistant. The researchers concluded that protease inhibitors do not reach the male reproductive organs.

12. Design an experiment to test the following hypothesis: "Eating chocolate causes zits." Include sample size, independent variable, dependent variable, the most important variables to standardize, and an experimental control.

13. Morgellons syndrome is a medical mystery. Patients experience sensations of stinging, biting, or crawling skin; they may also have rashes or sores that are slow to heal. Scientists have proposed several hypotheses to explain the symptoms: the patients may be imagining the disorder and creating the sores by picking at their skin; "Morgellons syndrome" may simply be a new name for a recognized skin disorder such as dermatitis or a bacterial infection; or exposure to toxins in the environment may cause the symptoms. If you had unlimited resources, what data might you collect to test each hypothesis?

14. Review "The Saccharin Scare" box on page 14. If you were investigating a possible saccharin–cancer link today, how would you improve on the design of the experiments conducted in the 1970s? Also, develop a hypothesis that could explain why more male than female rats developed tumors. Design an experiment that would help you test your hypothesis.

Pull It Together

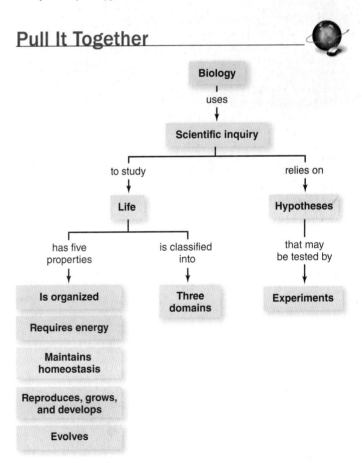

1. What are the elements of a controlled experiment?
2. Name and briefly describe the three domains of life.
3. What is the relationship between natural selection and evolution?
4. List the levels of biological organization, from atoms to the biosphere, and describe the relationships among them.

The Chemistry of Life

A tiny messenger molecule called nitric oxide can help relieve chest pain.

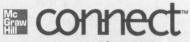

connect | BIOLOGY

Enhance your study of
this chapter with practice quizzes,
animations and videos, answer keys,
and downloadable study tools.
www.mhhe.com/hoefnagels

Just Say NO

AMONG BIOLOGICAL MOLECULES, NITRIC OXIDE—ABBREVIATED NO—HAS SOME ODD TRAITS. First, it is a gas. Second, it is tiny. NO consists of just two atoms, one of the element nitrogen and the other of oxygen. Third, NO apparently can slip freely into and out of cells. Most other molecules must either enter the cell through special channels or bind to receptors on receiving cells.

Before 1980, NO was known mostly for its harmful presence in smog, cigarette smoke, and acid rain. But later research identified NO's function in animals, plants, fungi, and bacteria. A few roles of NO include:

- *Blood pressure.* NO causes blood vessels to widen by relaxing muscles in the vessel walls. On a whole-body scale, NO therefore lowers blood pressure by giving the blood more room to move. This effect has important medical applications. For example, angina is chest pain that occurs when the heart muscle does not get enough oxygen. The drug nitroglycerine increases NO production in the body, relieving angina by widening the arteries that supply oxygen-rich blood to the heart.
- *Erectile function.* A mammal's penis consists of three chambers of spongy tissue that surround blood vessels. Upon sexual stimulation, nerves and other cells lining the insides of the blood vessels release NO. The NO then activates other chemicals that widen the blood vessels, allowing blood to fill the spongy tissue. The penis becomes erect. Some men with erectile dysfunction (impotence) cannot achieve erection because an enzyme interferes with the NO-activated chemical cascade. The pharmaceutical drugs Viagra (sildenafil), Cialis (tadalafil), and Levitra (vardenafil) treat erectile dysfunction by blocking the action of the enzyme. The NO-mediated reaction proceeds, producing an erection.
- *Plant defenses.* Many bacteria, fungi, and viruses are parasites on living plant cells. When a plant detects an attack, NO triggers a cascade of reactions that kill the tissues in a zone surrounding the initial site of infection. This early reaction, called the hypersensitive response, stops the invader before it spreads.

Life is made of chemicals, most of which are much larger than NO. Understanding biology is impossible without an introduction to chemistry. This chapter describes how nitrogen, oxygen, and other atoms come together to form the molecules of life.

UNIT 1

What's the Point?

Learning Outline

Learn How to Learn

Organize Your Time, and Don't Try to Cram

Get a calendar, and study the syllabus for every class you are taking. Write each due date in your calendar. Include homework assignments, quizzes, and exams, and add new dates as you learn them. Then, block out time before each due date to work on each task. Success comes much more easily if you take a steady pace instead of waiting until the last minute.

2.1 | Atoms Make Up All Matter

If you have ever touched a plant in a restaurant to see if it's fake, you know that we all have an intuitive sense of what life is made of. A living leaf feels moist and pliable; a fake one is dry and stiff. But what does chemistry tell us about the composition of life?

Your desk, your book, your body, your sandwich, a plastic plant—indeed, all objects in the universe, including life on Earth—are composed of matter and energy (figure 2.1). **Matter** is any material that takes up space, such as organisms, rocks, the oceans, and gases in the atmosphere. This chapter and the next

concentrate on the composition of living matter. Physicists define energy, on the other hand, as the ability to do work. In this context, *work* means moving matter. Heat, light, and chemical bonds are all forms of energy; chapter 4 discusses the energy of life in detail.

A. Elements Are Fundamental Types of Matter

The matter that makes up every object in the universe consists of one or more elements. A chemical **element** is a pure substance that cannot be broken down by chemical means into other substances. Examples include pure oxygen (O), carbon (C), nitrogen (N), sodium, (Na), and hydrogen (H).

Scientists had already noticed patterns in the chemical behavior of the elements by the mid-1800s, and several had proposed schemes for organizing the elements into categories. Nineteenth-century Russian chemist Dmitry Mendeleyev invented the **periodic table,** the chart that we still use today. The chart is "periodic" because the chemical properties of the elements repeat in each column of the table. Figure 2.2 illustrates an abbreviated periodic table, emphasizing the elements that make up organisms. (Appendix D contains a complete periodic table.)

Only about 25 elements are essential to life. Of these, the **bulk elements** are required in the largest amounts because they make up the vast majority of every living cell. The four most abundant bulk elements in life are carbon, hydrogen, oxygen, and nitrogen. **Minerals** are essential elements other than C, H, O, and N. Some minerals, including phosphorus (P), sodium (Na), magnesium (Mg), potassium (K), and calcium (Ca), are bulk elements. Others are **trace elements,** meaning they are required in small amounts.

Figure 2.1 Matter and Energy. Every object in the universe is composed of matter and energy.

Figure 2.2 The Periodic Table of Elements. Each element has a symbol, which can come from the element's English name (He for helium, for example) or from a name in another language (Na for sodium, which is *natrium* in Latin). Elements 58 through 71 and 90 through 103 are omitted for clarity; a complete periodic table appears in appendix D.

1																	2
H																	He
3	4											5	6	7	8	9	10
Li	Be											B	C	N	O	F	Ne
11	12											13	14	15	16	17	18
Na	Mg											Al	Si	P	S	Cl	Ar
19	20	21	22	23	24	25	26	27	28	29	30	31	32	33	34	35	36
K	Ca	Sc	Ti	V	Cr	Mn	Fe	Co	Ni	Cu	Zn	Ga	Ge	As	Se	Br	Kr
37	38	39	40	41	42	43	44	45	46	47	48	49	50	51	52	53	54
Rb	Sr	Y	Zr	Nb	Mo	Tc	Ru	Rh	Pd	Ag	Cd	In	Sn	Sb	Te	I	Xe
55	56	57	72	73	74	75	76	77	78	79	80	81	82	83	84	85	86
Cs	Ba	La	Hf	Ta	W	Re	Os	Ir	Pt	Au	Hg	Tl	Pb	Bi	Po	At	Rn
87	88	89	104	105	106	107	108	109	110	111	112	113	114	115	116		
Fr	Ra	Ac	Rf	Db	Sg	Bh	Hs	Mt	Ds	Rg	Uub	Uut	Uuq	Uup	Uuh		

- ☐ Bulk elements
- ☐ Trace elements
- ☐ Possibly essential trace elements

A person whose diet is deficient in any mineral can become ill or die. The thyroid gland, for example, requires the element iodine (I). If the diet does not supply enough iodine, the thyroid may become enlarged, forming a growth called a goiter. Similarly, blood requires iron (Fe) to carry oxygen to the body's tissues. An iron-poor diet can cause anemia.

B. Atoms Are Particles of Elements

An **atom** is the smallest possible "piece" of an element that retains the characteristics of the element. An atom is composed of three types of subatomic particles (**figure 2.3** and **table 2.1**). **Protons,** which carry a positive charge, and **neutrons,** which are uncharged, together form a central **nucleus.** Negatively charged **electrons** surround the nucleus. An electron is vanishingly small compared with a proton or a neutron.

For simplicity, most illustrations of atoms show the electrons closely hugging the nucleus. In reality, however, if the nucleus of a hydrogen atom were the size of a meatball, the electron belonging to that atom could be about 1 kilometer away from it! Thus, most of an atom's mass is concentrated in the nucleus, while the electron cloud occupies virtually all of its volume.

How can this electron cloud, which is mostly empty space, account for the solid "feel" of the objects in our world? The fact that the electrons are in constant motion helps explain this paradox. A good analogy is a ceiling fan. When the fan is not spinning, it is easy to move your hand between two blades. But when the fan is on, the rotating blades essentially form a solid disk.

Each element has a unique **atomic number,** the number of protons in the nucleus. Hydrogen, the simplest type of atom, has an atomic number of 1. In contrast, an atom of uranium has 92 protons. Elements are arranged sequentially in the periodic table by atomic number, which appears above each element's symbol (see figure 2.2).

When the number of protons equals the number of electrons, the atom is electrically neutral; that is, it has no net charge. An **ion** is an atom (or group of atoms) that has gained or lost electrons and therefore has a net negative or positive charge. Common positively charged ions, also called cations, include hydrogen (H^+), sodium (Na^+), and potassium (K^+). Negatively charged ions (anions) include hydroxide (OH^-) and chloride (Cl^-). Ions participate

Table 2.1   Types of Subatomic Particles

Particle	Charge	Mass	Location
Electron	–	0	Surrounding nucleus
Neutron	None	1	Nucleus
Proton	+	1	Nucleus

in many biological processes, including the transmission of messages in the nervous system. They also form ionic bonds, discussed in section 2.2. ▶ action potential, p. 532

C. Isotopes Have Different Numbers of Neutrons

An atom's **mass number** is the total number of protons and neutrons in its nucleus. Because neutrons and protons have the same mass (see table 2.1), subtracting the atomic number from the mass number therefore yields the number of neutrons in an atom.

All atoms of an element have the same atomic number but not necessarily the same number of neutrons. An **isotope** is any of these different forms of a single element. For example, carbon has three isotopes, designated ^{12}C (six neutrons), ^{13}C (seven neutrons), and ^{14}C (eight neutrons). The superscript in each isotope's symbol denotes the mass number.

Figure It Out

The most abundant isotope of iron (Fe) has a mass number of 56. Using the information in figure 2.2, how many neutrons are in each atom of ^{56}Fe?

Answer: 30.

Often one isotope of an element is very abundant, and others are rare. For example, about 99% of carbon isotopes are ^{12}C, and only 1% are ^{13}C or ^{14}C. An element's **atomic mass** (also called atomic weight) is the average mass of all isotopes. Because nearly all carbon atoms contain six neutrons, carbon's atomic mass is very close to 12 in the periodic table.

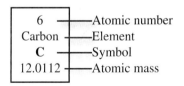

6	——Atomic number
Carbon	——Element
C	——Symbol
12.0112	——Atomic mass

Many of the known isotopes are unstable and **radioactive,** which means they emit energy as rays or particles when they break down into more stable forms. Every radioactive isotope has a characteristic half-life, which is the time it takes for half of the atoms in a sample to emit radiation, or "decay" to a different, more stable form. Scientists have determined the half-life of each radioactive isotope experimentally. Depending on the isotope, the half-life might range from a fraction of a second to millions

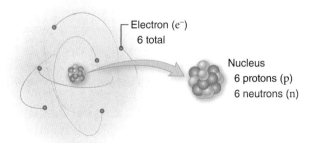

Figure 2.3 Atom Anatomy. The nucleus of an atom is made of protons and neutrons. A cloud of electrons surrounds the nucleus. This example has six protons, so it is a carbon atom.

Electron (e⁻)
6 total

Nucleus
6 protons (p)
6 neutrons (n)

Table 2.2   Selected Uses of Radioactive Isotopes

Application	Example(s)
Kill disease-causing organisms	Radiation from cobalt-60 sterilizes hospital equipment and kills microorganisms on food surfaces.
Tracers	Isotopes used as "tracers" allow biologists to understand life processes. The tracer emits radiation or particles that the researcher can track. • A plant biologist might expose a leaf to radioactive carbon-14, then track the isotope's progress during and after the chemical reactions of photosynthesis. Likewise, a biologist might learn how plants transport and use nutrients by supplying radioactive phosphorus-32 to roots. • A physician might give a patient a radioactive tracer, then track how the isotope moves in the body to search for tumors or examine a physiological process. (This is the basis of a PET scan, which uses fluorine-18 as a tracer.)
Radiometric dating	Archaeologists and paleontologists use the known half-lives of radioactive isotopes to determine the ages of artifacts and fossils (see chapter 12). Isotopes with relatively short half-lives are useful in dating younger materials, whereas those with long half-lives reveal the ages of more ancient items.
Cancer therapy	Directing radiation at a tumor kills cancer cells. Cobalt-60 targets cancer cells deep in the body, whereas phosphorus-32, which emits lower energy radiation, treats skin cancers. A patient being treated for an overactive thyroid gland (Graves disease) or thyroid cancer might ingest radioactive iodine-131, which accumulates in the thyroid and kills some of its cells.

Table 2.3   A Miniglossary of Matter

Term	Definition
Element	A fundamental type of substance
Atom	The smallest unit of an element that retains the characteristics of that element
Atomic number	The number of protons in an atom's nucleus
Mass number	The number of protons plus the number of neutrons in an atom's nucleus
Isotope	Any of the different forms of the same element, distinguished from one another by the number of neutrons in the nucleus
Atomic mass	The average mass of all isotopes of an element

or even billions of years (relatively large samples of isotopes are required to calculate the longest half-lives).

Because radioactive isotopes have the same chemical properties as stable isotopes, they have a variety of uses in science and medicine, some of which are listed in table 2.2. But the same properties that make radioactive isotopes useful can also make them dangerous. Exposure to excessive radiation can lead to radiation sickness, and radiation-induced mutations of a cell's DNA can cause cancer (see chapter 8). The lead-containing "bib" that a dentist places on your chest during mouth X-rays protects you from radiation.

Table 2.3 reviews the terminology of matter.

2.1 | Mastering Concepts

1. Which chemical elements do organisms require in large amounts?
2. Where in an atom are protons, neutrons, and electrons located?
3. What does an element's atomic number indicate?
4. What is the relationship between an atom's mass number and an element's atomic mass?
5. How are different isotopes of the same element different from one another?

2.2 | Chemical Bonds Link Atoms

Like all organisms, you are composed mostly of carbon, hydrogen, oxygen, and nitrogen atoms. But the arrangement of these atoms is not random. Instead, your atoms are organized into molecules (see figure 1.2). A **molecule** is two or more chemically joined atoms.

Some molecules, such as the gases hydrogen (H_2), oxygen (O_2), or nitrogen (N_2), are "diatomic," meaning that they consist of two atoms of the same element. More often, however, the elements in a molecule are different. A **compound** is a molecule composed of two or more different elements. Nitric oxide, described in the chapter opening essay, is a compound consisting of one nitrogen and one oxygen atom. Likewise, water (H_2O) is made of two atoms of hydrogen and one of oxygen. Many large biological compounds, including DNA and proteins, consist of tens of thousands of atoms.

A compound's characteristics can differ strikingly from those of its constituent elements. Consider table salt, sodium chloride. Sodium (Na) is a silvery, highly reactive solid metal, whereas chlorine (Cl) is a yellow, corrosive gas. But when equal numbers of these two atoms combine, the resulting compound forms the familiar white salt crystals that we sprinkle on food—an excellent example of an emergent property. Another example is methane, the main component of natural gas. Its components are carbon (a black sooty solid) and hydrogen (a light, combustible gas). ▶ emergent properties, p. 6

Scientists describe molecules by writing the symbols of their constituent elements and indicating the numbers of atoms of each element in one molecule as subscripts. For example, methane is written CH_4, which denotes one carbon atom attached to four hydrogen atoms. This representation of the atoms in a compound is termed a molecular formula. Table salt's formula is NaCl, that of water is H_2O, and that of the gas carbon dioxide is CO_2.

What forces hold together the atoms that make up each of these molecules? To understand the answer, we must first learn more about how electrons are arranged around the nucleus.

A. Electrons Determine Bonding

Electrons occupy distinct energetic regions around the nucleus. They are constantly in motion, so it is impossible to determine the exact location of any electron at any instant in time. Instead, chemists use the term **orbitals** to describe the most likely location for an electron relative to its nucleus. Each orbital can hold up to two electrons. Consequently, the more electrons in an atom, the more orbitals they occupy.

Electron orbitals exist in several energy levels; an **energy shell** is a group of orbitals that share the same level. The number of orbitals in each shell determines the number of electrons the shell can hold. The lowest energy shell, for example, contains just one orbital and thus holds up to two electrons. The next shell contains up to eight electrons in four orbitals.

Electrons occupy the lowest energy level available to them, starting with the innermost one. As each energy shell fills, any

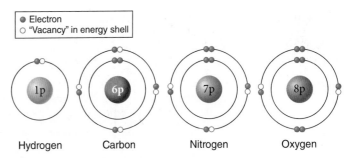

○ Electron
○ "Vacancy" in energy shell

Hydrogen   Carbon   Nitrogen   Oxygen

Figure 2.4 Energy Shells. Shown here are Bohr models of the most common atoms in organisms.

additional electrons must reside in higher energy shells. For example, hydrogen has only one electron in the lowest energy orbital, and helium has two. Carbon has six electrons; two occupy the lowest energy orbital, and four are in the next energy shell. Oxygen, with eight electrons total, has two electrons in the lowest energy orbital and six in three orbitals at the next higher energy level.

We can thus envision any atom's electrons as occupying a series of concentric energy shells, each having a higher energy level than the one inside it. In accordance with this view, electrons often are illustrated as dots moving in two-dimensional circles around a nucleus (**figure 2.4**). These depictions, called Bohr models, are useful for visualizing the interactions between atoms to form bonds. However, most orbitals are not spherical (**figure 2.5**). Bohr models therefore do not accurately portray the three-dimensional structure of atoms.

An atom's **valence shell** is its outermost occupied energy shell. Atoms are most stable when their valence shells are full. The gases helium (He) and neon (Ne), for example, are inert. Because their outermost shells are full, they exist in nature without combining with other atoms.

For most atoms, however, the valence shell is only partially filled. Such an atom will become most stable if its valence-shell "vacancies" fill. To arrive at exactly the right number, atoms share, steal, or donate electrons. The result is a **chemical bond:**

a. **Innermost shell:** up to 2 electrons in one spherical orbital (1*s*).

b. **Second shell:** up to 8 electrons in one spherical orbital (2*s*) and three dumbbell-shaped orbitals (2*p*).

Figure 2.5 Electron Orbitals. Orbitals depict the probability that an electron is in any given location. (a) The first (lowest) energy level contains up to two electrons in one orbital. (b) The second energy level has four orbitals, each containing up to two electrons.

Table 2.4   Chemical Bonds: A Summary

Type	Chemical Basis	Strength	Example
Covalent bond	Two atoms share pairs of electrons.	Strong	O—H bond within water molecule
Ionic bond	One atom donates one or more electrons to another atom, forming oppositely charged ions that attract each other.	Strong but breaks easily in water	Sodium chloride (NaCl)
Hydrogen bond	An atom with a partial negative charge attracts an atom with a partial positive charge. Hydrogen bonds form between adjacent molecules or between different parts of a large molecule.	Weak	Attraction between adjacent water molecules

an attractive force that holds atoms together (table 2.4). The remainder of this section describes three types of chemical bonds that are important in biology.

B. In a Covalent Bond, Atoms Share Electrons

A **covalent bond** forms when two atoms share electrons. The shared electrons travel around both nuclei, strongly connecting the atoms together. Most of the bonds in biological molecules are covalent.

Methane provides an excellent example of how atoms share electrons to fill their valence shells. A carbon atom has six electrons, two of which occupy its innermost shell. That leaves four electrons in its valence shell, which has a capacity of eight. Carbon therefore requires four more electrons to fill its outermost shell. A carbon atom can attain the stable eight-electron configuration by

sharing electrons with four hydrogen atoms, each of which has one electron in its only shell. The resulting molecule is methane, CH_4 (figure 2.6a). Figure 2.6b shows how oxygen and hydrogen form covalent bonds as they combine to produce water.

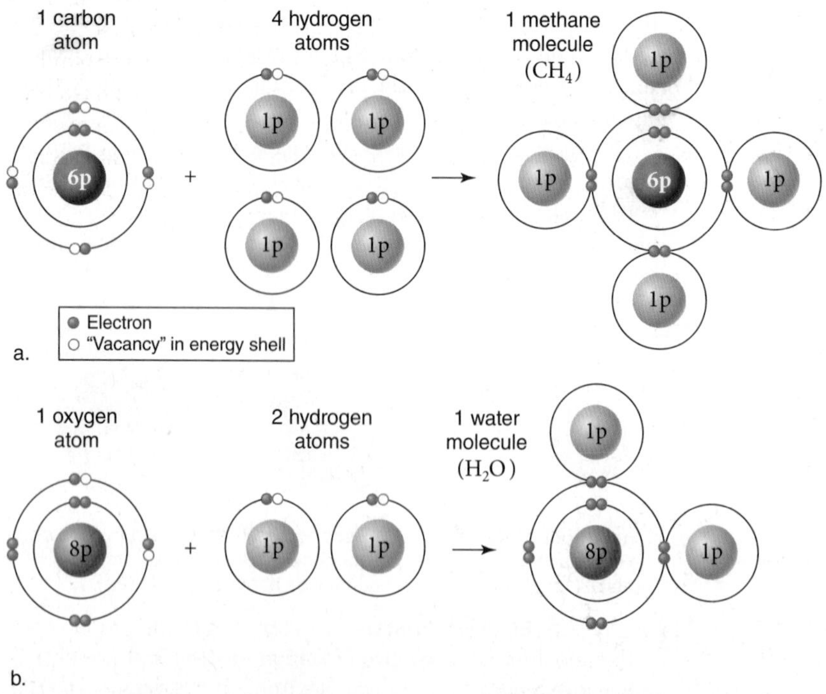

Figure It Out

Use the information in figure 2.2 to predict the number of covalent bonds that nitrogen (N) forms.

Answer: 3.

Covalent bonds are usually depicted as lines between the interacting atoms, with each line representing one bond. Each single bond contains two electrons, one from each atom. Atoms can also share two or three pairs of electrons, forming double and triple covalent bonds, respectively (figure 2.7). The diatomic molecule O_2, for example, has one double bond; a strong triple bond holds together the two atoms in N_2.

Covalent bonding means "sharing," but the partnership is not necessarily equal. **Electronegativity** is a measure of an atom's ability to attract electrons (figure 2.8). Oxygen, for example, strongly attracts electrons. Carbon and hydrogen have low electronegativity relative to oxygen. A **nonpolar covalent bond** is a "bipartisan" union in which both atoms exert approximately equal pull on their shared electrons. A bond between two atoms of the same element is nonpolar; after all, a bond between two identical atoms must be electrically balanced. H_2, N_2, and O_2 are all nonpolar molecules. Carbon and hydrogen atoms have similar electronegativity. A carbon–hydrogen bond is therefore also nonpolar.

A **polar covalent bond,** in contrast, is a lopsided union in which one nucleus exerts a much stronger pull on the shared electrons than does the other nucleus. Polar bonds form whenever a highly electronegative atom such as oxygen shares electrons—unequally—with a less electronegative partner such as carbon or hydrogen. Like a battery, a polar covalent bond has a positive end and a negative end.

Polar covalent bonds are critical to biology. As described in section 2.2D, they are responsible for hydrogen bonds, which in turn help define not only the unique properties of water (see section 2.3) but also the shapes of DNA and proteins (section 2.5).

Figure 2.6   Atoms Share Electrons in Covalent Bonds. (a) Methane (CH_4) contains four covalent bonds, formed when one carbon and four hydrogen atoms complete their outermost shells by sharing electrons. (b) A water molecule (H_2O) has two covalent bonds.

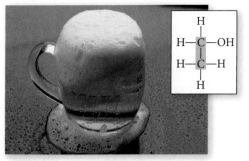

a. Ethanol

b. Ethylene

c. Acetylene

d. Caffeine

Figure 2.7 **Carbon Atoms Form Four Covalent Bonds.** (a) Ethanol is an alcohol built around two singly bonded carbon atoms. (b) Ethylene has two carbon atoms linked by a double bond. Ethylene is a plant hormone that triggers fruit to ripen. (c) Acetylene includes two carbons held by a triple bond. Tremendous heat energy is released when the triple bond in this flammable gas is broken. (d) Caffeine illustrates a double-ring structure of carbon bonded with nitrogen, oxygen, and hydrogen atoms.

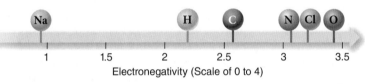

Electronegativity (Scale of 0 to 4)

Figure 2.8 **Unequal Attraction.** Atoms vary widely in their electronegativity, the ability to attract electrons.

C. In an Ionic Bond, One Atom Transfers Electrons to Another Atom

So far, we have seen covalent bonds in which atoms share electrons either equally (nonpolar) or unequally (polar). Is it possible for two atoms to have such different electronegativities that one actually takes one or more of its partner's electrons?

The answer is yes. Recall that an atom is most stable if its valence shell is full. The most electronegative atoms, such as chlorine (Cl), are usually those whose valence shells have only one "vacancy." Likewise, sodium (Na) and other weakly electronegative atoms have only one electron in their outermost shells. Neither chlorine nor sodium would benefit from sharing. Instead, sodium is most stable if it simply releases its extra electron to chlorine, which needs this "scrap" electron to complete its own valence shell.

An **ion** is an atom that has lost or gained electrons. The atom that has lost electrons carries a positive charge, whereas the one that has gained electrons acquires a negative charge. An **ionic bond** results from the electrical attraction between two ions with opposite charges. In general, such bonds form between an atom whose outermost shell is almost empty and one whose valence shell is nearly full. Thus, when sodium donates its electron to an atom of chlorine, the two atoms bond ionically to form NaCl (**figure 2.9**).

Figure 2.9 **Table Salt, an Ionically Bonded Molecule.** (a) A sodium atom (Na) can donate the one electron in its valence shell to a chlorine atom (Cl), which has seven electrons in its outermost shell. Notice that the valence shells of both atoms are now full. The resulting ions (Na$^+$ and Cl$^-$) form the compound sodium chloride (NaCl). (b) The ions that constitute NaCl occur in a repeating pattern that produces salt crystals.

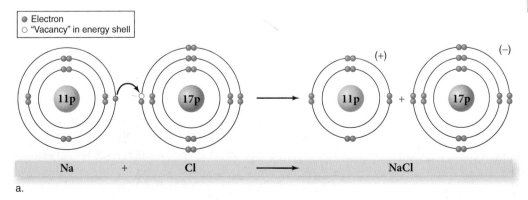

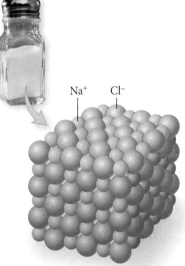

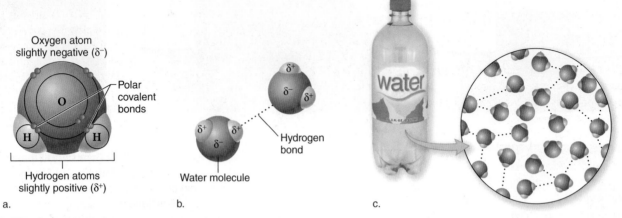

Figure 2.10 Hydrogen Bonds in Water. (a) In water's polar covalent bonds, oxygen attracts the shared electrons more strongly than does hydrogen. The O atom bears a partial negative charge (δ^-), and the H atoms carry partial positive charges (δ^+). (b) The hydrogen bond is the attraction between partial charges on adjacent molecules. (c) In liquid water, many molecules stick to one another with hydrogen bonds.

In NaCl, the most stable configuration of Na^+ and Cl^- is a three-dimensional crystal. Ionic bonds in crystals are strong, as demonstrated by the stability of the salt in your shaker. Those same crystals, however, dissolve when you stir them into water. As described in section 2.3, water molecules pull ionic bonds apart.

Nonpolar covalent bonds, polar covalent bonds, and ionic bonds represent points along a continuum. Two atoms of similar electronegativity share electrons equally in nonpolar covalent bonds. If one atom tugs at the shared electrons much more than the other, the covalent bond is polar. And if one atom is so electronegative that it rips electrons from another atom's valence shell, an ionic bond forms. Notice that the bond type depends on the *difference* in electronegativity, so the same element can participate in different types of bonds. Oxygen, for example, forms nonpolar bonds with itself (as in O_2) and polar bonds with hydrogen (as in H_2O).

Figure It Out

Potassium (K) has electronegativity of 0.82. Using the information in figure 2.8, what type of bond should form between K and Cl?

Answer: Ionic.

D. Partial Charges on Polar Molecules Create Hydrogen Bonds

When a covalent bond is polar, the negatively charged electrons spend more time around the nucleus of the more electronegative atom than around its partner. The "electron-hogging" atom therefore has a partial negative charge (written as "δ^-"), and the less-electronegative partner has an electron "deficit" and a partial positive charge (δ^+).

In a **hydrogen bond,** opposite partial charges on *adjacent molecules*—or within a single large molecule—attract each other. The name comes from the fact that the atom with the partial positive charge is always hydrogen. The atom with the partial negative charge, on the other hand, is a highly electronegative atom such as oxygen or nitrogen.

Water provides the simplest illustration of hydrogen bonds (figure 2.10). Each water molecule has a "boomerang" shape, owing to the oxygen atom's two pairs of unshared valence electrons. Moreover, the two O—H bonds in water are polar, with the nucleus of each oxygen atom attracting the shared electrons more strongly than do the hydrogen nuclei.

Each hydrogen atom in a water molecule therefore has a partial positive charge, which attracts the partial negative charge of the oxygen atom on an adjacent molecule. This attraction is the hydrogen bond. The partial charges on O and H, plus the bent shape, cause water molecules to stick to one another and to some other substances. (This slight stickiness is another example of an emergent property, because it arises from interactions between O and H.)

Hydrogen bonds are relatively weak compared with ionic and covalent bonds. In 1 second, the hydrogen bonds between one water molecule and its nearest neighbors form and re-form some 500 billion times. Even though hydrogen bonds are weak, they account for many of water's unusual characteristics—the subject of section 2.3. In addition, multiple hydrogen bonds help stabilize some large molecules, including proteins and DNA (see section 2.5).

2.2 | Mastering Concepts

1. How are atoms, molecules, and compounds related?
2. How does the number of valence electrons determine an atom's tendency to form bonds?
3. Explain how electronegativity differences between atoms result in nonpolar covalent bonds, polar covalent bonds, and ionic bonds.
4. What is the relationship between polar covalent bonds and hydrogen bonds?

2.3 | Water Is Essential to Life

Although water may seem to be a rather ordinary fluid, it is anything but. The tiny, three-atom water molecule has extraordinary properties that make it essential to all organisms, which explains why the search for life on other planets begins with the search for water. Indeed, life on Earth began in water, and for at least the first 3 billion years of life's history on Earth, all life was aquatic (see chapter 14). It was not until some 475 million years ago, when plants and fungi colonized land, that life could survive without being surrounded by water. Even now, terrestrial organisms still cannot live without it. This section explains some of the properties that make water central to biology.

A. Water Is Cohesive and Adhesive

Hydrogen bonds contribute to a property of water called **cohesion**—the tendency of water molecules to stick together. Without cohesion, water would evaporate instantly in most locations on Earth's surface. Cohesion also contributes to the observation that you can sometimes fill a glass so full that water is above the rim, yet it doesn't flow over the side unless disturbed.

This tendency of a liquid to hold together at its surface is called surface tension, and not all liquids exhibit it. Water has high surface tension because it is cohesive. At the boundary between water and air, the water molecules form hydrogen bonds with neighbors to their sides and below them in the liquid. These bonds tend to hold the surface molecules together, creating a thin "skin" that is strong enough to support small insects without breaking through (figure 2.11).

A related property of water is **adhesion,** the tendency to form hydrogen bonds with other substances. Both adhesion and cohesion are at work when water seemingly defies gravity as it rises in a small-diameter tube (figure 2.12), soaks into a paper towel, or moves from a plant's roots to its highest leaves. This movement depends upon cohesion of water within the plant's

Figure 2.12 **Defying Gravity.** Water adheres to the sides of this glass capillary tube, which has an interior diameter of just 0.5 millimeters. This adhesion, coupled with cohesion among the water molecules, causes the liquid to rise in the tube; the narrower the tube's diameter, the higher the water can rise.

conducting tubes. Water entering roots is drawn up through these tubes as water evaporates from leaf cells. Adhesion to the walls of the conducting tubes also helps lift water to the topmost leaves of trees. ▶ transpiration, p. 479

B. Many Substances Dissolve in Water

Another reason that water is vital to life is that it can dissolve a wide variety of chemicals. To illustrate this process, picture the slow disappearance of table salt as it dissolves in water. Although the salt crystals appear to vanish, the sodium and chloride ions remain. Water molecules surround each ion individually, separating them from one another (figure 2.13).

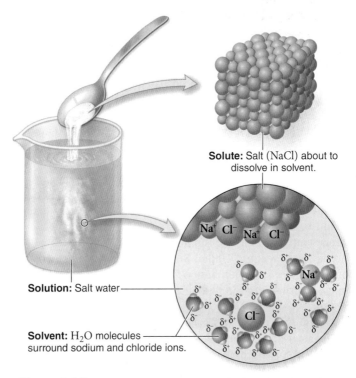

Solute: Salt (NaCl) about to dissolve in solvent.

Solution: Salt water

Na^+ Cl^- Na^+ Cl^-

Na^+

Cl^-

Solvent: H_2O molecules surround sodium and chloride ions.

Figure 2.13 **Solutions Are Mixtures of Molecules.** As salt crystals dissolve, polar water molecules surround each sodium and chloride ion.

Figure 2.11 **Running on Water.** A lightweight body and water-repellent legs allow this water strider to "skate" across a pond without breaking the water's surface tension.

In this example, water is a **solvent:** a chemical in which other substances, called **solutes,** dissolve. A **solution** consists of one or more solutes dissolved in a liquid solvent. In a so-called aqueous solution, water is the solvent. But not all solutions are aqueous. According to the rule "Like dissolves like," polar solvents such as water dissolve polar molecules; similarly, nonpolar solvents dissolve nonpolar substances.

Scientists divide chemicals into two categories, based on their affinity for water. **Hydrophilic** substances are either polar or charged, so they readily dissolve in water (the term literally means "water-loving"). Examples include sugar, salt, and ions. Electrolytes are ions in the body's fluids, and the salty taste of sweat illustrates water's ability to dissolve them. Sports drinks replace not only water but also sodium, potassium, magnesium, and calcium ions that are lost in perspiration during vigorous exercise. Electrolytes are essential to many processes, including heart and nerve function.

Not every substance, however, is water-soluble. Nonpolar molecules made mostly of carbon and hydrogen, such as fats and oils, are called **hydrophobic** ("water-fearing") because they do not dissolve in, or form hydrogen bonds with, water. This is why water alone will not remove grease from hands, dishes, or clothes. Dry cleaning companies use nonpolar solvents to remove oily spots from fabric. Detergents contain molecules that attract both water and fats, so they can dislodge greasy substances and carry the mess down the drain with the wastewater.

C. Water Regulates Temperature

Another unusual property of water is its ability to resist temperature changes. When molecules absorb energy, they move faster. Water's hydrogen bonds tend to counter this molecular movement; as a result, more heat is needed to raise water's temperature than is required for most other liquids, including alcohols. Because an organism's fluids are aqueous solutions, the same effect holds: an organism may encounter considerable heat before its body temperature becomes dangerously high. Likewise, the body cools slowly in cold temperatures.

At a global scale, water's resistance to temperature change explains why coastal climates tend to be mild. People living along the California coast have good weather year-round because the Pacific Ocean's steady temperature helps keep winters warm and summers cool. Far away from the ocean, in the central United States, winters are much colder and summers are much hotter. These differences in local climate contribute to the unique ecosystems that occur in each region (see chapter 38).

Hydrogen bonds also mean that a lot of heat is required to evaporate water. **Evaporation** is the conversion of a liquid into a vapor. When sweat evaporates from skin, individual water molecules break away from the liquid droplet and float into the atmosphere. Surface molecules must absorb energy to escape, and when they do, heat energy is removed from those that remain, drawing heat out of the body—an important part of the mechanism that regulates body temperature.

D. Water Expands as It Freezes

Water's unusual tendency to expand upon freezing also affects life. In liquid water, hydrogen bonds are constantly forming and breaking, and the water molecules are relatively close together. But in an ice crystal, the hydrogen bonds are stable, and the molecules are "locked" into roughly hexagonal shapes. Therefore, the less-dense ice floats on the surface of the denser liquid water below (figure 2.14). This characteristic benefits aquatic organisms. When the air temperature drops, a small amount of water freezes at the pond's surface. This solid cap of ice retains heat in the water below. If ice were to become denser upon freezing, it would sink to the bottom. The lake would then gradually turn to ice from the bottom up, entrapping the organisms that live there.

The formation of ice crystals inside cells, however, can be deadly. The expansion of ice inside a frozen cell can rupture the delicate outer membrane, killing the cell. How, then, do organisms survive in extremely cold weather? Mammals have thick layers of insulating fur and fat that help their bodies stay warm (figure 2.15). Ice fishes that live in the subfreezing waters of the

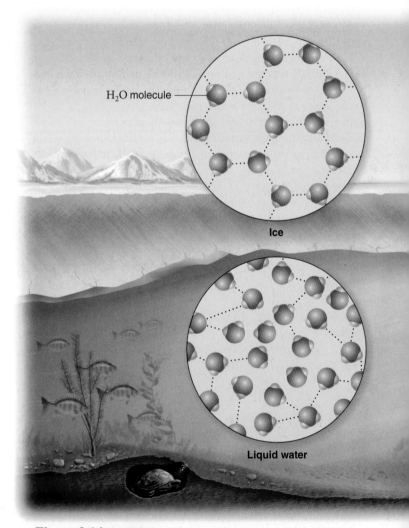

H₂O molecule

Ice

Liquid water

Figure 2.14 Ice Floats. Thanks to hydrogen bonds, ice crystals are less dense than liquid water. Ice therefore floats to the top of a freezing lake.

Figure 2.15 **Harp Seal Pup.** Thick blubber and fur help this young seal survive in an extremely cold habitat.

Antarctic have a different adaptation: they produce antifreeze chemicals that prevent their cells from freezing solid.

E. Water Participates in Life's Chemical Reactions

Life exists because of thousands of simultaneous chemical reactions. In a **chemical reaction,** two or more molecules "swap" their atoms to yield different molecules; that is, some chemical bonds break and new ones form. Chemists depict reactions as equations with the **reactants,** or starting materials, to the left of an arrow; the **products,** or results of the reaction, are listed to the right.

Consider what happens when the methane in natural gas burns inside a heater or gas oven:

$$CH_4 + 2O_2 \longrightarrow CO_2 + 2H_2O$$

$$\text{methane} + \text{oxygen} \longrightarrow \text{carbon dioxide} + \text{water}$$

In words, this says that one methane molecule combines with two oxygen molecules to produce a carbon dioxide molecule and two molecules of water. The bonds of the methane and oxygen molecules have broken, and new bonds formed in the products.

Note that the total number of atoms of each element is the same on either side of the equation: that is, one carbon, four hydrogens, and four oxygens. In chemical reactions, atoms are neither created nor destroyed.

Nearly all of life's chemical reactions occur in the watery solution that fills and bathes cells. Moreover, water is either a reactant in or a product of many of these reactions. In photosynthesis, for example, plants use the sun's energy to assemble food out of just two reactants: carbon dioxide and water (see chapter 5). Section 2.5 describes two other water-related reactions, hydrolysis and dehydration synthesis, that are vital to life.

2.3 | Mastering Concepts

1. How are cohesion and adhesion important to life?
2. What is the difference between hydrophilic and hydrophobic molecules?
3. How does water help an organism regulate its body temperature?
4. How does the density difference between ice and water affect life?
5. What happens in a chemical reaction?
6. How does water participate in the chemistry of life?

2.4 | Organisms Balance Acids and Bases

Life requires abundant water, but it is rarely chemically pure. Surprisingly, one of the most important substances dissolved in water is one of the simplest: H$^+$ ions. Each H$^+$ is a hydrogen atom stripped of its electron; in other words, it is simply a proton. But its simplicity belies its enormous effects on living systems. Too much or too little H$^+$ can ruin the shapes of critical molecules inside cells, rendering them nonfunctional.

One source of H$^+$ is pure water. At any time, about one in a million water molecules spontaneously breaks into two pieces, with one of the hydrogen atoms separating from the rest of the molecule. When this happens, the highly electronegative oxygen atom keeps the electron from the breakaway hydrogen atom. The result is one hydrogen ion (H$^+$) and one hydroxide ion (OH$^-$):

$$H_2O \longrightarrow H^+ + OH^-$$

In pure water, the number of hydrogen ions must exactly equal the number of hydroxide ions. A **neutral** solution likewise has exactly the same amount of H$^+$ as OH$^-$.

Some substances, however, alter this balance. An **acid** is a chemical that adds H$^+$ to a solution, making the concentration of H$^+$ ions exceed the concentration of OH$^-$ ions. Examples include hydrochloric acid (HCl), sulfuric acid (H$_2$SO$_4$), and sour foods such as vinegar and lemon juice. Adding sulfuric acid to pure water releases H$^+$ ions into the solution:

$$H_2SO_4 \longrightarrow 2H^+ + SO_4^{-2}$$

Because no OH$^-$ ions were added at the same time, the balance of H$^+$ to OH$^-$ skews toward extra H$^+$.

A **base** is the opposite of an acid: it makes the concentration of OH$^-$ ions exceed the concentration of H$^+$ ions. Bases work in one of two ways. They come apart to directly add OH$^-$ ions to the solution, or they absorb H$^+$ ions. Either way, the result is the same: the balance between H$^+$ and OH$^-$ shifts toward OH$^-$. Two common household bases are baking soda and sodium hydroxide (NaOH), an ingredient in oven and drain cleaners. When NaOH dissolves in water, it releases OH$^-$ into solution:

$$NaOH \longrightarrow Na^+ + OH^-$$

What happens if a person mixes an acid with a base? The acid releases protons, while the base either absorbs the H$^+$ or releases OH$^-$. Acids and bases therefore neutralize each other.

Both acids and bases are important in everyday life. The tart flavors of yogurt, sour cream, and spoiled milk come from acid-producing bacteria. Your stomach produces hydrochloric acid that kills microbes and activates enzymes that begin the digestion of food. Antacids contain bases that neutralize excess acid, relieving an upset stomach. In the environment, some air pollutants return to Earth as acid precipitation. The acidic rainfall kills plants and aquatic life, and it damages buildings and outdoor sculptures. ▶ acid deposition, p. 808

A. The pH Scale Expresses Acidity or Alkalinity

Scientists use a system of measurement called the **pH scale** to gauge how acidic or basic a solution is. The pH scale ranges from 0 to 14, with 7 representing a neutral solution such as pure water (figure 2.16). An acidic solution has a pH lower than 7, whereas an **alkaline,** or basic, solution has a pH greater than 7. Note that it is

a "reverse scale," in that the higher the H^+ concentration of a solution, the lower its pH. Thus, 0 represents a strongly acidic solution and 14 represents an extremely basic one (low H^+ concentration).

Each unit on the pH scale represents a 10-fold change in H^+ concentration. A solution with a pH of 4 is therefore 10 times more acidic than one with a pH of 5, and it is 100 times more acidic than one with a pH of 6.

All species have optimal pH requirements. Some organisms, such as the bacteria that cause ulcers in human stomachs, are adapted to low-pH environments. In contrast, the normal pH of human blood is 7.35 to 7.45. Extremely shallow breathing or kidney failure can cause the blood's pH to drop below 7. Vomiting, hyperventilating, or taking some types of alkaloid drugs, on the other hand, can raise the blood's pH above 7.8. Straying too far from the normal pH in either direction can be deadly.

Figure It Out

Flask A contains 100 milliliters of a solution with pH 5. After you add 100 ml of solution from Flask B, the pH rises to 7. What was the pH of the solution in Flask B?

Answer: 9.

B. Buffer Systems Regulate pH in Organisms

Maintaining the correct pH of body fluids is critical, yet organisms frequently encounter conditions that could alter their internal pH. How do they maintain homeostasis? The answer lies in **buffer systems,** pairs of weak acids and bases that resist pH changes.

Hydrochloric acid is a strong acid because it releases all of its H^+ when dissolved in water. As you can see in figure 2.16, the pH of pure HCl is 0. A weak acid, in contrast, does not release all of its H^+ into solution. An example is carbonic acid, H_2CO_3, which forms one part of the human body's buffer system:

$$H_2CO_3 \longleftrightarrow H^+ + HCO_3^-$$

carbonic acid   bicarbonate

The dual arrow indicates that the reaction can proceed in either direction, depending on the pH of the fluid. If a base removes H^+ from the solution, the reaction moves to the right to produce more H^+, restoring acidity. Alternatively, if an acid contributes H^+ to the solution, the reaction proceeds to the left and consumes the excess H^+. This action keeps the pH of the solution relatively constant. Carbonic acid is just one of several buffers that maintain the pH of blood at about 7.4.

2.4 | Mastering Concepts

1. How do acids and bases affect a solution's H^+ concentration?
2. How do the values of 0, 7, and 14 relate to the pH scale?
3. How do buffer systems regulate the pH of a fluid?

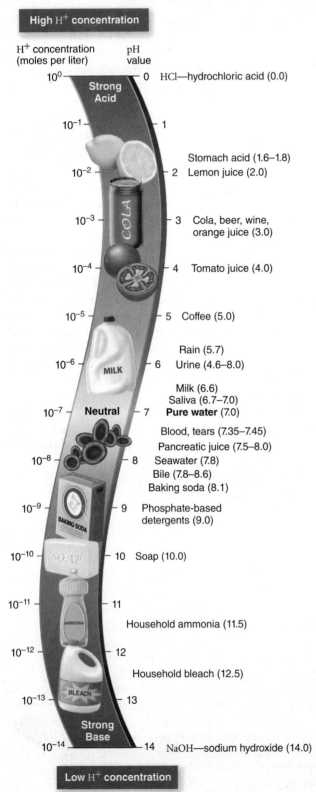

High H^+ concentration

H^+ concentration (moles per liter)	pH value	
10^0	0	HCl—hydrochloric acid (0.0)
		Strong Acid
10^{-1}	1	
10^{-2}	2	Stomach acid (1.6–1.8) Lemon juice (2.0)
10^{-3}	3	Cola, beer, wine, orange juice (3.0)
10^{-4}	4	Tomato juice (4.0)
10^{-5}	5	Coffee (5.0)
10^{-6}	6	Rain (5.7) Urine (4.6–8.0)
10^{-7}	7	Milk (6.6) Saliva (6.7–7.0) **Neutral** **Pure water (7.0)**
10^{-8}	8	Blood, tears (7.35–7.45) Pancreatic juice (7.5–8.0) Seawater (7.8) Bile (7.8–8.6) Baking soda (8.1)
10^{-9}	9	Phosphate-based detergents (9.0)
10^{-10}	10	Soap (10.0)
10^{-11}	11	
10^{-12}	12	Household ammonia (11.5)
10^{-13}	13	Household bleach (12.5)
10^{-14}	14	**Strong Base** NaOH—sodium hydroxide (14.0)

Low H^+ concentration

Figure 2.16 The pH Scale. The pH scale indicates the concentration of hydrogen ions (H^+). The lower the pH, the higher the concentration of free H^+ and the more acidic the solution. Conversely, the higher the pH, the more free hydroxide (OH^-) ions and the more alkaline (basic) the solution.

2.5 | Organic Molecules Generate Life's Form and Function

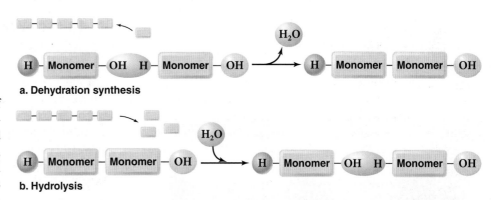

a. Dehydration synthesis

b. Hydrolysis

Figure 2.17 Opposite Reactions. (a) In dehydration synthesis, water is removed and a new covalent bond forms between two monomers. (b) In hydrolysis, water is added when the bond between monomers is broken.

Organisms are composed mostly of water and **organic molecules,** chemical compounds that contain both carbon and hydrogen. As you will learn later in this unit, plants and other autotrophs can produce all the organic molecules they require, whereas heterotrophs—including humans—must obtain them from food.

Life uses a tremendous variety of organic compounds. Organic molecules consisting almost entirely of carbon and hydrogen are called hydrocarbons; methane (CH_4) is the simplest example. Because a carbon atom forms four covalent bonds, however, this element can assemble into much more complex molecules, including long chains, intricate branches, and rings. Many organic compounds also include other elements, such as oxygen, nitrogen, phosphorus, or sulfur.

The four most abundant types of organic molecules in organisms are carbohydrates, lipids, proteins, and nucleic acids. Vitamins are also biologically important organic compounds, but they are required in smaller amounts. Vitamin deficiencies can cause illnesses such as scurvy (vitamin C), beriberi (vitamin B_1), and pellagra (vitamin B_3).

Proteins, nucleic acids, and some carbohydrates all share a property in common with one another: they are chains of small molecular subunits called **monomers.** Linked together, these monomers form **polymers,** just as a train is made of individual railcars.

How does your body produce new muscle proteins and other polymers? Cells use a chemical reaction called dehydration synthesis, also called a condensation reaction, to link the monomers together (**figure 2.17a**). In a **dehydration synthesis** reaction, a protein called an enzyme removes an —OH (hydroxyl group) from one molecule and a hydrogen atom from another, forming H_2O and a new covalent bond between the two smaller components. (The term *dehydration* means that water is lost). By repeating this reaction many times, cells can build extremely large polymers consisting of thousands of monomers.

The reverse reaction also occurs, breaking the covalent bonds that link monomers (figure 2.17b). In **hydrolysis,** enzymes use atoms from water to add a hydroxyl group to one molecule and a hydrogen atom to another (*hydrolysis* means "breaking with water.") Hydrolysis happens in your body when digestive enzymes in your stomach and intestines break down the proteins and other polymers in food.

Table 2.5 reviews the characteristics of the four major types of organic molecules in life. The rest of this section takes a closer look at each one.

Table 2.5 The Macromolecules of Life: A Summary

Type of Molecule	Chemical Structure	Function(s)
Carbohydrates		
Simple sugars	Monosaccharides and disaccharides	Provide quick energy
Complex carbohydrates (cellulose, chitin, starch, glycogen)	Polymers of monosaccharides	Support cells and organisms (cellulose, chitin); store energy (starch, glycogen)
Lipids		
Triglycerides (fats, oils)	Glycerol + 3 fatty acids	Store energy
Phospholipids	Glycerol + 2 fatty acids + phosphate group (see chapter 3)	Form major part of biological membranes
Sterols	Four fused rings, mostly of C and H	Stabilize animal membranes; sex hormones
Waxes	Fatty acids + other hydrocarbons or alcohols	Provide waterproofing
Proteins	Polymers of amino acids	Carry out nearly all the work of the cell (see table 2.6)
Nucleic acids (DNA, RNA)	Polymers of nucleotides	Store and use genetic information and transmit it to the next generation

Carbohydrates (starch); lipids

Proteins; lipids

Carbohydrates (cellulose)

Burning Question

What does it mean when food is "organic" or "natural?"

The word *organic* has multiple meanings. To a chemist, an organic compound contains carbon and hydrogen atoms. Chemically, all food is therefore organic. To a farmer or consumer, however, organic foods are produced according to a defined set of standards.

The U.S. Department of Agriculture (USDA) certifies crops as organically grown if the farmer did not apply pesticides (with few exceptions), petroleum-based fertilizers, or sewage sludge. Organically raised cows, pigs, and chickens cannot receive growth hormones or antibiotics, and they must have access to the outdoors and eat organic feed. In addition, no food labeled "organic" may be genetically engineered or treated with ionizing radiation.

Conventional and organically grown foods differ in several ways in their effects on human health and the environment. Many people believe that pesticide residues on food can damage the nervous system, cause cancer, or disrupt hormones. Pesticides and fertilizers can pollute waterways, and pesticides can kill nontarget organisms. Animals on conventional farms regularly consume antibiotics, which may select for antibiotic-resistant bacteria. Finally, whereas pesticides and fertilizers cost tremendous amounts of energy to produce, organic agriculture emphasizes soil and water conservation and the use of renewable resources.

A natural food may or may not be organic. The term *natural* refers to the way in which foods are processed, not how they are grown. Standards for what constitutes a natural food are fuzzy. The USDA specifies that meat and poultry labeled as natural cannot contain artificial ingredients or added color, but no such standards exist for other foods.

Submit your burning question to:
marielle_hoefnagels@mcgraw-hill.com

A. Carbohydrates Include Simple Sugars and Polysaccharides

Anyone following a "low-carb" diet can recite a list of the foods to avoid: potatoes, pasta, bread, cereal, sugary fruits, and sweets. All are rich in **carbohydrates,** organic molecules that consist of carbon, hydrogen, and oxygen, often in the proportion 1:2:1.

Carbohydrates are the simplest of the four main types of organic compounds, mostly because just a few monomers account for the most common types in cells. The two main groups of carbohydrates are simple sugars and complex carbohydrates.

Simple Sugars   **Monosaccharides,** the smallest carbohydrates, usually contain five or six carbon atoms (figure 2.18a). Monosaccharides with the same number of carbon atoms can differ from one another by how their atoms are bonded. For example, glucose (blood sugar) and fructose (fruit sugar) are both six-carbon monosaccharides with the molecular formula $C_6H_{12}O_6$, but their chemical structures differ.

A **disaccharide** ("two sugars") is two monosaccharides joined by dehydration synthesis. Figure 2.18b shows how sucrose (table sugar) forms when a molecule of glucose bonds to a molecule of fructose. Lactose, or milk sugar, is also a disaccharide.

Together, the sweet-tasting monosaccharides and disaccharides are called sugars, or simple carbohydrates. Their function in cells is to provide a ready source of energy, which is released when their bonds are broken (see chapter 6). Sugarcane sap and sugar beet roots contain abundant sucrose, which the plants use to fuel growth. The disaccharide maltose provides energy in sprouting seeds; beer brewers also use it to promote fermentation.

Oligosaccharides are carbohydrates of intermediate length, consisting of three to 100 monomers. A protein with an attached oligosaccharide is called a glycoprotein ("sugar protein"). Among other functions, glycoproteins on cell surfaces are important in immunity. For example, a person's blood type—A, B, AB, or O—refers to the combination of glycoproteins in the membranes of his or her red blood cells. A transfusion of the "wrong" blood type can trigger a harmful immune reaction. ▸ blood type, p. 605

Complex Carbohydrates   **Polysaccharides** ("many sugars"), also called complex carbohydrates, are huge molecules consisting of hundreds of monosaccharide monomers (figure 2.18c). The most common polysaccharides are cellulose, chitin, starch, and glycogen. They are all long chains of glucose, but they differ from one another by the orientation of the bonds that link the monomers.

Cellulose forms part of plant cell walls. Although it is the most common organic compound in nature, humans cannot digest it. Yet cellulose is an important component of the human diet, making up much of what nutrition labels refer to as "fiber." Cotton fibers, wood, and paper consist largely of cellulose. ▸ plant cell wall, p. 64

Like cellulose, chitin also supports cells. The cell walls of fungi contain chitin, as do the flexible exoskeletons of insects, spiders, and crustaceans. Chitin is the second most common polysaccharide in nature. It resembles a glucose polymer, but a

a. Monosaccharides: simple sugars composed of carbon, hydrogen, and oxygen in the proportions 1:2:1.

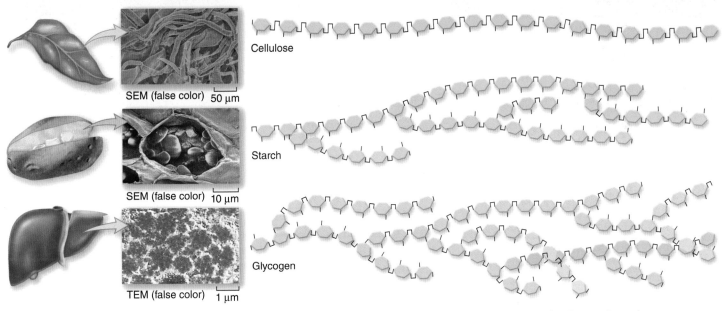

Ribose
$C_5H_{10}O_5$

Glucose
$C_6H_{12}O_6$

Fructose
$C_6H_{12}O_6$

b. Disaccharides: molecules composed of two monosaccharides joined by dehydration synthesis. Hydrolysis converts disaccharides into their component monosaccharides. (The structures of the molecules are simplified to emphasize the joining process.)

H_2O

Dehydration

Hydrolysis

H_2O

Glucose
$C_6H_{12}O_6$

Fructose
$C_6H_{12}O_6$

Sucrose
$C_{12}H_{22}O_{11}$

c. Polysaccharides: complex carbohydrates composed of long chains of simple sugars, usually glucose. Their chemical characteristics are determined by the orientation and location of the bonds between the monomers.

Cellulose

SEM (false color)   50 µm

Starch

SEM (false color)   10 µm

Glycogen

TEM (false color)   1 µm

Figure 2.18 Carbohydrates—Simple and Complex. (a) Monosaccharides are composed of single sugar molecules, such as glucose or fructose. (b) Disaccharides form by dehydration synthesis. In this example, glucose and fructose bond to form sucrose. (c) Polysaccharides are long chains of monosaccharides such as glucose. Different orientations of covalent bonds produce different characteristics in the polymers.

nitrogen-containing group replaces one hydroxyl group in each monomer. Because it is tough, flexible, and biodegradable, chitin is used in the manufacture of surgical thread.

Starch and glycogen have similar structures and functions. Both act as storage molecules that readily break down into their glucose monomers when cells need a burst of energy. Most plants store starch. Potatoes, rice, and wheat are all starchy, high-energy staples in the human diet. Glycogen occurs in animal and fungal cells. In humans, for example, skeletal muscle cells and the liver store energy as glycogen.

B. Lipids Are Hydrophobic and Energy-Rich

Lipids are organic compounds with one property in common: they do not dissolve in water. They are hydrophobic because they contain large areas dominated by nonpolar carbon–carbon and carbon–hydrogen bonds. Unlike carbohydrates, lipids are not polymers consisting of long chains of monomers. Instead, they have extremely diverse chemical structures.

This section discusses several groups of lipids: triglycerides, sterols, and waxes. Another important group, phospholipids, forms the majority of cell membranes; chapter 3 describes them. ▸ phospholipids, p. 54

Triglycerides   A **triglyceride** consists of three long hydrocarbon chains called **fatty acids** bonded to **glycerol,** a three-carbon molecule that forms the triglyceride's backbone. Although triglycerides do not consist of long strings of similar monomers, cells nevertheless use dehydration synthesis to produce them (**figure 2.19**). Each fatty acid has a **carboxyl group,** a carbon atom double-bonded to one oxygen and single-bonded to another oxygen

carrying a hydrogen atom. Enzymes link the —OH from one fatty acid to each of glycerol's three oxygen atoms, yielding three water molecules per triglyceride.

Many dieters try to avoid triglycerides, commonly known as fats. Red meat, butter, margarine, oil, cream, cheese, lard, fried foods, and chocolate are all examples of high-fat foods. Nutrition labels divide these fats into two groups: saturated and unsaturated. The degree of saturation is a measure of a fatty acid's hydrogen content. A **saturated** fatty acid contains all the hydrogens it possibly can. That is, single bonds connect all the carbons, and each carbon has two hydrogens (see the straight chains in figure 2.19). Animal fats are saturated and tend to be solid; bacon fat and butter are two examples. Most nutritionists recommend a diet low in saturated fats, citing their tendency to clog arteries.

A fatty acid is **unsaturated** if it has at least one double bond between carbon atoms. These double bonds cause kinks to form in the fatty acid "tails," producing an oily (liquid) consistency at room temperature. Olive oil, for example, is an unsaturated fat, as are most plant-derived lipids. These fats are healthier than are their saturated counterparts.

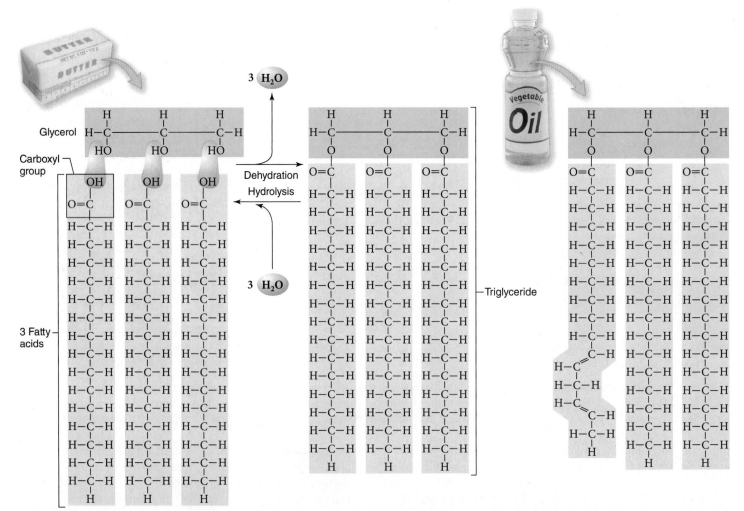

Figure 2.19 **Triglycerides.** A triglyceride consists of three fatty acids bonded to glycerol. In saturated fats such as butter, the fatty acid chains contain only single carbon–carbon bonds. In unsaturated fats, one or more double bonds bend the fatty acid tails, making the lipid more fluid. Vegetable oil is an unsaturated fat.

Apply It Now

Bad and Good Cholesterol

Cholesterol is not water-soluble, so it travels in the bloodstream encased in proteins. The resulting packets of cholesterol and protein are called lipoproteins, and they occur in two varieties.

Low-density lipoprotein (LDL) particles carry cholesterol to the arteries. Excess LDL cholesterol that does not enter cells may accumulate on the inner linings of blood vessels, impeding blood flow and triggering clot formation. High levels of LDL cholesterol (commonly called "bad cholesterol") increase the risk of heart disease.

In contrast, high-density lipoproteins (HDL) carry cholesterol away from the heart and to the liver, which removes it from the bloodstream. High levels of HDL cholesterol ("good cholesterol") promote heart health.

Physicians commonly prescribe statin drugs such as Lipitor to reduce LDL in patients with high cholesterol. These drugs block a liver enzyme that normally helps produce cholesterol, and they improve the liver's ability to get rid of LDL.

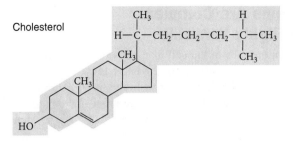

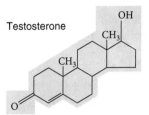

Figure 2.20 **Steroids.** All steroid molecules consist of four interconnected rings. Cholesterol and testosterone are two variations on this theme.

Food chemists have discovered how to turn vegetable oils into solid fats such as some brands of margarine, shortening, and peanut butter. A technique called partial hydrogenation adds hydrogen to the oil to solidify it—in essence, partially saturating a formerly unsaturated fat. One byproduct of this process is **trans fats,** which are unsaturated fats whose fatty acid tails are straight, not kinked. Trans fats are common in fast foods, fried foods, and many snack products, and they raise the risk of heart disease even more than saturated fats. Nutritionists therefore recommend a diet as low as possible in trans fats.

Despite their unhealthful reputation, fats and oils are vital to life. Fat is an excellent energy source, providing more than twice as much energy as equal weights of carbohydrate or protein. Animals must have dietary fat for growth; this requirement explains why human milk is rich in lipids, which fuel the brain's rapid growth during the first 2 years of life. Fats also slow digestion, and they are required for the use of some vitamins and minerals.

Fat cells aggregate as adipose tissue in animals. White adipose tissue forms most of

the fat in human adults, cushioning organs and helping to retain body heat as insulation. Brown adipose tissue releases heat energy that keeps infants and hibernating mammals warm.

Sterols **Sterols** are lipids that have four interconnected carbon rings. Vitamin D and cortisone are examples of sterols, as is cholesterol (figure 2.20). Cholesterol is a key part of animal cell membranes. In addition, animal cells use cholesterol as a starting material to make other lipids, including the sex hormones testosterone and estrogen. ▶ steroid hormones, p. 571

Although cholesterol is essential, an unhealthy diet can easily contribute to cholesterol levels that are too high, increasing the risk of cardiovascular disease (see the Apply It Now box on this page). Because saturated fats stimulate the liver to produce more cholesterol, it is important to limit dietary intake of both saturated fats and cholesterol.

Waxes **Waxes** are fatty acids combined with alcohols or other hydrocarbons, usually forming a stiff, water-repellent material. The waxy compartments of a honeycomb may hold pollen, honey, or larval bees (figure 2.21). In other species, waxes keep fur, feathers, leaves, fruits, and stems waterproof. Jojoba oil, used in cosmetics and shampoos, is unusual in that it is a liquid wax.

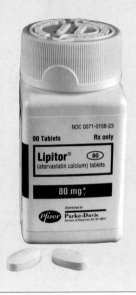

Figure 2.21 **A Waxy Nest.** Beeswax makes up this honeycomb.

C. Proteins Are Complex and Highly Versatile

Proteins do more jobs in the cell than any other type of biological molecule. **Table 2.6** lists a few of the thousands of kinds of proteins in the human body. Proteins literally control all the activities of life, so much so that illness or death can result if even one is missing or faulty. To name one example, the protein insulin controls the amount of sugar in the blood. The failure to produce insulin leads to one form of diabetes, an illness that can be deadly.

Amino Acid Structure and Bonding A

protein is a chain of monomers called **amino acids.** Each amino acid has a central carbon atom bonded to four other atoms or groups of atoms (**figure 2.22a**). One is a hydrogen atom; another is a carboxyl group; a third is an **amino group,** a nitrogen atom single-bonded to two hydrogen atoms (—NH$_2$); and the fourth is a side chain, or **R group,** which can be any of 20 chemical groups.

Organisms use 20 types of amino acids; figure 2.22 shows three of them. (Appendix E includes a complete set of amino acid structures). The R groups distinguish the amino acids from one another, and they have diverse chemical structures. An R group may be as simple as the lone hydrogen atom in glycine or as complex as the two rings of tryptophan. Some R groups are acidic or basic; some are strongly hydrophilic or hydrophobic.

Just as the 26 letters in our alphabet combine to form a nearly infinite number of words in many languages, mixing and matching the 20 amino acids gives rise to an endless diversity of unique proteins. This variety means that proteins have a seemingly limitless array of structures and functions.

The **peptide bond,** which forms by dehydration synthesis, is the covalent bond that links each amino acid to its neighbor (figure 2.22b). Two linked amino acids form a dipeptide; three form a tripeptide. Chains with fewer than 100 amino acids are peptides, and finally, those with 100 or more amino acids are **polypeptides.** A polypeptide is called a protein once it folds into its functional shape; a protein may consist of one or more polypeptide chains.

Where do the amino acids in your own proteins come from? Humans can synthesize most of them from scratch, but the eight "essential" amino acids must come from protein-rich foods such as meat, fish, dairy products, beans, and tofu. Digestive enzymes catalyze the hydrolysis reac-

Table 2.6 Protein Diversity in the Human Body

Protein(s)	Function or Location	Protein(s)	Function or Location
Actin, myosin, dystrophin	Muscle contraction	Growth factors	Promote cell division
Antibodies, cytokines	Immunity	Hemoglobin, myoglobin	Transport and store oxygen
Carbohydrases, lipases	Digestive enzymes*	Insulin, glucagon	Regulate blood glucose level
Casein	Milk protein	Keratin	Builds hair and fingernails
Collagen, elastin	Connective tissue	Transferrin	Transports iron in blood
Colony-stimulating factors	Blood cell formation	Tubulin, actin	Cell movements
DNA and RNA polymerase	Enzymes* required for DNA replication, gene expression	Tumor suppressors	Block cell division
Fibrin, thrombin	Blood clotting		

*Enzymes, discussed further in chapter 4, are proteins that speed chemical reactions. Without enzymes, most of the cell's reactions would proceed much too slowly to sustain life.

a.

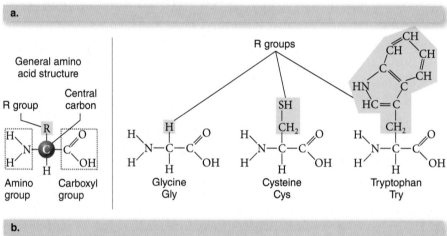

b.

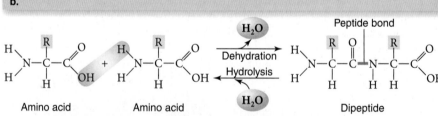

Figure 2.22 Amino Acids. (a) An amino acid is composed of an amino group, a carboxyl group, and one of 20 R groups attached to a central carbon atom. Three examples appear here. (b) A peptide bond forms by dehydration synthesis, joining two amino acids together.

tion that breaks peptide bonds and releases amino acids from proteins in food. The body then uses these monomers to build its own polypeptides.

Protein Folding   Unlike polysaccharides, most proteins do not exist as long chains inside cells. Instead, the peptide chain folds into a unique three-dimensional structure determined by the order and kinds of amino acids. Biologists describe the conformation of a protein at four levels (figure 2.23):

- **Primary (1°) structure:** The amino acid sequence of a polypeptide chain. This sequence determines all subsequent structural levels.

- **Secondary (2°) structure:** A "substructure" with a defined shape, resulting from hydrogen bonds between parts of the polypeptide. These interactions fold the chain of amino acids into coils, sheets, and loops. Each protein can have multiple types of secondary structure.

- **Tertiary (3°) structure:** The overall shape of a polypeptide, arising primarily through interactions between R groups and water. Inside a cell, water molecules surround each

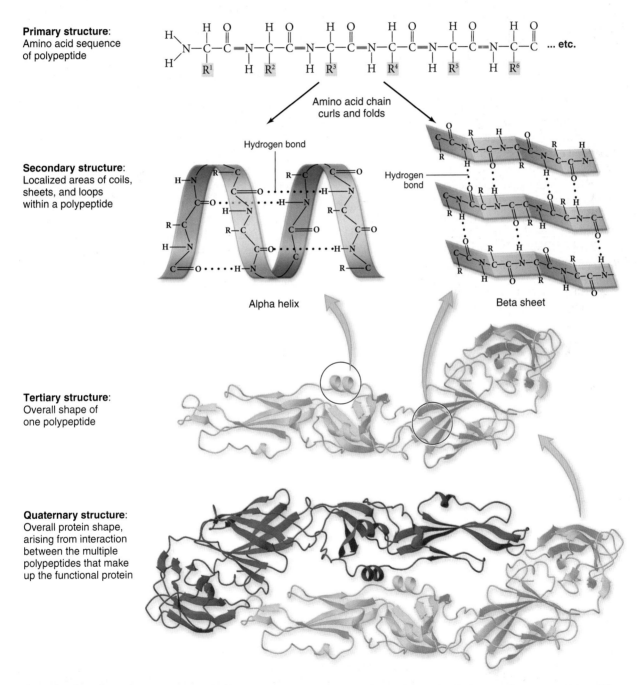

Figure 2.23 Four Levels of Protein Structure. The amino acid sequence of a polypeptide forms the primary structure, while hydrogen bonds create secondary structures such as helices and sheets. The tertiary structure is the overall three-dimensional shape of a protein. The interaction of multiple polypeptides forms the protein's quaternary structure.

polypeptide. The hydrophobic R groups move away from water toward the protein's interior. In addition, hydrogen bonds and (rarely) ionic bonds form between the peptide backbone and some R groups. Covalent bonds between sulfur atoms in some R groups further stabilize the structure. These disulfide bonds are abundant in structural proteins such as keratin, which forms hair, scales, beaks, feathers, wool, and hooves (figure 2.24).

- **Quaternary (4°) structure:** The shape arising from interactions between multiple polypeptide subunits of the same protein. The protein in figure 2.23 consists of two polypeptides; similarly, the oxygen-toting blood protein hemoglobin is composed of four polypeptide chains.

As detailed in chapter 7, an organism's genetic code specifies the amino acid sequence of each protein. A genetic mutation may therefore change a protein's primary structure. The protein's secondary, tertiary, and quaternary structures all depend upon the primary structure. Genetic mutations are often harmful because they result in misfolded, nonfunctional proteins.

Many biologists devote their careers to deducing protein structures, in part because the research has so many practical applications. Misfolded infectious proteins called prions, for instance, cause mad cow disease. Knowledge of protein structure can also aid in the treatment of infectious disease. If scientists can determine the shape of a protein unique to the organism that causes malaria, for example, they may be able to use that information to create effective new drugs with few side effects. Some consumer products also exploit protein shape. "Permanent wave" solutions and hair straighteners break disulfide bonds in keratin. The bonds return once the hair is in the desired conformation.

Denaturation: Loss of Function   A protein's function depends on its overall shape. A digestive enzyme, for example, is a protein that holds a large food molecule in just the right way to break the nutrient apart. An antibody protein binds to a very specific part of a molecule on the surface of a bacterium. Muscle proteins form long, aligned fibers that slide past one another, shortening their length to create muscle contractions.

Proteins are therefore vulnerable to conditions that alter their shapes. Heat, excessive salt, or the wrong pH can disrupt the hydrogen bonds that maintain the protein's secondary and tertiary structures. The protein is **denatured** if its structure is modified enough to destroy its function. As an example, consider what happens to an egg as it cooks. Proteins unfold in the heat, then clump and refold randomly as the once-clear egg protein turns solid white. Similarly, fish turns from translucent to opaque as it cooks.

Most denatured proteins will not renature; there is no way to uncook an egg. Gentle denaturation, however, is sometimes reversible. Edible gelatin, for example, is a protein derived from pig and cow collagen. Short chains of amino acids in powdered gelatin wrap around each other, forming minuscule "ropes." When a cook dissolves the powder in hot water, the ropes unwind. As the gelatin cools, some of the ropes re-form. Pockets of liquid trapped within the tangled strands create a jellylike texture in the finished product.

As different as carbohydrates, lipids, and proteins are, food chemists have discovered ways to use all three substances to make artificial sweeteners and fat substitutes. The Apply It Now box on page 39 describes how they do it.

a.   b.   c.

Figure 2.24   A Gallery of Keratin-Based Structures. Keratin forms (a) a bird's beak and feathers, (b) human hair, and (c) a ram's horns and fur.

Apply It Now

Sugar Substitutes and Fake Fats

Many weight-conscious people turn to artificial sweeteners and fat substitutes to cut calories without sacrificing their favorite foods. Chemically, how do these sugar and fat replacements compare with the real thing?

Artificial Sweeteners

Table sugar delivers about 4 Calories per gram. (As described in chapter 4, a nutritional Calorie—with a capital C—is a measure of energy that represents 1000 calories.) The use of artificial sweeteners reduces calorie intake, but not always because the additives are truly calorie-free. Instead, most are hundreds of times sweeter-tasting than sugar, so a tiny amount of artificial sweetener achieves the same effect as a teaspoon of sugar. A few popular artificial sweeteners include:

- **Saccharin** (sold as Sweet'n Low and Sugar Twin): This sweetener, which has only 1/32 of a Calorie per gram, was originally derived from coal tar in 1879. It consists of a double-ring structure that includes nitrogen and sulfur. (Saccharin's eventful history as a food additive is the topic of the Apply It Now box in chapter 1.)

Saccharin

- **Aspartame** (sold as NutraSweet and Equal): Surprisingly, aspartame's chemical structure does not resemble sugar. Instead, it consists of two amino acids, phenylalanine and aspartic acid. Like sugar, it delivers about 4 Calories per gram, but it is about 200 times sweeter than sugar, so less is needed.

- **Sucralose** (sold as Splenda): This sweetener is a close relative of sucrose, except that three chlorine (Cl) atoms replace three of sucrose's hydroxyl groups. Sucralose is about 600 times sweeter than sugar, but the body digests little if any of it, so it is virtually calorie-free.

Sucralose

- **Acesulfame-K** (sold as Sweet One and Sunett): Structurally similar to saccharin, acesulfame-K is about 200 times sweeter than sugar. However, it is essentially calorie-free because the body cannot absorb or use it as an energy source. (The "K" in the name stands for potassium, since acesulfame is sold as a potassium salt.)

Acesulfame-K (potassium salt)

Fat Substitutes

Because fat is so calorie-dense (about 9 Calories per gram), cutting fat is a quick way to trim calories from the diet. Excess dietary fat can be harmful, leading to weight gain and increasing the risk of heart disease and cancer. It is important to remember, however, that some dietary fat is essential for good health. Fat aids in the absorption of some vitamins and provides fatty acids that human bodies cannot produce. Fats also lend foods taste and consistency.

Fat substitutes are chemically diverse. The most common ones are based on carbohydrates, proteins, or even fats, and a careful reading of nutrition labels will reveal their presence in many processed foods.

- **Carbohydrate-based fat substitutes:** Modified food starches, dextrins, guar gum, pectin, and cellulose gels are all derived from polysaccharides, and they all mimic fat's "mouth feel" by absorbing water to form a gel. Depending on whether they are indigestible (cellulose) or digestible (starches), these fat substitutes deliver 0 to 4 Calories per gram. They cannot be used to fry foods.

- **Protein-based fat substitutes:** These food additives are derived from egg whites or whey (the watery part of milk). When ground into "microparticles," these proteins mimic fat's texture as they slide by each other in the mouth. Protein-based fat substitutes deliver about 4 Calories per gram, and they cannot be used in frying.

- **Fat-based fat substitute:** Olestra (marketed as Olean) is a hybrid molecule that combines a central sucrose molecule with six to eight fatty acids. Its chief advantage is that it tastes and behaves like fat—even for frying. Olestra is currently approved only for savory snacks such as chips. It is indigestible and calorie-free, but some people have expressed concern that olestra removes fat-soluble vitamins as it passes through the digestive tract. Others have publicized its reputed laxative properties. Most people, however, do not experience problems after eating small quantities of olestra.

Olestra

Sugar and fat substitutes can be useful for people who cannot—or do not wish to—eat much of the real thing. But nutritionists warn that these food additives should not take the place of a healthy diet and moderate eating habits.

Figure 2.25 Nucleotides. (a) A nucleotide consists of a sugar, one or more phosphate groups, and one of several nitrogenous bases. In DNA, the sugar is deoxyribose, whereas RNA nucleotides contain ribose. In addition, the base thymine appears only in DNA; uracil is only in RNA. (b) Dehydration synthesis joins two nucleotides together.

Phosphate group

Sugar (Deoxyribose)

Nitrogenous base (Guanine)

Adenine (A)   Cytosine (C)   Guanine (G)   Thymine (T)   Uracil (U)

a.

Dehydration
Hydrolysis
H_2O

b.

D. Nucleic Acids Store and Transmit Genetic Information

How does a cell "know" which amino acids to string together to form a particular protein? The answer is that each protein's primary structure is encoded in the sequence of a **nucleic acid,** a polymer consisting of monomers called nucleotides. Cells contain two types of nucleic acids, **deoxyribonucleic acid (DNA)** and **ribonucleic acid (RNA).**

Each **nucleotide** monomer consists of three components (figure 2.25a). At the center is a five-carbon sugar—ribose in RNA and deoxyribose in DNA. Attached to one of the sugar's carbon atoms is at least one phosphate group (PO_4). Attached to the opposite side of the sugar is a **nitrogenous base:** adenine (A), guanine (G), thymine (T), cytosine (C), or uracil (U). DNA contains A, C, G, and T, whereas RNA contains A, C, G, and U.

Dehydration synthesis links nucleotides together (figure 2.25b). In this reaction, a covalent bond forms between the sugar of one nucleotide and the phosphate group of its neighbor.

A DNA polymer is a double helix that resembles a spiral staircase. Alternating sugars and phosphates form the rails of the staircase, and nitrogenous bases form the rungs (figure 2.26). Hydrogen bonds between the bases hold the two strands of nucleotides together: A with T, C with G. The two strands are therefore complementary, or "opposites," of each other. Because of complementary base pairing, one strand of DNA contains the information for the other, providing a mechanism for the molecule to replicate.
▶ DNA replication, p. 154

DNA's main function is to store genetic information. Every organism inherits DNA from its parents (or parent, in the case of asexual reproduction). Slight changes in DNA from generation to generation, coupled with natural selection, account for many of the evolutionary changes that have occurred throughout life's history.

Unlike DNA, RNA is single-stranded (see figure 2.26). One function of RNA is to enable cells to use the protein-encoding information in DNA (see chapter 7). In addition, a modified RNA nucleotide, adenosine triphosphate (ATP), carries the energy that cells use in many biological functions. ▶ ATP, p. 76

2.5 | Mastering Concepts

1. What is the relationship between hydrolysis and dehydration synthesis?
2. Describe the monomers that form polysaccharides, proteins, and nucleic acids.
3. List examples of carbohydrates, lipids, proteins, and nucleic acids, and name the function of each.
4. What are the components of a triglyceride?
5. What is the significance of a protein's shape, and how can that shape be destroyed?
6. What are some differences between RNA and DNA?

Figure 2.26 Nucleic Acids: DNA and RNA. DNA consists of two strands of nucleotides entwined to form a double-helix shape. RNA is usually single-stranded.

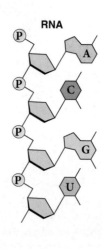

RNA

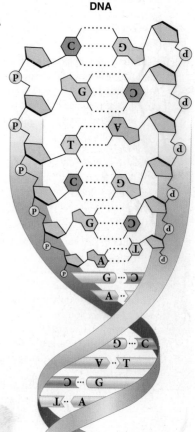

DNA

2.6 | Investigating Life: E. T. and the Origin of Life

Life's chemistry reflects life's unity: the same elements make up the same types of molecules in all organisms. This observation is consistent with evolution by common descent. After all, if one or a few ancient organisms gave rise to all of life's diversity, then contemporary life should reflect the chemistry of that ancestor.

Evolution accounts for the diversity of species on Earth, but it cannot explain how life started in the first place. We know that cells require carbohydrates, lipids, proteins, and nucleic acids. At the dawn of life on Earth, where did those first critical ingredients come from?

One hypothesis is that meteorites or comets carrying organic molecules seeded Earth with the precursors of life. Evidence for this explanation comes from a meteorite that fell to Earth near an Australian town called Murchison in 1969 (figure 2.27). Researchers discovered that the meteorite contained amino acids and other organic compounds. But an obvious question immediately arose: Did these molecules contaminate the meteorite after it fell, or did they really come from space?

The atoms that make up organic molecules can help answer this question. Carbon and nitrogen are two of the most abundant elements in life, and each has multiple isotopes (see section 2.1C). On Earth, 98.89% of C atoms are ^{12}C, with six protons and six neutrons. Just 1.11% of Earthly carbon atoms have seven neutrons (^{13}C). Likewise, 99.63% of nitrogen atoms on Earth are ^{14}N, and 0.37% are ^{15}N. But the heavier isotopes, ^{13}C and ^{15}N, are slightly more abundant in materials from outer space than they are on Earth. Scientists can use this small difference to distinguish between terrestrial and extraterrestrial materials.

Michael Engel, from the University of Oklahoma, and the University of Virginia's Stephen Macko tested the hypothesis that the Murchison meteorite's amino acids are extraterrestrial. They chemically extracted amino acids from a meteorite stone. Next, they measured the amounts of ^{14}N and ^{15}N in the amino acids. Engel and Macko predicted that the Murchison amino acids should be enriched in ^{15}N.

In another study, Zita Martins of Imperial College in London and an international group of colleagues analyzed the Murchison meteorite for the nucleotide bases that characterize RNA and DNA. They found uracil in the meteorite, along with xanthine, a base that today's cells need to produce thymine. Martins and her team then measured the amounts of ^{13}C and ^{12}C in the bases. Like Engel and Macko, the Martins group predicted that bases from the meteorite should contain more ^{13}C than do terrestrial organic molecules.

Table 2.7 shows the results of both studies. In this table, a positive number means that a sample contained more ^{15}N or ^{13}C than a known standard, and a negative number means that the sample contained less. As predicted, the amino acids and bases from the Murchison meteorite were enriched in both ^{15}N and ^{13}C relative to the same molecules on Earth. These results support the hypothesis that the Murchison meteorite carried amino acids, uracil, and xanthine to Earth.

Does this mean that life (or its key molecules) originally came from outer space? Not necessarily. As you will see in chapter 14, life's organic molecules may have arisen by chemical processes occurring entirely on Earth. We may never know how life started, but it is intriguing to think that some of its key ingredients may literally have fallen from the sky.

Table 2.7 Heavy Isotopes in Organic Molecules

Amino Acids from Murchison Meteorite	^{15}N (parts per thousand) Relative to Standard	Bases from Murchison Meteorite	^{13}C (parts per thousand) Relative to Standard
Glycine	+37	Uracil	+44.5
Alanine	+57	Xanthine	+37.7
Aspartic acid	+61		
Glutamic acid	+58		
Typical terrestrial organic compounds	−5 to +10	Typical terrestrial organic compounds	−110 to 0

Engel, M. H., and S. A. Macko. 1997. Isotopic evidence for extraterrestrial nonracemic amino acids in the Murchison meteorite. *Nature,* vol. 389, pages 265–268.

Martins, Zita, Oliver Botta, Marilyn L. Fogel, and six coauthors. 2008. Extraterrestrial nucleobases in the Murchison meteorite. *Earth and Planetary Science Letters,* vol. 270, pages 130–136.

2.6 | Mastering Concepts

1. What question were these researchers trying to answer?
2. Why are ^{15}N and ^{13}C called "heavy" isotopes? How are they different from ^{14}N and ^{12}C?
3. Both groups of researchers collected samples from the meteorite's interior. Why does the sample location matter?
4. How would the results have differed if the amino acids and bases were contaminants acquired after the meteorite fell to Earth?

Figure 2.27

Murchison Meteorite. When the Murchison meteorite struck Earth in 1969, it scattered rocks such as this one over a large area near Murchison, Australia.

Chapter Summary

2.1 | Atoms Make Up All Matter

- All **matter** can be broken down into pure substances called **elements**.

A. Elements Are Fundamental Types of Matter

- **Bulk elements** are essential to life in large quantities, and **trace elements** are required in smaller amounts. A **mineral** is any essential element other than C, H, O, or N.

B. Atoms Are Particles of Elements

- An **atom** is the smallest unit of an element. Positively charged **protons** and neutral **neutrons** form the **nucleus.** The negatively charged, much smaller **electrons** circle the nucleus.
- Elements are organized in the **periodic table** according to **atomic number** (the number of protons).
- An **ion** is an atom that gains or loses electrons.

C. Isotopes Have Different Numbers of Neutrons

- **Isotopes** of an element differ by the number of neutrons. A **radioactive** isotope is unstable.
- An element's **atomic mass** reflects the average **mass number** of all isotopes, weighted by the proportions in which they naturally occur.

2.2 | Chemical Bonds Link Atoms

- A **molecule** is two or more atoms joined together; if they are of different elements, the molecule is called a **compound.**

A. Electrons Determine Bonding

- Electrons move constantly; they are most likely to occur in volumes of space called **orbitals.** Orbitals are grouped into **energy shells.**
- An atom's tendency to fill its **valence shell** with electrons drives it to form **chemical bonds** with other atoms.

B. In a Covalent Bond, Atoms Share Electrons

- **Covalent bonds** form between atoms that can fill their valence shells by sharing one or more pairs of electrons.
- Atoms in a **nonpolar covalent bond** share electrons equally. **Electronegative** atoms in covalent bonds attract electrons away from less electronegative atoms, forming **polar covalent bonds.**

C. In an Ionic Bond, One Atom Transfers Electrons to Another Atom

- An **ionic bond** is an attraction between two oppositely charged ions, which form when one atom strips one or more electrons from another atom.

D. Partial Charges on Polar Molecules Create Hydrogen Bonds

- **Hydrogen bonds** result from the attraction between opposite partial charges on adjacent molecules or between oppositely charged parts of a large molecule.

2.3 | Water Is Essential to Life

A. Water Is Cohesive and Adhesive

- Water is **cohesive** and **adhesive,** sticking to itself and other materials.

B. Many Substances Dissolve in Water

- A **solution** consists of a **solute** dissolved in a **solvent.**
- Water dissolves **hydrophilic** (polar and charged) substances but not **hydrophobic** (nonpolar) substances.

C. Water Regulates Temperature

- Water helps regulate temperature in organisms because it resists both temperature change and **evaporation.**
- Large bodies of water help keep coastal climates mild.

D. Water Expands as It Freezes

- Ice is less dense than liquid water. As a result, ice floats on the surface of lakes and oceans.
- Ice crystals can damage living cells.

E. Water Participates in Life's Chemical Reactions

- In a **chemical reaction,** the **products** are different from the **reactants.**
- Most biochemical reactions occur in a watery solution.

2.4 | Organisms Balance Acids and Bases

- In pure water, the numbers of H^+ and OH^- in water are equal, and the solution is **neutral.** An **acid** adds H^+ to a solution, and a **base** adds OH^- or removes H^+.

A. The pH Scale Expresses Acidity or Alkalinity

- The **pH scale** measures H^+ concentration. Pure water has a pH of 7, acidic solutions have a pH below 7, and an **alkaline** solution has a pH between 7 and 14.

B. Buffer Systems Regulate pH in Organisms

- **Buffers** consist of weak acid–base pairs that maintain the pH ranges of body fluids.

2.5 | Organic Molecules Generate Life's Form and Function

- Many **organic molecules** consist of small subunits called **monomers,** which link together to form **polymers. Dehydration synthesis** is the chemical reaction that joins monomers together, releasing a water molecule.
- The **hydrolysis** reaction uses water to break polymers into monomers.

A. Carbohydrates Include Simple Sugars and Polysaccharides

- **Carbohydrates** consist of carbon, hydrogen, and oxygen in the proportions 1:2:1.
- **Monosaccharides** are single-molecule sugars such as glucose. Two bonded monosaccharides form a **disaccharide.** These simple sugars provide quick energy.
- **Oligosaccharides** are short chains of three to 100 monosaccharides.
- **Polysaccharides** are complex carbohydrates consisting of hundreds of monosaccharides. They provide support and store energy.

B. Lipids Are Hydrophobic and Energy-Rich

- **Lipids** are diverse hydrophobic compounds consisting mainly of carbon and hydrogen.
- **Triglycerides** (fats and oils) consist of **glycerol** and three **fatty acids,** which may be **saturated** (no double bonds) or **unsaturated** (at least one double bond). They store energy, slow digestion, cushion organs, and preserve body heat.
- **Sterols,** including cholesterol and sex hormones, are lipids consisting of four carbon- and hydrogen-rich rings.
- **Waxes** are hard, waterproof coverings made of fatty acids combined with other molecules.

C. Proteins Are Complex and Highly Versatile

- **Proteins** consist of **amino acids,** which join into **polypeptides** by forming **peptide bonds** through dehydration synthesis.
- A protein's three-dimensional shape is vital to its function. A **denatured** protein has a ruined shape.
- Proteins have a great variety of functions, participating in all the work of the cell.

D. Nucleic Acids Store and Transmit Genetic Information

- **Nucleic acids,** including **DNA** and **RNA,** are polymers consisting of **nucleotides.**
- DNA carries genetic information and transmits it from generation to generation. RNA copies the information, enabling the cell to make proteins.

2.6 | Investigating Life: E. T. and the Origin of Life

- Amino acids and nucleotide bases extracted from the Murchison meteorite suggest that early life's organic molecules may have come from outer space.

Multiple Choice Questions

1. The *mass number* of an atom represents the total number of
   a. electrons.
   b. protons.
   c. neutrons.
   d. protons + neutrons.

2. The atomic number of the element neon (Ne) is 10. How many electrons does a neutral atom of neon contain?
   a. 5
   b. 10
   c. 20
   d. It can't be determined from this information.

3. A *covalent bond* forms when
   a. electrons are present in a valence shell.
   b. a valence electron is removed from one atom and added to another.
   c. a pair of valence electrons is shared between two atoms.
   d. the electronegativity of one atom is greater than that of another atom.

4. The atomic number of silicon (Si) is 14. Use the idea of energy shells to predict the number of covalent bonds that Si could form.
   a. 2
   b. 3
   c. 4
   d. 8

5. An *ionic bond* forms when
   a. an electrical attraction occurs between two atoms of different charge.
   b. a nonpolar attraction is formed between two atoms.
   c. a valence electron is shared between two atoms.
   d. two atoms have similar electronegativity.

6. A hydrophilic substance is one that can
   a. form covalent bonds with hydrogen.
   b. dissolve in water.
   c. buffer a solution.
   d. mix with nonpolar solvents.

7. What type of chemical bond is being broken when methane is burned?
   a. Ionic
   b. Hydrogen
   c. Polar covalent
   d. Nonpolar covalent

8. What type of chemical bond forms during a dehydration synthesis reaction?
   a. Covalent
   b. Ionic
   c. Hydrogen
   d. Polymer

9. A sugar is an example of a _____ molecule, whereas DNA is a _____ .
   a. protein; nucleic acid
   b. nucleic acid; lipid
   c. lipid; protein
   d. carbohydrate; nucleic acid

10. The shape of a protein is determined by
    a. the sequence of amino acids.
    b. chemical bonds between amino acids.
    c. temperature and pH.
    d. All of the above are correct.

Write It Out

1. Define the following terms: *atom, element, molecule, compound, isotope,* and *ion.*

2. Consider the following atomic numbers: oxygen (O) = 8; fluorine (F) = 9; neon (Ne) = 10; magnesium (Mg) = 12. Build a Bohr model of each atom, and then predict how many bonds each atom should form.

3. Distinguish between nonpolar covalent bonds, polar covalent bonds, and ionic bonds.

4. If oxygen is highly electronegative, why is a covalent bond between two oxygen atoms considered nonpolar?

5. Can nonpolar molecules such as CH_4 participate in hydrogen bonds? Why or why not?

6. Define *solute, solvent,* and *solution.*

7. Explain why each of the following properties of water is essential to life: cohesion, adhesion, ability to dissolve solutes, resistance to temperature change.

8. Using your knowledge of the properties of water, explain the quote "Hydrogen bonds sank the *Titanic.*"

9. Why are buffer systems important in organisms?

10. Compare and contrast the chemical structures and functions of carbohydrates, lipids, proteins, and nucleic acids.

11. How is an amino acid's R group analogous to a nucleotide's nitrogenous base?

12. Pickles and several other foods are preserved in acids such as vinegar. Why is an acid a good preservative? (*Hint:* Consider the effect of acids on protein shape.)

13. Complete and explain the following analogy: a protein is to a knitted sweater as a denatured protein is to a ____.

14. A topping for ice cream contains fructose, hydrogenated soybean oil, salt, and cellulose. What types of chemicals are in it?

15. Using information in "Sugar Substitutes and Fake Fats" on page 39 and the amino acid structures in appendix E, draw the dipeptide called aspartame (NutraSweet).

Pull It Together

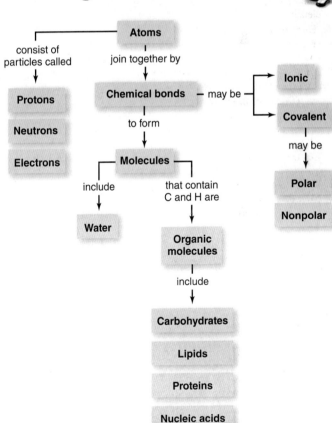

1. How do ions and isotopes fit into this concept map?

2. Add *covalent bonds, ionic bonds,* and *hydrogen bonds* to this concept map.

3. Besides water, what are other examples of molecules that are essential to life?

4. Add *monomers, polymers, dehydration synthesis,* and *hydrolysis* to this concept map.

Since 1983, the Komen Race for the Cure has raised money for the fight against breast cancer.

BIOLOGY

Enhance your study of
this chapter with practice quizzes,
animations and videos, answer keys,
and downloadable study tools.
www.mhhe.com/hoefnagels

Cancer Cells: A Tale of Two Drugs

CANCER IS A FAMILY OF DISEASES IN WHICH A PERSON'S OWN CELLS MULTIPLY OUT OF CONTROL. As described in chapter 8, the abnormal cells form tumors or uncontrolled populations of cells that invade nearby tissues and spread to other parts of the body. The illness can turn deadly if the cancerous cells destroy normal tissues or interfere with vital body functions.

Preventing and curing cancer are compelling reasons to study cell biology. For example, biologists have studied the structural and chemical differences between cancer cells and normal cells. Their findings have yielded spectacular new treatments that exploit some of these differences. Research into the chemical signals that control cell division has been especially fruitful, producing new drugs that inhibit only abnormal cells.

One example is Herceptin (trastuzumab). This drug, which was created to treat some forms of breast cancer, got its name from its target: HER2, a receptor protein on the surface of breast cells. The HER2 receptor binds to a molecule that stimulates the cell to divide. A normal breast cell has thousands of HER2 receptors, but cells of one form of breast cancer have many millions. Cells with too many HER2 receptors divide and spread rapidly. Herceptin prevents this by binding to HER2 receptors.

Another successful drug is Gleevec (imatinib), which treats some forms of leukemia and gastrointestinal cancers. Leukemia is a disease in which the body produces many abnormal white blood cells. In a type of leukemia called chronic myeloid leukemia, a genetic mutation causes cells to produce an abnormal protein. This protein prompts the body to produce cancerous cells. Gleevec blocks the protein, selectively slowing division of the abnormal cells without harming normal cells.

Both Herceptin and Gleevec took decades to develop. Each interferes with a feature that is unique to the cancer cells, producing fewer side effects than older treatments that destroy healthy cells, too. These drugs owe their success to generations of cell biologists who painstakingly documented the structures in and on cells. These cellular parts are the subject of this chapter.

Learning Outline

Learn How to Learn
Bite-Sized Pieces

Many students think they need to read a whole chapter in one sitting. Instead, try working through one topic at a time. Read just one section of the chapter, and compare it to your class notes. Think of each chapter as a meal: you eat a sandwich one bite at a time, so why not tackle biology the same way?

3.1 | Cells Are the Units of Life

A human, a hyacinth, a mushroom, and a bacterium appear to have little in common other than being alive. However, on a microscopic level, these organisms share many similarities. For example, all organisms consist of microscopic structures called **cells,** the smallest unit of life that can function independently. Within cells, highly coordinated biochemical activities carry out the basic functions of life. This chapter introduces the cell, and the chapters that follow delve into the cellular events that make life possible.

A. Simple Lenses Revealed the Cellular Basis of Life

The study of cells began in 1660, when English physicist Robert Hooke melted strands of spun glass to create lenses. He focused on bee stingers, fish scales, fly legs, feathers, and any type of insect he could hold still. When he looked at cork, which is bark from a type of oak tree, it appeared to be divided into little boxes, left by cells that were once alive. Hooke called these units "cells" because they looked like the cubicles (Latin, *cellae*) where monks studied and prayed. Although Hooke did not realize the significance of his observation, he was the first person to see the outlines of cells. His discovery initiated a new field of science, now called cell biology.

In 1673, Antony van Leeuwenhoek of Holland improved lenses further (**figure 3.1a**). He used only a single lens, but it produced a clearer and more highly magnified image than most two-lens microscopes then available. One of his first objects of study was tartar scraped from his own teeth, and his words best describe what he saw there:

> To my great surprise, I found that it contained many very small animalcules, the motions of which were very pleasing to behold. The motion of these little creatures, one among another, may be likened to that of a great number of gnats or flies disporting in the air.

Leeuwenhoek opened a vast new world to the human eye and mind (figure 3.1b). He viewed bacteria and protists that people hadn't known existed. He also described, with remarkable accuracy, microscopic parts of larger organisms, including human red blood cells and sperm. However, he failed to see the single-celled "animalcules" reproduce. He therefore perpetuated the idea of spontaneous generation, which suggested that life arises from nonliving matter or from nothing.

B. The Cell Theory Emerges

In the nineteenth century, more powerful microscopes with improved magnification and illumination revealed details of structures inside cells. In the early 1830s, Scottish surgeon Robert Brown noted a roughly circular object in cells from orchid plants. He saw the structure in every cell, then identified it in cells of a variety of organisms. He named it the "nucleus,"

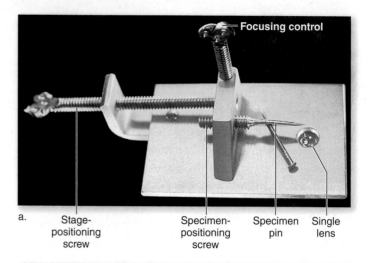

Focusing control

a. Stage-positioning screw — Specimen-positioning screw — Specimen pin — Single lens

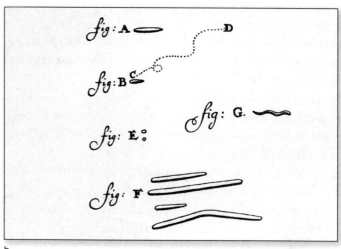

b.

Figure 3.1 **Early Microscope.** (a) Antony van Leeuwenhoek made many simple microscopes like this one. The object he was studying would have been at the tip of the specimen pin. (b) Leeuwenhoek's sketches were the first record of microorganisms.

a term that stuck. Soon microscopists distinguished the translucent, moving material that made up the rest of the cell, calling it the cytoplasm.

In 1839, German biologists Mathias J. Schleiden and Theodor Schwann proposed a new theory, based on many observations made with microscopes. Schleiden first noted that cells were the basic units of plants, and then Schwann compared animal cells to plant cells. After observing similarities in many different plant and animal cells, they concluded that cells were "elementary particles of organisms, the unit of structure and function." Schleiden and Schwann used their observations to formulate the **cell theory,** which originally had two main components: all organisms are made of one or more cells, and the cell is the fundamental unit of all life.

German physiologist Rudolf Virchow added a third component to the cell theory in 1855, when he proposed that all cells

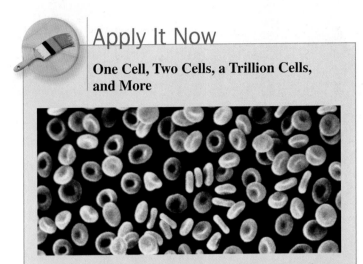

Apply It Now

One Cell, Two Cells, a Trillion Cells, and More

How many cells are in the human body? For adults, estimates range widely from about 10 trillion to 100 trillion, indicating that this question is harder to answer than it appears. First, the number changes throughout life. A child's growth comes from cell division that adds new cells, not from the expansion of existing ones. Second, no one has found a good way to count them all. Cells come in so many different shapes that it is hard to extrapolate from a small sample to the whole body. Also, new cells arise as old cells die, so a "true" count is a moving target.

Surprisingly, nonhuman cells vastly outnumber the body's own cells. Microbiologists estimate that the number of bacteria living in and on a typical human is *10 times* the number of human cells! Although some of these bacteria can cause disease, most exist harmlessly on the skin and in the mouth and gastrointestinal tract. These inconspicuous guests, which so vastly outnumber your own cells, also help extract nutrients from food and prevent disease.

come from preexisting cells. This idea contradicted spontaneous generation. When the French chemist and microbiologist Louis Pasteur finally disproved spontaneous generation in 1859, he provided additional evidence in support of cell theory.

The existence of cells is an undisputed fact, yet the cell theory is still evolving. For 150 years after its formulation, biologists focused on documenting the parts of a cell and the process of cell division. Since the discovery of DNA's structure and function in the 1950s, however, the cell theory has focused on the role of genetic information in dictating what happens inside cells. Modern cell theory therefore adds the ideas that all cells have the same basic chemical composition (chapter 2), use energy (chapters 4, 5, and 6), and contain DNA that is duplicated and passed on as each cell divides (chapters 7, 8, and 9).

Like any scientific theory, the cell theory is *potentially* falsifiable—yet many lines of evidence support each of its components, making it one of the most powerful ideas in biology.

C. Microscopes Magnify Cell Structures

Most cells are too small for the unaided human eye to see, so studying life at the cellular and molecular levels requires magnification. Cell biologists use a variety of microscopes to produce different types of images. This section describes several types; **figure 3.2** provides a sense of the size of objects that each can reveal.

Light Microscopes Light microscopes are ideal for generating true-color views of living or preserved cells. Because light must pass through an object to reveal its internal features, however, the specimens must be transparent or thinly sliced to generate a good image.

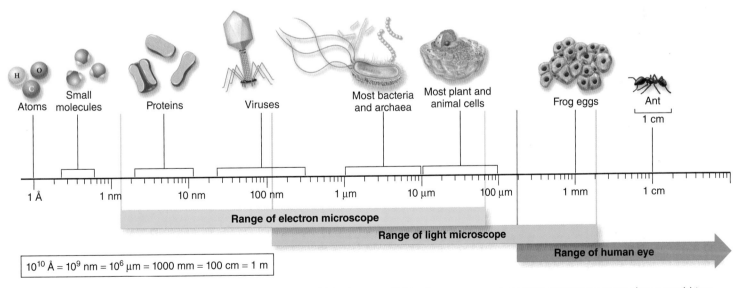

$$10^{10} \text{ Å} = 10^9 \text{ nm} = 10^6 \text{ μm} = 1000 \text{ mm} = 100 \text{ cm} = 1 \text{ m}$$

Figure 3.2 Ranges of Light and Electron Microscopes. Biologists use light microscopes and electron microscopes to view a world too small to see with the unaided eye. This illustration uses the metric system to measure size (see appendix C). Thanks to the overlapping capabilities of the different microscopes, we can visualize objects ranging in size from large molecules to entire cells.

a.

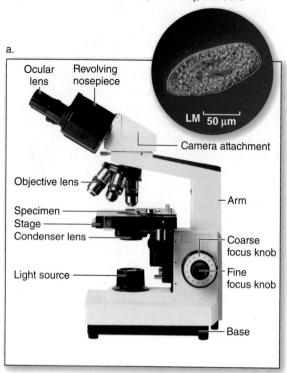

Ocular lens
Revolving nosepiece
Camera attachment
Objective lens
Arm
Specimen
Stage
Condenser lens
Coarse focus knob
Fine focus knob
Light source
Base

LM 50 μm

b.

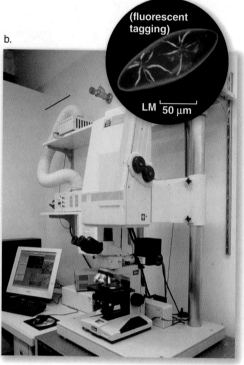

(fluorescent tagging)

LM 50 μm

c.

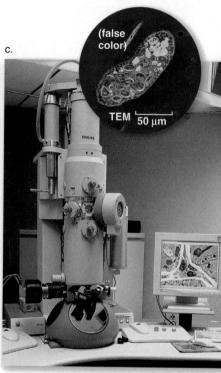

(false color)

TEM 50 μm

Two types of light microscopes are the compound microscope and the confocal microscope (figure 3.3a, b). A compound scope uses two or more lenses to focus visible light through a specimen; the most powerful ones can magnify up to 1600 times and resolve objects that are 200 nanometers apart. A confocal microscope enhances resolution by focusing white or laser light through a lens to the object. The image then passes through a pinhole. The result is a scan of highly focused light on one tiny part of the specimen. Computers can integrate multiple confocal images of specimens exposed to fluorescent dyes to produce spectacular three-dimensional peeks at living structures.

Transmission and Scanning Electron Microscopes

Instead of using light, the transmission electron microscope (TEM) sends a beam of electrons through a very thin slice of a specimen, using a magnetic field rather than a glass lens to focus the beam. The microscope translates the contrasts in electron transmission into a high-resolution, two-dimensional image that shows the internal features of the object (figure 3.3c). TEMs can magnify up to 50 million times and resolve objects less than 1 angstrom (10^{-10} meters) apart.

The scanning electron microscope (SEM) scans a beam of electrons over the surface of a metal-coated, three-dimensional specimen. Its images have lower resolution than the TEM; in SEM, the maximum magnification is about 250,000 times, and the resolution limit is 1 to 5 nanometers. The chief advantage of SEM is its ability to highlight crevices and textures on the surface of a specimen (figure 3.3d).

Both TEM and SEM provide much greater magnification and resolution than light microscopes. Nevertheless, they do have limitations. First, they are extremely expensive to build, op-

erate, and maintain. Second, electron microscopy normally requires that a specimen be killed, chemically fixed, and placed in a vacuum. These treatments can distort natural structures. Light microscopy, in contrast, allows an investigator to view living organisms. Third, unlike light microscopes, all images from electron microscopes are black and white, although artists often add false color to highlight specific objects in electron micrographs. (In this book, each photo taken through a microscope is tagged with the magnification and the type of microscope; the presence of false color is also noted where applicable.)

D. All Cells Have Features in Common

Microscopes and other tools clearly reveal that although cells can appear very different, they all have some of the same features. All cells, from the simplest to the most complex, have the following structures and molecules in common that allow them to reproduce, grow, respond to stimuli, and obtain energy:

- DNA, the cell's genetic information;
- RNA, which participates in the production of proteins (see chapter 7);
- **ribosomes,** structures that manufacture proteins;
- proteins that carry out all of the cell's work, from orchestrating reproduction to processing energy to regulating what enters and leaves the cell;
- **cytoplasm,** the fluid that occupies much of the volume of the cell; and
- a lipid-rich **cell membrane** (also called the plasma membrane) that forms a boundary between the cell and its environment (see section 3.3).

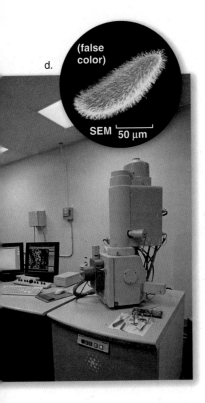

(false color)

SEM 50 μm

Figure 3.3 Light and Electron Microscopes Reveal Different Details. These photographs show four types of microscopes, along with sample images of a protist called *Paramecium*. (a) Compound light microscope. (b) Confocal microscope. (c) Transmission electron microscope. (d) Scanning electron microscope.

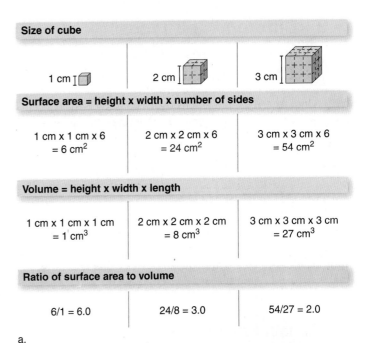

Size of cube

1 cm | 2 cm | 3 cm

Surface area = height x width x number of sides

| 1 cm x 1 cm x 6 = 6 cm² | 2 cm x 2 cm x 6 = 24 cm² | 3 cm x 3 cm x 6 = 54 cm² |

Volume = height x width x length

| 1 cm x 1 cm x 1 cm = 1 cm³ | 2 cm x 2 cm x 2 cm = 8 cm³ | 3 cm x 3 cm x 3 cm = 27 cm³ |

Ratio of surface area to volume

| 6/1 = 6.0 | 24/8 = 3.0 | 54/27 = 2.0 |

a.

One other feature common to nearly all cells is small size, typically less than 0.1 millimeter in diameter (see figure 3.2). Why so tiny? The answer is that nutrients, water, oxygen, carbon dioxide, and waste products enter or leave a cell through its surface. Each cell must have abundant surface area to accommodate these exchanges. As an object grows, however, its volume increases much faster than its surface area. Figure 3.4a illustrates this principle for a series of cubes, but the same applies to cells: small cell size maximizes the ratio of surface area to volume.

Figure It Out

For a cube 5 centimeters on each side, calculate the ratio of surface area to volume.

Answer: 1.2

Cells avoid surface area limitations in several ways. Nerve cells may be long (up to a meter or so), but they are also extremely thin, so the ratio of surface area to volume remains high. The flattened shape of a red blood cell maximizes its ability to carry oxygen, and the many microscopic extensions of an amoeba's membrane provide a large surface area for absorbing oxygen and capturing food (figure 3.4b). A transportation system that quickly circulates materials throughout the cell also helps.

The concept of surface area is everywhere in biology. A pine tree's pollen grains have extensions that maximize flotation on air currents; root hairs have tremendous surface area for absorbing water; the broad, flat leaves of plants maximize exposure to light; a fish's feathery gills absorb oxygen from water; a jackrabbit's enormous ears help the animal lose excess body heat in the desert air—the list goes on and on. Conversely, low surface areas minimize the exchange of materials or heat with the environment. A hibernating animal, for example, conserves warmth by tucking its limbs close to its body.

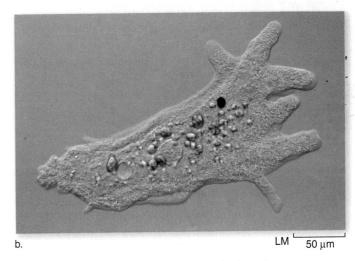

b. LM 50 μm

Figure 3.4 The Relationship Between Surface Area and Volume. (a) This simple example shows that smaller objects have more surface area *relative to their volume* than do larger objects with the same overall shape. (b) The membrane of this amoeba is highly folded, producing a large surface area relative to the cell's volume.

3.1 | Mastering Concepts

1. What is a cell?
2. How have microscopes contributed to the study of cells?
3. What are the main components of cell theory?
4. Describe the differences between light and electron microscopes.
5. Which molecules and structures occur in all cells?
6. Describe adaptations that increase the ratio of surface area to volume in cells.

3.2 | Different Cell Types Characterize Life's Three Domains

Until recently, biologists recognized just two types of cells, prokaryotic and eukaryotic. **Prokaryotes,** the simplest and most ancient forms of life, are organisms whose cells lack a nucleus (*pro* = before; *karyon* = kernel, referring to the nucleus). About 2.7 billion years ago, prokaryotes gave rise to **eukaryotes,** whose cells contain a nucleus and other membranous organelles (*eu* = true).

In 1977, however, microbiologist Carl Woese studied key molecules in many cell types and detected differences that suggested that some prokaryotes represented a completely different form of life. Biologists subsequently divided life into three domains: Bacteria, Archaea, and Eukarya (figure 3.5). This section describes them briefly.

A. Domain Bacteria Contains Earth's Most Abundant Organisms

Bacteria are the most abundant and diverse organisms on Earth. Some species, such as *Streptococcus* and *Escherichia coli*, can cause illnesses, but others living on your skin and inside your intestinal tract are essential for good health. Bacteria are also very valuable in research, food and beverage processing, and pharmaceutical production. In ecosystems, bacteria play critical roles as decomposers and producers.

Bacterial cells are structurally simple (figure 3.6a). The **nucleoid** is the area where the cell's circular DNA molecule congregates. Unlike a eukaryotic cell's nucleus, the bacterial nucleoid is not bounded by a membrane. Located near the DNA in the cytoplasm are the enzymes, RNA molecules, and ribosomes needed to produce the cell's proteins.

A rigid **cell wall** surrounds the cell membrane of most bacteria, protecting the cell and preventing it from bursting if it absorbs too much water. This wall also gives the cell its shape: usually rod-shaped, round, or spiral (figure 3.6b, c, d). Many antibiotic drugs, including penicillin, halt bacterial infection by interfering with the microorganism's ability to construct its protective cell wall. In some bacteria, polysaccharides on the cell wall form a capsule that adds protection or enables the cell to attach to surfaces.

Many bacteria can swim in fluids. **Flagella** (singular: flagellum) are tail-like appendages that enable these cells to move. One or more flagella are anchored in the cell wall and underlying cell membrane. Bacterial flagella rotate like a propeller, moving the cell forward or backward.

B. Domain Archaea Includes Prokaryotes with Unique Biochemistry

Archaean cells resemble bacterial cells in many ways (figure 3.7). Like bacteria, they are smaller than most eukaryotic cells, and they lack a membrane-bounded nucleus and other organelles. Most have cell walls, and flagella are also common. Because of these similarities, Woese first named his newly recognized group Archaebacteria.

The name later changed to Archaea when genetic sequences revealed that the resemblance to bacteria was only superficial. Archaea have their own domain because they build their cells out of biochemicals that are different from those in either bacteria or eukaryotes. Their phospholipids, cell walls, and flagella are all chemically unique. Their ribosomes, however, are more similar to those of eukaryotes than to those of bacteria. Archaea may therefore be the closest relatives of eukaryotes.

The first members of Archaea to be described were methanogens, microbes that use carbon dioxide and hydrogen from the environment to produce methane. Archaea subsequently became famous as "extremophiles" because scientists discovered many of them in habitats that are extremely hot, acidic, or salty. This characterization is somewhat misleading, however, because bacteria also occupy the same environments. Moreover, researchers have now discovered archaea in a variety of moderate habitats, including soil, swamps, rice paddies, oceans, and even the human mouth.

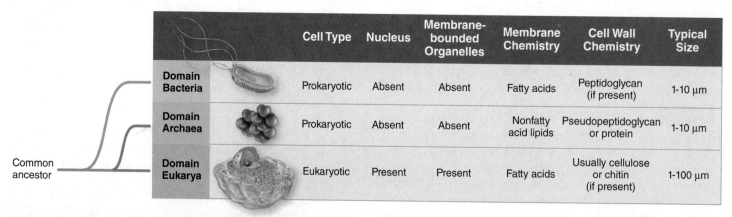

		Cell Type	Nucleus	Membrane-bounded Organelles	Membrane Chemistry	Cell Wall Chemistry	Typical Size
Common ancestor	Domain Bacteria	Prokaryotic	Absent	Absent	Fatty acids	Peptidoglycan (if present)	1-10 μm
	Domain Archaea	Prokaryotic	Absent	Absent	Nonfatty acid lipids	Pseudopeptidoglycan or protein	1-10 μm
	Domain Eukarya	Eukaryotic	Present	Present	Fatty acids	Usually cellulose or chitin (if present)	1-100 μm

Figure 3.5   The Three Domains of Life. Biologists distinguish domains Bacteria, Archaea, and Eukarya based on unique features of cell structure and biochemistry. The small evolutionary tree shows that archaea are the closest relatives of the eukaryotes.

Figure 3.6 Anatomy of a Bacterium. (A) Bacterial cells lack internal compartments. (b) Rod-shaped cells of *E. coli* inhabit human intestines. (c) Spherical *Staphylococcus aureus* cells cause infections that range from mild to deadly. (d) The corkscrew-shaped *Campylobacter jejuni* lives in the digestive tract of many animals.

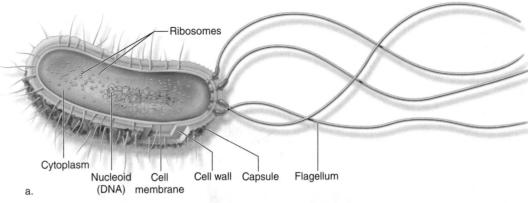

a.

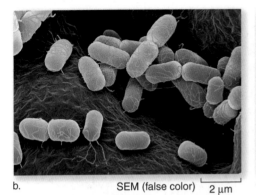

b.   SEM (false color)   2 μm

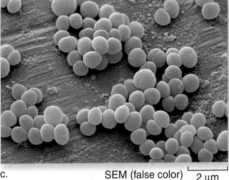

c.   SEM (false color)   2 μm

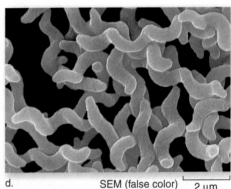

d.   SEM (false color)   2 μm

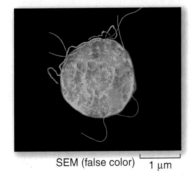

Figure 3.7 An Archaeon. *Sulfolobus acidocaldarius* thrives in hot springs, at a temperature of 80°C and a pH of 2.0. Note the prominent flagella.

SEM (false color)   1 μm

C. Domain Eukarya Contains Organisms with Complex Cells

An astonishing diversity of other organisms, including humans, belong to domain Eukarya. Our fellow animals are eukaryotes, as are yeasts, mushrooms, and other fungi. Plants are also eukaryotes, and so are one-celled protists such as *Amoeba* and *Paramecium*.

Burning Question

What is the smallest living organism?

Since the invention of microscopes, investigators have wondered just how small an organism can be and still sustain life. This seemingly simple question is hard to answer; *life* is hard to define.

Some people consider viruses alive because they share some, but not all, characteristics with cells (see chapter 15). Viruses are indeed miniscule: The smallest are less than 20 nanometers in diameter (see figure 3.2). Yet most biologists do not consider them alive, in part because viruses do not consist of cells or reproduce on their own.

50 nm
SEM
(false color)

Figure 3.A Nanobes. Alive or not?

Some scientists consider "nanobes" to be the world's smallest microorganisms, at about 20 to 150 nanometers long (figure 3.A). Other researchers are skeptical. These minuscule filaments are hard to analyze for hallmarks of life such as DNA, RNA, ribosomes, and protein. Their status remains controversial.

For now, the smallest certifiable living organisms are bacteria called mycoplasmas. Besides their small size (150 nanometers and larger), these microorganisms are unusual among bacteria because they lack cell walls. Biologists have studied mycoplasmas in detail for two reasons. First, some cause human disease such as urinary tract infections and pneumonia. Second, with only 482 genes, mycoplasmas have the smallest amount of genetic material of any known free-living cell. Studies on mycoplasmas are helping to reveal which genes are minimally required to sustain life.

Submit your burning question to:
marielle_hoefnagels@mcgraw-hill.com

Despite their great differences in external appearance, all eukaryotic organisms share many features on a cellular level. Figures 3.8 and 3.9 depict generalized animal and plant cells. Although both of the illustrated cells have many structures in common, there are some differences. Most notably, plant cells have chloroplasts and a cell wall, which animal cells lack.

One obvious feature that sets eukaryotic cells apart is their large size, typically 10 to 100 times greater than prokaryotic cells. The other main difference is that the cytoplasm of a eukaryotic cell is divided into **organelles** ("little organs"), compartments that carry out specialized functions. An elaborate system of internal membranes creates these compartments.

In general, organelles keep related biochemicals and structures close enough to make them function efficiently. At the same time, they keep potentially harmful substances away from other cell contents. Compartmentalization also saves energy be-

Figure 3.8 An Animal Cell. The large, generalized view shows the relative sizes and locations of a typical animal cell's components. The electron micrograph at right shows a human white blood cell with a prominent nucleus and many mitochondria.

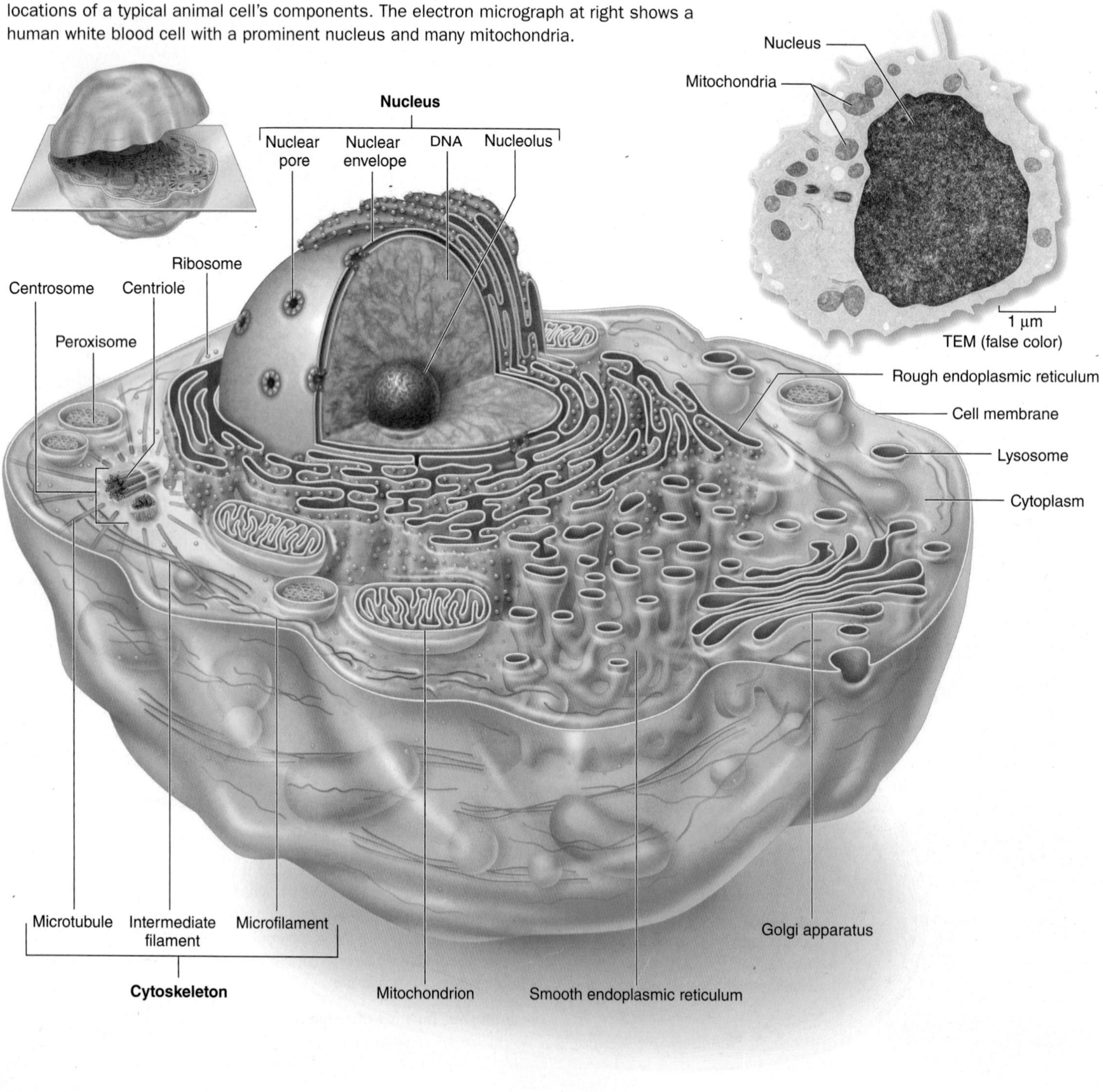

Nucleus

Mitochondria

1 μm

TEM (false color)

Nucleus

Nuclear pore   Nuclear envelope   DNA   Nucleolus

Ribosome

Centrosome   Centriole

Peroxisome

Rough endoplasmic reticulum

Cell membrane

Lysosome

Cytoplasm

Microtubule   Intermediate filament   Microfilament

Cytoskeleton

Mitochondrion

Smooth endoplasmic reticulum

Golgi apparatus

cause the cell maintains high concentrations of each biochemical only in certain organelles, not throughout the entire cell.

The rest of this chapter describes the structure of the eukaryotic cell in greater detail, and the illustrated table at the end of the chapter summarizes the functions of the eukaryotic organelles (see table 3.2 on page 66).

3.2 | Mastering Concepts

1. How do prokaryotic cells differ from eukaryotic cells?
2. How are bacteria and archaea similar to and different from each other?
3. How do organelles contribute to efficiency in eukaryotic cells?

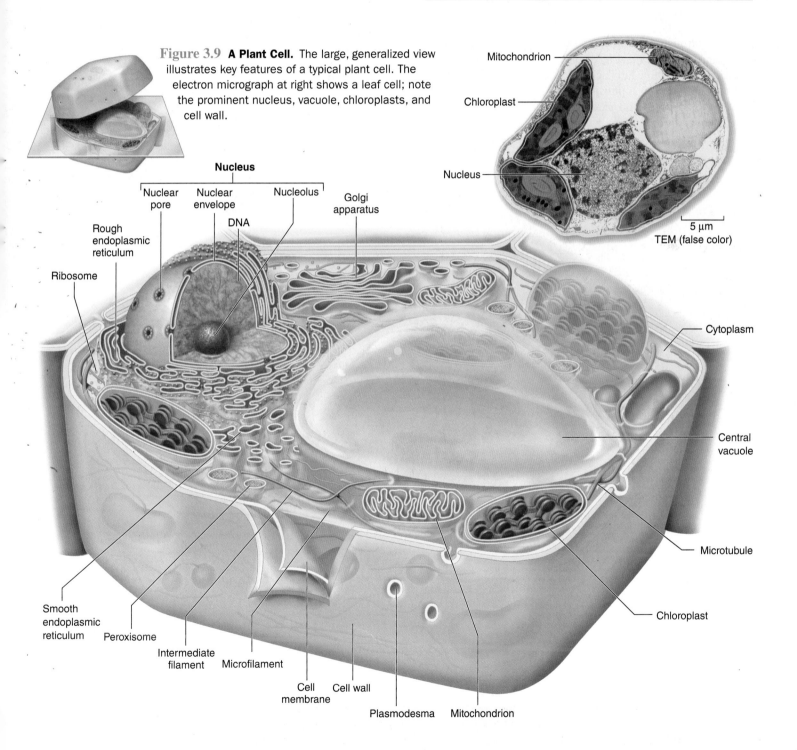

Figure 3.9 A Plant Cell. The large, generalized view illustrates key features of a typical plant cell. The electron micrograph at right shows a leaf cell; note the prominent nucleus, vacuole, chloroplasts, and cell wall.

Mitochondrion

Chloroplast

Nucleus

5 µm
TEM (false color)

Nucleus

Nuclear pore

Nuclear envelope

Nucleolus

Golgi apparatus

DNA

Rough endoplasmic reticulum

Ribosome

Cytoplasm

Central vacuole

Microtubule

Chloroplast

Smooth endoplasmic reticulum

Peroxisome

Intermediate filament

Microfilament

Cell membrane

Cell wall

Plasmodesma

Mitochondrion

3.3 | A Membrane Separates Each Cell from Its Surroundings

A cell membrane is one feature common to all cells. The membrane separates the cytoplasm from the cell's surroundings. The cell's surface also transports substances into and out of the cell (see chapter 4), and it receives and responds to external stimuli. Inside a eukaryotic cell, internal membranes enclose the organelles.

The cell membrane is composed of phospholipids, which are organic molecules that resemble triglycerides (**figure 3.10**). In a triglyceride, three fatty acids attach to a three-carbon glycerol molecule. But in a **phospholipid,** glycerol bonds to only two fatty acids; the third carbon binds to a phosphate group attached to additional atoms. ▸ triglycerides, p. 34

This chemical structure gives phospholipids unusual properties in water. The phosphate "head" end, with its polar covalent bonds, is attracted to water; that is, it is hydrophilic. The other end, consisting of two fatty acid "tails," is hydrophobic. In water, phospholipid molecules spontaneously arrange themselves into the most energy-efficient organization: a **phospholipid bilayer** (**figure 3.11**). In this two-layered, sandwichlike structure, the hydrophilic surfaces (the "bread" of the sandwich) are exposed to the watery medium outside and inside the cell. The hydrophobic tails face each other on the inside of the sandwich, like cheese between the bread slices. Unlike a sandwich, however, the bilayer forms a three-dimensional sphere, not a flat surface.

Thanks to the phospholipid bilayer, a biological membrane has selective permeability. The hydrophobic interior of the phospholipid bilayer prevents ions and polar molecules from passing freely into and out of a cell. The membrane does not, however, block lipids and small, nonpolar molecules such as O_2 and CO_2.

Cell membranes consist not only of phospholipid bilayers but also of sterols, proteins, and other molecules (**figure 3.12**). The cell membrane is often called a **fluid mosaic** because many of the proteins and phospholipids are free to move laterally

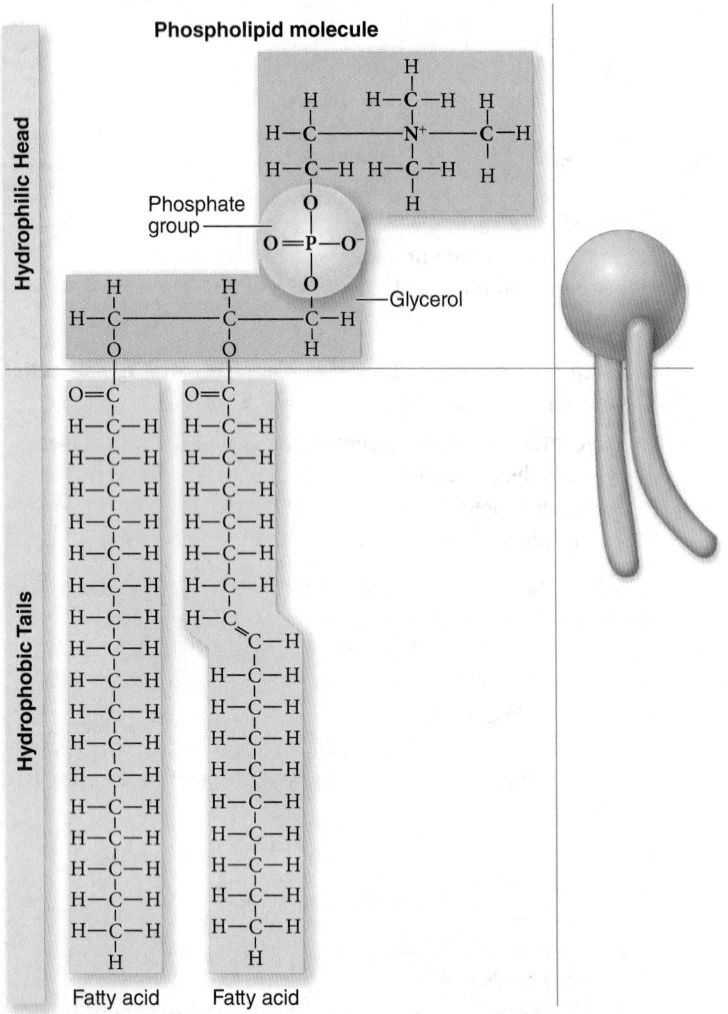

Figure 3.10 Membrane Phospholipids. A phospholipid molecule consists of a glycerol molecule attached to a hydrophilic phosphate "head" group and two hydrophobic fatty acid "tails." The drawing at right shows a simplified phospholipid structure.

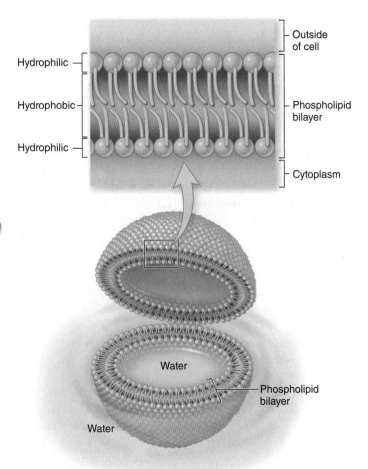

Figure 3.11 Phospholipid Bilayer. In water, phospholipids form a bilayer. The hydrophilic head groups are exposed to the water; the hydrophobic tails face each other, minimizing contact with water. A sphere of phospholipids forms the basis for the cell membrane.

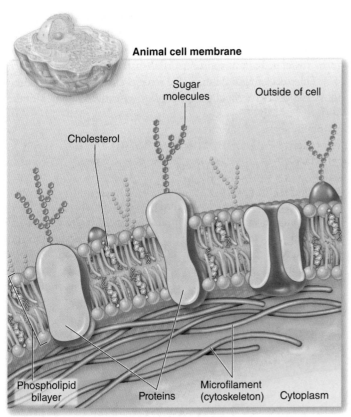

Animal cell membrane

Sugar molecules

Outside of cell

Cholesterol

Phospholipid bilayer

Proteins

Microfilament (cytoskeleton)

Cytoplasm

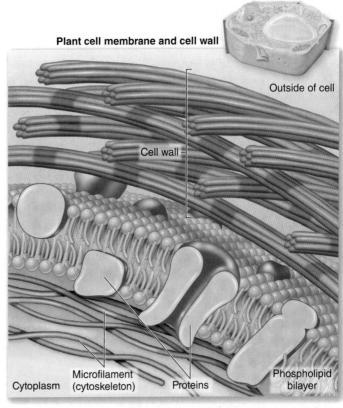

Plant cell membrane and cell wall

Outside of cell

Cell wall

Microfilament (cytoskeleton)

Cytoplasm

Proteins

Phospholipid bilayer

Figure 3.12 Anatomy of a Cell Membrane. The cell membrane is a "fluid mosaic" of proteins embedded in a phospholipid bilayer. Note that animal cell membranes contain cholesterol, but plant cell membranes do not. The outer face of the animal cell membrane also features carbohydrate (sugar) molecules linked to proteins. A cell wall of cellulose fibers surrounds plant cells.

within the bilayer. Sterols, including cholesterol in animal cell membranes, maintain the membrane's fluidity.

Whereas phospholipids and sterols provide the membrane's structure, proteins are especially important to its function. Researchers estimate that about one third of every organism's genome encodes membrane proteins. Some of the proteins lie completely within the phospholipid bilayer, whereas others extend out of one or both sides. Their functions include:

- **Transport proteins:** Transport proteins embedded in the phospholipid bilayer create passageways through which water-soluble molecules and ions pass into or out of the cell. Section 4.5 describes membrane transport in more detail.

- **Enzymes:** These proteins facilitate chemical reactions that otherwise would proceed too slowly to sustain life. (Not all enzymes, however, are associated with membranes.)
  ▸ enzymes, p. 78

- **Recognition proteins:** Carbohydrates attached to cell surface proteins serve as "name tags" that help the body recognize its own cells. The immune system attacks cells with unfamiliar surface molecules, which is why transplant recipients often reject donated organs. Surface structures also distinctively mark cells of different tissues in an individual, so a bone cell's surface is different from that of a nerve cell or a muscle cell.

- **Adhesion proteins:** These membrane proteins enable cells to stick to one another.

- **Receptor proteins:** Receptor proteins bind to molecules outside the cell and trigger a reaction inside the cell. The receptor protein HER2, described in this chapter's opening essay, receives signals that stimulate cell division.

Understanding membrane proteins is a vital part of human medicine, in part because at least half of all drugs bind to them. One example is omeprazole (Prilosec). This drug relieves heartburn and gastric reflux by blocking some of the transport proteins that pump hydrogen ions into the stomach. Another is the antidepressant drug fluoxetine (Prozac), which prevents receptors on brain cell surfaces from absorbing a mood-altering biochemical called serotonin.

3.3 | Mastering Concepts

1. Chemically, how is a phospholipid different from a triglyceride?
2. How does the chemical structure of phospholipids enable them to form a bilayer in water?
3. Where in the cell do phospholipid bilayers occur?
4. What are some functions of membrane proteins?

3.4 | Eukaryotic Organelles Divide Labor

In eukaryotic cells, organelles have specialized functions that carry out the work of the cell. If you think of a eukaryotic cell as a home, each organelle would be analogous to a room. For example, your kitchen, bathroom, and bedroom each hold unique items that suit the uses of those rooms. Likewise, each organelle has distinct sets of proteins and other molecules that fit the organelle's function. The "walls" of these cellular compartments are membranes, often intricately folded and studded with enzymes and other proteins.

Many of these internal membranes form a coordinated **endomembrane system,** which consists of several interacting organelles: the nuclear envelope, endoplasmic reticulum, Golgi apparatus, lysosomes, vacuoles, and cell membrane. As you will see, the organelles of the endomembrane system are connected by small "bubbles" of membrane that can pinch off of one organelle, travel within the cell, and fuse with another. These membranous spheres, which are also part of the endomembrane system, form **vesicles** that transport materials inside the cell.

This section will describe the structures and functions of the most important organelles, beginning with the endomembrane system.

A. The Nucleus, Endoplasmic Reticulum, and Golgi Interact to Secrete Substances

The organelles of the endomembrane system enable cells to produce, package, and release complex mixtures of biochemicals. This section focuses on each step involved in the production and secretion of one such mixture: milk (**figure 3.13**).

Special cells in the mammary glands of female mammals produce milk, which contains proteins, fats, carbohydrates, and

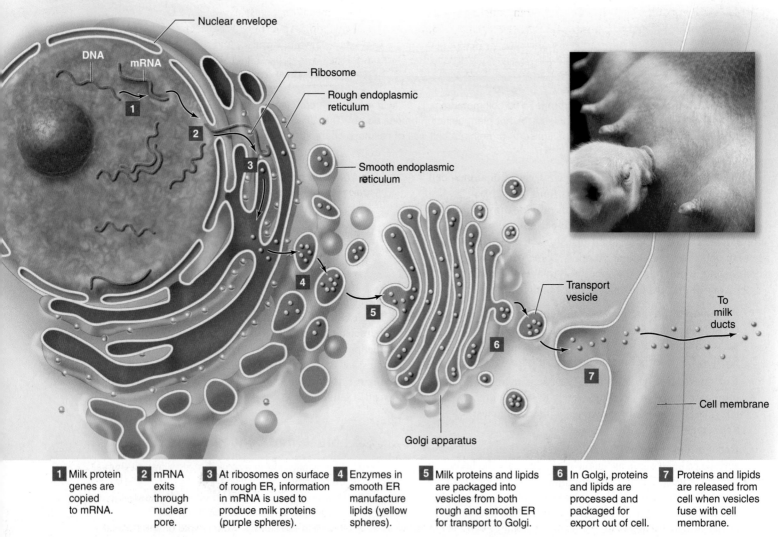

| **1** Milk protein genes are copied to mRNA. | **2** mRNA exits through nuclear pore. | **3** At ribosomes on surface of rough ER, information in mRNA is used to produce milk proteins (purple spheres). | **4** Enzymes in smooth ER manufacture lipids (yellow spheres). | **5** Milk proteins and lipids are packaged into vesicles from both rough and smooth ER for transport to Golgi. | **6** In Golgi, proteins and lipids are processed and packaged for export out of cell. | **7** Proteins and lipids are released from cell when vesicles fuse with cell membrane. |

Figure 3.13 Making Milk. Several organelles participate in the production and secretion of milk from a cell in a mammary gland; the numbers (1) through (7) indicate the order in which organelles participate in this process. The inset shows a piglet suckling from a sow.

water in a proportion ideal for development of a newborn. Human milk is rich in lipids, which the rapidly growing baby's nervous system requires. (Cows' milk contains a higher proportion of protein, better suited to a calf's rapid muscle growth.) Milk also contains calcium, potassium, and antibodies that help jumpstart the infant's immunity to disease.

The milk-producing cells of the mammary glands are dormant most of the time, but they undergo a burst of productivity shortly after the female gives birth. How do each cells' organelles work together to manufacture milk?

The Nucleus   The process of milk production and secretion begins in the **nucleus** (see figure 3.13, step 1), the most prominent organelle in most eukaryotic cells. The nucleus contains DNA, an informational molecule that specifies the "recipe" for every protein a cell can make (such as milk protein and enzymes required to synthesize carbohydrates and lipids). The cell copies the genes encoding these proteins into another nucleic acid, messenger RNA (mRNA).

The mRNA molecules exit the nucleus through **nuclear pores,** which are holes in the double-membrane **nuclear envelope** that separates the nucleus from the cytoplasm (figure 3.13, step 2, and figure 3.14). Nuclear pores are highly specialized channels composed of dozens of types of proteins. Traffic through the nuclear pores is busy, with millions of regulatory proteins entering and mRNA molecules leaving each minute.

Also inside the nucleus is the **nucleolus,** a dense spot that assembles the components of ribosomes. These ribosomal subunits leave the nucleus through the nuclear pores, and they come together in the cytoplasm to form complete ribosomes.

The Endoplasmic Reticulum and Golgi Apparatus

The remainder of the cell, between the nucleus and the cell membrane, is the cytoplasm. In all cells, the cytoplasm contains a watery mixture of ions, enzymes, RNA, and other dissolved substances. In eukaryotes, the cytoplasm also includes organelles and arrays of protein rods and tubules called the cytoskeleton (see section 3.5).

Once in the cytoplasm, mRNA coming from the nucleus binds to a ribosome, which manufactures proteins (see figure 3.13, step 3). Ribosomes that produce proteins for use inside the cell are free-floating in the cytoplasm. But many proteins are destined for the cell membrane or for secretion (in milk, for example). In that case, the entire complex of ribosome, mRNA, and partially made protein anchors to the surface of the **endoplasmic reticulum,** a network of sacs and tubules composed of membranes. (*Endoplasmic* means "within the cytoplasm," and *reticulum* means "network.")

The endoplasmic reticulum (ER) originates at the nuclear envelope and winds throughout the cell. Close to the nucleus, the membrane surface is studded with ribosomes making proteins that enter the inner compartment of the ER; these proteins are destined to be secreted from the cell. This section of the network

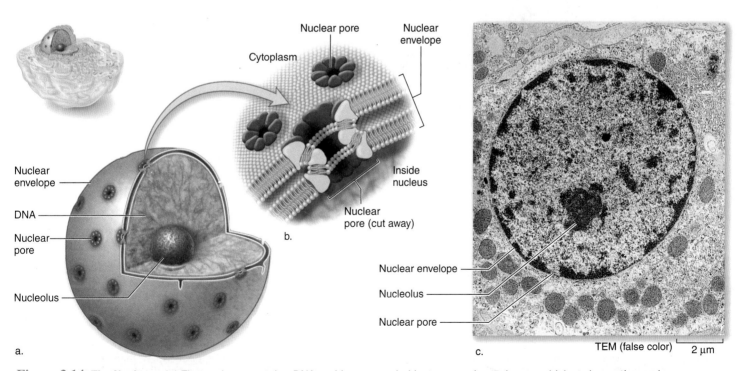

Figure 3.14 The Nucleus. (a) The nucleus contains DNA and is surrounded by two membrane layers, which make up the nuclear envelope. (b) Large pores in the nuclear envelope allow proteins to enter and mRNA molecules to leave the nucleus. (c) This transmission electron micrograph shows the nuclear envelope and nucleolus.

is called the **rough ER** because the ribosomes give these membranes a roughened appearance (figure 3.15).

Adjacent to the rough ER, a section of the network called the **smooth ER** synthesizes lipids—such as those that will end up in the milk—and other membrane components (see figure 3.13, step 4, and figure 3.15). The smooth ER also houses enzymes that detoxify drugs and poisons. In muscle cells, a specialized type of smooth ER stores and delivers the calcium ions required for muscle contraction. ▸ muscle function, p. 591

The lipids and proteins made by the ER exit the organelle in vesicles. A loaded transport vesicle pinches off from the tubular endings of the ER membrane (see figure 3.13, step 5) and takes its contents to the next stop in the production line, the **Golgi apparatus** (figure 3.16). This organelle is a stack of flat, membrane-enclosed sacs that functions as a processing center. Proteins from the ER pass through the series of Golgi sacs, where they complete their intricate folding and become functional (see figure 3.13, step 6). Enzymes in the Golgi apparatus also manufacture and attach carbohydrates to proteins or lipids, forming glycoproteins or glycolipids.

The Golgi apparatus sorts and packages materials into vesicles, which move toward the cell membrane. Some of the proteins it receives from the ER will become membrane surface proteins; other substances (such as milk protein and fat)

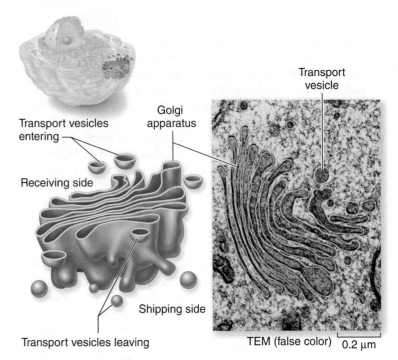

Figure 3.16 The Golgi Apparatus. The Golgi apparatus is composed of a series of flattened sacs, plus transport vesicles that deliver and remove materials. Proteins are sorted and processed as they move through the Golgi apparatus on their way to the cell surface or to a lysosome.

are packaged for secretion from the cell. In the production of milk, these vesicles fuse with the cell membrane and release the proteins outside the cell (see figure 3.13, step 7). The fat droplets stay suspended in the watery milk because they retain a layer of surrounding membrane when they leave the cell.

This entire process happens simultaneously in countless specialized cells lining the milk ducts of the breast, beginning shortly after a baby's birth. When the infant suckles, hormones released in the mother's body stimulate muscles surrounding balls of these cells to contract, squeezing milk into the ducts that lead to the nipple.

B. Lysosomes, Vacuoles, and Peroxisomes Are Cellular Digestion Centers

Besides producing molecules for export, eukaryotic cells also break down molecules in specialized compartments. All of these "digestion center" organelles are sacs surrounded by a single membrane.

Lysosomes   **Lysosomes** are organelles containing enzymes that dismantle and recycle food particles, captured bacteria, worn-out organelles, and debris (figure 3.17). They are so named because their enzymes lyse, or cut apart, their substrates.

The rough ER manufactures the enzymes inside lysosomes. The Golgi apparatus detects these enzymes by recognizing a

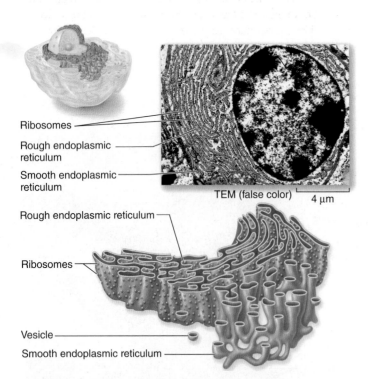

Figure 3.15 The Endoplasmic Reticulum, Rough and Smooth. The endoplasmic reticulum is a network of membranes extending from the nuclear envelope. Ribosomes dot the surface of the rough ER, giving it a "rough" appearance. The smooth ER is a series of interconnecting tubules and is the site for lipid production and other metabolic processes.

sugar attached to them, then packages them into vesicles that eventually become lysosomes. The lysosomes, in turn, fuse with transport vesicles carrying debris from outside or from within the cell. The enzymes inside the lysosome break down the large organic molecules into smaller subunits by hydrolysis, releasing them into the cytoplasm for the cell to use.

What keeps a lysosome from digesting the entire cell? The lysosome's membrane maintains the pH of the organelle's interior at about 4.8, much more acidic than the neutral pH of the rest of the cytoplasm. If one lysosome were to burst, the liberated enzymes would no longer be at their optimum pH, so they could not digest the rest of the cell. Nevertheless, a cell injured by extreme cold, heat, or another physical stress may initiate its own death by bursting all of its lysosomes at once. ▸ pH, p. 30

Some cells have more lysosomes than others. White blood cells, for example, have many lysosomes because these cells engulf and dispose of debris and bacteria. Liver cells also require many lysosomes to process cholesterol.

Malfunctioning lysosomes can cause illness. In Tay-Sachs disease, for example, a defective lysosomal enzyme allows a lipid to accumulate to toxic levels in nerve cells of the brain. The nervous system deteriorates, and an affected person eventually becomes unable to see, hear, or move. In the most severe forms of the illness, death usually occurs by age 5.

Vacuoles
Most plant cells lack lysosomes, but they do have an organelle that serves a similar function. In mature plant cells, the large central **vacuole** contains a watery solution of enzymes

that degrade and recycle molecules and organelles (see figure 3.9).

The vacuole also has other roles. Most of the growth of a plant cell comes from an increase in the volume of its vacuole. In some plant cells, the vacuole occupies up to 90% of the cell's volume. As the vacuole acquires water, it exerts pressure (called turgor pressure) against the cell membrane. Turgor pressure helps plants stay rigid and upright.

Besides water and enzymes, the vacuole also contains a variety of salts, sugars, and weak acids. Therefore, the pH of the vacuole's solution is usually somewhat acidic. In citrus fruits, the solution is very acidic, producing the tart taste of lemons and oranges. Water-soluble pigments also reside in the vacuole, producing blue, purple, and magenta colors in some leaves, flowers, and fruits.

Some protists have vacuoles, although their function is different from that in plants. The contractile vacuole in *Paramecium,* for example, pumps excess water out of the cell. In *Amoeba,* food vacuoles digest nutrients that the cell has engulfed.

Peroxisomes
All eukaryotic cells contain **peroxisomes,** organelles that contain several types of enzymes that dispose of toxic substances. Although they resemble lysosomes in size and function, peroxisomes originate at the ER (not the Golgi) and contain different enzymes.

Liver and kidney cells contain many peroxisomes that help dismantle toxins from the blood. Peroxisomes also break down fatty acids and produce cholesterol and some other lipids. In some peroxisomes, the concentration of enzymes reaches such

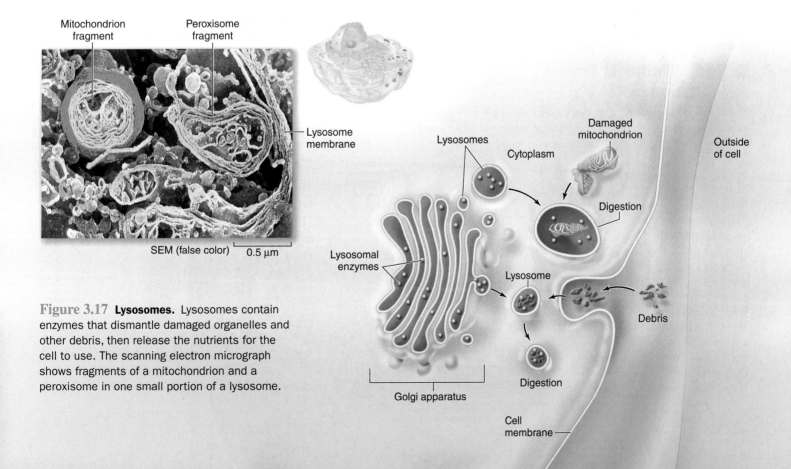

Figure 3.17 Lysosomes. Lysosomes contain enzymes that dismantle damaged organelles and other debris, then release the nutrients for the cell to use. The scanning electron micrograph shows fragments of a mitochondrion and a peroxisome in one small portion of a lysosome.

Mitochondrion fragment

Peroxisome fragment

Lysosome membrane

SEM (false color)   0.5 μm

Lysosomes

Cytoplasm

Damaged mitochondrion

Outside of cell

Digestion

Lysosomal enzymes

Lysosome

Debris

Golgi apparatus

Digestion

Cell membrane

Animal cell

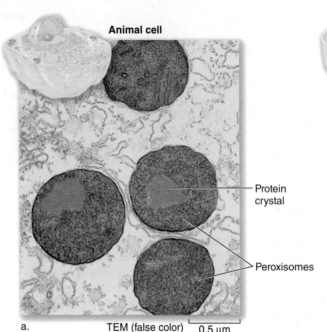

a.   TEM (false color)   0.5 μm

Protein crystal

Peroxisomes

Plant cell

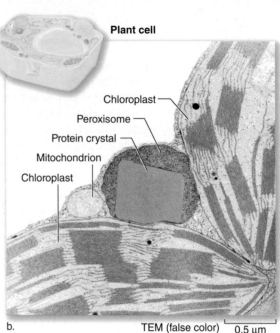

Chloroplast
Peroxisome
Protein crystal
Mitochondrion
Chloroplast

b.   TEM (false color)   0.5 μm

Figure 3.18
Peroxisomes.
Protein crystals give peroxisomes their characteristic appearance in (a) an animal cell and (b) a plant cell.

high levels that the proteins condense into easily recognized crystals (figure 3.18).

Peroxisomes protect cells from toxic byproducts. For instance, some of the reactions in the peroxisome (and other organelles) produce hydrogen peroxide, H_2O_2. This highly reactive compound can produce oxygen free radicals that can damage the cell. To counteract the free-radical buildup, peroxisomes contain an enzyme that detoxifies H_2O_2 and produces harmless water molecules in its place.

Abnormal peroxisomes can cause illness. In a disease called adrenoleukodystrophy (ALD), a faulty enzyme causes fatty acids to accumulate to toxic levels in the brain. The film *Lorenzo's Oil* depicted a boy with this disease and his parents' struggle to find treatment options.

C. Photosynthesis Occurs in Chloroplasts

Plants and many protists carry out photosynthesis, a process that uses energy from sunlight to produce glucose and other food molecules (see chapter 5). These nutrients sustain not only the

photosynthetic organisms but also the consumers (including humans) that eat them.

The **chloroplast** (figure 3.19) is the site of photosynthesis in eukaryotes. Each chloroplast contains multiple membrane layers. Two outer membrane layers enclose an enzyme-rich fluid called the stroma. Within the stroma is a third membrane system folded into flattened sacs called thylakoids, which are stacked and interconnected in structures called grana. Photosynthetic pigments such as chlorophyll are embedded in the thylakoids.

A chloroplast is one representative of a larger category of plant organelles called plastids. Some plastids synthesize lipid-soluble red, orange, and yellow carotenoid pigments, such as those found in carrots and ripe tomatoes. Plastids that assemble starch molecules are important in cells specialized for food storage, such as those in potatoes and corn kernels. Interestingly, any plastid can convert into any other type.

Unlike most other organelles, all plastids (including chloroplasts) contain their own DNA and ribosomes. The genetic material encodes proteins unique to plastid structure and function, including some of the enzymes required for photosynthesis.

Figure 3.19
Chloroplasts.
Photosynthesis occurs inside chloroplasts. Each chloroplast contains stacks of thylakoids that form the grana within the inner compartment, the stroma. Enzymes and light-harvesting pigments embedded in the membranes of the thylakoids convert sunlight to chemical energy.

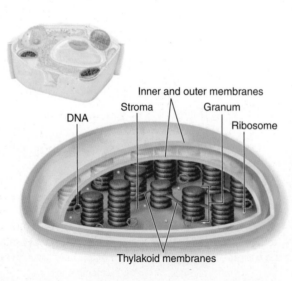

DNA
Stroma
Inner and outer membranes
Granum
Ribosome
Thylakoid membranes

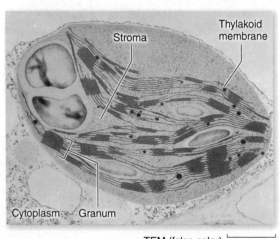

Stroma
Thylakoid membrane
Granum
Cytoplasm
Granum

TEM (false color)   1 μm

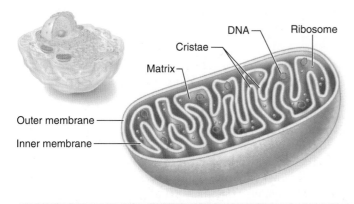

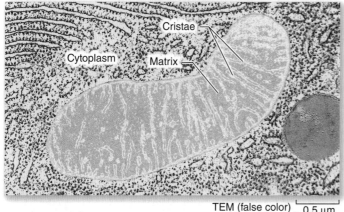

TEM (false color) | 0.5 μm

Figure 3.20 Mitochondria. Cellular respiration occurs inside mitochondria. Each mitochondrion contains a highly folded inner membrane, where many of the reactions of cellular respiration occur.

D. Mitochondria Extract Energy from Nutrients

Growth, cell division, protein production, secretion, and many chemical reactions in the cytoplasm all require a steady supply of energy. **Mitochondria** (singular: mitochondrion) are organelles that use a process called cellular respiration to extract this needed energy from food (see chapter 6). With the exception of a few types of protists, all eukaryotic cells have mitochondria.

A mitochondrion has two membrane layers: an outer membrane and an intricately folded inner membrane that encloses the mitochondrial matrix (**figure 3.20**). Within the matrix is DNA that encodes proteins essential for mitochondrial function; ribosomes occupy the matrix as well. **Cristae** are the folds of the inner membrane. The cristae add tremendous surface area to the inner membrane, which houses the enzymes that catalyze the reactions of cellular respiration.

In most mammals, mitochondria are inherited from the female parent only. (This is because the mitochondria in a sperm cell stay in the sperm's tail, which never enters the egg.) Mitochondrial DNA is therefore useful for tracking inheritance through female lines in a family. For the same reason, genetic mutations that cause defective mitochondria also pass only from mother to offspring. Mitochondrial illnesses are most serious when they affect the muscles or brain, because these energy-hungry organs depend on the functioning of many thousands of mitochondria in every cell.

The similarities between chloroplasts and mitochondria—both have their own DNA and ribosomes, and both are surrounded by double membranes—provide clues to the origin of eukaryotic cells, an event that apparently occurred about 2.7 billion years ago. According to the endosymbiosis theory, some ancient organism (or organisms) engulfed bacterial cells. Rather than digesting them as food, the host cells kept them on as partners: mitochondria and chloroplasts. The structures and genetic sequences of today's bacteria, mitochondria, and chloroplasts supply powerful evidence for this theory. ▸ endosymbiosis, p. 304

Organelles divide a cell's work, just as departments in a large store group related items together. Some specialty stores, however, sell only shoes or women's clothing. Likewise, cells can also have specialized functions (**figure 3.21**). For example, an active muscle cell is long and thin compared with a neuron, which produces extensions that touch adjacent nerve cells. A leaf cell is packed with chloroplasts. The protective epidermis of an onion, on the other hand, is dry and tough; because it forms underground, it lacks chloroplasts. Keep these specialized structures and functions in mind as you study cell processes throughout this book.

3.4 | Mastering Concepts

1. Which organelles interact to produce and secrete a complex substance such as milk?
2. What is the function of the nucleus and its contents?
3. Which organelles are the cell's "recycling centers"?
4. What are some functions of plastids?
5. Which organelle houses the reactions that extract chemical energy from nutrient molecules?
6. Which three organelles contain DNA?

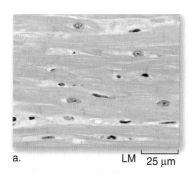

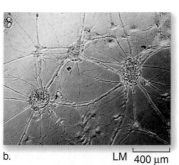

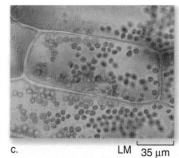

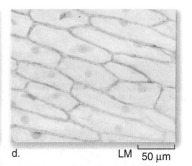

a. LM | 25 μm   b. LM | 400 μm   c. LM | 35 μm   d. LM | 50 μm

Figure 3.21 Specialized Cells. (a) Muscle cells in the heart contract in unison, thanks to electrical signals that pass rapidly from cell to cell. (b) These highly branched neurons are beginning to form a communication network. (c) The chloroplasts in these leaf cells carry out photosynthesis. (d) The cells of an onion's protective epidermis lack chloroplasts; the onion bulb forms underground, away from light.

3.5 | The Cytoskeleton Supports Eukaryotic Cells

The cytoplasm of a eukaryotic cell contains a **cytoskeleton,** an intricate network of protein "tracks" and tubules. The cytoskeleton is a structural framework with many functions. It is a transportation system, and it provides the structural support necessary to maintain the cell's characteristic three-dimensional shape (**figure 3.22**). It aids in cell division and helps connect cells to one another. The cytoskeleton also enables cells—or parts of a cell—to move.

Given the cytoskeleton's many functions, it is not surprising that defects can cause disease. For example, people with Duchenne muscular dystrophy lack a protein called dystrophin, part of the cytoskeleton in muscle cells. Without dystrophin, muscles—including those in the heart—degenerate. Another faulty cytoskeleton protein, ankyrin, causes a genetic disease of blood. In healthy red blood cells, the cytoskeleton maintains a concave disk shape. When ankyrin is missing, red blood cells are small, fragile, and misshapen, greatly reducing their ability to carry oxygen.

Injuries can also cause serious damage to the cytoskeleton. When a person suffers a strong blow to the head, such as in a fall or an auto accident, the cells that make up the brain can become stretched and distorted. The resulting damage to the cytoskeleton can trigger a chain reaction that ends with the death of the affected brain cells. People with such injuries may recover, but many die, lapse into a coma, or suffer from permanent disabilities.

The cytoskeleton includes three major components: microfilaments, intermediate filaments, and microtubules (**figure 3.23**). They are distinguished by protein type, diameter, and how they aggregate into larger structures. Other proteins connect these components to one another, creating an intricate meshwork.

The thinnest component of the cytoskeleton is the **microfilament,** a long rod composed of the protein actin. Each microfilament is only about 7 nanometers in diameter. Actin microfilament networks are part of nearly all eukaryotic cells. Muscle contraction, for example, relies on actin filaments and another protein, myosin. Microfilaments also provide strength for cells to survive stretching and compression, and they help to anchor one cell to another (see section 3.6). ▶ sliding filaments, p. 591

Intermediate filaments are so named because their 10-nanometer diameters are intermediate between those of

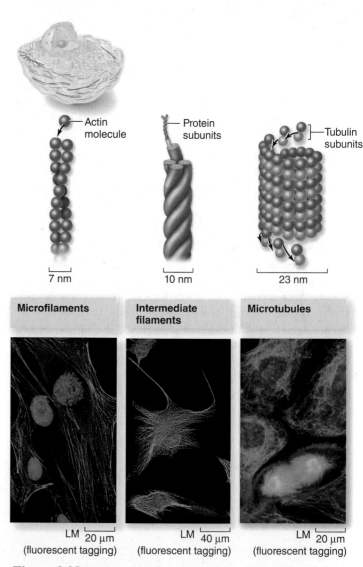

Figure 3.23 Elements of the Cytoskeleton. The cytoskeleton consists of three types of protein filaments, arranged in this figure from smallest to largest diameter. The photos show actin microfilaments (left, colored red), intermediate filaments (center, colored green), and microtubules (right, colored green).

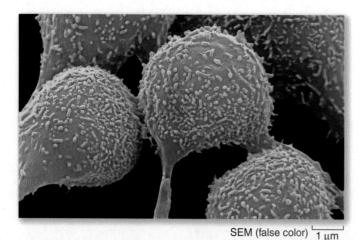

Figure 3.22 Cellular Architecture. A white blood cell's cytoskeleton enables it to produce long, thin extensions that reach out to "foreign" cell surfaces.

microfilaments and microtubules. Unlike the other components of the cytoskeleton, which consist of a single protein type, intermediate filaments are made of different proteins in each specialized cell type. They maintain a cell's shape by forming an internal scaffold in the cytoplasm and resisting mechanical stress. Intermediate filaments also help bind some cells together (see section 3.6).

A **microtubule** is composed of a protein called tubulin, assembled into a hollow tube 23 nanometers in diameter. The cell can change the length of a microtubule rapidly by adding or removing tubulin molecules.

Microtubules have many functions in eukaryotic cells. For example, chapter 8 describes how microtubules pull a cell's duplicated chromosomes apart during cell division. Microtubules also form a type of "trackway" along which organelles and proteins move within a cell. Some organisms, such as chameleons and squids, can change colors quickly by using this process to rearrange pigment molecules in their skin cells.

In animal cells, structures called **centrosomes** organize the microtubules. (Plants typically lack centrosomes and assemble microtubules at sites scattered throughout the cell.) The centrosome contains two centrioles, which are visible in figure 3.8. The centrioles apparently form the basis of structures called basal bodies, which in turn give rise to the extensions that enable some cells to move: cilia and flagella (**figure 3.24**).

Cilia are short and numerous, like a fringe. Some protists, such as the *Paramecium* in figure 3.3, have thousands of cilia that enable the cells to "swim" in water. In the human respiratory tract, coordinated movement of cilia sets up a wave that propels particles up and out; other cilia can move an egg cell through the female reproductive tract. ▶ ciliates, p. 363

Unlike cilia, flagella occur singly or in pairs, and a flagellum is much longer than a cilium. Flagella are more like tails, and their whiplike movement propels cells. Sperm cells in many species (including humans) have prominent flagella.

Cilia and flagella have the same internal structure. A basal body anchors the appendage inside the cell. The external portion is constructed of nine microtubule pairs surrounding two separate microtubules, forming a pattern described as "9 + 2." A protein called dynein connects the outer microtubule pairs and links them to the central pair, a little like a wheel. Dynein molecules shift in a way that slides adjacent microtubules against each other. This movement bends the cilium or flagellum. (The bacterial flagellum has a different structure.)

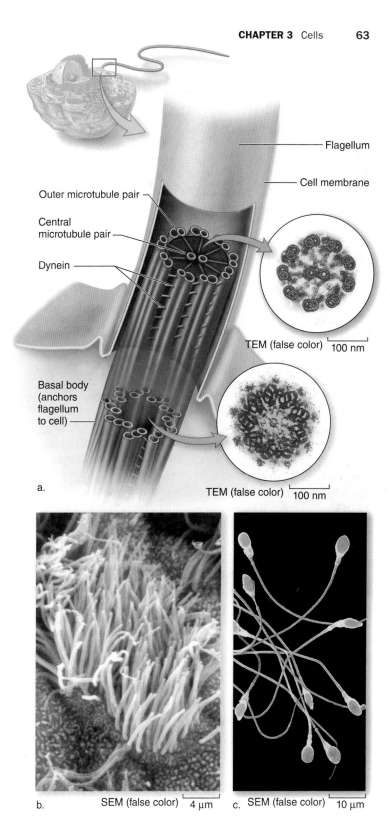

TEM (false color) 100 nm

TEM (false color) 100 nm

b. SEM (false color) 4 μm   c. SEM (false color) 10 μm

Figure 3.24 Microtubules Move Cells. (a) The proteins that form cilia and eukaryotic flagella have a characteristic "9 + 2" organization of microtubule "doublets." The basal body that gives rise to each cilium or flagellum, however, consists of a ring of microtubule "triplets." (b) These cilia line the human respiratory tract, where their coordinated movements propel dust particles upward so the person can expel them. (c) The flagella on mature human sperm cells enable them to swim.

3.5 | Mastering Concepts

1. What are some functions of the cytoskeleton?
2. What are the main components of the cytoskeleton?
3. How are cilia and flagella similar, and how are they different?

3.6 | Cells Stick Together and Communicate with One Another

So far, this chapter has described individual cells. But multicellular organisms, including plants and animals, are made of many cells that work together. How do these cells adhere to one another so that your body—or that of a plant—doesn't disintegrate in a heavy rain? Also, how do cells in direct contact with one another communicate to coordinate development and respond to the environment?

This section describes how the cells of plant and animal tissues stick together and how neighboring cells share signals. Table 3.1 lists the types of cell–cell connections for plants and animals, and table 3.2 (on page 66) summarizes the structures and functions of the main organelles in plant and animal cells.

A. Cell Walls Are Strong, Flexible, and Porous

Cell walls surround the cell membranes of nearly all bacteria, archaea, fungi, algae, and plants. But *cell wall* is a misleading term: it is not just a barrier that outlines the cell. Cell walls impart shape, regulate cell volume, prevent bursting when a cell takes in too much water, and interact with other molecules to help determine how a cell in a complex organism specializes. In plants, for example, a given cell may become a root, shoot, or leaf, depending on which cell walls it touches.

Table 3.1 Intercellular Junctions: A Summary

Type	Function	Example of Location
Plasmodesmata	Allow substances to move between plant cells	Plant cell walls
Tight junctions	Close the spaces between animal cells by fusing cell membranes	Cells in inner lining of small intestine
Anchoring (adhering) junctions	Spot weld adjacent animal cell membranes	Cells in outer skin layer
Gap junctions	Form channels between animal cells, allowing exchange of substances	Muscle cells in heart and digestive tract

Many materials may make up a cell wall. Bacterial cell walls, for example, are composed of peptidoglycan, whereas those of fungi contain chitin. Much of the plant cell wall consists of cellulose molecules aligned into microfibrils, which in turn aggregate and twist to form larger fibrils (figure 3.25). This

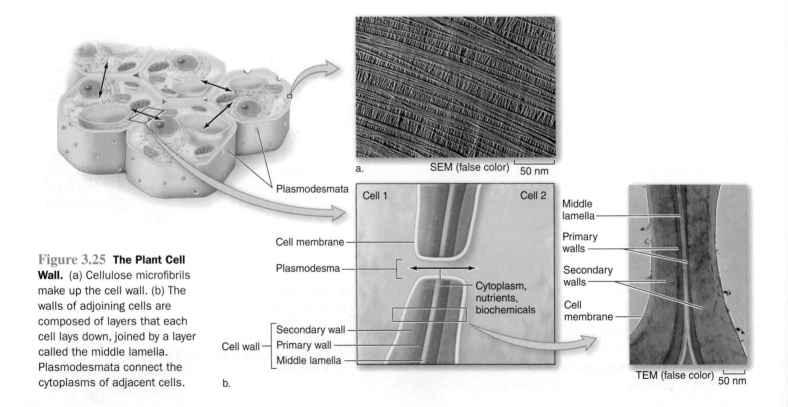

Figure 3.25 The Plant Cell Wall. (a) Cellulose microfibrils make up the cell wall. (b) The walls of adjoining cells are composed of layers that each cell lays down, joined by a layer called the middle lamella. Plasmodesmata connect the cytoplasms of adjacent cells.

a. SEM (false color) 50 nm

Plasmodesmata
Cell 1   Cell 2
Cell membrane
Plasmodesma
Cytoplasm, nutrients, biochemicals
Cell wall — Secondary wall / Primary wall / Middle lamella
b.

Middle lamella
Primary walls
Secondary walls
Cell membrane
TEM (false color) 50 nm

fibrous organization imparts great strength. Other molecules, including the polysaccharides hemicellulose and pectin, glue adjacent cells together and add strength and flexibility. Plant cell walls also contain glycoproteins, enzymes, and many other proteins. ▶ cellulose, p. 32

A plant cell secretes many of the components of its wall from the inside. The oldest layer of a cell wall is therefore on the exterior of the cell, and the newer layers hug the cell membrane. The region where adjacent cell walls meet is called the middle lamella.

Some cells have rigid secondary cell walls beneath the initial, more flexible primary one. The rigidity of the secondary cell wall comes from lignin, a polymer so complex that only a few types of organisms can decay it. Wood's toughness comes from secondary cell walls.

How do plant cells communicate with their neighbors through the wall? **Plasmodesmata** (singular: plasmodesma) are channels that connect adjacent cells. They are essentially "tunnels" in the cell wall, through which the cytoplasm, hormones, and some of the organelles of one plant cell can interact with those of another (see figure 3.25). Plasmodesmata are especially plentiful in parts of plants that conduct water or nutrients and in cells that secrete oils and nectars.

Plasmodesmata enable cell-to-cell communication and coordination of function within a plant. These tunnels, however, may also play a role in the spread of disease within a plant; viruses use them as conduits to pass from cell to cell.

B. Animal Cell Junctions Occur in Several Forms

Unlike plants and fungi, animal cells lack cell walls. Instead, many animal cells secrete a complex extracellular matrix that holds them together and coordinates many aspects of cellular life. In such tissues, cells are not in direct contact with one another. ▶ extracellular matrix, p. 514

In other tissues, however, the plasma membranes of adjacent cells directly connect to one another via several types of junctions (figure 3.26):

- A **tight junction** fuses cells together, forming an impermeable barrier between them. Proteins anchored in membranes connect to actin in the cytoskeleton and join cells into sheets, such as those lining the inside of the human digestive tract and the tubules of the kidneys. These connections allow the body to control where biochemicals move, since fluids cannot leak between the joined cells. Tight junctions create the "blood–brain barrier," densely packed cells that prevent many harmful substances from entering the brain. However, this barrier readily admits lipid-soluble drugs such as heroin and cocaine across its cell membranes, accounting for the rapid action of these drugs.
- An **anchoring (or adhering) junction** connects adjacent cells by linking their intermediate filaments in one spot,

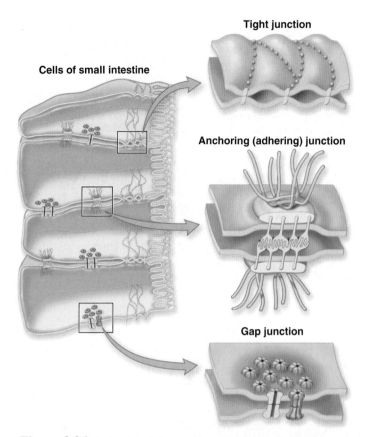

Figure 3.26 **Animal Cell Connections.** These cells illustrate all three types of animal cell junctions. Tight junctions fuse neighboring cell membranes, anchoring junctions form "spot welds," and gap junctions allow small molecules to move between adjacent cells.

somewhat like rivets or "spot welds." These junctions hold skin cells in place by anchoring them to the extracellular matrix.

- A **gap junction** is a protein channel that links the cytoplasm of adjacent cells, allowing exchange of ions, nutrients, and other small molecules. It is therefore analogous to plasmodesmata in plants. Gap junctions link heart muscle cells to one another, allowing groups of cells to contract together. Similarly, the muscle cells that line the digestive tract coordinate their contractions to propel food along its journey, courtesy of countless gap junctions.

3.6 | Mastering Concepts

1. What functions do cell walls provide?
2. What is the chemical composition of a plant cell wall?
3. What are plasmodesmata?
4. What are the three types of junctions that link cells in animals?

Table 3.2   Structures and Functions of Eukaryotic Organelles: A Summary

Organelle		Structure	Function(s)	Plant Cells?	Animal Cells?
Nucleus 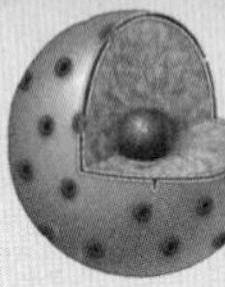		Perforated sac containing DNA, proteins, and RNA; surrounded by double membrane	Separates DNA from rest of cell; site of first step in protein synthesis; nucleolus produces ribosomal subunits	Yes	Yes
Ribosome 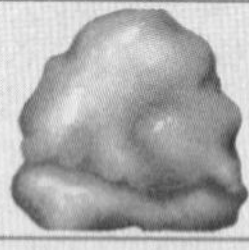		Two globular subunits composed of RNA and protein	Location of protein synthesis	Yes	Yes
Rough endoplasmic reticulum 		Membrane network studded with ribosomes	Produces proteins destined for secretion from the cell	Yes	Yes
Smooth endoplasmic reticulum		Membrane network lacking ribosomes	Synthesizes lipids; detoxifies drugs and poisons	Yes	Yes
Golgi apparatus 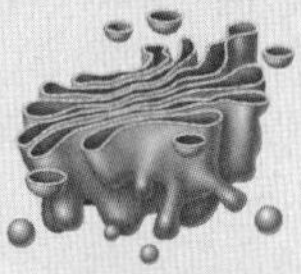		Stacks of flat, membranous sacs	Packages materials to be secreted; produces lysosomes	Yes	Yes
Lysosome 		Sac containing digestive enzymes; surrounded by single membrane	Dismantles and recycles components of food, debris, captured bacteria, and worn-out organelles	Rarely	Yes
Central vacuole 		Sac containing enzymes, acids, water-soluble pigments, and other solutes; surrounded by single membrane	Produces turgor pressure; recycles cell contents; contains pigments	Yes	No
Peroxisome 		Sac containing enzymes, often forming visible protein crystals; surrounded by single membrane	Disposes of toxins; breaks down fatty acids; eliminates hydrogen peroxide	Yes	Yes
Chloroplast 		Two membranes enclosing stacks of membrane sacs, which contain photosynthetic pigments and enzymes; contains DNA and ribosomes	Produces food (glucose) by photosynthesis	Yes	No
Mitochondrion 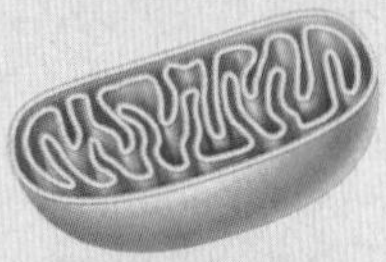		Two membranes; inner membrane is folded into enzyme-studded cristae; contains DNA and ribosomes	Releases energy from food by cellular respiration	Yes	Yes
Cytoskeleton 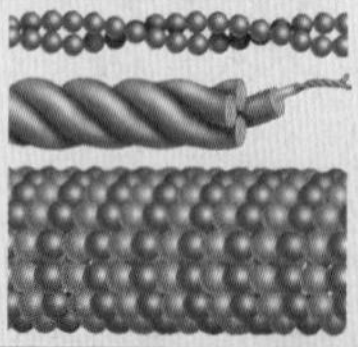		Network of protein filaments and tubules	Transports organelles within cell; maintains cell shape; basis for flagella/cilia; connects adjacent cells	Yes	Yes
Cell wall 		Porous barrier of cellulose and other substances (in plants)	Protects cell; provides shape; connects adjacent cells	Yes	No

3.7 | Investigating Life: Did the Cytoskeleton Begin in Bacteria?

Some of the most intriguing evolutionary puzzles arise when one group of organisms has a feature that its ancestors do not. The spots on a leopard's fur, the stinger on a scorpion's tail, and the spines of a cactus all prompt the question, "Where did those come from?"

On a much smaller scale, the cytoskeleton is one feature that distinguishes eukaryotic from prokaryotic cells. This supportive framework consists of actin, tubulin, and many other types of proteins. Because all eukaryotic cells have a cytoskeleton, these proteins must have been present in their last common (shared) ancestor. But the cytoskeleton is not present in prokaryotic cells, which were the first life forms. Where did this essential part of the eukaryotic cell come from?

Researchers are beginning to solve this mystery by questioning the assumption that prokaryotic cells lack a cytoskeleton. In the late 1990s, for example, researchers studying the bacterium *E. coli* discovered a protein with a structure similar to tubulin. Further study revealed that the protein is essential for cell division in both bacteria and archaea.

More recently, researchers have found two bacterial proteins similar to actin. Laura Jones, Rut Carballido-López, and Jeffery Errington at the University of Oxford discovered that the proteins form filaments that lie just inside the cell surface, helping the bacterium keep its shape. They used several tools, including altered genes, microscopy, and comparisons of molecules from several species, to learn about the proteins.

The scientists studied a bacterium called *Bacillus subtilis*. This soil-dwelling organism ordinarily forms rod-shaped cells, but the researchers noticed that when they experimentally "turned off" either of two genes, the cells had abnormal shapes. A gene is a length of DNA that specifies the amino acid sequence of one protein. If a gene is switched off, a cell cannot make the corresponding protein. When the scientists turned off the gene encoding a protein called MreB, the cells were the correct length but appeared abnormally inflated or rounded. When they turned off the gene encoding the protein Mbl instead, the cells were bent and twisted (figure 3.27a).

Another way to study protein function is to mutate the gene that encodes it. Mutating a gene usually ruins the function of the encoded protein. The researchers mutated both genes and found again that many of the mutant bacteria had abnormal shapes. These experiments provided additional evidence that *B. subtilis* requires both MreB and Mbl to form normal cells.

Next, the team used microscopes to see where in the *B. subtilis* cell the two proteins are located. An individual protein, however, is far too small for even the most powerful microscope to resolve. So the researchers used glowing fluorescent "tags" to reveal exactly where MreB and Mbl occurred. This method exploits the ability of antibodies to recognize specific molecules. Antibodies are a critical part of the human immune system,

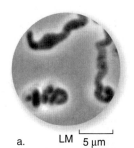 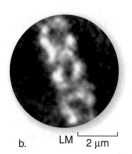

a. LM 5 μm   b. LM 2 μm

Figure 3.27 Cytoskeleton Precursor? (a) Mutated genes produced misshapen *Bacillus subtilis* cells. The cells are normally shaped like short rods. (b) Mbl protein tagged with a fluorescent label accumulates in a helix just inside the cell wall.

binding to cells or molecules that the body does not recognize as its own. Each antibody binds to just one type of molecule. The scientists created antibodies that had two critical abilities: they could bind to MreB or Mbl, and they could glow under ultraviolet light. The results were remarkable: each protein, marked with its fluorescent tag, formed a helix just inside the surface of the cell (figure 3.27b).

These experiments showed that MreB and Mbl together help determine cell shape in *B. subtilis*. Could the same proteins be at work in other microbes as well? The researchers searched the DNA sequences of dozens of other bacteria and archaea, looking for genes similar to the one encoding MreB. They found that most of the species with such a gene had cells shaped like rods, filaments, or corkscrews. Most species without it had spherical cells.

Finally, the scientists searched for proteins of similar structure in computerized databases containing amino acid sequences from other species. Both MreB and Mbl were similar to actin from both yeast and human cells. Although the sequences were not identical, they were similar at critical locations, suggesting related functions.

This research has brought biologists closer to answering the question of "Where did that come from?" for the cytoskeleton. The function, location, and amino acid sequences of MreB and Mbl in *B. subtilis* strongly suggest that the actin in eukaryotic cells has a prokaryotic counterpart with a function that persists to this day. At least some elements of the cytoskeleton apparently evolved before the two cell types diverged some 2.7 billion years ago.

Jones, Laura J. F., Rut Carballido-López, and Jeffery Errington. 2001. Control of cell shape in bacteria: helical, actin-like filaments in *Bacillus subtilis*. *Cell*, vol. 104, pages 913–922.

3.7 | Mastering Concepts

1. How did the researchers use multiple lines of evidence to answer their question?
2. How would the bacterial cells in figure 3.27b look different if Mbl occurred throughout the cytoplasm?

Chapter Summary

3.1 | Cells Are the Units of Life

- **Cells** are the microscopic components of all organisms.

A. Simple Lenses Revealed the Cellular Basis of Life

- The first person to see cells was Robert Hooke, who viewed cork with a crude lens in the late seventeenth century. Antony van Leeuwenhoek used a simple light microscope to view many cells.

B. The Cell Theory Emerges

- Schleiden, Schwann, and Virchow's formulation of the **cell theory** states that all life is composed of cells, that cells are the functional units of life, and that all cells come from preexisting cells.
- Contemporary cell biology focuses on the role of genetic information, the cell's chemical components, and the metabolic processes inside cells.

C. Microscopes Magnify Cell Structures

- Light microscopes and electron microscopes are essential tools for viewing the parts of a cell.

D. All Cells Have Features in Common

- All cells have DNA, RNA, **ribosomes** that build proteins, **cytoplasm,** and a **cell membrane** that is the interface between the cell and the outside environment.
- Complex cells also have specialized compartments called **organelles.**
- The surface area of a cell must be large relative to its volume.

3.2 | Different Cell Types Characterize Life's Three Domains

- Cells are **prokaryotic** (lacking a nucleus and other organelles) or **eukaryotic** (having a nucleus and other **organelles**). Prokaryotic cells include bacteria and archaea.

A. Domain Bacteria Contains Earth's Most Abundant Organisms

- Bacterial cells are structurally simple, but they are abundant and diverse. Most have a **cell wall** and one or more **flagella.** DNA occurs in an area called the **nucleoid.**

B. Domain Archaea Includes Prokaryotes with Unique Biochemistry

- Archaea share some characteristics with bacteria and eukaryotes but also have unique structures and chemistry.

C. Domain Eukarya Contains Organisms with Complex Cells

- Eukaryotic cells include those of protists, plants, fungi, and animals. Most eukaryotic cells are larger than prokaryotic cells.

3.3 | A Membrane Separates Each Cell from Its Surroundings

- A biological membrane consists of a **phospholipid bilayer** embedded with movable proteins and sterols, forming a **fluid mosaic.**
- Membrane proteins carry out a variety of functions.

3.4 | Eukaryotic Organelles Divide Labor

- The **endomembrane system** includes the nuclear envelope, endoplasmic reticulum, Golgi apparatus, lysosomes, vacuoles, cell membrane, and the **vesicles** that transport materials within cells.

A. The Nucleus, Endoplasmic Reticulum, and Golgi Interact to Secrete Substances

- A eukaryotic cell houses DNA in a **nucleus. Nuclear pores** allow the exchange of materials through the **nuclear envelope;** assembly of the ribosome's subunits occurs in the **nucleolus.**
- The **smooth** and **rough endoplasmic reticulum** and the **Golgi apparatus** work together to synthesize, store, transport, and release molecules.

B. Lysosomes, Vacuoles, and Peroxisomes Are Cellular Digestion Centers

- A eukaryotic cell degrades wastes and digests nutrients in **lysosomes.**
- In plants, a watery **vacuole** degrades wastes, exerts turgor pressure, and stores acids and pigments.
- **Peroxisomes** help digest fatty acids and detoxify many substances, including hydrogen peroxide.

C. Photosynthesis Occurs in Chloroplasts

- Cells of plants and algae have **chloroplasts,** organelles that use solar energy to make food.

D. Mitochondria Extract Energy from Nutrients

- Nearly all eukaryotic cells have **mitochondria.** The **cristae** (folds) of the inner mitochondrial membrane house the reactions of cellular respiration.

3.5 | The Cytoskeleton Supports Eukaryotic Cells

- The **cytoskeleton** is a network of protein rods and tubules that provides cells with form, support, and the ability to move.
- **Microfilaments** are the thinnest components of the cytoskeleton. They are composed of the protein actin.
- **Intermediate filaments** are intermediate in diameter between microtubules and microfilaments. They consist of various proteins, and they strengthen the cytoskeleton.
- **Microtubules** are hollow tubes that self-assemble from tubulin subunits. They form **cilia,** flagella, and the fibers that pull replicated chromosomes apart during cell division.

3.6 | Cells Stick Together and Communicate with One Another

A. Cell Walls Are Strong, Flexible, and Porous

- The cells of most organisms other than animals have cell walls, which provide protection and shape. Plant cell walls consist of cellulose fibrils connected by hemicellulose, pectin, and various proteins.
- **Plasmodesmata** are openings that extend between the cell walls of adjacent plant cells.

B. Animal Cell Junctions Occur in Several Forms

- Junctions connecting animal cells include **tight junctions, anchoring junctions,** and **gap junctions.** Tight junctions create a seal between adjacent cells. Anchoring junctions are "spot welds" that secure cells in place. Gap junctions allow adjacent cells to exchange signals and cytoplasmic material.

3.7 | Investigating Life: Did the Cytoskeleton Begin in Bacteria?

- Although the cytoskeleton occurs only in eukaryotic cells, bacteria do have actinlike proteins that help control cell shape.

Multiple Choice Questions

1. Why are cells considered to be the smallest unit of life?
   a. Because you need a microscope to see them
   b. Because a cell is the smallest thing that carries out all the functions of life
   c. Because they have an organized structure
   d. Because all cells have a nucleus with DNA

2. Which of the following is NOT a feature found in all cells?
   a. Proteins   c. Cell wall
   b. Ribosomes   d. Cell membrane

3. A cell membrane is said to be a *fluid mosaic* because
   a. there is water in the membrane.
   b. the membrane is made of lipids and proteins that can move.
   c. it forms a bilayer.
   d. transport proteins allow for the movement of water-soluble molecules.

4. One property that distinguishes cells in domain Bacteria from those in domain Eukarya is the presence of
   a. a cell wall.   c. flagella.
   b. DNA.   d. membranous organelles.

5. Which of the following organelles are associated with the job of cellular digestion?
   a. Lysosomes and peroxisomes
   b. Golgi apparatus and vesicles
   c. Nucleus and nucleolus
   d. Smooth and rough endoplasmic reticulum

6. Which of the following organelles does not belong to the endomembrane system?
   a. Golgi apparatus   c. Mitochondrion
   b. Endoplasmic reticulum   d. Lysosome

7. Within a single cell, which of the following is physically the smallest?
   a. Nuclear envelope   c. Cell membrane
   b. Phospholipid molecule   d. Mitochondrion

8. Which of the following organelles does not contain DNA?
   a. Nucleus
   b. Chloroplast
   c. Rough endoplasmic reticulum
   d. Mitochondrion

9. What cellular process is involved in the production of milk-specific mRNA molecules?
   a. Protein synthesis   c. Lipid synthesis
   b. Digestion   d. Cellular respiration

10. Anchoring junctions are stabilized by intermediate filaments. What else is required for a cell to form an anchoring junction?
    a. A cell wall   c. A receptor protein
    b. Extracellular matrix   d. Plasmodesmata

11. Which has a greater ratio of surface area to volume, a hippopotamus or a mouse? Which animal would lose heat faster in a cold environment and why?

12. What advantages does compartmentalization confer on a large cell?

13. List the chemicals that make up cell membranes.

14. Emulsifiers are common food additives. A typical emulsifier molecule has a hydrophilic end and a hydrophobic end. Draw a diagram explaining how an emulsifier can enable oil to mix with water.

15. Compare and contrast the phospholipid bilayer with two pieces of Velcro sticking to each other.

16. Choose one of the functions of membrane proteins mentioned in this chapter. If a person were born with a faulty version of a protein with that function, what symptoms might you predict?

17. One way to understand cell function is to compare the parts of a cell to the parts of a factory. For example, the Golgi apparatus would be analogous to the factory's shipping department. How would the other cell parts fit into this analogy?

18. Which three types of organelles in a eukaryotic cell are surrounded by a double membrane?

19. This chapter used the endomembrane system to illustrate the organelles involved in milk production. Once a baby drinks the milk, which organelles in the infant's cells extract the raw materials and potential energy in the milk to fuel growth?

20. Why does a muscle cell contain many mitochondria and a white blood cell (an immune cell that engulfs bacteria) contain many lysosomes?

21. List the components and functions of the cytoskeleton.

22. How do plant cells interact with their neighbors through the cell wall?

23. Describe how animal cells use junctions in different ways.

24. List examples of highly folded organelles with huge surface area.

Write It Out

1. How has improved technology enabled the expansion of the cell theory from Schleiden and Schwann's original formulation?

2. Until recently, biologists thought that there were only two types of cells. How has that view changed?

3. List the features that all cells share, then name three structures or activities found in eukaryotic cells but not in bacteria or archaea.

4. The simplest viruses consist only of a protein coat surrounding DNA or RNA. What does a virus have in common with cells? What does it lack that all cells have?

5. In what ways is a prokaryotic cell like a baseball stadium, but a eukaryotic cell is more like an office building?

6. Biologist J. Craig Venter has designed and built an artificial chromosome, based on the DNA in a bacterial cell. If he wants to use that DNA in an artificial cell, what other ingredients will he need? Can you foresee any benefits from, or ethical problems with, the ability to create artificial life? What do you need to know to be able to answer these questions?

7. Your friend claims that an ostrich egg is the largest single cell, but you are skeptical that one cell can be that large. If you had access to an ostrich egg and a high-quality light microscope, what features would you look for to resolve the argument?

8. Suppose you discover a new organism and that you have access to light and electron microscopes to examine its cells. List a specific question you could answer by using each type of microscope.

9. Why are large organisms made of numerous small cells instead of a few large ones?

10. A liver cell has a volume of 5000 μm^3. Its total membrane area, including the membranes of the organelles and the cell membrane, is 110,000 μm^2. A cell in the pancreas that manufactures digestive enzymes has a volume of 1000 μm^3 and a total membrane area of 13,000 μm^2. Which cell is probably more efficient in carrying out activities that require extensive membrane surfaces? Why?

Pull It Together

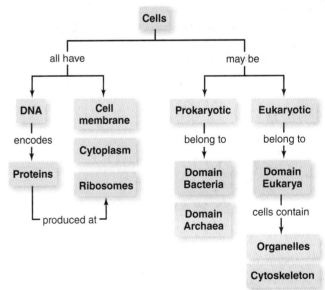

1. What are the functions of each structure in a eukaryotic cell?

2. How do a cell's organelles interact with one another?

3. Add the eukaryotic kingdoms of life to this concept map.

4. How do plant cells differ from animal cells?

5. How do plant and animal cells stick together and communicate with their neighbors?

6. Which cell types have a cell wall?

4

The Energy of Life

Riding a bicycle up a hill takes energy, which is provided by metabolic reactions inside cells.

connect |BIOLOGY

Enhance your study of
this chapter with practice quizzes,
animations and videos, answer keys,
and downloadable study tools.
www.mhhe.com/hoefnagels

Whole-Body Metabolism: Energy on an Organismal Level

"I WISH I HAD YOUR METABOLISM!" Perhaps you have overheard a calorie-counting friend make a similar comment to someone who stays slim on a diet of fattening foods.

In that context, the word *metabolism* means how fast a person burns food. But biochemists define metabolism as all of the chemical reactions that build and break down molecules within any cell. How are these two meanings related?

Interlocking networks of metabolic reactions supply the energy that every cell needs to stay alive. In humans, teams of metabolizing cells perform specialized functions such as digestion, muscle movement, hormone production, and countless other activities. It all takes a reliable energy supply—food, which we "burn" at an ever-changing rate.

Minimally, a body needs energy to maintain heartbeat, temperature, breathing, brain activity, and other basic life requirements. For an adult human male, the average energy use is 1750 Calories in 24 hours; for a female, 1450 Calories. These numbers do not include the energy required for physical activity or digestion, so the number of Calories needed to get through a day generally exceeds the minimum requirement.

Of course, these averages mask the fact that people have different metabolic rates. Age, sex, weight, and activity level all influence metabolic rates, as does body fat composition. All other things being equal, a person with the most lean tissue (muscle, nerve, liver, and kidney) will have the highest metabolic rate, because lean tissue consumes more energy than relatively inactive fat tissue. A thyroid hormone, thyroxine, also influences energy expenditure.

So why *do* some people gain weight? The simple answer is that if you eat more Calories than you spend, you gain weight. A typical food plan for a healthy, active adult includes 2000 Calories per day. By comparison, a single fast-food meal of a burger, large fries, and a chocolate shake may contain nearly 2500 Calories. These foods are inexpensive and easy to find, and it is hard to exercise enough to offset a steady diet of energy-rich foods. It is little wonder that Americans are facing an obesity epidemic.

On the other hand, if you eat fewer Calories than you spend, you slim down. This is why reducing caloric intake and exercising are two basic weight-loss recommendations. Unfortunately, metabolic differences make it difficult to translate this simple energy balance into a one-size-fits-all weight loss plan.

This chapter describes the fundamentals of metabolism, including how cells organize, regulate, and fuel the chemical reactions that sustain life.

UNIT 1

What's the Point?

Learning Outline

Learn How to Learn

Focus on Understanding, Not Memorizing

When you are learning the language of biology, be sure to concentrate on how each new term fits with the others. Are you studying multiple components of a complex system? Different steps in a process? The levels of a hierarchy? As you study, always make sure you can see how each part relates to the whole.

4.1 | All Cells Capture and Use Energy

You're running late. You overslept, you have no time for breakfast, and you have a full morning of classes. You rummage through your cupboard and find something called an "energy bar"—just what you need to get through the morning. But what is energy?

A. Energy Allows Cells to Do Life's Work

Physicists define **energy** as the ability to do work—that is, to move matter. This idea, abstract as it sounds, is fundamental to biology. Life depends on rearranging atoms and trafficking substances across membranes in precise ways. These intricate movements represent work, and they require energy.

Although it may seem strange to think of a "working" cell, all organisms do tremendous amounts of work on a microscopic scale. For example, a plant cell assembles glucose molecules into long cellulose fibers, moves ions across its membranes, and performs thousands of other tasks simultaneously. Likewise, a gazelle grazes on a plant's tissues to acquire energy that will enable it to do its own cellular work. A crocodile eats that gazelle for the same reason.

The total amount of energy in any object is the sum of energy's two forms: potential and kinetic (figure 4.1 and table 4.1). **Potential energy** is stored energy available to do work. A bicyclist at the top of a hill illustrates potential energy. Likewise, unburned gasoline—and that energy bar you grabbed—contains potential energy stored in the chemical bonds of its molecules. A chemical gradient is another form of potential energy (see section 4.5).

Kinetic energy is energy being used to do work; any moving object possesses kinetic energy. The bicyclist coasting down the hill in figure 4.1 demonstrates kinetic energy. Moving pistons, a rolling bus, and contracting muscles also have kinetic energy. Light and sound are other types of kinetic energy. Inside a cell, each molecule also has kinetic energy; in fact, all of the chemical reactions that sustain life rely on collisions between moving molecules, and many substances enter and leave cells by random motion alone.

Table 4.1 Examples of Energy in Biology

Type of Energy	Examples
Potential energy	Chemical energy (stored in bonds)
	Concentration gradient across a membrane
Kinetic energy	Light
	Sound
	Movement of atoms and molecules
	Muscle contraction

Calories are units used to measure energy. One **calorie** (cal) is the amount of energy required to raise the temperature of 1 gram of water from 14.5°C to 15.5°C. The most common unit for measuring the energy content of food, however, is the **kilocalorie** (kcal), which equals 1000 calories. (In nutrition, one food Calorie—with a capital C—is actually a kilocalorie.) A typical energy bar, for example, contains 240 kilocalories of potential energy stored in the chemical bonds of its ingredients: mostly carbohydrates, proteins, and fats.

B. The Laws of Thermodynamics Describe Energy Transfer

Thermodynamics is the study of energy transformations. The first and second laws of thermodynamics describe the energy conversions vital for life, as well as those that occur in the nonliving world. They apply to all energy transformations—gasoline combustion in a car's engine, a burning chunk of wood, or a cell breaking down glucose.

The **first law of thermodynamics** is the law of energy conservation. It states that energy cannot be created or destroyed, although energy can be converted to other forms. This means that the total amount of energy in the universe is constant.

Every aspect of life centers on converting energy from one form to another (figure 4.2). The most important energy transformations are photosynthesis and cellular respiration. In photosynthesis, plants and some microorganisms use carbon dioxide, water, and the kinetic energy in sunlight to assemble glucose molecules.

Figure 4.1 **Potential and Kinetic Energy.** Potential energy in the chemical bonds of food is converted to kinetic energy as muscles push the cyclist to the top of the hill. The potential energy of gravity provides a free ride by conversion to kinetic energy on the other side.

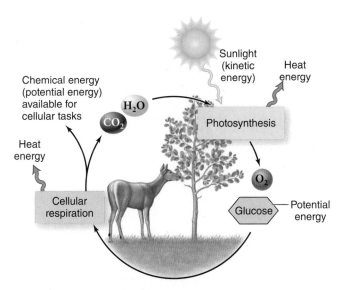

Endocytosis

Figure 4.2 Energy Can Take Many Forms. In photosynthesis, plants transform the kinetic energy in sunlight into the potential energy contained in the chemical bonds of glucose. Respiration, in turn, releases the potential energy in glucose. Heat energy is lost at every step along the way.

These carbohydrates contain potential energy in their chemical bonds. During cellular respiration, the energy-rich glucose molecules change back to carbon dioxide and water, liberating the energy necessary to power life. Cells translate the potential energy in glucose into the kinetic energy of molecular motion and use that burst of kinetic energy to do work.

Most organisms obtain energy from the sun, either directly through photosynthesis or indirectly by consuming other organisms. Even the potential energy in fossil fuels originated as solar energy. However, a few species of microorganisms can extract potential energy from the chemical bonds of inorganic chemicals, then synthesize organic compounds that are nutrients for them and the organisms that consume them.

The **second law of thermodynamics** states that all energy transformations are inefficient because every reaction loses some

energy to the surroundings as heat. If you eat your energy bar on the way to your first class, your cells can use the potential energy in its chemical bonds to make proteins, divide, or do other forms of work. According to the second law of thermodynamics, however, you will lose some energy as heat with every chemical reaction. This process is irreversible; the lost heat energy will not return to a useful form.

Heat energy results from random molecular movements. Because heat is disordered, and all energy eventually becomes heat, it follows that all energy transformations must head toward increasing disorder. **Entropy** is a measure of this randomness. In general, the more disordered a system is, the higher its entropy (figure 4.3).

Because organisms are highly organized, they may seem to defy the second law of thermodynamics—but only when considered alone as closed systems. Organisms can remain organized because they are *not* closed systems. They use incoming energy and matter from sources such as sunlight and food to maintain their organization and stay alive. The second law of thermodynamics implies that organisms can increase in complexity *as long as something else decreases in complexity by a greater amount.* Ultimately, life remains ordered and complex because the sun is constantly supplying energy to Earth. But the entropy of the universe as a whole, including the sun, is always increasing.

The ideas in this chapter and the two that follow describe how organisms acquire and use the energy they need to sustain life.

4.1 | Mastering Concepts

1. What are some examples of the "work" of a cell?
2. Give an example of how your body has both potential and kinetic energy.
3. What are the first and second laws of thermodynamics?
4. Why does the amount of entropy in the universe always increase?

Highly ordered	Highly disordered
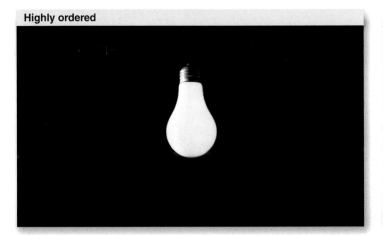	

Figure 4.3 Entropy. In an instant, a highly organized light bulb is transformed into broken glass and metal fragments. Entropy has irreversibly increased; no matter how many times you drop the glass and metal, the pieces will not arrange themselves back into a light bulb.

4.2 | Networks of Chemical Reactions Sustain Life

The number of chemical reactions occurring in even the simplest cell is staggering. Thousands of reactants and products form interlocking pathways that resemble complicated road maps.

The word **metabolism** encompasses all of these chemical reactions in cells, including those that build new molecules and those that break down existing ones. Each reaction rearranges atoms into new compounds, and each reaction either absorbs or releases energy. Digesting your morning energy bar and using its carbohydrates to fuel muscle movement are part of your metabolism. Photosynthesis and respiration are part of the metabolism of the grass under your feet as you hurry to class.

A. Chemical Reactions Absorb or Release Energy

Biologists group metabolic reactions into two categories based on energy requirements: endergonic and exergonic (**figure 4.4**). An **endergonic reaction** requires an input of energy to proceed (the prefix *end-* or *endo-* means "put into"). That is, the products contain more energy than the reactants. Typically, endergonic reactions build complex molecules from simpler components.

An example of an endergonic reaction is photosynthesis. Glucose ($C_6H_{12}O_6$), the product of photosynthesis, contains more potential energy than do carbon dioxide (CO_2) and water (H_2O), the reactants. The energy source that powers this reaction is sunlight.

In contrast, an **exergonic reaction** releases energy (*ex-* or *exo-* means "out of"). The products contain less energy than the reactants. Such reactions break large, complex molecules into their smaller, simpler components. Cellular respiration, the breakdown of glucose to carbon dioxide and water, is an example. The products, carbon dioxide and water, contain less energy than glucose.

What happens to the energy released in an exergonic reaction? According to the second law of thermodynamics, some is lost as heat; entropy always increases. But some of the energy released can be used to do work. For example, the cell may use the energy to form other bonds or to power other endergonic reactions. As we shall see, life's biochemistry is full of endergonic reactions that proceed at the expense of exergonic ones.

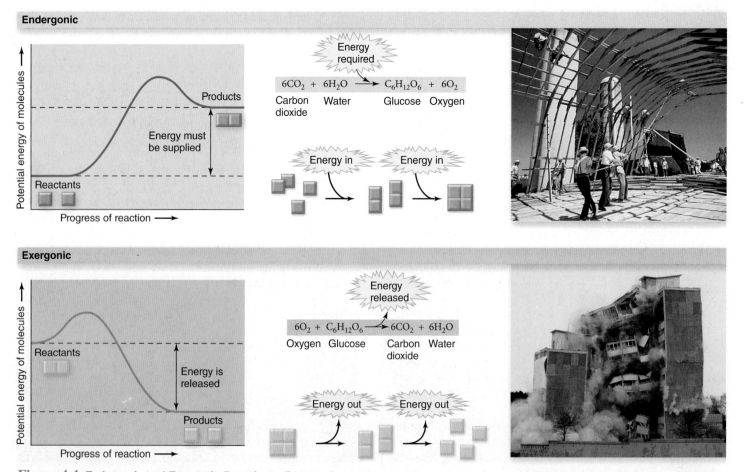

Figure 4.4 Endergonic and Exergonic Reactions. Endergonic reactions require an input of energy to build complex molecules from small components, like building a barn from bricks and boards. Exergonic reactions release energy by dismantling complex molecules.

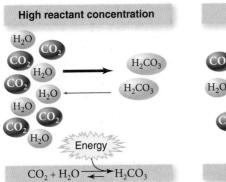

Figure 4.5
Chemical Equilibrium. At equilibrium, a reversible reaction is equally likely to proceed in either direction. By manipulating the concentrations of reactants and products, living cells can "drive" a reaction in one direction or the other.

High reactant concentration drives reaction forward.

At equilibrium, forward and reverse reactions are occurring at the same rate.

High product concentration drives reaction in reverse.

B. At Chemical Equilibrium, Reaction Rates Are in Balance

Most chemical reactions can proceed in both directions. That is, when enough product forms, some of it converts back to reactants. Arrows going in both directions between reactants and products indicate reversible reactions (figure 4.5). If reactants accumulate, the reaction is more likely to go forward; an excess of products, on the other hand, increases the chance that the reaction will proceed in reverse. At **chemical equilibrium,** the reaction goes in both directions at the same rate. Equilibrium does not necessarily mean that the *amounts* of products and reactants are equal; rather, their *rates of formation* are equal.

Cells must remain far from chemical equilibrium for their metabolic processes to occur. They do this by preventing the accumulation of products. For example, a metabolic pathway is a series of chemical reactions in which the product of one reaction is quickly consumed in the next reaction. The "disappearance" of the product prevents equilibrium, so the reactions keep moving in the direction that the cell requires.

C. Linked Oxidation and Reduction Reactions Form Electron Transport Chains

Electrons can carry energy. Most energy transformations in organisms occur in **oxidation–reduction ("redox") reactions,** which transfer energized electrons from one molecule to another.
▸ electrons, p. 21

Oxidation means the loss of electrons from a molecule, atom, or ion. Oxidation reactions, such as the breakdown of glucose to carbon dioxide and water, are exergonic; they release energy as they degrade complex molecules into simpler products. Conversely, **reduction** means a gain of electrons (plus any energy contained in the electrons). Reduction reactions are therefore endergonic; they require a net input of energy.

Oxidations and reductions occur simultaneously because electrons removed from one molecule during oxidation must join another molecule and reduce it. That is, if one molecule is reduced (gains electrons), then another must be oxidized (loses electrons).

Some proteins are electron-shuttling "specialists." Groups of these electron carriers often align in membranes. In an **electron transport chain,** each protein accepts an electron from the molecule before it and passes it to the next, like a bucket brigade (figure 4.6). Small amounts of energy are released at each step of an electron transport chain, and the cell uses this energy in other reactions. As you will see in chapters 5 and 6, electron transport chains play key roles in both photosynthesis and respiration.

4.2 | Mastering Concepts

1. What is metabolism on a cellular level?
2. Distinguish between endergonic and exergonic reactions.
3. What distinguishes a reaction that has reached chemical equilibrium?
4. What are oxidation and reduction, and why are they always linked?
5. What is an electron transport chain?

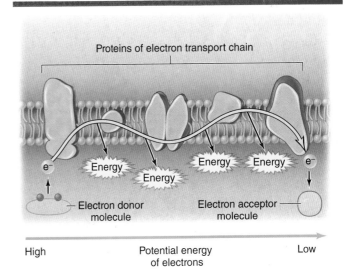

Figure 4.6 Electron Transport Chain. An electron donor transfers an electron to the first protein in the chain. This protein donates the electron to its neighbor, and so on, until the electron is transferred to a final electron acceptor. Both photosynthesis and respiration use electron transport chains.

4.3 | ATP Is Cellular Energy Currency

All cells contain a maze of interlocking chemical reactions—some releasing energy and others absorbing it. The covalent bonds of **adenosine triphosphate,** a molecule more commonly known as **ATP,** temporarily store much of the released energy. ATP holds energy released in exergonic reactions—such as the digestion of an energy bar—just long enough to power muscle contractions and all other endergonic reactions.

Recall from chapter 3 that ATP is a type of nucleotide (figure 4.7). Its components are the nitrogen-containing base adenine, the five-carbon sugar ribose, and three phosphate groups (PO_4). These phosphate groups place three negative charges very close to one another. This arrangement makes the molecule unstable, so it releases energy when the covalent bonds between the phosphates break. ▶ nucleotides, p. 40

In eukaryotic cells, organelles called mitochondria produce most of a cell's ATP. As you will see in chapter 6, a mitochondrion uses the potential energy in the bonds of one glucose molecule to generate dozens of ATP molecules in cellular respiration. Not surprisingly, the most energy-hungry cells, such as those in the muscles and brain, also contain the most mitochondria.

A. Coupled Reactions Release and Store Energy in ATP

All cells depend on the potential energy in ATP to power their activities. When a cell requires energy for an endergonic reaction, it "spends" ATP by removing the endmost phosphate group (figure 4.8). The products of this exergonic hydrolysis reaction are adenosine *di*phosphate (ADP, in which only two phosphate groups remain attached to ribose), the liberated phosphate group, and a burst of energy:

$$ATP + H_2O \longrightarrow ADP + (P) + energy$$

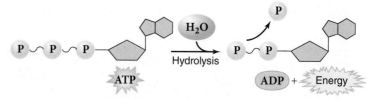

Figure 4.8 ATP Hydrolysis Releases Energy. Removing the endmost phosphate group of ATP yields ADP and a free phosphate group. The cell uses the released energy to do work.

In the reverse situation, energy can be temporarily stored by adding a phosphate to ADP, forming ATP and water:

$$ADP + (P) + energy \longrightarrow ATP + H_2O$$

The energy for this endergonic reaction comes from molecules broken down in other reactions, such as those in cellular respiration.

These reactions are fundamental to biology because ATP is the "go-between" that links endergonic to exergonic reactions. **Coupled reactions,** as their name implies, are simultaneous reactions in which one provides the energy that drives the other (figure 4.9). Cells couple the hydrolysis of ATP to endergonic reactions that occur at the same time. The ATP hydrolysis reaction drives the endergonic one, which does work or synthesizes new molecules.

As an example, consider the dehydration synthesis reaction that builds proteins from individual amino acids (see figure 2.22). This reaction is endergonic: it requires a net input of energy. When ATP provides energy, the reaction proceeds. ATP hydrolysis, an exergonic reaction, is therefore coupled to protein synthesis. ▶ dehydration synthesis and hydrolysis, p. 31

B. Transfer of Phosphate Completes the Energy Transaction

How does this coupling work? A cell uses ATP as an energy source by **phosphorylating** (transferring its phosphate group to)

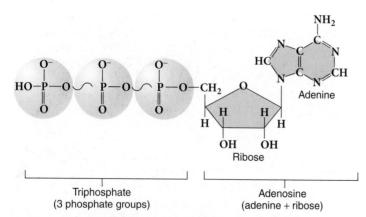

Figure 4.7 ATP. ATP is a nucleotide consisting of adenine, ribose, and three phosphate groups.

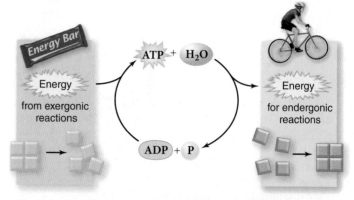

Figure 4.9 Coupled Reactions. Cells use ATP hydrolysis, an exergonic reaction, to fuel endergonic reactions. The cell regenerates ATP in other exergonic reactions, such as cellular respiration.

Apply It Now

Summer Light Show

Many organisms emit light in a process called bioluminescence. Some fishes, squids, and jellyfishes, for example, harbor pockets of glowing bacteria or protists in special light organs. The live-in microbes convert chemical energy into light energy.

Bioluminescence is much less common on land than in the sea, but one familiar example in many parts of the United States is the summertime glow of fireflies. Members of each of the more than 1900 species of fireflies use a distinctive repertoire of light signals to attract mates. Typically, flying males emit a pattern of flashes. Wingless females, called glowworms, usually are on leaves, where they emit light in response to the male's signals.

A series of chemical reactions generates the glow (figure 4.A). First, a molecule called *luciferin* reacts with ATP, yielding an intermediate compound and two phosphate groups. The enzyme *luciferase* then catalyzes a reaction of this intermediate with O_2 to yield *oxyluciferin* and a flash of light. Oxyluciferin is then reduced to luciferin, and the cycle starts over.

Although we understand the biochemistry of the firefly's glow, the ways that the animals use their bioluminescence remain mysterious. This is particularly true for the simultaneous flashing of fireflies in the same trees. When night falls, first one firefly, then another, then more begin flashing from the tree. Soon the tree twinkles like a Christmas tree. But then, order slowly descends. In small parts of the tree, the lights begin to blink on and off together. More fireflies join in. A half hour later, the entire tree seems to blink on and off every second. Biologists studying animal behavior are trying to determine just what the fireflies are doing—or saying—when they synchronize their signals.

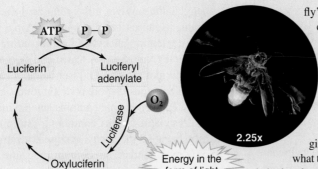

Figure 4.A

The Flash of the Firefly. ATP hydrolysis creates flashes of light as energy is transferred to a specialized molecule called luciferin.

---

another molecule. This transfer may have either of two effects. The presence of the phosphate may energize the target molecule, making it more likely to bond with other molecules. In this way, ATP fuels endergonic reactions. The other possible consequence of phosphorylation is a change in the shape of the target molecule. For example, adding phosphate can force a protein to take a different shape; removing phosphate returns the protein to its original form. The cell uses these changes to move substances throughout the cell. Muscle contraction is the large-scale effect of millions of small molecules changing shape in a coordinated way. ATP provides the energy.

ATP is sometimes described as energy "currency" for the cell. Just as you can use money to purchase a great variety of different products, all cells use ATP in many chemical reactions to do different kinds of work. Besides muscle contraction, other examples of jobs that require ATP include transporting substances across cell membranes, moving chromosomes during cell division, and synthesizing the large molecules that make up cells.

ATP is also analogous to a fully charged rechargeable battery. A full battery represents a versatile source of potential energy that can provide power to many types of electronic devices. Although a dead battery is no longer useful as an energy source, you can recharge a spent battery to restore its utility. Likewise, the cell can use respiration to reconstitute its pool of ATP.

C. ATP Represents Short-Term Energy Storage

Organisms require huge amounts of ATP. A typical adult human uses the equivalent of 2 billion ATP molecules a minute just to stay alive. Organisms recycle ATP at a furious pace, adding phosphate groups to ADP to reconstitute ATP, using the ATP to drive reactions, and turning over the entire supply every minute or so. If you ran out of ATP, you would die instantly.

Even though ATP is essential to life, cells do not stockpile it in large quantities. ATP's high-energy phosphate bonds make the molecule too unstable for long-term storage. Instead, cells store energy-rich molecules such as fats, starch, and glycogen. When ATP supplies run low, cells divert some of their lipid and carbohydrate reserves to the metabolic pathways of cellular respiration. This process soon produces additional ATP.

4.3 | Mastering Concepts

1. What is the molecular structure of ATP?
2. How does ATP hydrolysis supply energy for cellular functions?
3. Describe the relationships among endergonic reactions, ATP hydrolysis, and cellular respiration.

4.4 | Enzymes Speed Biochemical Reactions

Enzymes are among the most important of all biological molecules. An **enzyme** is an organic molecule that catalyzes (speeds up) a chemical reaction without being consumed. Most enzymes are proteins, although some are made of RNA.

Many of the cell's organelles, including mitochondria, chloroplasts, lysosomes, and peroxisomes, are specialized sacs of enzymes. Enzymes copy DNA, build proteins, digest food, recycle a cell's worn-out parts, and catalyze oxidation-reduction reactions, just to name a few of their jobs. Without enzymes, all of these reactions would proceed far too slowly to support life.

A. Enzymes Bring Reactants Together

Enzymes speed reactions by lowering the **energy of activation,** the amount of energy required to start a reaction (figure 4.10). Even exergonic reactions, which ultimately release energy, require an initial "kick" to get started. The enzyme brings reactants (also called substrates) into contact with one another, so that less energy is required for the reaction to proceed.

Most enzymes can catalyze only one or a few chemical reactions. An enzyme that dismantles a fatty acid, for example, cannot break down the starch in your energy bar. The key to this specificity lies in the shape of the enzyme's **active site,** the region to which the substrates bind (figure 4.11). The substrates fit like puzzle pieces into the active site. Once the reaction occurs, the enzyme releases the products. Note that the reaction does not consume or alter the enzyme. Instead, after the protein releases the products, its active site is empty and ready to pick up more substrate.

Enzymes are so critical to life that just one faulty or missing enzyme can have dramatic effects. Lactose intolerance is one example. People whose intestinal cells do not secrete an enzyme called lactase cannot digest milk sugar (lactose). Fortunately, a product called Lactaid can supply the missing enzyme. Phenylketonuria (PKU) is a much more serious disease. A PKU sufferer lacks an enzyme required to break down an amino acid called

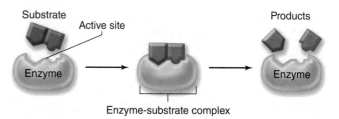

Figure 4.11 How Enzymes Work. An enzyme's active site has a specific shape that binds to one or more substrates. After the reaction is complete, the enzyme releases the product.

phenylalanine. When this amino acid accumulates in the bloodstream, it causes brain damage. People with PKU must avoid foods containing phenylalanine, including the artificial sweetener aspartame (NutraSweet). ▶ artificial sweeteners, p. 39

Enzymes also have household applications. Many detergents contain enzymes that break down food stains on clothing or dirty dishes. Raw pineapple contains an enzyme that breaks down protein. A gelatin dessert containing raw pineapple will fail to solidify because the fruit's enzymes destroy the gelatin protein. Some meat tenderizers contain the same enzyme, which breaks down muscle tissue and makes the meat easier to chew.

B. Enzymes Have Partners

Nonprotein "helpers" called **cofactors** are substances that must be present for an enzyme to catalyze a chemical reaction. Cofactors are often oxidized or reduced during the reaction, but, like enzymes, they are not consumed. Instead, they return to their original state when the reaction is complete.

Some cofactors are metals such as zinc, iron, and copper. Magnesium ions (Mg^{2+}), for example, help to stabilize many important enzymes. Other cofactors are organic molecules; an organic cofactor is called a coenzyme. The cell uses many water-soluble vitamins, including B_1, B_2, B_6, B_{12}, niacin, and folic acid, to produce coenzymes; vitamin C is a coenzyme itself. Diets lacking in vitamins can lead to reduced enzyme function and, eventually, diseases such as scurvy and beriberi.

C. Cells Control Reaction Rates

The intricate network of metabolic pathways may seem chaotic, but in reality it is just the opposite. Cells precisely control the rates of their chemical reactions. If they did not, some vital compounds would always be in short supply, and others might accumulate to wasteful (or even toxic) levels.

One way to regulate a metabolic pathway is by **negative feedback** (also called feedback inhibition), in which the product of a reaction inhibits the enzyme that controls its formation (figure 4.12). For example, the production of amino acids requires multiple steps. When an amino acid accumulates, it binds to an enzyme that acts early in the synthesis pathway. For a time, the synthesis of that amino acid stops. But when the level falls, the block on the enzyme lifts, and the cell can once again produce the amino acid.

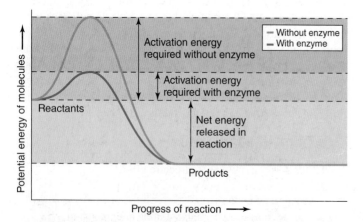

Figure 4.10 Less Energy Required. Enzymes speed chemical reactions by lowering the activation energy required to start the reaction.

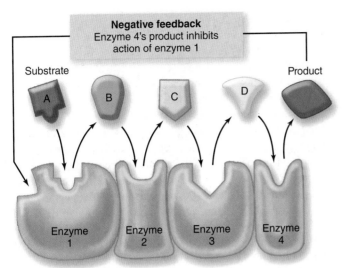

Figure 4.12 Negative Feedback. Histidine production requires several enzyme-catalyzed steps. When histidine accumulates, it inhibits the activity of the first enzyme in the pathway, temporarily halting further production.

Negative feedback works in two general ways to prevent too much of a substance from accumulating (figure 4.13). In **noncompetitive inhibition,** product molecules bind to the enzyme at a location other than the active site, in a way that alters the enzyme's shape so that it can no longer bind substrate. (Figure 4.12 shows an example.) Alternatively, in **competitive inhibition,** the product of a reaction binds to the enzyme's active site, preventing it from binding substrate. It is "competitive" because the product competes with the substrate to occupy the active site.

The opposite of negative feedback is **positive feedback,** in which a product activates the pathway leading to its own production. Blood clotting, for example, begins when a biochemical pathway synthesizes fibrin, a threadlike protein. The products of the later reactions in the clotting pathway stimulate the enzymes that catalyze fibrin production. Fibrin accumulates faster and faster, until there is enough to stem the blood flow. When the clot forms and the blood

flow stops, however, the clotting pathway shuts down—an example of negative feedback. Positive feedback is much rarer than negative feedback in organisms.

D. Many Factors Affect Enzyme Activity

Enzymes are very sensitive to conditions in the cell. An enzyme can become denatured and stop working if the pH changes or if the salt concentration becomes too high or too low. Temperature also greatly influences enzymes (figure 4.14). Enzyme action generally speeds up as the temperature climbs because reactants have more kinetic energy at higher temperatures. If it gets too hot, however, the enzyme rapidly denatures and can no longer function. ▸ denatured proteins, p. 38

Pharmaceutical drugs can also inhibit enzyme function. Many antibiotics, including triclosan (an ingredient in antibacterial soap), kill microorganisms—but not people—by inhibiting enzymes not present in our own cells. Aspirin relieves pain by binding to an enzyme that cells use to produce pain-related molecules called prostaglandins. Likewise, some poisons are also enzyme inhibitors. For example, a chemical called glyphosate (the active ingredient in an herbicide called Roundup®) competitively inhibits an enzyme found in plant cells but not in animals.

4.4 | Mastering Concepts

1. What do enzymes do in cells?
2. How does an enzyme lower a reaction's activation energy?
3. Distinguish between an enzyme and a coenzyme.
4. What are the roles of negative and positive feedback?
5. List three conditions that influence enzyme activity.

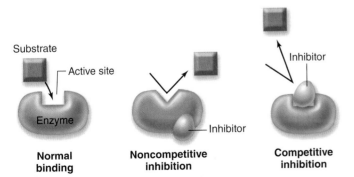

Figure 4.13 Enzyme Inhibitors. In noncompetitive inhibition, a substance binds to an enzyme in a place other than the active site, changing the shape of the protein. A competitive inhibitor physically blocks the active site of an enzyme.

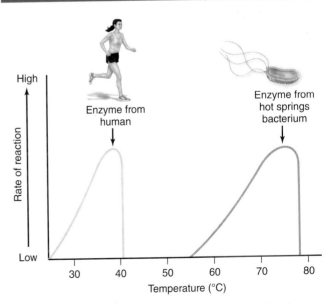

Figure 4.14 Temperature Matters. These graphs show how temperature affects the activity of enzymes from a human (left) and a bacterium that lives in hot springs (right). The microbes have heat-tolerant enzymes that denature only at very high temperatures.

4.5 | Membrane Transport May Release Energy or Cost Energy

The cell membrane is a busy place. Like a well-used border crossing between two countries, raw materials enter the cell and wastes exit in a continuous flow of traffic.

How do membranes regulate the traffic into and out of the cell? As described in chapter 3, a biological membrane is a phospholipid bilayer studded with proteins. This structure means that a membrane is "choosy," or selectively permeable. Some substances pass freely through the bilayer, but others—such as the sugar from a digested energy bar—require help from proteins.

Thanks to the regulation of membrane transport, the interior of a cell is chemically different from the outside. Concentrations of some dissolved substances (solutes) are higher inside the cell than outside, and others are lower. Likewise, the inside of each organelle in a eukaryotic cell may be chemically quite different from the solution in the rest of the cell.

The term *gradient* describes any such difference between two neighboring regions. In a **concentration gradient,** a solute is more concentrated in one region than in a neighboring region. For example, you can immediately see a concentration gradient when you first place a tea bag in a cup of hot water: near the tea bag, there are many more brown tea molecules than elsewhere in the cup (**figure 4.15**).

Over time, however, the brownish color spreads to create a uniform brew. This occurs because a concentration gradient dissipates *unless energy is expended to maintain it*. Random molecular motion always increases the amount of disorder (entropy). It costs energy to counter this tendency toward disorder. For the same reason, however, an existing concentration gradient represents a form of potential energy.

All forms of transport across membranes involve gradients. In the simplest types of transport, a gradient dissipates across the membrane. A substance moving from an area where it is more concentrated to an area where it is less concentrated is said to be "moving down" or "following" its concentration gradient. In other situations, a cell spends energy to maintain a concentration difference. This section describes three basic forms of traffic across the membrane; table 4.2 on page 82 provides a summary.

A. Passive Transport Does Not Require Energy Input

In **passive transport,** a substance moves across a membrane without the direct expenditure of energy. All forms of passive transport involve **diffusion,** the spontaneous movement of a substance from a region where it is more concentrated to a region where it is less concentrated (see figure 4.15). Because diffusion represents the dissipation of a chemical gradient—and the loss of potential energy—it does not require energy input.

How does any substance "know" which way to diffuse? The answer is, of course, that atoms and molecules know nothing. Diffusion occurs because all substances have kinetic energy; that is, they are in constant, random motion. To simplify the tea example, suppose each molecule can move randomly along one of 10 possible paths (in reality, the number of possible directions is infinite). Assume further that only one path leads back to the tea bag. Since nine of the 10 possibilities point away from the tea bag, the tea molecules tend to spread out; that is, they move down their concentration gradient.

If diffusion continues long enough, the gradient disappears. Diffusion *appears* to stop at that point, but the molecules do not stop moving. Instead, they continue to travel randomly back and forth at the same rate, so at equilibrium the concentration remains equal throughout the solution.

Simple Diffusion: No Proteins Required

Simple diffusion is a form of passive transport in which a substance moves down its concentration gradient without the use of a carrier molecule. Substances may enter or leave cells by simple diffusion only if they can pass freely through the membrane. Lipids and small, nonpolar molecules such as oxygen (O_2) and carbon dioxide (CO_2), for example, diffuse easily across the hydrophobic portion of a biological membrane.

If gradients dissipate without energy input, how can a cell use simple diffusion to acquire essential substances or get rid of toxic wastes? The answer is that the cell maintains the gradients, either by continually consuming the substances as they diffuse in or by producing more of the substances that diffuse out. For example, mitochondria consume O_2 as soon as it diffuses into the cell, maintaining the O_2 gradient that drives diffusion. Respiration also produces CO_2, which diffuses out because its concentration always remains higher in the cell than outside.

Osmosis: Diffusion of Water Across a Selectively Permeable Membrane

Two solutions of different concentrations may be separated by a selectively permeable membrane through which water, but not solutes, can pass. In that case, water will diffuse down its own gradient toward the side with the high solute concentration. **Osmosis** is this simple diffusion of water across a selectively permeable membrane (**figure 4.16**).

A human red blood cell demonstrates the effects of osmosis (**figure 4.17**). The cell's interior is normally **isotonic** to the surrounding blood plasma, which means that the plasma's solute concentration is the same as the inside of the cell (*iso-* means "equal", and *tonicity* is the ability of a solution to cause water movement). Water

Figure 4.15 A Gradient Dissipates.
Solute particles from a tea bag can move in any direction, with only a few paths leading back to the source. Eventually, the solutes are distributed uniformly throughout the cup.

Solvent

Solute

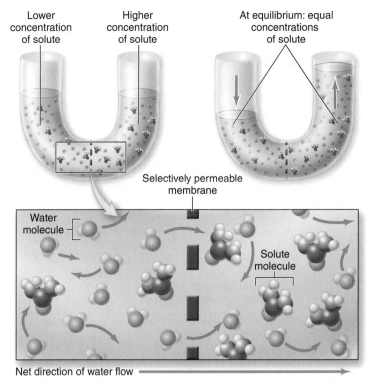

Net direction of water flow

Figure 4.16 Osmosis. The selectively permeable membrane dividing this U-shaped tube permits water but not solutes to pass. Water diffuses from the left side (low solute concentration) toward the right side (high solute concentration). At equilibrium, water flow is equal in both directions, and the solute concentrations will be equal on both sides of the membrane.

therefore moves into and out of the cell at equal rates. In a **hypotonic** environment, the solute concentration is lower than it is inside the cell (*hypo-* means "under," as in hypodermic). Water therefore moves by osmosis into a blood cell placed into hypotonic surroundings; since animal cells lack a cell wall, the membrane may even burst. Conversely,

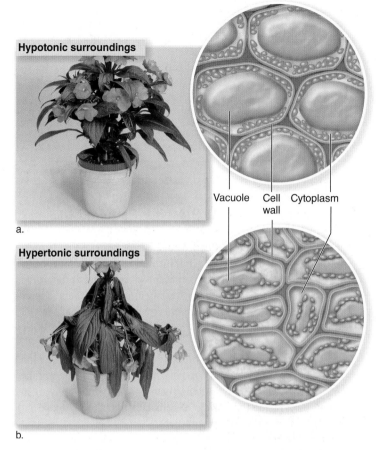

Hypotonic surroundings

a.

Hypertonic surroundings

b.

Vacuole Cell wall Cytoplasm

Figure 4.18 Plant Cells Keep Their Shapes by Regulating Diffusion. (a) The interior of a plant cell usually contains more concentrated solutes than its surroundings. Water enters the cell by osmosis, generating turgor pressure. (b) In a hypertonic environment, turgor pressure is low. The plant wilts.

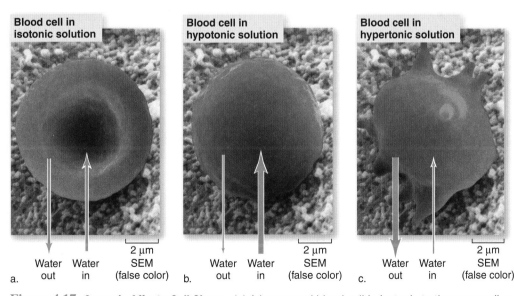

Blood cell in isotonic solution

Water out Water in 2 μm SEM (false color)

a.

Blood cell in hypotonic solution

Water out Water in 2 μm SEM (false color)

b.

Blood cell in hypertonic solution

Water out Water in 2 μm SEM (false color)

c.

Figure 4.17 Osmosis Affects Cell Shape. (a) A human red blood cell is isotonic to the surrounding plasma. Water enters and leaves the cell at the same rate, and the cell maintains its shape. (b) When the salt concentration of the plasma decreases, water flows into the cell faster than it leaves. The cell swells and may even burst. (c) In salty surroundings, the cell loses water and shrinks.

hypertonic surroundings have a higher concentration of solutes than the cell's cytoplasm (*hyper-* means "over," as in hyperactive). In a hypertonic environment, a cell shrivels and may die for lack of water.

Hypotonic and *hypertonic* are relative terms that can refer to the surrounding solution or to the solution inside the cell. The same solution might be hypertonic to one cell but hypotonic to another, depending on the solute concentrations inside the cells.

A plant's roots are often hypertonic to the soil, particularly after a heavy rain. Water rushes in, and the central vacuoles of the plant cells expand until the cell walls constrain their growth. **Turgor pressure** is the resulting force of water against the cell wall (**figure 4.18**). A limp, wilted piece of lettuce demonstrates the effect of lost turgor pressure. But the leaf becomes crisp again if placed in water, as individual cells expand like inflated balloons. Turgor pressure helps keep plants erect.

Figure It Out

A 0.9% salt solution is isotonic to human red blood cells. What will happen if you place a red blood cell in a 2.0% solution of salt water?

Answer: Water will leave the cell.

Facilitated Diffusion: Proteins Required   Ions and polar molecules cannot freely cross the hydrophobic part of the phospholipid bilayer; instead, transport proteins form "pores" that help these solutes cross. **Facilitated diffusion** is a form of passive transport in which a membrane protein assists the movement of a polar solute along its concentration gradient. Facilitated diffusion does not require energy expenditure because the solute moves from where it is more concentrated to where it is less concentrated.

Glucose moves into red blood cells via facilitated diffusion. This sugar is too hydrophilic to pass freely across the membrane, but glucose transporter proteins form channels that allow it in. Respiration inside the red blood cells consumes the glucose and maintains the concentration gradient.

Membrane proteins can enhance osmosis, too. Although membranes are somewhat permeable to water, osmosis can be slow. The cells of many organisms, including bacteria, plants, and animals, use membrane proteins called aquaporins to increase the rate of water flow. Kidney cells control the amount of water that enters urine by changing the number of aquaporins in their membranes.

B. Active Transport Requires Energy Input

Both simple diffusion and facilitated diffusion dissipate an existing concentration gradient. Often, however, a cell needs to do the opposite: create and maintain a concentration gradient. A plant's root cell, for example, may need to absorb nutrients from soil water that is much more dilute than the cell's interior. In **active transport,** a cell uses a transport protein to move a substance *against* its concentration gradient—from where it is less concentrated to where it is more concentrated (figure 4.19). Because a gradient represents a form of potential energy, the cell must

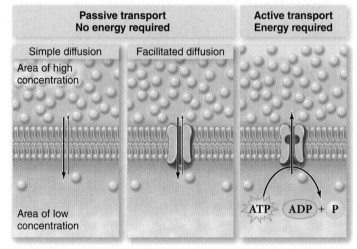

Figure 4.19 Passive and Active Transport Compared. Passive transport, which includes simple diffusion and facilitated diffusion, does not require direct energy input to move a substance down its concentration gradient. Active transport uses energy and proteins to move a substance against its concentration gradient.

Table 4.2   Movement Across Membranes: A Summary

Mechanism	Characteristics	Example
Passive transport	Net movement is down concentration gradient; does not require energy input.	
Simple diffusion	Substance to which the membrane is permeable moves across a membrane without the assistance of transport proteins.	Oxygen gas diffuses from lung into blood vessel.
Osmosis	Water diffuses across a selectively permeable membrane (note that protein channels called aquaporins enhance osmotic rate in many cells).	Water is reabsorbed into blood from kidney tubules.
Facilitated diffusion	Substances to which the membrane is not permeable move across the membrane with the assistance of transport proteins.	Glucose diffuses into red blood cells.
Active transport	Transport protein moves substance against its concentration gradient; requires energy input, often from ATP.	Salts are reabsorbed into blood from kidney tubules.
Transport using vesicles	Membranous vesicle carries materials into or out of a cell.	
Endocytosis	Membrane engulfs a substance and draws it into the cell.	White blood cells ingest bacteria.
Exocytosis	Vesicle fuses with cell membrane, releasing substances outside of the cell.	Neuron secretes neurotransmitters.

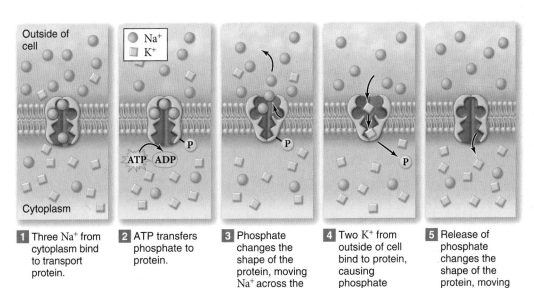

1. Three Na$^+$ from cytoplasm bind to transport protein.

2. ATP transfers phosphate to protein.

3. Phosphate changes the shape of the protein, moving Na$^+$ across the membrane.

4. Two K$^+$ from outside of cell bind to protein, causing phosphate release.

5. Release of phosphate changes the shape of the protein, moving K$^+$ into the cytoplasm.

Figure 4.20 The Sodium–Potassium Pump. This "pump" is a protein embedded in the cell membrane. It uses energy released in ATP hydrolysis to move potassium ions (K$^+$) into the cell and sodium ions (Na$^+$) out of the cell. In each case, the ions move from where they are less concentrated to where they are more concentrated.

expend energy to create it. Energy for active transport often comes from ATP.

Cells must contain high concentrations of K$^+$ and low concentrations of Na$^+$ to perform many functions. In animals, for example, sodium and potassium ion gradients are essential for nerve and muscle function (see chapters 25 and 28). One active transport system in the membranes of most animal cells is a protein called the **sodium–potassium pump** (figure 4.20), which uses ATP as an energy source to expel three sodium ions (Na$^+$) for every two potassium ions (K$^+$) it admits. Maintaining these ion gradients is costly: the million or more sodium–potassium pumps embedded in a cell's membrane use some 25% of the cell's ATP.

Concentration gradients are an important source of potential energy that cells can use to do work. For example, chapters 5 and 6 describe how cells establish concentration gradients of hydrogen ions (H$^+$) during photosynthesis and respiration. By controlling how and when H$^+$ diffuses back across the membrane, the chloroplast or mitochondrion can convert the potential energy stored in the gradient into another form of potential energy—chemical energy in the bonds of ATP.

C. Endocytosis and Exocytosis Use Vesicles to Transport Substances

Most molecules dissolved in water are small, and they can cross cell membranes by simple diffusion, facilitated diffusion, or active transport. Large particles, however, must enter and leave cells with the help of vesicles that form from cell membranes.

Endocytosis allows a cell to engulf fluids and large molecules and bring them into the cell (figure 4.21). The cytoskeleton deforms in a way that forms a small indentation in the cell membrane. The indentation becomes a "bubble" of membrane that closes in on itself, forming a vesicle that traps whatever was outside the membrane. ▶ cytoskeleton, p. 62

The two main forms of endocytosis are pinocytosis and phagocytosis. In pinocytosis, the cell engulfs small amounts of fluids and dissolved substances. In **phagocytosis,** the cell

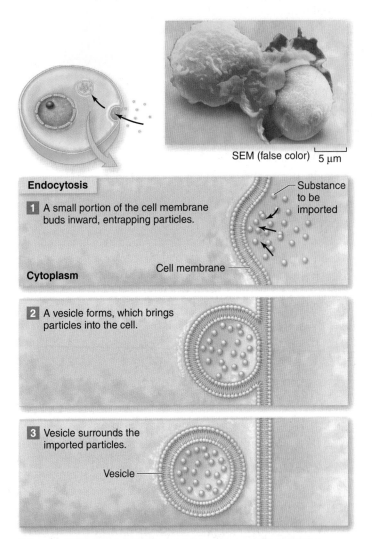

SEM (false color) 5 μm

Endocytosis

1. A small portion of the cell membrane buds inward, entrapping particles.

Substance to be imported

Cytoplasm

Cell membrane

2. A vesicle forms, which brings particles into the cell.

3. Vesicle surrounds the imported particles.

Vesicle

Figure 4.21 Endocytosis. Large particles enter a cell by endocytosis. The inset (top right) shows a white blood cell engulfing a yeast cell by phagocytosis, a form of endocytosis.

Burning Question

What causes headaches?

Headaches have many triggers. Muscle tension, nighttime teeth-grinding, stress and anxiety, bright lights, and some food ingredients are just a few examples.

Dehydration can also cause headaches. The body becomes dehydrated when a person does not drink enough fluids to compensate for water lost in urine, sweat, and breathing. Not just any fluid will do. In fact, the infamous "hangover headache" associated with a night of heavy drinking is likely a result of dehydration.

The cause-and-effect relationship between alcohol consumption and dehydration originates at the kidneys. Normally, when the concentration of solutes in blood is too high, the brain releases a hormone that stimulates the kidneys to conserve water (see chapter 32). As the kidneys return more water to the bloodstream, urine production declines.

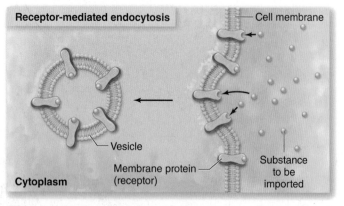

Alcohol interferes with the "water conservation" hormone. The kidneys produce more urine, and the body becomes dehydrated. A headache soon follows. To prevent or cure this painful side effect, experts recommend drinking more water, both with the alcohol and after the merriment ends.

Submit your burning question to:
marielle_hoefnagels@mcgraw-hill.com

Figure 4.23 Exocytosis. Cells package substances to be secreted into vesicles, which fuse with the cell membrane to release the materials.

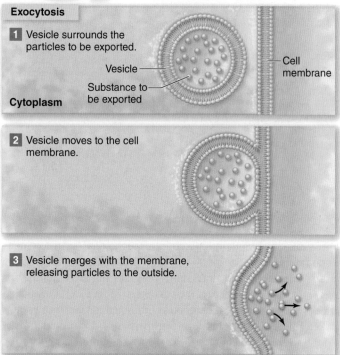

Exocytosis

1 Vesicle surrounds the particles to be exported.

Vesicle —
Substance to — be exported
Cytoplasm
— Cell membrane

2 Vesicle moves to the cell membrane.

3 Vesicle merges with the membrane, releasing particles to the outside.

captures and engulfs large particles, such as debris or even another cell. The vesicle then fuses with a lysosome, where hydrolytic enzymes dismantle the cargo. ▶ lysosomes, p. 58

When biologists first viewed endocytosis in white blood cells in the 1930s, they thought a cell would gulp in anything at its surface. They now recognize a more specific form of the process called receptor-mediated endocytosis (figure 4.22). A receptor protein on a cell's surface binds a biochemical; the cell membrane then indents, embracing the substance and drawing it

into the cell. Liver cells use receptor-mediated endocytosis to absorb cholesterol-toting proteins from the bloodstream.

Exocytosis, the opposite of endocytosis, uses vesicles to transport fluids and large particles out of cells (figure 4.23). Inside a cell, the Golgi apparatus produces vesicles filled with substances to be secreted. The vesicle moves to the cell membrane and joins with it, releasing the substance outside the membrane. For example, the tip of a neuron releases neurotransmitters by exocytosis; these chemicals then stimulate or inhibit neural impulses in a neighboring cell. The secretion of milk into milk ducts, depicted in figure 3.13, is another example.

Figure 4.22 Receptor-Mediated Endocytosis. The binding of a substance to a receptor protein triggers receptor-mediated endocytosis.

Receptor-mediated endocytosis
— Cell membrane
— Vesicle
Membrane protein (receptor)
Cytoplasm
Substance to be imported

4.5 | Mastering Concepts

1. What is diffusion?
2. What types of substances diffuse freely across a biological membrane?
3. How do differing concentrations of solutes in neighboring solutions drive osmosis?
4. Why does it cost energy for a cell to maintain a concentration gradient?
5. Distinguish between simple diffusion, facilitated diffusion, and active transport.
6. How do exocytosis and endocytosis use vesicles to transport materials across cell membranes?

4.6 | Investigating Life: Does Natural Selection Maintain Some Genetic Illnesses?

An individual enzyme or membrane protein may seem too small to be very important—until you consider that a single faulty one can cause serious illness. Cystic fibrosis, a disease that affects about 30,000 Americans, is one example. One in every 2500 babies born each year in the United States has cystic fibrosis. Each affected person lacks a protein called CFTR in his or her cell membranes.

CFTR, which stands for "*cystic fibrosis transmembrane conductance regulator*," is a membrane transport protein that moves chloride ions (Cl⁻) out of cells by active transport. As it does so, the solute concentration outside the cell increases, drawing water out by osmosis. Not surprisingly, CFTR occurs in tissues that secrete watery fluids such as mucus and perspiration.

One of the many locations where CFTR does its job is in cells lining the airspaces of the lungs. Water moving out of these cells thins the lung's mucus, which beating cilia clear away. Patients with cystic fibrosis, however, lack a working CFTR protein. The mucus in the lungs remains thick, making breathing difficult and creating an ideal breeding ground for bacteria. The patient eventually succumbs to chronic infections, often before age 30.

Cystic fibrosis may render patients too sick to have children or even take their lives before they are old enough to reproduce. So why hasn't natural selection eliminated this deadly disease from the human population? A possible answer to this evolutionary mystery may lie not in the lungs but in another place where CFTR proteins occur: the cells lining the digestive tract.

Some disease-causing bacteria that affect the digestive tract exploit CFTR. For example, bacteria that cause cholera *(Vibrio cholerae)* produce a toxin that overstimulates CFTR, triggering Cl⁻ and water to pour from the lining of the small intestine. The water and ions (along with many *Vibrio* cells) leave the body in watery diarrhea; the resulting dehydration can be deadly if left untreated. Researcher Sherif Gabriel and his colleagues at the University of North Carolina hypothesized that the abnormal CFTR protein associated with cystic fibrosis may actually increase resistance to cholera.

The team knew that everyone has two versions (alleles) of every gene, one inherited from each parent. To test their hypothesis, the researchers therefore bred three groups of mice. The animals in one group had two normal (functioning) CFTR gene copies. A second set of mice had two defective copies, and a third group had one normal and one defective copy.

The team then gave all the mice *Vibrio cholerae* toxin (via a feeding tube) and measured the amount of fluid produced in the small intestine (**figure 4.24**). As predicted, mice with two normal copies of the CFTR-encoding gene produced the most fluid, indicating they were highly vulnerable to cholera. Mice with two faulty genes resisted the toxin's effects, and those with two different copies lost intermediate amounts of fluid. The amount of faulty CFTR was therefore correlated with resistance to cholera.

This study helps explain how natural selection might maintain harmful alleles in populations. A person develops cystic fibrosis only if he or she receives a defective copy of the CFTR-encoding gene from both parents. On average, cystic fibrosis sufferers leave fewer offspring than healthy people do. But inheriting just one normal CFTR gene is enough to keep cystic fibrosis from developing. Evolutionary biologists suggest that, in some areas of the world, cholera resistance gives people with one faulty CFTR gene a reproductive edge over people with two copies of the normal gene. This advantage would occur wherever cholera threatens human populations.

From an evolutionary point of view, improved resistance to infectious disease apparently offsets losing some children to cystic fibrosis. A similar phenomenon occurs in human populations with a high frequency of the sickle cell trait. In that case, inheriting one copy of a faulty hemoglobin gene confers some resistance to another infectious disease, malaria.

Membrane proteins are among the most important components of cells. Studying their connection to genetic diseases such as cystic fibrosis may someday lead to a cure. At the same time, it is intriguing to consider that the defective gene that causes this terrible illness also has a flip side.

Gabriel, Sherif E., K. N. Brigman, B. H. Koller, et al. Oct. 7, 1994. Cystic fibrosis heterozygote resistance to cholera toxin in the cystic fibrosis mouse model. *Science,* vol. 266, pages 107–109.

4.6 | Mastering Concepts

1. What is the normal role of CFTR in humans, and how can faulty CFTR proteins cause cystic fibrosis?
2. Summarize the question Gabriel and his colleagues asked, and explain how their experiment helped answer the question.
3. How do you think the results in figure 4.24 would have been different if, before adding cholera toxin, the researcher had added a chemical that blocked the site at which the toxin binds to CFTR?

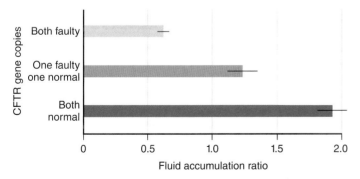

Figure 4.24 Cholera Toxin and CFTR. This graph shows the amount of fluid accumulated in the small intestines of mice after exposure to the cholera toxin. Mice with normal CFTR proteins lost the most fluid and were therefore the most susceptible to cholera. (Error bars represent standard errors; see Appendix B.)

Chapter Summary

4.1 | All Cells Capture and Use Energy

A. Energy Allows Cells to Do Life's Work
- **Energy** is the ability to do work. **Potential energy** is stored energy, and **kinetic energy** is action.
- Energy is measured in units called **calories.** One food Calorie is 1000 calories, or 1 **kilocalorie.**

B. The Laws of Thermodynamics Describe Energy Transfer
- The **first law of thermodynamics** states that energy cannot be created or destroyed but only converted to other forms.
- The **second law of thermodynamics** states that all energy transformations are inefficient because every reaction results in increased **entropy** (disorder) and the loss of usable energy as heat.

4.2 | Networks of Chemical Reactions Sustain Life

- **Metabolism** is the sum of the chemical reactions in a cell.

A. Chemical Reactions Absorb or Release Energy
- **Endergonic reactions** require energy input because the products have more energy than the reactants.
- Energy is released in **exergonic reactions,** in which the products have less energy than the reactants.

B. At Chemical Equilibrium, Reaction Rates Are in Balance
- At **chemical equilibrium,** a reaction proceeds in both directions at the same rate. Cells avoid chemical equilibrium by consuming reaction products, driving reactions forward.

C. Linked Oxidation and Reduction Reactions Form Electron Transport Chains
- Many energy transformations in organisms occur via **oxidation-reduction (redox) reactions. Oxidation** is the loss of electrons; **reduction** is the gain of electrons. Oxidation and reduction reactions occur simultaneously.
- In both photosynthesis and respiration, proteins shuttle electrons along **electron transport chains.**

4.3 | ATP Is Cellular Energy Currency

A. Coupled Reactions Release and Store Energy in ATP
- **ATP** stores energy in its high-energy phosphate bonds. Cellular respiration generates ATP.
- Many energy transformations involve **coupled reactions,** in which the cell uses the energy released in ATP hydrolysis to drive another reaction.

B. Transfer of Phosphate Completes the Energy Transaction
- **Phosphorylation** is the transfer of a phosphate group from ATP to another molecule, causing the recipient to become energized or to change shape.

C. ATP Represents Short-Term Energy Storage
- ATP is too unstable for long-term storage. Instead, cells store energy as fats and carbohydrates.

4.4 | Enzymes Speed Biochemical Reactions

A. Enzymes Bring Reactants Together
- **Enzymes** are organic molecules (usually proteins) that speed biochemical reactions by lowering the **energy of activation.**
- Substrate molecules fit into the enzyme's **active site.**

B. Enzymes Have Partners
- **Cofactors** are inorganic or organic substances that enzymes require to catalyze reactions. Like enzymes, cofactors are not consumed in the reaction.

C. Cells Control Reaction Rates
- In **negative feedback,** a reaction product temporarily shuts down its own synthesis whenever its levels rise. Negative feedback may occur by **competitive** or **noncompetitive inhibition.**
- In **positive feedback,** a product stimulates its own further production.

D. Many Factors Affect Enzyme Activity
- Enzymes have narrow ranges of conditions in which they function. Some poisons and drugs bind to essential enzymes.

4.5 | Membrane Transport May Release Energy or Cost Energy

- A **concentration gradient** is a difference in solute concentration between two neighboring regions, such as across a membrane. Gradients dissipate without energy input.

A. Passive Transport Does Not Require Energy Input
- All forms of **passive transport** involve **diffusion,** the dissipation of a chemical gradient by random molecular motion.
- In **simple diffusion,** a substance passes through a membrane along its concentration gradient without the aid of a transport protein.
- **Osmosis** is the simple diffusion of water across a selectively permeable membrane. Terms describing tonicity (**isotonic, hypotonic,** and **hypertonic**) predict whether cells will swell or shrink when the surroundings change. When plant cells lose too much water, the resulting loss of **turgor pressure** causes the plant to wilt.
- In **facilitated diffusion,** a membrane protein admits a substance along its concentration gradient without expending energy.

B. Active Transport Requires Energy Input
- In **active transport,** a carrier protein uses energy (ATP) to move a substance against its concentration gradient. In animals cells, the **sodium–potassium pump** uses active transport to exchange sodium ions for potassium ions.

C. Endocytosis and Exocytosis Use Vesicles to Transport Substances
- In **endocytosis,** a cell engulfs liquids or large particles. Pinocytosis brings in fluids; **phagocytosis** brings in solid particles.
- In **exocytosis,** vesicles inside the cell carry substances to the cell membrane, where they fuse with the membrane and release the cargo to the outside of the cell.

4.6 | Investigating Life: Does Natural Selection Maintain Some Genetic Illnesses?

- The faulty membrane protein that causes cystic fibrosis may help protect against cholera.

Multiple Choice Questions

1. Which of the following is the best example of potential energy in a cell?
   a. Cell division
   b. A molecule of glucose
   c. Movement of a flagellum
   d. Assembly of a cellulose fiber

2. An *endergonic* reaction is one that is characterized
   a. by a rapid release of energy.
   b. as needing an input of energy.
   c. by phosphorylation.
   d. as occurring spontaneously.

3. How do proteins contribute to the function of an electron transport chain?
   a. They become oxidized and reduced.
   b. They undergo osmosis.
   c. They are involved in the hydrolysis of electrons.
   d. They consume electrons.

4. Where in a molecule of ATP is the stored energy that is used by the cell?
   a. Within the nitrogenous base, adenine
   b. Within the five-carbon ribose sugar
   c. In the covalent bonds between the phosphate groups
   d. In the bond between adenine and ribose

5. What is the role of an enzyme in a cell?
   a. To speed up chemical reactions
   b. To become consumed during a chemical reaction
   c. To increase the energy required to make a reaction occur
   d. To provide energy to the cell

6. Which of the following is true regarding noncompetitive inhibition?
   a. The cofactors of an enzyme are altered.
   b. Excess product blocks the active site.
   c. The active site is unable to bind substrate.
   d. Excess product enhances enzyme activity.

7. The movement of water molecules during osmosis is due to
   a. simple diffusion.   c. pinocytosis.
   b. active transport.   d. endocytosis.

8. What would happen to a cell that was placed into a *hypertonic* environment?
   a. There would be no change.
   b. It would swell and burst.
   c. It would exhibit turgor pressure.
   d. It would shrink.

9. What type of transport occurs when a polar molecule moves from a region of high concentration to a region of low concentration?
   a. Simple diffusion   c. Active transport
   b. Facilitated diffusion   d. Phagocytosis

10. A concentration gradient is an example of
    a. oxidation-reduction.   c. entropy.
    b. potential energy.   d. equilibrium.

Write It Out

1. Cite everyday illustrations of the first and second laws of thermodynamics. How do the laws of thermodynamics underlie every organism's ability to function?
2. Some people claim that life's high degree of organization defies the second law of thermodynamics. What makes this statement false?
3. State the differences between endergonic and exergonic reactions.
4. What is chemical equilibrium?
5. Why are oxidation and reduction reactions linked?
6. Why is ATP called the cell's "energy currency"?
7. How does an enzyme speed a chemical reaction?
8. In what ways is an enzyme's function similar to engineers digging a tunnel through a mountain rather than building a road over the peak?
9. Why would a cell's fat-digesting enzymes not be able to digest an artificial fat such as Olestra (see chapter 2)?
10. Figure 4.14 shows the effect of temperature on enzyme activity. Draw similar curves that show the optimal pH for trypsin (an enzyme in the small intestine, pH 10), amylase (an enzyme in the saliva, pH 6.5) and pepsin (an enzyme in the stomach, pH 2).
11. When a person eats a fatty diet, excess cholesterol accumulates in the bloodstream. Cells then temporarily stop producing cholesterol. What phenomenon described in the chapter does this control illustrate?
12. Why does poking a hole in a cell's membrane kill the cell?

13. Explain the differences among diffusion, facilitated diffusion, active transport, and endocytosis.
14. Diffusion is an efficient means of transport only over small distances. How does this relate to a cell's surface-area-to-volume ratio (see chapter 3)?
15. In the kidneys, water moves by osmosis from tubules called nephrons to the bloodstream. Do you expect a nephron to be hypotonic, isotonic, or hypertonic relative to the blood? Explain.
16. A drop of a 5% salt (NaCl) solution is added to a leaf of the aquatic plant *Elodea*. When the leaf is viewed under a microscope, colorless regions appear at the edges of each cell as the cell membranes shrink from the cell walls. What is happening to these cells?
17. Seawater contains about 35 grams of salt per liter, whereas a liter of fresh water contains 0.5 g of salt or less. The blood of a fish has about 10 g of dissolved salt per liter. Cells in a fish's gills have transport proteins that pump salts across their membranes. In what direction would a saltwater fish pump ions? What about a freshwater fish?
18. Liver cells are packed with glucose. If the concentration of glucose in a liver cell is higher than in the surrounding fluid, what mechanism could the cell use to import even more glucose? Why would only this mode of transport work?

Pull It Together

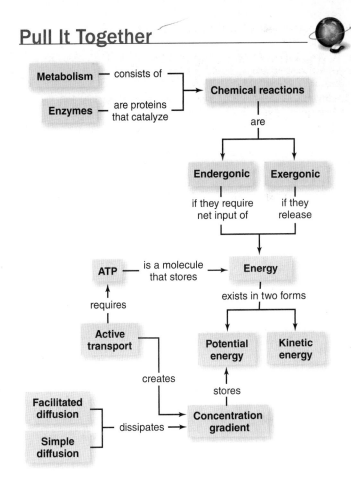

1. What types of molecules are ATP and enzymes?
2. What are some examples of potential energy and kinetic energy?
3. Add the terms *substrate, active site,* and *activation energy* to this concept map.
4. Where does passive transport fit on this concept map?

5 Photosynthesis

Photosynthetic algae help feed coral animals. These two brain coral colonies are competing for space off the coast of Roatan, Honduras.

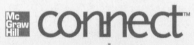

CONNECT

|BIOLOGY

Enhance your study of
this chapter with practice quizzes,
animations and videos, answer keys,
and downloadable study tools.
www.mhhe.com/hoefnagels

Is It Easier Being Green?

IF YOU COULD HAVE A PHOTOSYNTHETIC CHILD, WOULD YOU? Imagine the benefit. Like a houseplant, he could make his own food, free of charge, simply by sitting outside.

Of course, your child would look odd: his skin would be green, for starters, and he might move a little slowly, especially at night. He might even have skin flaps that capture extra sunlight. He wouldn't eat, so he would need another way to acquire essential minerals; perhaps his feet would grow rootlike extensions that would absorb water and nutrients from soil.

Maybe photosynthetic cows, pigs, and chickens—or pets such as dogs and cats—would be a better idea. Feed-free animals would be a commercial and environmental triumph, costing less to own and generating less waste than the animals we raise now.

Fortunately or unfortunately, scientists will probably never be able to create photosynthetic people, chickens, or pooches. Mammals and birds move, breathe, pump blood, and maintain high body temperatures. All of this activity would likely require energy beyond what photosynthesis alone could supply.

Some invertebrate animals, however, have adopted the "green" lifestyle by harboring live-in photosynthetic partners (see section 5.8). The closest to a true plant–animal hybrid is probably the sea slug *Elysia chlorotica,* a solar-powered mollusk with chloroplasts (photosynthetic organelles) in the cells lining its digestive tract. The chloroplasts come from algae in the slug's diet. As the animal grazes, it punctures the algal cells and discards everything but the chloroplasts, which migrate into the animal's cells. Light passes through the slug's skin and strikes the food-producing chloroplasts. Once its "solar panels" are in place, the animal may not eat again for months!

Perhaps the most famous animals to "farm" photosynthetic partners are corals. Inside the cells lining the coral animal's digestive cavity are tiny single-celled eukaryotes called dinoflagellates. These protists use the sun's energy to feed the coral animals. In exchange, the animals provide a home for the dinoflagellates.

Sometimes, however, the partners break up. Corals under stress sometimes expel their dinoflagellates, or the protists may leave on their own. The reef then turns white. The coral animals eventually die, endangering the entire reef ecosystem. Pollution, disease, shading, excessively warm water, and ultraviolet radiation all trigger coral bleaching. Biologists predict that global warming will only make this problem worse.

Corals and sea slugs are not the only animals whose lives depend on photosynthesis. Yours does, too, as you will learn in the next two chapters.

Learning Outline

Learn How to Learn
See What's Coming

Check out the Learning Outline at the beginning of each chapter. Each heading is a complete sentence that summarizes the most important idea of the section. Read through these statements before you start each chapter. That way, you can keep the "big picture" in mind while you study. You can also flip to the end of the chapter before you start to read; the chapter summary and concept map can provide a preview of what's to come.

5.1 | Life Depends on Photosynthesis

It is spring. A seed germinates, its tender roots and pale yellow stem extending rapidly in a race against time. For now, the seedling's sole energy source is food stored along with the embryonic plant in the seed itself. If its shoot does not reach light before its reserves run out, the seedling will die. But if it makes it, the shoot will quickly turn green and unfurl leaves that spread and catch the light. The seedling begins to feed itself, and an independent new life begins.

Organisms that can produce their own food underlie every ecosystem on Earth. It is not surprising, therefore, that if asked to designate the most important metabolic pathway, most biologists would not hesitate to cite **photosynthesis:** the process by which plants, algae, and some microorganisms harness solar energy and convert it into chemical energy. With the exception of deep-ocean hydrothermal vent communities, all life on this planet ultimately depends on photosynthesis.

A. Photosynthesis Builds Carbohydrates Out of Carbon Dioxide and Water

Most plants are easy to grow (compared with animals, anyway) because their needs are simple. Give a plant water, essential elements in soil, carbon dioxide, and light, and it will produce food and oxygen not only for itself but also for a host of consumers. How can plants do so much with such simple raw materials?

In photosynthesis, pigment molecules in plant cells capture energy from the sun (**figure 5.1**). In a series of chemical reactions, that energy is then used to build the carbohydrate glucose ($C_6H_{12}O_6$) from carbon dioxide (CO_2) molecules. The plant uses water in the process and releases oxygen gas (O_2) as a byproduct. The reactions of photosynthesis are summarized as follows:

$$6CO_2 + 6H_2O \xrightarrow{\text{light energy}} C_6H_{12}O_6 + 6O_2$$

Photosynthesis is an oxidation–reduction (redox) process. "Oxidation" means that electrons are removed from an atom or molecule; "reduction" means electrons are added. As you will soon see, photosynthesis strips electrons from the oxygen atoms in H_2O (i.e., the oxygen atoms are oxidized). These electrons are eventually used to reduce the carbon in CO_2. Because oxygen atoms attract electrons more strongly than do carbon atoms (as depicted in figure 2.8), moving electrons from oxygen to carbon requires energy input. The energy source for this endergonic reaction is, of course, sunlight. ▶ redox reactions, p. 75

Several fates await the glucose produced in photosynthesis. A plant's cells use about half of the glucose as fuel for their own cellular respiration, the metabolic pathway described in chapter 6. Roots, flowers, fruits, seeds, and other nonphotosynthetic plant parts could not grow without sugar shipments from green leaves and stems. Plants also combine glucose with other substances to manufacture additional compounds, including

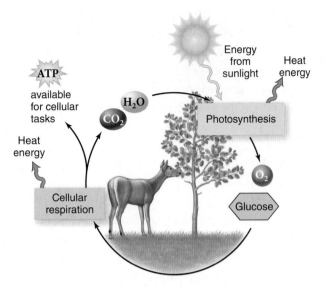

Figure 5.1   Cycle of Life. Photosynthesis produces glucose and O_2, which are the starting materials for cellular respiration.

amino acids and a host of economically important products such as rubber, medicines, and spices.

Plants also use glucose as a raw material to build a cellulose wall for each of its cells. Wood is the remains of dead cells (see chapter 21), and it is mostly made of cellulose. The timber in the world's forests therefore stores enormous amounts of carbon. So do vast deposits of coal and other fossil fuels, which are the remains of plants and other organisms that lived long ago. Burning wood or fossil fuels releases this stored carbon into the atmosphere as CO_2. As the amount of CO_2 in the atmosphere has increased, Earth's average temperature has risen. Living forests help reduce climate change by locking carbon in wood. ▶ global climate change, p. 809

If a plant produces more glucose than it immediately needs for respiration or building cell walls, it may store the excess as starch. Carbohydrate-rich tubers and grains, such as potatoes, rice, corn, and wheat, are all energy-storing plant organs. Some plants, including sugarcane and sugar beets, store energy as sucrose instead. Table sugar comes from these crops, just as sweet maple syrup comes from the sucrose-rich sap of a sugar maple tree. In addition, people use both starch (from corn kernels) and sugar (from sugarcane) to produce biofuels such as ethanol. ▶ biofuels, p. 376

Photosynthesis not only feeds plants, but it also provides the energy, raw materials, and oxygen for most other organisms on Earth (see figure 5.1). Animals, fungi, and other consumers eat the leaves, stems, roots, flowers, nectar, fruits, and seeds of the world's producers. Even the waste product of photosynthesis, O_2, is essential to much life on Earth. Some scientists consider tropical rainforests to be the world's "lungs," because they produce much of the oxygen that we breathe.

Because humans live on land, we are most familiar with the contribution that plants make to Earth's terrestrial ecosystems. In fact, however, more than half of the world's photosynthesis occurs

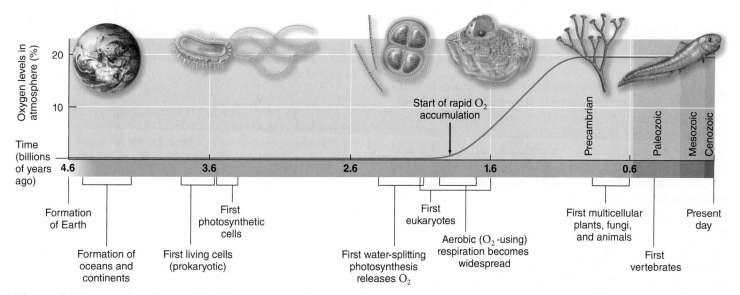

Figure 5.2 Oxygen Gas Changed the World. Billions of years ago, photosynthesis began to pump oxygen gas (O_2) into the atmosphere. As oxygen accumulated, life's diversity exploded. Brackets represent estimated dates of major events in life's history. (*Source:* National Academy of Sciences, *Teaching About Evolution and the Nature of Science,* 1998.)

in the vast oceans, courtesy of countless algae and bacteria. Several groups of bacteria are photosynthetic, some using pigments and metabolic pathways that are completely different from those in plants. For example, some photosynthetic microbes do not use water as an electron source or generate oxygen gas. (This chapter focuses on photosynthesis as it occurs in plants and algae.)

In short, Earth without photosynthesis would not long be a living world. If the sky were blackened by a nuclear holocaust, cataclysmic volcanic eruption, or massive meteor impact, the light intensity reaching Earth's surface would decline to about a tenth of its normal level. Photosynthetic organisms would die as they depleted their energy reserves faster than they could manufacture more food. Animals that normally ate these producers would go hungry, as would the animals that ate the herbivores. A year or even two might pass before enough life-giving light could penetrate the hazy atmosphere, but by then, it would be too late. The lethal chain reaction would already be well into motion, destroying food webs at their bases. No wonder biologists consider photosynthesis to be Earth's most important metabolic process.

B. The Evolution of Photosynthesis Changed Planet Earth

Most of today's organisms rely directly or indirectly on photosynthesis, so it may seem surprising that early life lacked the ability to capture sunlight. For the first billion or so years of life's history, however, all organisms were **heterotrophs,** meaning that they obtained carbon by consuming preexisting organic molecules. As these early heterotrophs oxidized the carbon compounds from their surroundings, they released CO_2 into the environment. But these organisms could not use the carbon in CO_2, so they faced extinction as soon as they depleted the organic compounds in their habitats.

Soon, however, some organisms developed a new talent: the ability to make their own food. **Autotrophs** use inorganic substances such as water and CO_2 to produce organic compounds. Most autotrophs use light as their energy source. This novel ability to convert light energy into chemical energy soon supported most other forms of life.

The evolution of photosynthesis some 3.5 billion years ago radically altered Earth in other ways as well. Photosynthesis by ancient cyanobacteria filled the atmosphere with a new waste product: oxygen gas (**figure 5.2**). The organisms that could use O_2 in respiration had an advantage: they could extract the most energy from food. Eventually, organisms using aerobic cellular respiration outcompeted most other life forms. With more energy available, life took on new shapes and sizes.

In addition, O_2 from photosynthesis reacted with free oxygen atoms to produce ozone (O_3). As ozone accumulated high in the atmosphere, it blocked harmful ultraviolet radiation from reaching the planet's surface, which prevented some genetic damage and allowed new varieties of life to arise. The new ozone layer therefore also helps explain the explosion in the diversity of life that followed the evolution of photosynthesis.

5.1 | Mastering Concepts

1. What is photosynthesis? Describe the reactants and products in words and in chemical symbols.
2. Why is photosynthesis essential to life on Earth?
3. What happens to the glucose that plants produce?
4. How is an autotroph different from a heterotroph?
5. How did the origin of photosynthesis alter Earth's atmosphere and the evolution of life?

5.2 | Sunlight Is the Energy Source for Photosynthesis

Each minute, the sun converts more than 100,000 kilograms of matter to energy, releasing much of it outward as waves of electromagnetic radiation. After an 8-minute journey, about two billionths of this energy reaches Earth's upper atmosphere. Of this, only about 1% is used for photosynthesis, yet this tiny fraction of the sun's power ultimately produces nearly 2 quadrillion kilograms of carbohydrates a year! Light may seem insubstantial, but it is a powerful force on Earth.

A. What Is Light?

Visible light is a small sliver of a much larger **electromagnetic spectrum,** the range of possible frequencies of radiation (**figure 5.3**). All electromagnetic radiation, including light, consists of **photons,** discrete packets of kinetic energy. A photon's **wavelength** is the distance it moves during a complete vibration. The shorter a photon's wavelength, the more energy it contains.

The sunlight that reaches Earth's surface consists of three main components of the electromagnetic spectrum: ultraviolet radiation, visible light, and infrared radiation. Of the three, ultraviolet radiation has the shortest wavelengths. Its high-energy photons damage DNA, causing sunburn and skin cancer. In the middle range of wavelengths is visible light, which provides the energy that powers photosynthesis; we perceive visible light of different wavelengths as distinct colors. Infrared radiation, with its longer wavelengths, contains too little energy per photon to be useful to organisms. Most of its energy is converted immediately to heat.

Table 5.1   Pigments of Photosynthesis

Pigment	Color(s)	Organisms
Major pigment		
Chlorophyll *a*	Blue-green	Plants, algae, cyanobacteria
Accessory pigments		
Chlorophyll *b*	Yellow-green	Plants, green algae
Carotenoids (carotenes and xanthophylls)	Red, orange, yellow	Plants, algae, bacteria, archaea

B. Photosynthetic Pigments Capture Light Energy

Plant cells contain several pigment molecules that capture light energy (see this chapter's Burning Question). The most abundant is **chlorophyll a,** a green photosynthetic pigment in plants, algae, and cyanobacteria. Photosynthetic organisms usually also have several types of **accessory pigments,** which are energy-capturing pigment molecules other than chlorophyll *a* (**table 5.1**). Chlorophyll *b* and carotenoids are accessory pigments in plants.

The photosynthetic pigments have distinct colors because they absorb only some wavelengths of visible light, while transmitting or reflecting others (**figure 5.4**). Chlorophylls *a* and *b* absorb red and blue wavelengths; they appear green because they reflect green light. Carotenoids, on the other hand, reflect longer wavelengths of light, so they appear red, orange, or yellow. (Carrots, tomatoes, lobster shells, and the flesh of salmon all owe

Figure 5.3 The Electromagnetic Spectrum. Sunlight reaching Earth consists of ultraviolet radiation, visible light, and infrared radiation, all of which is just a small part of a continuous spectrum of electromagnetic radiation. The shorter the wavelength, the more energy associated with the radiation.

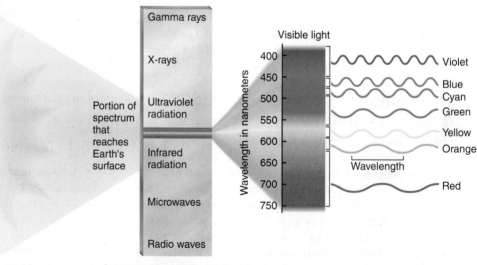

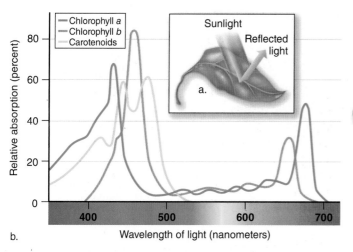

b.

Figure 5.4 Everything but Green. (a) Overall, a leaf reflects green and yellow wavelengths of light and absorbs the other wavelengths. (b) Each type of pigment absorbs some wavelengths of light and reflects others.

their distinctive colors to carotenoid pigments, which the animals must obtain from their diets.)

Only absorbed light is useful in photosynthesis. Accessory pigments absorb wavelengths that chlorophyll *a* cannot, so they extend the range of light wavelengths that a cell can harness. This is a little like the members of the same team on a quiz show, each contributing answers from a different area of expertise.

Figure It Out

If you could expose plants to just one wavelength of light at a time, would a wavelength of 300 nm, 450 nm, or 600 nm produce the highest photosynthetic rate?

Answer: 450 nm.

C. Chloroplasts Are the Sites of Photosynthesis

In plants, leaves are the main organs of photosynthesis. Their broad, flat surfaces expose abundant surface area to sunlight. But light is just one requirement for photosynthesis. Water is also essential; roots absorb this vital ingredient, which moves up stems and into the leaves. And plants also exchange CO_2 and O_2 with the atmosphere. How do these gases get into and out of leaves?

The answer is that CO_2 and O_2 enter and exit a plant through **stomata** (singular: stoma), tiny openings in the epidermis of a leaf or stem (figure 5.5). Stomata allow for gas exchange, but water evaporates through the same openings. When the plant loses too much water, pairs of specialized "guard cells" surrounding each stoma collapse against one another, closing the pores. Stomata therefore help balance the competing needs of gas exchange and water conservation. ▶ leaf epidermis, p. 461

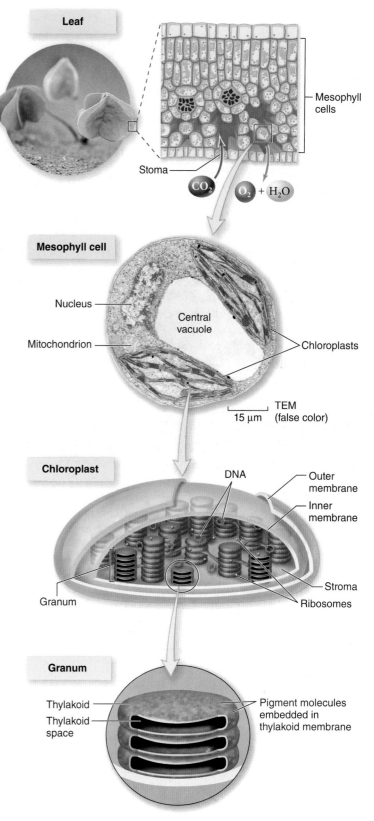

Figure 5.5 Leaf and Chloroplast Anatomy. Leaf mesophyll tissue consists of cells that contain many chloroplasts.

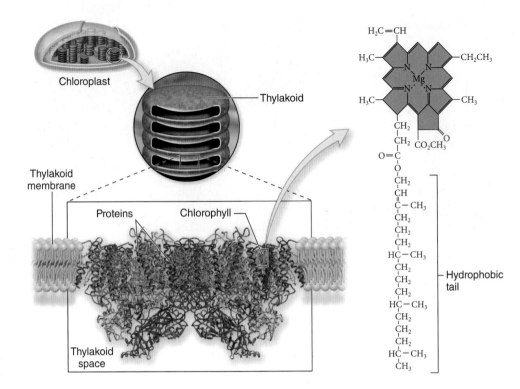

Figure 5.6 Photosystem. This diagram of a photosystem shows a complex grouping of proteins and pigments embedded in the chloroplast's thylakoid membrane.

Most photosynthesis occurs in cells filling the leaf's interior (see figure 5.5). **Mesophyll** is the collective term for these internal cells (*meso-* means "middle," and *-phyll* means "leaf"). In many plants, at least part of the mesophyll has a "spongy" texture, reflecting the air spaces that maximize gas exchange within the leaf.

Leaf mesophyll cells contain abundant **chloroplasts,** the organelles of photosynthesis in plants and algae. Most photosynthetic cells contain 40 to 200 chloroplasts, which add up to about 500,000 per square millimeter of leaf—an impressive array of solar energy collectors.

Each chloroplast contains tremendous surface area for the reactions of photosynthesis. Two membranes enclose the **stroma,** a gelatinous fluid containing ribosomes, DNA, and enzymes. (Be careful not to confuse the *stroma* with a *stoma,* or leaf pore). Suspended in the stroma of each chloroplast are between 10 and 100 **grana** (singular: granum), each composed of a stack of 10 to 20 disk-shaped thylakoids. Each **thylakoid,** in turn, consists of a membrane studded with photosynthetic pigments and enclosing a volume called the **thylakoid space.**

The pigments and proteins that participate in photosynthesis are grouped into photosystems in the thylakoid membrane (**figure 5.6**). One **photosystem** consists of chlorophyll *a* aggregated with other pigment molecules and the proteins that anchor the entire complex in the membrane.

Within each photosystem are some 300 chlorophyll molecules and 50 accessory pigments. Although all of the pigment molecules absorb light energy, only one chlorophyll *a* molecule

per photosystem actually uses the energy in photosynthetic reactions. The photosystem's **reaction center** is this chlorophyll *a* molecule and its associated proteins. All other pigments in the photosystem are called **antenna pigments** because they capture photon energy and funnel it to the reaction center. If the different pigments are like a quiz show team, then the reaction center is analogous to the one member who announces the team's answer to the show's moderator.

Why does only one chlorophyll molecule out of a few hundred actually participate in photosynthetic reactions? A single chlorophyll *a* molecule can absorb only a small amount of light energy. Several pigment molecules near each other capture much more energy because they can pass the energy on to the reaction center, freeing them to absorb other photons as they strike. Thus, the photosystem's organization greatly enhances the efficiency of photosynthesis.

5.2 | Mastering Concepts

1. What are the three main components of sunlight?
2. Describe the relationship among the chloroplast, stroma, grana, and thylakoids.
3. How does it benefit a photosynthetic organism to have more than one type of pigment?
4. How does the reaction center chlorophyll interact with the antenna pigments in a photosystem?

5.3 | Photosynthesis Occurs in Two Stages

 Inside a chloroplast, photosynthesis occurs in two stages: the light reactions and the carbon reactions. **Figure 5.7** summarizes the entire process, and sections 5.4 and 5.5 describe each part in greater detail.

The **light reactions** convert solar energy to chemical energy. (You can think of the light reactions as the "photo-" part of photosynthesis.) In the chloroplast's thylakoid membranes, pigment molecules in two linked photosystems capture kinetic energy from photons and store it as potential energy in the chemical bonds of two molecules: ATP and NADPH.

Recall from chapter 4 that **ATP** is a nucleotide that stores potential energy in the covalent bonds between its phosphate groups. ATP forms when a phosphate group is added to ADP (see figure 4.9). The other energy-rich product of the light reactions, **NADPH,** is a coenzyme that carries pairs of energized electrons. In photosynthesis, these electrons come from chlorophyll molecules. Once the light reactions are underway, chlorophyll, in turn, replaces its "lost" electrons by splitting water molecules, yielding O_2 as a waste product.
▶ coenzymes, p. 78

These two resources (energy and "loaded" electron carriers) set the stage for the second part of photosynthesis. The **carbon reactions** use ATP and the high-energy electrons in NADPH to reduce CO_2 to glucose molecules. (These reactions are the "-synthesis" part of photosynthesis.) The ATP and NADPH come from the light reactions, and the CO_2 comes from the atmosphere. Once inside the leaf, CO_2 diffuses into a mesophyll cell and across the chloroplast membrane into the stroma, where the carbon reactions occur.

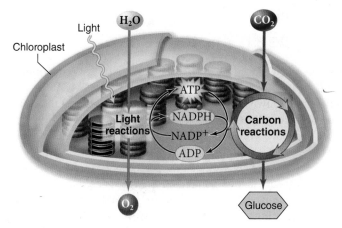

Figure 5.7 Overview of Photosynthesis. In the light reactions, pigment molecules capture sunlight energy and transfer it to molecules of ATP and NADPH. The carbon reactions use this energy to build glucose out of carbon dioxide.

Because the carbon reactions do not directly require light, they are sometimes called the "dark reactions" of photosynthesis. This term is misleading, however, because the carbon reactions can occur at any time of day or night, as long as ATP and NADPH are available. A more accurate alternative would be the "light-independent reactions."

5.3 | Mastering Concepts

1. What happens in each of the two main stages of photosynthesis?
2. Where in the chloroplast do the light reactions and the carbon reactions occur?

Burning Question

Why do leaves change colors in the fall?

Most leaves are green throughout a plant's growing season, although there are exceptions; some ornamental plants, for example, have yellow or purple leaves. The near-ubiquitous green color comes from chlorophyll *a,* the most abundant pigment in photosynthetic plant parts.

But the leaf also has other photosynthetic pigments. Carotenoids contribute brilliant yellow, orange, and red hues. Purple pigments, such as anthocyanins, are not photosynthetically active, but they do protect leaves from damage by ultraviolet radiation.

These accessory pigments are less abundant than chlorophyll, so they usually remain invisible to the naked eye during the growing season. As winter approaches, however, deciduous plants prepare to shed their leaves. The chlorophyll degrades, and the now "unmasked"

accessory pigments reveal their colors for a short time as a spectacular autumn display. These pigments soon disappear as well, and the dead leaves turn brown.

Springtime brings a flush of fresh, green leaves. The energy to produce the foliage comes from glucose the plant produced during the last growing season and stored as starch. The new leaves make food throughout the spring and summer, so the tree can grow—both above ground and below—and produce fruits and seeds.

As the days grow shorter and cooler in autumn, the cycle will continue, and the colorful pigments will again participate in one of nature's great disappearing acts.

Submit your burning question to:
marielle_hoefnagels@mcgraw-hill.com

5.4 | The Light Reactions Begin Photosynthesis

A plant placed in a dark closet literally starves. Without light, the plant cannot generate ATP or NADPH. And without these two critical energy and electron carriers, the plant cannot feed itself. Once its stored reserves are gone, the plant dies. The plant's life thus depends on the light reactions of photosynthesis, which occur in the membranes of chloroplasts.

We have already seen that the pigments and proteins of the chloroplast's thylakoid membranes are organized into photosystems (see figure 5.6). More specifically, the thylakoid membranes of algae and higher plants contain two types of photosystems, dubbed I and II. The two photosystems "specialize" in slightly different wavelengths of light. The reactive chlorophyll of photosystem I, called P700, absorbs light energy mostly at 700 nm. Photosystem II's reactive chlorophyll is called P680 because it absorbs wavelengths of 680 nm.

An electron transport chain connects the two photosystems. Recall from chapter 4 that an **electron transport chain** is a group of proteins that shuttle electrons like a bucket brigade, releasing energy with each step. As you will see, the electron transport chain that links photosystems I and II provides energy for ATP synthesis. A second electron transport chain ends in the production of NADPH.

Figure 5.8 depicts the arrangement of the photosystems and electron transport chains in the thylakoid membrane. Refer to this illustration as you work through the rest of this section.

A. Photosystem II Produces ATP

Photosynthesis begins in the cluster of pigment molecules of photosystem II. This may seem illogical, but the two photosystems were named as they were discovered. Photosystem II was discovered after photosystem I, but it functions first in the overall process.

Pigment molecules in photosystem II absorb light and transfer the energy to a chlorophyll *a* reaction center, where it boosts two electrons to an orbital with a higher energy level. The "excited" electrons, now packed with potential energy, are ejected from this chlorophyll *a* molecule and grabbed by the first protein in the electron transport chain that links the two photosystems. ▸ electron orbitals, p. 23

How does the chlorophyll *a* molecule replace these two electrons? They come from water (H_2O), which donates two electrons when it splits into oxygen gas and two protons (H^+). Chlorophyll *a* picks up the electrons, and the O_2 is a waste product that the plant releases to the environment.

Meanwhile, the chloroplast uses the potential energy in the electrons to create a proton gradient. As the electrons pass along the electron transport chain, the energy they lose drives the active transport of protons from the stroma into the thylakoid space. The resulting proton gradient between the stroma and the inside of the thylakoid represents a form of potential energy. ▸ active transport, p. 82

An enzyme complex called **ATP synthase** transforms the gradient's energy into chemical energy in the form of ATP (figure 5.9). A channel in ATP synthase allows protons trapped inside the thylakoid space to return to the chloroplast's stroma. As the gradient dissipates, energy is released. The ATP synthase enzyme uses this energy to add phosphate to ADP, generating ATP.

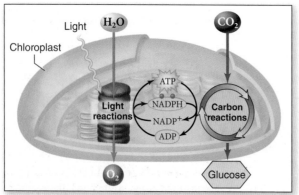

Figure 5.8 **Photosystems and Electron Transport Chains.** Chlorophyll molecules in photosystem II transfer light energy to electrons taken from water molecules, releasing oxygen. The energized electrons pass to photosystem I via an electron transport chain. Each transfer releases energy that is used to pump hydrogen ions into the thylakoid space; figure 5.9 shows how the resulting hydrogen gradient is used to generate ATP. In photosystem I, the electrons absorb more light energy and are transferred to $NADP^+$, creating the energy-rich NADPH.

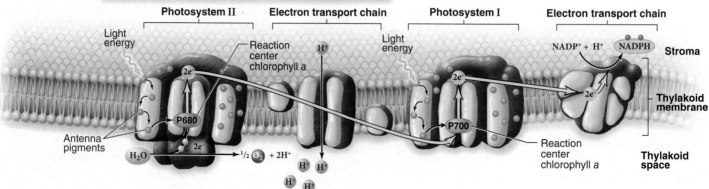

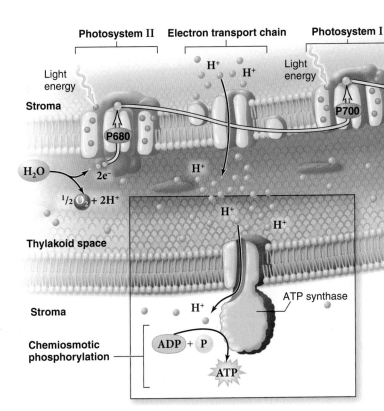

Photosystem II   Electron transport chain   Photosystem I   Electron transport chain

$NADP^+ + H^+$ NADPH

Light energy

Stroma

H^+ H^+

Light energy

P700

P680

H_2O $2e^-$

H^+

$1/2 O_2 + 2H^+$

Thylakoid space

H^+

H^+

Stroma

ATP synthase

H^+

Chemiosmotic phosphorylation

ADP + P

ATP

Figure 5.9 Making ATP. ATP synthase is a channel through which protons can escape from the thylakoid space. As they do, phosphate is added to ADP, producing ATP.

This mechanism is similar to using a dam to produce electricity. As water accumulates, tremendous pressure (a form of potential energy) builds on the face of the dam. That pressure is released by diverting water through a large pipe at the base of the dam, turning massive blades that spin an electric generator.

The coupling of ATP formation to the release of energy from a proton gradient is called **chemiosmotic phosphorylation** because it is the addition of a phosphate to ADP (phosphorylation) using energy from the movement of protons across a membrane (chemiosmosis). As described in chapter 6, the same process also occurs in cellular respiration: an electron transport chain provides the energy to create a proton gradient, and ATP synthase uses the gradient's potential energy to produce ATP.

B. Photosystem I Produces NADPH

Photosystem I functions much as photosystem II does. Photon energy strikes energy-absorbing molecules of chlorophyll *a*, which pass the energy to the reaction center. The reactive chlorophyll molecules eject electrons to an electron carrier molecule in a second electron transport chain. The boosted electrons in photosystem I are then replaced with electrons passing down the first electron transport chain from photosystem II.

Unlike in photosystem II, however, the second electron transport chain does not generate ATP, nor does it pass its electrons to yet another photosystem. Instead, the electrons reduce a molecule of $NADP^+$ to NADPH. This NADPH is the electron carrier that will reduce carbon dioxide in the carbon reactions, while the ATP generated in photosystem II will provide the energy.

5.4 | Mastering Concepts

1. Describe the events in photosystem II, beginning with light and ending with the production of ATP.
2. How do electrons pass from photosystem II to photosystem I?
3. How are the electrons from photosystem II replaced?
4. What happens in photosystem I?

Apply It Now

Weed Killers

No plant can survive for very long in the dark. One low-tech way to kill an unwanted plant, therefore, is to deprive it of light. Gardeners who want to convert a lawn into a garden, for example, might kill the grass by covering it with layers of newspaper for several weeks. The light reactions of photosynthesis cannot occur in the dark; the plants die.

Many herbicides also stop the light reactions. For example, DCMU (short for 3-(3,4-dichlorophenyl)-1,1-dimethylurea and known by the name diuron) blocks electron flow in photosystem II. Paraquat, noted for its use in destroying marijuana plants, diverts electrons from photosystem I.

Other herbicides take a different approach. Accessory pigments called carotenoids protect plants from damage caused by free radicals. Triazole herbicides kill plants by blocking carotenoid synthesis. No longer protected from free-radical damage, the cell's organelles are destroyed.

Still other weed killers exploit pathways not directly related to photosynthesis. For instance, glyphosate (Roundup) inhibits an enzyme that plants require for amino acid synthesis. Another herbicide, 2,4-D (short for 2,4-dichlorophenoxyacetic acid), mimics a plant hormone called auxin, as described in chapter 23.

5.5 | The Carbon Reactions Produce Carbohydrates

The carbon reactions, also called the Calvin cycle, occur in the chloroplast's stroma. The **Calvin cycle** is the metabolic pathway that uses NADPH and ATP to assemble CO_2 molecules into three-carbon carbohydrate molecules (**figure 5.10**). These products are eventually assembled into glucose and other sugars. (The pathway is named in honor of its discoverer, American biochemist Melvin Calvin.)

The first step of the Calvin cycle is **carbon fixation**—the initial incorporation of carbon from CO_2 into an organic compound. Specifically, CO_2 combines with **ribulose bisphosphate (RuBP),** a five-carbon sugar with two phosphate groups.

The enzyme that catalyzes this essential first reaction is RuBP carboxylase/oxygenase, also known as **rubisco.** As an essential component of every plant, rubisco is one of the most abundant and important proteins on Earth.

The six-carbon product of the initial reaction immediately breaks down into two three-carbon molecules called phosphoglycerate (PGA). Further steps in the cycle convert PGA to phosphoglyceraldehyde (PGAL), which is the carbohydrate product that leaves the Calvin cycle. The cell can use PGAL to build larger carbohydrate molecules such as glucose and sucrose. Some of the PGAL, however, is rearranged to form additional RuBP, perpetuating the cycle.

ATP and NADPH produced in the light reactions provide the potential energy and electrons necessary to reduce CO_2. As long as ATP and NADPH are plentiful, the Calvin cycle continuously "fixes" the carbon from CO_2 into small organic molecules, in both darkness and light.

5.5 | Mastering Concepts

1. What is the product of the carbon reactions?
2. What are the roles of rubisco, RuBP, ATP, and NADPH in the Calvin cycle?
3. What is the relationship between the light reactions and the carbon reactions?

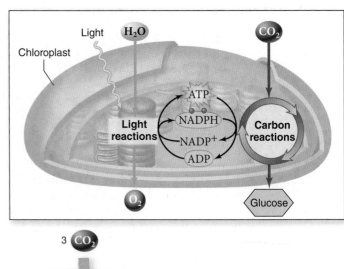

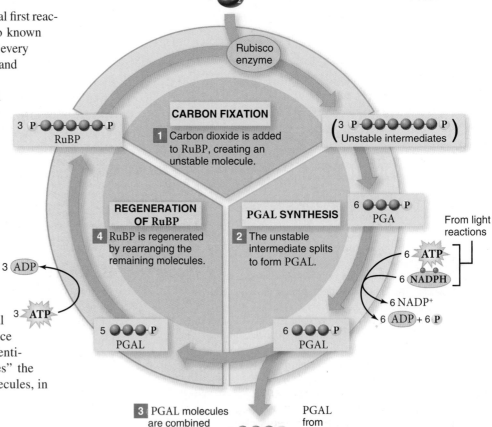

Figure 5.10 The Calvin Cycle. ATP and NADPH from the light reactions power the Calvin cycle, simplified here. The cycle generates a three-carbon molecule, PGAL, which is used to build glucose and other carbohydrates.

5.6 | C₃ Plants Use Only the Calvin Cycle to Fix Carbon

The Calvin cycle is also known as the **C₃ pathway** because a three-carbon molecule, PGA, is the first stable compound in the pathway. Although all plants use the Calvin cycle, C₃ plants use *only* this pathway to fix carbon from CO_2. About 95% of plant species are C₃, including cereals, peanuts, tobacco, spinach, sugar beets, soybeans, most trees, and some lawn grasses.

C₃ photosynthesis is obviously a successful adaptation, but it does have a weakness: inefficiency. All energy transformations are inefficient because some energy is always lost as heat (see section 4.1B). Photosynthesis therefore has a theoretical efficiency rate of about 30%. In reality, however, photosynthesis falls far short of that. On cloudy days, individual plants average from 0.1% to 3% photosynthetic efficiency.

How do plants waste so much solar energy? One contributing factor is a process that counteracts photosynthesis. In the metabolic pathway called **photorespiration,** the rubisco enzyme uses O_2 as a substrate instead of CO_2, starting a process that removes already-fixed carbon from the carbon reactions (**figure 5.11**).

A plant with open stomata minimizes the photorespiration rate. This is because CO_2 and O_2 compete for rubisco's active site; when stomata are open, CO_2 from the atmosphere enters the leaf, and O_2 produced in the light reactions diffuses out. But plants in hot, dry climates face a trade-off. If the stomata remain open too long, a plant may lose water, wilt, and die. If the plant instead closes its stomata, CO_2 runs low, and O_2 builds up in the leaves. Under those conditions, photorespiration becomes much more likely.

5.6 | Mastering Concepts

1. Why is the Calvin cycle also called the C₃ pathway?
2. How does photorespiration counter photosynthesis?
3. What conditions maximize photorespiration?

Figure 5.11 Photorespiration. (a) The Calvin cycle proceeds when rubisco reacts with CO_2. (b) Photorespiration occurs when the rubisco enzyme binds with O_2 instead of CO_2, forming one PGA molecule plus a two-carbon molecule called phosphoglycolate. Much of the phosphoglycolate is eventually liberated as CO_2.

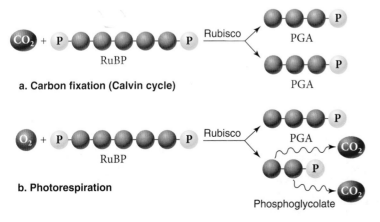

a. Carbon fixation (Calvin cycle)

b. Photorespiration

5.7 | The C₄ and CAM Pathways Save Carbon and Water

Photorespiration has a high cost. Plants may lose as much as 30% of their fixed carbon to this pathway, which has no known benefit. In hot climates, plants that minimize photorespiration may therefore have a significant competitive advantage. One way to improve efficiency is to ensure that rubisco always encounters high CO_2 concentrations. The C₄ and CAM pathways are two adaptations that do just that.

C₄ plants physically separate the light reactions and the carbon reactions into different cells. The light reactions occur in mesophyll cells, as does a carbon-fixation reaction called the C₄ pathway. In the **C₄ pathway,** CO_2 combines with a three-carbon molecule to form the four-carbon compound, oxaloacetate (hence the name C₄). The oxaloacetate is usually reduced to malate, another four-carbon molecule.

Malate then moves into adjacent **bundle-sheath cells** that surround the leaf veins. The CO_2 is liberated inside these cells, where the Calvin cycle fixes the carbon a second time. Meanwhile, at the cost of two ATP molecules, the three-carbon "ferry" returns to the mesophyll to pick up another CO_2.

C₄ plants owe their efficiency to the arrangement of cells in their leaves (**figure 5.12**). Unlike mesophyll cells, bundle-sheath cells are not exposed directly to atmospheric O_2. The rubisco in bundle-sheath cells is therefore much more likely to bind CO_2 instead of O_2, reducing photorespiration.

As a bonus, O_2 does not compete for the active site of the enzyme that first fixes CO_2 in the C₄ pathway. C₄ plants can therefore acquire the CO_2 they need with fewer, smaller stomata than C₃ plants. Since water loss occurs primarily through stomata, C₄ plants require about half as much water as C₃ plants.

About 1% of plants use the C₄ pathway. All are flowering plants growing in hot, open environments, including crabgrass and crop plants such as sugarcane and corn. C₄ plants are less abundant, however, in cooler, moister habitats. In those environments, the ATP cost of ferrying each CO_2 from a mesophyll cell to a bundle-sheath cell apparently exceeds the benefits of reduced photorespiration.

Figure 5.12 C₄ and C₄ Leaf Anatomy. In C₃ plants, the light reactions and the Calvin cycle occur in mesophyll cells. In C₄ plants, the light reactions occur in mesophyll, but the inner ring of bundle-sheath cells houses the Calvin cycle.

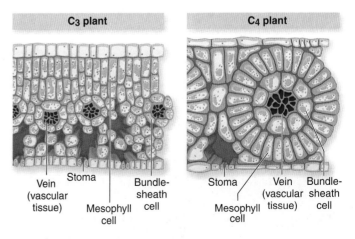

Another energy- and water-saving strategy, called crassulacean acid metabolism (CAM), occurs in desert plants in the Crassulaceae family. Plants that use the **CAM pathway** add a new twist: they open their stomata to fix CO_2 only at night, then fix it again in the Calvin cycle during the day. Unlike in C_4 plants, however, both fixation reactions occur in the same cell.

A CAM plant's stomata open at night, when the temperature drops and humidity rises. CO_2 diffuses in. Mesophyll cells incorporate the CO_2 into malate, which they store in large vacuoles. The stomata close during the heat of the day, but the stored malate moves from the vacuole to a chloroplast and releases its CO_2. The chloroplast then fixes the CO_2 in the Calvin cycle. The CAM pathway reduces photorespiration by generating high CO_2 concentrations inside chloroplasts.

About 3% to 4% of plant species, including pineapple and cacti, use the CAM pathway. All CAM plants are adapted to dry habitats. In cool environments, however, CAM plants cannot compete with C_3 plants. Their stomata are only open at night, so CAM plants have much less carbon available to their cells for growth and reproduction.

Figure 5.13 compares and contrasts C_3, C_4, and CAM plants.

5.7 | Mastering Concepts

1. Describe how a C_4 plant minimizes photorespiration.
2. How is the CAM pathway similar to C_4 metabolism, and how is it different?

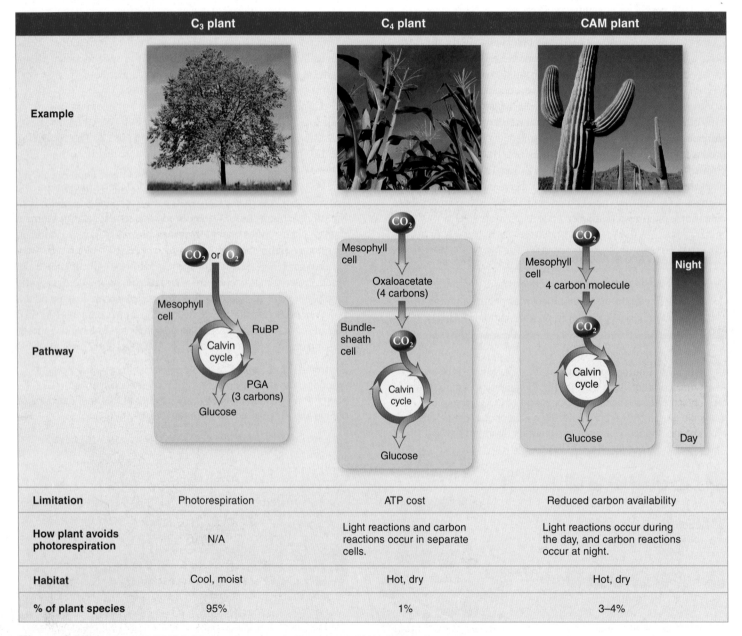

	C₃ plant	C₄ plant	CAM plant
Limitation	Photorespiration	ATP cost	Reduced carbon availability
How plant avoids photorespiration	N/A	Light reactions and carbon reactions occur in separate cells.	Light reactions occur during the day, and carbon reactions occur at night.
Habitat	Cool, moist	Hot, dry	Hot, dry
% of plant species	95%	1%	3–4%

Figure 5.13 C₃, C₄, and CAM Pathways Compared. The C_4 and CAM pathways are adaptations that minimize photorespiration.

5.8 | Investigating Life: Solar-Powered Sea Slugs

Most animals have an indirect relationship with photosynthesis. Plants and other autotrophs use the sun's energy in photosynthesis, and the food they make goes on to feed the animals.

But *Elysia chlorotica* is an unusual animal by all accounts (figure 5.14). This sea slug lives in salt marshes along the eastern coast of North America. As mentioned in this chapter's opening essay, *E. chlorotica* is solar-powered: it harbors chloroplasts in the lining of its gut.

These invertebrate animals do not inherit their solar panels from their parents; instead, they acquire the chloroplasts by eating algae called "water felt," or *Vaucheria litorea*. As a young sea slug grazes, it punctures the yellow-green filaments of the algae and sucks out the cell's contents. The animal digests most of the nutrients, but cells lining the slug's gut absorb the chloroplasts. The organelles stay there for the rest of the animal's life, carrying out photosynthesis as if they were still in the alga's cells. Like a plant, the solar-powered sea slug can live on sunlight and air.

A chloroplast requires a few thousand genes to carry out photosynthesis. Although chloroplasts contain their own DNA, these genes encode less than 10% of the required proteins. DNA in a plant cell's nucleus makes up the difference. But slugs are animals, and the nuclei inside their cells presumably lack these critically important genes. How can the chloroplasts operate inside their mollusk partners?

Mary E. Rumpho, of the University of Maine, collaborated with James R. Manhart, of Texas A&M University, to find out the answer. They considered two possibilities. Either the chloroplasts can work inside the host slug's digestive tract without the help of supplemental genes, or the slug's own cells provide the necessary proteins.

The researchers tested the first possibility by searching the chloroplast's DNA for genes that are essential for photosynthesis. They discovered that a gene called *psbO* was missing from the chloroplast. The *psbO* gene encodes a protein that is an essential part of photosystem II. Without *psbO*, photosynthesis is impossible. The researchers therefore rejected the hypothesis that the chloroplasts are autonomous.

That left the second possibility, which suggested that the slug's cells contain the DNA necessary to support the chloroplasts. The team looked for the *psbO* gene in the animal's DNA, and they found it (figure 5.15). Moreover, when they sequenced the *psbO* gene from the slug's genome, it was identical to the same gene in algae.

How could a gene required for photosynthesis have moved from a filamentous yellow-green alga to the genome of a sea slug? No one knows, but the researchers speculate that cells in a slug's digestive tract may have taken up fragments of algal DNA that spilled from partially eaten filaments.

Biologists do know that bacterial species often swap genes in a process called horizontal gene transfer. Rumpho and Manhart's study provides convincing evidence that horizontal gene transfer can and does occur between distantly related eukaryotes, too. Moreover, genetic evidence from many organisms suggests that horizontal gene transfer may have been extremely common throughout life's long history. As a result, many biologists are discarding the notion of a tidy evolutionary "tree" in favor of a messier, but perhaps more fascinating, evolutionary thicket.

Rumpho, Mary E., and seven colleagues, including James R. Manhart. 2008. Horizontal gene transfer of the algal nuclear gene *psbO* to the photosynthetic sea slug *Elysia chlorotica. Proceedings of the National Academy of Sciences,* vol. 105, pages 17867–17871.

5.8 | Mastering Concepts

1. Explain the most important finding of this study, and describe the evidence the researchers used to arrive at their conclusion.

2. The researchers also looked for the *psbO* gene in pufferfish (a vertebrate animal) and slime molds (a nonphotosynthetic protist). The gene was absent in both species. How was this finding important to the interpretation of the results of this study?

Figure 5.14 A Slug with Solar Panels. The leaflike body of the sea slug *Elysia chlorotica* is typically 2 to 3 centimeters long.

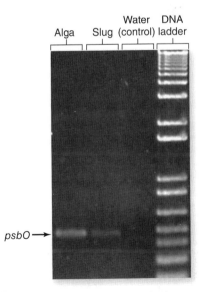

Figure 5.15 Photosynthesis Gene. Both algae and the "solar-powered" sea slug contain *psbO*, a gene required for photosynthesis. This electrophoresis gel sorts DNA fragments by size as they migrate from the top to the bottom of the gel. The "ladder" contains DNA pieces of known size, allowing the researchers to estimate the size of the DNA being studied.

Chapter Summary

5.1 | Life Depends on Photosynthesis

- **Photosynthesis** converts kinetic energy in light to potential energy in the covalent bonds of glucose. Plants, algae, and some bacteria are photosynthetic.

A. Photosynthesis Builds Carbohydrates Out of Carbon Dioxide and Water

- Photosynthesis is a redox reaction in which water is oxidized and CO_2 is reduced to glucose.
- Plants use glucose to generate ATP, grow, nourish nonphotosynthetic plant parts, and produce cellulose and many other biochemicals. Most store excess glucose as starch or sucrose.
- Most life ultimately depends on photosynthesis.

B. The Evolution of Photosynthesis Changed Planet Earth

- Before photosynthesis evolved, organisms were **heterotrophs** that relied on organic molecules as a carbon source. The first **autotrophs** developed the ability to produce their own organic molecules from atmospheric CO_2.
- Over billions of years, oxygen produced in photosynthesis changed Earth's climate and the history of life.

5.2 | Sunlight Is the Energy Source for Photosynthesis

A. What Is Light?

- Visible light is a small part of the **electromagnetic spectrum.**
- **Photons** move in waves. The longer the **wavelength,** the less kinetic energy per photon. Visible light occurs in a spectrum of colors representing different wavelengths.

B. Photosynthetic Pigments Capture Light Energy

- **Chlorophyll *a*** is the primary photosynthetic pigment in plants. **Accessory pigments** absorb wavelengths of light that chlorophyll *a* cannot absorb, extending the range of wavelengths useful for photosynthesis.

C. Chloroplasts Are the Sites of Photosynthesis

- Plants exchange gases with the environment through pores called **stomata.**
- Leaf **mesophyll** cells contain abundant **chloroplasts.**
- A chloroplast consists of a gelatinous matrix called the **stroma,** which contains stacks of **thylakoid** membranes called **grana.** Photosynthetic pigments are embedded in the thylakoid membranes, which enclose the **thylakoid space.**
- A **photosystem** consists of **antenna pigments** and a **reaction center.**

5.3 | Photosynthesis Occurs in Two Stages

- The **light reactions** of photosynthesis produce **ATP** and **NADPH;** these molecules provide energy and electrons for the glucose-producing **carbon reactions.**

5.4 | The Light Reactions Begin Photosynthesis

A. Photosystem II Produces ATP

- Photosystem II captures light energy and sends electrons from reactive chlorophyll *a* to an **electron transport chain** that joins photosystem II to photosystem I.
- Electrons from chlorophyll are replaced with electrons from water. O_2 is the waste product.
- The energy released in the electron transport chain drives the active transport of protons into the thylakoid space. The protons diffuse out through channels in **ATP synthase.** This movement powers the phosphorylation of ADP to ATP.
- The coupling of the proton gradient and ATP formation is called **chemiosmotic phosphorylation.**

B. Photosystem I Produces NADPH

- Photosystem I receives electrons from the electron transport chain and uses them to reduce $NADP^+$, producing NADPH. Light provides the energy.

5.5 | The Carbon Reactions Produce Carbohydrates

- The carbon reactions use energy from ATP and electrons from NADPH in **carbon fixation** reactions that incorporate CO_2 into organic compounds.
- In the **Calvin cycle, rubisco** catalyzes the reaction of CO_2 with **ribulose bisphosphate** (RuBP) to yield two molecules of PGA. These are converted to PGAL, the immediate carbohydrate product of photosynthesis. PGAL later becomes glucose.

5.6 | C₃ Plants Use Only the Calvin Cycle to Fix Carbon

- The Calvin cycle is also called the **C₃ pathway.** Most plant species are C₃ plants, which use only this pathway to fix carbon.
- **Photorespiration** wastes carbon and energy when rubisco reacts with O_2 instead of CO_2.

5.7 | The C₄ and CAM Pathways Save Carbon and Water

- The **C₄ pathway** reduces photorespiration by separating the light and carbon reactions into different cells. In mesophyll cells, CO_2 is fixed as a four-carbon molecule, which moves to a **bundle-sheath cell** and liberates CO_2 to be fixed again in the Calvin cycle.
- In the **CAM pathway,** desert plants such as cacti open their stomata and take in CO_2 at night, storing the fixed carbon in vacuoles. During the day, they split off CO_2 and fix it in chloroplasts in the same cells.

5.8 | Investigating Life: Solar-Powered Sea Slugs

- The sea slug *Elysia chlorotica* contains chloroplasts acquired from its food, a filamentous alga. The slug's DNA includes a gene required for photosynthesis.

Multiple Choice Questions

1. Where does the energy come from to drive photosynthesis?
   a. A chloroplast
   b. ATP
   c. The sun
   d. Glucose

2. Photosynthesis is an example of an _____ chemical reaction because _____.
   a. exergonic; energy is released by the reaction center pigment
   b. endergonic; light energy is used to build chemical bonds
   c. exergonic; light energy is captured by pigment molecules
   d. endergonic; the reactions occur inside a cell

3. The evolution of photosynthesis resulted in
   a. an increase in the amount of O_2 in the atmosphere.
   b. the initial appearance of heterotrophs.
   c. global warming.
   d. an increase in the amount of CO_2 in the atmosphere.

4. A plant appears green because
   a. it contains chloroplasts.
   b. chlorophyll *a* absorbs red and blue light.
   c. chlorophyll *a* absorbs ultraviolet light.
   d. Both a and c are correct.

5. Only high-energy light can penetrate the ocean and reach photosynthetic organisms in coral reefs. What color of light would you predict these organisms use?
   a. Red
   b. Yellow
   c. Blue
   d. Orange

6. Which part of the chloroplast is associated with the production of carbohydrates?
   a. The thylakoid
   b. The grana
   c. The thylakoid space
   d. The stroma

7. The ATP that is produced in the light reactions is used by the cell to
   a. reproduce and grow.
   b. build carbohydrate molecules.
   c. move electrons through the electron transport chain.
   d. split water into H^+ and O_2.

8. Can carbon fixation occur at night?
   a. Yes, because CO_2 can always enter a leaf.
   b. No, because a plant cell is not active at night.
   c. Yes, if there is a source of ATP and NADPH.
   d. No, photorespiration occurs at night.

9. What happens to the enzyme rubisco during photorespiration?
   a. The enzyme speeds up the formation of glucose.
   b. The enzyme's active site binds to O_2 instead of CO_2.
   c. It becomes denatured.
   d. The enzyme catalyzes the breakdown of glucose.

10. A plant that only opens its stomata at night is a
    a. C_2 plant.
    b. C_3 plant.
    c. C_4 plant.
    d. CAM plant.

Write It Out

1. Photosynthesis takes place in plants, algae, and some microbes. How does it affect a meat-eating animal?

2. What color would plants be if they absorbed all wavelengths of visible light? Why?

3. Define these terms and arrange them from smallest to largest:
   a. thylakoid membrane
   b. chloroplast
   c. reaction center
   d. photosystem
   e. electron transport chain

4. Determine whether each of the following molecules is involved in the light reactions, the carbon reactions, or both, and explain how: O_2, CO_2, carbohydrate, chlorophyll *a*, photons, NADPH, ATP, H_2O.

5. The light reactions described in this chapter are sometimes called "noncyclic photophosphorylation" because the electron transport chain that generates ATP does not return the electrons to chlorophyll. Some photosynthetic bacteria, however, generate ATP by cyclic phosphorylation. In these cells, light energy boosts chlorophyll's electrons, which pass through an electron transport chain and then return to the chlorophyll molecule. Would the light reactions of cyclic photophosphorylation produce ATP? NADPH? O_2? Explain your answers.

6. Of the many groups of photosynthetic bacteria, only cyanobacteria use chlorophyll *a*. How does this observation support the hypothesis that cyanobacteria gave rise to the chloroplasts of today's plants and algae?

7. One of the first investigators to explore photosynthesis was Flemish physician and alchemist Jan van Helmont. In the early 1600s, he grew willow trees in weighed amounts of soil, applied known amounts of water, and noted that in 5 years the trees gained more than 45 kg, but the soil had lost only a little weight. Because he had applied large amounts of water, van Helmont concluded (incorrectly) that plants grew solely by absorbing water. What is the actual source of the added biomass? Explain your answer.

8. One of the classic experiments in photosynthesis occurred in 1771, when Joseph Priestley found that if he placed a mouse in an enclosed container with a lit candle, the mouse would die. But if he also added a plant to the container, the mouse could live. Priestley concluded that plants "purify" air, allowing animals to breathe. What is the biological basis for this observation?

9. In 1941, biologists exposed photosynthesizing cells to water containing a heavy oxygen isotope, designated ^{18}O. The "labeled" isotope appears in the O_2 gas released in photosynthesis, showing that the oxygen came from the water. Where would the ^{18}O have ended up if the researchers had used ^{18}O-labeled CO_2 instead of H_2O?

10. How does photorespiration counteract photosynthesis?

11. When vegetables and flowers are grown in greenhouses in the winter, their growth rate greatly increases if the CO_2 concentration is raised to two or three times the level in the natural environment. What is the biological basis for the increased rate of growth?

12. Over the past decades, the CO_2 concentration in the atmosphere has increased.
    a. Predict the effect of increasing carbon dioxide concentrations on photorespiration.
    b. Scientists suggest that increasing CO_2 concentrations are leading to higher average global temperatures. If temperatures are increasing, does this change your answer to part (a)?

13. How is the CAM pathway adaptive in a desert habitat?

14. Explain how C_4 photosynthesis is based on a spatial arrangement of structures, whereas CAM photosynthesis is temporally based.

15. Explain why each of the following misconceptions about photosynthesis is false:
    a. Only plants are autotrophs.
    b. Plants do not need cellular respiration because they carry out photosynthesis.
    c. Chlorophyll is the only photosynthetic pigment.

Pull It Together

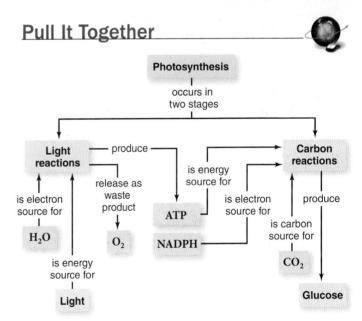

1. Where do electron transport chains fit into this concept map?

2. What specific event in the light reactions gives rise to the waste product, O_2?

3. How would you incorporate the Calvin cycle, rubisco, C_3 plants, C_4 plants, and CAM plants into this concept map?

4. Where do humans and other heterotrophs fit into this concept map?

5. Build another small concept map showing the relationships among the terms *chloroplast, stroma, grana, thylakoid, photosystem,* and *chlorophyll.*

6. What happens to the glucose produced in photosynthesis?

How Cells Release Energy

This African rock python is consuming a Thomson's gazelle.

Eating for Life

THE AFRICAN ROCK PYTHON LAY IN WAIT FOR THE LONE GAZELLE. When the gazelle came close, the snake moved suddenly, positioning the victim's head and holding it in place while swiftly entwining its 9-meter long body snugly around the mammal. Each time the gazelle exhaled, the snake squeezed, shutting down the victim's heart and lungs in less than a minute.

Thanks to the adaptations of its digestive system, the snake can swallow and digest a meal over half its own size. The reptile begins by opening its jaws at an angle of 130 degrees (compared with 30 degrees for the most gluttonous human) and places its mouth over the gazelle's head, using strong muscles to gradually envelop and push along the carcass. Saliva coats the prey, easing its journey to the snake's stomach. After several hours, the huge meal arrives at the stomach, and the remainder of the digestive tract readies itself for several weeks of dismantling the gazelle. Hydrochloric acid (HCl) builds up in the snake's stomach, lowering the pH sufficiently for the digestive enzymes to function, and the output of digestive enzymes in the intestines increases 60-fold.

Most organisms invest 10% to 23% of each meal's energy in digesting it and assimilating its nutrients. By comparison, the snake pays dearly for its meal, investing 32% in energy acquisition. Why the difference? The snake must use considerable muscle power to capture, subdue, and swallow its enormous prey. The reptile also expends energy in the rapid buildup of HCl and enzymes in its digestive tract.

As the gazelle passes through the snake's digestive system, it breaks into clumps of cells. These cells disintegrate, releasing proteins, carbohydrates, and lipids. After the snake digests these macromolecules, the component parts are small enough to enter the blood and move to the body's tissues. The animal's cells absorb these smaller nutrient molecules. Then, in cellular respiration, energy in the bonds of the food molecules is transferred to the high-energy phosphate bonds of ATP. Afterward, only a few chunks of hair and bone will remain to be eliminated.

In humans, snakes, and every other organism, nearly all activities depend on energy stored in ATP. Yet nothing eats ATP directly. This chapter describes how cells convert what we do eat—glucose and other food molecules—into those little ATP molecules that nothing can live without.

Learning Outline

Learn How to Learn

Don't Skip the Figures

As you read the narrative in the text, pay attention to the figures as well. Each one is trying to teach you something, but what is it? Sometimes, a figure summarizes the narrative and helps you see the chapter's "big picture." Other illustrations show the parts of a structure or the steps in a process; still others summarize a technique or help you classify information. Flip through this book and see if you can find examples of each type.

6.1 | Cells Use Energy in Food to Make ATP

No cell can survive without **ATP**—adenosine triphosphate. Without this energy carrier, you could not have developed from a fertilized egg into an adult. You could not breathe, chew, talk on the phone, circulate your blood, blink your eyes, walk, or listen to music. Without ATP, a plant could not take up soil nutrients, grow, or produce flowers, fruits, and seeds. A fungus could not acquire food or produce mushrooms. Like a car without gasoline, a cell without ATP simply dies. ▶ ATP, p. 76

ATP is essential because it powers nearly every activity that requires energy input in the cell: synthesis of DNA, RNA, proteins, carbohydrates, and lipids; active transport across the membranes surrounding cells and organelles; separation of duplicated chromosomes during cell division; movement of cilia and flagella; muscle contraction; and many others. This constant need for ATP explains the need for a steady food supply: all organisms use the potential energy stored in food to make ATP.

Where does the food come from in the first place? Chapter 5 explains the answer: In most ecosystems, plants and other autotrophs use photosynthesis to make organic molecules such as glucose ($C_6H_{12}O_6$) out of carbon dioxide (CO_2) and water (H_2O). Light supplies the energy. The glucose produced in photosynthesis feeds not only autotrophs but also all of the animals, fungi, and microbes that share the ecosystem (see figure 5.1).

All cells need ATP, but they don't all produce it in the same way. The pathways that generate ATP from food fall into three categories. In aerobic cellular respiration, the main subject of this chapter, a cell uses oxygen gas (O_2) and glucose to generate ATP. Plants, animals, and many microbes, especially those in O_2-rich environments, use aerobic respiration. The other two pathways, anaerobic respiration and fermentation, generate ATP from glucose without using O_2. Section 6.8 describes these two processes, both of which are most common in microorganisms.

The overall equation for **aerobic respiration** is essentially the reverse of photosynthesis:

$$\text{glucose} + \text{oxygen} \longrightarrow \text{carbon dioxide} + \text{water} + \text{ATP}$$
$$C_6H_{12}O_6 + 6O_2 \longrightarrow 6CO_2 + 6H_2O + 36ATP$$

This equation reveals that aerobic cellular respiration requires organisms to acquire O_2 and get rid of CO_2 (**figure 6.1**). These gases simply diffuse across the cell membranes of single-celled organisms, but more complex organisms have specialized organs of gas exchange such as gills or lungs. In humans and many other animals, O_2 from inhaled air diffuses into the bloodstream across the walls of microscopic air sacs in the lungs. The circulatory system carries the inhaled O_2 to cells, where gas exchange occurs. O_2 diffuses into the cell's mitochondria, the sites of respiration. Meanwhile, CO_2 diffuses out of the cells and into the bloodstream. After moving from the blood into the lungs, the CO_2 is exhaled.

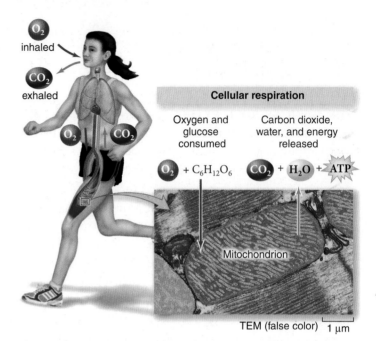

Figure 6.1 Breathing and Cellular Respiration Are Linked. The athlete breathes in O_2, which enters the bloodstream in the lungs and is distributed to all cells. There, in mitochondria, the O_2 participates in the reactions of cellular respiration. Energy-rich ATP is generated from potential energy in food; CO_2, a metabolic waste, is exhaled.

Many people mistakenly believe that plants do not use cellular respiration because they are photosynthetic. In fact, plants use O_2 to respire about half of the glucose they produce. Why do plants have a reputation for producing O_2, if they also consume it? The reason is that plants incorporate much of the remaining glucose into cellulose, starch, and other stored organic molecules. Therefore, they absorb much more CO_2 in photosynthesis than they release in respiration, and they release more O_2 than they consume.

The rest of this chapter describes how cells use the potential energy in food to generate ATP. Like photosynthesis, the journey entails several overlapping metabolic pathways and many different chemicals. But if we consider energy release in major stages, the logic emerges.

6.1 | Mastering Concepts

1. Why do all organisms need ATP?
2. What are the three general ways to generate ATP from food, and which organisms use each pathway?
3. What is the overall equation for cellular respiration?
4. How is cellular respiration related to breathing?
5. How can plants release more O_2 in photosynthesis than they consume in respiration?

6.2 | Cellular Respiration Includes Three Main Processes

The chemical reaction that generates ATP is straightforward: an enzyme tacks a phosphate group onto ADP, yielding ATP. As described in chapter 4, however, ATP synthesis requires an input of energy. The metabolic pathways of respiration harvest potential energy from food molecules and use it to make ATP. This section briefly introduces these pathways; later sections explain them in more detail.

Like photosynthesis, respiration is an oxidation–reduction reaction. The pathways of aerobic respiration oxidize (remove electrons from) glucose and reduce (add electrons to) O_2. Because of oxygen's strong attraction for electrons, this reaction is "easy," like riding a bike downhill. It therefore releases energy, which the cell traps in the bonds of ATP. ▶ redox reactions, p. 75

This reaction does not happen all at once. If a cell released all the potential energy in glucose's chemical bonds in one uncontrolled step, the sudden release of heat would destroy the cell; in effect, it would act like a tiny bomb. Rather, the chemical bonds and atoms in glucose are rearranged one step at a time, releasing a tiny bit of energy with each transformation. According to the second law of thermodynamics, some of this energy is lost as heat. But much of it is stored in the chemical bonds of ATP.

Biologists organize the intricate biochemical pathways of respiration into three main groups: glycolysis, the Krebs cycle, and electron transport (**figure 6.2**). In **glycolysis** (literally, "breaking sugar"), a six-carbon glucose molecule splits into two three-carbon **pyruvate** molecules. This process harvests energy in two forms. First, some of the electrons from glucose are transferred to an electron carrier molecule called **NADH** (nicotine adenine dinucleotide). Second, glycolysis generates two molecules of ATP.

Additional reactions, including the **Krebs cycle,** oxidize the pyruvate and release CO_2. Enzymes rearrange atoms and bonds in ways that transfer the pyruvate's potential energy and electrons to ATP, NADH, and another electron carrier molecule—**FADH$_2$** (flavin adenine dinucleotide).

By the time the Krebs cycle is complete, the carbon atoms that made up the glucose are gone—liberated as CO_2. The cell has generated a few molecules of ATP, but most of the potential energy from glucose now lingers in the high-energy electron carriers, NADH and FADH$_2$. The cell uses them to generate more ATP.

The **electron transport chain** transfers energy-rich electrons from NADH and FADH$_2$ through a series of membrane proteins. As electrons pass from carrier to carrier in the electron transport chain, the energy is used to create a gradient of hydrogen ions. (Recall from chapter 2 that a hydrogen ion is simply a hydrogen atom stripped of its electron, leaving just a proton.) The mitochondrion uses the potential energy stored in this proton gradient to generate ATP. An enzyme called **ATP synthase** forms a channel in the membrane, releasing the protons and using their

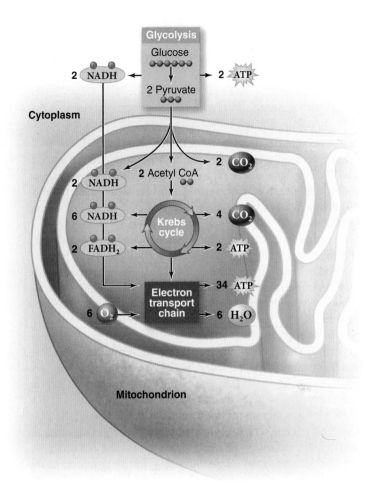

Figure 6.2 Overview of Aerobic Cellular Respiration. Glucose is broken down to carbon dioxide in three main stages: glycolysis, the Krebs cycle, and the electron transport chain. Along the way, energy is harvested as ATP. Except for glycolysis, these reactions occur inside the mitochondria of eukaryotic cells.

potential energy to add phosphate to ADP. (As described in section 5.4, the same enzyme generates ATP in the light reactions of photosynthesis.) In the meantime, the "spent" electrons are transferred to O_2, generating water as a waste product.

A common misconception is that any ATP-generating pathway in a cell is considered "respiration." In fact, however, all forms of respiration, aerobic and anaerobic, require an electron transport chain. As you will see in section 6.8, fermentation is not respiration because it generates ATP from glycolysis only.

6.2 | Mastering Concepts

1. Why do the reactions of respiration occur step-by-step instead of all at once?

2. What occurs in each of the three stages of cellular respiration?

6.3 | In Eukaryotic Cells, Mitochondria Produce Most ATP

Glycolysis always occurs in the cytoplasm, but the location of the other pathways in aerobic respiration depends on the cell type. In bacteria and archaea, the enzymes of the Krebs cycle are in the cytoplasm, and electron transport proteins are embedded in the cell membrane. The eukaryotic cells of protists, plants, fungi, and animals, however, contain organelles called **mitochondria** that house the other reactions of cellular respiration (figure 6.3).

A mitochondrion consists of an outer membrane and a highly folded inner membrane. **Cristae** are folds that greatly increase the surface area of the inner membrane. The **intermembrane compartment** is the area between the two membranes, and the mitochondrial **matrix** is the space enclosed within the inner membrane.

In a eukaryotic cell, the two pyruvate molecules produced in glycolysis cross both of the mitochondrial membranes and move into the matrix. Here, enzymes cleave pyruvate and carry out the Krebs cycle. Then, $FADH_2$ and NADH from glycolysis and the Krebs cycle move to the inner mitochondrial membrane, which is studded with electron transport proteins and ATP synthase. The inner membrane's cristae greatly increase the surface area on which the reactions of the electron transport chain can occur.

Electron transport chains and ATP synthase also occur in the thylakoid membranes of chloroplasts, which generate ATP in the light reactions of photosynthesis (see chapter 5). Similar enzymes also operate in the cell membranes of respiring bacteria and archaea, making ATP synthase one of the most highly conserved proteins over evolutionary time.

Mitochondria and chloroplasts also share another similarity: Both types of organelles contain DNA and ribosomes. Mitochondrial DNA encodes ATP synthase and most of the proteins of the electron transport chain. Not surprisingly, a person with abnormal versions of these genes may be very ill or even die. The worst mitochondrial diseases affect the muscular and nervous systems. Muscle and nerve cells are especially energy-hungry; each may contain as many as 10,000 mitochondria.

6.3 | Mastering Concepts

1. What are the parts of a mitochondrion?
2. Which respiratory reactions occur in each part of the mitochondrion?

Figure 6.3 **Anatomy of a Mitochondrion.** Eukaryotic cells, such as the ones that make up leaves, contain mitochondria that provide most of the cell's ATP. Each mitochondrion includes two membranes. The inner membrane encloses fluid called the mitochondrial matrix, and the space between the inner and outer membranes is the intermembrane compartment.

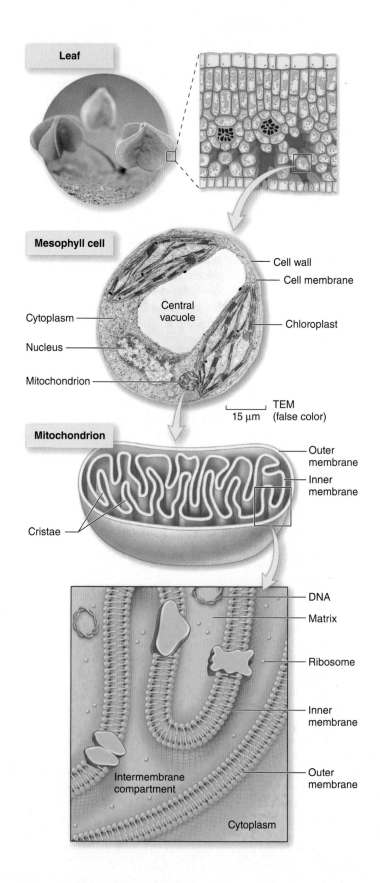

Leaf

Mesophyll cell

Cell wall
Cell membrane
Cytoplasm
Central vacuole
Chloroplast
Nucleus
Mitochondrion
15 µm   TEM (false color)

Mitochondrion

Outer membrane
Inner membrane
Cristae

DNA
Matrix
Ribosome
Inner membrane
Intermembrane compartment
Outer membrane
Cytoplasm

6.4 | Glycolysis Breaks Down Glucose to Pyruvate

Glycolysis is a more-or-less universal metabolic pathway that splits glucose into two three-carbon pyruvate molecules. The name of the pathway reflects its function: *glyco-* means sugar, and *-lysis* means to break.

The entire process of glycolysis requires 10 steps, all of which occur in the cell's cytoplasm (figure 6.4). None of the steps requires O_2, so cells can use glycolysis in both oxygen-rich and anaerobic environments.

The first five steps of glycolysis use ATP to "activate" glucose, redistributing energy in the molecule and splitting it in half. The rest of the pathway then stores some energy in two molecules of the electron carrier, NADH. Other steps extract more of the energy, regaining the two ATP molecules invested earlier plus producing two more. The net gain is therefore two NADHs and two ATPs per molecule of glucose.

Glycolysis produces ATP by **substrate-level phosphorylation,** which simply means that a high-energy "donor" molecule physically transfers a phosphate group to ADP, forming ATP. Unlike chemiosmotic phosphorylation, which is described in the section 6.5, this method of producing ATP does not require a proton gradient or the ATP synthase enzyme.

Glucose contains considerable bond energy, but cells recover only a small portion of it as ATP and NADH during glycolysis. Cells that carry out fermentation, such as yeasts that produce wine and beer, survive on this paltry ATP yield (see section 6.8). Yet the two pyruvate molecules still retain most of the potential energy of the original glucose molecule. As you will see, the pathways of aerobic respiration extract much more of that energy.

6.4 | Mastering Concepts

1. What are the starting materials of glycolysis?
2. How is substrate-level phosphorylation different from chemiosmotic phosphorylation?
3. What is the net gain of ATP and NADH for each glucose molecule undergoing glycolysis?

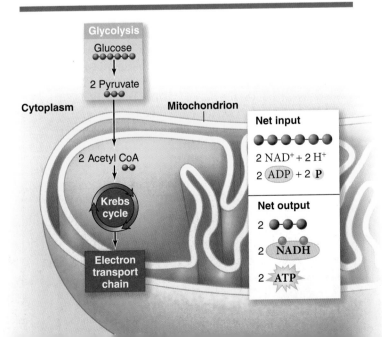

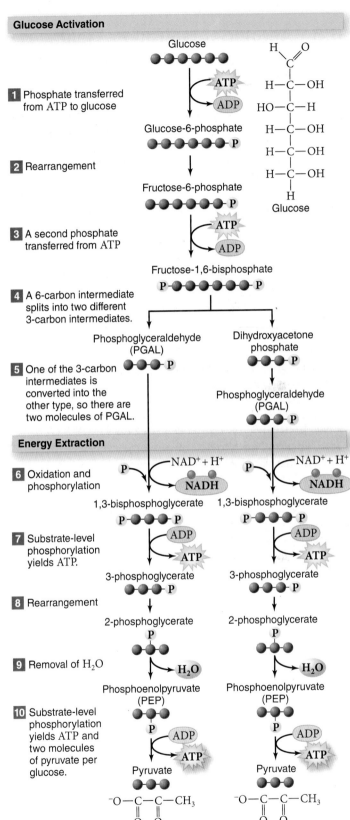

Figure 6.4 **Glycolysis.** In glycolysis, glucose splits into two molecules of pyruvate, producing a net yield of two ATPs and two NADHs. The mitochondrion illustrated at left shows an overview of glycolysis, and the right side of the figure shows the entire 10-step process. (Each gray sphere represents a carbon atom.)

6.5 | Aerobic Respiration Yields Much More ATP than Glycolysis Alone

Overall, aerobic cellular respiration taps much of the potential energy remaining in the pyruvate molecules that emerge from the pathways of glycolysis. The Krebs cycle and electron transport chain are the key ATP-generating processes. This section explains how they work.

A. Pyruvate Is Oxidized to Acetyl CoA

After glycolysis, pyruvate moves into the mitochondrial matrix, but it is not directly used in the Krebs cycle. Instead, a preliminary chemical reaction further oxidizes each pyruvate molecule (figure 6.5). First, a molecule of CO_2 is removed, and NAD^+ is reduced to NADH. The remaining two-carbon molecule, called an acetyl group, is transferred to a coenzyme to form acetyl coenzyme A (abbreviated acetyl CoA). **Acetyl CoA** is the compound that enters the Krebs cycle. ▸ coenzymes, p. 78

B. The Krebs Cycle Produces ATP and Electron Carriers

The Krebs cycle completes the oxidation of each acetyl group, releasing CO_2 (figure 6.6). The cycle begins when acetyl CoA sheds the coenzyme and combines with a four-carbon molecule called oxaloacetate. The resulting six-carbon molecule is called citrate, which is why the Krebs cycle is also known as the citric acid cycle.

The remaining steps in the Krebs cycle rearrange and oxidize citrate through several intermediates. Along the way, two carbon atoms are released as CO_2. In addition, some of the transformations transfer electrons to NADH and $FADH_2$; others produce ATP by substrate-level phosphorylation. Eventually, the molecules in the Krebs cycle re-create the original acceptor molecule, oxaloacetate. The cycle can now repeat.

The Krebs cycle turns twice per glucose molecule. Thus, the combined net output to this point (glycolysis, acetyl CoA formation, and the Krebs cycle) is four ATP molecules, 10 NADH molecules, two $FADH_2$ molecules, and six molecules of CO_2. Of course, this process does not capture all of the potential energy in glucose. According to the second law of thermodynamics, some is always lost as heat.

Besides continuing the breakdown of glucose, the Krebs cycle also has another function not directly related to respiration. The cell uses intermediate compounds formed in the Krebs cycle to manufacture other organic molecules, such as amino acids or fats. Section 6.7 explains that the reverse process also occurs; amino acids and fats can enter the Krebs cycle to generate energy from food sources other than carbohydrates.

C. The Electron Transport Chain Drives ATP Formation

The products generated in glycolysis, acetyl CoA formation, and the Krebs cycle are CO_2, ATP, NADH, and $FADH_2$. The cell ejects the CO_2 as waste and uses ATP to fuel essential processes. But what becomes of the NADH and $FADH_2$? An electron

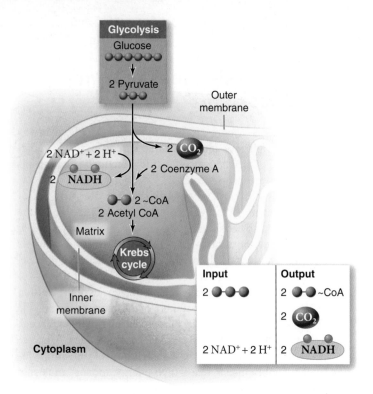

Figure 6.5 **Transition to the Mitochondria.** After pyruvate moves into a mitochondrion, it is oxidized to form a two-carbon acetyl group, CO_2, and NADH. The acetyl group joins with coenzyme A to form acetyl CoA. For every glucose molecule that entered glycolysis, two acetyl CoA molecules are now ready to enter the Krebs cycle.

transport chain in the inner mitochondrial membrane transfers the potential energy of these electron carriers to ATP as well.

The electron transport chain uses the energy from NADH and $FADH_2$ in stages (figure 6.7). The first protein in the chain accepts electrons from NADH; $FADH_2$ donates its electrons to the second protein. Subsequent proteins use some of the energy from the electrons to pump hydrogen ions (H^+) from the matrix into the intermembrane compartment. The electrons then pass to the next protein in the chain, and the next, and so on. In aerobic respiration, the final electron acceptor is O_2, which combines with hydrogen ions to form water. Breathing provides the O_2.

The electron transport chain therefore uses the energy in NADH and $FADH_2$ to establish a proton gradient across the inner mitochondrial membrane. As explained in chapter 4, a gradient represents a form of potential energy. The mitochondrion harvests this energy as ATP in the final stage of cellular respiration, with the help of the ATP synthase enzyme. In **chemiosmotic phosphorylation,** protons move down their gradient through ATP synthase channels back into the matrix, and ADP is phosphorylated to ATP. The ATP synthase enzyme therefore captures the potential energy of the proton gradient and saves it in a form the cell can use: ATP.

6.5 | Mastering Concepts

1. Pyruvate contains three carbon atoms; an acetyl group has only two. What happens to the other carbon atom?
2. How does the Krebs cycle generate CO_2, ATP, NADH, and $FADH_2$?
3. How do NADH and $FADH_2$ power ATP formation?
4. What is the role of O_2 in the electron transport chain?

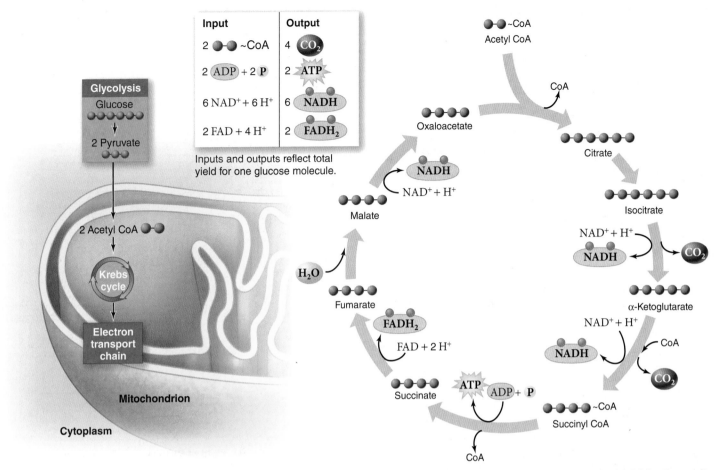

Figure 6.6 Krebs Cycle. In the mitochondrial matrix, acetyl CoA enters the Krebs cycle and is oxidized to two molecules of CO_2. Some of the energy is trapped as ATP, NADH, and $FADH_2$. The left half of this figure summarizes the inputs, outputs, and location of the Krebs cycle; the right half shows the entire cycle, step-by-step.

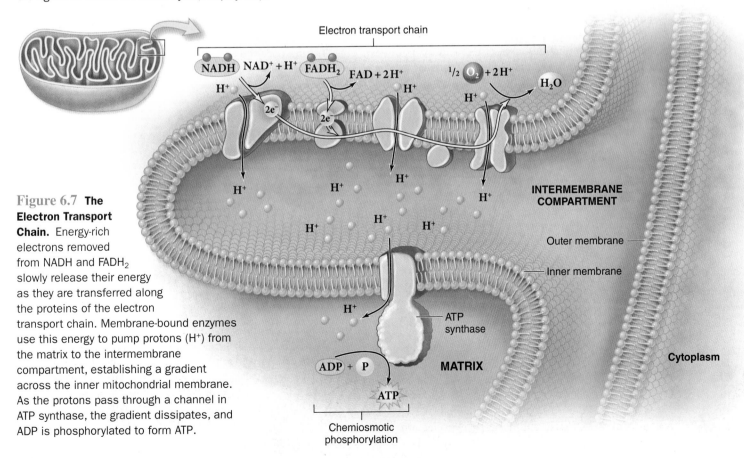

Figure 6.7 The Electron Transport Chain. Energy-rich electrons removed from NADH and $FADH_2$ slowly release their energy as they are transferred along the proteins of the electron transport chain. Membrane-bound enzymes use this energy to pump protons (H^+) from the matrix to the intermembrane compartment, establishing a gradient across the inner mitochondrial membrane. As the protons pass through a channel in ATP synthase, the gradient dissipates, and ADP is phosphorylated to form ATP.

6.6 | How Many ATPs Can One Glucose Molecule Yield?

 To estimate the yield of ATP produced from every glucose molecule that enters aerobic cellular respiration, we can add the maximum number of ATPs generated in glycolysis, the Krebs cycle, and the electron transport chain (figure 6.8).

Substrate-level phosphorylation yields two ATPs from glycolysis and two ATPs from the Krebs cycle (one ATP each from two turns of the cycle). These are the only steps that produce ATP directly. In addition, each glucose yields two NADH molecules from glycolysis, two NADHs from acetyl CoA production, and six NADHs and two $FADH_2$s from two turns of the Krebs cycle.

In theory, the ATP yield from electron transport is three ATPs per NADH and two ATPs per $FADH_2$. Electrons from the 10 NADHs from glycolysis and the Krebs cycle therefore yield up to 30 ATPs; electrons from the two $FADH_2$ molecules yield four more. Add the four ATPs from substrate-level phosphorylation, and the total is 38 ATPs per glucose. However, NADH from glycolysis must be shuttled into the mitochondrion, usually at a cost of one ATP for each NADH. This reduces the net theoretical production of ATPs to 36.

In reality, some protons leak across the inner mitochondrial membrane on their own, and the cell spends some energy to move pyruvate and ADP into the matrix. These "expenses" reduce the actual ATP yield to about 30 per glucose. The number of calories stored in 30 ATPs is about 32% of the total calories stored in the glucose bonds; the rest of the potential energy in glucose is lost as heat. This may seem wasteful, but for a biological process, it is reasonably efficient. To put this energy yield into perspective, an automobile uses only about 20% to 25% of the energy contained in gasoline's chemical bonds; the rest is lost as heat.

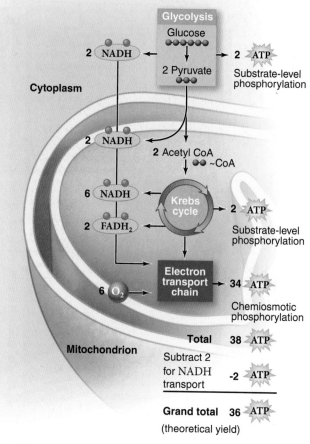

Figure 6.8 **Energy Yield of Respiration.** Breaking down glucose to carbon dioxide can yield as many as 36 ATPs, mostly from chemiosmotic phosphorylation at the electron transport chain.

6.6 | Mastering Concepts

1. Explain how to arrive at the estimate that each glucose molecule theoretically yields 36 ATPs.
2. How does the actual ATP yield compare to the theoretical yield?

Apply It Now

Some Poisons Inhibit Respiration

Many toxic chemicals kill by blocking one or more reactions in respiration. Poisons are therefore the tools of murderous villains—and of biochemists. The judicious use of poisons in isolated cells (or even isolated mitochondria) can reveal much about the chemistry of the Krebs cycle and electron transport. The following lists a few examples of chemicals that inhibit respiration:

Krebs cycle inhibitor:

- Arsenic interferes with several essential chemical reactions. For example, arsenic binds to part of a biochemical needed for the formation of acetyl CoA. It therefore blocks the Krebs cycle.

Electron transport inhibitors:

- Some mercury compounds are toxic because they stop an oxidation–reduction reaction early in the electron transport chain. Mercury is used in some thermometers, in fluorescent lights, and in many industrial applications.

- Cyanide blocks the final transfer of electrons to O_2. This highly poisonous compound has no household uses, but it is used in mining and some other industries.

- Carbon monoxide (CO) blocks electron transport at the same point as cyanide. This colorless, odorless gas is a byproduct of incomplete fuel combustion. CO from unvented heaters, stoves, and fireplaces can accumulate to deadly levels in homes. Car exhaust and cigarette smoke are other sources of CO.

Chemiosmosis inhibitors:

- The insecticide 2,4-dinitrophenol (DNP) kills by making the inner mitochondrial membrane permeable to protons, blocking formation of the proton gradient necessary to drive ATP synthesis.
- Oligomycin blocks the phosphorylation of ADP by inhibiting the part of the ATP synthase enzyme that lets the protons through. Oligomycin is mostly used in laboratory studies of respiration.

6.7 | Other Food Molecules Enter the Energy-Extracting Pathways

So far, we have focused on the complete oxidation of glucose. But food also includes starch, proteins, and lipids that contribute calories to the diet. These molecules also enter the energy pathways (figure 6.9).

The digestion of starch from potatoes, wheat, and other carbohydrate-rich food begins in the mouth and continues in the small intestine. Enzymes snip the long starch chains into individual glucose monomers, which generate ATP as described in this chapter. Another polysaccharide, glycogen, follows essentially the same path as starch. ▶ carbohydrates, p. 32

Proteins are digested into monomers called amino acids. The cell does not use most of these amino acids to produce ATP. Instead, most of them are incorporated into new proteins. When an organism depletes its immediate carbohydrate supplies, however, cells may use amino acids as an energy source. First, ammonia (NH_3) is stripped from the amino acid and excreted. The remainder of each molecule enters the energy pathways as pyruvate, acetyl CoA, or an intermediate of the Krebs cycle, depending on the amino acid. ▶ amino acids, p. 36

Meanwhile, enzymes in the small intestine digest fat molecules from food into glycerol and three fatty acids, which enter the bloodstream and move into the body's cells. Enzymes convert the glycerol to pyruvate, which then proceeds through the rest of cellular respiration as though it came directly from glucose. The fatty acids enter the mitochondria, where they are cut into many two-carbon pieces that are released as acetyl CoA. From here, the pathways continue as they would for glucose. ▶ lipids, p. 34

Fats contain more calories per gram than any other food molecule. A fat molecule has three fatty acids, each of which may contain 20 or more carbon atoms. A single fat molecule therefore yields 30 or so two-carbon acetyl CoA groups for the Krebs cycle. Conversely, the body can also store excess energy from either carbohydrates or fat by doing the reverse: diverting acetyl CoA away from the Krebs cycle and using the two-carbon fragments to build fat molecules. These lipids are stored in fat tissue that the body can use for energy if food becomes scarce.

6.7 | Mastering Concepts

1. At which points do digested polysaccharides, proteins, and fats enter the energy pathways?
2. How does the body store extra calories as fat?

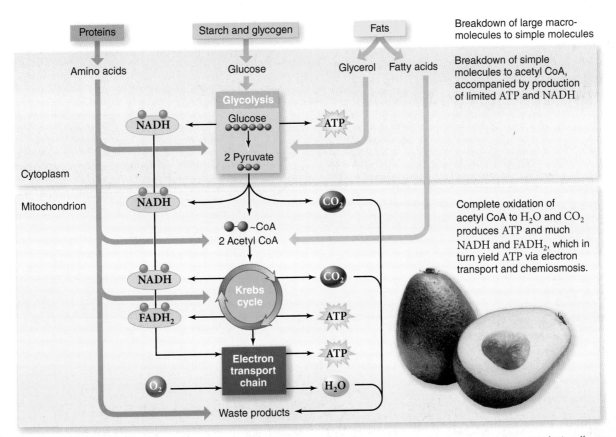

Figure 6.9 **Other Foods Enter the Energy Pathways.** Most cells use carbohydrates as a primary source of energy, but cells can also use amino acids and lipids to generate ATP.

6.8 | Some Energy Pathways Do Not Require Oxygen

 Most of the known organisms on Earth, including humans, use aerobic cellular respiration. Nevertheless, life thrives without O_2 in waterlogged soils, deep puncture wounds, sewage treatment plants, and your own digestive tract, to name just a few places. In the absence of O_2, the microbes in these habitats generate ATP using anaerobic metabolic pathways. Two examples are anaerobic respiration and fermentation (figure 6.10).

A. Anaerobic Respiration Uses an Electron Acceptor Other than O_2

Anaerobic respiration is essentially the same as aerobic respiration, except that an inorganic molecule other than O_2 is the electron acceptor at the end of the electron transport chain. Alternative electron acceptors include NO_3^- (nitrate), SO_4^{2-} (sulfate), and CO_2. The number of ATPs generated per molecule of glucose depends on the electron acceptor, but it is always lower than the ATP yield for aerobic respiration.

Many bacteria and archaea generate ATP by anaerobic respiration, and they play an important role in global nutrient cycles. For example, in waterlogged, oxygen-poor soils, bacteria that use NO_3^- as an electron acceptor begin a chain reaction that ends with the production of nitrogen gas (N_2). This gas drifts into the atmosphere, leaving the soil less fertile for plant growth. Bacteria that live in wetlands may use SO_4^{2-}, producing smelly hydrogen sulfide (H_2S) as a byproduct. And archaea living inside the intestines of cattle use CO_2 as an electron acceptor, generat-

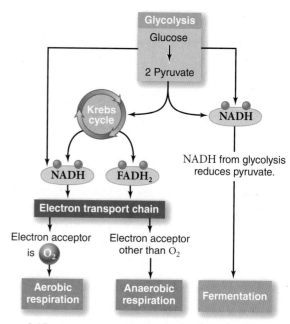

Figure 6.10 Alternative Metabolic Pathways. If O_2 is available, most organisms generate ATP in aerobic respiration. Two other pathways, anaerobic respiration and fermentation, can occur in the absence of O_2. Both alternatives yield less ATP than does aerobic respiration.

ing methane gas (CH_4). The methane, which the cattle emit as belches and flatulence, is an important greenhouse gas. ▶▶ carbon cycle, p. 774; nitrogen cycle, p. 775

? Burning Question

How do diet pills work?

Ads for diet pills are everywhere. Some are for weight-loss drugs that the U.S. Food and Drug Administration (FDA) has approved as safe and effective. Others are for dietary supplements that are not subject to FDA approval at all. All of these products, and their promises of effortless weight loss, may seem to be a dream come true. How do they work?

The FDA has approved three prescription weight-loss drugs. One is orlistat (Xenical); the over-the-counter drug Alli is a low-dose version of the same medicine. This drug interferes with lipase, the enzyme that digests fat in the small intestine. Undigested fat leaves the body in feces; orlistat therefore reduces calorie intake by reducing the body's absorption of high-energy fat molecules. The other two prescription weight-loss drugs are sibutramine (Meridia) and phentermine (Adipex-P). These medicines also reduce calorie intake, but in a different way: they suppress appetite.

All three prescription drugs can help a person lose weight but only if combined with exercise, a low-calorie diet, and behavior modification. Each also has side effects.

Dietary supplements greatly outnumber prescription weight-loss drugs, and with good reason. The FDA does not require the manufacturers of dietary supplements to show that the remedies are either safe or effective. Ads for "natural" supplements such as hoodia, green tea extract, and fucoxanthin make extraordinary promises of rapid weight loss, but the claims remain largely untested in scientific studies. The mechanism by which they work (if they work at all) usually remains unclear.

Unfortunately, some dietary supplements have serious side effects. Ephedra is one example. Before 2004, ephedra was marketed as a weight-loss aid and energy booster, but studies eventually linked it to fatal seizures, strokes, and heart attacks. The FDA therefore banned the sale of ephedra in the United States in 2004. An herb called bitter orange has taken its place in many "ephedra-free" weight-loss aids. But bitter orange has side effects that are similar to ephedra's, and its safety remains unknown.

Submit your burning question to:
marielle_hoefnagels@mcgraw-hill.com

B. Fermenters Acquire ATP Only from Glycolysis

Some microorganisms, including many inhabitants of your digestive tract, use fermentation. In these organisms, glycolysis still yields two ATPs, two NADHs, and two molecules of pyruvate per molecule of glucose. But the NADH does not donate its electrons to an electron transport chain, nor is the pyruvate further oxidized.

Instead, in **fermentation,** electrons from NADH reduce pyruvate. This process regenerates NAD$^+$ so that glycolysis can continue, but it generates no additional ATP. Fermentation is therefore far less efficient than respiration. Not surprisingly, fermentation is common among microorganisms that live in sugar-rich environments where food is essentially unlimited.

Figure It Out

Compare the number of molecules of ATP generated from 100 glucose molecules undergoing aerobic respiration versus fermentation.

Answer: 3600 (theoretical) yield for aerobic respiration; 200 for fermentation

Some microorganisms make their entire living by fermentation. An example is *Entamoeba histolytica,* a protist that causes a form of dysentery in humans. Others, including the gut-dwelling bacterium *Escherichia coli,* use O$_2$ when it is available but switch to fermentation when it is not. Most multicellular organisms, however, require too much energy to rely on fermentation exclusively.

Of the many fermentation pathways that exist, one of the most familiar produces ethanol (an alcohol). In **alcoholic fermentation,** pyruvate is first converted to acetaldehyde and CO$_2$, and then NADH reduces the acetaldehyde to produce NAD$^+$ and ethanol (figure 6.11a). Alcoholic fermentation produces the airy texture of breads; wine and champagne from grapes; the syrupy drink called mead from honey; and cider from apples. Fermenting grain—barley, rice, or corn—produces beer, sake, whisky, and other spirits.

In **lactic acid fermentation,** a cell uses NADH to reduce pyruvate, but in this case, the products are NAD$^+$ and the three-carbon compound lactic acid (figure 6.11b). The bacterium *Lactobacillus,* for example, ferments the lactose in milk, producing lactic acid that gives yogurt its sour taste. Bacteria can also ferment sugars in cabbage to produce the acids in sauerkraut, and the tangy taste of sourdough bread comes from yeasts and bacteria that use lactic acid fermentation.

The same pathway also occurs in human muscle cells. During vigorous exercise, muscles work so strenuously that they consume their available oxygen supply. In this "oxygen debt" condition (e.g., during the second half of a 100-meter dash), the muscle cells can acquire ATP only from glycolysis. The cells use lactic acid fermentation to generate NAD$^+$ so that glycolysis can continue. If too much lactic acid accumulates, however, the muscle fatigues and cramps. After the race, when the circulatory system catches up with the muscles' demand and O$_2$ is once again present, liver cells convert lactic acid back to pyruvate. Mitochondria then process the pyruvate as usual, by aerobic respiration.

6.8 | Mastering Concepts

1. What are some examples of alternative electron acceptors used in anaerobic respiration?
2. How many ATP molecules per glucose does fermentation produce?
3. What are two examples of fermentation pathways?

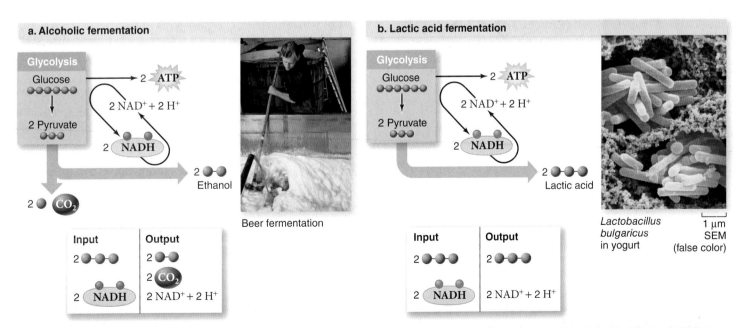

a. Alcoholic fermentation

Beer fermentation

b. Lactic acid fermentation

Lactobacillus bulgaricus in yogurt — 1 μm — SEM (false color)

Figure 6.11 Fermentation. In fermentation, ATP comes only from glycolysis. (a) Yeasts produce ethanol and carbon dioxide by alcoholic fermentation. The man in the photograph is stirring a large vat of fermenting beer. (b) Lactic acid fermentation occurs in some bacteria and, occasionally, in mammalian muscle cells. The photograph shows *Lactobacillus bulgaricus* bacteria in yogurt.

6.9 | Photosynthesis and Respiration Are Ancient Pathways

As you may have noticed, photosynthesis, glycolysis, and cellular respiration are intimately related (table 6.1 and figure 6.12). The carbohydrate product of photosynthesis—glucose—is the starting material for glycolysis. The O_2 released in photosynthesis becomes the final electron acceptor in aerobic respiration. CO_2 generated in respiration enters the carbon reactions in chloroplasts. Finally, photosynthesis splits water produced by aerobic respiration. Together, these energy reactions sustain life. How might they have arisen?

Glycolysis is probably the most ancient of the energy pathways because it occurs in virtually all cells. Glycolysis evolved when the atmosphere lacked or had very little O_2. These reactions enabled the earliest organisms to extract energy from simple organic compounds in the nonliving environment. Photosynthesis, in turn, may have evolved from glycolysis; some of the reactions of the Calvin cycle are the reverse of some of those of glycolysis. ▶ Calvin cycle, p. 98

The first photosynthetic organisms could not have been plants, because such complex organisms were not present on the early Earth. Rather, photosynthesis may have originated in an anaerobic cell that used hydrogen sulfide (H_2S) instead of water as an electron donor. These first photosynthetic microorganisms would have released sulfur, rather than O_2, into the environment. Eventually, changes in pigment molecules enabled some of these organisms to use water instead of H_2S as an electron source. Fossil evidence of cyanobacteria show that oxygen-generating photosynthesis arose at least 3.5 billion years ago. Once this pathway started, the accumulation of O_2 in the primitive atmosphere altered life on Earth forever (see section 5.1).

Later, in a process called endosymbiosis, a large "host" cell may have engulfed one of those ancient cyanobacteria and thereby transformed itself into a eukaryotic-like cell, complete with

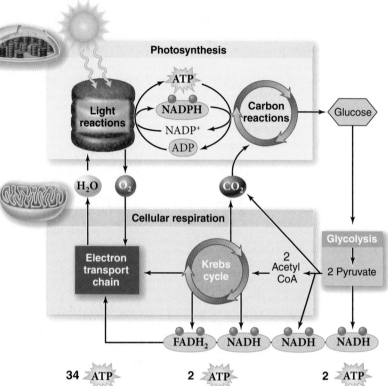

Figure 6.12 **The Energy Pathways and Cycles Connect Life.** An overview of metabolism illustrates how biological energy reactions are interrelated.

chloroplasts. Mitochondria might have evolved in a similar way, when larger cells engulfed bacteria capable of using O_2. The observation that both mitochondria and chloroplasts contain DNA and ribosomes lends support to this theory. ▶ endosymbiosis, p. 304

Afterward, different types of complex cells probably diverged, leading to the evolution of a great variety of eukaryotic organisms. Today, the interrelationships among photosynthesis, glycolysis, and aerobic respiration, along with the great similarities of these reactions in diverse species, demonstrate a unifying theme of biology: All types of organisms are related at the biochemical level.

Table 6.1 Photosynthesis and Respiration Compared

	Photosynthesis	Respiration
Food	Produced	Consumed
Energy	Stored as glucose	Released from glucose
Light	Required	Not required
H_2O	Consumed	Released
CO_2	Consumed	Released
O_2	Released	Consumed

6.9 | Mastering Concepts

1. Which energy pathway is probably the most ancient? What is the evidence?
2. Why must the first metabolic pathways have been anaerobic?
3. What is the evidence that photosynthesis may have evolved from glycolysis?

6.10 | Investigating Life: Plants' "Alternative" Lifestyles Yield Hot Sex

Think of an organism that feels warm. Did you think of yourself? A puppy? Your cat? Chances are you thought of a mammal or perhaps a bird, but certainly not a plant. Yet some plants, including *Philodendron,* do warm themselves (or at least their reproductive parts) to several degrees above ambient temperature (figure 6.13). How do they do it and, more important, what do they get out of it?

Philodendron flowers generate heat with a metabolic pathway involving the electron transport chain. As described in section 6.5, electrons from NADH and $FADH_2$ pass along a series of proteins embedded in the inner mitochondrial membrane. Along the way, the proteins pump H^+ into the space between the two mitochondrial membranes; ATP synthase uses the resulting proton gradient to generate ATP. The last protein in the electron transport chain dumps the electrons on O_2, yielding water as a waste product.

Plants and a few other types of organisms have another pathway, unimaginatively dubbed "alternative oxidase," that diverts electrons from the electron transport chain. NADH and $FADH_2$ still donate electrons to a protein in the chain, but that electron acceptor transfers them immediately to O_2 instead of to the next carrier.

The alternative oxidase pathway generates heat, but it does not help the mitochondrion produce ATP. So what does *Philodendron* gain by warming its flowers? One clue comes from the observation that the plant heats *just* its flowers and not its leaves, stem, or roots. Since flowers are reproductive parts, could the hot blooms somehow improve the plant's reproductive success?

In many plants, reproduction depends on animals that carry pollen from flower to flower. The plants may give away free meals of sweet nectar that lure pollinators such as insects, birds, and mammals. As the animal collects the offering, it brushes against the pollen-producing (male) flower parts. It then deposits the pollen on the female part of the next flower it visits.

Australian researcher Roger Seymour and his colleagues wondered whether heat from the flowers of *Philodendron solimoesense* helped the plant attract pollinators. They did a simple set of experiments to find out. First, they measured the temperature of *Philodendron* flowers. The central spike of the flower peaked at 40°C, about 15° above ambient temperature, while the floral

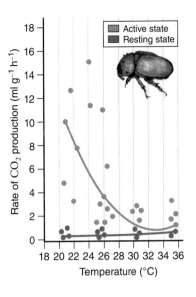

Figure 6.14 Energy Saver. Resting beetles respired at the same rate no matter what the temperature, but active beetles saved energy in warmer surroundings.

chamber was consistently at least a few degrees warmer than the surrounding air.

Next, the researchers turned their attention to beetles (*Cyclocephala colasi*) known to pollinate the flowers. The team used a device called a respirometer to measure the amount of CO_2 generated by active and resting beetles at a range of temperatures from 20°C to 35°C. Since respiration generates CO_2 as a waste product, the respirometer indirectly measures how much energy an organism uses. Resting beetles emitted approximately the same amount of CO_2 at all temperatures, but active ones (such as those that would visit flowers) produced only about one tenth as much CO_2 at 30°C as they did at 20°C (figure 6.14).

Finally, the researchers used their data to calculate the "energy-saving factor" attributed to floral heat. They concluded that the beetles used 2.0 to 4.8 times more energy at ambient temperature than at the temperature of the warmed flower, depending on time of night. The beetles therefore save energy simply by loitering on or near the flowers, energy that they can use to find food or lure mates even as they pollinate the plant. The hot flowers—courtesy of the seemingly wasteful alternative oxidase pathway—therefore enhance the reproductive success of both *Philodendron* and the beetles.

Seymour, Roger S., Craig R. White, and Marc Gibernau. November 20, 2003. Heat reward for insect pollinators. *Nature,* vol. 426, pages 243–244.

6.10 | Mastering Concepts

1. What hypothesis were the researchers testing, and what experiments did they design to help them test the hypothesis?

2. Suppose you hold one group of active beetles at 20°C and another group at 30°C. After several hours, you place each beetle in a device that measures how far the animal can fly at 20°C. Which group of beetles do you predict will fly farther?

Figure 6.13

Hot Bloom. The central part of this *Philodendron solimoesense* flower generates heat.

Chapter Summary

6.1 | Cells Use Energy in Food to Make ATP

- Every cell requires **ATP** to power reactions that require energy input.
- **Aerobic respiration** is a biochemical pathway that extracts energy from glucose in the presence of oxygen.
- The overall reaction for cellular respiration is

$$C_6H_{12}O_6 + 6O_2 \longrightarrow 6CO_2 + 6H_2O + 36ATP$$

- In humans and many other animals, the respiratory system acquires the oxygen that aerobic cellular respiration requires.
- Photosynthetic organisms such as plants also use aerobic respiration to generate ATP.

6.2 | Cellular Respiration Includes Three Main Processes

- In respiration, electrons stripped from glucose are used to reduce O_2.
- Nearly all cells use **glycolysis** as the first step in harvesting energy from glucose. The **Krebs cycle** and an **electron transport chain** follow.
- The electron transport chain establishes a proton gradient that powers phosphorylation of ADP to ATP by the enzyme **ATP synthase.**

6.3 | In Eukaryotic Cells, Mitochondria Produce Most ATP

- In eukaryotes, the Krebs cycle and electron transport chain occur in **mitochondria.** Each mitochondrion has two membranes.
- The inner membrane of a mitochondrion encloses the **matrix,** where the Krebs cycle occurs. **Cristae** are the folds of the inner membrane.
- The electron transport chain establishes a proton gradient in the **intermembrane compartment,** and ATP synthase spans the inner membrane.

6.4 | Glycolysis Breaks Down Glucose to Pyruvate

- In glycolysis, glucose is broken into three-carbon molecules of **pyruvate.**
- The reactions of glycolysis also add electrons to NAD^+, forming **NADH,** and they form two ATPs by **substrate-level phosphorylation** (phosphate transfer between organic compounds).

6.5 | Aerobic Respiration Yields Much More ATP than Glycolysis Alone

A. Pyruvate Is Oxidized to Acetyl CoA
- In the mitochondria, pyruvate is broken down into **acetyl CoA** in a coupled reaction that produces CO_2 and reduces NAD^+ to NADH.

B. The Krebs Cycle Produces ATP and Electron Carriers
- Acetyl CoA enters the Krebs cycle, a series of oxidation–reduction reactions that produces ATP, NADH, **FADH₂**, and CO_2. Substrate-level phosphorylation produces ATP in the Krebs cycle.

C. The Electron Transport Chain Drives ATP Formation
- Energy-rich electrons from NADH and $FADH_2$ fuel an electron transport chain. Electrons move through a series of carriers that release energy at each step. The terminal electron acceptor, O_2, is reduced to form water.
- The proteins of the electron transport chain pump protons from the mitochondrial matrix into the intermembrane compartment. As protons diffuse back into the matrix through channels in ATP synthase, their potential energy drives **chemiosmotic phosphorylation** of ADP to ATP.

6.6 | How Many ATPs Can One Glucose Molecule Yield?

- In the combined pathways of aerobic respiration, each glucose molecule theoretically yields 36 ATP molecules. The actual yield is about 30 ATP per glucose.

6.7 | Other Food Molecules Enter the Energy-Extracting Pathways

- Starch and other polysaccharides are digested to glucose before undergoing cellular respiration.
- Amino acids enter the energy pathways as pyruvate, acetyl CoA, or an intermediate of the Krebs cycle.
- Fatty acids enter as acetyl CoA, and glycerol enters as pyruvate.

6.8 | Some Energy Pathways Do Not Require Oxygen

A. Anaerobic Respiration Uses an Electron Acceptor Other than O₂
- In the absence of O_2, some organisms can use an alternative electron acceptor such as nitrate or sulfate.

B. Fermenters Acquire ATP Only from Glycolysis
- **Fermentation** pathways oxidize NADH to NAD^+, which is recycled to glycolysis, but these pathways do not produce additional ATP. **Alcoholic fermentation** reduces acetaldehyde to ethanol and loses carbon dioxide. **Lactic acid fermentation** reduces pyruvate to lactic acid.

6.9 | Photosynthesis and Respiration Are Ancient Pathways

- Photosynthesis and respiration are interrelated, with common intermediates and some reactions that mirror those of other pathways.
- Glycolysis may be the oldest energy pathway because it is present in nearly all cells; the other pathways are more specialized.
- Eukaryotes may have arisen in a process called endosymbiosis, when larger cells engulfed bacteria that were forerunners to mitochondria and chloroplasts.

6.10 | Investigating Life: Plants' "Alternative" Lifestyles Yield Hot Sex

- *Philodendron* plants use a modified respiratory pathway, creating a "heat reward" for their insect pollinators.

Multiple Choice Questions

1. Which of the following best describes *anaerobic* respiration?
   a. The production of ATP energy from glucose in the presence of oxygen
   b. The production of very little energy in the absence of oxygen
   c. The production of ATP energy from the sun in the presence of oxygen
   d. The production of ATP energy from glucose in the absence of oxygen

2. Which stage in cellular respiration directly requires the presence of O_2?
   a. Glycolysis   c. Electron transport
   b. The Krebs cycle   d. Both a and b are correct.

3. What is the role of ATP synthase?
   a. It uses ATP to make glucose.
   b. It uses a hydrogen ion gradient to make ATP.
   c. It uses ATP to make a hydrogen ion gradient.
   d. It synthesizes ATP directly from glucose.

4. How many ATP are made as a result of glycolysis?
   a. Two ATP are made.
   b. Four ATP are made.
   c. Two ATP are made, but two are consumed for a net gain of zero.
   d. Four ATP are made, but the net gain is two.

5. Which of the following molecules has the greatest amount of potential energy?
   a. Pyruvate   c. Glucose
   b. Acetyl CoA   d. CO_2

6. Which of the following molecules can be used to generate ATP energy?
   a. Carbohydrates   c. Fats
   b. Amino acids   d. All of the above are correct.

7. The difference between anaerobic and aerobic respiration is
   a. the amount of NADH that is produced.
   b. the electron carriers.
   c. the electron acceptors.
   d. the presence of $FADH_2$ instead of NADH.

8. Why is it important to regenerate NAD^+ during fermentation?
   a. It helps maintain the reactions of glycolysis.
   b. So it can transfer an electron to the electron transport chain
   c. To maintain the concentration of pyruvate in a cell
   d. To produce alcohol or lactic acid for the cell

9. Why is glycolysis considered to be the oldest metabolic reaction?
   a. Because glucose is a simple molecule
   b. Because it occurs in the absence of O_2
   c. Because it occurs in all cells
   d. Because it requires sunlight

10. What is *endosymbiosis*?
    a. A type of fermentation
    b. The transport of pyruvate into the matrix of the mitochondria
    c. A possible explanation for the origin of mitochondria
    d. The movement of electrons along the electron transport chain

Write It Out

1. How are breathing and cellular respiration similar? How are they different?

2. How do chemiosmotic phosphorylation and substrate-level phosphorylation each generate ATP? In which pathways does each occur?

3. How does aerobic respiration yield so much ATP from each glucose molecule, compared with glycolysis alone?

4. Cite a reaction or pathway that occurs in each of the following locations:
   a. cytoplasm
   b. mitochondrial matrix
   c. inner mitochondrial membrane
   d. intermembrane compartment

5. Health-food stores sell a product called "pyruvate plus," which supposedly boosts energy. Why is this product unnecessary? What would be a much less expensive substitute that would accomplish the same thing?

6. At what point does O_2 enter the energy pathways of aerobic respiration? What is the role of O_2? Why does respiration stop if a person cannot breathe? What happens if cellular respiration stops?

7. In a properly functioning mitochondrion, is the pH in the matrix lower than, higher than, or the same as the pH in the intermembrane compartment? If you apply one or more poisons described in this chapter's Apply It Now box, how does your answer change?

8. A chemical works as a disinfectant by poking holes in bacterial cell membranes. Why would this stop the cells from making ATP? Why would the inability to make ATP kill a cell?

9. Describe the energy pathways that are available for cells living in the absence of O_2.

10. Ben decides to bake bread. The recipe says to dissolve yeast in a mixture of sugar and hot water. Shortly after he does so, the mixture begins to bubble. What is happening? How would the outcome change if Ben forgets to add the sugar?

11. How are photosynthesis, glycolysis, and cellular respiration interrelated?

12. A student runs 5 kilometers each afternoon at a slow, leisurely pace. One day, she runs 2 km as fast as she can. Afterward she is winded and feels pain in her chest and leg muscles. She thought she was in great shape! What, in terms of energy metabolism, has she experienced?

13. Explain the fact that species as diverse as humans and yeasts use the same biochemical pathways to extract energy from nutrient molecules.

14. A seed is a plant embryo packaged with a food supply. Soaking a seed in water prompts the embryo to begin metabolizing its food supply to fuel its growth. Suppose that Anna has 50 soaked seeds. She boils half of them, killing their embryos, and lets them return to room temperature. She then places the dead seeds in one closed container and live seeds in another. If she later measures the temperature in the two containers, will they be different? Explain your answer.

15. Birds and mammals are endotherms: they maintain a constant internal body temperature no matter whether the environment is cold or hot (within limits, of course). An endotherm that gets too cold will increase its metabolic rate to generate heat. An ectothermic animal such as a reptile, on the other hand, allows its body temperature to fluctuate with the environment. If you own a pet rat and a pet snake of equal weight, which will require more food and why?

Pull It Together

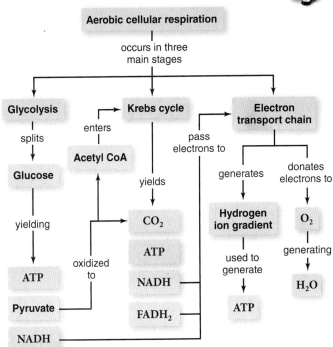

1. Where in the cell does each stage of respiration occur?
2. Where do the O_2 and glucose used in respiration come from?
3. How many ATP, NADH, CO_2, $FADH_2$, and H_2O molecules are produced at each stage of respiration?
4. What happens to the CO_2 and H_2O waste products of respiration?
5. What do cells do with the ATP they generate in respiration?
6. Where would photosynthesis, fermentation, anaerobic respiration, and ATP synthase fit into this concept map?

The "Who Am I" exhibition in London, England, celebrated the 10th anniversary of the sequencing of the entire human genome. Even with the complete sequence in hand, biologists have only just begun to explain how genes define human life.

connect™ |BIOLOGY

Enhance your study of this chapter with practice quizzes, animations and videos, answer keys, and downloadable study tools.
www.mhhe.com/hoefnagels

Learn How to Learn
Pause at the Checkpoints
As you read, get out a piece of paper and see if you can answer the "Figure It Out" and "Mastering Concepts" questions. If not, you may want to study a bit more before you move on.

The Human Genome Project Is Just the Beginning

WHEN THE HUMAN GENOME PROJECT FINISHED A DRAFT SEQUENCE OF HUMAN DNA IN MID-2000, NEWS STORIES SUGGESTED THAT PARENTS WOULD SOON BE ABLE TO SCREEN THEIR UNBORN CHILDREN FOR EYE COLOR, INTELLIGENCE, HEIGHT, AND SUSCEPTIBILITY TO HUNDREDS OF DISEASES. Talk of "designer babies" soon followed. The final sequence was completed in 2003, but we remain far from the ability to manipulate DNA to create the complex traits we might desire. ▶ DNA sequencing, p. 170

One misconception about the Human Genome Project is that the DNA sequence is a "blueprint for human life." A gene's nucleotide sequence encodes a protein, but just knowing a gene's sequence does not provide instant insight into everything needed to make a human. By itself, a DNA sequence does not explain how the cell turns each gene on and off, how the gene's encoded protein folds into its final shape, the function of the protein, or what happens if the gene mutates. Nor does it explain the function of the huge swaths of DNA that do not code for protein.

One goal of the Human Genome Project was to discover the basic set of genes that control human development and human life. All humans share these genes; the small variations within DNA sequences are what make each person unique. We now know, however, that only a 0.1% difference separates any two individuals. Researchers are actively investigating these differences, which will likely answer such important questions as why some people get cancer and others do not or why a medication helps some people but harms others.

A genome is an organism's entire set of DNA; an organism's proteome is all of the proteins that it expresses. Although an organism's genome changes little throughout its life, the combination of proteins that its cells produce depends on the cell type, the organism's stage of development, and the influence of other proteins and the environment. Proteomic studies are important for basic research into cell biology, but they also have medical applications. Understanding how the proteins produced in breast cancer cells differ from those in normal cells, for example, may reveal new targets for anticancer drugs.

We begin this genetics unit with a look at the intimate relationship between DNA and proteins. Subsequent chapters describe how cells copy DNA just before they divide and how cell division leads to the fascinating study of inheritance.

Learning Outline

7.1 | Experiments Identified the Genetic Material

The nucleic acid DNA is one of the most familiar molecules, the subject matter of movies and headlines (figure 7.1). Criminal trials hinge on DNA evidence; the idea of human cloning raises questions about the role of DNA in determining who we are; and DNA-based discoveries are yielding new diagnostic tests, medical treatments, and vaccines.

More important than DNA's role in society is its role in life itself. DNA is a biochemical with a remarkable function: it stores the information that each cell needs to produce proteins. These instructions make life possible. In fact, before a cell divides, it first makes an exact replica of its DNA. This process, described in chapter 8, copies the precious information that will enable the next generation of cells to live.

The recognition of DNA's role in life was a long time in coming. By the early 1900s, researchers had recognized the connection between inheritance and protein. For example, English physician Archibald Garrod noted that people with inherited "inborn errors of metabolism" lacked certain enzymes. Other researchers linked abnormal or missing enzymes to unusual eye color in fruit flies and nutritional deficiencies in bread mold. But how were enzyme deficiencies and inheritance linked? Experiments in bacteria would answer the question.

A. Bacteria Can Transfer Genetic Information

In 1928, English microbiologist Frederick Griffith contributed the first step in identifying DNA as the genetic material (figure 7.2).

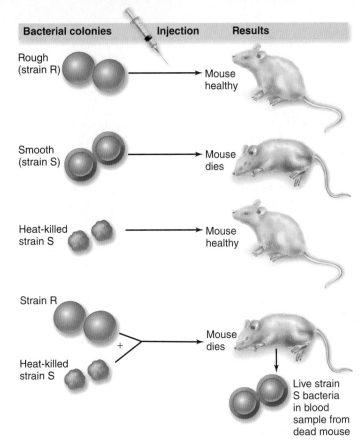

Figure 7.2 A Tale of Two Microbes. Griffith's experiments showed that a molecule in a lethal strain of bacteria (type S) could transform harmless type R bacteria into killers.

Griffith studied two strains of a bacterium, *Streptococcus pneumoniae*. Type R bacteria, named for their "rough" colonies, do not cause pneumonia when injected into mice. Type S ("smooth") bacteria, on the other hand, cause pneumonia. The smooth polysaccharide capsule that encases type S bacteria is apparently necessary for infection.

Griffith found that heat-killed type S bacteria did not cause pneumonia in mice. However, when he injected mice with a mixture of live type R bacteria plus heat-killed type S bacteria, neither of which could cause pneumonia alone, the mice died. Moreover, their bodies contained live type S bacteria encased in polysaccharide. Something in the heat-killed type S bacteria transformed the normally harmless type R strain into a killer.

How had the previously harmless bacteria acquired the ability to cause disease? In the 1940s, U.S. physicians Oswald Avery, Colin MacLeod, and Maclyn McCarty finally learned the identity of the "transforming principle." When the researchers treated the solution from the type S strain with a protein-destroying enzyme, the type R strain still changed into a killer. Therefore, a protein was not transmitting the killing trait. But when they treated the solution with a DNA-destroying enzyme, the type R bacteria remained harmless. The conclusion: DNA from type S cells altered the type R bacteria, enabling them to manufacture the smooth coat necessary to cause infection.

Figure 7.1 DNA—The Molecule in the Media. *Jurassic Park* was a 1993 blockbuster movie in which fictional scientists recreated dinosaurs. The dinosaur DNA came from blood found in ancient mosquitoes entombed in amber.

B. Hershey and Chase Confirmed the Genetic Role of DNA

At first, biologists hesitated to accept DNA as the biochemical of heredity. They knew more about proteins than about nucleic acids. They also thought that protein, with its 20 building blocks, could encode many more traits than DNA, which includes just four types of building blocks. In 1950, however, U.S. microbiologists Alfred Hershey and Martha Chase conclusively showed that DNA—not protein—is the genetic material.

Hershey and Chase used a very simple system. They infected the bacterium *Escherichia coli* with a bacteriophage, which is a virus that infects only bacteria. The virus consisted of a protein coat surrounding a DNA core. We now know that when the virus infects a bacterial cell, it injects its DNA, but the protein coat remains loosely attached to the bacterium. The viral DNA directs the bacterium to use its own energy and raw materials to manufacture more virus particles, which then burst from the cell. But much of this information was not available in 1950. In fact, Hershey and Chase wanted to know which part of the vi-

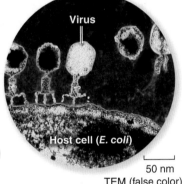

Figure 7.3 DNA's Role Is Confirmed. A bacteriophage is a virus that infects bacteria. Hershey and Chase used radioactive isotopes to distinguish the bacteriophage's protein coat from the DNA. They showed that the virus transfers DNA (not protein) to the bacterium, and this viral DNA caused bacterial cells to produce viruses.

Virus

Host cell (*E. coli*)

50 nm
TEM (false color)

rus controls its replication: the DNA or the protein coat. ▶▶ bacteriophages, p. 324; Focus on Model Organisms (*E. coli*), p. 346

To answer the question, the researchers "labeled" two batches of viruses, one with radioactive sulfur that marked protein and the other with radioactive phosphorus that marked DNA. They used each type of labeled virus to infect a separate batch of bacteria (figure 7.3). Then they agitated each mixture in a blender, which removed the unattached viruses and empty protein coats from the surfaces of the bacteria. They poured the mixtures into test tubes and spun them at high speed. The infected bacteria settled to the bottom of each test tube because they were heavier than the liberated viral protein coats.

Hershey and Chase examined the bacteria and the fluid in each tube. In the test tube containing sulfur-labeled viral proteins, the bacteria were not radioactive, but the fluid portion of the material in the tube was. In the other tube, where the virus contained DNA marked with radioactive phosphorus, the infected bacteria were radioactive, but the fluid was not.

The "blender experiments" therefore showed that the part of the virus that could enter the bacteria and direct them to mass-produce viruses was the part with the phosphorus label—namely, the DNA. The genetic material, therefore, was DNA and not protein.

7.1 | Mastering Concepts

1. How did Griffith's research, coupled with the work of Avery and his colleagues, demonstrate that DNA, not protein, is the genetic material?

2. How did the Hershey–Chase "blender experiments" confirm Griffith's results?

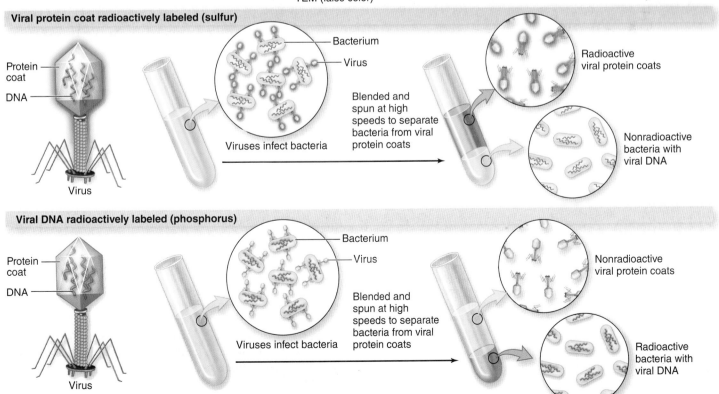

Viral protein coat radioactively labeled (sulfur)

Protein coat

DNA

Virus

Bacterium

Virus

Viruses infect bacteria

Blended and spun at high speeds to separate bacteria from viral protein coats

Radioactive viral protein coats

Nonradioactive bacteria with viral DNA

Viral DNA radioactively labeled (phosphorus)

Protein coat

DNA

Virus

Bacterium

Virus

Viruses infect bacteria

Blended and spun at high speeds to separate bacteria from viral protein coats

Nonradioactive viral protein coats

Radioactive bacteria with viral DNA

7.2 | DNA Is a Double Helix of Nucleotides

The early twentieth century also saw advances in the study of the structure of DNA. By 1929, biochemists had discovered the distinction between **RNA (ribonucleic acid)** and **DNA (deoxyribonucleic acid),** the two types of nucleic acid. Later, they determined that **nucleotides,** the building blocks of nucleic acids, include sugars, nitrogen-containing groups, and one or more phosphorus-containing components. Another important clue was the observation that DNA and RNA nucleotides always contain the same sugars and phosphates, but they may have any one of several different nitrogen-containing bases.

In the early 1950s, two lines of evidence together revealed DNA's chemical structure. Austrian-American biochemist Erwin Chargaff showed that DNA contains equal amounts of the bases adenine (A) and thymine (T) and equal amounts of the bases guanine (G) and cytosine (C). English physicist Maurice

Wilkins and chemist Rosalind Franklin bombarded DNA with X-rays, using a technique called X-ray diffraction to determine the three-dimensional shape of the molecule. The X-ray diffraction pattern revealed a regularly repeating structure of building blocks (figure 7.4a and b).

In 1953, U.S. biochemist James Watson and English physicist Francis Crick, working at the Cavendish laboratory in Cambridge in the United Kingdom, used these clues to build a ball-and-stick model of the DNA molecule. The now familiar double helix included equal amounts of G and C and of A and T, and it had the sleek symmetry revealed in the X-ray diffraction pattern. Watson, Crick, and Wilkins won the 1962 Nobel Prize in physiology or medicine for their discovery (figure 7.4c).

The DNA double helix resembles a twisted ladder (figure 7.5). The twin rails of the ladder, also called the sugar–phosphate "backbones," are alternating units of deoxyribose and phosphate joined with covalent bonds. The ladder's rungs are A–T and G–C base pairs joined by hydrogen bonds.

The A–T and G–C pairs arise from their chemical structures (figure 7.6). Adenine and guanine are purines, bases with a double ring structure. Cytosine and thymine are pyrimidines, which have a single ring. Each A–T pair is the same width as a C–G pair because each includes a purine and a pyrimidine.

The two strands of a DNA molecule are **complementary** to each other because the sequence of each strand defines the sequence of the other; that is, an A on one strand means a T on the opposite strand, and a G on one strand means a C on the other. Complementary base pairing is also the basis of gene function, as described later in this chapter.

Although the two chains of the DNA double helix are parallel to each other, they are oriented in opposite directions, like the northbound and southbound lanes of a highway. This head-to-tail ("antiparallel") arrangement is apparent when the carbon atoms

a. Rosalind Franklin

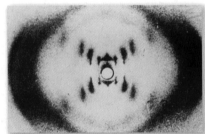
b. X–ray diffraction

c.

Figure 7.4 **Discovery of DNA's Structure.** (a) Rosalind Franklin produced (b) high-quality X-ray images of DNA that were crucial in the discovery of DNA's structure. (c) Maurice Wilkins, Francis Crick, and James Watson (first, third, and fifth from the left) shared the 1962 Nobel Prize in physiology or medicine for their now-famous discovery. Franklin had died in 1958, and by the rules of the award, she could not be included.

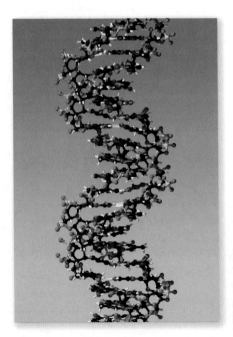

Figure 7.5 **DNA.** This model shows the three-dimensional shape of the familiar double helix.

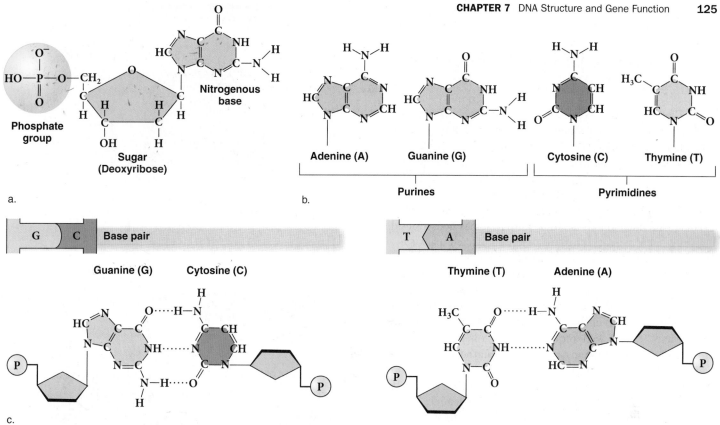

a.

b.

c.

Figure 7.6 Complementary Base Pairing. (a) Each nucleotide consists of the sugar deoxyribose, one or more phosphate groups, and a nitrogenous base. (b) Adenine and guanine are purines; cytosine and thymine are pyrimidines. (c) Purines pair with complementary pyrimidines; specifically, cytosine pairs with guanine, and adenine pairs with thymine.

in deoxyribose are numbered (figure 7.7). When the nucleotides are joined into a chain, opposite ends of the strand are designated "**3 prime**" (3′) and "**5 prime**" (5′). At the same end of the double helix, one chain therefore ends with the 3′ carbon, while the other chain ends with the 5′ carbon.

Figure It Out

Write the complementary DNA sequence of the following:
3′-ATCGGATCGCTACTG-5′

Answer: 5′-TAGCCTAGCGATGAC-3′

7.2 | Mastering Concepts

1. What are the components of DNA and its three-dimensional structure?
2. What evidence enabled Watson and Crick to decipher the structure of DNA?
3. Identify the 3′ and 5′ ends of a DNA strand.

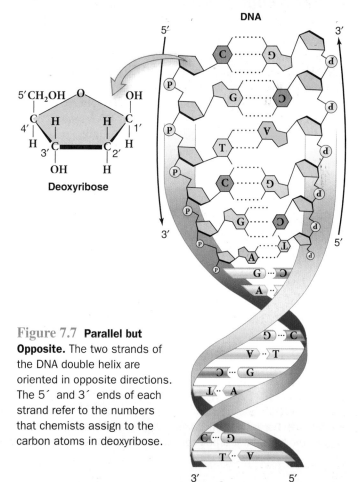

Figure 7.7 Parallel but Opposite. The two strands of the DNA double helix are oriented in opposite directions. The 5′ and 3′ ends of each strand refer to the numbers that chemists assign to the carbon atoms in deoxyribose.

7.3 | DNA Contains the "Recipes" for a Cell's Proteins

The amount of DNA in any cell is immense. In humans, for example, each pinpoint-sized nucleus contains some 6.4 billion base pairs of genetic information.

An organism's **genome** is all of the genetic material in its cells (**figure 7.8**). Genomes vary greatly in size and packaging. The genome of a bacterial cell mainly consists of one circular DNA molecule. In a eukaryotic cell, however, the majority of the genome is divided among multiple chromosomes housed inside the cell's nucleus; each **chromosome** is a discrete package of DNA and associated proteins. The mitochondria and chloroplasts of eukaryotic cells also contain DNA and therefore have their own genomes.

What does all of that DNA do? As described in more detail toward the end of this chapter, much of it has no known function. But some of it has a well-known role, which is to encode all of the cell's RNA and proteins. This section introduces the **gene,** which is a sequence of DNA nucleotides that codes for a specific protein or RNA molecule. Because many proteins are essential to life, each organism has many genes. The human genome, for example, includes 20,000 to 25,000 genes scattered on its 23 pairs of chromosomes.

A. Protein Synthesis Requires Transcription and Translation

In the 1940s, biologists working with the fungus *Neurospora crassa* deduced that each gene somehow controls the production of one protein. In the next decade, Watson and Crick described this relationship between nucleic acids and proteins as a flow of information they called the "central dogma."

Figure 7.9 illustrates the process of protein production. First, in **transcription,** a cell copies a gene's DNA sequence to a complementary RNA molecule. Then, in **translation,** the information in RNA is used to manufacture a protein by joining a

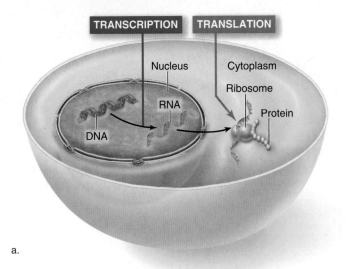

Figure 7.9 DNA to RNA to Protein. (a) The central dogma of biology states that information stored in DNA is copied to RNA (transcription), which is used to assemble proteins (translation). (b) DNA stores the information used to make proteins, just as a recipe stores the information needed to make brownies.

specific sequence of amino acids into a polypeptide chain. ▶ Focus on Model Organisms *(Neurospora),* p. 399

According to this model, a gene is therefore somewhat like a recipe in a cookbook. A recipe specifies the ingredients and instructions for assembling one dish, such as spaghetti sauce or brownies. Likewise, a protein-encoding gene contains the instructions for assembling a protein, amino acid by amino acid. A cookbook that contains many recipes is analogous to a chromosome, which is an array of genes. A person's entire collection of cookbooks, then, would be analogous to a genome.

To illustrate DNA's function with a concrete example, suppose a cell in a female mammal's breast is producing milk to feed an infant (see figure 3.13). One of the many proteins in milk is albumin. The steps below summarize the production of albumin, starting with its genetic "recipe":

1. Inside the nucleus, an enzyme first transcribes the albumin gene's DNA sequence to a complementary sequence of RNA.

2. After some modification, the RNA emerges from the nucleus and binds to a ribosome.

3. At the ribosome, amino acids are assembled in a specific order to produce the albumin protein.

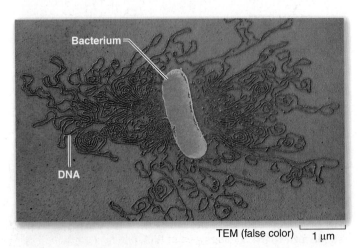

TEM (false color) 1 μm

Figure 7.8 Bacterial DNA. Genetic material bursts from this bacterium, illustrating just how much DNA is packed into a single cell.

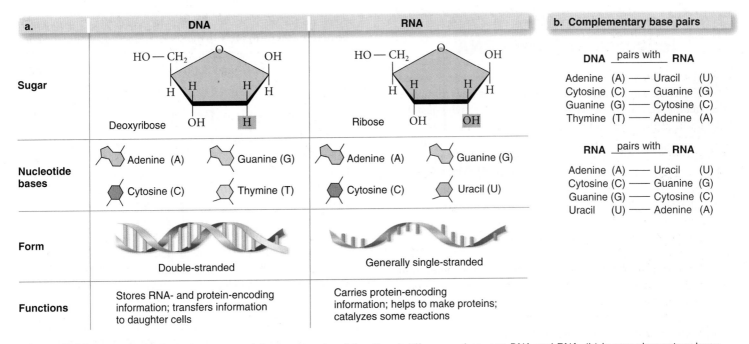

Figure 7.10 DNA and RNA. (a) Summary of the structural and functional differences between DNA and RNA. (b) In complementary base pairs, uracil in RNA behaves chemically like thymine in DNA.

The amino acid sequence in albumin is dictated by the sequence of nucleotides in the RNA molecule. The RNA, in turn, was transcribed from DNA. In this way, DNA provides the recipe for albumin and every other protein in the cell.

B. RNA Is an Intermediary Between DNA and a Polypeptide Chain

RNA is a multifunctional nucleic acid that differs from DNA in several ways (figure 7.10). First, its nucleotides contain the sugar ribose instead of deoxyribose. Second, RNA has the nitrogenous base uracil, which behaves similarly to thymine; that is, uracil binds with adenine in complementary base pairs. Third, unlike DNA, RNA can be single-stranded. Finally, RNA can catalyze chemical reactions, a role not known for DNA.

RNA is central to the flow of genetic information. Three types of RNA interact to synthesize proteins (table 7.1):

- **Messenger RNA (mRNA)** carries the information that specifies a protein. Each group of three mRNA bases in a row forms a **codon,** which is a genetic "code word" that corresponds to one amino acid.

- **Ribosomal RNA (rRNA)** combines with proteins to form a **ribosome,** the physical location of protein synthesis. Some rRNAs help to correctly align the ribosome and mRNA, and others catalyze formation of the bonds between amino acids in the developing protein.

- **Transfer RNA (tRNA)** molecules are "connectors" that bind an mRNA codon at one end and a specific amino acid at the other. Their role is to carry each amino acid to the ribosome at the correct spot along the mRNA molecule.

Table 7.1 Major Types of RNA

Molecule	Typical Number of Nucleotides	Function
mRNA	500–3000	Encodes amino acid sequence
rRNA	100–3000	Associates with proteins to form ribosomes, which structurally support and catalyze protein synthesis
tRNA	75–80	Binds mRNA codon on one end and an amino acid on the other, linking a gene's message to the amino acid sequence it encodes

The function of each type of RNA is further explained later in this chapter, beginning in section 7.4 with the first stage in protein production: transcription.

7.3 | Mastering Concepts

1. What is the relationship between a gene and a protein?
2. What are the two main stages in protein synthesis?
3. What are the three types of RNA, and how does each contribute to protein synthesis?

7.4 | Transcription Uses a DNA Template to Create RNA

Transcription produces an RNA copy of one gene. If a gene is analogous to a recipe for a protein, then transcription is like opening a cookbook to a particular page and copying just the

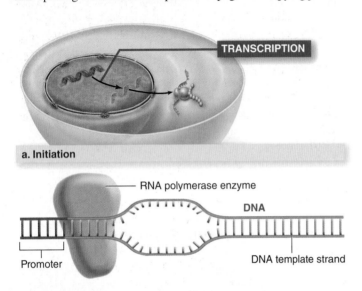

a. Initiation

b. Elongation

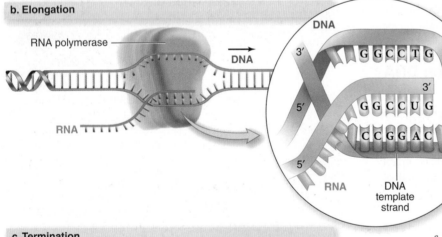

c. Termination

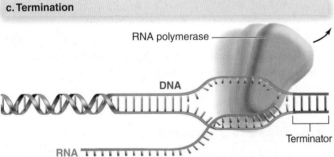

Figure 7.11 Transcription of RNA from DNA. Transcription occurs in three stages: initiation, elongation, and termination. (a) Initiation is the control point that determines which genes are transcribed and when. (b) RNA nucleotides are added during elongation. (c) A terminator sequence in the gene signals the end of transcription.

recipe for the dish you want to prepare. After the copy is made, the book can return safely to the shelf. Just as you would then use the instructions on the copy to make your meal, the cell uses the information in RNA—and not the DNA directly—to make each protein.

A. Transcription Occurs in Three Steps

Complementary base pairing underlies transcription, just as it does DNA replication (see chapter 8). In fact, transcription resembles DNA replication, with two main differences: (1) the product of transcription is RNA, not DNA; and (2) transcription copies just one gene from one DNA strand, rather than copying both strands of an entire chromosome.

In transcription, RNA nucleotide bases bind with exposed complementary bases on the **template strand,** which is the strand in the DNA molecule that is actually copied to RNA (figure 7.11). The process occurs in three stages:

1. **Initiation:** Enzymes unzip the DNA double helix, exposing the template strand. **RNA polymerase** (the enzyme that builds an RNA chain) binds to the **promoter,** a DNA sequence that signals the gene's start.

2. **Elongation:** RNA polymerase moves along the DNA template strand in a 3′-to-5′ direction, adding nucleotides only to the 3′-end of the growing RNA molecule.

3. **Termination:** A **terminator** sequence signals the end of the gene. Upon reaching the terminator sequence, the RNA polymerase enzyme separates from the DNA template and releases the newly synthesized RNA. The DNA molecule then resumes its usual double-helix shape.

As the RNA molecule is synthesized, it curls into a three-dimensional shape dictated by complementary base pairing within the molecule. The final shape determines whether the RNA functions as mRNA, tRNA, or rRNA.

The observation that the cell's DNA encodes all types of RNA—not just mRNA—has led to debate over the definition of the word *gene*. Originally, a gene was defined as any stretch of DNA that encodes one protein. More recently, however, biologists have expanded the definition to include any DNA sequence that is transcribed. The phrase *gene expression* can therefore mean the production of either a functional RNA molecule or a protein.

Figure It Out

Write the sequence of the mRNA molecule transcribed from the following DNA template sequence: 5′-TTACACTTGCAAC-3′

B. mRNA Is Altered in the Nucleus of Eukaryotic Cells

In bacteria and archaea, ribosomes may begin translating mRNA to a protein before transcription is even complete. In eukaryotic cells, however, the presence of the nuclear membrane prevents one mRNA from being simultaneously transcribed and translated. Moreover, in eukaryotes, mRNA is usually altered before it leaves the nucleus to be translated (figure 7.12).

5′ Cap and Poly A Tail   After transcription, a short sequence of modified nucleotides, called a cap, is added to the 5′ end of the mRNA molecule. At the 3′ end, 100 to 200 adenines are added, forming a "poly A tail." Together, the cap and poly A tail enhance translation by helping ribosomes attach to the 5′ end of the mRNA molecule. The length of the poly A tail may also determine how long an mRNA lasts before being degraded.

Intron Removal   In archaea and in eukaryotic cells, only part of an mRNA molecule is translated into an amino acid sequence. Figure 7.12 shows that an mRNA molecule consists of alternating sequences called introns and exons. **Introns** are portions of the mRNA that are removed before translation. The word *intron* is short for *intragenic regions*, where *intra-* means "within"

and *-genic* refers to the gene. Small catalytic RNAs and proteins remove the introns from the mRNA.

The remaining portions, the **exons,** are spliced together to form the mature mRNA that leaves the nucleus to be translated. (A tip for remembering this is that *exons* are the portions of an mRNA molecule that are actually *ex*pressed or that *ex*it the nucleus.)

The amount of genetic material devoted to introns can be immense. The average exon is 100 to 300 nucleotides long, whereas the average intron is about 1000 nucleotides long. Some mature mRNA molecules consist of 70 or more spliced-together exons; the cell therefore simply discards much of the RNA created in transcription. Although introns may seem wasteful, section 7.6 explains that this cutting and pasting is important in the regulation of gene expression.

7.4 | Mastering Concepts

1. What happens during each stage of transcription?
2. Where in the cell does transcription occur?
3. What is the role of RNA polymerase in transcription?
4. What are the roles of the promoter and terminator sequences in transcription?
5. How is mRNA modified before it leaves the nucleus of a eukaryotic cell?

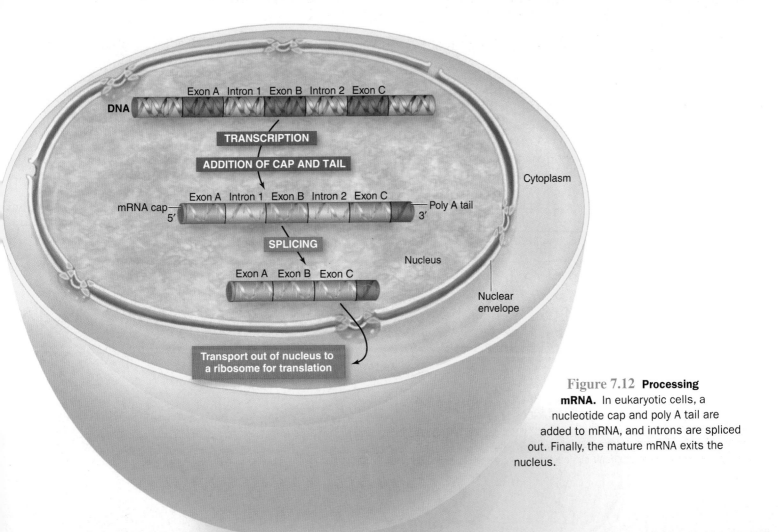

Figure 7.12 Processing mRNA. In eukaryotic cells, a nucleotide cap and poly A tail are added to mRNA, and introns are spliced out. Finally, the mature mRNA exits the nucleus.

7.5 | Translation Builds the Protein

Transcription copies the information encoded in a DNA base sequence into the complementary language of mRNA. Once transcription is complete and mRNA is processed, the cell is ready to translate the mRNA "message" into a sequence of amino acids. If mRNA is like a copy of a recipe, then translation is like preparing the dish.

A. The Genetic Code Links mRNA to Protein

The **genetic code** is the set of "rules" by which a cell uses the nucleotides in mRNA to assemble amino acids into a protein (figure 7.13). Each codon is a group of three mRNA bases corresponding either to one amino acid or to a "stop" signal.

In the 1960s, however, researchers did not yet understand exactly how the genetic code worked. One early question was the number of RNA bases that specify each amino acid. Researchers reasoned that RNA contains only four different nucleotides, so a genetic code with a one-to-one correspondence of mRNA bases to amino acids could specify only four different amino acids—far fewer than the 20 amino acids that make up biological proteins. A code consisting of two bases per codon could specify only 16 different amino acids. A code with three bases per codon, however, yields 64 different combinations, more than enough to specify the 20 amino acids in life. Experiments later confirmed the triplet nature of the genetic code.

A second, and more difficult, problem was to determine which codons correspond to which amino acids. In the 1960s, researchers answered this question by synthesizing mRNA molecules in the laboratory. They added these synthetic mRNAs to test tubes containing all the ingredients needed for translation, extracted from *E. coli* cells. Analyzing the resulting polypeptides allowed scientists to finish deciphering the genetic code in

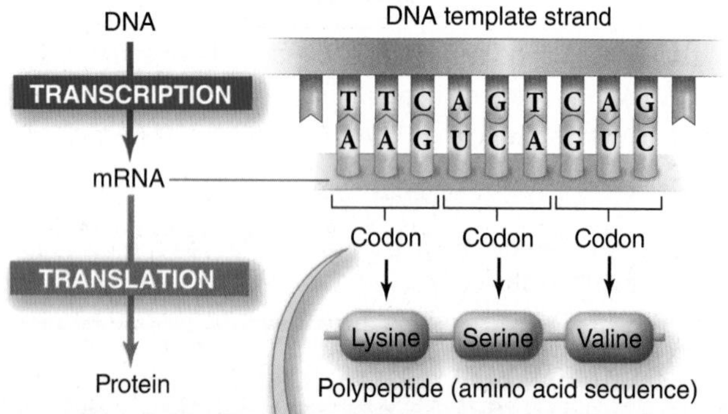

Figure 7.13 The Genetic Code. In translation, transfer RNA matches mRNA codons with amino acids as specified in the genetic code.

The Genetic Code					
Second letter of codon					
First letter of codon	U	C	A	G	Third letter of codon
U	UUU UUC — Phenylalanine (Phe; F) UUA UUG — Leucine (Leu; L)	UCU UCC UCA UCG — Serine (Ser; S)	UAU UAC — Tyrosine (Tyr; Y) UAA **Stop** UAG **Stop**	UGU UGC — Cysteine (Cys; C) UGA **Stop** UGG Tryptophan (Trp; W)	U C A G
C	CUU CUC CUA CUG — Leucine (Leu; L)	CCU CCC CCA CCG — Proline (Pro; P)	CAU CAC — Histidine (His; H) CAA CAG — Glutamine (Gln; Q)	CGU CGC CGA CGG — Arginine (Arg; R)	U C A G
A	AUU AUC — Isoleucine (Ile; I) AUA AUG **Start** Methionine (Met; M)	ACU ACC ACA ACG — Threonine (Thr; T)	AAU AAC — Asparagine (Asn; N) AAA AAG — Lysine (Lys; K)	AGU AGC — Serine (Ser; S) AGA AGG — Arginine (Arg; R)	U C A G
G	GUU GUC GUA GUG — Valine (Val; V)	GCU GCC GCA GCG — Alanine (Ala; A)	GAU GAC — Aspartic acid (Asp; D) GAA GAG — Glutamic acid (Glu; E)	GGU GGC GGA GGG — Glysine (Gly; G)	U C A G

less than a decade—a monumental task. Chemical analysis eventually showed that the genetic code also contains directions for starting and stopping translation. AUG is typically the first codon in mRNA, and the codons UGA, UAA, and UAG each signify "stop."

Nearly all species use the same mRNA codons to specify the same amino acids. Mitochondria and a handful of species use alternative codes that differ only slightly from the code in figure 7.13. The most logical explanation for this observation is that all life on Earth evolved from a common ancestor.

B. Translation Requires mRNA, tRNA, and Ribosomes

Translation—the actual construction of the protein—requires the following participants:

- **mRNA:** This product of transcription carries the genetic information that encodes a protein, with each three-base codon specifying one amino acid.

- **tRNA:** This "bilingual" molecule binds to an mRNA codon and to an amino acid (**figure 7.14**). The **anticodon** is a three-base loop that is complementary to one mRNA codon. The other end of the tRNA molecule forms a covalent bond to the amino acid corresponding to that codon. For example, a tRNA with the anticodon sequence AAG always picks up the amino acid phenylalanine.

- **Ribosome:** The ribosome, built of rRNA and proteins, anchors mRNA during translation. Each ribosome has one large and one small subunit that join at the initiation of protein synthesis (**figure 7.15**). Ribosomes may be free in the cytoplasm or attached to the rough endoplasmic reticulum (see figure 3.15).

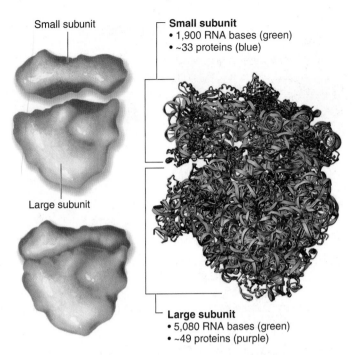

Small subunit
- 1,900 RNA bases (green)
- ~33 proteins (blue)

Large subunit
- 5,080 RNA bases (green)
- ~49 proteins (purple)

Figure 7.15 **The Ribosome.** A ribosome from a eukaryotic cell has two subunits containing a total of 82 proteins and four rRNA molecules.

C. Translation Occurs in Three Steps

The process of translation can be divided into three stages, during which mRNA, tRNA molecules, and ribosomes come together, link amino acids into a chain, and then dissociate again (**figure 7.16**).

1. **Initiation:** The leader sequence at the 5' end of the mRNA molecule bonds with a small ribosomal subunit. The first mRNA codon to specify an amino acid is usually AUG, which attracts a tRNA that carries the amino acid methionine. This methionine signifies the start of a polypeptide. A large ribosomal subunit attaches to the small subunit to complete initiation.

2. **Elongation:** A tRNA molecule carrying the second amino acid (glycine in figure 7.16) then binds to the second codon, GGA in this case. The two amino acids, methionine and glycine, align, and a covalent bond forms between them. With that peptide bond in place, the ribosome releases the first tRNA, which will pick up another methionine and be used again. ▶ peptide bond, p. 36

Next, the ribosome moves down the mRNA by one codon. A third tRNA enters, carrying its amino acid. This third amino acid aligns with the other two and forms a covalent bond to the second amino acid in the growing chain. The tRNA attached to glycine is released and

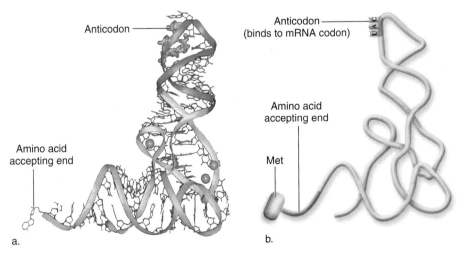

Figure 7.14 **Transfer RNA.** (a) Three-dimensional view of tRNA. (b) A simplified tRNA molecule shows the anticodon at one end and the amino acid-binding region at the opposite end. This particular tRNA is carrying the amino acid methionine.

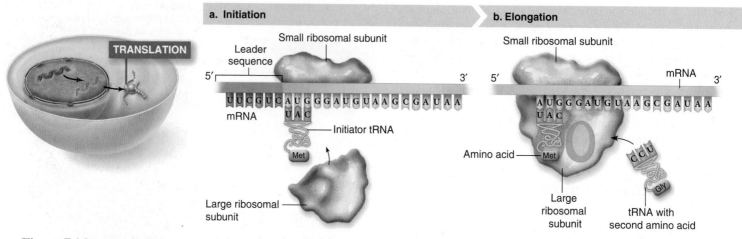

| a. Initiation | b. Elongation |

Figure 7.16 **Translation Creates the Protein.** (a) Initiation brings together the ribosomal subunits, mRNA, and an initiator tRNA. (b) As elongation begins, the anticodon of a tRNA molecule bearing a second amino acid forms hydrogen bonds with the second codon. The first amino acid forms a covalent bond with the second amino acid. Additional tRNAs bring subsequent amino acids encoded in the mRNA. (c) Termination occurs when a release factor protein binds to the stop codon. All components of the translation machine are liberated, and the completed polypeptide is released.

recycled. With the help of proteins called elongation factors, the polypeptide grows one amino acid at a time, as tRNAs continue to deliver their cargo.

3. **Termination:** Elongation halts at a "stop" codon (UGA, UAG, or UAA). No tRNA molecules correspond to these stop codons. Instead, proteins called release factors bind to the stop codon, prompting the release of the last tRNA from the ribosome. The ribosomal subunits separate from each other and are recycled, and the new polypeptide is released.

Protein synthesis can be very speedy. A plasma cell in the human immune system can manufacture 2000 identical antibody proteins per second. How can protein synthesis occur fast enough to meet all of a cell's needs? First, it is efficient. Transcription produces multiple copies of each mRNA, and dozens of ribosomes may simultaneously bind along the length of a single mRNA molecule (figure 7.17). Thanks to this "assembly line," a cell can make many copies of a protein from the same mRNA. Second, ribosomes zip along mRNA molecules, incorporating some 15 amino acids per second.

Figure It Out

If a DNA sequence is 5'-AAAGCAGTACTA-3', what would be the corresponding amino acid sequence?

Answer: Phe-Arg-His-Asp

D. Proteins Must Fold Correctly after Translation

The newly synthesized protein cannot do its job until it folds into its final shape (see figure 2.23). Some regions of the amino acid chain attract or repel other parts, contorting the polypeptide's overall shape. Enzymes catalyze the formation of chemical bonds, and "chaperone" proteins stabilize partially folded regions.

Proteins can fold incorrectly if the underlying DNA sequence is altered (see section 7.7), because the encoded protein

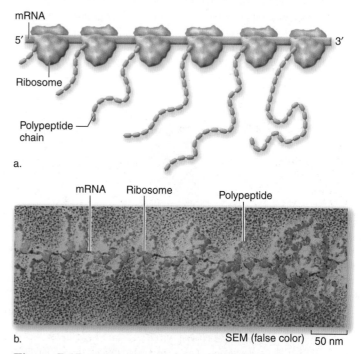

Figure 7.17 **Efficient Translation.** (a) Multiple ribosomes can simultaneously translate one mRNA. (b) This micrograph shows about two dozen ribosomes producing proteins from the same mRNA.

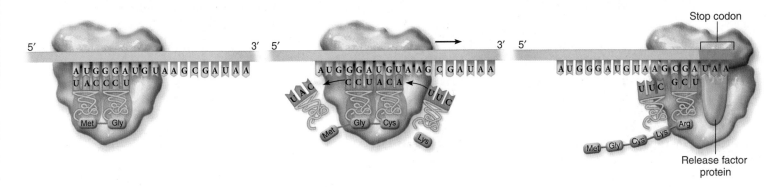

c. Termination

5′ ... 3′ 5′ ... 3′ 5′ ... Stop codon

Release factor protein

may have the wrong sequence of amino acids. Serious illness may result. In some forms of cystic fibrosis, for example, a membrane protein that normally controls the flow of chloride ions does not fold correctly into its final form.

Errors in protein folding can occur even if the underlying genetic sequence remains unchanged. Alzheimer disease, for example, is associated with a protein called amyloid that folds improperly and then forms an abnormal mass in brain cells. Likewise, mad cow disease and similar conditions in sheep and humans are caused by abnormal clumps of misfolded proteins called prions in nerve cells. ▶ prions, p. 332

In addition to folding, some proteins must be altered in other ways before they become functional. For example, insulin, which is 51 amino acids long, is initially translated as the 80-amino-acid polypeptide proinsulin. Enzymes cut proinsulin

to form insulin. A different type of modification occurs when polypeptides join to form larger protein molecules. The oxygen-carrying blood protein hemoglobin, for example, consists of four polypeptide chains (two alpha and two beta) encoded by separate genes.

7.5 | Mastering Concepts

1. How did researchers determine that the genetic code is a triplet and learn which codons specify which amino acids?
2. What happens in each stage of translation?
3. Where in the cell does translation occur?
4. How are polypeptides modified after translation?

Apply It Now

Some Poisons Disrupt Protein Synthesis

We learned in chapter 6 that some poisons kill by interfering with respiration. Here we list a few poisons that inhibit protein synthesis; a cell that cannot make proteins quickly dies.

- **Amanatin:** This toxin occurs in the "death cap mushroom," *Amanita phalloides* (pictured at right). Amanatin inhibits RNA polymerase, making transcription impossible.

- **Diphtheria toxin:** Bacteria called *Corynebacterium diphtheriae* secrete a toxin that causes a respiratory illness, diphtheria. The toxin inhibits an elongation factor, a protein that helps add amino acids to a polypeptide chain during translation.

- **Antibiotics:** Clindamycin, chloramphenicol, tetracyclines, and gentamicin are all antibiotics that bind to bacterial ribosomes. When its ribosomes are disrupted, a cell cannot make proteins, and it dies.

- **Ricin:** Derived from seeds of the castor bean plant (*Ricinus communis*), ricin is a potent natural poison that consists of

two parts. One part binds to a cell, and the other enters the cell and inhibits protein synthesis by an unknown mechanism. Interestingly, the part of the molecule that enters the cell is apparently more toxic to cancer cells than to normal cells, making ricin a potential cancer treatment.

- **Trichothecenes:** Fungi in genus *Fusarium* produce toxins called trichothecenes. During World War II, thousands of people died after eating bread made from moldy wheat, and many researchers believe trichothecenes were used as biological weapons during the Vietnam War. The mode of action is unclear, but the toxins seem to interfere somehow with ribosomes.

7.6 | Cells Regulate Gene Expression

Producing proteins costs tremendous amounts of energy. For example, an *E. coli* cell spends 90% of its ATP on protein synthesis. Transcription and translation require energy, as does the synthesis of nucleotides, enzymes, ribosomal proteins, and other molecules that participate in protein synthesis. Splicing out introns and making other modifications to the mRNA require still more energy. ▶ ATP, p. 76

Cells constantly produce essential proteins, such as the enzymes involved in the energy pathways described in chapters 5 and 6. Considering the high cost of making protein, however, it makes sense that cells save energy by not producing unneeded proteins.

Beyond energy savings, cells have many additional reasons to regulate gene expression. First, multicellular organisms consist of many types of specialized cells. Humans, for example, have at least 200 different cell types. If each cell contains the same complete set of genes, how does it acquire its unique function? The answer is that each type of cell expresses a different subset of genes. A hair follicle cell, for example, produces a lot of keratin (the protein that makes up hair). The same cell never makes the protein hemoglobin. Conversely, a red blood cell produces a lot of hemoglobin but leaves the keratin gene turned off. Overall, each cell's function is unique because it produces a unique combination of proteins.

Second, regulating gene expression gives cells flexibility to respond to changing conditions. For example, the enormous python pictured on page 104 ramps up its production of digestive enzymes shortly after it begins to swallow the gazelle. The genes encoding those enzymes turn off once the meal is gone. Likewise, specialized immune system cells churn out antibodies in response to an infection. Once the threat is gone, antibody production halts.

Third, an intricate set of genetic instructions orchestrates the growth and development of a multicellular organism. Early in an animal embryo's development, for example, protein signals "tell" cells whether they are at the head end of the body, the tail end, or somewhere in between. These signals, in turn, regulate the expression of unique combinations of genes that enable cells in each location to specialize. Similarly, in a flowering plant, genes that were silent early in the plant's life become active when external signals trigger flower formation.

All of these examples illustrate the important idea that cells produce many proteins only under certain conditions. This section describes some of the mechanisms that regulate gene expression in cells.

A. Operons Are Groups of Bacterial Genes That Share One Promoter

Soon after the discovery of DNA's structure, researchers began to unravel the controls of gene expression. In 1961, French biologists François Jacob and Jacques Monod described how and when *E. coli* bacteria produce the three enzymes that degrade the sugar lactose. The bacteria produce the proteins only when lactose is present in the cell's surroundings. What signals "tell" a simple bacterial cell to transcribe all three

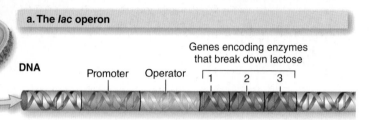

a. The *lac* operon

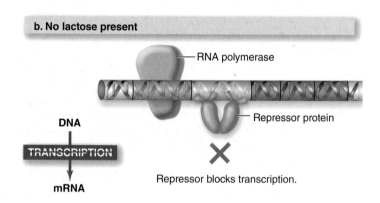

b. No lactose present

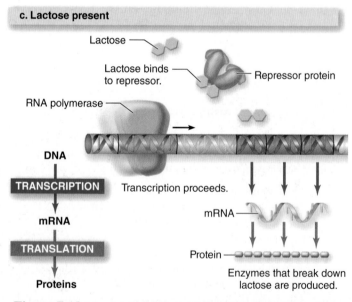

c. Lactose present

Figure 7.18 The *Lac* Operon. (a) One promoter controls the expression of three genes, which encode the enzymes that break down lactose. (b) In the absence of lactose, a repressor protein binds to the operator, preventing transcription of the genes. (c) If lactose is present, the sugar binds the repressor, changing the protein's shape and causing it to release the operator. Transcription proceeds, and the lactose can be digested.

genes at precisely the right time? ▶ Focus on Model Organisms *(E. coli)*, p. 346

Jacob and Monod showed that in *E. coli* and other bacteria, related genes are organized as operons. An **operon** is a group of genes plus a promoter and an operator that control the transcription of the entire group at once. The promoter, as described earlier, is the site to which RNA polymerase attaches to begin transcription. The **operator** is a DNA sequence located between the promoter and the protein-encoding regions. If a protein called a **repressor** binds to the operator, it prevents the transcription of the genes.

E. coli's **lac operon** consists of the three genes that encode lactose-degrading proteins plus the promoter and operator that control their transcription (figure 7.18). To understand how the *lac* operon works, first imagine an *E. coli* cell in an environment lacking lactose. Expressing the three genes would be a waste of energy. The repressor protein therefore binds to the operator, preventing RNA polymerase from transcribing the genes (figure 7.18b). The genes are effectively "off."

When lactose is present, however, a slightly altered form of the sugar attaches to the repressor, which changes its shape so that it detaches from the DNA. RNA polymerase is now free to transcribe the genes (figure 7.18c). After translation, the resulting enzymes enable the cell to absorb and degrade the sugar. Lactose, in a sense, causes its own dismantling.

Soon, geneticists discovered other groups of genes organized as operons. Some, like the *lac* operon, negatively control transcription by removing a block. Others produce factors that turn on transcription. As Jacob and Monod stated in 1961, "The genome contains not only a series of blueprints, but a coordinated program of protein synthesis and means of controlling its execution."

B. Eukaryotic Organisms Use Transcription Factors

In eukaryotic cells, groups of proteins called **transcription factors** bind DNA at specific sequences that regulate transcription. RNA polymerase cannot bind to a promoter or initiate transcription of a gene in the absence of transcription factors. A transcription factor may bind to a gene's promoter or to an **enhancer,** a regulatory DNA sequence that lies outside the promoter. An enhancer may be located near the gene (or even within it), but often they are thousands of base pairs away.

Figure 7.19 shows how transcription factors prepare a promoter to receive RNA polymerase. The first transcription factor to bind is attracted to a part of the promoter called the TATA box. This transcription factor attracts others, including proteins bound to an enhancer. Finally, RNA polymerase joins the complex, binding just in front of the start of the gene sequence. With RNA polymerase in place, transcription can begin.

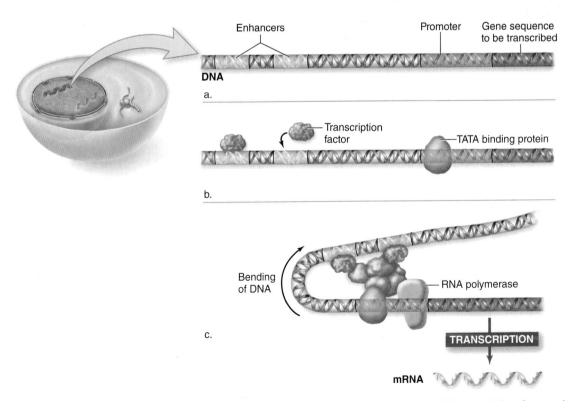

Figure 7.19 Transcription Factors. (a) Enhancers and promoters are regulatory DNA sequences. (b) Transcription factors, including TATA binding proteins, bind to these regulatory sequences. (c) DNA bends to bring the transcription factors together. RNA polymerase initiates transcription only if specific transcription factors are bound to a gene's promoter and enhancers.

Transcription factors respond to external stimuli that signal a gene to turn "on." Once a signaling molecule binds to the outside of a target cell, a series of chemical reactions occurs inside the cell. The last step in the series can be the activation or deactivation of a transcription factor. Because one transcription factor may influence many genes, one stimulus may trigger many simultaneous changes in the cell. One example is the series of events that follows the fusion of egg and sperm (see chapter 34).

The ability to digest lactose provides an excellent example of the importance of transcription factors and enhancers in gene regulation. All infants produce lactase, the enzyme that digests the lactose in milk. But many adults are lactose intolerant because their lactase-encoding gene remains turned off after infancy. Without the enzyme, lactose is indigestible. Some people, however, can continue to digest milk into adulthood. In these lactose-tolerant adults, an enhancer is modified in a way that promotes transcription of the lactase gene throughout life.

One gene can have multiple enhancers, each of which regulates transcription in a different cell type. For example, cells in the kidneys, lungs, and brain all produce a membrane protein called Duffy, as do red blood cells. Each cell type uses a different enhancer to control expression of Duffy. Most people from western Africa lack Duffy on their red blood cells, even though their other cells do produce the protein. The explanation is a change in the enhancer that controls Duffy production in just the red blood cells. The "Duffy-negative" blood cells apparently protect against malaria but increase a person's susceptibility to HIV.

Hundreds of transcription factors are known, and in humans, defects in them underlie some diseases, including cancers. This makes sense, because the signals that trigger cell division are proteins; expressing these genes too much or too little causes cells to divide out of control. In addition, some drugs interfere with transcription factors. The "abortion pill" RU486, for example, indirectly blocks the action of transcription factors needed for the development of an embryo. ▶ cancer, p. 162

C. Eukaryotic Cells Also Use Additional Regulatory Mechanisms

In addition to transcription factors, eukaryotic cells also have several other ways to control gene expression (figure 7.20).

DNA Availability Chromosomes must be unwound for genes to be expressed. In addition, a cell can "tag" unneeded DNA with methyl groups ($-CH_3$). Proteins inside the cell bind to the tagged DNA, preventing gene expression and signaling the cell to fold that section of DNA more tightly. Transcription factors and RNA polymerase cannot access highly compacted DNA, so these modifications turn off the genes.

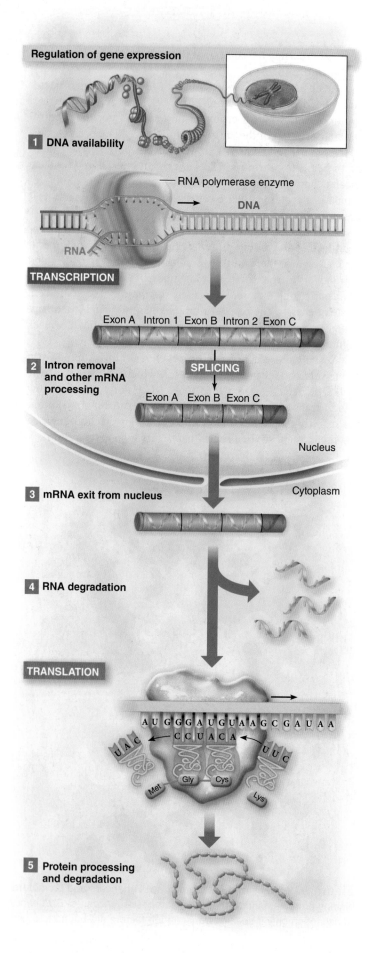

Regulation of gene expression

1 DNA availability

RNA polymerase enzyme

DNA

RNA

TRANSCRIPTION

Exon A Intron 1 Exon B Intron 2 Exon C

2 Intron removal and other mRNA processing

SPLICING

Exon A Exon B Exon C

Nucleus

Cytoplasm

3 mRNA exit from nucleus

4 RNA degradation

TRANSLATION

AUGGGAUGUAAGCGAUAA
CCUACA
UAC   UUC
Met   Gly   Cys   Lys

5 Protein processing and degradation

Figure 7.20 Regulation of Gene Expression. Eukaryotic cells have many ways to control whether each gene is turned on or off.

Burning Question

Is there a gay gene?

Despite periodic headlines about newly discovered genes "for" homosexuality, the reality is a bit more complex.

Linking a human behavior to one or more genes is difficult for several reasons. First, the question of a "gay gene" is somewhat misleading. Genes encode RNA and proteins, not behaviors, so any relationship between DNA and sexual behavior is necessarily indirect. Second, to establish a clear link to DNA, a researcher must be able to define and measure a behavior. This in itself is difficult, because people disagree about what it means to be homosexual. Third, an individual who possesses a gene version that contributes to a trait will not necessarily express the gene; many genes in each cell remain "off" at any given time. Fourth, multiple genes and a strong environmental influence are likely to be involved in anything as complex as sexual orientation.

Despite these complications, research has yielded some evidence of a biological component to homosexuality, at least in males. For example, a male homosexual's identical twin is much more likely to also be homosexual than is a nonidentical twin, indicating a strong genetic contribution. In addition, the more older brothers a male has, the more likely he is to be homosexual. This "birth order" effect occurs only for siblings with the same biological mother; having older stepbrothers does not increase the chance that a male is homosexual. Events before birth, not social interactions with brothers, are therefore apparently responsible for the effect.

Other research has produced ambiguous results. Anatomical studies of cadavers have revealed differences in the size of a particular brain structure between heterosexual and homosexual men, but the relative contribution of genes and environment to this structure is unknown. One study linked homosexuality in males, but not in females, to part of the X chromosome; a subsequent study did not support this conclusion.

So is there a gay gene? The short answer is no, because there is unlikely to be a single gene that "causes" homosexuality. At the same time, research suggests a genetic contribution to sexual orientation. Sorting out the complex interactions between multiple genes and the environment, however, remains a formidable challenge.

Submit your burning question to:
marielle_hoefnagels@mcgraw-hill.com

RNA Processing One gene can encode multiple proteins if different introns are removed from the mRNA. For example, one gene known to be expressed in the nervous system of fruit flies can theoretically be spliced into more than 38,000 different configurations!

mRNA Exit from Nucleus For a protein to be produced, mRNA must leave the nucleus and attach to a ribosome. If the mRNA fails to leave, the gene is silenced.

RNA Degradation Not all mRNA molecules are equally stable. Some are rapidly degraded, perhaps before they can be translated, whereas others are more stable.

Moreover, tiny RNA sequences called microRNAs can play a role in regulating gene expression. Each is only about 21 to 23 nucleotides long. A cell may produce a microRNA that is complementary to a coding mRNA. If the microRNA attaches to the mRNA, the resulting double-stranded RNA cannot be translated at a ribosome and is likely to be degraded.

Protein Processing and Degradation Some proteins, such as insulin, must be altered before they become functional. Dozens of modifications are possible, including the addition of sugars or an alteration in the protein's structure. Insulin, for example, is cut in two places after translation. If these modifications fail to occur, the protein cannot function.

In addition, to do its job, a protein must move from the ribosome to where the cell needs it. For example, a protein secreted in milk must be escorted to the Golgi apparatus and be packaged for export (see figure 3.13). A gene is effectively silenced if its product never moves to the correct destination.

Finally, like RNA, not all proteins are equally stable. Some are degraded shortly after they form, whereas others persist longer.

A human cell may express hundreds to thousands of genes at once. Unraveling the complex regulatory mechanisms that control the expression of each gene is an enormous challenge. As described in section 7.10, biologists now have the technology to begin navigating this regulatory maze. The payoff will be a much better understanding of cell biology, along with many new medical applications. The same research may also help scientists understand how external influences on gene expression contribute to complex traits, such as the one described in this chapter's Burning Question.

7.6 | Mastering Concepts

1. What are some reasons that cells regulate gene expression?
2. How do proteins determine whether a bacterial operon is expressed?
3. How do enhancers and transcription factors interact to regulate gene expression?
4. What are some other ways that a cell controls which genes are expressed?

7.7 | Mutations Change DNA Sequences

A **mutation** is a change in a cell's DNA sequence, either in a protein-coding gene or in noncoding DNA such as an enhancer. Many people think that mutations are always harmful, perhaps because some of them cause such dramatic changes (**figure 7.21**). Although some mutations do cause illness, they also provide the variation that makes life interesting (and makes evolution possible).

To continue the cookbook analogy introduced earlier, a mutation in a gene is similar to an error in a recipe. A small typographical error might be barely noticeable. A minor substitution of one ingredient for another might hurt (or improve) the flavor. But serious errors such as missing ingredients or truncated instructions are likely to ruin the dish.

A. Mutations Range from Silent to Devastating

A **point mutation** changes one or a few base pairs in a gene; larger-scale mutations may also occur. The mutation may be a single-base change, an insertion or deletion that shifts the codon "reading frame," or the expansion of repeated sequences (**table 7.2**). Some are not detectable except by DNA fingerprinting, while others may be lethal. ▶ DNA profiling, p. 172

Substitution Mutations A substitution mutation is the replacement of one DNA base with another. Such a mutation is "silent" if the mutated gene encodes the same protein as the original gene version. Silent mutations can occur because more than one codon encodes most amino acids.

Often, however, a substitution mutation changes a base triplet so that it specifies a different amino acid. This change is called a missense mutation. The substituted amino acid may drastically alter the protein's shape, changing its function. Sickle cell disease results from this type of mutation (**figure 7.22**).

Table 7.2 Types of Mutations

Type	Illustration
Original sequence	THE ONE BIG FLY HAD ONE RED EYE
Missense	TH**Q** ONE BIG FLY HAD ONE RED EYE
Nonsense	THE ONE BIG
Frameshift	THE ONE **Q**BI GFL YHA DON ERE DEY
Deletion of three letters	THE ONE BIG HAD ONE RED EYE
Duplication	THE ONE BIG FLY **FLY** HAD ONE RED EYE
Insertion	THE ONE BIG **WET** FLY HAD ONE RED EYE
Expanding repeat	Generation 1: THE ONE BIG FLY HAD ONE RED EYE Generation 2: THE ONE BIG FLY **FLY FLY** HAD ONE RED EYE Generation 3: THE ONE BIG FLY FLY **FLY FLY FLY FLY** HAD ONE RED EYE

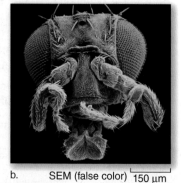

a.   SEM (false color) ⌐150 µm⌐   b.   SEM (false color) ⌐150 µm⌐

Figure 7.21 One Mutation Can Make a Big Difference.
Mutations in some genes can cause parts to form in the wrong places. (a) A normal fruit fly. (b) This fly has legs growing where antennae should be; it has a mutation that affects development.

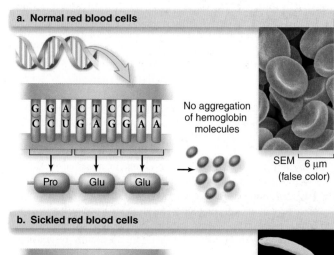
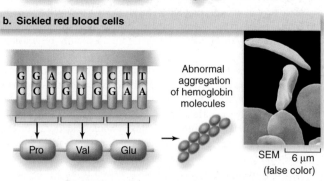

a. **Normal red blood cells**

G G A C T C C T T A
C C U G A G G A A

Pro — Glu — Glu

No aggregation of hemoglobin molecules

SEM ⌐6 µm⌐ (false color)

b. **Sickled red blood cells**

G G A C A C C T T A
C C U G U G G A A

Pro — Val — Glu

Abnormal aggregation of hemoglobin molecules

SEM ⌐6 µm⌐ (false color)

Figure 7.22 Sickle Cell Mutation. The most common form of sickle cell anemia results from a mutation in one of two hemoglobin genes. (a) Normal hemoglobin molecules do not aggregate, enabling the red blood cell to assume a rounded shape. (b) In sickle cell disease, a substitution mutation replaces one amino acid with a different one. As a result, hemoglobin molecules clump into long, curved rods that deform the red blood cell.

In other cases, called nonsense mutations, a base triplet specifying an amino acid changes into one that encodes a "stop" codon. This shortens the protein product, which can profoundly influence the organism. At least one of the mutations that give rise to cystic fibrosis, for example, is a nonsense mutation. Instead of the normal 1480 amino acids, the faulty protein has only 493 and therefore cannot function.

Figure It Out

Suppose that a substitution mutation replaces the first "A" in the following mRNA sequence with a "U":

5′-AAAGCAGUACUA-3′.

How many amino acids will be in the polypeptide chain?

Answer: Zero

Base Insertions and Deletions

One or more nucleotides can be added to or deleted from a gene. A **frameshift mutation** adds or deletes nucleotides in any number other than a multiple of three (figure 7.23). Because triplets of DNA bases specify amino acids, such a mutation disrupts the codon reading frame. It therefore also alters the sequence of amino acids and usually devastates a protein's function. Some mutations that cause cystic fibrosis result from the addition or deletion of just one or two nucleotides.

Even if a small insertion or deletion does not shift the reading frame, the effect might still be significant if the change drastically

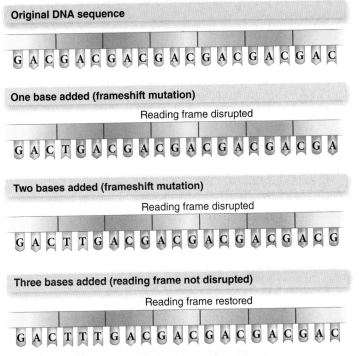

Figure 7.23 Frameshift Mutations. Insertions or deletions of one or two nucleotides dramatically alter a gene's codons. Adding or deleting three nucleotides restores the reading frame.

alters the protein's shape. The most common mutation that causes severe cystic fibrosis, for example, deletes only a single group of three nucleotides. The resulting protein lacks just one amino acid, but it cannot function.

Expanding Repeats

In an **expanding repeat mutation,** the number of copies of a three- or four-nucleotide sequence increases over several generations. With each generation, the symptoms begin earlier or become more severe (or both). Expanding genes underlie several inherited disorders, including fragile X syndrome and Huntington disease. In Huntington disease, expanded repeats of GTC cause extra glutamines (an amino acid) to be incorporated into the gene's protein product. The abnormal protein forms fibrous clumps in the nuclei of some brain cells, which causes the symptoms of uncontrollable movements and personality changes.

B. What Causes Mutations?

Some mutations occur spontaneously (without outside causes). A spontaneous substitution mutation usually originates as a DNA replication error. Replication errors can also cause insertions and deletions, especially in genes with repeated base sequences, such as GCG CGC . . . . It is as if the molecules that guide and carry out replication become "confused" by short, repeated sequences, as a proofreader scanning a manuscript might miss the spelling errors in the words "happpiness" and "bananana." ▶ DNA replication, p. 154

The average rate of replication errors for most genes is about 1 in 100,000 bases, but it varies among organisms and among genes. The larger a gene, the more likely it is to mutate. In addition, the more frequently DNA replicates, the more it mutates. Bacteria accumulate mutations faster than cells of complex organisms simply because their DNA replicates more often. Likewise, rapidly dividing skin cells tend to have more mutations than the nervous system's neurons, which divide slowly if at all.

Exposure to harmful chemicals or radiation may also damage DNA. A **mutagen** is any external agent that induces mutations. Examples include the ultraviolet radiation in sunlight, X-rays, radioactive fallout from atomic bomb tests and nuclear accidents, chemical weapons such as mustard gas, and chemicals in tobacco. The more contact a person has with mutagens, the higher the risk for cancer. Coating skin with sunscreen, wearing a lead "bib" during dental X-rays, and avoiding tobacco all lower cancer risk by reducing exposure to mutagenic chemicals and radiation.

Mutations may also occur during a specialized type of cell division called meiosis (see chapter 9). During an early stage of meiosis, paired chromosomes align and swap portions of their DNA. If genes are misaligned at a crossover point, they may lose or gain nucleotides.

Sometimes part of a chromosome can become inverted or fused with a different chromosome. Either event can cause mutations by bringing together gene segments that were not previously joined (see figure 9.14). When part of chromosome 9 breaks off and fuses with chromosome 22, for example, the resulting "Philadelphia chromosome" has a fused gene whose protein product causes a type of leukemia.

Movable DNA sequences are yet another source of mutations. A **transposable element,** or transposon for short, is a DNA sequence that can "jump" within the genome. A transposon can insert itself randomly into chromosomes. If it lands within a gene, the transposon can disrupt the gene's function; it can also leave a gap in the gene when it leaves.

C. Mutations May Pass to Future Generations

A **germline mutation** occurs in the cells that give rise to sperm and egg cells. Germline mutations are heritable because the mutated DNA will be passed down in at least some of the sex cells that the organism produces. As a result, every cell of the organism's affected offspring will carry the mutation as well. Such mutations may run in families for generations, or they can appear suddenly. For example, two healthy people of normal height may have a child with a form of dwarfism called achondroplasia. The child's achondroplasia arose from a new mutation that occurred by chance in the mother's or father's germ cell.

Most mutations, however, do not pass from generation to generation. A **somatic mutation** occurs in nonsex cells, such as those that make up the skin, intestinal tract, or lungs. All cells derived from the altered one will also carry the mutation, but the mutation does not pass to the organism's offspring. The children of a cigarette smoker with mutations that cause lung cancer, for example, do not inherit the parent's damaged genes.

D. Mutations Are Important

A mutation in a gene sometimes changes the structure of its encoded protein so much that the protein can no longer do its job. Inherited diseases, including cystic fibrosis and sickle cell anemia, stem from such DNA sequence changes. Some of the most harmful mutations affect the genes encoding the proteins that repair DNA. Additional mutations then rapidly accumulate in the cell's genetic material, which can kill the cell or lead to cancer. ▶ cancer, p. 162

Mutations are also extremely important because they produce genetic variability. They are the raw material for evolution because they create new **alleles,** or variants of genes. Except for identical twins, everyone has a different combination of alleles for the 25,000 or so genes in the human genome. The same is true for any genetically variable group of organisms.

Some of these new alleles are "neutral" and have no effect on an organism's fitness. Your reproductive success, for example, does not ordinarily depend on your eye color or your shoe size. As unit 3 explains, however, variation has important evolutionary consequences. In every species, individuals with some allele combinations reproduce more successfully than others. Natural selection "edits out" the less favorable allele combinations.

Homeotic genes illustrate the importance of mutations in evolution. These genes encode transcription factors that are expressed during the development of an embryo. If the transcription factors are faulty, the signals that control the formation of an organism's body parts become disrupted. The flies in figure 7.21 show what happens when homeotic genes are mutated. Having parts in the wrong places is, of course, usually harmful. But studies of many species reveal that mutations in homeotic genes have profoundly influenced animal evolution. Limb modifications such as arms, hooves, wings, and flippers trace their origins to homeotic mutations.

Mutations sometimes enhance an organism's reproductive success. Consider, for example, the antibiotic drugs that kill bacteria by targeting membrane proteins, enzymes, and other structures. Random mutations in bacterial DNA encode new versions of these targeted proteins. The descendants of some of the mutated cells become new strains that are unaffected by these antibiotics. The medical consequences are immense. Antibiotic-resistant bacteria have become more and more common, and many people now die of bacterial infections that once were easily treated with antibiotics.

Likewise, random mutations in viral genomes enable viruses to jump from other animals to humans. Evolving viruses have caused the global epidemics of HIV, influenza, and other diseases (see section 15.7).

Mutations can also be enormously useful in science and agriculture. Geneticists frequently induce mutations to learn how genes normally function. For example, biologists discovered how genes control flower formation by studying mutant *Arabidopsis* plants in which flower parts form in the wrong places. Plant breeders also induce mutations to create new varieties of many crop species (**figure 7.24**). Some kinds of rice, grapefruit, oats, lettuce, begonias, and many other plants owe their existence to breeders treating cells with radiation and then selecting interesting new varieties from the mutated individuals.

7.7 | Mastering Concepts

1. What is a mutation?
2. What are the types of mutations, and how does each alter the encoded protein?
3. What causes mutations?
4. What is the difference between a germline mutation and a somatic mutation?
5. How are mutations important?

a.   b.   c.

Figure 7.24   Useful Mutants. (a) Rio Red grapefruits and several varieties of (b) rice and (c) cotton are among the many plant varieties that have been created by using radiation to induce mutations.

BIOTECHNOLOGY

7.8 | The Human Genome Is Surprisingly Complex

The Human Genome Project revealed the complete sequence of human DNA in 2003. Chapter 8 explains how researchers determined the sequence of A, C, T, and G; for now, it is enough to know that scientists can study the structure and function of the entire human genome.

The Human Genome Project revealed some unexpected contradictions. For example, although our genome includes approximately 25,000 protein-encoding genes, our cells can produce some 400,000 different proteins. Furthermore, only about 1.5% of the human genome actually encodes protein.

How can so few genes specify so many proteins? Part of the answer lies in introns. By removing different combinations of introns from an mRNA molecule, a cell can produce several proteins from one gene—a departure from the old idea that each gene encodes exactly one protein. So far, no one understands exactly how a cell "decides" which introns to remove.

And what is the function of the 98.5% of our genome that does not encode proteins? Some of it is regulatory, such as the enhancers that control gene expression. In addition, much of our DNA is transcribed to rRNA, tRNA, and microRNA. Chromosomes also contain many **pseudogenes,** DNA sequences that are very similar to protein-encoding genes and that are transcribed but whose mRNA is not translated into protein. Pseudogenes may be remnants of old genes that once functioned in our nonhuman ancestors; eventually they mutated too far from the normal sequence to encode a working protein.

The human genome is also riddled with highly repetitive sequences that have no known function. The most abundant type of repeats are transposons (see section 7.7B), which were originally identified in corn by Barbara McClintock in the 1940s. These movable pieces of DNA make up about 45% of the human genome.

The genome also contains many tandem repeats (or "satellite DNAs"). These sequences consist of one or more bases repeated many times, such as CACACA or ATTCGATTCG. Huge clusters of tandem repeats, up to 100 million base pairs long, occur at the tips of chromosomes, among other places. Thousands of shorter tandem repeats (a few dozen to 30,000 base pairs long) also litter much of the rest of the genome. The exact number of repeats varies from person to person; DNA fingerprinting technology exploits these differences to match suspects with evidence left at the scene of a crime. Some expanding repeated sequences may cause illness such as Huntington disease. In other cases, the repeats appear to have no effect. ▸ DNA profiling, p. 172

7.8 | Mastering Concepts

1. How can the number of proteins encoded in DNA exceed the number of genes in the genome?
2. List some functions of the 98.5% of the human genome that does not specify protein.

BIOTECHNOLOGY

7.9 | Genetic Engineering Moves Genes Among Species

The fact that virtually all species use the same genetic code means that one type of organism can express a gene from another. Biologists take advantage of this fact by coaxing cells to take up **recombinant DNA,** which is genetic material that has been spliced together from multiple organisms. Scientists first accomplished this feat of "genetic engineering" in *E. coli* in the 1970s, but many other organisms have since been genetically modified. ▸ Focus on Model Organisms (*E. coli*), p. 346

A. Transgenic Organisms Contain DNA from Multiple Species

A **transgenic** organism is one that receives recombinant DNA. In the pharmaceutical industry, transgenic bacteria produce several dozen drugs, including human insulin to treat diabetes, blood clotting factors to treat hemophilia, immune system biochemicals, and fertility hormones. Other genetically modified bacteria produce the amino acid phenylalanine, which is part of the artificial sweetener aspartame. Still others degrade petroleum, pesticides, and other soil pollutants. Transgenic yeast cells produce a milk-curdling enzyme called chymosin used by many U.S. cheese producers. ▸ artificial sweeteners, p. 39

Transgenic crop plants may resist pests, survive harsh environmental conditions, or contain nutrients that they otherwise wouldn't. A large portion of the corn and soybeans grown in the United States is transgenic, containing genes encoding proteins that help the plants resist herbicide applications or fight off insect pests (see section 10.10). "Golden Rice" is a genetically engineered plant that contains genes from petunias and bacteria. The genes enable the rice plant to produce beta-carotene (a vitamin A precursor) and extra iron, making the rice grains gold in color and more nutritious.

Transgenic animals also have diverse applications. A glow-in-the-dark zebra fish was the first genetically modified house pet (figure 7.25). On a more practical note, a transgenic mouse

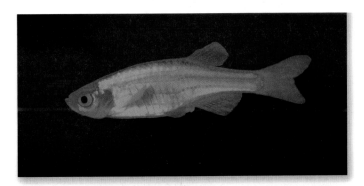

Figure 7.25 Transgenic Animal. Glow-in-the-dark zebra fish have been genetically altered to produce a fluorescent protein.

"model" for a human gene can reveal how a disease begins, enabling researchers to develop new treatments. Transgenic farm animals can secrete human proteins in their milk or semen, yielding abundant, pure supplies of otherwise rare substances that are useful as drugs.

B. Creating Transgenic Organisms Requires Cutting and Pasting DNA

The creation of a transgenic organism requires several stages, which are illustrated in **figure 7.26.**

1. Acquire Source DNA
The first step is to obtain DNA from a source cell, usually a bacterium, plant, or animal. The researcher may synthesize the DNA in the laboratory or extract it directly from the source cell. Extracting the DNA poses a problem, however, if the gene's source is a eukaryotic cell and the recipient will be a bacterium. Bacterial cells cannot remove introns from mRNA, so the DNA would encode a defective protein in bacteria.

Researchers therefore first isolate a mature mRNA molecule with the introns already removed. Then, they use an enzyme called **reverse transcriptase** to make a DNA copy of the mRNA. (As described in chapter 15, retroviruses such as HIV use this enzyme when they infect cells.) The resulting complementary DNA, or cDNA, encodes the eukaryotic protein but leaves out the introns.

2. Select a Cloning Vector
Next, the researcher chooses a **cloning vector,** a self-replicating genetic structure that will carry the source DNA into the recipient cell. (In molecular biology, "cloning" means to make many identical copies of a DNA sequence.) A common type of cloning vector is a **plasmid,** which is a small circle of double-stranded DNA separate from the cell's chromosome. Viruses are also used as vectors. They are altered so that they transport DNA but cannot cause disease.

3. Create Recombinant DNA
The next step is to create a recombinant plasmid. To create DNA fragments that can be spliced together, researchers use **restriction enzymes,** which are proteins that cut double-stranded DNA at a specific base

Figure 7.26 **Creating Transgenic Bacteria.** The first steps in creating a transgenic bacterium are to [1] isolate source DNA and [2] select a plasmid or other cloning vector. [3] Researchers use the same restriction enzyme to cut DNA from the donor cell and the plasmid. When the pieces are mixed, the "sticky ends" of the DNA fragments join, forming recombinant plasmids. [4] After the plasmid is delivered into a bacterium, it is mass-produced as the bacterium divides.

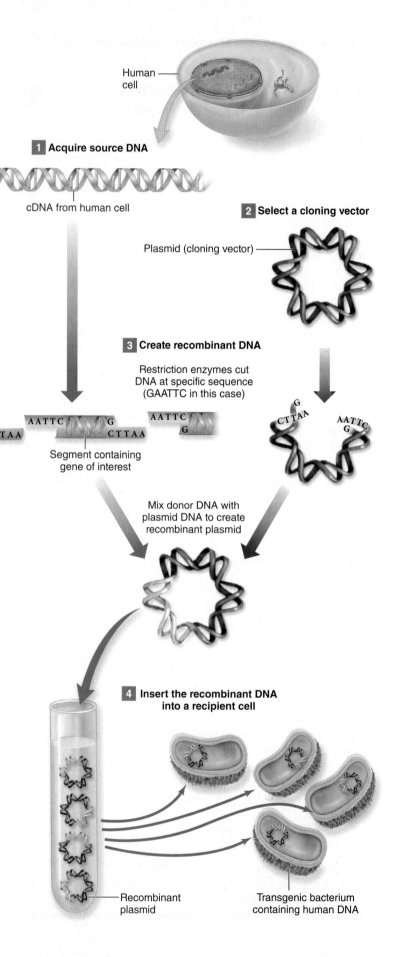

Human cell

1 Acquire source DNA

cDNA from human cell

2 Select a cloning vector

Plasmid (cloning vector)

3 Create recombinant DNA

Restriction enzymes cut DNA at specific sequence (GAATTC in this case)

Segment containing gene of interest

Mix donor DNA with plasmid DNA to create recombinant plasmid

4 Insert the recombinant DNA into a recipient cell

Recombinant plasmid

Transgenic bacterium containing human DNA

sequence. Some restriction enzymes generate single-stranded ends that "stick" to each other by complementary base pairing. The natural function of restriction enzymes is to protect bacteria by cutting up DNA from infecting viruses. Biologists, however, use them to cut and paste segments of DNA from different sources. When plasmid and donor DNA is cut with the same restriction enzyme and the fragments are mixed, the single-stranded sticky ends of some plasmids form base pairs with those of the donor DNA. Another enzyme, DNA ligase, seals the segments together.

4. Insert the Recombinant DNA into a Recipient
Cell Next, the researchers move the cloning vector with its recombinant DNA into a recipient cell. Zapping a bacterial cell with electricity opens temporary holes that admit naked DNA. Alternatively, "gene guns" shoot DNA-coated pellets directly into cells. DNA can also be packaged inside a fatty bubble called a liposome that fuses with the recipient cell's membrane, or it can be hitched to a virus that subsequently infects the recipient cell.

One tool for introducing new genes into plant cells is a bacterium called *Agrobacterium tumefaciens*. In nature, these bacteria enter the plant at a wound and inject a plasmid into the host's cells. The plasmid normally encodes proteins that stimulate the infected plant cells to divide rapidly, producing a tumorlike gall where the bacteria live. (The name of the plasmid, Ti, stands for "tumor inducing.")

As a first step in creating a transgenic plant, scientists can replace some of the Ti plasmid's natural genes with other DNA, such as a gene encoding a protein that confers herbicide resistance. They then allow the transgenic *Agrobacterium* to inject these modified plasmids into plant cells (figure 7.27).

All plants that grow from the infected cells should express the new herbicide-resistance gene.

Plants can grow from isolated cells, but most animals cannot. Therefore, biologists create transgenic animals by using viruses to introduce genes into a fertilized egg. The organism that develops will carry the foreign genes in every cell.

Regardless of whether the recipient of recombinant DNA is a bacterium, plant, or animal, the result is the same: When cells containing the DNA divide, all of their daughter cells also harbor the new genes. These transgenic organisms express their new genes just as they do their own, producing the desired protein along with all of the others that they normally make.

Although transgenic organisms have many practical uses, some people question whether their benefits outweigh their potential dangers. Some fear that ecological disaster could result if genetically modified organisms displace closely related species in the wild. Others worry that unfamiliar protein combinations in genetically modified crops could trigger food allergies. Still others object to the "unnatural" practice of combining genes from organisms that would never breed in nature.

7.9 | Mastering Concepts

1. What is recombinant DNA?
2. What are transgenic organisms, and how are they useful?
3. What are the steps in creating a recombinant plasmid?
4. How do bacteria, plant, and animal cells take up recombinant DNA?

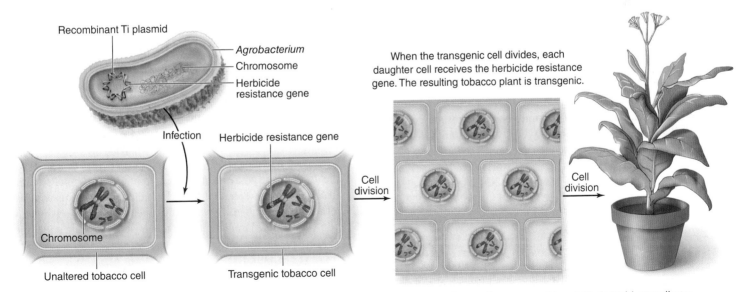

Figure 7.27 Creating a Transgenic Plant. This genetically modified *Agrobacterium* cell contains a recombinant Ti plasmid encoding a gene that confers herbicide resistance. The bacterium infects a tobacco plant cell, inserting the Ti plasmid into the plant cell's DNA. The transgenic plant cells can be grown into tobacco plants that express the herbicide resistance gene in every cell.

7.10 | Researchers Can Fix, Block, or Monitor Genes

Biotechnology holds out hope for an unprecedented ability to treat some genetic diseases. It is also an incredibly powerful tool for research. This section describes some applications.

A. Gene Therapy Repairs Faulty Genes

Thousands of diseases are caused by faulty genes: cancer, cystic fibrosis, sickle cell anemia, hemophilia, and Tay-Sachs disease are just a few examples. Most genetic illnesses currently have no cure, but **gene therapy** may someday provide one by replacing the faulty gene in a person's cells.

Gene therapy is challenging for several reasons. The new, therapeutic gene must be delivered directly to the cell type that needs correction. Viruses may be ideal for carrying DNA into target cells, but for gene therapy to be safe, the viruses must not alert the immune system. In addition, the gene therapy patient must express the repaired genes long enough for his or her health to improve.

Gene therapy trials in humans have proceeded very slowly since 1999, when 18-year-old Jesse Gelsinger received a massive infusion of viruses carrying a gene to correct an inborn error of metabolism. He died within days from an overwhelming immune system reaction. Gelsinger's death prompted a temporary halt to several gene therapy studies and stricter rules for conducting experiments. Nevertheless, gene therapy research and clinical trials continue.

B. Antisense RNA and Gene Knockouts Block Gene Expression

Sometimes it is useful to block gene expression, perhaps to silence a harmful gene or to learn a gene's normal function. Antisense and knockout technologies do this; both have potential applications in agriculture and health care.

Antisense technology exploits RNA's ability to form a double-stranded molecule. Ribosomes cannot translate double-stranded RNA, and cells normally destroy RNA in this form. Artificially adding an RNA sequence complementary to a messenger RNA therefore blocks a gene's expression (figure 7.28). This type of gene inactivation is called RNA interference, or RNAi. Theoretically, antisense RNA can suppress the activity of any gene, if the RNA can persist long enough and if it can be delivered to the appropriate tissue.

Knockout technology blocks a gene's function by replacing a normal copy of the gene with a disabled version; in a sense, it is therefore the opposite of gene therapy. In mice, researchers alter the DNA of isolated cells of a very early embryo. These genetically altered cells are then transferred to embryos and implanted into the uterus of a female mouse. Breeding the resulting offspring to each other then yields some mice with two copies of the knocked-out gene in every cell.

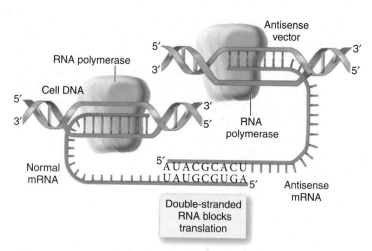

Figure 7.28 Silencing Gene Expression. Antisense RNA is a complementary nucleic acid sequence that binds with mRNA, preventing the mRNA from being translated into a protein.

Researchers can compare knockout organisms with their normal counterparts to learn the deleted gene's function. For example, researchers routinely knock out mouse genes to better understand the function of the corresponding genes in humans.

C. DNA Microarrays Help Monitor Gene Expression

A **DNA microarray,** also known as a "DNA chip," is a collection of short DNA fragments of known sequence placed in tens of thousands of defined spots on a small square of glass or other inert material (see figure 8.23). This tool can monitor gene expression in different cell types or in cells exposed to a variety of conditions.

As a simple example, suppose that a researcher wants to know how gene expression differs between kidney and liver cells. The first step would be to add fluorescent tags to cDNA copies of the mRNA in both types of cells. The cDNA would then be applied to a chip containing sequences representing the human genome. Dots of light would appear wherever the sample DNA is complementary to the chip's DNA. A computer then analyzes the pattern of colored dots on the square.

DNA microarrays promise to individualize medicine. A DNA chip for leukemia, for example, scans the genes expressed in a patient's white blood cells. The pattern reveals the cancer subtype, whether the person's cells will admit a particular drug, whether the drug will be safe and effective, and how the immune system is likely to respond to both the cancer and the drug. Chips with bacterial DNA help identify which microbes are causing an infection, allowing targeted treatment with the most appropriate antibiotics.

7.10 | Mastering Concepts

1. What is gene therapy, and why is it difficult to accomplish?
2. How do antisense RNA and gene knockouts silence genes?
3. How are DNA microarrays useful?

7.11 | Investigating Life: Clues to the Origin of Language

As you chat with your friends and study for your classes, you may take language for granted. Communication is not unique to humans, but a complex spoken language does set us apart from other organisms. Every human society has language. Without it, people could not transmit information from one generation to the next, so culture could not develop. Its importance to human evolutionary history is therefore incomparable. But how and when did such a crucial adaptation arise?

One clue emerged in the early 1990s, when scientists described a family with a high incidence of an unusual language disorder. Affected family members had difficulty controlling the movements of their mouth and face, so they could not pronounce sounds properly. They also had lower intelligence compared with unaffected individuals, and they had trouble applying simple rules of grammar.

Researchers traced the language disorder to one mutation in a single gene on chromosome 7. Further research revealed that the gene belongs to the large *f*orkhead *b*ox family of genes, abbreviated *FOX*. All members of the *FOX* family encode transcription factors, proteins that bind to DNA and control gene expression. The "language gene" on chromosome 7, eventually named *FOXP2*, is not solely responsible for language acquisition. But the fact that the gene encodes a transcription factor explains how it can simultaneously affect both muscle control and the brain.

To learn more about the evolution of *FOXP2*, scientists Wolfgang Enard, Svante Pääbo, and colleagues at Germany's Max Planck Institute and at the University of Oxford compared the sequences of the 715 amino acids that make up the FOXP2 protein in humans, several other primates, and mice (figure 7.29). Chimpanzees, gorillas, and the rhesus macaque monkey all have identical FOXP2 proteins; their version differs from the mouse's by only one amino acid. The human version differs from the mouse's by three amino acids.

This result showed that in the 70 million or so years since the mouse and primate lineages split, the FOXP2 protein changed by only one amino acid. Yet in the 5 or 6 million years since humans split from the rest of the primates, the *FOXP2* gene changed twice.

Initially, the new, human-specific *FOXP2* version would have been rare, as are all mutations. Today, however, nearly everyone has the same allele of *FOXP2*. The human-specific *FOXP2* allele evidently conferred such improved language skills that individuals with the allele consistently produced more offspring than those without it. That is, natural selection "fixed" the new, beneficial allele in the growing human population.

The research team used mathematical models to estimate that the original mutation happened within the past 200,000 years. A subsequent study, however, pushed that date back. The new research, published in 2007, showed that Neandertal DNA contains the same two changes observed in modern humans. These results suggested that the mutations occurred before modern humans and Neandertals split from their last common ancestor, some 300,000 to 400,000 years ago.

Species	Number of Differences Relative to Mouse Protein
 Mouse	N/A
 Rhesus monkey	1
 Gorilla	1
 Chimpanzee	1
 Human	3

Figure 7.29 FOXP2 Protein Compared. The mouse version of the FOXP2 protein differs from that of nonhuman primates by just one amino acid out of 715 in the protein. The human version has three differences when compared with that of the mouse.

The study of *FOXP2* is important because it helps us understand a critical period in human history. The gene changed after humans diverged from chimpanzees, and then individuals with the new, highly advantageous allele produced more offspring than those with any other version. By natural selection, the new allele quickly became fixed in the human population. Without those events, human communication and culture (including everything you chat about with your friends) might never have happened.

Enard, Wolfgang, Molly Przeworski, Simon E. Fisher, and five coauthors, including Svante Pääbo. August 22, 2002. Molecular evolution of *FOXP2*, a gene involved in speech and language. *Nature*, vol. 418, pages 869–872.

Krause, Johannes, Carles Lalueza-Fox, Ludovic Orlando, and 10 coauthors, including Svante Pääbo. November 6, 2007. The derived *FOXP2* variant of modern humans was shared with Neandertals. *Current Biology*, vol. 17, pages 1908–1912.

7.11 Mastering Concepts

1. What question about the *FOXP2* gene were the researchers trying to answer?
2. What insights could scientists gain by intentionally mutating the *FOXP2* gene in a developing human? Would such an experiment be ethical?

Chapter Summary

7.1 | Experiments Identified the Genetic Material

A. Bacteria Can Transfer Genetic Information

- Frederick Griffith determined that an unknown substance transmits a disease-causing trait between two types of bacteria.
- With the help of protein- and DNA-destroying enzymes, scientists subsequently showed that Griffith's "transforming principle" was DNA.

B. Hershey and Chase Confirmed the Genetic Role of DNA

- Using viruses that infect bacteria, Alfred Hershey and Martha Chase confirmed that the genetic material is DNA and not protein.

7.2 | DNA Is a Double Helix of Nucleotides

- Erwin Chargaff discovered that A and T, and G and C, occur in equal proportions in DNA. Maurice Wilkins and Rosalind Franklin provided X-ray diffraction data. James Watson and Francis Crick combined these clues to propose the double-helix structure of **DNA.**
- DNA is made of building blocks called **nucleotides.** The rungs of the DNA "ladder" consist of **complementary** base pairs (A with T; C with G).
- The two chains of the DNA double helix are antiparallel, with the **3′** end of one strand aligned with the **5′** end of the complementary strand.

7.3 | DNA Contains the "Recipes" for a Cell's Proteins

A. Protein Synthesis Requires Transcription and Translation

- A **gene** is a stretch of DNA that is transcribed to **RNA.** To produce a protein, a cell **transcribes** the gene's information to mRNA, which is **translated** into a sequence of amino acids.

B. RNA Is an Intermediary Between DNA and a Polypeptide Chain

- Three types of RNA (**mRNA, rRNA,** and **tRNA**) participate in gene expression.

7.4 | Transcription Uses a DNA Template to Create RNA

A. Transcription Occurs in Three Steps

- Transcription begins when **RNA polymerase** binds to a **promoter** on the DNA **template strand.** RNA polymerase then builds an RNA molecule. Transcription ends when RNA polymerase reaches a **terminator** sequence in the DNA.

B. mRNA Is Altered in the Nucleus of Eukaryotic Cells

- After transcription, the cell adds a cap and poly A tail to mRNA. **Introns** are cut out of RNA, and the remaining **exons** are spliced together.

7.5 | Translation Builds the Protein

A. The Genetic Code Links mRNA to Protein

- The correspondence between codons and amino acids is the **genetic code.**
- Each group of three consecutive mRNA bases is a **codon** that specifies one amino acid (or a stop codon).
- Experiments with synthetic mRNA enabled scientists to match each codon with its corresponding amino acid.

B. Translation Requires mRNA, tRNA, and Ribosomes

- mRNA carries a protein-encoding gene's information. rRNA associates with proteins to form **ribosomes,** which support and help catalyze protein synthesis.
- On one end, tRNA has an **anticodon** sequence complementary to an mRNA codon; the corresponding amino acid binds to the other end.

C. Translation Occurs in Three Steps

- Translation begins when mRNA joins with a small ribosomal subunit and a tRNA, usually carrying methionine. A large ribosomal subunit then joins the small one.
- In the elongation stage, a second tRNA binds to the next codon, and its amino acid bonds with the methionine that the first tRNA brought in. The ribosome moves down the mRNA as the chain grows.
- Upon reaching a "stop" codon, the ribosome is released, and the new polypeptide breaks free.

D. Proteins Must Fold Correctly After Translation

- Chaperone proteins help fold the polypeptide, which may be shortened or combined with others to form the finished protein.

7.6 | Cells Regulate Gene Expression

- Protein synthesis requires substantial energy input.

A. Operons Are Groups of Bacterial Genes That Share One Promoter

- In bacteria, **operons** coordinate expression of grouped genes whose encoded proteins participate in the same metabolic pathway. *E. coli*'s *lac* **operon** is a well-studied example. Transcription does not occur if a **repressor** protein binds to the **operator** sequence of the DNA.

B. Eukaryotic Organisms Use Transcription Factors

- In eukaryotic cells, proteins called **transcription factors** bind to promoters and **enhancers,** which are DNA sequences that regulate which genes a cell transcribes.

C. Eukaryotic Cells Also Use Additional Regulatory Mechanisms

- Other regulatory mechanisms include inactivating regions of a chromosome; alternative splicing; controls over mRNA stability and translation; and controls over protein folding and movement.

7.7 | Mutations Change DNA Sequences

- A **mutation** adds, deletes, alters, or moves nucleotides.

A. Mutations Range from Silent to Devastating

- A **point mutation** alters one or a few DNA bases. A substitution mutation may result in an mRNA that encodes the wrong amino acid or that substitutes a "stop" codon for an amino acid–coding codon. Substitution mutations can also be "silent."
- Inserting or deleting nucleotides (a **frameshift mutation**) may disrupt the reading frame of a gene, changing the amino acid sequence of the encoded protein.
- **Expanding repeat mutations** cause some inherited illnesses.

B. What Causes Mutations?

- A gene can mutate spontaneously, especially if it contains regions of repetitive DNA sequences. **Mutagens,** such as chemicals or radiation, induce mutations.
- Problems in meiosis can cause mutations if portions of chromosomes are deleted, inverted, or moved.

C. Mutations May Pass to Future Generations

- A **germline mutation** originates in cells that give rise to gametes and therefore appears in every cell of an offspring that inherits the mutation. A **somatic mutation,** which occurs in nonsex cells, affects a subset of cells in the body but does not affect the offspring.

D. Mutations Are Important

- Mutations create new alleles, which are the raw material for evolution.
- Induced mutations help scientists deduce gene function and help plant breeders produce new varieties of fruits and flowers.

7.8 | BIOTECHNOLOGY The Human Genome Sequence Is Surprisingly Complex

- Only 1.5% of the 3.2 billion base pairs of the human genome encode protein, yet those 25,000 or so genes specify hundreds of thousands of distinct proteins.
- Alternative splicing of introns explains how a limited number of genes can encode a larger number of proteins.
- The 98.5% of the human genome that does not encode protein encodes RNA, control sequences, **pseudogenes, transposable elements,** and other repeats.

7.9 | BIOTECHNOLOGY Genetic Engineering Moves Genes Among Species

A. Transgenic Organisms Contain DNA from Multiple Species
- **Transgenic** organisms are important in industry, research, and agriculture.

B. Creating Transgenic Organisms Requires Cutting and Pasting DNA
- **Restriction enzymes, cloning vectors** such as **plasmids,** and **reverse transcriptase** are tools that help researchers create **recombinant DNA** and introduce it to recipient cells.
- Several methods induce cells to take up recombinant DNA and become transgenic.

7.10 | BIOTECHNOLOGY Researchers Can Fix, Block, or Monitor Genes

A. Gene Therapy Repairs Faulty Genes
- **Gene therapy** requires placing a functional gene into cells expressing a faulty gene.

B. Antisense RNA and Gene Knockouts Block Gene Expression
- Silencing specific genes can treat illness or help researchers understand gene function.

C. DNA Microarrays Help Monitor Gene Expression
- **DNA microarrays** help researchers visualize the expression of many genes simultaneously.

7.11 | Investigating Life: Clues to the Origin of Language
- A family with a language disorder led researchers to discover a gene that is apparently involved in the acquisition of language.
- Comparing the human version of the gene with that in other primates and in mice suggests that the gene apparently began evolving rapidly soon after modern humans arose. Eventually one allele became fixed in the human population.

Multiple Choice Questions

1. A nucleotide is composed of all of the following EXCEPT a
   a. sugar.
   b. nitrogen-containing group.
   c. sulfur-containing group.
   d. phosphorus-containing group.

2. If one strand of DNA has the sequence ATTGTCC, then the sequence of the complementary strand would be
   a. TAACAGG.
   c. ACCTCGG.
   b. CGGAGTT.
   d. CCTGTTA.

3. Why did the DNA from the heat-killed type S cells *transform* the type R cells?
   a. Because it was mutated
   b. Because DNA determines which proteins are made in a cell
   c. Because the type R cells did not have DNA
   d. Because DNA is less complex than proteins

4. Choose the mRNA sequence that is complementary to the gene sequence GGACTTACG.
   a. CCTGAATGC
   c. GGTCAATCG
   b. AACUGGCUA
   d. CCUGAAUGC

5. The segments of eukaryotic mRNA that are translated into protein are called
   a. promoters.
   c. exons.
   b. introns.
   d. caps.

6. What is the job of the tRNA during translation?
   a. It carries amino acids to the mRNA.
   b. It triggers the formation of a covalent bond between amino acids.
   c. It binds to the small ribosomal subunit.
   d. It triggers the termination of the protein.

7. What could cause the *lac* operon to shut off after it has been activated?
   a. The binding of the sugar lactose to the promoter
   b. The inactivation of RNA polymerase by the addition of a modified sugar
   c. The rebinding of the repressor to the operator after all the lactose is degraded
   d. The binding of the repressor to the promoter

8. What might happen if you changed one nucleotide in a codon?
   a. The protein would stop being made.
   b. The protein would have the wrong amino acid sequence.
   c. There would be no effect on the protein.
   d. All of the above are possible.

9. Are mutations harmful?
   a. Yes, because the DNA is damaged.
   b. No, because changes in the DNA result in better alleles.
   c. Yes, because mutated proteins don't function.
   d. It depends on how the mutation affects the protein's function.

10. Which biotechnology would you use to temporarily stop a gene from being expressed?
    a. Antisense RNA
    c. Microarray
    b. Gene knockout
    d. Transgenic

Write It Out

1. Describe the three-dimensional structure of DNA.
2. How would the results of the Hershey–Chase experiment have differed if protein were the genetic material?
3. Write the complementary DNA sequence of each of the following base sequences:
   a. TCGAGAATCTCGATT
   b. CCGTATAGCCGGTAC
   c. ATCGGATCGCTACTG
4. Put the following in order from smallest to largest: nucleotide, genome, nitrogenous base, gene, nucleus, cell, codon, chromosome.
5. What is the function of DNA?
6. List the differences between RNA and DNA.
7. Define and distinguish between transcription and translation. Where in a eukaryotic cell does each process occur?
8. This chapter compared a chromosome to a cookbook and a gene to a recipe. List the ways that chromosomes and genes are UNLIKE cookbooks and recipes.
9. Some people compare DNA to a blueprint stored in the office of a construction company. Explain how this analogy would extend to transcription and translation.
10. List the three major types of RNA and their functions.
11. List the sequences of the mRNA molecules transcribed from the following template DNA sequences:
    a. TTACACTTGCTTGAGAGTT
    b. GGAATACGTCTAGCTAGCA
12. Given the following partial mRNA sequences, reconstruct the corresponding DNA template sequences:
    a. GUGGCGUAUUCUUUUCCGGGUAGG
    b. AGGAAAACCCCUCUUAUUAUAGAU

13. Refer to the figure to answer these questions:

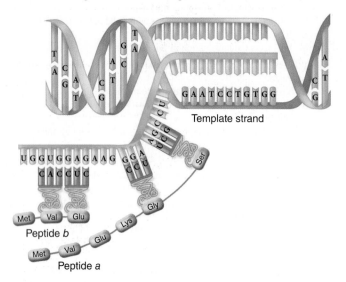

Template strand

Peptide b

Peptide a

a. Add labels for mRNA (including the 5′ and 3′ ends) and tRNA. In addition, draw in the RNA polymerase enzyme and the ribosomes, including arrows indicating the direction of movement for each.

b. What are the next three amino acids to be added to peptide *b?*

c. Fill in the nucleotides in the mRNA complementary to the template DNA strand.

d. What is the sequence of the DNA complementary to the template strand (as much as can be determined from the figure)?

e. Does this figure show the entire peptide that this gene encodes? How can you tell?

f. What might happen to peptide *b* after its release from the ribosome?

g. Does this figure depict a prokaryotic or a eukaryotic cell? How can you tell?

14. Consult the genetic code to write codon changes that could account for the following changes in amino acid sequence.

a. tryptophan to arginine

b. glycine to valine

c. tyrosine to histidine

15. Titin is a muscle protein whose gene has the largest known coding sequence: 80,781 DNA bases. How many amino acids long is titin?

16. If a protein is 1259 amino acids long, what is the minimum size of the gene that encodes the protein? Why might the gene be longer than the minimum?

17. On the television program *The X Files,* Agent Scully discovers an extraterrestrial life form that has a triplet genetic code but with five different bases instead of the four of earthly inhabitants. How many different amino acids can this code specify?

18. A mouse's genome has 1500 olfactory genes encoding proteins that enable the animal to detect odors. In each olfactory sensory neuron, only one of these genes is expressed; the others remain "off." List all of the ways that a mouse cell might silence the unneeded genes.

19. The genome of the human immunodeficiency virus (HIV) includes nine genes. Two of the genes encode four different proteins each. How is this possible?

20. The shape of a finch's beak reflects the expression of a gene that encodes a protein called calmodulin. A cactus finch has a long, pointy beak; its cells express the gene more than a ground finch, which has a short, deep beak. When researchers boosted gene expression in a ground finch embryo, the bird's upper beak was longer than normal. Develop a hypothesis that explains this finding.

21. If a gene is like a cake recipe, then a mutation is like a cake recipe containing an error. List the major types of mutations, and describe an analogous error in a cake recipe.

22. A protein-encoding region of a gene has the following DNA sequence:
G T A G C G T C A C A A A C A A A T C A G C T C
Determine how each of the following mutations alters the amino acid sequence:

a. substitution of a T for the C in the 10th position

b. substitution of a G for the C in the 19th position

c. insertion of a T between the 4th and 5th DNA bases

d. insertion of a GTA between the 12th and 13th DNA bases

e. deletion of the first DNA nucleotide

23. Explain how a mutation in a protein-encoding gene, an enhancer, or a gene encoding a transcription factor can all have the same effect on an organism.

24. How can a mutation alter the sequence of DNA bases in a gene but not produce a noticeable change in the gene's polypeptide product? How can a mutation alter the amino acid sequence of a polypeptide yet not alter the organism?

25. Parkinson disease causes rigidity, tremors, and other motor symptoms. Only 2% of cases are inherited, and these tend to have an early onset of symptoms. Some inherited cases result from mutations in a gene that encodes the protein parkin, which has 12 exons. Indicate whether each of the following mutations in the parkin gene would result in a smaller protein, a larger protein, or no change in the size of the protein.

a. deletion of exon 3

b. deletion of six consecutive nucleotides in exon 1

c. duplication of exon 5

d. disruption of the splice site between exon 8 and intron 8

e. deletion of intron 2

26. In a disorder called gyrate atrophy, cells in the retina begin to degenerate in late adolescence, causing night blindness that progresses to total blindness. The cause is a mutation in the gene that encodes an enzyme, ornithine aminotransferase (OAT). Researchers sequenced the *OAT* gene for five patients with the following results:

- Patient A: A change in codon 209 of UAU to UAA
- Patient B: A change in codon 299 of UAC to UAG
- Patient C: A change in codon 426 of CGA to UGA
- Patient D: A two-nucleotide deletion at codons 64 and 65 that results in a UGA codon at position 79
- Patient E: Exon 6, including 1071 nucleotides, is entirely deleted.

a. Which patient(s) have a frameshift mutation?

b. How many amino acids is patient E missing?

c. Which patient(s) will produce a shortened protein?

27. Researchers use computer algorithms that search DNA sequences for indications of specialized functions. Explain the significance of detecting the following sequences:

a. a promoter

b. a sequence of 75 to 80 nucleotides that folds into a shape resembling a backwards letter "L"

c. a gene with a sequence very similar to that of a known protein-encoding gene but that is not translated into protein

d. RNAs with poly A tails

28. How do researchers create recombinant DNA and transgenic organisms, and what are some applications of this technology?

29. Transgenic crops often require fewer herbicides and insecticides than conventional crops. In that respect, they could be considered environmentally friendly. Use the Internet to research the question of why some environmental groups oppose transgenic technology.

30. Define *gene therapy, antisense RNA, gene knockout,* and *DNA microarray.*
31. Which biotechnology might be able to accomplish the following goals? More than one answer may be possible.
    a. Silence the HIV genes integrated into the chromosomes of people with HIV infection (which leads to AIDS).
    b. Create bacteria that produce human growth hormone, used to treat extremely short stature.
32. Explain the ethical issues that gene therapy presents.
33. Many patients waste precious time taking anticancer drugs that are ineffective or too toxic. How might DNA microarray technology refine the treatment of cancer?
34. A young zebra finch must learn to sing. Researchers used a modified virus to deliver a "mirror image" of the *FOXP2* gene to the brain of a young finch. With the *FOXP2* gene silenced, the bird's song-learning ability was impaired. Why did the treatment silence the gene? How does this experiment relate to the study of human language?
35. Choose an experiment mentioned in the chapter and analyze how it follows the scientific method.
36. Give an example from the chapter of different types of experiments used to address the same hypothesis. Why might this be necessary?

Pull It Together

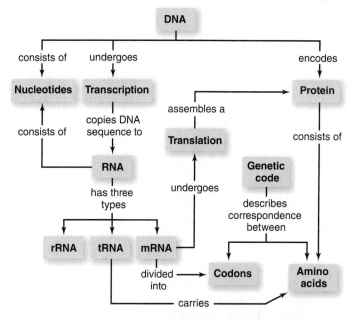

1. Why is protein production essential to cell function?
2. Where do promoters, terminators, stop codons, transcription factors, RNA polymerase, and enhancers fit into this concept map?
3. How do transgenic organisms fit into this concept map?
4. Use the concept map to explain why a mutation in DNA sometimes causes protein function to change.

DNA Replication, Mitosis, and the Cell Cycle

Carbon Copies. Tabouli, left, plays with her sibling, Baba Ganoush. In an effort to publicize its pet gene banking and cloning services, a private company produced the kittens by cloning a one-year-old cat.

connect

|BIOLOGY

Enhance your study of this chapter with practice quizzes, animations and videos, answer keys, and downloadable study tools.
www.mhhe.com/hoefnagels

Learn How to Learn

Explain It, Right or Wrong

As you work through the multiple choice questions at the end of each chapter, make sure you can explain why each correct choice is right. You can also test your understanding by taking the time to explain why each of the other choices is wrong.

The Clones Are Here

IMAGINE BEING ABLE TO GROW A NEW INDIVIDUAL, GENETICALLY IDENTICAL TO YOURSELF, FROM A BIT OF SKIN OR THE ROOT OF A HAIR. Although humans cannot reproduce in this way, many organisms do the equivalent. They develop parts of themselves into genetically identical individuals—clones—that then detach and live independently.

Cloning, or asexual reproduction, has been a part of life since the first cell arose billions of years ago. Long before sex evolved, each individual simply reproduced by itself, without a partner to contribute half the offspring's genetic information.

In its simplest form, asexual reproduction consists of the division of a single cell. In bacteria, archaea, and single-celled eukaryotes such as *Amoeba,* the cell's DNA replicates, and then the cell splits into two identical, individual organisms. Although the details of cell division differ, the result is the same: one individual becomes two.

Most multicellular eukaryotes (plants, fungi, animals, and some types of protists) reproduce sexually, but at least some organisms in each kingdom also use asexual reproduction. This strategy is especially common in plants and fungi. Hobbyists and commercial plant growers clone everything from fruit trees to African violets by cultivating cuttings from a parent plant's stems, leaves, and roots. Many fungi also are phenomenal breeders, asexually producing countless microscopic spores on bread, cheese, and every other imaginable food supply.

Asexual reproduction is much less common in animals. Sponges, coral animals, hydra, and jellyfishes "bud" genetically identical clones that break away from the parent. Mammals, however, normally reproduce only sexually. That is why biologists attracted worldwide attention in the 1990s by creating a lamb called Dolly, the first clone of an adult mammal (see section 8.7).

All eukaryotes rely on a process called mitosis, which enables cells to copy themselves faithfully. Without mitosis, no eukaryote could grow, replace dead cells, or repair injuries. This chapter describes how and when eukaryotic cells divide—and explains how they die.

Learning Outline

8.1 | Cells Divide and Cells Die

Your cells are too small to see without a microscope, so it is hard to appreciate just how many you lose as you sleep, work, and play. Each minute, for example, you shed tens of thousands of dead skin cells. If you did not have a way to replace these building blocks, your body would literally wear away. Instead, cells in your deep skin layers divide and replace the ones you lose. Each new cell lives an average of about 35 days, so you will gradually replace your entire skin in the next month or so—without even noticing!

Cell division produces a continuous supply of replacement cells, both in your skin and everywhere else in your body. But cell division has other functions as well. No living organism can reproduce without cell division, and the growth and development of a multicellular organism also require the production of new cells.

This chapter explores the opposing but coordinated forces of cell division and cell death, and considers what happens if either process goes wrong. We begin by exploring cell division's role in reproduction, growth, and development.

A. Sexual Life Cycles Include Mitosis, Meiosis, and Fertilization

Organisms must reproduce—generate other individuals like themselves—for a species to persist. The most straightforward and ancient way for a single-celled organism to reproduce is **asexually,** by replicating its genetic material and splitting the contents of one cell into two. Except for the occasional mutation, asexual reproduction generates genetically identical offspring. Most bacteria and archaea, for example, reproduce by binary fission, the simplest type of asexual cell division. Many protists and multicellular eukaryotes also reproduce asexually, as described in the opening essay for this chapter. ▶ binary fission, p. 343

Sexual reproduction, in contrast, is the production of offspring whose genetic makeup comes from two parents. Each parent contributes a sex cell, and the fusion of these cells signals the start of the next generation. Because sexual reproduction mixes up and recombines traits, the offspring are genetically different from each other.

Figure 8.1 illustrates how two types of cell division, meiosis and mitosis, interact in a sexual life cycle. In humans and many other species, the male parent provides sperm cells, and a female produces egg cells. **Meiosis,** described in chapter 9, is the specialized type of cell division that gives rise to these sex cells (collectively called **gametes**). Meiosis produces cells that are genetically different from one another. This variation among gametes explains why the offspring of two parents usually look different from one another.

Fertilization is the union of the sperm and the egg cell, producing the first cell of the new offspring. Immediately after fertilization, the other type of cell division—mitotic—takes over.

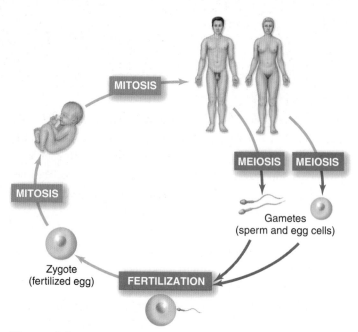

Figure 8.1 Sexual Reproduction. In the life cycle of humans and many other organisms, adults produce gametes by meiosis. Fertilization unites sperm and egg, and mitotic cell division accounts for the growth of the new offspring.

Mitosis divides a eukaryotic cell's genetic information into two identical daughter cells. Mitotic cell division explains how you grew from a single cell into an adult (**figure 8.2**), how you repair damage after an injury, and how you replace the cells that you lose every day. Likewise, mitotic cell division accounts for the growth and development of plants, mushrooms, and other multicellular eukaryotes.

Each of the trillions of cells in your body retains the genetic information that was present in the fertilized egg. Inspired by the astonishing precision with which this occurs, geneticist Herman J. Müller wrote in 1947:

> *In a sense we contain ourselves, wrapped up within ourselves, trillions of times repeated.*

This quotation eloquently expresses the powerful idea that every cell in the body results from countless rounds of cell division, each time forming two genetically identical cells from one.

B. Cell Death Is Part of Life

The development of a multicellular organism requires more than just cell division. Cells also die in predictable ways, carving distinctive structures. **Apoptosis** is cell death that is a normal part of development. Like cell division, it is a precise, tightly regulated sequence of events (see section 8.6). Apoptosis is therefore also called "programmed cell death."

During early development, both cell division and apoptosis shape new structures. For example, the feet of both chickens and

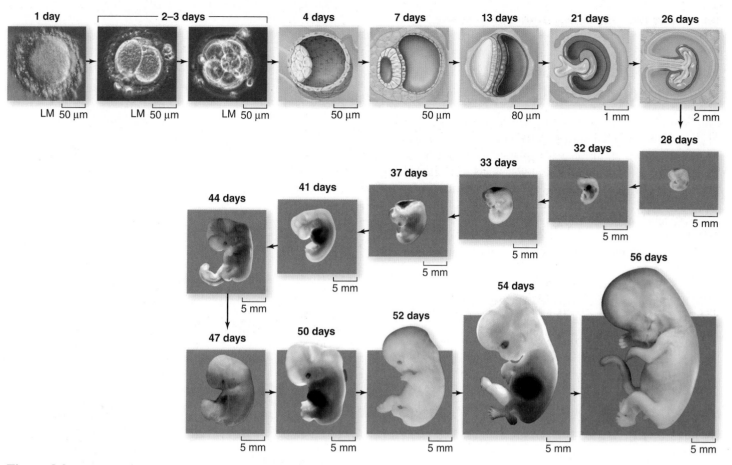

Figure 8.2 Human Growth and Development. These photos show the development of a human fetus from a zygote. Mitotic division produces the cells that build the body.

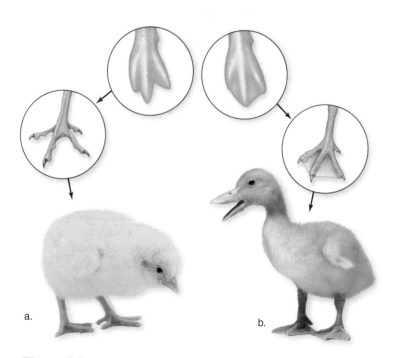

Figure 8.3 Apoptosis Carves Toes. (a) A developing chicken foot undergoes extensive apoptosis; the toes take shape as selected cells die. (b) A duck's foot retains webbing between the digits.

ducks start out as webbed paddles when the birds are embryos. The webs of tissue remain in the duck's foot throughout life. In the chicken, however, individual toes form as cells between the digits die (figure 8.3). Likewise, cells in the tail of a tadpole die as the young frog develops into an adult.

Throughout an animal's life, cell division and cell death are in balance, so tissue neither overgrows nor shrinks. Cell division compensates for the death of skin and blood cells, a little like adding new snow (cell division) to a snowman that is melting (apoptosis). Both cell division and apoptosis also help protect the organism. For example, cells divide to heal a scraped knee; apoptosis peels away sunburnt skin cells that might otherwise become cancerous.

Before learning more about mitosis and apoptosis, it is important to understand how the genetic material in a eukaryotic cell copies itself and condenses in preparation for cell division. Sections 8.2 and 8.3 explain these events.

8.1 | Mastering Concepts

1. Explain the roles of mitotic cell division, meiosis, and fertilization in the human life cycle.
2. Why are both cell division and apoptosis necessary for the development of an organism?

8.2 | DNA Replication Precedes Cell Division

Before any cell divides—by binary fission, mitotically, or meiotically—it must first duplicate its entire **genome,** which consists of all of the cell's genetic material. Chapter 7 describes the cell's genome as a set of "cookbooks," each containing "recipes" (genes) that encode tens of thousands of proteins. In DNA replication, the cell copies all of this information, letter by letter. Without a full set of instructions, a new cell will die.

Recall from figure 7.7 that DNA is a double-stranded nucleic acid. Each strand of the double helix is composed of nucleotides. Hydrogen bonds between the nitrogenous bases of the nucleotides hold the two strands together. The base adenine (A) pairs with its complement, thymine (T); similarly, cytosine (C) forms complementary base pairs with guanine (G).

Figure It Out

Write the complementary strand for the following DNA sequence:
5′-TCAATACCGATTAT-3′

Answer: 3′-AGTTATGGCTAATA-5′

When Watson and Crick reported DNA's structure, they understood that they had uncovered the key to DNA replication. Their paper ends with the tantalizing statement, "It has not escaped our notice that the specific pairing we have postulated immediately suggests a possible copying mechanism for the genetic material." They envisioned DNA unwinding, exposing unpaired bases that would attract their complements, and neatly knitting two double helices from one. This route to replication, which turned out to be essentially correct, is called semiconservative because each DNA double helix conserves half of the original molecule (**figure 8.4**).

DNA does not, however, replicate by itself. Instead, an army of enzymes copies DNA just before a cell divides (**figure 8.5**). Enzymes called helicases unwind and "unzip" the DNA, while binding proteins prevent the two single strands from rejoining each other. Other enzymes then guide the assembly of new DNA strands. **DNA polymerase** is the enzyme that adds new DNA nucleotides that are complementary to the bases on each exposed strand. As the new DNA strands grow, hydrogen bonds form between the complementary bases.

Curiously, DNA polymerase can only add nucleotides to an existing strand. A primase enzyme therefore must build a short complementary piece of RNA, called an RNA primer, at the start of each DNA segment to be replicated. The RNA primer attracts the DNA polymerase enzyme. Once the new strand of DNA is in place, another enzyme removes each RNA primer

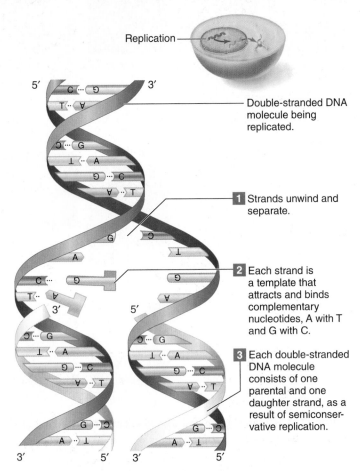

Replication

Double-stranded DNA molecule being replicated.

1 Strands unwind and separate.

2 Each strand is a template that attracts and binds complementary nucleotides, A with T and G with C.

3 Each double-stranded DNA molecule consists of one parental and one daughter strand, as a result of semiconservative replication.

Figure 8.4 DNA Replication Is Semiconservative. In this simplified view of DNA replication, [1] DNA strands unwind and separate. [2] New nucleotides form complementary base pairs with each exposed strand. [3] The process ends with two identical double-stranded DNA molecules. Note that the RNA primers and enzymes that participate in DNA replication are not shown here.

and replaces it with the correct DNA nucleotides. Enzymes called **ligases** form covalent bonds between the resulting DNA segments.

Furthermore, like the RNA polymerase enzyme described in section 7.4, DNA polymerase can add new nucleotides only to the exposed 3′ end—never the 5′ end—of a growing strand. Replication therefore proceeds continuously on only one new DNA strand, called the leading strand. On the other strand, called the lagging strand, replication occurs in short 5′ to 3′ pieces. These short pieces, each preceded by its own RNA primer, are called Okazaki fragments.

Enzymes copy the DNA simultaneously at hundreds of points, called origins of replication, on each chromosome. Copying proceeds in both directions at once from each origin. This arrangement is similar to the way that hurried office workers might split a lengthy report into short pieces and then divide

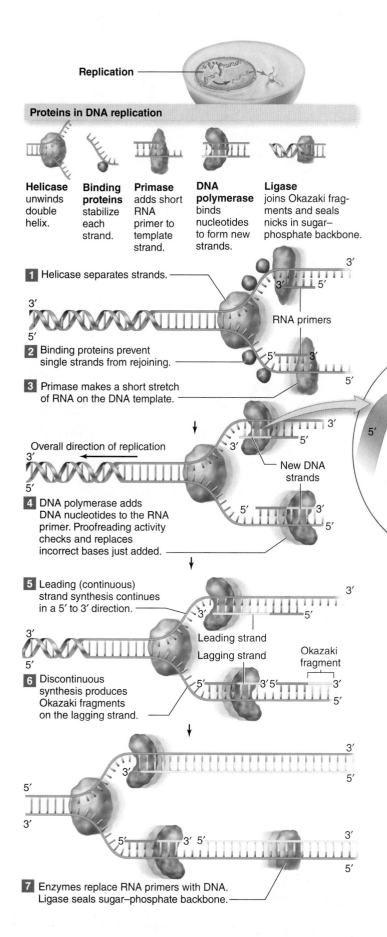

Proteins in DNA replication

Helicase unwinds double helix.

Binding proteins stabilize each strand.

Primase adds short RNA primer to template strand.

DNA polymerase binds nucleotides to form new strands.

Ligase joins Okazaki fragments and seals nicks in sugar–phosphate backbone.

1 Helicase separates strands.

RNA primers

2 Binding proteins prevent single strands from rejoining.

3 Primase makes a short stretch of RNA on the DNA template.

Overall direction of replication

New DNA strands

4 DNA polymerase adds DNA nucleotides to the RNA primer. Proofreading activity checks and replaces incorrect bases just added.

5 Leading (continuous) strand synthesis continues in a 5′ to 3′ direction.

Leading strand

Lagging strand

Okazaki fragment

6 Discontinuous synthesis produces Okazaki fragments on the lagging strand.

G A C C G G A C

C T G G C C U G G

New DNA strand

7 Enzymes replace RNA primers with DNA. Ligase seals sugar–phosphate backbone.

the sections among many copy machines operating at the same time. Thanks to this division of labor, copying the billions of DNA nucleotides in a human cell's 46 chromosomes takes only 8 to 10 hours.

DNA replication is incredibly accurate. DNA polymerase "proofreads" as it goes, discarding mismatched nucleotides and inserting correct ones. After proofreading, DNA polymerase incorrectly incorporates only about 1 in a billion nucleotides. Other repair enzymes help ensure the accuracy of DNA replication by cutting out and replacing incorrect nucleotides.

Nevertheless, mistakes occasionally remain. The result is a **mutation,** which is any change in a cell's DNA sequence. To extend the cooking analogy, a mutation is similar to a mistake in one of the recipes in a cookbook. Section 7.7 describes the many ways that a mutation can affect the life of a cell.

Overall, DNA replication requires a great deal of energy because a large, organized nucleic acid contains much more potential energy than do many individual nucleotides. Energy is required to synthesize nucleotides and to create the covalent bonds that join them together in the new strands of DNA. Many of the enzymes that participate in DNA replication, including helicase and ligase, also require energy in the form of ATP to catalyze their reactions.
▶ ATP, p. 76

8.2 | Mastering Concepts

1. Why does DNA replicate?
2. What is semiconservative replication?
3. What are the steps of DNA replication?
4. What is the role of RNA primers in DNA replication?
5. What happens if DNA polymerase fails to correct an error?

Figure 8.5 DNA Replication Requires Many Proteins. Although the proteins that participate in DNA replication are depicted separately for clarity, in reality they form a single cluster that moves rapidly along a DNA molecule. Many such clusters operate simultaneously in a cell that is preparing to divide.

8.3 | Replicated Chromosomes Condense as a Cell Prepares to Divide

Binary fission, the type of cell division that occurs in bacteria and archaea, is relatively uncomplicated because the genetic material in these cells consists of a single circular DNA molecule (see figure 16.7). In a eukaryotic cell, however, distributing the DNA into daughter cells is a bit more complex because the genetic information consists of multiple chromosomes housed inside the cell's nucleus.

A **chromosome** is a single molecule of DNA and its associated proteins. Each species has a characteristic number of chromosomes. A mosquito's cell has 6 chromosomes; grasshoppers, rice plants, and pine trees all have 24; humans have 46; dogs and chickens have 78; a carp has 104. Each of these numbers is even because sexually reproducing organisms inherit one set of chromosomes from each parent. Human sperm and egg cells, for example, each contain 23 chromosomes; fertilization therefore yields an offspring with 46 chromosomes in every cell.

With so much genetic material, a eukaryotic cell must balance two needs. On the one hand, the cell must have access to the information in its DNA. On the other hand, if the cell is to divide, it must package its DNA into a portable form that can easily move into the two daughter cells (figure 8.6). To understand how cells maintain this balance, we must take a closer look at the structure of the chromosome.

Eukaryotic chromosomes consist of **chromatin,** which is a collective term for all of the cell's DNA and its associated proteins. These proteins include the many enzymes that help replicate the DNA and transcribe it to a sequence of RNA (see chapter 7). Others serve as scaffolds around which DNA entwines, helping to pack the DNA efficiently inside the cell.

To illustrate the importance of DNA packing, consider this fact: Stretched end to end, the DNA in one human cell would

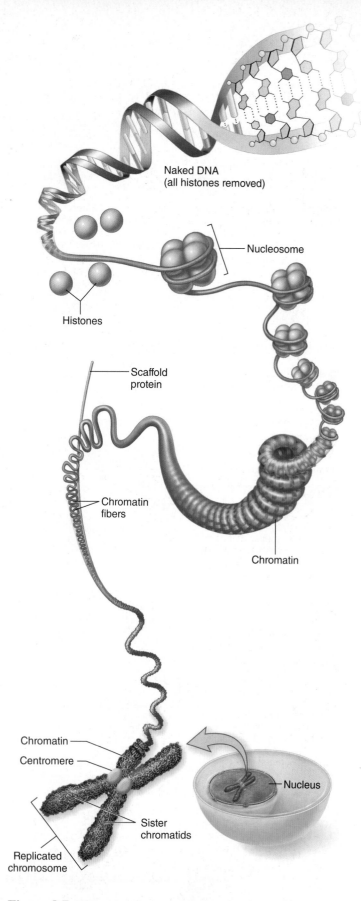

Naked DNA (all histones removed)

Nucleosome

Histones

Scaffold protein

Chromatin fibers

Chromatin

Chromatin

Centromere

Sister chromatids

Replicated chromosome

Nucleus

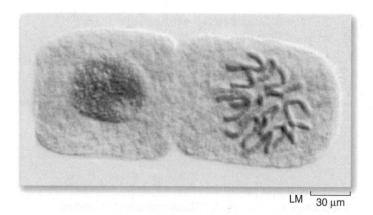

LM   30 μm

Figure 8.6 Two Views of DNA. In the cell on the left, DNA is loosely packed in the nucleus and available for DNA replication and protein synthesis. Before a cell divides, however, the DNA winds into the compact, portable chromosomes visible in the cell on the right.

Figure 8.7 Parts of a Chromosome. DNA replicates just before a cell divides, and then the chromatin condenses into this familiar, compact form. The two genetically identical chromatids of a replicated chromosome attach at the centromere.

form a thread some 2 meters long. If the DNA bases of all 46 human chromosomes were typed as A, C, T, and G, the several billion letters would fill 4000 books of 500 pages each! How can a cell only 100 microns in diameter contain so much material?

The explanation is that chromatin is organized into units called **nucleosomes,** each consisting of a stretch of DNA wrapped around eight proteins (histones). A continuous thread of DNA connects nucleosomes like beads on a string (figure 8.7). When the cell is not dividing, chromatin is barely visible because the nucleosomes are loosely packed together. The information in the DNA is therefore accessible for the cell to produce the enzymes and other proteins that it needs for all of its metabolic activities. DNA replication in preparation for cell division also requires that the cell's DNA be unwound.

The chromosome's appearance begins to change shortly after DNA replication. The nucleosomes gradually fold into progressively larger structures, eventually taking on the familiar, dense, compact shapes that are easily visible in the microscope. DNA packing is somewhat similar to winding a very long length of thread around a wooden spool. Just as spooled thread occupies less space and is easier to move than a disorganized wad, condensed DNA is much easier for the cell to manage than is unwound chromatin.

Once condensed, a chromosome has readily identifiable parts (table 8.1 and figure 8.7). A replicated chromosome consists of two **chromatids,** each with a DNA sequence identical to the other. These paired chromatids are therefore called "sister chromatids." The **centromere** is a small section of DNA and associated proteins that attaches the sister chromatids to each other. It often appears as a constriction in a replicated chromosome. As a cell's genetic material divides, the centromere splits, and the sister chromatids move apart. At that point, each chromatid becomes an individual chromosome in its own right.

8.3 | Mastering Concepts

1. What is the relationship between chromosomes and chromatin?
2. How does DNA interact with histones?
3. What are the main parts of a chromosome?

Table 8.1 Miniglossary of Chromosome Terms

Term	Definition
Chromatin	Collective term for all of the DNA and associated proteins in a cell's nucleus
Chromosome	A discrete, continuous molecule of DNA wrapped around protein. Eukaryotic cells contain multiple linear chromosomes, whereas bacterial cells each contain one circular chromosome.
Chromatid	One of two identical attached copies that make up a replicated chromosome
Centromere	A small part of a chromosome that attaches sister chromatids to each other

8.4 | Mitotic Division Generates Exact Cell Copies

Suppose you scrape your leg while sliding into second base during a softball game. At first, the wound bleeds, but the blood soon clots and forms a scab. Underneath the dried crust, cells of the immune system clear away trapped dirt and dead cells. At the same time, undamaged skin cells bordering the wound begin to divide repeatedly, producing fresh, new daughter cells that eventually fill the damaged area.

Those actively dividing skin cells illustrate the **cell cycle,** which describes the events that occur from one cell division until the next. Biologists divide the cell cycle into stages (figure 8.8). During **interphase,** the cell is not dividing, but protein synthesis, DNA replication, and many other events occur. Immediately following interphase is mitosis, during which the contents of the nucleus divide. **Cytokinesis** is the splitting of the cell itself. After cytokinesis is complete, the daughter cells enter interphase, and the cell cycle begins anew.

Mitotic cell division occurs some 300 million times per minute in your body, replacing cells lost to abrasion or cell death. In each case, the products are two daughter cells, each receiving complete, identical genetic instructions and the molecules and organelles they need to maintain metabolism.

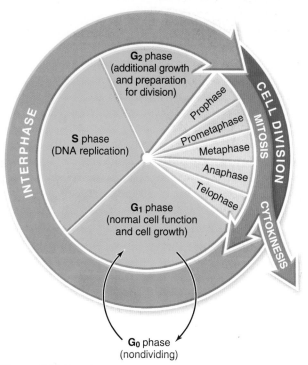

Figure 8.8 The Cell Cycle. Interphase includes gap phases (G$_1$ and G$_2$), when the cell grows and some organelles duplicate. Nondividing cells can leave G$_1$ and enter G$_0$ indefinitely. During the synthesis phase (S) of interphase, DNA replicates. Mitosis divides the replicated genetic material between two nuclei. Cytokinesis then splits the cytoplasm in two, producing two identical daughter cells.

A. Interphase Is a Time of Great Activity

Biologists once mistakenly described interphase as a time when the cell is at rest. The chromatin is unwound and therefore barely visible, so the cell appears inactive. However, interphase is actually a very active time. The cell carries out its basic functions, from muscle contraction to insulin production to bone formation. DNA replication also occurs during this stage.

Interphase is divided into "gap" phases (designated G_1, G_0, and G_2), separated by a "synthesis" (S) phase. During G_1, the cell grows, carries out basic functions, and produces molecules needed to build new organelles and other components it will require if it divides. A cell in G_1 is exquisitely sensitive to external signals that "tell" it whether it should divide, stop to repair damaged DNA, die, or enter a nondividing stage called G_0.

In G_0, a cell continues to function, but it does not replicate its DNA or divide. At any given time, most cells in the human body are in G_0, a stage that is usually reversible. Nerve cells in the brain, however, are permanently in G_0, which explains both why the brain does not grow after it reaches its adult size and why brain damage is often irreparable.

During **S phase,** enzymes replicate the cell's genetic material and repair damaged DNA (see section 8.2). As S phase begins, each chromosome includes one DNA molecule. By the end of S phase, each chromosome consists of two identical, attached DNA molecules—the sister chromatids—although they are not yet visible with a light microscope.

Another event that occurs during S phase is the duplication of the centrosome in an animal cell. **Centrosomes** are structures that organize the proteins that will move the chromosomes during mitosis. Each centrosome includes a cloud of proteins enclosing a pair of barrel-shaped centrioles. Most plant cells lack distinct centrosomes. Instead, the organization of these proteins occurs at many locations throughout the cell.

In G_2, the cell continues to grow but also prepares to divide, producing the proteins that will coordinate the movements of the chromosomes during mitosis. The DNA winds more tightly around its histone proteins, and this start of chromosome condensation signals impending mitosis. Interphase has ended.

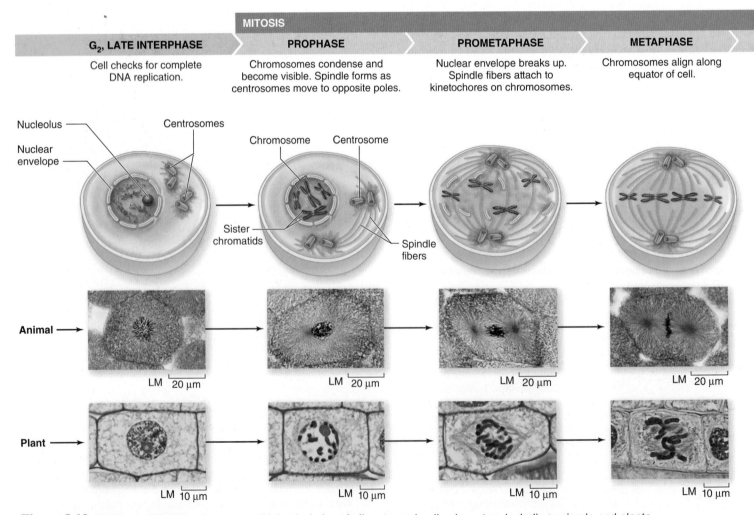

G_2, LATE INTERPHASE	PROPHASE	PROMETAPHASE	METAPHASE
Cell checks for complete DNA replication.	Chromosomes condense and become visible. Spindle forms as centrosomes move to opposite poles.	Nuclear envelope breaks up. Spindle fibers attach to kinetochores on chromosomes.	Chromosomes align along equator of cell.

Figure 8.10 **Stages of Mitosis.** Mitotic cell division includes similar stages in all eukaryotes, including animals and plants.

Figure It Out

A cell that has completed interphase contains __ times as much DNA as a cell at the start of interphase.

Answer: Two

B. Chromosomes Divide During Mitosis

Overall, mitosis separates the genetic material that replicated during S phase. For the chromosomes to be evenly distributed, they must line up in a way that enables them to split equally into two sets that are then pulled to opposite poles of the cell. The **mitotic spindle** is the array of proteins that coordinate these chromosome movements (**figure 8.9**). In animal cells, the centrosomes organize the microtubules that make up the spindle. ▶ microtubules, p. 63

Mitosis is a continuous process, but biologists divide it into stages for ease of understanding. **Figure 8.10** summarizes the key events of mitosis; you may find it helpful to consult this figure as you read on.

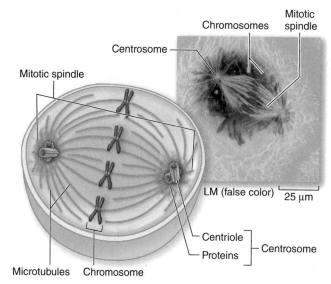

Figure 8.9 The Spindle Aligns Chromosomes. The mitotic spindle consists of microtubules that form the fibers that grow outward from two centrosomes. The spindle fibers push and pull to align the chromosomes. The inset shows the spindle in an amphibian cell during metaphase.

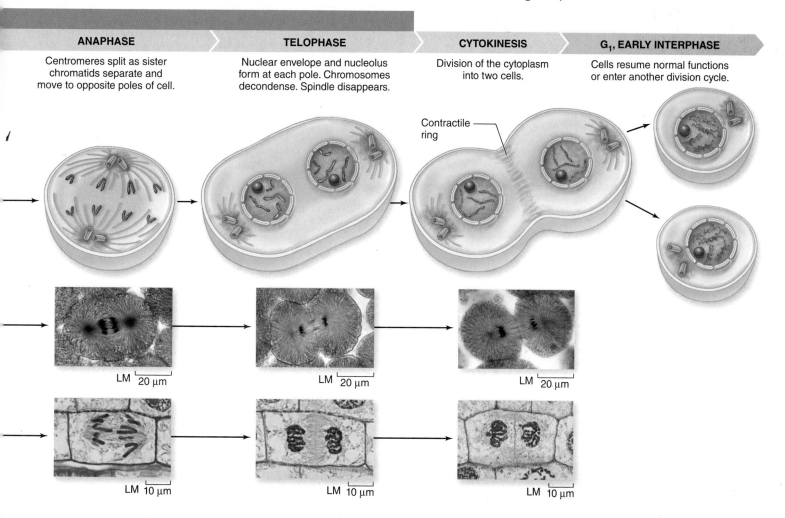

ANAPHASE	TELOPHASE	CYTOKINESIS	G₁, EARLY INTERPHASE
Centromeres split as sister chromatids separate and move to opposite poles of cell.	Nuclear envelope and nucleolus form at each pole. Chromosomes decondense. Spindle disappears.	Division of the cytoplasm into two cells.	Cells resume normal functions or enter another division cycle.

During **prophase,** DNA coils very tightly, as illustrated in figure 8.7, shortening and thickening the chromosomes. As the chromosomes condense, they become visible when stained and viewed under a microscope. For now, the chromosomes remain randomly arranged in the nucleus. Also during prophase, the two centrosomes migrate toward opposite poles of the cell, and the mitotic spindle begins to form. The nucleolus (the dark area in the nucleus) also disappears.

Prometaphase occurs immediately after the formation of the spindle. The nuclear envelope and associated endoplasmic reticulum break into small pieces, enabling the spindle fibers to reach the chromosomes. Meanwhile, proteins called **kineto-chores** begin to assemble on each centromere; these proteins attach the chromosomes to the spindle.

As **metaphase** begins, the mitotic spindle aligns the chromosomes down the center, or equator, of the cell. This alignment ensures that each cell will contain one sister chromatid from each duplicated chromosome.

In **anaphase,** the centromeres split as the mitotic spindle pulls the sister chromatids apart. Some microtubules in the spindle shorten, "reeling in" the chromosomes to opposite sides of the cell. Other spindle fibers lengthen in a way that moves the poles farther apart, stretching the dividing cell.

Telophase, the final stage of mitosis, essentially reverses the events of prophase and prometaphase. The mitotic spindle disassembles, and the chromosomes begin to unwind. In addition, a nuclear envelope and nucleolus form at each end of the stretched-out cell. As telophase ends, the division of the genetic material is complete, and the cell contains two nuclei—but not for long.

C. The Cytoplasm Splits in Cytokinesis

In cytokinesis, organelles and macromolecules are distributed into the two forming daughter cells, which then physically separate. The process differs somewhat between animal and plant cells (**figure 8.11**).

In an animal cell, the first sign of cytokinesis is the **cleavage furrow,** a slight indentation around the middle of the cell. This indentation results from a contractile ring of actin and myosin proteins that forms beneath the cell membrane. The proteins contract like a drawstring, separating the daughter cells.

Unlike animal cells, plant cells are surrounded by cell walls. A dividing plant cell must therefore construct a new wall that separates the two daughter cells. The first sign of cell wall construction is a line, called a **cell plate,** between the forming cells. Vesicles from the Golgi apparatus travel along microtubules, delivering structural materials such as cellulose fibers, other polysaccharides, and proteins to the midline of the dividing cell. The layer of cellulose fibers embedded in surrounding material makes a strong, rigid wall that gives a plant cell its shape. ▶ cell wall, p. 64

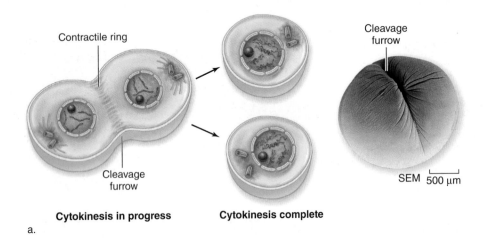

Contractile ring

Cleavage furrow

Cleavage furrow

SEM 500 μm

Cytokinesis in progress   **Cytokinesis complete**

a.

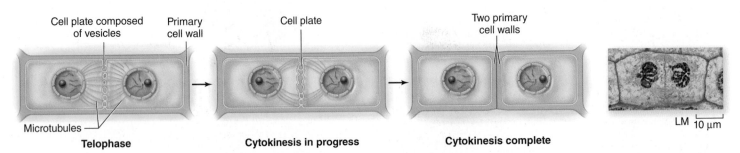

Cell plate composed of vesicles   Primary cell wall

Cell plate

Two primary cell walls

Microtubules

LM 10 μm

Telophase   **Cytokinesis in progress**   **Cytokinesis complete**

b.

Figure 8.11 Cytokinesis. (a) In an animal cell, the first sign of cytokinesis is an indentation called a cleavage furrow, which is formed by a contractile ring consisting of actin and myosin proteins. (b) In plant cells, the cell plate is the first stage in the formation of a new cell wall.

Table 8.2 Miniglossary of Cell Division Terms

Term	Definition
Interphase	Stage of the cell cycle in which chromosomes replicate and the cell grows
G_0 phase	Gap stage of interphase in which the cell functions but does not divide
G_1 phase	Gap stage of interphase in which the cell grows and carries out its functions
G_2 phase	Gap stage of interphase in which the cell produces and stores membrane components and spindle proteins
S phase	Synthesis stage of interphase when DNA replicates
Mitosis	Division of a cell's chromosomes into two identical nuclei
Prophase	Stage of mitosis when chromosomes condense and the mitotic spindle begins to form
Prometaphase	Stage of mitosis when the nuclear membrane breaks up and spindle fibers attach to kinetochores
Metaphase	Stage of mitosis when chromosomes are aligned down the center of the cell
Anaphase	Stage of mitosis when the spindle pulls sister chromatids toward opposite poles of the cell
Telophase	Stage of mitosis when chromosomes arrive at opposite poles and nuclear envelopes form
Centrosome	Structure that organizes the microtubules that make up the mitotic spindle
Mitotic spindle	Part of the cytoskeleton that moves chromosomes during mitosis
Kinetochore	Protein to which the mitotic spindle attaches on a chromosome's centromere
Cytokinesis	Distribution of cytoplasm to daughter cells following division of a cell's chromosomes
Cleavage furrow	Indentation in cell membrane of an animal cell undergoing cytokinesis
Cell plate	Material that forms the beginnings of the cell wall in a plant cell undergoing cytokinesis

Although cytokinesis typically follows mitosis, there are exceptions. Some types of green algae and slime molds, for example, exist as enormous cells containing thousands of nuclei, the products of many rounds of mitosis without cytokinesis. ▸ slime molds, p. 360

Table 8.2 summarizes some of the vocabulary related to the cell cycle.

8.4 | Mastering Concepts

1. What are the three main events of the cell cycle?
2. What happens during interphase?
3. How does the mitotic spindle form, and what is its function?
4. What happens during each phase of mitosis?
5. Distinguish between mitosis and cytokinesis.

Burning Question

What are the galls that form on plants?

Galls are abnormal growths that often form on the leaves and stems of plants. The growths may be smooth and perfectly round, as in the stem galls shown here. They may also cause grotesque deformities on stems, leaves, flowers, roots, and other plant parts.

Many organisms cause plants to form galls, including fungi, bacteria, and even parasitic plants. The most common galls, however, are traced to a distinctive group of wasps. A female gall wasp lays an egg in the vein of a stem or leaf. When the egg hatches and develops into a larva, it secretes chemicals that stimulate the plant's cells to divide. The resulting gall does not usually hurt or help the tree, but it does form a protective shell that houses and feeds the young wasp until adulthood.

Submit your burning question to:
marielle_hoefnagels@mcgraw-hill.com

8.5 | Cancer Arises When Cells Divide out of Control

Some cells divide more or less constantly. The cells at the tips of a plant's roots, for example, may continue to divide throughout the growing season, exploring the soil for water and nutrients. Likewise, stem cells in your bone marrow constantly produce new blood cells. On the other hand, the skin cells bordering a wound quit dividing once healing is complete; brain cells rarely divide once they are mature. How do any of these cells "know" what to do?

A. Chemical Signals Regulate Cell Division

Two groups of proteins guide a cell's progress through the cell cycle. The concentrations of proteins called cyclins fluctuate in predictable ways during each stage. For example, cyclin E peaks between G_1 and S phases of interphase, whereas cyclin B is essentially absent at that time but has its highest concentration between G_2 and mitosis. Proteins that bind to each cyclin, in turn, translate these fluctuations into action by activating the transcription factors that stimulate entry into the next stage of the cell cycle. ▶ regulation of gene expression, p. 134

The interactions between these signaling proteins contribute to several "checkpoints" that ensure that a cell does not enter one stage until the previous stage is complete. That is, a cell that fails to "pass" a checkpoint correctly will not undergo the change in cyclin concentrations that allows it to progress to the next stage. These checkpoints are therefore somewhat like the guards that check passports at border crossings, denying entry to travelers without proper documentation.

Figure 8.12 illustrates a few cell cycle checkpoints:

- The G_1 checkpoint screens for DNA damage. If the DNA is damaged, a protein called p53 promotes the expression of genes encoding DNA repair enzymes. Badly damaged DNA prompts p53 to trigger apoptosis, and the cell dies.

- Several S phase checkpoints ensure that DNA replication occurs properly. If the cell does not have enough nucleotides to complete replication or if a DNA molecule breaks, the cell cycle may pause or stop at this point.

- The G_2 checkpoint is the last one before the cell begins mitosis. If the cell does not contain two full sets of identical DNA or if the spindle-making machinery is not in place, the cell cycle may be delayed. Alternatively, the p53 protein may trigger apoptosis.

- The metaphase checkpoint ensures that all chromosomes are aligned and that the spindle fibers attach correctly to the chromosomes. If everything checks out, the cell proceeds to anaphase.

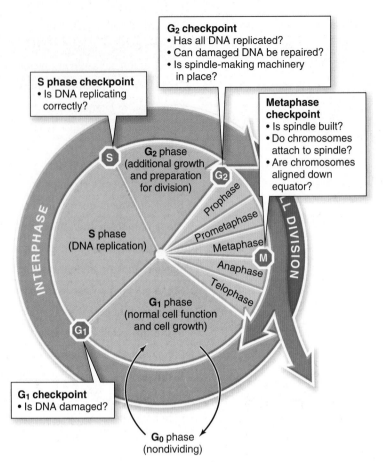

G$_2$ checkpoint
- Has all DNA replicated?
- Can damaged DNA be repaired?
- Is spindle-making machinery in place?

S phase checkpoint
- Is DNA replicating correctly?

Metaphase checkpoint
- Is spindle built?
- Do chromosomes attach to spindle?
- Are chromosomes aligned down equator?

G$_1$ checkpoint
- Is DNA damaged?

INTERPHASE

G$_2$ phase (additional growth and preparation for division)

S phase (DNA replication)

Prophase
Prometaphase
Metaphase
Anaphase
Telophase

CELL DIVISION

G$_1$ phase (normal cell function and cell growth)

G$_0$ phase (nondividing)

Figure 8.12 Cell Cycle Control Checkpoints. These checkpoints ensure that the cell completes each stage of the cell cycle correctly before proceeding to the next.

Precise timing of the chemical signals that regulate the cell cycle is essential. Too little cell division, and an injury may go unrepaired; too much, and an abnormal growth forms. An understanding of these signals may therefore help reveal how diseases such as cancer arise.

B. Cancer Cells Break Through Cell Cycle Controls

What happens when the body loses control over the balance between cell division and cell death? Sometimes, a **tumor**—an abnormal mass of tissue—forms. Biologists classify tumors into two groups (figure 8.13). **Benign tumors** are usually slow-growing and harmless, unless they become large enough to disrupt nearby tissues or organs. A tough capsule surrounding the tumor prevents it from invading nearby tissues or spreading to other parts of the body. Warts and moles are examples of benign tumors of the skin.

In contrast, a **malignant tumor** invades adjacent tissue. Because it lacks a surrounding capsule, a malignant tumor is likely to **metastasize,** meaning that its cells can break away from the

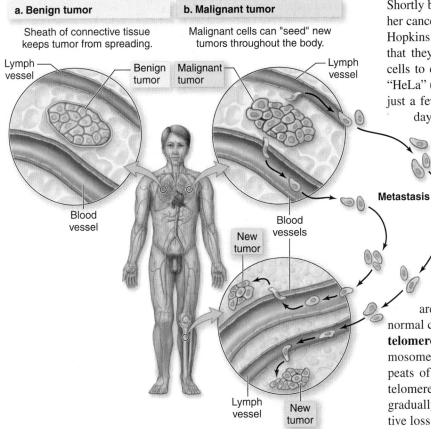

a. Benign tumor	b. Malignant tumor
Sheath of connective tissue keeps tumor from spreading.	Malignant cells can "seed" new tumors throughout the body.

Lymph vessel
Benign tumor
Malignant tumor
Lymph vessel
Blood vessel
Metastasis
New tumor
Blood vessels
Lymph vessel
New tumor

Figure 8.13 Benign and Malignant Tumors. (a) A capsule of connective tissue keeps a benign tumor from invading adjacent tissues. (b) A malignant tumor is not encapsulated and therefore can spread throughout the body in blood and lymph.

original mass and travel in the bloodstream or lymphatic system to colonize other areas of the body. **Cancer** is a class of diseases characterized by malignant cells.

Cancer begins when a single cell accumulates genetic mutations that cause it to break through its death and division controls. As the cell continues to divide, a tumor develops. All tumors grow slowly at first, because only a few cells are dividing. However, not all tumors continue to grow at the same rate. In one study, for example, researchers measured how long it took for tumors in lung cancer patients to double in size. For patients with the fastest-growing tumors, the doubling time was about 68 days; the slowest-growing masses took about 225 days to double. In general, the slower a tumor's growth rate, the better the patient's prognosis.

C. Cancer Cells Differ from Normal Cells in Many Ways

Given sufficient nutrients and space, cancer cells can divide uncontrollably and eternally. The cervical cancer cells of a woman named Henrietta Lacks vividly illustrate these characteristics.

Shortly before Lacks died in 1951, researchers removed some of her cancer cells and began to grow them in a laboratory at Johns Hopkins University. Lacks's cells grew so well, dividing so often, that they quickly became a favorite of cell biologists seeking cells to culture that would divide indefinitely. Still used today, "HeLa" (for *H*enrietta *La*cks) cells replicate so vigorously that if just a few of them contaminate a culture of other cells, within days they completely take over.

In addition to uncontrolled division, cancer cells have other unique characteristics as well. First, a cancer cell looks different from a normal cell (**figure 8.14**). It is rounder, its cell membrane is more fluid, and it may lose some of the specialized features of its parent cells. Some cancer cells have multiple nuclei. These visible differences allow pathologists to detect cancerous cells by examining tissue under a microscope.

Second, unlike normal cells, many cancer cells are essentially immortal, ignoring the "clock" that limits normal cells to 50 or so divisions. This cellular clock resides in **telomeres,** the noncoding DNA at the tips of eukaryotic chromosomes. Telomeres consist of hundreds to thousands of repeats of a specific DNA sequence. At each cell division, the telomeres lose nucleotides from their ends, so the chromosomes gradually become shorter. After about 50 divisions, the cumulative loss of telomere DNA signals division to cease in a normal cell. Cells that produce an enzyme called **telomerase,** however, can continually add DNA to chromosome tips. Their telomeres stay long, which enables them to divide beyond the 50-or-so division limit. Cancer cells have high levels of telomerase; inactivating this enzyme could therefore have tremendous medical benefits.

A third difference between normal cells and cancer cells lies in **growth factors,** proteins that stimulate cell division. These

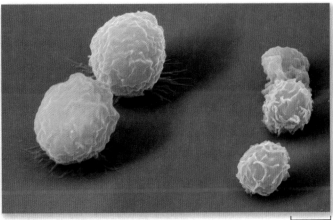

SEM (false color) |— 6 μm

Figure 8.14 Cancer Cells Are Abnormal. The two cancerous leukemia cells on the left are larger than the normal marrow cells on the right.

proteins bind to receptors on a receiving cell's membrane, and then a cascade of chemical reactions inside the cell initiates division. At a wound site, for example, a protein called epidermal growth factor stimulates cells to divide and produce new skin underneath a scab. Normal cells stop dividing once external growth factors are depleted. Many cancer cells, however, divide even in the absence of growth factors.

Fourth, normal cells growing in culture exhibit **contact inhibition,** meaning that they stop dividing when they touch one another in a one-cell-thick layer. Cancer cells lack contact inhibition, so they tend to pile up in culture. In addition, normal cells divide only when attached to a solid surface, a property called anchorage dependence. The observation that cancer cells lack anchorage dependence helps explain how metastasis occurs.

Cancer cells have other unique features as well. For example, a normal cell dies (undergoes apoptosis) when badly damaged, but many cancer cells do not. In addition, cancer cells send signals that stimulate the development of new blood vessels. In this way, a tumor builds its own blood supply that delivers nutrients and removes wastes. Disrupting these signals may lead to new cancer treatments (see section 8.9).

D. Inheritance and Environment Both Can Cause Cancer

Proteins control both the cell cycle and apoptosis. Genes encode proteins, so genetic mutations (changes in genes) play a key role in causing cancer. So far, researchers know of hundreds of genes that contribute to cancer. Two classes of cancer-related genes, oncogenes and tumor suppressor genes, appear to be in a perpetual "tug of war" in determining whether cancer develops (**figure 8.15**).

Oncogenes are mutated variants of genes that normally stimulate cell division (*onkos* is the Greek word for "mass" or "lump"). The normal versions of these genes, called proto-oncogenes, encode many types of proteins, from the receptors that bind growth factors outside the cell to any of the participants in the series of reactions that trigger cell division. If the protein is abnormally active or expressed at too high a concentration, the cell cycle will be accelerated, and cancer may develop. Oncogenes cause some cancers of the cervix, bladder, and liver.

Recall from section 8.3 that our cells contain 23 pairs of chromosomes, with one member of each pair coming from each parent. Oncogenes are especially dangerous because only one of the two versions in a cell needs to be damaged for cancer to develop. The oncogene's abnormal protein is an "accelerator" that overrides the normal protein encoded by the proto-oncogene.

Tumor suppressor genes encode proteins that normally block cancer development; that is, they promote apoptosis or prevent cell division. Inactivating, deleting, or mutating these genes therefore eliminates crucial limits on cell division. One example of a tumor suppressor gene is *p53,* which encodes a protein that participates in the cell cycle control checkpoints described earlier. Mutations in *p53* apparently cause about half of all human cancers. *BRCA1,* a gene associated with some types of breast cancer, is another example of a tumor suppressor gene. Unlike oncogenes, usually both of a cell's versions of a tumor

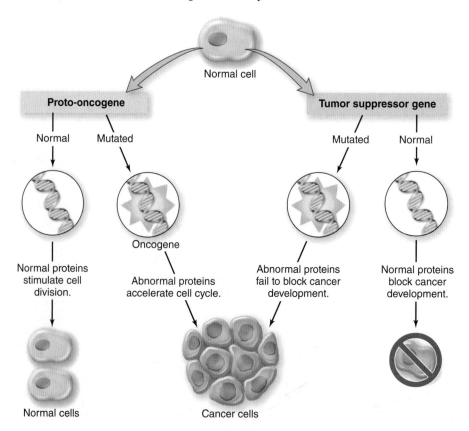

Figure 8.15 **Cancer-Related Genes.**
Oncogenes and tumor suppressor genes both influence the cell cycle. When proto-oncogenes are mutated, they form oncogenes that accelerate cell division. Tumor suppressor genes normally inhibit cell division, but when these genes are mutated, cancer can develop.

suppressor gene must be damaged for cancer to develop. That is, as long as one tumor-suppressor gene is functioning, the cell continues to produce the protective proteins.

The more oncogenes or mutated tumor suppressor genes in a person's cells, the higher the probability of cancer. Where do these mutations come from? Sometimes, a person inherits mutated DNA from one or both parents. The mutations may run in families, or they may have arisen spontaneously in a parent's sperm- or egg-producing cells. Often, however, people acquire the cancer-causing mutations throughout their lifetimes.

Figure 8.16 depicts some choices a person can make to reduce cancer risk. Some of these strategies are straightforward. For example, UV radiation and many chemicals in tobacco are mutagens, which means they damage DNA. Reducing sun exposure and avoiding tobacco therefore directly reduce cancer risk. Likewise, condoms can help prevent infection with cancer-causing viruses that are sexually transmitted. ▶ mutagens, p. 139

Other risk factors illustrated in figure 8.16 are less obvious. Obesity, for example, greatly increases the risk of death from cancers of the breast, cervix, uterus, and ovaries in women; obese men have an elevated risk of dying from prostate cancer. High-calorie foods that are rich in animal fats and low in fiber, coupled with a lack of exercise, contribute to high body weight. But scientists remain uncertain why obesity itself is a risk factor for cancer. Perhaps fat tissue secretes hormones that contribute to metastasis, or maybe obesity reduces immune system function. Research into the cancer–obesity connection is increasingly important as obesity rates continue to climb.

One thing is clear: an enormous variety of illnesses are grouped under the category of "cancer," and each is associated with a unique suite of risk factors. It therefore pays to be skeptical of claims that any one product can miraculously fight cancer. A healthy lifestyle remains the best way to reduce cancer risk.

E. Cancer Treatments Remove or Kill Abnormal Cells

Medical professionals describe the spread of cancer cells as a series of stages. In one system used to classify colon cancer, for example, a stage I cancer has started invading tissue layers adjacent to the tumor's origin, but cancerous cells remain confined to the colon. At stage II, the tumor has spread to tissues around the colon but has not yet reached nearby lymph nodes. Stage III cancers have spread to organs and lymph nodes near the cancer's origin, and stage IV cancers have spread to distant sites. The names and criteria for each stage vary among cancers. In general, however, the lower the stage, the better the prospect for successful treatment.

Physicians use many techniques to estimate the stage of a patient's cancer. For example, X-rays, CAT scans, MRIs, PET scans, ultrasound, and other imaging tests are noninvasive ways to detect and measure tumors inside the body. A physician can also use an endoscope to inspect the inside of some organs, such as the esophagus or intestines. The same tool can also collect a biopsy sample; pathologists then use microscopes to search the tissue for suspicious cells. Blood tests can reveal more clues, including an abnormal number of white blood cells or a high level of a "tumor marker" such as prostate-specific antigen (PSA). Combining many such lines of evidence helps medical professionals diagnose cancer and determine the stage, which in turn helps guide treatment decisions.

Traditional cancer treatments include surgical tumor removal, drugs (chemotherapy), and radiation. Chemotherapy drugs, usually delivered intravenously, are intended to stop cancer cells anywhere in the body from dividing. Radiation therapy uses directed streams of energy from radioactive isotopes to kill tumor cells in limited areas. ▶ isotopes, p. 21

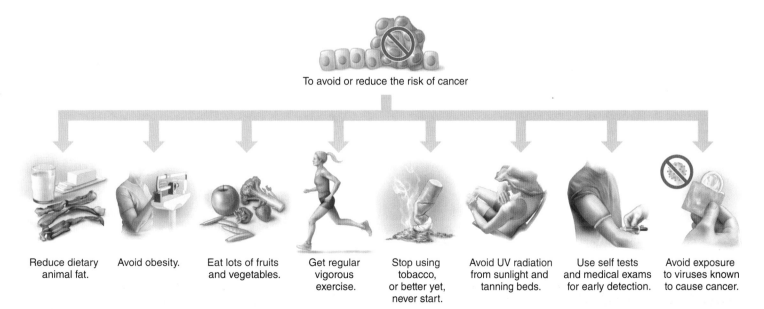

To avoid or reduce the risk of cancer

| Reduce dietary animal fat. | Avoid obesity. | Eat lots of fruits and vegetables. | Get regular vigorous exercise. | Stop using tobacco, or better yet, never start. | Avoid UV radiation from sunlight and tanning beds. | Use self tests and medical exams for early detection. | Avoid exposure to viruses known to cause cancer. |

Figure 8.16 Cancer Risk. Many aspects of a person's lifestyle influence the risk of cancer.

Burning Question

What causes skin cancer?

Cancer has many forms, some inherited and others caused by radiation or harmful chemicals. Exposure to ultraviolet radiation from the sun or from tanning beds, for example, increases the risk of skin cancer because UV radiation damages DNA. If genetic mutations occur in genes encoding proteins that control the pace of cell division, cells may begin dividing out of control, forming a malignant tumor on the skin.

How might a person determine whether a mole, sore, or growth on the skin is cancerous? The abnormal skin may vary widely in appearance, and only a physician can tell for sure. Nevertheless, most cancers have a few features in common. "ABCD" is a shortcut for remembering these four characteristics:

- **Asymmetry:** Each half of the area looks different from the other.
- **Borders:** The borders of the growth are irregular, not smooth.
- **Color:** The color varies within a patch of skin, from tan to dark brown to black. Other colors, including red, may also appear.
- **Diameter:** The diameter of a cancerous area is usually greater than 6 mm, which is about equal to the size of a pencil eraser.

Skin cancer is the most common form of cancer in the United States. Several types of skin cancer exist, including basal cell carcinoma, squamous cell carcinoma, and melanoma. Basal cell carcinoma is the most common, but melanoma causes the most deaths because the cancerous cells quickly spread to other parts of the body.

The highest risk for skin cancer occurs among people who have light-colored skin and eyes, and who spend a lot of time in the sun. Avoiding exposure to UV radiation, both in the sun and in tanning beds, can help minimize this risk. Sunscreen is a must when outdoors. In addition, medical professionals recommend that people pay attention to changes in their skin. Carcinomas and melanomas are treatable if detected early.

Submit your burning question to:
marielle_hoefnagels@mcgraw-hill.com

Chemotherapy and radiation are relatively "blunt tools" that target rapidly dividing cells, whether cancerous or not. Examples of cells that divide frequently include those in the bone marrow, digestive tract, and hair follicles. The death of these cells account for the most notorious side effects of cancer treatment: fatigue, nausea, and hair loss. Fortunately, the healthy cells usually return after the treatment ends. Some patients, especially those who receive high doses of chemotherapy or radiation, also have bone marrow transplants to speed the replacement of healthy blood cells.

Basic research into the cell cycle has yielded new cancer treatments with fewer side effects. For example, drugs that target a cancer cell's unique molecules have been very successful in treating some forms of breast cancer and leukemia (see the opening essay for chapter 3). Drugs called angiogenesis inhibitors block a tumor's ability to recruit blood vessels, starving the cancer cells of their support system. In the future, cancer patients may receive gene therapy treatments that replace faulty genes with functional copies. ▶ gene therapy, p. 144

The success of any cancer treatment depends on many factors, including the type of cancer and the stage in which it is detected. Surgery can cure cancers that have not spread or that have only invaded local lymph nodes. Once cancer metastasizes, however, it becomes more difficult to treat. As cancer cells spread, their DNA often mutates. Treatment that shrank the original tumor may have no effect on this new, changed growth. Also, a treatment that kills 99.9% of a tumor's cells can still leave millions of cells to divide and regrow (see section 8.9).

8.5 | Mastering Concepts

1. What prevents normal cells from dividing when they are not supposed to?
2. What happens at cell cycle checkpoints?
3. What is the difference between a benign and a malignant tumor?
4. How do cancer cells differ from normal cells?
5. What is the relationship between mutations and cancer?
6. How does a person acquire the mutations associated with cancer?
7. Distinguish among the treatments for cancer.

8.6 | Apoptosis Is Programmed Cell Death

Development relies on a balance between cell division and programmed cell death, or apoptosis (see figure 8.3). Apoptosis is different from necrosis, which is the "accidental" cell death that follows a cut or bruise. Whereas necrosis is sudden, traumatic, and disorderly, apoptosis results from a precisely coordinated series of events that dismantle a cell.

The process begins when a "death receptor" protein on a doomed cell's membrane receives a signal to die (**figure 8.17**). Within seconds, apoptosis-specific "executioner" proteins begin to cut apart the cell's proteins and destroy the cell. Immune system cells descend, and the cell is soon gone.

Apoptosis has two main functions in animals. First, apoptosis eliminates excess cells, carving out functional structures such as fingers, toes, nostrils, and ears as an animal grows. The second function of apoptosis is to weed out aging or defective cells that otherwise might harm the organism. A good example is the peeling skin that follows a sunburn. Sunlight contains UV radiation that can cause cancer by damaging the DNA in skin cells. Apoptosis helps protect against skin cancer by eliminating severely damaged cells, which die and simply peel away.

Plant cells die, too, but not in precisely the way that animal cells meet their programmed fate. Instead, plant cells are digested by enzymes in their own vacuoles; when the vacuole bursts, the cell dies. Plants also use a form of cell death to kill cells infected by fungi or bacteria, limiting the spread of the pathogen.

8.6 | Mastering Concepts

1. What events happen in a cell undergoing apoptosis?
2. Describe two functions of apoptosis.

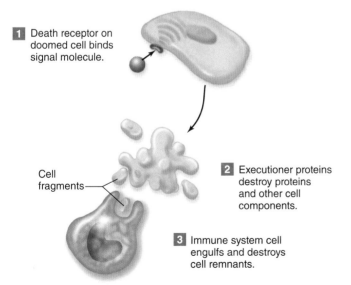

1 Death receptor on doomed cell binds signal molecule.

Cell fragments

2 Executioner proteins destroy proteins and other cell components.

3 Immune system cell engulfs and destroys cell remnants.

Figure 8.17 Death of a Cell. [1] Soon after a death receptor receives the bad news, [2] enzymes trigger apoptosis and destroy the cell. [3] Immune system cells mop up the debris.

8.7 | Stem Cells and Cloning Present Ethical Dilemmas

The public debates over stem cells and cloning combine science, philosophy, religion, and politics in ways that few other modern issues do (**figure 8.18**). What is the biology behind the headlines?

A. Stem Cells Divide to Form Multiple Cell Types

A human develops from a single fertilized egg into an embryo and then a fetus—and eventually into an infant, child, and adult—thanks to mitotic cell division. As development continues, more and more cells become permanently specialized into muscle, skin, liver, brain, and other cell types. All contain the same DNA, but some genes become irreversibly "turned off" in specialized cells. Once committed to a fate, a mature cell rarely reverts to another type. ▶ regulation of gene expression, p. 134

Animal development therefore relies on stem cells. In general, a **stem cell** is any undifferentiated cell that can give rise to specialized cell types. When a stem cell divides mitotically to yield two daughter cells, one remains a stem cell, able to divide again. The other specializes.

a.

b.

Figure 8.18 Stem Cell Controversy. Debates over stem cells often pit (a) people who advocate the use of embryonic stem cells in medicine against (b) people with moral objections.

Animals have two general categories of stem cells: embryonic and adult (**figure 8.19**). **Embryonic stem cells** give rise to all cell types in the body (including adult stem cells) and are therefore called "totipotent"; *toti-* comes from the Latin word for "entire." **Adult stem cells** are more differentiated and produce a limited subset of cell types. For example, stem cells in the skin replace cells lost through wear and tear, and stem cells in the bone marrow produce all of the cell types that make up blood. Adult stem cells are "pluripotent"; *pluri-* means "many" in Latin.

Stem cells are important in biological and medical research. The study of these cells lends insight into how animals develop and grow. In addition, with the correct combination of chemical signals, medical researchers should theoretically be able to coax stem cells to divide in the laboratory and produce blood cells, neurons, or any other cell type. Many people believe that stem cells hold special promise as treatments for neurological disorders such as Parkinson disease and spinal cord injuries, since neurons ordinarily do not divide to replace injured or diseased tissue. Stem cell therapies may also conquer diabetes, heart disease, and many other illnesses that are currently incurable.

Moreover, biologists may be able to observe specialized cells to learn how a disease such as diabetes develops from its start, rather than seeing just the end stages of the illness in a patient. Their observations may lead to new drugs or other treatments that target the early stages of the disease.

Stem cells may improve drug testing as well. Pharmaceutical companies currently evaluate drug safety and efficacy primarily in whole organisms, such as mice and rats. The ability to test on just kidney or brain cells, for example, would allow researchers to predict with much more precision the likely side effects of a new drug. It might also reduce the need for laboratory animals.

Both embryonic and adult stem cells have advantages and disadvantages for medical use. Embryonic stem cells are extremely versatile, but a patient's immune system would probably reject tissues derived from another individual's cells. In addition, research on embryonic stem cells is controversial because of their origin. In fertility clinics, technicians fertilize eggs *in vitro*, and only a few of the resulting embryos are ever implanted into a woman's uterus. Researchers destroy some of the "spare" embryos at 4 to 5 days old to harvest the stem cells. (The other embryos are either stored for possible later implantation or discarded.) Many people consider it unethical to use human embryos in medical research, even if those embryos would otherwise have been thrown away.

Biologists are therefore investigating adult stem cells in skin, bone marrow, the lining of the small intestine, and other locations in the body. A patient's immune system would not reject tissues derived from his or her own adult stem cells. These stem cells are less abundant, however, and they usually give rise to only some cell types.

New laboratory techniques may eliminate some of these drawbacks. Researchers have discovered how to manipulate gene expression in a way that induces adult stem cells to behave like embryonic stem cells. This technique could allow differentiated cells taken from an adult to be turned into stem cells, which could then be coaxed to develop into any other cell type. Time will tell how useful these so-called "induced pluripotent stem cells" will be or whether they will match the medical potential of embryonic stem cells.

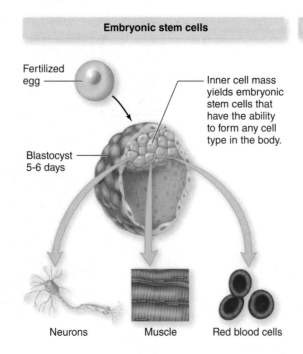

Figure 8.19 Stem Cells.
(a) Human embryonic stem cells are derived from the inner cell mass of a blastocyst, a stage in development that occurs several days after fertilization. Shortly thereafter, cells begin to specialize. (b) The adult body also contains stem cells, but they may not have the potential to develop into as many different cell types as do embryonic stem cells.

Embryonic stem cells

Adult stem cells

Fertilized egg

Inner cell mass yields embryonic stem cells that have the ability to form any cell type in the body.

Blastocyst 5-6 days

Neurons   Muscle   Red blood cells

a.

Bone marrow

Blood–forming stem cell (produces blood and immune system cells)

Stromal stem cell (produces bone and fat cells)

b.

B. Cloning Creates Identical Copies of an Organism

Cloning is another current topic related to mitotic cell division. Unlike many other organisms, mammals do not naturally clone themselves. In 1996, however, researcher Ian Wilmut and his colleagues in Scotland used a new procedure to create Dolly the sheep, the first clone of an adult mammal.

The researchers used a technique called **somatic cell nuclear transfer** to create Dolly (figure 8.20). First, they removed the nucleus of a cell taken from a donor sheep's mammary gland. (The name of the cloning technique derives in part from the fact that mammary glands consist of *somatic* cells, which are body cells that do not give rise to sperm or eggs.) They then transferred this "donor" nucleus to a sheep's egg cell whose own nucleus had been removed. The resulting cell divided mitotically to form an embryo, which the researchers implanted in a surrogate mother's uterus. The embryo then developed into a lamb, named Dolly (after country singer Dolly Parton).

Scientifically, this achievement was remarkable because it showed that the DNA from a differentiated somatic cell (in this case, from a mammary gland) could "turn back the clock" and revert to an undifferentiated state. Mitotic cell division then produced every cell in Dolly's body.

Dolly appeared normal, and she gave birth to six healthy lambs (via sexual reproduction). But she had arthritis in her hind legs, and she died of a lung infection in 2003 at age 6. Normally, sheep of Dolly's breed live 11 or 12 years, and some people have speculated that Dolly's relatively early death may have been related to the already-shortened telomeres she "inherited" from the 6-year-old sheep from which she was cloned (see section 8.5C). Nevertheless, there is no evidence that Dolly's short lifespan was related to her being a clone.

Since Dolly's birth, scientists have used somatic cell nuclear transfer to clone other mammals as well, including cats, dogs, mice, bulls, and a champion horse that had been castrated (and could therefore not reproduce). Cloning may even help rescue endangered species or recover extinct species. For example, an extinct mountain goat was cloned from preserved DNA, but the animal died shortly after birth.

Many people wonder whether humans can and should be cloned. Reproductive cloning, as achieved with Dolly, could help infertile couples have children. Scientists could also use cloned human embryos as a source of stem cells, which could be used to grow "customized" artificial organs that the patient's immune system would not reject. This application of cloning is called therapeutic cloning.

Despite the potential benefits, however, human cloning carries unresolved ethical questions. For example, most clones die early in development, presumably because the gene regulation mechanisms in an adult cell's nucleus are somehow incompatible with those in the egg cell. Even the tiny percentage of clones that make it to birth often have abnormalities. This difficulty emphasizes the ethical issues surrounding human reproductive cloning. In addition, therapeutic cloning still requires the destruction of an embryo to harvest the stem cells. As we have already seen, many people question the practice of creating human embryos only to destroy them. Finally, both reproductive and therapeutic cloning require unfertilized human eggs. The removal of eggs from a woman's ovaries is costly and poses medical risks.

8.7 | Mastering Concepts

1. Describe the differences between embryonic, adult, and induced pluripotent stem cells.
2. What are the potential medical benefits of stem cells?
3. Why is the cloning technique called somatic cell nuclear transfer?
4. Summarize the steps scientists use to clone an adult mammal.

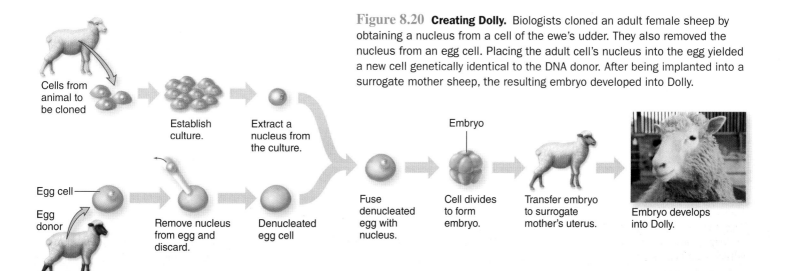

Figure 8.20 Creating Dolly. Biologists cloned an adult female sheep by obtaining a nucleus from a cell of the ewe's udder. They also removed the nucleus from an egg cell. Placing the adult cell's nucleus into the egg yielded a new cell genetically identical to the DNA donor. After being implanted into a surrogate mother sheep, the resulting embryo developed into Dolly.

Cells from animal to be cloned

Establish culture.

Extract a nucleus from the culture.

Egg cell

Egg donor

Remove nucleus from egg and discard.

Denucleated egg cell

Fuse denucleated egg with nucleus.

Embryo

Cell divides to form embryo.

Transfer embryo to surrogate mother's uterus.

Embryo develops into Dolly.

8.8 | Several Technologies Use DNA Replication Enzymes

Chapter 7 describes how the scientific understanding of the relationship between DNA and protein led to the ability to create transgenic organisms. Likewise, knowledge of DNA replication has led to its own share of useful technologies.

A. DNA Sequencing Reveals the Order of Bases

The uses of DNA sequences, from individual genes to entire genomes, seem endless. Researchers can apply sequence information to everything from identifying viruses, to analyzing the DNA in a malignant tumor, to predicting protein sequences, to determining evolutionary relationships. How do investigators get the DNA sequence information they need? ▶ Human Genome Project, p. 141

Modern DNA sequencing instruments use a highly automated version of a basic technique Frederick Sanger developed in 1977 (figure 8.21). Sanger's chain terminator technique uses the DNA polymerase enzyme to generate a series of DNA fragments that are complementary to the DNA being sequenced. Included in the reaction mixes are low concentrations of specially modified nucleotides. Each time DNA polymerase incorporates one of these modified nucleotides, the new DNA chain stops growing. The result is a group of fragments that differ in length from one another by one end base. Once a collection of such pieces is generated, a technique called electrophoresis can be used to separate the fragments by size. Reading the end bases in order by size reveals the sequence of the complement, and deriving the original DNA sequence is then easy.

Sanger used gel electrophoresis and radioactive labels to sort and visualize the fragments. Researchers today use a slightly different technique to sort the fragments by size, and fluorescent labels mark each of the four base types. The data appear as a sequential readout of the wavelengths of the fluorescence from the labels (figure 8.22).

DNA microarrays offer a way to sequence DNA on a smaller scale. A **DNA microarray** (also called a DNA chip) is a small glass square to which are attached short DNA fragments of known sequence (figure 8.23). In one version, the 4096 possible six-base combinations of DNA are immobilized on a microchip measuring about 1 square centimeter. Copies of an unknown DNA segment incorporating a fluorescent label are then also placed on the microchip. The copies stick to all six-base strands on the chip whose sequences are complementary to any part of the unknown DNA segment's sequence. Under laser light, the bound sequences fluoresce. Because the researcher (or computer program) knows which strands occupy which positions on the microchip, a scan of the chip reveals which six-base sequences are bound to the labeled DNA. Then, software aligns the

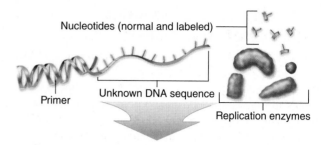

1 Four solutions contain unknown DNA sequence, primers, normal nucleotides (**A**,**C**,**G**, and **T**), labeled nucleotides, replication enzymes, and a small amount of "terminator" nucleotide.

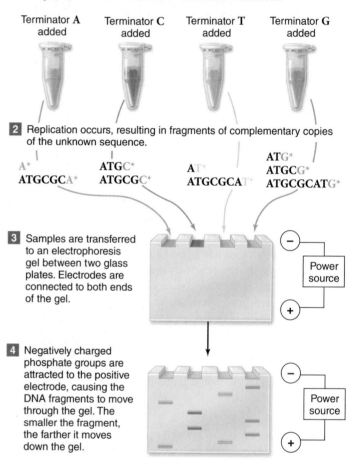

Terminator **A** added   Terminator **C** added   Terminator **T** added   Terminator **G** added

2 Replication occurs, resulting in fragments of complementary copies of the unknown sequence.

A*   ATGC*   A T*   ATG*
ATGCGCA*   ATGCGC*   ATGCGCA T   ATGCG*
                                                            ATGCGCATG*

3 Samples are transferred to an electrophoresis gel between two glass plates. Electrodes are connected to both ends of the gel.

Power source

4 Negatively charged phosphate groups are attracted to the positive electrode, causing the DNA fragments to move through the gel. The smaller the fragment, the farther it moves down the gel.

Power source

5 The fragments are read off by size, and the original sequence can be deduced.

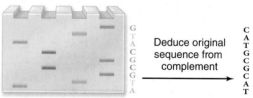

Deduce original sequence from complement

G T A C G C G T A → C A T G C G C A T

Figure 8.21 **Determining the Sequence of DNA.** In the Sanger method of DNA sequencing, complementary copies of an unknown DNA sequence are terminated early because of the addition of chemically modified "terminator" nucleotides. Placing the fragments in order by size reveals the sequence.

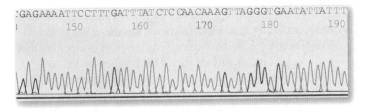

Figure 8.22 New Way to Read DNA. A computerized readout of a DNA sequence, made possible by fluorescent labels.

identified fragments by their overlaps. This reconstructs the complement of the entire unknown sequence.

Microarrays are increasingly common in research because they offer a fast, inexpensive way to generate sequence data. One day, they may also appear in doctors' offices. If medical professionals can quickly determine a patient's DNA sequence at key locations, they may be able to pinpoint the cause of a patient's cancer or customize drug treatments to minimize side effects.

B. PCR Replicates DNA in a Test Tube

The **polymerase chain reaction (PCR)** taps into a cell's DNA copying machinery to rapidly produce millions of copies of a DNA sequence of interest (**figure 8.24**). But instead of occurring inside a living cell, PCR replicates the selected sequence of DNA in a test tube. The requirements include the following:

- A target DNA sequence to be replicated.
- *Taq* polymerase, a heat-stable DNA polymerase produced by *Thermus aquaticus,* a bacterium that inhabits hot springs. (Other heat-tolerant polymerases can also be used.)
- Two types of laboratory-made DNA primers, each around 20 bases long, that are complementary to sequences known to occur at each end of the target sequence. The primers are necessary because DNA polymerase can only attach nucleotides to an existing strand.
- A supply of the four types of DNA nucleotides.

The DNA replication reactions occur in an automated device called a thermal cycler that controls key temperature changes. In the first step of PCR, heat separates the two strands of the target DNA. Next, the temperature is lowered, and the short primers attach to the separated target strands by complementary base pairing. *Taq* DNA polymerase adds nucleotides to the primers and builds sequences complementary to the target sequence. The newly synthesized strands then act as templates in the next round of replication, which is initiated immediately by raising the temperature to separate the strands once more. The number of pieces of DNA doubles with every round of PCR, so that after just 20 cycles, a million-fold increase occurs in the number of copies of the target sequence.

PCR is an extremely powerful, versatile, and useful tool that has a wide variety of applications. Lab workers routinely use it whenever they need to mass-produce a particular sequence of

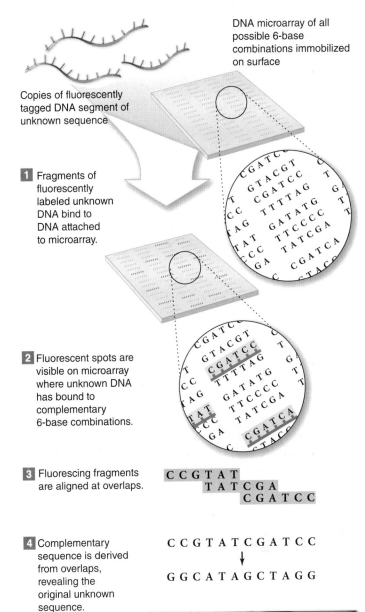

DNA microarray of all possible 6-base combinations immobilized on surface

Copies of fluorescently tagged DNA segment of unknown sequence

1 Fragments of fluorescently labeled unknown DNA bind to DNA attached to microarray.

2 Fluorescent spots are visible on microarray where unknown DNA has bound to complementary 6-base combinations.

3 Fluorescing fragments are aligned at overlaps.

```
C C G T A T
    T A T C G A
        C G A T C C
```

4 Complementary sequence is derived from overlaps, revealing the original unknown sequence.

```
C C G T A T C G A T C C
          ↓
G G C A T A G C T A G G
```

Figure 8.23 Sequencing on a Chip. A labeled DNA segment of unknown sequence binds to short, known DNA sequences immobilized on a small glass microchip. Identifying areas of overlap among the bound sequences reveals the unknown DNA sequence.

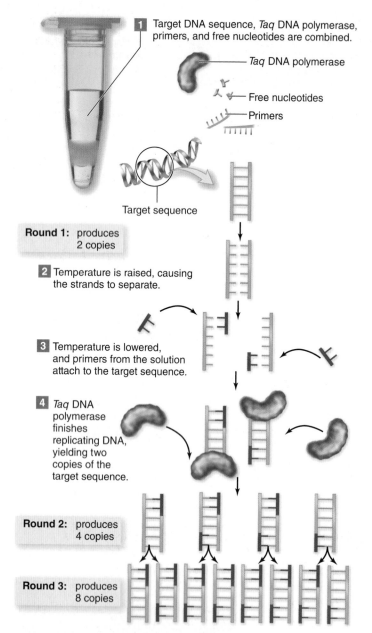

1 Target DNA sequence, *Taq* DNA polymerase, primers, and free nucleotides are combined.

Taq DNA polymerase

Free nucleotides

Primers

Target sequence

Round 1:   produces 2 copies

2 Temperature is raised, causing the strands to separate.

3 Temperature is lowered, and primers from the solution attach to the target sequence.

4 *Taq* DNA polymerase finishes replicating DNA, yielding two copies of the target sequence.

Round 2:   produces 4 copies

Round 3:   produces 8 copies

Figure 8.24 Polymerase Chain Reaction. In PCR, primers bracket a DNA sequence of interest. A heat-stable *Taq* DNA polymerase uses these primers, and plenty of nucleotides, to build up millions of copies of the target sequence.

DNA for analysis. PCR's greatest strength is that it works on tiny samples. Thanks to PCR, trace amounts of DNA extracted from a single hair or a few skin cells left at a crime scene can yield enough genetic material for DNA profiling. In forensics, PCR is used to amplify the DNA needed to establish genetic relationships, identify remains, convict criminals, and exonerate the falsely accused.

In agriculture, veterinary medicine, environmental science, and human health care, PCR can be used to amplify the nucleic acids of microorganisms, viruses, and other parasites. Medical workers can use PCR to identify known disease-causing genes in a cell's genome. Evolutionary biologists use PCR to amplify DNA from long-dead plants and animals. The list goes on and on.

PCR's greatest weakness, ironically, is its extreme sensitivity. A blood sample contaminated by leftover DNA from a previous run or by a stray eyelash dropped from the person running the reaction can yield a false result.

C. DNA Profiling Has Many Applications

On average, each person's DNA sequence differs from that of a nonrelative by just one nucleotide out of 1000. Finding these small differences by sequencing and comparing entire genomes would be time-consuming, tedious, costly, and impractical. Instead, **DNA profiling** uses just the most variable parts of the genome to detect genetic differences between individuals.

One approach to DNA profiling examines **short tandem repeats (STRs),** which are sequences of a few nucleotides that are repeated in noncoding regions of DNA. People within a population have different numbers of these repeats (**figure 8.25**). That is, for a given STR site, one individual might have five instances of the repeated nucleotides, whereas another person might have four or seven. To generate a DNA profile, a technician extracts DNA from a person's cells and uses PCR to amplify the DNA at each of 13 STR sites. The technician can then use electrophoresis to determine the number of repeats at each site.

As an example of how a DNA profile might aid a criminal investigation, suppose that a hair is found on a murder victim's body. Does the hair belong to the victim, the alleged murderer, or another person? To find out, technicians extract DNA from the hair, from cells of the victim, and from the cells of one or more suspects. Statistics suggest that the chance that any two unrelated individuals have the same pattern at all 13 STR markers is one in 250 trillion. A hair that matches the victim's DNA profile would be useless as evidence. A match between the hair's DNA and that of a suspect, however, would support a conviction. Conversely, a suspect can use dissimilar DNA profiles as evidence of his or her innocence. Since 1989, DNA analysis of stored evidence has proved the innocence of hundreds of people serving time in prison for violent crimes they did not commit.

In addition to DNA extracted from the nucleus, analysis of the DNA in mitochondria is also sometimes useful. Mitochondrial DNA typically contains only about 16,500 base pairs, so it is far shorter than nuclear DNA. But because each cell contains multiple mitochondria, each of which contains many DNA molecules, mitochondria can often yield useful information even when nuclear DNA is badly degraded. Investigators extract mitochondrial DNA from hair, bones, and teeth, then use PCR to amplify the variable regions for analysis. ▶ mitochondria, p. 61

Because everyone inherits mitochondria only from his or her mother, this technique cannot distinguish among siblings or many other close family members. It is very useful, however, for verifying the relationship between woman and child. For

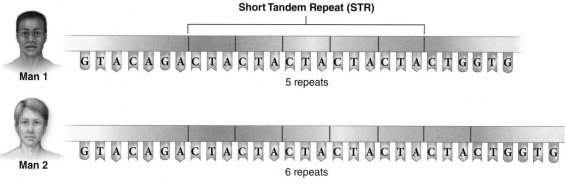

Short Tandem Repeat (STR)

Man 1
G T A C A G A C T A C T A C T A C T A C T A C T G G T G
5 repeats

Man 2
G T A C A G A C T A C T A C T A C T A C T A C T A C T G G T G
6 repeats

Man 3
G T A C A G A C T A C T A C T A C T A C T A C T A C T A C G T G G T G
7 repeats

Figure 8.25 Genetic Differences. Modern DNA profiling techniques detect differences in the number of short tandem repeats (STRs) in parts of the human genome that are known to be genetically variable.

example, children who were kidnapped during infancy can be matched to their biological mothers or grandmothers. Likewise, only males have a Y chromosome. STRs on the Y chromosome can therefore verify a father–son relationship.

DNA profiling is regularly featured in news stories about violent crimes and paternity claims. Another notable example was the effort to identify the remains of the thousands of people who died in the terrorist attacks of September 11, 2001. This chapter's Apply It Now box briefly describes this massive project.

8.8 | Mastering Concepts

1. How do researchers use the Sanger method and DNA microarrays to deduce a DNA sequence?
2. How do target DNA, primers, nucleotides, and *Taq* DNA polymerase interact in PCR?
3. Why is PCR useful?
4. What are STRs, and how are they used in DNA profiling?
5. Why does mitochondrial DNA provide different information from nuclear DNA?

Apply It Now

Identifying Victims of the Terrorist Attacks of September 11, 2001

Until September 11, 2001, the most challenging application of DNA profiling had been identifying plane-crash victims, a grim task eased by having lists of passengers. The terrorist attacks on the World Trade Center provided a staggeringly more complex situation, for several reasons: the high number of casualties, the condition of the remains, and the lack of a list of who was actually in the buildings.

In the days following September 11, somber researchers at Myriad Genetics, Inc., in Salt Lake City, Utah, who usually analyze DNA for breast cancer genes, received frozen DNA from soft tissue recovered from the disaster site. The laboratory also received cheek scrapings from relatives of the missing and tissue from the victims' toothbrushes, razors, and hairbrushes. The workers used PCR to determine the numbers of copies of STRs at 13 locations in the genome. If the STR pattern of a sample from the disaster scene matched DNA from a victim's toothbrush, identification was fairly certain.

DNA extracted from tooth and bone bits was sent to Celera Genomics Corporation in Rockville, Maryland. Here, DNA sequences were analyzed from mitochondria, which can survive incineration.

Overall, the disaster yielded more than a million DNA samples, which the labs used to identify about 850 of the more than 2700 people reported missing. It was a very distressing experience

for the technicians and researchers whose jobs had suddenly shifted from detecting breast cancer and sequencing genomes to helping identify victims of a terrorist attack.

8.9 | Investigating Life: Cutting off a Tumor's Supply Lines in the War on Cancer

When Charles Darwin proposed natural selection as a mechanism of evolutionary change, he envisioned selective forces operating on tortoises, flowering plants, and other whole organisms. But the power of natural selection extends to a much smaller scale, including the individual cells that make up a tumor. The advance, retreat, resurgence, and death of these renegade cells command dramatic headlines in the war on cancer.

Our weapons against cancer include powerful chemotherapy drugs, but drug-resistant tumor cells are a significant barrier to successful treatment. Rapidly dividing tumor cells develop resistance to drugs because frequent cell division produces abundant opportunities for mutations. An alternative cancer-fighting strategy, therefore, might be to launch an indirect attack on a tumor's slower growing support tissues instead.

Any tumor larger than 1 or 2 cubic millimeters needs a blood supply to carry nutrients, oxygen, and wastes. Blood travels throughout the body in vessels lined with endothelial tissue. For a blood vessel to grow, its endothelial cells must divide, which happens only rarely in adults. Cancer cells, however, secrete molecules that stimulate endothelial cells to divide and form new blood vessels. This sprouting of new "supply lines" is called angiogenesis.

Angiogenesis inhibitors are cancer-fighting drugs that stop blood vessel growth. One example is endostatin, a 184-amino acid protein that keeps endothelial cells from dividing without affecting other cells in the body. It should therefore choke off a tumor's supply lines without toxic side effects. The drug became the focus of intense media attention when cancer researchers Thomas Boehm, Judah Folkman, and their colleagues at the Dana Farber Cancer Center and Harvard Medical School reported that endostatin suppressed tumor development in mice—and that the cancer cells did not develop resistance to it.

To test endostatin, the researchers first induced cancer in 6-week-old male mice by injecting each animal with one of three types of cancer cells. After tumors developed, the researchers injected some of the mice with endostatin, while control mice received a placebo (injections of saline solution). Injections continued until tumors in endostatin-treated mice were barely detectable. When the tumors regrew, the researchers repeated the injections.

The results were astounding: the tumors never developed resistance (**figure 8.26**). With each dose of endostatin, the tumors shriveled. Standard chemotherapy drugs might temporarily shrink a tumor or delay its development by slowing cell division, but resistant cells soon caused the tumor to bounce back. But after two to six treatments with endostatin, the tumors never grew back, and the mice remained healthy.

The results of subsequent clinical trials with human cancer patients, however, were mixed. Endostatin shrank tumors in a handful of people without side effects. But the drug was ineffective in most patients, and its U.S. manufacturer eventually stopped making it, citing high production costs. (No one knows

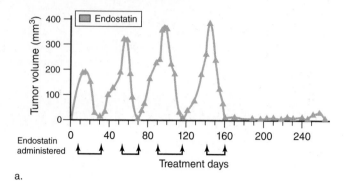

a.

b.

Figure 8.26 No Resistance. (a) Endostatin repeatedly shrank tumors in mice, and the tumors never developed resistance to the drug. (b) A traditional chemotherapy drug, cyclophosphamide, delayed but did not prevent tumor growth in mice.

why it worked so much better in mice than in people, but such disparities are common in cancer research.)

What does endostatin have to do with evolution? The logic behind its use as an anticancer drug relies on natural selection. Because DNA may mutate every time it replicates, rapidly dividing cancer cells are genetically different from one another. A conventional chemotherapy drug may kill most cancer cells in a tumor, but a few have mutations that let them survive. These cells divide; over time, the entire tumor is resistant to the drug. Unlike other drugs, however, endostatin does not target the tumor; instead, it affects a blood vessel's endothelial cells. These cells rarely divide and therefore accumulate mutations slowly, reducing the chance that they will become resistant to endostatin.

This may seem comforting, but evolution will not stand still for our convenience. New mutations may still enable tumor cells to inactivate or break down endostatin. Understanding natural selection helps researchers know what to look for—and perhaps even launch new offensives in the war on cancer.

Boehm, Thomas, Judah Folkman, Timothy Browder, and Michael S. O'Reilly. November 27, 1997. Antiangiogenic therapy of experimental cancer does not induce acquired drug resistance. *Nature*, vol. 390, pages 404–407.

8.9 | Mastering Concepts

1. Why doesn't endostatin select for drug-resistant cancer cells, as other chemotherapy drugs do?

2. Suppose you learn of a study in which ginger slowed tumor growth in mice for 30 days. What questions would you ask before deciding whether to recommend that a cancer-stricken relative eat more ginger?

Chapter Summary

8.1 | Cells Divide and Cells Die

A. Sexual Life Cycles Include Mitosis, Meiosis, and Fertilization

- In **sexual reproduction,** two parents produce genetically variable **gametes** by **meiosis. Fertilization** produces the first cell of the new offspring.
- **Mitotic** cell division produces identical eukaryotic cells used in growth, tissue repair, and **asexual reproduction.**

B. Cell Death Is Part of Life

- **Apoptosis** is programmed cell death that occurs during the normal development of an organism.

8.2 | DNA Replication Precedes Cell Division

- A dividing cell must first duplicate its **genome.**
- To replicate, DNA unwinds, and the hydrogen bonds between the two strands break. **DNA polymerase** adds DNA nucleotides to a short RNA primer. **Ligase** seals the sugar–phosphate backbone after the RNA primer is replaced with DNA.
- Replication proceeds only in a 5′ to 3′ direction, so the process is discontinuous in short stretches on one strand.
- Enzymes repair damaged DNA and correct errors made in replication, but **mutations** occasionally remain.

8.3 | Replicated Chromosomes Condense as a Cell Prepares to Divide

- A **chromosome** consists of **chromatin** (DNA plus protein). In eukaryotic cells, chromatin is organized into **nucleosomes,** which enable the cell to pack a lot of DNA into a small space.
- Once replicated, a chromosome consists of two identical sister **chromatids** attached at a section of DNA called a **centromere.**

8.4 | Mitotic Division Generates Exact Cell Copies

- The **cell cycle** is a sequence of events in which a cell is preparing to divide (interphase), dividing its genetic material (mitosis), or dividing its cytoplasm (cytokinesis).

A. Interphase Is a Time of Great Activity

- **Interphase** includes gap periods, G_1 and G_2, when the cell grows and produces molecules required for cell function and division. DNA replicates during the synthesis period **(S).** A cell that is not dividing is in G_0.

B. Chromosomes Divide During Mitosis

- The microtubules of the **mitotic spindle** move a cell's chromosomes during mitosis. In animal cells, spindle proteins arise from paired **centrosomes.**
- **Mitosis** consists of five overlapping stages. In **prophase,** the chromosomes condense, the nucleolus disassembles, and the spindle forms. In **prometaphase,** the nuclear envelope breaks up, and spindle fibers attach to **kinetochores.** In **metaphase,** spindle fibers align replicated chromosomes along the cell's equator. In **anaphase,** the chromatids of each replicated chromosome separate, sending a complete set of genetic instructions to each end of the cell. In **telophase,** the spindle breaks down, and nuclear envelopes form.

C. The Cytoplasm Splits in Cytokinesis

- **Cytokinesis** is the physical separation of the two daughter cells. When an animal cell undergoes cytokinesis, a **cleavage furrow** forms, and a contractile ring draws the two cells apart. In a plant cell, a **cell plate** between the daughter cells marks the site where a new cell wall will form.

8.5 | Cancer Arises When Cells Divide out of Control

A. Chemical Signals Regulate Cell Division

- Cell cycle control checkpoints ensure that each stage of the cell cycle is complete before the next begins. The cell may pause briefly to repair errors. If damage is too great to repair, the checkpoints may trigger apoptosis.

B. Cancer Cells Break Through Cell Cycle Controls

- **Tumors** can result from excess cell division or deficient apoptosis. A **benign** tumor does not spread, but a **malignant** tumor invades nearby tissues and **metastasizes** if it reaches the bloodstream or lymph.
- **Cancer** is a family of diseases characterized by malignant cells.

C. Cancer Cells Differ from Normal Cells in Many Ways

- A cancer cell divides uncontrollably and has a distinctive appearance compared to a healthy cell.
- When **telomeres** become very short, division ceases. An enzyme called **telomerase** adds DNA to telomeres in some cells. Cancer cells produce telomerase, so they retain long telomeres and divide continually.
- Normal cells divide only in response to the presence of external signals called **growth factors.** Cancer cells may continue to divide even after growth factors are depleted.
- A cancer cell lacks **contact inhibition** and anchorage dependence, may not undergo apoptosis, and secretes chemicals that stimulate the growth of blood vessels.

D. Inheritance and Environment Both Can Cause Cancer

- A mutated proto-oncogene is an **oncogene.** Oncogenes speed cell division, and mutated **tumor suppressor genes** fail to stop excess cell division.
- Mutations in cancer-related genes may be inherited or acquired during a person's lifetime. Several lifestyle factors influence cancer risk.

E. Cancer Treatments Remove or Kill Abnormal Cells

- Surgery, chemotherapy, and radiation are the most common cancer treatments. The chance of successful treatment depends on the type of cancer and the stage at which it is detected. Gene therapy may provide new treatment options in the future.

8.6 | Apoptosis Is Programmed Cell Death

- Apoptosis shapes structures and protects an organism by killing cells that could become cancerous.
- After a cell receives a signal to die, enzymes destroy the cell's components. Immune system cells dispose of the remains.

8.7 | BIOTECHNOLOGY Stem Cells and Cloning Present Ethical Dilemmas

A. Stem Cells Divide to Form Multiple Cell Types

- **Embryonic stem cells** give rise to all cells in the body; **adult stem cells** produce only a limited subset of cell types.
- Induced pluripotent stem cells may eliminate some of the ethical issues associated with embryonic stem cells.

B. Cloning Creates Identical Copies of an Organism

- Researchers use a technique called **somatic cell nuclear transfer** to clone adult mammals.
- Human reproductive and therapeutic cloning have potential medical applications, but they also involve ethical dilemmas.

8.8 | BIOTECHNOLOGY Several Techniques Use DNA Replication Enzymes

A. DNA Sequencing Reveals the Order of Bases

- The Sanger method uses modified nucleotides to generate DNA fragments of various lengths. Sorting the fragments by size reveals the DNA sequence.

- **DNA microarrays** contain all possible combinations of short nucleotide sequences. DNA of unknown sequence sticks to some of the chip-bound DNA, then computers reconstruct the original sequence.

B. PCR Replicates DNA in a Test Tube

- In the **polymerase chain reaction (PCR),** heat causes DNA to separate into two strands, then a primer provides DNA polymerase with a starting point for replication. Repeated cycles of heating and cooling allow for rapid amplification of the target DNA sequence.
- PCR finds many applications in research, forensics, medicine, agriculture, and other fields.

C. DNA Profiling Has Many Applications

- **DNA profiling** detects genetic differences among individuals who vary in **short tandem repeats (STRs).**
- Investigators can use known frequencies of variation in STR sites to calculate the probability that two DNA samples come from the same individual.
- Analysis of mitochondrial DNA and Y chromosomes can verify maternal and father–son relationships.

8.9 | Investigating Life: Cutting off a Tumor's Supply Lines in the War on Cancer

- Natural selection occurs inside tumors. As chemotherapy drugs eliminate susceptible cells, resistant ones survive and divide to regrow the tumor.
- Endostatin starves tumors by stopping the growth of blood vessels. Because endothelial cells in blood vessels divide much more slowly than tumor cells, they are much less likely to become resistant to endostatin.

Multiple Choice Questions

1. What would happen to DNA replication if the helicase enzyme did not function?
   a. Replication could occur, but there would be errors.
   b. The new DNA strands would not be held together by covalent bonds.
   c. Replication would occur in a single direction.
   d. Replication would not occur at all.

2. Why is a chromosome sometimes composed of *two* chromatids?
   a. Because the nucleosomes are folded
   b. Because the chromosome contains the entire genome of a cell
   c. Because the DNA has replicated
   d. Because the cell has two parents

3. How is G_0 different from G_1?
   a. In G_0 the cell is replicating its DNA.
   b. A G_1 cell is getting ready to divide, but a G_0 cell has already divided.
   c. G_0 occurs at the end of interphase.
   d. A G_1 cell can continue to divide, but a G_0 cell does not.

4. What would happen to an animal cell that repeatedly underwent interphase and mitosis but not cytokinesis?
   a. The number of nuclei in the cell would increase over time.
   b. The amount of DNA in the cell would decrease over time.
   c. The cell would enter G_0.
   d. The cell would not form a new cell wall.

5. Why is the metaphase checkpoint so important?
   a. Because it determines how quickly a cell goes through mitosis
   b. Because it ensures that the chromosomes will be properly separated
   c. Because it ensures that cytokinesis will proceed properly
   d. Because it can trigger apoptosis

6. Predict how excess telomerase activity would affect a cell.
   a. It would cause the telomeres of the chromosomes to rapidly shrink.
   b. It would reduce the number of chromosomes in the cell.
   c. It would increase the number of times the cell could divide.
   d. It would inhibit growth of the organism.

7. What is an *oncogene*?
   a. A gene that normally slows the cell cycle
   b. A gene that regulates cell death
   c. An abnormal version of a gene that can accelerate the cell cycle
   d. A gene that responds to the signals associated with contact inhibition

8. What is the role of enzymes in apoptosis?
   a. They kill the cell by destroying its proteins.
   b. They function as the "death receptor" on the surface of the cell.
   c. They are part of the immune response that eliminates the cells.
   d. They cause the cell to swell and burst.

9. Why is PCR useful?
   a. Because it replicates all the DNA in a cell
   b. Because it can create large amounts of DNA from small amounts
   c. Because it uses a heat-tolerant, *Taq* polymerase
   d. Because it occurs in an automated device

10. In 1920, a woman claiming to be a surviving member of the royal family of Tsar Nicholas II of Russia appeared in Europe. What modern method of DNA profiling would you use to verify this person's claims?
    a. Somatic cell nuclear transfer
    b. Analysis of the sequence of the woman's entire genome
    c. Analysis of mitochondrial DNA sequences
    d. Comparison of STR site frequencies to population databases

Write It Out

1. If a cell contains all the genetic material it needs to synthesize protein (see chapter 7), why must the DNA also replicate?

2. State the functions of each of the following participants in DNA replication: primer, DNA polymerase, ligase, and helicase.

3. A person with deficient DNA repair may have an increased cancer risk, and his or her chromosomes cannot heal breaks. The person is, nevertheless, alive. How long would an individual lacking DNA polymerase be likely to survive?

4. Explain the relationships among chromatin, chromosome, chromatid, and centromere.

5. Obtain a rubber band and twist it as many times as you can. What happens to the overall shape of the rubber band? How is this similar to what happens to chromosomes as a cell prepares to divide? How is it different?

6. Biologists once thought interphase was a time of cellular rest, but it is not. What happens during interphase?

7. Why is G_1 a crucial time in the life of a cell?

8. Does a cell in G_1 contain more, less, or the same amount of DNA as a cell in G_2? Explain your answer.

9. Describe what will happen to a cell if interphase happens, but mitosis does not.

10. If a drug prevents microtubules from forming, at which stage of mitosis would a cell become "stuck"?

11. Cytochalasin B is a drug that blocks cytokinesis by disrupting the actin and myosin microfilaments in the contractile ring. What effect would this drug have on cell division? Explain your answer.

12. How do biochemicals from outside the cell control the cell cycle? What signals inside the cell affect cell division?

9.2 | Diploid Cells Contain Two Homologous Sets of Chromosomes

Before exploring sexual reproduction further, a quick look at a cell's chromosomes is in order. Recall from chapters 7 and 8 that a **chromosome** is a single molecule of DNA and its associated proteins.

A sexually reproducing organism consists mostly of **diploid cells** (abbreviated 2*n*), which contain two full sets of chromosomes; one set is inherited from each parent. Each diploid human cell, for example, contains 46 chromosomes (**figure 9.3**). The photo in figure 9.3 illustrates a **karyotype,** a size-ordered chart of all the chromosomes in a cell. Notice that the 46 chromosomes are arranged in 23 pairs; your mother and your father each contributed one member of each pair.

Of the 23 chromosome pairs in a human cell, 22 pairs are **autosomes**—chromosomes that are the same for both sexes. The remaining pair is made up of the two **sex chromosomes,** which determine whether an individual is female or male. Females have two X chromosomes, whereas males have one X and one Y chromosome.

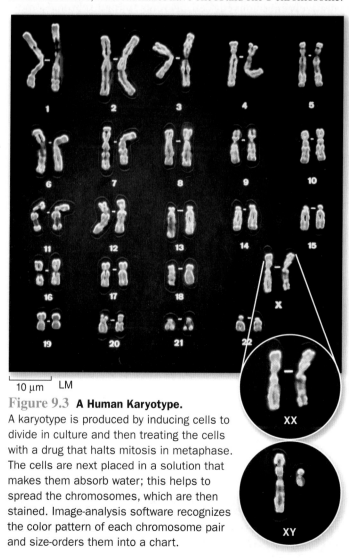

10 μm   LM

Figure 9.3 A Human Karyotype.
A karyotype is produced by inducing cells to divide in culture and then treating the cells with a drug that halts mitosis in metaphase. The cells are next placed in a solution that makes them absorb water; this helps to spread the chromosomes, which are then stained. Image-analysis software recognizes the color pattern of each chromosome pair and size-orders them into a chart.

The two members of most chromosome pairs are homologous to each other. A **homologous pair** of chromosomes is a matching pair of chromosomes that look alike and have the same sequence of genes. (The word *homologous* means "having the same basic structure.") The physical similarities between any two homologous chromosomes are evident in figure 9.3: they share the same size, centromere position, and pattern of light- and dark-staining bands. The karyotype does not, however, show that the two members of a homologous pair of chromosomes also carry the same sequence of genes. Chromosome 21, for example, includes 367 genes, always in the same order.

Homologous chromosomes, however, are not identical—after all, nobody has two identical parents! Instead, the two homologs differ in the combination of **alleles,** or versions, of the genes they carry (**figure 9.4**). As described in chapter 7, each allele of a gene encodes a different version of the same protein. A chromosome typically carries exactly one allele of each gene, so a person inherits one allele per gene from each parent. Depending on the parents' chromosomes, the two alleles may be identical or different. Overall, however, the members of each homologous pair of chromosomes are at least slightly different from each other.

Unlike the autosome pairs, the X and Y chromosomes are not homologous to each other. X is much larger than Y, and its genes are completely different. Nevertheless, in males, the sex chromosomes behave as homologous chromosomes during meiosis.

9.2 | Mastering Concepts

1. What are autosomes and sex chromosomes?
2. What is a karyotype?
3. How are the members of a homologous pair similar and different?

Figure 9.4 Homologous Chromosomes. On these homologous chromosomes, both alleles for gene *A* are the same, as are those for gene *D*. The two alleles for gene *B*, however, are different.

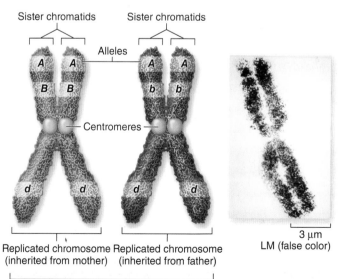

3 μm
LM (false color)

Replicated chromosome (inherited from mother)   Replicated chromosome (inherited from father)

Homologous pair of chromosomes

9.3 | Meiosis Is Essential in Sexual Reproduction

Sexually reproducing species range from humans to ferns to the mold that grows on bread. This section describes some of the features that all sexual life cycles share.

A. Gametes Are Haploid Sex Cells

Sexual reproduction poses a practical problem: maintaining the correct chromosome number. We have already seen that most cells in the human body contain 46 chromosomes. If a baby arises from the union of a man's sperm and a woman's egg, then why does a human baby not have 92 chromosomes per cell (46 from each parent)? And if that offspring later reproduced, wouldn't cells in the next generation have 184 chromosomes?

In fact, the normal chromosome number does not double with each generation. The explanation is that the special cells required for sexual reproduction, sperm cells and egg cells, are not diploid. Rather, they are **haploid cells** (abbreviated n); that is, they contain only one full set of genetic information instead of two.

These haploid cells, called **gametes,** are sex cells that combine to form a new offspring. **Fertilization** merges the gametes from two parents, creating a new cell: the diploid **zygote,** which is the first cell of the new organism (**figure 9.5**). The zygote has two full sets of chromosomes, one set from each parent. In most species, including plants and animals, the zygote begins dividing mitotically shortly after fertilization.

Thus, the life of a sexually reproducing, multicellular organism requires two ways to package DNA into daughter cells. **Mitosis,** described in chapter 8, divides a eukaryotic cell's chromosomes into two identical daughter cells. Mitotic cell division produces the cells needed for growth, development, and tissue repair. **Meiosis,** the subject of this chapter, forms genetically variable gametes that each contain half the number of chromosomes as the organism's diploid cells.

B. Specialized Germ Cells Undergo Meiosis

Only some cells can undergo meiosis and produce gametes. In humans and other animals, these specialized diploid cells, called **germ cells,** occur only in the ovaries and testes. Plants don't have the same reproductive organs as animals, but they do have specialized gamete-producing cells in flowers and other reproductive parts.

The rest of the body's diploid cells, called **somatic cells,** do not participate directly in reproduction. Leaf cells, root cells, skin cells, muscle cells, and neurons are examples of somatic cells. Most somatic cells can divide mitotically, but they do not undergo meiosis.

To make sense of this, consider your own life. It began when a small, swimming sperm cell carrying 23 chromosomes from your father wriggled toward your mother's comparatively enormous egg cell, also containing 23 chromosomes. You were conceived when the sperm fertilized the egg cell. At that moment, you were a one-celled zygote, with 46 chromosomes. That first cell then began dividing, generating identical copies of itself to form an embryo, then a fetus, infant, child, and eventually an adult (see figure 8.2). Once you reached reproductive maturity, germ cells in your testes or ovaries produced haploid gametes of your own, perpetuating the cycle.

The human life cycle is of course most familiar to us, and many animals reproduce in essentially the same way. Gametes are the only haploid cells in our life cycle; all other cells are diploid. Sexual reproduction, however, can take many other forms as well. In some organisms, including plants, both the haploid and the diploid stages are multicellular. Section 9.8 describes the life cycle of a sexually reproducing plant in more detail.

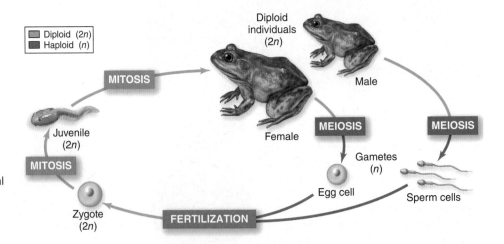

Figure 9.5 **Sexual Reproduction.** All sexual life cycles include meiosis and fertilization; mitotic cell division enables the organism to grow.

C. Meiosis Halves the Chromosome Number and Scrambles Alleles

No matter the species, meiosis has two main outcomes. First, the resulting gametes contain half the number of chromosomes as the rest of the body's cells. They therefore ensure that the chromosome number does not double with every generation. The second function of meiosis is to scramble genetic information, so that two parents can generate offspring that are genetically different from both the parents and from one another. As described in section 9.1, genetic variability is one of the evolutionary advantages of sex.

Although meiosis has unique functions, many of the events are similar to those of mitosis. As you work through the stages of meiosis, it may therefore help to think of what you already know about mitotic cell division. For example, a cell dividing mitotically undergoes interphase, followed by the overlapping phases of mitosis and then cytokinesis (see figure 8.8). Similarly, interphase occurs just before meiosis; the names of the phases of meiosis are similar to those in mitosis; and cytokinesis occurs after the genetic material is distributed.

Despite these similarities, meiosis has two unique outcomes, highlighted in figure 9.6. First, meiosis includes two divisions, which create four haploid cells from one diploid germ cell. Second, meiosis shuffles genetic information, setting the stage for each haploid nucleus to receive a unique mixture of alleles.

Sections 9.4 and 9.5 explain in more detail how meiosis simultaneously halves the chromosome number and produces genetically variable nuclei. We then turn to problems that can occur in meiosis and describe how humans package haploid nuclei into individual sperm or egg cells.

9.3 | Mastering Concepts

1. What is the difference between somatic cells and germ cells?
2. How do haploid and diploid nuclei differ?
3. What are the roles of meiosis, gamete formation, and fertilization in a sexual life cycle?
4. What is a zygote?

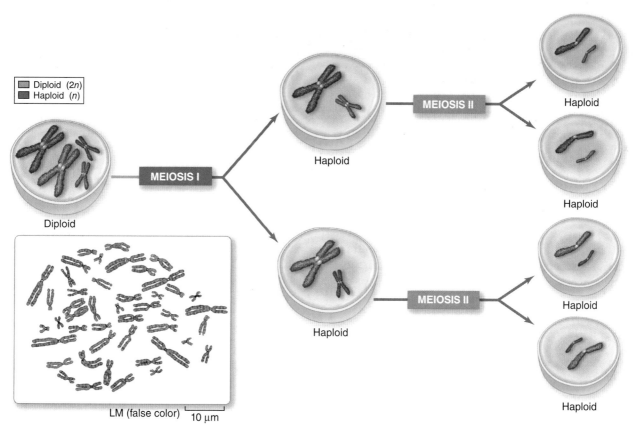

Figure 9.6 **Summary of Meiosis.** In meiosis, a diploid nucleus gives rise to four haploid nuclei. The figure is simplified in the sense that the diploid cell contains only two pairs of homologous chromosomes. In reality, a diploid human cell contains 23 pairs of homologous chromosomes (inset). The figure also omits the effects of crossing over (see figure 9.8).

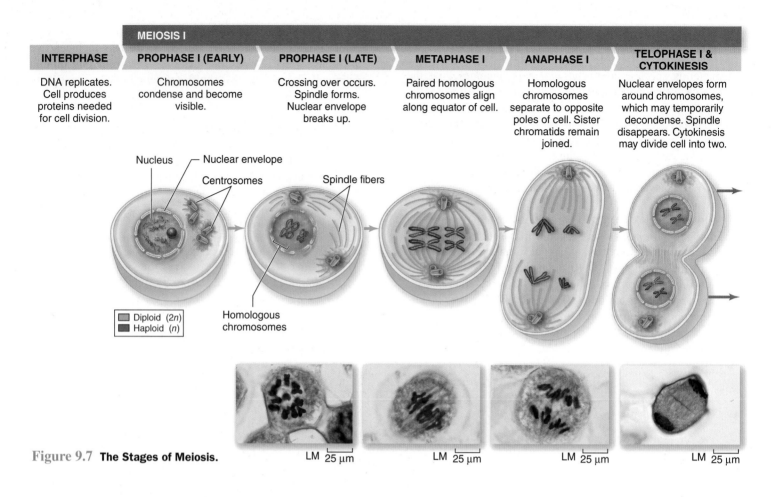

MEIOSIS I					
INTERPHASE	PROPHASE I (EARLY)	PROPHASE I (LATE)	METAPHASE I	ANAPHASE I	TELOPHASE I & CYTOKINESIS
DNA replicates. Cell produces proteins needed for cell division.	Chromosomes condense and become visible.	Crossing over occurs. Spindle forms. Nuclear envelope breaks up.	Paired homologous chromosomes align along equator of cell.	Homologous chromosomes separate to opposite poles of cell. Sister chromatids remain joined.	Nuclear envelopes form around chromosomes, which may temporarily decondense. Spindle disappears. Cytokinesis may divide cell into two.

Nucleus
Nuclear envelope
Centrosomes
Spindle fibers

□ Diploid (2n)
■ Haploid (n)

Homologous chromosomes

LM 25 μm   LM 25 μm   LM 25 μm   LM 25 μm

Figure 9.7 **The Stages of Meiosis.**

9.4 | In Meiosis, DNA Replicates Once, but the Nucleus Divides Twice

Before meiosis occurs, a germ cell first undergoes interphase. The cell grows during G_1 of interphase and synthesizes the molecules necessary for division. All of the cell's DNA replicates during S phase, after which each of the cell's chromosomes consists of two identical sister chromatids attached at a centromere. The cell also produces proteins and other enzymes necessary to divide the cell. Finally, in G_2, the chromatin begins to condense, and the cell produces the spindle proteins that will move the chromosomes. ▸ DNA replication, p. 154

The germ cell is now ready for meiosis to begin. The key to learning the events of meiosis is to pay careful attention to the movements of the cell's homologous chromosome pairs. During the first half of meiosis, called meiosis I, each chromosome physically aligns with its homolog. The homologous pairs split into two cells toward the end of meiosis I. Meiosis II then partitions the genetic material into four haploid nuclei. Figure 9.7 diagrams the entire process; you may find it helpful to refer to it as you read the rest of this section.

A. In Meiosis I, Homologous Chromosomes Pair Up and Separate

Homologous pairs of chromosomes find each other and then split up during the first meiotic division.

Prophase I   During prophase I (that is, the prophase of meiosis I), replicated chromosomes condense. A spindle begins to form from microtubules assembled at the centrosomes, and spindle attachment points called kinetochores assemble on each centromere. The nuclear envelope breaks up, allowing the spindle fibers to reach the chromosomes.

The events described so far resemble those of prophase of mitosis, but something unique happens during prophase I of meiosis: The homologous chromosomes line up next to one another. That is, chromosome 1 aligns with its homolog, X aligns with Y, and so forth. (Mules are sterile because their germ cells cannot complete this stage, as described in this chapter's Burning Question.) Section 9.5 describes how this arrangement allows for a gene-shuffling mechanism called crossing over.

Metaphase I   In metaphase I, the paired homologs align down the center of the cell. Each member of a homologous pair attaches to a spindle fiber stretching to one pole. The stage is now set for the homologous pairs to be separated.

Anaphase I, Telophase I, and Cytokinesis   Homologous pairs separate in anaphase I, although the sister chromatids that make up each chromosome remain joined together. The chromosomes complete their movement to opposite poles in telophase I. In most species, cytokinesis occurs after telophase I, splitting the original germ cell into two.

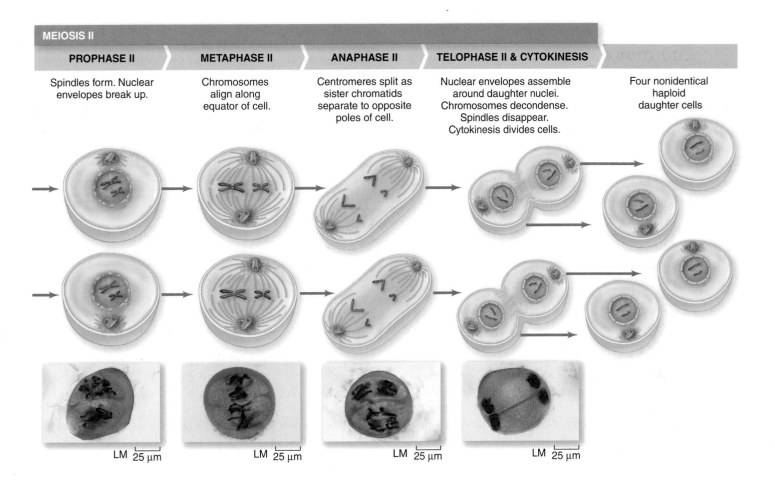

MEIOSIS II

PROPHASE II	METAPHASE II	ANAPHASE II	TELOPHASE II & CYTOKINESIS	
Spindles form. Nuclear envelopes break up.	Chromosomes align along equator of cell.	Centromeres split as sister chromatids separate to opposite poles of cell.	Nuclear envelopes assemble around daughter nuclei. Chromosomes decondense. Spindles disappear. Cytokinesis divides cells.	Four nonidentical haploid daughter cells

LM 25 μm   LM 25 μm   LM 25 μm   LM 25 μm

Burning Question

If mules are sterile, then how are they produced?

A mule is the hybrid offspring of a mating between a male donkey and a female horse. The opposite cross (female donkey with male horse) yields a hybrid called a hinny.

Mules and hinnies may be male or female, but they are usually sterile. Why?

A peek at the parents' chromosomes reveals the answer. Donkeys have 31 pairs of chromosomes, whereas horses have 32 pairs. When gametes from horse and donkey unite, the resulting hybrid zygote has 63 chromosomes (31+32). The zygote divides mitotically to yield the cells that make up the mule or hinny.

These hybrid cells cannot undergo meiosis for two reasons. First, they have an odd number of chromosomes, which disrupts meiosis because at least one chromosome lacks a homologous partner. Second, donkeys and horses have slightly different chromosome structures, so the hybrid's parental chromosomes cannot align properly during prophase I. The result: an inability to produce sperm and egg cells. The only way to produce more mules and hinnies is to again mate horses with donkeys. ▶ hybrid infertility, p. 276

Submit your burning question to:
marielle_hoefnagels@mcgraw-hill.com

B. Meiosis II Yields Four Haploid Cells

A second interphase precedes meiosis II in many species. During this time, the chromosomes unfold into very thin threads. The cell manufactures proteins, but the genetic material does not replicate a second time.

Meiosis II strongly resembles mitosis. The process begins with prophase II, when the chromosomes again condense and become visible. In metaphase II, the chromosomes align down the center of each cell. In anaphase II, the centromeres split, and the separated sister chromatids move to opposite poles. In telophase II, nuclear envelopes form around the separated sets of chromosomes. Cytokinesis then separates the nuclei into individual cells. The overall result: one diploid germ cell has divided into four haploid cells.

Figure It Out

A cell that is entering prophase I contains __ times as much DNA as one daughter cell at the end of meiosis.

Answer: Four

9.4 | Mastering Concepts

1. What happens during interphase of meiosis?
2. How do the events of meiosis I and meiosis II produce four haploid cells from one diploid germ cell?

9.5 | Meiosis Generates Enormous Variability

By creating new combinations of alleles, meiosis generates astounding genetic variety among the offspring from just two parents. Three mechanisms account for this diversity: crossing over, independent assortment, and random fertilization.

A. Crossing Over Shuffles Genes

Crossing over is a process in which two homologous chromosomes exchange genetic material (**figure 9.8**). During prophase I, the homologs align themselves precisely, gene by gene, in a process called synapsis. The chromosomes are attached at a few points along their lengths, called chiasmata (singular: chiasma), where the homologs exchange chromosomal material.

As an example, consider what takes place in your own diploid germ cells. You inherited one member of each homologous pair from your mother; the other came from your father. Crossing over means that pieces of these homologous chromosomes physically change places. New combinations arise whenever the homologous chromosomes carry different alleles of one or more genes.

Suppose, for instance, that one chromosome carries genes that dictate hair color, eye color, and finger length. Perhaps the version you inherited from your father combines alleles that specify blond hair, blue eyes, and short fingers. The homolog from your mother is different; its alleles dictate black hair, brown eyes, and long fingers. Now, suppose that crossing over occurs between the homologous chromosomes. Afterward, one recombinant chromatid might combine alleles for blond hair, brown eyes, and long fingers; another would specify black hair, blue eyes, and short fingers. The two chromatids that did not form chiasmata would remain unchanged. Even though all of the alleles in your ova-

ries or testes came from your parents, half of the chromatids now contain allele *combinations* that are new.

The result of crossing over is four genetically different chromatids in place of two pairs of identical chromatids. As meiosis continues in the germ cell, each chromatid will end up in a separate haploid cell. Thus, crossing over ensures that each haploid cell will be genetically different from the others.

B. Chromosome Pairs Align Randomly During Metaphase I

A look at figure 9.7 reveals a second way that meiosis creates genetic variability. At metaphase I, the paired chromosomes line up at the cell's center; each red chromosome from one parent is attached (for the moment) to its blue homolog from the other parent. Examine the orientation of these chromosomes. Notice that the blue chromosome is "on top" in the pair on the left, whereas the red chromosome occupies that position in the pair on the right. In anaphase I, the chromosomes separate, and the resulting nuclei have a mixture of paternal and maternal genetic material.

The next time a germ cell in the same individual undergoes meiosis, the orientation of the chromosomes may be the same, or it may not be. The alignment of chromosomes at metaphase I is a random process, and all possible combinations are equally probable.

The number of possible arrangements is related to the number of chromosomes, according to the formula 2^n (where n is the haploid number). For two pairs of homologs, each resulting gamete may have any of four (2^2) unique chromosome configurations. For three pairs of homologs, as shown in **figure 9.9**[, eight (2^3) unique configurations can occur in the gametes. Extending this formula to humans, with 23 chromosome pairs, each gamete contains one of 8,388,608 (2^{23}) possible chromosome combinations—all equally likely.

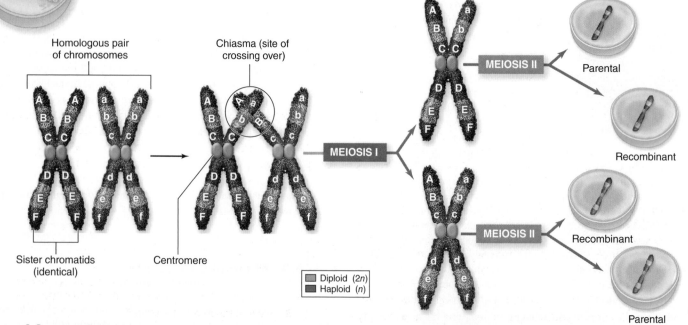

	Diploid (2n)
	Haploid (n)

Figure 9.8 Crossing Over. Crossing over between homologous chromosomes generates genetic diversity by mixing up parental traits, creating recombinant chromatids. The capital and lowercase letters represent different alleles of six genes.

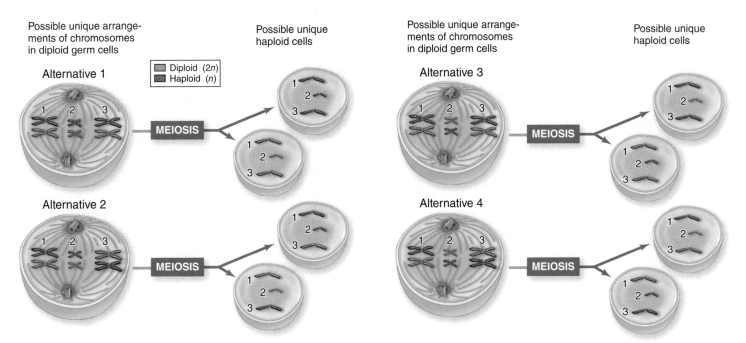

Possible unique arrange-
ments of chromosomes
in diploid germ cells

Possible unique
haploid cells

Possible unique arrange-
ments of chromosomes
in diploid germ cells

Possible unique
haploid cells

Alternative 1

☐ Diploid (2n)
☐ Haploid (n)

MEIOSIS

Alternative 2

MEIOSIS

Alternative 3

MEIOSIS

Alternative 4

MEIOSIS

Figure 9.9 Many Possibilities. A germ cell containing three homologous pairs of chromosomes can generate eight genetically different gametes. The number of unique gametes skyrockets when one considers all 23 chromosome pairs (in humans), plus the effects of crossing over.

Monozygotic (identical) twins

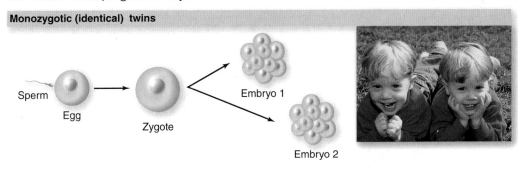

Sperm

Egg

Zygote

Embryo 1

Embryo 2

Dizygotic (fraternal) twins

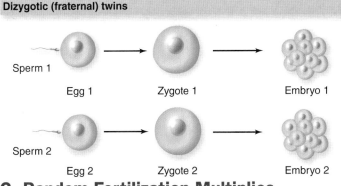

Sperm 1

Egg 1

Zygote 1

Embryo 1

Sperm 2

Egg 2

Zygote 2

Embryo 2

Figure 9.10 Two Origins for Twins. Monozygotic twins are genetically identical because they come from the same zygote. Dizygotic, or fraternal, twins are no more alike than nontwin siblings because they start as two different zygotes.

C. Random Fertilization Multiplies the Diversity

We have already seen that every germ cell undergoing meiosis is likely to produce haploid nuclei with different combinations of chromosomes. Furthermore, it takes two to reproduce. In one mating, any of a woman's 8,388,608 possible egg cells can combine with any of the 8,388,608 possible sperm cells of a partner. One couple could therefore theoretically create more than 70 trillion ($8,388,608^2$) genetically unique individuals! And this enormous number is an underestimate, because it does not take into account the additional variation from crossing over.

With so much potential variability, the chance of two parents producing genetically identical individuals seems exceedingly small. How do the parents of identical twins defy the odds? The answer is that identical twins result from just one fertilization

event. The resulting zygote or embryo splits in half, creating separate, identical babies (figure 9.10). Identical twins are called "monozygotic" because they derive from one zygote. In contrast, nonidentical (fraternal) twins occur when two sperm cells fertilize two separate egg cells. The twins are therefore called "dizygotic." See this chapter's Apply It Now box on page 195 for more on multiple births.

9.5 | Mastering Concepts

1. How does crossing over shuffle genes?
2. Explain how events in metaphase I enable a human to produce over 8 million genetically different gametes.
3. What is random fertilization?
4. How are identical twins different from fraternal twins?

9.6 | Mitosis and Meiosis Have Different Functions: A Summary

Mitosis and meiosis are both mechanisms that divide a eukaryotic cell's genetic material (figure 9.11). The two processes share many events, as revealed by the similar names of the stages. The cell copies its DNA during an interphase stage that precedes both mitosis and meiosis, after which the chromosomes condense. Moreover, spindle fibers orchestrate the movements of the chromosomes in both mitosis and meiosis.

However, the two processes also differ in many ways:

- Mitosis occurs in somatic cells throughout the body, and it occurs throughout the life cycle. In contrast, meiosis occurs only in germ cells and only at some stages of life (see section 9.8).
- Homologous chromosomes do not align with each other during mitosis, as they do in meiosis. This alignment allows for crossing over, which also occurs only in meiosis.

- Mitosis yields identical daughter cells for growth, repair, and asexual reproduction. Thanks to crossing over and the random orientation of chromosome pairs during metaphase I, meiotic division generates genetically variable daughter cells. Organisms use these variable cells in sexual reproduction.
- Following mitosis, cytokinesis occurs once for every DNA replication event. The product of mitotic division is therefore two daughter cells. In meiosis, cytokinesis occurs twice, although the DNA has replicated only once. One cell therefore yields four daughter cells.
- After mitosis, the chromosome number in the daughter cells is the same as in a parent cell. In contrast, only diploid cells divide meiotically, producing four haploid daughter cells.

9.6 | Mastering Concepts

1. In what ways are mitosis and meiosis similar?
2. In what ways are mitosis and meiosis different?

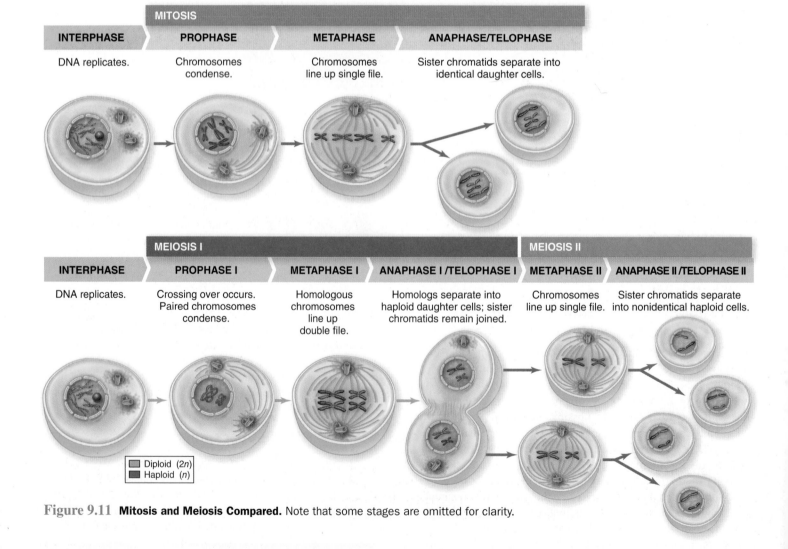

Figure 9.11 Mitosis and Meiosis Compared. Note that some stages are omitted for clarity.

9.7 | Errors Sometimes Occur in Meiosis

Considering the number of separate events that occur during meiosis, it is not surprising that things occasionally take a wrong turn. The result can be gametes with extra or missing chromosomes. Even small chromosomal abnormalities can have devastating effects on health.

A. Polyploidy Means Extra Chromosome Sets

An error in meiosis, such as the failure of the spindle to form properly, can produce a **polyploid cell** with one or more complete sets of extra chromosomes (*polyploid* means "many sets"). For example, if a sperm with the normal 23 chromosomes fertilizes an abnormal egg cell with two full sets (46), the resulting zygote will have three copies of each chromosome (69 total), a type of polyploidy called triploidy. Most human polyploids cease developing as embryos or fetuses.

In contrast to humans, about 30% of flowering plant species tolerate polyploidy well, and many crop plants are polyploids. The durum wheat in pasta is tetraploid (it has four sets of seven chromosomes), and the wheat species in bread is hexaploid, with six sets of seven chromosomes. Polyploidy is an important force in plant evolution; section 9.9 describes one example.

B. Nondisjunction Results in Extra or Missing Chromosomes

Some gametes have just one extra or missing chromosome. The cause of the abnormality is an error called **nondisjunction,** which occurs when chromosomes fail to separate at either the first or the second meiotic division (figure 9.12). The result is a sperm or egg cell with two copies of a particular chromosome or none at all, rather than the normal one copy. When such a gamete fuses with another at fertilization, the resulting zygote has either 45 or 47 chromosomes instead of the normal 46.

Most embryos with incorrect chromosome numbers cease developing before birth; they account for about 50% of all spontaneous abortions (miscarriages) that occur early in a pregnancy. Extra genetic material, however, causes fewer problems than missing material. This is why most children with the wrong number of chromosomes have an extra one—a trisomy—rather than a missing one. ▸ when a pregnancy ends, p. 711

Following is a look at some syndromes in humans resulting from too many or too few chromosomes.

Extra Autosomes: Trisomy 21, 18, or 13
A person with trisomy 21, the most common cause of Down syndrome, has

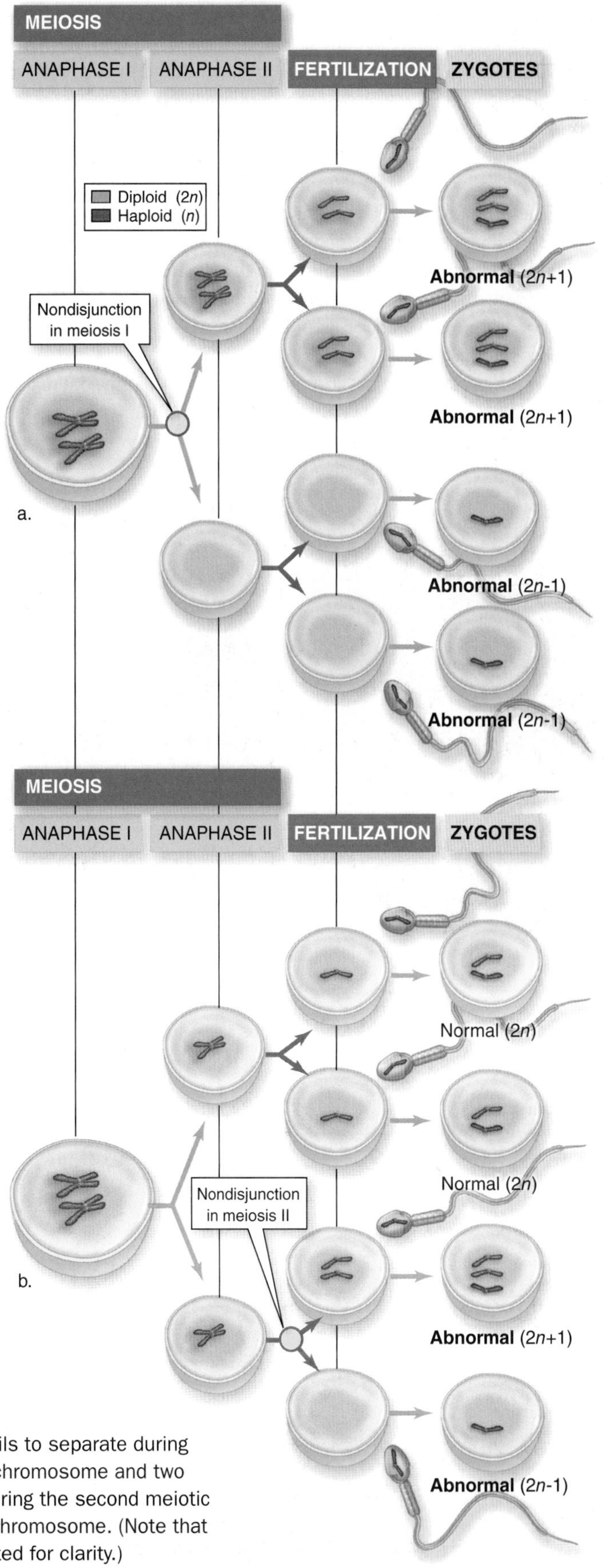

Figure 9.12 Nondisjunction. (a) A homologous pair of chromosomes fails to separate during the first division of meiosis. The result: two nuclei with two copies of the chromosome and two nuclei that lack the chromosome. (b) Sister chromatids fail to separate during the second meiotic division. One nucleus has an extra chromosome, and one is missing the chromosome. (Note that all chromosomes other than the ones undergoing nondisjunction are omitted for clarity.)

three copies of chromosome 21 (figure 9.13). An affected person has distinctive facial features and a unique pattern of hand creases. Intelligence varies greatly; some children have profound mental impairment, whereas others learn well. Many affected children die before their first birthdays, often because of congenital heart defects. People with Down syndrome also have an above-average risk for leukemia and Alzheimer disease.

The probability of giving birth to a child with trisomy 21 increases dramatically as a woman ages. For women younger than 30, the chances of conceiving a child with the syndrome are 1 in 3000. For a woman of 48, the incidence jumps to 1 in 9. An increased likelihood of nondisjunction in older females may account for this age association.

Trisomy 21 is the most common autosomal trisomy, but that is only because the fetus is most likely to remain viable. Trisomies 18 and 13 are the next most common, but few infants with these genetic abnormalities survive infancy. Trisomies undoubtedly occur with other chromosomes, but the embryos fail to develop.

Extra or Missing Sex Chromosomes: XXX, XXY, XYY, and XO Nondisjunction can produce a gamete that contains two X or Y chromosomes instead of only one. Fertilization then produces a zygote with too many sex chromosomes: XXX, XXY, or XYY. A gamete may also lack a sex chromosome altogether. If one gamete contains an X chromosome and the other gamete has neither X nor Y, the resulting zygote is XO. Interestingly, medical researchers have never reported a person with one Y and no X chromosome. When a zygote lacks an X chromosome, so much genetic material is missing that it probably cannot sustain more than a few cell divisions. Table 9.1 summarizes some of the sex chromosome abnormalities.

C. Smaller-Scale Chromosome Abnormalities Also Occur

Parts of a chromosome may be deleted, duplicated, inverted, or even moved to a new location (figure 9.14). Because each chromosome includes hundreds or thousands of genes, even small changes in a chromosome's structure can affect an organism.

A chromosomal **deletion** results in the loss of one or more genes. Cri du chat syndrome (French for "cat's cry"), for example, is associated with deletion of several genes on chromosome 5. The illness is named for the odd cry of an affected child, similar to the mewing of a cat. The gene deletion also causes severe mental retardation and developmental delay.

In the opposite situation, a **duplication** produces multiple copies of part of a chromosome. Fragile X syndrome, for example, results from repeated copies of a three-base sequence (CGG) on the X chromosome. The disorder can produce a range of symptoms, including mental retardation. The number of repeats can range from fewer than 10 to more than 200. Individuals with the most copies of the repeat are the most severely affected.

The duplication of entire genes sometimes plays an important role in evolution. If one copy of the original gene continues to do its old job, then a mutation in a "spare" copy will not be harmful. Although these mutations often ruin the gene, they can also lead to new functions. As just one example, biologists have studied a gene that was originally required for the secretion of calcium in tooth enamel in vertebrates. Mutations in duplicate genes created new functions, including the production of calcium-rich breast milk.

In an **inversion,** part of a chromosome flips and reinserts, changing the gene sequence. Unless inversions break genes, they are usually less harmful than deletions, because all the genes are still present. Fertility problems can arise, however, if an adult has an inversion in one chromosome but its homolog is normal. During crossing over, the inverted chromosome and its noninverted partner may twist around each other in a way that generates chromosomes with deletions or duplications. Because the gametes will have extra or missing genes, the result may be a miscarriage or birth defects.

a.

LM 10 μm

Trisomy 21

b.

Figure 9.13 **Trisomy 21.** (a) A normal human karyotype reveals 46 chromosomes, in 23 pairs. (b) A child with three copies of chromosome 21 has Down syndrome.

Table 9.1   Examples of Sex Chromosome Abnormalities

Chromosomes	Name of Condition	Incidence	Symptoms
XXX	Triplo-X	1 in every 1000 to 2000 females	Symptoms are tallness, menstrual irregularities, and a normal-range IQ that is slightly lower than that of other family members. A woman with triplo-X may produce some egg cells bearing two X chromosomes, which increases her risk of giving birth to triplo-X daughters or XXY sons.
XXY	Klinefelter or XXY syndrome	1 in every 500 to 1000 males	The syndrome varies greatly. Often, however, affected individuals are sexually underdeveloped, with rudimentary testes and prostate glands and no pubic or facial hair. They also have very long limbs and large hands and feet, and they may develop breast tissue. Individuals with XXY syndrome may be slow to learn, but they are usually not mentally retarded unless they have more than two X chromosomes, which is rare.
XYY	Jacobs or XYY syndrome	One in every 1000 males	The vast majority of XYY males are apparently normal, although they may be very tall. They may also have acne and problems with speech and reading.
XO	Turner syndrome	1 in every 2000 females	Young women with the missing chromosome are short and sexually undeveloped. They are usually of normal intelligence. Although women with Turner syndrome are infertile, treatment with hormone supplements can promote growth and sexual development.

Figure 9.14 Chromosomal Abnormalities. Portions of a chromosome can be (a) deleted, (b) duplicated, or (c) inverted. (d) In translocation, two nonhomologous chromosomes exchange parts. The micrograph shows a portion of chromosome 5 (larger pair) that has switched places with part of chromosome 14 (smaller pair).

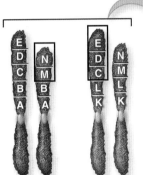
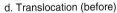
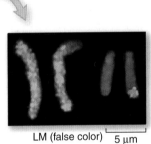

Normal   a. Deletion   b. Duplication   c. Inversion   d. Translocation (before)   Translocation (after)

LM (false color)   5 μm

In a **translocation,** nonhomologous chromosomes exchange parts. Translocations often break genes, and sometimes the result is leukemia or other cancers. In about 95% of people with chronic myelogenous leukemia, for example, part of one gene from chromosome 9 fuses with a gene on chromosome 22. The combined gene encodes a protein that speeds cell division and suppresses normal cell death (apoptosis), causing leukemia, a form of cancer in which blood cells divide out of control. ▶ apoptosis, p. 167

If no genes are broken in a translocation, then the person has the normal amount of genetic material; it is simply rearranged. Such a person is healthy but may have fertility problems. Some sperm or egg cells will receive one of the translocated chromosomes but not the other, causing a genetic imbalance—some genes are duplicated, and others are deleted. The consequences depend on which genes the rearrangement disrupts.

9.7 | Mastering Concepts

1. What is polyploidy?
2. How can nondisjunction during meiosis lead to gametes with extra or missing chromosomes?
3. How can deletions, duplications, inversions, and translocations cause illness?
4. How do inversions and translocations cause fertility problems?

9.8 | Haploid Nuclei Are Packaged into Gametes

The events of meiosis explain how a diploid germ cell produces four genetically different haploid nuclei. The same process occurs in both sexes, yet sperm and egg cells typically look very different from each other (figure 9.15). Usually, a sperm is lightweight and can swim; an egg cell is huge by comparison and packed with nutrients and organelles. How do males and females package those haploid nuclei into such different-looking gametes?

A. In Humans, Gametes Form in Testes and Ovaries

The formation and specialization of sperm cells is called spermatogenesis (figure 9.16). Inside the testes, spermatogonia are diploid stem cells that divide mitotically to produce two kinds of cells: more spermatogonia and specialized germ cells called primary spermatocytes. It is these germ cells that undergo meiosis. During interphase, primary spermatocytes accumulate cytoplasm and replicate their DNA. The first meiotic division yields two equal-sized haploid cells called secondary spermatocytes.

Each secondary spermatocyte then completes its second meiotic division. The products are four equal-sized spermatids, each of which specializes into a mature, tadpole-shaped sperm cell. The entire process, from spermatogonium to sperm, takes about 74 days.

In comparison to a sperm cell, an egg cell is massive. The female produces these large cells by unequally packaging the

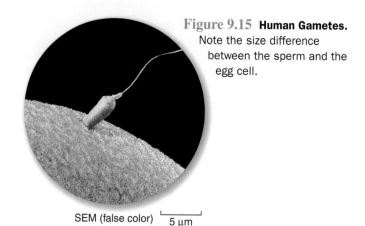

Figure 9.15 Human Gametes. Note the size difference between the sperm and the egg cell.

SEM (false color) 5 μm

cytoplasm from the two meiotic divisions. The egg cell gets most of the cytoplasm, and the other products of meiosis are tiny.

The formation of egg cells is called oogenesis (figure 9.17). It occurs in the ovaries and begins with a diploid stem cell, an oogonium. This cell can divide mitotically to produce more oogonia or a germ cell called a primary oocyte. In meiosis I, the primary oocyte divides into a small haploid cell with very little cytoplasm, called a polar body, and a much larger haploid cell called a secondary oocyte. In meiosis II, the secondary oocyte divides unequally to produce another polar body and the mature egg cell, or ovum, which contains a large amount of cytoplasm. The tiny polar bodies normally play no further role in reproduction.

Chapter 34 explores human reproduction and development in more detail.

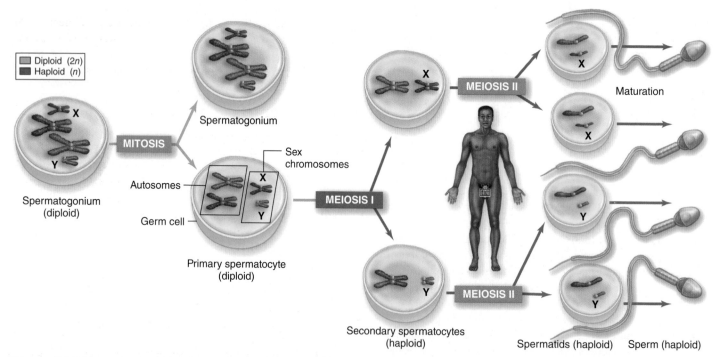

Figure 9.16 Sperm Formation (Spermatogenesis). In humans, diploid primary spermatocytes undergo meiosis, yielding four equal-sized, haploid sperm. Of the normal 23 pairs of chromosomes, only one pair of autosomes and one pair of sex chromosomes are shown.

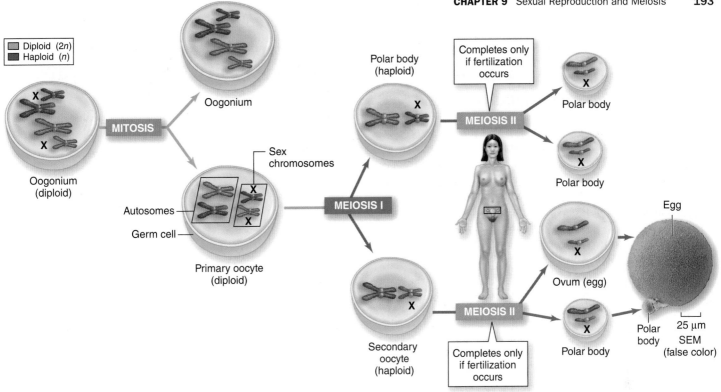

Figure 9.17 **Ovum Formation (Oogenesis).** Diploid primary oocytes undergo meiosis. Meiotic division in females allocates most of the cytoplasm to one large egg cell. The body discards the other products of meiosis, called polar bodies, which contain the other sets of chromosomes. Of the normal 23 pairs of chromosomes, only one pair of autosomes and one pair of sex chromosomes are shown.

B. In Plants, Gametophytes Produce Gametes

Plant life cycles include an **alternation of generations** between multicellular haploid and diploid individuals (figure 9.18). Germ cells in the diploid plant, or sporophyte, undergo meiosis to produce haploid cells called spores. The spores germinate, dividing mitotically to produce a multicellular haploid plant called a ga-

metophyte. The gametophyte, in turn, produces sperm or egg cells by mitotic cell division. Sperm fertilizes egg to form a diploid zygote, which divides mitotically and develops into a sporophyte. The cycle begins anew.

In mosses and ferns, the gametophytes are small green plants that are visible with the unaided eye. In flowering plants, however, the gametophyte is microscopic and relies on the sporophyte for nutrition. The egg-producing female gametophyte, for example, is buried deep within a flower.

Some plants produce swimming sperm cells. In mosses and ferns, the male gametes use flagella to swim in a film of water to the stationary egg cell. The sperm cells of conifers and flowering plants, however, do not swim. Instead, these plants produce pollen grains—male gametophytes—that travel in wind or on animals to reach female plant parts. Pollen germination delivers sperm cells directly to the stationary egg cell. Chapters 18 and 23 further describe plant reproduction.

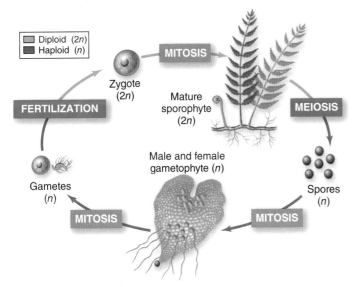

Figure 9.18 **Plant Reproduction.** Plant life cycles include an alternation of multicellular haploid and diploid generations.

9.8 | Mastering Concepts

1. What are the stages of sperm development in humans?
2. What are the stages of development of an egg cell in humans?
3. How does gamete production in plants differ from that in animals?

9.9 | Investigating Life: A New Species Is Born, but Who's the Daddy?

The title of Charles Darwin's book *On the Origin of Species* conveys a central question in biology: Where do new types of organisms come from? All species descend from a common ancestor, so each must have arisen from a preexisting one. But how? Unfortunately, since most new species formed long ago, details about the exact events usually remain lost to history.

Sometimes, however, science catches a lucky break, as in the case of a flowering plant called goat's beard, or *Tragopogon*. This weedy plant is a type of dandelion. Europeans introduced three *Tragopogon* species to North America in the 1900s, cultivating the plants for their edible roots. The seeds ride the wind on a parachute-like crown of fluff, and the plants have spread widely across the continent. The most common of the three introduced plants is *T. dubius*, whereas two rarer species are *T. pratensis* and *T. porrifolius*.

A brief lesson on plant reproduction will help clarify why *Tragopogon* is so important to biologists who study how species form. Plants in the three introduced species are diploid, and each individual has two parents, just as you do. Pollen carries haploid sperm cells to a female flower part containing an egg cell. After fertilization, the diploid zygote develops into an embryo, which is packaged along with a food supply into a seed.

Errors occasionally occur during gamete production, however, and a plant may produce sex cells containing two full sets of chromosomes instead of just one. If a diploid sperm cell fertilizes a diploid egg cell, the resulting zygote has four sets of chromosomes; in other words, it is tetraploid.

Species	Diploid or Tetraploid	Number of Chromosomes	Parents
T. dubius	Diploid	12	—
T. pratensis	Diploid	12	—
T. porrifolius	Diploid	12	—
T. mirus	Tetraploid	24	*T. dubius* and *T. porrifolius*
T. miscellus	Tetraploid	24	*T. dubius* and *T. pratensis*

Figure 9.19 ***Tragopogon* Species Origins.** Genetic studies have revealed which "parents" hybridized to produce the two tetraploid species. But which species contributed the pollen and which contributed the egg?

This type of "mistake" sometimes occurs in *Tragopogon*. In fact, biologists have known since the 1950s that the union of diploid gametes gave rise to two brand-new tetraploid species, named *T. mirus* and *T. miscellus* (**figure 9.19**). These tetraploid plants are considered full-fledged species because they can mate among themselves but not with the "parental" diploid plants.

Because *Tragopogon* did not exist in North America before the 1900s, *T. mirus* and *T. miscellus* must have arisen in just half

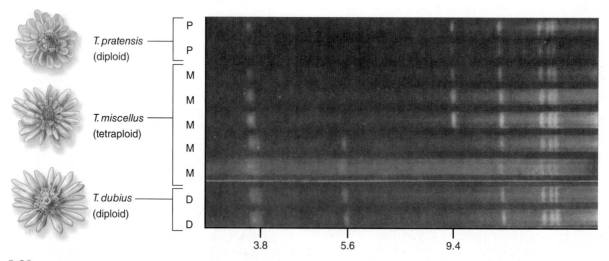

Figure 9.20   Chloroplast DNA. The Soltises purified chloroplast DNA from multiple individuals of *T. pratensis*, *T. miscellus*, and *T. dubius*. Because chloroplast DNA comes from only the egg, a matching pattern on the electrophoresis gel reveals the identity of the mother. The tetraploid species *T. miscellus* may have *T. pratensis* or *T. dubius* as a maternal parent, indicating that this hybrid has arisen more than once.

a century. To learn more about how the two species formed, Washington State University biologists Douglas and Pamela Soltis studied DNA extracted from chloroplasts, the organelles of photosynthesis. Each plant inherits chloroplast DNA from just one parent, the female. The egg contributes cytoplasm, containing mitochondria and chloroplasts, to the zygote. The comparatively tiny sperm cell contains few organelles.

Chloroplast DNA can therefore answer a question about the short history of North American *Tragopogon:* Which diploid species was the father and which was the mother of each tetraploid species? To find out, the Soltises collected seeds from 39 natural populations of the five *Tragopogon* species. They allowed the seeds to germinate in large trays of soil, then isolated the chloroplast DNA from each plant's leaves. The researchers then treated the DNA with 18 restriction enzymes, each of which cut the DNA at a different sequence. Electrophoresis separated the fragments, and stains made the bands of DNA visible. ▶ DNA profiling, p. 172

The Soltises knew that different patterns of DNA fragments would reflect underlying differences in chloroplast DNA sequences. As expected, the gels revealed a unique fragment pattern for each diploid *Tragopogon* species. When Soltis and Soltis compared these genetic "fingerprints" to those of the tetraploid hybrids, it was clear that *T. porrifolius* was the female parent of *T. mirus;* the male parent was *T. dubius.* The same species also fathered most populations of *T. miscellus,* with *T. pratensis* being the female parent (**figure 9.20**). Interestingly, however, two samples of *T. miscellus* had chloroplast DNA like that of *T. dubius. Tragopogon miscellus* has therefore arisen more than once.

The story of goat's beard is important because it shows that new species do not always arise gradually; instead, a sudden genetic change can instantly separate a brand new species from its parents. The observation that this process has occurred more than once in 50 years, at least for *Tragopogon,* is tantalizing. How many times has it happened in life's history? We will probably never know. Nearly 150 years after *On the Origin of Species,* however, our window on evolution is clearer than ever.

Soltis, Douglas E., and Pamela S. Soltis. 1989. Allopolyploid speciation in *Tragopogon:* Insights from chloroplast DNA. *American Journal of Botany,* vol. 76, pages 1119–1124.

9.9 | Mastering Concepts

1. How did the researchers use chloroplast DNA to learn about the evolutionary history of tetraploid *Tragopogon* species?

2. The Soltises suggest that because *T. dubius* is so much more common than the other two diploid species, its pollen is also the most abundant. How would you test the hypothesis that the most common plant is most likely to be the father of a tetraploid hybrid?

Apply It Now

Multiple Births

As you learned in section 9.5, twins can be fraternal (no more alike genetically than siblings born singly) or identical. But how do triplets, quadruplets, and higher-order multiple births arise?

Triplets come about in several ways. The least common route is for a single embryo to split and develop into three genetically identical babies (monozygotic triplets). Alternatively, if three sperm fertilize three separate egg cells, the triplets will all be fraternal (trizygotic). Most commonly, however, an embryo splits and forms two identical babies, and another embryo develops into an additional, nonidentical baby.

Identical quadruplets are exceedingly rare, occurring perhaps once in 11 million deliveries. Monozygotic quintuplets are even more unusual, with only one set ever known to have been born. Just as for triplets, higher-order multiples usually are combinations of identical and fraternal siblings.

Multiple births have become more common since the 1980s for two reasons. First, older women are more likely to have multiple births, and childbearing among these women has become more common. Second, treatment for infertility has increased. Some fertility drugs stimulate a woman to release more than one egg cell. If sperm fertilize all of them, a multiple birth could result.

Another infertility therapy is *in vitro* fertilization, in which sperm fertilize egg cells harvested from a woman's ovaries in the lab. One or more embryos judged most likely to result in a live birth are then implanted into the woman's uterus. Multiple births often result. A notable example occurred in 2009, when a woman gave birth to octuplets conceived by *in vitro* fertilization. The children represent only the second full set of octuplets born alive in U.S. history.

Chapter Summary

9.1 | Why Sex?

- **Asexual reproduction** is reproduction without sex. **Sexual reproduction** mixes traits from two parents as it produces new individuals.
- **Conjugation** is a form of gene transfer that occurs in some microorganisms.
- Asexual reproduction can be successful in a stable environment, but a changing environment selects for sexual reproduction.

9.2 | Diploid Cells Contain Two Homologous Sets of Chromosomes

- **Diploid cells** have two full sets of **chromosomes,** one from each parent.
- In humans, the **sex chromosomes** (X and Y) determine whether an individual is male or female. The 22 **homologous pairs** of **autosomes** do not determine sex.
- Homologous chromosomes share the same size, banding pattern, and centromere location, but they differ in the **alleles** they carry.

9.3 | Meiosis Is Essential in Sexual Reproduction

A. Gametes Are Haploid Sex Cells
- In sexual life cycles, **meiosis** halves the genetic material to produce **haploid gametes.**
- **Fertilization** occurs when gametes fuse, forming the diploid **zygote.** **Mitotic cell division** produces the body's cells.

B. Specialized Germ Cells Undergo Meiosis
- **Somatic** cells do not participate in reproduction. In contrast, diploid **germ cells** undergo meiosis to produce haploid sex cells.
- In most animals, gametes are the only haploid cells.

C. Meiosis Halves the Chromosome Number and Scrambles Alleles
- Meiosis shares some similarities with mitosis, but the unique events of meiosis ensure that gametes are haploid and genetically variable.

9.4 | In Meiosis, DNA Replicates Once, but the Nucleus Divides Twice

- Interphase happens before meiosis.

A. In Meiosis I, Homologous Chromosomes Pair Up and Separate
- Homologous pairs of chromosomes align during prophase I, then split apart during anaphase I.

B. Meiosis II Yields Four Haploid Cells
- In meiosis II, the two products of meiosis I divide to yield four haploid cells.

9.5 | Meiosis Generates Enormous Variability

A. Crossing Over Shuffles Genes
- **Crossing over,** which occurs in prophase I, produces variability when portions of homologous chromosomes switch places. After crossing over, the chromatids carry new combinations of parental alleles.

B. Chromosome Pairs Align Randomly During Metaphase I
- Every possible orientation of homologous pairs of chromosomes at metaphase I is equally likely. As a result, each human gamete contains one of over 8 million possible unique combinations of paternal and maternal chromosomes.

C. Random Fertilization Multiplies the Diversity
- Because any sperm can fertilize any egg cell, a human couple can produce over 70 trillion genetically different offspring.
- Identical (monozygotic) twins arise when a zygote splits into two embryos. Fraternal (dizygotic) twins develop from separate zygotes.

9.6 | Mitosis and Meiosis Have Different Functions: A Summary

- Mitotic division produces identical copies of a cell and occurs throughout life.
- Meiosis produces genetically different haploid cells. It occurs only in specialized cells and only during some parts of the life cycle.

9.7 | Errors Sometimes Occur in Meiosis

A. Polyploidy Means Extra Chromosome Sets
- **Polyploid** cells have one or more extra sets of chromosomes.

B. Nondisjunction Results in Extra or Missing Chromosomes
- **Nondisjunction** is the failure of chromosomes to separate in meiosis, and it causes gametes to have incorrect chromosome numbers. A sex chromosome abnormality is typically less severe than an incorrect number of autosomes.

C. Smaller-Scale Chromosome Abnormalities Also Occur
- Chromosomal rearrangements can **delete** or **duplicate** genes. An **inversion** flips gene order, possibly disrupting vital genes. In a **translocation,** two nonhomologs exchange parts. Some translocations cause cancer.

9.8 | Haploid Nuclei Are Packaged into Gametes

A. In Humans, Gametes Form in Testes and Ovaries
- Spermatogenesis begins in the testes with diploid spermatogonia, which divide mitotically to produce primary spermatocytes (the germ cells). After meiosis I, the cells are haploid secondary spermatocytes. In meiosis II, these cells divide, each yielding two spermatids. The spermatids differentiate along the male reproductive tract, becoming sperm.
- In oogenesis, diploid oogonia divide mitotically, yielding germ cells called primary oocytes. In meiosis I, the primary oocyte divides, distributing cytoplasm to one large secondary oocyte and a much smaller polar body. In meiosis II, the secondary oocyte divides, yielding the large egg cell and another small polar body. Oogenesis occurs in the ovaries.

B. In Plants, Gametophytes Produce Gametes
- In plants, sexual reproduction involves an **alternation of generations** with multicellular haploid and diploid phases. Meiosis occurs in the sporophyte to yield haploid spores, which develop into the haploid gametophyte generation. The gametophyte produces gametes by mitotic cell division.

9.9 | Investigating Life: A New Species Is Born, but Who's the Daddy?

- Analysis of chloroplast DNA has revealed the parental origin of two species of *Tragopogon* plants that have arisen since the 1950s.

Multiple Choice Questions

1. The unique feature of sex is
   a. the ability of a cell to divide.
   b. the production of offspring.
   c. the ability to generate new genetic combinations.
   d. All of the above are correct.

2. A __ cell is ___ and a gamete is ___.
   a. sperm; diploid; haploid   c. germ; haploid; diploid
   b. somatic; diploid; haploid   d. somatic; haploid; diploid

3. Fertilization results in the formation of a
   a. diploid zygote.   c. diploid somatic cell.
   b. haploid gamete.   d. haploid zygote.

4. What is the relationship between homologous chromosomes?
   a. They are exact copies.
   b. They carry the same genes but in different order.
   c. They came from a single parent.
   d. They carry different versions of the same genes.

5. Crossing over occurs during which phase of meiosis?
   a. Prophase I   c. Metaphase II
   b. Metaphase I   d. Anaphase II

6. Which of the following is *not* a mechanism that contributes to diversity?
   a. Random fertilization   c. Cytokinesis
   b. Crossing over   d. Independent assortment

7. Nondisjunction is most likely due to an error at which stage of meiosis?
   a. Prophase II   c. Anaphase I or II
   b. Metaphase I or II   d. Telophase I

8. A translocation occurs when
   a. a part of the chromosome flips and reinserts into the same chromosome.
   b. nonhomologous chromosomes exchange DNA.
   c. a section of a chromosome is lost.
   d. multiple copies of a gene become incorporated into a chromosome.

9. Down syndrome results from which of the following?
   a. An extra chromosome 21
   b. An extra X chromosome
   c. The absence of a Y chromosome
   d. The absence of chromosome 18

10. Why can a gametophyte produce gametes by mitosis?
    a. Because a plant's gametes are diploid
    b. Because the gametophyte's cells are already haploid
    c. Because the gametes will go through meiosis later
    d. Because spores function like germ cells

Write It Out

1. Distinguish between asexual reproduction and sexual reproduction.

2. What is the evidence that sexual reproduction has been successful over evolutionary time?

3. Some fungi reproduce asexually while nutrients are abundant but switch to sexual reproduction when conditions are not as good. Explain this observation.

4. Sketch the relationship between mitosis, meiosis, and fertilization in a sexual life cycle.

5. What is the difference between haploid and diploid cells? Are your skin cells haploid or diploid? What about germ cells? Gametes?

6. Many male veterans of the Vietnam War claim that their children born years later have birth defects caused by a contaminant in the herbicide Agent Orange used as a defoliant in the conflict. What types of cells would the chemical have to have affected in these men to cause birth defects years later? Explain your answer.

7. How are mitosis and meiosis different?

8. Huntington disease is caused by a faulty gene on chromosome 4. Before researchers discovered the actual gene in 1993, people at risk for the disease were tested for the presence of a nearby "marker" on chromosome 4. The marker was not the gene itself, but it reliably predicted who would develop the disease later in life. Sequences farther away from the disease gene, however, were not good predictors. How do the events of crossing over explain this observation? Create a sketch to accompany your answer.

9. Draw all possible metaphase I chromosomal arrangements for a cell with a diploid number of 8. How many unique gametes are possible for this species?

10. A dog has 39 pairs of chromosomes. Considering only the orientation of homologous chromosomes during metaphase I, how many genetically different puppies are possible from the mating of two dogs? Is this number an underestimate or an overestimate? Why?

11. What is the difference between monozygotic and dizygotic twins?

12. Is it possible for a boy–girl pair of twins to be genetically identical? Why or why not?

13. List some examples of chromosomal abnormalities, and explain how each relates to an error in meiosis.

14. Define the following terms: *crossing over, synapsis, gamete, autosome, nondisjunction,* and *homologous pair.*

15. How does spermatogenesis differ from oogenesis, and how are the processes similar?

16. Describe how a plant life cycle may include a multicellular haploid and a diploid phase.

Pull It Together

1. Fit the following terms into this concept map: *chromatid, centromere, nondisjunction, fertilization,* and *mitosis.*

2. What happens in meiosis I and meiosis II?

3. What two processes in meiosis I generate genetic variation among gametes?

4. Why must diploid organisms produce haploid gametes?

5. Where do the two sets of homologous chromosomes in a diploid cell come from?

10

Patterns of Inheritance

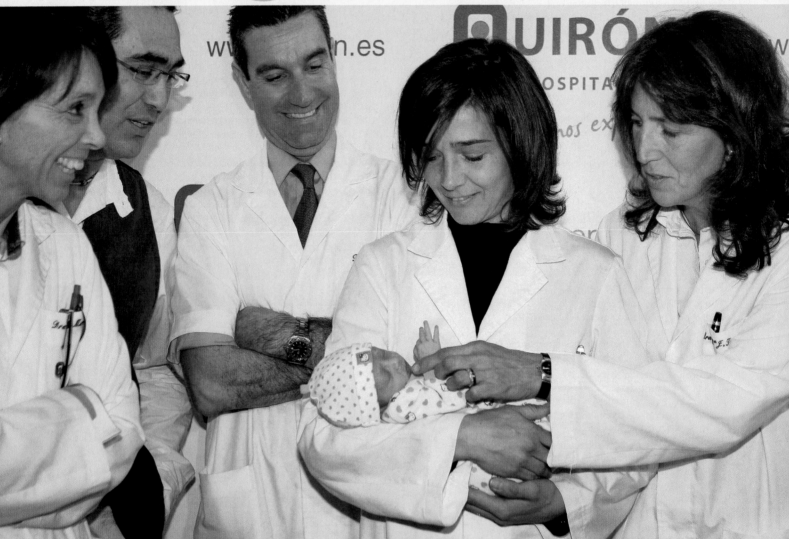

A medical team admires a healthy newborn child after she was born without a genetic disease called facioscapulohumeral muscular dystrophy (FSHD), which runs in the baby's family. While still an embryo conceived by *in vitro* fertilization, her cells were screened and found to be free of the disease-causing allele.

McGraw Hill **connect** |BIOLOGY

Enhance your study of this chapter with practice quizzes, animations and videos, answer keys, and downloadable study tools.
www.mhhe.com/hoefnagels

Learn How to Learn
Be a Good Problem Solver

This chapter is about the principles of inheritance, and you will find many genetics problems among its pages. The guide at the end of this chapter shows a systematic, step-by-step approach to solving three of the most common types of genetics problems. Keep using the guide until you feel comfortable solving any problem type.

From Mendel to Medical Genetics

INTEREST IN HEREDITY IS PROBABLY AS OLD AS HUMANKIND ITSELF. But of all the people who have studied inheritance, one nineteenth-century investigator, Gregor Mendel, made the most lasting impression on what would become the science of genetics.

Mendel was born in 1822 and spent his early childhood in a small village in what is now the Czech Republic, where he learned how to tend fruit trees. After finishing school, Mendel became a priest at a monastery where he could teach and do research in natural science. The young man eagerly learned how to artificially pollinate crop plants to control their breeding. The monastery sent him to earn a college degree at the University of Vienna, where courses in the sciences and statistics fueled his interest in plant breeding. Mendel began to think about experiments to address a compelling question for plant breeders: Why did some traits disappear, only to reappear a generation later?

From 1857 to 1863, Mendel crossed and cataloged some 24,034 plants through several generations. He observed consistent ratios of traits in the offspring and deduced that the plants transmitted distinct units, or "elementen" (now called genes). Mendel described his work to the Brno Medical Society in 1865 and published it in the organization's journal the next year.

Interestingly, Charles Darwin puzzled over natural selection and evolution at the same time that Mendel was tending his plants. No one knew it at the time, but each scientist was exploring genetic variation from a different point of view. Mendel focused on the fate of specific traits from generation to generation; Darwin studied larger-scale shifts in variation within populations. Thanks to another century of biological research, we now know that all variation traces to mutations in DNA. That insight ties together the ideas of Mendel, Darwin, and many other scientists. The so-called "modern evolutionary synthesis" integrates genetic variation, inheritance, and natural selection to explain evolutionary changes in populations. We take up this idea again in unit 3.

Biology has made great strides since Mendel and Darwin's time. Today, genetics and DNA are familiar to nearly everyone, and the entire set of genetic instructions to build a person—the human genome—has been deciphered. The infant pictured at left shows how far medical genetics has come: a medical team confirmed that the baby was free of a genetic disease while she was just a microscopic embryo. Even so, every family grappling with inherited illness encounters the same principles of heredity that Mendel derived in his experiments with peas. Our look at genetics begins the traditional way, with Gregor Mendel, but we can now appreciate his genius in light of what we know about DNA.

UNIT 2

What's the Point?

Learning Outline

10.1 Chromosomes Are Packets of Genetic Information: A Review

10.2 Mendel's Experiments Uncovered Basic Laws of Inheritance
- A. Why Peas?
- B. Dominant Alleles Appear to Mask Recessive Alleles
- C. For Each Gene, a Cell's Two Alleles May Be Identical or Different
- D. Every Generation Has a Name

10.3 The Two Alleles of Each Gene End Up in Different Gametes
- A. Monohybrid Crosses Track the Inheritance of One Gene
- B. Meiosis Explains Mendel's Law of Segregation

10.4 Genes on Different Chromosomes Are Inherited Independently
- A. Dihybrid Crosses Track the Inheritance of Two Genes at Once
- B. Meiosis Explains Mendel's Law of Independent Assortment

10.5 Genes on the Same Chromosome May Be Inherited Together
- A. Genes on the Same Chromosome Are Linked
- B. Studies of Linked Genes Have Yielded Chromosome Maps

10.6 Gene Expression Can Appear to Alter Mendelian Ratios
- A. Incomplete Dominance and Codominance Add Phenotype Classes
- B. Some Inheritance Patterns Are Especially Difficult to Interpret

10.7 Sex-Linked Genes Have Unique Inheritance Patterns
- A. X and Y Chromosomes Determine Sex in Humans
- B. X-Linked Recessive Disorders Affect More Males Than Females
- C. X Inactivation Prevents "Double Dosing" of Proteins

10.8 Pedigrees Show Modes of Inheritance

10.9 Most Traits Are Influenced by the Environment and Multiple Genes
- A. The Environment Can Alter the Phenotype
- B. Polygenic Traits Depend on More Than One Gene

10.10 Investigating Life: Heredity and the Hungry Hordes

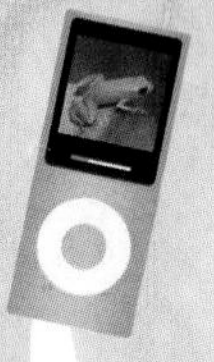

10.1 | Chromosomes Are Packets of Genetic Information: A Review

A healthy young couple, both with family histories of cystic fibrosis, visits a genetic counselor before deciding whether to have children. The counselor suggests genetic tests, which reveal that both the man and the woman are carriers of cystic fibrosis. The counselor tells the couple that each of their future children has a 25% chance of inheriting this serious illness. How does the counselor arrive at that one-in-four chance? This chapter will explain the answer.

First, however, it may be useful to review some concepts from previous chapters in this unit. Chapter 7 explained that cells contain DNA, a molecule that encodes all of the information needed to sustain life. Human DNA includes some 25,000 genes. A **gene** is a portion of DNA whose sequence of nucleotides (A, C, G, and T) encodes a protein. When a gene's nucleotide sequence mutates, the encoded protein may also change. Each gene can therefore exist as one or more **alleles,** or alternative forms, each arising from a different mutation.

The DNA in the nucleus of a eukaryotic cell is divided among multiple **chromosomes,** which are long strands of DNA associated with proteins. Recall that a **diploid cell** contains two sets of chromosomes, with one set inherited from each parent. The human genome consists of 23 pairs of chromosomes (**figure 10.1a**). Of these, 22 pairs are **autosomes,** which are the chromosomes that are the same for both sexes. The single pair of **sex chromosomes** determines whether a person is male or female: a person with two X chromosomes is female, whereas a male has one X and one Y.

With the exception of X and Y, the chromosome pairs are homologous (figure 10.1b). As described in chapter 9, the two members of a **homologous pair** of chromosomes look alike and have the same sequence of genes in the same positions. (A gene's *locus* is its physical place on the chromosome.) But the two homologs may or may not carry the same alleles. Since each homolog comes from a different parent, each person inherits two alleles for each gene in the human genome.

An analogy may help clarify the relationships among these terms. If each chromosome is like a cookbook, then the human genome is a "library" that consists of 46 such volumes, arranged in 23 pairs of similar books. The entire cookbook library includes about 25,000 recipes, each analogous to one gene.

The two alleles for each gene, then, are comparable to two of the many ways to prepare brownies; some recipes include nuts, for example, whereas others use different types of chocolate. The two "brownie recipes" in a cell may be exactly the same, slightly different, or very different from each other. Furthermore, with the exception of identical twins, everyone inherits a unique combination of alleles for all of the genes in the human genome.

Another important idea to review from chapter 9 is the role of meiosis and fertilization in a sexual life cycle (see figure 9.5). **Meiosis** is a specialized form of cell division that occurs in diploid germ cells and gives rise to **haploid cells,** each containing just one

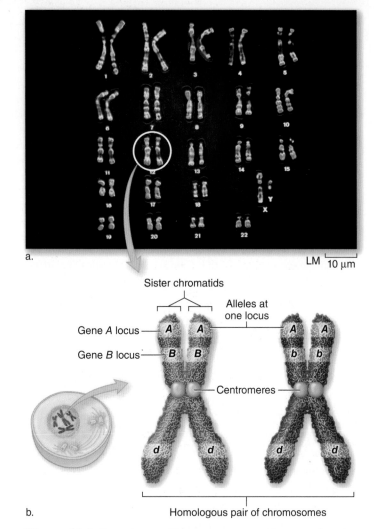

a.   LM ⎸10 μm

Figure 10.1 Homologous Chromosomes. (a) A human diploid cell contains 23 pairs of chromosomes. (b) Each chromosome has one allele for each gene. For the chromosome pair in this figure, both alleles for gene *A* are identical. The same is true for gene *D*, but the chromosomes carry different alleles for gene *B*.

set of chromosomes. In humans, these haploid cells are **gametes**—sperm or egg cells. **Fertilization** unites the gametes from two parents, producing the first cell of the next generation. Gametes are the cells that convey chromosomes from one generation to the next, so they play a critical part in the study of inheritance.

No one can examine a gamete and say for sure which allele it carries for every gene. As we shall see in this chapter, however, for some traits, we can use knowledge of a person's characteristics and family history to say that a gamete has a 100% chance, 50% chance, or 0% chance of carrying a specific allele. With this information for both parents, it is simple to calculate the probability that a child will inherit the allele.

10.1 | Mastering Concepts

1. Describe the relationships among chromosomes, DNA, genes, and alleles.
2. How do meiosis, fertilization, diploid cells, and haploid cells interact in a sexual life cycle?

10.2 | Mendel's Experiments Uncovered Basic Laws of Inheritance

Gregor Mendel, the nineteenth-century researcher who discovered the basic principles of genetics (see the chapter opening essay), knew nothing about DNA, genes, chromosomes, or meiosis. But he nevertheless discovered how to calculate the probabilities of inheritance, at least for some traits. This section explains how he used careful observations of pea plants to draw his conclusions.

A. Why Peas?

As Mendel discovered, the pea plant *(Pisum sativum)* is a good choice for studying heredity. Pea plants are easy to grow, develop quickly, and produce many offspring. Moreover, peas have many traits that appear in two easily distinguishable forms. For example, seeds may be round or wrinkled, yellow or green. Pods may be smooth or may conform to the shape of the peas inside. Stems may be tall or short.

Pea plants also have another advantage for studies of inheritance: it is easy to control which plants mate with which (**figure 10.2**). An investigator can take pollen from the male flower parts of one plant and apply it to the female part of the same plant (self-fertilization) or another plant (cross-fertilization). The resulting offspring are seeds that develop inside pods; each pea represents a genetically unique offspring, analogous to you and your siblings. Traits such as seed color or seed shape are evident right away; for other characteristics, such as plant height or flower color, the investigator must sow the seeds and observe each plant that develops.

B. Dominant Alleles Appear to Mask Recessive Alleles

Mendel's first experiments with peas dealt with single traits that have two expressions, such as yellow or green seed colors. He set up all possible combinations of crosses: yellow with yellow, green with green, and yellow with green (**figure 10.3**).

Mendel noted that some plants were always **true-breeding;** that is, self-fertilization always produced offspring identical to the parent plant. Plants derived from green seeds, for example, always produced green seeds when self-fertilized. But crosses involving yellow-seeded plants were more variable. Sometimes these plants were true-breeding, but in other cases, the offspring included a mix of yellow and green seeds. Sometimes the green trait vanished in one generation, only to reappear in the next.

Mendel noticed a similar mode of inheritance when he studied other pea plant characteristics: one trait seemed to obscure the other. Mendel called the masking trait *dominant;* the trait being masked is called *recessive.* The yellow-seed trait, for example, is dominant to the green trait.

Although Mendel referred to *traits* as dominant or recessive, modern biologists reserve these terms for *alleles.* A **dominant allele** is one that exerts its effects whenever it is present; a **recessive allele** is one whose effect is masked if a dominant allele is also present. When a gene has only two alleles, it is common to symbolize the dominant allele with a capital letter (such as *Y* for yellow) and the recessive allele with the corresponding lowercase letter (*y* for green).

The "dominance" of an allele may seem to imply that it "dominates" in the population as a whole. The most common allele, however, is not always the dominant one. In humans, the allele that causes a form of dwarfism called achondroplasia is dominant, but it is very rare—as is the dominant allele that causes Huntington disease. Conversely, blue eyes are the norm in people of northern European origin, but the alleles that produce this eye color are recessive.

The term *dominant* also conjures images of a bully that forces a weak, recessive allele into submission. After all, the recessive allele seems to hide when a dominant allele is present, emerging from its hiding place only if the dominant allele is

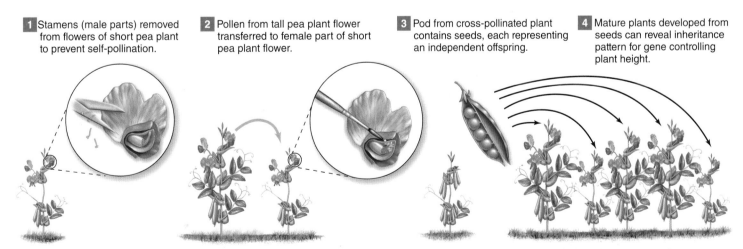

1 Stamens (male parts) removed from flowers of short pea plant to prevent self-pollination.

2 Pollen from tall pea plant flower transferred to female part of short pea plant flower.

3 Pod from cross-pollinated plant contains seeds, each representing an independent offspring.

4 Mature plants developed from seeds can reveal inheritance pattern for gene controlling plant height.

Figure 10.2 Mendel's Experimental Approach for Breeding Peas. One of the advantages of working with pea plants is that an investigator can easily control which plants breed with each other. Gregor Mendel used this technique to set up carefully designed crosses of pea plants, so he could observe the appearance of traits in the next generation.

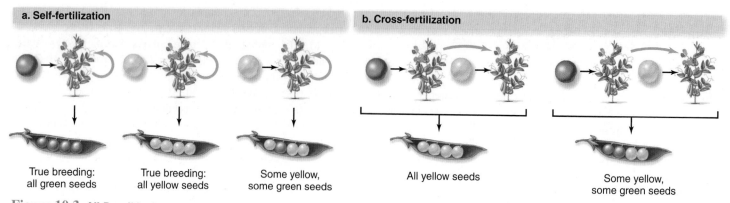

Figure 10.3 All Possible Crosses. (a) When Mendel self-fertilized green-seeded pea plants, the pods contained only green seeds. Self-fertilizing yellow-seeded plants, however, sometimes yielded only yellow seeds; other times, the green trait appeared among the offspring. (b) A cross between a yellow-seeded plant and a green-seeded plant could produce either all yellow seeds or a mixture of offspring.

absent. How does the recessive allele "know" what to do? In fact, alleles cannot hide, emerge, or know anything. A recessive allele remains a part of the cell's DNA, regardless of the presence of a dominant allele. It only seems to hide because it codes for a nonfunctional protein. If a dominant allele is also present, the organism usually has enough of the functional protein to maintain its normal appearance (although section 10.6 describes some exceptions). It is only when both alleles are recessive that the lack of the functional protein becomes noticeable. This chapter's first Burning Question, on page 203, describes a health-related consequence of a nonfunctional protein encoded by a recessive allele.

C. For Each Gene, a Cell's Two Alleles May Be Identical or Different

Mendel chose traits encoded by genes with only two alleles, but some genes have hundreds of forms. Regardless of the number of possibilities, however, a diploid cell can have only two alleles for each gene. After all, each diploid individual has inherited one set of chromosomes from each parent, and each chromosome carries only one allele per gene.

For a given gene, a diploid cell's two alleles may be identical or different. The **genotype** expresses the genetic makeup of an individual, and it is written as a pair of letters representing the alleles. An individual that is **homozygous** for a gene has two identical alleles, meaning that both parents contributed the same gene version. If both of the alleles are dominant, the individual's genotype is homozygous dominant (written as *YY,* for example). If both alleles are recessive, the individual is homozygous recessive *(yy)*. An individual with a **heterozygous** genotype, on the other hand, has two different alleles for the gene *(Yy)*. That is, the two parents each contributed different genetic information.

The organism's genotype is distinct from its **phenotype,** or observable characteristics. Flower color, seed color, and stem length are examples of pea plant phenotypes that Mendel

studied. Your own phenotype includes not only your height, eye color, shoe size, number of fingers and toes, skin color, and hair texture but also other characteristics that are not readily visible, such as your blood type or the specific shape of your hemoglobin proteins. As described in section 10.9, most phenotypes result from a complex interaction between genes and environment. Mendel, however, chose traits controlled exclusively by genes.

Mendel's observation that only some yellow-seeded pea plants were true-breeding arises from the two possible genotypes for the yellow phenotype (homozygous dominant and heterozygous). All homozygous plants are true-breeding because all of their gametes contain the same allele. Heterozygous plants, however, are not true-breeding because they may pass on either the dominant or the recessive allele.

Today, biologists use additional terms to describe organisms. A **wild-type** allele, genotype, or phenotype is the most common form or expression of a gene in a population. Wild-type fruit flies, for example, have two antennae and one pair of wings. A **mutant** allele, genotype, or phenotype is a variant that arises when a gene undergoes a mutation. Mutant phenotypes for fruit flies include having multiple pairs of wings or having legs instead of antennae growing out of the head (see figure 7.21).

D. Every Generation Has a Name

Part of Mendel's genius was that he kept careful tallies of the offspring from countless crosses, which required a systematic accounting of multiple generations of plants. Today's biologists still use Mendel's system of standardized names to keep track of inheritance patterns.

The purebred **P generation** (for "parental") is the first set of individuals being mated; the **F_1 generation,** or first filial generation, is the offspring from the P generation (*filial* derives from the Latin word for "child"). The **F_2 generation** is the offspring of the F_1 plants, and so on. Although these terms are applicable only to

Table 10.1 Miniglossary of Genetic Terms

Term	Definition
Generations	
P	The parental generation
F_1	The first filial generation; offspring of P generation
F_2	The second filial generation; offspring of F_1 generation
Chromosomes and genes	
Chromosome	A continuous molecule of DNA plus associated proteins
Gene	A sequence of DNA that encodes a protein
Locus	The physical location of a gene on a chromosome
Allele	One of the alternative forms of a specific gene
Dominant and recessive	
Dominant allele	An allele that is expressed if present in the genotype
Recessive allele	An allele whose expression is masked by a dominant allele
Genotypes and phenotypes	
Genotype	An individual's allele combination for a particular gene
Homozygous	Possessing identical alleles of one gene
Heterozygous	Possessing different alleles of one gene
Phenotype	An observable characteristic
True breeding	Homozygous; self-fertilization yields offspring identical to self for a given trait
Wild type	The most common phenotype, genotype, or allele in a population
Mutant	A phenotype, genotype, or allele resulting from a mutation in a gene

Burning Question

Why does diet soda have a warning label?

Foods containing the artificial sweetener aspartame carry a warning label that says "Phenylketonurics: contains phenylalanine." Only people with a metabolic disorder called phenylketonuria (abbreviated PKU) need to heed this warning.

Aspartame contains an amino acid called phenylalanine. In most people, an enzyme converts phenylalanine into another amino acid. A mutated allele of the gene encoding this enzyme, however, results in the production of an abnormal, nonfunctional enzyme.

People who have just one copy of this recessive allele are healthy because the cell has enough of the normal enzyme, thanks to the dominant allele. The recessive allele seems to "vanish," just as in Mendel's pea plants. Individuals who inherit two copies of the recessive allele, however, cannot produce the normal enzyme. The disease symptoms appear when phenylalanine accumulates to toxic levels, causing mental retardation and other problems. Avoiding foods containing phenylalanine helps minimize the effects of the disease—hence the warning.

Submit your burning question to:
marielle_hoefnagels@
mcgraw-hill.com

lab crosses, they are analogous to human family relationships. If you consider your grandparents the P generation, your parents are the F_1 generation, and you and your siblings are the F_2 generation.

Table 10.1 summarizes the important terms encountered so far. The remainder of the chapter uses this basic vocabulary to integrate Mendel's findings with what biologists now know about genes, chromosomes, and reproduction.

10.2 | Mastering Concepts

1. Why did Gregor Mendel choose pea plants as his experimental organism?
2. Distinguish between dominant and recessive; heterozygous and homozygous; phenotype and genotype; wild type and mutant.
3. Define the P, F_1, and F_2 generations.

10.3 | The Two Alleles of Each Gene End Up in Different Gametes

Mendel used a systematic series of crosses to deduce the rules of inheritance, beginning with single genes.

A. Monohybrid Crosses Track the Inheritance of One Gene

Mendel began with a P generation consisting of true-breeding plants derived from yellow seeds *(YY)* and true-breeding green-seeded plants *(yy)*. The F_1 offspring produced in this cross had yellow seeds (genotype *Yy*). The green trait therefore seemed to disappear in the F_1 generation.

Next, he used the F_1 plants to set up a **monohybrid cross:** a mating between two individuals that are both heterozygous for the same gene. The resulting F_2 generation had both yellow and green phenotypes, in a ratio of 3:1; that is, for every three yellow seeds, Mendel observed one green seed.

A diagram called a **Punnett square** uses the genotypes of the parents to reveal which allele combinations the offspring may inherit. The Punnett square in figure 10.4, for example, shows how the green phenotype reappeared in the F_2 generation. In a monohybrid cross, both parents are heterozygous *(Yy)* for the seed color gene. Each therefore produces some gametes carrying the *Y* allele and some gametes carrying *y*. All three possible genotypes may therefore appear in the F_2 generation, in the ratio 1 *YY:* 2 *Yy:* 1 *yy.* The corresponding phenotypic ratio is three yellow seeds to one green seed, or 3:1. Mendel saw similar results for all seven traits that he studied (table 10.2).

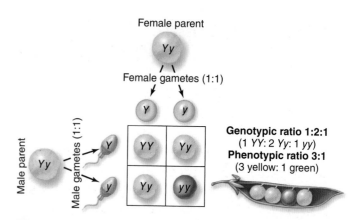

Figure 10.4 Punnett Square. This diagram depicts Mendel's monohybrid cross of two heterozygous yellow-seeded *(Yy)* pea plants. The two possible types of female gametes are listed along the top of the square; the two possible male gametes are listed on the left-hand side. Each compartment within the square contains the genotype and phenotype that results when the corresponding gametes join.

Figure It Out

If Mendel mated a true-breeding tall plant with a heterozygous tall plant, what percent of the offspring would also be tall?

Answer: 100%

Mendel could tally the plants with each phenotype, but he also needed to keep track of each genotype. He knew that the green-seeded plants were always homozygous recessive *(yy)*. But what was the genotype of each yellow seed, *YY* or *Yy*? He had no way to tell just by looking, so he set up breeding experiments called test crosses to distinguish between the two possibilities. A **test cross** is a mating between an individual of unknown

Table 10.2   Mendel's Monohybrid Crosses

Experiment	Total	Plants Expressing Dominant Allele	Plants Expressing Recessive Allele	Ratio*
1. Seed form	7324	5474 round *(R)*	1850 wrinkled *(r)*	2.96:1
2. Seed color	8023	6022 yellow *(Y)*	2001 green *(y)*	3.01:1
3. Pod form	1181	882 inflated *(V)*	299 restricted *(v)*	2.95:1
4. Pod color	580	428 green *(G)*	152 yellow *(g)*	2.82:1
5. Flower position	858	651 axial *(F)*	207 terminal *(f)*	3.14:1
6. Seed coat color	929	705 gray *(A)*	224 white *(a)*	3.15:1
7. Stem length	1064	787 tall *(L)*	277 short *(l)*	2.84:1

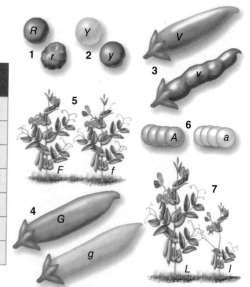

* Each ratio deviates slightly from the expected 3:1 because inheritance reflects the rules of probability. Repeating each experiment would likely yield slightly different ratios, each very close to 3:1.

If plant is homozygous dominant (YY):

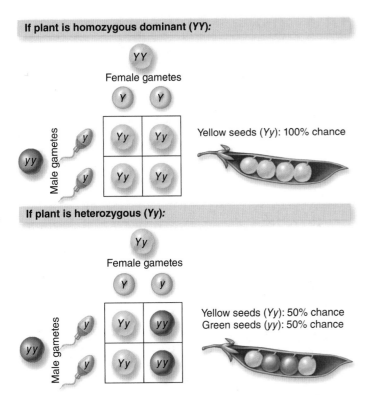

Female gametes

Yellow seeds (Yy): 100% chance

If plant is heterozygous (Yy):

Female gametes

Yellow seeds (Yy): 50% chance
Green seeds (yy): 50% chance

Figure 10.5 **Test cross.** A yellow-seeded pea plant may be homozygous dominant *(YY)* or heterozygous *(Yy)*. To determine the genotype, the plant is mated with a homozygous recessive *(yy)* plant. If the unknown plant is *YY,* all offspring of the test cross will share its phenotype; if the unknown plant is *Yy,* about half the offspring are likely to produce green seeds.

genotype and a homozygous recessive individual (figure 10.5). If a yellow-seeded plant crossed with a *yy* plant produced only yellow seeds, Mendel knew the unknown genotype was *YY;* if the cross produced seeds of both colors, he knew it must be *Yy.*

B. Meiosis Explains Mendel's Law of Segregation

All of Mendel's breeding experiments and calculations added up to a brilliant description of basic genetic principles. Without any knowledge of chromosomes or genes, Mendel used his data to conclude that genes occur in alternative versions (which we now call alleles). He further determined that each individual inherits two alleles for each gene and that these alleles may be the same or different. Finally, he deduced his **law of segregation,** which states that the two alleles of each gene are packaged into separate gametes; that is, they "segregate," or move apart from each other, during gamete formation. (In science, a *law* is a statement about a phenomenon that is invariable, at least as far as anyone knows. Unlike a theory, a law does not necessarily explain the phenomenon.)

Mendel's law of segregation makes perfect sense in light of what we now know about reproduction. During meiosis I, homologous pairs of chromosomes separate and move to opposite poles of the cell. A plant of genotype *Yy* therefore produces equal numbers of gametes carrying *Y* or *y,* whereas a *YY* plant produces only *Y* gametes (figure 10.6). When gametes from the two parents meet at fertilization, they combine at random. About 50% of the time, both gametes carry *Y;* the other 50% of the time, one contributes *Y* and the other; *y.*

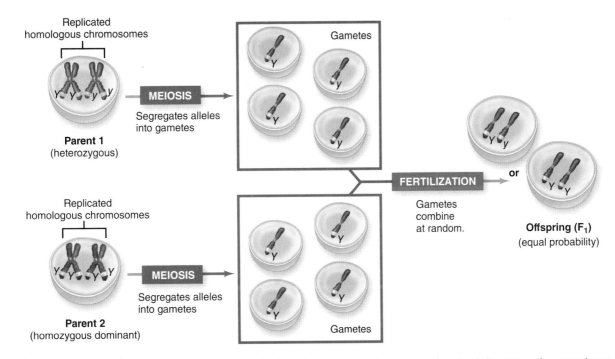

Figure 10.6 **Mendel's Law of Segregation.** During meiosis, homologous pairs of chromosomes (and the genes they carry) segregate from one another and are packaged into separate gametes. At fertilization, gametes combine at random to form the next generation (in this figure, red and blue denote different parental origins of the chromosomes).

Mother: healthy carrier
Female gametes

Healthy noncarrier (*FF*): 25% chance
Healthy carrier (*Ff*): 50% chance
Affected (*ff*): 25% chance

Figure 10.7 **Mendel's Law Applied to Humans.** This Punnett square shows the possible results of a mating between two carriers of cystic fibrosis. The two brothers in the photo have cystic fibrosis; they are undergoing treatment to reduce the buildup of sticky mucus in their lungs.

This principle of inheritance applies to all diploid species, including humans. Return for a moment to the couple and their genetic counselor introduced in section 10.1. Cystic fibrosis arises when a person has two recessive alleles for a particular gene on chromosome 7 (see section 4.6). Genetic testing revealed that the man and the woman are both carriers. In genetic terms, this means that although neither has the disease, both are heterozygous for the gene that causes cystic fibrosis. Just as in Mendel's monohybrid crosses, each of their children has a 25% chance of inheriting two recessive alleles (**figure 10.7**). Each child also has a 50% chance of being a carrier (heterozygous) and a 25% chance of inheriting two dominant alleles.

Note that Punnett squares, including the one in figure 10.7, show the *probabilities* that apply to each offspring. That is, if the couple has four children, there will not necessarily be exactly one with genotype *FF,* two with *Ff,* and one with *ff.* Similarly, the chance of tossing a fair coin and seeing "heads" is 50%, but two tosses will not necessarily yield one head and one tail. If you toss the coin 1000 times, however, you will likely approach the expected 1:1 ratio of heads to tails. As Mendel discovered, pea plants are ideal for genetics studies in part because they produce many offspring in each generation.

10.3 | Mastering Concepts

1. What is a monohybrid cross, and what are the genotypic and phenotypic ratios expected in the offspring of the cross?
2. How are Punnett squares helpful in following inheritance of single genes?
3. What is a test cross, and why is it useful?
4. How does the law of segregation reflect the events of meiosis?

10.4 | Genes on Different Chromosomes Are Inherited Independently

Mendel's law of segregation arose from his studies of the inheritance of single traits. He next asked himself whether the same law would apply if he followed two different characters at the same time. Would one trait influence the inheritance of the other, or would each trait follow its own independent inheritance pattern?

Mendel therefore began another set of breeding experiments in which he simultaneously examined the inheritance of two characteristics of peas: shape and color. A pea's shape may be round or wrinkled (determined by the *R* gene, with the dominant allele specifying round shape). At the same time, its color may be yellow or green (determined by the *Y* gene, with the dominant allele specifying yellow).

A. Dihybrid Crosses Track the Inheritance of Two Genes at Once

As he did before, Mendel began with a P generation consisting of true-breeding parents (**figure 10.8a**). He crossed plants grown from wrinkled, green seeds (homozygous recessive for both genes, *rr yy*) with plants derived from round, yellow seeds (homozygous dominant for genes *R* and *Y,* denoted *RR YY*). All F_1 offspring were heterozygous for both genes (*Rr Yy*) and therefore had round, yellow seeds.

Next, Mendel crossed F_1 plants with each other (figure 10.8b). A **dihybrid cross** is a mating between individuals that are each heterozygous for two genes. Each *Rr Yy* individual in the F_1 generation produced equal numbers of gametes of four different types: *R Y, R y, r Y,* and *r y.* After Mendel completed the crosses, he found four phenotypes in the F_2 generation, reflecting all possible combinations of seed shape and color. The Punnett square predicts that the four phenotypes will occur in a ratio of 9:3:3:1. That is, nine of 16 offspring should have round, yellow seeds; three should have round, green seeds; three should have wrinkled, yellow seeds; and just one should have wrinkled, green seeds. This prediction almost exactly matches Mendel's results.

Figure It Out

In a cross between an *Rr Yy* plant and an *rr yy* plant, what proportion of the offspring should be homozygous recessive for both seed shape and seed color?

Answer: 25%.

B. Meiosis Explains Mendel's Law of Independent Assortment

Based on the results of the dihybrid cross, Mendel proposed what we now know as the **law of independent assortment.** It states that during gamete formation, the segregation of the alleles for

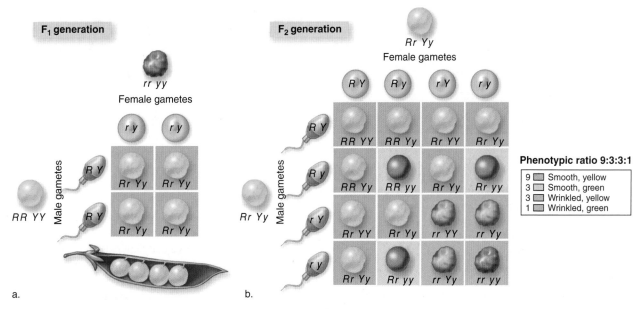

Figure 10.8 **Plotting a Dihybrid Cross.** (a) In the parental generation, one parent is homozygous dominant for both genes; the other is homozygous recessive. The F₁ generation is therefore heterozygous for both. (b) When plants from the F₁ generation are self-fertilized, phenotypes occur in a distinctive ratio in the F₂ generation.

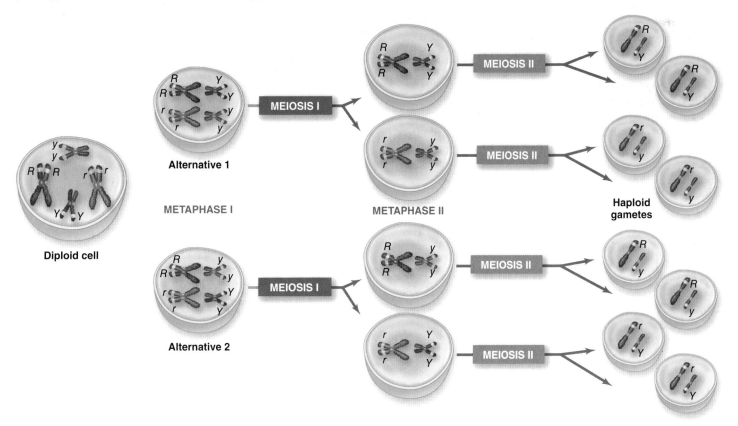

Figure 10.9 **Mendel's Law of Independent Assortment.** Homologous chromosome pairs align at random during metaphase I of meiosis. The exact allele combination in a gamete depends on which chromosomes happen to be packaged together. An individual of genotype *Rr Yy* therefore produces approximately equal numbers of four types of gametes: *RY, ry, Ry,* and *rY.*

one gene does not influence the alleles for another gene (provided the genes are on separate chromosomes). That is, alleles for two different genes are randomly packaged into gametes with respect to each other. With this second set of experiments, Mendel had again inferred a principle of inheritance based on meiosis (**figure 10.9**).

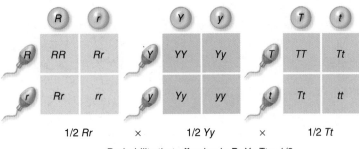

1/2 Rr　　×　　1/2 Yy　　×　　1/2 Tt

Probability that offspring is *Rr Yy Tt* = 1/8

Figure 10.10　The Product Rule. What is the chance that two parents that are heterozygous for three genes *(Rr Yy Tt)* will give rise to an offspring with that same genotype? To find out, multiply the individual probabilities for each gene.

Punnett squares become cumbersome when analyzing more than two genes. A Punnett square for three genes has 64 boxes; for four genes, 256 boxes. An easier way to predict genotypes and phenotypes is to use the rules of probability on which Punnett squares are based. The **product rule** states that the chance that two independent events will both occur (for example, an offspring inheriting specific alleles for two genes) equals the product of the individual chances that each event will occur.

The product rule can predict the chance of obtaining wrinkled, green seeds *(rr yy)* from dihybrid *(Rr Yy)* parents. Consider the dihybrid individual one gene at a time. The probability that two *Rr* plants will produce *rr* offspring is 25%, or ¼. Similarly, the chance of two *Yy* plants producing a *yy* individual is ¼. According to the product rule, the chance of dihybrid parents *(Rr Yy)* producing homozygous recessive *(rr yy)* offspring is therefore ¼ multiplied by ¼, or ¹⁄₁₆. Now consult the 16-box Punnett square for Mendel's dihybrid cross (see figure 10.8). As expected, only one of the 16 boxes contains *rr yy*. **Figure 10.10** applies the product rule to three traits.

Interestingly, Mendel found some trait combinations for which a dihybrid cross did not yield the expected phenotypic ratio. Mendel could not explain this result. No one could, until Thomas Hunt Morgan's work led to the chromosomal theory of inheritance. As you will see in section 10.5, the law of independent assortment does not apply to genes that are close together on the same chromosome.

10.4 | Mastering Concepts

1. What is a dihybrid cross, and what is the phenotypic ratio expected in the offspring of the cross?
2. How does the law of independent assortment reflect the events of meiosis?
3. How can the product rule be used to predict the results of crosses in which multiple genes are studied simultaneously?

10.5 | Genes on the Same Chromosome May Be Inherited Together

Biologists did not appreciate the significance of Gregor Mendel's findings during his lifetime, but his careful observations laid the foundation for modern genetics. In 1900, three botanists working independently each rediscovered the principles of inheritance. They eventually found the paper that Mendel had published in 1866, and other scientists demonstrated Mendel's ratios again and again in several species. At about the same time, advances in microscopy were allowing chromosomes to be observed and described for the first time.

It soon became apparent that what Mendel called "elementen" (later renamed "genes") had much in common with chromosomes. Both genes and chromosomes, for example, come in pairs. In addition, alleles of a gene are packaged into separate gametes, as are the members of a homologous pair of chromosomes. Finally, both genes and chromosomes are inherited in random combinations.

As biologists cataloged traits and the chromosomes that transmit them in several species, it soon became clear that the number of traits far exceeded the number of chromosomes. Fruit flies, for example, have only four pairs of chromosomes, but they have dozens of different bristle patterns, body colors, eye colors, wing shapes, and other characteristics. How might a few chromosomes control so many traits? The answer: each chromosome carries many genes.

A. Genes on the Same Chromosome Are Linked

Linked genes are carried on the same chromosome; they are therefore inherited together. Unlike genes on different chromosomes, they do not assort independently during meiosis. The seven traits that Mendel followed in his pea plants were all transmitted on separate chromosomes. Had the same chromosome carried these genes, Mendel's dihybrid crosses would have generated markedly different results.

The inheritance pattern of linked genes was first noticed in the early 1900s, when William Bateson and R. C. Punnett observed offspring ratios in pea plants that were different from the ratios predicted by Mendel's laws. The researchers crossed true-breeding plants that had crimson flowers and long pollen grains (genotype *PP LL*) with true-breeding plants that had red flowers and round pollen grains (genotype *pp ll*). Then they crossed the heterozygous F_1 plants, of genotype *Pp Ll*, with each other.

Surprisingly, the F_2 generation did not show the expected phenotypic ratio for an independently assorting dihybrid cross (**figure 10.11**). Two types of F_2 offspring—those with the same phenotypes as the P generation—were more abundant than predicted. The other two classes of offspring, with a mix of phenotypes (genotypes *pp L_* and *P_ ll*) were less common. Bateson and Punnett hypothesized that this pattern reflected two genes located on the same chromosome, as depicted in figure 10.11b.

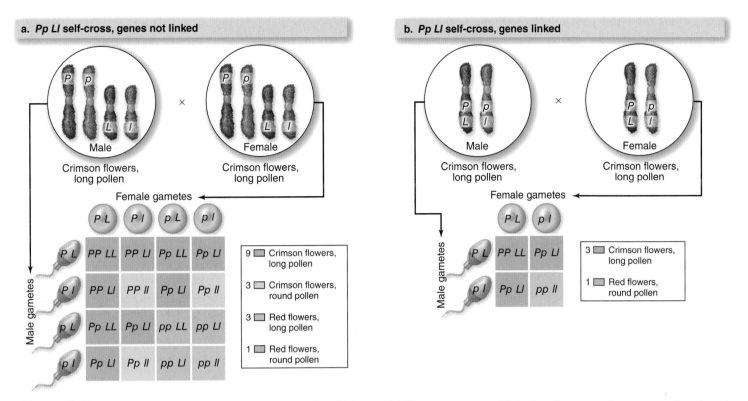

a. Pp Ll self-cross, genes not linked

Male — Crimson flowers, long pollen × Female — Crimson flowers, long pollen

Female gametes: P L, P l, p L, p l

Male gametes: P L, P l, p L, p l

	P L	P l	p L	p l
P L	PP LL	PP Ll	Pp LL	Pp Ll
P l	PP Ll	PP ll	Pp Ll	Pp ll
p L	Pp LL	Pp Ll	pp LL	pp Ll
p l	Pp Ll	Pp ll	pp Ll	pp ll

9 ▭ Crimson flowers, long pollen
3 ▭ Crimson flowers, round pollen
3 ▭ Red flowers, long pollen
1 ▭ Red flowers, round pollen

b. Pp Ll self-cross, genes linked

Male — Crimson flowers, long pollen × Female — Crimson flowers, long pollen

Female gametes: P L, p l

Male gametes: P L, p l

	P L	p l
P L	PP LL	Pp Ll
p l	Pp Ll	pp ll

3 ▭ Crimson flowers, long pollen
1 ▭ Red flowers, round pollen

Figure 10.11 Gene Linkage Changes the Results of a Dihybrid Cross. (a) When genes are not linked on the same chromosome, they assort independently. The gametes then represent all possible allele combinations, and the cross yields the expected phenotypic ratio. (b) If genes are linked on the same chromosome, only two allele combinations are expected in the gametes, and the number of phenotypes is reduced.

While Bateson and Punnett were studying linkage in peas, Thomas Hunt Morgan at Columbia University was breeding the fruit fly *Drosophila melanogaster* and studying the inheritance of pairs of traits. The data began to indicate four **linkage groups,** collections of genes that tended to be inherited together. Within each linkage group, dihybrid crosses did not produce the proportions of offspring that Mendel's law of independent assortment predicts. Because the number of linkage groups was the same as the number of homologous pairs of chromosomes, scientists eventually realized that each

linkage group was simply a set of genes transmitted together on the same chromosome. ▶ Focus on Model Organisms *(Drosophila)*, p. 425

Nevertheless, the researchers did sometimes see offspring with trait combinations not seen in either parent. How could this occur? The answer turned out to involve yet another event in meiosis: **crossing over,** an exchange of genetic material between homologous chromosomes during prophase I (see figure 9.8). After crossing over, no two chromatids in a homologous pair of chromosomes are identical (figure 10.12).

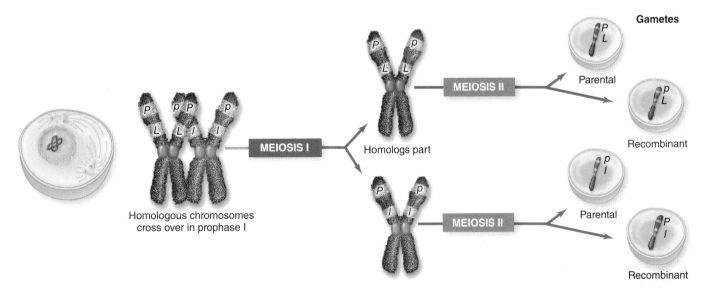

Homologous chromosomes cross over in prophase I → MEIOSIS I → Homologs part → MEIOSIS II → Gametes: Parental (P L), Recombinant (p L); Parental (p l), Recombinant (P l)

Figure 10.12 Crossing Over. Linkage between two alleles is interrupted if crossing over occurs at a point between the two genes. As a result, some gametes contain recombinant arrangements of the alleles.

Crossing over is a random process, and it might or might not occur between two linked genes. For any pair of genes on the same chromosome, then, most offspring inherit a **parental chromatid,** which retains the allele combination from each parent. But whenever crossing over happens between the two genes, some of the offspring inherit a **recombinant chromatid** with a mix of maternal and paternal alleles.

B. Studies of Linked Genes Have Yielded Chromosome Maps

Morgan wondered why some crosses produced a higher proportion of recombinant offspring than others. Might the differences reflect the physical relationships between the genes on the chromosome? Alfred Sturtevant, Morgan's undergraduate assistant, explored this idea. In 1911, Sturtevant proposed that the farther apart two alleles are on the same chromosome, the more likely crossing over is to separate them—simply because more space separates the genes (figure 10.13a).

Sturtevant's idea became the basis for mapping genes on chromosomes. By determining the percentage of recombinant offspring, investigators can infer how far apart the genes are on one chromosome. Crossing over frequently separates alleles on opposite ends of the same chromosome, so recombinant offspring occur frequently. In contrast, a crossover would rarely separate alleles lying very close together on the chromosome, and the proportion of recombinant offspring would be small. Geneticists use this correlation between crossover frequency and the distance between genes to construct **linkage maps,** which are diagrams of gene order and spacing on chromosomes.

In 1913, Sturtevant published the first genetic linkage map, depicting the order of five genes on the X chromosome of the fruit fly (figure 10.13b). Researchers then rapidly mapped genes on all four fruit fly chromosomes. Linkage maps for human chromosomes followed over the next half century.

At one time, phenotypes were the basis for linkage maps. Now, however, genetic marker technology can associate a known, detectable DNA sequence with a specific phenotype. The marker does not have to be part of the gene that controls the phenotype; the two must simply be located close enough together that the presence of the marker correlates strongly with the phenotype.

Genetic markers are useful tools for predicting the chance that a person will develop a particular inherited illness, even if the actual disease-causing gene is unknown. The first such use of genetic markers occurred in the 1980s. An extensive study of a small Venezuelan village with a high incidence of Huntington disease revealed that all family members with the disorder also carried a unique marker on chromosome 4. Healthy relatives did not have this sequence. Like a flag, the marker indicates the presence of the disease-causing allele to which it is closely linked. The actual gene was not identified until the 1990s, but in the meantime, researchers could identify who was likely to carry the Huntington disease–causing allele long before the symptoms developed.

The linkage maps of the mid-twentieth century provided the rough drafts to which DNA sequence information was added later in the century. Today, entire genomes are routinely sequenced using powerful computers that assemble many short overlapping DNA sequences. ▶ DNA sequencing, p. 170

10.5 | Mastering Concepts

1. How do patterns of inheritance differ for unlinked versus linked pairs of genes?
2. What is the difference between recombinant and parental chromatids, and how do they arise?
3. How do biologists use crossover frequencies to map genes on chromosomes?

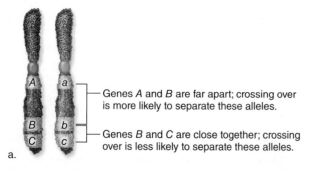

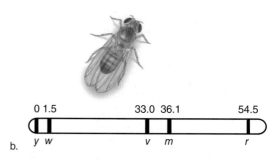

Genes *A* and *B* are far apart; crossing over is more likely to separate these alleles.

Genes *B* and *C* are close together; crossing over is less likely to separate these alleles.

a.

0 1.5   33.0 36.1   54.5

y w   *v   m*   *r*

b.

Figure 10.13 **Breaking Linkage.** (a) Crossing over is more likely to separate the alleles of genes *A* and *B* (or *A* and *C*) than to separate the alleles of genes *B* and *C,* because there is more room for an exchange to occur. (b) A linkage map of a fruit fly chromosome, showing the locations of five genes. The numbers represent crossover frequencies relative to the leftmost gene, *y.*

10.6 | Gene Expression Can Appear to Alter Mendelian Ratios

Mendel's crosses yielded easily distinguishable offspring. A pea is either yellow or green, round or wrinkled; a plant is either tall or short. Often, however, offspring traits do not occur in the proportions that Punnett squares or probabilities predict. It may appear that Mendel's laws do not apply—but they do. The underlying genotypic ratios are there, but the nature of the phenotype, other genes, or the environment alter how traits appear. Sections 10.6 and 10.7 describe some situations that may alter Mendelian ratios.

A. Incomplete Dominance and Codominance Add Phenotype Classes

For the traits Mendel studied, one allele is completely dominant and the other is completely recessive. The phenotype of a heterozygote is therefore identical to that of a homozygous dominant individual. For many genes, however, heterozygous offspring do not share the phenotype of either parent. As you will see, biologists apply unique notation to alleles for these traits. Designating one allele with a capital letter and the other in lowercase does not work, because neither allele is dominant over the other.

When a gene shows **incomplete dominance,** the heterozygote has a third phenotype that is intermediate between those of the two homozygotes. For example, a red-flowered snapdragon of genotype r_1r_1 crossed with a white-flowered r_2r_2 plant gives rise to an r_1r_2 plant with pink flowers (**figure 10.14**). The single copy of allele r_1 in the pink heterozygote directs less pigment production than the two copies in a red-flowered r_1r_1 plant. Although the red color seems to be "diluted" in the all-pink F_1 generation, the Punnett square shows that crossing two of these pink plants can yield F_2 offspring with red, white, or pink flowers.

In **codominance,** two different alleles are expressed together in the phenotype. For example, a person's ABO blood type is determined by the I gene, which has three possible alleles: I^A, I^B, and i (**figure 10.15**). The I gene encodes an enzyme that inserts either an "A" or a "B" molecule onto the surfaces of red blood cells. Allele i is recessive, so a person with genotype ii produces neither molecule A nor molecule B and therefore has type O blood. A person who produces only molecule A (genotype I^AI^A or

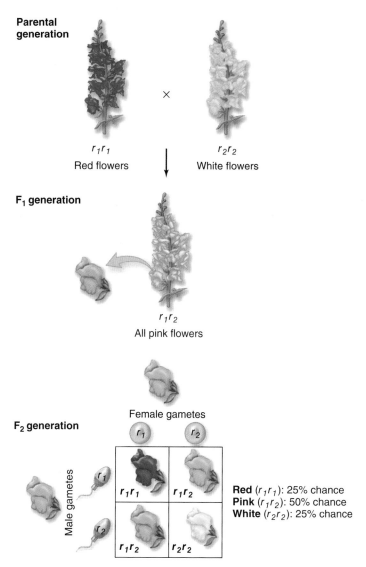

Figure 10.14 Incomplete Dominance. In snapdragons, a cross between a plant with red flowers (r_1r_1) and a plant with white flowers (r_2r_2) produces a heterozygous plant with pink flowers (r_1r_2). The red and white phenotypes reappear in the F_2 generation.

Genotypes	Phenotypes	
	Surface molecules	ABO blood type
I^AI^A I^Ai	Only A	Type A 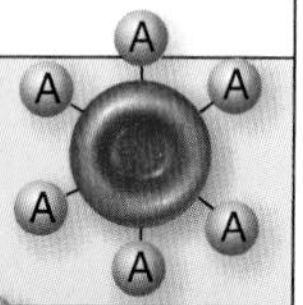
I^BI^B I^Bi	Only B	Type B 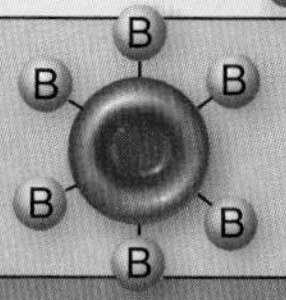
I^AI^B	Both A and B	Type AB 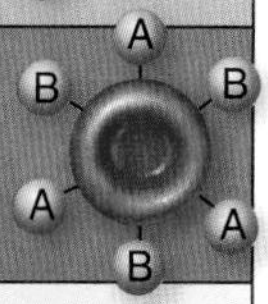
ii	None	Type O 

Figure 10.15 Codominance. The I^A and I^B alleles of the I gene are codominant, meaning that both are expressed in a heterozygote. Allele i is recessive.

$I^A i$) has type A blood; likewise, someone with only molecule B (genotype $I^B I^B$ or $I^B i$) has type B blood. But genotype $I^A I^B$ yields type AB blood. The I^A and I^B alleles are codominant because both are equally expressed when both are present.

What is the difference between the recessive i allele and the codominant I^A and I^B alleles? Recall that a recessive allele encodes a nonfunctional protein. Both alleles I^A and I^B code for functional proteins, so people of blood type AB have both molecules A and B on the surfaces of their red blood cells. Neither allele is silent.

The ABO blood type example also illustrates another condition that can alter phenotypic ratios: a gene with three or more possible alleles can yield many phenotypes. In the ABO blood type system, two codominant alleles (I^A and I^B) and one recessive allele (i) produce six possible genotypes and four phenotypes (blood types A, B, AB, and O).

Figure It Out

A woman with type AB blood has children with a man who has type O blood. What is the probability that a child they conceive will have type B blood?

Answer: 50%

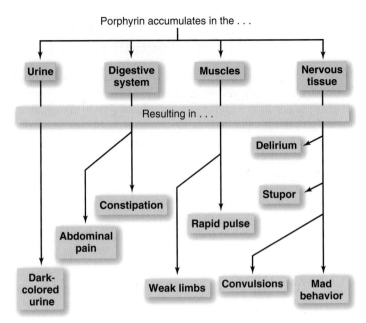

Figure 10.16 Pleiotropy. Complex networks of enzymes normally convert molecules called porphyrins into other compounds. In a disorder called porphyria, at least one of the enzymes is nonfunctional, so porphyrins build up. The resulting effects on the body are widespread.

B. Some Inheritance Patterns Are Especially Difficult to Interpret

Some conditions are especially difficult to trace through families. For example, one gene may influence the phenotype in many ways; conversely, multiple genes may contribute to one phenotype. Although the basic rules of inheritance apply to each gene, the patterns of phenotypes that appear in grandparents, parents, and siblings may be hard to interpret.

Pleiotropy   In **pleiotropy,** one gene has multiple effects on the phenotype. Pleiotropy arises when one protein is important in different biochemical pathways or affects more than one body part or process. For example, a single connective tissue protein abnormality causes Marfan syndrome, with symptoms including long limbs, spindly fingers, a caved-in chest, a weakened aorta, and lens dislocation. Volleyball player and Olympic silver medalist Flo Hyman had the disease; she died in 1986 during a game in Japan. Abraham Lincoln may also have had Marfan syndrome. Like achondroplasia and Huntington disease, Marfan syndrome is an example of an uncommon disorder that is caused by a dominant allele. ▶ connective tissue, p. 515

Another famous example of pleiotropy is a disorder that is believed to have afflicted Britain's King George III in the late 1700s. Starting at age 50, he experienced recurrent bouts of a long list of symptoms. In the twentieth century, researchers realized that one of George's symptoms, dark-colored urine, may have

indicated an illness known as porphyria variegata (figure 10.16). Lack of a liver enzyme results in buildup of compounds called porphyrins (hence the name, porphyria). Biologists still do not understand how this condition causes the symptoms. Careful study of George's relatives revealed that some of them had different subsets of the symptoms, which were originally thought to be unrelated diseases.

Although Mendelian rules of inheritance apply to pleiotropic genes, the conditions they cause can be difficult to trace through families because individuals with different subsets of symptoms may appear to have different disorders.

Protein Interactions   With thousands of genes active in a cell, it is not surprising that many proteins can interact to contribute to a single phenotype. These interactions can complicate the analysis of inheritance patterns because mutations in different genes can produce similar or identical phenotypes.

Some interactions occur among multiple proteins involved in the same biochemical pathway. For example, blood clot formation requires 11 separate biochemical reactions. A different gene encodes each enzyme in the pathway, and clotting disorders may result from abnormalities in any of these genes. The phenotypes are the same (poor blood clotting), but the genotypes differ. Porphyria likewise has several forms because eight enzymes participate in the conversion of porphyrins to other compounds. Mutations in any of the genes encoding these enzymes can cause different forms of porphyria.

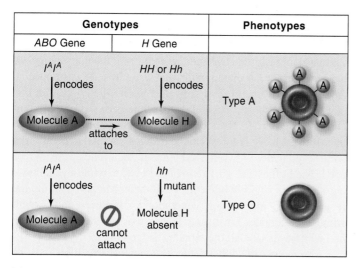

Genotypes		Phenotypes
ABO Gene	*H* Gene	
$I^A I^A$ ↓ encodes Molecule A ····· attaches to → Molecule H	*HH* or *Hh* ↓ encodes Molecule H	Type A
$I^A I^A$ ↓ encodes Molecule A ⊘ cannot attach	*hh* ↓ mutant Molecule H absent	Type O

Figure 10.17 Epistasis. The *H* gene encodes a protein that normally attaches molecule A or B to the red blood cells of a person with alleles I^A or I^B. But if the genotype for the *H* gene is *hh*, then the molecule has nothing to attach to. The person's blood will test as type O, even though allele I^A or I^B is still present.

Another type of interaction is **epistasis,** which occurs when one gene's product affects the expression of another gene. As a simple example, male pattern baldness (a genetic condition) is associated with gradual hair loss, which hides the effects of the allele for a "widow's peak" hairline.

Epistasis is also responsible for some ABO blood type inconsistencies between parents and their children (figure 10.17). Most people with alleles I^A or I^B have blood type A, B, or AB. However, a protein must physically link the A and B molecules to the blood cell's surface. If a gene called *H* is mutated, then part of that protein is missing, and A and B cannot attach.

As a result, a person with the extremely rare genotype *hh* has blood that always tests as type O, even though he or she may have a genotype indicating type A, B, or AB blood. Confusion over paternity might arise if two parents have ABO genotypes indicating that type O offspring are not possible, yet they have a child with type O blood. Moreover, a person with genotype *hh* can only receive blood from other *hh* individuals. The explanation is that the intact H protein in normal blood can induce an immune reaction in a person whose blood lacks that protein.

10.6 | Mastering Concepts

1. How do incomplete dominance and codominance increase the number of phenotypes?
2. What is pleiotropy?
3. How can the same phenotype stem from many different genotypes?
4. How can epistasis decrease the number of phenotypes observed in a population?

10.7 | Sex-Linked Genes Have Unique Inheritance Patterns

Huntington disease, cystic fibrosis, and other conditions controlled by genes that are on autosomes affect both sexes equally. A few conditions, including red–green color blindness and hemophilia, however, occur much more frequently in males than females. Phenotypes that affect one sex more than the other are **sex-linked;** that is, the alleles controlling them are on the X or Y chromosomes.

A. X and Y Chromosomes Determine Sex in Humans

In many species, including humans, the sexes have equal numbers of autosomes but differ in the types of sex chromosomes they have. Females have two X chromosomes, whereas males have one X and one much smaller Y chromosome (figure 10.18). People who seek to increase the odds of conceiving a boy or a girl can exploit this size difference between X and Y chromosomes, as described in this chapter's Apply It Now box.

The Y chromosome plays the largest role in human sex determination. All human embryos start with rudimentary

Figure 10.18 The Sperm Determines the Sex of the Baby. In humans, each egg contains 23 chromosomes, one of which is a single X chromosome. A sperm cell's 23 chromosomes include either an X or a Y chromosome. If a Y-bearing sperm cell fertilizes an egg, the baby will be a male (XY). If an X-bearing sperm cell fertilizes an egg, the baby will be a female (XX).

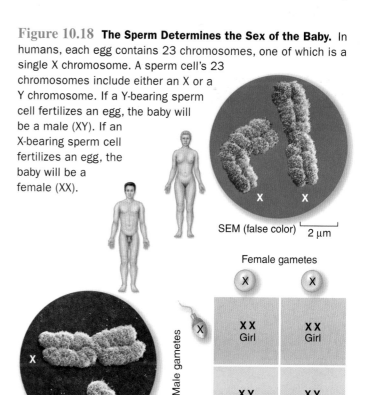

SEM (false color) ⊢ 2 μm

SEM (false color) ⊢ 2 μm

	Female gametes	
	X	X
X (Male gametes)	**X X** Girl	**X X** Girl
Y (Male gametes)	**X Y** Boy	**X Y** Boy

Girl (XX): 50% chance
Boy (XY): 50% chance

female structures (see figure 34.18). An embryo having a working copy of a Y-chromosome gene called *SRY* (for *sex-determining region of the Y*) develops into a male. *SRY* encodes a protein that switches on other genes that direct the undeveloped testes to secrete the male sex hormone testosterone. Cascades of other gene activities follow, promoting the development of male sex organs. The SRY protein also turns on a gene encoding a protein that dismantles embryonic female structures.

Despite this critical role, the Y chromosome carries fewer than 100 genes. Scientists therefore know of very few **Y-linked** disorders; most involve defects in sperm production.

The human X chromosome, on the other hand, carries more than 1000 protein-encoding genes, most of which have nothing to do with sex determination. Most human sex-linked traits are therefore **X-linked;** that is, they are controlled by genes on the X chromosome.

B. X-Linked Recessive Disorders Affect More Males Than Females

Thomas Hunt Morgan was the first to unravel the unusual inheritance patterns associated with genes on the X chromosome (figure 10.19). The eyes of fruit flies are normally red, but one day Morgan discovered a male with white eyes. To study the inheritance of this odd phenotype, he created true-breeding lines of flies with each eye color. When he mated a parental generation of white-eyed males with red-eyed females, the F_1 flies were all red-eyed. The F_2 flies had a 3:1 ratio of red-eyed to white-eyed flies—but all the flies with white eyes were male. Morgan also did the reverse cross, mating a P generation of red-eyed males with white-eyed females. That time, all the males of the F_1 generation had white eyes, and all the females had red eyes.

Morgan reasoned that the recessive white-eye allele must be on the X chromosome. Because X and Y carry different genes, a

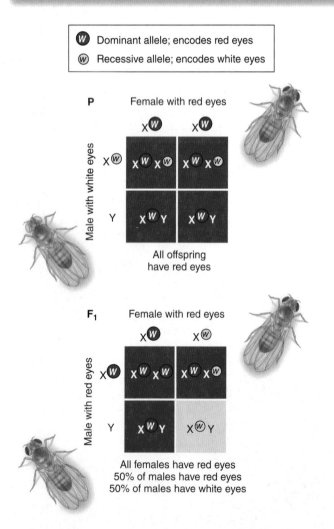

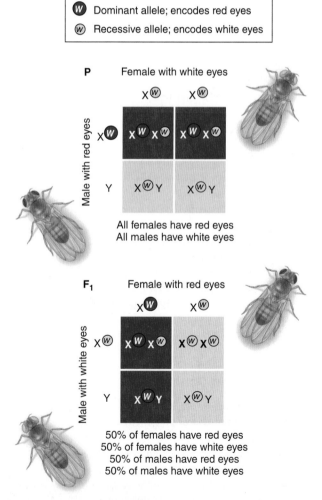

Figure 10.19 Fly Eyes. Thomas Hunt Morgan first described the unusual inheritance patterns of X-linked traits. In the crosses shown, *W* represents the dominant allele, which encodes red eyes, and *w* is the recessive allele for white eyes. (a) A cross between a true-breeding white-eyed male with a red-eyed female. (b) A cross between a true-breeding red-eyed male and a white-eyed female. Had this eye color gene been on an autosome, the outcome of the cross shown in (a) would have been the same as that in (b).

male fly can never "mask" the white-eye allele on his X chromosome with a corresponding dominant allele. A female, however, has two X chromosomes. She will express the white-eye phenotype only if *both* of her X chromosomes carry the recessive eye color allele. X-linked inheritance therefore explained why white eyes appeared more frequently in males than in females.

Recessive alleles cause most X-linked disorders in humans, although a few are associated with dominant alleles (**table 10.3**). As in fruit flies, a human female inherits an X chromosome from both parents; a male inherits his X chromosome from his mother (see figure 10.18). A female therefore exhibits an X-linked recessive disorder only if she inherits the recessive alleles from both parents. A male, in contrast, expresses every allele on his single X chromosome, whether dominant or recessive.

Figure 10.20 shows the inheritance of hemophilia A, a disorder with an X-linked recessive mode of inheritance. In hemophilia, a protein called a clotting factor is missing or defective. Blood therefore clots very slowly, and bleeding is excessive. The heterozygous female in the Punnett square does not exhibit symptoms because her dominant allele encodes a functional blood-clotting protein. When she has children with a normal male, however, each son has a 50% chance of being affected, and each daughter has a 50% chance of being a carrier.

Figure It Out

A woman who is heterozygous for the X-linked gene associated with color blindness marries a color-blind man. What is the probability that their first child will be a son who is color-blind?

Answer: 25%.

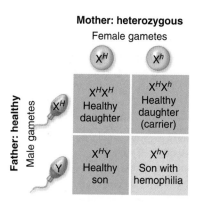

Figure 10.20

Hemophilia A. This Punnett square depicts a cross between a heterozygous female (a "carrier" of hemophilia A) and a normal male, the most common way to transmit any X-linked recessive allele.

Healthy daughter, noncarrier ($X^H X^H$): 25% chance
Healthy daughter, carrier ($X^H X^h$): 25% chance
Healthy son ($X^H Y$): 25% chance
Affected son ($X^h Y$): 25% chance

C. X Inactivation Prevents "Double Dosing" of Proteins

Relative to males, female mammals have a "double dose" of every gene on the X chromosome. Cells balance this inequality by **X inactivation,** in which a cell shuts off all but one X chromosome in each cell. This process happens early in the embryonic development of a mammal.

X inactivation is directly observable because a turned-off X chromosome absorbs a stain much more readily than an active X chromosome does; the inactivated X forms a Barr body that is visible with a microscope. A normal male cell has no Barr bodies because the single X chromosome remains active.

Table 10.3 Some X-Linked Disorders in Humans

Disorder	Genetic Explanation	Characteristics
X-linked recessive inheritance		
Duchenne muscular dystrophy	Mutant allele for gene encoding dystrophin	Rapid muscle degeneration early in life
Fragile X syndrome	Unstable region of X chromosome has unusually high number of CCG repeats.	Most common form of inherited mental retardation
Hemophilia A	Mutant allele for gene encoding blood clotting protein (factor VIII)	Uncontrolled bleeding, easy bruising
Red–green color blindness	Mutant alleles for genes encoding receptors for red or green (or both) wavelengths of light	Reduced ability to distinguish between red and green
Rett syndrome	Mutant allele for gene required for development of nerve cells	Severe developmental disorders. Almost all affected children are female; affected male embryos cease development before birth.
X-linked dominant inheritance		
Extra hairiness (congenital generalized hypertrichosis; some forms)	Mechanism unknown	Many more hair follicles than normal
Hypophosphatemic rickets (some forms)	Mutant allele for gene involved in phosphorus absorption	Low blood phosphorus level causes defective bones
Retinitis pigmentosa (some forms)	Mutant allele for cell-signaling protein; mechanism unknown	Defects in retina cause partial blindness

Apply It Now

Choosing the sex of your baby

Scientists have used the size difference between the X and Y chromosomes to develop technologies that may help people choose the sex of their babies. Technologies for sex selection include:

- **MicroSort:** Chromosomes in a sperm sample are stained with a fluorescent dye. A sperm cell with an X chromosome absorbs more dye and therefore glows more brightly than one with a Y chromosome.

- **Ericsson method:** Sperm swim through a container holding a thick fluid. The Y-carrying sperm are lighter and should be faster swimmers than the heavier X-carrying sperm. Therefore, the Y-carrying sperm cells reach the bottom of the container faster than their slower counterparts. Removing the sperm from the top or the bottom of the container should produce a sample enriched in X- or Y-carrying sperm cells, respectively.

- **Spin method:** A sperm sample is placed in test tube, which is placed into a centrifuge. The centrifuge spins the test tube so that the heavier (X-containing) sperm fall to the bottom, whereas the lighter sperm remain near the top.

After Y-carrying sperm are separated from X-carrying sperm, the woman is inseminated with the desired fraction of the sperm. Alternatively, if a woman is using *in vitro* fertilization, the egg is fertilized in the laboratory with the sperm most likely to give a baby of the desired sex. None of the methods, however, works all the time; at best, the sperm samples are enriched in sperm that favor one sex over the other.

Which X chromosome becomes inactivated—the one inherited from the father or the one from the mother—is a random event. As a result, a female expresses the paternal X chromosome alleles in some cells and the maternal alleles in others. Moreover, when a cell with an inactivated X chromosome divides mitotically, all of the daughter cells have the same X chromosome inactivated. Because the inactivation occurs early in development, females have patches of tissue that differ in their expression of X-linked alleles. **Figure 10.21** shows how inactivation of an X-linked coat color gene causes the distinctive appearance of calico and tortoiseshell cats, which are always female (except for rare XXY males).

X chromosome inactivation also explains another interesting observation: X-linked dominant disorders are typically less severe in females than in males. Thanks to X chromosome inactivation, a female who is heterozygous for an X-linked gene will express a dominant disease-causing allele in only some of her cells. As a result, she experiences less severe symptoms than an affected male, who expresses the dominant allele in every cell.

10.7 | Mastering Concepts

1. What determines a person's sex?
2. What is the role of the *SRY* gene in sex determination?
3. Why do males and females express recessive X-linked alleles differently?
4. Why does X inactivation occur in female mammals?

a.

b.

Figure 10.21 X Inactivation. In cats, the X chromosome carries a coat color gene with alleles for black or orange coloration. Calico and tortoiseshell cats are heterozygous for this gene; one of the two X chromosomes is inactivated in each colored patch. (a) X inactivation happened early in the development of this cat, producing large patches of cells with the same color. (A different gene accounts for the white background). (b) X inactivation occurred later in this cat, producing smaller patches. Can you see why calico cat clones do *not* share identical fur color patterns?

10.8 | Pedigrees Show Modes of Inheritance

Although Gregor Mendel did not study human genetics, our species nevertheless has some "Mendelian traits": those determined by single genes with alleles that are either dominant or recessive. For these traits, how do we know whether each phenotype is associated with a dominant or a recessive allele?

We have already seen in chapter 9 that karyotypes are useful for diagnosing chromosomal abnormalities. The same tool is useless in determining the inheritance of single-gene disorders, however, because different alleles look exactly alike under a microscope. To discover the inheritance pattern of an individual gene, researchers therefore track its incidence over multiple generations, much as Mendel did with his pea plants more than a century ago.

Genes on autosomes exhibit two modes of inheritance: autosomal dominant and autosomal recessive (table 10.4). An **autosomal dominant** disorder is expressed in heterozygotes and therefore typically appears in every generation. Because the allele is dominant and located on an autosome, one or both of the affected individual's parents must also have the disorder (unless the disease-causing allele arose by a new mutation). If a generation arises in which no individuals inherit the allele by chance, transmission of the disorder stops in that family.

Inheriting an **autosomal recessive** disorder requires that a person receive the disease-causing allele from both parents. Each parent must therefore have at least one copy of the allele, either because they are homozygous recessive and have the disease or because they are heterozygotes ("carriers"). If both parents are carriers, autosomal recessive conditions may seem to disappear in one generation, only to reappear in the next.

X-linked recessive conditions such as hemophilia produce unique inheritance patterns. Such disorders mostly affect males in a family; females are rarely affected, but many are heterozygous "carriers" for the disease-causing allele. Each child of a carrier has a 50% chance of inheriting the recessive allele from the mother. If a son receives the recessive allele, he will have the condition. Any daughter who inherits the recessive allele from the mother, however, will be a carrier—unless she also inherits the same allele from her father.

Table 10.4 Some Autosomal Dominant and Autosomal Recessive Disorders in Humans

Disorder	Genetic Explanation	Characteristics
Autosomal dominant inheritance		
Achondroplasia	Mutant allele of gene on chromosome 4 causes deficiency of receptor protein for growth factor.	Dwarfism with short limbs; head and trunk sizes are normal
Familial hypercholesterolemia	Mutant allele of gene on chromosome 2 encodes faulty cholesterol-binding protein.	High cholesterol, heart disease
Huntington disease	Mutant allele of gene on chromosome 4 encodes protein with extra amino acids that cause it to misfold and form clumps in brain cells.	Progressive uncontrollable movements and personality changes, beginning in middle age
Marfan syndrome	Mutant allele of gene on chromosome 15 causes connective tissue disorder.	Long limbs, sunken chest, lens dislocation, spindly fingers, weakened aorta
Neurofibromatosis (type 1)	Mutant allele of gene on chromosome 17 encodes faulty cell signaling protein.	Brown skin marks (café-au-lait spots), benign tumors beneath skin
Polydactyly	Multiple genes on multiple chromosomes; mechanism is unknown.	Extra fingers or toes or both
Autosomal recessive inheritance		
Albinism	Mutant allele of gene on chromosome 11 encodes faulty gene in biochemical pathway required for pigment production.	Lack of pigmentation in skin, hair, and eyes
Cystic fibrosis	Mutant allele of gene on chromosome 7 encodes faulty chloride channel protein.	Lung infections and congestion, infertility, poor fat digestion, poor weight gain, salty sweat
Phenylketonuria (PKU)	Mutant allele of gene on chromosome 12 causes enzyme deficiency in biochemical pathway that breaks down the amino acid phenylalanine.	Buildup of phenylalanine and related compounds causes mental retardation
Tay-Sachs disease	Mutant allele of gene on chromosome 15 causes deficiency of lysosome enzyme.	Buildup of byproducts causes nervous system degeneration

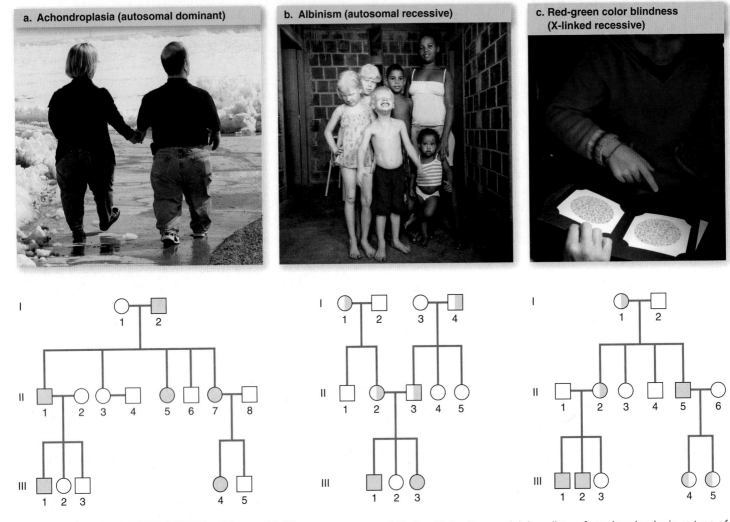

Figure 10.22 **Pedigrees Reveal Mode of Inheritance.** (a) A pedigree for achondroplasia, a type of dwarfism with an autosomal dominant mode of inheritance. (b) A pedigree for albinism, which has an autosomal recessive mode of inheritance. (c) A pedigree for red–green color blindness, a disorder with an X-linked recessive mode of inheritance.

Pedigree charts depicting family relationships and phenotypes are useful for determining the mode of inheritance (figure 10.22). In a pedigree chart, squares indicate males and circles denote females. Colored shapes indicate individuals with the disorder, and half-filled shapes represent known carriers. Horizontal lines connect parents. Siblings connect to their parents by vertical lines and to each other by an elevated horizontal line.

Figure 10.22 shows typical pedigrees for genes carried on autosomes and the X chromosome. In studying the differences among these three pedigrees, notice especially the patterns of the colored shapes. For example, affected individuals appear in every generation for autosomal dominant traits, whereas recessive conditions often skip generations. Also, note that males are most commonly affected with X-linked recessive disorders.

Despite their utility, pedigrees can be difficult to construct and interpret for several reasons. People sometimes hesitate to supply information because they are embarrassed or want to protect their privacy. Adoption, children born out of wedlock, serial marriages and the resulting blended families, and assisted reproductive technologies complicate tracing family ties. In addition, many people cannot trace their families back far enough to reveal modes of inheritance.

10.8 | Mastering Concepts

1. How are pedigrees helpful in determining a disorder's mode of inheritance?

2. How do the pedigrees differ for autosomal dominant, autosomal recessive, and X-linked recessive conditions?

10.9 | Most Traits Are Influenced by the Environment and Multiple Genes

Mendel's data were clear enough for him to infer principles of inheritance because he observed characteristics determined by single genes with two easily distinguished alleles. Moreover, the traits he selected are unaffected by environmental conditions. A genetic counselor can likewise be confident in telling two cystic fibrosis carriers that each of their children has a 25% probability of getting the disease.

But the counselor cannot calculate the probability that the child will be an alcoholic, have depression, be a genius, or wear size 9 shoes. The reason is that multiple genes and the environment control these and most other traits.

A. The Environment Can Alter the Phenotype

The external environment can profoundly affect gene expression; that is, a gene may be active in one circumstance but inactive in another. As a simple example, temperature influences the quantity of pigment molecules in the fur of some animals. Siamese cats and Himalayan rabbits have dark ears, noses, feet, and tails because these parts are colder than the animals' abdomens (**figure 10.23**). In crocodiles, the incubation temperature of the egg determines whether the baby will develop as a male or a female. Different combinations of genes are activated at each temperature, greatly altering the phenotype of the offspring.

In humans, fetal alcohol syndrome is an example of the effect of environment on phenotype: prenatal exposure to alcohol can cause a developing baby to develop facial abnormalities or epilepsy. Likewise, personal circumstances ranging from hormone levels to childhood experiences to diet influence a person's susceptibility to depression, alcoholism, and type II diabetes.

These three diseases have a genetic component as well, but sorting out the relative contributions of "nature" and "nurture" is difficult. Studies of twins are often helpful, as are careful observations of everything from family composition to brain structures. The Burning Question in chapter 7, for example, explains some strategies that researchers use to explore the genetic connection to homosexuality. Discovering the exact mechanism by which genes interact with external stimuli is a very active area of research.

Even human diseases with simple, single-gene inheritance patterns can have an environmental component. Cystic fibrosis, for example, is a single-gene disorder. Because cystic fibrosis patients are very susceptible to infection, however, the course of the illness depends on which infectious agents a person encounters.

Burning Question

Is male baldness really from the female side of the family?

Male pattern baldness is the distinctive hair loss that many men (and some women) experience as they enter their 20s, 30s, and 40s. The baldness spreads outward from the temples and crown of the head in a characteristic pattern.

Two conditions are required for male pattern baldness to develop. First, hormones called androgens must be present in larger than normal amounts. Testosterone and dihydrotestosterone (DHT) are androgens; they bind to and enter hair follicle cells, interacting with the DNA to stop growth of the hair follicle. Second, the individual must have a genetic predisposition for the condition.

Many people believe that a man inherits the tendency for baldness from his mother's side of the family, but multiple genes apparently control this trait, and at least some of the genes reside on autosomes. However, researchers have found that one of the most important factors controlling male pattern baldness is the structure of an androgen receptor protein. The X chromosome carries the gene encoding this critical protein. Therefore, men inherit at least part of the tendency for baldness from their mothers.

Submit your burning question to:
marielle_hoefnagels@mcgraw-hill.com

B. Polygenic Traits Depend on More Than One Gene

Unlike cystic fibrosis, most inherited traits are **polygenic;** that is, the phenotype reflects the activities of more than one gene. Male pattern baldness is one example of a condition influenced by multiple genes, as described in the Burning Question on this page. Eye color is another example of a polygenic trait. The production of pigments in the eye's iris is influenced by multiple enzymes, which are encoded by multiple genes.

Figure 10.23 Environment Affects Phenotype. In Siamese cats, the gene encoding an enzyme required for pigment production is mutated. The heat-sensitive enzyme is active in cool areas, such as the paws, ears, snout, and tail. But the enzyme is inactive at body temperature, so the cat's fur remains light-colored where the skin is warmer.

Figure 10.24 Height Is Polygenic and Environmental.
(a) These students from the University of Connecticut at Storrs lined up by height in 1920, demonstrating the characteristic bell-shaped distribution of polygenic traits. (b) A similar photo taken in 1997 reveals more tall students than lineups photographed early in the twentieth century, illustrating the influence of improved health care and nutrition. In 1920, the tallest student was 6'2"; in 1997, the tallest was 6'5".

To complicate matters, the environment often profoundly affects the expression of both single-gene and polygenic traits. For example, in plants, polygenic traits typically include flower color, the density of leaf pores (stomata), and crop yield. But these traits do not remain static throughout a plant's life. Soil pH can affect flower color, CO_2 concentration can change the number of stomata per square centimeter, and nutrient and water availability greatly influence crop production. ▸ stomata, p. 462

When the frequencies of all the phenotypes associated with a polygenic trait are plotted on a graph, they form a characteristic bell-shaped curve. **Figure 10.24** shows a bell curve for height, which is a product of genetics, childhood nutrition, and health care. **Figure 10.25** shows a continuum of gene expression for skin color, a trait that is affected by genes and exposure to sunlight. Body weight and intelligence are other traits that are both polygenic and influenced by the environment.

10.9 | Mastering Concepts

1. How can the environment affect a phenotype?
2. What is a polygenic trait?

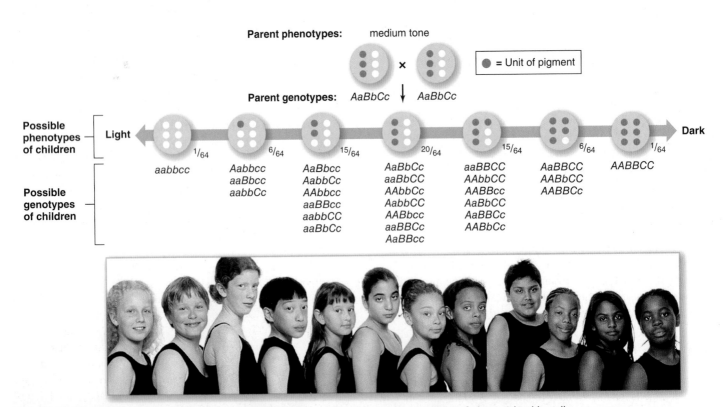

Figure 10.25 Skin Color Is Polygenic. Multiple genes interact to determine the quantity of pigment in skin cells.

10.10 | Investigating Life: Heredity and the Hungry Hordes

Agriculture provides a steady food supply, but not just to humans. Hungry insects and other animals can devastate crops by eating the leaves, roots, seeds, and fruits of the plants we grow for food or fiber. Farmers continually seek new ways to kill these competitors, but each tactic selects for new adaptations in the insects. It is a long-standing, and seemingly unavoidable, evolutionary arms race.

The larvae of butterflies and moths are especially voracious. A good example is the pink bollworm (*Pectinophora gossypiella;* figure 10.26). The adults of this species are moths that lay eggs on cotton bolls. When the eggs hatch, the pink caterpillars tunnel into the boll and eat the seeds, damaging the cotton fibers.

One tool that keeps bollworms and other caterpillars at bay is a soil bacterium called *Bacillus thuringiensis,* abbreviated Bt. This microbe produces a toxic protein that pokes holes in a caterpillar's intestinal tract, leaving the animal vulnerable to infection and unable to digest food. Bt does not affect humans, because we lack the specific molecule to which the toxin binds. Bt is among the few insecticides that organic farmers can spray on their plants.

Bt is not perfect. Some caterpillars, such as those that tunnel inside an ear of corn, escape death because they never eat the Bt sprayed on a plant's surface. In the 1990s, however, biotechnology solved this problem when scientists inserted the bacterial gene for the Bt toxin into plant cells. Every cell in these so-called Bt plants produces the toxin. Nibbling on any hidden or exposed plant part spells death to the caterpillar. ▶ transgenic organisms, p. 141

Genetically modified Bt corn and cotton are enormously popular with farmers in the United States, greatly increasing the selection pressure for Bt resistance in insect populations. When Bt was available only as a spray, insect populations encountered it infrequently. But genetically modified crops produce the toxin throughout their lives, so caterpillars are exposed from the moment they begin feeding. Large fields of Bt plants are therefore likely to eliminate most of the susceptible individuals, leaving resistant caterpillars to produce the next generation.

To combat this selective pressure, farmers growing Bt crops must agree to surround each field with a buffer strip planted with a conventional (non-Bt) variety of the same crop—a refuge. The

Figure 10.26 Hungry Caterpillar. Pink bollworms cost cotton producers tens of millions of dollars each year.

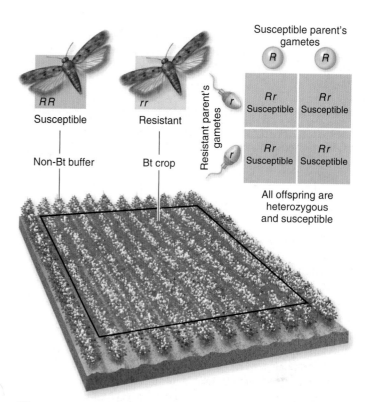

Figure 10.27 Bt Crops Require Buffers. This Punnett square shows the logic of using buffer strips around Bt crops to help keep the incidence of recessive alleles low.

success of the refuge strategy relies on the assumption that most Bt resistance alleles are recessive. A homozygous recessive, resistant moth emerging from the Bt crop has a good chance of encountering a susceptible mate from the buffer strip (figure 10.27). All of their heterozygous offspring are susceptible and therefore will die if they eat the Bt plants. A ready pool of susceptible mates from the refuge should therefore keep the recessive allele rare.

How can scientists test whether refuges really do keep recessive alleles rare in the real world? One solution is to find a way to measure a bollworm's genotype directly. Biologists Shai Morin, Bruce Tabashnik, and their colleagues at the University of Arizona tackled this problem by studying Bt resistance in several populations of pink bollworms. One population, reared for decades in a lab without exposure to Bt, was susceptible. Other groups, originally collected in Arizona and Texas, were artificially selected for resistance by feeding them Bt-laced meals and allowing the survivors to reproduce. ▶ artificial selection, p. 233

Laboratory experiments confirmed that alleles conferring Bt resistance are indeed recessive (figure 10.28). In addition, from previous studies, the researchers knew that Bt resistance comes from changes in a protein called cadherin, which is the target molecule to which the toxin binds. The recessive allele encodes a defective version of cadherin, rendering the Bt toxin harmless.

To learn more about the resistance alleles, the researchers extracted DNA from resistant pink bollworms, determined the cadherin gene sequence for each, and compared it with the gene from susceptible insects. The results suggested that bollworm

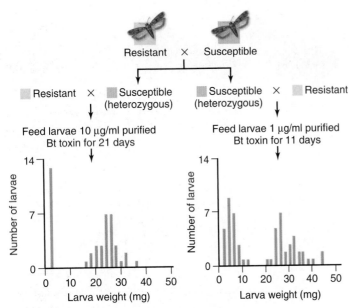

Figure 10.28 Bt Resistance Is Recessive. When a heterozygous, susceptible insect is bred with a resistant mate, half the offspring should be resistant and half should be susceptible. Tests with Bt toxin show that this is indeed the case: half of the insects thrived in the presence of Bt toxin, while the susceptible ones died or were very small.

populations harbor three unique resistance alleles, each encoding a different variation on the cadherin protein's shape. An insect with two such resistance alleles is immune to the toxin. Furthermore, the unique DNA sequence associated with each allele is a useful "tag" for identifying Bt resistance in pink bollworms. Researchers can simply extract an insect's DNA and test for the presence of the recessive alleles.

As a result, we can now spy on the pink bollworm's evolution as it happens. The stakes are high. If the resistance alleles become very common, Bt may become useless as a control measure, forcing growers to switch back to broad-spectrum pesticides that kill many beneficial insects. Careful monitoring will be crucial as we continue to wage war against the insects that will forever compete for our crops.

Morin, Shai, Robert W. Biggs, Mark S. Sisterson, and 10 other authors (including Bruce E. Tabashnik). April 29, 2003. Three cadherin alleles associated with resistance to *Bacillus thuringiensis* in pink bollworm. *Proceedings of the National Academy of Sciences*, vol. 100, pages 5004–5009.

10.10 | Mastering Concepts

1. Explain the logic of planting non-Bt-crop buffer strips around fields planted with Bt crops.

2. How did researchers in this study use a breeding experiment to demonstrate that Bt resistance alleles in pink bollworms are recessive?

3. What do you predict will happen to the incidence of resistance alleles in pink bollworm populations if farmers choose not to plant the required refuge?

Chapter Summary

10.1 | Chromosomes Are Packets of Genetic Information: A Review

- Each **gene** encodes a protein; mutations in genes create **alleles.**
- A **chromosome** is a continuous molecule of DNA with associated proteins. A **diploid** human cell contains 22 **homologous pairs** of **autosomes** and one pair of **sex chromosomes.**
- **Meiosis** is a type of cell division that gives rise to **haploid gametes. Fertilization** unites gametes and restores the diploid number.

10.2 | Mendel's Experiments Uncovered Basic Laws of Inheritance

A. Why Peas?
- Gregor Mendel studied inheritance patterns in pea plants because they are easy to grow, develop quickly, and produce abundant offspring. It is also easy to control crosses between pea plants.

B. Dominant Alleles Appear to Mask Recessive Alleles
- An individual is **true breeding** for a trait if self-fertilization always yields offspring identical to the parent for that trait.
- A **dominant** allele is always expressed if it is present; a **recessive** allele is masked by a dominant allele. Dominant does not always mean "most common."

C. For Each Gene, a Cell's Two Alleles May Be Identical or Different
- The combination of alleles for a gene is the individual's **genotype.** A **heterozygote** has two different alleles for a gene. A **homozygous** recessive individual has two recessive alleles, and a homozygous dominant individual has two dominant alleles.
- A **phenotype** is any observable characteristic of an organism.
- A **wild-type** allele is the most common in a population. A change in a gene is a **mutation** and may result in a **mutant** phenotype.

D. Every Generation Has a Name
- In genetic crosses, the purebred parental generation is designated **P;** the next generation is the first filial generation, or F_1; and the next is the second filial generation, or F_2.

10.3 | The Two Alleles of Each Gene End Up in Different Gametes

A. Monohybrid Crosses Track the Inheritance of One Gene
- A **monohybrid cross** is a mating between two individuals that are heterozygous for the same gene.
- **Punnett squares** are useful for calculating the probability of each possible genotype and phenotype among the offspring of a genetic cross.
- A **test cross** reveals an unknown genotype by breeding the individual to a homozygous recessive individual.

B. Meiosis Explains Mendel's Law of Segregation
- Mendel's **law of segregation** states that the two alleles of the same gene separate into different gametes. Each individual receives one allele of each gene from each parent.

10.4 | Genes on Different Chromosomes Are Inherited Independently

A. Dihybrid Crosses Track the Inheritance of Two Genes at Once
- A **dihybrid cross** is a mating between individuals that are heterozygous for two genes.

B. Meiosis Explains Mendel's Law of Independent Assortment
- According to Mendel's **law of independent assortment,** the inheritance of one gene does not affect the inheritance of another gene on a different chromosome. This law reflects meiosis, in which homologous pairs of

chromosomes (and the genes they carry) align randomly during metaphase I.

- The **product rule** is an alternative to Punnett squares for following inheritance of two or more traits at a time.

10.5 | Genes on the Same Chromosome May Be Inherited Together

A. Genes on the Same Chromosome Are Linked

- **Linked genes** are located on the same chromosome. Dihybrid crosses for pairs of linked genes produce genotypic and phenotypic ratios that are different from those predicted by the law of independent assortment.
- **Linkage groups** are collections of genes that are often inherited together because they are on the same chromosome.
- The farther apart two linked genes are on a chromosome, the more likely **crossing over** is to separate their alleles. If crossing over occurs between two alleles, some offspring will have **recombinant** genotypes; otherwise, offspring will have **parental** genotypes for the two genes.

B. Studies of Linked Genes Have Yielded Chromosome Maps

- Breeding studies reveal the crossover frequencies used to create **linkage maps**—diagrams that show the order of genes on a chromosome.

10.6 | Gene Expression Can Appear to Alter Mendelian Ratios

A. Incomplete Dominance and Codominance Add Phenotype Classes

- Heterozygotes for alleles with **incomplete dominance** have phenotypes intermediate between those of the two homozygotes. **Codominant** alleles are both expressed in a heterozygote.

B. Some Inheritance Patterns Are Especially Difficult to Interpret

- A **pleiotropic** gene has multiple effects on the body and therefore produces many different phenotypes.
- Many biochemical reactions require multiple proteins. Mutated genes encoding any of the proteins can stop the pathway, producing the same phenotype.
- In **epistasis,** one gene masks the effect of another.

10.7 | Sex-Linked Genes Have Unique Inheritance Patterns

- Genes controlling **sex-linked** traits are located on the X or Y chromosomes.

A. X and Y Chromosomes Determine Sex in Humans

- In humans, the male has X and Y sex chromosomes, and the female has two X chromosomes. The *SRY* gene on the Y chromosome controls other genes that stimulate development of male structures and suppress development of female structures.

B. X-Linked Recessive Disorders Affect More Males Than Females

- An **X-linked** gene passes from mother to son because the male inherits his X chromosome from his mother and his Y chromosome from his father. Scientists know of many more X-linked disorders than Y-linked disorders.

C. X Inactivation Prevents "Double Dosing" of Proteins

- **X inactivation** shuts off all but one X chromosome in the cells of female mammals, equalizing the number of active X-linked genes in each sex.

10.8 | Pedigrees Show Modes of Inheritance

- **Pedigrees** trace phenotypes in families and reveal modes of inheritance.
- An **autosomal dominant** disorder affects both sexes and can be inherited from one affected parent. An **autosomal recessive** disorder can appear in either sex, must be inherited from both parents (who are either carriers or are affected), and can skip generations. X-linked recessive disorders affect mostly males.

10.9 | Most Traits Are Influenced by the Environment and Multiple Genes

A. The Environment Can Alter the Phenotype

- Unlike traits that Mendel studied, most traits have environmental as well as genetic influences.

B. Polygenic Traits Depend on More Than One Gene

- A **polygenic trait** varies continuously in its expression, and the frequencies of the phenotypes form a bell-shaped curve.

10.10 | Investigating Life: Heredity and the Hungry Hordes

- Researchers have developed a way to test for the presence of recessive alleles in insect pests of cotton. This genetic test allows researchers to monitor caterpillars for resistance to Bt, an insecticidal toxin produced in genetically modified cotton.

Multiple Choice Questions

1. Which of the following is a difference between an autosome and a sex chromosome?
   a. An autosome has more DNA.
   b. A sex chromosome is only present in the germ cells.
   c. Only autosomes can be diploid.
   d. There are more autosomes than sex chromosomes in a cell.

2. According to Mendel, if an individual is *heterozygous* for a gene, the phenotype will correspond to that of
   a. the recessive trait alone.
   b. the dominant trait alone.
   c. a blend of the dominant and recessive traits.
   d. a wild-type trait.

3. If an individual is *homozygous* for a gene, then the genotype will contain
   a. only the recessive allele.
   b. only the dominant allele.
   c. both a dominant and a recessive allele.
   d. Either a or b could be true.

4. What can you conclude if all of the many offspring of a test cross show the phenotype associated with the dominant allele?
   a. One parent was homozygous dominant.
   b. One parent was heterozygous.
   c. The offspring are all homozygous dominant.
   d. Both b and c are correct.

5. Which of the following is a possible gamete for an individual with the genotype *PP rr*?
   a. *PP*   c. *pr*
   b. *Pr*   d. *rr*

6. Use the product rule to determine the chance of obtaining an offspring with the genotype *Rr Yy* from a dihybrid cross between parents with the genotype *Rr Yy*.
   a. 1/2   c. 1/8
   b. 1/4   d. 1/16

7. Recombination is most likely to occur between
   a. closely linked genes.
   b. genes on nonhomologous chromosomes.
   c. linked genes that are far apart.
   d. parental chromosomes.

8. How does incomplete dominance affect the phenotype of a heterozygote?
   a. It results in a blend of the dominant and recessive phenotypes.
   b. It results in the expression of only the recessive phenotype.
   c. The dominant phenotype is still expressed, but only in patches.
   d. The trait is not observed in the individual.

9. White cats have at least one dominant allele of the *W* gene, which masks the expression of all other genes that contribute to coat color. The *W* gene is an example of
   a. pleiotropy.
   b. independent assortment.
   c. codominance.
   d. epistasis.

10. Suppose a woman is a symptomless carrier of a recessive X-linked disease. If her husband has the disease, what is the chance that their first child is a girl who also has the disease?
    a. 100%
    b. 50%
    c. 25%
    d. 0

11. How does X inactivation contribute to genetic diversity?
    a. It controls the number and kind of genes expressed on an X chromosome.
    b. It determines which X chromosome is expressed in a male.
    c. It allows for expression of either the maternal or paternal X in different cells.
    d. It enhances expression of Y-linked genes in males.

12. What is a polygenic trait?
    a. A trait that reflects expression of both dominant and recessive alleles
    b. A trait that reflects the influence of the environment
    c. A trait that reflects the expression of many alleles of the same gene
    d. A trait that reflects the expression of many different genes

Write It Out

1. What advantages do pea plants and fruit flies have for studies of inheritance? Why aren't humans equally suitable?

2. Some people compare a homologous pair of chromosomes to a pair of shoes. Explain the similarity. How would you extend the analogy to the sex chromosomes for females and for males?

3. In an attempt to breed winter barley that is resistant to barley mild mosaic virus, agricultural researchers cross a susceptible domesticated strain with a resistant wild strain. The F_1 plants are all susceptible, but when the F_1 plants are crossed with each other, some of the F_2 individuals are resistant. Is the resistance allele recessive or dominant? How do you know?

4. Given the relationship between genes, alleles, and proteins, how can a recessive allele appear to "hide" in a heterozygote?

5. Chapter 8 explained the roles of oncogenes and tumor suppressor genes in cancer. Both types of genes encode proteins that regulate the cell cycle. Determine the function of each type of protein, then explain why oncogenes are dominant and mutant alleles of tumor suppressor genes are recessive.

6. Many plants are polyploid (see chapter 9); that is, they have more than two sets of chromosomes. How would having four (rather than two) copies of a chromosome more effectively mask expression of a recessive allele?

7. Springer spaniels often suffer from canine phospho-fructokinase (PFK) deficiency. The dogs lack an enzyme that is crucial in extracting energy from glucose molecules. Affected pups have extremely weak muscles and die within weeks. A DNA test is available to identify male and female dogs that are carriers. Why would breeders wish to identify carriers if these dogs are not affected?

8. How did Mendel use evidence from monohybrid and dihybrid crosses to deduce his laws of segregation and independent assortment? How do these laws relate to meiosis?

9. In a dihybrid cross, the predicted phenotype ratio is 9:3:3:1. The "9" represents the proportion of plants expressing at least one dominant allele for both traits. How would you use test crosses to determine whether these plants are homozygous dominant or heterozygous for one or both genes?

10. A white woman with fair skin, blond hair, and blue eyes marries a black man with dark brown skin, hair, and eyes. They have fraternal twins. One twin has blond hair, brown eyes, and light skin, and the other has dark hair, brown eyes, and dark skin. What Mendelian law does this real-life case illustrate?

11. The radish has nine groups of traits. Within each group, dihybrid crosses do not yield a 9:3:3:1 phenotypic ratio. Instead, such crosses yield an overabundance of phenotypes like those of the parents. What does this information reveal about the chromosomes of this plant?

12. How does gene linkage interfere with Mendel's law of independent assortment? Why doesn't the inheritance pattern of linked genes disprove Mendel's law?

13. How does crossing over "unlink" genes?

14. If two different but linked genes are located very far apart on a chromosome, how may the inheritance pattern create the appearance of independent assortment?

15. Explain how each of the following appears to disrupt Mendelian ratios: incomplete dominance, codominance, pleiotropy, epistasis.

16. Suppose a single trait is controlled by a gene with four codominant alleles. A person can inherit any combination of two of the four alleles. How many phenotypes are possible for this trait?

17. What is the role of the Y chromosome in human sex determination?

18. Do you agree with the statement that all alleles on the Y chromosome are dominant? Why or why not?

19. Suppose a fetus has X and Y chromosomes but lacks receptors for the protein encoded by the *SRY* gene. Will the fetus develop as a male or as a female? Explain your answer.

20. How are X-linked genes inherited differently in male and female humans?

21. What does X inactivation accomplish?

22. Rett syndrome is a severe X-linked disorder that affects mostly female children. How does X inactivation explain this observation?

23. The cells of a track runner are collected before an Olympic competition. Technicians examining the cells discover two Barr bodies in each nucleus. How is this unusual? What would you expect to find if you constructed a karyotype of the runner's chromosomes?

24. A family has an X-linked dominant form of congenital generalized hypertrichosis (excessive hairiness). Although the allele is dominant, males are more severely affected than females. Moreover, the women in the family often have asymmetrical, hairy patches on their bodies. How does X chromosome inactivation explain this observation?

25. Why are male calico cats rare?

26. In the following pedigree, is the disorder's mode of inheritance autosomal dominant, autosomal recessive, or X-linked recessive? Explain your reasoning.

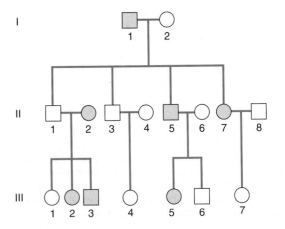

27. Pedigree charts can sometimes be difficult to construct and interpret. People may refuse to supply information, and adoption or serial marriages can produce blended families. Artificial insemination may involve anonymous sperm donors. Many traits are strongly influenced by the environment. How does each of these factors complicate the use of pedigrees?

28. Explain the following "equation":

Genotype + Environment = Phenotype

29. Mitochondria and chloroplasts contain DNA that encodes some proteins essential to life. These organelles are inherited only via the female parent's egg cell. Do you expect these genes to follow Mendelian laws of inheritance? Explain your answer.

Genetics Problems

See pages 226 and 227 for step-by-step guides to solving genetics problems.

1. Holstein cattle suffer from the condition citrullinemia, in which homozygous recessive calves die within a week of birth because they cannot break down ammonia that is produced when amino acids are metabolized. If a cow that is heterozygous for the citrullinemia gene is inseminated by a bull that is homozygous dominant, what is the probability that a calf inherits citrullinemia?

2. Wild-type canaries are yellow. A dominant mutant allele of the color gene, designated W, causes white feathers. Inheriting two dominant alleles is lethal to the embryo. If a yellow canary is crossed to a white canary, what is the probability that an offspring will be yellow? What is the probability that it will be white?

3. In humans, more than 100 forms of deafness are inherited as recessive alleles on many different chromosomes. Suppose that a woman who is heterozygous for a deafness gene on one chromosome has a child with a man who is heterozygous for a deafness gene on a different chromosome. Does the child face the general population risk of inheriting either form of deafness or the 25% chance that Mendelian ratios predict for a monohybrid cross? Explain your answer.

4. A man and a woman each have dark eyes, dark hair, and freckles. The genes for these traits are on separate chromosomes. The woman is heterozygous for each of these genes, but the man is homozygous. The dominance relationships of the alleles are as follows:

B = dark eyes; b = blue eyes
H = dark hair; h = blond hair
F = freckles; f = no freckles

a. What is the probability that their child will share the parents' phenotype?
b. What is the probability that the child will share the same genotype as the mother? As the father?
Use the product rule or a Punnett square to obtain your answers. Which method do you think is easier?

5. Genes J, K, and L are on the same chromosome. The crossover frequency between J and K is 19%, the crossover frequency between K and L is 2%, and the crossover frequency between J and L is 21%. Use this information to create a linkage map for the chromosome.

6. A particular gene in dogs contributes to coat color. The two alleles exhibit incomplete dominance. Dogs with genotype mm have normal pigmentation; genotype Mm leads to "dilute" pigmentation; genotype MM produces an all-white dog. If a breeder mates a normal dog with a white dog, what will be the genotypes and phenotypes of the puppies? If two Mm dogs are mated, what is the probability that a puppy will be all white?

7. Three babies are born in the hospital on the same day. Baby X has type AB blood; Baby Y has type B blood; Baby Z has type O blood.

Use the information in the table below to determine which baby belongs to which couple. (Assume that all individuals are homozygous dominant for the H gene.)

Couple	Mother	Blood type	Father	Blood type
1	Abby	B	Seth	AB
2	Carol	A	Sam	A
3	Nancy	AB	Bill	O

8. Consider a woman whose brother has hemophilia A but whose parents are healthy. What is the chance that she has inherited the hemophilia allele? What is the chance that the woman will conceive a son with hemophilia?

Pull It Together

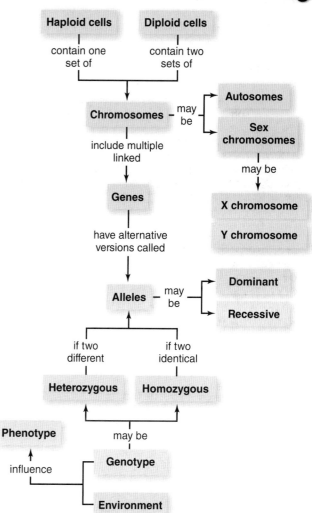

1. Which cells in the human body are haploid? Which cells are diploid?
2. What is the difference between a genotype and a phenotype?
3. What is the difference between a dominant and a recessive allele?
4. Add *meiosis, gametes, mutations, incomplete dominance, codominance, pleiotropy,* and *epistasis* to this concept map.

How to Solve a Genetics Problem: One Gene

Sample problem: Phenylketonuria (PKU) is an autosomal recessive disorder. If a man with PKU marries a woman who is a symptomless carrier, what is the probability that their first child will be born with PKU?

1. **Write a key.** Pick ONE letter to represent the gene in your problem. Use the capital form of your letter to symbolize the dominant allele; use the lowercase letter to symbolize the recessive allele.
   Sample: The dominant allele is *K;* the recessive allele is *k.*

2. **Summarize the problem's information.** Make a table listing the phenotypes and genotypes of both parents.
   Sample:

	Male	**Female**
Phenotype	Has PKU	No PKU (carrier)
Genotype	*kk*	*Kk*

3. **Sketch the parental chromosomes and gametes.** Use the genotypes in your table to draw the alleles onto chromosomes. Then draw short arrows to show the homologous chromosomes moving into separate gametes for each parent.

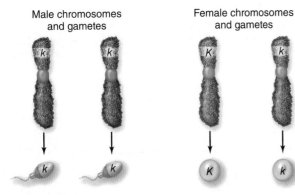

Male chromosomes and gametes

Female chromosomes and gametes

4. **Make a Punnett square.** Arrange the gametes you sketched in step #3 along the edges of the square, and fill in the genotypes of the offspring.

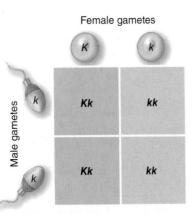

5. **Calculate the genotypic ratio.** Count the number of squares that contain each offspring genotype.
   Sample: 2 *Kk;* 2 *kk*

6. **Calculate the phenotypic ratio.** Count the number of squares that contain each offspring phenotype.
   Sample: 2 PKU carriers; 2 PKU sufferers

7. **Calculate the probability of each phenotype.** Divide each number in step #6 by 4 (the total number of squares) and multiply by 100.
   Sample: 50% probability that a child will be a carrier; 50% probability that a child will have PKU

How to Solve a Genetics Problem: Two Genes

Sample problem: A student collects pollen (male sex cells) from a pea plant that is homozygous recessive for the genes controlling seed form and seed color. She uses the pollen to fertilize a plant that is heterozygous for both genes. What is the probability that an offspring plant has the same genotype and phenotype as the male parent? Assume the genes are not linked.

1. **Write a key.** Pick ONE letter to represent each of the genes in your problem. Use the capital form of your letter to symbolize the dominant allele; use the lowercase letter to symbolize the recessive allele.
   Sample: For seed form, the dominant allele (round) is *R;* the recessive allele (wrinkled) is *r;* for seed color, the dominant allele (yellow) is *Y;* the recessive allele (green) is *y.*

2. **Summarize the problem's information.** Make a table listing the phenotypes and genotypes of both parents.
   Sample:

	Male	**Female**
Phenotype	Wrinkled, green	Round, yellow
Genotype	*rr yy*	*Rr Yy*

3. **Sketch the parental chromosomes and gametes.** Use the genotypes in your table to draw the alleles onto two sets of chromosomes, one for each parent. The law of independent assortment means that you need to draw all possible configurations. So redraw the chromosomes, this time switching the order of the alleles in one pair. Then draw short arrows to show the chromosomes separating, and sketch the four possible gametes for each parent. Depending on the parents' genotypes, some of the gametes produced by a parent may have the same genotype.

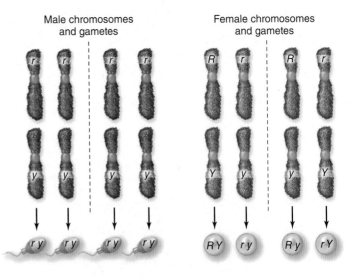

Male chromosomes and gametes

Female chromosomes and gametes

4. **Make a Punnett square.** Arrange the gametes you sketched in step #3 along the edges of the square, and fill in the genotypes of the offspring.

Female gametes

	R Y	*r y*	*R y*	*r Y*
r y	**Rr Yy**	**rr yy**	**Rr yy**	**rr Yy**
r y	**Rr Yy**	**rr yy**	**Rr yy**	**rr Yy**
r y	**Rr Yy**	**rr yy**	**Rr yy**	**rr Yy**
r y	**Rr Yy**	**rr yy**	**Rr yy**	**rr Yy**

Male gametes

5. **Calculate the genotypic ratio.** Count the number of squares that contain each offspring genotype.
Sample: 4 *Rr Yy*; 4 *rr yy*, 4 *Rr yy*, 4 *rr Yy*

6. **Calculate the phenotypic ratio.** Count the number of squares that correspond to each possible phenotype combination.
Sample: 4 round, yellow; 4 wrinkled, green; 4 round, green; 4 wrinkled, yellow

7. **Calculate the probability of each phenotype.** Divide each number in step #6 by 16 (the total number of squares) and multiply by 100.
Sample: 25% probability that an offspring has the same genotype and phenotype as the male parent.

NOTE: The product rule is a simpler way to solve the same problem and eliminates the need for a large Punnett square. To use the product rule in this case, first calculate the probability that the parents ($rr \times Rr$) produce an offspring with genotype rr (½, or 50%). Then calculate the chance that $yy \times Yy$ parents produce a yy offspring (½, or 50%). Multiply the two probabilities to calculate the probability that both events occur simultaneously: ½ × ½ = ¼, or 25%. See section 10.4 for more on the product rule.

How to Solve a Genetics Problem: X-Linked Gene

Sample problem: Hemophilia is caused by an X-linked recessive allele. If a man who has hemophilia marries a healthy woman who is not a carrier, what is the chance that their child will have hemophilia?

1. **Write a key.** Pick ONE letter to represent the gene in your problem. Use the capital form of your letter to symbolize the dominant allele; use the lowercase letter to symbolize the recessive allele.
Sample: The dominant allele is *H;* the recessive allele is *h*. Because these alleles are on the X chromosome, inheritance will differ between males and females. It is therefore best to designate the chromosomes and alleles together as X^H and X^h.

2. **Summarize the problem's information.** Make a table listing the phenotypes and genotypes of both parents.

Sample:

	Male	**Female**
Phenotype	Has hemophilia	Healthy
Genotype	$X^h Y$	$X^H X^H$

3. **Sketch the parental chromosomes and gametes.** Use the genotypes in your table to draw the alleles onto chromosomes. Then draw short arrows to show the chromosomes moving into separate gametes for each parent.

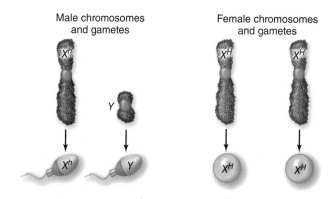

Male chromosomes and gametes   Female chromosomes and gametes

4. **Make a Punnett square.** Arrange the gametes you sketched in step #3 along the edges of the square; fill in the genotypes of the offspring.

Female gametes

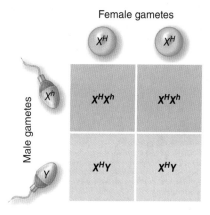

	X^H	X^H
X^h	$X^H X^h$	$X^H X^h$
Y	$X^H Y$	$X^H Y$

Male gametes

5. **Calculate the genotypic ratio.** Count the number of squares that contain each offspring genotype.
Sample: 2 $X^H X^h$; 2 $X^H Y$

6. **Calculate the phenotypic ratio.** Count the number of squares that correspond to each possible phenotype.
Sample: 2 female carriers; 2 healthy males

7. **Calculate the probability of each phenotype.** Divide each number in step #6 by 4 (the total number of squares) and multiply by 100.
Sample: 50% probability that a child will be a female carrier; 50% probability that a child will be a healthy male. No child, male or female, will have hemophilia.

11 The Forces of Evolutionary Change

Behind Bars. Extreme crowding in this Russian prison promotes the spread of antibiotic-resistant tuberculosis and other infectious diseases.

BIOLOGY

Enhance your study of
this chapter with practice quizzes,
animations and videos, answer keys,
and downloadable study tools.
www.mhhe.com/hoefnagels

The Rise of Antibiotic Resistance

Do you owe your life to antibiotics? Even if you have never had a serious infection, chances are that one or more of your ancestors did, and antibiotics may have saved their lives. Had it not been for these "wonder drugs," you may never have been born.

Many antibiotics are naturally occurring chemicals. Soil fungi and bacteria secrete these compounds into their surroundings, giving them an edge against microbial competitors. Biologists discovered antibiotics in the early 1900s, but it took decades for chemists to figure out how to mass-produce them. Once that occurred, antibiotics revolutionized medical care in the twentieth century and enabled people to survive many once-deadly bacterial infections.

Unfortunately, the miracle of antibiotics is under threat, and many infections that once were easily treated are reemerging as killers. Ironically, the overuse and misuse of antibiotics is partly responsible for the problem. Physicians sometimes prescribe antibiotics for viral infections, even though the drugs kill only bacteria; moreover, many patients fail to take the drugs as directed. Agricultural practices contribute to the problem as well. Producers of cattle, chickens, and other animals use antibiotics to treat and prevent disease, even adding small amounts to the animals' feed to promote growth. Farmers also spray antibiotics on fruit- and vegetable-producing plants to treat bacterial infections.

By saturating the environment with antibiotics, we have profoundly affected the evolution of bacteria. Microbes that can defeat the drugs are becoming increasingly common, and the explanation is simple. Antibiotics kill susceptible bacteria and leave the resistant ones alone. The survivors multiply, producing a new generation of antibiotic-resistant bacteria. This is an example of natural selection in action, and the public health consequences are both widespread and severe. ▸ antibiotics, p. 342

Antibiotic-resistant bacteria appeared just four years after these drugs entered medical practice in the late 1940s, and researchers responded by discovering new drugs. But the microbes kept pace. Today, 40% of hospital *Staphylococcus* infections resist all antibiotics but one. These so-called "superbugs" are called MRSA, which is short for methicillin-resistant *Staphylococcus aureus*. Worse, some laboratory strains are resistant to all antibiotics. Researchers fear that the discovery of new antibiotics will not keep up with the global proliferation of resistant strains.

Natural selection is just one mechanism of evolution, a process that is ongoing in every species. This chapter explains how evolution occurs.

Learning Outline

Learn How to Learn

Practice Your Recall

Here's an old-fashioned study tip that still works. When you finish reading a passage, close the book and write what you remember—in your own words. In this chapter, for example, you will learn about several forces of evolutionary change. After you read about them, can you list and describe them without peeking at your book? Try it and find out!

11.1 | Evolutionary Thought Has Evolved for Centuries

Scientific reasoning has profoundly changed human thinking about our own origins. Just 250 years ago, most people believed Earth was about 6000 years old. A century later, scientists accepted evidence that Earth is much older (millions of years old or more) but still believed that a creator made all life on Earth in its present form. Contemporary scientists, using evidence from many fields of research, now accept evolution as the best explanation for life's diversity.

But what *is* evolution? A simple definition of **evolution** is descent with modification. "Descent" implies inheritance; "modification" refers to changes in heritable traits from generation to generation. For example, we see evolution at work in the many types of cats—large and small, striped and spotted—that have descended from an ancestral cat species.

Evolution has another, more specific, definition as well. Recall from unit 2 that each gene can have several different versions, or alleles, in a population. We have also seen that a **population** consists of interbreeding members of the same species (see figure 1.2). Biologists say that evolution occurs in a population when some alleles become more common, and others less common, from one generation to the next. A more precise definition of evolution, then, is genetic change in a population over multiple generations. These changes in allele frequencies account for both short- and long-term changes in the history of life on Earth.

As the chapters in this unit will repeatedly demonstrate, evolution occurs everywhere and is obvious in many ways. It serves as such a compelling conceptual framework for many observations about life that geneticist Theodosius Dobzhansky gave this title to a much-quoted article he wrote: "Nothing in Biology Makes Sense Except in the Light of Evolution."

Evolution does not, however, answer one question that fascinates many people: how did life begin in the first place? Because little evidence remains from life's ancient origin, this question is difficult to answer scientifically; chapter 14 describes some of what we do know.

A. Many Explanations Have Been Proposed for Life's Diversity

People have tried to explain the diversity of life for a very long time (figure 11.1). In ancient Greece, Aristotle (384–322 BCE) recognized that all organisms are related in a hierarchy of simple to complex forms, but he believed that all members of a species were created identical to one another in form and capacity. This idea influenced scientific thinking for nearly 2000 years.

Several other ideas were also considered fundamental principles of science well into the 1800s. Among them was the concept of a "special creation," the sudden appearance of organisms on Earth. People believed that this creative event was planned and purposeful, that species were fixed and unchangeable, and that Earth was relatively young. The idea of a special creation also implied that there could be no extinctions.

What About Fossils?   Scientists struggled to reconcile these beliefs with compelling evidence that species could in fact change. Fossils, which had been discovered at least as early as 500 BCE, were at first thought to be oddly shaped crystals or faulty attempts at life that arose spontaneously in rocks. But by the mid-1700s, the increasingly obvious connection between organisms and fossils argued against these ideas.

To explain how fossils came to be yet not deny the role of a creator, scientists suggested that fossils represented organisms killed during the biblical flood. Yet some of the fossils depicted organisms not seen before. Because people believed that species created by God could not become extinct, these fossils presented a paradox. The conflict between ideology and observation widened as geologists discovered that different rock layers revealed different groups of extinct fossilized organisms.

Aristotle Individuals in a species are basically identical and species are unchanging.

Hutton Changes in nature are gradual; uniformitarianism.

Lamarck New species come from existing species through environmental forces.

Darwin & Wallace Individuals in a population are different; species arise through the process of natural selection.

350 BCE   AD 1749   1785   1798   1809   1830   1859

Buffon Species change as they spread from their original location.

Cuvier Species reappear after catastrophes; fossils represent extinctions.

Lyell All changes in nature are gradual; renewed uniformitarianism.

Figure 11.1 Early Evolutionary Thought. Many scientists made significant contributions over the years, developing the foundation that Charles Darwin and Alfred Russel Wallace used to describe natural selection as the mechanism for evolution.

Newer rock
layers

Older rock
layers

Figure 11.2 Rock Layers Reveal Earth's History. Layers of sedimentary rock formed from sand, mud, and gravel that were deposited in ancient seas. Such sediment layers are visible along the Grand Canyon. The rock layers on the bottom are older than those on top. Rock strata sometimes contain fossil evidence of organisms that lived (and died) when the layer was formed, providing clues about when the organism existed.

New Ideas from Geology   In 1749, French naturalist Georges-Louis Buffon (1707–1788) became one of the first to openly suggest that closely related species arose from a common ancestor and were changing—a radical idea at the time. By moving the discussion into the public arena, he made possible a new consideration of evolution and its causes from a scientific point of view.

Meanwhile, in the 1700s and 1800s, much of the study of nature focused on geology. In 1785, physician James Hutton (1726–1797) proposed the theory of **uniformitarianism,** which suggested that the processes of erosion and sedimentation that act in modern times have also occurred in the past, producing profound changes in Earth over time. On the other side, Georges Cuvier (1769–1832) was convinced of **catastrophism,** the theory that a series of brief, violent, global upheavals such as enormous floods, volcanic eruptions, and earthquakes were responsible for most geological formations.

Cuvier also used his knowledge of anatomy to identify fossils and to describe the similarities among organisms. He was the first to recognize the **principle of superposition**—the idea that lower layers of rock (and the fossils they contain) are older than those above them (figure 11.2). Although he had to accept that some species must have become extinct, he refused to believe that they were not originally formed through creation. He argued that catastrophes would destroy most of the organisms in an area, but then new life would arrive from surrounding areas.

Early Ideas About the Origin of Species   Once fossils were recognized as evidence of extinct life, it became clear that species could in fact change. Still, no one had proposed how this might happen. Then, in 1809, French taxonomist Jean Baptiste de Lamarck (1744–1829) proposed the first scientifically testable evolutionary theory. He reasoned that organisms that used one part of their body repeatedly would increase their abilities, very much like weight lifters developing strong arms. Conversely, disuse would weaken an organ until it disappeared. Lamarck surmised (incorrectly) that these changes would pass to future generations.

Geologist Charles Lyell (1797–1875) renewed the argument for uniformitarianism in 1830. He suggested that natural processes are slow and steady, and that Earth is much older than 6000 years—perhaps millions or hundreds of millions of years old. One obvious conclusion from his contribution is that gradual changes in some organisms could be represented in successive fossil layers. Lyell was so persuasive that many scientists began to reject catastrophism in favor of the idea of gradual geologic change.

B. Charles Darwin's Voyage Provided a Wealth of Evidence

With these new theories and ideas, people were beginning to accept the concept of evolution but could not understand how it could result in the formation of new species. Ultimately, Charles Darwin (1809–1882) recognized their application to the changing diversity of life on Earth. Darwin was the son of a physician and the grandson of noted physician and poet Erasmus Darwin, who had anticipated evolutionary theory by writing in 1796 that all animals arose from a single "living filament."

Young Charles Darwin attended Cambridge University in England and, at the urging of his family, completed studies to enter the clergy. Meanwhile, he also followed his own interests. He joined geological field trips and met several eminent geology professors.

Eventually, Darwin was offered a position as ship's naturalist aboard the HMS *Beagle.* Before the ship set sail for its 5-year voyage in 1831 (figure 11.3), the botany professor who had arranged Darwin's position gave the young man the first volume of Lyell's *Principles of Geology.* Darwin picked up the second and third volumes in South America. By the time he finished reading Lyell's works, Darwin was an avid proponent of uniformitarianism.

He recorded his observations as the ship journeyed around the coast of South America. He noted forces that uplifted new land, such as earthquakes and volcanoes, and the constant erosion that wore it down. He marveled at forest plant fossils interspersed with sea sediments and at shell fossils in a mountain cave. Darwin tried to reconstruct the past from contemporary observations and wondered how each fossil had arrived where he found it.

He was particularly aware of similarities and differences among organisms. If there had been a single special creation, then why was one sort of animal or plant created to live on a mountaintop in one part of the world, yet another type was on mountains elsewhere? Even more puzzling was the resemblance between organisms living in similar habitats in different parts of the world. We now know that such species have undergone convergent evolution. That is, two species that live on opposite sides of the planet may nevertheless share characteristics because they evolved in similar environmental conditions. ▶ convergent evolution, p. 263

In the fourth year of the voyage, the HMS *Beagle* spent a month in the Galápagos Islands, off the coast of Ecuador. The notes and samples Darwin brought back would form the seed of his theory of evolution by natural selection.

C. *On the Origin of Species* Proposed Natural Selection as an Evolutionary Mechanism

Toward the end of the voyage, Darwin began to assimilate all he had seen and recorded. Pondering the great variety of organisms in South America and their relationships to fossils and geology, he began to think that these were clues to how species originate.

Descent with Modification   Darwin returned to England in 1836, and by 1837 he began assembling his notes in earnest. In March 1837, Darwin consulted ornithologist (bird expert) John Gould about the finches the *Beagle* brought back from the Galápagos Islands. Gould could tell from bill structures that some of the birds ate small seeds, whereas others ate large seeds, fruits, or insects. In all, he described 14 distinct types of finches, each different from the birds on the mainland yet sharing some features.

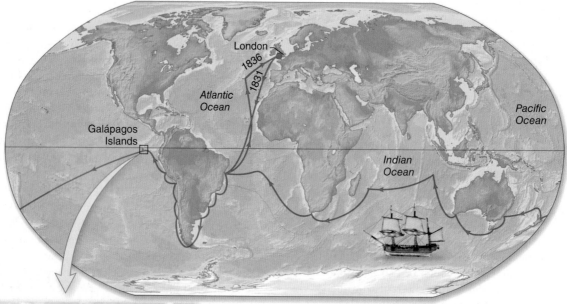

Figure 11.3 The Voyage of the *Beagle*. Darwin observed life and geology throughout the world during the journey of the HMS *Beagle*. Many of Darwin's ideas about natural selection and evolution had their origins in the observations he made on the Galápagos Islands (inset).

Darwin thought that the different varieties of finches on the Galápagos had probably descended from a single ancestral type of finch that had flown to the islands and, finding a relatively unoccupied new habitat, flourished. Over the next 3 million years, the finch population gradually branched in several directions. Different groups ate insects, fruits, and seeds of different sizes, depending on the resources each island offered. Darwin also noted changes in other species, including Galápagos tortoises. He coined the phrase "descent with modification" to describe gradual changes from an ancestral type.

Malthus's Ideas on Populations   In September 1838, Darwin read a work that helped him understand the diversity of finches on the Galápagos Islands. Economist and theologian Thomas Malthus's *Essay on the Principle of Population,* written 40 years earlier, stated that food availability, disease, and war limit the size of a human population. Wouldn't other organisms face similar limitations? If so, then individuals that could not obtain essential resources would die.

The insight Malthus provided was that individual members of a population were not all the same, as Aristotle had taught. Instead, individuals better able to obtain resources were more likely to survive and reproduce. This would explain the observation that more individuals are produced in a generation than survive; they do not all obtain enough vital resources to live. Over time, environmental challenges would eliminate the more poorly equipped variants, and gradually, the population would change.

The Concept of Natural Selection   Darwin used the term *natural selection* to describe "this preservation of favourable variations and the rejection of injurious variations." Biologists later modified the definition to add modern genetics terminology. We now say that **natural selection** occurs when environmental factors cause the differential reproductive success of individuals with particular genotypes.

Darwin got the idea of natural selection from thinking about artificial selection (also called selective breeding). In **artificial selection,** a human chooses one or a few desired traits, such as milk production or seed size, and then allows only the individuals that best express those qualities to reproduce (**figure 11.4**). Artificial selection is responsible not only for agriculturally important varieties of animals and plants but also for the many breeds of domesticated cats and dogs (see this chapter's Apply It Now box on page 234). Darwin himself raised pigeons and developed several new breeds by artificial selection.

How did natural selection apply to the diversity of finch species on the Galápagos? Originally, some finches flew from the mainland to one island. Eventually that island population outgrew the supply of small seeds, and birds that could eat nothing else starved. But finches that could eat other things, perhaps because of an inherited difference in bill structure, survived and reproduced. Since their food was plentiful, these once-unusual birds gradually came to make up more of the population.

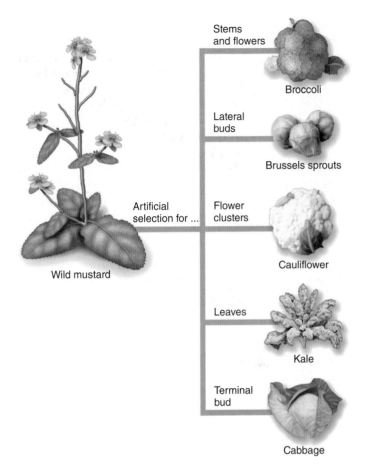

Figure 11.4 Artificial Selection. By selecting for different traits, plant breeders used the same type of wild mustard to create these five vegetable varieties.

Darwin further realized that he could extend this idea to multiple islands, each of which had a slightly different habitat and therefore selected for different varieties of finches. A new species might arise when a population adapted to so many new conditions that its members could no longer breed with the original group (see chapter 13). In a similar way, new species have evolved throughout the history of life as populations have adapted to different resources. All species are therefore ultimately united by common ancestry.

Publication of *On the Origin of Species*   Darwin described his theory of evolution by natural selection in a 35-page sketch in 1842; two years later, he developed a 230-page analysis. He did not publish either account. Instead, he continued to work on his ideas until 1858, when he received a manuscript entitled *On the Tendency of Varieties to Depart Indefinitely from the Original Type* from British naturalist Alfred Russel Wallace (1823–1913). Wallace had observed the diverse insects, birds, and mammals of South America and southeast Asia, and his manuscript independently proposed that natural selection was the driving force of evolution.

Apply It Now

Dogs Are Products of Artificial Selection

People have been breeding dogs for thousands of years, beginning with domesticated wolves. Dog fanciers now recognize hundreds of breeds, each the product of artificial selection for a different trait that originally occurred as natural genetic variation. Bloodhounds, for example, are selected for their keen sense of smell. Border collies herd livestock (or anything else that moves), and the sleek greyhound is bred for speed.

Behind the carefully bred traits, however, lurk small gene pools and extensive inbreeding, which may harm the health of purebred show animals (table 11.A). Dog breeders can select for desired characteristics, but they can't always avoid the hereditary health problems associated with each breed.

Figure 11.A shows two examples. The tiny jaws and massive teeth of bulldogs cause dental and breathing problems, sinusitis, and their notorious "dog breath." The runny, sad eyes of the basset hound can be quite painful if they become clogged with dirt and mucus. Short legs make the dog prone to arthritis, the long abdomen promotes back injuries, and the characteristic floppy ears often hide ear infections.

Table 11.A   Purebred Plights

Breed	Health Problem(s)
Cocker spaniel	Nervousness, ear infections, hernias, kidney problems
Collie	Blindness, bald spots, seizures
Dalmatian	Deafness
German shepherd	Hip dysplasia
Golden retriever	Lymphatic cancer, muscular dystrophy, skin allergies, hip dysplasia, absence of one testicle
Great Dane	Heart failure, bone cancer
Labrador retriever	Dwarfism, blindness
Shar-pei	Skin disorders

Bulldog

Basset hound

Figure 11.A
Artificial Selection in Dogs. The bulldog was selected for its flattened face and fierce demeanor, and the mournful expression of the basset hound accompanies its heightened sense of smell. These traits originally occurred as natural genetic variation in their wolf ancestors.

Darwin submitted his own paper, along with Wallace's, to the Linnaean Society meeting later that year. In 1859, Darwin finally published the 490-page *On the Origin of Species by Means of Natural Selection, or Preservation of Favoured Races in the Struggle for Life*. It would form the underpinning of modern life science.

Table 11.1 summarizes Darwin's main arguments in support of natural selection. He observed that individuals in a species are different from one another and that at least some of this variation is heritable. If more individuals are born than can survive, then competition will determine which ones live long enough to reproduce. Darwin realized that those with the most adaptive traits would be

Table 11.1   The Logic of Natural Selection: A Summary

Observations of nature

1. Organisms are varied, and some variations are inherited. Within a species, no two individuals (except identical siblings) are exactly alike.

2. More individuals are born than survive to reproduce.

3. Individuals compete with one another for the limited resources that enable them to survive.

Inferences from observations

1. Within populations, the inherited characteristics of some individuals make them more likely to survive and produce fertile offspring.

2. Because of the environment's selection against nonadaptive traits, only individuals with adaptive traits live long enough to transmit their genes to the next generation. Over time, natural selection can change the characteristics of populations, even giving rise to new species.

most likely to reproduce and pass those traits to the next generation. Over many generations, natural selection coupled with environmental change or a new habitat could change a population's characteristics or even give rise to new species. Darwin's own observations of ants, pigeons, orchids, and many other organisms provided abundant support for his ideas.

Some members of the scientific community happily embraced Darwin's efforts. Upon reading *On the Origin of Species,* his friend Thomas Henry Huxley remarked, "How stupid of me not to have thought of that." Others, however, were less appreciative. People in some religious denominations perceived a clash with their beliefs that all life arose from separate special creations, that species did not change, and that nature is harmonious and purposeful. Perhaps most disturbing to many people was the idea that humans were just one more species competing for resources.

D. Evolutionary Theory Continues to Expand

Although Charles Darwin's arguments were fundamentally sound, he could not explain all that he saw. For instance, he did not understand the source of variation within a population, nor did he know how heritable traits were passed from generation to generation. Ironically, Austrian monk Gregor Mendel was solving the puzzle of inheritance at the same time that Darwin was pondering natural selection (see the opening essay for chapter 10). Mendel's work, however, remained obscure until after Darwin's death.

Since Darwin's time, scientists have learned much more about genes, chromosomes, and the origin and inheritance of genetic variation (figure 11.5). In the 1930s, scientists finally recognized the connection between natural selection and genetics. They unified these ideas into the **modern evolutionary synthesis,** which suggests that genetic mutations create heritable

variation and that this variation is the raw material upon which natural selection acts.

After the discovery of DNA's structure in the 1950s, the picture became clearer still. We now know that mutations are changes in DNA sequence (described in chapter 7) and that mutations occur at random in all organisms. Sexual reproduction amplifies this variability by shuffling and reshuffling parental alleles to produce genetically different offspring (see chapter 9).

Today, overwhelming evidence supports the theory of evolution by natural selection; chapter 12 describes some of the data in detail. Contemporary biologists therefore accept evolution as the best explanation for the fact that diverse organisms use the same genetic code, the same chemical reactions to extract energy from nutrients, and many of the same (or very similar) enzymes and other proteins. Descent from a common ancestor explains both this great unity of life and the spectacular diversity of organisms today. Coupled with a wide variety of changing habitats and enormous amounts of time, the result of natural selection is a planet packed with millions of variations of the same underlying biochemical theme.

11.1 | Mastering Concepts

1. What is evolution?
2. How did people think that species arose and diversified before Charles Darwin published his theory of evolution by natural selection?
3. What did Darwin observe that led him to develop his ideas about the origin of species?
4. How is artificial selection different from natural selection?
5. How did Darwin's ideas challenge prevailing beliefs about life's diversity and the status of humans?
6. What is the modern evolutionary synthesis?

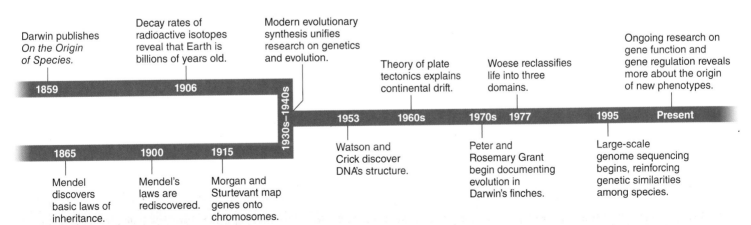

Figure 11.5 Evolutionary Theory Since Darwin. Charles Darwin and Gregor Mendel laid a foundation for evolutionary theory, but many scientists since that time have added to our understanding of how evolution works.

11.2 | Natural Selection Molds Evolution

Biological evolution includes large-scale, so-called "macroevolutionary" events such as the appearance of new species of finches. **Microevolution,** the relatively short-term genetic changes within a population or species, occurs on a much smaller scale. Note that evolution involves changes in allele frequencies. Because an individual's alleles do not change, *evolution can occur in populations but not in individuals.*

Natural selection is the most famous, and often the most important, mechanism of microevolution. This section explains the basic requirements for natural selection to occur. The rest of this chapter describes natural selection and several others forces of evolutionary change in more detail.

A. Adaptations Enhance Reproductive Success

As Darwin knew, organisms of the same species are different from one another, and every population produces more individuals than resources can support (figure 11.6). Some members of any population will not survive to reproduce. A struggle for existence is therefore inevitable.

The variation inherent in each species means that some individuals in each population are better than others at obtaining nutrients and water, avoiding predators, tolerating temperature changes, attracting mates, and reproducing. The heritable traits conferring these advantages are **adaptations**—features that provide a selective advantage because they improve an organism's ability to survive and reproduce.

The word *adaptation* can be confusing because it has multiple meanings. For example, a student might say, "I have adapted well to college life," but short-term changes in an individual do not constitute evolution. Adaptations in the evolutionary sense include only those structures, behaviors, or physiological processes that are heritable and that contribute to reproductive success.

In any population, individuals with the best adaptations are most likely to reproduce and pass their advantage to their offspring. Because of this "differential reproductive success," a population changes over time, with the best available adaptations to the existing environment becoming more common with each generation (figure 11.7).

Natural selection requires preexisting genetic diversity. Ultimately, this diversity arises largely by chance. Nevertheless, it is important to realize that natural selection itself is not a random process. Instead, it selectively eliminates most of the individuals that are least able to compete for resources or cope with the prevailing environment.

One additional important note is that not all variation within a population is subject to natural selection. Some features are "selectively neutral," meaning that they neither increase nor decrease reproductive success. As a simple example, the shape of a person's earlobes has little to do with how many offspring he or she produces. Earlobe shape differences therefore represent neutral variation in the human population.

Other types of neutral variation do not change an individual's appearance at all. For example, a mutation may occur in a stretch of DNA that does not encode a protein. In addition, thanks to redundancy in the genetic code, a mutation in a gene may not change the amino acid sequence of the encoded protein. Because such mutations are "silent," they may persist in the population without being subject to natural selection.

Figure 11.6
Requirements for Natural Selection.
(a) This basketball player and referee illustrate genetic variation within the human population. (b) Dandelions produce many offspring, but few survive.

a.

b.

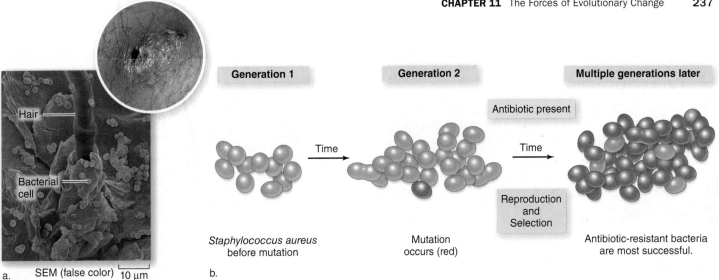

Figure 11.7 Natural Selection. (a) *Staphylococcus aureus* bacteria cause serious skin infections and other illnesses. (b) Natural selection requires preexisting variation. The presence of an antibiotic strongly selects for those bacteria that are immune to the drug.

B. Natural Selection Eliminates Phenotypes

Recall from chapter 10 that an individual's phenotype is its observable properties, most of which arise from a combination of environmental influences and the action of multiple genes. By "weeding out" individuals with poorly adapted phenotypes, natural selection indirectly changes allele frequencies in the population.

The phenotype that is the "best" depends entirely on the time and place; a trait that is adaptive in one set of circumstances may become a liability in another. Consider the finches on the tiny Galápagos Island of Daphne Major. In a very dry season in the early 1980s, birds with large beaks were more likely to survive because they could eat the large, tough seeds that remained after the small seeds were depleted (figure 11.8). In 1983, however, many small seeds accumulated following 8 months of extremely heavy rainfall. Over the next 2 years, small-beaked finches, which could easily eat the tiny seeds, came to predominate in the population. Constantly changing conditions mean that evolution never stops. (Learn more about the finches on Daphne Major in section 12.7.)

C. Natural Selection Does Not Have a Goal

Because most species have become more complex over life's long evolutionary history, many people erroneously believe that natural selection leads to ever more "perfect" organisms or that evolution works toward some long-term goal. Explanations that use the words *need* or *in order to* typically reflect this misconception. For example, a person might say, "The beaks in figure 11.8 grew because the finches needed to eat the large, tough seeds" or "The beaks grew in order to give the finches the ability to eat the largest seeds."

Both explanations are incorrect because evolution does not have a goal. How could it? No known mechanism allows the environment to tell DNA how to mutate and generate the alleles needed to confront future conditions. Nor does natural selection strive for perfection; if it did, the vast majority of species in life's history would still exist. Instead, most are extinct.

Several factors combine to prevent natural selection from producing all of the traits that a species might find useful. First,

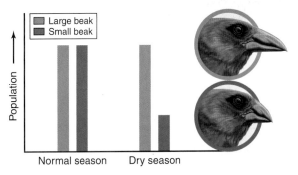

Figure 11.8 Finch Beak Shape Reflects Natural Selection. Since the early 1970s, Princeton University researchers Peter and Rosemary Grant have meticulously measured changes in the finch populations of the Galápagos. The Grants captured medium ground finches *(Geospiza fortis)* and monitored their beak sizes. Within a population, the average beak size (an inherited trait) can change appreciably in as short a time as a year. Following a very dry season, when seeds were sparse, the small, easy-to-crack ones were eaten rapidly. Birds with large beaks survived because only they were strong enough to open the large, tough-to-crack seeds.

a.

2 cm

b.

Figure 11.9 **Extinction.** (a) Sea scorpions once thrived worldwide. These animals became extinct some 250 million years ago, during one of Earth's several mass extinction events. (b) These petrified trees in the Arizona desert are from a family of trees that is now extinct in the northern hemisphere. More than 200 million years ago, this region was in the tropics. As the continents shifted and the climate changed, the trees died out, leaving only these fossils.

every genome has limited potential, imposed by its evolutionary history. The structure of the human skeleton, for example, will not allow for the sudden appearance of wheels, no matter how useful they might be on paved roads. Second, no population contains every allele needed to confront every possible change in the environment. If the right alleles aren't available at the right time, an environmental change may wipe out a species (**figure 11.9**). Third, disasters such as floods and volcanic eruptions can indiscriminately eliminate the best allele combinations, simply by chance (see section 11.6). And finally, some harmful genetic

a.

b.

Figure 11.10 **Fitness Is Reproductive Success.** One key to fitness is living long enough to reproduce. (a) For a redwood, that may take centuries. (b) In an annual plant, it may take just a few weeks.

traits are out of natural selection's reach, such as diseases that appear only after reproductive age.

D. What Does "Survival of the Fittest" Really Mean?

Natural selection is often called "the survival of the fittest," but this phrase is not entirely accurate or complete. In everyday language, the "fittest" individual is the one in the best physical shape: the strongest, fastest, or biggest. Physical fitness, however, is not the key to natural selection (although it may play a part).

Rather, in an evolutionary sense, **fitness** refers to an organism's genetic contribution to the next generation. A large, quick, burly elk scores zero on the evolutionary fitness scale if poor eyesight makes it vulnerable to an early death in the jaws of a wolf. On the other hand, a mayfly that dies in the act of producing thousands of offspring is highly fit.

Paradoxically, natural selection promotes any trait that increases fitness, even if the trait virtually guarantees an individual's death. For example, a male praying mantis may not resist if the female begins to eat his head during copulation. The male's passive behavior is adaptive because the extra food the female obtains in this way will enhance the chance of survival for their young. Likewise, section 31.5 describes a male spider that somersaults his abdomen into the jaws of his mate during copulation. Unlike the mantis, his body is too small to offer the female a nutritional benefit. Instead, the male's suicidal behavior prolongs copulation, so he sires the most possible offspring. As in the case of the mantis, the male spider does not survive—but his alleles will.

These examples illustrate an important point: *by itself, survival is not enough*. Because successful reproduction is the only way for an organism to perpetuate its genes, fitness depends on the ability to survive just long enough to reproduce. Plants that germinate, grow, flower, produce seeds, and die within just a few weeks may have fitness equal to a redwood that lives for centuries (figure 11.10).

Many adaptations contribute to an organism's overall fitness. The ability to overcome poor weather conditions, combat parasites and other disease-causing organisms, evade predators, and compete for resources all enhance an organism's chances of reaching reproductive age. At that point, the ability to attract mates (or pollinators, in the case of many flowering plants) affects the number of offspring an organism produces.

Fitness includes not only the total number of offspring produced but also the proportion that reach reproductive age. Some organisms have few offspring but invest large amounts of energy in each one. Others produce thousands of young but invest minimally in each. Section 36.4 further describes this evolutionary trade-off between "quality" and "quantity."

Some people wonder whether evolution could produce one species that could thrive and reproduce in every habitat on Earth, outcompeting other species across the globe. This chapter's Burning Question discusses this possibility.

? Burning Question

Why doesn't natural selection produce one superorganism?

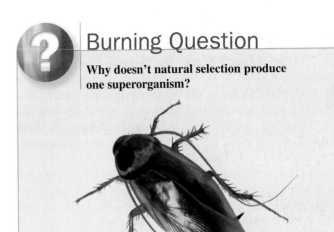

Natural selection cannot produce one "perfect" organism that is supremely adapted to every possible habitat on Earth. The simple reason is that the adaptations that seem "perfect" in one habitat would be completely wrong in another. To take an extreme example, a trout's adaptations that work so well in a cold mountain stream are useless in the sands of the Sahara. The variety of habitats on Earth—from oceans, to freshwater, to the tundra, prairie, desert, and forest—is just too great for one species to be able to live everywhere.

Some people believe that evolution actually has produced a superorganism: humans. True, we have the intelligence, dexterity, and cultural background to occupy every continent on Earth. But humans can only visit Earth's waters and the highest mountains, areas where other organisms thrive, for brief periods. And few organisms can live in the extreme heat of Earth's interior.

Nevertheless, organisms that can survive in a wide variety of habitats may threaten Earth's ecosystems. Weeds such as cheat grass and dandelions, along with animal pests such as cockroaches and rats, crowd out native species and appear to be able to live anywhere. Yet even they can't withstand conditions that are too hot, too cold, too dry, or too wet. Because different habitats have such different constraints, it seems unlikely that any organism will ever evolve with the combination of traits that would enable it to live everywhere.

Submit your burning question to:
marielle_hoefnagels@mcgraw-hill.com

11.2 | Mastering Concepts

1. What is an adaptation, and how do adaptations become more common within a population?
2. What is the role of genetic variation in natural selection?
3. How can natural selection favor different phenotypes at different times?
4. Why doesn't natural selection produce perfectly adapted organisms?
5. What is evolutionary fitness?

11.3 | Evolution Is Inevitable in Real Populations

Biologists can detect evolution by examining a population's **gene pool**—its entire collection of genes and their alleles. As we have already seen, evolution is a change in **allele frequencies;** one allele's frequency is calculated as the number of copies of that allele, divided by the total number of alleles in the population. Suppose, for example, that a gene has two possible alleles, *A* and *a*. In a population of 100 individuals, the gene has 200 alleles (assuming the individuals are diploid). If 160 of those alleles are *a*, then the frequency of allele *a* is 160/200, or 0.8.

As this section describes, we can sometimes use known allele frequencies to derive **genotype frequencies.** Each genotype's frequency is the number of individuals with that genotype, divided by the total size of the population. For example, if 64 of the 100 individuals in our population are homozygous recessive, then the frequency of genotype *aa* is 64/100, or 0.64.

Allele and genotype frequencies determine the genetic characteristics of a population. Most people in Sweden, for example, might have alleles conferring blond hair; a population of Asians would have very few, if any, such alleles. Instead, their gene pool would contain mostly alleles conferring darker hair. If Swedes migrate to Asia and interbreed with the locals (or vice versa), allele frequencies change.

These shifting allele frequencies in populations are the small steps of change that collectively drive evolution. Given the large number of genes in any organism and the many factors that can alter allele frequencies (including but not limited to natural selection), evolution is not only possible but unavoidable. This section explains why.

A. At Hardy–Weinberg Equilibrium, Allele Frequencies Do Not Change

Hardy–Weinberg equilibrium is the highly unlikely situation in which allele frequencies and genotype frequencies do not change from one generation to the next. It occurs only in populations that meet the following assumptions: (1) mutations do not occur, so no new alleles arise; (2) individuals mate at random; (3) individuals do not migrate into or out of the population; (4) the population is infinitely large, or at least large enough to eliminate random changes in allele frequencies (genetic drift); and (5) natural selection does not occur.

Hardy–Weinberg equilibrium is named after mathematician Godfrey H. Hardy and physician Wilhelm Weinberg. In 1908, they independently proposed that the expression $p + q = 1$ could represent the frequency of both of the alleles $(p + q)$ for a gene in a population of diploid organisms, if only two alleles exist for that gene.

For example, suppose that in a population of ferrets, the frequency for the dark fur allele *(D)* is 0.6; the frequency of the alternative allele *d,* which confers tan fur, is 0.4 (figure 11.11). The two frequencies add up to 1 because the two alleles represent all the possibilities in the population. Knowing these frequencies is important because changes in allele frequencies are the direct measure of evolution.

At Hardy–Weinberg equilibrium, we can use allele frequencies to calculate genotype frequencies. If *p* represents the frequency of the allele *D,* then the proportion of the population with genotype *DD* equals p^2, which in the ferret example equals 0.6×0.6, or 0.36. Likewise, the proportion of population members with genotype *dd* equals 0.4×0.4 (q^2), or 0.16. To calculate the frequency of the heterozygous class, multiply *pq* by 2 ($2 \times 0.6 \times 0.4 = 0.48$). That is, 48% of the ferrets in the population have genotype *Dd* and dark fur.

Figure It Out

In a species of ladybug, one gene controls whether the beetle has spots or not, with the allele conferring spots being dominant. Suppose you find a swarm consisting of 1000 ladybugs; 250 of them lack spots. What are the frequencies of the dominant and recessive alleles?

Answer: $q^2 = 0.25$, so $q = 0.5$. Therefore, $p = 0.5$ as well.

Since the homozygotes and the heterozygotes account for all possible genotypes in the population, the sum of their frequencies must add up to one:

$$p^2 + 2pq + q^2 = 1$$

If conditions of Hardy–Weinberg equilibrium are met, allele and genotype frequencies will not change in future generations. When the ferrets produce gametes, the proportion of *D* alleles will equal that of the homozygous dominant class (0.36) plus one half of the gametes from the heterozygotes (one half of 0.48, or 0.24). Therefore, the proportion of *D* alleles would be $p = 0.36 + 0.24 = 0.60$. If $p = 0.6$, then $q = 1 - p$, or 0.4. We are back at the beginning: in the next generation, the same proportion of ferrets will have dark fur. Hardy–Weinberg equilibrium persists, and evolution is not occurring.

Figure It Out

Mr. Morton's backyard is home to a population of 200 squirrels. He has discovered that 98 of the squirrels are homozygous dominant for a particular gene. About how many squirrels should he expect to be heterozygous for the gene if all of the assumptions of Hardy–Weinberg are met?

Answer: 84.

B. In Reality, Allele Frequencies Always Change

The theoretical ferret population in figure 11.11 is at Hardy–Weinberg equilibrium and therefore does not evolve. A real

Allele Frequencies

p = frequency of D (dominant allele) = dark fur = 0.6
q = frequency of d (recessive allele) = tan fur = 0.4

Algebraic Expression	What It Means
$p + q = 1$ $(0.6 + 0.4 = 1)$	Frequency of all dominant alleles (p) plus frequency of all recessive alleles (q) for this gene.
$p^2 + 2pq + q^2 = 1$ $(0.36 + 0.48 + 0.16 = 1)$	The frequencies of homozygous dominant individuals (p^2) plus heterozygotes ($2pq$) plus homozygous recessives (q^2) add up to all of the individuals in the population.

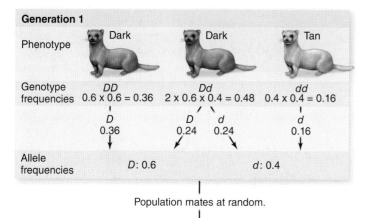

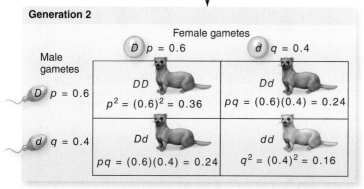

population, however, violates some or all of the assumptions of Hardy–Weinberg equilibrium. In reality, the frequency of an allele in a population may change when

- mutation introduces new alleles;
- individuals remain in closed groups, mating among themselves rather than with the larger population (nonrandom mating);
- individuals migrate among populations;
- allele frequencies change due to chance (genetic drift);
- some phenotypes are better adapted to the environment than others (natural selection).

All of these events are common. If you think about the human population, for example, it is clear that we do not choose our mates at random. Migration alters allele frequencies as we move and mix. We can sometimes counter natural selection by medically correcting phenotypes that would otherwise prevent some of us from having children, but natural selection still acts to reduce the frequency of alleles that cause deadly childhood illnesses.

Finally, no population is infinitely large. Allele frequencies therefore always change over multiple generations. In other words, evolution is inevitable.

Even though its assumptions do not apply to real populations, the concept of Hardy–Weinberg equilibrium is nevertheless important. The reason is that it serves as a basis of comparison to reveal when microevolution is occurring. Additional studies can then reveal which mechanism of evolution is acting on the population. Sections 11.4 through 11.6 describe these mechanisms of microevolution in more detail.

11.3 | Mastering Concepts

1. What are the five conditions required for Hardy–Weinberg equilibrium?
2. Why is the concept of Hardy–Weinberg equilibrium important?
3. Explain the components and meaning of the equation $p^2 + 2pq + q^2 = 1$.
4. Why doesn't Hardy–Weinberg equilibrium occur in real populations?

Figure 11.11 Hardy–Weinberg Equilibrium. At Hardy–Weinberg equilibrium, allele frequencies remain constant from one generation to the next; evolution does not occur. This figure depicts the random mating that underlies Hardy–Weinberg equilibrium. The sample population of ferrets shown in the figure contains only 25 individuals; one of the assumptions of Hardy–Weinberg equilibrium, however, is that a population must be very large.

11.4 | Natural Selection Can Shape Populations in Many Ways

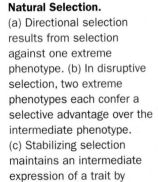

Of all the mechanisms by which a population can evolve, natural selection is probably the most important. Natural selection changes the genetic makeup of a population by favoring the alleles that contribute to reproductive success and selecting against those that are harmful.

Natural selection, however, does not eliminate alleles directly. Instead, individuals with the "best" phenotypes are most likely to pass their alleles to the next generation; those with poorly suited phenotypes are less likely to survive long enough to reproduce. Three modes of natural selection—directional, disruptive, and stabilizing—are distinguished by their effects on the phenotypes in a population (figure 11.12).

In **directional selection,** one extreme phenotype is fittest, and the environment selects against the others. A change in tree trunk color from light to dark, for example, may select for dark-winged moths and against light-winged individuals. The rise of antibiotic resistance among bacteria, described in this chapter's opening essay, also reflects directional selection, as does the increase in pesticide-resistant insects (see section 10.10). The fittest phenotype may initially be rare, but its frequency increases over multiple generations as the environment changes—for example, after exposure to the antibiotic or insecticide.

In **disruptive selection** (sometimes called diversifying selection), two or more extreme phenotypes are fitter than the intermediate phenotype. Consider, for example, a population of marine snails that live among tan rocks encrusted with white barnacles. The white snails near the barnacles are camouflaged, and the tan ones on the bare rocks likewise blend in. The snails that are not white or tan, or that lie against the oppositely colored background, are more often seen and eaten by predatory shorebirds.

In a third form of natural selection, called **stabilizing selection** (or normalizing selection), extreme phenotypes are less fit than the optimal intermediate phenotype. Human birth weight illustrates this tendency to stabilize. Very small or very large newborns are less likely to survive than babies of intermediate weight. By eliminating all but the individuals with the optimal phenotype, stabilizing selection tends to reduce the variation in a population. It is therefore most common in stable, unchanging environments.

These three models of natural selection might seem to suggest that, for each trait, only one or a few beneficial alleles ought to persist in the population. The harmful alleles should gradually become less common until they disappear, while the others become "fixed" in the population.

For some genes, however, natural selection maintains a **balanced polymorphism,** in which multiple alleles of a gene persist indefinitely in the population at more or less constant frequencies. (*Polymorphism* means "multiple forms"). In a balanced polymorphism, even harmful alleles may remain in the population. This situation seems contrary to natural selection; how can it occur?

One circumstance that can maintain a balanced polymorphism is a **heterozygote advantage,** which occurs when an

Figure 11.12 Types of Natural Selection.
(a) Directional selection results from selection against one extreme phenotype. (b) In disruptive selection, two extreme phenotypes each confer a selective advantage over the intermediate phenotype. (c) Stabilizing selection maintains an intermediate expression of a trait by selecting against extreme variants.

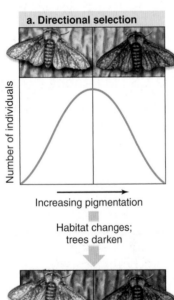
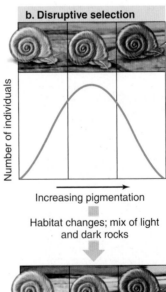
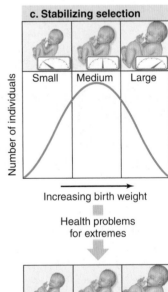

a. Directional selection

Number of individuals
Increasing pigmentation

Habitat changes; trees darken

b. Disruptive selection

Number of individuals
Increasing pigmentation

Habitat changes; mix of light and dark rocks

c. Stabilizing selection

Number of individuals
Small Medium Large
Increasing birth weight

Health problems for extremes

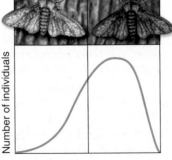
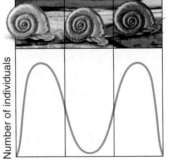
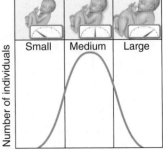

Number of individuals

Number of individuals

Number of individuals
Small Medium Large

Table 11.2 Examples of Heterozygote Advantage

Illness with Balanced Polymorphism	Heterozygote Advantage	Possible Explanation
Cystic fibrosis	Protection against diarrheal diseases such as cholera (see section 4.6)	Carriers have too few functional chloride channels in intestinal cells, blocking toxin
G6PD deficiency	Protection against malaria	Red blood cells inhospitable to malaria parasite
Phenylketonuria (PKU)	Protection against miscarriage induced by ochratoxin A, a fungal toxin	Excess amino acid (phenylalanine) in carriers inactivates toxin
Sickle cell disease	Protection against malaria	Red blood cells inhospitable to malaria parasite
Tay-Sachs disease	Protection against tuberculosis	Unknown
Noninsulin-dependent diabetes mellitus	Protection against starvation	Tendency to gain weight protects against starvation during famine

individual with two different alleles for a gene (a heterozygote) has greater fitness than those whose two alleles are identical (homozygotes). Heterozygotes can maintain a harmful recessive allele in a population, even if homozygous recessive individuals have greatly reduced fitness.

Table 11.2 lists several examples of balanced polymorphisms in which heterozygote advantage plays a role. The best documented example is sickle cell disease. The disease-causing allele encodes an abnormal form of hemoglobin. The abnormal hemoglobin proteins do not fold properly; instead, they form chains that bend a red blood cell into a characteristic sickle shape (see figure 7.22). In a person who is homozygous recessive for the sickle cell allele, all the red blood cells are affected. Symptoms include anemia, joint pain, a swollen spleen, and frequent, severe infections; the person may not live long enough to reproduce. On the other hand, a person who is heterozygous for the sickle cell allele is only mildly affected. Some of his or her red blood cells may take abnormal shapes, but the resulting mild anemia is not usually harmful. Still, why doesn't natural selection eliminate the sickle cell allele from the population?

The answer is that heterozygotes also have a reproductive edge over people who are homozygous for the *normal* hemoglobin allele. Specifically, heterozygotes are resistant to a severe infectious disease, malaria. When a mosquito carrying cells of a protist called *Plasmodium* feeds on a human with normal hemoglobin, the parasite enters the red blood cells. Eventually, infected blood cells burst, and the parasite travels throughout the body. But the sickled red blood cells of an infected carrier halt the parasite's spread. ▶ malaria, p. 364

The heterozygote advantage for the sickle cell allele is regional (figure 11.13). In malaria-free areas, anemia is the predominant selective force, and heterozygotes have no advantage. The sickle cell allele is therefore rare. Wherever malaria rages, however, the two opposing selective forces—anemia and malaria—maintain a balanced polymorphism for the sickle cell trait. In these areas, sickle cell carriers remain healthiest; they are resistant to malaria but not ill from sickle cell disease. The frequency of the

sickle cell allele remains high because carriers have more children than people who are homozygous for either allele.

Unfortunately, as described in chapter 10, two carriers have a 25% chance of producing a child who is homozygous for the sickle cell allele. These children pay the evolutionary price for the genetic protection against malaria. (Cystic fibrosis, the topic of section 4.6, provides another example of a balanced polymorphism.)

11.4 | Mastering Concepts

1. Distinguish among directional, disruptive, and stabilizing selection.
2. How can natural selection maintain harmful alleles in a population?

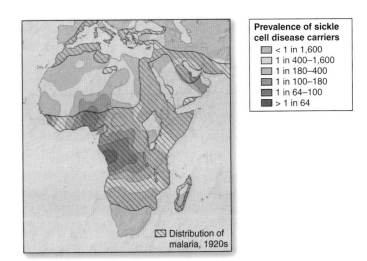

Prevalence of sickle cell disease carriers
< 1 in 1,600
1 in 400–1,600
1 in 180–400
1 in 100–180
1 in 64–100
> 1 in 64

Distribution of malaria, 1920s

Figure 11.13 Heterozygote Advantage. The distribution of people with one copy of the sickle cell allele overlaps closely with the areas where malaria is prevalent. Sickle cell anemia and malaria are opposing selective forces that maintain a balanced polymorphism for the two alleles of the hemoglobin gene.

11.5 | Sexual Selection Directly Influences Reproductive Success

In many species, the sexes look alike. The difference between a male and a female house cat, for example, is not immediately obvious. In some species, however, natural selection can maintain a **sexual dimorphism,** which is a difference in appearance between males and females. One sex may be much larger or more colorful than the other, or one sex may have distinctive structures such as horns or antlers.

Some of these sexually dimorphic features may seem to violate natural selection. For example, the vivid red feathers of male cardinals make the birds much more obvious to predators than their brown female counterparts, and the extravagant tail of a peacock is both brightly colored and too long to permit flight. How can natural selection allow for traits that apparently reduce survival?

The answer is that a special form of natural selection is at work. **Sexual selection** is a type of natural selection resulting from variation in the ability to obtain mates (figure 11.14). If female cardinals prefer bright red males, then showy plumage directly increases a male's chances of reproducing. Because the brightest males get the most chances to reproduce, alleles that confer red plumage are common in the population.

Sexual selection has two forms. In **intrasexual selection,** the members of one sex compete among themselves for access to the opposite sex; mate choice plays no part in deciding the winner. Male bighorn sheep, for example, use their horns to battle for the right to mate with multiple females. The strongest rams are therefore the most likely to pass on their alleles. In **intersexual selection,** members of one sex choose their mates from among multiple members of the opposite sex.

Why do males usually show the greatest effects of sexual selection? In most (but not all) animal species, females spend more time and energy rearing each offspring than do males. Because of this high investment in reproduction, females tend to be selective about their mates. Males are typically less choosy and must compete for access to females.

The evolutionary origin of the males' elaborate ornaments remains an open question. One possibility is that long tail feathers and bright colors are costly to produce and maintain; they are therefore indirect advertisements of good health or disease resistance. Likewise, the ability to win fights with competing males is also an indicator of good genes. A female who instinctively chooses a high-quality male will increase not only his fitness but also her own.

11.5 | Mastering Concepts

1. How does sexual selection promote traits that would seem to decrease fitness?
2. What is the difference between intrasexual selection and intersexual selection?

Figure 11.14 Examples of Sexual Selection. (a) Female weaver birds select mates based on the male's nest-building ability. (b) The male bird-of-paradise displays bright plumes that attract females. (c) Two bighorn sheep prepare to butt heads in the Rocky Mountains. (d) A female long-horned, wood-boring beetle chooses her mate by evaluating the male's territory size. A better territory means more food for offspring.

11.6 | Evolution Occurs in Several Additional Ways

Natural selection is responsible for adaptations that enhance survival and reproduction, but it is not the only mechanism of evolution. This section describes four more ways that a population can evolve: mutation, genetic drift, nonrandom mating, and gene flow. All occur frequently, and each can, by itself, disrupt Hardy–Weinberg equilibrium. The changes in allele frequencies that constitute microevolution therefore occur nearly all the time.

A. Mutation Fuels Evolution

A change in an organism's DNA sequence introduces a new allele to a population. The new variant may be harmful, neutral, or beneficial, depending on how the mutation affects the sequence of the encoded protein. ▶ mutations, p. 139

Mutations are the raw material for evolution because genes contribute to phenotypes, and natural selection acts on phenotypes. For example, random mutations in bacterial DNA may change the shapes of key proteins in the cell's ribosomes or cell wall. Exposure to antibiotics selects for some of the new phenotypes if they happen to make the cell resistant to the drug. In that case, the mutations will pass to the next generation.

A common misconception is that a mutation produces a novel adaptation precisely when a population "needs" it to confront a new environmental challenge. For example, many people mistakenly believe that antibiotics *create* resistance; that is, that resistance arises in bacteria *in response* to exposure to the drugs.

In reality, genes do not "know" when to mutate; the chance that a mutation will occur is independent of whether a new phenotype would benefit the organism. The only way antibiotic resistance arises is if some bacteria happen to have a mutation that confers antibiotic resistance *before* exposure to the drug. The drug creates a situation in which these variants can flourish. That trait will then become more common within the population by natural selection. If no bacteria start out resistant, the drug kills the entire population. Because bacterial populations are often enormous, however, it is likely that at least a few individuals carry such a mutation.

The rate at which mutations occur varies, both among different genes and within a gene. The average rate is around one DNA sequence change per 10^9 base pairs. At first, this number may seem too low to pose a significant force in evolution. Each genome, however, has an enormous number of base pairs, and a large number of cell divisions occur throughout life. During the human lifespan, for example, about 175 new mutations are estimated to occur in each person's genome.

A mutation affects evolution only if subsequent generations inherit it. In asexually reproducing organisms such as bacteria, each mutated cell gives rise to mutant offspring (if the mutation does not prevent reproduction). In a multicellular organism, however, a mutation can pass to the next generation only if it arises in a germ cell (i.e., one that will give rise to gametes; see chapter 9). For example, a cigarette smoker with lung cancer will not pass any smoking-induced mutations to her children, because her egg cells will not contain the altered DNA.

B. Genetic Drift Occurs by Chance

Genetic drift is a change in allele frequencies that occurs purely by chance. Unlike mutation, which increases diversity, genetic drift tends to eliminate alleles from a population.

All forms of genetic drift are rooted in sampling error, which occurs when a sample does not match the larger group from which it is drawn (**figure 11.15**). Suppose, for example, that one allele of a gene occurs at a very low frequency in a population. If, by chance, none of the individuals carrying the rare allele happens to reproduce, that variant will disappear from the population. Even if some do reproduce, the allele still might not pass to the next generation. After all, the events of meiosis ensure that each allele has only a 50% chance of passing to each offspring. A rare allele can therefore vanish from a population—not because it reduces fitness but simply by chance.

Allele frequencies for any gene will fluctuate at random from generation to generation, but significant sampling errors are most likely to affect small populations. Hardy–Weinberg equilibrium requires populations to be very large (approaching infinity)

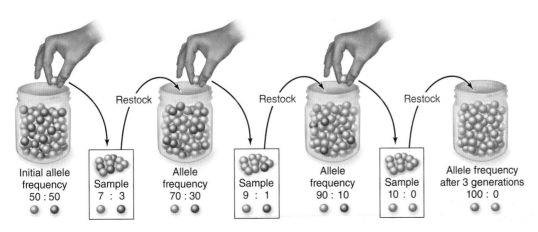

Figure 11.15 Sampling Error. Ten marbles are drawn at random from the jar on the left. Even though 50% of the marbles in the jar are blue, the random sample contains only 30% blue marbles. In the next "generation," 30% of the marbles are blue, but the random sample contains just one blue marble. The third sample contains no blue marbles and therefore eliminates the blue "allele" from the population, purely by chance.

Initial allele frequency
50 : 50

Sample
7 : 3

Restock

Allele frequency
70 : 30

Sample
9 : 1

Restock

Allele frequency
90 : 10

Sample
10 : 0

Restock

Allele frequency after 3 generations
100 : 0

to minimize the effects of these random changes in allele frequencies. In reality, of course, many populations are small, so genetic drift can be an important evolutionary force.

Random chance can eliminate rare alleles, but genetic drift can also operate in other ways. As described below, sampling errors can occur when a small population separates from a larger one or when a large population is reduced to a very small size.

The Founder Effect

One cause of genetic drift is the **founder effect,** which occurs when a small group of individuals leaves its home population and establishes a new, isolated settlement. The small group's "allele sample" may not represent the allele frequencies of the original population. Some traits that were rare in the original population may therefore be more frequent in the new population. Likewise, other traits will be less common or may even disappear.

The Amish people of Pennsylvania provide a famous example of the founder effect. About 200 followers of the Amish denomination immigrated to North America from Switzerland in

the 1700s. One couple, who immigrated in 1744, happened to carry the recessive allele associated with Ellis–van Creveld syndrome (**figure 11.16**). This allele is extremely rare in the population at large. Intermarriage among the Amish, however, has kept the disease's incidence high in this subgroup more than two centuries after the original immigrants arrived.

The Bottleneck Effect

Genetic drift also may result from a population **bottleneck,** which occurs when a population drops rapidly over a short period, causing the loss of many alleles that were present in the larger ancestral population (**figure 11.17**). Even if the few remaining individuals mate and eventually restore the population's numbers, the loss of genetic diversity is permanent.

Cheetahs are currently undergoing a population bottleneck. Until 10,000 years ago, these cats were common in many areas. Today, just two isolated populations live in South and East Africa, numbering only a few thousand animals. Inbreeding has made the South African cheetahs so genetically alike that even unrelated animals can accept skin grafts from each other. Researchers attribute the genetic uniformity of cheetahs to two bottlenecks: one that occurred at the end of the most recent ice age, when habitats changed drastically, and another when humans slaughtered many cheetahs during the nineteenth century.

North American species have undergone severe bottlenecks as well. Bison, for example, nearly went extinct because of overhunting in the 1800s. And habitat loss has caused the population of greater prairie chickens to plummet from about 100 million in 1900 to a several hundred today. The loss of genetic diversity in cheetahs, bison, and prairie chickens presents a potential disaster: a single change in the environment might doom them all.

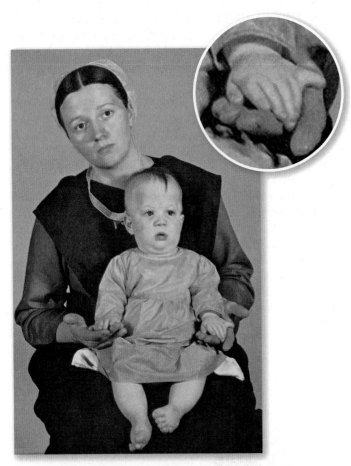

Figure 11.16 The Founder Effect. This Amish child from Lancaster County, Pennsylvania, has inherited Ellis–van Creveld syndrome, a condition characterized by short-limbed dwarfism, extra fingers, and other symptoms. This autosomal recessive disorder occurs in 7% of the people of this Amish community but is extremely rare elsewhere.

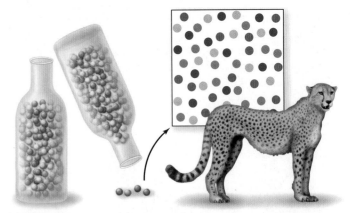

Original cheetah population contains 25 different alleles of a particular gene.

Cheetah population is drastically reduced.

Repopulation occurs. Only four different alleles remain.

Figure 11.17 The Bottleneck Effect. A population bottleneck occurs when a population shrinks rapidly, eliminating some alleles at random. In this figure, each color represents a different allele in the cheetah population.

C. Nonrandom Mating Concentrates Alleles Locally

To achieve and maintain Hardy–Weinberg equilibrium, a population must have completely random mating, in which each individual has an equal chance of mating with any other member of the population.

In reality, mating is rarely random. Many factors influence mate choice, including geographical restrictions, access to the opposite sex, and behavior. Most species also exhibit some form of preference in mate choice, including sexual selection (figure 11.18). Humans, for example, give great consideration to choosing mates; it is hardly a random process. For many people, cultural factors such as social position and religion are especially important in choosing partners.

The practice of artificial selection is another way to reduce random mating. Humans select those animals or plants that have a desired trait and then prevent them from mating with those lacking that trait. The result is a wide variety of subpopulations (such as different breeds of dogs) that humans maintain by selective breeding.

D. Gene Flow Moves Alleles Between Populations

Under Hardy–Weinberg equilibrium, no alleles ever leave or enter a population. In reality, **gene flow** moves alleles among populations. Migration is one common way that gene flow occurs (figure 11.19). Departing members of a population take their alleles with them. Likewise, new members entering and interbreeding with an existing population may add new alleles. But gene flow does not require the movement of entire individuals. Wind

Figure 11.18 **Nonrandom Mating.** If mating among toads were random, all individuals would have an equal chance of reproducing. Instead, a male mates only if his song attracts a willing female. This male toad inflates his throat pouch and generates a call.

Figure 11.19 **Gene Flow.** Immigration is one way for new alleles to enter a population.

can carry a plant's pollen for miles, for example, spreading one individual's alleles to a new population.

Gene flow characterizes large cities, which defy Hardy–Weinberg equilibrium by their very existence. Waves of immigration built the population of New York City, for example. The original Dutch settlers of the 1600s lacked many of the alleles present in today's metropolis; immigrants from other parts of Europe, Africa, Central and South America, and Asia introduced them. Within the city, pockets of ethnicity illustrate nonrandom mating as people have children with mates who are most like themselves.

Gene flow can counteract the effects of both natural selection and genetic drift. At one time, isolated European populations had unique allele frequencies for many genetic diseases. Geographical barriers, such as mountain ranges and large bodies of water, historically restricted migration and kept the gene pools separate. Highways, trains, and airplanes, however, have eliminated physical barriers to migration. Eventually, gene flow should make these regional differences disappear.

We have now seen how natural selection, mutation, genetic drift, nonrandom mating, and gene flow disrupt Hardy–Weinberg equilibrium and bring about evolutionary change. Figure 11.20, on page 248, summarizes their effects on populations.

11.6 | Mastering Concepts

1. What are some ways that mutations affect an organism's phenotype?
2. Under what conditions does a mutation in one organism pass to subsequent generations?
3. How does sampling error cause genetic drift?
4. What is the difference between the founder effect and a population bottleneck?
5. How do nonrandom mating and gene flow disrupt Hardy–Weinberg equilibrium?

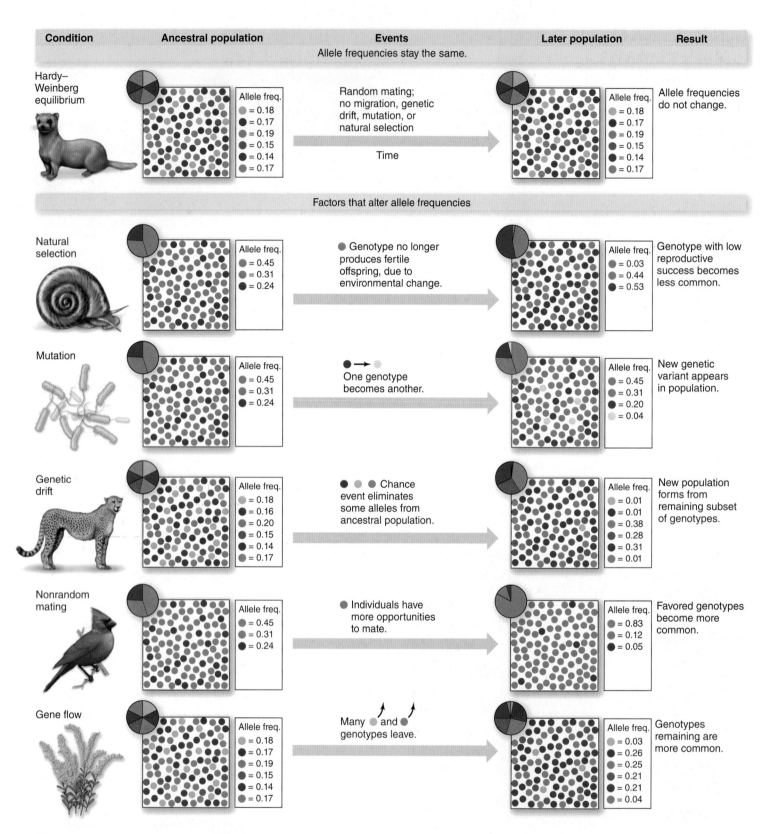

Figure 11.20 Mechanisms of Evolution: A Summary. At Hardy–Weinberg equilibrium, allele frequencies remain unchanged. Each mechanism of evolution—natural selection, mutation, genetic drift, nonrandom mating, and gene flow—change allele frequencies. In these images, different colored dots represent the alleles for a particular gene. The left image of each pair shows the allele frequencies in the ancestral population; the right image shows the frequencies after evolutionary change has occurred.

11.7 | Investigating Life: Size Matters in Fishing Frenzy

Studying the mechanisms of evolution helps us to understand life's history, but it also has practical consequences. A good example of natural selection is unfolding in fisheries worldwide. The selective force stems from a surprisingly mundane source—fishing regulations—but it affects everything from restaurant menus, to coastal economies, to the future of the ocean ecosystem.

The past several decades have seen devastating declines in the numbers of large predatory fishes such as swordfish, marlin, and sharks, as well as smaller animals including tuna, cod, and flounder. From a biological point of view, the reason for the fisheries decline is simple: the animals' death rate exceeds their reproductive rate. Industrial-scale fishing is the culprit. Since the 1950s, fishing fleets have employed larger ships and improved technologies in pursuit of their prey.

Fishing regulations usually allow the harvest of only those fish that exceed some minimum size. This strategy is logical, because the smallest fish are most likely to be juveniles. Protecting the youngsters should permit the population to recover from the harvest of adult fish. Yet these regulations also have predictable evolutionary side effects. If humans selectively harvest the largest individuals, fish that are small at maturity are the most likely to survive long enough to reproduce. Large fish may become more scarce over many generations. The same policy should also select for slow-growing fish, since they would be last to exceed the minimum allowed size.

Fish ecologists David Conover and Stephan Munch of the State University of New York tested these predictions by studying small coastal fish called Atlantic silversides (*Menidia menidia*). Conover and Munch set up their experiment by randomly dividing a large, captive population of Atlantic silversides into six tanks, each containing about 1100 juvenile fish. After about 6 months, the researchers assigned two tanks to each of the following treatments:

- **Large harvested:** Remove the largest 90% of the fish from two of the tanks, leaving the smallest 10% to reproduce. This treatment simulated fishing policies that protect all fish below a minimum size.

- **Small harvested:** Remove the smallest 90% of the silversides.

- **Random harvested (control):** Remove 90% of the fish, without size bias.

After the harvests, the 100 or so survivors in each tank reproduced, and their descendants were reared in identical conditions until it was again time to harvest 90% of each population. The researchers repeated the treatments over four generations.

Predictably, both the total harvest weight and the weight of the average caught fish were initially highest for the large-harvested fish. Over four generations of size-biased fish removal, however, the small-harvested treatment favored both large size and rapid growth; the opposite was true in the large-harvested population (figure 11.21). The researchers concluded that the three treatments imposed different selective forces that changed the genetic structure of the populations.

Conover and Munch's experiment is more than a straightforward demonstration of natural selection in action; it also has economic and ecological applications. Revised regulations that protect the smallest and the largest fish would spare the juveniles that are critical to a species' future reproduction while also selecting for fast-growing fish. Imposing a maximum size limit would also have ecological benefits, restoring the feeding patterns and other "ecosystem services" of the largest fish. This wide range of implications beautifully illustrates the powerful ideas that spring from understanding one fundamental idea: natural selection.

Conover, David O. and Stephan B. Munch. 2002. Sustaining fisheries yields over evolutionary time scales. *Science,* vol. 297, pages 94–96.

11.7 | Mastering Concepts

1. What hypothesis did Conover and Munch test?
2. How is a population reared in a tank different from a population in the "real world"?
3. The heritability of a trait ranges from 0 (entirely under environmental control) to 1.0 (100% controlled by genes). In Atlantic silversides, the heritability of body size is about 0.2. How would the results of this experiment differ if heritability of body size were higher? What if it approached 0?

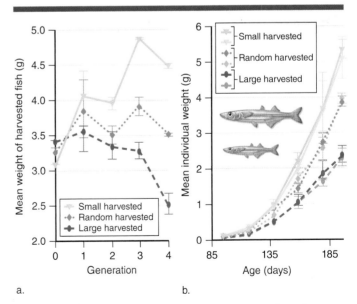

a.   b.

Figure 11.21 Size Matters in Fish Harvests. (a) Over four generations, the average per-fish weight in the large-harvested population was much less than the average for the small-harvested population. (b) Small-harvested fish grow faster than large-harvested fish.

Chapter Summary

11.1 | Evolutionary Thought Has Evolved for Centuries

- Biological **evolution** is change in allele frequencies in populations. Evolution has occurred in the past and continues in the present.

A. Many Explanations Have Been Proposed for Life's Diversity

- Early attempts to explain life's diversity relied on belief in a creator.
- Geology laid the groundwork for evolutionary thought. Some people explained the distribution of rock strata with the idea of **catastrophism** (a series of floods). The more gradual **uniformitarianism** (continual remolding of Earth's surface) became widely accepted.
- The **principle of superposition** states that lower rock strata are older than those above, suggesting an evolutionary sequence for fossils within them.
- Lamarck was the first to propose a testable mechanism of evolution, but it was erroneously based on **inheritance of acquired characteristics.**

B. Charles Darwin's Voyage Provided a Wealth of Evidence

- During the voyage of the HMS *Beagle,* Darwin observed the distribution of organisms in diverse habitats and their relationships to geological formations. After much thought and consideration of input from other scientists, he developed his theory of the origin of species by means of **natural selection.**

C. *On the Origin of Species* Proposed Natural Selection as an Evolutionary Mechanism

- Natural selection is based on multiple observations: individuals vary for inherited traits; many more offspring are born than survive; and life is a struggle to acquire limited resources. The environment eliminates poorly adapted individuals, so only those with the best adaptations reproduce.
- **Artificial selection** is based on similar requirements, except that a human breeder takes the place of the environment.
- *On the Origin of Species* offered abundant evidence for descent with modification. However, people who believed that Earth is young and that humans are unique had difficulty accepting Darwin's ideas.

D. Evolutionary Theory Continues to Expand

- The **modern evolutionary synthesis** unifies ideas about DNA, mutations, inheritance, and natural selection.

11.2 | Natural Selection Molds Evolution

- Natural selection is one mechanism of **microevolution,** the small-scale genetic changes within a species.

A. Adaptations Enhance Reproductive Success

- Individuals with the best **adaptations** to the current environment are most likely to leave fertile offspring, and therefore their alleles become more common in the population over time.
- Natural selection requires variation, which arises ultimately from random mutations.

B. Natural Selection Eliminates Phenotypes

- Natural selection weeds out some phenotypes, causing changes in allele frequencies over multiple generations.

C. Natural Selection Does Not Have a Goal

- Natural selection does not work toward a goal, nor can it achieve perfectly adapted organisms.

D. What Does "Survival of the Fittest" Really Mean?

- Organisms with the highest evolutionary **fitness** are the ones that have the greatest reproductive success. Many traits contribute to an organism's fitness.

11.3 | Evolution Is Inevitable in Real Populations

- A **gene pool** includes the alleles for all of the genes in a population. Calculations of **allele frequencies** and **genotype frequencies** allow biologists to detect whether evolution has occurred.

A. At Hardy–Weinberg Equilibrium, Allele Frequencies Do Not Change

- We can calculate the proportion of genotypes and phenotypes in a population by inserting known allele frequencies into an algebraic equation: $p^2 + 2pq + q^2 = 1$.
- If a population meets all assumptions of **Hardy–Weinberg equilibrium,** evolution does not occur because allele frequencies do not change from generation to generation.

B. In Reality, Allele Frequencies Always Change

- The conditions for Hardy–Weinberg equilibrium do not occur together in natural populations, suggesting that allele frequencies always change from one generation to the next.

11.4 | Natural Selection Can Shape a Population in Many Ways

- In **directional selection,** one extreme phenotype becomes more prevalent in a population.
- In **disruptive selection,** multiple extreme phenotypes survive at the expense of intermediate forms.
- In **stabilizing selection,** an intermediate phenotype has an advantage over individuals with extreme phenotypes.
- In **balanced polymorphism,** natural selection indefinitely maintains more than one allele for a gene.
- Harmful recessive alleles may remain in a population because of a **heterozygote advantage,** in which carriers have a reproductive advantage over homozygotes.

11.5 | Sexual Selection Directly Influences Reproductive Success

- **Sexual dimorphisms** differentiate the sexes. They result from **sexual selection,** a form of natural selection in which inherited traits—even those that seem nonadaptive—make an individual more likely to mate.
- **Intrasexual selection** is competition that does not involve a choice by the opposite sex; **intersexual selection** reflects mate choice by members of the opposite sex.

11.6 | Evolution Occurs in Several Additional Ways

A. Mutation Fuels Evolution

- Mutation alters allele frequencies by changing one allele into another, sometimes providing new phenotypes for natural selection to act on. Many mutations do not pass to the next generation.

B. Genetic Drift Occurs by Chance

- In **genetic drift,** allele frequencies change purely by chance events, especially in small populations. The **founder effect** and population **bottlenecks** are forms of genetic drift.

C. Nonrandom Mating Concentrates Alleles Locally

- Nonrandom mating causes some alleles to concentrate in subpopulations.

D. Gene Flow Moves Alleles Between Populations

- Allele movement between populations, as by migration, is **gene flow.**

11.7 | Investigating Life: Size Matters in Fishing Frenzy

- Fishing regulations that spare only the smallest fish in a population select for small, slow-growing individuals. Studies of Atlantic silversides suggest that protecting the largest fish as well would increase fishery productivity in the long run.

Multiple Choice Questions

1. The idea that geological processes have occurred in the past as they are occurring today is characterized as
   a. catastrophism.
   b. uniformitarianism.
   c. the principle of superposition.
   d. inheritance of acquired characteristics.

2. How are artificial selection and natural selection similar?
   a. They both rely on human intervention.
   b. They both work toward a predetermined goal.
   c. They both select for specific traits within a population.
   d. They are processes that only affect animals.

3. Microevolution applies to changes that occur
   a. only within small populations of organisms.
   b. in small regions of DNA.
   c. to the allele frequencies in a population or species.
   d. to small cells such as bacteria.

4. How does natural selection influence a gene pool?
   a. It selects for alleles that will be useful in the future.
   b. It triggers mutations, leading to increased diversity.
   c. It alters the frequency of alleles within an individual.
   d. It alters the frequency of alleles within a population.

5. Which of the following is the best definition of evolutionary fitness?
   a. The ability to increase the number of alleles in a gene pool
   b. The ability to survive for a long time
   c. The ability of an individual to adapt to a changing environment
   d. The ability to produce many offspring

6. Selection that favors two or more extreme phenotypes is called
   a. directional selection.   c. disruptive selection.
   b. stabilizing selection.   d. normalizing selection.

7. Huntington disease is caused by a lethal dominant allele that usually affects individuals late in life. How can natural selection explain why the disease-causing allele remains in the human population?
   a. The heterozygous individual might have some advantage.
   b. The disease does not interfere with the ability to reproduce.
   c. New mutations generate the allele.
   d. Both a and b are correct.

8. Assume a population is 36% homozygous dominant and 16% homozygous recessive. What percentage of the population is heterozygous, based on Hardy–Weinberg equilibrium?
   a. 64%   c. 48%
   b. 52%   d. 26%

9. Darwin observed that different types of organisms were found on either side of a geographic barrier. The barrier was preventing
   a. gene flow.   c. sexual selection.
   b. genetic drift.   d. mutation.

10. Which of the following processes is nonrandom?
    a. A population bottleneck   c. The founder effect
    b. Natural selection   d. Mutation

Write It Out

1. How did James Hutton, Georges Cuvier, Georges-Louis Buffon, Jean Baptiste de Lamarck, Charles Lyell, and Thomas Malthus influence Charles Darwin's thinking?

2. Your boss is a plant breeder who asks you to develop a sweeter variety of apple. How would you use artificial selection to achieve this goal? How might natural selection produce the same result? Which process would occur faster?

3. How does variation arise in an asexually reproducing population? A sexually reproducing population?

4. What happens to a population if conditions change and no individuals have the allele combinations required to survive and produce offspring?

5. Many articles about the rise of antibiotic-resistant bacteria claim that overuse of antibiotics creates resistant strains. How is this incorrect?

6. Explain how harmful recessive alleles can persist in populations, even though they prevent homozygous individuals from reproducing.

7. Fraggles are mythical, mouselike creatures that live underground beneath a large vegetable garden. Of the 100 Fraggles in this population, 84 have green fur, and 16 have gray fur. A dominant allele *F* confers green fur, and a recessive allele *f* confers gray fur. Assuming Hardy–Weinberg equilibrium is operating, answer the following questions. (a) What is the frequency of the gray allele *f*? (b) What is the frequency of the green allele *F*? (c) How many Fraggles are heterozygotes *(Ff)*? (d) How many Fraggles are homozygous recessive *(ff)*? (e) How many Fraggles are homozygous dominant *(FF)*?

8. One spring, a dust storm blankets the usually green garden of the Fraggles in gray. The green Fraggles therefore become visible to the Gorgs, who tend the gardens and try to kill the Fraggles to protect their crops. The gray Fraggles, however, blend easily into the dusty background. How might this event affect microevolution in the Fraggles? What mode of natural selection does this represent?

9. Describe examples of directional, disruptive, and stabilizing selection other than those mentioned in the chapter.

10. How does sexual selection maintain sexual dimorphism?

11. One of the most striking examples of convergent evolution occurs among animals that permanently live in caves. Among other shared traits, many of these species are eyeless. Develop a hypothesis to explain how an eyeless species could evolve from an ancestral population with eyes. How does your hypothesis compare to how Lamarck might have explained it?

Pull It Together

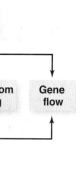

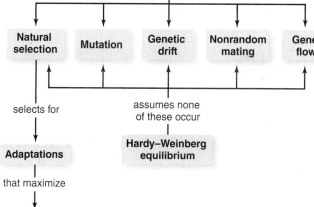

1. What is the biological definition of evolution?

2. Add the terms *genotype, phenotype, allele frequencies, founder effect, bottleneck effect,* and *sexual selection* to this concept map.

3. How does each mechanism of evolution change allele frequencies in a population?

4. Describe the three modes of natural selection.

The theropod dinosaur *Protarchaeopteryx* lived about 145 MYA in China. It had feathers that resembled down on its body and tail but also larger barbed feathers at the end of the tail and perhaps elsewhere. The model shows what the animal might have looked like, based on evidence from fossils.

| BIOLOGY

Enhance your study of
this chapter with practice quizzes,
animations and videos, answer keys,
and downloadable study tools.
www.mhhe.com/hoefnagels

Are Birds Dinosaurs?

LIKE A PUZZLE WITH MANY PIECES MISSING, THE HISTORY OF LIFE HAS MANY POSSIBLE INTERPRETATIONS. But as more information surfaces, the pieces fit together, and the story becomes clearer. Investigating the relationship between birds and dinosaurs is one such story of life.

The idea that birds are closely related to small, meat-eating dinosaurs called theropods began with the discovery of a 150-million-year-old fossil of a theropod with feathers in Bavaria. It was 1860, the year after Darwin published *On the Origin of Species*. The animal, named *Archaeopteryx lithographica,* was the size of a blue jay, with a mix of features seen in birds and nonavian reptiles: wings and feathers, a toothed jaw, and a long, bony tail.

In 1870, Thomas Henry Huxley reported that he had identified 35 features that only ostriches and theropod skeletons shared, which he interpreted as evidence of a close relationship. Others dismissed Huxley's ideas, largely because dinosaurs were not known to have flown. By 1916, the hypothesis seemed even less likely, when Gerhard Heilmann, a Danish physician and fossil collector, noted that theropods lack clavicles, which form a bird's wishbone. Fossils of theropods with clavicles had simply not yet been discovered.

In the 1960s, however, Yale University paleontologist John Ostrom strengthened the bird–dinosaur link with an exhaustive comparison of bones discovered in the 1920s, including the clavicles of theropods. Since then, the similarities between dinosaurs and birds have added up. Many of the structures that make flight possible were present in dinosaurs that lived millions of years before birds. For example, some theropods had hollow bones, an upright stance in which they stood on their toes, a horizontal back, long arms, and a short tail. Rare finds of dinosaur nests reveal behavioral and reproductive similarities to birds.

For years, all that was missing was evidence of feathered dinosaurs. In 1996, however, researchers discovered fossils of *Sinosauropteryx* in China. This turkey-sized animal had a fringe of downlike structures along its neck, back, and flanks. Then, in 1998, a bonanza fossil find introduced *Protarchaeopteryx* and *Caudipteryx,* both even more feathery than *Sinosauropteryx.*

Still another piece of the puzzle fell into place in 2007, when scientists revealed amino acid sequences of collagen protein molecules extracted from *Tyrannosaurus rex* fossils. Of all the vertebrate species studied, the *T. rex* collagen sequence was more similar to that of a chicken than it was to other reptiles, including alligators and lizards.

The surprising bird–dinosaur link illustrates the changeable nature of science and reinforces the importance of evidence. Every chapter of this textbook contains a section titled Investigating Life that explains how biologists test hypotheses about evolution. This chapter summarizes the main types of evidence they use.

UNIT 3

What's the Point?

Learning Outline

 Learn How to Learn

Make a Chart

One way to organize all of the information in a chapter is to make a summary chart or matrix. The chart's contents will depend on the chapter. For this chapter, for example, you might write the following headings along the top of a piece of paper: "Type of Evidence," "Definition," "How It Works," "What It Tells Us," and "Example." Then you would list the lines of evidence for evolution along the left edge of the chart. Start filling in your chart, using the book at first to find the information you need. Later, you should be able to recreate your chart from memory.

12.1 | Clues to Evolution Lie in the Earth, Body Structures, and Molecules

The millions of species alive today did not just pop into existence all at once; they are the result of continuing evolutionary change in organisms that lived billions of years in the past. Many types of clues enable us to hypothesize about how modern species evolved from extinct ancestors and to understand the relationships among organisms that live today. (See this chapter's Burning Question.)

Traditionally, evidence for evolution came mostly from **paleontology,** the study of fossil remains or other clues to past life (figure 12.1). The discovery of many new types of fossils in the early 1800s created the climate that allowed Charles Darwin to make his tremendous breakthrough (see chapter 11). As people recognized that fossils must represent a history of life, scientists developed theories to explain that history. The geographical locations of fossils and modern species provided additional clues.

At Darwin's time, some scientists suspected that Earth was hundreds of millions of years old, but no one knew the exact age. We now know that Earth's history is about 4.6 billion years long, a duration that most people find nearly unimaginable. Scientists describe the events along life's long evolutionary path in the context of the **geologic timescale,** which divides Earth's history into a series of eras defined by major geological or biological events such as mass extinctions (figure 12.2). ▶ mass extinctions, p. 284

Fossils and biogeographical studies provided the original evidence for evolution, revealing when species most likely diverged from common ancestors in the context of other events happening on Earth. Comparisons of embryonic development and anatomical structures provided additional supporting data. An entirely new type of evidence emerged in the 1960s and 1970s, when scientists began analyzing the sequences of DNA, proteins, and other biological molecules. Since then, the explosion of molecular data has revealed in unprecedented detail how species are related to one another.

Chapter 11 explained how natural selection and other mechanisms of evolution account for changes within species. This chapter examines the different approaches to studying evolution in both living and extinct species. Chapter 13 explains how new species form and become extinct, ending with a description of how scientists assemble diverse clues into hypotheses of the relationships among species. Chapter 14 continues on this theme by offering a brief history of life on Earth.

12.1 | Mastering Concepts

1. What is the geologic timescale?
2. What types of information provide the clues that scientists use in investigating evolutionary relationships?

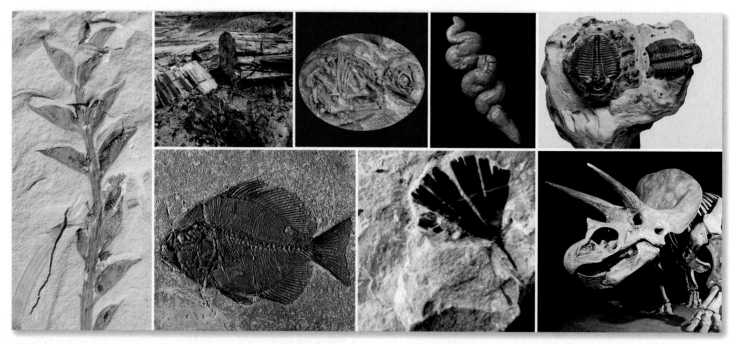

Figure 12.1 Fossil Evidence Is Diverse. Plants and animals have left a rich fossil record spanning hundreds of millions of years. Clockwise from left: *Archaefructus,* an early flowering plant that lived in China 138 million years ago; petrified wood; a 190-million-year-old dinosaur embryo encased in its egg; fossilized feces of an unidentified turtle from the Miocene epoch (between 5 million and 24 million years ago); trilobites, which were arthropods that became extinct some 250 million years ago; a skull of *Triceratops,* a dinosaur that lived in North America until 65 million years ago; a *Ginkgo* leaf; an exceptionally well-preserved fish.

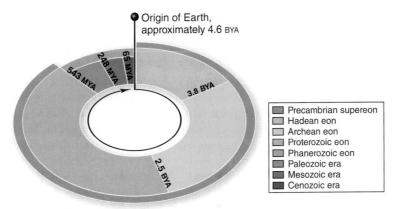

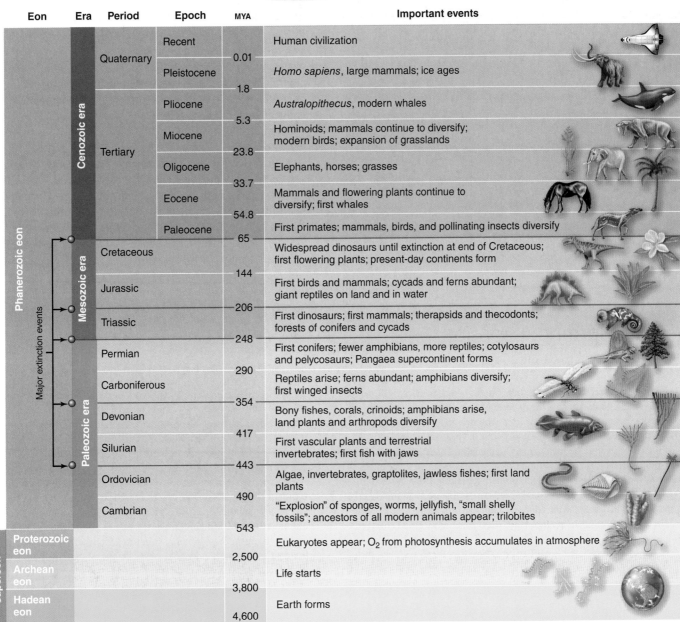

Figure 12.2 The Geologic Timescale. Scientists divide Earth's 4.6-billion-year history into five eras and multiple periods and epochs. Notice that the current era, the Cenozoic, is but a flicker of time compared with all of Earth's existence. By contrast, the Archean and Proterozoic eras are by far the longest, yet we know the least about them. (MYA = million years ago.)

12.2 | Fossils Record Evolution

 A **fossil** is any evidence of an organism from more than 10,000 years ago. Fossils come in all sizes, documenting the evolutionary history of everything from microorganisms to dinosaurs to humans. These remains give us our only direct evidence of organisms that preceded human history. They occur all over the world and represent all major groups of organisms, revealing much about the geological past. For example, the abun-

Figure 12.3 Big Change. Ammonite fossils such as these are common in land-locked Oklahoma, indicating that what is now the central United States was once covered by an ocean.

dant remains of extinct marine animals called ammonites in Oklahoma indicate that a vast, shallow ocean once submerged what is now the central United States (**figure 12.3**).

Fossils do more than simply provide a collection of ancient remains of plants, animals, and microbes. They also allow researchers to test predictions about evolution. Section 20.17 describes an excellent example: the discovery of *Tiktaalik*, an extinct animal with characteristics of both fishes and amphibians. Based on many lines of evidence showing the close relationship between these two groups, biologists had long predicted the existence of such an animal. *Tiktaalik*, described in 2006, finally provided direct fossil evidence of the connection.

A. Fossils Form in Many Ways

Evidence of past life comes in many forms (**figure 12.4**). Often, a fossil forms when an organism dies, becomes buried in sediments, and then is chemically altered. For example, coal, oil, and natural gas (also called fossil fuels) are the decomposed remains of plants and other organisms preserved by compression. Alternatively, minerals can replace the organic matter left by a decaying organism, which literally turns to stone. Petrified wood forms in this way.

Impression fossils form when an organism presses against soft sediment, which then hardens, leaving an outline or imprint of the body. An impression fossil may also be evidence of an animal's movements, such as footprints. If the imprint later fills with mud that hardens into rock, the resulting cast is a rocky replica of the ancient organism. Teeth, bones, arthropod exoskeletons, and tree trunks often leave casts.

Burning Question

Is evolution really testable?

Some people believe that evolution cannot be tested because it happened in the past. Although no experiment can recreate the conditions that led to today's diversity of life, evolution is testable. In fact, its validity has been verified repeatedly over the past 150 years.

All scientific theories, including evolution, not only explain existing data but also predict future observations. Some other explanations for life's diversity, including intelligent design, are unscientific because they do not offer testable predictions; that is, no future observation can disprove the idea that an intelligent creator designed life on Earth. One strength of evolutionary theory, in contrast, is its ability to predict future discoveries.

For example, if vertebrate life started in the water and then moved onto land, an evolutionary biologist might predict that fishes should appear in the fossil record before reptiles or mammals. A newly found fossil that contradicts the prediction would require investigators to form a new hypothesis. On the other hand, fossils that confirm the prediction lend additional weight to the theory's validity.

Evaluating common descent is somewhat similar to solving a crime. The perpetrator may leave footprints, tire tracks, fingerprints,

and DNA at a crime scene. Detectives would develop hypotheses about possible suspects by piecing together the physical and biological evidence. Innocent suspects can exonerate themselves by providing additional information, such as a verified alibi, that contradicts the hypothesis. Ultimately, the best explanation is the one that is consistent with all available evidence.

A mountain of evidence supports the idea of common descent. Extinct organisms have left traces of their existence, both as fossils and as the genetic legacy that all current organisms have inherited. The distribution of life on Earth offers other clues, as does the study of everything from anatomical structures to protein sequences. Laboratory experiments and field observations of natural populations likewise suggest likely mechanisms for evolutionary change and have even documented the emergence of new species. No other scientifically testable hypothesis explains and unifies all of these observations as well as common descent.

Submit your burning question to:
marielle_hoefnagels@mcgraw-hill.com

a. Compression

Leaf sinks.

Fine sediment covers leaf.

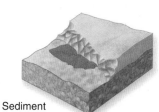

Sediment compresses, forming sedimentary rock.

b. Petrifaction

Animal dies, decays, and is buried.

Water containing dissolved minerals seeps through.

Organic matter replaced by minerals "turns to stone."

c. Impression

Animal dies, making impression in mud.

Animal decays away.

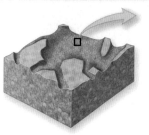

Mud hardens to rock.

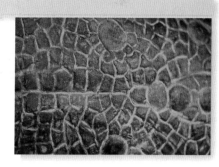

d. Cast

Animal dies and sinks into soft sediment.

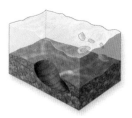

Animal decays away.

Imprint fills with mud.

Mud hardens to rock.

e. Intact preservation

Oozing sap traps an insect.

Figure 12.4 How Fossils Form. (a) A compression fossil of a leaf preserves part of the plant. (b) Most fossils of human ancestors consist of mineralized bones and teeth, usually found in fragments. (c) An impression fossil reveals an imprint of anatomical details, such as the scaly skin of a dinosaur. (d) This horn coral is a cast. Once-living material dissolved and was replaced by mud that hardened into rock. (e) Fossils can also be preserved intact in tree resin, which hardens to form amber.

Rarely, a whole organism is preserved intact. Fossil pollen grains in waterlogged lake sediments provide clues to long-ago climates. Sticky tree resin entraps plants and animals, hardening them in translucent amber tombs. The La Brea Tar Pits in Los Angeles have preserved more than 660 species of animals and plants that became stuck in the gooey tar thousands of years ago.

The most striking fossils have formed when sudden catastrophes, such as mud slides and floods, rapidly buried organisms in an oxygen-poor environment. Decomposition and tissue damage were minimal in the absence of oxygen; scavengers could not reach the dead. In those conditions, even delicate organisms have left detailed anatomical portraits (**figure 12.5**).

B. The Fossil Record Is Often Incomplete

Occasionally, researchers find fossils that document, step-by-step, the evolution of one species into another (see section 17.6). Usually, however, the fossil record is incomplete, meaning that some of the features marking the transition from one group to another are not recorded in fossils.

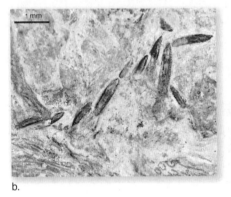

Figure 12.5 A Field of Dinosaur Embryos. (a) This plain in present-day Argentina was a valley about 89 MYA. Thousands of dinosaur eggs containing about-to-hatch babies were buried in mud when streams overflowed their banks. The newborns would have been only 40 centimeters long but would have attained an adult length of over 13 meters. (b) Preserved embryonic teeth were so tiny that a mouthful of 32 of them would fit into an "o" in this sentence. (c) The skin from some of the embryos was remarkably well preserved.

Several explanations account for this partial history. First, the vast majority of organisms never leave a fossil trace. Soft-bodied organisms, for example, are much less likely to be preserved than are those with teeth, bones, or shells. Organisms that decompose or are eaten after death, rather than being buried in sediments, are also unlikely to fossilize. Second, erosion or the movements of Earth's continental plates have destroyed many fossils that did form. Third, scientists are unlikely to ever discover the many fossils that must be buried deep in the Earth or submerged under water.

C. The Age of a Fossil Can Be Estimated in Two Ways

Scientists use two general approaches to estimate when a fossilized organism lived: relative dating and absolute dating.

Relative Dating   **Relative dating** places a fossil into a sequence of events without assigning it a specific age. It is usually based on the principle of superposition, with lower rock strata presumed to be older than higher layers (see figure 11.2). The farther down a fossil is, therefore, the longer ago the organism it represents lived—a little like a memo at the bottom of a pile of papers on a desk being older than one nearer the top. Relative dating therefore places fossils in order from "oldest" to "most recent."

Absolute Dating and Radioactive Decay   Researchers use **absolute dating** to assign an age to a fossil by testing either the fossil itself or the sediments above and below the fossil. Either way, the dates usually are expressed in relation to the present. For example, scientists studying fossilized *Archaeopteryx* showed that this animal lived about 150 million years ago (MYA). Although the term *absolute dating* seems to imply pinpoint accuracy, absolute dating techniques typically return a range of likely dates.

Radiometric dating is a type of absolute dating that uses radioactive isotopes as a "clock." Recall from chapter 2 that each isotope of an element has a different number of neutrons. Some isotopes are naturally unstable, which causes them to emit radiation as they radioactively decay. Each radioactive isotope decays at a characteristic and unchangeable rate, called its half-life. The **half-life** is the time it takes for half of the atoms in a sample of a radioactive substance to decay. If an isotope's half-life is 1 year, for example, 50% of the radioactive atoms in a sample will have decayed in a year. In another year, half of the remaining radioactive atoms will decay, leaving 25%, and so on. If we measure the amount of a radioactive isotope in a sample, we can use the isotope's known half-life to deduce when the fossil formed. ▶ isotopes, p. 21

One radioactive isotope often used to assign dates to fossils is carbon-14 (^{14}C; **figure 12.6**), which forms in the atmosphere when cosmic rays from space bombard nitrogen gas. Carbon-14 has a half-life of 5730 years; it decays to the more stable nitrogen-14 (^{14}N). Organisms accumulate ^{14}C during photosynthesis or by eating organic matter. One in every trillion carbon atoms present in

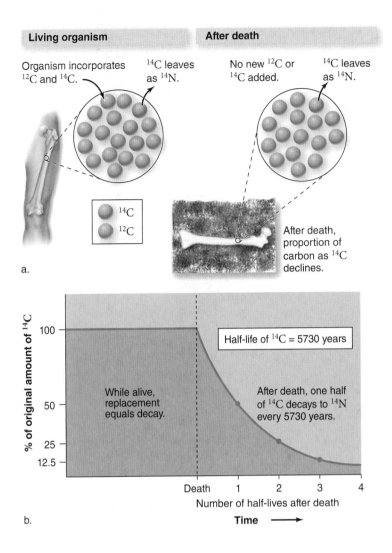

a.

c.

Figure 12.6 Carbon-14 Dating. (a) Living organisms accumulate radioactive carbon-14 (^{14}C) by photosynthesis or eating other organisms. During life, ^{14}C is replaced as fast as it decays to nitrogen-14 (^{14}N). After death, no new ^{14}C enters the body, so the proportion of ^{14}C to ^{12}C declines. (b) During one half-life, 50% of the remaining radioactive atoms in a sample decay. (c) Measuring the proportion of ^{14}C to ^{12}C allows scientists to determine how long ago a fossilized organism—such as this woolly mammoth—died.

living tissue is ^{14}C; most of the rest are ^{12}C, a nonradioactive (stable) isotope. When an organism dies, however, its intake of carbon, including ^{14}C, stops. As the body's ^{14}C decays without being replenished, the ratio of ^{14}C to ^{12}C decreases. This ratio is then used to determine when death occurred, up to about 40,000 years ago.

For example, radioactive carbon dating determined the age of fossils of vultures that once lived in the Grand Canyon. The birds' remains have about one fourth the ^{14}C-to-^{12}C ratio of a living organism. Therefore, about two half-lives, or about 11,460 years, passed since the animals died. It took 5730 years for half of the ^{14}C to decay, and another 5730 years for half of what was left to decay to ^{14}N.

Figure It Out

Kennewick Man is a human whose remains were found in 1996 in Washington state. Radiometric dating of a bone fragment suggests that he lived about 9300 years ago. At the time scientists dated the bone, about what percent of the original amount of ^{14}C remained in his bones?

Answer: Approximately 30%

Another widely used radioactive isotope, potassium-40 (^{40}K), decays to argon-40 (^{40}Ar) with a half-life of 1.3 billion years, so it is valuable in dating very old rocks containing traces of both isotopes. Chemical analyses can detect the accumulation of ^{40}Ar in amounts corresponding to fossils that are about 300,000 years old or older.

One limitation of ^{14}C and potassium–argon dating is that they leave a gap, resulting from the different half-lives of the radioactive isotopes. To cover the missing years, researchers use isotopes with intermediate half-lives or turn to other techniques, such as tree-ring comparisons. ▶ tree rings, p. 469

12.2 | Mastering Concepts

1. What are some of the ways that fossils form?
2. Why will the fossil record always be incomplete?
3. Distinguish between relative and absolute dating of fossils.
4. How does radiometric dating work?

12.3 | Biogeography Considers Species' Geographical Locations

Geographical barriers greatly influence the origin of species (see chapter 13). It is therefore not surprising that the studies of geography and biology overlap in one field, **biogeography,** the study of the distribution of species across the planet.

A. The Theory of Plate Tectonics Explains Earth's Shifting Continents

Despite the occasional volcanic eruptions and earthquakes, Earth's geological history might seem rather uneventful. Not so. Fossils tell the story of ancient seafloors rising all the way to Earth's "ceiling": the Himalayan Mountains. Littering the Kali Gandaki River in the mountains of Nepal are countless fossilized ammonites, large mollusks similar to the chambered nautilus. How did fossils of marine animals end up more than 3600 meters above sea level?

The answer is that Earth's continents are in motion, an idea called "continental drift" (**figure 12.7**). According to the theory of **plate tectonics,** Earth's surface consists of several rigid layers, called tectonic plates, that move in response to forces acting deep within the planet. In some areas where plates come together, mountain ranges form as the plates become wrinkled and distorted. Long ago, the Indo-Australian plate (which includes the subcontinent of India) moved slowly north and eventually collided with the Eurasian plate. The mighty Himalayas—once an ancient seafloor—rose at the boundary, lifting the marine fossils toward the sky.

In other places, called subduction zones, one plate dives beneath another, forming a deep trench. Meanwhile, new plate material forms at areas where plates move apart and molten rock seeps to Earth's surface at the seam. As a result, wide oceans now separate continents that were once joined together. This slow-motion dance of the continental plates has dramatically affected life's history as oceans shifted, land bridges formed and disappeared, and mountain ranges emerged.

It may seem hard to imagine that Earth's continents have not always been located where they are now. But a wealth of evidence, including the distribution of some key fossils, indicate that the continents were once united (**figure 12.8**). Deep-sea probes that measure seafloor spreading, along with the locations of the world's earthquake-prone and volcanic "hot spots," reveal that the continents continue to move today.

B. Species Distributions Reveal Evolutionary Events

Biogeographical studies have shed light on past evolutionary events. The rest of this section describes three examples.

The Case of the Missing Marsupials   Marsupials are pouched mammals such as kangaroos, koalas, and sugar gliders.

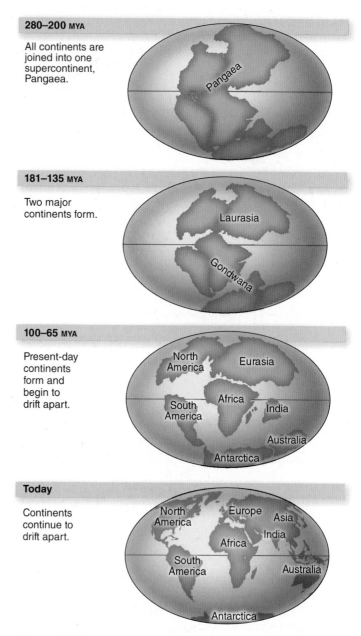

280–200 MYA

All continents are joined into one supercontinent, Pangaea.

Pangaea

181–135 MYA

Two major continents form.

Laurasia

Gondwana

100–65 MYA

Present-day continents form and begin to drift apart.

North America   Eurasia

Africa

South America   India

Australia

Antarctica

Today

Continents continue to drift apart.

North America   Europe   Asia

Africa   India

South America

Australia

Antarctica

Figure 12.7 A Changing World. The distribution of continents on Earth has changed with time, due to shifting tectonic plates.

The young are born tiny, hairless, blind, and helpless. As soon as they are born, they crawl along the mother's fur to tiny, milk-secreting nipples inside her pouch.

Marsupials were once the most widespread land mammals on Earth. About 110 MYA, however, a second group, the placental mammals, had evolved. The young of placental mammals develop within the female's body, nourished in the uterus by the placenta. Baby placental mammals are born more fully developed than are marsupials, giving them a better chance of survival after birth. Because of this reproductive advantage, placental mammals soon displaced marsupials on most continents, including North America. ▶ mammals, p. 445

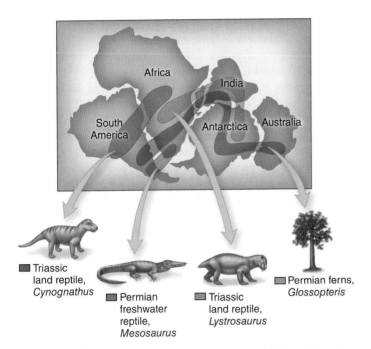

Figure 12.8 Fossils Tell the Tale. Continental drift explains the modern-day distributions of fossils representing life in the southern hemisphere at the time of Pangaea.

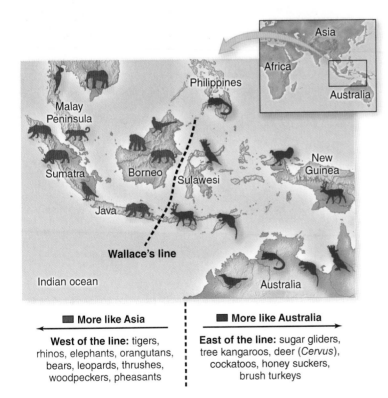

Figure 12.9 Wallace's Line. As Alfred Russel Wallace traveled around the Malay archipelago, he noticed distinct patterns of animal life on either side of an imaginary boundary, which eventually came to be called Wallace's line.

Nevertheless, fossil evidence suggests that marsupials were diverse and abundant in South America until about 1 or 2 MYA, long after their counterparts on most other continents had disappeared. The reason is that water separated South America from North America until about 3 MYA. But sediments eroded from both continents eventually created a new land bridge, the isthmus of Panama, which permitted migration between North and South America. The resulting invasion of the placental mammals eventually spelled extinction for most South American marsupials.

Australia's marsupials remained isolated from competition from placental mammals for much longer. Until about 140 MYA, Antarctica and Australia were part of Gondwana (see figure 12.7). Then, about 60 or 70 MYA, Australia separated from Antarctica and began drifting toward Eurasia. The isolation from the other continents meant marsupials remained free of competition from placentals. In fact, Australia remains unique in that most of its native mammals are still marsupials.

Island Biogeography A smaller-scale application of biogeography is the study of species on island chains. Hawaii, the Galápagos, and other island groups house unique groups of species that appear closely related to those on the nearest mainland. The explanation is that the only organisms that can colonize a newly formed volcanic island are those that can swim, fly, or raft from an inhabited location. The exchange of organisms between an isolated island and the mainland is probably quite rare. Those organisms that do successfully colonize the island encounter conditions that are different from those on the mainland. With limited migration between the island and the ancestral populations, the relocated organisms have evolved and diversified into new species, a process called adaptive radiation. Just as they did for the marsupials, geographical barriers have influenced the course of evolution on island chains. ▶ adaptive radiation, p. 282

Wallace's Line Biogeography figured prominently in the early history of evolutionary thought. Alfred Russel Wallace, the British naturalist who independently discovered natural selection along with Charles Darwin, had noticed unique assemblages of birds and mammals on either side of an imaginary line in the Malay Archipelago (figure 12.9). The explanation for what came to be called "Wallace's line" turned out to be a deep-water trench that has separated the islands on either side of the line, even as sea levels rose and fell over tens of millions of years. The watery barrier prevented the migration of most species, so evolution produced a unique variety of organisms on each side of Wallace's line.

12.3 | Mastering Concepts

1. How have the positions of Earth's continents changed over the past 200 million years?
2. How do biogeographical observations help biologists interpret evolutionary history?

12.4 | Anatomical Comparisons May Reveal Common Descent

Many clues to the past come from the present. As unit 1 explains, all life is made of cells, and eukaryotic cells are very similar in the structure and function of their membranes and organelles. On a molecular scale, cells share many similarities in their enzymes, signaling proteins, and metabolic pathways. Unit 2 describes another set of common features: the relationship between DNA and proteins, and the mechanisms of inheritance. We now turn to the whole-body scale, where comparisons of anatomy and physiology reveal still more commonalities among modern species.

A. Homologous Structures Have a Shared Evolutionary Origin

Two structures are termed **homologous** if the similarities between them reflect common ancestry. (Recall from chapter 9 that two chromosomes are homologous if they have the same genes, though not necessarily the same alleles, arranged in the same order.) Homologous genes, chromosomes, anatomical structures, or other features are similar in their configuration, position, or developmental path.

The organization of the vertebrate skeleton illustrates homology. All vertebrate skeletons support the body, are made of the same materials, and consist of many of the same parts. Amphibians, birds and other reptiles, and mammals typically have four limbs, and the numbers and positions of the bones that make up those appendages are strikingly similar (**figure 12.10**). The simplest explanation is that modern vertebrates descended from a common ancestor that originated this skeletal organization. Each group gradually modified the skeleton as species adapted to different environments.

Note that homologous structures share a common evolutionary origin, but they may not have the same function. The middle ear bones of mammals, for example, originated as bones that sup-

ported the jaws of primitive fishes, and they still exist as such in some vertebrates. These bones are homologous and reveal our shared ancestry with fishes. Likewise, the forelimbs pictured in figure 12.10 have varying functions.

Homology is a powerful tool for discovering evolutionary relationships. For example, as described in section 12.6, a newly sequenced gene or genome can be compared with homologous genes from other species to infer how closely related any two species are. Likewise, fossilized structures are often compared with homologous parts in known species. Comparative studies can also provide clues to why humans have some of the features we do (see this chapter's Apply It Now box on page 268).

B. Vestigial Structures Have Lost Their Functions

Evolution is not a perfect process. As environmental changes select against some structures, others persist even if they are not used. A **vestigial** structure has no apparent function in one species, yet it is homologous to a functional organ in another. (Darwin compared vestigial structures to silent letters in a word, such as the "g" in *night;* they are not pronounced, but they offer clues to the word's origin.) In some whales and snakes, tiny leg bones are vestigial, retained from vertebrate ancestors that used legs to walk on land (**figure 12.11**).

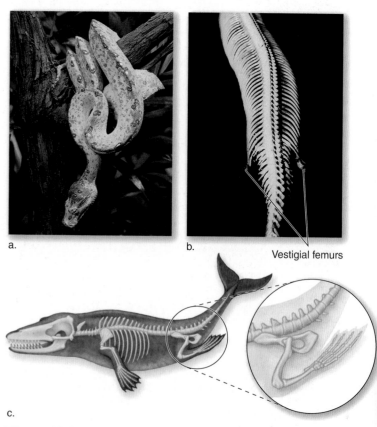

a.

b.

Vestigial femurs

c.

Figure 12.11 Vestigial Structures. (a) Some snakes have tiny femurs (leg bones) that are vestigial but (b) detectable only in the skeleton. (c) Some whales likewise have a vestigial pelvis and hind limbs.

Figure 12.10 Homologous Limbs. Although their forelimbs have different functions, all of these vertebrates have skeletons that are similarly organized and composed of the same type of tissue.

Humans have several vestigial organs. The tiny muscles that make hairs stand on end helped our furry ancestors conserve heat or show aggression; in us, they apparently serve only as the basis of goose bumps. Human embryos have tails, which usually disintegrate long before birth; in other vertebrates, tails persist into adulthood. Above our ears, a trio of muscles (which most of us can't use) helps other mammals move their ears in a way that improves hearing. Each vestigial structure links us to other animals that still use these features.

C. Convergent Evolution Produces Superficial Similarities

Some anatomical parts have similar functions and appear superficially similar among different species, but they are not homologous. Rather, they are **analogous,** meaning that the structures evolved independently. Flight, for example, evolved independently in birds and in insects. The bird's wing is a modification of vertebrate limb bones, whereas the insect's wing is an outgrowth of the exoskeleton that covers its body (figure 12.12). The wings have the same function—flight—and enhance fitness in the face of similar environmental challenges. The differences in structure, however, indicate they do not have a common developmental pathway. They are analogous, not homologous.

Analogous structures are often the product of **convergent evolution,** which produces similar adaptations in organisms that

Figure 12.12 **Analogous Structures.** The wings of birds and butterflies both function in flight, but they are analogous structures because they are not made of the same materials, nor are they organized in the same way. They are not inherited from a recent common ancestor.

do not share the same evolutionary lineage. The loss of pigmentation and eyes in diverse cave animals provides a striking example of convergent evolution (figure 12.13). Likewise, the similarities between sharks and dolphins illustrate the power of selective forces in shaping organisms. A shark is a fish, whereas a dolphin is a mammal that evolved from terrestrial ancestors. The two animals are not closely related. Nevertheless, their marine habitat and predatory lifestyle have selected for many shared adaptations, including the streamlined body and the shape and locations of the fins or flippers.

12.4 | Mastering Concepts

1. What can homologies reveal about evolution?
2. What is a vestigial structure? What are some examples of vestigial structures in humans and other animals?
3. What is convergent evolution?

a.

b.

c.

Figure 12.13 **Convergent Evolution.** (a) The blind cave salamander (*Proteus anguinus*) and (b) the blind cave isopod (*Titanethes albus*) occur in Slovenia. (c) This cave crayfish from Florida (*Troglocambarus maclanei*) also lacks eyes and pigment.

12.5 | Embryonic Development Patterns Provide Evolutionary Clues

Because related organisms share many physical traits, they must also share the processes that produce those traits. Developmental biologists study how the adult body takes shape from its single-celled beginning. Careful comparisons of developing body parts can be enlightening. As just one example, **figure 12.14** shows that similar-looking skulls can develop in different ways, depending on how each part grows in proportion to the others.

By comparing embryos at different stages, it should be possible to deduce some of the steps that have led to differences among species. Historically, however, the field of comparative embryology has seen some controversy. In 1874, German naturalist Ernst Haeckel published nearly identical drawings of embryos from different vertebrate species, including fishes,

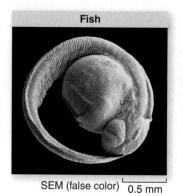
Fish

SEM (false color) 0.5 mm

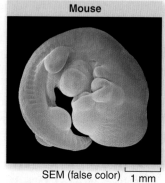
Mouse

SEM (false color) 1 mm

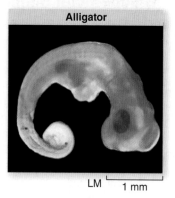
Alligator

LM 1 mm

Figure 12.15 Embryo Resemblances. Vertebrate embryos appear alike early in development. As development continues, parts grow at different rates in different species, and the embryos begin to look less similar.
Source: Fish © Dr. Richard Kessel/Visuals Unlimited, Inc.

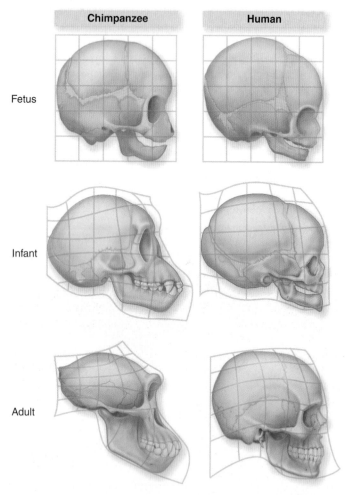
Chimpanzee   Human

Fetus

Infant

Adult

Figure 12.14 Same Parts, Different Proportions. As fetuses, humans and chimpanzees have very similar skulls, but the two species follow different developmental pathways. The grids superimposed on each image show that in an adult human, the brain is larger and the jaw is smaller than in chimpanzees.

reptiles, birds, and mammals. Shortly thereafter, Haeckel admitted that he altered some details to make the embryos look more alike. He also failed to represent the relative sizes of each structure. That is, although embryos might start out with the same basic parts, different growth rates (such as the very fast growth of the head in the primate fetus) distinguish species as development proceeds.

Haeckel's fudging fueled skepticism about the true degree of similarity among vertebrate embryos. Years later, however, developmental biologists photographed embryos and fetuses of a variety of vertebrate species throughout development (**figure 12.15**). The data show that there really are similarities in embryonic structures, supporting the concept of common ancestry.

More recently, the discovery of genes that contribute to development has spawned the field of evolutionary developmental biology (or "evo-devo" for short). Recall from chapter 7 that a gene is a region of DNA that encodes a protein. Some genes encode proteins that dictate how an organism will develop. One goal of research in evolutionary developmental biology is to identify these genes and determine how mutations can give rise to new body forms.

A basic question in developmental biology, for example, is how a clump of identical cells transforms into a body with a distinct head, tail, segments, and limbs. The fruit fly illustrates one possible answer (**figure 12.16**). One end of a newly produced fruit fly egg contains a protein called bicoid. This "bicoid end" will develop into the head end of the fly; in contrast, the rear end contains no bicoid at all. Bicoid therefore distinguishes the front from the rear. Along with other proteins, bicoid stimulates the development of other protein gradients that help divide the body

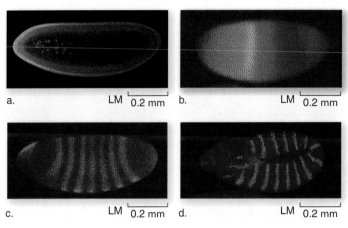

Figure 12.16 Early Development in Fruit Flies. Genes are expressed unequally along the length of a fly embryo, setting up distinct environments that profoundly influence development. (a) A protein called bicoid distinguishes the front and rear ends of a fertilized egg. The colors represent different concentrations of bicoid. (b) Bicoid stimulates the synthesis of two other proteins, one shown in red and one in green. (c) A half hour later, another gene directs production of a protein that divides the embryo into seven stripes. (d) Yet another gene controls the division of each section into two.

into segments, eventually giving rise to legs, wings, and other specialized body parts.

Not surprisingly, mutations in the genes encoding proteins that regulate development can produce dramatic new phenotypes. **Homeotic** is a general term describing any gene that, when mutated, leads to organisms with structures in abnormal or unusual places (such as the fruit fly with legs growing out of its head, shown in figure 7.21). Many homeotic genes encode proteins that regulate the expression of other genes.

Homeotic genes occur in all animal phyla studied to date, as well as in plants and fungi, and they provide important clues about development. The genes that prompt limb formation in mouse and chicken embryos, for example, do so by turning on in cells at specific points along the length of the embryo. In snake embryos, these genes turn on all along the body, rather than at defined locations. In the absence of a unique developmental signal telling the embryo where to start sprouting limbs, the snake's legs remain mere buds (see figure 12.11).

Researchers have also discovered that new phenotypes can come from mutations in DNA that does not encode proteins. Recall from chapter 7 that eukaryotic cells require proteins called transcription factors for gene expression. The transcription factors bind to areas of DNA called enhancers, signaling a gene to turn on. **Figure 12.17** shows that mutations in these enhancers can cause new features such as wing spots to develop in fruit flies. Perhaps more interesting is the observation that changes in other enhancers can cause the loss of a trait (in this case, pigmentation), even though the gene encoding the pigment remains intact.

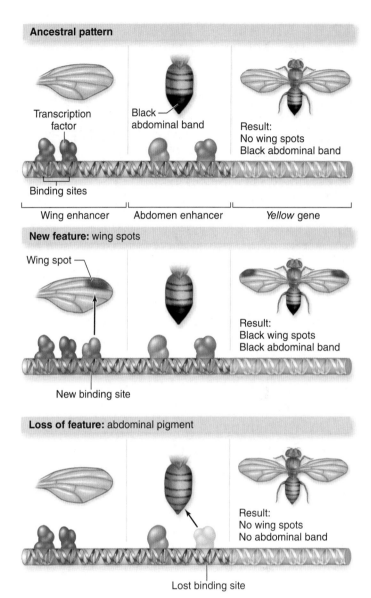

Figure 12.17 Gene Regulation. In fruit flies, a gene called *yellow* is required for the production of black pigment. Different combinations of transcription factors bound to enhancers determine whether *yellow* is activated in a fly's wings and abdomen.

These examples only scratch the surface of the types of information about evolution that biologists can learn by studying development. The relatively new field of evolutionary developmental biology is sure to yield much more insight into evolution in the future.

12.5 | Mastering Concepts

1. How does the study of embryonic development reveal clues to a shared evolutionary history?
2. Why are evolutionary biologists interested in how genes influence development?

12.6 | Molecules Reveal Relatedness

The evidence for evolution described so far in this chapter is compelling, but it is only the beginning. Since the 1970s, biologists have compiled a wealth of additional data by comparing the molecules inside the cells of diverse organisms. The results have not only confirmed many previous studies but have also added unprecedented detail to our ability to detect and measure the pace of evolutionary change.

The molecules that are most useful to evolutionary biologists are nucleic acids (DNA and RNA) and proteins. As described in chapter 2, nucleic acids are long chains of subunits called nucleotides, whereas proteins are composed of amino acids. Cells use the information in nucleic acids to produce proteins.

This intimate relationship between nucleic acids and proteins is itself a powerful argument for common ancestry. All species use the same genetic code in making proteins, and cells use the same 20 amino acids. The fact that biologists can move DNA among species to create transgenic organisms is a practical reminder of the universal genetic code. ▶ transgenic organisms, p. 141

A. Comparing DNA and Protein Sequences May Reveal Close Relationships

To study molecular evolution, biologists compare nucleotide and amino acid sequences among species. It is highly unlikely that two unrelated species would evolve precisely the same DNA and protein sequences by chance. It is more likely that the similarities were inherited from a common ancestor and that differences arose by mutation after the species diverged from the ancestral type. Thus, an underlying assumption of molecular evolution studies is that the greater the similarity between two modern species, the closer their evolutionary relationship. One advantage of molecular comparisons is that they are less subjective than deciding whether two anatomical structures are homologous or analogous.

DNA   The ability to rapidly sequence DNA has led to an explosion of information. The Human Genome Project has drawn attention to whole-genome comparisons among species. But DNA differences can also be assessed for a few bases, for one gene, or for families of genes with related structures or functions. Biologists routinely locate a gene in one organism, then scan huge databases to study its function in other species. ▶ DNA sequencing, p. 170

Biologists have also sequenced noncoding DNA, including transposons and pseudogenes (see section 7.8). Neither transposons nor pseudogenes have any known function, yet their sequences are similar for closely related species. The best explanation for these observations is common descent.

The recent explosion of genetic information has also helped explain how evolution works. We now know that cells may add new functions by acquiring DNA from other organisms (horizontal gene transfer; see section 5.8) and by duplicating entire genes (figure 12.18). And we have already seen that the study of gene regulation is especially fruitful in showing how the timing and location

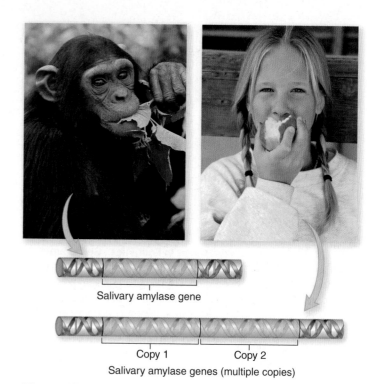

Salivary amylase gene

Copy 1   Copy 2

Salivary amylase genes (multiple copies)

Figure 12.18 **Gene Duplication.** Chimpanzees consume little starch compared to humans. This dietary difference apparently selected for additional copies of the gene encoding amylase, the starch-digesting enzyme.

of gene expression can differ among closely related species (see figure 12.17). The list of additional applications is endless, ranging from studies pinpointing the origin of disease-causing microorganisms (see sections 15.7 and 16.6) to monitoring the evolution of pesticide-resistant insects (see section 10.10).

Proteins   Like DNA, protein sequences also often support fossil and anatomical evidence of evolutionary relationships. One study, for example, found seven of twenty proteins to be identical in humans and chimps, our closest relatives. Many other proteins have only minor sequence differences from one species to another. The keratin of sheep's wool, for example, is virtually identical to that of human hair. The similarity reflects the shared evolutionary history of all mammals.

Keratin is useful in studies of mammals, but other species lack this protein. To study broader groups of organisms, biologists use proteins that are present in all species. One example is cytochrome *c*, which is part of the electron transport chain in respiration (see section 6.5C). Figure 12.19 shows that the more closely related two species are, the more alike is their cytochrome *c* amino acid sequence.

B. Molecular Clocks Help Assign Dates to Evolutionary Events

A biological molecule such as DNA can act as a "clock." If biologists know the mutation rate for a gene, plus the number of differences in the DNA sequences for that gene in two species, they can use the DNA as a **molecular clock** to estimate the time when the organisms diverged from a common ancestor (figure 12.20).

Cytochrome c Evolution	
Organism	Number of amino acid differences from humans
Chimpanzee	0
Rhesus monkey	1
Rabbit	9
Cow	10
Pigeon	12
Bullfrog	20
Fruit fly	24
Wheat	37
Yeast	42

Figure 12.19
Cytochrome c Comparison. The more recent the shared ancestor with humans, the fewer the differences in the amino acid sequence for the respiratory protein cytochrome c.

For example, many human and chimpanzee genes differ in about 4% to 6% of their nucleotides, and substitutions occur at an estimated rate of 1% per 1 million years. Therefore, about 4 million to 6 million years have passed since the two species diverged.

Molecular clock studies are not quite as straightforward as glancing at a wristwatch. DNA replication errors occur in different regions of a chromosome at different rates, and some genes affect phenotypes more than others. Therefore, to make a large-scale tree that incorporates all life, researchers consider slowly evolving DNA sequences common to all organisms. The genes encoding ribosomal RNA fit the bill. Carl Woese used these genes to deduce the existence of the three domains (Archaea, Bacteria, and Eukarya; see figure 1.8).

Another useful molecular clock measures mutations in mitochondrial DNA (mtDNA). Mitochondria contain only a small amount of genetic material. But mtDNA is especially valuable in tracking recent evolutionary events in closely related organisms because its molecular clock "ticks" 5 to 10 times faster than the nuclear DNA clock. Moreover, mtDNA can sometimes be extracted from extinct organisms and museum specimens, even if the nuclear DNA is badly degraded. ▶ mitochondria, p. 61

Mitochondrial DNA presents unique possibilities because it is inherited from mothers only, in the egg. Unlike nuclear DNA, mitochondrial DNA does not participate in crossing over during meiosis. It is therefore easier to trace long-term evolutionary events with mitochondrial DNA than with nuclear DNA. For example, many studies show that people from Africa have the most diverse mtDNA sequences. Because mutations take time to accumulate, these results suggest that Africans have existed longer than other modern populations. The idea that early humans originated in Africa and then migrated to the other continents is called the single origin ("out of Africa") hypothesis. ▶▶ crossing over, p. 186; human evolution, p. 312

The Y chromosome, which passes only from father to son, has also been useful in tracking human migration. Because the Y chromosome does not exchange genetic material with the X chromosome during meiosis, it accumulates changes much more slowly than do the other chromosomes. Researchers have combined studies of the Y chromosome with mtDNA data to generate increasingly accurate portraits of human history.

12.6 | Mastering Concepts

1. How does analysis of DNA and proteins support other evidence for evolution?
2. What is the basis of using a molecular clock to determine when two species diverged from a common ancestor?
3. What is an advantage of using mtDNA instead of nuclear DNA in tracing evolution?

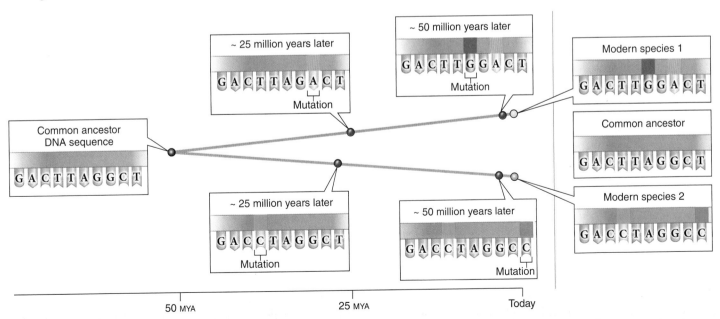

Figure 12.20 Molecular Clock. If a sequence of DNA accumulates mutations at a regular rate, then the number of sequence changes can act as a "clock" that tells how much time has passed since two species last shared a common ancestor.

12.7 | Investigating Life: Darwin's Finches Reveal Ongoing Evolution

Fossils, anatomical structures, and molecules reveal past evolution, but Darwin's finches on the Galápagos Islands unmistakably show that we can also watch species evolve in real time, in the real world. These birds received their name not only because Charles Darwin collected them during *Beagle*'s voyage but also because subsequent research on the birds has taught scientists so much about natural selection and evolution.

Princeton University ecologists Peter and Rosemary Grant, along with their students, have conducted the most famous studies of these birds. Since the 1970s, they have captured, weighed, measured, banded, and then released nearly every finch on two small islands; they have done the same for each bird's descendants. They also began collecting DNA from the birds in the 1990s.

Throughout the decades they have also measured rainfall, plant cover, seed diversity, seed abundance, and seed size on the islands.

The main goal of their massive project has been to learn more about how the 14 species of Darwin's finches diverged from a common ancestor over the past 3 million years. But the long-term data on individual birds, coupled with ecological information, also enable the Grants to analyze the island's short-term "natural experiments." That is, whenever conditions change dramatically, the team can document evolutionary changes by comparing multiple generations before and after the event.

This section focuses on just one of the Grants' many studies on the island of Daphne Major. Two permanent residents of Daphne Major are the medium ground finch, *Geospiza fortis*, and the slightly larger cactus finch, *Geospiza scandens* (figure 12.21). The beaks of the two birds reflect their different diets. The medium ground finch uses its crushing bill to eat seeds of various sizes, whereas the cactus finch uses a probing bill to dine on the pollen, nectar, and medium-sized seeds of the *Opuntia* cactus.

Apply It Now

An Evolutionary View of the Hiccups

It happens to everyone: we eat or drink too fast, and then the hiccups begin. Typically, they last for several minutes and stop on their own. Hiccup triggers include everything from consuming too much alcohol, to vigorous laughing or sobbing, to tumors in the chest, to more than a dozen other possible causes.

Whatever the trigger, scientists know what happens during a hiccup: An involuntary muscle spasm causes us to inhale sharply. At the same time, the epiglottis blocks the airway to the lungs, producing the classic "hic" sound. The events are familiar, yet the very existence of hiccups poses a mystery. The quick intake of air apparently doesn't prevent or solve any known problem. So why do we get the hiccups at all?

Scientists recently found two clues in the anatomy of our vertebrate ancestors (figure 12.A). Hiccups result from irritation of one or both phrenic nerves, which emerge from the spinal cord near the base of the skull and extend through the chest cavity to the diaphragm. Stimulation by the phrenic nerve triggers contraction of the diaphragm, which normally controls breathing. The distance between the neck and the diaphragm offers many opportunities for irritation of the phrenic nerve and, thus, hiccups. A more practical arrangement would be for the phrenic nerve to emerge from the spinal cord nearer the diaphragm. But we inherited our existing anatomy from our fishy ancestors, whose gills are near where the phrenic nerve emerges.

The second clue to the origin of hiccups comes from a close examination of young amphibians. Tadpoles have both lungs and gills, so they can gulp air or extract oxygen from water. In the latter case, they pump mouthfuls of water across their gills; at the same time, the epiglottis closes to keep water out of the lungs.

The tadpole's breathing action almost exactly matches what happens in a human hiccup. Scientists therefore surmise that the hiccup is an accident of evolution—a vestigial remnant of our shared evolutionary history with the vertebrates that came before us.

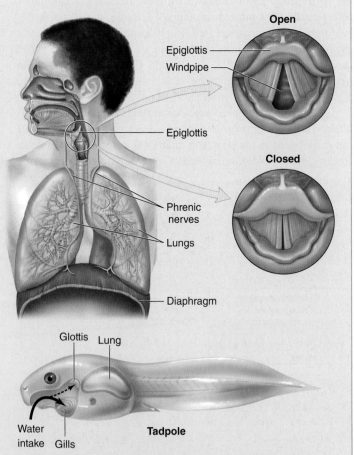

Figure 12.A **Hiccup Origins.** Amphibian tadpoles may help explain why humans get the hiccups.

Geospiza fortis,
the medium ground finch

Geospiza scandens,
the cactus finch

Figure 12.21 **Two of Darwin's Finches.**

In late 1982 to mid-1983, a severe El Niño event triggered heavy rains for 8 months (**figure 12.22a**). The downpour changed the assortment of seeds on the island. Large, hard seeds dominated before the rainy year, but smaller, softer ones became much more abundant in the years that followed (figure 12.22b).

The change in food source was a selective force on the medium ground finches' population for several years after 1983. Finches with small bills, which easily handle small seeds, had more food available than larger beaked birds. Since they could eat the most, they could raise the most offspring. As a result, average beak width declined between 1984 and 1987.

What about the cactus finches, *G. scandens*? The El Niño event did not prompt a shift in the size of their beaks because their diets did not change; they simply kept eating *Opuntia* seeds.

This research beautifully complements a famous earlier study documenting a severe drought in 1977 that wiped out most plants on the island. In the resulting food shortage, finches rapidly devoured the small, easy-to-eat seeds. Afterward, only birds with beaks that could crack the largest, toughest seeds survived long enough to reproduce. They passed the "large beak" alleles to their offspring, and data from subsequent years clearly show a shift toward larger, stronger bills (see figure 11.8).

The biotechnology revolution has enabled the Grants and their colleagues to delve even deeper into the evolution of Darwin's finches. For example, in 2004, researchers reported that they had discovered a gene that influences the development of a finch chick's beak. Species with the widest beaks expressed the gene earlier and at a higher level than species with narrower beaks. To confirm the gene's role in beak development, the researchers increased its expression in a chicken embryo, and the experimental bird responded by growing a wide, deep, finchlike beak.

The data that the Grants have collected in the Galápagos Islands clearly reveal that heritable traits really do change measurably in response to short-term climatic changes, even in a natural setting. It is exciting to consider that technology now allows us to link the measurable features of generations of Darwin's finches

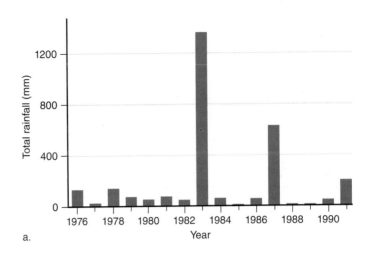

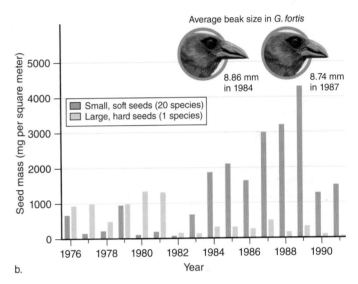

Figure 12.22 **Evolution in Action.** (a) Rainfall was heavy on the Galápagos island of Daphne Major in 1983. (b) After 1983, small, soft seeds became much more abundant than large, hard seeds. Average beak width in *Geospiza fortis*, the medium ground finch, declined during the same period.

to the genes that dictate their development. In the future, it may be possible to track the frequencies of specific alleles over multiple generations, taking us on yet another step along the journey that Darwin himself started some 150 years ago.

Grant, B. Rosemary, and Peter R. Grant. February 22, 1993. Evolution of Darwin's finches caused by a rare climatic event. *Proceedings of the Royal Society of London* B., vol. 251, pages 111–117.

12.7 | **Mastering Concepts**

1. What hypothesis are the Grants testing, and what observations are they using to test it?
2. Would careful examination of finch fossils or DNA be a better way to address the questions of short-term evolution that the Grants are studying?

Chapter Summary

12.1 | Clues to Evolution Lie in the Earth, Body Structures, and Molecules

- The **geologic timescale** divides Earth's history into eras defined by major events such as mass extinctions.
- Evidence for evolutionary relationships comes from **paleontology** (the study of past life) and from comparing anatomical and biochemical characteristics of species.

12.2 | Fossils Record Evolution

- **Fossils** are the remains of ancient organisms.

A. Fossils Form in Many Ways

- Fossils may form when mineral replaces tissue gradually, or they may be indirect evidence such as footprints or feces. Rarely, organisms are preserved whole.

B. The Fossil Record Is Often Incomplete

- Many organisms that lived in the past have not left fossil evidence.

C. The Age of a Fossil Can Be Estimated in Two Ways

- The position of a fossil in the context of others provides a **relative date.**
- **Radiometric dating** uses the ratio of a radioactive isotope to its breakdown product to give an **absolute date,** which is a range of time when an organism lived. Radioactive isotopes vary in their **half-lives;** some isotopes are therefore useful for dating ancient fossils, whereas others are best for determining the ages of younger objects.

12.3 | Biogeography Considers Species' Geographical Locations

- **Biogeography** is the study of the distribution of species on Earth.

A. The Theory of Plate Tectonics Explains Earth's Shifting Continents

- The **plate tectonics** theory indicates that forces deep inside Earth have moved the continents throughout much of life's history, creating and eliminating geographical barriers.

B. Species Distributions Reveal Evolutionary Events

- Biogeography provides insight into large- and small-scale evolutionary events.

12.4 | Anatomical Comparisons May Reveal Common Descent

A. Homologous Structures Have a Shared Evolutionary Origin

- **Homologous** anatomical structures and molecules have similarities that indicate they were inherited from a shared ancestor, although they may differ in function.

B. Vestigial Structures Have Lost Their Functions

- **Vestigial** structures have no function in an organism but are homologous to functioning structures in related species.

C. Convergent Evolution Produces Superficial Similarities

- **Analogous** structures are similar in function but do not reflect shared ancestry. **Convergent evolution** can produce analogous structures, such as the wings of birds and insects.

12.5 | Embryonic Development Patterns Provide Evolutionary Clues

- Embryos of different species reflect the actions of genes retained from ancestors.
- Evolutionary developmental biology combines the study of development with the study of DNA sequences. Many genes, including **homeotic genes,** influence the development of new phenotypes.

12.6 | Molecules Reveal Relatedness

A. Comparing DNA and Protein Sequences May Reveal Close Relationships

- Molecular sequences contain so much information that it is unlikely that similarities happened by chance; descent from a shared ancestor is more likely.
- DNA sequence comparisons provide an indication of the relationships among species, as can the amino acid sequences of proteins.

B. Molecular Clocks Help Assign Dates to Evolutionary Events

- A **molecular clock** compares DNA sequences to estimate the time when two species diverged from a common ancestor.

12.7 | Investigating Life: Darwin's Finches Reveal Ongoing Evolution

- Peter and Rosemary Grant's classic studies of Darwin's finches on the Galápagos Islands show how evolution unfolds in real time.
- A short-term change in climate on the island of Daphne Major triggered changes in food availability. As a result, the average beak size in a species of seed-eating birds declined.

Multiple Choice Questions

1. You are living in which of the following geological eras?
   a. Archean
   b. Mesozoic
   c. Paleozoic
   d. Cenozoic

2. Which form of a fossil would be most likely to provide material for DNA analysis?
   a. Petrifaction
   b. Impression
   c. Intact preservation
   d. Cast

3. Suppose biologists find a fossilized lizard bone containing one eighth of the amount of ^{14}C present in the atmosphere. How long ago did the lizard die?
   a. 5730 years ago
   b. 11,460 years ago
   c. 17,190 years ago
   d. 22,920 years ago

4. The distributions of some species can be explained by
   a. plate tectonics.
   b. changes in ocean levels.
   c. the distance between islands and the nearest mainland.
   d. All of the above are correct.

5. The wing of a bird and the wing of a fly are
   a. homologous structures.
   b. vestigial structures.
   c. analogous structures.
   d. convergent structures.

6. What term is used to describe the tiny, nonfunctional eyes of certain species of cavefish?
   a. homeotic
   b. vestigial
   c. analogous
   d. fossilized

7. As described in chapter 7, the genetic code's redundancies mean that a genetic mutation does not necessarily change the amino acid sequence of a protein. For a given gene, which molecule would you expect to change the most over evolutionary time?
   a. The DNA sequence should change more than the protein sequence.
   b. The protein sequence should change more than the DNA sequence.
   c. The two should have exactly the same number of changes.
   d. The answer depends on the gene.

8. How does the activity of a homeotic gene relate to evolution?
    a. Homeotic genes serve as a marker for convergent evolution.
    b. Organisms with similar homeotic genes have the same vestigial structures.
    c. Mutations in homeotic genes can produce new body plans.
    d. The presence of homeotic genes helps to identify a fossil as that of an animal.
9. Which of the following would be most useful for comparing ALL known groups of organisms?
    a. DNA encoding ribosomal RNA
    b. DNA encoding the keratin protein
    c. Mitochondrial DNA
    d. Y chromosome DNA
10. How does a molecular clock work?
    a. It uses mutation rates to estimate when two species diverged.
    b. It measures the loss of ^{14}C isotopes in a sample.
    c. It estimates the time required for a population to divide into two species.
    d. All of the above are correct.

Write It Out

1. What types of information are used to hypothesize how species are related to one another by descent from shared ancestors? Give an example of how multiple types of evidence can support one another.
2. Describe six types of fossils and how they form. What present environmental conditions might preserve today's organisms to form the fossils of the future?
3. Why is the fossil record useful, even if it doesn't represent every type of organism that ever lived?
4. Index fossils represent organisms that were widespread but lived during relatively short periods of time. How might index fossils be useful in relative dating?
5. Why can't scientists use potassium–argon dating on a preserved woolly mammoth that died 20,000 years ago?
6. How do molecular sequences provide different information than relative and absolute dating?
7. How have geological events such as continental movements and the emergence of new volcanic islands influenced the history of life on Earth?
8. Why is it important for evolutionary biologists to be able to distinguish between homologous and analogous anatomical structures?
9. Provide examples of homologous and analogous structures other than those mentioned in this chapter.
10. The giant anteater lives in South America, the scaly anteater lives in tropical parts of Asia and Africa, and the spiny anteater lives in Australia. The three animals are not at all closely related, but they resemble one another closely. They have long, sticky tongues that they use to eat ants, no teeth, large salivary glands, and long, bald snouts. Which phenomenon do these animals illustrate?
11. How did the discovery of homeotic genes help launch the new field of "evo-devo"?
12. Suggest a type of genetic change that could have a drastic effect on the evolution of a species.
13. How do biologists use sequences of proteins and genes to infer evolutionary relationships?

14. Scientists have extracted flu viruses from people who died in the 1918 influenza epidemic. How could genetic sequences from these viruses be useful today?
15. Why is the DNA sequence of one gene a less accurate indicator of the evolutionary relationship between two species than a comparison of large portions of the two genomes?
16. Some genes are more alike between human and chimp than other genes are from person to person. Does this mean that chimps are humans or that humans with different alleles are different species? What other explanation fits the facts?
17. If scientists could extract DNA from dinosaur fossils, how could they use the sequence to learn more about the origin of birds?
18. Birds normally lack teeth, but scientists have discovered a mutant chicken embryo with a mouth full of teeth resembling those of alligators. They subsequently found that a mutation in one gene was responsible for the toothy bird. How does this finding provide evidence for evolution?
19. Why are molecular clocks useful, and what are their limitations?
20. Evolutionary biologists often try to assign an approximate date when two organisms last shared a common ancestor. Why do you think that molecular evidence often yields an earlier date than fossil evidence?
21. Genetic anthropology combines the study of DNA with physical evidence such as fossils to reveal the history of the human species. Use the Internet to research the goals and methods of the Genographic Project, HapMap, or the Human Genome Diversity Project. What are the benefits of these projects? Can you see any ethical issues arising from this type of research?
22. Search the Internet for a copy of the Declaration of Independence. Look for words that were spelled differently in 1776 compared to today's spelling. How are changes in language similar to evolutionary changes? How are they different?

Pull It Together

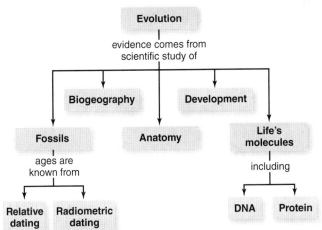

1. What are some examples of each line of evidence for evolution?
2. Add the following terms to this concept map: *homologous structures, vestigial structures, homeotic genes,* and *molecular clock.*

Going Extinct? The 'I'iwi is a honeycreeper native to the Hawaiian islands. Like many island species, this scarlet bird is endangered by habitat loss, introduced predators, and disease.

Enhance your study of this chapter with practice quizzes, animations and videos, answer keys, and downloadable study tools.
www.mhhe.com/hoefnagels

Learn How to Learn
Don't Neglect the Boxes
You may be tempted to skip the chapter opening essays and boxed readings because they're not "required." Read them anyway. The contents should help you remember and visualize the material you are trying to learn. And who knows? You may even find them interesting.

Islands Provide Windows on Speciation and Extinction

OVER BILLIONS OF YEARS, MANY NEW SPECIES HAVE APPEARED ON EARTH IN A PROCESS CALLED SPECIATION. Yet most species that have ever lived are now extinct. The opposing, ongoing processes of speciation and extinction have defined life's history.

Islands provide ideal opportunities to study speciation. Their small land areas house populations that are relatively easy to monitor. In addition, few organisms can travel vast distances across oceans to reach isolated islands. The descendants of those that have done so have diversified and exploited multiple habitats. Islands located far from a mainland therefore often have groups of closely related species found nowhere else.

Many examples illustrate the spectacular diversity that islands may host. For example, about 800 species of *Drosophila* flies and their close relatives inhabit the Hawaiian Islands. Twenty-eight species of plants called silverswords thrive in every imaginable Hawaiian island habitat, yet all apparently descended from one ancestor that colonized Hawaii long ago. The same islands are also home to birds called honeycreepers, each with a bill adapted to a different food source.

At the same time, island species are especially vulnerable to extinction. Many factors conspire against them. A single hurricane, fire, or flood may destroy a small island population, as can the extinction of a prey species or the introduction of a new predator. Even random fluctuations in birth and death rates can doom a small population.

A volcanic island called Mauritius, in the Indian Ocean, illustrates how humans increase the extinction risk that island populations face. Until the sixteenth century, Mauritius teemed with tall forests, colorful birds, scurrying insects, and basking reptiles. One inhabitant was the large, flightless dodo bird, described by one scientist as "a magnificently overweight pigeon." When European sailors arrived in the 1500s, however, the men ate dodo meat, and their pet monkeys and pigs ate dodo eggs. Rats and mice swam ashore from ships and attacked native insects and reptiles. The sailors' Indian myna birds inhabited nests of the native echo parakeet, while imported plants crowded the seedlings of native trees. By the mid-1600s, only 11 of the original 33 species of birds remained. The dodo was exterminated by 1681, the first of many recorded extinctions caused by human activities.

Both speciation and extinction have likely been a part of evolution since life began. This chapter explains how these processes occur.

Learning Outline

13.1 | The Definition of "Species" Has Evolved over Time

Throughout the history of life, the types of organisms have changed: new species have appeared, and existing ones have gone extinct. The term **macroevolution** describes these large, complex changes in life's panorama. Macroevolutionary events tend to span very long periods, whereas the microevolutionary processes described in chapter 11 happen so rapidly that we can sometimes observe them over just a few years. Nevertheless, the many small changes that accumulate in a population by microevolution eventually lead to large-scale macroevolution.

Evolution has produced an obvious diversity of life. A bacterium, for example, is clearly distinct from a tree or a bird (figure 13.1a). At the same time, some organisms are more closely related than others, as the three cats in figure 13.1b illustrate. To make sense of these observations, biologists recognize the importance of grouping similar individuals into **species**—that is, distinct types of organisms. This task requires agreement on what the word *species* means. Perhaps surprisingly, the definition has changed considerably over time and is still the topic of vigorous debate among biologists.

A. Linnaeus Devised the Binomial Naming System

Swedish botanist Carolus Linnaeus (1707–1778) was not the first to ponder what constitutes a species, but his contributions last to this day. Linnaeus defined species as "all examples of creatures that were alike in minute detail of body structure." Importantly, he was the first investigator to give every species a two-word

SEM (false color) 3 μm

Figure 13.1 **Distinctive Species.** (a) Bacteria, a tree, and a bird are about as dissimilar as three types of organisms can be. (b) These three species of felines have many characteristics in common.

name. Each name combines the broader classification *genus* (plural: genera) with a second word that designates the species. The scientific name for humans, for example, is *Homo sapiens.*

Linnaeus also devised a hierarchical system for classifying species. He grouped similar genera into orders, classes, and kingdoms; scientists now use additional categories, as described in Section 13.6. Linnaeus's classifications organized the great diversity of life and helped scientists communicate with one another.

His system did not, however, consider the role of evolutionary relationships. Linnaeus thought that each species was created separately and could not change. Therefore, species could not appear or disappear, nor were they related to one another.

Charles Darwin (1809–1882) finally connected species diversity to evolution, writing that "our classifications will come to be, as far as they can be so made, genealogies." As the theory of evolution by natural selection became widely accepted in the nineteenth and twentieth centuries, scientists no longer viewed classifications merely as ways to organize life. They considered them to be hypotheses about the evolutionary history of life. We pick up this topic again in section 13.6.

B. Ernst Mayr Developed the Biological Species Concept

In the 1940s, Harvard biologist Ernst Mayr amended the work of Linnaeus and Darwin by considering reproduction and genetics. Mayr defined a **biological species** as a population, or group of populations, whose members can interbreed and produce fertile offspring. **Speciation,** the formation of new species, occurs when members of a population can no longer successfully interbreed.

How might this happen? A new species can form if a population somehow becomes divided. Recall from chapter 11 that a population's **gene pool** is its entire collection of genes and their alleles. An intact, interbreeding population shares a common gene pool. After a population splits in two, however, microevolutionary changes such as natural selection, genetic drift, and gene flow can lead to genetic divergence between the groups. With the accumulation of enough differences in their separate gene pools, the two groups can no longer produce fertile offspring even if they come into contact once again. In this way, microevolution becomes macroevolution.

The biological species definition does not rely on physical appearance, so it is much less subjective than Linnaeus's observations. Under the system of Linnaeus, it would be impossible to determine whether two similar-looking butterflies belong to different species (figure 13.2). Using Mayr's definition, however, if the two groups can produce fertile offspring together, they share a gene pool and therefore belong to one species.

Nevertheless, Mayr's species definition raises several difficulties. First, the biological species concept cannot apply to asexually reproducing organisms, such as bacteria, archaea, and many fungi and protists. Second, it is likewise impossible to apply the biological species definition to extinct organisms known

Figure 13.2 How Many Species? Linnaeus would have categorized these butterflies based on their physical appearance. Mayr's biological species definition, however, provides an objective rule for determining whether each group really is a separate species.

only from fossils. Third, some types of organisms have the *potential* to interbreed in captivity, but they do not do so in nature. Fourth, for some species, reproductive isolation is not absolute. Many closely related species of plants, for example, sometimes produce fertile offspring together.

As a result, the biological species concept does not always help biologists decide whether two individuals belong to the same or different species. DNA sequence analysis has helped to fill in some of these gaps. Biologists working with bacteria and archaea, for example, use a stretch of DNA that encodes ribosomal RNA to define species. If the DNA sequences of two specimens are more than 97% identical, they are considered to be the same species. These genetic sequences, however, still present some ambiguity because they cannot reveal whether genetically similar organisms currently share a gene pool.

Despite these difficulties, reproductive isolation is the most common criterion used to define species. The rest of this chapter therefore uses the biological species concept to describe how speciation occurs.

13.1 | Mastering Concepts

1. What is the relationship between macroevolution and microevolution?
2. How does the biological species concept differ from Linnaeus's definition?
3. What are some of the challenges in defining species?

13.2 | Reproductive Barriers Cause Species to Diverge

In keeping with Mayr's biological species concept, a new species forms when one portion of a population can no longer breed and produce fertile offspring with the rest of the population. That is, the two separate groups no longer share a common gene pool, and each begins to follow its own, independent evolutionary path.

Two parts of a population can become reproductively isolated in many ways, because successful reproduction requires so many complex events. Any interruption in courtship, fertilization, embryo formation, or offspring development can be a reproductive barrier.

Biologists divide the many mechanisms of reproductive isolation into two broad groups: prezygotic and postzygotic. Prezygotic reproductive barriers occur before the formation of the zygote, or fertilized egg; postzygotic barriers reduce the fitness of a hybrid offspring. (A hybrid, in this case, is the offspring of individuals from two different species.) Figure 13.3 summarizes the reproductive barriers, which are described in more detail below.

A. Prezygotic Barriers Prevent Fertilization

Mechanisms of **prezygotic reproductive isolation** affect the ability of two species to combine gametes and form a zygote. These reproductive barriers include:

- **Ecological (or habitat) isolation:** A difference in habitat preference can separate two populations in the same geographic area. For example, one species of ladybird beetles eats one type of plant, while a closely related species eats a different plant. The two species never occur on the same host plant, although they interbreed freely in the laboratory. The different habitat preferences are the reproductive barrier that keeps the gene pools of the two species separate.

- **Temporal isolation:** Two species that share a habitat will not mate if they are active at different times of day or reach reproductive maturity at different times of year. Among field crickets in Virginia, for example, one species takes much longer to develop than another, so the adults of the two species may not meet until late in the season, if at all. Similarly, the different emergence schedules for 13- and 17-year periodical cicadas suggest that temporal isolation prevents interbreeding between these two species.

- **Behavioral isolation:** Behavioral differences may prevent two closely related species from mating. The males of two species of tree frogs, for instance, use distinct calls to attract mates. Female frogs choose males of their own species based on the unique calls. Likewise, sexual selection in many birds is based on intricate mating dances. Any variation in the ritual from one group to another could prevent them from being attracted to one another.

- **Mechanical isolation:** In many animal species, male and female parts fit together almost like a key in a lock. Any change in the shape of the gamete-delivering or -receiving structures may prevent groups from interbreeding. In plants, males and females do not copulate, but mechanical barriers still apply. For example, although two species of sage plants in California can interbreed, in practice they rarely do because they use different pollinators. One species has flowers that accommodate small to medium bees, whereas the other attracts large bees. The different pollinators effectively isolate the gene pools of the two plant species. Section 13.7 explores a similar reproductive barrier in another pair of plant species. ▸ pollination, p. 493

- **Gametic isolation:** If a sperm cannot fertilize an egg cell, then no reproduction will occur. For example, many marine organisms, such as sea urchins, simply release sperm and egg cells into the water. Each species's gametes display unique surface molecules that enable an egg to recognize sperm of the same species. In the absence of a "match," fertilization will not occur, and the gene pools will remain separate.

B. Postzygotic Barriers Prevent Viable or Fertile Offspring

Individuals of two different species may produce a hybrid zygote. Even then, **postzygotic reproductive isolation** may keep the species separate by selecting against the hybrid offspring, effectively preventing genetic exchange between the populations. Collectively, these postzygotic barriers are sometimes called hybrid incompatibility.

Postzygotic reproductive barriers include:

- **Hybrid inviability:** A hybrid embryo may die before reaching reproductive maturity, typically because the genes of its parents are incompatible. For example, two species of eucalyptus trees coexist in California forests, but hybrid offspring are rare. Either the hybrid seeds fail to germinate, or the seedlings die soon after sprouting.

- **Hybrid infertility (sterility):** Some hybrids are infertile. A familiar example is the mule, a hybrid offspring of a female horse and a male donkey. Mules are infertile because a horse's egg has one more chromosome than a donkey's sperm cell. The animal can grow and develop, but meiosis does not occur in the mule's germ cells because the chromosomes are not homologous. Similarly, a liger is the hybrid offspring of a male lion and a female tiger; ligers are usually sterile. ▸▸ homologous chromosomes, p. 181; mules, p. 185

- **Hybrid breakdown:** Some species produce hybrid offspring that are fertile. When the hybrids reproduce, however,

Barrier	Description	Example	Illustration
PREZYGOTIC REPRODUCTIVE ISOLATION			
Ecological (habitat) isolation	Different environments	Ladybugs feed on different plants	
Temporal isolation	Active or fertile at different times	Field crickets mature at different rates	
Behavioral isolation	Different activities	Frog mating calls differ	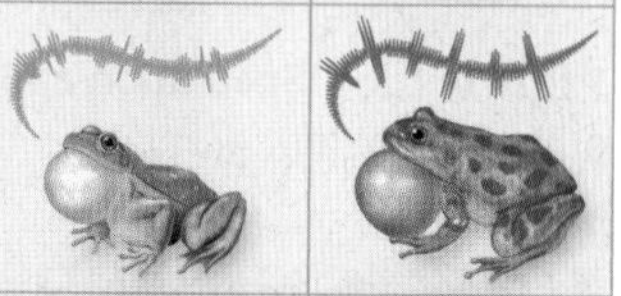
Mechanical isolation	Mating organs or pollinators incompatible	Sage species use different pollinators	
Gametic isolation	Gametes cannot unite	Sea urchin gametes incompatible	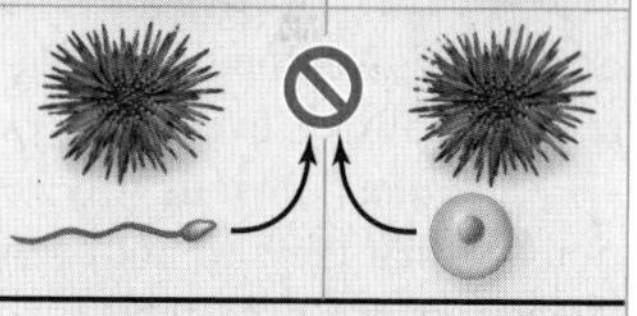
POSTZYGOTIC REPRODUCTIVE ISOLATION			
Hybrid inviability	Hybrid offspring fail to reach maturity	Hybrid eucalyptus seeds and seedlings not viable	
Hybrid infertility (sterility)	Hybrid offspring unable to reproduce	Lion-tiger cross (liger) infertile	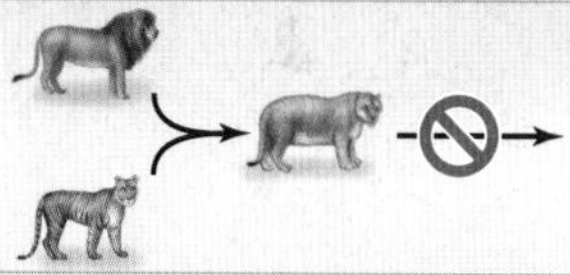
Hybrid breakdown	Second generation hybrid offspring have reduced fitness	Offspring of hybrid mosquitoes have abnormal genitalia	

Figure 13.3 Reproductive Barriers: A Summary.

their offspring may have abnormalities that reduce their fitness. Some second-generation hybrid offspring of the mosquito species *Aedes aegypti* and *Aedes mascarensis*, for example, have abnormal genitalia that make mating difficult.

Successful hybridization is rare in animals, but it frequently occurs in plants (see section 9.9). One of the problems with introducing nonnative species of plants into a region is the potential production of hybrids that displace the native plants. On the other hand, many of our food crops are the result of hybridization. The

tangelo, for example, is the hybrid offspring of a tangerine and a pomelo (grapefruit).

13.2 | Mastering Concepts

1. How do reproductive barriers lead to speciation?
2. Name five modes of prezygotic reproductive isolation.
3. What are three ways that postzygotic reproductive isolation may occur?

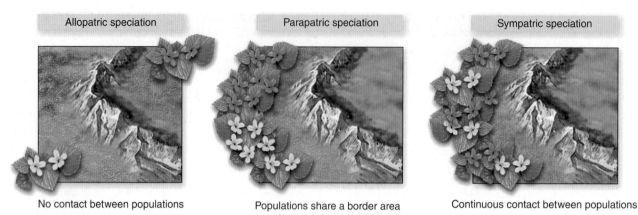

| Allopatric speciation | Parapatric speciation | Sympatric speciation |

No contact between populations   Populations share a border area   Continuous contact between populations

Figure 13.4 **Speciation and Geography.** The three modes of speciation are distinguished based on whether populations are separated by a physical barrier or mingle within a shared border area.

13.3 | Spatial Patterns Define Three Types of Speciation

Reproductive barriers keep related species apart, but how do these barriers arise in the first place? More specifically, how could two populations of the same species evolve along different pathways, eventually yielding two species?

The most obvious way is to physically separate the populations so that they do not exchange genes. Natural selection and genetic drift would then act independently in the two populations. Eventually, the genetic differences between the populations would give rise to one or more reproductive barriers.

Yet speciation can also separate populations that have physical contact with each other. They may divide into species even though they inhabit neighboring regions or even share a habitat. Biologists recognize these different circumstances by dividing the geographic setting of speciation into three categories: allopatric, parapatric, and sympatric (**figure 13.4**).

A. Allopatric Speciation Reflects a Geographic Barrier

In **allopatric speciation,** a geographic barrier physically separates a population into two groups that cannot interbreed (*allo-* means "other," and *patria* means "fatherland"). The isolation may result from rivers, deserts, glaciers, changes in sea level, the formation or destruction of mountains or bodies of water, or the appearance of any other physical barrier to gene exchange.

If the two groups cannot contact each other, gene flow between them stops, and the forces of microevolution act independently in each group. The result may be one or more reproductive

Figure 13.5 **Allopatric Speciation in Pupfish.** Devil's Hole is a pool in a limestone cavern east of Death Valley National Park. The pupfish species that lives there is a product of allopatric speciation.

barriers. When the descendants of the original two populations can no longer interbreed, one species has branched into two.

The Devil's Hole pupfish, which inhabits a warm spring at the base of a mountain near Death Valley, California, illustrates one way that allopatric speciation occurs (**figure 13.5**). The spring was isolated from other bodies of water about 50,000 years ago, preventing genetic exchange between the fish trapped in the spring and those in the original population. In that time, the gene pool has shifted sufficiently so that a Devil's Hole pupfish cannot mate with fish from another spring. It has become a distinct species.

In addition to isolated springs, island archipelagos also offer opportunities for allopatric speciation. For example, 11 subspecies of tortoise (*Geochelone nigra*) occupy the various Galápagos islands. Researchers have compared the DNA collected from 161 individuals and used the sequences to reconstruct many of the events that occurred after tortoises first colonized the islands a couple of million years ago (**figure 13.6**).

According to the DNA analysis, a few newcomers from the South American mainland first colonized either San Cristóbal or Española. As the population grew, the tortoises migrated to nearby islands, where they encountered new habitats that selected for different adaptations, especially in shell shape. Dry islands with sparse vegetation have selected for notched shells that enable the tortoises to reach for higher food sources. On islands with lush, low-growing vegetation, the tortoises have domed shells.

Although many of the subspecies look distinctly different from one another, the tortoises can interbreed. They are not yet separate species, but the genetic similarities among tortoises on each island suggest that gene flow from island to island was historically rare. The Galápagos tortoises illustrate an ongoing process of allopatric speciation.

Allopatric speciation has been considered the most common mechanism because the evidence for it is the most abundant and obvious. The diversification of plant and animal species on island archipelagos such as Hawaii and the Galápagos provide many striking examples. So do the world's fishes: of the 29,000 or so known species of fishes, 36% live in freshwater habitats, although these places account for only 1% of Earth's surface. Compared with the vast interconnected oceans, the countless lakes, ponds, streams, and rivers provide diverse habitats and ample barriers to genetic exchange.

B. Parapatric Speciation Occurs in Neighboring Regions

In **parapatric speciation,** part of a population enters a new habitat bordering the range of the parent species (*para-* means "alongside"). Most individuals mate within their own populations, although gene flow may still occur among individuals that venture into the shared border zone.

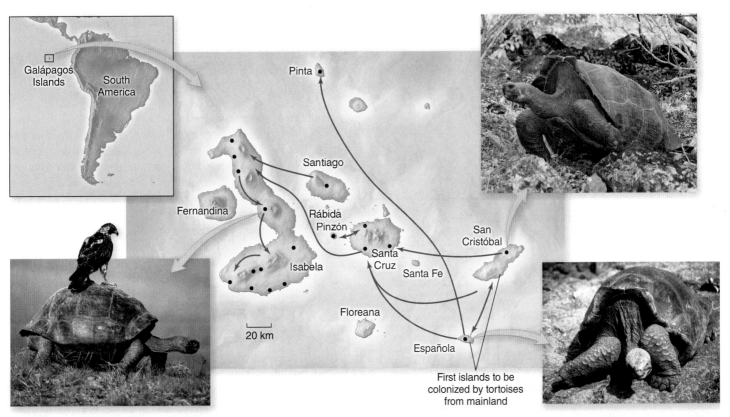

Figure 13.6 Allopatric Speciation in Tortoises. Descendants of the first tortoises to arrive on the Galápagos have colonized most of the islands, evolving into many subspecies. Unique habitats have selected for different sets of adaptations in the tortoises.

Despite this limited gene flow, parapatric speciation can occur if the two habitats are different enough to drive disruptive selection. As described in section 11.4, disruptive selection occurs when individuals with intermediate forms have lower fitness than those at either extreme. Parapatric speciation might begin if different features confer high fitness in the two neighboring habitats. Disruptive selection would counteract gene flow by eliminating individuals not well suited for either habitat. Over time, the two populations would become genetically (and reproductively) isolated.

Consider the little greenbul (*Andropadus virens),* a small green bird that lives in the tropical rain forest of Cameroon, West Africa (**figure 13.7**). The birds also inhabit the isolated patches of forest that characterize the transitional areas (called ecotones) between rain forest and grassland. In one study, researchers captured birds from six tropical rain forest sites and six ecotone sites. The birds in the ecotone patches had greater weight, deeper bills, and longer legs and wings than their rain forest counterparts.

What forces might have selected for these differences? The researchers speculated that longer wings increase fitness in the ecotone by improving flight, since the little greenbuls are vulnerable to predation in the open grassland areas separating the patches of forest. Presumably, the foods available in the ecotone and rain forest habitats account for the difference in bill depth.

The little greenbuls from the ecotones can still mate with those from the rain forest, so speciation has not yet occurred. Nevertheless, the researchers concluded that the forces of natural selection are greater than the gene flow between the two populations, gradually taking the groups farther apart. We are likely seeing speciation in action.

C. Sympatric Speciation Occurs in a Shared Habitat

In **sympatric speciation,** populations diverge genetically while living in the same physical area (*sym-* means "together"). Among evolutionary biologists, the idea of sympatric speciation can be controversial. After all, how can a new species arise in the midst of an existing population?

Often sympatric speciation reflects the fact that a habitat that appears uniform actually consists of many microenvironments. Fishes called cichlids, for example, have diversified into many species within the same lakes in Africa. **Figure 13.8** shows two cichlids in Cameroon's tiny Lake Ejagham. Although the lake is only 18 meters deep and has a surface area of less than half a square kilometer, it has distinct ecological zones. The lake bottom is muddy near the center, whereas leaves and twigs from nearby trees cover the sandy bottom near the shore.

Two closely related cichlids each specialize in a different zone. The larger fish consume insects and other invertebrates that feed on leaves near the shore. The smaller cichlids specialize on tiny floating prey in the deep offshore region. The fish also prefer to breed in different habitats. The two forms therefore typically remain reproductively isolated, and researchers surmise that the cichlids are undergoing sympatric speciation.

In plants, a common mechanism of sympatric speciation is **polyploidy,** which occurs when the number of sets of chromosomes increases. Many major crops, including wheat, corn, sugar cane, potatoes, and coffee, are derived from polyploid plants.
▸ polyploidy, p. 189

Polyploid organisms sometimes arise when gametes from two different species fuse. For example, section 9.9 describes how new polyploid species of *Tragopogon* wildflowers arose after plants in this genus were introduced to Washington state in the 1950s. Cotton plants provide another example (**figure 13.9**). An Old World species of wild cotton has 26 large chromosomes, whereas one from Central and South America has 26 small chromosomes. Plant breeders crossed the two species to create a hybrid with 26 chromosomes, then applied a drug called colchicine that caused the chromosome number to double. The resulting cotton plant, *Gossypium hirsutum,* is a polyploid with 52 chromosomes: 26 large and 26 small. Farmers around the world cultivate this species to harvest cotton for cloth.

Polyploidy can also occur when meiosis fails. A diploid individual may occasionally produce diploid sex cells. Self-fertilization will produce tetraploid offspring (with four sets of chromosomes). This organism represents an "instant species" because it is reproductively isolated from its diploid ancestors. After all, when the tetraploid plant's gametes (each containing two sets of chromosomes) combine with haploid gametes from the

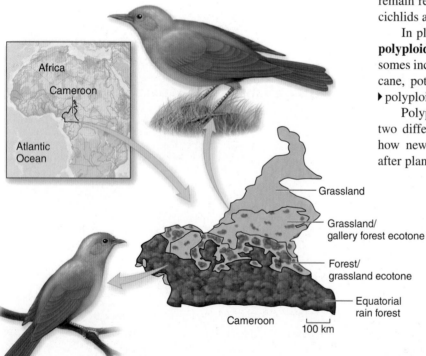

Figure 13.7   Parapatric Speciation. The little greenbul lives in the rain forest and in ecotones, which are patches of forest in the border zone with grasslands. Birds from the ecotone are larger than their rain forest counterparts.

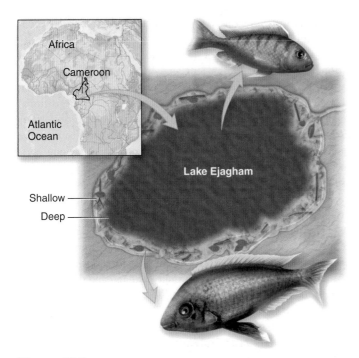

Figure 13.8 **Sympatric Speciation.** Cichlids in the deep waters of Lake Ejagham have smaller bodies than do shallow-water fish.

parent species, each offspring inherits three sets of chromosomes. This hybrid is infertile because it cannot complete meiosis: three chromosome sets cannot form the required homologous pairs.

Nearly half of all flowering plant species are natural polyploids, as are about 95% of ferns. Clearly, this form of reproductive isolation is extremely important in plant evolution. In animals, however, polyploidy is rare, possibly because the extra "dose" of chromosomes is usually fatal.

D. Determining the Type of Speciation May Be Difficult

Biologists sometimes debate whether a speciation event is allopatric, parapatric, or sympatric. One reason for the disagreement is that the definitions represent three points along a continuum, from complete reproductive isolation to continuous intermingling. Another difficulty is that we may not be able to detect the barriers that are important to other species. Is an apparently uniform patch of forest a potential setting for sympatric speciation, because it appears the same throughout? Our perspective may differ from that of a soil-dwelling insect, whose habitat differs greatly from the environment in the treetops.

The problem of perspective also leads to debate over the size of the geographic barrier needed to separate two populations, which depends on the distance over which a species can spread its gametes. A plant with windblown pollen or a fungus producing lightweight spores encounters few barriers to gene flow. Pollen and spores can travel thousands of miles in the upper atmosphere. On the other hand, a desert pupfish cannot migrate out of its aquatic habitat, so the isolation of its pool instantly creates an insurmountable geographic barrier. The same circumstance would not deter gene flow in species that walk or fly between pools.

13.3 | Mastering Concepts

1. Distinguish among allopatric, parapatric, and sympatric speciation, and provide examples of each.
2. How can polyploidy contribute to sympatric speciation?
3. Why is it sometimes difficult to determine whether speciation is allopatric, parapatric, or sympatric?

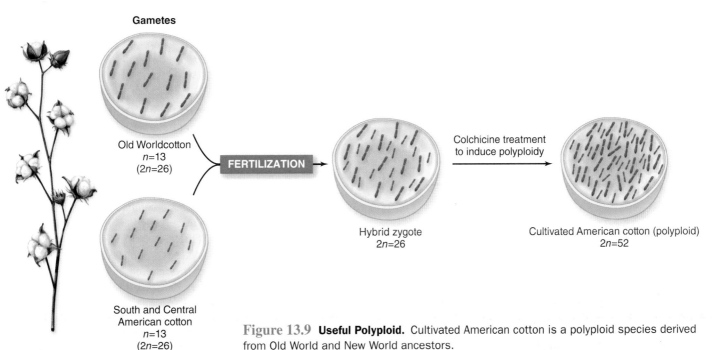

Figure 13.9 **Useful Polyploid.** Cultivated American cotton is a polyploid species derived from Old World and New World ancestors.

13.4 | Speciation May Be Gradual or Occur in Bursts

On a global level, Earth has seen times of relatively little change in the living landscape but also periods of rapid speciation and times of mass extinctions. Even within a short time, one species can evolve rapidly, while another hardly changes at all (this chapter's Burning Question on page 293 explains this disparity). As described in this section, speciation can happen quickly, gradually, or at any rate in between.

A. Gradualism and Punctuated Equilibrium Are Two Models of Speciation

Darwin envisioned one species gradually transforming into another through a series of intermediate stages. The pace as Darwin saw it was slow, although not necessarily constant. This idea, which became known as **gradualism,** held that evolution proceeds in small, incremental changes over many generations (**figure 13.10a**).

If the gradualism model is correct, "slow and steady" evolutionary change should be evident in the fossil record. In fact, however, many steps in species formation did not leave fossil evidence, and so we do not know of many intermediate or transitional forms between species. Much of the fossil record instead suggests the opposite: that a new species may appear relatively suddenly and evolve quickly, followed by long periods over which the species changed little.

What accounts for the "missing" transitional forms? One explanation is that the fossil record is incomplete, for many reasons: poor preservation of biological material, natural forces that destroyed fossils, and the simple fact that we haven't discovered every fossil on Earth. Chapter 12 explores these reasons in more detail.

Another explanation for the absence of some predicted transitional forms is that such "missing links" may have been too rare to leave many fossils. After all, periods of rapid biological changes would not leave much fossil evidence of transitional forms. In 1972, paleontologists Stephen Jay Gould and Niles Eldredge coined the term **punctuated equilibrium** to describe relatively brief bursts of rapid evolution interrupting long periods of little change (see figure 13.10b). The fossil record, they argued, lacks some transitional forms because they never existed in a particular location or because there were simply too few organisms to leave fossils.

The punctuated equilibrium model fits well with the concept of allopatric speciation. Imagine the isolated population of desert pupfish that, over time, became genetically distinct from its ancestral population. If the climate changed and the pool containing the new species rejoined its "old" pool, the fossil record might show that a new fish species suddenly appeared with its ancestors. Unless paleontologists were lucky enough to recover fossils from the formerly isolated pool, there would be no apparent transition between the two species. Afterward, unless the environment changed again, no new selective forces would drive the formation of new species. Thus, a period of stability would ensue.

The fossil record supports both punctuated equilibrium and gradualism. Microscopic protists such as foraminiferans and diatoms, for example, have evolved gradually. Vast populations of these asexually reproducing organisms span the oceans. Since isolated populations rarely form, it is perhaps not surprising that speciation has occurred only gradually among those protists. On the other hand, the fossils of animals such as bryozoans, mollusks, and mammals all reveal many examples of rapid evolution followed by periods of stability.

B. Bursts of Speciation Occur During Adaptive Radiation

Speciation can happen in rapid bursts during an **adaptive radiation,** in which a population inhabiting a patchy or heterogeneous environment gives rise to multiple specialized forms in a relatively short time. (In this context, the term *radiation* means spreading outward from a central source.)

In adaptive radiation, speciation typically occurs in response to the availability of new resources. For example, a few individuals might colonize a new, isolated habitat such as a mountaintop or island. Multiple food sources, such as plants with different-sized seeds, would simultaneously select for different adaptations. Over time, multiple species would develop.

Adaptive radiation is especially common in island groups. One stunning example occurred in the Caribbean islands of Cuba, Jamaica, Puerto Rico, and Hispaniola (**figure 13.11**). Adaptive radiation occurred separately on each island, where at least

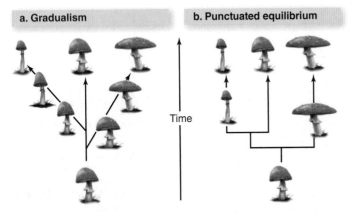

a. Gradualism	b. Punctuated equilibrium

Time

Figure 13.10 Evolution—Both Gradual and Dramatic. (a) In gradualism, three species arise from one ancestor by way of small, incremental steps. (b) Punctuated equilibrium produces the same result, except that the new species arise in rapid bursts followed by periods of little change.

Upper trunk/canopy

Large toe pads, can change color

Cuba—*Anolis porcatus*
Hispaniola—*A. chlorocyanus*
Jamaica—*A. grahami*
Puerto Rico—*A. evermanni*

A. evermanni

Midtrunk

Long forelimbs, vertically flattened body

Cuba—*Anolis loysiana*
Hispaniola—*A. distichus*
Jamaica—none found
Puerto Rico—none found

A. loysiana

Lower trunk/ground

Stocky body, long hind limbs

Cuba—*Anolis sagrei*
Hispaniola—*A. cybotes*
Jamaica—*A. lineatopus*
Puerto Rico—*A. cristatellus*

A. cristatellus

Grass/bush

Slender body, very long tail

Cuba—*Anolis alutaceus*
Hispaniola—*A. olssoni*
Jamaica—none found
Puerto Rico—*A. pulchellus*

A. pulchellus

Figure 13.11 Adaptive Radiation. More than 150 anole lizard species have evolved on four Caribbean islands. Each island has a unique collection of species, yet the adaptations to each habitat are similar—a striking example of convergent evolution (see section 12.4C).

150 species of small lizards called anoles have adapted in very similar ways to different parts of their habitats. Other examples of adaptive radiation occur on Hawaii (as described in the chapter opening essay), the Galápagos, and the Malay archipelago, where Alfred Russel Wallace collected a multitude of beetles and butterflies.

As another path to adaptive radiation, some members of a population may inherit a key adaptation that gives them an advantage. The oldest flowering plant fossil, for example, is about 125 million years old. The descendants of the first flowering plants diversified rapidly, and all of today's major lineages were already in place 100 million years ago. Hundreds of thousands of flowering plant species now inhabit Earth. The new adaptation—the flower—apparently unleashed an entirely new set of options for reproduction, prompting rapid diversification.

A third type of adaptive radiation occurs when some members of a population have a combination of adaptations that enable them to survive a major environmental change. After the poorly suited organisms perish, the survivors diversify as they exploit the new resources in the changed environment. Mammals, for example, underwent an enormous adaptive radiation when the extinctions of the nonavian dinosaurs opened up many new habitats (**figure 13.12**).

13.4 | Mastering Concepts

1. Describe the theories of gradualism and punctuated equilibrium. How can the fossil record support both?
2. What are three ways that adaptive radiation can occur?

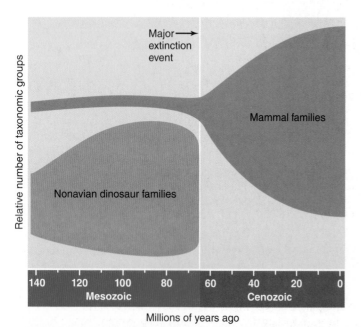

Figure 13.12 **Speciation Following a Mass Extinction.** With many ecological niches vacated when most dinosaurs became extinct about 65 million years ago, mammals flourished.

13.5 | Extinction Marks the End of the Line

A species goes **extinct** when all of its members have died. If speciation is the birth of a species, extinction represents its death.

A. Many Factors Can Combine to Put a Species at Risk

Many factors can cause extinction, but all amount to a failure to adapt to environmental change. Any species will eventually vanish if its gene pool does not contain the "right" alleles necessary for individuals to produce fertile offspring and sustain the population.

The change that wipes out a species may be habitat loss, new predators, or new diseases. Extinction may also be a matter of bad luck: sometimes no individual of a species survives a catastrophe such as a volcanic eruption or asteroid impact.

When faced with a shifting environment, what is the chance that a species will become extinct? The answer depends in part on how fast conditions change. Species such as elephants, which reach reproductive maturity at 30 years, are much less likely to survive a sudden change than are mice, which produce three generations a year.

In addition, the smaller the initial size of a population, the less likely it is to endure a major challenge. Small populations experience fewer genetic mutations, which are the ultimate source of new adaptations (see section 11.6). Low genetic diversity within a population, in turn, poses two problems. First, the population may contain too few beneficial alleles to withstand a new challenge. An emerging disease, for example, may wipe out most individuals, leaving too few survivors to maintain the population's size. Second, inbreeding in a small population tends to bring together lethal recessive alleles, which can weaken the organisms and make them less able to survive and reproduce.

B. Extinction Rates Have Varied over Time

Biologists distinguish between two different types of extinction events. The **background extinction rate** results from the gradual loss of species as populations shrink in the face of new challenges. Paleontologists have used the fossil record to calculate that most species exist from 1 million to 10 million years before becoming extinct. Thus, the background rate is roughly 0.1 to 1.0 extinctions per year per million species. Most extinctions overall occur as part of this background rate.

Earth has also witnessed several periods of **mass extinctions,** when a great number of species disappeared over relatively short expanses of time (**figure 13.13**). Mass extinctions have had a great influence on Earth's history because they have periodically opened vast new habitats for adaptive radiation to occur.

Paleontologists study clues in Earth's sediments to understand the catastrophic events that contribute to mass extinctions. Two theories have emerged to explain these events, although several processes have probably contributed to mass extinctions.

Apply It Now

Recent Species Extinctions

Species extinctions have occurred throughout life's long history. They continue today, often accelerated by human activities. Overharvesting, for example, caused the extinction of the passenger pigeon (pictured above) and many other species. Habitat loss to agriculture, urbanization, damming, or pollution also takes a toll. Introduced plants and animals can deplete native species by competing with or preying on them. Many biologists estimate that we are currently experiencing a sixth global mass extinction; global climate change may make the effects even worse (see chapter 39).

The table in this box lists a few species of vertebrate animals that have disappeared during the past few centuries. This list is far from complete; many more species of animals and plants have become extinct during the same time. Countless others are threatened or endangered, meaning that they are at risk for extinction.

Name	Cause of Extinction	Former Location
Fishes:		
Chinese paddlefish (*Psephurus gladius*)	Habitat destruction	China
Las Vegas dace (*Rhinichthys deaconi*)	Habitat destruction	North America
Amphibians:		
Palestinian painted frog (*Discoglossus nigriventer*)	Habitat destruction	Israel
Southern day frog (*Taudactylus diurnus*)	Undetermined	Australia
Reptiles:		
Yunnan box turtle (*Cuora yunnanensis*)	Habitat destruction, overharvesting	China
Martinique lizard (*Leiocephalus herminieri*)	Undetermined	Martinique
Birds:		
Dodo (*Raphus cucullatus*)	Habitat destruction, overharvesting	Mauritius
Moa (*Megalapteryx diderius*)	Overharvesting	New Zealand
Laysan honeycreeper (*Himatione sanguinea*)	Habitat destruction	Hawaii
Black mamo (*Drepanis funerea*)	Habitat destruction, introduced predators	Hawaii
Passenger pigeon (*Ectopistes migratorius*)	Overharvesting	North America
Great auk (*Alca impennis*)	Overharvesting	North Atlantic
Mammals:		
Quagga (*Equus quagga quagga*)	Overharvesting	South Africa
Steller's sea cow (*Hydrodamalis gigas*)	Overharvesting	Bering Sea
Bali tiger (*Panthera tigris balica*)	Habitat destruction, overharvesting	Indonesia
Javan tiger (*Panthera tigris sondaica*)	Habitat destruction, overharvesting	Indonesia
Caspian tiger (*Panthera tigris virgata*)	Habitat destruction, overharvesting	Central Asia

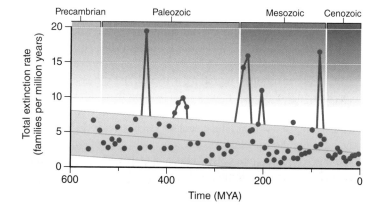

Figure 13.13 Background and Mass Extinctions. This graph shows extinction rates for marine animals over the past 600 million years. The shaded area near the bottom of the graph estimates the background extinction rate; peaks show five mass extinctions.

The first explanation, called the **impact theory,** suggests that meteorites or comets have crashed to Earth, sending dust, soot, and other debris into the sky, blocking sunlight and setting into motion a deadly chain reaction. Without sunlight, plants died. The animals that ate plants, and the animals that ate those animals, then perished. Evidence for the impact theory of the extinction at the end of the Cretaceous period includes centimeter-thick layers of earth that are rich in iridium, an element rare on Earth but common in meteorites (figure 13.14).

A second theory is that movements of Earth's crust may explain some mass extinctions. The crust, or uppermost layer of the planet's surface, is divided into many pieces, called tectonic plates. During Earth's history, movement of tectonic plates caused continents to drift apart, then come back together. Oceans mixed and separated. The result was dramatic environmental change that profoundly affected life. Organisms that had thrived in their habitats had new competitors for limited resources. Climates changed as continents moved toward or away from the poles. Colliding continents also altered shorelines, causing shallow coastal areas packed with life to disappear. ▶ plate tectonics, p. 260

The role of *Homo sapiens* in causing extinctions is evident today. For many types of organisms, ecologists are documenting an alarming increase in background extinction rates to 20 to 200 extinctions per million species per year. Habitat loss and habitat fragmentation, pollution, introduced species, and overharvesting combine to imperil many species (see chapter 39). This chapter's Apply It Now box, on page 285, lists a few of the many vertebrate species that have recently become extinct. Only time will tell if we are on the brink of another mass extinction or are just at a peak in the many ebbs and flows of biodiversity that have characterized life on Earth.

13.5 | Mastering Concepts

1. What factors can cause or hasten extinction?
2. Distinguish between background extinction and mass extinctions.
3. How have humans influenced extinctions?

Figure 13.14 Impact Theory Evidence. This distinctive layer of rock marks the end of the Cretaceous period.

13.6 | Biological Classification Systems Are Based on Common Descent

Darwin proposed that evolution occurs in a branched fashion, with each species giving rise to other species as populations occupy and adapt to new habitats. As described in chapter 12, ample evidence has shown him to be correct.

The goal of modern classification systems is to reflect this shared evolutionary history. **Systematics,** the study of classification, therefore incorporates two interrelated specialties: taxonomy and phylogenetics. **Taxonomy** is the science of describing, naming, and classifying species; **phylogenetics** is the study of evolutionary relationships among species. This section describes how biologists apply the evidence for evolution to the monumental task of organizing life's diversity into groups.

A. The Taxonomic Hierarchy Organizes Species into Groups

Carolus Linnaeus, the sixteenth-century biologist introduced at the start of this chapter, made a lasting contribution to systematics. He devised a way to organize life into a hierarchical classification scheme that assigned a consistent, scientific name to each type of organism.

Linnaeus's idea is the basis of the taxonomic hierarchy used today. Biologists organize life into nested groups of taxonomic levels, based on similarities (figure 13.15). The three domains—Archaea, Bacteria, and Eukarya—are the most inclusive levels. Each domain is divided into kingdoms, which in turn are divided into phyla, then classes, orders, families, genera, and species. A **taxon** (plural: taxa) is a group at any rank; that is, domain Eukarya is a taxon, as is the order Liliales and the species *Aloe vera*. Some disciplines also use additional ranks, such as superfamilies and subspecies.

The more features two organisms have in common, the more taxonomic levels they share. A human, a squid, and a fly are all members of the animal kingdom, but their many differences place them in separate phyla. A human, rat, and pig are more closely related—all belong to the same kingdom, phylum, and class (Mammalia). A human, orangutan, and chimpanzee are even more closely related, sharing the same kingdom, phylum, class, order, and family (Order Primates, Family Hominidae). As humans, our full classification is Eukarya-Animalia-Chordata-Mammalia-Primates-Hominidae-*Homo*-*Homo sapiens.*

The taxonomic hierarchy is useful, but it has a flaw. The system includes eight main levels, ranging from domain to species, which might give the incorrect impression that evolution took eight "steps" to produce each modern species. In fact, however, the taxonomic ranks are not equivalent across kingdoms. As just one example, the family Felidae contains 37 species of cats, whereas the family Orchidaceae contains about 22,000 species of

Taxonomic group	*Aloe vera* plant found in:	Number of species
Domain	Eukarya	Several million
Kingdom	Plantae	~375,000
Phylum	Anthophyta	~235,000
Class	Liliopsida	~65,000
Order	Liliales	~1200
Family	Asphodelaceae	785
Genus	*Aloe*	500
Species	*Aloe vera*	1

Figure 13.15 Taxonomic Hierarchy. Life is divided into three domains, which are subdivided into kingdoms. The kingdoms, in turn, contain numerous smaller categories. This diagram shows the complete classification for the plant *Aloe vera*.

orchids. The taxon "family" therefore does not reflect the same species diversity in plants as it does in animals. The Linnaean system is an imperfect reflection of evolutionary history, but it remains in widespread use because biologists have not agreed on a better approach.

B. A Cladistics Approach Is Based on Shared Derived Traits

Biologists illustrate life's diversity in the form of evolutionary trees, also called **phylogenies,** which depict species's relation-

ships based on descent from shared ancestors. Systematists have long constructed phylogenetic trees, which are hypotheses about evolutionary relationships. (Charles Darwin included an example in *On the Origin of Species*.)

Biologists use multiple lines of evidence to construct phylogenetic trees. Anatomical features of fossils and existing organisms are useful, as are behaviors, physiological adaptations, and molecular sequences. In the past, systematists constructed phylogenetic tree diagrams by comparing as many characteristics as possible among species. Those organisms with the most characteristics in common would be neighbors on the tree's branches.

Basing a tree entirely on similarities, however, can be misleading. As just one example, many types of cave animals are eyeless and lack pigments. But these resemblances do not mean that the fish and snail species that occupy caves are closely related to one another; instead, they are the product of convergent evolution. Insofar as the goal of a classification system is to group related organisms together, simply attending to similarities might therefore lead to an incorrect classification.

A cladistics approach solves this problem. Widely adopted beginning in the 1990s, **cladistics** is a phylogenetic system that defines groups by distinguishing between ancestral and derived characters. **Ancestral characters** are inherited attributes that resemble those of the ancestor of a group; an organism with **derived characters** has features that are different from those found in the group's ancestor.

Cladistics therefore builds on the concept of homology, described in chapter 12. Homologous structures are inherited from a common ancestor; examples include the forelimbs of birds and mammals (see figure 12.10). We have already seen that natural selection can modify homologous structures (and genes) into a wide variety of new forms. In a cladistics analysis, an ancestral character has changed little from its state in a group's ancestor; a derived character, on the other hand, has changed more.

How does a researcher know which characters are ancestral and which are derived? They choose an **outgroup** consisting of comparator organisms that are not part of the group being studied. For example, in a cladistic analysis of land vertebrates, an appropriate outgroup might be lungfishes. Features that are present in lungfishes and land vertebrates are assumed to be ancestral. A segmented backbone and two eyes are therefore ancestral features. Derived features would include four limbs, feathers, fur, or other characteristics that appear only in some land vertebrates but not in lungfish.

In a cladistics approach, biologists use shared derived characters to define groups. A **clade,** also called a **monophyletic** group, is a group of organisms consisting of a common ancestor and all of its descendants; in other words, it is a group of species united by a single evolutionary pathway. For example, birds form a clade because they all descended from the same group of reptiles. A clade may contain any number of species, as long as all of its members share an ancestor that organisms outside the clade do not share.

C. Cladograms Depict Hypothesized Evolutionary Relationships

The result of a cladistics analysis is a **cladogram,** a treelike diagram built using shared derived characteristics (**figure 13.16**). The emphasis in a cladogram is not similarities but rather historical relationships. To emphasize this point, imagine a lizard, a crocodile, and a chicken. Which resembles the lizard more closely: the crocodile or the chicken? Clearly, the most *similar* animals are the lizard and the crocodile. But these resemblances are only superficial. The shared derived characters tell a more complete story of evolutionary history. Based on the evidence, crocodiles are more closely related to birds than they are to lizards.

Figure It Out

How many clades are represented in the phylogenetic tree in figure 13.16?

Answer: 6

All cladograms have features in common. The tips of the branches represent the taxa in the group being studied. Existing species, such as birds and turtles, are at the tips of longer branches in figure 13.16; the nonavian dinosaurs are extinct and therefore occupy a shorter branch. Each node in a cladogram indicates where two groups arose from a common ancestor. A branching pattern of lines therefore represents populations that diverge genetically, splitting off to form a new species.

A common mistake in interpreting cladograms is to incorrectly assume that a taxon must be closely related to both groups that appear next to it on the tree. In figure 13.16, for example, mammals are adjacent to both turtles and amphibians. Does this mean that rabbits are as closely related to frogs as they are to turtles? To find out, look at the relative amount of time that has passed since mammals last shared a common ancestor with each group. Because the common ancestor of mammals and turtles existed more recently, these groups are more closely related than are mammals and amphibians.

Cladograms depict nested hierarchies of evolutionary relationships. Each clade can rotate on its branch without changing the meaning of the cladogram. Moreover, biologists can use either diagonal lines or square brackets when drawing a cladogram. As a result, many equivalent cladograms can tell the same evolutionary story (**figure 13.17**).

Figure 13.18 demonstrates how to construct a cladogram using the derived characters that were important in the evolution of birds. The first step is to select the most relevant traits, such as feathers and lightweight bones. Next, the researcher collects the data and makes a chart showing which species have which traits. Then, in constructing the tree, species sharing the

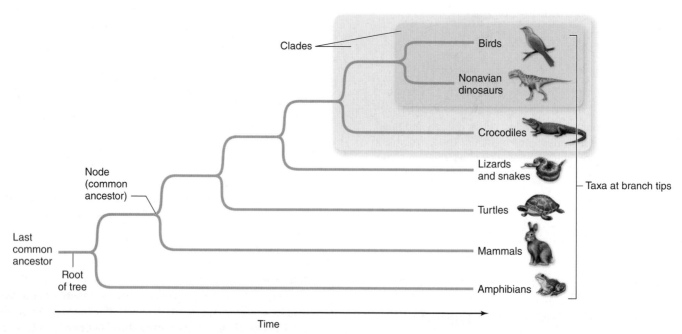

Figure 13.16 Reading a Cladogram. Groups at the tips of branches are linked in a nested hierarchy. The more recently any two groups shared a common ancestor, the more closely related they are.

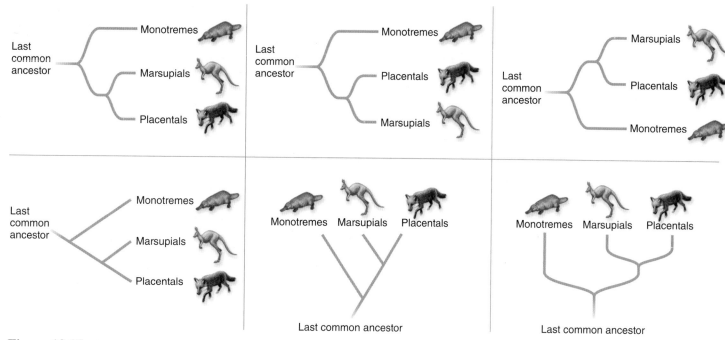

Figure 13.17 Different Cladograms, Same Relationships. All of these cladograms depict the same information about evolutionary relationships among mammals.

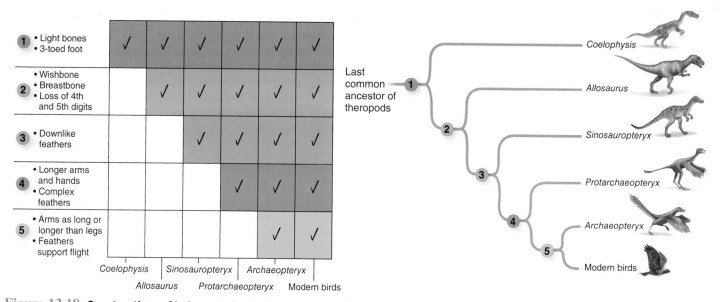

Figure 13.18 Constructing a Cladogram. To build a cladogram, tally up the derived characters that species share. Pairs that share the greatest number of derived characters are hypothesized to be most closely related. *Source:* Cladogram data based on Kevin Padian, June 25, 1998, When is a bird not a bird? *Nature*, vol. 393, page 729.

most derived characters occupy the branches farthest from the root. The resulting cladogram shows the hypothesized relationship between modern birds and their close relatives, the non-avian dinosaurs.

The cladogram in figure 13.18 is based on the physical features of not only existing birds but also the fossils of extinct dinosaurs.

Molecular sequences are also extremely useful because they add many more characters on which to build cladograms and deduce evolutionary relationships. Figure 13.19 shows a simple cladogram based on mutations in a 10-nucleotide sequence of DNA. (In reality, cladograms are typically based on far longer DNA sequences.) Scientists can analyze DNA not only from living species but also

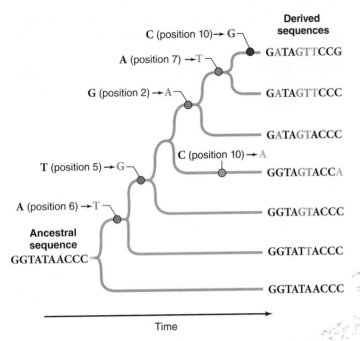

Figure 13.19 Cladogram Based on DNA. Mutations in DNA produce unique sequence variations, characters that can be used to build cladograms.

from long-dead organisms preserved in museums, amber, and permafrost (see section 18.6).

Comparing molecular sequence differences among species also provides a way to assign approximate dates to the branching points of a cladogram. If researchers can estimate the rate at which a selected stretch of DNA accumulates mutations, they can use "molecular clocks" to estimate how much time has passed since two species diverged from a common ancestor. ▶ molecular clocks, p. 266

A problem with cladistics is that the analysis becomes enormously complicated when many species and derived characters are included. Mathematically, many trees can accommodate the same data set; for instance, just 10 taxa can be arranged into millions of possible trees. Sorting through all the possibilities requires tremendous computing power. How do researchers settle on the "best" tree? One strategy is to select the most **parsimonious** tree, which is the one that requires the fewest steps to construct. The most parsimonious tree therefore invokes the fewest evolutionary changes needed to explain the data.

All phylogenetic trees are based on limited and sometimes ambiguous information. They are therefore not peeks into the past but rather tools that researchers can use to construct hypotheses about the relationships among different types of organisms. These investigators can then add other approaches to test the hypotheses.

D. Many Traditional Groups Are Not Monophyletic

Contemporary scientists using a cladistics approach typically assign names only to clades and not to groups that reflect incomplete clades or that combine portions of multiple clades. Many familiar groups of species, however, are not monophyletic. Instead, these groups may be paraphyletic or polyphyletic (figure 13.20).

For example, according to the traditional Linnaean classification system, class Reptilia includes turtles, lizards, snakes, crocodiles, and the extinct dinosaurs, but it excludes

Figure 13.20 Paraphyletic and Polyphyletic Groups. The Linnaean class Reptilia is paraphyletic; it excludes birds and therefore does not form a clade. A group containing only endothermic animals is polyphyletic because it excludes the most recent common ancestor of these animals.

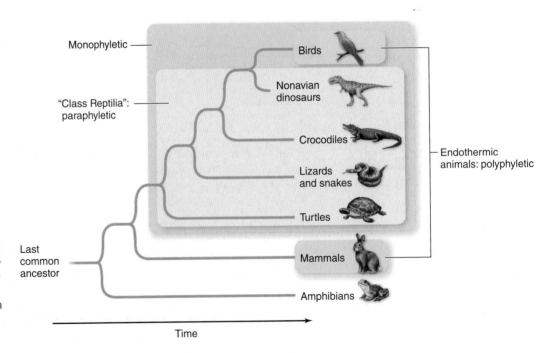

birds. The cladogram, however, places birds in the same clade with the reptiles based on their many shared derived characteristics. Most biologists therefore now consider birds to be reptiles, so they make a distinction between birds and "non-avian" dinosaurs.

The traditional class Reptilia is considered **paraphyletic** because it contains a common ancestor and some, but not all, of its descendants. The kingdom Protista is also paraphyletic (figure 13.21). Protists include mostly single-celled eukaryotes that do not fit into any of the three eukaryotic kingdoms (plants, fungi, and animals). Yet all three of these groups share a common eukaryotic ancestor with the protists. Biologists are currently struggling to divide kingdom Protista into monophyletic groups, an immense task.

Polyphyletic groups exclude the most recent common ancestor shared by all members of the group. For example, a group consisting of endothermic (formerly called "warm-blooded") animals includes only birds and mammals. This group is polyphyletic because it excludes the most recent common ancestor of birds and mammals, which was an ectotherm (formerly called "cold-blooded"). Likewise, the term *algae* reflects a polyphyletic grouping of many unrelated species of aquatic organisms that carry out photosynthesis. In general, polyphyletic groups reflect characteristics—such as ectothermy or photosynthesis— that have evolved independently in multiple species.

Systematists try to avoid paraphyletic and polyphyletic groups, which do not reflect a shared evolutionary history. Nevertheless, many such group names remain in everyday usage.

Table 13.1 summarizes some of the terminology used in systematics.

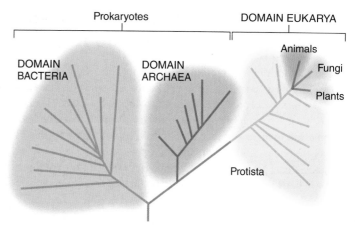

Figure 13.21 Paraphyletic Protists. Although domains Bacteria, Archaea, and Eukarya are all monophyletic, the group of eukaryotes called protists is paraphyletic because it excludes plants, fungi, and animals.

13.6 | Mastering Concepts

1. Describe the taxonomic hierarchy.
2. What is the advantage of a cladistics approach over a more traditional approach to phylogeny?
3. Distinguish between ancestral and derived characters.
4. What sorts of evidence do biologists use in a cladistics analysis?
5. How is a cladogram constructed?
6. How do paraphyletic and polyphyletic groups differ from monophyletic groups?

Table 13.1 Miniglossary of Systematics Terms

Term	Definition
Ancestral characters	Features present in the common ancestor of a clade
Clade	Group of organisms consisting of a common ancestor and all of its descendants; a monophyletic group
Cladistics	Phylogenetic system that groups organisms by characteristics that best indicate shared ancestry
Cladogram	Phylogenetic tree built on shared derived characters
Derived characters	Features of an organism that are different from those found in a clade's ancestors
Monophyletic group	Group of organisms consisting of a common ancestor and all of its descendants; a clade
Outgroup	Comparator organism outside the group being studied; useful for identifying ancestral traits
Paraphyletic group	Group of organisms consisting of a common ancestor and some, but not all, of its descendants
Phylogenetic tree	Diagram depicting hypothesized evolutionary relationships
Polyphyletic group	Group of species that excludes the most recent common ancestor
Systematics	The combined study of taxonomy and evolutionary relationships among organisms

13.7 | Investigating Life: Birds Do It, Bees Do It

The origin of biological diversity has long intrigued biologists, especially after Charles Darwin published *On the Origin of Species*. According to Mayr's biological species concept, the formation of a reproductive barrier that divides a population is the event that signals the birth of a new species. It is easy to understand how mountains, glaciers, and other physical obstacles can prevent interbreeding. But how can a new reproductive barrier arise within a single population, creating two species from one?

Plants offer an ideal opportunity to answer this question. Many species of flowering plants depend on animals such as birds or bees as pollen carriers. Closely related plant species may employ different pollinators, thus creating a reproductive barrier. Consider two species of wildflowers native to the western United States (**figure 13.22**): the purple monkeyflower *(Mimulus lewisii)* and the scarlet monkeyflower *(Mimulus cardi-*

Figure 13.22 Wild-Type and Mutant *Mimulus*. By selective breeding, researchers created *Mimulus* plants that resembled their wild-type counterparts in every respect—except flower color. They used the wild-type and mutant plants to test hypotheses about reproductive barriers in *Mimulus*.

nalis). Bumblebees pollinate *M. lewisii*, whereas hummingbirds prefer the red-flowered *M. cardinalis.* If, by chance, a hummingbird carries *M. cardinalis* pollen to *M. lewisii*, the resulting hybrid offspring are viable. Such cross-breeding rarely happens in the wild, however, thanks to the more-or-less exclusive relationship between each plant and its pollinators.

Because the two monkeyflowers are so closely related, it is reasonable to suppose that, in the past, one ancestral *Mimulus* species gave rise to both types. The reproductive barrier that separates them may have arisen after the slow, constant buildup of mutations in many genes. Alternatively, a mutation in one or a few "major" genes (those that control flower form or color, for example) may have had the same effect. We know that a single mutation can dramatically alter an organism's appearance (see the fly with a homeotic mutation in figure 7.21). Can a single mutation in a plant species also create a reproductive barrier by attracting a new pollinator?

Biologist H.D. "Toby" Bradshaw, of the University of Washington, and Michigan State University's Douglas Schemske studied the two *Mimulus* species to find out. They knew that these plants have a gene locus (area of a chromosome) that controls the concentration of yellowish orange pigments called carotenoids in the flower petals. The locus, which may consist of one or a few genes, is called *YUP* (for *y*ellow *up*per petal). In the wild-type purple monkeyflower, the dominant allele prevents carotenoids from forming, leaving the pinkish-purple anthocyanin pigments to provide the petal color. Wild-type scarlet monkeyflowers have two copies of the recessive *yup* allele. In these plants, carotenoid pigments accumulate, turning the flowers red in the presence of anthocyanins (**table 13.2**).

Bradshaw and Schemske used a breeding trick to create plants that had different-colored flowers from their parents. They began by crossing the two species with each other; the resulting hybrid offspring were heterozygous for the *YUP* locus. They then repeatedly backcrossed the hybrids with the wild-type parents, selecting each time for the desired allele. After four generations of backcrossing, the researchers had produced two new plant lineages: *M. lewisii*

Table 13.2 Wild-Type Monkeyflower Characteristics

Species	Flower Color	Genotype at Flower Color Locus	Pollinator
Purple monkeyflower, *Mimulus lewisii*	Pink-purple	YUP/YUP or YUP/yup	Bumblebee
Scarlet monkeyflower, *Mimulus cardinalis*	Red	yup/yup	Hummingbird

that were homozygous for the recessive *yup* allele and *M. cardinalis* plants that expressed the dominant *YUP* allele. Each was 97% identical to its parent strain but with one obvious difference—the flower colors. The new variety of *Mimulus lewisii* had pale yellow-orange petals, while the new *M. cardinalis* plants had dark pink blooms (see figure 13.22). The different-colored plants mimicked the effect of a mutation in the *YUP* allele.

To test the hypothesis that a change in the *YUP* locus could prompt a pollinator shift, Bradshaw and Schemske planted all four *Mimulus* varieties (the two wild-type species and their mutant siblings) at a California location where both species normally occur. For about a week and a half, the researchers observed the plants from dawn until evening, recording the animal species that visited each flower. They found that the substituted *YUP* alleles did indeed alter pollinator preferences (table 13.3). The mutant yellow-orange *M. lewisii* flowers attracted fewer bees but more hummingbirds than did their wild-type purple counterparts. *M. cardinalis* drew far more bumblebee visits with its new dark pink petals than did the wild-type red plants, while hummingbird visits stayed about the same.

This experiment supports the hypothesis that a change in just one gene locus may have jump-started speciation in *Mimulus*. Additional mutations, coupled with natural selection, eventually sculpted the flower shapes and petal positions that define the purple and scarlet monkeyflower species. It is tempting to extend this finding to other species. Can a single mutation modify an animal's mating ritual just enough to create a new reproductive barrier? Or cause a flower to open its petals at a different time of day? Or enable an animal to exploit a new food source, extending its habitat? This experiment cannot answer these questions. Every now and then, however, small mutations may fuel the birth of an entirely new species.

Bradshaw, H. D., Jr., and Douglas W. Schemske. November 13, 2003. Allele substitution at a flower colour locus produces a pollinator shift in monkeyflowers. *Nature*, vol. 426, pages 176–178.

Table 13.3 Pollinator Visits

		Number of Visits (10⁻³ visits/flower/hour)	
		Bumblebees	Hummingbirds
M. lewisii			
	Wild-type (pink-purple)	15.4	0.0212
	Mutant (yellow-orange)	2.63	1.44
M. cardinalis			
	Wild-type (red)	0.148	189
	Mutant (dark pink)	10.9	168

13.7 | Mastering Concepts

1. What hypothesis were the investigators testing, and why did they choose these two plant species as experimental subjects?
2. If biologists could mutate the gene that controls carotenoid concentration, they could generate pairs of plants that differ only in that one gene, improving on the 97% similarity they achieved by backcrossing hybrids with wild-type plants. Explain why this improvement would help the researchers test their hypothesis.

Burning Question

Why does evolution occur rapidly in some species but slowly in others?

Even though the same basic forces of microevolution operate on all populations, species evolve at different rates. This may seem like a paradox until you think of all the different ways that a population's gene pool can change over multiple generations (see chapter 11). In general, evolution occurs most quickly in:

- genes with high mutation rates
- small populations, which are most susceptible to genetic drift
- species with a high rate of polyploidy
- species that reproduce rapidly, such as bacteria and insects
- times of rapid environmental change
- populations with high rates of immigration or emigration
- species that reproduce sexually

Biologists have much to learn about evolution rates. One unsolved mystery is how a species's DNA can evolve rapidly, even as external appearances remain unchanged. Such is the case for the tuatara, a New Zealand reptile. When researchers compared DNA extracted from modern tuataras and from 8750-year-old fossils, they measured the fastest known molecular evolution rate. Yet the tuataras themselves have changed little. The solution to the mystery awaits further research.

Submit your burning question to:
marielle_hoefnagels@mcgraw-hill.com

Chapter Summary

13.1 | The Definition of "Species" Has Evolved over Time

- **Macroevolution** refers to large-scale changes in life's diversity, including the appearance of new **species** and higher taxonomic levels.

A. Linnaeus Devised the Binomial Naming System

- Linnaeus's species designations and classifications helped scientists communicate. Darwin added evolutionary meaning.

B. Ernst Mayr Developed the Biological Species Concept

- Mayr added the requirement for reproductive isolation to define **biological species.**
- **Speciation** is the formation of a new species, which occurs when a population's **gene pool** is divided and each part takes its own evolutionary course.

13.2 | Reproductive Barriers Cause Species to Diverge

A. Prezygotic Barriers Prevent Fertilization

- **Prezygotic reproductive isolation** occurs before or during fertilization. It includes obstacles to mating such as space, time, and behavior; mechanical mismatches between male and female; and molecular mismatches between gametes.

B. Postzygotic Barriers Prevent Viable or Fertile Offspring

- **Postzygotic reproductive isolation** results in offspring that die early in development, are infertile, or produce a second generation of offspring with abnormalities.

13.3 | Spatial Patterns Define Three Types of Speciation

A. Allopatric Speciation Reflects a Geographic Barrier

- **Allopatric speciation** occurs when a geographic barrier separates a population. The two populations then diverge genetically to the point that their members can no longer produce fertile offspring together.

B. Parapatric Speciation Occurs in Neighboring Regions

- **Parapatric speciation** occurs when two populations live in neighboring areas but share a border zone. Genetic divergence between the two groups exceeds gene flow, driving speciation.

C. Sympatric Speciation Occurs in a Shared Habitat

- **Sympatric speciation** enables populations that occupy the same area to diverge, often via major genetic changes such as **polyploidy.**

D. Determining the Type of Speciation May Be Difficult

- The distinction between allopatric, parapatric, and sympatric speciation is not always straightforward, partly because it is difficult to define the size of a geographic barrier.

13.4 | Speciation May Be Gradual or Occur in Bursts

A. Gradualism and Punctuated Equilibrium Are Two Models of Speciation

- Evolutionary change occurs at many rates, from slow and steady **gradualism** to the periodic bursts that characterize **punctuated equilibrium.**

B. Bursts of Speciation Occur During Adaptive Radiation

- In **adaptive radiation,** an ancestral species rapidly branches into several new species, reflecting either the availability of new resources or the appearance of a particularly beneficial adaptation.

13.5 | Extinction Marks the End of the Line

- **Extinction** is the disappearance of a species.

A. Many Factors Can Combine to Put a Species at Risk

- A slow reproductive rate and low genetic diversity make species vulnerable to extinction, especially during rapid environmental change.

B. Extinction Rates Have Varied over Time

- The **background extinction rate** reflects ongoing losses of species on a local scale.
- Historically, **mass extinctions** have resulted from global changes such as continental drift. The **impact theory** suggests that a meteorite or comet changed Earth's climate and caused a mass extinction at the end of the Cretaceous period.
- Human activities are increasing the extinction rate.

13.6 | Biological Classification Systems Are Based on Common Descent

- The study of **systematics** includes **taxonomy** (the science of classification) and **phylogenetics** (the study of species relationships).

A. The Taxonomic Hierarchy Organizes Species into Groups

- Biologists use a taxonomic hierarchy to classify life's diversity, with **taxa** ranging from domain to species.

B. A Cladistics Approach Is Based on Shared Derived Traits

- Cladistics defines groups based on **ancestral** and **derived characters.** A group of species related by common descent is a **clade,** or **monophyletic** group.
- An **outgroup** helps researchers detect ancestral characters.

C. Cladograms Depict Hypothesized Evolutionary Relationships

- A **cladogram** shows evolutionary relationships as a branching hierarchy.
- The most **parsimonious** cladogram is the simplest tree that fits the data.

D. Many Traditional Groups Are Not Monophyletic

- **Paraphyletic** and **polyphyletic** group names often remain in common usage, but they do not reflect complete evolutionary relationships.

13.7 | Investigating Life: Birds Do It, Bees Do It

- Researchers have studied *Mimulus* plants to determine how the reproductive barrier that separates two species may have arisen.
- A single mutation in the gene that controls petal color can induce flowers to attract new pollinators.

Multiple Choice Questions

1. The term *macroevolution* refers to
   a. large-scale changes in the diversity of organisms.
   b. large-scale changes in the DNA of organisms.
   c. evolutionary changes that affect larger organisms.
   d. evolutionary changes that can be observed.

2. The biological species concept defines species based on
   a. external appearance.
   b. the number of adaptations to the same habitat.
   c. ability to interbreed.
   d. DNA and protein sequences.

3. Which reproductive barrier applies to two species that cannot interbreed because they are active at different times of day?
   a. Temporal isolation   c. Mechanical isolation
   b. Gametic isolation   d. Ecological isolation

4. How can infertility occur in a hybrid whose parents have different numbers of chromosomes?
   a. The difference prevents mitotic cell division.
   b. The cells of the hybrid cannot grow, so the embryo dies.
   c. Meiosis is blocked, so gametes cannot form.
   d. Mitosis is altered, so the gametes are inviable.

5. Agriculture and urban development often divide wildlife habitat into isolated fragments. What type of speciation might occur as a result of this human activity?
   a. Sympatric speciation
   b. Allopatric speciation
   c. Parapatric speciation
   d. Paraphyletic speciation

6. Adaptive radiation can best be defined as
   a. the evolution of a phenotype that adapts a population to a new environment.
   b. an adaptation that specifically helps individuals find new environments.
   c. migration of a population to a better environment.
   d. speciation in response to a new or changing environment.

7. Why is a species with a small population more likely than a large population to undergo an extinction?
   a. Because they cannot produce enough offspring
   b. Because there is less genetic diversity
   c. Because the individuals are isolated from one another
   d. Because they take too long to produce offspring

8. All organisms within a clade
   a. belong to the same species.
   b. share identical DNA sequences with a common ancestor.
   c. share one or more characteristics derived through evolution from a common ancestor.
   d. share a common developmental pathway.

9. How does DNA contribute to our understanding of phylogenies?
   a. It allows scientists to determine whether speciation was parapatric or sympatric.
   b. It allows scientists to estimate when species last shared a common ancestor.
   c. It demonstrates whether gradualism or punctuated equilibrium occurred.
   d. It identifies extinctions.

10. Kingdom Protista contains many lineages but does not include all descendants of their common ancestor. Kingdom Protista is therefore
    a. homophyletic.
    b. paraphyletic.
    c. polyphyletic.
    d. a clade.

Write It Out

1. What did Linnaeus, Darwin, and Mayr contribute to the meaning of the term *species*?
2. What type of reproductive barrier applies to each of these scenarios?
   a. Humans introduced apple trees to North America in the 1800s. Insects called hawthorn flies, which feed and mate on hawthorn plants, quickly discovered the new fruits. Some flies preferred the taste of apples to their native host plants. Because these flies mate where they eat, this difference in food preference quickly led to a reproductive barrier.
   b. Water buffalo and cattle can mate, but the embryos die early in development.
   c. Eastern and western meadowlarks are difficult to distinguish between based on size, shape, and color. Yet their calls are distinct, a difference that presumably helps the birds identify potential mates.
   d. The shells of two species of snails in the genus *Bradybaena* spiral in different directions. The snails' genital openings are therefore not aligned, so they cannot mate.
   e. A surface water fish species rarely encounters a closely related fish species that lives in caves.
3. If the apple-feeding flies from question 2 form a different species from their hawthorn-feeding relatives, which type of speciation has occurred?
4. Examine the lizards and their characteristics shown in figure 13.11. Propose an advantage for each of the variations, given the environment. How could each of these species have arisen from a common ancestor?

5. How does natural selection predict a gradualistic mode of evolution? Does the presence of fossils that are consistent with punctuated equilibrium mean that natural selection does not occur?
6. Why do species become extinct? Choose a species that has recently become extinct and describe some possible evolutionary consequences to other species that interacted with that species before its extinction.
7. What information would you need to determine what the background extinction rate was hundreds of millions of years ago? How might you determine the current extinction rate?
8. Examine the following cladogram, which shows the relationships among a fossilized tree (*Hymenaea protera*), three living relatives, and six other organisms:

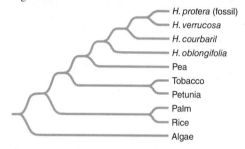

   a. Which organism is *H. protera*'s closest relative?
   b. Is *H. protera* more closely related to tobacco or to palm?
   c. Which organism depicted is ancestral to all the others?
   d. Redraw the tree so that *H. protera* is next to tobacco without changing the evolutionary relationships among any of the species.
9. Figure 18.3 summarizes the hypothesized evolutionary relationships among living plants. Are the gymnosperms monophyletic, polyphyletic, or paraphyletic? Explain your answer.
10. Figure 20.2 shows a phylogenetic tree for animals. How many nodes are in the tree? Suggest three ways to rearrange the tree without changing the evolutionary relationships among the animals.

Pull It Together

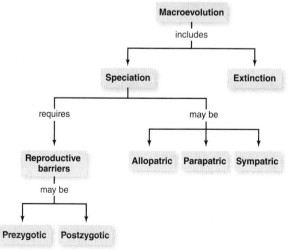

1. How do reproductive barriers relate to the biological species concept?
2. Distinguish between pre- and postzygotic reproductive barriers.
3. Describe allopatric, parapatric, and sympatric speciation and give an example of each.
4. Add *gradualism* and *punctuated equilibrium* to this concept map.
5. How do species become extinct?

The Origin and History of Life

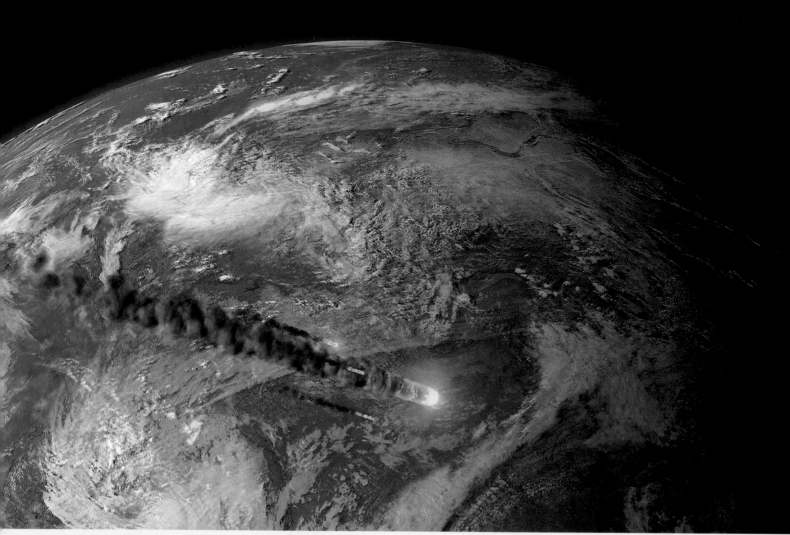

Bombardment. Meteoroids enter Earth's atmosphere every day, as they have done for eons. Some scientists hypothesize that ancient objects from space carried the organic molecules needed for life to begin on Earth.

Enhance your study of
this chapter with practice quizzes,
animations and videos, answer keys,
and downloadable study tools.
www.mhhe.com/hoefnagels

Life from Space

IN A 1908 BOOK ENTITLED *WORLDS IN THE MAKING*, SWEDISH PHYSICAL CHEMIST SVANTE ARRHENIUS SUGGESTED THAT LIFE CAME TO EARTH FROM THE COSMOS. He later broadened the idea, calling it "panspermia" and proposing that life-carrying spores arrived on interstellar dust, comets, asteroids, and meteorites. Later versions of panspermia proposed that instead of spores, which are protected cells, only the organic chemicals required for life arrived from space.

What is the evidence for panspermia? Modern proponents point to several intriguing clues:

- Some microorganisms can survive under extreme conditions, which would be necessary to endure the high radiation and cold temperatures of space during a journey that could take millions of years. A bacterium called *Deinococcus radiodurans,* for example, tolerates a thousand times the radiation level that a person can; it even lives in nuclear reactors! Microbes also live in pockets of water within the ice of Antarctic lakes, surroundings not unlike the icy insides of a comet. In addition, researchers have revived some bacteria after a dormancy lasting millions of years.
- Meteorites that have fallen to Earth from space sometimes contain organic compounds such as amino acids (see section 2.6).
- The surface of one of Jupiter's moons, Europa, is covered with a thick layer of ice. The frozen surface may conceal an ocean of liquid water, a prerequisite for life. ▶ essential water, p. 27
- Canyons, shorelines, and other physical features of Mars leave little doubt that liquid water once flowed on the red planet. The Martian climate is now too cold for liquid water, but ice—and the possibility of life—remains. In 2002, the orbiting *Mars Odyssey* spacecraft remotely detected vast deposits of ice below the planet's surface. Then in 2004, another orbiter, the *Mars Express,* found high concentrations of water vapor and methane in the Martian atmosphere, leading to speculation that methane-producing microbes might live in liquid water beneath Mars's surface. Moreover, the Mars lander *Phoenix* "tasted" Martian soil and verified the presence of subsurface ice in 2008.

Although this sparse evidence is enough to keep the debate alive, panspermia is not widely accepted, in part because it sidesteps the question of life's ultimate origin in the universe. Instead, most scientists accept that life probably arose from simple chemical substances on Earth. This process and its astounding aftermath are the subjects of this chapter.

UNIT 3

What's the Point?

Learning Outline

Learn How to Learn
Write Your Own Test Questions
Have you ever tried putting yourself in your instructor's place by writing your own multiple-choice test questions? It's a great way to pull the pieces of a chapter together. The easiest questions to write are based on definitions and vocabulary, but those will not always be the most useful. Try to think of questions that integrate multiple ideas or that apply the concepts in a chapter. Write 10 questions, and then let a classmate answer them. You'll probably both learn something new.

14.1 | Life's Origin Remains Mysterious

Reconstructing life's start is like reading all the chapters of a novel except the first. A reader can get some idea of the events and setting of the opening chapter from clues throughout the novel. Similarly, scattered clues from life through the ages reflect events that may have led to the origin of life.

Scientists describe the origin and history of life in the context of the **geologic timescale,** which divides time into eons, eras, periods, and epochs defined by major geological or biological events. **Figure 14.1** shows a simplified geologic timescale; see figure 12.2 for a complete version.

The study of life's origin begins with astronomy and geology. Earth and the solar system's other planets formed about 4.6 BYA (billion years ago) as solid matter condensed out of a vast expanse of dust and gas swirling around the early Sun. The red-hot ball that became Earth cooled enough to form a crust by about 4.2 to 4.1 BYA, when the surface temperature ranged from 500°C to 1000°C, and atmospheric pressure was 10 times what it is now.

The geological evidence paints a chaotic picture of this Hadean eon, including volcanic eruptions, earthquakes, and ultraviolet radiation. Analysis of craters on other objects in the solar system suggests that comets, meteorites, and possibly asteroids bombarded Earth's surface during its first 500 million to 600 million years (figure 14.2). These impacts repeatedly boiled off the seas and vaporized rocks to carve the features of the fledgling world.

Still, protected pockets of the environment probably existed where organic molecules could aggregate and perhaps interact. So harsh and unsettled was the early environment of Earth that organized groups of chemicals may have formed many times and at many places, only to be torn apart by heat, debris from space, or radiation. We can't know.

At some point, however, probably during the Archean eon, an entity arose that could survive, thrive, reproduce, and diversify. The clues from geology and paleontology suggest that from 4.2 to 3.85 BYA, simple cells (or their precursors) arose. This section describes some of the major steps in the chemical evolution that eventually led to the first cell.

Figure 14.2 Early Earth. Intense bombardment by meteorites marked Earth's first 500 million to 600 million years.

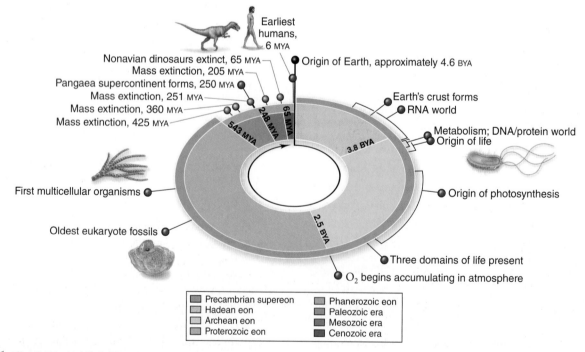

Earliest humans, 6 MYA

Origin of Earth, approximately 4.6 BYA

Nonavian dinosaurs extinct, 65 MYA
Mass extinction, 205 MYA
Pangaea supercontinent forms, 250 MYA
Mass extinction, 251 MYA
Mass extinction, 360 MYA
Mass extinction, 425 MYA

Earth's crust forms
RNA world

Metabolism; DNA/protein world
Origin of life

543 MYA
248 MYA
65 MYA
3.8 BYA

First multicellular organisms

Origin of photosynthesis

Oldest eukaryote fossils

2.5 BYA

Three domains of life present

O_2 begins accumulating in atmosphere

☐ Precambrian supereon	☐ Phanerozoic eon
☐ Hadean eon	☐ Paleozoic era
☐ Archean eon	☐ Mesozoic era
☐ Proterozoic eon	☐ Cenozoic era

Figure 14.1 Highlights in Life's History. In this simplified geologic timescale, the size of each eon and era is proportional to its length in years. (BYA = billion years ago; MYA = million years ago)

A. The First Organic Molecules May Have Formed in a Chemical "Soup"

Early Earth was not only different geologically from today's planet, but it was also different chemically. The atmosphere today contains gases such as nitrogen (N_2), oxygen (O_2), carbon dioxide (CO_2), and water vapor (H_2O). What might it have been like 4 BYA?

Russian chemist Alex I. Oparin hypothesized in his 1938 book, *The Origin of Life,* that a hydrogen-rich, or reducing, atmosphere was necessary for organic molecules to form on Earth. Oparin thought that this long-ago atmosphere included methane (CH_4), ammonia (NH_3), water, and hydrogen (H_2), similar to the atmospheres of the outer planets today. Due to the extreme conditions on Earth immediately after cooling, little O_2 would have been available, because most of the newly formed chemicals were highly reactive with oxygen. Without O_2, Oparin suggested, chemical reactions that form amino acids and nucleotides could have occurred.

Miller's Experiment   In 1953, Stanley Miller, a graduate student in chemistry at the University of Chicago, and his mentor, Harold Urey, decided to test whether Oparin's atmosphere could indeed

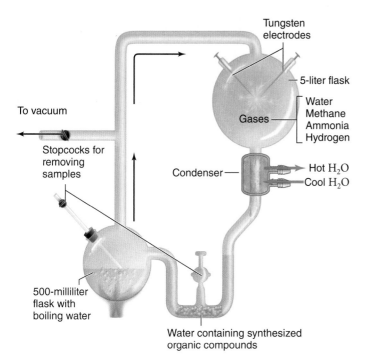

Tungsten electrodes

5-liter flask

To vacuum

Gases

Water
Methane
Ammonia
Hydrogen

Stopcocks for removing samples

Condenser

Hot H_2O
Cool H_2O

500-milliliter flask with boiling water

Water containing synthesized organic compounds

"You have to define 'simulate.' One has to reconstruct an historical event—how did it happen on the primitive earth? You don't even need to argue if you can construct exactly how it happened, but what you can do is to go through a plausible process, from an initial atmosphere, through something that is capable of self-replication. When you do a good prebiotic experiment, you see biological material—amino acids, purines, pyrimidines, and sugars—just fall out. That is telling us something."
—Stanley Miller

Figure 14.3 **The First Prebiotic Simulation.** When Stanley Miller passed an electrical spark through heated gases, the mixture generated amino acids and other organic molecules.

Burning Question

Does new life spring from inorganic molecules now, as it did in the past?

It is intriguing to think of the possibility that new life could be forming from nonliving matter now, just as it did long ago in Earth's history. Although theoretically possible, scientists have never seen life emerging from a collection of simple chemicals. Such a finding would be a major blow to the cell theory, which says that cells come only from preexisting cells (see section 3.1B).

The emergence of new life from simple molecules, however, is improbable today. One reason is that when Earth was young, no life existed, so the first simple cells encountered no competition. Now, however, life thrives nearly everywhere on Earth (see chapter 16's Burning Question). Perhaps new life *is* forming, but before it has a chance to become established, a hungry microbe gobbles it up. Such an event would be extremely difficult to detect.

A second reason is that the chemical and physical environment on the young Earth was nothing like that of today's world. The conditions that allowed new life to develop billions of years ago simply no longer exist.

So does the ancient chemical origin for life on Earth violate the cell theory? The answer is no, because the cell theory applies to today's circumstances, which are very different from those on the early Earth. Scientists have never observed the formation of life from nonliving matter—but that does not mean that it did not happen in the distant past.

Submit your burning question to:
marielle_hoefnagels@mcgraw-hill.com

give rise to organic molecules. Miller built a sterile glass enclosure to contain Oparin's four gases, through which he passed electric discharges to simulate lightning (figure 14.3). He condensed the gases in a narrow tube and passed them over an electric heater, a laboratory version of a volcano. ▸ organic molecules, p. 31

After a few failures and adjustments, Miller saw the condensed liquid turn yellowish. Chemical analysis showed that he had made glycine, the simplest amino acid in organisms. When he let the brew cook a full week, the solution turned varying shades of red, pink, and yellow-brown; he subsequently found a few more amino acids, some of which are found in life. (In fact, the solution contained even more chemicals than Miller realized; a 2008 reanalysis of material saved from one of Miller's experiments revealed 22 amino acids.)

A prestigious journal published the original work, which Urey gallantly refused to put his name on. The 25-year-old Miller made headlines in newspapers and magazines reporting (incorrectly) that he had created "life in a test tube."

Life is far more than just a few amino acids, but "the Miller experiment" went down in history as the first **prebiotic simulation,** an attempt to re-create chemical conditions on Earth before life arose. Miller and many others later extended his results by altering conditions or using different starting materials. For example, methane and ammonia could form clouds of hydrogen cyanide (HCN), which produced amino acids in the presence of ultraviolet light and water. Prebiotic "soups" that included phosphates yielded nucleotides, including the biological energy molecule ATP. Other experiments produced carbohydrates and phospholipids similar to those in biological membranes.

The experiment has survived criticisms that Earth's early atmosphere actually contained abundant CO_2, a gas not present in Miller's original setup. Organic molecules still form, even with an adjusted gas mixture.

Hydrothermal Vents as a Model   More recent prebiotic simulations mimic deep-sea hydrothermal vents. Here, in a zone where hot water meets cold water, chemical mixtures could have encountered a rich brew of minerals spewed from Earth's interior. One laboratory version combines mineral-rich lava with seawater containing dissolved CO_2; under high temperature and pressure, simple organic compounds form. In another vent model, nitrogen compounds and water mix with an iron-containing mineral under high temperature and pressure. The iron catalyzes reactions that produce ammonia—one of the components of the original Miller experiment. ▶ hydrothermal vents, p. 783

The Possible Role of Clays   Once the organic building blocks of macromolecules were present, they had to have linked into chains (polymerized). This may have happened on hot clays or other minerals that provided ample, dry surfaces.

Clays may have played an important role in early organic chemistry for at least three reasons. First, clays consist of sheetlike minerals. Their flat surfaces can therefore form templates on which chemical building blocks could have linked to build larger molecules. Second, some types of clay also contain minerals that can release electrons, providing energy to form chemical bonds. Third, these minerals may also have acted as catalysts to speed chemical reactions.

Prebiotic simulations demonstrate that the first RNA molecules could have formed on clay surfaces (figure 14.4). Not only do the positive charges on clay's surface attract and hold negatively charged RNA nucleotides, but clays also promote formation of the covalent bonds that link the nucleotides into chains. They even attract other nucleotides to form a complementary strand. About 4 BYA, clays might have been fringed with an ever-increasing variety of growing polymers. Some of these might have become the macromolecules that would eventually build cells.

B. Some Investigators Suggest an "RNA World"

Life requires an informational molecule. That molecule may have been RNA, or something like it, because RNA is the most versatile molecule that we know of. It stores genetic information and uses it to manufacture proteins. RNA can also catalyze chemical reactions and duplicate on its own. As Stanley Miller summed it

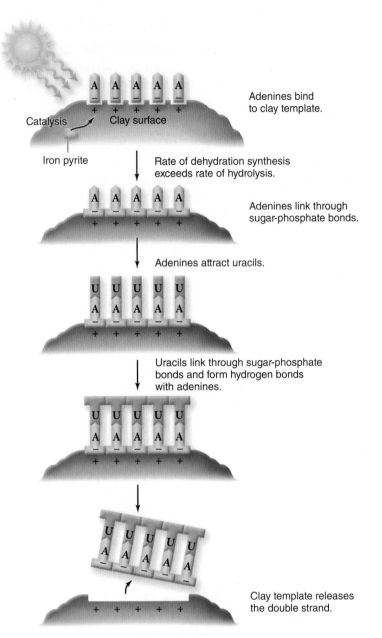

Figure 14.4 A Possible Role for Clay. Chains of nucleotides may have formed on clay templates. In this hypothesized scenario, iron pyrite ("fool's gold") was the catalyst for polymer formation, and sunlight provided the energy.

up, "The origin of life is the origin of evolution, which requires replication, mutation, and selection. Replication is the hard part. Once a genetic material could replicate, life would have just taken off."

Perhaps pieces of RNA on clay surfaces continued to form and accumulate, growing longer, becoming more complex in sequence, and changing as replication errors led to mutations. Some members of this accumulating community of molecules would have been more stable than others, leading to an early form of natural selection. The term **"RNA world"** has come to

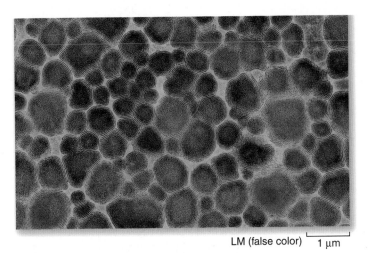

LM (false color) |— 1 μm

Figure 14.5 Liposomes. Each of these bubbles, or liposomes, contains a watery solution and has a phospholipid bilayer boundary similar to a cell membrane.

describe how self-replicating RNA may have been the first independent precursor to life on Earth.

At some point, RNA might have begun encoding proteins, just short chains of amino acids at first. An RNA molecule may eventually have grown long enough to encode the enzyme reverse transcriptase, which copies RNA to DNA. With DNA, the chemical blueprints of life found a much more stable home. Protein enzymes eventually took over some of the functions of catalytic RNAs.

C. Membranes Enclosed the Molecules

Meanwhile, lipids would have been entering the picture. Under the right temperature and pH conditions, and with the necessary precursors, phospholipids could have formed membranelike structures, some of which left evidence in ancient sediments. Laboratory experiments show that pieces of membrane can indeed grow on structural supports and break free, forming a bubble called a liposome (figure 14.5).
▸ phospholipids, p. 54

Perhaps an ancient liposome enclosed a collection of nucleic acids and proteins to form a cell-like assemblage (figure 14.6). Carl Woese, who described the domain Archaea, gave the term **progenotes** to these hypothetical, ancient aggregates of RNA, DNA, proteins, and lipids. Also called protocells or protobionts, these were precursors of cells but they were not nearly as complex.

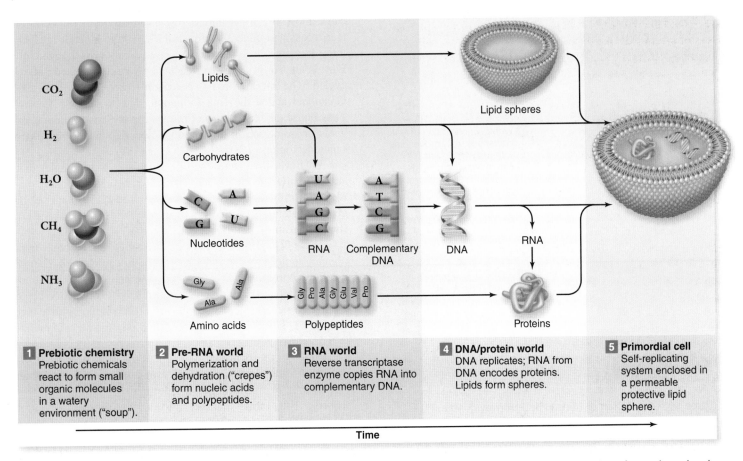

1 Prebiotic chemistry	**2 Pre-RNA world**	**3 RNA world**	**4 DNA/protein world**	**5 Primordial cell**
Prebiotic chemicals react to form small organic molecules in a watery environment ("soup").	Polymerization and dehydration ("crepes") form nucleic acids and polypeptides.	Reverse transcriptase enzyme copies RNA into complementary DNA.	DNA replicates; RNA from DNA encodes proteins. Lipids form spheres.	Self-replicating system enclosed in a permeable protective lipid sphere.

Time

Figure 14.6 Pathway to a Cell. The steps leading to the origin of life on Earth may have started with the formation of organic molecules from simple precursors. However it originated, the first cell would have contained self-replicating molecules enclosed in a phospholipid bilayer membrane.

D. The Origin of Metabolism Would Have Involved Early Enzymes

The next step in the origin of life was the evolution of metabolism. As described in unit 1, metabolism is the ability to acquire and use energy to maintain the organization necessary for life. In a simple sense, this means converting one kind of molecule to another.

The capacity of nucleic acids to mutate may have enabled progenotes to become increasingly self-sufficient, giving rise eventually to the reaction pathways of metabolism (**figure 14.7**). Imagine a progenote that fed on a molecule, nutrient A, that was abundant in its environment. As the progenote reproduced, it began to use up nutrient A. The ancient seas, however, probably held more than one genetic variety of the progenote. One type might have had an enzyme that could convert another nutrient, B, into the original nutrient A. The second progenote would have

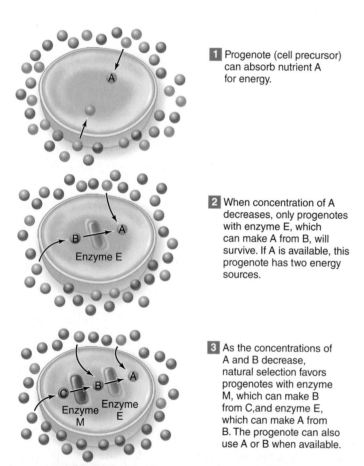

1 Progenote (cell precursor) can absorb nutrient A for energy.

2 When concentration of A decreases, only progenotes with enzyme E, which can make A from B, will survive. If A is available, this progenote has two energy sources.

3 As the concentrations of A and B decrease, natural selection favors progenotes with enzyme M, which can make B from C, and enzyme E, which can make A from B. The progenote can also use A or B when available.

Figure 14.7 Evolution of Metabolic Pathways. As early cells developed the ability to use more substances as nutrients, metabolic pathways may have emerged. Natural selection would have favored cells that could use the greatest variety of nutrients.

had a nutritional advantage because it could extract energy from two food sources, B and A. ▶ enzymes, p. 78

Soon the first type of progenote, totally dependent on nutrient A, would die out as its food vanished. For a while, the progenote that could convert nutrient B to A would flourish. In time, however, nutrient B would also become scarce. Meanwhile, perhaps a new type of progenote would have arisen with an additional enzyme that converted nutrient C to B (and then, using the first enzyme, B to A). In time, this progenote would flourish; then another, and yet another. Over time, an enzyme-catalyzed sequence of connected chemical reactions, a biochemical pathway linking D to C to B to A, would arise. More pathways would form. As intermediates of one pathway spawned others, metabolism evolved.

No one knows what these first metabolic pathways were or how they eventually led to respiration, photosynthesis, and thousands of other chemical reactions that support life today. Despite intriguing similarities between glycolysis and some photosynthetic reactions (see section 6.9), these early stages of metabolism may not have left enough evidence for us ever to understand their origins.

E. Early Life Changed Earth Forever

Unfortunately, direct evidence of the first life is likely gone because most of Earth's initial crust has been destroyed. Erosion tears rocks and minerals into particles, only to be built up again into sediments, heated and compressed. Seafloor is dragged into Earth's interior at deep-sea trenches, where it is melted and recycled. The oldest rocks that remain today, from an area of Greenland called the Isua formation, date to about 3.85 BYA. They house the oldest hints of life: quartz crystals containing organic deposits rich in the carbon isotopes found in organisms. ▶ plate tectonics, p. 260

Whatever they were, the first organisms were simpler than any cell known today. Several types of early cells probably prevailed for millions of years, competing for resources and sharing genetic material. Eventually, a type of cell arose that was the last shared ancestor of all life on Earth today.

These first cells lived in the absence of O_2 and probably used organic molecules as a source of both carbon and energy. Another source of carbon, however, was the CO_2 in the atmosphere. Photosynthetic organisms eventually evolved that could use light for energy and atmospheric CO_2 as a carbon source (see chapter 5). These cells no longer relied on organic compounds in their surroundings for food.

Photosynthesis probably originated in cells that used hydrogen sulfide (H_2S) instead of water as an electron donor. These first photosynthetic microorganisms would have released sulfur, rather than O_2, into the environment. Eventually, changes in pigment molecules enabled some of these organisms to use H_2O instead of H_2S as an electron donor. Cells using this new form of photosynthesis released O_2 as a waste product.

Figure 14.8 Stromatolites. Cyanobacteria build distinctive formations in Shark Bay, Western Australia.

Some of the oldest known fossils are from 3.7-billion-year-old rock in Warrawoona, Australia, and Swaziland, South Africa. The fossils strongly resemble large formations of cyanobacteria called stromatolites (figure 14.8). These ancient cyanobacteria, along with many others, would have changed the composition of Earth's atmosphere by consuming CO_2 and releasing O_2 over millions of years during the Archean and Proterozoic eons (see figure 5.2).

Extensive iron deposits dating to about 2 billion years ago provide evidence of the changing atmosphere. As O_2 from photosynthesis built up in the oceans, iron that was previously dissolved in seawater reacted with the O_2 and sank to the bottom of the sea, producing distinctive layers of iron-rich sediments.

The evolution of photosynthesis forever altered life on Earth. Photosynthetic organisms formed the base of new food chains. In addition, natural selection began to favor aerobic organisms that could use O_2 in metabolism, while anaerobic species would persist in pockets of the environment away from oxygen. Ozone (O_3) also formed from O_2 high in the atmosphere, blocking the sun's damaging ultraviolet radiation. The overall result was an explosion of new life that eventually gave rise to today's microbes, plants, fungi, and animals (figure 14.9).

14.1 | Mastering Concepts

1. How were conditions on Earth before life began different from current conditions?
2. What types of information can prebiotic simulations provide?
3. Why is RNA likely to have been pivotal in life's beginnings?
4. What is a progenote?
5. How might metabolic pathways have originated and evolved?
6. About when did the first cells probably originate?
7. How did early life change Earth?

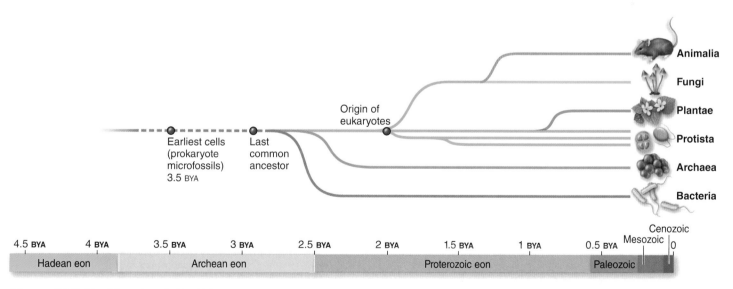

Figure 14.9 The Diversity of Life. This evolutionary tree shows the relationships among the two domains of prokaryotes (Bacteria and Archaea) and the eukaryotic protists, plants, fungi, and animals. Note that life has increased in complexity, but simpler organisms such as bacteria and archaea still flourish.

14.2 | Complex Cells and Multicellularity Arose over a Billion Years Ago

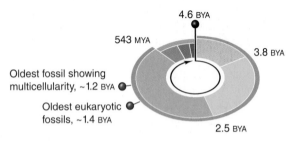

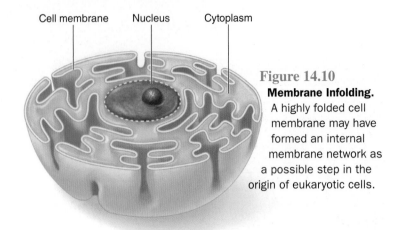

Figure 14.10
Membrane Infolding. A highly folded cell membrane may have formed an internal membrane network as a possible step in the origin of eukaryotic cells.

Until this point, we have considered the origin of prokaryotic cells. Fossil evidence shows that eukaryotic cells emerged during the Proterozoic era, at least 1.9 to 1.4 BYA. Australian fossils consisting of organic residue 1.69 billion years old are chemically similar to eukaryotic membrane components and may have come from a very early unicellular eukaryote.

Recall from chapter 3 that prokaryotic cells are structurally simple compared with compartmentalized eukaryotic cells. We may never know the origin of the nuclear envelope, endoplasmic reticulum, Golgi apparatus, and other membranes within the eukaryotic cell. The membranes of these organelles consist of phospholipids and proteins, as does the cell's outer membrane. Perhaps the outer membrane of an ancient cell repeatedly folded in on itself, eventually pinching off inside the cell to form a complex internal network of organelles (**figure 14.10**). Unfortunately, that hypothesis is difficult or impossible to test, so we can only speculate about that aspect of eukaryotic cell evolution.

Some details, however, are becoming clear. Section 3.7 describes the discovery of cytoskeleton-like proteins in bacterial

cells. In addition, the endosymbiont theory may explain the origin of two types of membrane-bounded organelles.

A. Endosymbiosis Explains the Origin of Mitochondria and Chloroplasts

The **endosymbiont theory** proposes that mitochondria and chloroplasts originated as free-living bacteria that were engulfed by other prokaryotic cells (**figure 14.11**). (This process would be similar to endocytosis, pictured in figure 4.21.) The term *endosymbiont* derives from *endo-*, meaning "inside," and *symbiont*, meaning "to live together." Biologist Lynn Margulis at the University of Massachusetts at Amherst proposed this theory in the late 1960s.

The evidence supporting the idea that mitochondria and chloroplasts originated as independent organisms includes:

- similarities in size, shape, and membrane structure between the organelles and some types of bacteria;
- the double membrane surrounding mitochondria and chloroplasts, a presumed relic of the original engulfing event;

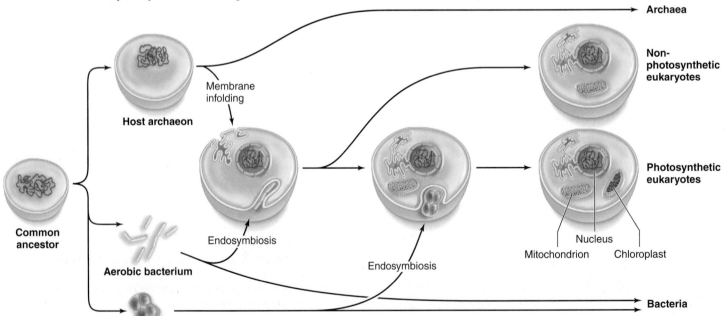

Figure 14.11 The Endosymbiont Theory. Mitochondria and chloroplasts may have originated from an ancient union of bacterial cells with archaean cells.

- the observation that mitochondria and chloroplasts are not assembled in cells but instead divide, as do bacterial cells;

- the similarity between the photosynthetic pigments in chloroplasts and those in cyanobacteria;

- the observation that mitochondria and chloroplasts contain DNA, RNA, and ribosomes, which are similar to those in bacterial cells; and

- DNA sequence analysis, which shows a close relationship between mitochondria and aerobic bacteria (proteobacteria), and between chloroplasts and cyanobacteria.

Mitochondria must have come first, because virtually all eukaryotes have these organelles. Chloroplasts came later, in the lineage that eventually gave rise to photosynthetic protists and plants.

After the ancient endosymbiosis events, many genes moved from the DNA of the organelles to the nuclei of the host cells. These genetic changes made the captured microorganisms unable to live on their own outside their hosts. Over time, they came to depend on one another for survival. The result of this biological interdependency, according to the endosymbiont theory, is the compartmentalized cells of modern eukaryotes.

Endosymbiosis has been a potent force in eukaryote evolution. In fact, the chloroplasts of some types of photosynthetic protists apparently derive from a secondary endosymbiosis—that is, of a eukaryotic cell engulfing a eukaryotic red or green alga (figure 14.12; see also figure 17.2). In these species, three or four membranes surround the chloroplasts; some of their cells even retain remnants of the engulfed cell's nucleus.

B. Multicellularity May Also Have Its Origin in Cooperation

Another critical step leading to the evolution of plants, fungi, and animals was the origin of multicellularity, which occurred about 1.2 BYA. The earliest fossils of multicellular life are from a red alga that lived about 1.25 BYA to 950 MYA (million years ago) in Canada (figure 14.13). Abundant fossil evidence of

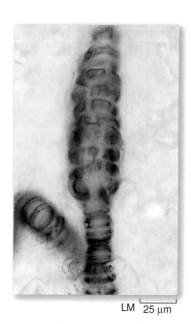

Figure 14.13 Early Multicellularity. This 1.2-billion-year-old fossil of *Bangiomorpha pubescens* is among the oldest evidence of multicellularity.

LM 25 μm

multicellular algae, dating from a billion years ago, also comes from eastern Russia.

Exactly how life developed from the single-celled to the many-celled is a mystery. Perhaps many individual cells came together, joined, and took on specialized tasks to form a multicellular organism. The life cycle of modern-day protists called slime molds illustrates how this may have occurred. Alternatively, a single-celled organism may have divided, and the daughter cells may have remained stuck together rather than separating. After many rounds of cell division, these cells may have begun expressing different subsets of their DNA. The result would be a multicellular organism with specialized cells—similar to the way in which modern animals and plants develop from a single fertilized egg cell. ▶ slime molds, p. 360

Whichever way it happened, the origin of multicellularity ushered in the possibility of specialized cells, which allowed for new features such as attachment to a surface or an upright orientation. The resulting explosion in the variety of body sizes and forms introduced new evolutionary possibilities and opened new habitats for other organisms. The rest of this chapter describes the diversification of multicellular life.

14.2 | Mastering Concepts

1. When did eukaryotic cells first appear?
2. How might the endoplasmic reticulum, nuclear envelope, and other internal membranes have arisen in eukaryotic cells?
3. What is the evidence that mitochondria and chloroplasts descend from simpler cells engulfed long ago?
4. When did the first multicellular organisms appear in the fossil record?
5. What are two ways that multicellular organisms may have originated?

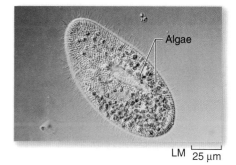

LM 25 μm

Figure 14.12 Secondary Endosymbiosis. This protist, *Paramecium bursaria*, has engulfed eukaryotic algae, which live symbiotically inside its cytoplasm. This dual organism, and many others in similar partnerships, provide evidence for the endosymbiosis theory.

14.3 | Life's Diversity Exploded in the Past 500 Million Years

It would take many thousands of pages to capture all of the events that passed from the rise of the first cells to life today—if we even knew them. This section highlights a few key events in the history of multicellular, eukaryotic life.

As you read through this section, keep in mind what you have already learned about evolution, speciation, and extinction. Species have come and gone in response to major environmental changes such as rising and falling temperatures, the increasing concentration of O_2 in the atmosphere, the shifting continents, and the advance and retreat of glaciers. Short-term catastrophes such as floods, volcanic eruptions, and meteorite impacts have also taken their toll on some species and created new opportunities for others. ▶ mass extinctions, p. 284

A. The Strange Ediacarans Flourished Late in the Precambrian

The "Precambrian" is an informal name for the eventful 4 billion years that preceded the Cambrian period of the Paleozoic era. During the Precambrian, life originated, reproduced, and diversified into many forms. In addition, photosynthesis evolved, O_2 accumulated in Earth's atmosphere, and eukaryotes arose, as did the first multicellular algae and animals.

Perhaps the most famous Precambrian residents were the mysterious Ediacaran organisms, which left no known modern descendants (**figure 14.14**). One example, *Dickinsonia,* could reach 1 meter in diameter but was less than 3 millimeters thick. Biologists have interpreted these fossils as everything from worms to ferns to fungi. The Ediacarans vanished from the fossil record about 544 MYA. (In 2004, geologists named the last portion of the Precambrian, from 543 to about 600 MYA, the Ediacaran period in honor of these strange creatures.)

B. Paleozoic Plants and Animals Emerged onto Land

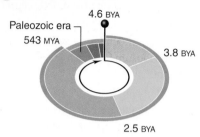

Cambrian Period (543 to 490 MYA)   Fossils of all major phyla of animals appeared within a few million years of one another in the Cambrian seas, a spectacular period of diversification sometimes called the "Cambrian explosion." During this time, remnants of the Ediacaran world coexisted with abundant red and green algae, sponges, jellyfishes, and worms. Most notable were the earliest known organisms with hard parts, such as insectlike trilobites, nautiloids, scorpion-like eurypterids, and brachiopods, which resembled clams. Many of these invertebrates left remnants identified only as "small, shelly fossils." The early Cambrian seas were also home to diverse wormlike, armored animals, some of which would die out.

The Burgess Shale from the Canadian province of British Columbia preserves a glimpse of life from this time. A mid-Cambrian mud slide buried enormous numbers of organisms, including animals with skeletons and soft-bodied invertebrates not seen elsewhere. The Burgess Shale animals were abundant, diverse, and preserved in exquisite detail (**figure 14.15**).

Ordovician Period (490 to 443 MYA)   During the Ordovician period, the seas continued to support huge communities of algae and invertebrates such as sponges, corals, snails, clams, and cephalopods. The first vertebrates to leave fossil evidence, jawless fishes called ostracoderms, appeared at this time.

Figure 14.14
Ediacarans Were . . . Different. (a) *Dickinsonia* was an Ediacaran organism with segments, two different ends, and internal features that paleontologists have interpreted as a simple circulatory or digestive system. But just what it was remains unclear. (b) No one knows what type of organism *Spriggina* was, either.

a.   1 cm

b.   0.5 cm

Fern or animal?

Figure 14.15 **Cambrian Life.** *Opabinia* is one of many strange animals whose fossils have been discovered in the Burgess Shale. Its body was segmented, and its head sported stalked eyes and a long, flexible snout.

Fossilized spores indicate that life had ventured onto land, in the form of primitive plants that may have resembled modern liverworts (see section 18.2).

The Ordovician period ended with a mass extinction that killed huge numbers of marine invertebrates. Apparently the supercontinent of Gondwana drifted toward the South Pole, causing temperatures to cool. Sea levels dropped as glaciers accumulated, destroying shoreline habitats.

Silurian Period (443 to 417 MYA) The first plants with specialized water- and mineral-conducting tissues, the vascular plants, evolved during the Silurian (figure 14.16). These plants were larger than their ancestors, so they provided additional food and shelter for animals. The first terrestrial animals to leave fossils resembled scorpions, which may have preyed upon other small animals exploring the land. Fungi likely colonized land at the same time.

Figure 14.16 **Plants Settle the Land.** *Cooksonia*, which lived during the Silurian, is one of the earliest known land plants.

Aquatic life also continued to change. Fishes with jaws arose, as did the first freshwater fishes, but their jawless counterparts were still widespread during the Silurian. The oceans also contained abundant corals, trilobites, and mollusks.

Devonian Period (417 to 354 MYA) The Devonian period was the "Age of Fishes." The seas continued to support more life than did the land. The now prevalent invertebrates were joined by fishes with skeletons of cartilage or bone. Corals and animals called crinoids that resembled flowers were abundant.

The fresh waters of the Devonian were home to the lobe-finned fishes (figure 14.17a). These animals had fleshy, powerful fins and could obtain O_2 through both gills and primitive lunglike structures. Toward the end of the Devonian period, about 360 MYA, the first amphibians appeared (figure 14.17b and c). *Acanthostega* had a fin on its tail like a fish and used its powerful tail to move underwater, but it also had hips, paddlelike legs, and toes. Preserved footprints indicate that the animal could venture briefly onto land. A contemporary of *Acanthostega,* called

a. *Eusthenopteron*

b. *Acanthostega*

c. *Ichthyostega*

Figure 14.17 **From Water to a Land.** (a) *Eusthenopteron* was a lobe-finned fish; the bones of its fins closely resemble those of a tetrapod's limbs. (b) *Acanthostega* stayed mostly in the water but had legs and other adaptations that permitted it to spend short periods on land. (c) *Ichthyostega* could spend longer periods on land because its rib cage was stronger. It retained the skull shape and finned tail of its ancestors.

Ichthyostega, had more powerful legs and a rib cage strong enough to support the animal's weight on land, yet it had a skull shape and finned tail reminiscent of fish ancestors.

Many fossils indicate that by this time, plants were diversifying to ferns, horsetails, and seed plants. Scorpions, millipedes, and other invertebrates lived on the land.

A mass extinction of marine life marks the end of the Devonian period. Warm-water invertebrates and jawless fishes were hit especially hard, and jawed fishes called placoderms became extinct. Life on land, however, was largely spared. The cause of this mass extinction remains unknown.

Carboniferous Period (354 to 290 MYA)

The amphibians flourished from about 350 to 300 MYA, giving this period the name the "Age of Amphibians." These animals spent time on land, but they had to return to the water to wet their skins and lay eggs. Although today's amphibians are small, some of their ancient relatives were huge—up to 9 meters long!

During the Carboniferous, some amphibians arose that coated their eggs with a hard shell. These animals branched from the other amphibians, eventually giving rise to reptiles. The first vertebrates capable of living totally on land, the primitive reptiles, appeared about 300 MYA.

The swamps of this time had fernlike plants and conifers, which towered to 40 meters (**figure 14.18**). The air was alive with the sounds of grasshoppers, crickets, and giant dragonflies with 75-centimeter wingspans. Land snails and other invertebrates flourished in the sediments. By the end of the period, many of the plants had died, buried beneath the swamps to form coal beds during the coming millennia. In fact, the term *Carboniferous* means "coal-bearing," in reference to those long-buried remains; see this chapter's Apply It Now box on page 311 for more on coal.

Meanwhile, in the oceans, the bony fishes and sharks were beginning to resemble modern forms, and protists called foraminiferans were abundant. Bryozoans and brachiopods were plentiful, but trilobites were becoming less common.

Permian Period (290 to 248 MYA)

During the Permian, seed plants called gymnosperms became more prominent. Reptiles were also becoming more prevalent. The reptile introduced a new adaptation, the amniote egg, in which an embryo could develop completely on dry land (see figure 20.38). Amniote eggs persist today in reptiles, birds, and a few mammals.

The Permian period foreshadowed the dawn of the dinosaur age. Cotylosaurs were early Permian reptiles that gave rise to the dinosaurs and all other reptiles. They coexisted with their immediate descendants, the pelycosaurs, or sailed lizards.

The Permian period ended with what paleontologists call "the mother of mass extinctions." It affected marine life the most, wiping out more than 90% of species in shallow areas of the sea. On the land, many types of insects, amphibians, and reptiles disappeared, paving the way for the age of dinosaurs.

Paleontologists hypothesize that the Permian extinctions were partly the result of a drop in sea level, which dried out coastline

Figure 14.18 Carboniferous Forest Life. (a) About 300 MYA, lush forests dominated the landscape. The fernlike plants in the foreground are ancient seed-bearing plants. Other plants and trees, also extinct, gave rise to modern club mosses and ground pines. (b) These forests were eventually preserved in massive coal beds containing the remains of Carboniferous plants, such as this fossilized fern frond.

a.

b.

communities. Carbon dioxide from oxidation of organic molecules accumulated in the atmosphere, raising global temperature and depleting the sea's dissolved O_2. The loss of O_2, in turn, may have favored bacteria that produce toxic hydrogen sulfide (H_2S) as a waste. A long series of volcanic eruptions, beginning 255 MYA and lasting a few million years, further altered global climate. Finally, sea level rose again, drowning coastline communities.

C. Reptiles and Flowering Plants Thrived During the Mesozoic Era

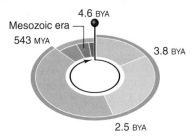

Figure 14.19 **Marine Reptile.** Plesiosaurs were enormous carnivorous reptiles that swam in the Jurassic seas.

Triassic Period (248 to 206 MYA)   During the Triassic period, small reptiles called thecodonts flourished. Thecodonts shared the forest of cycads, ginkgos, and conifers with other animals called therapsids, which were the ancestors of mammals.

At the close of the Triassic period, yet another mass extinction affected life in the oceans and on land. Many marine animals were wiped out, as were many reptiles and amphibians on land. As a result, much larger animals began to infiltrate a wide range of habitats. These new, well-adapted animals were the dinosaurs, and they would dominate for the next 120 million years.

Jurassic Period (206 to 144 MYA)   By the Jurassic period, giant reptiles were everywhere (figure 14.19). Ichthyosaurs, plesiosaurs, and giant marine crocodiles swam in the seas alongside sharks and rays, feasting on fish, squid, and ammonites. Apatosaurs and stegosaurs roamed the land. Carnivores, such as allosaurs, preyed on the herbivores. Pterosaurs glided through the air, as did *Protoarchaeopteryx* and then *Archaeopteryx*—the first birds (see chapter 12's opening essay).

tall ferns and conifers, ginkgos, club mosses, and horsetails. The first frogs and the first true mammals, which were no larger than rats, appeared as well.

Cretaceous Period (144 to 65 MYA)   The Cretaceous period was a time of great biological change. By around 100 MYA, flowering plants had spread in spectacular diversity; many modern insects arose at about the same time. Marine reptiles hunted mollusks and fish, and birds and pterosaurs roamed the skies. Duck-billed maiasaurs traveled in groups of thousands in what is now Montana. Huge herds of apatosaurs migrated from the plains of Alberta to the Arctic, northern Europe, and Asia, which were joined as one continent at the time. Near the end of the Cretaceous, *Tyrannosaurus* roamed in what is now western North America, and *Triceratops* was so widespread that some paleontologists call it the "cockroach of the Cretaceous."

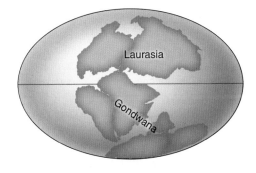

At the same time, the first flowering plants (angiosperms) appeared on land. The forests, however, still consisted largely of

The reign of the giant reptiles ended about 65 MYA, with the extinction of ichthyosaurs, plesiosaurs, mososaurs, pterosaurs, and nonavian dinosaurs. Ammonites vanished too, as did many types of foraminiferans, sea urchins, and bony fishes. In all, nearly 75% of species perished. The mass extinction opened up habitats for many species that survived, including flowering plants, mollusks, amphibians, some smaller reptiles, birds, and mammals. Many of these groups, including the mammals that gave rise to our own species, subsequently flourished.

The mass extinction that ended the Mesozoic era coincides with an asteroid impact near the Yucatán peninsula. The asteroid, which was about 10 kilometers in diameter, left a debris- and clay-filled crater offshore and a huge semicircle of sinkholes onshore (figure 13.14 shows the distinctive iridium layer, an important piece of evidence for the impact theory). Biologists estimate that photosynthesis was almost nonexistent for 3 years, as debris that was thrown into the sky circulated in the atmosphere, blocking sunlight. Plankton, which provide microscopic food for many larger marine dwellers, died as well, causing a devastating chain reaction.

Figure It Out

Suppose that a 100-meter track represents Earth's 4.6 billion-year history. How close to the end of the track would you mark the Permian and Cretaceous extinctions?

Answer: 5.4 m (Permian) and 1.4 m (Cretaceous)

D. Mammals Diversified During the Cenozoic Era

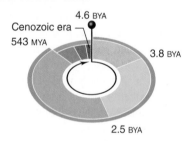

Tertiary Period (65 to 1.8 MYA)   The dawn of the Cenozoic era was a time of great adaptive radiation for mammals, according to the fossil record (see figure 13.12). Within just 1.6 million years during the Tertiary period, 15 of the 18 modern orders of placental mammals arose (**figure 14.20**). ▶ adaptive radiation, p. 282

At the start of the Tertiary period, diverse hoofed mammals grazed the grassy Americas. Many may have been marsupials (pouched mammals) or egg-laying monotremes, ancestors of the platypus. Then placental mammals appeared, and fossil evidence indicates that they rapidly dominated the mammals.

Geology and the resulting climate changes molded the comings and goings of species throughout the Cenozoic. The era began with the formation of new mountains and coastlines as tectonic plates shifted. The wet warmth of the Paleocene epoch, which opened up many habitats for mammals, continued into the

Figure 14.20 Old Bat. This fossilized bat (a placental mammal) dates from the Eocene epoch, tens of millions of years ago.

Eocene, providing widespread forests and woodlands. Grasslands began to replace the forests by the end of the Eocene, when the temperature and humidity dropped. Extinctions of some mammals paralleled the changing plant populations as the forests diminished, but grazing mammals thrived throughout the remaining three epochs of the Tertiary period.

Quaternary Period (1.8 MYA to Present)   The Pleistocene epoch accounts for all but the last 10,000 years of the Quaternary period. During several Pleistocene ice ages, huge glaciers covered about 30% of Earth's surface and then withdrew again.

Many organisms of the time were similar to those that are familiar now, including flowering plants, insects, birds, and mammals. Some Pleistocene species, however, are extinct (**figure 14.21**). The woolly mammoths, mastodons, saber-toothed cats, and other large mammals that once roamed North America, Asia, and Europe are known only from their fossils and the occasional DNA fragment (see section 18.6).

Our species, *Homo sapiens* ("the wise human"), probably first appeared about 200,000 years ago in Africa and had migrated throughout most of the world by about 10,000 years

Figure 14.21 **American Mastodon.** These enormous mammals lived from about 3.7 million years ago until they went extinct about 10,000 years ago.

ago. Other species of *Homo*, including Neandertals, vanished during the Pleistocene, leaving *Homo sapiens* as the sole human species.

But the history of our species began long before that time, some 60 MYA, when the order Primates arose. We pick up the story of human evolution in more detail in section 14.4.

14.3 | Mastering Concepts

1. When did the Ediacarans live, and what were they like?
2. What types of organisms flourished in the Cambrian?
3. How did Paleozoic life diversify during the Ordovician, Silurian, Devonian, Carboniferous, and Permian periods?
4. How did the Paleozoic era end?
5. Which organisms came and went during the Mesozoic era?
6. How did the Mesozoic era end?
7. Which new organisms arose during the Cenozoic era?

Apply It Now

Coal's Costs

The relationship between ancient plants and today's electronic gadgetry may seem remote, but it is not. Gadgets require electricity, and nearly half of the electricity in the United States comes from coal-fired power plants. Each lump of coal, in turn, comes from the long-dead remains of the Carboniferous forests.

The ferns and other plants in these forests produced their own food by photosynthesis. That is, they used water, energy from the sun, and CO_2 from the atmosphere to build the sugars and complex molecules that made up their bodies. Many of these leaves and stems did not decompose when the plants died; instead, they were buried under sediments. Heat and pressure eventually transformed the plant matter into coal. When we burn the coal, we release the potential energy and CO_2 that ancient plants trapped in photosynthesis hundreds of millions of years ago.

Burning coal to generate electricity is relatively inexpensive, compared to other energy sources. Coal powered the Industrial Revolution and continues to be economically important today. This fuel does not, however, come without costs.

First, coal mining has environmental and health consequences. Extracting coal from underground deposits can destroy soil, plants, and wildlife habitat. Water quality can suffer as exposed mountainsides erode away. And although coal mining in developed countries is safer than in times past, miners still risk injury and death in mine collapses and explosions.

Second, burning coal releases CO_2 into the atmosphere. This heat-trapping gas plays a large role in global climate change. Coal combus-

tion also releases potentially harmful heavy metals such as mercury, which can accumulate in body tissues. Moreover, nitrogen- and sulfur-rich gases from coal-fired power plants contribute to acid deposition. Chapter 39 describes these environmental issues in more detail.

Feeding the human population's voracious demand for energy requires enormous amounts of coal and other fossil fuels. As you consider the costs of coal, keep in mind that no energy source is free, not even renewable ones such as sunlight and wind; all require equipment, roads, power lines, and maintenance. The best way to avoid these costs is to reduce the overall demand for energy.

14.4 | Fossils and DNA Tell the Human Evolution Story

In many ways, humans are Earth's dominant species. We are not the most numerous—more microbes occupy one person's intestinal tract than there are people on Earth. In the short time of human existence, however, we have colonized most continents, altered Earth's surface, eliminated many species, and changed many others to fit our needs. Where did we come from?

A. Humans Are Primates

If you watch the monkeys or apes in a zoo for a few minutes, it is almost impossible to ignore how similar they seem to humans. Young ones scramble about and play. They sniff and handle food. Babies cling to their mothers. Adults gather in small groups or sit quietly, snoozing or staring into space.

It is no surprise that we see ourselves reflected in the behaviors of monkeys and apes. All **primates**—including monkeys, apes, and humans—share a suite of physical characteristics (**figure 14.22**). First, primates have grasping hands with opposable thumbs that can bend inward to touch the pads of the fingers. Some primates also have grasping feet with opposable big toes. Second, a primate's fingers and toes have flat nails instead of claws. Third, eyes set in the front of the skull give primates binocular vision with overlapping fields of sight that produce excellent depth perception. Fourth, the primate brain is large by comparison with body size.

Compared with many other groups of mammals, primate anatomy is unusually versatile. For instance, bat wings are useful for flight but not much else; likewise, horse hooves are best for fast running. In contrast, primates have multipurpose fingers and

Table 14.1   Miniglossary of Primate Terminology

Term	Animals Included in Group*
Primates	Prosimians, monkeys (simians), and apes
Hominoids	"Lesser apes" (gibbons) and great apes
Hominids	"Great apes" (orangutans, gorillas, chimpanzees, humans)
Hominines	Gorillas, chimpanzees, humans
Hominins	Extinct and modern humans

*Each group also contains extinct representatives known only from fossils.

toes that are useful not only for locomotion but also for grasping and manipulating small objects. Primate limbs are similarly versatile.

The Primate Lineage   The primate lineage contains three main groups: prosimians, simians, and hominoids. **Prosimian** is an informal umbrella term for lemurs, aye-ayes, lorises, tarsiers, and bush babies. **Simian** is the corresponding term for monkeys, both Old World (native to Africa and Asia) and New World (native to South and Central America). The **hominoids** are apes, including humans.

Hominoids are further divided into two groups. One contains the gibbons, or "lesser apes." The other, the **hominids,** contains all of the "great apes": orangutans, gorillas, chimpanzees (including bonobos), and humans. Orangutans, however, are not as closely related to the other great apes. **Hominines** include only gorillas, chimpanzees, and humans, and **hominins** are extinct

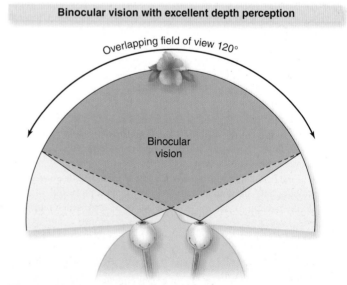

Flat nails and grasping hands with opposable thumbs

Opposable thumb   Flat fingernails

Tarsier   Gorilla   *Homo sapiens*

Large brains

Human brain   Chimpanzee brain

Binocular vision with excellent depth perception

Overlapping field of view 120°

Binocular vision

Figure 14.22 Primate Characteristics. Primates share several physical characteristics, including opposable thumbs, excellent depth perception, and large brains relative to body size.

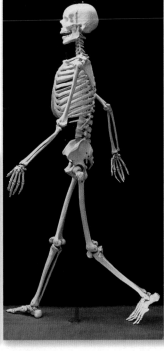

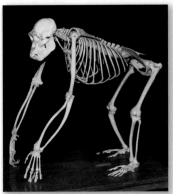

Figure 14.23 Clues from Bones. The skeleton on the left, from a chimpanzee, shares many similarities with the human skeleton on the right.

and modern humans. Table 14.1 summarizes the groups of primates.

Of the many sources of information indicating how these groups are related, paleontologists have used one the longest: the physical characteristics of skeletons (figure 14.23). Most human fossils consist of bones and teeth. Comparing these remains with existing primates reveals surprisingly detailed information about locomotion and diet. For this reason, knowledge of primate skeletal anatomy is essential to interpreting human fossils.

Primate Locomotion Among the most important characteristics in hominoid skeletons are adaptations related to locomotion. Brachiation is swinging from one arm to the other while the body dangles below. Many hominoids move through the treetops in this way; in contrast, monkeys run on all fours along the tops of branches. Orangutans spend most of their lives in trees and move by brachiation when they are in treetops. Gibbons, the most superbly acrobatic hominoids, have long arms and hands. The size and opposability of the thumb are reduced, but their arms connect to the shoulders by ball-and-socket joints that allow free movement of the arms in 360 degrees. In addition, a long collarbone acts as a brace and keeps the shoulder from collapsing toward the chest.

Heavier-bodied chimpanzees and gorillas don't brachiate as much as gibbons and orangutans, but like humans, they can do so. Humans seldom brachiate, with the exception of small, light-bodied children playing on schoolyard "monkey bars." Adult human arms are too weak to support the heavy torso and legs.

Chimpanzees and gorillas move by knuckle-walking, a behavioral modification that allows an animal to run rapidly on the ground on all fours, with their weight resting on the knuckles.

The proportionately longer arms of chimps and gorillas are an adaptation to knuckle-walking.

One important feature distinguishes humans from the other great apes: bipedalism, or the ability to walk upright on two legs. Adaptations to bipedalism include relatively short arms and longer, stronger leg bones. Foot bones form firm supports for walking, with the big toe fixed in place and not opposable. The bowl-shaped pelvis supports most of the weight of the body, and lumbar vertebrae are robust enough to bear some body weight.

Bipedalism is also reflected in the bones of the head. The foramen magnum is the large hole in the skull where the spinal cord leaves the brain. In modern humans, this hole is tucked beneath the skull. In gorillas and chimps, the foramen magnum is located somewhat closer to the rear of the skull; in animals that run on all fours, such as horses and dogs, the hole is at the back of the skull.

Dietary Adaptations in Primates Other skeletal characteristics, including the size and shape of the teeth, are related to diet (figure 14.24). Upper and lower molar teeth have ridges that fit together, much as the teeth of gears intermesh. Food caught between these surfaces is ground, crushed, and mashed. The size of these teeth is an adaptation that reflects the toughness of the diet.

As you examine the skulls in figure 14.24, notice the differences in the sagittal crests. This bony ridge runs lengthwise along the top of the skull and is an attachment point for muscles. A prominent sagittal crest indicates particularly strong jaws, another clue to an animal's diet.

Other important features in primate skulls include the size of the jaw bones, the prominence of the ridge of bone above the eye, the degree to which the jaw protrudes, and the shape of the curve of the tooth row. All of these characteristics allow paleoanthropologists, the scientists who study human fossils, to identify hominid species.

B. Molecular Evidence Documents Primate Relationships

Fossil evidence and anatomical similarities were once the only lines of evidence that paleoanthropologists could analyze in tracing the course of human evolution. Around 1960, however, scientists began to use molecular sequences to investigate relationships among primates. Studies of blood proteins and DNA presented a new picture of primate evolution, as it became clear that humans are a species of great ape. One of the astounding findings of these molecular studies was that the genes of humans and chimpanzees are 99% identical.

Other research further eroded the distinctions between humans and other great apes. Previously, humans had been placed in a separate group, supposedly characterized by upright walking, tool-making, and language. Then, in the 1970s, chimpanzees and wild gorillas were observed to use tools, and captive great apes learned to use sign language to communicate with their trainers. The only characteristic that now remains unique to *Homo* is bipedal locomotion.

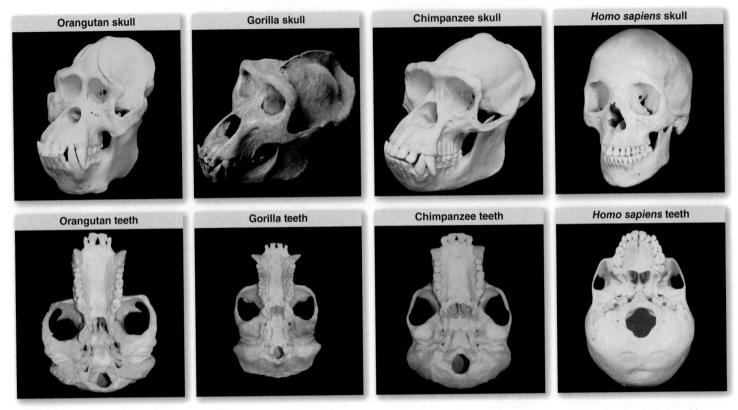

Figure 14.24 Skulls and Teeth. The skulls and teeth of an orangutan, gorilla, chimpanzee, and human reveal details about diet and jaw strength. Biologists compare fossils of extinct species to bones of existing primates to learn how our ancestors lived. The top row of photos shows side views; the bottom row shows skulls from underneath, revealing the hole through which the spinal cord enters the brain.

As the new molecular evidence continued to pour in, scientists began to view primate relationships differently. Figure 14.25 shows an evolutionary tree that takes into account both the anatomical characteristics and the molecular data. As you examine this cladogram, keep in mind two important concepts. First, note that humans are not descended from other groups of modern apes. Instead, all living humans and chimpanzees share a common ancestor and diverged from that ancestor perhaps 7 MYA. Second, note that gibbons, orangutans, gorillas, and chimpanzees are not "less evolved" than humans. All living species are on an equal evolutionary footing, although some may belong to older lineages.

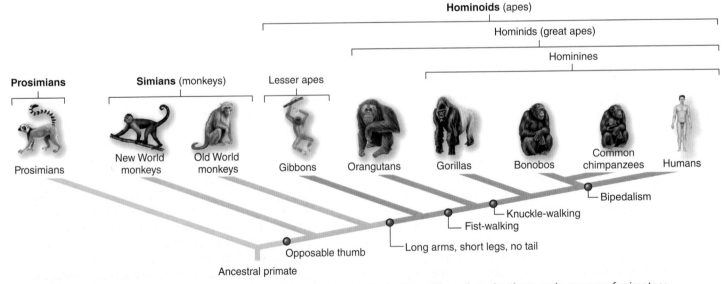

Figure 14.25 Primate Lineages. This cladogram shows the physical traits that differentiate the three main groups of primates: prosimians, simians, and hominoids. Molecular data support this hypothesis of the evolutionary relationships among primates.

C. Hominin Evolution Is Partially Recorded in Fossils

Even though DNA and proteins provide overwhelming evidence of the relationships between living primates, these molecules deteriorate with time. Scientists therefore cannot usually use molecular data to establish relationships of prehistoric hominins. For this, we must turn to studies of fossilized remains.

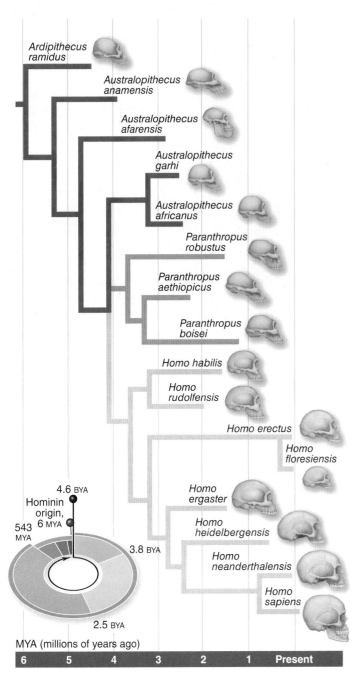

MYA (millions of years ago)

| 6 | 5 | 4 | 3 | 2 | 1 | Present |

Figure 14.26 Human Family Tree. Fossilized remains place our close relatives into three main groups (australopiths, *Paranthropus*, and *Homo*), as indicated by the branch colors. Skull sizes are approximately to scale and range from about 400 cm³ for *A. afarensis* to about 1450 cm³ for modern humans. The evolutionary relationships in this tree are hypothesized.

To interpret fossils from the human family tree, paleoanthropologists compare details of the ancient skeletal features with modern primates and try to reconstruct as much information about diet and lifestyle as they can. So far, fossil hominins in the human family tree have fallen into the following groups (**figure 14.26**):

- ***Ardipithecus.*** In 2009, researchers revealed a detailed analysis of *Ardipithecus ramidus,* a human ancestor discovered in 1994 in Ethiopia. *Ardipithecus* (or "Ardi" for short) dates to about 4.4 MYA and is the oldest representative of the human lineage discovered to date. The 1994 discovery, coupled with dozens of other fossils from the same species, gives several clues about Ardi's life. The teeth indicate that Ardi was an omnivore. The pelvis supported both upright walking and powerful climbing, as did the feet, which had opposable big toes. The flexible hands had long, grasping fingers with which Ardi could have carried objects while walking upright.

- **Australopiths.** Fossils of four or five species of extinct small apes have been assigned to the genus *Australopithecus,* meaning "Southern ape-man." The downward position of the foramen magnum indicates that these apes walked upright. A trail of fossilized footprints is additional evidence of their hominin heritage (**figure 14.27**). *Australopithecus afarensis* (including the famous "Lucy" fossil in **figure 14.28**) and *A. africanus* are members of this group, which dates from about 4 to 2.5 MYA.

- ***Paranthropus.*** This extinct group, whose name literally means "beside humans," is characterized by extremely large teeth, protruding jaws, and skulls that have a sagittal crest. All of these specializations probably relate to the large jaw muscles needed to crush tough plants or crack nuts. Most researchers hypothesize that *Paranthropus* descended from *Australopithecus. Paranthropus aethiopicus, P. boisei,* and *P. robustus* are members of this group, which dates from about 3 to 1.5 MYA. *Paranthropus* seems to be an evolutionary dead end that gave rise to no other group.

- ***Homo.*** Fossils in this group are associated with stones thought to have been tools. *Homo* species tend to have larger bodies and larger brains than do australopiths. All members of genus *Homo* are considered humans, and *Homo habilis, H. ergaster,* and *H. erectus* belong to the cluster of extinct species that are called "early *Homo.*" These species lived from about 2.5 MYA to about 1 MYA and gave rise to "recent *Homo.*" Recent species of *Homo* have smaller teeth, lighter jaws, larger braincases, less protruding jaws, and lighter brow ridges. Their fossils are associated with evidence of culture. Recent *Homo* species are *H. heidelbergensis, H. neanderthalensis, H. floresiensis* (the hobbit-like humans whose bones were discovered in Indonesia in 2003), and *H. sapiens.* The only human species alive today is *Homo sapiens.*

Figure 14.27 Fossil Footprints. The so-called Laetoli footprints, which date to about 3.7 MYA, are preserved in volcanic ash. They were discovered in Tanzania in 1978.

Figure 14.28 Early Human. This reconstruction of the *Australopithecus afarensis* specimen called "Lucy" was made from the casting of bones discovered in 1974.

One interesting trend in human evolution has been an adaptive radiation of species, followed by extinctions. Fossil evidence shows that about 1.8 MYA, as many as five species of hominins lived together in Africa. About 200,000 years ago, three species of recent *Homo* coexisted in Europe. Today, however, all except *Homo sapiens* are extinct.

What happened to the other *Homo* species? No one knows, but many anthropologists wonder whether *H. sapiens* contributed to their extinction. Scientists had speculated that Neandertals disappeared after interbreeding with *H. sapiens,* a hypothesis called "extinction through absorption." But an analysis of Neandertal DNA has led most scientists to reject that hypothesis.

D. Environmental Changes Have Spurred Hominin Evolution

What provoked the hominid ancestors of humans to abandon brachiation in favor of bipedal, upright walking? What allowed the large brains that are characteristic of recent *Homo* to develop? To find these answers, we have to consider a related question: Where did hominids evolve?

Charles Darwin was one of the first to speculate that humans evolved in Africa. About 12 MYA, tectonic movements caused a period of great mountain building. The continental plates beneath India and the Himalayan region collided and ground together, heaving up the Himalayas. The resulting climatic shift had enormous ecological consequences. Cooler temperatures reduced the thick tropical forests that had covered much of Europe, India, the Middle East, and East Africa. Open plains appeared, bringing new opportunities for species that could live there. These included less competition for food in the treetops, a new assortment of foods, and a different group of predators. Experts speculate that one type of small ape moved out of the trees and began life on the African savannas.

Perhaps at first this species alternated between running on all fours and bipedal walking. On open plains, however, there are advantages to bipedal walking, especially the elevated vantage point for sensing danger and spotting food and friends. This environment would have selected for apes with the best skeletal adaptations to bipedalism, and the trait would have been preserved and honed in the plains. Bipedalism also freed hominin hands to carry objects and use the tools that are so characteristic of *Homo* species.

No one knows what might have spurred the evolution of the large brain that characterizes humans. Some experts relate the development of a large brain to tool use; others relate it to life in social groups and language.

E. Migration and Culture Have Changed *Homo sapiens*

Mitochondrial DNA sampled from people around the world has revealed a compelling portrait of human migration out of Africa (**figure 14.29**). Asia, Australia, and Europe all were colonized at least 40,000 years ago, but it took somewhat longer for humans to reach the Americas.

As humans spread throughout the world, new habitats selected for different adaptations. Near the equator, for example, sunlight is much more intense than at higher latitudes. One component of sunlight is ultraviolet (UV) radiation, which is both harmful and beneficial. On the one hand, UV radiation damages DNA and causes skin cancer. On the other hand, some UV wavelengths help the skin produce vitamin D, which is essential to bone development and overall health.

These two counteracting selective pressures help explain why skin pigmentation is strongly correlated with the amount of ultraviolet radiation striking the Earth (see section 24.6). One pigment that contributes to skin color, melanin, blocks UV radiation. Intense UV radiation selects for alleles that confer abundant melanin. People whose ancestry is near the equator, such as in Africa and Australia, therefore tend to have very dark brown skin. In northern Europe and other areas with weak sunlight, however, heavily pigmented skin would block so much UV light that indigenous people would suffer from vitamin D deficiency. Northern latitudes are therefore correlated with a high frequency of alleles conferring pale, pinkish skin.

No matter where people roamed, one byproduct of the large human brain was **culture**: the knowledge, beliefs, and behaviors that we transmit from generation to generation. Among the earliest signs of culture are cave art from about 14,000 years ago, which indicates that our ancestors had developed fine hand coordination and could use symbols.

By 10,000 years ago, depending on which native plants and animals were available for early farmers to domesticate, agriculture began to replace a hunter–gatherer lifestyle in many places. Agriculture meant increased food production, which profoundly changed societies. Freed from the necessity of producing their own food, specialized groups of political leaders, soldiers, weapon-makers, religious leaders, scientists, engineers, artists, writers, and many other types of workers arose. These new occupations meant improved transportation and communication, better technologies, and the ability to explore the world for new lands and new resources.

Undoubtedly, humans are a special species. We can modify the environment much more than, for example, a slime mold or an earthworm can. We can also alter natural selection, in our own species and in others. Finally, culture allows each generation to build on information accumulated in the past. The knowledge, beliefs, and behaviors that shape each culture are constantly modified within a person's lifetime, in stark contrast to the millions of years required for biological evolution. Our species is therefore extremely responsive to short-term changes.

Despite our unique set of features, however, we are a species, descended from ancestors with which we share many characteristics. It is intriguing to think about where the human species is headed, which species will vanish, and how life will continue to diversify in the next 500 million years.

14.4 | Mastering Concepts

1. Name and describe the three groups of contemporary primates. To which group do humans belong?
2. What can skeletal anatomy and DNA sequences in existing primates tell us about the study of human evolution?
3. What are the three groups of hominins in the human family tree, and which still exist today?
4. Which conditions may have contributed to the evolution of humans?

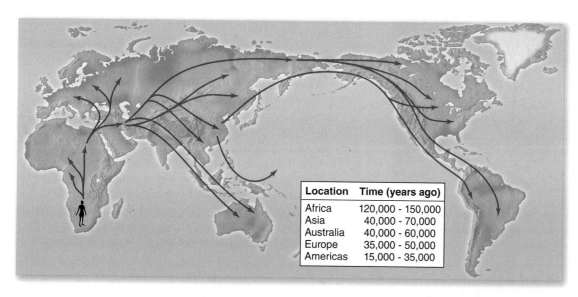

Location	Time (years ago)
Africa	120,000 - 150,000
Asia	40,000 - 70,000
Australia	40,000 - 60,000
Europe	35,000 - 50,000
Americas	15,000 - 35,000

Figure 14.29 Out of Africa. Researchers used mitochondrial DNA sequences to deduce approximately when humans originally settled each continent after migrating out of Africa.

14.5 | Investigating Life: What Makes Us Human?

Perhaps no scientific issue is more tantalizing and entangled with philosophy than the question of what makes us human. One way to look for answers is to study the similarities and differences between humans and chimpanzees, our closest living relatives on the evolutionary tree (**figure 14.30**). A team of scientists has sequenced the 3 billion or so DNA nucleotides that make up the chimpanzee genome and has begun comparing it with our own.

The story begins with the Chimpanzee Sequencing and Analysis Consortium, a group of 67 researchers in the United States, Europe, and Israel who collaborated to determine the genetic sequence of one chimpanzee *(Pan troglodytes)*. The subject was a captive male chimp named Clint, who lived at a primate research center in Atlanta until he died of heart failure in 2004. (For comparison, the Human Genome Project, completed in 2003, used DNA from many anonymous donors.)

The scientists used the "whole-genome shotgun approach" to sequence the genome. They first isolated DNA from Clint's blood cells, then broke the genetic material into many small fragments. Next, they inserted each fragment into a separate plasmid, which is a small ring of DNA (see figure 7.26). Each plasmid that carried chimp DNA was placed into a different bacterial cell. The researchers allowed the bacteria to replicate on culture plates, producing countless copies of the plasmid. Then, when it was a fragment's "turn" to be processed, a technician simply retrieved a sample of the bacteria, extracted the plasmid, and determined the sequence of A, C, G, and T (see section 8.8A). Finally, the

scientists used powerful computers to assemble the sequences from tens of millions of fragments. ▶ DNA sequencing, p. 170

Long before this project began, the startling genetic similarities between humans and chimps were well known. The chromosomes of the two species, for example, are extremely similar (**figure 14.31**). Although chimpanzees do have one more pair of chromosomes than do humans, a close look at figure 14.31 reveals why: our chromosome 2 formed when two smaller chromosomes fused, an event that occurred after the human and chimpanzee lineages split.

The complete DNA sequences for both species, however, reveal exactly how much we have in common: The two genomes are 96% alike, with the differences concentrated in the noncoding regions. The coding regions (the sequences that specify proteins) are 99% alike.

Although sequencing the chimp and human genomes is a noteworthy accomplishment in itself, the work has just begun. Scientists must still scrutinize both genomes, not only to identify the 25,000 or so coding regions and their regulatory sequences but also to determine which sequences correspond to previously discovered genes. They must also sequence the genomes of additional individuals to locate the variable regions, and they must determine which gene variants (alleles) contribute to which traits.

It all represents an enormous investment of time, money, and energy. Biologists consider the effort worthwhile, however, because it will reveal an unprecedented view of human biology and evolution. Here is a sampling of questions we may soon be able to answer:

- **Which genes define humans?** Previously, scientists could only compare the human genome with those of bacteria such as *E. coli*, plants such as *Arabidopsis*, and animals such as nematodes *(Caenorhabditis elegans)*, fruit flies *(Drosophila melanogaster)*, and mice *(Mus musculus)*. Those comparisons gave important insights into the traits that are common to all animals, all eukaryotes, or all cells. With the chimp genome complete, we can finally determine precisely how human DNA *differs* from that of our closest relative. To answer this question, scientists are searching the two genomes for regions that have been duplicated, inserted, deleted, and otherwise changed since humans and chimps last shared a common ancestor.

- **What accounts for bipedalism, large brains, complex language, and other uniquely human features?** Considering how genetically similar humans and chimps are, we have strikingly different phenotypes. At least two hypotheses could explain this curious observation. Some scientists suggest that human and chimp proteins are essentially the same but that we turn the genes encoding those proteins on and off at different times. For example, section 7.11 illustrates how a small change in just one transcription factor can affect a person's ability to use language. A competing hypothesis is that after humans and chimps

Figure 14.30 Close Relatives. Humans share 99% of our protein-encoding genes with chimpanzees. What are the differences that make us human?

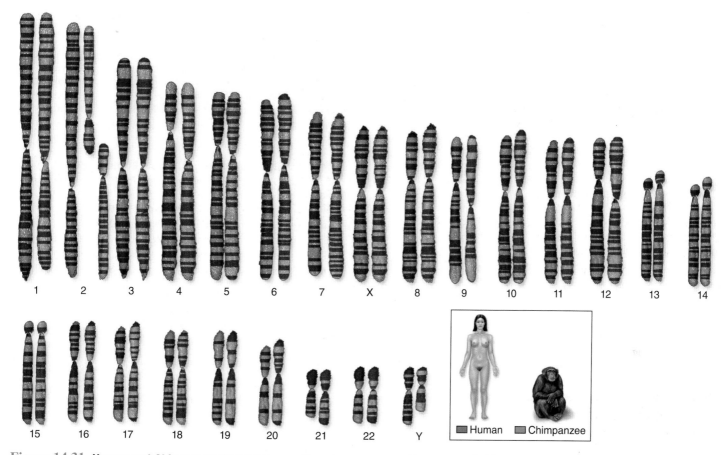

1 2 3 4 5 6 7 X 8 9 10 11 12 13 14

15 16 17 18 19 20 21 22 Y

■ Human ■ Chimpanzee

Figure 14.31 Human and Chimpanzee Chromosomes. The chromosomes of humans and chimpanzees look virtually identical, and molecular studies have indicated extremely high sequence similarity between the two species.

diverged, changes in the human lineage caused some previously functional genes to stop working. Such "degenerate" genes may make us less muscular and less hairy than chimpanzees. Comparing the chimp and human genotypes will help biologists test these hypotheses and yield insights into human evolution.

- **Why do humans and chimps have different diseases?** Our two species are very closely related, so it is reasonable to suppose that we suffer from the same illnesses. Yet the number of differences is surprising. For example, humans are susceptible to Alzheimer disease and carcinomas (cancers of the epithelium), but chimps are not. HIV progresses to AIDS in humans but not in chimps. At least some of the differences will certainly lie in our genes, and their discovery may yield new disease cures.

Whatever the source of our humanity, it is revealed partly in our ability to make reasoned decisions and in our compassion for others. Chimpanzees are endangered in the wild, where they succumb to habitat loss, hunting, the pet trade, and biomedical research. The Chimpanzee Sequencing and Analysis Consortium's paper advocates the protection of chimpanzees in the wild, and it ends with this statement: "We hope that elaborating how few differences separate our species will broaden recognition of our duty to these extraordinary primates that stand as our siblings in the family of life."

Chimpanzee Sequencing and Analysis Consortium. September 1, 2005. Initial sequence of the chimpanzee genome and comparison with the human genome. *Nature*, vol. 437, pages 69–87.

Additional reference: Olson, Maynard V. and Ajit Varki. January 2003. Sequencing the chimpanzee genome: insights into human evolution and disease. *Nature Reviews Genetics*, vol. 4, pages 20–28.

14.5 | Mastering Concepts

1. What information can researchers gain by comparing the human and chimpanzee genome sequences?

2. Comparing the chimpanzee and human genomes does not reveal which species has the "ancestral" form of each gene (the version present in the last common ancestor of humans and chimpanzees). How could genome sequences from orangutans or gorillas help scientists resolve which gene variants are ancestral?

Chapter Summary

14.1 | Life's Origin Remains Mysterious

- The solar system formed about 4.6 BYA, and life first left evidence on Earth by about 3.7 BYA. The **geologic timescale** describes these and many other events in life's history.

A. The First Organic Molecules May Have Formed in a Chemical "Soup"

- **Prebiotic simulations** combine simple inorganic chemicals to form life's organic building blocks, including amino acids and nucleotides.
- These monomers may have linked together to form polymers on hot clay or mineral surfaces.

B. Some Investigators Suggest an "RNA World"

- The **RNA world** theory proposes that RNA preceded formation of the first cells. Proteins provided enzymes and structural features. Reverse transcriptase could have copied RNA's information into DNA.

C. Membranes Enclosed the Molecules

- Phospholipid sheets that formed bubbles around proteins and nucleic acids may have formed cell precursors, or **progenotes.**

D. The Origin of Metabolism Would Have Involved Early Enzymes

- Metabolic pathways may have originated when progenotes mutated in ways that enabled them to use alternative or additional nutrients.

E. Early Life Changed Earth Forever

- Early organisms permanently changed the physical and chemical conditions in which life continued to evolve.

14.2 | Complex Cells and Multicellularity Arose over a Billion Years Ago

- The internal membranes of eukaryotic cells may have formed when the outer membrane folded in on itself repeatedly.

A. Endosymbiosis Explains the Origin of Mitochondria and Chloroplasts

- The **endosymbiont theory** proposes that chloroplasts and mitochondria originated as free-living bacteria that were engulfed by larger cells

B. Multicellularity May Also Have Its Origin in Cooperation

- The evolution of multicellularity, which occurred about 1.2 BYA, is poorly understood.

14.3 | Life's Diversity Exploded in the Past 500 Million Years

A. The Strange Ediacarans Flourished Late in the Precambrian

- The Ediacarans were soft, flat organisms that were completely unlike modern species. They lived during the late Precambrian and early Cambrian periods.

B. Paleozoic Plants and Animals Emerged onto Land

- The Cambrian explosion introduced many species, notably those with hard parts. Amphibian-like animals ventured onto land about 360 MYA, followed by reptiles, birds, and mammals. Invertebrates, ferns, and forests flourished.

C. Reptiles and Flowering Plants Thrived During the Mesozoic Era

- Dinosaurs prevailed throughout the Mesozoic era, when forests were largely cycads, ginkgos, and conifers. In the middle of the era, flowering plants became prevalent. When the nonavian dinosaurs died out 65 MYA, resources opened up for mammals.

D. Mammals Diversified During the Cenozoic Era

- Mammals diversified during the Tertiary period. Humans arose during the Pleistocene epoch of the Quaternary period. Repeated ice ages and the extinction of many large mammals also occurred during the Pleistocene.

14.4 | Fossils and DNA Tell the Human Evolution Story

A. Humans Are Primates

- **Primates** have grasping hands, opposable thumbs, binocular vision, large brains, and flat nails. The three groups of primates are **prosimians, simians,** and **hominoids** (apes).
- **Hominids** are the "great apes," whereas **hominines** are gorillas, chimpanzees, and humans. **Hominins** are extinct and modern humans.
- Fossil bones and teeth reveal how extinct species moved and what they ate.

B. Molecular Evidence Documents Primate Relationships

- Protein and DNA analysis has altered how scientists draw the human family tree.

C. Hominin Evolution Is Partially Recorded in Fossils

- Three groups of hominins are australopiths, *Paranthropus,* and *Homo.*

D. Environmental Changes Have Spurred Hominin Evolution

- Millions of years ago, new mountain ranges arose, causing climate shifts. Savannas replaced tropical forests, and apes—the ancestors of humans—moved from the trees to the savanna.

E. Migration and Culture Have Changed *Homo sapiens*

- After migrating out of Africa, humans encountered new habitats that selected for new allele combinations.
- Humans owe our success to language and **culture.**

14.5 | Investigating Life: What Makes Us Human?

- Scientists completed the chimpanzee genome project in 2005. Comparing the chimp and human genomes will lead to unprecedented insight into the genes, phenotypes, and diseases that are unique to humans.

Multiple Choice Questions

1. The first important step leading to life on Earth was most likely the formation of
   a. membrane-enclosed structures.
   b. O_2 gas.
   c. organic molecules.
   d. CO_2 gas.

2. What must be true for natural selection to occur in an "RNA world"?
   a. RNA must encode proteins.
   b. RNA molecules must undergo mutations.
   c. RNA molecules must replicate.
   d Both b and c are correct.

3. Photosynthetic cells affected early Earth by
   a. adding O_2 to the atmosphere.
   b. increasing the amount of hydrogen sulfide in the early oceans.
   c. depleting the ozone layer.
   d. changing the pH of the early oceans.

4. The endosymbiont theory explains the origin of
   a. cellular life.
   b. the nuclear membrane in eukaryotes.
   c. mitochondria and chloroplasts in eukaryotes.
   d. the plasma membrane of prokaryotes.

5. Which of the following provides the strongest support for the idea that mitochondria were once independent organisms?
   a. Their size is similar to that of prokaryotes.
   b. They are surrounded by a membrane.
   c. They are shaped like a prokaryote.
   d. They have their own DNA and ribosomes.

6. How have mass extinctions contributed to the evolution of life on Earth?
   a. Extinctions created new selective pressures.
   b. Extinctions eliminated only the abundant species.
   c. Extinctions allowed for adaptive radiation.
   d. Both a and c are correct.

7. Which of the following represents the correct order of appearance, from earliest to most recent?
   a. Fishes, reptiles, Ediacarans, flowering plants
   b. Ediacarans, fishes, reptiles, flowering plants
   c. Ediacarans, flowering plants, fishes, reptiles
   d. Flowering plants, reptiles, fishes, Ediacarans

8. Primates share all of the following characteristics except
   a. opposable thumbs.   c. bipedalism.
   b. excellent depth perception.   d. flat fingernails.

9. Evidence suggests that the closest relatives to humans are
   a. monkeys.   c. gorillas.
   b. orangutans.   d. chimpanzees.

10. Which group of hominines is the oldest?
   a. Australopiths   c. *Homo*
   b. *Paranthropus*   d. Simians

10. What can scientists learn by comparing the fossilized skeletons of extinct primates with the bones of modern species?

11. The foramen magnum in *Australopithecus africanus* is closer to the front of the skull than in gorillas. What does this observation indicate about *A. africanus*?

12. Use the Internet to learn about *National Geographic*'s Genographic Project. What are the main objectives and components of the project, and how are researchers using the information they gather to learn more about human evolution?

13. In what ways has culture been an important factor in human evolution?

14. At one time, several species of *Homo* existed at the same time. Propose at least two hypotheses that might explain why only *Homo sapiens* remains.

15. How do you predict that a scientist would respond to a question about whether humans "evolved from monkeys"?

16. The video game "Spore" invites players to design creatures and guide them through "five stages of evolution." Search the Internet for information about "Spore," then describe how evolution in this game is similar to, and different from, the evolution of life on Earth.

Write It Out

1. Panspermia suggests that life on Earth developed from primitive "spores" that arrived from outer space. Suppose it did so on multiple planets, eventually giving rise to humans (on Earth) and *Star Trek*'s Vulcans (on planet Vulcan). In that case, would humans be more closely related to Vulcans or to Earth species such as crabgrass and bread mold? Given your answer, does it seem likely that *Star Trek*'s Mr. Spock could have had one Vulcan and one human parent?

2. About how soon after Earth formed did life first appear, according to fossil evidence? How much longer did it take for multicellular organisms to colonize land?

3. Review the structures of nucleic acids and proteins in chapter 2. What chemical elements had to have been in primordial "soup" to generate these organic molecules?

4. How might the first metabolic pathway have arisen?

5. The amoeba *Pelomyxa palustris* is a single-celled eukaryote with no mitochondria, but it contains symbiotic bacteria that can live in the presence of O_2. How does this observation support the endosymbiont theory?

6. The antibiotic streptomycin kills bacterial cells but not eukaryotic cells; diphtheria toxin kills eukaryotic cells but not bacteria. Which of these two substances do you predict would kill mitochondria and chloroplasts? Explain your answer

7. List the major events of Precambrian time and of the Paleozoic, Mesozoic, and Cenozoic eras.

8. Distinguish between the terms *primate, hominid, hominin,* and *Homo*.

9. Search the Internet for information about *Homo floresiensis*. What types of evidence are paleontologists using to place *H. floresiensis* on the human family tree?

Pull It Together

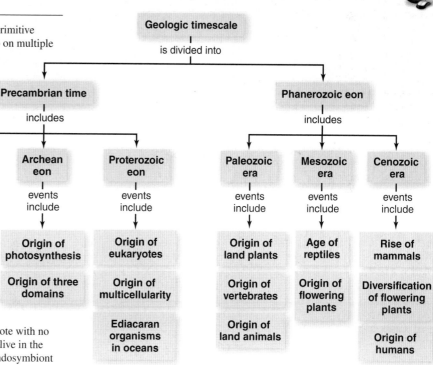

1. What events mark the transitions between eras in the geologic timescale?

2. How has the emergence of new species changed Earth's history?

3. Create an additional concept map that depicts the evolution of humans.

4. What scientific evidence supports human evolution?

Chapter

15 Viruses

Flu Protection. Protective masks help prevent the spread of influenza among passengers on this tram in western Ukraine.

Mc Graw Hill **connect** |BIOLOGY

Enhance your study of
this chapter with practice quizzes,
animations and videos, answer keys,
and downloadable study tools.
www.mhhe.com/hoefnagels

From the Birds: Influenza

MOST VIRUSES INFECT ONLY A FEW, CLOSELY RELATED SPECIES, BUT SOME—INCLUDING THOSE THAT CAUSE INFLUENZA—JUMP AMONG MORE DISTANT RELATIVES SUCH AS BIRDS, PIGS, AND HUMANS.

Influenza, known as "flu" for short, probably began thousands of years ago in China, where the influenza virus moved from wild to domesticated ducks. In the seventeenth century, the Chinese brought domesticated ducks to live among rice paddies, where they were close to people, pigs, and chickens.

Pigs contribute to influenza epidemics because cells that line a pig's throat carry receptors that bind to both the avian (bird) and human versions of flu viruses. Avian and human flu viruses infecting pigs commonly exchange segments of their genomes, generating new strains. These new viral varieties sometimes cause serious epidemics because most humans lack immunity against viruses that normally infect birds or pigs.

The flu pandemic of 1918 killed more people in the United States than World Wars I and II, the Korean War, and the Vietnam War combined. Unlike modern flu outbreaks, which typically kill children and the elderly, most victims of the 1918 flu were 20 to 40 years old. Researchers are trying to determine why this flu was so deadly by studying viruses in lung tissue from victims preserved in graves in the Alaskan permafrost.

Flu pandemics also occurred in 1957 and 1968. In 1997, epidemiologists feared yet another pandemic when a new flu variant appeared to jump from birds directly to humans. A 3-year-old boy in Hong Kong died of a fierce flu caused by a virus never before seen in humans but known in birds. When two other Hong Kong children fell ill with the chicken virus, panic set in. To avert an epidemic, the government killed every chicken in Hong Kong. Fortunately, the outbreak never went beyond 18 people, six of whom died.

Sporadic outbreaks of avian flu in Asia have occurred annually since 2003, infecting dozens of people each time. The virus is common in birds, and the mortality rate in humans is high, ranging from 32% to 70%. So far, however, the avian influenza virus has not acquired the genes necessary to spread directly from person to person. Outbreaks in humans are therefore limited. On the other hand, the swine flu virus that raced around the world in 2009 prompted health officials to declare the first influenza pandemic of the 21st century. Fortunately, the fatality rate was less than 1%.

Flu viruses evolve rapidly. Epidemiologists therefore watch carefully for new variants that are both deadly and easily transmitted among humans. A new virus with both qualities might trigger an outbreak rivaling the disastrous 1918 flu pandemic.

Viruses infect every type of organism, not just humans, birds, and pigs. This chapter explains what viruses are, how they replicate and cause disease, and why they have proven so hard to defeat.

Learning Outline

Learn How to Learn
Take the Best Possible Notes

Some students take notes only on what they consider "important" during a lecture. Others write down words but not diagrams. Both strategies risk losing vital information and connections between ideas that could help in later learning. Instead, write down as much as you can during lecture, including sketches of the diagrams. It will be much easier to study later if you have a complete picture of what happened in every class.

15.1 | Viruses Are Infectious Particles of Genetic Information and Protein

Smallpox, influenza, the common cold, rabies, polio, chickenpox, warts, mononucleosis, AIDS—this diverse list includes illnesses that range from merely inconvenient to deadly. All have one thing in common: they are infectious diseases caused by viruses.

Many people mistakenly lump viruses and bacteria together as "germs." Viruses, however, are not bacteria. In fact, they are not even cells. A **virus** is a small, infectious agent that is simply genetic information enclosed in a protein coat. The 2000 or so known species of viruses therefore straddle the boundary between the chemical and the biological.

A. Viruses Are Smaller and Simpler Than Cells

A virus is much smaller than a cell (see figure 3.2). At about 10 μm (microns) in diameter, an average human cell is perhaps one-tenth the diameter of a human hair. A bacterium is about one-tenth again as small, at about 1 μm (1000 nm) long. The average virus, with a diameter of about 80 nm, is more than 12 times smaller than a bacterium.

A virus does not have a nucleus, organelles, ribosomes, or even cytoplasm. Only a few types of viruses contain enzymes. All viruses share two features:

- **Genetic information.** All viruses contain genetic material (either DNA or RNA) that carries instructions to make their molecular components. The major criterion for classifying viruses is whether the genetic material is DNA or RNA (**table 15.1**). Either type of nucleic acid may be single- or double-stranded. ▶ nucleic acids, p. 40

- **Protein coat.** The **capsid,** or protein coat, surrounds the genetic material. The capsid's shape determines a virus's overall form, which is another characteristic used in classification (**figure 15.1**). Many viruses are spherical or icosahedral (a 20-faced shape built of triangular sections). Others are rod-shaped, oval, or filamentous.

Some viruses have other features besides genetic material and a protein coat. For example, some have a lipid-rich **envelope,** a layer of membrane outside the protein coat. The envelope may include embedded proteins that help a virus invade a host cell. An example of an enveloped virus is the human immunodeficiency virus (HIV), which causes acquired immunodeficiency syndrome (AIDS). The influenza virus also has an envelope. The presence or absence of an envelope is another criterion for virus classification. ▶ cell membrane, p. 54

Despite having relatively few components, a virus's overall structure can be quite intricate and complex. For example, some **bacteriophages,** which are viruses that infect bacteria,

Table 15.1 Some Viruses That Infect Humans

Genetic Material	Virus (Disease)
DNA	Variola major (smallpox)
	Herpesviruses (oral and genital herpes; chickenpox)
	Epstein–Barr virus (mononucleosis, Burkitt lymphoma)
	Papillomaviruses (warts, cervical cancer)
	Hepatitis B virus
RNA	Human immunodeficiency virus (AIDS)
	Poliovirus
	Influenza viruses
	Measles virus
	Mumps virus
	Rabies virus
	Ebola virus
	Rhinovirus (common cold)
	West Nile virus
	Hepatitis A and C viruses

have parts that resemble tails, legs, and spikes. These viruses look like the spacecrafts once used to land on the moon (see figure 15.1b).

B. A Virus's Host Range Consists of the Organisms It Infects

The **host range** of a virus is the kinds of organisms or cells that it can infect. A virus can enter only a cell that has a specific target attachment molecule, or receptor, on its surface. Virtually all species of animals, fungi, plants, protists, and bacteria get viral infections, but a bacteriophage cannot attack human cells because our cell surfaces lack the correct target molecules.

Some target molecules occur on a very small subset of cells in an organism, whereas others are more widespread. HIV, for example, infects only human helper T cells (a type of white blood cell). The rabies virus, on the other hand, can infect any mammal, including humans, skunks, foxes, raccoons, bats, and dogs. All of these animals have the target molecule that the rabies virus uses to recognize a potential host.

The **reservoir** of a virus is the site where it exists in nature. For many viruses that infect humans, the reservoir is a host animal that may or may not show symptoms of infection. A reservoir animal acts as a continual source of the virus to other host species. Examples of reservoirs for viruses that cause diseases in humans

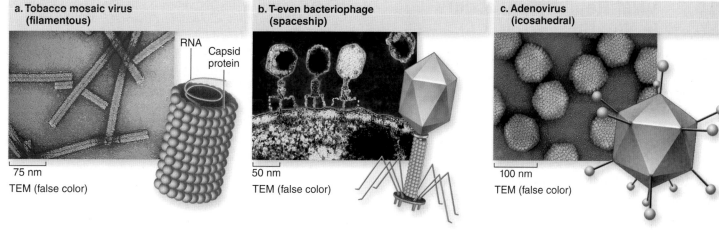

a. Tobacco mosaic virus (filamentous)

RNA
Capsid protein

75 nm
TEM (false color)

b. T-even bacteriophage (spaceship)

50 nm
TEM (false color)

c. Adenovirus (icosahedral)

100 nm
TEM (false color)

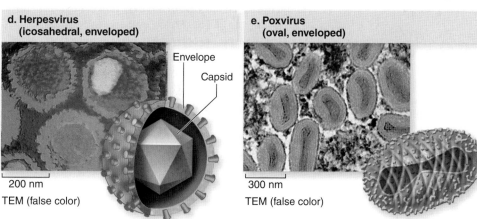

d. Herpesvirus (icosahedral, enveloped)

Envelope
Capsid

200 nm
TEM (false color)

e. Poxvirus (oval, enveloped)

300 nm
TEM (false color)

Figure 15.1 Viruses of Many Shapes and Sizes. Each type of virus has a characteristic structure, visible only with an electron microscope. (a) The capsid of the filamentous tobacco mosaic virus (TMV) consists of repeated protein subunits. (b) T-even viruses look like tiny spaceships. (c) Adenoviruses cause respiratory infections similar to the common cold. (d) Herpesviruses cause cold sores and rashes. (e) A poxvirus causes smallpox.

include wild birds (avian influenza and West Nile encephalitis), rodents (hantavirus pulmonary syndrome), mosquitoes (yellow fever), and raccoons (rabies).

C. Are Viruses Alive?

Most biologists do not consider a virus to be alive because it does not metabolize, respond to stimuli, or reproduce on its own. Instead, a virus must enter a living host cell to manufacture more of itself. ▶ what is life? p. 4

Nevertheless, viruses do have some features in common with life, including genetic material. Both DNA and RNA can mutate, which means that viruses evolve just as life does. Each time a virus replicates inside a host cell, random mutations occur. The genetic variability among the new viruses is subject to natural selection. That is, some variants are better than others at infecting and replicating in host cells. Many mutant viruses die out, but others pass their successful gene versions to the next generation. Over time, natural selection shapes the genetic composition of each viral population.

Although viruses evolve, their extreme genetic and structural diversity suggests that they do not share a single common ancestor. As a result, viruses are not part of the taxonomic hierarchy that biologists use to classify life. Scientists group viruses into species, genera, and families based on the type of nucleic acid, the structure of the virus, how it replicates, and the type of disease it causes. Although some virus families have been assigned to orders, most have not. Biologists do not recognize classes, phyla, kingdoms, or domains for viruses. ▶ taxonomic hierarchy, p. 286

15.1 | Mastering Concepts

1. How are viruses similar to and different from bacteria and eukaryotic cells?
2. What features do all viruses share?
3. What determines a virus's host range?
4. Why is it important to know the reservoir species for a virus that causes disease in humans?
5. How do viruses evolve?
6. What criteria do biologists use to classify viruses?

15.2 | Viral Replication Occurs in Five Stages

The production of new viruses is very different from cell division. When a cell divides, it doubles all of its components and splits in two. Virus production, on the other hand, more closely resembles the assembly of cars in a factory.

A cell infected with one virus may produce hundreds of new viral particles. Whatever the host species or cell type, the same basic processes occur during a viral infection (figure 15.2):

1. **Attachment:** A virus attaches to a host cell by adhering to a receptor molecule on the cell's surface. Generally, the virus can attach only to a cell within which it can reproduce. HIV cannot infect skin cells, for example, because its receptors occur only on helper T cells.

2. **Penetration:** The viral genetic material can enter the cell in several ways. Animal cells engulf virus particles and bring them into the cytoplasm via endocytosis. Viruses that infect plants often enter their host cells by hitching a ride on the mouthparts of herbivorous insects. Many bacteriophages inject their genetic material through a hole in the cell wall, somewhat like a syringe. ▶ endocytosis, p. 83

3. **Synthesis:** The host cell produces multiple copies of the viral genome; mutations during this stage are the raw material for viral evolution. In addition, the information encoded in viral DNA is used to produce the virus's proteins (see chapter 7). The host cell provides all of the resources required for the production of new viruses: ATP, tRNA, ribosomes, nucleotides, amino acids, and enzymes.

4. **Assembly:** The subunits of the capsid join, and then genetic information is packed into each protein coat. Enveloped viruses such as HIV are not complete until they bud from the host cell, acquiring their outer coverings from the host cell membrane.

5. **Release:** Once the virus particles are assembled, they are ready to leave the cell. Some bacteriophages induce production of an enzyme that breaks down the host's cell wall, killing the cell as it releases the viruses. HIV and herpesviruses, on the other hand, bud from the host cell by exocytosis. The cell may die as enveloped viruses carry off segments of the cell membrane. ▶ exocytosis, p. 84

The amount of time between initial infection and cell death varies. Bacteriophages need as little as a half hour to infect a cell and replicate. At the other extreme, for some animal viruses, years may elapse between initial attachment and the final burst of viral particles.

15.2 | Mastering Concepts

1. Describe the five steps in viral replication.
2. What is the source of energy and raw materials for the synthesis of viruses in a host cell?

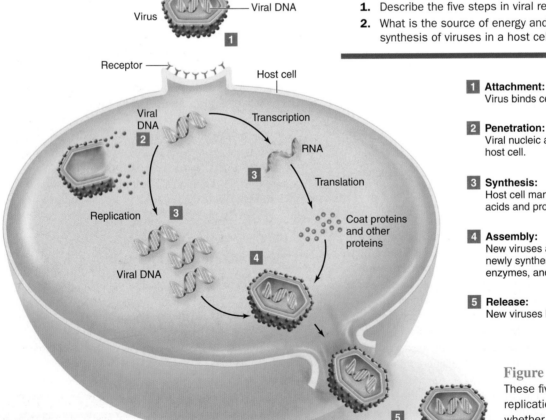

1 Attachment:
Virus binds cell surface receptor.

2 Penetration:
Viral nucleic acid is released inside host cell.

3 Synthesis:
Host cell manufactures viral nucleic acids and proteins.

4 Assembly:
New viruses are assembled from newly synthesized coat proteins, enzymes, and nucleic acids.

5 Release:
New viruses leave the host cell.

Figure 15.2 Viral Replication. These five basic steps of viral replication apply to any virus, whether the host cell is prokaryotic or eukaryotic.

15.3 | Cell Death May Be Immediate or Delayed

Following attachment to the host cell and penetration of the viral genetic material, viruses may or may not immediately cause cell death. Bacteriophages, the viruses that infect bacteria, can do either. The two viral replication strategies in bacteriophages are called lytic and lysogenic infections (figure 15.3).

A. Some Viruses Kill Cells Immediately

In a **lytic infection,** a virus enters a cell, immediately replicates, and causes the host cell to burst (lyse) as it releases a flood of new viruses. The newly released viruses infect other cells, repeating the process until all of the bacteria in a culture are dead.

Some researchers have investigated the possibility of using lytic bacteriophages to treat bacterial infections in people. "Phage therapy" would have two main advantages over antibiotics (drugs that kill bacteria).

First, unlike drugs, viruses evolve along with their bacterial hosts, and they keep killing until all host cells are dead. Bacterial populations are therefore unlikely to acquire resistance to the phages.

Second, each bacteriophage targets only one or a few strains of bacteria, so the treatment is tailored to the infection. Paradoxically, phage therapy's main weakness is related to this strength. Medical personnel must first identify the exact strain of bacteria causing infection before beginning phage therapy. This delay could be deadly.

B. Viral DNA Can "Hide" in a Cell

In a **lysogenic infection,** the genetic material of a virus is replicated along with the host cell's chromosome, but the cell is not immediately destroyed. At some point, however, the virus reverts to a lytic cycle, releasing new viruses and killing the cell.

Many lysogenic viruses use enzymes to cut the host cell DNA and join their own DNA with the host's. A **prophage** is the DNA of a lysogenic bacteriophage that is inserted into the host chromosome. Other lysogenic viruses maintain their DNA apart from the chromosome. Either way, however, when the cell divides, the viral genes replicate, too.

During a lysogenic stage, the viral DNA does not damage the host cell. Only a few viral proteins are produced, most functioning as a "switch" that determines whether the virus should become lytic. At some signal, such as stress from DNA damage or cell starvation, these viral proteins trigger a lytic infection cycle that kills the cell and releases new viruses that infect other cells. The next generation of viruses may enter a lytic or lysogenic replication cycle, depending on the condition of the host cells.

Much of what biologists know about viral replication comes from research on bacteriophages. This chapter's Focus on Model Organisms describes some of what scientists have learned from studies of one such bacteriophage, called lambda.

15.3 | Mastering Concepts

1. What is a lytic viral infection?
2. How is a lysogenic viral infection similar to and different from a lytic cycle?

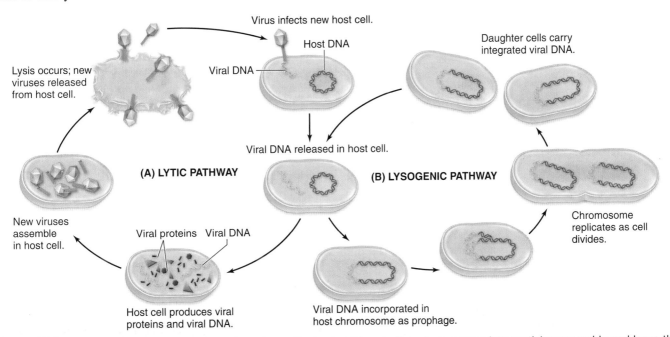

Figure 15.3 Lysis and Lysogeny. (a) In the lytic pathway, the host cell bursts (lyses) when new virus particles assemble and leave the cell. (b) In lysogeny, viral DNA replicates along with the cell, but new viruses are not immediately produced. An environmental change may trigger a lysogenic virus to switch to the lytic pathway.

15.4 | Effects of a Viral Infection May Be Mild or Severe

A person can acquire a viral infection by inhaling the respiratory droplets of an ill person or by ingesting food or water contaminated with viruses. Some viruses, such as HIV and hepatitis B, enter a person's bloodstream via a blood transfusion, sexual contact, or the use of contaminated needles. Some viral infections are symptomless, but often, the infected person becomes ill.

A. Symptoms Result from Cell Death and the Immune Response

Once a viral infection is established, the death of infected cells produces a wide range of symptoms that reflect the types of host cells destroyed. If enough cells die, the disease may be severe.

To illustrate how symptoms develop, consider the flu. Influenza viruses infect cells lining the human airway. As cells produce and release new viruses, the infection spreads rapidly in the lungs, trachea, throat, and nose. The dead and damaged cells in the airway cause the respiratory symptoms of influenza, including cough and sore throat.

What about the other symptoms, such as fever and body aches? These classic signs of influenza do not directly involve the respiratory system, where the infected cells are located. Instead, these whole-body symptoms result from the immune system's response to the viral infection. Cells of the immune system release signaling molecules called cytokines, which, in turn, induce fever; a high body temperature speeds other immune responses. Cytokines also trigger inflammation, which causes body ache and fatigue. These (and other) immune reactions usually defeat the influenza virus, but in the meantime, they can also make the host rather miserable! Chapter 33 describes the human immune system in more detail.

B. Some Animal Viruses Linger for Years

Like a lysogenic bacteriophage, a virus infecting an animal cell may remain dormant as a cell divides. A **latent** infection does not produce disease symptoms, yet the viral genetic information is inside the cell.

Some animal viruses remain latent until conditions make it possible, or necessary, to replicate. An example is herpes simplex virus type I, which causes cold sores on the lips. After initial infection, the viral DNA remains in host cells indefinitely. When a cell becomes stressed or damaged, new viruses are assembled and leave the cell to infect other cells. Cold sores, which reflect the localized death of these cells, periodically recur at the site of the original infection.

HIV is another virus that can remain latent inside a human cell (figure 15.4). HIV belongs to a family of viruses called retroviruses, all of which have an RNA genome. The virus attaches to and penetrates an immune system component called a helper T cell. The virus's reverse transcriptase enzyme

Focus on Model Organisms

Bacteriophage Lambda

Although viruses are not organisms, they have nevertheless contributed enormously to the scientific understanding of life. This box focuses on a virus that has never made you sick because it does not infect human cells. Rather, it kills *Escherichia coli* bacteria that live in the intestines of humans and other mammals. As with all viruses, bacteriophage lambda (Greek letter λ) injects its genetic material into its host cell, which subsequently turns into a virus-making factory. In so doing, phage lambda has revealed many processes fundamental to all life.

Phage lambda's double-stranded DNA genome of 50,000 base pairs contains all the information it takes to make more phages, and its 60 or so genes must turn on and off in proper sequence for the new viruses to form properly. Biologists learn the functions of viral proteins by studying viral mutants—phages with missing proteins. These studies have revealed that phage lambda's proteins fall into three general groups:

- **Capsid proteins:** Phage lambda's protein coat consists of a nearly spherical head (enclosing the viral DNA) and a tube-shaped tail with proteins that bind to *E. coli*'s surface (figure 15.A). DNA enters the host cell through the tail.

- **Regulatory proteins:** Some of phage lambda's proteins bind to viral DNA and either promote or prevent transcription of particular genes. Other regulatory proteins stop the virus from entering the lysogenic pathway or prevent the replication of other viruses that may have also infected the cell.

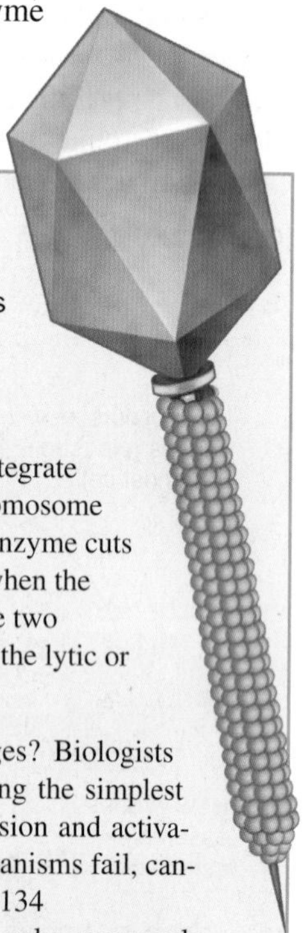

Figure 15.A Bacteriophage Lambda. This virus, deadly to the bacterium *E. coli*, is harmless to humans.

- **Enzymes:** A protein called integrase helps integrate phage lambda's DNA into the host cell's chromosome when it enters the lysogenic phase. Another enzyme cuts the viral DNA back out of the chromosome when the lytic phase begins. The balance between these two enzymes determines whether the virus enters the lytic or lysogenic phase of the life cycle.

Why study gene regulation in bacteriophages? Biologists can learn a lot about complex systems by studying the simplest models, since similar mechanisms of gene repression and activation also work in our own cells. When those mechanisms fail, cancer may result. ▶ regulation of gene expression, p. 134

Phage lambda has taught scientists about not only gene regulation but also recombination and protein folding, both of which are fundamental life processes. In addition, this virus has become something of a laboratory workhorse. Phage lambda is unusual because it can complete its life cycle even if a large amount of foreign DNA is inserted into its genome. This discovery has made phage lambda an extremely important tool for ferrying recombinant DNA into *E. coli*. ▶▶ protein folding, p. 37; transgenic organisms, p. 141

transcribes the viral RNA to DNA. The DNA then inserts itself into the host cell's DNA. Shortly after infection, many HIV particles are produced and released by budding. A strong immune response soon greatly reduces virus production, but infected cells in the lymph nodes continue to release small numbers of viruses. Infected individuals have almost no symptoms, yet HIV is present in their bloodstreams and can be transmitted to others. This phase, called clinical latency, can persist for years. ▶ T cells, p. 679

Throughout this latent period, immune function appears normal, but the number of helper T cells gradually declines. Eventually, the loss of T cells leaves the body unable to defend itself from infections or cancer. AIDS is the result. This chapter's Apply It Now box explains how some HIV-fighting treatments interfere with viral replication.

Because latent viruses persist by signaling their host cells to divide continuously, some cause cancer. A latent infection by some strains of human papillomavirus, which causes genital warts, can lead to cervical cancer (see this chapter's Burning Question box). Epstein–Barr virus is another example. More than 80% of the human population carries this virus, which infects B cells of the immune system. A person who is initially exposed to the virus may develop mononucleosis. The virus later maintains a latent infection in B cells. In a few people, especially those with weakened immune systems, the virus eventually causes a form of cancer called Burkitt lymphoma. ▶ cancer, p. 162

C. Drugs and Vaccines Help Fight Viral Infections

Halting a viral infection is a challenge, in part because viruses invade living cells. Some antiviral drugs interfere with enzymes or other proteins that are unique to viruses, but overall, researchers have developed few medicines that inhibit viruses without killing infected host cells. Many viral diseases therefore remain incurable.

Antiviral drug development is complicated by the genetic variability of many viruses. Consider the common cold. Many different cold viruses exist, and their genomes mutate rapidly. As a result, a different virus strain is responsible every time you get the sniffles. Developing drugs that work against all of these variations has so far proved impossible. Even if a drug inactivated 99.99% of cold-causing viruses, the remaining 0.01% would be resistant. These viruses would replicate, and natural selection would rapidly render the drug ineffective.

Vaccination remains our most potent weapon against many viral diseases. A **vaccine** "teaches" the immune system to recognize one or more molecular components of a virus without actually exposing the person to the disease. Some vaccines confer immunity for years, whereas others must be repeated annually. The influenza vaccine is an example of the latter. Flu viruses mutate rapidly, so this year's vaccine is likely to be ineffective against next year's strains. ▶ vaccines, p. 684

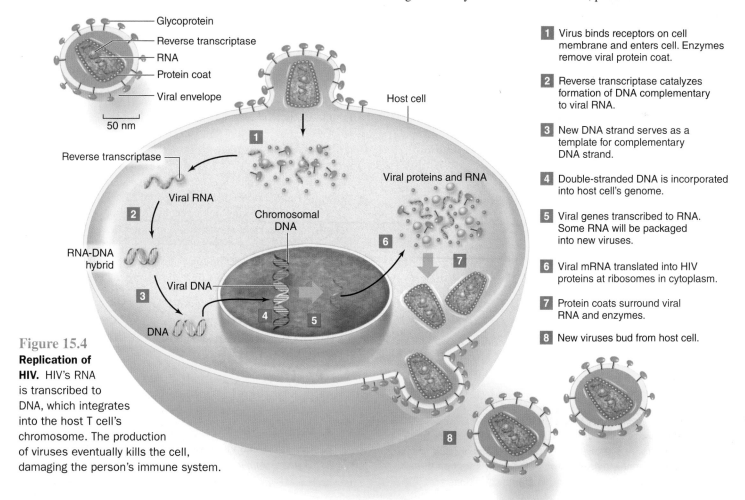

Figure 15.4 Replication of HIV. HIV's RNA is transcribed to DNA, which integrates into the host T cell's chromosome. The production of viruses eventually kills the cell, damaging the person's immune system.

Apply It Now

Anti-HIV Drugs

Decades of painstaking research on HIV at the molecular level has enabled biologists to develop medications that slow the replication of this deadly virus. Here are some ways that these drugs work:

- **Keep viruses out of host cells.** Entry inhibitors block molecules that HIV uses to recognize and enter a host cell. The drug enfuvirtide (Fuzeon) is an example.
- **Inhibit replication of viral genetic information.** Azidothymidine (AZT) stops reverse transcriptase from making a DNA copy of HIV's RNA genome.
- **Inhibit viral DNA integration into host DNA.** Integrase inhibitors are a new class of drugs that keep viral DNA from inserting into the host cell's chromosome.
- **Inhibit assembly of new viruses inside infected host cells.** Protease inhibitors prevent viral enzymes called proteases from cleaving the proteins that make up HIV's protein coat. The protein coat cannot form properly, so fewer mature viruses leave infected cells, and the rate of infection slows.

- **Inhibit host cell reproduction.** Cellular inhibitors stop the replication of infected host cells. Hydroxyurea, for example, keeps T cells from dividing, reducing the number of HIV particles produced. Not surprisingly, this drug should be used with caution in AIDS patients with low T-cell counts.

Many HIV-positive patients take several drugs in combination, a strategy that helps guard against the evolution of drug-resistant viruses. The rationale behind this strategy lies with HIV's high mutation rate. The reverse transcriptase enzyme does not proofread while it copies viral RNA to DNA, so mutations in HIV's genome are extremely common. Most of the resulting mutant viruses are defective and unable to replicate. By chance, however, some new variants may encode proteins that resist an anti-HIV drug. Because the other drugs are still likely to inhibit the new variant, a multidrug regimen lowers the chance that resistant viruses will survive and spread.

Thanks to successful global vaccination programs, smallpox has vanished from human populations, and polio is nearly defeated. Childhood vaccinations have greatly reduced the incidence of measles, mumps, and many other potentially serious illnesses. Unfortunately, researchers have been unable to develop vaccines against many deadly viruses, including HIV. High mutation rates in HIV's genetic material make this virus a moving target.

To produce viruses in a laboratory for vaccine manufacture or testing, scientists must inoculate host cells and wait for the cells to produce the desired number of viruses. The choice of host depends on the virus; bacteria, cultured animal cells, live animals, and live plants are all candidates. Fertilized chicken eggs, for example, provide the host cells needed to produce the raw materials of the influenza vaccine (**figure 15.5**).

The antibiotic drugs that kill bacteria never work against viral infections. The reason is that viruses lack the cell walls, ribosomes, and enzymes targeted by antibiotics. Although antibiotics are useless against viruses, many patients demand that physicians prescribe them for viral infections. This behavior selects for antibiotic-resistant bacteria, an enormous and growing public health problem. ▶ antibiotic resistance, p. 229

Nevertheless, a viral infection can sometimes promote bacterial growth. For example, patients sometimes develop sinus infections as a complication of influenza or the common cold. Physicians may prescribe antibiotics to treat these secondary bacterial infections, but the drugs will not affect the underlying virus.

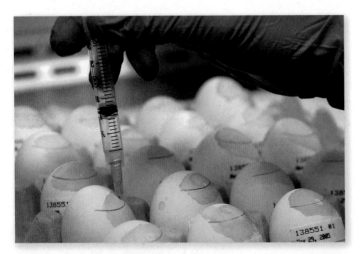

Figure 15.5 Viruses from Eggs. Influenza viruses replicate inside fertilized chicken eggs in a laboratory that produces the flu vaccine. People with egg allergies should therefore consult a physician before receiving a flu shot. (The yellow spots on the eggs are iodine, which helps maintain sterile conditions.)

15.4 | Mastering Concepts

1. How can a person acquire and transmit a viral infection?
2. How do symptoms of a viral infection develop?
3. What is a latent animal virus?
4. Describe how HIV replicates in host cells.
5. How are some latent viral infections linked to cancer?
6. How are viral infections treated and prevented?

15.5 | Viruses Cause Diseases in Plants

 Like all organisms, plants can have viral infections. The first virus ever discovered was tobacco mosaic virus (see figure 15.1a), which affects not only tobacco but also tomatoes, peppers, and more than 120 other plant species.

To infect a plant cell, a virus must penetrate waxy outer leaf layers and thick cell walls. Most viral infections spread when plant-eating insects such as leafhoppers and aphids move virus-infested fluid from plant to plant on their mouthparts. ▶ plant cell wall, p. 64

Once inside a plant, viruses multiply at the initial site of infection. The killed plant cells often appear as small dead spots on the leaves. Over time, the viruses spread from cell to cell through plasmodesmata (bridges of cytoplasm between plant cells). They can also move throughout a plant by entering the vascular tissues that distribute sap. Depending on the location and extent of the viral infection, symptoms may include blotchy, mottled leaves or abnormal growth. A few symptoms, such as the streaking of some flower petals, appear beautiful to us (**figure 15.6**).

Although plants do not have the same forms of immunity as do animals, they can fight off viral infections. For example, in a process called "posttranscriptional gene silencing," a plant cell degrades viral mRNA, which prevents the production of new viruses. Researchers are learning more about the role of posttranscriptional gene silencing in the defense against viruses in both plants and animals.

Figure 15.6 Sick Plants. Cucumber mosaic virus causes a characteristic mottling (spotting) of squash leaves, A virus has also caused the streaking on the petals of these tulips.

15.5 | Mastering Concepts

1. How do viruses enter plant cells and spread within a plant?
2. What are some symptoms of a viral infection in plants?

? Burning Question

Can a person get cancer by having sex?

Cancer usually has nothing to do with sex, but there is an exception. Each year, cervical cancer affects about 10,000 women in the United States and 510,000 women worldwide. Most cases are caused by exposure to a sexually transmitted virus.

The cancer-causing virus is called human papillomavirus, abbreviated HPV. About 100 strains of this virus exist, some causing harmless ailments such as warts on the soles of the feet. Thirty or so strains infect the genitals and are passed among male and female sexual partners. Usually the infection goes unnoticed. But some strains of HPV cause genital warts, and the virus can also cause cervical cancer in women.

What are the best ways to prevent cervical cancer? One important strategy is to have regular Pap tests, which detect abnormal, precancerous cells in the cervix. Another way to prevent cervical cancer is to reduce the chance of exposure to HPV. Abstaining from sex, limiting the number of

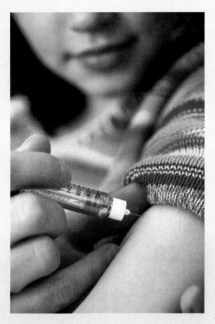

sexual partners, and using a condom are all commonsense ways to help prevent the spread of HPV.

In addition, a vaccine can help prevent cervical cancer. Approved for use in the United States in 2006, the human papillomavirus vaccine prevents infection by four strains of HPV. Two of these strains are associated with 70% of cervical cancer cases; the other two cause 90% of genital warts.

Although the vaccine is safe and effective in girls and women between 9 and 26 years old, health professionals recommend vaccination at age 11 or 12. Vaccinating girls before they become sexually active makes sense because the vaccine does not work in people who have already been exposed to the virus. Thanks to the new vaccine, epidemiologists predict that the incidence of cervical cancer will decline dramatically in the future; time will tell if they are correct.

Submit your burning question to:
marielle_hoefnagels@mcgraw-hill.com

15.6 | Viroids and Prions Are Other Noncellular Infectious Agents

The idea that something as simple as a virus can cause devastating illness may seem amazing. Yet some infectious agents are even simpler than viruses.

A. A Viroid Is an Infectious RNA Molecule

A **viroid** is a highly wound circle of RNA that lacks a protein coat; it is simply naked RNA that can infect a cell. The RNA coils tightly, bonding with itself to form double-stranded RNA. This configuration helps prevent degradation by host cell enzymes.

Although viroid RNA does not encode protein, it can nevertheless cause severe disease in many important crop plants, including tomatoes (**figure 15.7**). Apparently the viroid's RNA interferes with the plant's ability to produce one or more essential proteins.

Transmission of the viroid usually occurs in seeds or pollen. Alternatively, viroids can spread in many of the same ways as viruses do: wind causes contaminated plants to rub against their uninfected neighbors, insects feed on multiple plants, and farm equipment spreads contaminated plant sap throughout a field.

B. A Prion Is an Infectious Protein

Another type of infectious agent is a **prion,** which stands for "<u>pro</u>teinaceous <u>in</u>fectious particle." A prion protein (PrP for short) is a normal cellular protein that can exist in multiple three-dimensional shapes, at least one of which is abnormal and can cause disease. Upon contact with an abnormal form of PrP, a normal prion protein switches to the abnormal PrP configuration. The change triggers another round of protein refolding, and so on. As a result of this chain reaction, masses of abnormal prion proteins accumulate inside cells. ▶ protein folding, p. 37

The misshapen prion proteins cause brain cells to die. The brain eventually becomes riddled with holes, like a sponge. Diseases caused by prions are therefore called spongiform

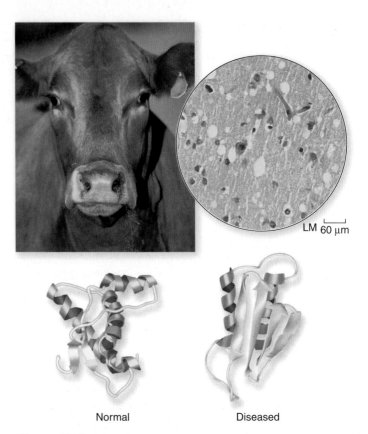

Normal   Diseased

Figure 15.8 **Prion Disease.** The holes and clumps of protein fibrils are evident in the brain of a cow that died of bovine spongiform encephalopathy, also known as "mad cow disease." Below are models of healthy and abnormal prion proteins.

encephalopathies. "Mad cow disease," or bovine spongiform encephalopathy (BSE), is one example (**figure 15.8**). Affected cattle grow fearful, then aggressive, and then have difficulty standing. The animals rapidly lose weight and die.

Animals acquire these diseases by ingesting an infected animal or receiving a transplant of infected tissue. Because of mad cow disease, governments now ban the practice of feeding cattle the processed remains of other cattle.

Prions cause a few human diseases, all of them extremely rare. One is kuru, which is associated with cannibalism. Creutzfeldt–Jakob disease (CJD) is another example; this illness may occur spontaneously, be acquired in medical procedures using tainted biological materials, or be inherited as a gene that encodes the abnormal prion protein. Another heritable prion disease is fatal familial insomnia.

Prions are extremely hardy; heat, radiation, and chemical treatments that destroy bacteria and viruses have no effect on prions. Luckily, commonsense precautions that keep brains and spinal cords out of food and medical products can prevent transmission of prion diseases.

15.6 | Mastering Concepts

1. How are viroids and prions different from viruses?
2. How do viroids and prions cause disease?
3. What is the best way to avoid prion diseases?

Figure 15.7 **Viroids Infect Plants.** The plant on the left has a viroid-caused disease called "tomato bunchy top"; the one on the right is healthy.

Viroid
(circular RNA)

15.7 | Investigating Life: Scientific Detectives Follow HIV's Trail

Anyone who watches crime dramas on TV knows that detectives collect many types of evidence when trying to match a suspect to a crime. Do footprints found at the scene match the suspect's shoes? Do skin cells found under the victim's fingernails contain DNA matching the suspect's? Was the suspect in the victim's neighborhood at the time of the crime?

Epidemiologists use similar tactics when they test hypotheses about the evolution of viruses. For example, scientists have suggested that the ancestor of HIV is a virus called simian immunodeficiency virus (SIV). Many strains of SIV exist. They typically cause symptomless infections in monkeys and apes, although some can cause an AIDS-like disease in chimpanzees.

Researchers have tested hypotheses about the origin of HIV by asking five independent questions:

1. Do the genomes of HIV and SIV consist of the same type of genetic material and have the same order, number, and types of genes?

2. Do the viral genes (and their encoded proteins) share similar sequences?

3. Is SIV common enough in its natural host that it has a chance of spreading to other hosts, including humans?

4. Do SIV and HIV occur in the same geographic region?

5. Is there a plausible transmission pathway for SIV to spread from its original host to humans?

The two major strains of HIV, called HIV-1 and HIV-2, apparently arose independently. HIV-2 is the milder of the two types. By the late 1980s and early 1990s, scientists had used the five criteria to identify a strain called SIVsm as the source of HIV-2. (The "sm" stands for sooty mangabey, a monkey that is SIVsm's primary host.)

The origin of HIV-1 remained unknown, but scientists suspected it arose from a different source, SIVcpz (for chimpanzee). University of Alabama medical researcher Beatrice Hahn, along with a multinational team of scientists, wanted to learn more about HIV-1's evolutionary history. They started with frozen tissue collected in 1985 after the death of a chimpanzee named Marilyn, who was born in Africa around 1960 and exported to the United States as an infant. Marilyn had never been used in AIDS research, yet she tested positive for antibodies against HIV-1. Hahn and her colleagues used the polymerase chain reaction (PCR) to amplify viral DNA from Marilyn's tissue. They found sequences that were similar to both HIV-1 and SIV but not identical to either. ▶ polymerase chain reaction, p. 171

Marilyn's virus was only the fourth SIV strain ever found in chimpanzees. Two of the other three SIV-infected chimps belonged to the same subspecies (the central chimpanzee) as Marilyn. The third was an eastern chimpanzee born in the Democratic Republic of Congo. These four SIV strains provided a critical clue to HIV's origin (figure 15.9). Amino acid sequences of a viral protein suggested that all known subgroups of HIV-1 (M, N, and O) were closely related to SIV from central chimpanzees.

The researchers thus used the epidemiological criteria to demonstrate the origin of HIV-1. Clearly, SIVcpz and HIV-1 share genetic similarities (criteria 1 and 2). Moreover, the habitat of central chimpanzees overlaps with the region of Africa where HIV-1 occurs (criterion 4), and humans who hunt chimps for meat provide a likely transmission route (criterion 5).

Only the third criterion remained questionable: If SIV is common in chimpanzees, why had scientists found it in only four animals out of more than 1500 tested? Hahn and her coauthors suggested that this low infection rate was a predictable consequence of studying chimps that were either wild-caught as infants or born in captivity. Compared with wild chimps, these captive animals would have fewer opportunities to acquire the virus. Indeed, a subsequent study revealed SIVcpz infection rates as high as 18% among wild chimpanzees in Gombe National Park, Tanzania. SIVcpz is therefore more common than previously thought.

Even when television cops nab the criminal, some details of the crime remain obscure. Similarly, many questions remain about HIV-1's origin from SIV. How and when did the virus jump from chimpanzees to humans? Why do HIV-1 subgroups N and O remain rare, whereas subgroup M is causing the global AIDS epidemic? Can we use HIV-1 and HIV-2 to learn about the potential for other viruses to emerge as human pathogens? The scientific detectives continue to search for the answers.

Gao, Feng, Elizabeth Bailes, David L. Robertson, and nine coauthors, including Beatrice Hahn. 1999. Origin of HIV-1 in the chimpanzee *Pan troglodytes troglodytes*. *Nature*, vol. 397, pages 436-441.

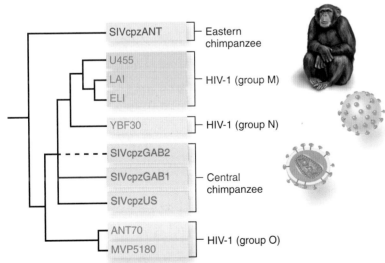

Figure 15.9 HIV-1 from SIV. Amino acid sequences of one viral protein reveal that the eastern chimpanzee's strain of SIV (blue) is distinct from all other SIVcpz strains (red) and from all known HIV-1 subgroups (green).

15.7 | Mastering Concepts

1. What hypothesis was Hahn's team investigating, and how did they test their hypothesis?

2. How could researchers use a similar strategy to study the origin of a new influenza virus?

Chapter Summary

15.1 | Viruses Are Infectious Particles of Genetic Information and Protein

A. Viruses Are Smaller and Simpler Than Cells
- A **virus** is a nucleic acid (DNA or RNA) in a **capsid.** A membranous **envelope** surrounds some viruses.
- A virus must infect a living cell to reproduce.
- Many viruses, including some **bacteriophages,** have relatively complex structures.

B. A Virus's Host Range Consists of the Organisms It Infects
- The species that a virus infects constitute its **host range.** The **reservoir** of a virus is its natural habitat, often an animal in which it does not cause symptoms.

C. Are Viruses Alive?
- Viruses are intracellular parasites that most biologists do not consider to be alive.
- Many viral genomes mutate rapidly.
- Scientists classify viruses based on the type of genetic material, the shape of the capsid, the presence or absence of an envelope, the replication strategy, and the type of disease.

15.2 | Viral Replication Occurs in Five Stages
- After a virus infects a cell, its host manufactures many copies of the viral proteins and nucleic acids, then assembles these components into new viruses.
- The five stages of viral replication within a host cell are attachment, penetration, synthesis, assembly, and release.

15.3 | Cell Death May Be Immediate or Delayed

A. Some Viruses Kill Cells Immediately
- In a **lytic infection,** new viruses are immediately manufactured, assembled, and released.

B. Viral DNA Can "Hide" in a Cell
- In a **lysogenic infection,** the virus's nucleic acid replicates along with that of a dividing cell without causing symptoms. The viral DNA may integrate as a **prophage** into the host chromosome.

15.4 | Effects of a Viral Infection May Be Mild or Severe

A. Symptoms Result from Cell Death and the Immune Response
- The effects of a virus depend on the cell types it infects. Viruses cause disease by killing infected cells and by stimulating immune responses.

B. Some Animal Viruses Linger for Years
- HIV and some other viruses remain **latent,** or hidden, inside animal cells. Some latent viruses are associated with cancer.

C. Drugs and Vaccines Help Fight Viral Infections
- Antiviral drugs and **vaccines** combat some viral infections.
- Antibiotics, drugs that kill bacteria, are ineffective against viruses.

15.5 | Viruses Cause Diseases in Plants
- Viruses infect plant cells, then spread via plasmodesmata.

15.6 | Viroids and Proteins Are Other Noncellular Infectious Agents

A. A Viroid Is an Infectious RNA Molecule
- **Viroids** are naked RNA molecules that infect plants.

B. A Prion Is an Infectious Protein
- A **prion** protein can take multiple shapes, at least one of which can cause diseases such as transmissible spongiform encephalopathies. Treatments that destroy other infectious agents have no effect on prions.

15.7 | Investigating Life: Scientific Detectives Follow HIV's Trail
- Scientists have tested hypotheses about HIV's origin by collecting multiple types of information.
- The subgroups of HIV-1 evolved independently from simian immunodeficiency viruses (SIV) infecting chimpanzees.

Multiple Choice Questions

1. Which of the following is NOT a feature associated with viruses?
   a. Cytoplasm
   b. Genetic information
   c. Protein coat
   d. Membrane

2. Which of the following is the largest?
   a. HIV
   b. RNA molecule
   c. *E. coli* cell
   d. human T cell

3. Why are viruses not considered to be alive?
   a. Because they depend on host cells to reproduce
   b. Because they do not evolve
   c. Because they do not contain genetic information
   d. Both a and c are correct.

4. At which stage in viral replication does the genetic information enter the host cell?
   a. Penetration
   b. Synthesis
   c. Assembly
   d. Release

5. A virus can jump from one host species to a different host species because
   a. viruses are small and numerous.
   b. mutations in the virus let it recognize a new host surface molecule.
   c. the host cell provides ATP, ribosomes, nucleotides, and amino acids.
   d. viruses leave the host cell by exocytosis.

6. Which step in viral replication would be the best target to prevent the spread of a virus?
   a. Attachment
   b. Penetration
   c. Synthesis
   d. Release

7. What occurs during a lysogenic infection?
   a. Viral particles attach but do not penetrate a host cell.
   b. Viral particles fill a host cell and cause it to burst.
   c. Viral genetic material replicates in a host cell without causing symptoms.
   d. The viral prophage DNA is packaged into a capsid.

8. The severity of the symptoms associated with a viral infection is related to
   a. the response of the immune system.
   b. the number of viruses released.
   c. the number and types of cells that become infected.
   d. both a and c.

9. What is a prion?
   a. A highly wound circle of RNA
   b. A virus that has not yet acquired its envelope
   c. A protein that can alter the shape of a second protein
   d. The protein associated with a latent virus

10. What property of a virus makes it useful for making transgenic organisms?
   a. The ability to replicate inside a host cell
   b. The insertion of viral genes into the host's DNA
   c. The ability of the host cell to release viruses
   d. The ability of viral genes to undergo rapid mutation

Write It Out

1. Your biology lab instructor gives you a Petri dish of agar covered with visible colonies. Your lab partner says the colonies are viruses, but you disagree. How do you know the colonies are bacteria?

2. Why is it inaccurate to refer to the "growth" of viruses?

3. With a diameter of about 600 nm, mimiviruses are enormous compared with other viruses. The mimivirus genome consists of about 1.2 million base pairs and encodes more than 1000 genes—more than some bacteria. If you encountered a mimivirus-like object in your research, what sorts of studies could you carry out to determine whether the object was a virus or a bacterium?

4. Raccoons account for the most reported rabies cases in the eastern United States, whereas most cases of rabies in the central and western United States involve bats, foxes, and skunks. Develop a hypothesis that explains this observation. How would you test your hypothesis? Can you think of a way to reduce the incidence of rabies in wild animals?

5. Rhinoviruses replicate in the mucus-producing cells in a person's nose, throat, and lungs, causing the common cold. Papillomaviruses, which infect skin cells, cause growths called warts. HIV infects T cells and causes AIDS. How do these three types of viruses "know" which human cells to infect?

6. Imagine a hybrid virus with the capsid of virus X and the DNA of virus Y. Will a host cell infected with this hybrid virus produce virus X, virus Y, a mix of virus X and Y, or hybrid viruses? Explain your answer.

7. Chapter 7 compared chromosomes to cookbooks and genes to recipes. How could you incorporate viruses into this analogy?

8. Describe how a virus's envelope can include lipid molecules that are also part of host cell membranes.

9. Why are lytic viruses better suited as agents of "phage therapy" than are lysogenic viruses? How would you test whether such a treatment would be effective? Would you be willing to take a "viral antibiotic"?

10. Search the Internet for information about the flu vaccine. Why is the flu shot administered annually when many other vaccines last for years? Is it possible for the flu vaccine to give a person influenza?

11. Why do antibiotics such as penicillin kill bacteria but leave viruses unharmed?

12. Several anti-HIV drugs are already on the market. List some reasons that we might need even more new drugs to fight HIV in the future.

13. Gene therapy aims to replace faulty or disease-causing genes with healthy DNA sequences. How might you use a virus to deliver these new genes into a cell?

14. No one knows how or when viruses originated. Some scientists suggest that viruses were once pieces of cellular DNA that eventually became independent. Why is this hypothesis hard to test?

15. How is a biological virus similar to and different from a computer virus?

16. The National Center for Biotechnology Information maintains an online list of viruses for which genome sequence data are available. Choose one of these viruses and describe some discoveries that have come from research on the virus you chose.

Pull It Together

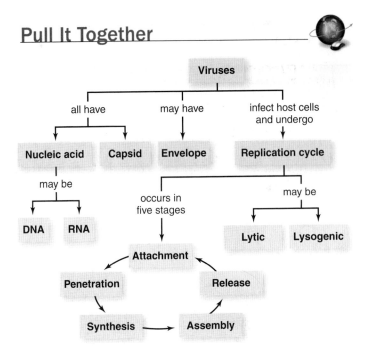

1. What are the four types and configurations of nucleic acids in viruses?
2. How is a virus similar to and different from a bacterium, a viroid, and a prion?
3. What events occur in each of the five stages of viral replication?
4. Distinguish between lytic and lysogenic infections.

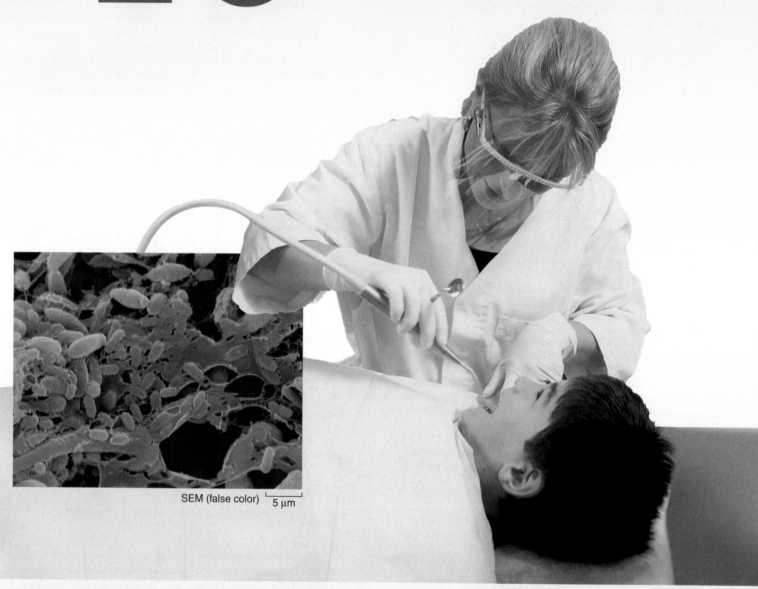

Chapter

16 Bacteria and Archaea

SEM (false color) | 5 μm

The plaque that accumulates on teeth is a familiar example of a bacterial biofilm.

 **connect** |BIOLOGY

Enhance your study of
this chapter with practice quizzes,
animations and videos, answer keys,
and downloadable study tools.
www.mhhe.com/hoefnagels

Bacterial Biofilms: "Mob Mentality" on a Microscopic Scale

BACTERIA ARE ONE-CELLED ORGANISMS, SO IT IS TEMPTING TO THINK OF THEM AS RUGGED INDIVIDU-ALISTS. Instead, however, they often build complex communities in which cells of the same species communicate with one another, protect one another, and even form structures with specialized functions. These organized aggregations of bacterial cells are called biofilms.

A biofilm forms when bacteria settle and reproduce on a solid surface. Once the cells reach a critical density, they express genes that trigger the secretion of a sticky slime made of polysaccharides. As cells continue to divide, they pile up into mushroom-shaped structures that maximize the biofilm's surface area. Occasionally, cells released from the biofilm colonize new habitats, starting the process anew.

Microbiologists are still learning about the cues that trigger biofilm formation. For example, bacteria use signaling molecules to detect the density of cells around them. These "quorum-sensing" signals allow the cells in a biofilm to coordinate their activities, somewhat like a multicellular organism. Thus, within the first 6 hours of biofilm formation, the bacteria turn off genes encoding the proteins that form the flagellum: after all, success in a sedentary lifestyle does not require the ability to swim. In contrast, genes encoding proteins that build attachment structures called pili are activated.

Interest in biofilms extends far beyond idle curiosity about bacterial life. These microbial mats degrade sewage in the trickling filters of community wastewater treatment plants, corrode water pipes, help mine copper ore, and coat the surfaces of plants' roots. Much of the attention that biofilms receive, however, is medical. Persistent biofilms can form on medical implants such as heart valves and catheters. They can also form dental plaque, induce gum disease, and colonize the thick mucus that accumulates in the lungs of cystic fibrosis patients. Bacteria in biofilms are much more resistant to immune defenses and antibiotic treatment than are individual cells.

As microbiologists learn more about biofilm formation, they may be able to develop new treatments to combat medically important bacteria. For example, it may be possible to disrupt biofilm formation by silencing quorum-sensing signals. Conversely, learning how to trigger biofilm formation in beneficial soil bacteria may enhance efforts to clean up toxic wastes.

The more scientists learn about prokaryotes—bacteria and archaea—the more we can appreciate the amazing capabilities of Earth's simplest organisms. As this chapter explains, microbes are much more than just "germs."

UNIT 4

What's the Point?

Learning Outline

Learn How to Learn

Skipping Class?

Attending lectures is important, but you may need to skip class once in a while. How will you find out what you missed? If your instructor does not provide complete lecture notes, you may be able to copy them from a friend. Whenever you borrow someone else's notes, it's a good idea to compare them with the assigned reading to make sure they are complete and accurate. You might also want to check with the instructor if you have lingering questions about what you missed.

16.1 | Prokaryotes Are a Biological Success Story

The microscopic world of life may be invisible to the naked eye, but its importance is immense. Chapter 15 described viruses, tiny infectious particles that straddle the line between life and nonlife. The remaining chapters in this unit focus squarely on life, beginning with the prokaryotes.

A **prokaryote** is a single-celled organism that lacks a nucleus and membrane-bounded organelles. DNA sequences and

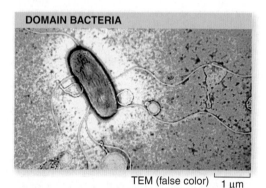

DOMAIN BACTERIA

TEM (false color)   1 μm

DOMAIN ARCHAEA

SEM (false color)   1 μm

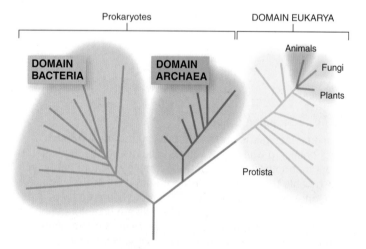

Prokaryotes   DOMAIN EUKARYA

Animals

DOMAIN
BACTERIA   Fungi

DOMAIN
ARCHAEA   Plants

Protista

Figure 16.1 Diversity of Prokaryotic Life. Domains Bacteria and Archaea form two of the three main branches of life.

other lines of evidence suggest the existence of two prokaryotic domains: **Bacteria** and **Archaea** (figure 16.1).

Microbiologists have probably discovered just a tiny fraction of prokaryotic life on Earth; new microbes turn up every time biologists examine a spoonful of soil or a milliliter of water. The total number of species in both domains may be anywhere between 100,000 and 10,000,000; no one knows.

Although we have much to learn of their diversity, it is clear that prokaryotic cells have had a huge influence on Earth's natural history. The earliest known fossils closely resemble today's bacteria, suggesting that the first cells were prokaryotic. Ancient photosynthetic microbes also contributed oxygen gas (O_2) to Earth's atmosphere, creating a protective ozone layer and paving the way for aerobic respiration. Along the road of evolution, bacteria probably gave rise to the chloroplasts and mitochondria of eukaryotic cells. ▶▶aerobic respiration, p. 106; endosymbiosis, p. 304

The reign of the prokaryotes continues today. Virtually no place on Earth is free of bacteria and archaea; their cells live within rocks and ice, high in the atmosphere, far below the ocean's surface, in thermal vents, nuclear reactors, hot springs, animal intestines, plant roots, and practically everywhere else. Many species prefer hot, cold, acidic, alkaline, or salty habitats that humans consider "extreme." This chapter's Burning Question, on page 345, describes some of the few places where microbes do not live.

Most of what we know about the biology of bacteria and archaea comes from the relatively few species that microbiologists can culture in the laboratory. However, these microorganisms do not represent the natural diversity of the prokaryotes. Microbial ecologists therefore extract genetic sequences directly from the environment, without culturing the cells first. These bits of DNA have revealed much about microbes that we could never have learned from cultured cells alone.

Before embarking on a tour of prokaryote biology and ecology, it is worth noting that the term *prokaryote* has become somewhat controversial among microbiologists. The reason is that the word falsely implies a close evolutionary relationship between bacteria and archaea, despite strong evidence that archaea are actually more closely related to eukaryotes. Nevertheless, many biologists continue to use the term as a handy shortcut for describing all cells that lack nuclei.

This chapter offers a taste of the microbial world, beginning with a tour of prokaryotic cells. A sampling of bacterial and archaean diversity comes next, and the chapter ends by explaining why microbes are essential to all other life.

16.1 | Mastering Concepts

1. What are two domains that contain prokaryotes?
2. List several ways that prokaryotes have influenced evolution.
3. In what habitats do bacteria and archaea live?
4. Why is the term *prokaryote* controversial?

16.2 | Prokaryote Classification Traditionally Relies on Visible Features

At about 1 to 10 μm long, a typical prokaryotic cell is 10 to 100 times smaller than most eukaryotic cells (see figure 3.2). Bacteria and archaea also lack the membrane-bounded organelles that characterize eukaryotic cells (see table 3.2). How do microbiologists classify the diversity of life within these two domains, given the tiny cell sizes and scarcity of distinctive internal structures?

The answer to that question has evolved over time. For hundreds of years, biologists classified microbes based on close scrutiny of their cells and metabolism. This section describes some of the most important features; table 16.1 illustrates how some of these characteristics differ between bacteria and archaea.

A. Microscopes Reveal Cell Structures

Viewing cells with a microscope is an essential step in identifying bacteria and archaea. Light microscopes (and sometimes electron microscopes) reveal the internal and external features unique to each species. Figure 16.2 illustrates a typical bacterial cell; in reading through this section, remember that a given cell may have some or all of the structures pictured.

Internal Structures   Like the cells of other organisms, all bacteria and archaea are bounded by a cell membrane that encloses cytoplasm, DNA, and ribosomes. A prokaryotic cell's DNA typically consists of one circular chromosome. The **nucleoid** is the region where this DNA is located, along with some RNA and a few proteins. Unlike the nucleus of a eukaryotic cell, a membranous envelope does not surround the nucleoid.

The cells of many bacteria and archaea also contain one or more **plasmids,** circles of DNA apart from the chromosome. The genes on a plasmid may encode the proteins necessary to replicate and transfer the plasmid to another cell. Other genes may provide the ability to resist a drug or toxin, cause disease, or alter the cell's metabolism. Recombinant DNA technology uses plasmids to ferry genes from one kind of cell to another. ▸ transgenic organisms, p. 141

Table 16.1   Bacteria and Archaea Compared

	Bacteria	Archaea
Similarities		
Predominantly unicellular	Yes	Yes
Cell size	1–10 μm	1–10 μm
Nucleus and other membrane-bounded organelles	No	No
Circular chromosome	Yes	Yes
Able to grow at temperatures above 80°C	Yes (some)	Yes (some)
Nitrogen fixation	Yes (some)	Yes (some)
Differences		
Cell wall composition	Peptidoglycan	Pseudo-peptidoglycan, protein
Membrane composition	Based on fatty acids	Based on nonfatty acid lipids (isoprenes)
Use chlorophyll in photosynthesis	Yes (some)	No
Generate methane as byproduct of metabolism	No	Yes (some)
Sensitive to streptomycin	Yes	No
Introns within genes	No	Sometimes

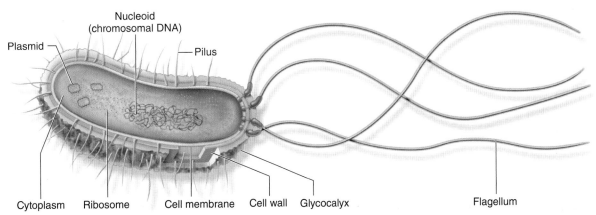

Figure 16.2 Prokaryotic Cell. This diagram shows the internal and external structures that are typical of a prokaryotic cell.

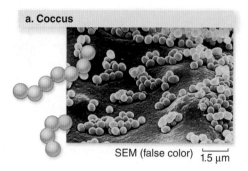

a. Coccus

SEM (false color) ⊢⊣ 1.5 μm

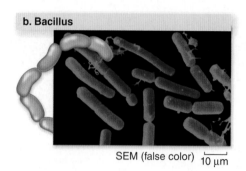

b. Bacillus

SEM (false color) ⊢⊣ 10 μm

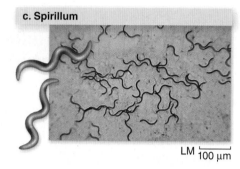

c. Spirillum

LM ⊢⊣ 100 μm

Figure 16.3 Cell Shapes. (a) Cocci such as these *Micrococcus* cells are spherical. (b) A bacillus is rod-shaped; *Bacillus megaterium* is an example. (c) A spirillum is spiral-shaped. These bacteria are called *Spirillum volutans*.

The **ribosomes** are structures where proteins are assembled, a process described in chapter 7. Bacterial, archaean, and eukaryotic ribosomes are all structurally different from one another. Some antibiotics, such as streptomycin, kill bacteria without harming eukaryotic host cells by exploiting this difference. This chapter's Apply It Now box describes more examples of how antibiotics work.

External Structures   The **cell wall** is a rigid barrier that surrounds the cells of most bacteria and archaea. Bacterial cell walls contain **peptidoglycan,** a complex polysaccharide that does not occur in the cell walls of archaea. The antibiotic penicillin inhibits the reproduction of bacteria (but not archaea) by interfering with the final steps in peptidoglycan synthesis.

The wall gives a cell its shape (**figure 16.3**). Three of the most common forms are **coccus** (spherical), **bacillus** (rod-shaped), and **spirillum** (spiral or corkscrew shaped). In addition, the arrangement of the cells in pairs, clusters (*staphylo-*), or chains (*strepto-*) is sometimes important in classification. The disease-causing bacterium *Staphylococcus,* for example, forms grapelike clusters of spherical cells.

The **Gram stain** reaction distinguishes between two types of cell walls (**figure 16.4**). The walls of gram-positive bacteria are made primarily of a thick layer of peptidoglycan. After the staining procedure is complete, gram-positive cells appear purple. In contrast, gram-negative cells have much thinner cell walls and stain pink. Medical technicians often use Gram-staining as a first step in identifying bacteria that cause infections. The distinction is important because gram-positive and gram-negative bacteria are susceptible to different antibiotic drugs.

Besides the thin inner layer of peptidoglycan, the cell walls of gram-negative bacteria have an outer membrane of lipid, polysaccharide, and protein. Parts of this outer layer trigger a strong immune response, including fever and inflammation (see chapter 33). The outer membrane also causes the toxic effects of many medically important gram-negative bacteria, such as *Salmonella.*

Many prokaryotic cells have distinctive structures outside the cell wall (**figure 16.5**). A **glycocalyx,** also sometimes called a capsule or slime layer, is a sticky layer composed of proteins or polysaccharides that may surround the cell wall. The glycocalyx has many functions, including attachment, resistance to drying, and protection from immune system cells. It also plays a role in the formation of biofilms, as described in the chapter opening essay.

Some cells have **pili** (singular: pilus), which are short, hairlike projections made of protein (figure 16.5b). Attachment pili enable cells to adhere to objects. The bacterium that causes cholera, for example, uses pili to attach to a human's intestinal wall. Other projections, called sex pili, aid in the transfer of DNA from cell to cell, as described in section 16.3.

Not all prokaryotes can move, but many can. In a response called **taxis,** cells move toward or away from an external stimulus such as food, a toxin, oxygen, or light. Cells that can move have a **flagellum,** which is a whiplike extension that rotates like a propeller. The bacterium in figure 16.5c has many flagella; other cells have one or a few. (Some eukaryotic cells also have flagella, but they are not homologous to those on bacterial or archaean cells.)

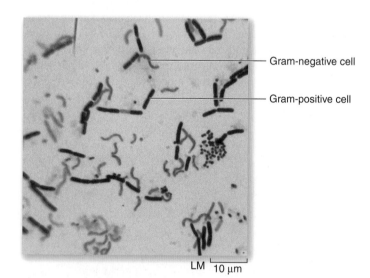

Gram-negative cell

Gram-positive cell

LM ⊢⊣ 10 μm

Figure 16.4 Gram Stain. The Gram stain procedure distinguishes bacteria on the basis of cell wall structure. This stained smear contains both gram-positive and gram-negative bacteria.

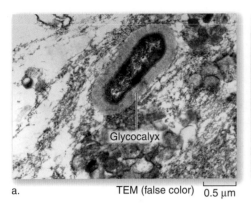

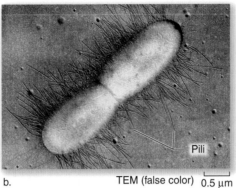

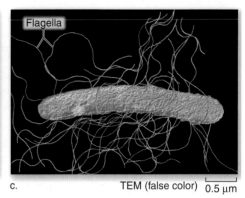

a.   TEM (false color) 0.5 μm   b.   TEM (false color) 0.5 μm   c.   TEM (false color) 0.5 μm

Figure 16.5 External Structures of Prokaryotic Cells. (a) A glycocalyx enables a cell such as this *Bacteroides* to adhere to a surface. (b) Pili attach cells to objects, surfaces, and other cells. This is *Escherichia coli*. (c) The numerous flagella on this *Proteus mirabilis* enable it to move.

Endospores   Two genera of gram-positive bacteria produce **endospores,** which are dormant, thick-walled structures that can survive harsh conditions (**figure 16.6**). An endospore can withstand boiling, drying, ultraviolet radiation, and disinfectants. Once environmental conditions improve, the endospore germinates and develops into a normal cell.

One spore-forming soilborne bacterium is *Clostridium botulinum.* Food canning processes typically include a high-pressure heat sterilization treatment to destroy endospores of this species. If any endospores survive, they can germinate inside the can, producing cells that thrive in the absence of oxygen. The cells produce a toxin that causes botulism, a severe (and sometimes deadly) form of food poisoning. Green beans, corn, and other vegetables that are improperly home-canned are the most frequent sources of foodborne botulism.

Another spore-forming bacterium is *Bacillus anthracis.* This organism, ordinarily found in soil, can cause a deadly disease when inhaled. Cultures of *B. anthracis* can be dried to induce formation of endospores, then ground into a fine powder that re-

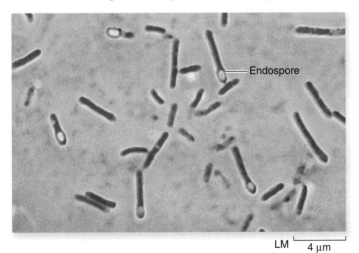

LM 4 μm

Figure 16.6 Endospores. *Clostridium botulinum* bacteria survive environmental extremes by forming thick-walled endospores.

mains infectious for decades. This property makes anthrax a potential biological weapon. In late 2001, anthrax-tainted mail killed five people in the United States. The case remained unsolved until 2008, when distinctive DNA sequences in the powder implicated a military scientist in the crime.

B. Metabolic Pathways May Be Useful in Classification

Over billions of years, bacteria and archaea have developed a tremendous diversity of chemical reactions that allow them to metabolize everything from organic matter to metal. One way to group microorganisms is to examine some of these key metabolic pathways.

The methods by which organisms acquire carbon and energy form one basis for classification. **Autotrophs,** for example, acquire carbon from inorganic sources such as carbon dioxide (CO_2); plants and algae are familiar autotrophs. **Heterotrophs,** on the other hand, get carbon by consuming organic molecules produced by other organisms. *Escherichia coli,* a notorious intestinal bacterium, is a heterotroph. The organism's energy source is also important. **Phototrophs** derive energy from the sun; **chemotrophs** oxidize inorganic or organic chemicals.

By combining these terms for carbon and energy sources, a biologist can describe how a microbe fits into the environment. Plants and cyanobacteria, for example, are photoautotrophs; they use sunlight (*photo-*) for energy and CO_2 (*auto-*) for carbon, as described in chapter 5. Many disease-causing bacteria are chemoheterotrophs because they use organic molecules from their hosts as sources of both carbon and energy.

In addition, oxygen requirements are often important in classification. **Obligate aerobes** require O_2 for generating ATP in respiration (see chapter 6). For **obligate anaerobes,** O_2 is toxic, and they live in habitats that lack it. *Clostridium tetani,* the bacterium that causes tetanus, is one example. **Facultative anaerobes,** which include the intestinal microbes *E. coli* and *Salmonella,* can live either with or without O_2.

Apply It Now

Antibiotics and Other Germ Killers

When a person develops a bacterial infection, a physician may prescribe antibiotics. These drugs typically inhibit structures and functions present in bacterial, but not host, cells. Some mechanisms of action include:

- **Inhibiting cell wall synthesis:** Penicillin is an antibiotic that interferes with cell wall formation. A bacterium that cannot make a rigid cell wall will burst and die.

- **Disrupting cell membranes:** All life depends on an intact membrane that regulates what enters and leaves a cell. Polymyxin antibiotics exploit differences between bacterial and eukaryotic cell membranes. ▶ membranes, p. 54

- **Inhibiting protein synthesis:** No organism can survive if it cannot make proteins. The antibiotics streptomycin, chloramphenicol, and erythromycin bind to different parts of bacterial ribosomes, but all have the same effect—they kill bacteria without killing us. ▶ ribosomes, p. 131

- **Inhibiting transcription:** Gene expression requires RNA synthesis. Rifamycin antibiotics prevent RNA synthesis in bacteria by binding to a bacterial form of RNA polymerase. ▶ transcription, p. 128

- **Inhibiting metabolic enzymes:** Theoretically, antibiotics could block any bacterial metabolic pathway that does not occur in host cells. Sulfanilamide, for example, mimics the substrate of a bacterial enzyme that participates in an essential chain of chemical reactions. ▶ enzymes, p. 78

Unfortunately, the misuse of antibiotics in medicine and agriculture has selected for antibiotic-resistant bacteria. How does this occur?

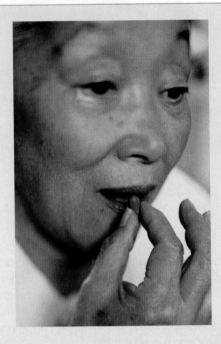

Bacterial DNA accumulates mutations each time a cell divides; cells may also acquire new genes, as described in section 16.3. Therefore, new bacterial strains that are resistant to one or more classes of antibiotics occasionally emerge. Each time we use an antibiotic drug, we kill the susceptible strains and select for the resistant ones. The increasing incidence of antibiotic-resistant bacteria is a huge and growing public health problem (see section 16.6).

C. Molecular Data Reveal Evolutionary Relationships

The traditional classification criteria based on cell structures and metabolism almost certainly group together organisms that are only distantly related to one another. They remain useful, however, because they are based on characteristics that are relatively easy to observe using a microscope and well-defined laboratory tests.

Molecular data such as DNA sequences are harder to obtain and analyze, but they have triggered a revolution in microbial taxonomy and evolutionary biology. Once microbiologists began to analyze the DNA sequences encoding ribosomal RNA (rRNA), for example, they ended up realigning all organisms into three domains (see figure 16.1). The prokaryotes, previously believed to belong to just one kingdom, were divided into two domains—Archaea and Bacteria. In addition, studies of DNA extracted from soil, water, and other habitats have revealed many new species of microbes. ▶ constructing a cladogram, p. 288

DNA sequence data have brought scientists closer to a classification system that reflects evolutionary relationships. Nevertheless, microbiologists are far from understanding how the dozens of phyla of bacteria and archaea are related to one another. Phylogenetic trees depicting their diversity therefore remain too preliminary to include in this chapter. With additional data filling in the gaps, evolutionary relationships should become clearer in the future.

16.2 | Mastering Concepts

1. What are the three most common cell shapes among microbes?
2. What are plasmids, and how are they important?
3. What does the Gram stain reveal about a cell?
4. What are the functions of a glycocalyx, pili, flagella, and endospores?
5. What terms do microbiologists use to describe carbon sources, energy sources, and oxygen requirements?
6. How are molecular data changing microbial taxonomy?

16.3 | Prokaryotes Transmit DNA Vertically and Horizontally

Like all organisms, bacteria and archaea transmit DNA from generation to generation as they reproduce, a process sometimes called **vertical gene transfer.** Eukaryotic cells divide by mitosis or meiosis, the subjects of chapters 8 and 9. In prokaryotes, reproduction occurs by **binary fission,** an asexual process that replicates DNA and distributes it and other cell parts into two daughter cells (**figure 16.7**).

As the prokaryotic cell prepares to divide, the DNA replicates. The chromosome and its duplicate are attached to the inner surface of the cell. The cell membrane grows between the two DNA molecules, separating them. Then the cell membrane dips inward, pinching off two daughter cells from the original one. Formation of cell walls completes the binary fission process. ▶ DNA replication, p. 154

In optimal conditions, some bacterial cells can divide every 20 minutes. This rapid reproduction explains how the oral microbes that survive your nightly tooth-brushing regimen produce countless descendants (and the notoriously unpleasant "morning breath") as you sleep.

Binary fission is a form of asexual reproduction, which usually means that genetic diversity arises only from random mutations in a cell's DNA. But bacteria and archaea can also acquire new genetic material from other sources. In **horizontal gene transfer,** a cell receives DNA from another cell that is not its ancestor.

Horizontal gene transfer occurs in three ways (**figure 16.8**). First, a cell may take up naked DNA without physically contacting

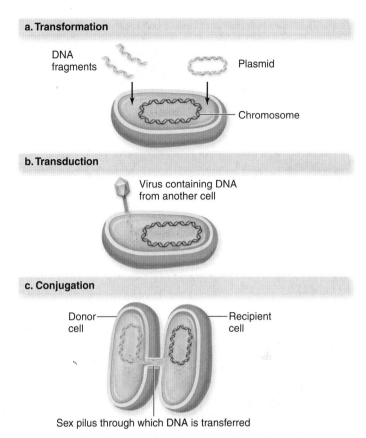

Figure 16.8 Horizontal Gene Transfer. Three forms of horizontal gene transfer include (a) transformation, (b) transduction, and (c) conjugation.

another cell. For example, a dying cell may release its genetic material as it bursts; **transformation** occurs when other cells absorb stray bits of its DNA. Second, viruses sometimes mistakenly package host cell DNA along with their own. In **transduction,** a virus transfers this combined DNA to a bacterial cell. Third, in **conjugation,** one cell receives DNA via direct contact with another cell. A **sex pilus** is the appendage through which DNA passes from donor to recipient.

Horizontal gene transfer has profound implications in fields as diverse as origin-of-life research, medicine, systematics, and biotechnology. Both transduction and conjugation, for example, move antibiotic-resistance genes among bacteria, a serious and growing public health problem. Yet at the same time, biologists take advantage of horizontal gene transfer when they send new genes into bacteria in recombinant DNA technology. ▶ antibiotic resistance, p. 229

16.3 | Mastering Concepts

1. What are the events of binary fission?
2. Distinguish between vertical and horizontal gene transfer.
3. What are the sources of genetic variation in bacteria and archaea?

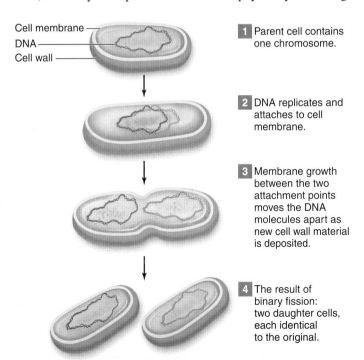

1 Parent cell contains one chromosome.
2 DNA replicates and attaches to cell membrane.
3 Membrane growth between the two attachment points moves the DNA molecules apart as new cell wall material is deposited.
4 The result of binary fission: two daughter cells, each identical to the original.

Figure 16.7 Vertical Gene Transfer. In binary fission, the cell grows and then indents as the DNA replicates, separating one cell into two.

16.4 | Prokaryotes Include Two Domains with Enormous Diversity

For many decades, the tendency to lump together all prokaryotic organisms hid much of the diversity in the microbial world. We now know of so many species of bacteria and archaea in so many habitats that it would take many books to describe them all—and many more species remain undiscovered. This section contains a small sampling of this extraordinary diversity.

A. Domain Bacteria Includes Many Familiar Groups

Scientists have identified 23 phyla within domain Bacteria, but the evolutionary relationships among them remain unclear. Table 16.2 lists a few of the main groups, and figure 16.9 illustrates two examples.

Phylum Proteobacteria is a group of gram-negative bacteria that exemplify the overall diversity within the domain. Some proteobacteria, including the purple sulfur bacteria, carry out photosynthesis. Others play important roles in nitrogen or sulfur cycling, whereas still others form a medically important group that includes enteric bacteria and vibrios. *Helicobacter,* the bacterium that causes ulcers in humans, is a proteobacterium, as are the intestinal bacteria *E. coli* and *Salmonella.* (*E. coli* is the subject of this chapter's Focus on Model Organisms).

Cyanobacteria is a second phylum in domain Bacteria. Billions of years ago, these autotrophs were the first to produce O_2

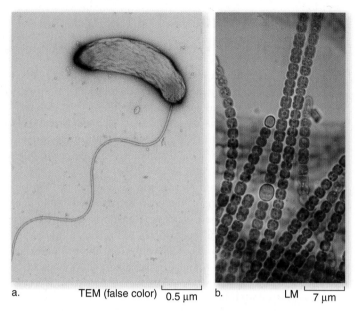

a.   TEM (false color) ⊢0.5 μm⊣   b.   LM ⊢7 μm⊣

Figure 16.9 Two Types of Bacteria. (a) *Vibrio cholera,* a proteobacterium. (b) Filaments of *Anabaena,* a cyanobacterium.

as a byproduct of photosynthesis. They also gave rise to the chloroplasts inside the cells of land plants and green algae. Cyanobacteria remain important in ecosystems, forming the base of aquatic food chains and participating in symbiotic relationships with fungi on land (see section 19.7).

A third phylum, Spirochaetes, contains some medically important bacteria. These spiral-shaped organisms include *Borrelia burgdorferi,* a bacterium that can cause Lyme disease when

Table 16.2   Selected Phyla in Domain Bacteria

Phylum	Features	Example(s)
Proteobacteria	Largest group of gram-negative bacteria	
Purple sulfur bacteria	Bacterial photosynthesis using H_2S (not H_2O) as electron donor	*Chromatium vinosum*
Enteric bacteria	Rod-shaped, facultative anaerobes in animal intestinal tracts	*Escherichia coli, Salmonella* species (cause gastrointestinal disease)
Vibrios	Comma-shaped, facultative anaerobes common in aquatic environments	*Vibrio cholerae* (causes cholera)
Cyanobacteria	Photosynthetic; some fix nitrogen; free-living or symbiotic with plants, fungi (in lichens), or protists	*Nostoc, Anabaena*
Spirochaetes	Spiral-shaped; some pathogens of animals	*Borrelia burgdorferi* (causes Lyme disease), *Treponema pallidum* (causes syphilis)
Firmicutes	Gram-positive bacteria with a low proportion of G+C in their DNA; aerobic or anaerobic; rods or cocci	*Bacillus anthracis* (causes anthrax), *Clostridium tetani* (causes tetanus), *Staphylococcus, Streptococcus*
Actinobacteria	Filamentous gram-positive bacteria with a high proportion of G+C in their DNA	*Streptomyces*
Chlamydiae	Grow only inside host cells; cell walls lack peptidoglycan	*Chlamydia*

Burning Question

Are there areas on Earth where no life exists?

Sand, bare rock, and polar ice may seem devoid of life, but they are not. Scientists using microscopes and molecular tools have discovered microbes living in the hottest, coldest, wettest, driest, saltiest, highest, most radioactive, and most pressurized places on the planet. Bacteria and archaea colonize every imaginable habitat, including places where no other organism can survive.

There are a few places, however, that humans keep artificially microbe-free for the sake of our own health. For example, people in many professions use autoclaves, radiation, and filters to sterilize everything from surgical tools, to medicines and bandages, to processed foods. Artificial sterilization eliminates microbes that could otherwise cause infections, food poisoning, or other illnesses.

Our own bodies are home to many, many microbes, both inside and out. Yet we manage to keep many of our internal fluids and tissues germ-free, including the sinuses, muscles, brain and spinal cord, ovaries and testes, blood, cerebrospinal fluid, urine in kidneys and the bladder, and semen before it enters the urethra. These areas are among the few places where microbes do not ordinarily live; if a bacterial infection does occur, the resulting illness can be deadly.

Submit your burning question to:
marielle_hoefnagels@mcgraw-hill.com

Figure 16.10 Extremophiles. Archaea such as *Sulfolobus* thrive in boiling mud pools. This is Krafla caldera in Iceland.

habitats such as boiling hot springs, whereas the halophiles prefer salt concentrations of up to 30%, and the acidophiles tolerate pH as low as 1.0. As more archaea are discovered in moderate environments such as soil or the open ocean, however, formal taxonomic descriptions become more important.

Microbiologists now tentatively divide domain Archaea into three phyla. One phylum, Euryarchaeota, contains archaea that live in stagnant waters and the anaerobic intestinal tracts of many animals, generating large quantities of methane gas. The same phylum also includes halophilic archaea that carry out photosynthesis in very salty habitats such as seawater, evaporating ponds, and salt flats.

A second phylum, Crenarchaeota, includes species that thrive in acidic hot springs or at hydrothermal vents on the ocean floor. The same phylum also contains a wide variety of soil and water microorganisms with moderate temperature requirements.

Other thermophiles are classified into a third phylum, Korarchaeota, known mostly from genes extracted from their habitats. They seem to be most closely related to the Crenarchaeota.

The importance of archaea in ecosystems is slowly becoming clearer as scientists decipher more about their roles in global carbon, nitrogen, and sulfur cycles. Many live in ocean waters and sediments, a hard-to-explore habitat in which the role of archaea is particularly poorly understood. Their immense numbers, however, suggest that archaea are critical players in ocean ecology.

transmitted to humans in a tick's bite. *Treponema pallidum* causes the sexually transmitted disease syphilis.

Phylum Firmicutes contains gram-positive bacteria with a unique genetic signature: they have a low proportion of guanine and cytosine (G+C) in their DNA. Some of the firmicutes form endospores; examples include *Bacillus anthracis* (the cause of anthrax) and *Clostridium tetani* (the cause of tetanus). Others include *Staphylococcus* and *Streptococcus,* two genera of bacteria that cause infectious disease in humans.

The actinobacteria form another phylum of gram-positive bacteria. These filamentous, soil-dwelling microbes are also medically important: they are the source of infection-fighting antibiotics, including streptomycin.

Bacteria in phylum Chlamydiae are unusual in that they lack cell walls. They cannot generate ATP on their own, so they must live inside a host cell. When they infect cells lining the human genital tract, they cause a sexually transmitted disease called chlamydia.

B. Many, But Not All, Archaea Are "Extremophiles"

Archaea are often collectively described as "extremophiles" because scientists originally found them in habitats that lacked oxygen or that were extremely hot, acidic, or salty (figure 16.10). At first, the organisms were informally divided into groups based on habitat. The thermophiles, for example, live in

16.4 | Mastering Concepts

1. In what ways are bacteria and archaea similar and different?
2. What are some examples of phyla within domain Bacteria?
3. What are the three phyla in Archaea?

Focus on Model Organisms

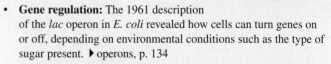

Escherichia coli

Of all the model organisms profiled in this unit, *Escherichia coli*, commonly called *E. coli*, may be the best understood (**figure 16.A**). Dr. Theodor Escherich (1857 to 1911) discovered this normal resident of the human intestinal tract in 1885, and it quickly became a popular lab organism. Biologists have studied every aspect of this bacterium for over a century, and its contributions to biology are enormous. The following are a few highlights:

- **DNA is genetic material:** In 1950, Hershey and Chase used virus-infected *E. coli* cells to demonstrate that DNA, not protein, is the genetic material. ▶ Hershey and Chase, p. 123

- **Genetic exchange:** In the 1940s and 1950s, biologists studying *E. coli* discovered conjugation and transduction (see section 16.3). Both phenomena have important implications for bacterial evolution and transgenic technology.

- **DNA replication:** In the late 1950s, Matthew Meselson and Franklin Stahl used *E. coli* in their famous experiments that demonstrated how DNA copies itself. ▶ DNA replication, p. 154

Figure 16.A
Escherichia coli.

- **Gene regulation:** The 1961 description of the *lac* operon in *E. coli* revealed how cells can turn genes on or off, depending on environmental conditions such as the type of sugar present. ▶ operons, p. 134

- **Transgenic technology:** In 1973, *E. coli* became the first organism to receive a gene from another species. The researchers used the newly discovered restriction enzyme *Eco*R1, derived from *E. coli,* to produce the recombinant plasmids. Since then, *E. coli* cells have received countless genes from many other species. ▶ transgenic organisms, p. 141

- **Gene function:** The *E. coli* genome sequence was completed in 1997, and many scientists are working to describe the functions of hundreds of *E. coli* genes. Many genes in other organisms, including humans, will no doubt have similar DNA sequences and functions. Such studies yield new insights into disease and cell function.

16.5 | Bacteria and Archaea Are Important to Human Life

Many people think of microbes as harmful "germs" that cause disease. Indeed, most of the familiar examples of bacteria listed in the previous section are pathogens. Although some bacteria do make people sick, most microbes do not harm us at all. This section describes some of the ways that bacteria and archaea affect our lives.

A. Microbes Form Vital Links in Ecosystems

Although it may seem hard to believe that one-celled organisms can be essential, the truth is that all other species would die without bacteria and archaea. For example, microbes play essential roles in the global carbon cycle. They decompose organic matter in soil and water, releasing CO_2. Other microorganisms absorb CO_2 in photosynthesis. And all kinds of microbes, both heterotrophs and autotrophs, are eaten by countless organisms in every imaginable habitat. Chapter 37 explains these community interactions in more detail. ▶ carbon cycle, p. 774

Another essential process is **nitrogen fixation,** the chemical reactions in which prokaryotes convert atmospheric nitrogen gas (N_2) to ammonia (NH_3). The element nitrogen is a component of protein, DNA, and many other organic molecules. The only organisms that can use N_2 directly are a few species of bacteria and

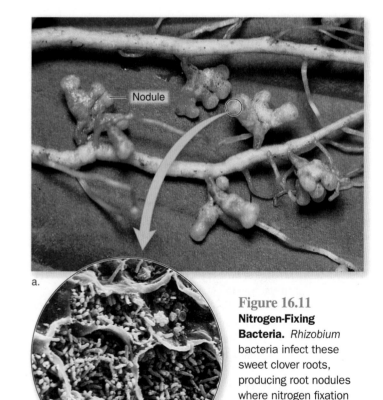

a.

b.   SEM ⌐ 4 µm

Figure 16.11
Nitrogen-Fixing Bacteria. *Rhizobium* bacteria infect these sweet clover roots, producing root nodules where nitrogen fixation occurs. The inset shows a cross section of a root nodule, revealing bacteria inside the plant's cells.

archaea. Ultimately, if not for nitrogen fixers releasing NH_3 that plants and other organisms can absorb, most of Earth's nitrogen would be locked in the atmosphere. The nitrogen cycle—and therefore all life—would eventually cease without these crucial microbes. ▶ nitrogen cycle, p. 775

Some nitrogen-fixing bacteria live in soil or water. Others, such as those in the genus *Rhizobium,* induce the formation of nodules in the roots of clover and other host plants in the legume family (**figure 16.11**). Inside the nodules, *Rhizobium* cells share the nitrogen that they fix with their hosts; in exchange, the bacteria receive nutrients and protection.

B. Bacteria and Archaea Live in and on Us

No matter how hard you scrub, it is impossible to escape the fact that you are a habitat for microorganisms. A menagerie of microbes lives on the skin and in the mouth, large intestine, urogenital tract, and upper respiratory tract (**figure 16.12**). These microscopic companions are beneficial because they help crowd out disease-causing bacteria.

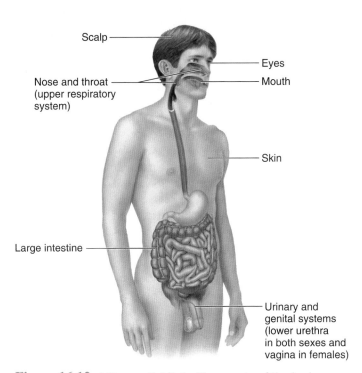

Scalp

Eyes

Nose and throat (upper respiratory system)

Mouth

Skin

Large intestine

Urinary and genital systems (lower urethra in both sexes and vagina in females)

Figure 16.12 A Human Habitat. Many parts of the body house thriving populations of microorganisms. The blood, nervous system, urinary bladder, and some other areas normally remain sterile.

Most people never notice these invisible residents unless something disrupts their personal microbial community. Suppose, for example, that your cat scratches your leg and the wound becomes infected. If you take antibiotics to fight the infection, the drug will probably also kill off some of the normal microbes in your body. As they die, harmful ones can take their place. The resulting microbial imbalance in the intestines or genital tract causes unpleasant side effects such as diarrhea or a vaginal yeast infection. These problems subside in time as the normal microbes divide and restore their populations.

Although most bacteria in and on the human body are harmless, some cause disease. (So far, no archaea are linked to human illnesses.) To cause an infection, bacteria must first enter the body. Animal bites transmit some bacteria, as can sexual activity. A person can also inhale air containing respiratory droplets from a sick coworker or ingest bacteria in contaminated food or water. Bacteria can also enter the body through open wounds.

Once inside the host, pili or slime capsules attach the pathogens to host cells. As the invaders multiply, disease symptoms may develop. Some symptoms result from damage caused by the bacteria themselves. The cells may produce enzymes that break down host tissues, for example, or they may release toxins that harm the host's circulatory, digestive, or nervous system. ▶ enzymes, p. 78

To illustrate bacterial toxins, consider *E. coli,* a normal inhabitant of animal intestines. Sometimes, cattle droppings containing *E. coli* contaminate water, milk, raw fruits and vegetables, hamburger, and other foods. Most strains of *E. coli* are harmless, but strain O157:H7 multiplies inside the body and produces a toxin that can cause belly pain, bloody diarrhea, and, in some cases, life-threatening kidney failure. Outbreaks of *E. coli* strain O157:H7 have led to widely publicized recalls of everything from raw spinach to ground beef to unpasteurized apple juice.

Raw eggs and other foods contaminated with animal feces also may contain *Salmonella,* a close relative of *E. coli. Salmonella, Staphylococcus aureus,* and *Bacillus cereus* are all examples of microbes that thrive in foods that have been improperly refrigerated or inadequately cooked. The toxins they produce in the food—not infection with the bacteria themselves—produce the vomiting and diarrhea associated with food poisoning.

Besides the effects of toxins, a bacterial infection may also trigger an immune reaction. As a result, the classic signs and symptoms of a bacterial infection develop: fever, swollen lymph nodes, pain, and nausea, among others (see chapter 33). Eventually, strong immune defenses usually defeat a bacterial infection.

But the arms race extends to the bacterial side as well. Natural selection favors bacteria that can evade the immune system by hiding inside host cells or forming protective biofilms (see the chapter opening essay). Most pathogens also spread efficiently to new hosts. Bacteria exit the body in many ways: in respiratory droplets, feces, vaginal discharge, or semen, for example. Blood-feeding animals such as ticks and mosquitoes also transmit some pathogenic bacteria.

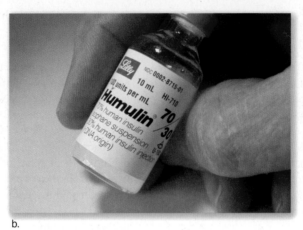

a. b. c.

Figure 16.13 Bacteria at Work. (a) Bacteria of genus *Lactococcus* manufacture cheddar cheese from fermenting milk. (b) Transgenic bacteria produce many drugs, including human insulin. (c) Raw sewage is sprayed on a trickling filter at a municipal wastewater treatment plant. Bacterial biofilms on the filter degrade the organic matter in the sewage.

C. Humans Put Many Prokaryotes to Work

Humans have exploited the metabolic talents of microbes for centuries, long before we could see cells under a microscope (figure 16.13). Scientists continue to invent new ways to use the microbial world.

For instance, many foods are the products of bacterial metabolism. Vinegar, sauerkraut, sourdough bread, pickles, olives, yogurt, and cheese are just a few examples; organic acids released in fermentation produce the tart flavors of these foods. ▶ fermentation, p. 115

Microbes also have many industrial applications. They help produce enormous quantities of vitamins and useful chemicals such as ethanol and acetone. Transgenic bacteria mass-produce human proteins, including insulin and blood-clotting factors. In addition, heat-, acid-, and salt-tolerant enzymes isolated from microbes have found their way into dishwashing detergents and other household products. A heat-tolerant enzyme also transformed modern biology by improving the efficiency of PCR, the polymerase chain reaction. ▶ polymerase chain reaction, p. 171

Water and waste treatment also use bacteria and archaea. Sewage treatment plants in most communities, for example, rely on biofilms consisting of countless microbes that degrade organic wastes. And a technique called bioremediation uses microorganisms to metabolize and detoxify pollutants such as petroleum spills, polychlorinated biphenyls (PCBs), and mercury.

16.5 | Mastering Concepts

1. In what ways are bacteria and archaea essential to eukaryotic life?
2. How are the microbes that colonize your body beneficial?
3. What adaptations enable pathogenic bacteria to enter the body and cause disease?
4. What are some practical uses of bacteria and archaea?

16.6 | Investigating Life: A Bacterial Genome Solves Two Mysteries

DNA has astonishing power as a tool for detecting evolutionary change. Genetic material from organisms that lived hundreds of thousands of years ago can provide clues about life in the distant past (see section 18.6). The same molecule can also teach us about events that occurred within the past 30 years.

Consider, for example, a bacterium called *Staphylococcus aureus*. Most people harbor this common gram-positive organism without becoming ill. If the bacteria breach the skin's defenses or enter the bloodstream, however, they can cause everything from painful boils to blood infection and death. Toxic shock syndrome, scalded skin syndrome, and damage to the heart's inner lining are a few of the potentially life-threatening consequences of a "staph" infection.

This variety of symptoms stems from *S. aureus*'s genetic diversity. Researcher James Musser and his colleagues at the National Institute of Allergy and Infectious Diseases wanted to learn more about the relationship between genetic variation, pathogenicity, and evolution in *S. aureus*. His team studied 37 strains representing the range of known genetic variability in *S. aureus* isolated in the United States, Europe, Canada, and Japan.

The design of Musser's study was simple. The researchers constructed identical microarrays loaded with short DNA sequences from a reference strain of *S. aureus* whose genome was sequenced in 2001. They washed the microarrays with DNA extracted from each of the test strains. If a bacterium's nucleotide sequence matched a spot on the microarray, its DNA stuck there; if not, it simply washed away. The more matches, the more genetically similar each test strain was to the reference. Moreover, the researchers could use the microarray patterns to deduce how similar the strains were to one another. ▶▶ DNA sequencing, p. 170; microarrays, p. 144

Musser and his team were especially interested in the origin of *S. aureus* strains that resist treatment with antibiotics. One such group is called methicillin-resistant *S. aureus*, abbreviated MRSA. These bacteria, which first appeared in the 1960s, earned the public's attention in the 1990s when they began causing frequent infections in hospital patients.

Scientists have offered two competing explanations for the origin of MRSA. One hypothesis is that MRSA evolved many times through horizontal transfer of the gene that confers antibiotic resistance. An alternative hypothesis is that the resistance gene arose by random mutation only once in a formerly susceptible cell. With replication and additional mutations, the descendants of that cell gave rise to multiple resistant strains.

The researchers tested both hypotheses by assembling their microarray data into a treelike diagram showing clusters of the most closely related organisms. The study included 12 MRSA strains. If a single lineage (tree "branch") had contained all of the resistant strains, it would have supported the hypothesis that all MRSA originated from a single ancestor. Instead, the resistant strains were scattered throughout the tree diagram, supporting the multiple-origin hypothesis (**figure 16.14**).

A related, hotly debated question is the origin of the toxic shock syndrome epidemic of 1980 (**figure 16.15**). Most cases were linked with the use of high-absorbency tampons among previously healthy young women. When technicians cultured bacteria from the vaginas of the ill women, they found *S. aureus* 90% of the time.

Why the sudden epidemic in 1980? Epidemiologists have proposed two hypotheses. One possibility is that the increasing use of ultra-high-absorbency tampons at that time created a new type of environment; many strains of *S. aureus* may have independently exploited this change. An alternative explanation is that just one cell mutated and became a highly successful "superpathogen" whose descendants swept across multiple continents.

The researchers' tree diagram again led to the answer (see figure 16.14). The 11 strains from toxic shock syndrome patients were not genetically identical, as would be expected if all were clones of the same superpathogen. The team therefore rejected the second hypothesis in favor of the first, concluding that multiple *S. aureus* strains together caused the epidemic.

DNA evidence is most famous for its role in solving crimes, but it is equally valuable to the "detectives" who study infectious disease. Understanding how bacteria have evolved in the recent past gives scientists a head start in anticipating future epidemics. Given the enormous adaptability of bacterial populations, we probably will need all the help we can get.

Fitzgerald, J. Ross and four coauthors, including James M. Musser. July 17, 2001. Evolutionary genomics of *Staphylococcus aureus*: Insights into the origin of methicillin-resistant strains and the toxic shock syndrome epidemic. *Proceedings of the National Academy of Sciences*, vol. 98, pages 8821–8826.

16.6 | Mastering Concepts

1. How did Musser's team use their data set to help answer two different questions?
2. Propose a change to figure 16.14 that would support the hypothesis that all strains associated with toxic shock syndrome arose from one "superpathogen."

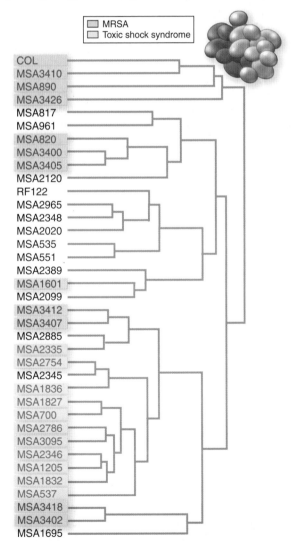

Figure 16.14 *Staphylococcus aureus* **Relationships.** This tree helped reveal the origins of MRSA and toxic shock syndrome.

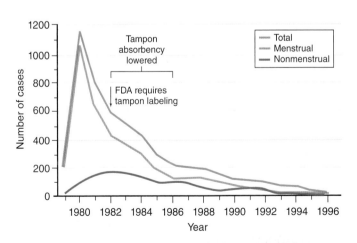

Figure 16.15 **Toxic Shock Syndrome Epidemic.** Ultra-high-absorbency tampons were linked to a spike in toxic shock syndrome cases in 1980.

Chapter Summary

16.1 | Prokaryotes Are a Biological Success Story

- **Prokaryotes** (domains **Bacteria** and **Archaea**) were the first organisms, and they may have given rise to mitochondria and chloroplasts in eukaryotic cells more than a billion years ago.
- Bacteria and archaea are very abundant and diverse, and they occupy a great variety of habitats.

16.2 | Prokaryote Classification Traditionally Relies on Visible Features

A. Microscopes Reveal Cell Structures

- Like other organisms, prokaryotic cells contain DNA and **ribosomes,** and they are bounded by a cell membrane.
- The chromosome, along with some RNA and proteins, is located in an area called the **nucleoid. Plasmids** are circles of DNA apart from the chromosome.
- A **cell wall** surrounds most prokaryotic cell membranes and confers the cell's shape: a spherical **coccus,** rod-shaped **bacillus,** or spiral-shaped **spirillum.**
- **Gram-staining** reveals differences in cell wall architecture. Gram-positive bacteria have a thick **peptidoglycan** layer; gram-negative bacteria have an outer membrane surrounding the thinner peptidoglycan cell wall.
- A **glycocalyx** outside the cell wall provides attachment to surfaces or protection from host immune system cells.
- **Pili** are projections that allow cells to adhere to surfaces or transfer DNA to other cells.
- **Flagella** rotate to allow a motile cell to move toward or away from a stimulus, a process called **taxis.**
- Some bacteria survive harsh conditions by forming protective **endospores.**

B. Metabolic Pathways May Be Useful in Classification

- **Autotrophs** acquire carbon from inorganic sources, and **heterotrophs** obtain carbon from other organisms. A **phototroph** derives energy from the sun, and a **chemotroph** acquires energy by oxidizing organic or inorganic chemicals.
- **Obligate aerobes** require oxygen, **facultative anaerobes** can live whether or not oxygen is present, and **obligate anaerobes** cannot function in the presence of oxygen.

C. Molecular Data Reveal Evolutionary Relationships

- Microbiologists are using molecular data to reconsider the traditional classification of bacteria and archaea.

16.3 | Prokaryotes Transmit DNA Vertically and Horizontally

- Cell division occurs by **binary fission** in prokaryotes. Binary fission is a form of **vertical gene transfer,** in which DNA is transferred to the next generation.
- Prokaryotes can also acquire new DNA directly from the environment (**transformation**), a virus (**transduction**), or another cell via a **sex pilus** (**conjugation**). All three are routes of **horizontal gene transfer.**

16.4 | Prokaryotes Include Two Domains with Enormous Diversity

A. Domain Bacteria Includes Many Familiar Groups

- Proteobacteria, cyanobacteria, spirochetes, firmicutes, actinobacteria, and chlamydias are a few examples of diversity within domain Bacteria.

B. Many, But Not All, Archaea Are "Extremophiles"

- Domain Archaea contains methanogens, halophiles, thermophiles, and many organisms that thrive in moderate conditions.

16.5 | Bacteria and Archaea Are Important to Human Life

A. Microbes Form Vital Links in Ecosystems

- All life depends on the bacteria and archaea that contribute gases to the atmosphere, recycle organic matter, and **fix nitrogen.**

B. Bacteria and Archaea Live in and on Us

- Pathogenic bacteria adhere to host cells and colonize tissues. Disease symptoms often result from the immune system's reaction to the infection.
- Toxins produced by bacteria, plus the ability to spread to new hosts, also contribute to disease.

C. Humans Put Many Prokaryotes to Work

- Bacteria and archaea are used in the manufacture of many foods, drugs, and other chemicals, in sewage treatment, and in bioremediation.

16.6 | Investigating Life: A Bacterial Genome Solves Two Mysteries

- Molecular studies revealed that *Staphylococcus aureus* strains resistant to multiple antibiotics arose independently several times, as did those that caused the toxic shock syndrome epidemic of 1980.

Multiple Choice Questions

1. A *prokaryotic* cell is one that
   a. lacks DNA.
   b. has membrane-bounded organelles.
   c. lacks a nucleus.
   d. lacks a plasma membrane.

2. What is the cell wall of a gram-negative cell made of?
   a. Pectin   c. Glycocalyx
   b. Peptidoglycan   d. Plasmids

3. The external structure that attaches the prokaryote to its environment is a
   a. pilus.   c. plasmid.
   b. flagellum.   d. endospore.

4. Which of these is a distinguishing characteristic between the domains Bacteria and Archaea?
   a. Their size
   b. The chemical composition of the cell wall and cell membrane
   c. Their ability to grow at high temperatures
   d. The presence of membrane-bounded organelles

5. What name is used to describe prokaryotic cells that are rod-shaped?
   a. Spirilla   c. Cocci
   b. Bacilli   d. Both a and b are correct.

6. Which form of prokaryote would be the most challenging to isolate and culture?
   a. A facultative anaerobe   c. An obligate anaerobe
   b. A photoautotroph   d. Both a and b are correct.

7. Which of the following is the best explanation of why binary fission can occur without a spindle structure like that found in mitotic cells?
   a. The cell is small, so there is less material to divide.
   b. There is only one chromosome, and it attaches to the membrane.
   c. The prokaryotic DNA does not need to replicate.
   d. The DNA is transferred through the sex pilus.

8. Which of the following requires the participation of a virus?
   a. Conjugation   c. Transduction
   b. Transformation   d. Transfection

9. What property of a prokaryote makes it useful for bioremediation?
   a. The ability to reproduce quickly
   b. The ability to live in extreme environments
   c. The diversity of metabolic pathways found in various prokaryotes
   d. All of the above are correct.

10. Which of the following processes occurs ONLY in prokaryotes?
    a. Nitrogen fixation
    b. Photosynthesis
    c. Asexual reproduction
    d. All of the above are correct.

Write It Out

1. Give five examples that illustrate how bacteria and archaea are important to other types of organisms.

2. How can the polymerase chain reaction (PCR; see chapter 8) be useful in studying microbes that do not survive in laboratory culture?

3. If you were developing a new "broad-spectrum" antibiotic to kill a wide variety of bacteria, which cell structures and pathways would you target? Which of those targets also occur in eukaryotic cells, and why is that important? How would your strategy change if you were designing a new "narrow-spectrum" antibiotic active against only a few types of bacteria?

4. Distinguish between the following pairs of terms: (a) a phototroph and a chemotroph; (b) a gram-negative and a gram-positive bacterium; (c) an autotroph and a heterotroph; (d) an obligate anaerobe and a facultative anaerobe; and (e) transformation and transduction.

5. Scientists are studying bacteria discovered on the seafloor. In the absence of light, these bacteria can use organic molecules to generate electricity. NASA hopes they may one day be useful in producing some of the electricity on spacecraft, using human waste as fuel. What mode of nutrition do these bacteria probably use?

6. How do prokaryotes reproduce, and what are three ways they can acquire genes other than by vertical gene transfer?

7. List the ways that binary fission is similar to and different from mitosis.

8. Why did the discovery of archaea generate interest in searching for cells on other planets?

9. What adaptations in pathogenic bacteria enable them to cause disease?

10. How have humans harnessed the metabolic diversity of bacteria and archaea for industrial purposes?

11. A young child develops a very high fever and an extremely painful sore throat. Knowing that the child could have an infection with a strain of *Streptococcus* that could be deadly, the physician seeks a precise diagnosis. What three approaches might the doctor (or a laboratory) use to tell whether this infection is viral or bacterial and, if the latter, to identify the bacterium?

12. Stomach ulcers, once thought to be entirely a product of spicy food or high stress, are now known to be caused by bacteria (*Helicobacter pylori*). How has ulcer treatment changed because of this new knowledge?

13. Researchers can raise mice in the complete absence of microorganisms. When these animals are subsequently exposed to pathogenic microbes, they are more likely to develop disease than are mice that have been raised in a normal (nonsterile) environment. Explain this finding.

14. Botox is a toxin produced by the bacterium *Clostridium botulinum*. When ingested with tainted food, Botox can kill by paralyzing muscles needed for breathing and heartbeat. Physicians inject small quantities of diluted Botox into facial muscles to paralyze them and reduce the appearance of wrinkles. Some people have expressed concern about a trend in which people come together for "Botox parties" at hair salons and other nonmedical settings. What are the risks of getting injections in such a setting?

15. Use the Internet to learn about the goals of the Human Microbiome Project. Why is it important to study the microbes that live in and on humans?

16. Use the Internet to learn how the federal government investigates outbreaks of foodborne illnesses such as *E. coli* O157:H7 and *Salmonella*. What can you do to protect yourself from food poisoning?

17. Use the Internet to learn about sexually transmitted diseases (STDs). Which of the most common STDs are caused by bacteria?

Choose one STD caused by bacteria to study in more depth; describe how the pathogen spreads and its symptoms, treatment, and prevention.

18. Impetigo, MRSA, and other bacterial skin infections spread easily among people living in close quarters such as dorms. To prevent the spread of skin infections, health experts recommend washing hands, showering after participating in sports, disinfecting surfaces, covering open wounds with bandages, not sharing personal items such as razors, and washing athletic clothing after use. How does each strategy help prevent infection?

19. If you worked for a school confronting an outbreak of *S. aureus*, how would you determine whether the strains were MRSA? What measures would you recommend to control the outbreak?

20. Genome sequencing projects are complete or in progress for many prokaryotes other than *E. coli*. Use the Internet to choose one of these bacteria or archaea, and describe some new discoveries that have come from research on the organism you chose.

21. *Mycobacterium tuberculosis* causes most cases of tuberculosis. Recently, strains of this bacterium that are resistant to all known antibiotic drugs have become increasingly common. Explain how this change occurred; use the terms *mutation, DNA,* and *natural selection* in your answer.

22. This book depicts the diversity of life with an evolutionary tree (see, for example, figure 16.1). Some biologists, however, are beginning to show life's evolutionary history with a more weblike diagram that accounts for genes moving among multiple branches and does not assume a single common root. What are some arguments in favor of each approach?

Pull It Together

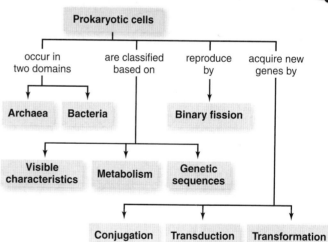

1. How are archaea different from bacteria?

2. What are the most common cell shapes for bacteria?

3. Add autotrophs, heterotrophs, phototrophs, and chemotrophs to this concept map.

4. Where do obligate aerobes, obligate anaerobes, and facultative anaerobes fit on this map?

5. Which terms on this concept map refer to types of horizontal gene transfer?

6. Create a new concept map that includes the internal and external parts of a prokaryotic cell.

7. What are the cell features and metabolic criteria by which biologists classify microbes?

Chapter

17 Protists

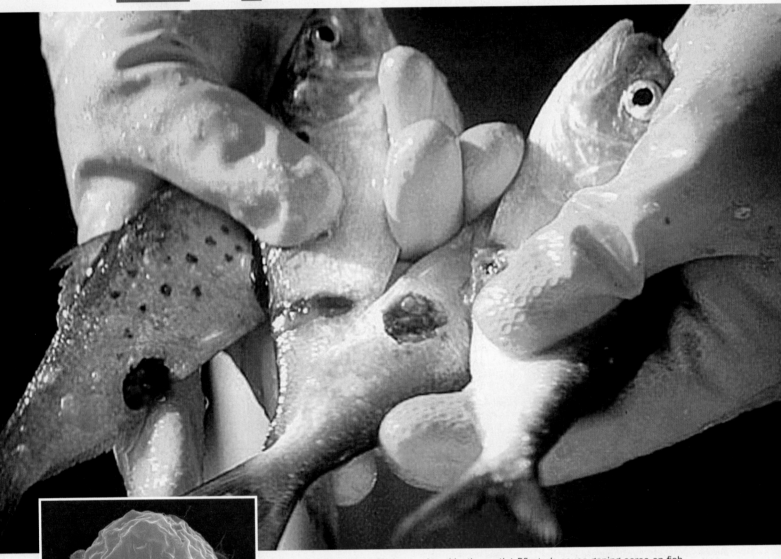

Pfiesteria Fish Lesions. Toxins produced by the protist *Pfiesteria* cause gaping sores on fish.

SEM (false color) 5 μm

Mc Graw Hill **connect** |BIOLOGY

Enhance your study of
this chapter with practice quizzes,
animations and videos, answer keys,
and downloadable study tools.
www.mhhe.com/hoefnagels

Science Lessons from a Killer Cell

OF THE MANY THOUSANDS OF SPECIES OF PROTISTS, THE VAST MAJORITY ARE HARMLESS TO ANIMALS. One exception, however, is *Pfiesteria,* which lives in water off the east coast of the United States. *Pfiesteria* belongs to a group of protists called dinoflagellates. Typically, *Pfiesteria* is nontoxic, feeding on algae and bacteria in coastal waters. But a few of the stages in its life cycle can produce extremely potent toxins that kill fish and accumulate in shellfish, making them poisonous to humans.

Pfiesteria produces at least two toxins. One is a powerful neurotoxin, and the other causes disintegration of the skin of a fish. Although the toxins break down rapidly, the infected fish often develop open sores. Many fish die. In humans, *Pfiesteria* toxins can produce rashes, open sores, fatigue, erratic heartbeat, breathing difficulty, personality changes, and extreme memory loss.

The on-again, off-again nature of the toxins has made them very difficult for scientists to study. Controversy erupted in 2002, when scientists reported that a species called *Pfiesteria shumwayae* did not produce toxins. In making their case, the researchers used several experimental strategies. They exposed fish to fluid in which *P. shumwayae* had grown (after first removing the cells). They also tried to extract toxic chemicals from water in which *P. shumwayae* grew. They searched the microbe's genome for genes known to encode toxins in other organisms. Finally, they grew fish and *P. shumwayae* cells in separate halves of two-part tanks, divided by a membrane through which toxins, but not cells, could pass. None of the experiments yielded evidence of toxins.

Later in the same year, however, other researchers refuted these results, saying that the first group raised the *Pfiesteria* cells under conditions that suppressed toxin production (the cells must grow with live fish to produce toxins). This second group used the same form of *P. shumwayae* as the first group, but they also included two control forms: one known to make toxins and the other known to be nontoxic. The results indicated that, under some conditions, *P. shumwayae* did indeed produce toxins that harmed fish.

These conflicting studies illustrate two important features of science. First, the conditions under which scientists conduct experiments can greatly affect their conclusions. Second, communication is vital. When the first group published their results, they described their *Pfiesteria* forms and experimental techniques in detail. The second group used this information to identify conditions known to suppress toxin production. The careful reporting of materials and methods enables scientists to evaluate one another's work.

This chapter's content illustrates a third feature of science: adaptability to new information. Protists were once considered a single kingdom, but we now know that the protists are much too diverse to fit into one neat category. This chapter presents a sampling of that diversity.

Learning Outline

Learn How to Learn

What's the Point of Rewriting Your Notes?

Your notes are your record of what happened in class, so why should you rewrite them after lecture is over? One answer is that the abbreviations and shorthand that make perfect sense while you take notes will become increasingly mysterious as time goes by. Rewriting the information in complete sentences not only reinforces learning but also makes your notes much easier to study before an exam.

17.1 | Protists Lie at the Crossroads Between Simple and Complex Organisms

Our tour of life's diversity began with viruses, infectious particles that lie on the border between the living and the nonliving. Chapter 16 then described the two domains of prokaryotes, the Bacteria and the Archaea. The ancestors of today's prokaryotes played pivotal roles in the evolution of life, inventing countless metabolic pathways and releasing oxygen into Earth's atmosphere.

About 2 billion years ago, the prokaryotes gave rise to a new, more complex cell type: the eukaryote. In contrast to a prokaryote, a **eukaryotic cell** has a nucleus and other membrane-bounded organelles, such as mitochondria and chloroplasts. Figures 3.8 and 3.9 clearly show the compartmentalization and division of labor typical of eukaryotic cells, and section 3.2 offers a more complete description of the distinctions among the cell types. We now embark on a tour of the protists, the simplest eukaryotes.

A. What Is a Protist?

Until recently, biologists recognized four eukaryotic kingdoms: Protista, Plantae, Fungi, and Animalia. The plants, fungi, and animals are distinguished based on their characteristics. Loosely defined, plants are multicellular eukaryotes that carry out photosynthesis; fungi are mostly multicellular eukaryotes that obtain food by external digestion; and animals are multicellular eukaryotes that obtain food by ingestion.

Kingdom Protista, in contrast, was defined by *exclusion*. An organism was designated a **protist** if it was a eukaryote that did not fit the description of a plant, fungus, or animal. Kingdom Protista was, in effect, a convenient but artificial "none of the above" category (**figure 17.1**). Not surprisingly, the nearly 100,000 named species of protists are metabolically diverse, displaying great variety in size, nutrition, locomotion, reproduction, and cell surfaces.

Textbooks and taxonomists have traditionally considered the protists in terms of the more familiar organisms that they resemble: the plantlike algae, funguslike slime molds, and animal-like protozoa. Modern systematists, however, try to group organisms based on evolutionary relationships. DNA sequences provide the most objective measure of relatedness. Based on these new molecular data, the former Kingdom Protista has shattered into dozens of groups whose relationships to one another remain uncertain. Whether those groups become full-fledged kingdoms, subkingdom "supergroups," phyla, or some other taxonomic designation remains to be seen. ▶ systematics, p. 286

Because the classification of protists is in transition and many of the new groupings are not universally accepted, this chapter uses the traditional approach to classification. Section 17.5, however, revisits modern trends in protist taxonomy.

B. Protists Are Important in Many Ways

The metabolic diversity among protists means they have an astonishingly wide variety of functions and roles in human life. In ecosystems, the autotrophic ("self-feeding") algae carry out photosynthesis, producing much of the O_2 in Earth's atmosphere and supporting food webs in oceans, lakes, rivers, and ponds. In addition, algae living among the threads of fungi on rocks and tree bark form lichens, the first step in building soil from bare rock. ▶▶ food webs, p. 768; lichens, p. 403

Medically important protists include parasites that infect plant and animal hosts (including humans). These heterotrophic organisms can contaminate food or water or be transmitted by insects. The costs amount to billions of dollars, incalculable suffering, and millions of deaths annually.

Protists also have found their way into diverse industrial applications. Some algae, for example, help make chocolate smooth and creamy, whereas others help make paints reflective. The distinctive fossils of still other species point the way to petroleum reserves. Conversely, biologists may one day be able to reduce our reliance on petroleum by harnessing protist photosynthesis and reproduction in "bioreactors" that use sunlight and CO_2 to produce oil and other biofuels.

C. Protists Have a Lengthy Evolutionary History

Evolutionary biologists are especially interested in protists. One reason is that the cells of today's protists may retain clues to important milestones in eukaryote history. The endosymbiosis theory, for example, suggests that early eukaryotes originally obtained mitochondria and chloroplasts by engulfing other cells.

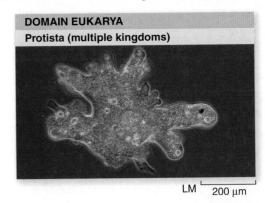

DOMAIN EUKARYA
Protista (multiple kingdoms)

LM   200 μm

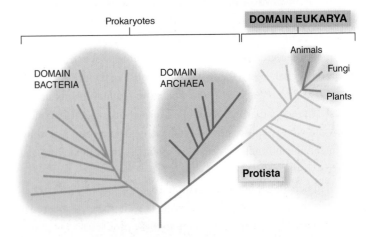

Figure 17.1 Protists at the Crossroads. "Kingdom" Protista consists of many lineages, each of which may eventually be considered its own kingdom. Plants, fungi, and animals trace their ancestry to protists, living or extinct.

Nearly all eukaryotes have mitochondria, so the cells that became these organelles were probably engulfed first. Biologists can learn more about the origin of mitochondria by studying protists whose mitochondrial genomes are very much like those of bacteria. ▶ endosymbiosis, p. 304

Chloroplasts have an especially colorful evolutionary history (figure 17.2). Biologists studying chloroplast DNA, membrane structure, and photosynthetic pigments suggest that the chloroplasts of red algae, green algae, and land plants all arose from cyanobacteria engulfed by some ancient eukaryotic cell. The chloroplasts in these species are surrounded by two membranes. In other photosynthetic protists, three or more membranes surround each chloroplast. The evidence suggests that these organelles originated long ago when red algae or green algae were themselves engulfed by other cells—an event called secondary endosymbiosis. No one knows exactly how many times, or when, these endosymbiosis events happened. We are just beginning to unravel the events surrounding the origins of all the different types of chloroplasts.

Another important milestone in eukaryote history is the origin of multicellularity. Biologists do not know how unicellular eukaryotes adopted a multicellular lifestyle. The fossil record is essentially silent on the matter, mostly because the first multicellular organisms lacked hard parts that fossilize readily. We do know, however, that multicellularity arose independently in multiple lineages. After all, genetic evidence clearly suggests that plants, fungi, and animals arose from different lineages of unicellular protists. Biologists hope to gain insight into the steps that led to multicellularity by studying protists that form colonies in which individuals interact as they move and obtain food. ▶ multicellularity, p. 305

A second reason that protists are important to studies of evolutionary biology is that these simple eukaryotes shed light on the evolutionary history of plants, fungi, and animals. The genetic and cellular similarities shared by green algae and plants, for example, illustrate the strong evolutionary connection between these organisms. Likewise, DNA evidence suggests that heterotrophic protists called choanoflagellates are the closest existing relatives to animals. These organisms bear an uncanny resemblance to specialized cells in sponges, which are the simplest animals. ▶ sponges, p. 415

Clearly, depicting the diversity of the protists in just one chapter is a challenge. In exploring these organisms, remember that the examples in each section represent just a small sampling of some of the most intriguing members of this uniquely variable group.

17.1 | Mastering Concepts

1. What features define the protists?
2. Describe examples of how protists are important.
3. Why are evolutionary biologists interested in protists?

Figure 17.2 Proposed Origin of Mitochondria and Chloroplasts. In primary endosymbiosis, an early eukaryote engulfs a bacterium, which subsequently becomes a mitochondrion or chloroplast. In a secondary endosymbiosis, one eukaryote engulfs another eukaryote, forming a chloroplast with more than two membranes.

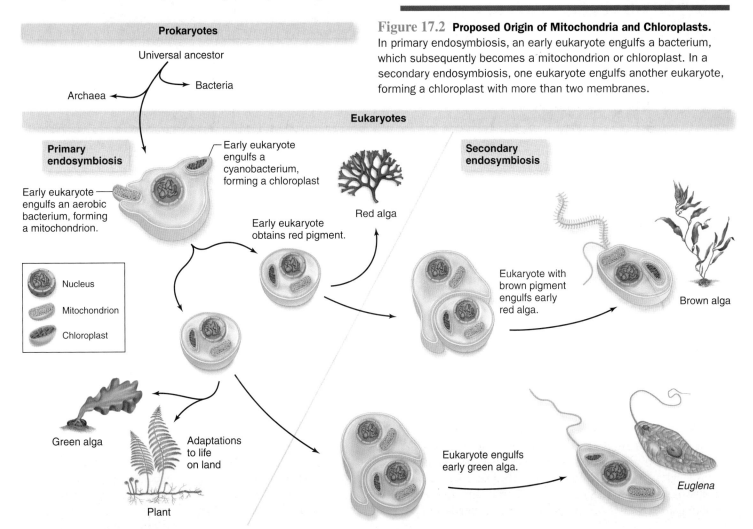

17.2 | Many Protists Are Photosynthetic

Most people probably think of algae as pond scum, but **algae** is a general term that refers to any photosynthetic protist that lives in water. Although the cyanobacteria were traditionally called "blue-green algae," most biologists now reserve the term *algae* for eukaryotes.

The cells of algae contain chloroplasts that house a rainbow of yellow, gold, brown, red, and green photosynthetic pigments. These organelles use light energy and CO_2 to produce carbohydrates and other organic molecules that support freshwater and marine food webs. They also release O_2 as a waste product.

Algae may be single-celled, colonial, filamentous, or multicellular. Some of the more complex species produce differentiated tissues. Although the body forms may resemble those of plants, algae are considered protists because they lack the distinctive reproductive structures that define plants. This section describes the major types of algae.

A. Euglenoids Are Heterotrophs and Autotrophs

The **euglenoids** are unicellular protists with elongated cells (figure 17.3). Most have a long, whiplike flagellum used in locomotion and a short flagellum that does not extend from the cell. Supporting the cell membrane is a pellicle, a protective layer made of rigid or elastic protein strips. An eyespot helps the cell orient toward light.

Most euglenoids inhabit fresh water. About one-third of the species are photosynthetic, and the rest feed on organic compounds suspended in the water. But these metabolic roles are not always fixed. Photosynthetic euglenoids such as *Euglena,* for example, may occasionally feed on organic matter. In darkness, their cells become entirely heterotrophic, although photosynthesis resumes once light returns.

Figure 17.3 *Euglena.* The pond-dwelling *Euglena* has a flagellum and chloroplasts, but it can also ingest food particles.

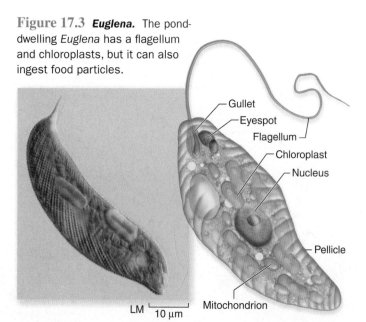

- Gullet
- Eyespot
- Flagellum
- Chloroplast
- Nucleus
- Pellicle
- Mitochondrion

LM |— 10 µm

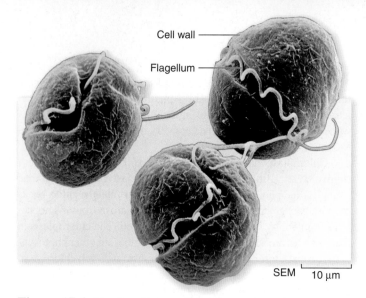

Cell wall ——
Flagellum ——

SEM |— 10 µm

Figure 17.4 **Dinoflagellates.** *Gymnodinium* sp. exhibits classic dinoflagellate structure. Note the flagella and cellulose plates that make up the cell wall.

B. Dinoflagellates Are "Whirling Cells"

The marine protists known as **dinoflagellates** are characterized by two flagella of different lengths (figure 17.4). One of the flagella beats in a way that propels the cell with a whirling motion (the Greek *dinein* means "to whirl"). In addition, many dinoflagellates have cell walls that consist of overlapping cellulose plates.

Dinoflagellates are a major component of plankton, the microscopic food that supports the ocean's vast food webs. About half the species are photosynthetic. Several species live within the tissues of jellyfishes, corals, sea anemones, or giant clams, providing carbohydrates to their host animals. Many other species of dinoflagellates are predators or parasites. Some are bioluminescent, producing flashing lights in tropical waters.

A red tide is a sudden population explosion, or "bloom," of dinoflagellates that turn the water red, orange, or brown (figure 17.5). Because some dinoflagellates are colorless, however, many biologists now prefer the term *harmful algal bloom* over *red tide*. Usually these blooms occur in response to a boost in the nutrient content of the water, as described in this chapter's Burning Question.

A bloom of dinoflagellates can be deadly. The chapter opening essay described *Pfiesteria,* a toxin-producing dinoflagellate that kills fish and causes neurological symptoms in humans. The toxins of other species may become concentrated in the tissues of clams, scallops, oysters, and mussels. A person who eats tainted

Figure 17.5 **Harmful Algal Bloom.** Huge populations of dinoflagellates cause red tides.

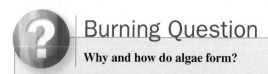

Burning Question

Why and how do algae form?

Figure 17.A Algal Bloom.

Algae are common aquatic organisms, but they are often inconspicuous. Sometimes, however, their populations grow so large that they seem to take over; ponds and poorly maintained swimming pools can turn bright green with algae (**figure 17.A**). This population explosion occurs when nutrients and sunlight are abundant.

Algal blooms are normal in some ecosystems, such as in many ponds. A bloom where water is normally clear, however, usually indicates that nutrients from sewage, fertilizer, or animal waste are polluting the waterway. The use of lawn fertilizers is a common cause of algal blooms in ponds in residential settings. Algal blooms induced by nutrient pollution can devastate aquatic life.
▶ eutrophication, p. 776

Submit your burning question to:
marielle_hoefnagels@mcgraw-hill.com

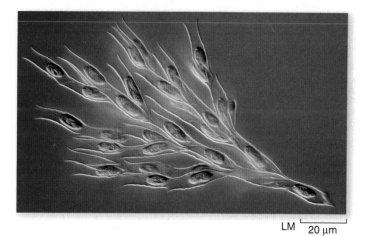

LM ⌐ 20 μm

Figure 17.6 Golden Algae. *Dinobryon* is a colonial golden alga.

cellular, but filamentous and colonial forms also exist. *Dinobryon* is an example of a freshwater genus of golden algae; in these protists, individual vase-shaped cells stack end to end to produce branched or unbranched chains.

Like *Euglena,* golden algae can act as autotrophs or heterotrophs. In some aquatic ecosystems, photosynthesis by golden algae provides a significant source of food for zooplankton. When light or nutrient supplies dwindle, however, many golden algae can consume bacteria or other protists.

Some golden algae can be harmful. In nutrient-rich streams and lakes, blooms of golden algae can release toxins that trigger fish kills (although humans are apparently unaffected).

Diatoms   **Diatoms** are unicellular algae with two-part silica cell walls that confer a variety of ornate shapes (**figure 17.7**). These protists occupy just about every moist habitat on Earth, from oceans to streams to fish tanks to damp soil. Their populations are sensitive to water pH, salinity, and other environmental conditions. Biologists therefore periodically sample freshwater diatom populations as indirect indicators of environmental quality.

shellfish may develop the numb mouth, lips, face, and limbs that are characteristic of paralytic shellfish poisoning. In this context, the familiar advice "never eat shellfish in a month without an R" makes sense, because toxic algae blooms are most frequent from May through August. In reality, however, most commercially harvested shellfish undergo testing and are safe to eat year-round.

C. Golden Algae, Diatoms, and Brown Algae Contain Yellowish Pigments

Three groups of algae contain a yellowish photosynthetic pigment called fucoxanthin in addition to chlorophylls *a* and *c*. This accessory pigment gives these organisms a golden, olive green, or brown hue.

Golden Algae   The **golden algae** are named for their color (**figure 17.6**). Their cells usually have two flagella. Most are uni-

Figure 17.7 Diatoms. The "glass houses" (silica cell walls) of these photosynthetic protists exhibit a dazzling variety of forms.

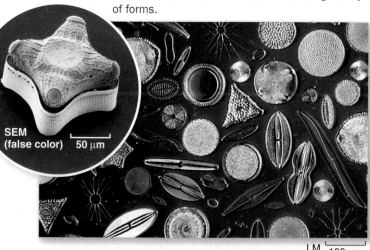

SEM (false color) ⌐ 50 μm

LM ⌐ 100 μm

Although diatoms occur nearly everywhere, most live in oceans. Their tiny photosynthetic cells can reach huge densities, removing CO_2 from the atmosphere and providing food for zooplankton. Over millions of years, the glassy shells of diatoms have accumulated on the ocean floor, forming deposits over 200 meters thick in some regions. The abrasive shells mined from these deposits are used in swimming pool filters, polishes, toothpaste, and many other products. Diatoms also impart the distinct reflective quality of paints used in roadway signs and license plates.

Few microorganisms fossilize as well as diatoms, and biologists know of some 35,000 extinct species. Section 17.6 describes how diatoms have left a complete record of the formation of a new species in a Wyoming lake.

Brown Algae   The **brown algae** are the most complex and largest protists. Although they are multicellular, they resemble the golden algae in their pigments and their reproductive cells, each of which bears two flagella.

Brown algae live in marine habitats all over the world. The Sargasso Sea in the northern Atlantic Ocean, for example, is named after floating masses of brown algae called *Sargassum.* The kelps, which are the largest of the brown algae, produce enormous underwater forests that provide food and habitat for many animals (figure 17.8). Some kelps exceed 30 meters in length.

Humans consume several species of kelp. *Laminaria digitata,* for example, is an ingredient in many Asian dishes. Algin, a chemical extracted from the cell walls of brown algae, is used as an emulsifying, thickening, and stabilizing agent in products including ice cream, candies, chocolate, salad dressings, sauces, soft drinks, beer, cough syrup, toothpaste, cosmetics, polishes, latex paint, and paper.

Figure 17.8 A Giant Kelp. A holdfast organ anchors a brown alga to a substrate, while the gas-filled bladders add buoyancy. The leaflike blades expand the surface area for photosynthesis.

Blade

Bladder

Holdfast

Figure 17.9 A Red Alga. *Bossiella* is a coralline red alga that secretes calcium carbonate and helps build coral reefs.

D. Red Algae Can Live in Deep Water

Most **red algae** are relatively large (figure 17.9), although some are microscopic. These marine organisms are somewhat similar to green algae in that they store carbohydrates as a modified form of starch, have cell walls containing cellulose, and produce chlorophyll *a.*

Red algae, however, can live in water exceeding 200 meters in depth, thanks to reddish and bluish photosynthetic pigments that absorb wavelengths of light that chlorophyll *a* cannot capture. All light becomes dimmer with increasing depth, but the wavelengths do not dissipate equally. The pigments in red algae can use some of the wavelengths that persist in deep water.

Figure It Out

Consult figure 5.4 to see the light absorption spectrum for chlorophyll *a.* Predict the most likely range of wavelengths absorbed by the photosynthetic pigments unique to red algae.

Answer: Between 500 and 650 nm

Humans use red algae in many ways. Agar, for example, is a polysaccharide in the cell walls of some species. This jellylike substance is used as a culture medium for microorganisms, an inert ingredient in medications, a gel in canned meats, and a thickener in ice cream and yogurt. Another useful product is carrageenan, a polysaccharide that emulsifies fats in chocolate bars and stabilizes paints, cosmetics, and creamy foods. A red alga called nori is used for wrapping sushi.

E. Green Algae Are the Closest Relatives of Land Plants

The **green algae** are the most plantlike of the protists. They use chlorophyll *a* and *b* as photosynthetic pigments, use starch as a storage carbohydrate, and have cell walls containing cellulose. Like plants, many green algae also have life cycles that feature an **alternation of generations,** in which a multicellular haploid (gametophyte) phase is followed by a diploid (sporophyte) phase (figure 17.10).

The habitats and body forms of green algae are diverse (figure 17.11). Most live in fresh water or in moist habitats on land, although some live in symbiotic relationships with fungi,

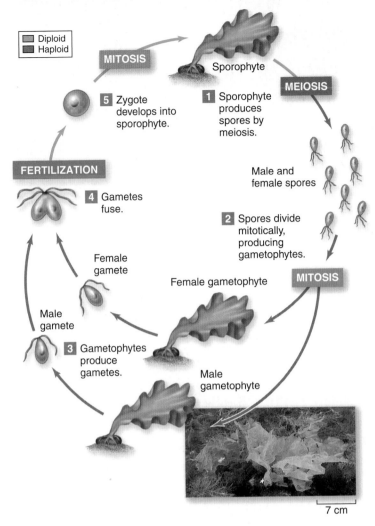

Figure 17.10 Life Cycle of a Green Alga. In the sea lettuce *Ulva*, a sporophyte undergoes meiosis and produces haploid spores, which develop into gametophytes. The gametophytes produce gametes by mitotic cell division. Fertilization yields a diploid zygote that matures to form a sporophyte. The cycle begins anew.

forming lichens. Green algae range in size from the smallest eukaryote *(Micromonas),* only 1 μm in diameter, to sea lettuce *(Ulva),* exceeding 1 meter in length. Green algae may be unicellular, filamentous, colonial, or multicellular. The multicellular species may have rootlike and stemlike structures, but they are far less specialized than plants. ▶ lichens, p. 403

One well-studied green alga is *Chlamydomonas,* a unicellular organism that reproduces asexually and sexually. Scientists study these algae to learn about the evolution of sex, how an individual's sex is determined, and how cells of opposite sexes recognize each other. A classroom favorite is the colonial green alga, *Volvox.* Hundreds to thousands of *Volvox* cells form hollow balls; the cells move their flagella in coordinated waves to move the sphere. New colonies remain within the parental ball of cells until they burst free.

Other green algae include the geometrically shaped desmids *(Micrasterias),* mermaid's wineglass *(Acetabularia),* the ribbonlike *Spirogyra,* and the tubular *Codium.* One species, *Chlorella,* is being considered as a food and oxygen source on prolonged space flights. An advantage of *Chlorella* over plants is that green algae can multiply very quickly, as evidenced by the rapid "greening" of a poorly maintained aquarium or swimming pool.

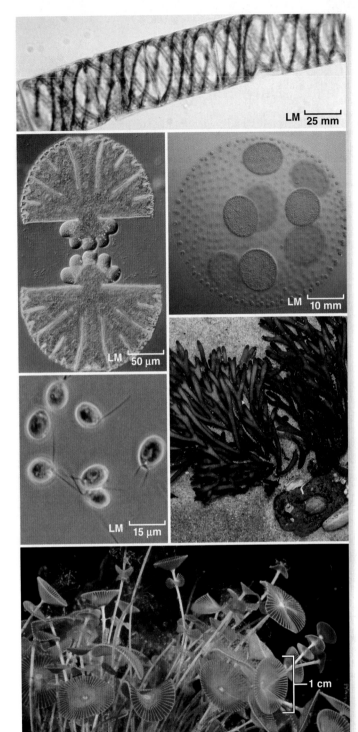

Figure 17.11 Gallery of Green Algae. Green algae have a variety of body forms, from solitary microscopic cells to complex multicellular forms. Clockwise from top: *Spirogyra, Volvox, Codium, Acetabularia, Chlamydomonas, Micrasterias.*

17.2 | Mastering Concepts

1. What mode of nutrition do the algae use?
2. Describe several criteria for classifying the algae.
3. List and describe the characteristics of the major groups of algae.

17.3 | Some Heterotrophic Protists Were Once Classified as Fungi

Slime molds and water molds are protists that resemble fungi in some ways: they are heterotrophic, and some produce filamentous feeding structures similar to those in fungi. They also commonly occur alongside fungi in many habitats. Nevertheless, molecular evidence clearly indicates that they are not closely related to fungi.

A. Slime Molds Are Unicellular and Multicellular

The slime molds are informally divided into two groups whose relationship to each other remains unclear. Both types live in damp habitats such as forest floors. In addition, each type of organism exists as single, amoeboid cells and as large masses that behave as one multicellular organism. The major difference between the two types is reflected in their names: plasmodial and cellular slime molds.

The feeding stage of a **plasmodial slime mold** consists of a plasmodium, which is a mass of thousands of diploid nuclei enclosed by a single cell membrane. This structure gives these organisms their other common name, the "acellular" slime molds. The plasmodium may be a conspicuous, slimy, bright yellow or orange mass up to 25 cm in diameter (**figure 17.12**). It migrates along the forest floor, engulfing bacteria and other microorganisms on leaves, debris, and rotting logs.

In times of drought or food shortages, the plasmodium halts and forms fruiting bodies, which produce thick-walled reproductive cells called spores. When favorable conditions return, the spores germinate and form haploid cells. Two of these cells may fuse, forming a diploid zygote nucleus that divides repeatedly by mitosis, forming a new multinucleate plasmodium. Scientists use the plasmodial slime mold *Physarum polycephalum* to study cell division and the movements of cytoplasm inside a cell.

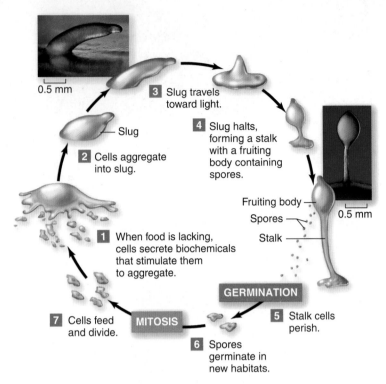

3 Slug travels toward light.

Slug

2 Cells aggregate into slug.

4 Slug halts, forming a stalk with a fruiting body containing spores.

Fruiting body
Spores
Stalk

0.5 mm

1 When food is lacking, cells secrete biochemicals that stimulate them to aggregate.

GERMINATION

MITOSIS

7 Cells feed and divide.

5 Stalk cells perish.

6 Spores germinate in new habitats.

Figure 17.13 **Life Cycle of the Cellular Slime Mold** *Dictyostelium discoideum.* The amoeboid cells of this species feed on bacteria. Starvation stimulates the cells to aggregate into a multicellular "slug," which crawls to a new habitat. Only the asexual portion of the life cycle is shown, although a sexual phase may also occur.

In contrast to the plasmodial slime molds, individual cells of a **cellular slime mold** retain their membranes throughout the life cycle. The cells exist as haploid amoebae, engulfing bacteria and other microorganisms in fresh water, moist soil, and decaying vegetation.

When food becomes scarce, the amoebae secrete chemical attractants that stimulate the neighboring cells to aggregate into a sluglike structure (**figure 17.13**). The "slug" moves toward light, stops, and forms a stalk topped by a fruiting body that produces haploid spores. The cells of the stalk perish, but the spores survive; water, soil animals, or birds spread them to new habitats. The spores then germinate to form haploid amoebae, and the cycle begins anew. Sexual reproduction also occurs in some conditions. *Dictyostelium discoideum* is a cellular slime mold used in many scientific studies (see this chapter's Focus on Model Organisms).

B. Water Molds Are Decomposers and Parasites

The **water molds,** or oomycetes, are decomposers or parasites of plants and animals in moist environments (**figure 17.14**). The filaments of water molds secrete digestive enzymes into their surroundings and absorb the nutrients. In addition, like fungi called chytrids, water molds produce swimming spores that aid their dispersal in water and wet soil. Despite the similarities, however, water molds are unlike fungi in many ways. The filaments of water molds are diploid, for example, whereas most fungal filaments are haploid. Also, fungi have cell walls containing chitin, but water mold cell walls contain cellulose.

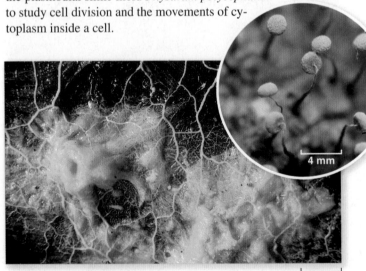

4 mm

1 cm

Figure 17.12 **Plasmodial Slime Mold.** Streaming masses of the plasmodial slime mold *Physarum* ooze across dead logs, leaf litter, and other organic matter. The inset shows fruiting bodies.

a.   b.

Figure 17.14 Water Molds. (a) *Phytophthora infestans* is a water mold that causes late blight of potatoes and was responsible for the Irish potato famine in the mid-1840s. (b) *Saprolegnia* infects a dead insect.

The best-known water molds are those that ruin crops, causing such diseases as downy mildew of grapes and lettuce. In the 1870s, downy mildew of grapes, caused by *Plasmopara viticola,* nearly destroyed the French wine industry. The water mold *Phytophthora infestans,* which means "plant destroyer," causes late blight of potato. This disease caused the devastating Irish potato famine from 1845 to 1847, during which more than a million people starved and millions more emigrated from Ireland. The Irish potato famine followed several rainy seasons, which fostered the rapid spread of the plant disease. A newly discovered relative of *P. infestans,* called *Phytophthora ramorum,* causes a tree disease called sudden oak death. Another well-known water mold is *Saprolegnia,* a protist that forms cottony masses on fishes and other aquatic organisms that are weakened or dead.

17.3 | Mastering Concepts

1. What mode of nutrition do the slime molds and water molds use?
2. Compare and contrast the plasmodial and cellular slime molds.
3. What has been the role of water molds in the environment and history?

Focus on Model Organisms

Dictyostelium discoideum

Slime molds have such an unappealing name that it may seem hard to imagine why anyone would study them. But *Dictyostelium discoideum* is an unusual organism, one that straddles the boundary between the unicellular and the multicellular. Its feeding phase consists of individual amoeba-like cells that move independently, feeding on bacteria by phagocytosis. When the food runs out, cells begin to aggregate into a multicelled structure that migrates toward light. The cells differentiate into a base, stalk, and spores; only the spores survive to colonize a new habitat.

Dictyostelium discoideum is useful as a model because, like other model organisms, it is easy to grow in the laboratory and has a short generation time. In addition, its cells are readily accessible to microscopy and genetic studies. As a result, *D. discoideum* (affectionately called "Dicty" by its researchers) remains a useful and fascinating organism. Discoveries by researchers working on *D. discoideum* include:

- **Cell movement.** A *Dictyostelium* cell eats by producing extensions that engulf and absorb food particles by phagocytosis. Scientists have discovered that this movement is possible because proteins such as actin and myosin move rapidly within the cell. These same proteins produce muscle movement in animals.
  ▶ muscle movement, p. 590

- **Cytokinesis.** Researchers observing cell division in *D. discoideum* have discovered that the protein myosin is also required for cytokinesis (the physical division of one cell into two).
  ▶ cytokinesis, p. 157

- **Chemotaxis.** Starving Dicty cells move toward one another and form a multicellular "slug." This movement toward a chemical stimulus, called chemotaxis, requires membrane proteins that not only detect the signals from other Dicty cells but also transmit the information to the inside of the cell. Similar signal transduction systems occur in many organisms. ▶ cell membrane, p. 54

- **Cell differentiation.** When individual Dicty cells come together, chemical signals presumably determine which cells will become stalk cells (and die) and which will become spore cells (and survive). Researchers have discovered a sterol-like compound that induces the differentiation of stalk cells. Such research may help answer questions about the origin of multicellularity.
  ▶ sterols, p. 35

17.4 | Protozoa Are Diverse Heterotrophic Protists

Finding a list of characteristics that unite the diverse **protozoa** is difficult. Most are unicellular, and the vast majority are heterotrophs, but several autotrophic species exist. They move by flagella, cilia, or pseudopodia. Some are free-living, and others are obligate parasites. Most are asexual, but sexual reproduction occurs in many species.

This chapter describes four groups of distantly related protozoa that are defined by locomotion and morphology. New molecular techniques are redefining the protozoa, but until the newer system of classification is better defined and more widely accepted, these four groups remain practical for general biology, education, and medicine.

A. Several Flagellated Protozoa Cause Disease

The **flagellated protozoa** are unicellular organisms with one or more flagella. Most are free-living in fresh water, the ocean, and soil. The euglenoids and dinoflagellates, groups already described with the algae, are flagellates. This section turns to a few of the heterotrophic species (**figure 17.15**).

One example of a flagellated protozoan is *Trichonympha,* a protist that lives in the intestines of termites. The cells of *Trichonympha,* in turn, harbor bacteria that digest cellulose. This bacterium-within-protist living organization enables termites to "digest" wood. Exposing termites to high oxygen or high temperature kills the symbionts. The insects soon die, with guts full of undigested wood.

Some parasitic flagellated protozoa cause disease in humans. For example, *Trichomonas vaginalis* resides in the urogenital tracts of both men and women. It is sexually transmitted and causes a form of vaginitis in females. *Giardia intestinalis* (also known as *Giardia lamblia*) causes "hiker's diarrhea," or giardia-sis. People ingest the cysts of the organism in contaminated water. As *Giardia* cells divide in the small intestine, they impair the host's ability to absorb nutrients, resulting in diarrhea and cramping.

Another group of disease-causing flagellates are the **trypanosomes.** These whip-shaped parasites invade the bloodstream and, in some cases, the brain. Insects transmit trypanosomes to humans. Tsetse flies, for example, carry *Trypanosoma brucei,* the organism that causes African sleeping sickness. In South and Central America, kissing bugs transmit *Trypanosoma cruzi* to humans from rodents, armadillos, and dogs. The resulting illness, called Chagas disease, kills 45,000 people annually. The sand fly transmits a related parasite, *Leishmania.*

B. Amoeboid Protozoa Produce Pseudopodia

The **amoeboid protozoa** produce cytoplasmic extensions known as pseudopodia (Latin, meaning "false feet"), which are important in locomotion and capturing food via phagocytosis. The most studied species is *Amoeba proteus,* a common freshwater microbe that engulfs bacteria, algae, and other protists in its pseudopodia (**figure 17.16**). The human digestive tract may be invaded by another species, *Entamoeba histolytica,* which can cause amoebic dysentery. ▶ phagocytosis, p. 83

The **foraminiferans,** or forams, are an ancient group of mostly marine amoeboid protozoa. They have complex, brilliantly colored shells made primarily of calcium carbonate (**figure 17.17a**). Their populations are immense: about one-third of the ocean floor is made of the shells of the marine foram *Globigerina.* England's White Cliffs of Dover, among other limestone and chalk deposits, are made largely of the tests (shells) of forams and other marine organisms that have been lifted out of the water by geologic forces. Paleontologists studying extinct forams have learned which species correlate with oil and gas deposits. The shells are also useful in dating rock strata.

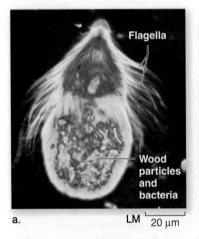

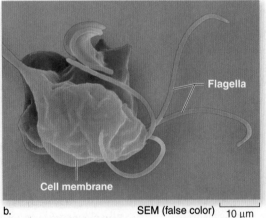

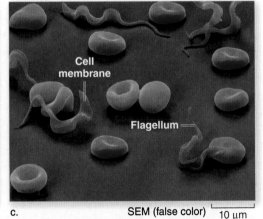

a. LM ⊢ 20 μm   b. SEM (false color) ⊢ 10 μm   c. SEM (false color) ⊢ 10 μm

Figure 17.15 Flagellated Protozoa. (a) *Trichonympha* lives inside the intestines of termites. Note the fringe of flagella. (b) *Trichomonas vaginalis* causes the sexually transmitted disease trichomoniasis. These organisms have multiple flagella. (c) *Trypanosoma brucei* causes African sleeping sickness. Each cell features a single flagellum.

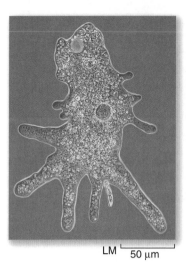

Figure 17.16 "False Feet." Pseudopodia are temporary projections from the cells of amoeboid organisms. A pseudopod enables the organism to move or take in food. This organism, *Amoeba proteus*, is consuming *Euglena*, another protist (small green cell at bottom center).

LM 50 μm

The **radiolarians** are among the oldest protozoa. They are planktonic organisms with intricate tests made of silica (**figure 17.17b**); pseudopodia extend through holes in the shells. "Radiolarian ooze" is sediment consisting of large numbers of their tests. On the ocean floor, radiolarian ooze can be as thick as 4000 meters.

C. Ciliates Are Common Protozoa with Complex Cells

The **ciliates** are complex, mostly unicellular protists that are characterized by abundant hairlike cilia (**figure 17.18**). The cilia have multiple functions. Waves of moving cilia propel the organism through the water. Cilia also sweep bacteria, algae, and other ciliates into the cell's gullet. ▶ cilia, p. 63

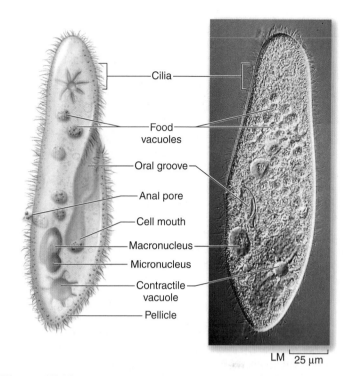

LM 25 μm

Figure 17.18 Anatomy of *Paramecium*. Structures in *Paramecium* include a micronucleus and a macronucleus; an oral groove with a "mouth" into which cilia wave food; food vacuoles for storage; contractile vacuoles that maintain solute concentrations; and an anal pore, which releases wastes.

Ciliate cells have other distinctive features as well. A food vacuole surrounds and transports each captured meal inside the cell, and a permanent anal pore releases the wastes. In freshwater habitats, water may enter the cell by osmosis. An organelle called

LM 200 μm

a.

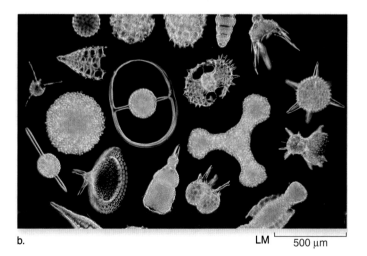

b.

LM 500 μm

Figure 17.17 Foraminiferans and Radiolarians.
(a) Foraminiferans such as *Globigerina* form vast amounts of ocean sediment. The White Cliffs of Dover are composed of mostly foraminiferan tests. (b) Radiolarians have intricate silicon-based shells. © Dr. Richard Kessel & Dr. Gene Shih/Visuals Unlimited.

a contractile vacuole helps maintain water balance by pumping the excess fluid out of the cell. ▶ osmosis, p. 80

In addition, many ciliates have two types of nuclei, a small micronucleus and a larger macronucleus. The DNA in the micronucleus is passed on during sexual reproduction, whereas the genes in the macronucleus have metabolic and developmental functions.

The habitats of ciliates are diverse. Most are free-living, motile cells such as *Paramecium* and its predator, *Didinium.* Several ciliate species, such as *Stentor,* are sessile or attached forms living on a variety of substrates. Nearly one-third of ciliates are symbiotic, living in the bodies of crustaceans, mollusks, and vertebrates. Some inhabit the stomachs of cattle, where they house bacteria that break down the cellulose in grass. Others are parasites. *Ichthyophthirius multifilis,* for example, causes a common epidermal disease called "ich" in freshwater fish.

D. Apicomplexans Include Nonmotile Animal Parasites

The **apicomplexans** are nonmotile, spore-forming, internal parasites of animals. The name *apicomplexa* comes from the apical complex, a cluster of microtubules and organelles at one end of the cell. This structure, visible only with an electron microscope, apparently helps the parasite attach to and invade host cells.

Apicomplexans include several organisms that cause illness. This chapter's Apply It Now box describes *Cryptosporidium,* a genus containing several species that cause waterborne disease.

Another example is *Toxoplasma gondii,* a protist that infects cats and other mammals. A person who handles feces from infected cats can accidentally ingest *Toxoplasma* cysts. The resulting infection may remain symptomless or develop into an illness called toxoplasmosis, especially in people with weakened immune systems. In the most severe cases, the parasite can damage the brain and eyes. The infection can also pass to a fetus, which is why pregnant women should avoid cat litter boxes.

Malaria is another example of an illness caused by an apicomplexan. From 300 million to 500 million people a year suffer from this disease, with more than 90% of them in sub-Saharan Africa. Globally, 2 million to 3 million people die each year from malaria. Among very ill children, the death rate is 50%. Even survivors may retain dormant forms of the parasite, becoming ill again months or years after the initial infection.

Four species of *Plasmodium* cause mosquito-borne malaria in humans. (*Plasmodium,* a genus of apicomplexans, is not to be confused with the plasmodium produced by some slime molds.) The life cycle is complex, involving many stages in multiple hosts and including both asexual and sexual reproduction. Figure 17.19 shows a simplified version.

A cycle of malaria begins when an infected female mosquito of any of 60 *Anopheles* species feeds on human blood. The insect's saliva transmits small haploid cells called sporozoites to the human host. The sporozoites travel in the bloodstream and enter the liver cells, where they multiply rapidly. Eventually the cells emerge as merozoites. Some of the merozoites infect red blood cells, feeding on hemoglobin and reproducing. Every 48 to

Apply It Now

Don't Drink That Water

The sign says, "Please shower before entering pool." What is the purpose of that request? Isn't showering a wasteful prelude to taking a refreshing plunge?

The truth is that the preswim shower is an important public health measure. Washing thoroughly with soap and hot water helps eliminate harmful microbes before they have a chance to contaminate the pool's water.

Cryptosporidium, or "crypto," is one example of a contagious microorganism that spreads easily in water. This apicomplexan protist lives in the intestinal tracts of infected humans, entering the body through the mouth and exiting in feces. It produces tough-walled cysts that can survive for days in the chlorinated water of public pools, water parks, splash pads, and other places where people gather to play in the water.

Even tiny amounts of feces, invisible to the unaided eye, can contaminate water with *Cryptosporidium* cysts, triggering an outbreak. Swimmers who accidentally swallow the tainted water typically become ill within a week, as cells released from the cysts invade the lining of the digestive tract. Symptoms include diarrhea, cramps, fever, vomiting, and dehydration.

Crypto outbreaks periodically occur in communities throughout the United States. One notable episode occurred in Milwaukee, Wisconsin, in 1993. A malfunctioning water treatment plant distributed contami-

nated water to hundreds of thousands of households, sickening about 25% of Milwaukee's population and killing 100 people.

Scattered small outbreaks also have occurred in pools and other recreational water venues, where the victims are often young children who ingest the water. Simple preventive measures include keeping children with diarrhea out of the water, washing hands thoroughly after using the toilet or changing diapers, and—as the sign says—showering before entering the pool.

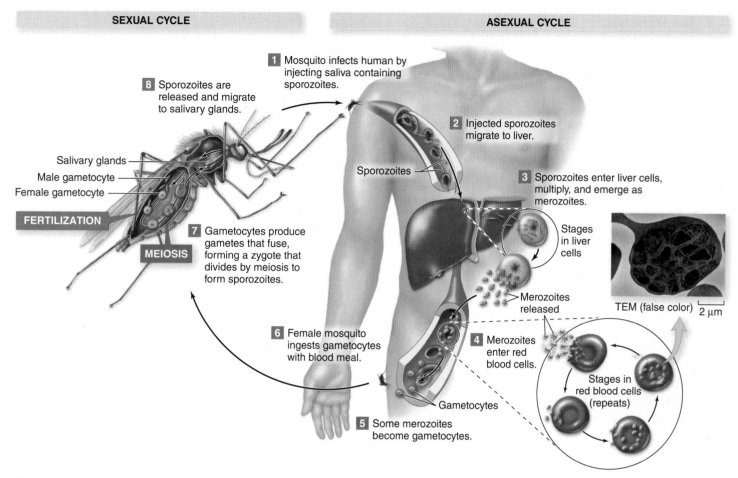

SEXUAL CYCLE

ASEXUAL CYCLE

1 Mosquito infects human by injecting saliva containing sporozoites.

8 Sporozoites are released and migrate to salivary glands.

2 Injected sporozoites migrate to liver.

Sporozoites

3 Sporozoites enter liver cells, multiply, and emerge as merozoites.

Salivary glands
Male gametocyte
Female gametocyte

Stages in liver cells

FERTILIZATION

7 Gametocytes produce gametes that fuse, forming a zygote that divides by meiosis to form sporozoites.

MEIOSIS

Merozoites released

TEM (false color) 2 μm

6 Female mosquito ingests gametocytes with blood meal.

4 Merozoites enter red blood cells.

Stages in red blood cells (repeats)

Gametocytes

5 Some merozoites become gametocytes.

Figure 17.19 *Plasmodium* **Life Cycle.** A mosquito transmits sporozoites of *Plasmodium,* the protist that causes malaria. The sporozoites migrate in the bloodstream to a person's liver. The sporozoites then divide to form merozoites, which may continue to cycle in red blood cells for some time. At some point, gametocytes form from merozoites. Mosquitoes ingest the gametocytes, starting the cycle anew.

72 hours they burst from the host cells and infect other red blood cells. This release is often synchronized throughout the victim's body, causing the characteristic recurrent chills and fever of malaria.

Other merozoites become specialized as male and female sexual forms called gametocytes. When mosquitoes ingest the gametocytes from an infected person's blood, the cells unite in the insect's stomach. After several additional steps, sporozoites form. These move to the mosquito's salivary glands, ready to enter a new host when the insect seeks its next blood meal.

Despite decades of research, malaria continues to be the world's most significant infectious disease. No effective vaccine has been developed, and nations troubled by poverty and civil unrest struggle to distribute drugs to prevent or treat malaria. Moreover, *Plasmodium* continues to develop resistance to drugs that were effective in the past. Malaria prevention efforts therefore focus on repelling and killing mosquitoes (even as the use of insecticides has selected for resistance in *Anopheles* mosquitoes). The careful use of antimalarial drugs, alone and in combination, helps prevent and treat malaria and reduces selection for resistant parasites.

Not everyone is equally susceptible to malaria. People with one copy of the recessive sickle cell allele are much less likely to contract malaria than are people with two dominant alleles. In areas of the world where malaria is endemic, human populations have a relatively high incidence of the sickle cell allele. In malaria-free areas, the sickle cell allele is much rarer. This pattern illustrates the selective force that malaria exerts on the human population. ▶▶ sickle cell trait, p. 138; balanced polymorphism, p. 242

17.4 | Mastering Concepts

1. What mode of nutrition do protozoa use?
2. What are the characteristics of each major group of protozoa?
3. List three diseases caused by flagellated protozoa.
4. Compare and contrast amoebae, foraminiferans, and radiolarians.
5. How do ciliates move and eat?
6. What are the distinguishing characteristics of apicomplexans?

17.5 | Protist Classification Is Changing Rapidly

This chapter illustrates some of the difficulties in classifying the protists. For example, the euglenoids and dinoflagellates could easily fall into either of two groups: the algae (because they photosynthesize) and the flagellated protozoa (because they have flagella and may be heterotrophic). Likewise, the water molds are traditionally grouped with slime molds because they share a habitat with fungi, but water molds are actually closely related to brown algae. Clearly, the traditional scheme groups unrelated organisms.

New research based on genetic sequences is helping to assign each species into a lineage with its closest relatives. Nevertheless, biologists have not yet firmly established the number of kingdoms, the rank of each taxonomic group, their names, or the evolutionary relationships among them.

One scheme organizes the eukaryotes into "supergroups," some of which are summarized in table 17.1. Several of the supergroups listed in this table unite organisms once thought to be dissimilar. The **alveolates,** for example, are eukaryotes that have a series of flattened sacs, or alveoli, just beneath the cell membrane. The alveolates include dinoflagellates, ciliates, apicomplexans, foraminiferans, and radiolarians.

Another new grouping is the **stramenopiles,** which include water molds, diatoms, brown algae, and golden algae. The word *stramenopile* means "flagellum-hair" (*stramen* = straw or flagellum, and *pilos* = hair). At some point in their life cycles, stramenopiles produce cells with two flagella, one of which is covered with tubular hairs. In addition, the photosynthetic members of this group produce the yellowish accessory pigment fucoxanthin.

The placement of many other protists remains unresolved. Protistan classification will continue to evolve as research reveals new molecular sequences, but it will likely remain a "work in progress" for years to come.

Finally, note that two of the supergroups, the Archaeplastida and the Opisthokonta, include three other eukaryotic kingdoms (the plants, fungi, and animals). These kingdoms are the subject of the remaining chapters in this unit.

17.5 | Mastering Concepts

1. How have molecular sequences changed protist classification?
2. What features unite some of the major lineages of protists?

Table 17.1 Proposed Eukaryotic "Supergroups": A Summary

Supergroup		Distinguishing Features	Examples
Excavata		Unicellular flagellated protists; many lack mitochondria	*Trichomonas, Trichonympha, Giardia*
Euglenozoa		Unicellular flagellated protists; most have two flagella (only one may emerge from the cell); photosynthetic or parasitic; chloroplasts derived from secondary endosymbiosis	*Euglena*, trypanosomes
Archaeplastida		Photosynthetic eukaryotes; chloroplasts derived from primary endosymbiosis	Red algae, green algae, land plants
Alveolata		Flattened sacs (alveoli) beneath cell membrane; chloroplasts (if present) derived from secondary endosymbiosis	Dinoflagellates, apicomplexans, ciliates
Stramenopila		Motile cells have two flagella, one of which has tubular hairs; fucoxanthin accessory pigment in photosynthetic forms; chloroplasts derived from secondary endosymbiosis	Water molds, diatoms, brown algae, golden algae
Rhizaria		Amoeboid movement; many produce shells	Radiolarians, foraminiferans
Amoebozoa		Amoeboid movement via pseudopodia; feed by phagocytosis; slime molds form spores	*Amoeba, Dictyostelium, Physarum*
Opisthokonta		Motile cells have one flagellum	Choanoflagellates, animals, fungi

17.6 | Investigating Life: Glassy Fossils Reveal the Birth of a Species

Gazing at a serene lake may not conjure images of evolution—nor do most people give a thought to the muddy sediments under the water's surface. But it might be a good idea to think again, especially at Yellowstone Lake in Wyoming. Preserved in its muck is a fossil record that documents speciation, the transition of one species into another (see chapter 13).

The story begins about 14,000 years ago, at the end of the last ice age, when retreating glaciers exposed Yellowstone Lake. Each year, warm summer weather triggers the explosive growth of diatoms and other aquatic life. Meanwhile, soil particles and pollen wash or blow into the lake from surrounding land, mixing with wastes and the remains of aquatic organisms. These materials combine to form sediments that rain onto the lake bottom at a rate of 0.6 to 0.7 mm per year.

The glassy fossils of diatoms provide a continuous, 14,000-year record of evolution in Yellowstone Lake. A research team led by Edward C. Theriot from the University of Texas and Sherilyn C. Fritz of the University of Nebraska studied both modern and fossilized diatoms in the lake. They wanted to learn whether the evolution of the diatoms was related to environmental change.

To find out, they collected an 8.5-meter-long sediment core from the lake floor. In the lab, they took small sediment samples at regular intervals along the core and examined the diatoms with a compound microscope. They were especially interested in two species: *Stephanodiscus yellowstonensis,* named after its home lake, and *S. niagarae,* which once occupied Yellowstone Lake but does not live there today.

Theriot and Fritz carefully examined any diatom resembling *S. yellowstonensis* or *S. niagarae.* They measured the diameter of the silica wall, the number of ridges radiating from the center of each wall, and the number of spines. They compared each with modern *S. niagarae* diatoms collected from nearby lakes and with two other *Stephanodiscus* species.

The researchers also used carbon-14 (^{14}C) analysis to determine the age of the sediments, which ranged from nearly 14,000 years old at the deepest part to less than 1000 years old at the shallowest. Finally, the team analyzed pollen from the core. Pollen grains that enter the lake become part of the sediments, recording changes in the nearby plant community. Because biologists know which plants thrive in different conditions, pollen analysis gives clues to climatic changes. ▶ radiometric dating, p. 258

The pollen record revealed that a major shift in the plant community began about 12,000 years ago (figure 17.20). Grasses and scattered trees surrounded the lake until about 12,600 years ago, but after the climate began to warm, the forests expanded. Tree pollen became increasingly abundant from around 12,600 to 10,700 years ago, after which the plant community changed little.

Meanwhile, for reasons that remain unclear, the diatoms were evolving. Until about 12,400 years ago, *S. niagarae* occupied the lake. The species then began to change, with many "transitional"

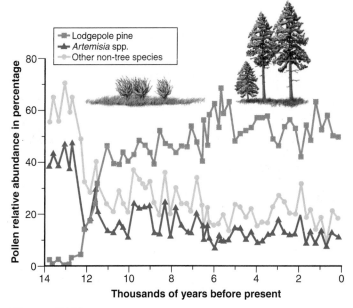

Figure 17.20 Changing Climate. As the climate warmed some 12,000 years ago, the plant community changed as well.

forms whose spine numbers were intermediate between *S. niagarae* and *S. yellowstonensis.* By 10,000 years ago, the transition was complete: *S. niagarae* was gone, and the less-spiny *S. yellowstonensis* had taken its place (figure 17.21). One species had evolved into another, leaving fossils behind to tell the tale.

The diatoms at Yellowstone Lake are unusual because their record of speciation is coupled with well-documented evidence of climatic change. Biologists can thank the decay-resistant silica shells of diatoms—and a continuous, gentle rain of sediment—for this extraordinarily detailed portrait of evolution.

Theriot, Edward C., Sherilyn C. Fritz, Cathy Whitlock, and Daniel J. Conley. 2006. Late Quaternary rapid morphological evolution of an endemic diatom in Yellowstone Lake, Wyoming. *Paleobiology,* vol. 32, pages 38–54.

17.6 | Mastering Concepts

1. What were the researchers asking about Yellowstone Lake, and how did they answer their question?
2. Spines increase a cell's surface area, so the spiniest diatoms should be the most buoyant. What selective force might explain the observed shift in spine number? How might you test your hypothesis in a lab?

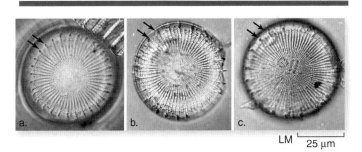

Figure 17.21 Diatoms in Transition. (a) Modern *Stephanodiscus niagarae.* (b) Transitional form from 11,730 years ago. (c) Modern S. *yellowstonensis.* Arrows point to spines.

Chapter Summary

17.1 | Protists Lie at the Crossroads Between Simple and Complex Organisms

A. What Is a Protist?
- **Protists** are **eukaryotes** that are not plants, fungi, or animals. The protists are very diverse in anatomy, size, and ecological role.
- Classification of protists is changing as molecular sequence data are considered along with traditional traits. Kingdom Protista is being divided into multiple new groups.

B. Protists Are Important in Many Ways
- Photosynthetic protists support food webs and release oxygen, whereas parasitic protists cause disease in plants and animals. Some protists have industrial applications.

C. Protists Have a Lengthy Evolutionary History
- Some protists may resemble early eukaryotes; others may provide clues to the origins of multicellularity and the ancestors of the plant, fungi, and animal lineages.

17.2 | Many Protists Are Photosynthetic
- The photosynthetic protists are known as **algae.** They range in size from microscopic to dozens of meters long.

A. Euglenoids Are Heterotrophs and Autotrophs
- **Euglenoids** are unicellular, elongated flagellates that commonly inhabit fresh water.

B. Dinoflagellates Are "Whirling Cells"
- The **dinoflagellates** have two different-sized flagella at right angles that generate a whirling movement. Many have outer plates made of cellulose. Dinoflagellates cause red tides.

C. Golden Algae, Diatoms, and Brown Algae Contain Yellowish Pigments
- The **golden algae** are photosynthetic but can consume other microorganisms when light or nutrient supplies decline.
- **Diatoms** are microscopic phytoplankton with intricate silica shells.
- **Brown algae** are large, multicellular seaweeds such as kelp.

D. Red Algae Can Live in Deep Water
- **Red algae** contain unique reddish and bluish photosynthetic pigments that expand their photosynthetic range.

E. Green Algae Are the Closest Relatives of Plants
- **Green algae** store carbohydrates as starch and use the same pigments as plants. Many have **alternation of generations.** Unlike plants, green algae lack true roots, stems, and leaves, and they have less specialized cells. The green algae have diverse body forms, ranging from microscopic, one-celled organisms to large, multicellular seaweeds.

17.3 | Some Heterotrophic Protists Were Once Classified as Fungi

A. Slime Molds Are Unicellular and Multicellular
- **Plasmodial slime molds** form plasmodia, brightly colored masses containing thousands of diploid nuclei. A plasmodium feeds by engulfing other cells.
- In **cellular slime molds,** individual cells retain their separate cell membranes throughout the life cycle.

B. Water Molds Are Decomposers and Parasites
- **Water molds** are filamentous heterotrophs that live in moist or wet environments. Some are pathogens of plants, and some decompose organic matter.

17.4 | Protozoa Are Diverse Heterotrophic Protists
- Most **protozoa** are heterotrophs, and most have motile cells.

A. Several Flagellated Protozoa Cause Disease
- Several species of **flagellated protozoa,** such as *Giardia, Trichomonas, Trypanosoma* (the **trypanosomes**), and *Leishmania,* cause disease in humans.

B. Amoeboid Protozoa Produce Pseudopodia
- **Amoeboid protozoa** move by means of "false feet," or pseudopodia. This group includes amoebae, **foraminiferans,** and **radiolarians.**

C. Ciliates Are Common Protozoa with Complex Cells
- **Ciliates** have complex cells with cilia, food vacuoles, contractile vacuoles, and two types of nuclei.

D. Apicomplexans Include Nonmotile Animal Parasites
- **Apicomplexans** are obligate parasites of animals. They have an apical complex of organelles that helps them attach to or penetrate host cells. Malaria and toxoplasmosis are diseases caused by apicomplexans.

17.5 | Protist Classification Is Changing Rapidly
- The traditional means of classifying protists is giving way to a newer scheme based on molecular evidence.
- **Stramenopiles** have cells that produce two dissimilar flagella. This group includes water molds, golden algae, brown algae, and diatoms.
- **Alveolates** have flattened sacs beneath the cell membrane and include the dinoflagellates, ciliates, and apicomplexans.
- The relationships among most lineages of protists remain unclear.

17.6 | Investigating Life: Glassy Fossils Reveal the Birth of a Species
- Diatoms in Wyoming's Yellowstone Lake have left a continuous fossil record that spans more than 10,000 years.
- Researchers studying the fossils have documented the formation of a new species, *S. yellowstonensis,* from its direct ancestor, *S. niagarae.* The period over which speciation happened coincided with a period of climatic change.

Multiple Choice Questions

1. Which of the following is NOT a defining characteristic of the protists?
   a. Unicellular
   b. Cells contain membrane-bounded organelles
   c. Cells contain a nucleus
   d. Eukaryotic

2. The toxic bloom associated with a red tide is the product of which type of protist?
   a. Euglenoids
   b. Diatoms
   c. Red algae
   d. Dinoflagellates

3. The different colors associated with the various phyla of photosynthetic protist result from
   a. variations in the cell walls of these organisms.
   b. the presence of different photosynthetic pigments.
   c. differences in the habitats occupied by these cells.
   d. the presence of silica or polysaccharides on the outside of the cell.

4. How is a plasmodial slime mold different from a cellular slime mold?
   a. The ability to form resistant spores
   b. The use of photosynthetic pigments
   c. The composition of their cell walls
   d. The ability to form a single cell with multiple nuclei

5. What property distinguishes an alga from a protozoan?
   a. The presence of a flagellum
   b. The ability to live in water
   c. The ability to use photosynthesis
   d. Both a and b are correct.

6. What form of motility is associated with amoebae?
   a. Flagella
   b. Cilia
   c. Pseudopodia
   d. Apical complex

7. Which of the following diseases is not caused by a flagellated protozoan?
   a. Vaginitis
   b. Hiker's diarrhea
   c. Malaria
   d. African sleeping sickness

8. Protozoans use cilia to
   a. collect food.
   b. move through the environment.
   c. attach to a surface.
   d. both a and b.

9. Which statement about malaria is false?
   a. Mosquitoes transmit malaria to humans.
   b. The protist that causes malaria infects red blood cells.
   c. *Plasmodium* is increasingly resistant to antimalarial drugs.
   d. Carriers of the sickle cell trait are extremely susceptible to malaria.

10. Why is the classification of protists based on DNA sequences considered useful?
    a. Because only protists have DNA
    b. Because genetic sequences have confirmed the traditional categories of protists
    c. Because DNA reveals evolutionary relationships, even among organisms that look different
    d. All of the above are correct.

Write It Out

1. Explain why evolutionary biologists are interested in choanoflagellates, green algae, and organisms with mitochondria whose genomes resemble those of bacteria.

2. Name an organism that uses each form of locomotion: cilia; flagella; amoeboid motion.

3. Name at least one type of organism that has each of the following structures: cyst; pseudopod; plasmodium; silica cell wall; apical complex.

4. Name the protist that causes each of the following diseases: malaria; toxoplasmosis; grape downy mildew; paralytic shellfish poisoning; amoebic dysentery; leishmaniasis; late blight of potato; hiker's diarrhea; African sleeping sickness; Chagas disease.

5. Describe the relationship between nutrient pollution and harmful algal blooms. Why might harmful algal blooms be more frequent in summer? What steps could coastal communities take to prevent nutrient pollution?

6. Explain why the fossil record for diatoms is much more complete than that of other protists, such as amoebae and slime molds.

7. A diatom's silica cell wall is rigid and somewhat resembles a shoebox with a lid. How might these protists divide? Draw a diagram of the process.

8. How is it adaptive for a red alga to have pigments other than chlorophyll?

9. Study the spherical *Volvox* colony in figure 17.11. The cells in the colony have eyespots and flagella, and they are connected to one another by thin strands of cytoplasm (these features are difficult to see in the photo). Propose some ways that *Volvox*'s colonial lifestyle might be adaptive in its freshwater habitat.

10. What adaptations enable cellular slime molds to survive a temporarily harsh environment?

11. Develop a hypothesis to explain why the plasmodium of *Physarum polycephalum* is bright yellow. How would you test your hypothesis?

12. Search the Internet for a news story about *Physarum polycephalum* using a simple maze to find the shortest distance between two food sources. Why would this behavior be adaptive? Do you think this experiment shows that slime molds are intelligent? Explain.

13. How does rainy weather foster the spread of *Phytophthora,* the protist that causes late blight of potato?

14. Use the Internet to learn about the sexually transmitted disease trichomoniasis. Describe how the pathogen spreads and the symptoms, treatment, and prevention of the disease.

15. The flagellated protist *Giardia* does not have true mitochondria, but in 2005, researchers reported that it does have tiny organelles called mitosomes. Chapter 6 describes the structure and function of the mitochondrion. What features would you look for to test the hypothesis that mitosomes are derived from mitochondria?

16. Insecticides such as DDT kill mosquitoes that transmit malaria. The use of DDT, however, causes reproductive problems in eagles, ospreys, and other fish-eating birds. The insecticide is therefore banned for use in the United States. In countries where malaria still rages, what information would you need to help you decide whether to allow the use of DDT in the fight against malaria?

17. How might global climate change affect the distribution of malaria? Explain your answer.

18. Give three examples of protists for which the classifications have recently changed. In each case, what was the justification for the old category, and what is the justification for the change?

19. Suppose someone hands you a microscope and a single-celled organism. Create a flowchart that you could use to identify the specimen.

Pull It Together

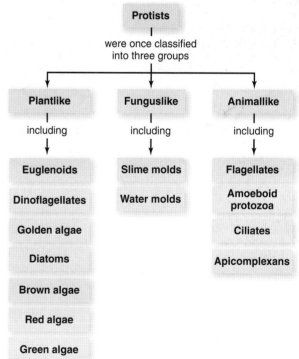

Protists

were once classified into three groups

Plantlike → including → Euglenoids, Dinoflagellates, Golden algae, Diatoms, Brown algae, Red algae, Green algae

Funguslike → including → Slime molds, Water molds

Animallike → including → Flagellates, Amoeboid protozoa, Ciliates, Apicomplexans

1. What features do all protists have in common?

2. How have molecular data changed the classification system depicted in this concept map?

3. What are the two groups of slime molds?

4. Name examples of disease-causing organisms in each group of protozoa.

5. Which organisms in this concept map are heterotrophs, and which are autotrophs?

Sphagnum, the source of peat moss, grows in bogs. This man is cutting peat for fuel.

Peat Moss, Pot Scrubbers, Drugs, and More: The Many Uses of Plants

MOST PEOPLE KNOW WHY PLANTS ARE ESSENTIAL TO ANIMAL LIFE: VEGETATION PROVIDES FOOD AND HABITAT, AND PHOTOSYNTHESIS PRODUCES OXYGEN GAS. Yet plants serve us in many unexpected ways as well. Here are some interesting examples from the four main plant lineages.

- **Peat moss:** Gardeners and houseplant lovers recognize peat moss as a major ingredient in potting mixes. Peat comes from partially decomposed sphagnum moss harvested from enormous bogs. The dried moss is unusually spongy, absorbing 20 times its weight in water. When mixed with soil, peat slowly releases water to plant roots. People also burn peat as cooking fuel or to generate electricity.

- **Horsetail:** *Equisetum,* or the horsetail, is a seedless plant related to ferns. Some horsetails are called "scouring rushes" because their stems and leaves contain abrasive silica particles. Native Americans used horsetails to polish bows and arrows, and early colonists and pioneers used them to scrub pots and pans. Some people believe that horsetails, taken as a dietary supplement, can strengthen fingernails and prevent osteoporosis.

- **Pacific yew:** *Taxus brevifolia,* the Pacific yew, is a conifer that contains the compound paclitaxel in its bark. Paclitaxel has anticancer properties, particularly for treating breast cancer. But Pacific yews are slow-growing trees, and harvesting them for their bark would mean their extinction. Fortunately, paclitaxel is now synthesized in the laboratory and marketed under the trade name Taxol.

- **Cotton:** You probably own many garments made of cotton, a cloth that comes from flowering plants in genus *Gossypium.* Cotton seeds develop in a dense web of cellulose fibers, which textile manufacturers spin into threads that make up T-shirts, blue jeans, underwear, towels, sheets, and many other cloth products. The seeds themselves also produce cooking oil. About three-fourths of the U.S. cotton crop is genetically modified to produce its own insecticides (see section 10.10).

Peat moss, horsetail, the Pacific yew, and cotton are just four of the many plants that humans use, and they represent only a tiny percentage of the diverse kingdom Plantae. This chapter highlights some of the history and diversity of these essential organisms.

Learning Outline

Learn How to Learn

Take Notes on Your Reading

Many classes require reading assignments. Taking notes as you read should help you not only retain information but also identify what you don't understand. Before you take notes, read through the assigned pages once; otherwise, you may have trouble distinguishing between main points and minor details. Then read it again. This time, pause after each section and write the most important ideas in your own words. What if you can't remember it or don't understand it well enough to summarize the passage? Read it again, and if that doesn't work, ask for help with the points that aren't clear.

18.1 | Plants Have Changed the World

If you glance at your surroundings in almost any outdoor setting, the plants are the first things you see. Grasses, trees, shrubs, ferns, or mosses exist nearly everywhere, at least on land (**table 18.1** and **figure 18.1**). Members of kingdom **Plantae** dominate habitats from moist bogs to parched deserts. They are so familiar that it is difficult to imagine a world without plants.

Plants are autotrophs: they use sunlight as an energy source to assemble CO_2 and H_2O into sugars (chapter 5 describes the reactions of photosynthesis). The sugars, in turn, provide the energy and raw materials that maintain and build a plant's body. Moreover, the chemical reactions of photosynthesis release oxygen gas, O_2, as a waste product. Animals and other organisms that use aerobic respiration need this gas; in addition, O_2 in the atmosphere helps form the ozone layer that protects life from the sun's harmful ultraviolet radiation.

Today's plants transform the landscape, but their effect on the evolution of life is even more dramatic. Once plants settled the land some 450 million years ago, they set into motion a

Figure 18.1 **Plants Are Everywhere.** Nearly every ecosystem on land is dominated by plants.

Table 18.1 Phyla of Plants

Phylum	Examples	Number of Existing Species
Nonvascular plants		
Marchantiophyta	Liverworts	9000
Anthocerotophyta	Hornworts	100
Bryophyta	True mosses	15,000
Seedless vascular plants		
Lycopodiophyta	Club mosses, spike mosses	1200
Pteridophyta	Whisk ferns, true ferns, horsetails	11,500
Seed plants		
Cycadophyta	Cycads	130
Ginkgophyta	Ginkgo	1
Pinophyta	Pines, firs, and other conifers	630
Gnetophyta	Gnetophytes	80
Magnoliophyta	All flowering plants, including roses, grasses, fruit trees, and oaks	>260,000

complex series of changes that would profoundly affect both the living and nonliving worlds. The explosion of photosynthetic activity from plants altered the atmosphere, lowering CO_2 levels and raising O_2 content.

As plants gradually expanded from the water's edge to the world's driest habitats, they formed the bases of intricate food webs, providing diverse habitats for many types of animals, fungi, and microbes. Herbivores consume living leaves, stems, roots, seeds, and fruits. Dead leaves accumulating on the soil surface feed countless soil microorganisms, insects, and worms. When washed into streams and rivers, this leaf litter supports a spectacular assortment of fishes and other aquatic animals.

Of course, plants remain essential to life today. From a human perspective, farms and forests provide the foods we eat, the paper we read, the lumber we use to build our homes, many of the clothes we wear, and some of the fuel we burn. (This chapter's Burning Question introduces biofuels derived from plants). The list goes on and on. It is amazing to think that plants do so much with such modest raw materials: sunlight, water, minerals, and CO_2.

A. Green Algae Are the Closest Relatives of Plants

All plants, from mosses to maple trees, are multicellular organisms with eukaryotic cells. With the exception of a few parasitic species, plants are autotrophs. A careful reading of section 17.2, however, will reveal that protists called algae have the same combination of traits. Which of the many lineages of algae, then, gave rise to plants?

The answer is that green algae and plants apparently share the most recent common ancestor. About 480 to 470 million years ago, or perhaps earlier, one group of green algae related to today's **charophytes** likely gave rise to plants (**figure 18.2**).

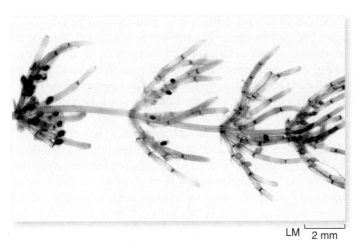

Figure 18.2 Charophyte. This green alga, called *Chara*, may resemble the ancestors of land plants.

Evidence for this evolutionary connection includes chemical and structural similarities. For example, the chloroplasts of plants and green algae contain the same photosynthetic pigments. In addition, like green algae, plants have cellulose-rich cell walls and use starch as a nutrient reserve. Similar DNA sequences offer additional evidence of a close relationship. ▶▶ green algae, p. 358; polysaccharides, p. 32

Nevertheless, the body forms of algae are quite different from those of plants, in part because water presents selective forces that are far different from those in the terrestrial landscape. Consider the aquatic habitat. Light, water, minerals, and dissolved gases surround the whole body of a submerged green alga, and the buoyancy of water provides physical support. In sexual reproduction, an alga simply releases gametes into the water, and the current carries the sex cells to another individual.

On land, the water and minerals essential for plant growth are in the soil, and only the aboveground part of the plant is exposed to light. Air not only provides much less physical support than does water, but it also dries out the plant's aerial tissues. Furthermore, the dispersal of gametes for sexual reproduction becomes more complicated on dry land.

These conditions have selected for unique adaptations in the body forms and reproductive strategies of plants. As described in the next section, biologists use some of these features to organize land plants into four main groups (**figure 18.3**): the bryophytes, seedless vascular plants, gymnosperms, and angiosperms.

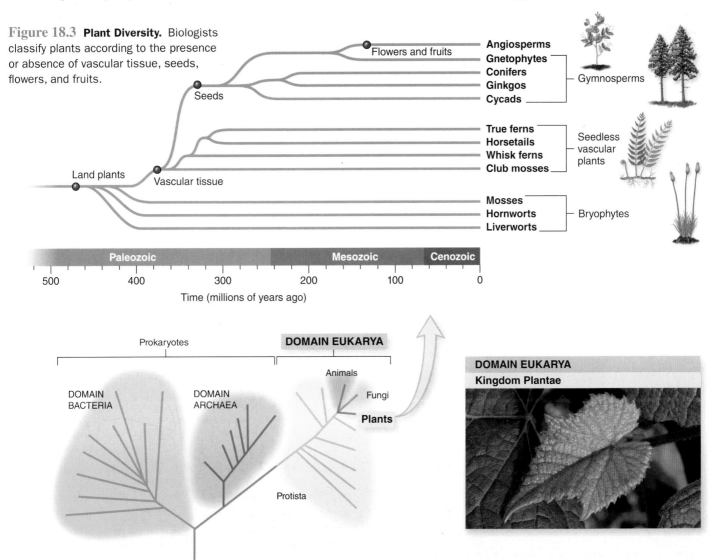

Figure 18.3 Plant Diversity. Biologists classify plants according to the presence or absence of vascular tissue, seeds, flowers, and fruits.

B. Plants Are Adapted to Life on Land

Figure 18.4 illustrates many of the adaptations that enable plants to grow upright, retain moisture, survive, and reproduce without being immersed in water. Refer to this figure frequently while reading the rest of this section.

Obtaining Resources A suite of adaptations enables plants to acquire light, CO_2, water, and minerals. Most plants, for example, have aboveground stems that support multiple leaves. The extensive surface area of the leaves maximizes exposure to sunlight and CO_2. Below the ground surface, highly branched root systems not only anchor the plant in the soil but also absorb water and minerals.

A plant that dries out will not survive. One water-conserving adaptation is the **cuticle,** a waxy coating that minimizes water loss from the aerial parts of a plant. Dry habitats such as deserts select for extra-thick cuticles; plants in moist habitats typically have thin cuticles.

The waxy cuticle is impermeable not only to water but also to gases such as CO_2 and O_2. How do plants exchange these gases with the atmosphere? The answer is that the cuticle is interrupted by many **stomata,** which are pores in the epidermis of stems and leaves. Two guard cells surround each stoma and control whether the pore is open or closed. As gases diffuse through open stomata, water also escapes from the plant's tissues. Plants close the stomata in dry weather, minimizing water loss. ▸ plant epidermis, p. 461

Internal Transportation and Support The division of labor between the above- and belowground parts of a plant poses a problem: roots need the food produced at the leaves, whereas leaves and stems need water and minerals from soil. In small plants called bryophytes, cell-to-cell diffusion meets these needs. Other plants, however, have **vascular tissue,** a collection of tubes that transport sugar, water, and minerals throughout the plant.

The two types of vascular tissue are xylem and phloem. **Xylem** conducts water and dissolved minerals from the roots to the leaves. **Phloem** transports sugars produced in photosynthesis to the roots and other nongreen parts of the plant. This internal transportation system has supported the evolution of specialized roots, stems, and leaves, many of which have adaptations that enable plants to exploit extremely dry habitats.

In addition, xylem is rich in **lignin,** a complex polymer that strengthens cell walls. The additional support from lignin means that vascular plants can grow tall and form branches, important adaptations in the intense competition for sunlight.

Reproduction Plants have a life cycle called **alternation of generations,** in which a multicellular diploid stage alternates with a multicellular haploid stage (**figure 18.5**). The **sporophyte** (diploid) generation develops from a zygote that forms when gametes come together at fertilization. In a mature sporophyte, some cells undergo meiosis and produce haploid spores, which divide mitotically to form the gametophyte. The haploid **gametophyte** produces gametes by mitotic cell division; these sex cells fuse at

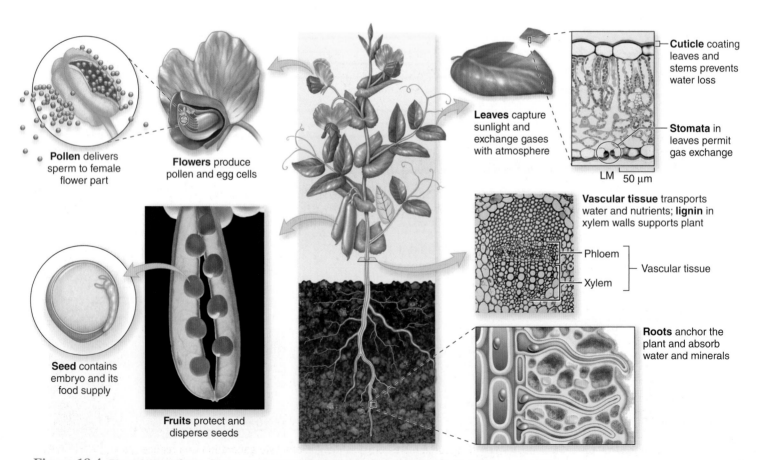

Figure 18.4 Plant Adaptations. Pea plants have many features that support life on land.

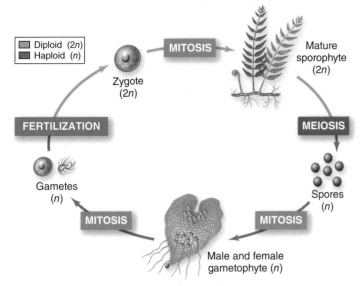

Figure 18.5 **Alternation of Generations.** Unlike animals, plants have multicellular haploid and diploid generations. The haploid, or gametophyte, generation follows a diploid sporophyte generation.

fertilization, starting the cycle anew. (Read more about mitosis and meiosis in chapters 8 and 9).

A prominent trend among land plants is a change in the relative sizes and independence of the gametophyte and sporophyte generations (figure 18.6). In a moss, for example, the brown sporophyte depends on the green gametophyte for nutrition. In more complex plants, the sporophyte generation is photosynthetic and much larger than the gametophytes. Ferns, pines, and flowering plants have gametophytes that range from microscopic to barely visible with the unaided eye. Keep this evolutionary trend in mind as you study the plant life cycles in this chapter.

Plant reproduction has other variations as well. The swimming sperm cells of mosses and ferns require a film of free water to reach an egg, limiting the distance over which gametes can spread. Gymnosperms and angiosperms can reproduce over far greater distances, thanks to pollen. **Pollen** consists of the male gametophytes of seed plants; each pollen grain produces sperm. In **pollination,** wind or animals deliver pollen directly to female plant parts, eliminating the need for free water in sexual reproduction.

Gymnosperms and angiosperms also share another reproductive adaptation: seeds. A **seed** is a dormant plant embryo, packaged together with a food supply inside a tough outer coat that prevents desiccation. The food supply sustains the young plant between the time the seed germinates and when the seedling begins photosynthesis.

The origin of pollen and seeds was a significant event in the evolution of plants. The spores of mosses and ferns—the seedless plants—take little energy to produce, but they are short-lived and tend to remain relatively close to the mother plant. The gymnosperms and angiosperms, in contrast, can disperse their gametes and seeds over long distances and in dry conditions. Moreover, seeds can remain dormant for years, germinating when conditions are favorable. Pollen and seeds therefore give gymnosperms and angiosperms a competitive edge in many habitats.

Two additional reproductive adaptations occur only in the angiosperms: flowers and fruits (see figure 18.4). **Flowers** are reproductive structures that produce pollen and egg cells. After fertilization, parts of the flower develop into a **fruit** that contains the seeds. Flowers and fruits help angiosperms protect and disperse both their pollen and their offspring. These adaptations are spectacularly successful; angiosperms far outnumber all other plants, both in numbers and species diversity.

Biologists classify plants based on the presence or absence of transport tissues, seeds, flowers, and fruits. The next four sections of this chapter describe the diversity of plants. Unit 5 delves more deeply into the anatomy, reproduction, and development of angiosperms.

18.1 | Mastering Concepts

1. How have plants changed the landscape, and how are they vital to life today?
2. What evidence suggests that plants evolved from green algae?
3. What are the functions of the cuticle and stomata?
4. How does vascular tissue adapt plants to land?
5. Describe the reproductive adaptations of plants.
6. What features differentiate the four major groups of plants?

	Bryophytes	Seedless Vascular Plants	Gymnosperms	Angiosperms
GAMETOPHYTE		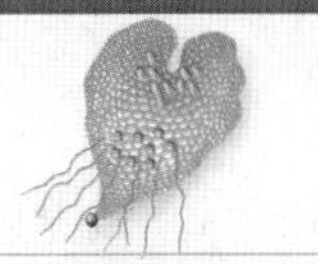	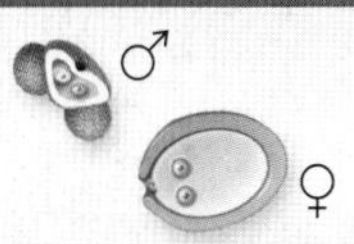	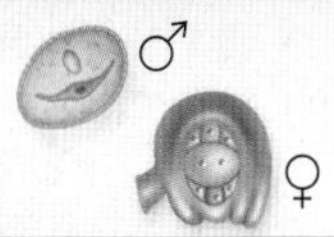
Size relative to sporophyte?	Varies	Small	Microscopic	Microscopic
Depends on sporophyte for nutrition?	No	No	Yes	Yes
SPOROPHYTE				
Size relative to gametophyte?	Varies	Large	Large	Large
Depends on gametophyte for nutrition?	Yes	No	No	No

Figure 18.6 **Changes in the Generations.** As plants became more complex, the gametophyte generation was reduced to just a few cells that depend on the sporophyte for nutrition.

18.2 | Bryophytes Are the Simplest Plants

- Angiosperms
- Gymnosperms
- **Seedless vascular plants**
- **Bryophytes**

The earliest plants, which probably resembled modern bryophytes, emerged onto land during the Ordovician period some 450 million years ago. All **bryophytes** are seedless plants that lack vascular tissue, but evidence suggests that they do not form a single clade (a group that includes one common ancestor and all of its descendants). This section describes them together, however, because they share some important features. ▸ cladistics, p. 287

A. Bryophytes Lack Vascular Tissue

Without vascular tissue and lignin, bryophytes lack the physical support to grow very large. Bryophytes are therefore small, compact plants. Their diminutive size means that each cell can absorb minerals and water directly from its surroundings. Materials move from cell to cell within the plant by diffusion and osmosis, not within specialized transport tissues.

Although bryophytes lack true leaves and roots, many have structures that are similar to these organs. For example, photosynthesis occurs at flattened leaflike areas. In addition, hairlike extensions called rhizoids cover a bryophyte's lower surface, anchoring the plant to its substrate.

Without true roots, bryophytes cannot tap distant sources of water when conditions turn dry. Many species are therefore restricted to moist, shady habitats that are unlikely to dry out. Others tolerate periods of desiccation by entering dormancy. When moisture returns, so does their metabolism.

a.

b.

c.

Figure 18.7 A Gallery of Bryophytes. (a) The gametophyte of this liverwort resembles small leaves. The umbrella-shaped structures produce sperm and egg cells. (b) In the hornwort *Anthoceros,* the tapered hornlike structures are sporophytes, below which the flat gametophytes are visible. (c) Short sporophytes topped with dark capsules peek above the gametophytes of *Sphagnum* moss.

? Burning Question

What are biofuels?

Petroleum and other fossil fuels consist of organisms buried millions of years ago. Burning gasoline and other fossil fuels releases their ancient carbon into the atmosphere. Biofuels are plant-based substitutes for fossil fuels. These alternative fuels have attracted attention in recent years, in part because they can help decrease reliance on foreign petroleum. Replacing fossil fuels with biofuels should also help reduce CO_2 emissions and the associated problem of global climate change. ▸ global climate change, p. 809

Two types of biofuels are biodiesel and ethanol. As the name suggests, biodiesel is a diesel fuel substitute. Currently, most biodiesel comes from oil extracted from crushed soybeans or canola seeds. To avoid driving up the price of these food crops, researchers are looking for economical, nonfood sources of biodiesel. Examples include everything from green algae to the seeds of a plant called the jatropha tree. This long-lived tree uses little water and tolerates poor soil, so it does not compete with food plants for rich farmland.

Ethanol, the other main biofuel, is a gasoline substitute. Corn kernels are the main source of ethanol in the United States. Starch extracted from the corn kernel is enzymatically digested to sugar, which is then fermented into ethanol. Sugarcane, the main source of ethanol in Brazil, is a more economical alternative in the tropics. Its tissues are high in sugar, not starch, so biofuel manufacturers can omit the costly enzymes from the ethanol production process. ▸ fermentation, p. 115

Researchers are also searching for nonfood sources of sugar to use in ethanol production. The inedible stems of corn or of prairie grasses such as switchgrass would be ideal; bacterial and fungal enzymes easily break the cellulose in the plant cell walls into simple sugars. One problem, however, is that the stems also contain lignin, a complex molecule that interferes with cellulose extraction. So far, the heat and acid treatment needed to eliminate the lignin is too costly and inefficient to make cellulose-derived ethanol economical.

Biofuels are promising, but it is important to realize that they are not exactly "carbon-neutral." Most plants used in biofuel production require fertilizers and pesticides—both of which come from fossil fuels.

Submit your burning question to:
marielle_hoefnagels@mcgraw-hill.com

Biologists classify the 24,000 or so species of bryophytes into three phyla (figure 18.7):

- **Liverworts** (phylum Marchantiophyta) have flattened leaflike structures. The diverse liverworts may be the bryophytes most closely related to ancestral land plants.
- **Hornworts** (phylum Anthocerotophyta) are the smallest group of bryophytes, with only about 100 species. They are named for their sporophytes, which are shaped like tapered horns.
- **Mosses** (phylum Bryophyta) are the closest living relatives to the vascular plants. The gametophytes resemble short "stems" with many "leaves." The brown or green sporophyte looks nothing like the gametophyte.

Nonvascular plants play important roles in ecosystems. For example, mosses can survive on bare rock or in a very thin layer of soil. As their tissues die, they contribute organic matter that helps build soil. Larger plants subsequently colonize the soil, triggering other changes in the ecosystem. A similar process occurs in forest canopies, where bryophytes

Gemma cup   Gemma   1 mm

Figure 18.8 **Asexual Reproduction.** The gametophytes of liverworts such as this *Lunularia* can produce fragments called gemmae. Raindrops splash the gemmae from the cups, and then each gemma develops into an identical new plant.

living on tree limbs help build an organic soil that sustains entire communities of tree-dwelling organisms.

B. Bryophytes Have a Conspicuous Gametophyte

Most reproduction in bryophytes occurs asexually. For example, mosses and liverworts produce gemmae, small pieces of tissue that detach and grow into new plants (figure 18.8).

Bryophytes also reproduce sexually (figure 18.9). Gametes form by mitosis in separate sperm- and egg-producing structures on the leafy gametophyte. Sperm swim to the egg cell in a film of water that coats the plants. The sporophyte generation begins at fertilization, with the formation of the diploid zygote. This cell divides mitotically, producing a stalk attached to the gametophyte. At the tip of the stalk, specialized cells inside a sporangium undergo meiosis and produce haploid spores. After the spores are released, they germinate, giving rise to new haploid gametophytes.

Bryophytes produce large numbers of spores, which offer temporary protection against drying out. Thanks to their tough walls, bryophyte spores are preserved in the fossil record. In fact, the earliest evidence of plants on land is a fossilized spore resembling those of today's liverworts.

18.2 | Mastering Concepts

1. Describe the three main groups of bryophytes.
2. Why do mosses usually live in moist, shady habitats?
3. How do bryophytes reproduce asexually and sexually?

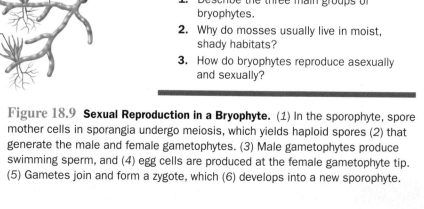

Figure 18.9 **Sexual Reproduction in a Bryophyte.** (*1*) In the sporophyte, spore mother cells in sporangia undergo meiosis, which yields haploid spores (*2*) that generate the male and female gametophytes. (*3*) Male gametophytes produce swimming sperm, and (*4*) egg cells are produced at the female gametophyte tip. (*5*) Gametes join and form a zygote, which (*6*) develops into a new sporophyte.

18.3 | Seedless Vascular Plants Have Xylem and Phloem But No Seeds

Angiosperms
Gymnosperms
Seedless vascular plants
Bryophytes

Most nonvascular plants are small and easily overlooked. Not so for the **seedless vascular plants,** the 12,000 species that have xylem and phloem but do not produce seeds. These plants have much larger representatives than their bryophyte counterparts.

A. Seedless Vascular Plants Include Ferns and Their Close Relatives

The earliest species of seedless vascular plants are extinct, but fossil evidence suggests that they originated early in the Devonian period, about 400 million years ago. The club mosses (not to be confused with the true mosses, which are bryophytes), are descendants of these early vascular plants. The other seedless

vascular plants, including the ferns, first appeared about 50 million years later.

As described in section 18.1, vascular tissue enabled plants to grow much larger than nonvascular plants, both in height and in girth. This increase in size was adaptive because taller plants have the edge over their shorter neighbors in the competition for sunlight. Larger plants also triggered evolutionary changes in other organisms by providing new habitats and more diverse food sources for arthropods, vertebrates, and other land animals.

Unlike the bryophytes, the seedless vascular plants typically have true roots, stems, and leaves. In many species, the leaves and roots arise from underground stems called rhizomes. Rhizomes sometimes also store carbohydrates that provide energy for the growth of new leaves and roots.

The seedless vascular plants include four lineages grouped into two phyla (figure 18.10):

- **Club mosses** (phylum Lycopodiophyta) are small plants in genus *Lycopodium*. These plants have simple leaves that resemble scales or needles, and the name reflects their club-shaped reproductive structures. Their close relatives are the spike mosses *(Selaginella).* Collectively, club mosses and spike mosses are sometimes called lycopods.

Figure 18.10 A Gallery of Seedless Vascular Plants. (a) The club moss *Lycopodium obscurum* produces upright stems. (b) The spike moss, *Selaginella martensii,* has scalelike foliage. (c) Yellowish spore-producing structures are visible on this whisk fern, *Psilotum nudum.* (d) *Equisetum fluvatile* is a horsetail. (e) The narrow beech fern *(Phegopteris connectilis)* is a true fern.

- **Whisk ferns** (phylum Pteridophyta) are simple plants that have rhizomes but not roots. Most species have no obvious leaves. Their name comes from the highly branched stems of *Psilotum,* which resemble whisk brooms.

- **Horsetails** (phylum Pteridophyta) grow along streams or at the borders of forests. The only living genus of horsetails, *Equisetum,* includes plants with branched rhizomes that give rise to green aerial stems bearing spores at their tips. Horsetails are also called scouring rushes, reflecting the abrasive silica particles in their tissues (see the chapter opening essay).

- **True ferns** (phylum Pteridophyta) make up the largest group of seedless vascular plants, with about 11,000 species. The fronds, or leaves, of ferns are their most obvious feature; some species are popular as ornamental plants. Ferns were especially widespread and abundant during the Carboniferous period, when their huge fronds dominated warm, moist forests. Their remains form most coal deposits. ▶ coal, p. 311

Most seedless vascular plants live on land, where their roots and rhizomes help stabilize soil and prevent erosion. Some species of ferns and horsetails are especially adept at colonizing disturbed soils such as road cuts. But not all species are terrestrial. The tiny fern *Azolla* lives in water, where its leaves house cyanobacteria that fix nitrogen. In Asia, rice farmers cultivate *Azolla* to help fertilize their crops. ▶ nitrogen fixation, p. 346

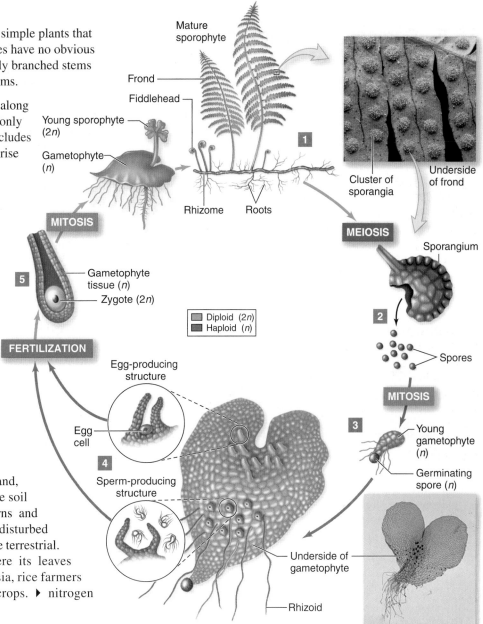

Figure 18.11 **Sexual Reproduction in a True Fern.** (*1*) Sporangia on sporophyte fronds house spore mother cells that (*2*) produce spores through meiosis. (*3*) A haploid spore develops into a gametophyte, which (*4*) produces egg cells and swimming sperm. (*5*) These gametes join and produce a zygote, which develops into the sporophyte.

B. Seedless Vascular Plants Have a Conspicuous Sporophyte and Swimming Sperm

Figure 18.11 illustrates the life cycle of a fern. The sporophyte produces haploid spores by meiosis in collections of sporangia on the underside of each frond. Once shed, the spores germinate and develop into tiny, heart-shaped gametophytes that produce gametes by mitotic cell division. The swimming sperm require a film of free water to reach the egg cell. The gametes fuse, forming a zygote. This diploid cell divides mitotically and forms the sporophyte, which quickly dwarfs the gametophyte.

Many seedless vascular plants live in shady, moist habitats. Like bryophytes, these plants produce swimming sperm and therefore cannot reproduce sexually in the absence of free water.

18.3 | Mastering Concepts

1. Describe the four groups of seedless vascular plants.
2. How do seedless vascular plants reproduce?
3. How are seedless vascular plants similar to and different from bryophytes?

18.4 | Gymnosperms Are "Naked Seed" Plants

Bryophytes and seedless vascular plants dominated Earth's vegetation for more than 150 million years. But during the Permian period, about 300 million years ago, plants with pollen and seeds appeared. The new reproductive adaptations allowed these plants to outcompete the seedless plants in many habitats.

The first seed plants were gymnosperms. The term **gymnosperm** derives from the Greek words *gymnos,* meaning "naked," and *sperma,* meaning "seed." The seeds of these plants are "naked" because they are not enclosed in fruits.

A. Gymnosperms Include Conifers and Three Related Groups

Living gymnosperms are remarkably diverse in reproductive structures and leaf types. The sporophytes of most gymnosperms are woody trees or shrubs, although a few species are more vine-like. Leaf shapes range from tiny reduced scales to needles, flat blades, and large fernlike leaves. The 800 or so species of gymnosperms group into four phyla (figure 18.12):

- **Cycads** (phylum Cycadophyta) live primarily in tropical and subtropical regions. They have palmlike leaves, and they produce large cones. Many cycads are planted as ornamentals, but only two species are native to the United States. Cycads dominated Mesozoic era landscapes; today,

many species are near extinction because of their slow growth, low reproductive rates, and shrinking habitats.

- The **ginkgo** (phylum Ginkgophyta), also called the maidenhair tree, has distinctive, fan-shaped leaves that have remained virtually unchanged for 80 million years. Only one species exists: *Ginkgo biloba*. It no longer grows wild in nature, but it is a popular cultivated tree. Ginkgos have male and female organs on separate plants; landscapers avoid planting female ginkgo trees because the fleshy seeds produce a foul odor. Although clinical trials have not conclusively supported its medical benefits, some people believe that extracts of *Ginkgo biloba* leaves may improve memory and concentration.

- **Conifers** (phylum Pinophyta) such as pine trees are by far the most familiar gymnosperms. These plants often have needlelike or scalelike leaves, and they produce egg cells and pollen in cones. Conifers are commonly called "evergreens" because most retain their leaves all year, unlike deciduous trees. This term is somewhat misleading, however, because conifers do shed their needles. They just do it a few needles at a time, turning over their entire needle supply every few years.

- **Gnetophytes** (phylum Gnetophyta) include some of the most distinctive (if not bizarre) of all seed plants. Botanists have struggled with the classification of these plants. Some details of their life history have led to speculation that gnetophytes are closely related to the flowering plants. Molecular evidence, however, places these puzzling plants with the conifers. *Ephedra* is a gnetophyte, as is *Welwitschia,* a slow-growing plant that lives in African deserts. Mature *Welwitschia* plants have a single pair of large, strap-shaped leaves that persist throughout the life of the plant.

a.   b.   c.   d.

Figure 18.12 A Gallery of Gymnosperms. (a) Cycads are ancient seed plants with cones that form in the center of a crown of large leaves. A seed cone is shown here. (b) The leaves of *Ginkgo biloba* turn yellow in the fall. The lower photo shows the fleshy seed. (c) This pinyon pine is an example of a conifer. The seed cone has woody scales. (d) *Ephedra* is a gnetophyte with cones that resemble tiny flowers.

B. Conifers Produce Pollen and Seeds in Cones

Pines illustrate the gymnosperm life cycle (figure 18.13). In these plants, **cones** are the organs that bear the reproductive structures. Large female cones bear two sporangia, called **ovules,** on the upper surface of each scale. Through meiosis, each ovule produces four haploid structures called megaspores, only one of which develops into a female gametophyte. Over many months, the female gametophyte undergoes mitosis and gives rise to two to six egg cells. At the same time, small male cones bear sporangia on thin, delicate scales. Through meiosis, these sporangia produce microspores, which eventually become wind-blown pollen grains (immature male gametophytes). Pollination occurs when pollen grains settle between the scales of female cones and adhere to drops of a sticky secretion.

The pollen grain germinates, giving rise to a pollen tube that grows through the ovule toward the egg cell. Two haploid sperm nuclei develop inside the pollen tube; one sperm nucleus fertilizes the haploid egg cell, and the other disintegrates. The resulting zygote is the first cell of the sporophyte generation. The whole process is so slow that fertilization occurs about 15 months after pollination.

Within the ovule, the haploid tissue of the female gametophyte nourishes the developing diploid embryo. Following a period of metabolic activity, the embryo becomes dormant, and the ovule develops a tough, protective seed coat. The seed may remain in the cone for another year. Eventually, however, the seed is shed and dispersed by wind or animals. If conditions are favorable, the seed germinates, giving rise to a new tree.

18.4 | Mastering Concepts

1. What are the characteristics of gymnosperms?
2. What are the four groups of gymnosperms?
3. What is the role of cones in conifer reproduction?
4. What happens during and after pollination in gymnosperms?

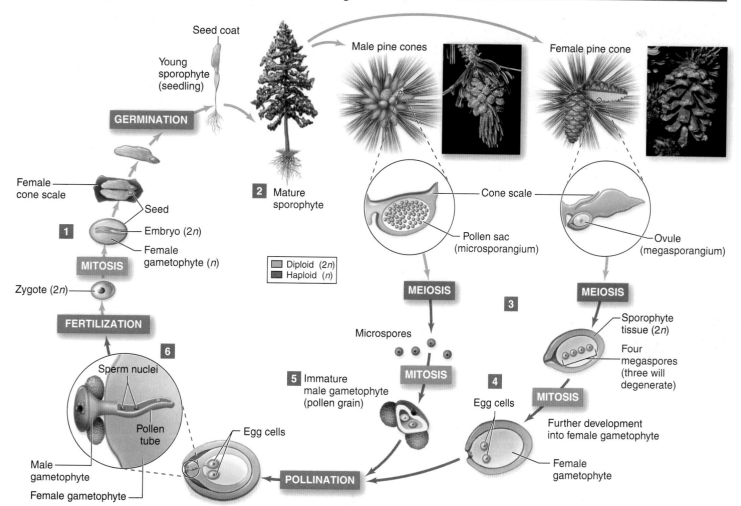

Figure 18.13 Sexual Reproduction in Pines. (*1*) A pine tree's seed contains the embryo, or young diploid sporophyte. (*2*) The seed germinates, and the sporophyte grows into a mature, cone-producing tree. (*3*) Cells in the male and female cone scales undergo meiosis, producing spores that develop into haploid gametophytes that consist of just a few cells each. (*4*) On the female cones, each scale has two ovules (only one is shown), each of which yields an egg-producing gametophyte. (*5*) The male cones produce pollen, the male gametophytes. (*6*) A pollen grain delivers a sperm nucleus to an egg cell via a pollen tube. More than a year later, each fertilized egg (zygote) completes its development into a seed.

18.5 | Angiosperms Produce Seeds in Fruits

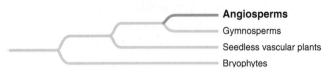

- **Angiosperms**
- Gymnosperms
- Seedless vascular plants
- Bryophytes

The bryophytes, seedless vascular plants, and gymnosperms make up less than 5% of all modern plant species. The other 95% are **angiosperms,** or flowering plants (phylum Magnoliophyta). Examples include apple trees, corn, roses, petunias, lilies, grasses, and many other familiar plants, including those we grow for our own food.

Many people think of flowers only as decorations for the human habitat. Although many ornamental plants are the products of selective breeding for their spectacular blooms, not all flowers are showy, sweet-smelling beauties (figure 18.14). The flowers of wind-pollinated plants such as grasses, for example, are plain and easily overlooked.

Fossil evidence places the origin of angiosperms in the early Cretaceous period, at least 130 million years ago. By 100 million years ago, all of today's major lineages of angiosperms were in place. Biologists have long puzzled about the sudden appearance and rapid diversification of the flowering plants. The adaptive radiation of the angiosperms coincided with the diversification of vertebrates and arthropods on land, but no one has established a definite cause-and-effect link. ▸ adaptive radiation, p. 282

A. Most Angiosperms Are Eudicots or Monocots

Biologists are still working to sort out the evolutionary relationships among the angiosperms; figure 18.15 shows one hypothesis.

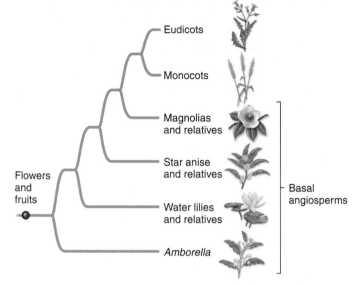

Figure 18.15 Angiosperm Phylogeny. This phylogenetic tree depicts the proposed evolutionary relationships among the eudicots, monocots, and basal angiosperms.

The two largest clades are the eudicots and the monocots. The **eudicots** have two cotyledons (the first leaf structures to arise in the embryo), and their pollen grains feature three or more pores. About 175,000 species exist. The diverse eudicots include roses, daisies, sunflowers, oaks, tomatoes, beans, and many others. *Arabidopsis thaliana,* the subject of this chapter's Focus on Model Organisms, is a eudicot.

Monocots are named for their single cotyledon; in addition, their pollen grains have just one pore. (Monocots and eudicots also differ by other characteristics, further described in chapter 21.) Examples of the 70,000 species of monocots are the orchids, lilies, grasses, bananas, and ginger. The grasses include

a.   b.   c.   d.

Figure 18.14 A Gallery of Angiosperms. The angiosperms exhibit an astonishing variety of flowers and fruits. (a) In bananas (genus *Musa*), flowers occur in clusters. Yellow flowers and green developing fruits are visible in this photograph. (b) Red maple (*Acer rubrum*) produces small, bright red flowers that develop into winged fruits after fertilization. (c) Passion vines are tropical plants with very showy flowers. (d) Cattails (*Typha latifolia*) are familiar wetland plants. The brown cylindrical "tails" are actually spikes of tiny brown flowers.

not only lawn plants but also sugarcane and grains such as rice, wheat, barley, and corn (see this chapter's Apply It Now box).

The other 3% or so of the flowering plants form a paraphyletic group informally called the basal angiosperms. Examples of basal angiosperms are magnolias, nutmeg, avocados, black pepper, water lilies, and star anise.

B. Flowers and Fruits Are Unique to the Angiosperm Life Cycle

The angiosperm life cycle is similar to that of gymnosperms in some ways (figure 18.16). For example, the sporophyte is the only conspicuous generation, and both types of plants produce pollen and seeds.

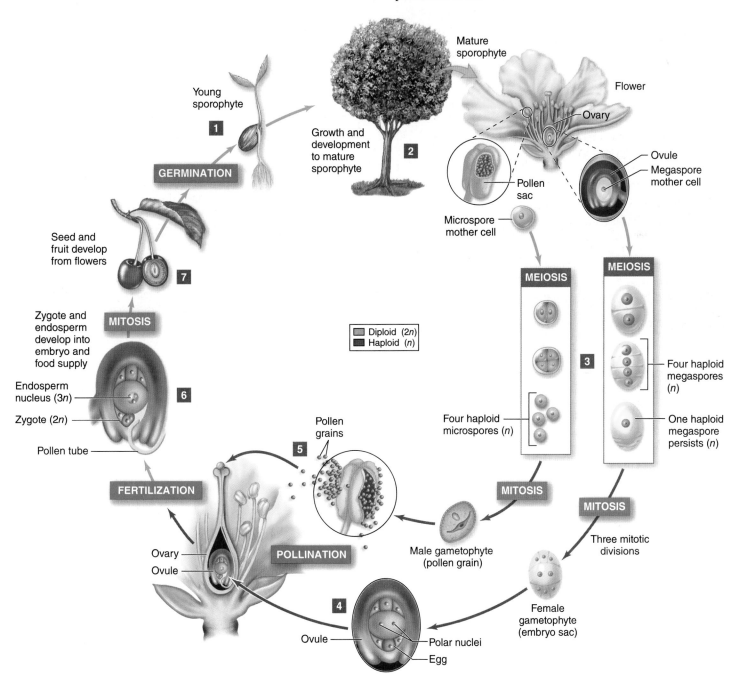

Figure 18.16 Sexual Reproduction in Angiosperms. (*1*) An angiosperm's seed germinates, and (*2*) the sporophyte grows into a mature tree. (*3*) Cells in the pollen sac and ovary undergo meiosis, producing spores that develop into haploid male and female gametophytes. (*4*) Inside the ovary, the female gametophyte includes one egg cell and two polar nuclei. (*5*) The pollen sac produces pollen, the male gametophytes. (*6*) A pollen grain delivers two sperm nuclei; one fertilizes the egg and the other fertilizes the polar nuclei, forming triploid tissue that develops into the endosperm. (*7*) The fruit develops from the ovary wall.

Apply It Now

Corn, Corn, Everywhere

Today's corn plant *(Zea mays)* is a product of thousands of years of artificial selection. Its ancestor, a wild grass called teosinte, had small, loosely packed kernels, each contained in its own husk. Farmers cultivated this monocot in Mexico at least 7000 years ago. The native people who settled North America brought it with them; somewhere along the line, farmers selected for the familiar cob and kernels that we now associate with corn.

Even if you don't eat the kernels off the cob, you likely have much more corn in your diet than you imagine. Besides fresh corn, frozen corn, canned corn, corn meal, corn oil, and corn starch, some of the many unexpected fates of corn include:

- baking powder and confectioners ("powdered") sugar, which often contain corn starch.
- corn syrup, a liquid sugar derived from corn starch, that sweetens soft drinks, ketchup, and many other foods.
- vanilla extract, which is often made with corn syrup.
- dextrin and maltodextrins, polysaccharides that thicken syrups and provide texture to low-fat foods (among many other uses). These food additives are produced by the partial hydrolysis of corn starch. ▶ hydrolysis, p. 31
- the simple sugars dextrose (glucose) and fructose.
- margarine, which often contains corn oil.
- the grain alcohol in bourbon whiskey. Ethanol from corn is also a biofuel (see this chapter's Burning Question).

The corn plant has also played a major role in the history of biology. In the 1940s, Barbara Mc-Clintock used multicolored Indian corn in her discovery of transposons, "jumping genes" that move from one location to another within a genome. She won the 1983 Nobel Prize for physiology or medicine for this work, which has led to new ways to mutate genes and, in turn, new insights into gene function. ▶ transposons, p. 140

Yet the life cycles differ in important ways. Most obviously, the reproductive organs in angiosperms are flowers, not cones. Another difference is that angiosperm seeds develop inside the flower's ovary following fertilization; other floral parts develop into a fruit that houses the seeds. Thus, an angiosperm's seeds are not "naked" like those of gymnosperms.

One other unique feature of the angiosperm life cycle is double fertilization. Each pollen grain produces two sperm nuclei, one of which fertilizes an egg. The other fertilizes a pair of polar nuclei in the female gametophyte. The resulting triploid nucleus develops into the endosperm, a tissue that supplies nutrients to the germinating seedling.

In many angiosperms, the main food storage molecule in the seed is starch. For example, the endosperm of wheat and other grains is starchy; bakers grind these seeds into flour to make bread and other baked goods. In other species, the endosperm is rich in oil. Coconuts and castor seeds are two sources of useful oils derived from endosperm.

C. Animals Often Participate in Angiosperm Reproduction

Angiosperms are also unique for the degree to which many of their life cycles depend on animals. Arthropods and vertebrates alike have two main roles in plant reproduction: pollination and the dispersal of seeds and fruits.

Pollination is at the heart of angiosperm sexual reproduction, and many species rely on animal "couriers" that unwittingly carry pollen from flower to flower. Of course, animals do not pollinate plants as an act of charity; they usually visit flowers in search of food. Bright colors and alluring scents signal the availability of a sweet reward such as nectar.

The relationship between angiosperms and their pollinators is sometimes so tight that one cannot reproduce or survive without the other. This situation can lead to **coevolution,** in which a genetic change in one species selects for subsequent change in another species. For example, an alteration in the shape or color of a flower can select for new adaptations in its pollinator (see section 13.7). The reverse can also occur: a change in the curve of a hummingbird's beak, for example, can select for corresponding modifications in the flower that it pollinates.

After pollination, one or more seeds develop inside a fruit. The main functions of the fruit are to protect and disperse the seeds. Some fruits, such as those of dandelions, have "parachutes" that promote wind dispersal. Others, however, spread only with the help of animals. Some have burrs that cling to animal fur. Others are sweet and fleshy, attracting animals that eat the fruits and later spit out or discard the seeds in their feces. ▶ seed dispersal, p. 497

Chapter 23 considers angiosperm reproduction in more detail.

18.5 | Mastering Concepts

1. What are the two largest clades of angiosperms?
2. In what ways are the life cycles of angiosperms similar to and different from those of conifers?
3. What is the relationship between flowers and fruits?
4. How do animals participate in angiosperm reproduction?

Focus on Model Organisms

Arabidopsis thaliana

The choice of a "star" model organism for plants is simple: *Arabidopsis thaliana* easily beats out all others (figure 18.A). This tiny angiosperm, a mustard relative, has become a staple of plant biology laboratories because of its small size, easy cultivation, and prolific reproduction. In addition, biologists can use a bacterium called *Agrobacterium tumefaciens* to carry new genes into *Arabidopsis* cells.

The amount of attention devoted to this plant may seem extravagant, considering its insignificance as a commercial plant. Because angiosperms are closely related to one another, however, discoveries in *Arabidopsis* will likely apply directly to economically important crop plants. For example, genetic and molecular studies of *Arabidopsis* can help researchers identify genes that enable plants to grow on poor soil. Understanding these genes may help plant scientists develop improved varieties of food crops such as rice, barley, wheat, and corn.

Moreover, many genes in *Arabidopsis* have counterparts in humans and other organisms, so research on this plant can have far-reaching applications. For example, scientists have discovered that *Arabidopsis* has at least 139 genes that correspond to human disease genes. Studies of *Arabidopsis* can therefore help us understand illnesses from colitis to Alzheimer disease to arthritis.

A few of the important discoveries resulting from work on *Arabidopsis* include:

- **Control of gene expression:** Each cell type in a multicellular organism turns on a different combination of genes. One way that cells regulate gene expression is to attach methyl groups to unneeded DNA; another is to produce small pieces of RNA that bind to genes that have already been transcribed. Researchers have studied both processes in *Arabidopsis,* in part because problems in gene expression cause some types of cancer in humans. ▶ regulation of gene expression, p. 134

- **Genome duplication:** The *Arabidopsis* genome sequencing project was completed in 2000. Analysis of the DNA suggests that the entire genome has duplicated two or three times, fueling speculation that all plants are polyploids. This finding may help shed light on the evolutionary history of plants. ▶ polyploids, p. 189

- **Disease resistance:** Some plants construct a sort of "fire break" around the spot where a bacterium or fungus has entered. Small areas of surrounding plant tissue die, and this zone of dead cells prevents the invader from spreading throughout the plant. Study of this response in *Arabidopsis* has led to new insights into how genes regulate this form of programmed cell death.
  ▶ apoptosis, p. 167

- **Response to the environment:** A plant cannot avoid extremes of temperature, light availability, and salinity by moving to a better location. Instead, it must adjust its physiology. *Arabidopsis* has genes that control its response when the weather turns cold, a finding that could help crop plants survive freezing.

- **Hormones:** Ethylene is a gas that helps control fruit ripening and plant senescence (aging). Mutant *Arabidopsis* plants that do not respond to ethylene have helped researchers find ethylene receptor proteins. Researchers have also discovered that the ethylene response requires copper, which the plant transports using a protein similar to one that transports copper in humans. (When faulty in humans, this protein causes Menkes disease.) ▶ ethylene, p. 500

- **Circadian rhythms:** Circadian processes occur in 24-hour cycles. In *Arabidopsis,* for example, proteins encoded by clock genes ensure that the expression of genes needed for photosynthesis peak at around noon. The same genes are repressed at night. Pigments called phytochromes "reset" the clock each day. ▶ phytochrome, p. 503

- **Flowering:** Angiosperms delay flowering until they reach reproductive maturity. How do they "know" when the time comes, and how do they build flower parts in the right places? *Arabidopsis* research has revealed genes that control the timing of flowering, the differentiation of cells that give rise to flowers, the development of individual flower parts, and the development of ovules inside the flower. For example, some genetic mutations induce flowers to develop into shoots; others promote early flowering.

Figure 18.A *Arabidopsis thaliana.*

18.6 | Investigating Life: Genetic Messages from the Dead Tell Tales of Ancient Ecosystems

Psychics claim to be able to communicate with the spirits of the dead. Although scientists do not assert the same spiritual connection, they can bring back the genetic remnants of species that lived and died long ago.

When an organism dies, its DNA usually degrades rapidly. But in some special cases, the genetic material remains intact indefinitely. Freezing is one way to preserve DNA. An ideal source of diverse ancient DNA is a landscape that once teemed with life but that has since become permanently frozen.

One such example is the land bridge, called Beringia, that once connected present-day northeastern Siberia to Alaska. During the last ice age, which ended about 10,000 years ago, much of Earth's water was locked in polar ice caps and glaciers. As a result, global sea levels were low, exposing the Beringian land bridge. At the same time, Beringia's dry climate kept the area free of the glaciers that scoured most of North America during the ice age. Giant mammals such as mammoth, bear, bison, and large cats roamed the grassy Beringian landscape.

Now that the ice age has ended, the land bridge is under the sea, and much of the rest of Beringia is permanently frozen land in Siberia, Alaska, and the Yukon. Could DNA trapped in frozen Siberian soil reveal an ancient ecosystem's organisms? Danish researchers Eske Willerslev, Anders Hansen, and their colleagues decided to find out.

To study the ancient remains, the scientists drove metal cylinders deep into the permafrost and removed long, thin rods (called "cores") of ice, soil, and organic material (figure 18.17). The deepest holes yield the oldest deposits, because new sediments accumulate over old ones. Radiometric dating (a technique that

uses radioactive decay of isotopes as a type of "clock"), pollen analysis, and other techniques helped them estimate the age of each layer of material in the cores. ▶ radiometric dating, p. 258

Willerslev and colleagues tried to extract DNA from eight sediment samples ranging in age from modern to 1.5 million to 2 million years old. Wherever DNA was present, they used the polymerase chain reaction to amplify two sets of genes. One target sequence was part of the gene encoding the protein rubisco, which is essential for photosynthesis (see chapter 5). This gene occurs in chloroplasts, so its presence indicates plant material in a sample. The other three target DNA sequences were fragments of genes found in mitochondria of vertebrate animals. (Vertebrates are animals with bony backbones, such as fishes, reptiles, and mammals). ▶ polymerase chain reaction, p. 171

The results showed that sediments from 300,000 to 400,000 years old contained plant DNA. When the scientists sequenced the DNA and compared it to genes cataloged in GenBank, a public database of nucleotide sequences, they found that the ancient sediments harbored tremendous diversity: 11 classes, 23 orders, and 28 families representing mosses, herbs, shrubs, and trees.

The oldest mitochondrial DNA was much younger, about 20,000 to 30,000 years old. Some came from grazers that still exist today, such as the horse, lemming, hare, musk ox, and reindeer, but the researchers also found genes from extinct mammoth and bison. (Previous researchers studying fossils of these huge herbivores had deposited the gene sequences in GenBank.) The presence of genes from animals that vanished long ago suggests that the sediment DNA was authentic and not simply a modern contaminant.

These gene fragments are important, but not for *Jurassic Park*-style reasons. After all, it is not possible to bring back the mammoth from just a tiny piece of its mitochondrial DNA. Rather, the study is significant because the long-buried sediments can help scientists reconstruct ancient ecosystems. The chloroplast DNA, for example, reveals that herbs (grasses and other nonwoody green plants) dominated the Beringian landscape 300,000 or so years ago but lost ground to shrubs over time (figure 18.18). The most

Figure 18.17 Core Sample. Cores of frozen sediment taken from permafrost may contain ancient DNA.

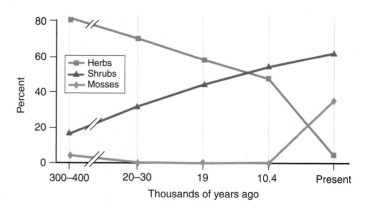

Figure 18.18 Communities Change. Chloroplast DNA isolated from sediment samples shows that grasses and other herbs have become much less common, while shrubs have become more common, over the past 400,000 or so years.

dramatic decline of grasses occurred in the past 10,000 years, a time that coincided with the extinction of the mammoth and bison. Did one event cause the other? Or did the end of the ice age cause both? What role did increasing human populations play? These questions remain unanswered for now.

Willerslev's team hopes their work will inspire others to extract DNA from ancient sediments around the world. The resulting patchwork of gene fragments, pieced together, will reveal valuable information about the changes that shaped ancient ecosystems—without the help of a psychic.

Willerslev, E., A. J. Hansen, J. Binladen, et al. May 2, 2003. Diverse plant and animal genetic records from Holocene and Pleistocene sediments. *Science,* vol. 300, pages 791–795.

18.6 | Mastering Concepts

1. Why do researchers collect DNA from permafrost?
2. What are some alternative hypotheses for why the researchers failed to recover any DNA from sediments that were more than 300,000 to 400,000 years old? How would you test your hypotheses?

Plant Characteristics: A Summary

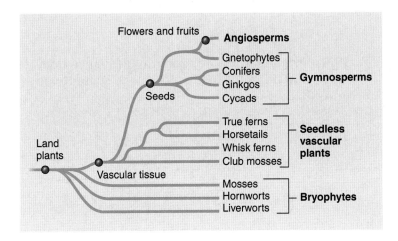

■ **Angiosperms (flowering plants)**
- >260,000 species
- Monocots, eudicots, basal angiosperms
- Independent sporophyte
- Pollen and egg cells develop in flowers
- Usually pollinated by wind or animals
- Seeds develop inside fruits

■ **Pines and other gymnosperms**
- ~830 species
- Cydads, ginkgos, conifers, gnetophytes
- Independent sporophyte
- Pollen and seeds usually develop on cone scales
- Usually wind-pollinated

■ **Ferns and other seedless vascular plants**
- ~12,000 species
- Club mosses, whisk ferns, horsetails, true ferns
- Independent sporophyte

■ **Mosses and other bryophytes (nonvascular plants)**
- ~24,000 species
- Liverworts, hornworts, mosses
- Sporophyte depends on gametophyte for nutrition

Group	Swimming Sperm	Vascular Tissue	Pollen	Seeds	Flowers	Fruits
Bryophytes	Yes	No	No	No	No	No
Seedless vascular plants	Yes	Yes	No	No	No	No
Gymnosperms	No	Yes	Yes	Yes	No	No
Angiosperms	No	Yes	Yes	Yes	Yes	Yes

Chapter Summary

18.1 | Plants Have Changed the World

- Members of kingdom **Plantae** provide food and habitat for other organisms, remove CO_2 from the atmosphere, and produce O_2. Humans rely on plants for food, lumber, clothing, paper, and many other resources.

A. Green Algae Are the Closest Relatives of Plants

- Plants evolved about 480 million years ago from green algae closely related to today's **charophytes.**
- Like many green algae, plants are multicellular, eukaryotic autotrophs that have cellulose cell walls and use starch as a carbohydrate reserve. Green algae and plants also use the same photosynthetic pigments. Unlike green algae, plants are adapted to conditions on land.

B. Plants Are Adapted to Life on Land

- Adaptations that enable plants to obtain and conserve resources include roots, leaves, a waterproof **cuticle,** and **stomata.**
- **Vascular tissue** (**xylem** and **phloem**) transports water, minerals, food, and other substances within a plant body. **Lignin** strengthens cell walls, providing physical support.
- Plant life cycles have an **alternation of generations,** with multicellular **sporophyte** (diploid) and **gametophyte** (haploid) phases. The sporophyte produces haploid spores by meiosis; the gametophyte produces haploid sperm and egg cells by mitosis. Fertilization restores the diploid number.
- In the simplest plants, the gametophyte generation is most prominent; in more complex plants, the sporophyte dominates.
- In gymnosperms and angiosperms, reproductive adaptations include **pollen** and **seeds. Pollination** delivers sperm to egg. The resulting zygote develops into an embryo, which is packaged along with a food supply into a seed. Angiosperms also produce **flowers** and **fruits.**
- Plants are classified by presence or absence of vascular tissue, seeds, flowers, and fruits.

18.2 | Bryophytes Are the Simplest Plants

A. Bryophytes Lack Vascular Tissue

- **Bryophytes** are small green plants lacking vascular tissue, leaves, roots, and stems.
- The three groups of bryophytes are **liverworts, hornworts,** and **mosses.**

B. Bryophytes Have a Conspicuous Gametophyte

- In bryophytes, the gametophyte stage is dominant. Many bryophytes reproduce asexually by fragmentation of the gametophyte. Sperm require free water to swim to egg cells.

18.3 | Seedless Vascular Plants Have Xylem and Phloem But No Seeds

A. Seedless Vascular Plants Include Ferns and Their Close Relatives

- **Seedless vascular plants** have vascular tissue but lack seeds. This group includes **club mosses, whisk ferns, horsetails,** and **true ferns.**

B. Seedless Vascular Plants Have a Conspicuous Sporophyte and Swimming Sperm

- The diploid sporophyte generation is the most obvious stage of a fern life cycle, but the haploid gametophyte forms a tiny separate plant.
- In the sexual life cycle of ferns, collections of sporangia appear on the undersides of fronds. Meiosis occurs in the sporangia and yields haploid spores, which germinate in soil and develop into gametophytes. The gametophytes produce egg cells and swimming sperm.

18.4 | Gymnosperms Are "Naked Seed" Plants

A. Gymnosperms Include Conifers and Three Related Groups

- **Gymnosperms** are vascular plants with seeds that are not enclosed in fruits. The four groups of gymnosperms are **cycads, ginkgos, conifers,** and **gnetophytes.**

B. Conifers Produce Pollen and Seeds in Cones

- In pines (a type of conifer), **cones** house the reproductive structures. Male cones release pollen, and female cones produce egg cells inside **ovules.** Pollen germination yields a pollen tube through which a sperm nucleus travels to the egg cell. After fertilization, the resulting embryo remains dormant in a seed until germination.
- In conifers, the sperm do not require free water to swim to the egg cell. Instead, most gymnosperms rely on wind to spread pollen.

18.5 | Angiosperms Produce Seeds in Fruits

A. Most Angiosperms Are Eudicots or Monocots

- **Angiosperms** are vascular plants that produce flowers. These reproductive structures produce pollen and egg cells.
- The two largest clades of angiosperms are **eudicots** and **monocots.**

B. Flowers and Fruits Are Unique to the Angiosperm Life Cycle

- Like gymnosperms, angiosperms use pollen to transport sperm to egg cells; both groups also produce seeds. Only angiosperms, however, have flowers. After pollination, the flower parts develop into the fruit, which protects the seeds.

C. Animals Often Participate in Angiosperm Reproduction

- Many animal-pollinated angiosperms have **coevolved** with their pollinators.
- The fruits of angiosperms aid in dispersal, usually by wind or animals.

18.6 | Investigating Life: Genetic Messages from the Dead Tell Tales of Ancient Ecosystems

- Researchers have extracted DNA from plants and animals buried long ago in frozen sediments. The DNA evidence reveals changes in the landscape over hundreds of thousands of years.

Multiple Choice Questions

1. Which of the following is NOT a property common to both land plants and green algae?
   a. Photosynthesis
   b. Starch as a storage form of energy
   c. Cellulose cell walls
   d. The presence of a cuticle and stomata

2. In the alternation of generations in plants, the gametophyte is _____ and produces gametes by _____.
   a. haploid; mitosis
   b. diploid; mitosis
   c. haploid; meiosis
   d. diploid; meiosis

3. Which group of bryophytes is most like its green algae ancestor?
   a. Mosses
   b. Liverworts
   c. Hornworts
   d. Both a and c are correct.

4. When moss spores are released from a sporangium, they form
   a. a sperm cell.
   b. a gametophyte.
   c. a sporophyte.
   d. a gemma.

5. How does the presence of vascular tissue (xylem and phloem) affect a plant?
   a. It reduces the plant's dependence on a moist environment.
   b. It allows specialization of roots, leaves, and stems.
   c. It allows for the growth of larger plants.
   d. All of the above are correct.

6. Why do many ferns require a shady, moist habitat?
   a. They lack vascular tissue.
   b. They use swimming sperm for sexual reproduction.
   c. They lack a water-retaining cuticle layer.
   d. The production of spores within the sporangium requires moisture.

7. Which type of plant produces seeds enclosed in fruits?
   a. Bryophyte
   b. Gymnosperm
   c. Angiosperm
   d. True fern

8. Reproduction in a pine tree is associated with
   a. male and female cones.
   b. windblown pollen.
   c. the formation of pollen tubes.
   d. All of the above are correct.

9. What is a key adaptation that is unique to the angiosperms?
   a. Dominant gametophyte generation
   b. Pollen grains that use pollen tubes to fertilize egg cells
   c. The use of flowers as reproductive structures
   d. Both a and b are correct.

10. How does the concept of coevolution apply to the angiosperms?
    a. Angiosperms coexist with simpler land plants such as bryophytes.
    b. Evolutionary changes to the environment have altered the diversity of fruits.
    c. The shape of a flower selects for the type of pollinator that can visit it.
    d. The type of pollinator selects for the form of seed dispersal.

Write It Out

1. What characteristics do all land plants have in common?

2. As described in chapter 17, the term *algae* encompasses many separate lineages of organisms. What evidence did biologists use to determine which types of algae are most closely related to land plants?

3. How are terrestrial habitats different from aquatic habitats? List the adaptations that enable plants to obtain resources, transport materials, and reproduce; explain how each adaptation contributes to a plant's reproductive success on land.

4. A few plant species, such as duckweed and water lilies, either float on or are submerged in water. These aquatic plants apparently evolved from terrestrial plants. Which adaptations do you expect to be minimized in these aquatic plants?

5. What are the four major groups of plants and the characteristics that distinguish them?

6. Give at least two explanations for the observation that bryophytes are much smaller than most vascular plants. How can increased height be adaptive? In what circumstances is small size adaptive?

7. Which adaptations enabled seedless vascular plants to invade new habitats?

8. A fern plant can produce as many as 50 million spores a year. How are these spores similar to and different from seeds? In a fern population that is neither shrinking nor growing, approximately what proportion of these spores is likely to survive long enough to reproduce? What factors might determine whether an individual spore successfully produces a new fern plant?

9. Some websites for gardeners warn against the "foul-smelling fruits" that female ginkgo trees produce. In what way is this statement incorrect?

10. How do the adaptations of gymnosperms and angiosperms enable them to live in drier habitats than bryophytes and seedless vascular plants?

11. How do angiosperms differ from gymnosperms? How are the two groups of plants similar?

12. Some viroids (infectious RNA molecules that cause diseases in plants) move to new plant hosts in pollen. Which of the four main groups of plants could "catch" viroids in this way? How could you test the hypothesis that a particular viroid spreads in pollen and not via an insect vector or direct plant-to-plant contact?

13. The immature fruit of the opium poppy produces many chemicals that affect animal nervous systems. In what way might these chemicals benefit the plant?

14. Scientists hypothesize that a geological catastrophe, such as an asteroid impact, killed the dinosaurs. If this is true, how might angiosperms have survived such a catastrophe?

15. Some angiosperm species have exclusive relationships with just one species of pollinator. How would this relationship benefit the plant? What are the risks to the plant?

16. Compare and contrast the life cycles of the four groups of plants. How does each group represent a variation on the common theme of alternation of generations?

17. Use the Internet to learn more about a well-studied plant species other than *Arabidopsis thaliana* and corn. How has the plant you chose contributed to knowledge of general biology and plant biology?

18. Make a list of everything you own or use that comes from plants or plant parts. Which of the four main groups of plants are represented in your list?

19. Suppose you and a friend are hiking and you see an unfamiliar plant. What observations would you make in trying to determine which type of plant it is?

20. People often move plants from one part of the world to another. Sometimes, an introduced plant species can become invasive, taking over native plant populations. The U.S. Department of Agriculture manages the National Invasive Species Information Center, whose website maintains a list of invasive plants. Which plant species are considered invasive in your home state? Why are those species harmful? Should invasive plants be eradicated? How?

21. Human activities and natural phenomena can drive plant species to extinction. The U.S. Department of Agriculture Natural Resources Conservation Service maintains lists of threatened and endangered plant species. What are some examples of threatened or endangered species in your area? What are the most important threats to those species? What are the potential consequences of a plant species extinction? What steps should we take to save threatened and endangered plants?

22. What are the pros and cons of pursuing biofuels as alternatives to fossil fuels? In your opinion, do the pros outweigh the cons, or vice versa? Justify your answer.

Pull It Together

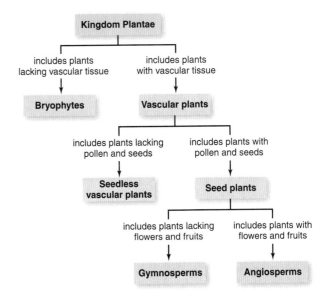

1. What are the main groups of plants within the bryophytes, seedless vascular plants, gymnosperms, and angiosperms?

2. For each of the four main groups of plants, describe the gametophyte and sporophyte generations.

3. How do bryophytes and seedless vascular plants reproduce if they lack pollen and seeds?

4. What is the relationship between pollen and seeds?

Mushrooms Galore. This man is harvesting white button mushrooms in a "grow room."

The Mushroom Mystique

FOR MANY PEOPLE, THE WORD *FUNGI* CONJURES IMAGES OF MUSHROOMS, ALONG WITH EMOTIONS THAT RANGE FROM DELIGHT TO REVULSION. Enthusiasts know that mushrooms can be delectable, yet fairy tales associate toadstools with dank, lonely forests teeming with witches and evil spirits.

Mushroom imagery is common in everyday language. A phrase such as "mushrooming private enterprise in China," for example, evokes the amazing growth rates of fungi. When warm weather follows a rain, mushrooms seem to pop up overnight on lawns and around trees. This extraordinarily rapid growth adds to their mystique.

How can mushrooms appear so suddenly? Like an apple on a tree, a mushroom is a reproductive structure attached to a much larger organism. Most of the fungal body, however, is underground. Fungi produce filaments that feed by penetrating their food, absorbing nutrients as they go. Sometimes, two threads unite, and the combined organism prepares to reproduce. When the temperature and moisture are just right, the cells of tiny "premushrooms" absorb water, expanding rapidly. Mushrooms erupt out of the ground and shed reproductive cells called spores. Afterward, the mushrooms wilt, vanishing as quickly as they appeared.

Picking and eating wild mushrooms can be extremely dangerous. One of the deadliest mushrooms is the death angel *(Amanita virosa)*. The poisons can be deceptive. Initial symptoms of nausea and diarrhea are followed by a lag of a day or two, during which the mushroom's toxins quietly destroy the liver, kidney, and other organs. By the time symptoms return, the damage is irreversible, and an emergency organ transplant is the only hope for survival. Unfortunately, there is no simple way to tell poisonous species from safe ones.

Commercial mushroom growers cultivate many edible fungi. The familiar white "button mushroom" *(Agaricus brunnescens)* grows on a compost of horse manure and straw. Shiitakes *(Lentinus edodes)* live on hardwood logs or compressed sawdust; oyster mushrooms *(Pleurotus* species) prefer wheat or rice straw. Hobbyists can purchase kits to grow these species at home.

Mushrooms represent only a small subset of kingdom Fungi. Diverse fungi spoil food and cause Dutch elm disease, chestnut blight, athlete's foot, diaper rash, ringworm, and yeast infections. On the other hand, they are essential decomposers in ecosystems, and they help humans manufacture cheese, bread, alcoholic beverages, soy sauce, antibiotics, dyes, and many other useful substances. Moreover, research on yeasts and other simple fungi has helped biologists learn about genetics and basic cell processes. This chapter opens the door to this fascinating group.

UNIT 4

What's the Point?

Learning Outline

Learn How to Learn
Use All Your Resources

Textbooks come with online quizzes, animations, and other resources that can help you learn biology. After you have studied the material in a chapter, test yourself by taking an online quiz. Take note of the questions for which you are unsure of the answers, and remember to go back to study those topics again. Also, check for animations that take you through complex processes one step at a time. Sometimes, the motion of an animation can help you understand what's happening more easily than studying a static image.

19.1 | Fungi Are Essential Decomposers

The members of kingdom **Fungi** live nearly everywhere—in soil, in and on plants and animals, in water, even in animal dung (figure 19.1). Microscopic fungi infect the cells of protists, while massive fungi extend enormous distances. For example, a single underground fungus extends over 890 hectares in an Oregon forest. Mycologists (biologists who study fungi) have identified more than 80,000 species of fungi, but 1.5 million or so are thought to exist (table 19.1).

Many people know that fungi can cause disease and turn foods moldy, so these organisms have a rather unsavory reputation. The chapter opening essay described why this reputation is undeserved. Fungi benefit humans directly in many ways: some are edible, others aid in food and beverage production, and still others are useful in biological research. In addition, fungi are vitally important in ecosystems. Many fungi secrete digestive enzymes that break down dead plants and animals, releasing inorganic nutrients and recycling them to plants. These decomposers are, in a sense, the garbage processors of the planet. Other fungi help plants absorb minerals or fight disease.

Table 19.1   Phyla of Fungi

Group	Examples	Number of Existing Species
Chytridiomycota	Parasite of frog skin	1000
Zygomycota	Black bread mold	1000
Glomeromycota	Arbuscular mycorrhizal fungi	200
Ascomycota	Morels, truffles	More than 50,000
Basidiomycota	Mushrooms, puffballs	30,000

A. Fungi Are Eukaryotic Heterotrophs That Digest Food Externally

The evolutionary history of fungi remains unclear. The hypothetical common ancestor of all fungi is an aquatic, single-celled, flagellated protist resembling contemporary fungi called chytrids. The identity of the closest living relative to the fungi, however, remains controversial.

Biologists do know, however, that fungi are more closely related to animals than to plants. This finding may surprise those who notice the superficial similarities between plants and fungi. Unlike plants, however, fungi cannot carry out photosynthesis. Moreover, fungi share many chemical and metabolic features with animals.

Fungi have a unique combination of characteristics:

- The cells of fungi are eukaryotic, as are the cells of protists, plants, and animals.

- Fungi are heterotrophs, as are animals, but these two groups acquire food in different ways. Animals ingest their food and digest it internally; fungi secrete enzymes that break down organic matter outside their bodies. The fungus then absorbs the nutrients.

- Fungal cell walls are composed primarily of the modified carbohydrate chitin. This tough, flexible molecule also forms the exoskeletons of some animals. ▶ carbohydrates, p. 32

- The storage carbohydrate of fungi is glycogen, the same as for animals (see figure 2.17).

- Most fungi are multicellular, but **yeasts** are unicellular (see figure 19.1a). The distinction between yeasts and multicellular fungi, however, is not absolute; some species can switch between uni- and multicellular phases.

- Fungi have unique reproductive cycles. In most species, the only diploid cell is the zygote, which forms when the cells of two different individuals unite. The zygote undergoes meiosis and yields haploid nuclei, which then divide mitotically as the organism grows. Some fungi remain

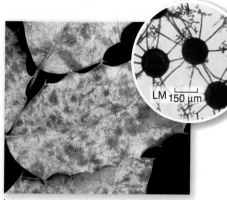

a.   SEM (false color) |—| 5 μm   b.

LM 150 μm

c.   d.

Figure 19.1 **Fungi Range Greatly in Size.** (a) These microscopic yeast cells are reproducing by budding. (b) Powdery mildew fungi live on leaf surfaces and produce microscopic reproductive structures (inset). (c) This moth was killed by a parasitic fungus, which has produced eerie yellow spikes. (d) *Armillaria ostoyae* mushrooms erupt from this decaying tree.

Burning Question

Why does food get moldy?

Fungal spores are everywhere. They germinate and grow into colonies only if they land on a surface that provides enough food and moisture. Fresh bread, cheese, and fruit are all perfect for fungal growth.

Perhaps a better question would be: why *don't* some foods get moldy? Humans have devised many ways to preserve food. Refrigeration dramatically slows the rate of fungal growth. Salt and sugar, in sufficiently high concentrations, also retard mold growth by limiting the fungus's ability to take up water by osmosis. Dried foods are preserved in the same way. Cooking

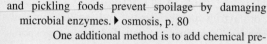

and pickling foods prevent spoilage by damaging microbial enzymes. ▸ osmosis, p. 80

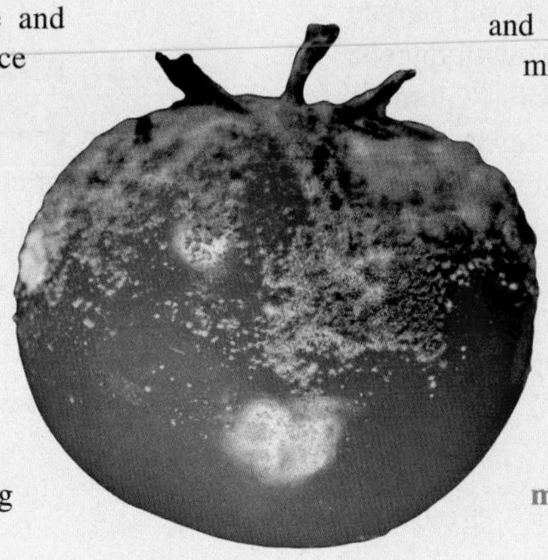

One additional method is to add chemical preservatives to foods. Organic acids such as sodium benzoate are common food additives that inhibit mold growth by disrupting fungal cell membranes. Many processed foods are so laden with preservatives that their shelf lives extend for years, a remarkable accomplishment in a world full of hungry microbes. ▸ membranes, p. 54

Submit your burning question to:
marielle_hoefnagels@mcgraw-hill.com

---

haploid throughout most of the life cycle, but others have a **dikaryotic** stage in which cells retain separate nuclei from two parents. Only fungi have dikaryotic cells. (Chapters 8 and 9 describe mitosis and meiosis in detail.)

The body of a fungus is much more extensive than just a mushroom or the visible fuzz on a moldy piece of food (figure 19.2). Instead, fungi usually consist of an enormous number of microscopic, threadlike filaments called **hyphae** (singular: hypha). These filaments branch rapidly within a food source, growing and absorbing nutrients at their tips. A **mycelium** is a mass of

aggregated hyphae that may form visible strands in soil or decaying wood.

While the feeding hyphae remain hidden in the food, the reproductive structures emerge at the surface. Most fungi produce abundant **spores,** which are microscopic reproductive cells (see this chapter's Burning Question). Spores that land on a suitable habitat can germinate and give rise to feeding hyphae, starting a new colony.

Spores can be asexually or sexually produced. **Conidia** are asexual spores (figure 19.3); the greenish or black powder on moldy food consists entirely of conidia. The production of sexual spores can be considerably more complex. In most fungal species, hyphae aggregate to form a **fruiting body,** a specialized sexual spore-producing organ such as a mushroom, puffball, or truffle.

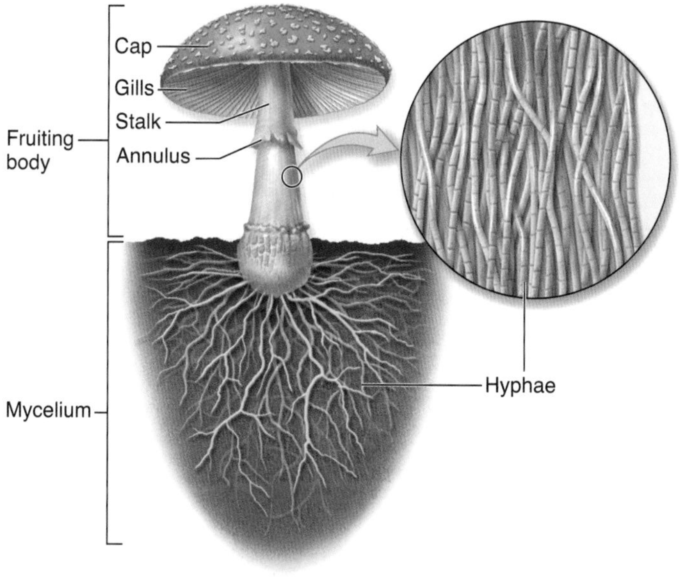

Cap

Gills

Stalk

Annulus

Fruiting body

Mycelium

Hyphae

Figure 19.2 The Fungal Body. A mushroom arises from hyphae penetrating the fungus's food source. The mushroom itself is composed of hyphae that are tightly aligned to form a solid structure.

SEM (false color)   50 μm

Figure 19.3 Conidia. This fungus, a common mold, is producing abundant asexual spores.

B. Fungal Classification Is Traditionally Based on Reproductive Structures

Mycologists classify fungi into five phyla based on the presence and types of sexual structures (**figure 19.4**). The chytridiomycetes, or chytrids, produce sexual and asexual spores with flagella. Zygomycetes produce thick-walled sexual zygospores. Glomeromycetes do not reproduce sexually at all; instead, they have large, distinctive asexual spores. Ascomycetes produce sexual spores in characteristic sacs, and basidiomycetes release sexual spores from club-shaped structures.

Biologists are still studying the evolutionary relationships among the chytrids, zygomycetes, and glomeromycetes. The ascomycetes and basidiomycetes, however, are clearly sister groups, as figure 19.4 indicates. These are the most complex fungi. Their fruiting bodies, which include morels, truffles, mushrooms, and puffballs, are usually visible with the unaided eye.

19.1 | Mastering Concepts

1. In what ways are fungi important in ecosystems?
2. What characteristics define the fungi?
3. What evidence suggests that fungi are more closely related to animals than to plants?
4. Describe the major structures of a fungus.
5. Which features distinguish the five phyla of fungi?

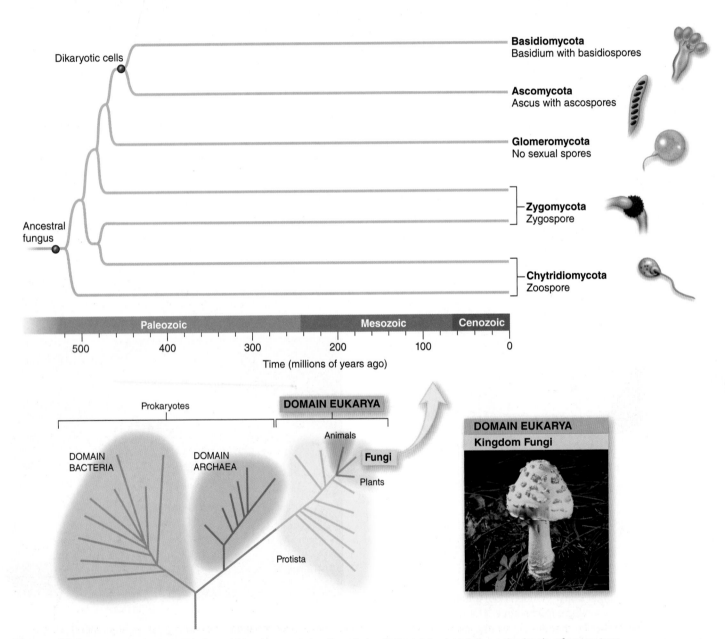

Figure 19.4 **Fungal Diversity.** The fungal kingdom contains five phyla, distinguished mainly on the basis of spore type.

19.2 | Chytridiomycetes Produce Swimming Spores

Basidiomycetes
Ascomycetes
Glomeromycetes
Zygomycetes
Chytridiomycetes

The 1000 or so species of **chytridiomycetes** (phylum Chytridiomycota) may provide a glimpse of what the earliest fungi were like. Their body forms vary from single cells to slender hyphae, and they differ from other fungi in that they produce **zoospores**—motile spores—each with a single flagellum (figure 19.5). Biologists have yet to discover the details of most chytrid life cycles.

These microscopic fungi are powerful decomposers, secreting enzymes that degrade cellulose, chitin, and keratin. One ecosystem where resident chytrids are particularly valuable is a ruminant's digestive tract. There, anaerobic chytrids start digesting the cellulose in a cow's grassy meal, paving the way for bacteria to continue the process.

Chytrids also contribute to the ongoing worldwide decline in amphibian populations. The chytrid *Batrachochytrium dendrobatidis* causes a lethal disease called cutaneous chytridiomycosis in frogs (figure 19.6). The fungus feeds on keratin in the frog's skin and coats the host's legs and undersides, impairing the frog's ability to breathe through its skin. The fungus spreads to new hosts by releasing zoospores into the water.
▶ amphibians, p. 440

The chytrids and zygomycetes are sometimes called the basal fungi because of their simple structures and position close to the base of the fungal family tree. Each group was once considered to be monophyletic, but molecular data have argued against

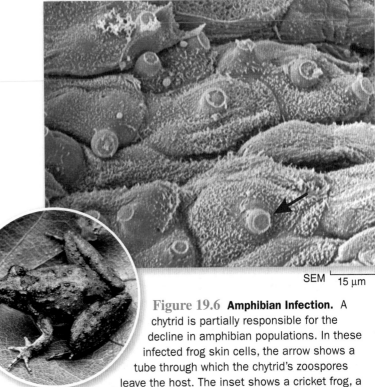

SEM | 15 µm

Figure 19.6 Amphibian Infection. A chytrid is partially responsible for the decline in amphibian populations. In these infected frog skin cells, the arrow shows a tube through which the chytrid's zoospores leave the host. The inset shows a cricket frog, a species that is often infected with the chytrid.

that hypothesis. Instead, each phylum apparently contains representatives from multiple lineages, and fungi once considered to be chytrids have since been reclassified as zygomycetes.

19.2 | Mastering Concepts

1. What are zoospores, and how do they adapt chytrids to moist environments?
2. What do chytrids eat?

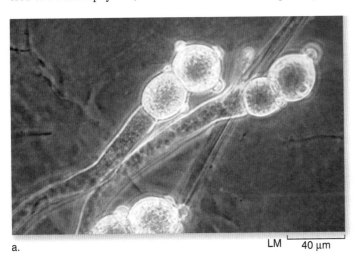

a.   LM | 40 µm

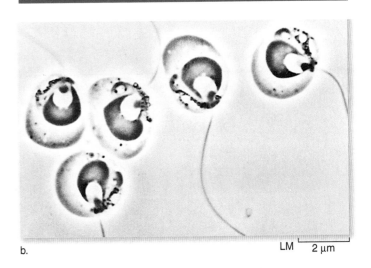

b.   LM | 2 µm

Figure 19.5 Chytrids. (a) Hyphae and spore sacs of the chytrid *Allomyces*. Flagellated zoospores will emerge from the rounded sacs at the tips of the hyphae. (b) Zoospores from the chytrid *Blastocladiella*.

19.3 | Zygomycetes Are Fast Growing and Prolific

The 1000 or so species of **zygomycetes** (phylum Zygomycota) account for only about 1% of identified fungi, but they include some familiar organisms. The black mold, *Rhizopus stolonifer,* which grows on bread, fruits, and vegetables, is a zygomycete. In addition to forming black fuzz on refrigerated leftovers, zygomycetes occur on decaying plant and animal matter in soil. Some are parasites of insects, and others must pass through the digestive system of an herbivore to complete their life cycles.

The zygomycetes are known for their spectacular growth rates, but they take their name from their mode of sexual reproduction: *zygon* is Greek for "yoke," implying the joining of two parts. As illustrated in **figure 19.7**, this "joining" occurs when two hyphae fuse. Then their haploid nuclei merge into a new structure, a diploid **zygospore** protected within a distinctive spiny, dark wall. After the merger, the cells from which they arose appear as two vacated areas that hug the zygosporangium.

The diploid zygospore nucleus undergoes meiosis, and a haploid hypha emerges. The hypha immediately produces a spore sac, which breaks open and releases numerous haploid spores; as the products of meiosis, they are genetically variable. Each spore then gives rise to a hypha that grows into a haploid mycelium.

In zygomycetes, asexual spores are much more common than zygospores. These conidia are the products of mitosis, so the spores produced by each hypha are genetically identical. The abundant haploid conidia spread easily to new habitats; **figure 19.8** shows two of the more unusual dispersal mechanisms. Spore germination produces a hypha, which quickly grows and branches as it digests its food. Within days, the new mycelium sprouts its own spore sacs, and the cycle begins anew.

19.3 | Mastering Concepts

1. Where do zygomycetes occur?
2. How do zygomycetes reproduce asexually and sexually?
3. How does the zygospore fit into the zygomycete life cycle?

Figure 19.7 **Zygomycete Reproduction.** (*1*) Haploid nuclei from cells of different mating types merge, (*2*) yielding a diploid zygote. (*3*) After meiosis, the zygospore germinates and (*4*) produces a sac filled with haploid spores. (*5*) As these spores germinate, they give rise to haploid hyphae that start the cycle anew. (*6*) In asexual reproduction, spore sacs release haploid conidia that give rise to hyphae, which, in turn, produce more spores.

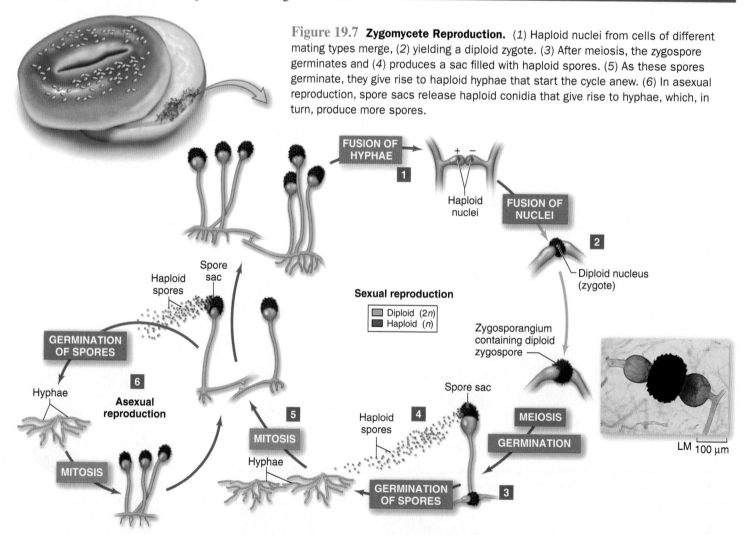

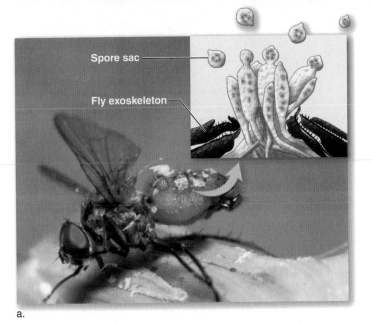

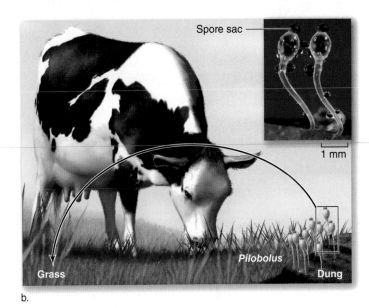

a.

b.

Figure 19.8 Zygomycete Spore Dispersal. (a) Spores of *Entomophthora muscae* produce hyphae that penetrate a fly's exoskeleton. Within days, the fly is dead. The fungus emerges, producing a shower of spores *(inset)* that infect additional flies. (b) On the dung of an herbivore, *Pilobolus* grows a stalk topped with a black sac containing thousands of spores *(inset)*. The spore sac is explosively launched from the stalk. Spores that land on the grass are eaten by cattle, which pass the intact spores in their dung.

19.4 | Glomeromycetes Colonize Living Plant Roots

The **glomeromycetes** (phylum Glomeromycota) form the smallest group of fungi, with only about 200 known species. Their place in kingdom Fungi is still uncertain. These organisms were once classified as zygomycetes, but molecular evidence suggests they form a clade that may be a sister group to the ascomycetes and basidiomycetes.

Glomeromycetes have some unusual features. First, they live only in association with living plant roots. This partnership with plants is called a **mycorrhiza** (literally, "fungus-roots"). Some of the oldest known plant fossils show evidence of mycorrhizae, indicating that plants and fungi moved onto land together hundreds of millions of years ago.

In a mycorrhiza, fungal hyphae colonize plant roots in a way that benefits both partners. The hyphae absorb water and minerals from the soil and share these resources with the plant; in return, the fungus gains carbohydrates that the plant produces in photosynthesis. The glomeromycetes are so dependent on their plant hosts that they will not grow on agar in laboratory culture.

Glomeromycetes form a type of mycorrhiza in which the exchange of materials occurs at structures called arbuscules; the fungus produces the arbuscules inside the root cells (figure 19.9). Ascomycetes and basidiomycetes form a different type of mycorrhiza; section 19.7 describes these partnerships in more detail.

Another notable feature of the glomeromycetes is their unusually large asexual spores (figure 19.10); some of the largest are visible with the unaided eye. At the same time, the glomeromycetes are not known to produce sexual spores. Because biologists classify most fungi based on their sexual structures, the absence of a sexual life cycle is partly responsible for the difficulty in placing these fungi in the phylogenetic tree.

19.4 | Mastering Concepts

1. What are the distinctive features of glomeromycetes?
2. Describe a mycorrhiza.

Figure 19.9 Glomeromycete Structures. This microscopic structure, which is inside a root cell, is an arbuscule produced by a mycorrhizal glomeromycete.

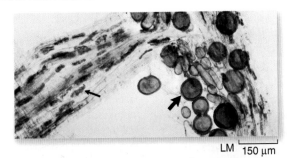

Figure 19.10 Big Spores. The arbuscules (small arrow) and spores (large arrow) of a glomeromycete fill this root.

19.5 | Ascomycetes Are the Sac Fungi

- Basidiomycetes
- **Ascomycetes**
- Glomeromycetes
- Zygomycetes
- Chytridiomycetes

With more than 50,000 species, the **ascomycetes** (phylum Ascomycota) make up the largest group of fungi. Their lifestyles vary widely. Some colonize living insects, others decompose organic matter, and still others form partnerships with photosynthetic organisms (see section 19.7). A few are carnivores (figure 19.11).

Some ascomycetes are pests. They cause most fungal diseases of plants, including Dutch elm disease and chestnut blight. A few species cause skin infections such as athlete's foot and other human diseases. Many of the common molds that ravage flood-damaged homes are ascomycetes (figure 19.12).

Nevertheless, ascomycetes also benefit humans. Truffles and morels are prized for their delicious flavors (figure 19.13). The common mold *Penicillium* is famous for secreting penicillin, the first antibiotic discovered. *Penicillium* species also lend their sharp flavors to Roquefort cheese. Fermentation by yeasts such as *Saccharomyces* is essential for baking bread, brewing beer, and making wine. *Aspergillus oryzae* ferments soybean pulp to produce soy sauce. Cyclosporine, a drug that suppresses the immune systems of organ transplant recipients, comes from an ascomycete. The red bread mold, *Neurospora crassa,* has contributed to biological research (see the Focus on Model Organisms box). ▶ fermentation, p. 115

Ascomycetes can produce enormous numbers of asexual and sexual spores (figure 19.14). In sexual reproduction, the hyphae of compatible mating types fuse, but the individual nuclei from the two parents do not immediately merge. The resulting cell is dikaryotic; when the cell divides, the two nuclei undergo mitosis separately.

The fruiting body consists of tightly woven hyphae, often in a cuplike or bottlelike shape, that form a fertile layer of dikaryotic cells. Eventually, the two nuclei in each dikaryotic

Figure 19.12 Mold Everywhere. A woman retrieves a photo album from a home that was flooded. Orange and black mold colonies cover the walls and ceiling.

cell fuse, forming a diploid zygote. The zygote immediately undergoes meiosis and produces four haploid nuclei, each of which usually divides once by mitosis. The result is eight haploid **ascospores,** so named because they form in a saclike **ascus** (plural: asci). After dispersal by wind, water, or animals, ascospore germination yields a new haploid individual.

19.5 | Mastering Concepts

1. List examples of ascomycetes that are important to humans.
2. Describe asexual and sexual reproduction in an ascomycete.
3. How do dikaryotic cells fit into the ascomycete life cycle?
4. How does an ascus come to contain eight ascospores?

Figure 19.11 A Carnivorous Fungus. Threadlike loops of the fungus *Arthrobotrys anchonia* constrict around a nematode worm. The fungus will thread its hyphae into its prey, releasing digestive enzymes and eating it from within.

SEM   15 μm

a.   b.

Figure 19.13 Edible Ascomycetes. (a) Truffles are familiar ascomycetes; they produce asci on internal folds of tissue. (b) Each "pit" in a morel's cap produces thousands of asci. This morel is about 10 cm tall.

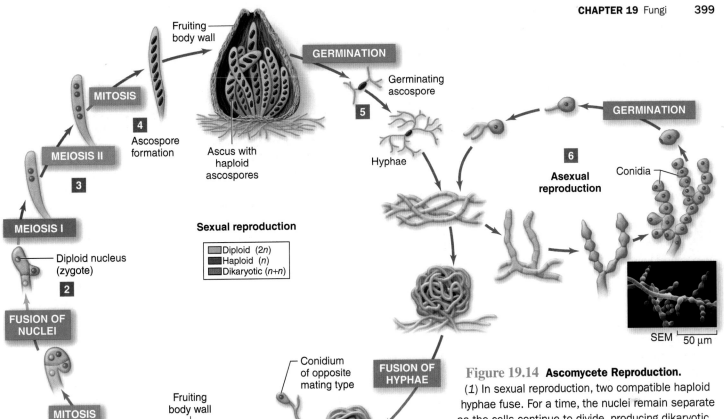

Sexual reproduction

☐ Diploid (2n)
■ Haploid (n)
■ Dikaryotic (n+n)

Figure 19.14 **Ascomycete Reproduction.**
(1) In sexual reproduction, two compatible haploid hyphae fuse. For a time, the nuclei remain separate as the cells continue to divide, producing dikaryotic cells. (2) Eventually the nuclei fuse, producing a diploid zygote. (3) This cell undergoes meiosis. (4) A subsequent mitotic division usually yields eight haploid spores inside an ascus. (5) Once released, each ascospore germinates and gives rise to new haploid hyphae, beginning the cycle again. (6) Many ascomycetes reproduce asexually by producing haploid conidia on haploid hyphae.

Focus on Model Organisms

Neurospora crassa

Why would biologists study *Neurospora crassa* (figure 19.A), a filamentous ascomycete commonly known as red bread mold? First, *N. crassa* can tell us about other ecologically, industrially, and medically important fungi. Second, *N. crassa* has the traits that all model organisms share: it is small and easy to grow, and it has a rapid reproductive cycle. Third, this fungus has special advantages for tracing inheritance patterns. Because its hyphae are haploid, *N. crassa* expresses every recessive or mutated allele in its DNA (in diploid organisms, dominant alleles can mask the presence of their recessive counterparts).

A full accounting of *Neurospora*'s contributions to basic biology could literally fill a book. The following list, however, details some ways this fungus has enhanced our understanding of life:

- **Crossing over.** In the 1920s and 1930s, *Neurospora* provided the first clear evidence of crossing over, which occurs during prophase I of meiosis. *Neurospora*'s narrow, tube-shaped asci are especially well suited to such studies because they retain the products of meiosis in the original order in which the cells divided (see figure 19.14). ▶ crossing over, p. 186

- **One-gene/one-enzyme hypothesis.** In the 1940s, George Beadle, Edward Tatum, and their colleagues isolated *Neurospora* mutants with unusual nutritional requirements. Their observation that each mutant had a single enzyme deficiency provided convincing

Figure 19.A **Neurospora crassa.** This photo shows the asci from a single squashed fruiting body. Note that each ascus contains exactly eight spores.

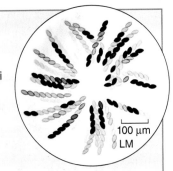

evidence for the idea that one gene encodes each protein in a cell, a topic described in detail in chapter 7.

- **Circadian rhythms.** *Neurospora* cultures produce conidia at roughly 24-hour intervals, a phenomenon called a circadian rhythm (*circa,* "about"; *dies,* "day"). Mutants that produce conidia constantly or not at all have yielded insights into regulation of the so-called clock-controlled genes.

- **Gene regulation.** Some genes in a cell are "silent" at any given time. *Neurospora* has helped reveal some of the ways cells regulate gene expression. For example, cells can "tag" DNA with chemical groups that prevent transcription, or they may destroy RNA after transcription. ▶ regulation of gene expression, p. 134

19.6 | Basidiomycetes Are the Familiar Club Fungi

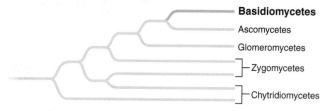

Basidiomycetes
Ascomycetes
Glomeromycetes
Zygomycetes
Chytridiomycetes

The 30,000 or so species of **basidiomycetes** (phylum Basidiomycota) include familiar representatives such as mushrooms, puffballs, stinkhorns, shelf fungi, and bird's nest fungi (figure 19.15). These organisms spread their spores in many ways. Wind carries the spores of puffballs and mushrooms; the putrid odor of the stinkhorn's slimy spore mass attracts flies, which carry the spores on their feet; raindrops splash the spore-laden "eggs" out of a bird's nest fungus.

Basidiomycetes play many roles in human life. Some mushrooms are edible, some are deadly, and others are hallucinogenic (see this chapter's Apply It Now box). Basidiomycetes called smuts and rusts are plant pathogens, causing serious diseases of cereal crops such as corn and wheat. In forests, wood decay fungi play a vital role in Earth's carbon cycle. Unfortunately, their talent for degrading the cellulose and lignin in fallen logs also makes them serious pests in another context: they cause dry rot in wooden wall studs and other building materials.

Basidiomycetes can reproduce asexually, but the sexual portion of the life cycle is usually the most prominent. As figure 19.16 shows, the fusion of two haploid hyphae creates a dikaryotic mycelium. This mycelium typically grows unseen within its food source. When environmental conditions are favorable, however, one or more mushrooms emerge. Lining each mushroom's gills are numerous dikaryotic, club-shaped cells called **basidia** (singular: basidium). Inside each basidium, the haploid nuclei fuse, giving rise to a diploid zygote. The zygote immediately undergoes meiosis, yielding four haploid nuclei. Each nucleus migrates into a **basidiospore,** which germinates after dispersal. The cycle begins anew.

Sometimes, a circle of mushrooms emerges from the ground all at once. The growth pattern of the underground mycelium explains this phenomenon. Hyphae extend outward in all directions from a colony's center. Mushrooms poke up at the margins of the mycelium, creating the "fairy rings" of folklore.

19.6 | Mastering Concepts

1. What are some familiar basidiomycetes?
2. How are basidiomycetes different from ascomycetes?
3. Describe the sexual life cycle of basidiomycetes.
4. How does a "fairy ring" form?

a.   2.5 cm   b.

c.   1 cm   d.   0.5 cm

Figure 19.15 **A Gallery of Basidiomycetes.** (a) Puffballs (genus *Lycoperdon*). (b) A stinkhorn (*Mutinus caninus*). (c) Turkey tail bracket fungi (*Trametes versicolor*). (d) Bird's nest fungi (order Nidulariales).

Apply It Now

Fungi and Human Health

Although most fungi are harmless (or even beneficial), some can threaten human health by causing mild to fatal infections, allergic reactions, or poisonings.

Infection: Fungi that degrade the protein keratin can infect our skin, hair, and nails, causing ringworm, athlete's foot, and other irritating diseases. Some fungi, such as *Candida albicans,* normally inhabit the mucous membranes of the mouth, intestines, or vagina. In a yeast infection, populations of these fungi grow out of control. Yeast infections commonly occur when the body's normal microbial community is disrupted, such as when a person uses antibiotics to fight a bacterial infection. People with weakened immune systems are also vulnerable to yeast infections.

Other fungi enter body tissues via wounds or the lungs. People who inhale dust containing spores of a fungus called *Coccidioides immitis* may develop a potentially fatal disease called valley fever (figure 19.B). Similarly, *Histoplasma capsulatum* inhabits bird droppings; inhaling the spores may result in a typically mild infection called histoplasmosis.

Allergic reactions: People in many parts of the United States receive "mold spore counts" along with their weather reports, and resi-

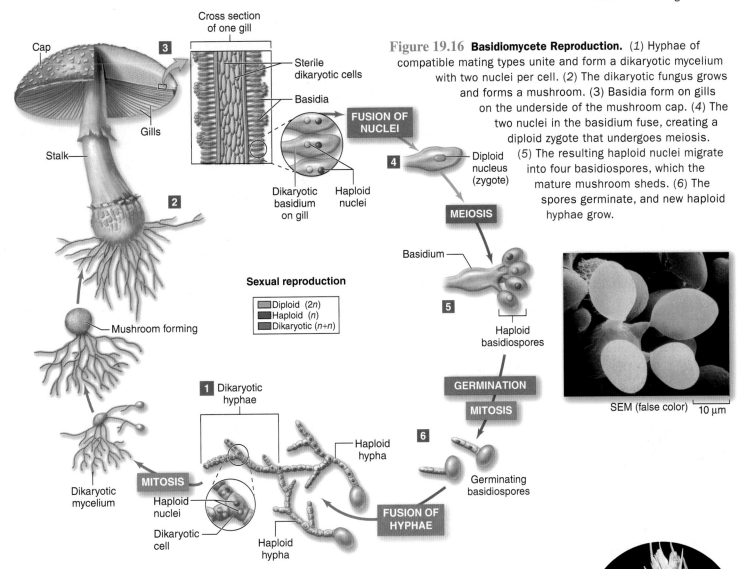

Cross section of one gill

3

Cap

Sterile dikaryotic cells

Basidia

Gills

Stalk

FUSION OF NUCLEI

2

Dikaryotic basidium on gill

Haploid nuclei

4

Diploid nucleus (zygote)

MEIOSIS

Mushroom forming

Sexual reproduction

Diploid (2n)
Haploid (n)
Dikaryotic (n+n)

Basidium

5

Haploid basidiospores

GERMINATION

MITOSIS

1 Dikaryotic hyphae

Figure 19.16 Basidiomycete Reproduction. (1) Hyphae of compatible mating types unite and form a dikaryotic mycelium with two nuclei per cell. (2) The dikaryotic fungus grows and forms a mushroom. (3) Basidia form on gills on the underside of the mushroom cap. (4) The two nuclei in the basidium fuse, creating a diploid zygote that undergoes meiosis. (5) The resulting haploid nuclei migrate into four basidiospores, which the mature mushroom sheds. (6) The spores germinate, and new haploid hyphae grow.

SEM (false color)   10 μm

6

Germinating basidiospores

Haploid hypha

Dikaryotic mycelium

MITOSIS

Haploid nuclei

Dikaryotic cell

Haploid hypha

FUSION OF HYPHAE

---

dents with mold allergies avoid going outside when the counts are high. Common allergenic fungi include species of *Alternaria* and *Aspergillus,* both of which produce abundant asexual spores. They can produce enormous spore masses in mold-infested buildings. ▶ allergies, p. 686

Toxicity: Some people report symptoms such as headaches and memory loss after living or working in moldy buildings. *Stachybotrys atra,* the toxic "black mold," is a commonly cited culprit, but proof that fungal toxins cause these symptoms is lacking.

Poisonous mushrooms such as those described in the chapter opening essay are the most famous toxic fungi, but others produce potent chemicals as well. *Claviceps purpurea* is an ascomycete that causes a

Figure 19.B *Coccidioides immitis.* This microscopic thick-walled structure, called a spherule, produces hundreds of fungal spores in infected tissues.

LM   7 μm

Figure 19.C Ergot. These black ergots are masses of hyphae of the ascomycete *Claviceps purpurea.* If the ergots are mixed in with grain used to mill flour, people who eat the flour may become ill.

plant disease called ergot (figure 19.C). People who eat bread made from ergot-contaminated grain can develop convulsions, gangrene, and psychotic delusions. Some of the women in Salem, Massachusetts, who were burned at the stake as witches in the late 1690s may have been suffering from ergotism; their uncontrollable movements were attributed to demons. The drug lysergic acid diethylamide (LSD) comes from lysergic acid, a chemical found in ergots.

Another toxin-producing fungus is the ascomycete *Aspergillus flavus.* Hyphae of *A. flavus* feed on nuts, grains, or other stored crops and release aflatoxin, a potent cancer-causing compound that can also be deadly at high doses. Acute toxicity is rare in humans, but in 2006, dozens of dogs died of liver failure after eating aflatoxin-tainted pet food.

19.7 | Fungi Interact with Other Organisms

Fungi are decomposers and parasites, but they also form symbiotic relationships with living organisms in which both partners benefit (a situation called a mutualism). Here we profile a few interesting examples. ▶ mutualism, p. 764

A. Endophytes Live in Aerial Plant Parts

Endophytes are fungi that live between the cells of a plant's leaves and stems without triggering disease symptoms (*endo-* means inside, and *-phyte* means plant). Every known plant, from mosses to angiosperms, harbors endophytes (**figure 19.17**). The ubiquity of endophytes has led some researchers to comment that "all plants are part fungi."

Some of these internal fungi neither help nor harm the plant. Others secrete substances that help defend the plants against herbivores. In grasses such as fescue, for example, endophytes can produce toxins that sicken grazing animals. Section 19.8 describes research showing that endophytes can also defend plants against other fungi.

B. Mycorrhizal Fungi Live on or in Roots

Mycorrhizae are associations between fungi and living roots that were introduced in section 19.4. About 80% of all land plants, including grasses, shrubs, and trees, form mycorrhizae. Some plants, such as orchids, cannot live without their fungal associates. Other plant species depend less on the fungi, especially in nutrient-rich soils.

Figure 19.17 **Endophytes.** Hyphae of an endophytic fungus grow out of a tiny piece of leaf tissue.

Figure 19.18 **Ectomycorrhizae.** (a) The creamy white root tips of this buckthorn tree are colonized by an ectomycorrhizal fungus. (b) Hyphae of a mycorrhizal fungus wrap around a host's root tip.

Glomeromycetes form the most common types of mycorrhizae. The fungi pierce the host plant's root cells and produce highly branched arbuscules through which the partners exchange materials (see figure 19.9). Hyphae also extend into the soil, absorbing water and minerals. Basidiomycetes and ascomycetes form a different type of mycorrhiza. In ectomycorrhizae, the fungal hyphae wrap around root tips and reach into the surrounding soil (**figure 19.18**). The filaments may extend into the root, but they never penetrate individual cells.

Many edible fungi depend on their ectomycorrhizal relationships with live tree roots; these mushrooms are difficult to cultivate commercially. The popularity and high price of these wild delicacies have lured many mushroom pickers into the woods. Scientists are debating whether the wild mushroom trade will harm populations of fungi—and the trees that rely on them—in the long term.

C. Some Ants Cultivate Fungi

The leaf-cutter ants of Central and South America and the southern United States cultivate the basidiomycete *Lepiota* in special underground chambers. Into each chamber, the ants transport a paste made from saliva and the disks that they cut from green leaves (**figure 19.19**). The fungi grow on the paste. Adult and larval ants eat the hyphae so quickly that mushrooms never get a chance to form.

The ants and their fungal gardens constitute a mutualistic partnership; both the ants and the fungi have a steady food supply. Both benefit. But how do the ants keep out competitors or pathogens? Biologists once thought that the ants simply ate all interlopers, but then they discovered a third partner. *Streptomyces* bacteria coat parts of the ants' exoskeletons and secrete a potent antibiotic that kills an ascomycete that attacks the cultivated fungus.

Figure 19.19 **Ants at Work.** Leafcutter ants use the vegetation they harvest to farm fungi—their food source.

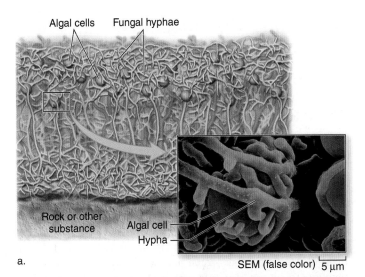

a.

SEM (false color) 5 μm

D. Lichens Are Distinctive Dual Organisms

A **lichen** forms when a fungus, either an ascomycete or a basidiomycete, harbors green algae or cyanobacteria among its hyphae (figure 19.20). The photosynthetic partner contributes carbohydrates; the fungus absorbs water and essential minerals.

Lichens are sometimes called "dual organisms" because the two species—the fungus and its autotrophic partner—appear to be one individual when viewed with the unaided eye. The body forms can vary widely. Many lichens are colorful, flattened crusts, but others form upright structures that resemble mosses or miniature shrubs. Still others are long, scraggly growths that dangle from tree branches.

Lichens are important ecologically because they secrete acids that break down rock, a first step in the soil-building process. Many also harbor nitrogen-fixing cyanobacteria, which make usable nitrogen available to plants. In addition, some animals, including caribou, eat lichens. ▶ nitrogen fixation, p. 346

Just about any stable surface, from tree bark to boulders to soil, can support lichens. They live everywhere from the driest deserts to the wettest tropical rain forests to the frozen Arctic. Lichens survive dehydration by suspending their metabolism, only to revive when moisture returns. One type of habitat, however, is hostile to lichens: polluted areas. Lichens absorb toxins but cannot excrete them. Toxin buildup hampers photosynthesis, and the lichen dies. Disappearance of native lichens is a sign that pollution is disturbing the environment; scientists therefore use lichens to monitor air quality.

b.

19.7 | Mastering Concepts

1. What are endophytes?
2. Which types of fungi form arbuscular mycorrhizae and ectomycorrhizae?
3. How do ants, plants, fungi, and bacteria interact in leaf-cutter ant colonies?
4. How do fungi and autotrophs interact in a lichen?
5. How do scientists use lichens to monitor pollution?

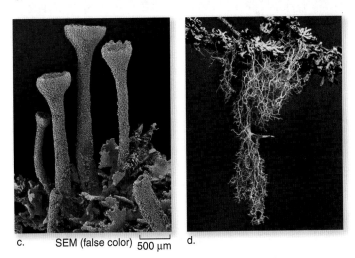

c. SEM (false color) 500 μm   d.

Figure 19.20 **Anatomy of a Lichen.** (a) Cross section of a lichen encrusting a rock; the fungal hyphae wrap tightly around their photosynthetic "partner" cells. (b) The lichens living on this rock look like single organisms when viewed at a large scale. (c) The tiny stalked cups of this trumpet lichen are fruiting bodies. (d) Grey tube lichens and greenish beard lichens colonize a tree branch.

19.8 | Investigating Life: The Battle for Position in Cacao Tree Leaves

Chocolate has many friends; its delectable taste and healthful antioxidants make it a favorite food. But chocolate also has its detractors, beyond low-fat diet enthusiasts. Most of its enemies are microbial pathogens of the cacao tree (*Theobroma cacao;* **figure 19.21**), the source of cocoa. Diseases such as frosty pod and witches' broom, for example, are the work of protists and fungi.

All organisms, including plants, defend themselves against disease. Waxy coverings on leaves and stems prevent the entry of many pathogens, and plant cells produce an array of noxious chemicals that deter those microbes that manage to enter. Besides these innate protections, some plants also enlist endophytes in the war against predators and pathogens.

Some plants pass endophytes to their offspring in seeds, but the cacao tree acquires its resident fungi from the environment. A brand-new leaf is endophyte-free, but fungal spores soon arrive in wind or rain and germinate on the young foliage. The hyphae penetrate the waxy leaf cuticle and set up shop within the plant's tissues, absorbing water and nutrients that leak out of the leaf veins. Oddly, the endophytes do not trigger the plant's defenses. This raises an interesting question: What does the tree gain from allowing plant-eating fungi to live inside its tissues?

University of Arizona biologist A. Elizabeth Arnold, along with a research team at the Smithsonian Tropical Research Institute in Panama, wondered whether the endophytes help protect cacao trees from disease. To find out, they collected seeds from cacao plants and soaked them briefly in dilute bleach to kill any fungal spores sticking to the seed coats. The researchers needed endophyte-free plants to test their hypothesis, so they planted the seeds in a greenhouse and protected them from fungal spores that might blow in from outside. Once the trees were 100 days old, the team sampled the leaves to ensure that the plants were endophyte-free.

Figure 19.21 The Cacao Tree. The pods of *Theobroma cacao* are the source of chocolate.

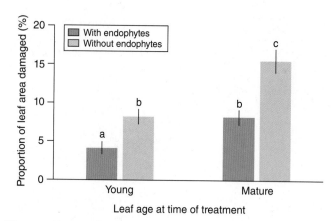

Figure 19.22 Helpful Partners. Endophytic fungi protected young and mature cacao leaves from damage by the pathogen *Phytophthora*.

Next, they sprayed a mixture of spores from several species of endophyte fungi on some of the leaves, so each tree had some treated leaves and some that remained free of the fungi. A couple of weeks later, once the fungi were thriving in the sprayed leaves, the researchers were ready to test their hypothesis. They inoculated both endophyte-treated and control leaves with spores of a protist called *Phytophthora,* which causes black pod disease of cacao.

Fifteen days later, Arnold and her team counted the number of dead leaves on each tree and measured the *Phytophthora*-damaged area on surviving leaves. These combined measures showed that plant parts without endophytes lost twice as much leaf area as those with the resident fungi (**figure 19.22**).

No one is sure what the endophytes gain from the relationship, other than the nutrients they absorb from inside the leaf. One hypothesis is that their biggest gain comes after their home leaf ages, dies, and falls to the ground. Like devoted fans who camp overnight to buy prime tickets for a concert, perhaps the endophytes colonize a living leaf to get "first dibs" on the dead tissue.

Win–win relationships, such as the one between the cacao tree and its resident fungi, are common in life. Scientists are trying to develop some of these cooperative organisms into biological control agents that might prevent disease in cacao plants without the use of harmful chemicals. Our endophyte allies are already living quietly in the stems and leaves of our crop plants. Perhaps one day farmers will enlist these friendly fungi in a more aggressive fight to preserve our chocolate fix.

Arnold, A. Elizabeth, Luis Carlos Mejia, Damond Kyllo, et al. December 23, 2003. Fungal endophytes limit pathogen damage in a tropical tree. *Proceedings of the National Academy of Sciences,* U.S.A., vol. 100, pages 15649–15654.

19.8 | Mastering Concepts

1. How did the researchers test their hypothesis that endophytes help prevent plant disease?
2. This study used a mix of endophyte species. How would you design an experiment to determine whether one species of endophyte or some combination of multiple species protects the leaves against pathogens?

Fungi Characteristics: A Summary

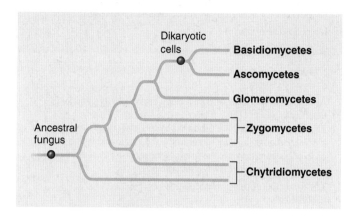

- **Basidiomycota**
  - ~30,000 species
  - Basidiospores on basidia
  - Mushrooms, stinkhorns, puffballs
  - Ectomycorrhizae, some lichens

- **Ascomycetes**
  - ~50,000 species
  - Ascospores inside ascus
  - Asexual reproduction common
  - Ectomycorrhizae, most lichens
  - Truffles, morels, most yeasts, Dutch elm disease, *Penicillium*

- **Glomeromycetes**
  - ~200 species
  - No sexual spores; large asexual spores
  - Most are obligate parasites
  - Arbuscular mycorrhizae

- **Chytridiomycetes**
  - ~1,000 species; paraphyletic group
  - Spores have single flagellum
  - Mostly aquatic
  - Decomposers and parasites on many organisms, including amphibians

- **Zygomycetes**
  - ~1,000 species; paraphyletic group
  - Zygospores
  - Black bread mold
  - Asexual reproduction more common than sexual reproduction
  - Decomposers and parasites

Phylum	Sexual Spore Type	Dikaryotic Cells	Complex Fruiting Body
Chytridiomycota	Zoospore	No	Absent
Zygomycota	Zygospore	No	Absent
Glomeromycota	None	No	Absent
Ascomycota	Ascospores in ascus	Yes	Present
Basidiomycota	Basidiospores on basidium	Yes	Present

 # Chapter Summary

19.1 | Fungi Are Essential Decomposers

- Fungi are widespread and profoundly affect ecosystems by feeding on living and dead organic material.

A. Fungi Are Eukaryotic Heterotrophs That Digest Food Externally

- Fungi are more closely related to animals than to plants.
- Fungal characteristics include heterotrophy, chitin cell walls, and glycogen. Fungi have both asexual and sexual reproduction, and some have unique **dikaryotic** cells with two genetically different nuclei.
- A fungal body includes a **mycelium** built of threads called **hyphae,** which may form a **fruiting body.** Some fungi occur both as filamentous hyphae and as unicellular **yeasts.**
- Fungi reproduce using asexual and sexual **spores. Conidia** are asexually produced spores.

B. Fungal Classification Is Traditionally Based on Reproductive Structures

- The five main groups of fungi are chytridiomycetes, zygomycetes, glomeromycetes, ascomycetes, and basidiomycetes.
- Molecular evidence indicates that ascomycetes and basidiomycetes are sister groups.

19.2 | Chytridiomycetes Produce Swimming Spores

- **Chytridiomycetes** are microscopic fungi that produce flagellated, motile **zoospores.**

- Chytrids decompose major biological carbohydrates such as cellulose, chitin, and keratin.

19.3 | Zygomycetes Are Fast Growing and Prolific

- **Zygomycetes** reproduce rapidly via asexually produced, haploid, thin-walled spores.
- Sexual reproduction occurs when hyphae of different mating types fuse their nuclei into a **zygospore.** The zygospore undergoes meiosis, then generates a spore sac. Spore germination yields haploid hyphae, continuing the cycle.

19.4 | Glomeromycetes Colonize Living Plant Roots

- **Glomeromycetes** form **mycorrhizae,** which are mutually beneficial associations between fungi and roots. Mycorrhizal plants have extra surface area for absorbing nutrients, and the fungus acquires carbohydrates from its plant host.
- Glomeromycetes have no known sexual stage.

19.5 | Ascomycetes Are the Sac Fungi

- **Ascomycetes** include important plant pathogens, but they also have many uses in industry.
- Hyphae are haploid for most of the life cycle. Asexual reproduction typically occurs via conidia.

- In sexual reproduction, haploid hyphae join, producing a brief dikaryotic stage. After the nuclei fuse, meiosis yields haploid **ascospores** in saclike **asci.**

19.6 | Basidiomycetes Are the Familiar Club Fungi

- **Basidiomycetes** include mushrooms and other familiar fungi. Wood decay fungi are important in ecosystems but also rot building materials. Smuts and rusts are important plant pathogens.
- Basidiomycetes are dikaryotic for most of the life cycle. Asexual reproduction occurs by budding, by fragmentation, or with asexual spores.
- Sexual reproduction occurs when haploid nuclei in **basidia** along the gills of mushrooms fuse. The resulting zygote then undergoes meiosis, producing haploid **basidiospores.** Shortly after spore germination, hyphae fuse, regenerating the dikaryotic state.

19.7 | Fungi Interact with Other Organisms

A. Endophytes Live in Aerial Plant Parts
- **Endophytes** are fungi that colonize plant leaves and stems without triggering disease symptoms.

B. Mycorrhizal Fungi Live on or in Roots
- Glomeromycetes form arbuscular mycorrhizae, whereas ascomycetes and basidiomycetes form ectomycorrhizae.

C. Some Ants Cultivate Fungi
- Leaf-cutter ants cultivate basidiomycetes on a nutritious paste made from leaf disks, and bacteria on the ants kill a parasitic ascomycete.

D. Lichens Are Distinctive Dual Organisms
- A **lichen** is a compound organism that consists of a fungus in intimate association with a cyanobacterium or a green alga.
- Lichens are useful air pollution indicators.

19.8 | Investigating Life: The Battle for Position in Cacao Tree Leaves

- Experimental evidence indicates that endophytes protect cacao tree leaves from *Phytophthora* pathogens.

Multiple Choice Questions

1. How are fungi similar to animals?
   a. They have a cell wall made of chitin.
   b. They store energy in the form of glycogen.
   c. They are predominantly haploid organisms.
   d. Both a and b are correct.

2. Fungi get nutrients through their
   a. fruiting body.
   b. hyphae.
   c. spores.
   d. All of the above are correct.

3. Fungi are considered _____ because they get their carbon and energy from _____.
   a. autotrophs; photosynthesis
   b. autotrophs; organic matter
   c. heterotrophs; organic matter
   d. heterotrophs; photosynthesis

4. Which phylum of fungus is most commonly eaten by humans?
   a. Chytridiomycetes
   b. Glomeromycetes
   c. Basidiomycetes
   d. Zygomycetes

5. Why is the zygospore diploid?
   a. Because it forms through meiosis
   b. Because it forms when two haploid nuclei fuse
   c. Because it forms by mitosis of the compatible mating types
   d. Because it forms from a multicellular stage in the life cycle

6. If a cell is *dikaryotic,* it
   a. contains two nuclei.
   b. is diploid.
   c. functions as a spore.
   d. is a zygospore.

7. Are the eight ascospores of an ascomycete typically genetically identical?
   a. Yes, because they are produced by mitosis.
   b. Yes, because they are the product of a single dikaryotic cell.
   c. No, because they are the products of meiosis, then mitosis.
   d. No, because they undergo mutations.

8. The tissue that forms a mushroom is
   a. haploid.
   b. diploid.
   c. dikaryotic.
   d. All of the above are correct.

9. What is a lichen?
   a. A type of photosynthetic fungus
   b. A combination of a fungus and an alga or cyanobacterium
   c. A combination of two phyla of fungi
   d. A combination of a fungus and a root

10. How does a mycorrhizal fungus differ from a free-living basidiomycete?
    a. It gains its nutrients from a living plant.
    b. It has no sexual stage in its life cycle.
    c. It is not edible.
    d. All of the above are correct.

Write It Out

1. Give examples of fungi that are important economically, ecologically, and as food for humans.
2. Suppose you are digging in forest soil and find a stringy, whitish structure. If you had access to a microscope, what features would you look for to determine whether the object is a plant root or a fungus?
3. What evidence suggests that fungi are more closely related to animals than to plants?
4. Which characteristics do all fungi share? Which characteristics distinguish each phylum of fungi?
5. Penicillin, an antibiotic derived from a fungus, kills bacteria by destroying their cell walls. Why doesn't penicillin harm fungal cell walls?
6. Review figure 18.5, which shows the alternation of generations in plants. Compare and contrast the life cycles of zygomycetes, ascomycetes, and basidiomycetes to the basic plant life cycle.
7. Sketch each of the following structures: mycelium; ascus with ascospores; zoospore; zygospore; basidium with basidiospores.
8. Distinguish between the following pairs of terms: yeast and mycelium; ascomycete and basidiomycete; mycorrhiza and lichen; zoospore and zygospore; diploid and dikaryotic.
9. Hyphae and arbuscules are highly branched structures. How does their extensive surface area contribute to their functions?
10. Other than shared DNA sequences, what characteristics place ascomycetes and basidiomycetes together as sister groups?
11. An ascomycete called *Cryphonectria parasitica* causes chestnut blight, a disease that has killed most of the chestnut trees in the United States.

If you were a plant breeder trying to bring back the chestnut, how would you use artificial selection to create disease-resistant varieties of the tree?

12. Use the Internet to find examples of chytrids, zygomycetes, ascomycetes, and basidiomycetes that cause diseases in plants or animals. How does each fungus infect a host and spread to new hosts? What can humans do to fight each disease? What are the costs and benefits of doing so?

13. Describe the relationships that connect:
    a. leaves and endophytes.
    b. roots and mycorrhizal fungi.
    c. leaf-cutter ants, their fungal food, bacteria, and pathogenic fungi.
    d. the fungal and photosynthetic partners in a lichen.

14. Some endophytes produce compounds that fight bacteria, making them potential new sources of antibiotics. Describe an experimental method that would allow you to screen endophytes for antibacterial activity.

15. Figure 19.22 shows that endophytes were associated with a small but statistically significant reduction in cacao leaf disease. In what ways might this seemingly small difference be important to the plant?

16. Describe how experiments might show that:
    a. chytrids are killing amphibians.
    b. fungi benefit more from a lichen relationship than do algae.
    c. bacteria help leaf-cutting ants cultivate one fungus while killing another.
    d. overharvesting of mycorrhizal basidiomycetes harms forest health.
    e. mold on a building's walls can cause illness in the building's occupants.

17. Genome sequencing projects are complete or in progress for many fungi other than *Neurospora crassa*. Search the Internet for a list of these fungi, choose one, and describe some discoveries that have come from research on the species you chose.

18. Use the Internet to learn how Alexander Fleming discovered penicillin. Many years passed before other scientists developed penicillin into a drug. How does the story of penicillin, from discovery to drug development, illustrate the process of science?

19. White nose syndrome is an illness that weakens and kills bats roosting in caves. Affected animals have a white fungus around their muzzle and on their wings, but no one knows if the fungus causes the illness or simply infects bats that are weakened by some other disease. White nose syndrome was discovered in 2006 and has since spread to several states in the northeastern United States. Search the Internet for the latest information about white nose syndrome, then propose a testable hypothesis to explain how the fungus might have entered bat populations. What information would you need to determine whether the fungus actually causes the disease? How can people help prevent the spread of the fungus from cave to cave? If the disease continues to spread, how might the widespread death of bats affect ecosystems?

Pull It Together

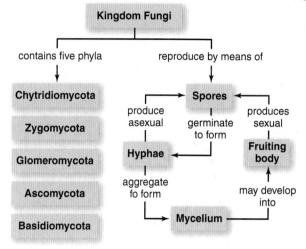

1. Which spore types are produced by each group of fungi?
2. Why are fungi important to other species?
3. How do dikaryotic cells fit into the sexual life cycle of fungi?

20

Animals

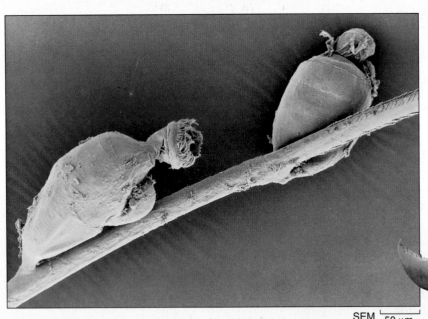

SEM ⊢ 50 µm ⊣

Home Sweet Home. Scientists discovered *Symbion pandora* amid the sensory hairs on a lobster's external skeleton.

BIOLOGY

Enhance your study of
this chapter with practice quizzes,
animations and videos, answer keys,
and downloadable study tools.
www.mhhe.com/hoefnagels

Life on a Lobster's Lips

Discovering and describing never-before-seen organisms is nothing new to biologists. Finding *Symbion pandora* was unusual, however, not only because of its habitat on the mouthparts of the Norwegian lobster, *Nephrops norvegicus,* but also because it did not fit into any known animal phylum. Danish researchers reported discovering the animal in late 1995.

The researcher who codiscovered *S. pandora,* Reinhardt Kristensen, says that many animals await our detection and description. In 1983, Kristensen discovered animals in another phylum, the Loricifera, living between grains of sand. Amazingly, even as other scientists are exploring the remotest reaches of Earth looking for new life, both of these phyla were discovered "hiding in plain sight." The lobster that hosts *S. pandora* lives in a busy Scandinavian shipping lane.

Among the smallest of animals, *S. pandora* measures less than a millimeter. Its saclike body adheres to a lobster's mouthparts. This unusual animal was assigned to a new phylum, Cycliophora, which means "wheel bearer" in Greek. The "wheel" refers to the ciliated ring that forms the animal's mouth, which opens into a funnel-shaped structure atop the animal's body.

Symbion pandora's life is full of competition and frequent relocation. Living on lobster mouthparts means ready access to nutrients: as a lobster crushes food in its jaws, *S. pandora* sweeps stray particles into its own ciliated mouth ring. But food and oxygen availability vary among the bristles surrounding the lobster's mouth, and competition for the prime dining spots is fierce.

In this competition, however, victory is fleeting. Like all arthropods, lobsters periodically shed their external skeletons. When they do, what was once prime real estate becomes worthless. Fortunately, *S. pandora* is adapted to this occasional upheaval. Its complex life cycle includes short swimming stages in addition to the lengthier immobile (sessile) stages. As soon as the lobster's new exoskeleton is complete, the animals resettle, renewing the competition for the best locations.

What exactly is *S. pandora?* DNA sequence comparisons suggest it may be related to a group of animals called rotifers, but additional research is needed to determine how these curious organisms fit into the animal evolutionary tree. In the meantime, it's intriguing to think of how much we do not know about Animalia— the most familiar of biological kingdoms.

Learning Outline

Learn How to Learn
Think While You Search the Internet

Your instructor may give you assignments that require you to find news or information on the Internet. You can use a search engine, but how do you know if the sites you find are credible? The Internet is full of misinformation, so you must evaluate every site you visit. Collaborative online encyclopedias such as Wikipedia are unreliable because anyone can change any article. For other sites, ask the following questions. Are you looking at someone's personal page? Is there an educational, governmental, nonprofit, or commercial sponsor? Is the author reputable? Is the information up-to-date? Does the page contain facts or opinions? Are the facts backed up with documentation? Taking the time to find the answers to these questions will help ensure that the information you find is valid.

20.1 | Animals Live Nearly Everywhere

Think of any animal. There's a good chance that the example that popped into your head was a mammal such as a dog, cat, horse, or cow. Although it makes sense that we think first of our most familiar companions, the 5800 species of mammals represent only a tiny subset of organisms in kingdom **Animalia.**

Biologists have described about 1,200,000 animal species, distributed among 37 phyla. This chapter considers nine of the phyla in detail (**table 20.1**); the Burning Questions box on page 431 briefly describes three others. The vast majority of animals are **invertebrates** (animals without backbones). The phylum Arthropoda alone, for example, includes more than 1 million identified species of invertebrates such as insects, crustaceans, and spiders. Only about 57,000 known animal species are **vertebrates** (animals with backbones) such as mammals and birds.

The diversity of animals is astonishing. Animals live in us, on us, and around us. They are extremely diverse in size, habitat, body form, and intelligence. Whales are immense; roundworms can be microscopic. Bighorn sheep scale mountaintops; crabs scuttle on the deep ocean floor. Earthworms are squishy; clams surround themselves in heavy armor. Sponges are witless; humans, chimps, and dolphins are clever. This chapter explores some of this amazing variety.

Table 20.1   Nine Phyla of Animals

Phylum	Examples	Number of Existing Species
Porifera	Sponges	5000
Cnidaria	Hydras, jellyfishes, corals, sea anemones	11,000
Platyhelminthes (flatworms)	*Planaria*, tapeworms, flukes	25,000
Mollusca	Bivalves, chitons, snails, slugs, squids, octopuses	112,000
Annelida	Earthworms, leeches, polychaetes	15,000
Nematoda (roundworms)	Pinworms, hookworms, *C. elegans*	80,000
Arthropoda	Horseshoe crabs, spiders, scorpions, crustaceans, insects	More than 1,000,000
Echinodermata	Sea stars, sea urchins, sand dollars	7000
Chordata	Tunicates, lancelets, fishes, amphibians, reptiles, mammals	60,000

A. The First Animals Likely Evolved from Protists

The animal you thought of a moment ago probably lives on land, since terrestrial animals are the ones we see most often. Nevertheless, only 10 of the 37 known phyla include species that live on land, and no phylum contains only terrestrial animals. All of today's animals clearly have their origins in aquatic ancestors.

The first animals, which arose about 570 million years ago (MYA), may have been related to aquatic protists called choanoflagellates (**figure 20.1**), organisms that strongly resemble collar cells in sponges. Although no one knows exactly what the first animal looked like, the Ediacaran organisms that thrived during the Precambrian left some of the oldest animal fossils ever found (see figure 14.14). Animal life diversified spectacularly during the Cambrian period, which ended about 490 MYA. Most of today's phyla of animals, including sponges, jellyfishes, arthropods, mollusks, and many types of worms, originated in the Cambrian seas. Their fossils are exceptionally abundant in an area of British Columbia called the Burgess Shale (see figure 14.15). ▶ Cambrian explosion, p. 306

Aquatic animals were already diverse by the time plants and fungi colonized the land about 475 MYA. Arthropods, vertebrates, and other animals soon followed, diversifying as they adapted to new food sources and habitats.

B. Animals Share Several Characteristics

All animals, from sponges to chimpanzees, share a combination of features. First, they are multicellular organisms with eukaryotic cells, as are plants, fungi, and some protists. Unlike the cells of plants and fungi, however, animal cells lack cell walls. ▶ animal cell, p. 52

a.   Solitary   Colonial   b.

Figure 20.1 Animal Ancestor? (a) An immediate animal ancestor may have resembled a choanoflagellate. (b) Whatever the ancestor was, it eventually gave rise to all modern animals, including monarch butterflies.

Second, all animals are heterotrophs, obtaining both carbon and energy from organic compounds produced by other organisms. Most animals ingest their food, break it down in a digestive tract, absorb the nutrients, and eject the indigestible wastes.

Third, animal development is unlike that of any other type of organism. After fertilization, the diploid zygote (the first cell of the new organism) divides rapidly. The early animal embryo begins as a solid ball of cells that quickly hollows out to form a **blastula,** a sphere of cells surrounding a fluid-filled cavity. (Embryonic stem cells, the source of so much controversy, come from embryos at this stage of development.) No other organisms go through a blastula stage of development. ▶ stem cells, p. 167

Fourth, animal cells secrete and bind to a nonliving substance called the extracellular matrix. This complex mixture of proteins and other substances enables some cells to move, others to assemble into sheets, and yet others to embed in supportive surroundings, such as bone or shell. ▶ extracellular matrix, p. 514

C. Biologists Classify Animals Based on Organization, Morphology, and Development

Figure 20.2 compiles the nine animal phyla described in this chapter into a phylogenetic tree. This section explains the features that biologists use to construct the tree. Section 20.1D lists traits that are sometimes important in describing animals but that do not represent branching points on the tree.

Cell and Tissue Organization   The first major branching point separates animals into two clades. The simplest animals, the sponges, have several specialized cell types, but the cells do not interact to provide specific functions as they would in a true tissue. The other clade contains **eumetazoans,** animals with true tissues. In complex animals, multiple tissue types interact to form organs such as a heart, brain, or kidney. The organs, in turn, interact in systems that circulate and distribute blood, dispose of

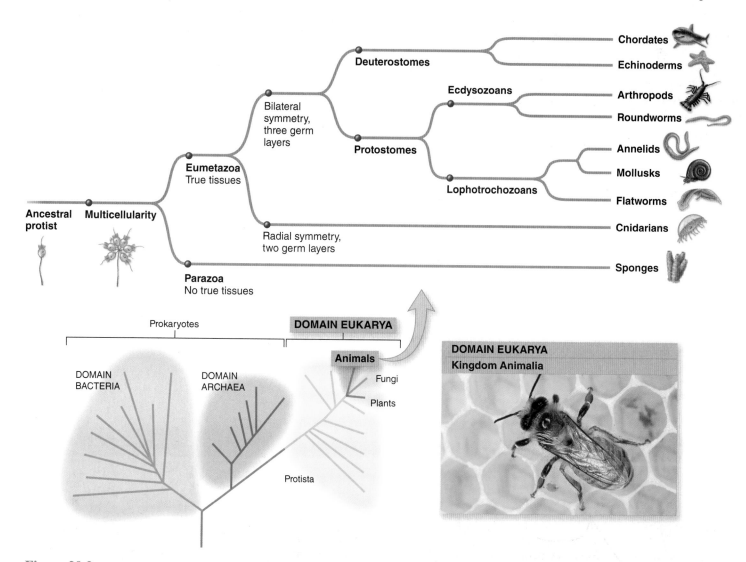

Figure 20.2 Animal Diversity. Biologists classify animals based on body form, developmental characteristics, and DNA sequences.

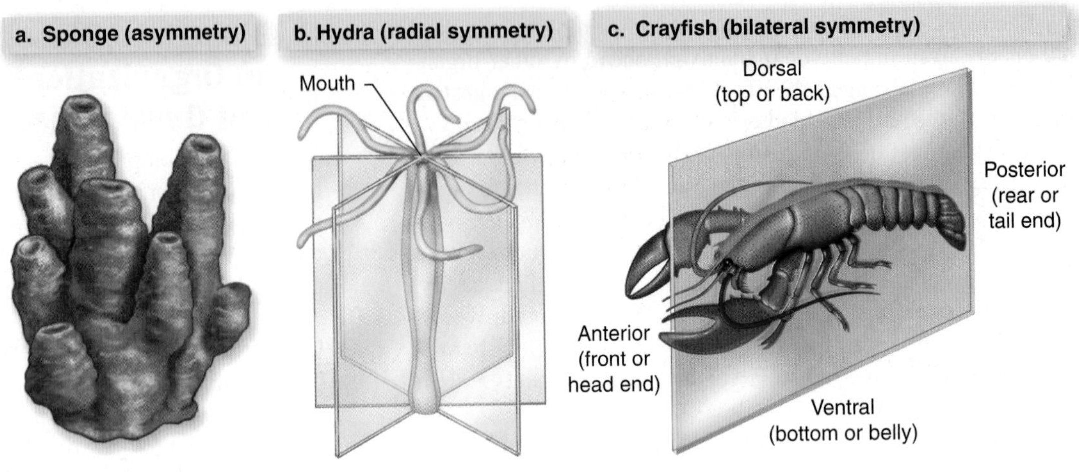

a. Sponge (asymmetry)

b. Hydra (radial symmetry)

Mouth

c. Crayfish (bilateral symmetry)

Dorsal
(top or back)

Posterior
(rear or
tail end)

Anterior
(front or
head end)

Ventral
(bottom or belly)

Figure 20.3 Types of Symmetry. (a) Many sponges are asymmetrical. (b) A hydra has radial symmetry. (c) A crayfish has bilateral symmetry. Animals with bilateral symmetry have a front (anterior) and rear (posterior) end, and a dorsal (back or top) and ventral (bottom or belly) side.

wastes, and carry out other functions. In general, the larger and more active an animal, the more complex and specialized its organ systems are.

Body Symmetry and Cephalization

Body symmetry is another major criterion used in animal classification (figure 20.3). Many sponges are asymmetrical; that is, they lack symmetry. Other sponges, hydras, jellyfishes, adult sea stars, and their close relatives have **radial symmetry,** in which any plane passing through the body from the mouth to the opposite end divides the body into mirror images. All other animals have **bilateral symmetry,** in which only one plane divides the animal into mirror images. Bilaterally symmetrical animals such as crayfish and humans have head (anterior) and tail (posterior) ends, and they typically move through their environment head first. This behavior is correlated with **cephalization,** the tendency to concentrate sensory organs and a brain at the head end of the animal.

Embryonic Development: Two or Three Germ Layers

Early embryos give other clues to evolutionary relationships (figure 20.4). In eumetazoans, the blastula folds in on itself to generate the **gastrula,** which is composed of two or three tissue layers called primary germ layers. The gastrulas of jellyfishes and their relatives have two germ layers: **ectoderm** to the outside, and **endoderm** to the inside. All other eumetazoans have a

third germ layer, **mesoderm,** that forms between the ectoderm and endoderm.

These germ layers give rise to all of the body's tissues and organs. Ectoderm develops into the skin and nervous system, whereas endoderm becomes the digestive tract and the organs derived from it. Mesoderm gives rise to the muscles, reproductive system, and many other specialized structures.

Embryonic Development: Protostomes and Deuterostomes

Biologists traditionally divide bilaterally symmetrical animals into two clades, based in part on events that occur after the embryo has begun to fold into a gastrula. As development proceeds, the inner cell layer of the gastrula fuses with the opposite side of the embryo, forming a tube with two openings. This cylinder of endoderm will develop into the animal's digestive tract.

In most **protostomes,** the first indentation to form develops into the mouth, and the anus develops from the second opening. (*Protostome* literally means "mouth first"). In **deuterostomes,** such as echinoderms and chordates, the first indentation becomes the anus, and the mouth develops from the second opening. (*Deuterostome* means "mouth second"). We now know that some animals classified as protostomes do not conform to the "mouth first" pattern. Nevertheless, DNA sequences support their placement in the protostome clade.

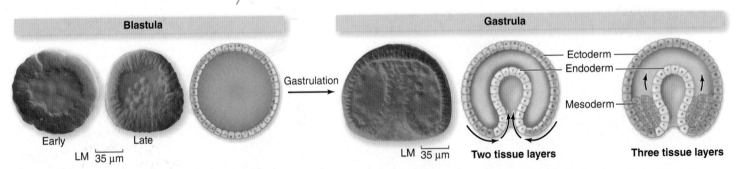

Blastula

Early   Late

LM 35 μm

Gastrulation

Gastrula

LM 35 μm

Ectoderm
Endoderm

Mesoderm

Two tissue layers

Three tissue layers

Figure 20.4 Two or Three Primary Germ Layers. During early development of eumetazoans, a fluid-filled ball of cells called a blastula folds in on itself and forms the gastrula. (The three photos show a sea urchin's blastula and gastrula). Animals with two primary germ layers have an outer ectoderm and an inner endoderm layer. In other animals, mesoderm forms between the ectoderm and endoderm.

a. Coelom

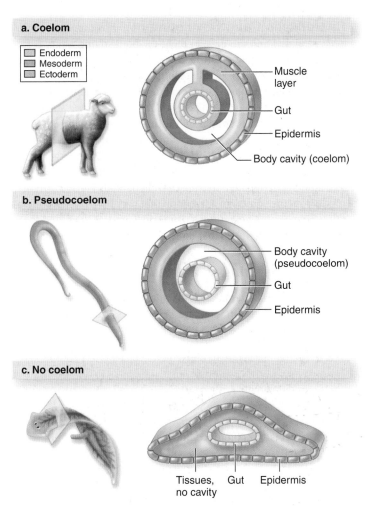

Endoderm
Mesoderm
Ectoderm

Muscle layer

Gut

Epidermis

Body cavity (coelom)

b. Pseudocoelom

Body cavity (pseudocoelom)

Gut

Epidermis

c. No coelom

Tissues, no cavity   Gut   Epidermis

Figure 20.5 Body Cavities. (a) Like many other animals, a sheep has a coelom. (b) A roundworm has a pseudocoelom. (c) A flatworm lacks a coelom. Note that these drawings are abstractions. In the sheep, for example, the internal organs grow into the coelom, greatly distorting the cavity's shape.

D. Biologists Also Consider Additional Characteristics

Other features that describe animals include the body cavity, the organization of the digestive tract, segmentation, and patterns of reproduction and development.

Body Cavity (Coelom) Biologists once classified bilaterally symmetrical animals based on the presence or absence of a coelom (figure 20.5). The **coelom** is a fluid-filled body cavity that forms completely within the mesoderm. Animals that have a coelom include earthworms, snails, insects, sea stars, and chordates. In contrast, roundworms have a body cavity called a **pseudocoelom** ("false coelom") that is lined partly with mesoderm and partly with endoderm. Flatworms lack a coelom, although evidence suggests their ancestors did have body cavities.

The coelom's chief advantage is flexibility. As internal organs such as the heart, lungs, liver, and intestines develop, they push into the coelom. The fluid of the coelom cushions the organs, protects them, and enables them to shift as the animal bends and moves.

In many animals, the coelom or pseudocoelom serves as a hydrostatic skeleton that provides support and movement. In a **hydrostatic skeleton,** muscles push against a constrained fluid. The earthworm, for example, burrows through soil by alternately contracting and relaxing muscles surrounding its coelom. Jellyfishes, flatworms, and other invertebrates lacking a coelom or pseudocoelom also have hydrostatic skeletons. In these animals, muscles act on other fluid-filled compartments in the body.

Digestive Tract A sponge lacks a digestive tract; instead, the animal has a simple opening through which filtered water leaves the body. In other animals, the digestive tract may be incomplete or complete (figure 20.6). Cnidarians and flatworms have an **incomplete digestive tract,** in which the mouth both

a. Incomplete digestive tract

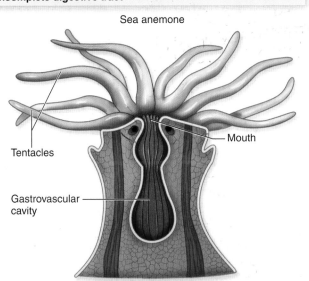

Sea anemone

Mouth

Tentacles

Gastrovascular cavity

b. Complete digestive tract

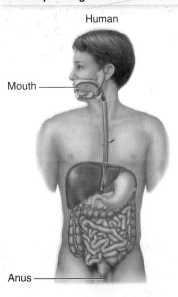

Human

Mouth

Anus

Figure 20.6 Digestive Tracts. (a) An animal with an incomplete digestive tract takes in food and ejects wastes through its mouth. (b) A complete digestive tract has one-way flow from mouth to anus.

Figure 20.7 **Segmentation.** This millipede illustrates segmentation—the division of the body into repeated parts.

takes in food and ejects wastes. In these animals, digestion occurs in the **gastrovascular cavity,** which secretes digestive enzymes and distributes nutrients to all parts of the animal's body.

In animals with a **complete digestive tract,** food passes in one direction from mouth to anus. A complete digestive tract allows for increased efficiency and specialization. Cells near the mouth can secrete digestive enzymes into the tract, "downstream" cells can absorb nutrients, and those near the anus can help eject wastes.

Segmentation   **Segmentation** is the division of an animal body into repeated parts (**figure 20.7**). In centipedes, millipedes, and earthworms, the segments are clearly visible. Insects and chordates also have segmented bodies, although the subdivisions may be less obvious.

Segmentation adds to the body's flexibility, and it enormously increases the potential for the development of specialized body parts. Activating different combinations of genes in each segment can create regions with unique functions. Antennae can form on an insect's head, for example, while wings or legs sprout from other segments.

Reproduction and Development   Most animals reproduce sexually, and the development of the resulting embryo follows either of two paths (**figure 20.8**). Animals that undergo **direct development** have no larval stage; at hatching or birth, they already resemble adults. A newborn elephant or newly hatched cricket, for example, looks like a smaller version of the adult.

In contrast, an animal with **indirect development** may spend part of its life as a **larva,** which is an immature stage that does not resemble the adult. The larvae eventually undergo **metamorphosis,** in which their bodies change greatly as they mature into adults. Larvae often live in different habitats and eat different foods from the adults, an adaptation that may help reduce competition between the generations. Butterflies, mosquitoes, and frogs are examples of animals that undergo indirect development.

20.1 | Mastering Concepts

1. What characteristics do all animals share?
2. When and in what habitat did animals likely originate, and when did today's major groups of animals arise?
3. What evidence do biologists use to construct the animal phylogenetic tree?
4. What are the events of early embryonic development in an animal?
5. Which animals have true tissues, and which have organs?
6. What is the difference between radial and bilateral symmetry?
7. Distinguish between a coelom and a pseudocoelom.
8. What are the two main types of digestive tract?
9. What advantages does segmentation confer?
10. How does direct development differ from indirect development?

Figure 20.8 **Animal Development.** (a) The young of animals that undergo direct development resemble the adult. (b) In indirect development, metamorphosis transforms a larva into a very different-looking adult.

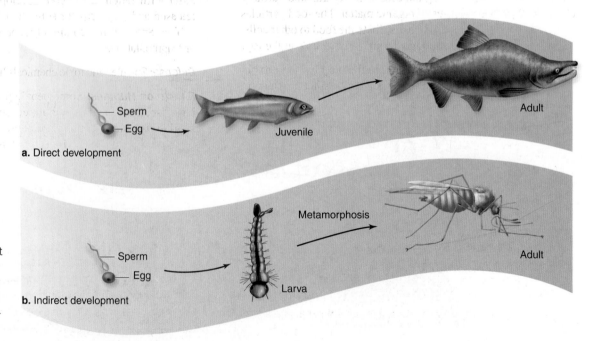

a. Direct development

Sperm
Egg

Juvenile

Adult

b. Indirect development

Sperm
Egg

Larva

Metamorphosis

Adult

20.2 | Sponges Are Simple Animals That Lack Differentiated Tissues

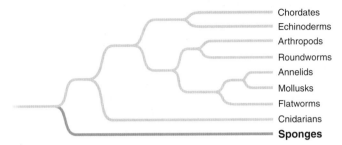

- Chordates
- Echinoderms
- Arthropods
- Roundworms
- Annelids
- Mollusks
- Flatworms
- Cnidarians
- **Sponges**

The **sponges** belong to phylum Porifera, which means "pore-bearers"—an apt description of these simple animals that lack true tissues and organs (**figure 20.9**).

Habitat: Aquatic. Most are marine, although some live in fresh water.

Body Structure: Sponges have hollow, porous bodies that are either asymmetrical or radially symmetrical (**figure 20.10**). The body wall has an outer layer of flattened cells and an inner layer of flagellated "collar cells" (the cells that resemble choanoflagellates). Sandwiched between these layers is a noncellular, jellylike matrix, the mesohyl, that gives the sponge body its shape. Embedded in the mesohyl are many types of cells, including amoebocytes that digest food, store and transport nutrients, divide, or secrete skeletal components.

Feeding: Sponges are filter feeders. Movement of flagella on the collar cells produces a current of water, which flows through the body wall and into the sponge's central cavity. (Contractile cells in the body wall can close the pores when the water contains too much sediment). Cells lining the central cavity trap and partially digest bacteria and particles of organic matter. The food particles then pass to amoebocytes, which distribute the food to other cells. Water and wastes exit the sponge through a large hole at the top.

Support and Movement: Embedded in the mesohyl are protein fibers and spicules, which are sharp slivers of silica or calcium

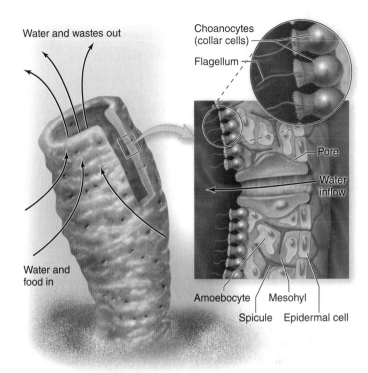

Figure 20.10 Sponge Anatomy. Water enters a sponge through pores in the body wall, and flagellated collar cells capture suspended food particles. Water and wastes exit through the large hole at the top. Amoebocytes have multiple functions, including secreting the spicules that make up the sponge's skeleton.

carbonate (see figure 20.10). Although some sponges can move very slowly, these animals are generally considered sessile, meaning they remain anchored to their substrate.

Reproduction and Development: Sponges are hermaphrodites; each animal releases sperm into water currents and retains eggs. After fertilization, the zygote develops into a blastula, which is released and may drift for some time before settling into a new habitat. Sponges also commonly reproduce asexually by budding or fragmentation.

Defense: Spicules and toxic chemicals help sponges deter predators.

Effects on Humans: Some people use natural sponges in bathing. Also, the chemicals that protect sponges from predators may yield useful anticancer and antimicrobial drugs. Collecting wild sponges, however, can harm ecosystems.

Figure 20.9 Porous Animals. Sponges are simple animals that lack true tissues.

20.2 | Mastering Concepts

1. How is a sponge's body different from that of other animals?
2. How do sponges feed?
3. What are two of the cell types in a sponge?
4. What is the function of spicules?
5. How do sponges reproduce sexually and asexually?
6. In what ways are sponges important?

20.3 | Cnidarians Are Radially Symmetrical, Aquatic Animals

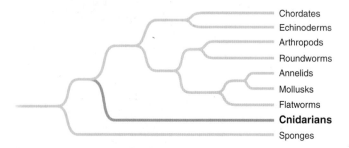

Phylum Cnidaria takes its name from the Greek word for "nettle," a stinging plant. The **cnidarians** all share the ability to sting predators and prey. These diverse aquatic animals include several familiar examples (**figure 20.11**).

Habitat: Aquatic (mostly marine).

Body Structure: Cnidarians are radially symmetrical, and their bodies can take either of two forms (**figure 20.12**). A **polyp** consists of a sessile stalk with tentacles on one end, whereas a **medusa** is a free-swimming, bell-shaped, "tentacles-down" body form typical of jellyfishes. In both body forms, two tissue layers sandwich a jellylike, noncellular substance called mesoglea. Tentacles surround the mouth. **Cnidocytes** embedded in the tentacles act as tiny harpoons that either inject venom or entangle the prey. One opening, the mouth, leads to the dead-end gastrovascular cavity.

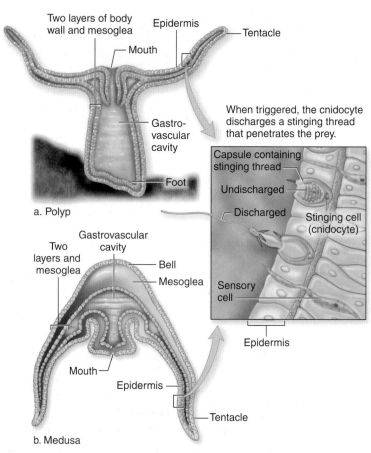

Figure 20.12 Cnidarian Anatomy. Cnidarians have two body forms: (a) the polyp and (b) the medusa. Stinging cells in the epidermis aid in defense and prey capture.

Figure 20.11 Diversity of Cnidarians. (a) A hydra is a tiny freshwater cnidarian. (b, c) Sea anemones have abundant tentacles that catch and sting fish. (d) Tiny coral animals build magnificent reefs, such as the one in (e). (f) The sea nettle jellyfish has tentacles dangling from its bell.

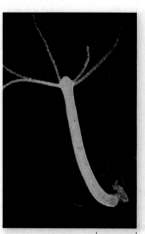

a.   250 µm   b.   c.   d.   e.

Diversity: Corals and sea anemones belong to a clade of cnidarians that exist exclusively as sessile polyps. A second clade contains hydras and the jellyfishes.

Feeding: Cnidarians are carnivores that use their tentacles to grab and sting passing prey and stuff it into their gastrovascular cavities. Cells lining the digestive tract secrete enzymes that digest the food. After absorbing the nutrients, the animal ejects indigestible matter through the mouth.

Support and Movement: In all cnidarians, the mesoglea acts as a hydrostatic skeleton. These animals can swim and move their tentacles courtesy of specialized nerve and muscle cells in the epidermis (outer body wall). Groups of linked neurons, called nerve nets, coordinate the contraction of muscle cells. As the cells contract, they force water out of the bell of a jellyfish, propelling the animal through the water. The same mechanism enables a sea anemone to stuff food into its gastrovascular cavity.

Some cnidarians also secrete calcium carbonate exoskeletons. Over many generations, the exoskeletons secreted by countless coral animals have built magnificent coral reefs. ▶ coral reefs, p. 796

Reproduction and Development: Cnidarians reproduce sexually and asexually (**figure 20.13**). In the moon jelly *Aurelia*, for example, male and female medusae release gametes into the water. After fertilization, a cilia-fringed larva develops, attaches to a surface, and metamorphoses into the polyp form. The polyp then buds asexually, generating a colony of additional polyps and, eventually, medusae. The cycle begins anew.

Defense: Stinging cnidocytes are the main defense against predators.

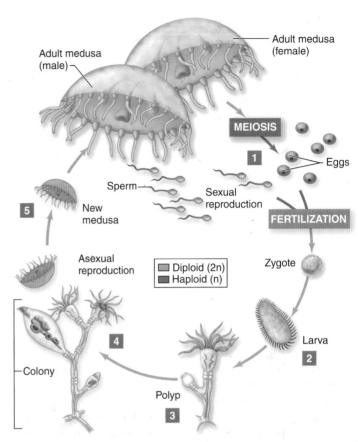

Figure 20.13 Cnidarian Reproduction. In the moon jelly *Aurelia*, (1) male and female medusae release sperm and egg cells. After fertilization, (2) a larva develops. The larva attaches to a surface and becomes (3) a polyp, which reproduces asexually to form (4) a colony. (5) New medusae form asexually from the polyp colony.

Effects on Humans: Huge swarms of jellyfish are becoming increasingly common, presenting a nuisance for tourist destinations and fisheries. Jellyfish stings may cause skin irritation or cramps; a few species have toxins that can be lethal on contact. On the positive side, coral reefs house many commercially important species of fishes and other animals, and they protect coastlines from erosion. As they build their calcium carbonate reefs, corals remove carbon from the atmosphere. A molecule that protects corals from damaging ultraviolet radiation is being developed into a sunscreen for human use.

20.3 | Mastering Concepts

1. What features do all cnidarians share?
2. What are some examples of cnidarians?
3. What is the difference between a polyp and a medusa?
4. How do cnidarians feed, move, and reproduce?
5. In what ways are cnidarians important?

f.

20.4 | Flatworms Have Bilateral Symmetry and Incomplete Digestive Tracts

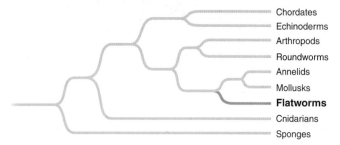

Chordates
Echinoderms
Arthropods
Roundworms
Annelids
Mollusks
Flatworms
Cnidarians
Sponges

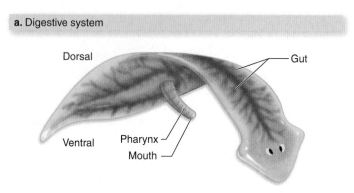

a. Digestive system

Dorsal — Gut — Pharynx — Mouth — Ventral

Figure 20.15 Flatworm Anatomy. (a) The pharynx is a muscular tube that opens into the incomplete digestive tract.

Phylum Platyhelminthes includes the **flatworms.** (*Platy* means "flat"; *helminth* means "worm"). Some of these animals are surprisingly beautiful, whereas others look downright scary.

Habitat: Free-living (usually in aquatic ecosystems) or parasitic on other animals.

Body Structure: As the name suggests, flatworms have flattened bodies, which enable each cell to independently exchange CO_2 and O_2 with the environment. Unlike the sponges and cnidarians, flatworms have bilateral symmetry, with a simple brain and sensory structures concentrated at the head end.

Diversity: Figure 20.14 shows the three main classes of flatworms, including the **free-living flatworms** (such as a marine flatworm and a planarian) and the parasitic **flukes** and **tapeworms.**

Feeding: Free-living flatworms usually are predators or scavengers. A muscular, tubelike pharynx takes in food and ejects wastes from the highly branched gut (**figure 20.15a**). A fluke feeds on blood and other host tissues, using its pharynx to pull food into the gastrovascular cavity. A tapeworm lacks a mouth and digestive system entirely; instead, it attaches to the host's intestine by a scolex (see figure 20.14c) and absorbs food through its body wall.

Circulation and Respiration: Flatworms do not have specialized circulatory or respiratory systems. CO_2 and O_2 simply diffuse through the body wall.

Excretion: Protonephridia are structures that maintain internal water balance and excrete nitrogenous wastes. Each consists of numerous flame cells, so named because their shape and beating cilia resemble a flickering candle (figure 20.15b). The tubules of protonephridia open to pores on the body surface.

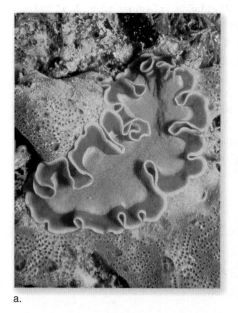

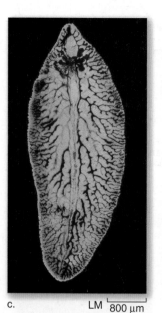

a. b. LM 200 µm c. LM 800 µm d. SEM (false color) 1 mm

Figure 20.14 Flatworm Diversity. (a) A marine flatworm and (b) a planarian are two examples of free-living flatworms. (c) The liver fluke, *Fasciola hepatica*, is a parasite that infects sheep, cattle, and humans. (d) Tapeworms are internal parasites with suckers that latch onto the host's intestines. © Dr. Richard Kissel & Dr. Gene Shih/Visuals Unlimited.

b. Excretory system

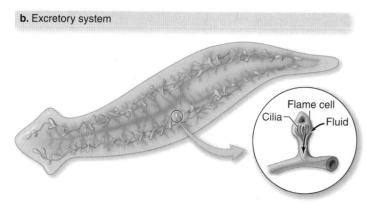

Flame cell
Cilia
Fluid

c. Nervous system

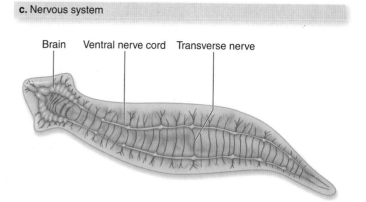

Brain   Ventral nerve cord   Transverse nerve

Figure 20.15 Flatworm Anatomy. (b) Flame cells form the excretory system. (c) A brain and ladderlike nerve cords make up the nervous system.

Nervous System: Some flatworms have nerve cords and concentrations of nerve cell bodies in their anterior ends, forming a simple brain (figure 20.15c). Others have a nerve net. Chemosensory (taste and smell) cells in some freshwater planarians cluster on ear-shaped structures; eyespots detect light but do not form images.

Support and Movement: Loosely packed cells surrounded by fluid inside the body act as a hydrostatic skeleton. Flatworms creep or swim by contracting muscles in a rolling motion. Many use cilia to glide along mucous secretions.

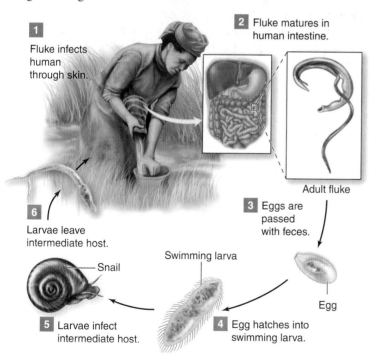

1 Fluke infects human through skin.

2 Fluke matures in human intestine.

Adult fluke

3 Eggs are passed with feces.

Egg

4 Egg hatches into swimming larva.

5 Larvae infect intermediate host.

Swimming larva

Snail

6 Larvae leave intermediate host.

Figure 20.16 Life Cycle of a Blood Fluke. (*1*) Humans acquire the blood fluke, *Schistosoma,* when contaminated water penetrates the skin. (*2*) The worms mature inside the host, (*3*) and eggs leave the body in urine or feces. (*4*) The eggs hatch in water. (*5*) Larvae infect an intermediate host, usually a snail. (*6*) Another organism may serve as a second intermediate before the fluke moves to its final vertebrate host.

Reproduction and Development: Many flatworms reproduce asexually. Free-living species, for example, may simply pinch in half and regenerate the missing parts asexually.

Blood flukes mate inside their vertebrate host (**figure 20.16**). The fertilized eggs leave the host's body in urine or feces and hatch into larvae upon reaching water. The larvae colonize other animals such as snails before infecting a new vertebrate host.

Tapeworms have a different strategy. The posterior end of a tapeworm's body consists of repeated reproductive organs called proglottids, each with a complete set of male and female reproductive structures. Proglottids containing fertilized eggs break off of the worm and leave the host in feces. When a new host swallows proglottids in contaminated water, the eggs hatch. The resulting larvae can migrate to the host's muscles; this is how people acquire tapeworm infections by eating undercooked fish, beef, or pork.

Defense: When inside the host, parasitic flatworms are safe from predators; free-living forms secrete a protective mucus.

Effects on Humans: Worldwide, infections with blood flukes, lung flukes, liver flukes, and tapeworms affect hundreds of millions of people and countless domesticated and wild animals.

20.4 | Mastering Concepts

1. What features do all flatworms share?
2. How does the body shape of a flatworm enhance gas exchange with the environment?
3. Describe the three classes of flatworms.
4. How do free-living flatworms, tapeworms, and flukes eat?
5. What types of specialized cells and tissues do flatworms have in their excretory and nervous systems?
6. How do flatworms move and reproduce?
7. In what ways are flatworms important?

20.5 | Mollusks Are Soft, Unsegmented Animals

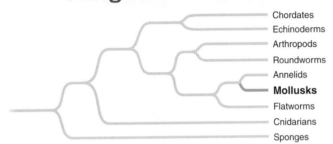

Chordates
Echinoderms
Arthropods
Roundworms
Annelids
Mollusks
Flatworms
Cnidarians
Sponges

Mollusks form a fascinating phylum that contains many familiar animals (**figure 20.17**). Clams and snails are mollusks, as is the largest known invertebrate—the colossal squid, which may be up to 14 meters long. The octopus, the most intelligent invertebrate, is also a mollusk.

Habitat: Terrestrial, marine, and freshwater.

Body Structure: The word *mollusk* comes from the Latin word for "soft," reflecting the fleshy bodies in this phylum of animals. Mollusks have several structures in common (**figure 20.18**). One is the **mantle,** a dorsal fold of tissue that secretes a shell in most species. A muscular **foot** provides movement, and an area called the **visceral mass** contains the digestive and reproductive organs. Many mollusks have a **radula,** a tonguelike strap with teeth made of chitin (a tough polysaccharide). They use the radula to scrape food into their mouths.

a.

b.

c.

d.

e.

f.

Figure 20.17 Mollusk Diversity. (a) The sea cradle chiton lives along the west coast of the United States. (b) The sea scallop is a bivalve. (c) The land snail and (d) the slug are gastropods. (e) This octopus is a cephalopod, as is (f) this squid.

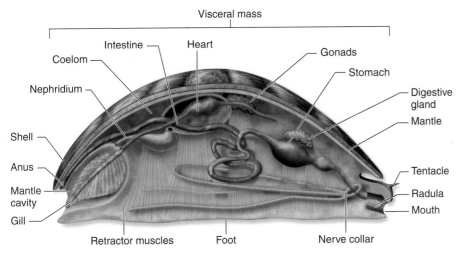

Figure 20.18 Mollusk Anatomy. In this generalized mollusk, the digestive, circulatory, excretory, and reproductive systems are all contained within the visceral mass. The muscular foot provides locomotion, and the mantle secretes a protective shell.

Diversity: The four largest classes of mollusks are the chitons, bivalves, gastropods, and cephalopods. **Chitons** are marine animals with eight flat shells that overlap like shingles. **Bivalves,** such as oysters, clams, scallops, and mussels, have two-part, hinged shells. **Gastropods** ("stomach-foot") are snails, slugs, sea slugs (nudibranchs), and limpets. Their name comes from the broad, flat foot on which they crawl. The **cephalopods** include marine animals such as octopuses, squids, and nautiluses. *Cephalopod* means "head-foot," a reference to the arms connected to the head of the animal.

Feeding: Chitons scrape algae and other cells off of rocks, whereas bivalves filter organic particles and small organisms out of the water. Cephalopods are active predators of fast-moving prey such as fishes. Most slugs and snails are herbivores; as described in section 5.8, some sea slugs remove chloroplasts from the algae they eat and carry out photosynthesis on their own.

Circulatory System: Most mollusks have an open circulatory system, in which a heart pumps blood to tissues throughout the body cavity instead of within vessels. Cephalopods, however, have a closed circulatory system, with blood confined to vessels. Their blood vessels efficiently exchange O_2 and CO_2 with respiring muscle cells, supporting a high metabolic rate.

Respiratory System: Aquatic mollusks have gills that exchange O_2 and CO_2 with the environment. Terrestrial snails and slugs have a lung derived from a space called the mantle cavity.

Excretory System: An excretory organ called a nephridium ("little kidney") filters blood and produces urine.

Nervous System: The molluscan nervous system varies from simple and ladderlike to complex and cephalized. An octopus's nervous system includes a brain, a highly developed visual system, nerve cords that extend into each of its eight arms, and an excellent sense of touch.

Support and Movement: All mollusks have a hydrostatic skeleton, and most have a supportive internal or external shell. The animal moves when muscles act on the constrained fluid of the coelom. The speed and type of locomotion, however, vary among the mollusks. On land, snails and slugs glide on a trail of mucus. Bivalves use the muscular foot to burrow into sediments. Cephalopods are the speediest invertebrates; they move by "jet propulsion." A structure called a siphon sucks water into their

mantles. When muscles in the mantle contract, the water shoots out and propels the animal.

Reproduction and Development: Many bivalves shed gametes into the water, where external fertilization occurs. Gastropods and cephalopods, on the other hand, mate and fertilize eggs internally. Development may be indirect or direct. In many marine mollusks, a ciliated, pear-shaped larva settles to the bottom of the sea and develops into an adult. In cephalopods and snails, however, the hatchlings resemble adults. The development of snails is especially unusual. The body twists inside its shell as it develops, so that in the adult animal, the anus empties digestive wastes near the head.

Defense: The hard shells of bivalves and snails protect against many predators, but in cephalopods, the shell is internal or absent. These mollusks have other defenses. Squids and octopuses, for example, can change their color and shape to match their background with uncanny camouflage. When alarmed, cephalopods can squirt a melanin-pigmented "ink" that cloaks their speedy escape.

Effects on Humans: Mollusks have diverse effects on human life, health, and environmental quality. We harvest pearls from oysters, and we eat clams, mussels, oysters, snails, squids, and octopuses. Bivalves can become poisonous, however, if they accumulate pollutants or toxins produced by protists called dinoflagellates. The cuttlebones that supply calcium to pet birds come from the thick internal shells of cephalopods called cuttlefish. Snails and slugs are voracious consumers of garden and crop plants. Some aquatic snails host parasitic worms (see figure 20.16), and cone snails can kill humans with a venomous "harpoon." Zebra mussels have disrupted the ecology of the Great Lakes since their introduction in the 1980s. ▶▶ invasive species, p. 812; dinoflagellates, p. 356

20.5 | Mastering Concepts

1. What structures do mollusks have?
2. What are the four largest classes of mollusks, and where do they live?
3. How do mollusks feed, move, excrete metabolic wastes, reproduce, and protect themselves?
4. In what ways are mollusks important?

20.6 | Annelids Are Segmented Worms

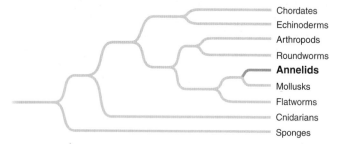

Earthworms and other segmented worms are **annelids.** The name of the phylum, Annelida, derives from the Latin word *annulus* ("little ring"), a reference to the segments that characterize these animals.

Habitat: Terrestrial, freshwater, and marine.

Body Structure: The most obvious characteristic of annelids is the repeated segments that make up their bodies. They also have a complete digestive tract, and a coelom separates the gut from the body wall.

Diversity: Biologists recognize two main classes of annelids (figure 20.19). One class contains annelids with a saddlelike thickening called a clitellum near the head end. Included in this class are the **oligochaetes,** such as earthworms. *Oligo-* means "few," a reference to the small number of bristles (setae) on the side of each segment. **Leeches** belong to the same class. Their segments lack bristles, but they have suckers and superficial rings called annuli within each segment.

The other class of annelids, the **polychaetes,** contains the marine segmented worms. Most polychaetes have pairs of fleshy, paddlelike appendages called parapodia that they use in locomotion. The name *polychaete* comes from the many bristles embedded in the parapodia.

Feeding: Annelids use diverse feeding strategies. Earthworms ingest soil, digest the organic matter, and eliminate the indigestible particles as castings. Many leeches suck blood from vertebrates, but most eat small organisms such as arthropods, snails, or other annelids. Polychaetes are often filter feeders or deposit feeders, although some are predators with formidable jaws.

Circulatory System: All annelids except leeches have a closed circulatory system. Earthworms have multiple heart-like aortic arches that pump blood throughout the body (**figure 20.20**).

Respiratory System: Some polychaetes have feathery gills, but oligochaetes and leeches exchange gases by diffusion across the body wall. Because the CO_2 and O_2 must dissolve in water, these animals must remain moist. This requirement explains why an earthworm will die if its body surface dries out.

Excretory System: The nephridium draws in fluid from the coelom, returns some ions and other substances to the blood, and discharges the waste-laden fluid outside the body through a pore (see figure 20.20).

Nervous System: The nervous system includes a simple "brain," a mass of nerve cells at the head end of the animal. These cells connect around the digestive tract to a ventral nerve cord, with lateral nerves running through each segment.

Support and Movement: Circular and longitudinal muscles push against the coelom as the worm crawls, burrows, or swims.

Figure 20.19 Annelid Diversity. (a) The earthworm (*Lumbricus terrestris*), is an oligochaete. (b) A leech has visible segments. (c) *Nereis is* a polychaete; the fleshy paddles are parapodia. (d) Feathery plumes extend from a colony of polychaetes called featherduster worms. The plumes are organs of feeding and respiration.

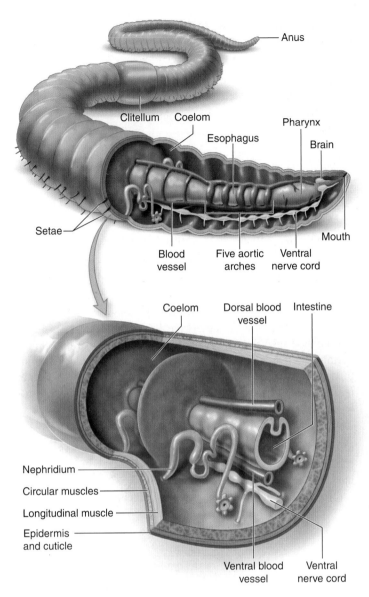

Figure 20.20 Earthworm Anatomy. Earthworms have the body segmentation that is typical of annelids. The upper illustration shows the overall organization of the worm's internal organs; the lower illustration provides a detailed view of one segment.

Leeches crawl, inchworm-style, by using the suckers at each end of the body. Parapodia can act as paddles that help polychaetes walk, swim, and dig.

Reproduction and Development: Leeches and oligochaetes are hermaphrodites, which means that each individual has the reproductive organs of both sexes. Two individuals copulate, each discharging sperm that its partner temporarily stores. In these animals, juveniles resemble the adults. In contrast, polychaetes have separate sexes, external fertilization, and indirect development.

Defense: Many annelids avoid predation by burrowing underground or in sediments. Polychaetes called tubeworms construct

tough tubes of chitin into which they can retract, and some also have powerful jaws.

Effects on Humans: Earthworms aerate and fertilize soil. Worm farms raise oligochaetes for sale as fishing bait or as soil conditioners (see this section's Apply It Now box). In medicine, a blood-thinning chemical from leeches can stimulate circulation in surgically reattached digits and ears, and physicians sometimes apply leeches to remove excess blood that accumulates after damage to the nervous system.

Apply It Now

Start Your Own Worm Farm

You may have heard of cow manure being used as a fertilizer, but what about worm castings? These pinhead-sized droppings are rich in the water-soluble nutrients that plants need. You can buy plant fertilizers made of worm castings, but if you produce food scraps in your kitchen, you can easily build a small-scale worm farm of your own.

To start the farm, acquire two stackable plastic tubs, one nesting inside the another. Drill holes in the bottom of the upper tub, which will contain the worms. The lower tub will collect the waste fluids that drain through the holes; like the castings, this liquid is an excellent fertilizer.

Purchase live worms locally or online. Red wigglers (*Eisenia fetida*) work best because they eat a lot and reproduce rapidly. Earthworms are not a good choice; they burrow deep into soil and are therefore poorly suited for life in a shallow container.

Create bedding for the worms by shredding some black and white newsprint or computer paper. Toss in a couple of handfuls of garden soil and some crushed, cooked egg shells. After you moisten the mixture, add your worms and some food. Worms can eat vegetable scraps, most fruits, stale bread, leftover pasta, coffee grounds, tea leaves, paper, and cardboard. Avoid meat and dairy products, which will produce unpleasant odors. In addition, onions and garlic repel worms, and citrus fruits cause the bins to become too acidic. Keep your farm in a dark, quiet location at room temperature.

When the bedding and food have decayed, it is time to harvest the castings. Drill holes into a third tub and prepare a fresh bedding mixture. Place the new bin on top of the one containing the worms. If the bottom of the tub is resting lightly on the old bedding, the worms will crawl through the holes into the new tub. After about two weeks, the lower tub will contain nothing but castings, which you can add to your plants.

20.6 | Mastering Concepts

1. What features do all annelids share?
2. What features distinguish the two classes of annelids, and where do members of each group live?
3. How do annelids feed, exchange gases, excrete wastes, move, reproduce, and defend themselves?
4. In what ways are annelids important?

20.7 | Nematodes Are Unsegmented, Cylindrical Worms

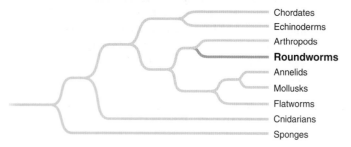

- Chordates
- Echinoderms
- Arthropods
- **Roundworms**
- Annelids
- Mollusks
- Flatworms
- Cnidarians
- Sponges

Most **roundworms** (phylum Nematoda) are barely visible to the unaided eye, but they are extremely abundant in every known habitat, from the oceans to the tropics to the North and South Poles.

Habitat: Some nematodes parasitize plants or animals, but most are free-living in soil or in the sediments of freshwater or marine ecosystems. Their abundance is astounding. A square meter of soil can yield millions of microscopic nematodes, representing dozens of species.

Body Structure: Nematodes are unsegmented, cylindrical worms with tapered ends and a complete digestive tract (**figure 20.21**). The name of the phylum reflects this long, thin shape; the Greek word *nema* means "thread." Most nematodes are microscopic, but the giant intestinal roundworm (*Ascaris*) can reach 40 cm long.

Diversity: Biologists have described more than 80,000 species of nematodes, and hundreds of thousands (or even millions) more may remain undescribed. Most classification relies on genetic sequences, and biologists have yet to agree on the number or composition of classes within the phylum.

Feeding: Free-living nematodes eat fungi, bacteria, plants, or insect larvae. Parasitic nematodes may suck blood or consume digested food in the host's intestines.

Circulation and Respiration: Nematodes lack specialized circulatory or respiratory organs; fluid in the pseudocoelom distributes nutrients, O_2, and CO_2 throughout the body (see figure 20.5).

Excretory System: Specialized cells maintain salt balance and remove nitrogenous wastes from the body cavity. An excretory pore opens to the outside of the body.

Nervous System: Nematodes have an anterior brain and two nerve cords along the length of the body. Bristles and other sensory structures on the body surface enable the worms to detect touch and chemicals.

Support and Movement: The pseudocoelom acts as a hydrostatic skeleton. Nematodes are limited to back-and-forth, thrashing motions because only longitudinal (lengthwise) muscles act on the pseudocoelom. As a result, a nematode can neither crawl nor lift its body above its substrate.

Reproduction and Development: Most species have separate sexes. Females produce large numbers of tough eggs that survive drying and exposure to damaging chemicals. The juveniles undergo direct development.

Defense: An external layer of tissue secretes a tough cuticle, which may protect nematodes. (Like arthropods, nematodes molt; that is, they shed the cuticle and grow a new one several times during their lives). Also, parasitic roundworms are protected from predators while inside their hosts. Many nematodes also have the remarkable ability to survive extreme heat, cold, or drying by entering a state of suspended animation; life resumes when favorable conditions return.

Effects on Humans: The most familiar nematodes are those that cause us harm, including intestinal parasites such as pinworms, hookworms, and *Ascaris*. Other parasitic roundworms, including *Trichinella*, live in the muscle tissue of humans and pigs and are transmitted by eating undercooked pork. Still other nematodes cause elephantiasis (**figure 20.22**)

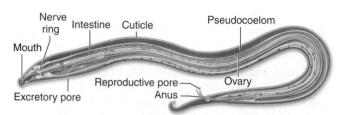

LM   15 µm

Figure 20.21 Nematode Anatomy. A free-living nematode is only about 1 mm long, but its body contains a complete digestive tract, a simple nervous system, and reproductive organs.

Nerve ring   Intestine   Cuticle   Pseudocoelom
Mouth
Excretory pore   Reproductive pore   Ovary   Anus

Figure 20.22 Elephantiasis. The nematode *Wuchereria bancrofti* (inset) inhabits lymph nodes. Excess fluid accumulates in the legs and feet, causing an illness called elephantiasis.

or African river blindness, a disease transmitted by the bite of a black fly. Moreover, heartworms are nematodes that infect the hearts, lungs, and blood vessels of dogs and cats. Nematodes also profoundly affect agriculture. Some are plant pathogens that cause diseases in and spread viruses among important food crops; others aid farmers by attacking insect pests. Finally, biologists use the nematode *Caenorhabditis elegans* in scientific research (see the Focus on Model Organisms box on this page).

20.7 | Mastering Concepts

1. What features do all roundworms share?
2. What are some examples of roundworms?
3. Compare and contrast the roundworm body structure with those of a flatworm and an annelid.
4. What evidence places roundworms in a clade with arthropods?
5. How do nematodes feed, excrete metabolic wastes, move, and reproduce?
6. In what ways are roundworms important?

Focus on Model Organisms

Caenorhabditis elegans and *Drosophila melanogaster*

Two invertebrates, a roundworm and an arthropod, share the spotlight in this box. Both have the characteristics common to all model organisms: small size, easy cultivation in the lab, and rapid life cycles. Each has provided crucial insights into life's workings.

The nematode: *Caenorhabditis elegans*

This roundworm, a soil inhabitant, is arguably the best understood of all animals. This was the first animal to have its genome sequenced (in 1998), revealing about 18,000 genes. A small sampling of the contributions derived from research on *C. elegans* includes:

- **Animal development:** An adult *C. elegans* consists of only about 1000 cells. Because the worm is transparent, biologists can watch each organ form, cell by cell, as the animal develops from a zygote into an adult. Eventually, researchers hope to understand every gene's contribution to the development of this worm.

- **Apoptosis:** Programmed cell death, or apoptosis, is the planned "suicide" of cells as a normal part of development. Researchers observing *C. elegans* development know exactly which cells will die at each stage. Learning about genes that promote apoptosis may lead researchers to a better understanding of cancer, a family of diseases in which cell division is unregulated. ▶▶ cancer, p. 162; apoptosis, p. 167

- **Muscle function:** The first *C. elegans* gene to be cloned, *unc-54*, revealed the amino acid sequence of one part of myosin, a protein required for muscle contraction. ▶ myosin, p. 590

- **Drug development:** Nematodes provide a good forum for preliminary testing of new pharmaceutical drugs. For example, researchers might identify a *C. elegans* mutant lacking a functional insulin gene, then test new diabetes drugs for the ability to replace the function of the missing gene. ▶ diabetes, p. 576

- **Aging:** Worms with mutations in some genes have life spans that are twice as long as normal. The selective destruction of neurons can also expand or reduce the life span, depending on which cells are destroyed. Insights on aging in *C. elegans* may eventually help increase the human life span.

The fruit fly: *Drosophila melanogaster*

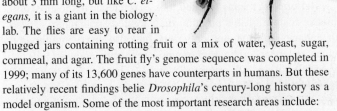

Drosophila melanogaster is only about 3 mm long, but like *C. elegans,* it is a giant in the biology lab. The flies are easy to rear in plugged jars containing rotting fruit or a mix of water, yeast, sugar, cornmeal, and agar. The fruit fly's genome sequence was completed in 1999; many of its 13,600 genes have counterparts in humans. But these relatively recent findings belie *Drosophila's* century-long history as a model organism. Some of the most important research areas include:

- **Heredity:** In the early 1900s, Thomas Hunt Morgan and his colleagues used *Drosophila* to show that chromosomes carry the information of heredity. Studies on mutant flies with different-colored eyes led to the discovery of sex-linked traits. Morgan's group also demonstrated that genes located on the same chromosome are often inherited together. In the process, they discovered crossing over. ▶▶ crossing over, p. 186; linked genes, p. 208; sex linkage, p. 213

- **Human disease:** The similarity of some *Drosophila* genes to those in the human genome has led to important insights into muscular dystrophy, cancer, and many other diseases. For example, researchers have studied the fly version of the human *p53* gene, which induces damaged cells to commit suicide (apoptosis). When that gene is faulty, the cell may continue to divide uncontrollably. The result: cancer.

- **Animal development:** Homeotic genes are "master switch" genes that regulate the overall development of the body, including segmentation and wing placement. Researchers discovered these genes in mutant flies with dramatic abnormalities, such as legs growing in place of antennae on the fly's head (see figure 7.21). Later, researchers discovered comparable genes in many organisms, including mice, leading to new insights into mammalian development. ▶ the mouse as model organism, p. 445

- **Circadian rhythms:** The expression of some genes in bacteria, plants, fungi, and animals cycles throughout a 24-hour day. How do the rhythmically expressed genes "know" what time it is? In *Drosophila*, clock genes called *period* and *timeless* encode proteins that turn off their own expression, much like a thermostat turns off a heater when the temperature is too high. This "master clock" controls the animal's other daily cycles of hormone secretion and behavior.

20.8 | Arthropods Have Exoskeletons and Jointed Appendages

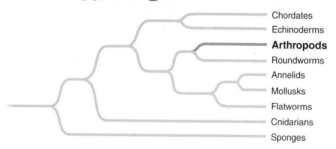

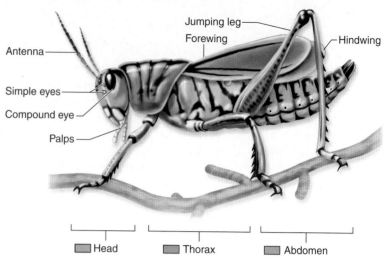

Figure 20.23 **Arthropod Features.** These segmented animals have an exoskeleton and jointed appendages.

If diversity and sheer numbers are the measure of biological success, then the phylum Arthropoda certainly is the most successful group of animals. More than 1,000,000 species of **arthropods** have been recorded already, and biologists speculate that this number could double.

Arthropoda means "jointed foot," a reference to the most distinctive feature of this phylum: their jointed appendages. These appendages include not only feet but also legs, mouthparts, wings, antennae, copulatory organs, ornaments, and weapons.

In addition, all arthropods have a versatile, lightweight **exoskeleton** made mostly of chitin, protein, and (sometimes) calcium salts. Thin, flexible areas create the moveable joints between body segments and within appendages. An exoskeleton has a drawback, though—to grow, an animal must molt and secrete a bigger one, leaving the animal vulnerable while its new exoskeleton is still soft. Molting is one feature that arthropods share with nematodes, another extraordinarily successful group in the ecdysozoan clade.

A. Arthropods Have Complex Organ Systems

Habitat: Terrestrial, freshwater, and marine.

Body Structure: In addition to the distinctive exoskeleton and jointed appendages, arthropod bodies are segmented. In many arthropods, the segments group into three major body regions: head, thorax, and abdomen (**figure 20.23**). The coelom is reduced to small fluid-filled cavities surrounding the reproductive and excretory organs.

Feeding: Arthropods eat almost everything imaginable, including dead organic matter, plant parts, and other animals.

Circulatory System: Arthropods have open circulatory systems (see figure 29.2). A tubelike heart propels the blood, which circulates around the animal's organs (**figure 20.24**).

Respiratory System: In most land arthropods, the body wall is perforated with holes (spiracles) that open into a series of branching tubes called tracheae, transporting oxygen and carbon dioxide to and from tissues. In contrast, aquatic arthropods have gills, and spiders and scorpions have stacked plates called book lungs.

Excretory System: Insects, spiders, and other terrestrial arthropods have organs called Malpighian tubules that collect and remove nitrogenous wastes while reabsorbing water. (The "green gland" in crayfish and lobsters has a similar function.) The tubules deposit dry, nitrogen-rich waste into the posterior end of the digestive tract. The animal ejects the waste, together with undigested food, through its anus.

Nervous System: Thanks to a nervous system with a brain and ventral nerve cords, many arthropods are active, fast, and sensitive to their environment. Consider, for example, the speed with which a fly can detect and react to a swatter. Their sensory systems can detect light, sound, touch, vibrations, air currents, and chemical signals. All of these clues help arthropods find food, identify mates, and escape predation. ▸ firefly bioluminescence, p. 77

Support and Movement: The tough exoskeleton protects the animal and gives it its shape. Internal muscles span the joints between body segments and within appendages. The resulting hollow lever systems can generate precise, forceful movements as the animal crawls, jumps, swims, or flies.

Reproduction and Development: Most arthropods have separate sexes. The male commonly produces a packet of sperm,

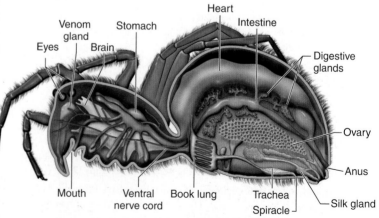

Figure 20.24 **Arthropod Organs.** The internal anatomy of a spider reveals complex organ systems.

Figure 20.25 Trilobites. These extinct marine arthropods have three distinct regions along the length of the body: a long central lobe plus flanking right and left lobes.

which the female takes into her genital opening. The sperm fertilize her eggs internally. In most species, the female then lays the fertilized eggs, although mites and scorpions bear live young. Ants and bees tend their young, but in most other arthropods, parental care is minimal.

Arthropod development may be direct or indirect. Butterflies and houseflies, for example, undergo indirect development, changing dramatically during a metamorphosis from larva to pupa (cocoon) to adult. Others, such as crickets, change only gradually from molt to molt.

Defense: Besides the protective exoskeleton, many arthropods can bite, sting, pinch, make noises, or emit foul odors or toxins that deter predators. Some have excellent camouflage that enables them to blend into their surroundings. Others have defensive behaviors; they may jump, run, roll into a ball, dig into soil, or fly away when threatened. Some moths unfurl wings with dramatic eyespots that startle or confuse predators.

Effects on Humans: Arthropods intersect with human society in about every way imaginable. Mosquitoes, flies, fleas, and ticks transmit infectious diseases as they consume human blood. Bees and scorpions sting, termites chew wood in our homes, and many insect larvae destroy crops. Yet entire industries rely on

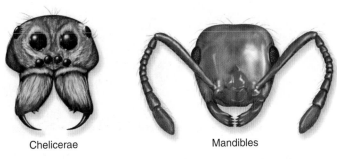

Chelicerae   Mandibles

Figure 20.26 Arthropod Mouthparts. Some arthropods, such as spiders, have chelicerae. Others, such as ants, have mandibles.

arthropods—consider beeswax, honey, silk, and delicacies such as shrimp, crabs, and lobsters. Insects pollinate many plants, and spiders eat crop pests. The fruit fly, *Drosophila melanogaster,* is important in biological research. On a much smaller scale, dust mites eat flakes of skin that we shed as we move about our homes, while follicle mites inhabit our pores (see the Apply It Now box in this section).

B. Arthropods Are the Most Diverse Animals

Phylum Arthropoda is divided into five subphyla. One contains the 17,000 species of extinct **trilobites** (figure 20.25). The other four subphyla are classified based on mouthpart shape (figure 20.26). Spiders, scorpions, and other **chelicerates** have grasping, clawlike mouthparts called chelicerae. The three subphyla of **mandibulates** have chewing, jawlike mouthparts termed mandibles.

Chelicerates: Spiders and Their Relatives

Habitat: Terrestrial and aquatic.

Body Structure: Most chelicerates have two major body regions: an abdomen and a fused head and thorax (figure 20.27). They also have chelicerae and four or more pairs of walking legs.

a.   b.   c.   d.

Figure 20.27 Chelicerate Arthropods. Chelicerates include (a) horseshoe crabs, (b) ticks, (c) spiders, and (d) scorpions.

a.   b.

Figure 20.28 Myriapods. (a) The millipede has two pairs of legs per segment. (b) The centipede has one pair per segment.

Diversity: The two most familiar groups of chelicerates are horseshoe crabs and arachnids. **Horseshoe crabs** are primitive-looking animals whose name refers to the hard, horseshoe-shaped exoskeleton, which covers a wide abdomen and a long tailpiece. The four species of horseshoe crabs are not true crabs, which are crustaceans. The more than 100,000 species of **arachnids** are eight-legged arthropods, including mites and ticks; spiders; harvestmen ("daddy longlegs"); and scorpions.

Special Features: Technicians use a compound extracted from the blood of horseshoe crabs to test medical supplies for bacteria. Spiders make "silk" and use it to produce webs, tunnels, egg cases, and spiderling nurseries. Some spiderlings use silk driftlines to float to a new habitat.

Mandibulates: Millipedes and Centipedes

About 13,000 species of **millipedes** and **centipedes** make up a group called the myriapods (**figure 20.28**).

Habitat: Terrestrial.

Body Structure: The head features mandibles and one pair of antennae. The rest of the body is divided into repeating subunits, each with one or two pairs of appendages.

Special Features: Centipedes subdue their prey with venom; the bite of a centipede can therefore be very painful.

Apply It Now

Your Tiny Companions

Even when you think you are alone, you aren't; your body may host a diverse assortment of arthropods. Head lice and body lice are biting insects that cause skin irritation. Ticks latch onto the skin and suck your blood, sometimes transmitting the bacteria that cause Lyme disease. The tiny larvae of chigger mites produce saliva that digests small areas of skin tissue, causing intense itching.

Lice, ticks, and chiggers are hard to ignore, but you may never notice one inconspicuous companion: the follicle mite, *Demodex* (figure 20.A). This arachnid, which is less than half a millimeter long, lives in hair follicles and nearby oil glands, where it eats skin secretions and dead skin cells. *Demodex* mites are by no means rare. Nearly everyone has them, and each follicle may house up to 25 of the tiny animals. (If you would like to see your own follicle mites, carefully remove an eyebrow hair or eyelash and examine it with a compound microscope.) Luckily, the infestation is typically symptomless, although occasionally the mites may cause a rash.

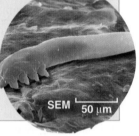

Figure 20.A Follicle Mites. Tiny *Demodex* mites live in skin pores and hair follicles.

Mandibulates: Crustaceans

Habitat: Mostly aquatic. Isopods, commonly known as pill bugs or "roly-polies," are the only terrestrial examples.

Body Structure: **Crustaceans** have mandibles, two pairs of antennae, two or three major body segments, and branched appendages (**figure 20.29**).

☐ Abdomen
☐ Cephalothorax

a.

b.

Figure 20.29 Crustaceans. (a) A lobster is a crustacean, as is (b) the horn-eyed ghost crab.

Figure 20.30 Insects. (a) This molting cicada is an insect, as is (b) this beetle and (c) this dragonfly.

Diversity: This group of about 52,000 species contains many familiar animals, including crabs, shrimp, and lobsters. Smaller crustaceans include terrestrial isopods, brine shrimp, water fleas (*Daphnia*), copepods, and barnacles.

Mandibulates: Insects

Insects colonized land shortly after plants, about 475 MYA, and they diversified rapidly in one of life's great adaptive radiations. Scientists know of well over 1 million species, with many more awaiting formal description. Why did so many species of insects evolve? Biologists surmise that the answer relates to the high reproductive rates of insects. In addition, mutations in homeotic genes can modify the body segments of insects into seemingly unlimited variations; some biologists liken the insect body plan to the versatility of a Swiss army knife. ▶ homeotic genes, p. 265

Habitat: Most insects are terrestrial, but some species live or reproduce in fresh water. The ocean, high altitudes, and extremely cold habitats are about the only places that are nearly devoid of insects.

Body Structure: All members of this group have mandibles and one pair of antennae. The body is divided into a head, thorax, and abdomen. Insects have six legs and usually two pairs of wings.

Diversity: Insect diversity almost defies description (figure 20.30 and table 20.2). They range in size from wingless soil-dwellers less than 1 mm long to fist-sized beetles, foot-long walking sticks, and flying insects with foot-wide wingspans. Some extinct dragonflies were even larger—one had a wingspan of about 75 cm!

Special Features: The evolution of flight was an important event in life's history. Insects were the first animals to fly, using their wings to disperse to new habitats, escape predators, court mates, and find food. Moreover, many of today's flowering plants

Table 20.2 Some Major Orders of Insects

Order	Examples
Thysanura	Silverfish
Ephemeroptera	Mayflies
Odonata	Dragonflies and damselflies
Orthoptera	Roaches, crickets, grasshoppers
Phthiraptera	Lice
Hemiptera	Cicadas, aphids
Coleoptera	Beetles
Hymenoptera	Ants, wasps, bees
Lepidoptera	Moths, butterflies
Diptera	True flies
Siphonaptera	Fleas

evolved in conjunction with flying insects, trading nectar for rapid, efficient pollination services. ▶ pollination, p. 493

20.8 | Mastering Concepts

1. What features distinguish the arthropods?
2. What are the main body regions of an arthropod?
3. How do arthropods use their jointed appendages?
4. Describe how arthropods feed, respire, excrete metabolic wastes, sense their environment, move, reproduce, and defend themselves.
5. What is the function of the exoskeleton?
6. In what ways are arthropods important?
7. How are chelicerates different from mandibulates?
8. Give an example of an animal in each subphylum of arthropods.

20.9 | Echinoderm Adults Have Five-Part, Radial Symmetry

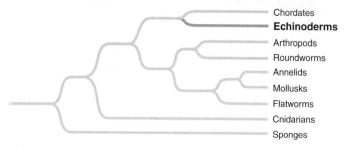

Chordates
Echinoderms
Arthropods
Roundworms
Annelids
Mollusks
Flatworms
Cnidarians
Sponges

The **echinoderms** (phylum Echinodermata) include some of the most colorful and distinctive sea animals. Their name literally means "spiny skin," a feature clearly visible in the sea urchins.

Habitat: Marine.

Body Structure: Adult echinoderms have radial symmetry, with the body divided into five parts (**figure 20.31**). This body form is most obvious in the adult sea stars and brittle stars, which usually have five arms. The other echinoderms lack arms, but close inspection reveals the five-part symmetry.

Another unique feature of echinoderms is the **water vascular system,** a series of enclosed, water-filled canals that end in hollow tube feet (**figure 20.32**). Coordinated muscle contractions

a.

b.

c.

d.

Figure 20.31 Echinoderm Diversity. (a) A sea star. (b) A sea cucumber. (c) A sea urchin. (d) A sand dollar.

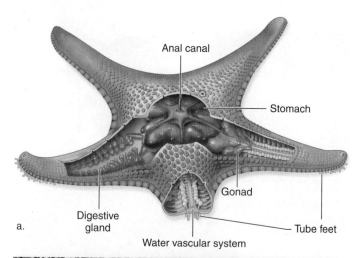

Anal canal

Stomach

Digestive gland

Gonad

Water vascular system

Tube feet

a.

b.

Figure 20.32 Echinoderm Anatomy. (a) The echinoderm water vascular system connects to tube feet. (b) The tube feet of this sea cucumber help the animal move along the ocean floor, acquire oxygen, and feel its surroundings.

extend and retract each foot, which can bend from side to side or create a suction-cup effect when applied to a hard surface.

Diversity: The five most common and familiar classes of echinoderms are sea lilies; sea stars; brittle stars; sea urchins and sand dollars; and sea cucumbers.

Feeding: Some echinoderms are predators. A sea star, for example, attaches its tube feet to a bivalve's shell and steadily pulls until the prey's muscles tire and the shell opens. The sea star then everts its stomach through its mouth, secretes digestive enzymes into the bivalve, and absorbs the liquefied food. Other echinoderms, such as sea cucumbers, eat dead organic matter; sea urchins scrape algae from rocks.

Circulatory and Respiratory Systems: Tube feet function like gills, absorbing oxygen from the water. Respiration may also occur by means of tiny skin gills (in sea stars) or by a network of passageways called a respiratory tree (in sea cucumbers).

Excretory System: The water vascular system helps flush metabolic wastes from echinoderm bodies.

Nervous System: Echinoderms lack heads and brains. Nerves extending down the arms and tentacles connect with a central nerve ring surrounding the gut. Tube feet can function as sense organs.

Burning Question

Are there really only nine kinds of animals?

The nine phyla described in this chapter represent a diverse cross-section of the 37 known animal phyla. But it would be a mistake to conclude that these nine groups contain the only important or interesting animals. The chapter opener described a tenth phylum, Cycliophora. Here are three additional invertebrate phyla, each representing microscopic animals with unusual features (figure 20.B):

Phylum Placozoa ("flat animals"): This phylum contains just one named species, *Trichoplax adhaerens*. An adult consists of a few thousand cells that are differentiated into just four cell types. The transparent, asymmetrical body resembles a microscopic sandwich, with an upper surface, a lower surface, and a connecting layer in between. Cilia enable the animal to glide or flow along a solid surface, like a slime mold. Biologists first discovered placozoans living in aquaria, and the phylogenetic position of these strange animals has been debated ever since. *Trichoplax* has the smallest genome of any animal, and its body plan is even simpler than that of a sponge. Yet genetic sequences suggest the placozoans are eumetazoans. ▶ slime molds, p. 360

Phylum Rotifera ("wheel bearer"): Like placozoans, the rotifers have tiny, transparent, unsegmented bodies. But the 2000 or so species of rotifers are considerably more complex, with bilateral symmetry and a complete digestive tract. These animals are named for the wheel-like tufts of cilia that sweep particles of decomposing organic matter into the mouth. Tiny hard "jaws" then grind the food. Rotifers inhabit fresh

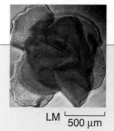

LM |—| 500 μm   LM |—| 150 μm   SEM |—| 300 μm (false color)

Figure 20.B Other Phyla. (a) *Trichoplax*. (b) Rotifer. (c) Tardigrade.

water and moist soil, and they can survive drying for years by suspending their metabolism. As with the placozoans, the phylogenetic position of rotifers remains controversial. The cuticle and pseudocoelom suggest a close relationship to nematodes, but molecular data place rotifers in a clade with annelids, mollusks, and flatworms.

Phylum Tardigrada ("slow walker"): These charismatic little animals are commonly called "water bears," owing to their aquatic habitat and overall shape. Their segmented bodies have eight legs, each ending in a claw. Covering the body is a cuticle of chitin, and the animal molts as it grows. These features place the tardigrades in a clade with arthropods and nematodes. More than 1000 species have been collected from habitats all over the world, from the poles to the tropics. Like rotifers, tardigrades can enter suspended animation, a state in which they can survive extreme cold or heat, desiccation, the vacuum of space, high pressure, or radiation doses that would kill a human.

Submit your burning question to:
marielle_hoefnagels@mcgraw-hill.com

Support and Movement: The wavelike pumping of water in and out of the tube feet allows echinoderms to glide slowly while maintaining a firm grip on the substrate, a clear advantage in a wave-pounded environment. Sand dollars and some sea cucumbers burrow into soft sediments, and some brittle stars can swim by using their appendages as oars. In addition, echinoderms have a unique, collagen-rich tissue that can rapidly interchange between soft and hard, allowing the animal to squeeze into tight spaces or, alternatively, stiffen to aid in feeding or to defend against predation.

Reproduction and Development: Echinoderms usually reproduce sexually, with male and female gametes from separate individuals

combining in the sea. The larvae start out with bilateral symmetry (figure 20.33), but then a group of cells assembles into a five-sided disc that turns inside out and consumes the remainder of the larva. The animal is now a tiny replica of the adult form.

Defense: Besides spines, the skin of echinoderms may also be equipped with small pincers that deter predators. In addition, sea stars have internal skeletal plates that are strong but lightweight. In sea urchins and sand dollars, these plates fuse into a protective shell. The soft-bodied sea cucumbers often produce poisonous chemicals, and many echinoderms can regenerate severed body parts.

Effects on Humans: People harvest some species of sea urchins for their gonads, an ingredient (uni) in Japanese sushi. Parts of some sea cucumbers are also edible. On the other hand, the crown-of-thorns sea star, *Acanthaster planci*, can cause painful wounds in humans who touch them. These animals eat corals and have wiped out large areas of Australia's Great Barrier Reef.

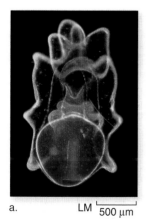
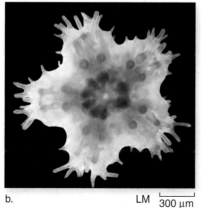

a. LM |—| 500 μm   b. LM |—| 300 μm

Figure 20.33 Echinoderm Metamorphosis. (a) This bilaterally symmetrical sea star larva does not resemble the adult organism. (b) As development proceeds, the five-part radial symmetry becomes apparent.

20.9 | Mastering Concepts

1. What characteristics distinguish the echinoderms?
2. Where do echinoderms live?
3. What is a water vascular system?
4. What are some examples of echinoderms?
5. How do echinoderms eat, respire, excrete metabolic wastes, sense their environment, move, reproduce, and defend themselves?
6. In what ways are echinoderms important?

20.10 | Most Chordates Are Vertebrates

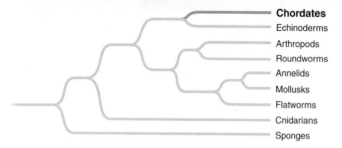

Chordates
Echinoderms
Arthropods
Roundworms
Annelids
Mollusks
Flatworms
Cnidarians
Sponges

Many people find phylum Chordata to be the most interesting of all, at least in part because it contains humans and many of the animals that we eat, keep as pets, and enjoy observing in zoos and in the wild. The **chordates** are a diverse group of at least 60,000 species (**table 20.3**). From the tiniest tadpole to fearsome sharks and lumbering elephants, chordates are dazzling in their variety of forms.

No one knows what the common ancestor of chordates was, but it was certainly an aquatic animal, and it apparently arose along with most other animal phyla during the Cambrian explosion. Like all other animals, the chordates began as invertebrates.

Table 20.3 Major Taxonomic Groups in Phylum Chordata

Group	Examples	Number of Existing Species
Tunicates (subphylum Urochordata)	Sea squirt	3000
Lancelets (subphylum Cephalochordata)	Amphioxus	30
Hagfishes (class Myxini)	Slime hag	60
Fishes (multiple classes)	Lamprey, shark, salmon, lungfish, coelacanth	30,000
Amphibians (class Amphibia)	Frog, salamander, caecilian	6000
Reptiles (class Reptilia and class Aves)	Turtle, lizard, snake, tuatara, crocodile, chicken, ostrich	8000 (nonavian reptiles); 9000 to 10,000 (birds)
Mammals (class Mammalia)	Platypus, kangaroo, dog, whale, human	5800

The most familiar chordates, however, are the vertebrates—fishes, amphibians, nonavian reptiles, birds, and mammals. All of these animals have an internal skeleton (endoskeleton) that includes a segmented backbone. ▸ Paleozoic life, p. 306

A. Four Features Distinguish Chordates

What sets Chordata apart from the other phyla? Every chordate has the following four features at some point during its life (**figure 20.34**):

1. **Notochord:** The notochord is a flexible rod that extends dorsally (along the back) down the length of a chordate's body. In most vertebrates, the notochord does not persist into adulthood but rather is replaced by the backbone that surrounds the spinal cord.

2. **Dorsal, hollow nerve cord:** The dorsal, hollow nerve cord is parallel to the notochord. In many chordates, the nerve cord develops into the spinal cord and enlarges at the anterior end, forming a brain. Many animals described earlier in this chapter have ventral nerve cords that are solid, not tubular. The dorsal position of the nerve cord is unique to the chordates. ▸ central nervous system, p. 540

3. **Pharyngeal pouches or slits:** In most chordate embryos, pouches or slits form in the pharynx, the muscular tube that begins at the back of the mouth. These structures have multiple functions. Invertebrate chordates feed by filtering food particles out of water that passes through the openings. In vertebrates, the pouches develop into gills, the middle ear cavity, or other structures.

4. **Postanal tail:** A muscular tail extends past the anus in all chordate embryos. In humans, chimpanzees, and gorillas, the tail normally withers away before birth, leaving only the tailbone as a vestige. (Rarely, a human baby is born with a tail, which is removed in minor surgery.) In fishes, salamanders, lizards, cats, and many other species, adults retain the tail.

(1) Notochord (2) Dorsal nerve cord (3) Pharyngeal slit (4) Postanal tail

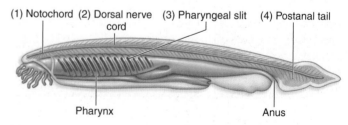

Pharynx

Anus

Figure 20.34 Defining Characteristics of a Chordate. At some time in its life, every chordate has each of these four features.

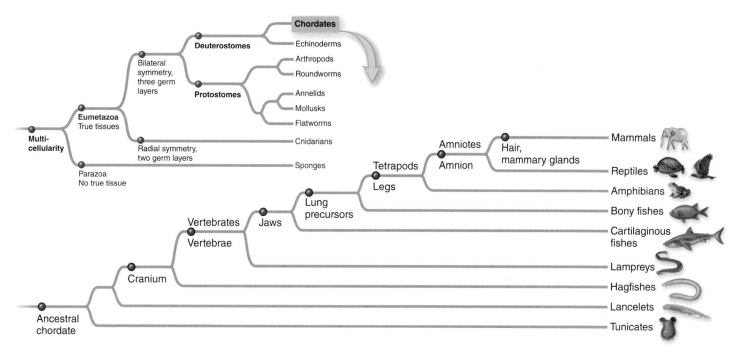

Figure 20.35 Chordate Diversity. Chordates include several groups of animals, including the invertebrate tunicates and lancelets and the better-known fishes, amphibians, reptiles (including birds), and mammals.

B. Biologists Use Many Features to Classify Chordates

Figure 20.35 depicts the evolutionary relationships within phylum Chordata. If you have previously studied the chordates, you may be surprised at the absence of a separate branch for birds in the phylogenetic tree. At one time, the nonavian reptiles such as snakes, lizards, turtles, and crocodiles were considered to form a separate clade from the birds. We now know, from molecular, fossil, and anatomical evidence, that birds are a type of reptile. Figure 20.35 reflects this new understanding.

This section explains the features that determine the branches on the chordate evolutionary tree, and the rest of this chapter considers each chordate group separately.

Cranium: In most chordates, a bony or cartilage-rich **cranium** surrounds and protects the brain (figure 20.36). Hagfishes and vertebrates form two clades of **craniates,** animals that have a cranium.

Vertebrae: Vertebrates are chordates that have a vertebral column, or backbone, composed of cartilage or bone (see figure 20.36). **Vertebrae** protect the spinal cord and provide attachment points for muscles, giving the animal a greater range of movement.

Jaws   **Jaws** are the bones that frame the entrance to the mouth. The development of hinged jaws from gill supports, shown in figure 20.37, greatly expanded the ways that animals could feed. In many vertebrate species, teeth or a beak are attached to the jaw.

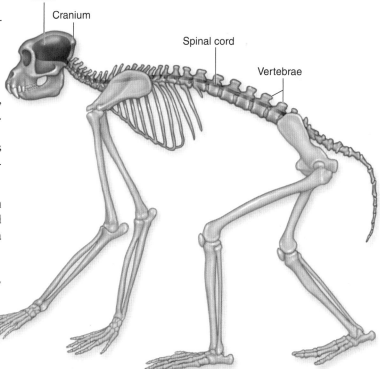

Figure 20.36 Skull and Backbone. A cranium and vertebrae are two of the characteristics that define the different groups of chordates.

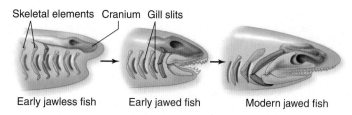

Skeletal elements Cranium Gill slits

Early jawless fish   Early jawed fish   Modern jawed fish

Figure 20.37 Origin of Jaws. Jaws developed from skeletal elements that supported gill slits near the mouth of the fish. The two elements closest to the mouth of the jawless fish were lost over time.

These features enhance the ability of the animal to grasp prey or gather small food items.

Lungs   Most fishes have gills that absorb O_2 from water and release CO_2. In contrast, most air-breathing vertebrates have internal saclike **lungs** as the organs of respiration. Lungs are homologous to the swim bladders of bony fishes. Both structures apparently arose from simple outgrowths of the esophagus. Originally, these sacs allowed fishes to gulp air; in the ancestors of terrestrial vertebrates, however, they developed into air-breathing lungs. (Section 30.1 shows gills and lungs in more detail).

Limbs   **Tetrapods** are vertebrates with two pairs of limbs that enable the animals to walk on land (*tetrapod* means "four legs"). Amphibians, reptiles (including birds), and mammals are all tetrapods. Some animals classified as tetrapods, however, have fewer than four limbs. Snakes, for example, lack

limbs entirely. The limbs of whales, dolphins, and sea lions are either modified into flippers or are too small to project from the body. Anatomical and molecular evidence, however, clearly links all of these animals to tetrapod ancestors (see figure 12.11 and section 20.17).

Amnion   The eggs of fishes and amphibians must remain moist, or the embryos inside will die. In contrast, reptiles (including birds) and mammals form a clade of **amniotes** that can breed in arid environments, in part because of the evolution of the **amniotic egg** (figure 20.38). Its leathery or hard outer layer surrounds a yolk that nourishes the developing embryo and enables it to survive outside of water. Also inside the egg are several membranes (the amnion, chorion, and allantois) that cushion the embryo, provide for gas exchange, and store metabolic wastes. These membranes are homologous to the protective structures that surround a developing fetus in the uterus of a female mammal.

Body Coverings   Fish scales derive from bone, and amphibians have naked, unscaly skin. In the three groups of amniotes, the body coverings all are composed of the same protein—keratin (figure 20.39). Nonavian reptiles such as snakes and crocodiles have dry, tough scales all over their bodies. Birds have similar scales on their legs; feathers cover the rest of the body. Mammals have fur or hair.

Additional Characteristics Distinguish Some Chordate Groups   The regulation of body temperature is an additional characteristic that is important in animal biology. The body

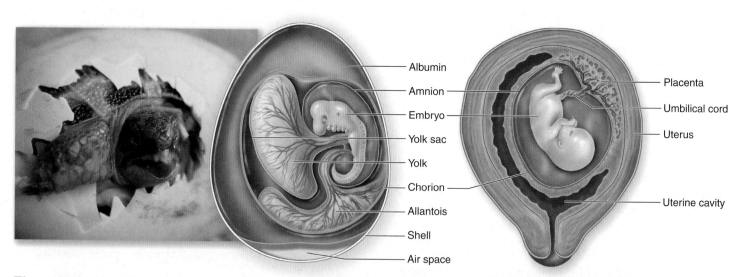

Albumin
Amnion
Embryo
Yolk sac
Yolk
Chorion
Allantois
Shell
Air space

Placenta
Umbilical cord
Uterus
Uterine cavity

Figure 20.38 The Amnion. The amnion is a sac that encloses the developing embryo of a reptile or mammal. In an amniotic egg, the embryo is encased in a hard, protective shell, and it is supported internally by three membranes—the amnion, allantois, and chorion. Placental mammals also enclose embryos in an amnion.

Figure 20.39 Body Coverings. Vertebrate body coverings include (clockwise from upper left) the bony scales of a fish; the dry, keratin-rich scales of a snake (a reptile); the fur of a mammal; the feathers of a bird; and the naked, unscaly skin of an amphibian.

temperature of an **ectotherm** tends to fluctuate with the environment; these animals have no internal mechanism that keeps their temperature within a narrow range. Invertebrates, fishes, amphibians, and nonavian reptiles are ectotherms. Many behaviors, such as basking in the sun or burrowing into the ground during the hottest part of the day, help an ectotherm adjust its temperature.

Birds and mammals are **endotherms,** which means they maintain their body temperature by using heat generated from their own metabolism. Endothermy requires an enormous amount of energy, which explains why birds and mammals must eat so much more food than ectotherms of the same size. Section 32.6 explores the origin of fur and feathers, which help endothermic animals conserve body heat.

The number of chambers in the heart is also important (see figure 29.3). Fishes have a two-chambered heart, with one atrium and one ventricle. In amphibians and most nonavian reptiles, the heart has two atria and one ventricle. Four chambers (two atria and two ventricles) make up the hearts of crocodiles, birds, and mammals. The more heart chambers, the better the separation of oxygen-rich blood from oxygen-poor blood, and the greater the efficiency with which blood delivers oxygen—a necessity in an energy-hungry endothermic animal. ▶ vertebrate circulatory systems, p. 603

In the first half of this chapter, the emphasis was on differences in the internal organ systems, which highlight the dramatic transitions between the simplest animals and the most complex. The rest of this chapter takes a slightly different approach. Most chordates have complex respiratory, digestive, excretory, and nervous systems. So we focus here on the evolutionary transitions between the main groups of chordates and on the diversity of animals within the phylum.

20.10 | Mastering Concepts

1. What are the four defining characteristics of chordates?
2. Which chordates are craniates, and which of those are also vertebrates?
3. How did the origin of jaws, lungs, limbs, and the amnion affect the course of vertebrate evolution?
4. How do the body coverings of fishes, amphibians, nonavian reptiles, birds, and mammals differ?
5. What is the difference between an ectotherm and an endotherm?
6. How does the number of heart chambers affect the efficiency of oxygen delivery to body tissues?

20.11 | Tunicates and Lancelets Have Neither Cranium nor Backbone

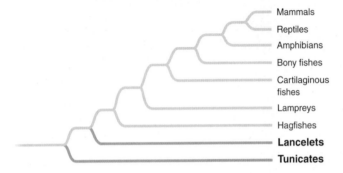

- Mammals
- Reptiles
- Amphibians
- Bony fishes
- Cartilaginous fishes
- Lampreys
- Hagfishes
- **Lancelets**
- **Tunicates**

Tunicates and lancelets form two subphyla of invertebrates that probably would attract little attention if not for one fact: they are the modern organisms that most resemble the ancestral chordates. Biologists continue to debate which group is more closely related to vertebrates.

The 3000 species of **tunicates** take their name from the tunic, a protective, flexible body covering that the epidermis secretes. The best studied tunicates, the ascidians, are marine animals that resemble a bag with two siphons (figure 20.40). Cilia pull water in through one siphon. As the water moves across gill slits in the pharynx, oxygen diffuses into nearby blood vessels, and carbon dioxide diffuses out. Mucus covering the gill slits traps suspended food particles, and the water exits through the other siphon. These animals are also called sea squirts because they can forcibly eject water from their siphons if disturbed.

The free-swimming tunicate larva resembles a tadpole, and it has all four chordate characteristics. Once it settles headfirst onto a solid surface, however, the tail and notochord disappear, and the nerve cord shrinks to nearly nothing. The adults, which are usually sessile, retain only the pharyngeal slits. Neither adult nor larva is segmented.

Some invasive tunicate species are a major nuisance in coastal areas. Huge colonies of rapidly reproducing sea squirts coat dock pilings and boats, and they smother desirable shellfish species.

Lancelets (also called amphioxus) resemble small, eyeless fishes with translucent bodies (figure 20.41). They live in shallow seas, with their tails buried in sediment and their mouths extending into the water. Cilia on the gills, coupled with the mucus secreted by the pharynx, trap and move food particles into the digestive tract. The filtered water passes out through the gill slits, leaving the body through a separate pore. Blood distributes nutrients to body cells, but gas exchange occurs directly across the skin. To reproduce, males and females release gametes into the water; the larvae, like the adults, resemble tiny fish.

Biologists have identified some 30 species of lancelets. These animals clearly display all four major chordate characteristics, as well as inklings of the organ systems that appear in the vertebrates. Like vertebrates, lancelets have segmented blocks of muscles. Furthermore, the lancelet nervous system consists of a nerve cord with a slight swelling at the head end, plus sensory receptors on the body. The simple lancelet brain appears to share some of the same divisions as the more elaborate vertebrate brain. However, these animals lack the sophisticated sensory organs, complex brains, and mobility of vertebrates.

20.11 | Mastering Concepts

1. How do tunicates use their siphons in feeding and gas exchange?
2. What is the relationship among tunicates, lancelets, and the vertebrate chordates?

Figure 20.40 Tunicates. On the surface, a tunicate looks as simple as a sponge, but its internal anatomy reveals a complete digestive tract, a heart, and other complex organs. The photo at right shows invasive tunicates along the Washington coast.

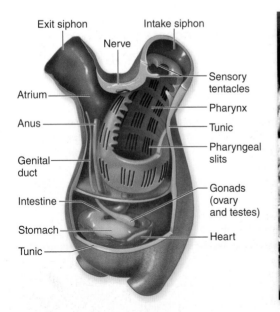

Exit siphon — Nerve — Intake siphon — Sensory tentacles — Pharynx — Tunic — Pharyngeal slits — Gonads (ovary and testes) — Heart — Atrium — Anus — Genital duct — Intestine — Stomach — Tunic

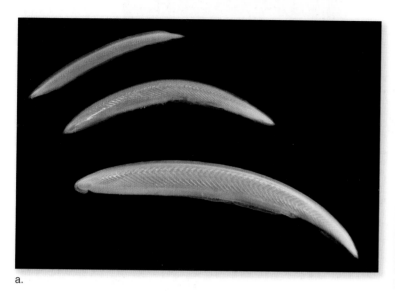

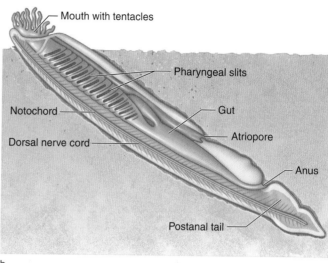

a.

b.

Figure 20.41 Lancelets. (a) The faint stripes on these lancelets are formed by blocks of muscle. The largest lancelets are about 8 cm long. (b) This lancelet is in its suspension-feeding posture, with its head sticking out of the substrate and its tail buried.

20.12 | Hagfishes Have a Cranium but Lack a Backbone

The common ancestor of the craniates may have resembled a **hagfish.** The long, slender hagfish looks something like an eel (**figure 20.42**), and hagfish skin is even marketed as "eel skin" in boots and wallets. But hagfishes are not eels. In a hagfish, cartilage makes up the notochord and supports the tail. But because vertebrae do not surround the nerve cord, hagfishes are not vertebrates. On the other hand, an eel is a true fish—a vertebrate.

Hagfishes live in cold ocean waters, eating marine invertebrates such as shrimp and worms or using their tongues to scavenge the soft tissues of dead or near-dead animals. Tentacles near the mouth help locate food.

These animals have some unusual abilities. They can slide their flexible bodies in and out of knots to pull on food, escape predation, or clean themselves. Hagfishes are also called "slime hags," in recognition of the glands that release copious amounts of a sticky white slime when the animal is disturbed.

Scientists once classified the 60 or so species of hagfishes with other jawless fishes (see section 20.13), but molecular evidence and the fact that hagfishes lack vertebrae justify their place in a separate subphylum.

20.12 | Mastering Concepts

1. How is a hagfish different from a true fish?
2. How do hagfishes eat and defend themselves?

Figure 20.42 Hagfishes. These marine animals secrete a sticky slime and scavenge dead animals, including whales.

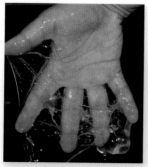

20.13 | Fishes Are Aquatic Vertebrates with Gills and Fins

- Mammals
- Reptiles
- Amphibians
- **Bony fishes**
- **Cartilaginous fishes**
- **Lampreys**
- Hagfishes
- Lancelets
- Tunicates

Fishes are the most diverse and abundant of the vertebrates, with more than 30,000 known species that vary greatly in size, shape, and color. They occupy nearly all types of water, from fresh to salty, from clear to murky, and from frigid to warm, although they cannot tolerate hot springs.

Fishes play important roles in their aquatic habitats. They graze on algae, scavenge dead organic matter, or prey on other animals, eating everything from mosquito larvae and other small invertebrates to one another. Tuna and many other fish species are also an important source of dietary protein for people (and their pets) on every continent. Angling for trout, bass, salmon, and other fishes remains a popular sport. Fishes also inspire a wide range of emotions, from an intense fear of sharks to the peace and tranquility that come from watching tropical fish in a home aquarium.

A. Fishes Changed the Course of Vertebrate Evolution

Fishes originated some 500 MYA from an unknown ancestor. Several features arose in this group that would have profound effects on vertebrate evolution. A segmented backbone, with its multiple muscle attachment points, expanded the range of motion. Jaws opened new feeding opportunities, which in turn selected for a more complex brain that could develop a hunting strategy or plan an escape route.

Two of the adaptations that enabled vertebrates to thrive on land originated in fishes: lungs and limbs. Lungs developed in a few species of fishes, and the air-breathing descendants of these animals eventually colonized the land. No fish has true limbs, but some fishes have fins with stronger bones and more flesh than the delicate, swimming fins of other fishes. These robust fins may have enabled tetrapod ancestors to move along the sediments of their shallow water homes. Whatever their original selective advantage, those fins eventually evolved into the limbs that define tetrapods.

B. Fishes May or May Not Have Jaws

Biologists recognize two groups of jawless fishes that are only distantly related to one another. The jawed fishes diversified into three main clades.

Figure 20.43
Jawless Fish.
A lamprey has a distinctive sucker mouth.

Jawless Fishes *Lampreys.* The modern jawless fishes include **lampreys** (figure 20.43). Lampreys are the simplest organisms to have cartilage around the nerve cord, so they are most like the first true vertebrates. They spend most of their lives as larvae, straining food from the water column. The adults feed on small invertebrates, although some species use their suckers to consume the blood of fish. Over the past century, some lampreys have ventured beyond their natural Lake Ontario range into the other Great Lakes, where they have been largely responsible for the decline in populations of lake trout and whitefish. ▶ invasive species, p. 812

Ostracoderms (Extinct). The earliest vertebrates to leave fossil evidence were jawless fishes called **ostracoderms.** These animals were filter feeders that lived on the ocean floor. They had notochords and brains, and their bodies were encased in bony, armorlike plates. They lived during the Cambrian period, but they vanished when fishes with biting jaws diversified during the Silurian period. Their phylogenetic placement remains controversial, but they may have been more closely related to the jawed fishes than are the lampreys.

Jawed Fishes *Placoderms (Extinct).* Around the time that ostracoderms lived, another group of fishes arose: the **placoderms.** Like the ostracoderms, the placoderms had a notochord and armor. They also had another feature: fearsome jaws, suggesting a predatory lifestyle.

Cartilaginous Fishes. The **cartilaginous fishes** include sharks, skates, and rays (figure 20.44). This clade arose some 450 MYA and has persisted even as the placoderms became extinct. As the name implies, their skeletons are made of cartilage. Although some sharks feed on plankton, the carnivorous species are notorious for their ability to detect blood in the water. They also have a **lateral line,** which is a sense organ extending along both sides of the fish. The lateral line is a series of canals that detect vibration in nearby water, helping the animal to find prey and escape predation. Some species of cartilaginous fishes must swim continuously to keep water flowing over their gills.

a.

b.

Figure 20.44 Cartilaginous Fishes. This group includes (a) skates and rays and (b) sharks.

Bony Fishes. The **bony fishes** form a clade that includes 96% of existing fish species. They have skeletons of bony tissue reinforced with mineral deposits of calcium phosphate (**figure 20.45**). Like sharks, bony fishes have a lateral line system. Unlike cartilaginous fishes, however, the bony fishes have a hinged gill covering that can direct water over the gills, eliminating the need for constant swimming. In addition, by expanding or contracting the volume of its swim bladder, a bony fish can adjust its buoyancy.

Bony fishes are divided into two classes (**figure 20.46**):

- The **ray-finned fishes** include nearly all familiar fishes: eels, minnows, catfish, trout, tuna, salmon, and many others. Their diversity and abundance are testament to their superb adaptations to a watery world.

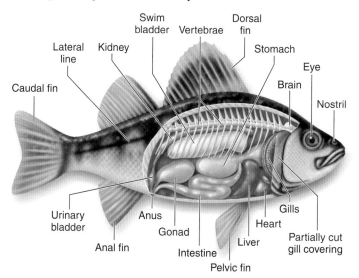

Figure 20.45 Anatomy of a Bony Fish. This illustration shows some of a fish's adaptations to life in water, including fins, a swim bladder, and gills.

a.

b.

c.

Figure 20.46 Bony Fishes. Most fishes belong to this group, which includes (a) ray-finned fishes, (b) lungfishes, and (c) coelacanths.

- The **lobe-finned fishes** are the bony fishes most closely related to the tetrapods, based on the anatomical structure of their fleshy paired fins. This group includes the lungfishes and the coelacanths. **Lungfishes** have lungs that are homologous to those of tetrapods. During droughts, a lungfish burrows into the mud beneath stagnant water, gulping air and temporarily slowing their metabolism. **Coelacanths** are called "living fossils"; they originated during the Devonian and remain the oldest existing lineage of vertebrates with jaws.

20.13 | Mastering Concepts

1. What features do all fishes share?
2. Which structures in fishes were the precursors to limbs?
3. What are the two types of jawless fishes? Which is extinct and which still exists?
4. What are the three groups of jawed fishes?
5. How were placoderms different from ostracoderms?
6. What are the major types of cartilaginous and bony fishes?

20.14 | Amphibians Lead a Double Life on Land and in Water

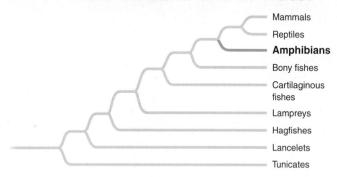

Mammals
Reptiles
Amphibians
Bony fishes
Cartilaginous fishes
Lampreys
Hagfishes
Lancelets
Tunicates

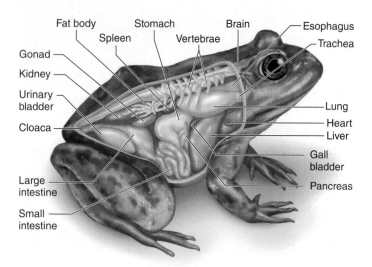

Figure 20.47 **Anatomy of a Frog.** Although amphibians have adaptations to life on land, their small lungs, thin skin, and reproductive requirements tie them to fresh water.

The word **amphibian** is Greek for "double-life," referring to the ability of these tetrapod vertebrates to live in fresh water and on land. Amphibians are probably the least familiar vertebrates because most people do not eat them or keep them as pets. Nevertheless, they are important in ecosystems, controlling both algae and populations of insects that transmit human disease. Their chorus of mating calls, ranging from squeaks to grunts and croaks, adds ambience to springtime evenings in many areas. Scientists are also studying toxins in amphibian skin as possible painkilling drugs.

Many biologists are concerned about dramatic declines in amphibian populations worldwide. The destruction of wetlands and forests devastates amphibians' breeding areas, and infection with a chytrid fungus has killed many of them outright (see sections 19.2 and 39.6). Collecting animals for the pet trade also takes a toll, as described in this section's Apply It Now box.

In addition, amphibians have porous skin that makes them especially vulnerable to pollution. They are therefore useful as indicators of environmental quality.

A. Amphibians Were the First Tetrapods

It is easy to envision how gulping lungfishes might have foreshadowed the lungs of amphibians or how the fleshy fins of lobe-finned fishes might have become legs. New fossil finds continue to fill in the pieces of the fish–amphibian transition, which occurred about 375 MYA (see figure 14.17 and section 20.17). ▶ Paleozoic life, p. 306

Life on land provided amphibian ancestors with space, protection, food, and plentiful oxygen, but it also presented new challenges. The animals faced wider swings in temperature, and delicate gills collapsed without the buoyancy of water. The new habitat therefore selected for new adaptations (figure 20.47). Lungs improved, and circulatory systems (including a three-chambered heart) grew more complex and powerful. The skeleton became denser and better able to withstand the force of gravity. Natural selection also favored acute hearing and sight, with tear glands and eyelids keeping eyes moist.

Yet amphibians retain a strong link to the water. Amphibian eggs, which lack protective shells and membranes, will die if they dry out. Also, the larvae (tadpoles) respire through external gills, which require water. Although most adults have lungs, many supplement oxygen intake by gas exchange through their skin, which must therefore remain moist.

B. Amphibians Include Three Main Lineages

The amphibians are grouped into three orders: frogs and toads; salamanders and newts; and caecilians (figure 20.48).

Figure 20.48 **Amphibian Diversity.** (a) This tree frog is native to tropical forests. (b) Salamanders are small amphibians with tails and four delicate legs. (c) Caecilians are legless amphibians that resemble earthworms or snakes.

Apply It Now

Wild-Caught Pets

Wild animals may seem to make fascinating and unique pets, but owning one is not usually a good idea. Many arguments justify laws against owning animals caught in the wild:

- **Population pressures:** Many amphibians and other wild animals are already endangered by habitat loss and pollution. Collecting for the pet trade only adds to the threat of extinction. The harvest of exotic coral reef fishes, for example, depletes their populations and damages the entire reef ecosystem. Trading in wild-caught parrots and macaws likewise threatens their native populations. Huge numbers of North American box turtles are captured for the pet trade each year. These animals have low reproductive rates, so it is difficult for the wild populations to recover. In addition, long life spans make owning a box turtle a major commitment. Unfortunately, many wild-caught animals die before reaching pet owners.

- **Threats to native species:** Virtually every animal in the pet industry is an introduced species, including birds, ferrets, gerbils, sugar gliders, snakes, lizards, amphibians, fishes, and invertebrates. If an exotic pet escapes or if its owner releases it, the animal may prey on, compete with, or spread disease to native organisms. Goldfish, for example, have become a nuisance species in lakes of the Pacific Northwest.

- **Physical dangers:** Large animals such as crocodiles and large cats represent a physical threat to their owners, becoming more dangerous as they grow.

- **Disease:** Wild-caught animals can carry disease. In 2003, monkeypox spread among people who kept prairie dogs as pets. The prairie dogs caught the disease from African rodents imported into the United States. Wild birds can also carry diseases that can infect native birds, poultry, and people. Likewise, hamsters, turtles, and iguanas can carry *Salmonella*.

Frogs and Toads   Most amphibian species are **frogs** or toads. Adults have large mouths, and they are "neckless"; that is, their heads are fused to their trunks. Their bodies lack both tails and scales.

In most species, the female frog lays her eggs directly in the water as a male clasps her back and releases sperm. The fertilized eggs hatch into legless, aquatic tadpoles. Most tadpoles feed on algae, although some species are omnivores. As they mature, tadpoles typically undergo a dramatic change in body form—a metamorphosis. They develop legs and lungs, lose the tail, and acquire carnivorous tastes.

Frogs and toads have remarkable adaptations that help them avoid predation. Whereas some have convincing camouflage, others display vibrant colors that warn of toxins secreted from glands in the skin. The poison dart frogs of Central and South America are famous for their toxins, which they acquire by eating ants and other prey. (Captive frogs fed diets lacking these toxins therefore eventually become nonpoisonous.) Frogs and toads also escape predation by jumping away, playing dead, and inflating their mouths so that a snake cannot swallow them.

Salamanders and Newts   **Salamanders** and newts have tails and four legs, so they resemble lizards. Unlike lizards, however, their skin lacks scales, their digits lack claws, and they always live near water. In most salamanders, the male deposits a

sperm packet inside the female's body; the female subsequently lays the fertilized eggs in water. Free-swimming larvae with a finlike tail hatch from the eggs. Both adults and young are carnivores, eating arthropods, worms, snails, fish, and other salamanders. In some groups of salamanders, the adults have larval features. The North American mudpuppy, for example, swims on pond bottoms and retains external gills.

Caecilians   The limbless **caecilians** resemble giant earthworms. Most species burrow under the soil in tropical forests, but a few inhabit shallow freshwater ponds. They are the only amphibians in which a male uses a sex organ to deliver sperm inside a female's body. Caecilians are carnivores, eating insects and worms.

20.14 | Mastering Concepts

1. What features do all amphibians share?
2. What role does water play in amphibian gas exchange and reproduction?
3. What features distinguish the three major orders of amphibians?
4. What are some reasons for the recent decline in amphibian populations worldwide?

20.15 | Reptiles Were the First Vertebrates to Thrive on Dry Land

- Mammals
- **Reptiles**
- Amphibians
- Bony fishes
- Cartilaginous fishes
- Lampreys
- Hagfishes
- Lancelets
- Tunicates

The changeability of scientific knowledge is evident in any modern discussion of **reptiles** and **birds.** The word *reptile* traditionally referred only to snakes, lizards, crocodiles, and other amniotes with dry, scaly skin. Birds had feathery body coverings and were considered a separate lineage. But that point of view has changed. We now know that birds form one of several clades of reptiles (**figure 20.49**). As a consequence, modern use of the term *reptile* must include both the nonavian reptiles and the birds.

Reptiles evolved from amphibians during the Carboniferous period, between 363 and 290 MYA. They dominated animal life during the Mesozoic era, until their decline beginning 65 MYA. Although many reptile species survived to the present day, many others are known only from fossils. The extinct groups include the terrestrial dinosaurs, the marine ichthyosaurs and plesiosaurs, and the flying pterosaurs. Of these, the dinosaurs especially capture the imagination (see the Burning Question in this section).
▶ Mesozoic life, p. 309

Unlike their amphibian ancestors, most reptiles have adaptations that enable them to live and reproduce on dry land (**figure 20.50**). Tough scales cover the skin, and the kidneys excrete only small amounts of water. In addition, internal fertilization

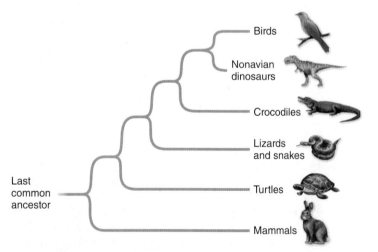

- Birds
- Nonavian dinosaurs
- Crocodiles
- Lizards and snakes
- Turtles
- Mammals

Last common ancestor

Figure 20.49 **Amniote Phylogeny.** Evidence suggests that birds form one of several clades of reptiles.

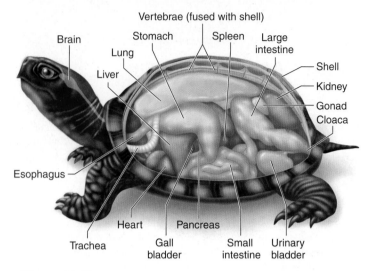

Figure 20.50 **Anatomy of a Reptile.** The skin, lungs, heart, and reproductive system of a tortoise adapt this animal to life on land.

meant that reptiles no longer deposited sperm in water. Their amniotic eggs are adapted to dry conditions (see figure 20.38). Finally, well-developed lungs and enhanced circulation increased their respiratory capacity beyond that of their aquatic ancestors.

? Burning Question

What were dinosaurs?

Dinosaurs were terrestrial reptiles that lived during the Mesozoic era, between 245 and 65 MYA. They ranged in stature from chicken-sized to truly gargantuan—the largest could have peeked into the sixth story of a modern-day building.

Biologists divide the hundreds of dinosaur species into two main lineages. One clade includes the beaked plant-eaters such as *Stegosaurus* and *Triceratops*. The other contains long-necked herbivores such as *Apatosaurus* and theropods, the only carnivorous dinosaurs; *Tyrannosaurus rex* belonged to this group, as did the ancestors of modern birds.

Many people mistakenly believe that all reptiles (or even mammals) that lived during the Mesozoic era were dinosaurs. In reality, lizards, snakes, crocodiles, and other terrestrial reptiles lived alongside the dinosaurs. The Mesozoic also saw marine reptiles such as plesiosaurs and flying reptiles such as pterodactyls, but they were not dinosaurs either; all dinosaurs were terrestrial. Nor did all dinosaurs live at the same time. As some species appeared, others went extinct throughout the Mesozoic.

Despite movie plots suggesting the contrary, humans and dinosaurs never coexisted. The last of the dinosaurs were gone by the time our own mammalian lineage—the primates—was just getting started. Tens of millions of years later, humans finally roamed the Earth and found the fossils that prove that these huge reptiles once existed.

Submit your burning question to
marielle_hoefnagels@mcgraw-hill.com

A. Nonavian Reptiles Include Four Main Groups

The orders of nonavian reptiles include turtles and tortoises; lizards and snakes; tuataras; and crocodilians (**figure 20.51**). Like fishes and amphibians, all of the nonavian reptiles are ectothermic.

Movies depict snakes, alligators, and crocodiles as terrifying killers, but most reptiles are inconspicuous animals that cannot harm people. Yet they are an important link in ecosystems, controlling populations of rodents and insects while providing food for owls and other predatory birds. In addition, some people keep snakes, lizards, or turtles as pets. In some parts of the world, reptiles have an even greater economic effect; the skins of farm-raised snakes and crocodiles are the raw material for boots, belts, and wallets, and some restaurants serve alligator meat.

Turtles and Tortoises   With a reputation for being "slow and steady," **turtles** and tortoises may not seem to have a recipe for success. Yet they have persisted in marine, freshwater, and terrestrial habitats since the Triassic period. Nowadays, habitat destruction threatens many land turtles, as does the pet trade.

The turtle's trademark feature is its shell, made of bony plates and a covering derived from the animal's epidermis. The

a.
b.
c.
d.
e.

Figure 20.51 Reptile Diversity. Nonavian reptiles include (a) turtles and tortoises, (b) lizards, (c) snakes, (d) tuataras, and (e) crocodiles and alligators.

shell's plates are fused to the animal's vertebrae and ribs, so it forms an integral part of the skeleton.

The largest member of this group is the giant leatherback sea turtle. These migratory animals are endangered worldwide because they are hunted for food, become trapped in fishing nets, swallow garbage, and ingest toxic chemicals. The destruction of nesting areas also takes its toll.

Lizards and Snakes   Almost 95% of nonavian reptile species are **snakes** or **lizards.** These animals are similar to each other, except that lizards usually have legs and snakes do not. Also, snakes never have external ear openings or moveable eyelids, and most lizards lack the distinctive forked tongue that characterizes snakes.

Several adaptations help lizards and snakes feed and avoid becoming food. A lizard might detach its tail as it escapes a predator. Camouflage and the ability to hold perfectly still enable snakes to surprise their prey, then subdue it by injecting venom or wrapping coils around the victim and squeezing the life out of it. Snakes then demonstrate a unique skill; they can unhinge their jaws, allowing them to swallow animals much larger than they are (see the opening figure for chapter 6).

Tuataras   **Tuataras** are reptiles that closely resemble lizards. The order containing tuataras once contained many species, all but two of which are now extinct. A captive breeding program may help restore populations of these endangered animals, which are native only to a few islands near New Zealand.

Crocodilians   The **crocodilians** (crocodiles, alligators, and their relatives) are carnivores that live in or near water. Their horizontally held heads have eyes on top and nostrils at the end of the elongated snout. Heavy scales cover their bodies; four legs project from the sides. Unlike most other nonavian reptiles, they have a four-chambered heart.

These reptiles look somewhat primitive, and indeed they are ancient animals, dating back to 230 MYA. Yet they have acute senses and complex behaviors, comparable to those of birds. For example, crocodilians lay eggs in nests, which the adults guard. The adults also care for the hatchlings, which stay with their mothers while they are young and call to the adults when they are in danger. Adults mark their territories by smacking their heads or jaws on the water surface. Like many birds and mammals, they even have dominance hierarchies.

B. Birds Are Warm, Feathered Reptiles

The behavioral similarities between crocodilians and birds are unsurprising in light of evolutionary history. Birds, dinosaurs, and crocodilians all belonged to a reptilian group called archosaurs, of which only the birds and crocodilians survive today.

Chickens and crocodiles look quite different, so what evidence suggests that birds really are reptiles? As described in chapters 12 and 13, *Archaeopteryx* and fossils of feathered reptiles provide important clues to the evolutionary history of birds, as do the skeletal similarities between birds and nonavian reptiles.

The bird's amniotic egg, with its hard calcium-rich shell, is another clue, as are the keratin-rich scales on a bird's legs.

Of course, birds have unique features that set them apart from other reptiles. Most birds can fly, although ostriches and some other species are flightless (figure 20.52). Anatomical adaptations to flight include a tapered body with a streamlined profile. Their lightweight bones are hollow, with internal struts that add support. The powerful four-chambered heart and unique lungs supply the oxygen that supports the high metabolic demands of flight (see figures 29.3 and 30.5). Wings provide lift in the air, while highly developed muscles power flight.

Birds are the only modern animals that have **feathers,** epidermal structures that provide insulation and enable birds to fly (figure 20.53). Feathers are also important in mating behavior, as anyone who has watched a peacock show off his plumage can attest. Like a snake's scales, a feather is built of the protein keratin. Different-sized and different-shaped feathers serve distinct functions, such as flight or insulation.

Like mammals and unlike all other reptiles, birds are endothermic. Paleontologists debate whether dinosaurs, the ancestors of birds, were endotherms or ectotherms. The posture, body size, habitats, growth rate, bone structure, and other features of dinosaurs seem to argue for the endothermic point of view, but the question remains open (see section 32.6).

Today, birds are a part of everyday human life. U.S. consumers eat hundreds of millions of chickens and turkeys every year, along with countless chicken eggs. We keep parrots, finches, canaries, and many other caged birds as pets, and we use bird feathers in everything from hats to down blankets. Songbirds enrich the lives of many birdwatchers, and hunters pursue wild turkeys, doves, and ducks. Birds are important in ecosystems as

a.   b.

Figure 20.52 **Two Groups of Birds.** (a) A kingfisher is a carinate, a bird with a keeled breastbone that supports powerful flight muscles. (b) This running ostrich is an example of a ratite, a flightless bird.

well. Some pollinate plants and disperse fruits and seeds, whereas others eat rodents, insects, and other vermin. But birds can also be pests. Starlings and pigeons are a nuisance in cities, fouling buildings and sidewalks with their droppings and speeding the rusting of bridges. Moreover, ducks and other domesticated birds transmit bird flu and other diseases to humans.

20.15 | Mastering Concepts

1. What features do all reptiles share?
2. How do scales and amniotic eggs adapt reptiles to land?
3. What features characterize each of the orders of nonavian reptiles?
4. What characteristics place the birds within the reptiles, and how are birds unique?
5. What are the functions of feathers?
6. What adaptations enable birds to fly?

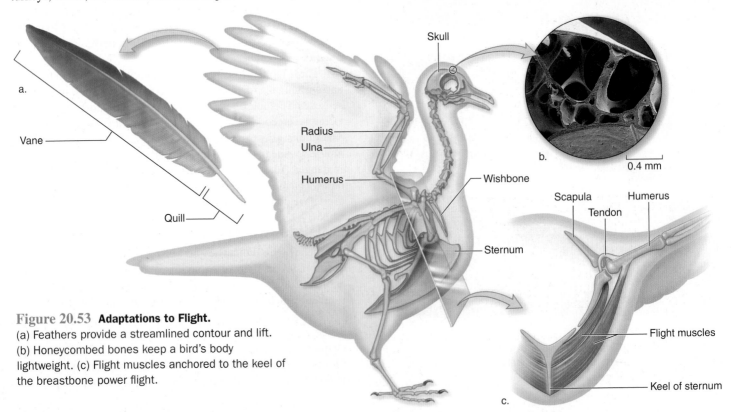

Figure 20.53 **Adaptations to Flight.**
(a) Feathers provide a streamlined contour and lift.
(b) Honeycombed bones keep a bird's body lightweight. (c) Flight muscles anchored to the keel of the breastbone power flight.

20.16 | Mammals Are Warm, Furry Milk-Drinkers

Mammals
Reptiles
Amphibians
Bony fishes
Cartilaginous fishes
Lampreys
Hagfishes
Lancelets
Tunicates

Mammals are by far the most familiar vertebrates, not only because we *are* them but also because we surround ourselves with them. We keep dogs, cats, rabbits, gerbils, hamsters, and ferrets as pets. Farmers raise many types of mammals for their meat or milk, including cows, pigs, and goats. Lambs and adult sheep provide meat and wool, and leather from cows makes up everything from upholstery to shoes. Horses, oxen, mules, and dogs are important work animals. Many people enjoy hunting deer for food and sport, and trappers kill mink, fox, beaver, and other mammals for their fur.

Mammals also play important roles in ecosystems. Coyotes, wolves, and foxes keep populations of herbivorous deer, rodents, rabbits, and other mammals in check. Some bats eat countless insects, and others pollinate plants. On the other hand, pests such as rats, mice, and skunks thrive alongside human populations. Some transmit diseases, including hantavirus and rabies.

A. Mammals Share a Common Ancestor with Reptiles

Mammals arose before the end of the Triassic period (about 200 MYA). DNA sequence similarities and the existence of egg-laying mammals suggest that mammals and reptiles share a common ancestor. Fossil evidence also provides clues to their origin. Mammals and their immediate ancestors are **synapsids;** that is, they have a single hole on each side of the skull, behind the eye orbits. (Birds and other reptiles have two such openings on each side.) One group of synapsids, called therapsids, had

Focus on Model Organisms

Mus musculus, the mouse

The history of the mouse, *Mus musculus,* as the stereotypical lab animal dates back to the early 1900s. Researchers discovered that the mouse's small size made it an excellent research animal. Also, mice are famous for their prolific breeding. They reach sexual maturity at the age of about 4 weeks, are sexually receptive every few days, and give birth to litters of 1 to 10 pups after a gestation of only about 3 weeks. Over a life span of 1.5 to 3 years, a single pair of mice can produce hundreds of offspring.

Researchers have benefited from biotechnology in their studies of mice. Transgenic mice have been available since the 1980s (see chapter 7). These mice are modified in countless ways, including altered susceptibility to human diseases. Another biotechnology, cloning, was applied to mice in 1998, making possible the production of genetically identical animals ideal for testing new disease treatments. ▶ cloning, p. 169

The mouse genome sequence was completed in 2002, revealing about 30,000 genes divided among 20 chromosomes. Not surprisingly, more than 99% of mouse genes have counterparts in the human genome. This similarity has made possible some of the following ways in which *Mus musculus* has contributed to biological research:

- **Immune function:** In the 1930s, the discovery that mice reject transplants from all but their very close relatives led to the discovery of the major histocompatibility complex (MHC). Since that time, biologists have discovered an array of genes related to immune function (see chapter 33).

- **Human disease:** Mice have been used to study human disease since the 1930s, when researchers discovered that mice could contract yellow fever. Vaccines were subsequently tested in mice. The availability of transgenic mice has opened new possibilities for research on the cause and treatment of human disease, including muscular dystrophy, Alzheimer disease, obesity, Parkinson disease, cancer, and HIV/AIDS. ▶ cancer, p. 162; ▶ HIV, p. 328

- **X chromosome inactivation:** In the 1960s, biologist Mary Lyon proposed that in female mammals, one of the two X chromosomes is inactivated early in embryonic development. This phenomenon, which is now often illustrated using calico cats, was first proved in mice with mottled coats. ▶ X inactivation, p. 215

- **Stem cells:** These undifferentiated cells, which can be derived from embryos or adults, can specialize into many other cell types. Mouse stem cell research has shown great promise in treating spinal cord injuries and many other ailments. ▶ stem cells, p. 167

Figure 20.54 **Monotremes.** (a) Duck-billed platypus. (b) Echidna.

characteristics of both mammals and reptiles; mammals are the only remaining members of the therapsid clade.

Most mammals were small until after the mass extinction that occurred 65 MYA. The loss of so many reptile species paved the way for the rapid diversification of many larger species of mammals. At the same time, the rise of flowering plants provided new food sources and habitats for mammals and for the arthropods that mammals eat. ▶▶ adaptive radiation, p. 282; Cenozoic life, p. 310

The word *mammal* derives from the Latin *mammae* for "breast," and refers to the milk-secreting **mammary glands** of the female. Infant mammals are nourished by their mother's milk. Mammals are also distinguished from other vertebrates by hair. Hair is composed of keratin and helps conserve body heat. Even whales and dolphins have hair at birth, but they lose it as they mature into their streamlined shapes.

Because fur and mammary glands do not leave fossils, ancient mammals are distinguished from reptilian fossils in other ways. A mammal has three middle ear bones compared with the reptilian one or two, and the lower jaw consists of one bone, compared with the reptile's several. Mammalian teeth are distinctive, too, with four types: molars, premolars, canines, and incisors. In reptiles, all the teeth are the same.

A few other features also characterize living mammals. First, like birds, mammals have a four-chambered heart, which evolved independently in the two groups. Second, the outer layer of the mammalian brain is very well developed, enabling mammals to learn, remember, and even think. As a result, many mammals

have unparalleled ability to plan and purposefully respond to stimuli. Third, only mammals have a dome-shaped muscular diaphragm that separates the chest cavity from the abdominal cavity. Contracting the muscles of the diaphragm draws air into the lungs. Birds and other reptiles use different mechanisms to ventilate their lungs.

B. Mammals Lay Eggs or Bear Live Young

Biologists divide mammals into two subclasses, one containing egg-laying mammals (monotremes) and the other containing live-bearing mammals. The latter subclass is further subdivided into two clades, the marsupials and the placental mammals.

The **monotremes** are mammals that lay eggs, such as the duck-billed platypus and echidnas of Australia and New Guinea (figure 20.54). The name of this group comes from their distinctive anatomy: the urinary, digestive, and reproductive tracts share a single ("mono") opening to the outside of the body. Reptiles have a similar anatomy.

The amniotic eggs in this group form another direct link to the reptiles. When the young hatch from the eggs, they are helpless, tiny, hairless, and blind. A newborn monotreme crawls along its mother's fur until it reaches milk-secreting pores in the skin. After a few months of suckling, the young leave the safety of their burrows and begin hunting their own food.

Marsupials, such as kangaroos and opossums, give birth to tiny, immature young about 4 to 5 weeks after conception (figure 20.55). In many marsupials, the babies crawl from the mother's vagina to a **marsupium,** or pouch, where they suckle milk and continue developing. Some species, however, have poorly developed pouches or none at all. In these marsupials, the nipples are exposed, and only the mother's body protects the suckling young.

The **placental mammals,** also called eutherians, are the most diverse of the three groups (figure 20.56). Placental mammals have much longer pregnancies than do marsupials. Females carry their young inside the uterus, where a **placenta** connects the maternal and fetal circulatory systems (see figure 20.38). The placenta nourishes and

a.   b.

Figure 20.55 **Marsupials.** (a) Kangaroo. (b) Opossum.

a.   b.

c.

Figure 20.56 Placental Mammals. (a) A human baby derives nourishment from his mother's milk. (b) Dolphins are aquatic mammals. (c) Bats are flying mammals.

removes wastes from the developing offspring. (The mouse, *Mus musculus,* is a placental mammal and the topic of this chapter's Focus on Model Organisms.)

Placental mammals diversified and displaced most marsupials early in the Cenozoic era, which began 65 MYA. Australian marsupials, however, continued to diversify long after marsupials on other continents had gone extinct; section 12.3 describes how continental drift explains this biogeographical pattern.

Bats, rodents, cetaceans (whales and dolphins), carnivores (dogs and cats), hoofed mammals, elephants, and many other mammals are eutherians. Humans are placental mammals as well. We belong to an order, Primates, that arose some 60 MYA. Section 14.4 describes how this group gave rise to the characteristics and abilities that distinguish our own species.

20.16 | Mastering Concepts

1. Which characteristics define a mammal?
2. What evidence suggests that mammals share a common ancestor with reptiles?
3. How do monotremes, marsupials, and placental mammals differ in how they reproduce?

20.17 | Investigating Life: Limbs Gained and Limbs Lost

Every now and then, a spectacular fossil grabs headlines worldwide. That is exactly what happened—twice—in April 2006, when fossils shed light on two important questions in vertebrate evolutionary biology: how did tetrapods gain their limbs, and how did snakes lose them?

The idea that early amphibians crawled onto land some 375 MYA is intriguing, especially since the descendants of those early colonists are today's amphibians, reptiles, and mammals. Fossils discovered over the past half-century, including *Acanthostega* and *Ichthyostega,* have clarified the fish-amphibian transition (see figure 14.17). Still, some details of this fascinating event remain poorly understood.

Scientists Edward T. (Ted) Daeschler of Philadelphia's Academy of Natural Sciences, Neil Shubin of the University of Chicago, and Harvard University's Farish Jenkins added new insights when they published back-to-back papers in the journal *Nature* in April 2006. The articles described fossils of an extinct animal, *Tiktaalik roseae,* that the researchers had unearthed in Arctic Canada (figure 20.57). *Tiktaalik* either crawled or paddled in shallow tropical streams about 380 MYA, during the late Devonian period. (Although today's Canada is anything but tropical, the entire North American continent was near the equator during the Devonian.)

Scientists jokingly call the animal a "fishapod" because of its uncanny mix of fish and tetrapod characteristics. Like a fish,

a.

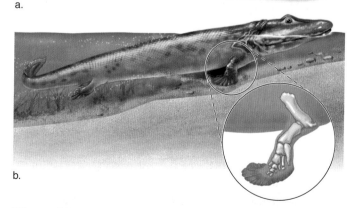

b.

Figure 20.57 Fossil "Fishapod." (a) This photo shows a portion of the *Tiktaalik* fossil discovered in 2006. (b) *Tiktaalik*'s limbs clearly contained bones, yet they were fringed with fins— one clue to the animal's aquatic heritage.

Tiktaalik had scales and gills. Like a tetrapod, it had lungs, and its ribs were robust enough to support its body. It could also move its head independently of its shoulders, something that a fish cannot do. But the appendages got the most attention. *Tiktaalik* had moveable wrist bones that were sturdy enough to support the animal in shallow water or on short excursions to land. Although the bones were clearly limblike, the "limbs" were fringed with fins, not toes.

The *Tiktaalik* fossils caught the world's eye because they were extraordinarily complete and exquisitely preserved. Scientifically, *Tiktaalik* is important for two reasons. First, it adds to our knowledge about tetrapod evolution. Second, it highlights the predictive power of evolutionary biology. The researchers did not simply stumble on *Tiktaalik* by accident. Instead, they were looking for a fossil representing the fish–amphibian transition, based on previous knowledge of how, when, and where vertebrates moved onto land hundreds of millions of years ago. Finding *Tiktaalik* confirmed the prediction.

Fossils have answered the question of how tetrapods got their limbs, but until recently, the same was not true for another important issue in vertebrate evolution: how did snakes *lose* their limbs? Molecular and anatomical information, including vestigial legs in some snakes (see figure 12.11), clearly indicated that snakes evolved from lizards. Yet the precise four-legged ancestor remained elusive.

Scientists proposed two competing hypotheses to explain the origin of snakes. Noting that snakes resemble two existing groups of burrowing lizards, some scientists suggested that snakes evolved on land. Opposing scientists, citing skull and jaw similarities, contend that snakes descend from mosasaurs, marine reptiles that thrived during the Cretaceous period.

Argentinian paleontologist Sebastián Apesteguía, along with Brazilian colleague Hussam Zaher, added a critical clue to the debate over snake origins in April 2006, when they reported finding three fossilized snakes in the Patagonia region of Argentina (**figure 20.58**). The snakes, which they named *Najash rionegrina*, lived during the Upper Cretaceous period, about 90 MYA.

Najash is different from other ancient snakes for at least two reasons. First, it is the first snake ever found to have not only functional legs and a pelvis but also a sacrum—a bone connecting the pelvis to the spine. (In other fossil snakes with limbs, the pelvis is "free floating" and not connected to the backbone.) The sacrum is important because lizards and other tetrapods have the same bone. *Najash* is therefore more primitive than any snake ever found (see figure 20.58b). Second, both the fossil's features and the rock where it was found suggest that it was terrestrial, not marine. Taken together, these two pieces of evidence seem to settle the matter: snakes originated on land.

Spectacular fossils such as *Tiktaalik* and *Najash* spark a flurry of excitement that obscures the countless hours of tedious, labor-intensive work needed to interpret fossils. Researchers scrutinize the scales, limbs, skull, jaw, teeth, vertebrae, and other parts of each new find to glean every possible piece of information about the animal and how it lived. The analysis may confirm previous knowledge, as in the case of *Tiktaalik*. Or, as with *Najash*, it may help settle a long-standing argument. Either way, fossils contribute immeasurably to our understanding of life's long history.

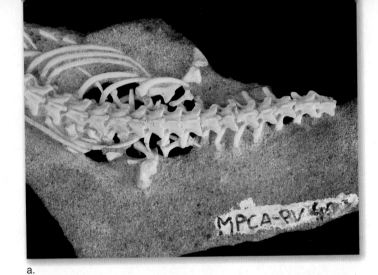

a.

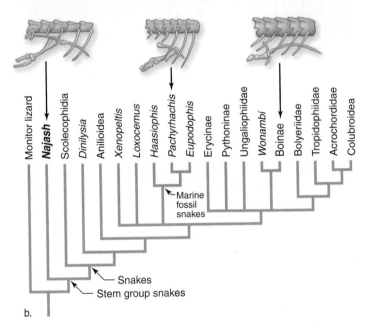

b.

Figure 20.58 *Najash*, **the Fossil Snake.** (a) *Najash* had hindlimbs, a pelvis, and a sacrum. (b) This phylogenetic tree places *Najash* as one of the most primitive snakes, whereas the three marine fossil snakes with legs (*Pachyrhachis*, *Haasiophis*, and *Eupodophis*) belong to a more derived group.

Apesteguía, Sebastián, and Hussam Zaher. April 20, 2006. A Cretaceous terrestrial snake with robust hindlimbs and a sacrum. *Nature*, vol. 440, pages 1037–1040.

Daeschler, Edward B., Neil H. Shubin, and Farish A. Jenkins, Jr. April 6, 2006. A Devonian tetrapod-like fish and the evolution of the tetrapod body plan. *Nature*, vol. 440, pages 757–763.

Shubin, Neil H., Edward B. Daeschler, and Farish A. Jenkins, Jr. April 6, 2006. A pectoral fin of *Tiktaalik roseae* and the origin of the tetrapod limb. *Nature*, vol. 440, pages 764–771.

20.17 | Mastering Concepts

1. How might the ability to crawl on land for short periods have enhanced the reproductive fitness of *Tiktaalik*?

2. How might the loss of hind limbs enhance the reproductive fitness of a burrowing animal such as *Najash*?

Animal Characteristics: A Summary

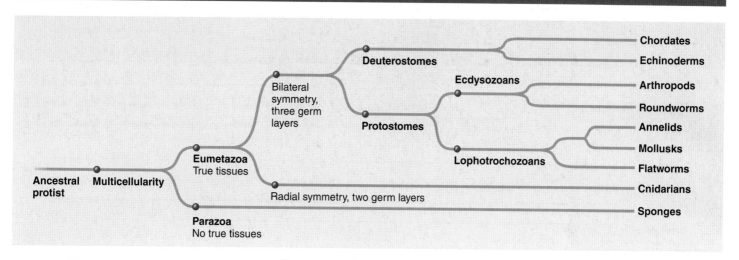

Ancestral protist — Multicellularity

Eumetazoa True tissues

Parazoa No true tissues

Bilateral symmetry, three germ layers

Radial symmetry, two germ layers

Deuterostomes

Protostomes

Ecdysozoans

Lophotrochozoans

Chordates
Echinoderms
Arthropods
Roundworms
Annelids
Mollusks
Flatworms
Cnidarians
Sponges

■ **Chordates**
- ~60,000 species
- Notochord, dorsal hollow nerve cord, pharyngeal pouches, postanal tail
- Fishes, amphibians, reptiles (including birds), mammals

■ **Echinoderms**
- ~7000 species
- Five-part symmetry
- Water vascular system
- Sea stars, sea urchins, sand dollars

■ **Arthropods**
- >1,000,000 species
- Jointed appendages; exoskeleton
- Trilobites, spiders and other arachnids, crustaceans, insects, millipedes, centipedes

■ **Roundworms**
- ~80,000 species
- Unsegmented, cylindrical worms
- Free-living or parasitic

■ **Annelids**
- ~15,000 species
- Segmented worms
- Earthworms, polychaetes, leeches

■ **Mollusks**
- ~112,000 species
- Mantle secretes shell
- Chitons, bivalves, cephalopods, gastropods

■ **Flatworms**
- ~25,000 species
- Flat bodies
- Marine flatworms, planarians, flukes, tapeworms
- Free-living or parasitic

■ **Cnidarians**
- ~11,000 species
- Hydra, jellyfish, coral
- Medusa or polyp forms
- Stinging cells

■ **Sponges**
- ~5000 species
- Porous bodies
- Filter feeders

Phylum	Level of Organization	Symmetry	Cephalization	Coelom	Digestive Tract	Segmentation
Porifera (sponges)	Cellular	Asymmetrical or radial	No	No	Absent	No
Cnidaria	Tissue	Radial	No	No	Incomplete	No
Platyhelminthes (flatworms)	Organ system	Bilateral	Yes	No	Incomplete (when present)	No
Mollusca	Organ system	Bilateral	Yes	Yes	Complete	No
Annelida	Organ system	Bilateral	Yes	Yes	Complete	Yes
Nematoda (roundworms)	Organ system	Bilateral	Yes	Pseudocoelom	Complete	No
Arthropoda	Organ system	Bilateral	Yes	Yes	Complete	Yes
Echinodermata	Organ system	Bilateral larvae; radial adults	No	Yes	Complete	No
Chordata	Organ system	Bilateral	Yes	Yes	Complete	Yes (usually)

Chordate Characteristics: A Summary

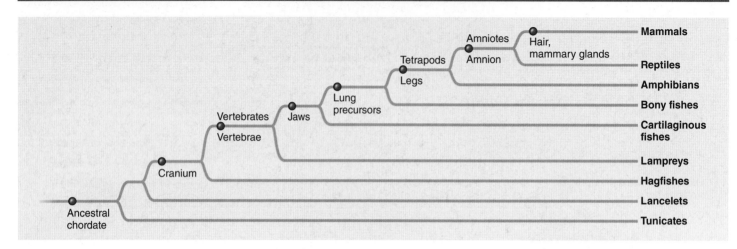

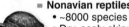

■ **Mammals**
- ~5800 species
- Hair/fur and mammary glands
- Amnion surrounds developing embryo
- Monotremes, marsupials, placentals

■ **Birds**
- ~10,000 species
- Feathers and hollow bones
- Amniotic eggs
- Carinates, ratites

■ **Nonavian reptiles**
- ~8000 species
- Dry, scaly skin; amniotic eggs
- Turtles, lizards, snakes, tuataras, crocodilians

■ **Amphibians**
- ~6000 species
- Scale-less tetrapods
- Can live on land but reproduce in water
- Frogs, salamanders, caecilians

■ **Fishes**
- ~30,000 species
- Scale-covered bodies with gills and fins
- Jawed fishes include cartilaginous fishes (sharks and rays) and bony fishes

■ **Hagfishes**
- ~60 species
- Wormlike bodies
- Marine carnivores
- "Slime hag"

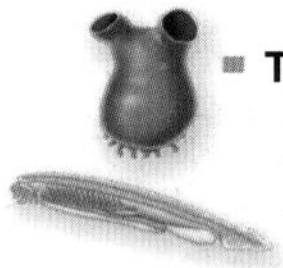

■ **Tunicates and lancelets**
- ~3000 species (tunicates)
- ~30 species (lancelets)
- Invertebrate filter feeders
- Sea squirts and amphioxus

Subphylum/ Class	Cranium	Vertebrae	Jaws	Skeleton	Lungs	Limbs	Amnion	Body Temperature Regulation	Number of Heart Chambers
Tunicates and lancelets	No	No	No	Absent	No	No	No	Ectotherm	–
Hagfishes	Yes	No	No	Cartilage	No	No	No	Ectotherm	–
Fishes	Yes	Yes	Yes (usually)	Cartilage or bone	No (usually)	No	No	Ectotherm	2
Amphibians	Yes	Yes	Yes	Bone	Yes	Yes	No	Ectotherm	3
Nonavian reptiles	Yes	Yes	Yes	Bone	Yes	Yes	Yes	Ectotherm	3 (4 in crocodilians)
Birds	Yes	Yes	Yes	Bone	Yes	Yes	Yes	Endotherm	4
Mammals	Yes	Yes	Yes	Bone	Yes	Yes	Yes	Endotherm	4

Chapter Summary

20.1 | Animals Live Nearly Everywhere

- Of the 1,200,000 or so species in kingdom **Animalia,** most are **invertebrates;** only one phylum contains **vertebrates,** which have a segmented backbone of cartilage or bone.

A. The First Animals Likely Evolved from Protists

- The immediate ancestor of animals was likely a choanoflagellate.
- The earliest fossil evidence of animals is from about 570 MYA. These animals lived in water, and existing animal diversity strongly reflects this aquatic heritage.

B. Animals Share Several Characteristics

- Animals are multicellular, eukaryotic heterotrophs whose cells secrete extracellular matrix but do not have cell walls. Most digest their food internally.
- The **blastula** is a stage in embryonic development that is unique to animals.

C. Biologists Classify Animals Based on Organization, Morphology, and Development

- Animal bodies exhibit degrees of organization into tissues, organs, and organ systems. **Eumetazoans** are animals with true tissues.
- In most phyla, body symmetry is **radial** or **bilateral. Cephalization** is correlated with bilateral symmetry.

- An animal zygote divides mitotically to form a blastula and then usually a **gastrula.** In some animals, the gastrula has two tissue layers (**ectoderm** and **endoderm**). In others, a third layer (**mesoderm**) forms between the other two.
- Bilaterally symmetrical animals are **protostomes** if the gastrula's first indentation forms into the mouth. In **deuterostomes,** the first indentation develops into the anus.

D. Biologists Also Consider Additional Characteristics
- Biologists also describe animals based on the presence or absence of a body cavity (**coelom** or **pseudocoelom**). The body cavity can act as a **hydrostatic skeleton.**
- Digestive tracts are **incomplete** or **complete;** a **gastrovascular cavity** is an incomplete digestive tract.
- **Segmentation** improves flexibility and increases the potential for specialized body parts.
- Most animals reproduce sexually. The resulting embryo may undergo **direct development** or **indirect development,** in which **metamorphosis** transforms a **larva** into an adult.

20.2 | Sponges Are Simple Animals That Lack Differentiated Tissues
- **Sponges** are asymmetrical or radially symmetrical. Their porous bodies filter small food particles out of water. Although they lack tissues, sponges have specialized cell types, including collar cells and amoebocytes.
- A sponge's skeleton consists of spicules or organic fibers (or both).
- Sponges reproduce sexually and may bud asexually.

20.3 | Cnidarians Are Radially Symmetrical, Aquatic Animals
- **Cnidarians** are mostly marine animals that capture prey with tentacles and stinging **cnidocytes.** They digest food in a gastrovascular cavity. Cnidarians move by contracting muscle cells that act on a hydrostatic skeleton.
- A cnidarian body form is a **polyp** or a **medusa.**
- Examples of cnidarians include corals, hydras, and jellyfishes.
- In species that alternate between polyp and medusa forms, sexual reproduction produces a larva, which attaches and becomes a polyp. The polyp reproduces asexually to generate a colony that yields medusae.

20.4 | Flatworms Have Bilateral Symmetry and Incomplete Digestive Tracts
- **Flatworms** are unsegmented animals that lack a coelom. The flat shape allows individual cells to exchange gases with their environment.
- These animals include **free-living flatworms** such as planarians; **flukes** and **tapeworms** are parasitic.
- Flatworms lack circulatory and respiratory systems, but protonephridia maintain water balance. They have simple nervous systems and hydrostatic skeletons. They reproduce asexually and sexually, and most are hermaphrodites.

20.5 | Mollusks Are Soft, Unsegmented Animals
- **Mollusks** have bilateral symmetry and a complete digestive tract. The main groups of mollusks are **chitons, bivalves, gastropods,** and **cephalopods.**
- The mollusk body includes a **mantle,** a muscular **foot,** and a **visceral mass.** Most mollusks have a shell, and many have a tonguelike **radula.** They are filter feeders, herbivores, or predators.
- Cephalopods have complex sensory and nervous systems.
- Sexes are usually separate in the mollusks; fertilization may be internal or external.

20.6 | Annelids Are Segmented Worms
- **Annelid** bodies consist of repeated segments. Annelids include **oligochaetes, leeches,** and **polychaetes,** and they feed in diverse ways.
- Organ systems include a complete digestive tract, a closed circulatory system, and respiratory, excretory, and nervous systems. The coelom acts as a hydrostatic skeleton.
- Leeches and oligochaetes are hermaphrodites; polychaetes have separate sexes.

20.7 | Nematodes Are Unsegmented, Cylindrical Worms
- **Roundworms** are unsegmented worms that molt periodically. They include parasitic and free-living species in soil and aquatic sediments.
- Nematodes have diverse diets and complete digestive tracts. The pseudocoelom is a hydrostatic skeleton. Most have separate sexes and undergo direct development.

20.8 | Arthropods Have Exoskeletons and Jointed Appendages
- **Arthropods** are segmented animals with jointed appendages and a chitin-rich **exoskeleton.**

A. Arthropods Have Complex Organ Systems
- Arthropods exhibit great diversity in feeding, respiratory systems, excretory systems, nervous systems, and reproduction. They have open circulatory systems.

B. Arthropods Are the Most Diverse Animals
- The extinct **trilobites** were arthropods, as are the **chelicerates** (**horseshoe crabs** and **arachnids**). **Mandibulate** arthropods include **centipedes** and **millipedes; crustaceans;** and **insects.**
- Insects are by far the most diverse arthropods.

20.9 | Echinoderm Adults Have Five-Part, Radial Symmetry
- **Echinoderms** are spiny-skinned marine animals whose adults have radial symmetry. Echinoderms are deuterostomes, as are the chordates.
- Echinoderms move by using tube feet, which are part of the **water vascular system.** This network of canals also aids in circulation and gas exchange.
- Most echinoderms reproduce sexually. The bilaterally symmetrical larvae look very different from the adults.

20.10 | Most Chordates Are Vertebrates

A. Four Features Distinguish Chordates
- **Chordates** share four characteristics: a **notochord,** a **dorsal hollow nerve cord, pharyngeal pouches** or slits in the pharynx, and a **postanal tail.**

B. Biologists Use Many Features to Classify Chordates
- The absence of a **cranium** distinguishes tunicates and lancelets from the **craniates** (hagfishes and vertebrates).
- Other features that distinguish chordates from one another include **vertebrae, jaws, lungs,** and the presence of limbs in **tetrapods.**
- Reptiles and mammals are **amniotes;** in these animals, the **amniotic egg** or amnion protects the developing embryo. The type of body covering is also important in classifying chordates.
- **Ectotherms** use the environment to regulate body temperature, whereas **endotherms** use heat from metabolism to maintain body temperature.
- Fishes have two heart chambers, amphibians and most nonavian reptiles have three, and birds and mammals have four.

20.11 | Tunicates and Lancelets Have Neither Cranium nor Backbone

- **Tunicates** obtain food and oxygen with a siphon system. The tunicate larva has all four chordate characteristics, but adults retain only the pharyngeal slits.
- **Lancelets** resemble eyeless fishes; adults have all four chordate characteristics.

20.12 | Hagfishes Have a Cranium but Lack a Backbone

- **Hagfishes** have a cranium of cartilage, but they are not vertebrates. They secrete slime and lack jaws.

20.13 | Fishes Are Aquatic Vertebrates with Gills and Fins

- **Fishes** are abundant and diverse vertebrates.

A. Fishes Changed the Course of Vertebrate Evolution

- Adaptations in fishes that allowed vertebrates to move onto land include lungs and fleshy, paired fins that were later modified as limbs. Jaws and a vertebral column also originated in fishes.

B. Fishes May or May Not Have Jaws

- Jawless fishes include the extinct **ostracoderms** and the modern **lampreys.**
- **Placoderms** are an extinct class of fishes with jaws.
- The **cartilaginous fishes** include skates, rays, and sharks. Sharks detect vibrations from prey with a **lateral line** system.
- The **bony fishes** account for 96% of existing fish species. Bony fishes have lateral line systems and swim bladders, which enable them to control their buoyancy. The two groups of bony fishes are the **ray-finned fishes** and **lobe-finned fishes,** which are further subdivided into **lungfishes** and **coelacanths.**

20.14 | Amphibians Lead a Double Life on Land and in Water

A. Amphibians Were the First Tetrapods

- **Amphibians** breed in water and must keep their skin moist to breathe. Adaptations to life on land include a sturdy skeleton, lungs, limbs, and a three-chambered heart.

B. Amphibians Include Three Main Lineages

- Amphibians include **frogs** and toads; **salamanders** and newts; and **caecilians.**

20.15 | Reptiles Were the First Vertebrates to Thrive on Dry Land

- **Reptiles** (including **birds**) have efficient excretory, respiratory, and circulatory systems. Internal fertilization and amniotic eggs permit reproduction on dry land.

A. Nonavian Reptiles Include Four Main Groups

- Nonavian reptiles include **turtles** and tortoises; **lizards** and **snakes; tuataras;** and **crocodilians.**

B. Birds Are Warm, Feathered Reptiles

- Birds retain scales and egg-laying from reptilian ancestors. Honeycombed bones, streamlined bodies, and **feathers** are adaptations that enable flight.
- Like mammals but unlike the other reptiles, birds are endothermic.

20.16 | Mammals Are Warm, Furry Milk-Drinkers

A. Mammals Share a Common Ancestor with Reptiles

- **Mammals** have fur, secrete milk from **mammary glands,** and have distinctive teeth and highly developed brains. They are **synapsids.**

B. Mammals Lay Eggs or Bear Live Young

- **Monotremes** are mammals that hatch from an egg. The young of **marsupial** mammals are born after a short pregnancy and often develop inside the mother's **marsupium. Placental mammals** have longer pregnancies; the young are nourished by the **placenta** in the mother's uterus.

20.17 | Investigating Life: Limbs Gained and Limbs Lost

- *Tiktaalik roseae* is an extinct "fishapod." Its fossils clearly reflect the transition between fishes and amphibians, which occurred about 375 MYA.
- Another fossil, *Najash rionegrina,* has helped researchers understand the origin of snakes. The features of *Najash* suggest that snakes arose on land from burrowing ancestors.

Multiple Choice Questions

1. Which animal phylum has bilateral symmetry but an incomplete digestive tract?
   - a. Porifera
   - b. Platyhelminthes
   - c. Nematoda
   - d. Cnidaria

2. Nematodes and arthropods are grouped together in the ecdysozoan clade because they
   - a. have a radula.
   - b. molt.
   - c. have a pseudocoelom.
   - d. have only two germ layers.

3. What is a key characteristic of all arthropods?
   - a. Six legs
   - b. Pseudocoelom
   - c. Hydrostatic skeleton
   - d. Exoskeleton

4. How is the body structure of an annelid different from an arthropod?
   - a. Annelids lack jointed appendages.
   - b. Annelids have a complete digestive tract.
   - c. Annelids have cephalization.
   - d. Annelids are bilateral.

5. When an earthworm is moving through the soil, its muscles are pushing against its
   - a. digestive tract.
   - b. coelom.
   - c. bony skeleton.
   - d. cnidocyte.

6. Arthropods molt because
   - a. they undergo indirect development.
   - b. the exoskeleton becomes damaged over time.
   - c. the exoskeleton prevents the organism from growing.
   - d. they have an open circulatory system.

7. Which of the following characteristics is associated with echinoderms?
   - a. Protostome
   - b. Two germ layers (ectoderm and endoderm)
   - c. Bilateral symmetry
   - d. All of the above are correct.

8. What is a *water vascular system*?
   - a. A network of canals that enables echinoderms to move
   - b. The circulatory system of an echinoderm
   - c. The excretory system of an echinoderm
   - d. All of the above are correct.

9. What is a notochord?
   - a. The spine of a chordate animal
   - b. One segment in a chordate's backbone
   - c. A type of germ layer
   - d. A fibrous rod that runs down the back of a chordate

10. Since a tunicate is considered to be a chordate, it must have a(n)
    - a. cranium.
    - b. notochord.
    - c. amniotic egg.
    - d. lung.

11. Lobe-finned fishes are important because they
    a. were the first vertebrates.   c. are closely related to tetrapods.
    b. were the earliest animals.   d. lack jaws.

12. Why is it important for amphibians to live in a moist habitat?
    a. To protect their scales   c. To maintain their buoyancy
    b. To maintain their eggs   d. Both a and b are correct.

13. How do reptiles and mammals differ from amphibians?
    a. Only reptiles and mammals are amniotes.
    b. Only reptiles and mammals are tetrapods.
    c. Only reptiles and mammals have lungs.
    d. All of the above are correct.

14. What is the evolutionary link between birds and crocodilians?
    a. Four-chambered hearts   c. Scales
    b. Egg laying   d. All of the above are correct.

15. Since a whale is a mammal, it must
    a. have scales.   c. produce milk.
    b. have gills.   d. All of the above are correct.

Write It Out

1. Compare the nine major animal phyla in the order in which the chapter presents them, listing the new features for each group.

2. You are visiting the aquarium, and your companion points at an animal in a tank. None of the signs shows a picture of the animal or its name. What criteria would you use to assign the animal to a phylum?

3. List the criteria used to distinguish: (a) animals from other organisms; (b) vertebrates from invertebrates; (c) protostomes from deuterostomes; (d) ectotherms from endotherms; (e) a tapeworm, a nematode, a slug, an earthworm, a snake, and a caecilian.

4. Distinguish between: (a) radial and bilateral symmetry; (b) blastula and gastrula; (c) direct and indirect development; (d) complete and incomplete digestive tract; (e) coelom and pseudocoelom.

5. Compare and contrast feeding in a sponge and a sea urchin.

6. Segmented animals occur in multiple phyla. How might segmentation benefit an animal? If segmentation is adaptive, why do unsegmented animals still exist?

7. Make a chart showing the characteristics of each subphylum of arthropods. Suppose you could examine a fossilized arthropod; use your chart to describe how you would assign the fossil to a subphylum.

8. What is the evidence for the surprisingly close relationship between echinoderms and chordates?

9. What are the four distinguishing characteristics of chordates?

10. How do tunicates and lancelets differ from fishes and tetrapods?

11. Draw from memory a phylogenetic tree that traces the evolutionary history of vertebrates. Include the features that mark each branching point in your tree.

12. If you found an eel-like animal at the beach, what features would you look for in deciding whether you had a hagfish or an eel (a true fish)?

13. List five adaptations that enable: (a) fishes to live in water; (b) amphibians to live on land; (c) snakes to live in the desert; (d) birds to fly.

14. Tunicates and lancelets have not left fossil evidence because their bodies lack hard parts. The hair and mammary glands that distinguish mammals are not hard either, yet mammals have a rich fossil record. Explain this difference.

15. How are a fish's and a bird's skeletons similar in structure and function?

16. Fishes are adapted to life in water, and tetrapods to life on land. Cite two criteria for assessing which group has been more successful.

17. Summarize the evidence for the idea that birds are reptiles. How does the changing placement of birds in the vertebrate family tree illustrate the scientific process? Why does this type of research matter?

18. If you found a fossil and were not sure whether it was from a reptile or a mammal, how might you tell the difference?

19. How are fishes, amphibians, nonavian reptiles, birds, and mammals important to humans? How are they important in ecosystems?

20. Among plants and vertebrates, some species are well-adapted to dry land, whereas others require moisture to reproduce. Compare and contrast the adaptations in plants and vertebrates that have allowed each group to breed in increasingly dry habitats.

21. Give three examples of interactions between animals classified in different phyla.

22. Explain how a sessile or slow-moving lifestyle, such as that of sponges, sea cucumbers, and tunicates, might select for bright colors and an arsenal of toxic chemicals.

23. Biologists have speculated that the first feathered dinosaurs may have used feathers in mating displays, to conserve body heat, or in flight. How might you use fossils, living species, or both to test each hypothesis?

24. Search the Internet to answer this question: Other than *Caenorhabditis elegans*, *Drosophila melanogaster*, and *Mus musculus*, what are some examples of invertebrate or vertebrate animals that have contributed to scientists' knowledge of general biology and animal biology? What other genomes of animals have scientists sequenced, and what are some resulting discoveries?

25. Invasive animal species are disrupting ecosystems around the world. Search the Internet for a list of invasive animal species. Which phyla are represented in the list? What harm do invasive species do? How important is it to try to eradicate invasive species?

Pull It Together

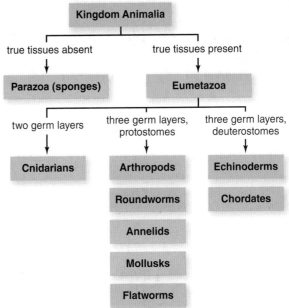

1. Which animals are cephalized?
2. Which animals have an incomplete digestive tract?
3. Which animals have segmented bodies?
4. Which animals have a coelom or pseudocoelom?
5. Draw a concept map that summarizes the chordates, including both invertebrates and vertebrates.

Cash Crop. Women harvest coca leaves in Peru.

Candy, Herbs, and Drugs: Plants Are Chemical Factories

FOOD, BEVERAGES, PAPER, LUMBER, TEXTILES, OILS, ROPE...MOST PEOPLE KNOW THAT THESE PRODUCTS COME FROM PLANTS. The cells of roots, stems, leaves, flowers, and fruits produce less conspicuous resources as well: a vast array of potent chemicals that people use every day.

Consider, for example, the refreshing flavor of peppermint. You may associate peppermint with candy, but the flavor originally comes from a plant. Adorning the leaves of peppermint plants are microscopic hairs that produce and store chemicals called terpenes. Menthol, the most abundant terpene in peppermint, produces a cooling sensation when eaten or inhaled. Spearmint, sage, basil, lavender, rosemary, thyme, and oregano are other aromatic herbs in the mint family; they produce different terpenes, each with its own distinctive scent.

Like menthol, hashish is also a leaf hair extract. Hashish is purified resin from the leaves and flowers of *Cannabis sativa*. A chemical called tetrahydrocannabinol (THC) produces the narcotic effect of hashish and marijuana. THC is part of a class of chemical compounds called phenolics.

Alkaloids are another class of powerful plant-derived compounds. Most people are familiar with alkaloids; capsaicin, for example, makes chili peppers taste "hot." Other examples include caffeine (from coffee beans), nicotine (from tobacco leaves), morphine and codeine (from opium poppy flowers), and cocaine (from coca leaves). Alkaloids can also have medicinal value. The antimalarial medicine quinine, for example, comes from the bark of Cinchona trees, and vincristine is an antileukemia drug from the leaves of a periwinkle, *Catharanthus roseus*.

Terpenes, phenolics, alkaloids, and many other potent chemicals are called secondary metabolites because they do not participate directly in photosynthesis, respiration, reproduction, or other essential processes. So why do plant cells invest energy to make them?

The answer lies in natural selection. Many secondary metabolites are bad-tasting or toxic to disease-causing organisms and herbivores. Quite simply, plants that defend themselves with these weapons retain more leaf area than those that don't. The more leaf area, the greater the potential for photosynthesis and, ultimately, the more resources available to produce offspring.

A chemical arsenal that repels or poisons hungry animals is just one defense against predation. As you'll see in this chapter, a plant's growth pattern and structural defenses play important roles as well.

UNIT 5

What's the Point?

Learning Outline

Learn How to Learn

Interpreting Images from Microscopes

Any photo taken through a microscope should include information that can help you interpret what you see. First, read the caption and labels so you know what you are looking at—usually an organ, a tissue, or an individual cell. Then study the scale bar and estimate the size of the image. Finally, check whether a light microscope (LM), scanning electron microscope (SEM), or transmission electron microscope (TEM) was used to create the image. Note that stains and false colors are often added to emphasize the most important features.

21.1 | Vegetative Plant Parts Include Stems, Leaves, and Roots

Imagine a rose bush that produces a bounty of sweet-smelling flowers. Biologists would divide your plant into two sets of parts. The gorgeous flowers, which will eventually give rise to fruits called rose hips, are the reproductive parts of the plant. Chapter 23 describes flowers and fruits in detail. The **vegetative,** or nonreproductive, parts are the roots, stems, and leaves.

This chapter and the next focus on the **anatomy** (form) and **physiology** (function) of a plant's vegetative organs. As you will see, a plant's **organs** are composed of multiple **tissues,** groups of cells that interact to provide a specific function. This section begins our tour of the plant body with an overview of its major organs.

A plant's vegetative organs work together (**figure 21.1**). The **shoot** is the aboveground part of the plant. The shoot's **stem** supports the **leaves,** which produce carbohydrates such as sucrose by photosynthesis. A large portion of this sugar moves down the stem and nourishes the **roots,** which are usually below ground. Root cells depend completely on the shoots to provide energy for their metabolism. At the same time, however, roots anchor the plant and absorb water and minerals that move via the stem to the leaves.

A close look at a stem reveals that it consists of alternating nodes and internodes. A **node** is a point at which one or more leaves attach to the stem. **Internodes** are the stem areas between the nodes. Along with at least one leaf, each node features an **axillary bud,** an undeveloped shoot that forms in the angle between the stem and leaf stalk. Although a bud has the potential to elongate to form a branch or flower, many remain dormant.

Informally, biologists divide plants into two categories based on the characteristics of the stem. A **herbaceous plant** has a green, soft stem at maturity. The chili plant in figure 21.1 is herbaceous, as are grasses, daisies, dandelions, and radishes. The stems of **woody plants** such as elm and cedar trees are made of tough wood covered with bark.

Natural selection ensures that roots, stems, and leaves do not always look exactly like those in figure 21.1. Depending on the habitat, plants may contend with everything from hungry animals to extreme drought to continuous flooding to frozen winters. These selective forces have sculpted the vegetative plant body into a tremendous diversity of forms.

Specialized root functions, for example, include storage, gas exchange, support, and even photosynthesis (**figure 21.2**). Beet and carrot roots stockpile starch, and desert plant roots may store water. In oxygen-poor habitats such as swamps, specialized roots grow up into the air, allowing oxygen to diffuse in. Orchids produce aerial roots that carry out photosynthesis. Thick buttress roots at the base of a tree provide support, as do prop roots that arise from a corn plant's stem.

Prop roots are examples of **adventitious roots,** which arise from any plant part other than the roots. If you have ever rooted a houseplant by placing a stem cutting in water, you have seen adventitious roots.

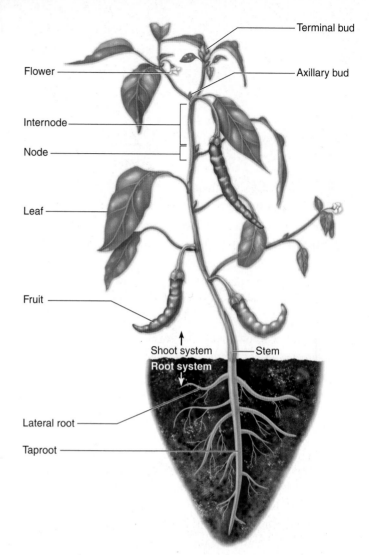

Figure 21.1 Parts of a Flowering Plant. A plant consists of a root system and a shoot system. Roots, stems, and leaves are vegetative organs; flowers and fruits are reproductive structures.

Stems may also have specialized functions (**figure 21.3**). The stems of climbing plants may form tendrils that coil around objects, maximizing exposure of the leaves to the sun. Stolons are horizontal stems that sprout from an existing plant and grow along the soil surface, asexually forming adventitious roots and new shoots at their nodes. Rhizomes are thickened underground stems that also produce shoots and adventitious roots. Tubers, such as potatoes, are swollen regions of stolons or rhizomes that store starch, whereas the succulent, fleshy stems of cacti stockpile water. Still other stems are protective; some types of thorns, such as those on hawthorn plants, are modified branches.

In addition to carrying out photosynthesis, leaves can have many other functions as well. Onion bulbs, for example, are collections of the fleshy bases of leaves that store nutrients. Cotyledons are embryonic leaves; in some species, they store carbohydrates that supply energy for seed germination. Cactus spines are modified leaves that deter predators. The sepals and petals of a flower are also modified leaves. And in a few carnivorous plant species, the leaves attract, capture, and digest prey, as described in chapter 22. ▶ flower parts, p. 493

a.

b.

c.

Figure 21.2 Modified Roots. (a) The roots of mangrove trees form pneumatophores, specialized structures that poke out of waterlogged soil and absorb oxygen from the air. (b) Aerial roots of orchids can carry out photosynthesis. (c) Adventitious prop roots on corn arise from the stem and support the plant.

Artificial selection has also modified stems, leaves, roots, flowers, and fruits to suit human needs. Table 21.1 shows a variety of plants that farmers cultivate for their edible parts. Most people divide these foods into two categories: fruits and vegetables. This chapter's Burning Question, on page 462, explains more about these overlapping terms.

The plants in table 21.1 are all angiosperms, or flowering plants. As described in chapter 18, the two largest clades of angiosperms are eudicots and monocots. Eudicots include everything from chili peppers to green beans to sunflowers to elm trees. Monocots include orchids, lawn grasses, corn, rice, wheat, and bamboo. The rest of this chapter describes the structural similarities and differences between the plants in these two clades.

21.1 | Mastering Concepts

1. How do stems, leaves, and roots support one another?
2. What is the relationship between the node and the internode of a stem?
3. Give examples of roots, stems, and leaves with specialized functions.

Figure 21.3 Modified Stems. (a) Some tendrils are stems that coil around objects, supporting and anchoring plants. This is a grapevine. (b) Iris leaves and roots arise from rhizomes. (c) The stem of the fishhook barrel cactus is the primary organ of photosynthesis, and it is highly modified to store water. Its thorns are modified leaves, not stems. (d) The thorns on this honey locust are outgrowths of the woody stem.

Table 21.1 Many Parts Are Edible

Plant Part*	Edible Example(s)
Vegetative	
Axillary (lateral) bud	Brussels sprout
Terminal bud	Cabbage
Bark	Cinnamon
Bulb	Onion, garlic
Leaf blade	Basil, parsley, spinach, chives, kale
Leaf petiole	Celery, rhubarb
Rhizome	Ginger "root"
Root	Carrot, parsnip, beet, sweet potato
Seedling	Alfalfa sprout, bean sprout
Stem	Asparagus, kohlrabi, sugar cane
Tuber	Potato
Reproductive	
Flower bud	Broccoli, cauliflower, artichoke
Style and stigma (flower parts)	Saffron
Seeds and fruits	Apple, raspberry, tomato, string bean, olive, cucumber, pumpkin, corn kernel, bell pepper, black peppercorn, rice, wheat, walnut, peanut, pea, nutmeg

*Why no wood? Because it's mostly empty cell walls made of cellulose and lignin, neither of which we can digest.

21.2 | Plants Have Flexible Growth Patterns, Thanks to Meristems

Consider the plight of a green plant. Rooted in place, it seems vulnerable and defenseless against drought, flood, wind, fire, and hungry herbivores. Yet plants dominate nearly every habitat on land. How do they do it? Some plant defenses come from their arsenal of secondary metabolites, as described at the start of this chapter. But in many ways, plants owe their success to modular growth.

A. Plants Grow by Adding New Modules

To understand modular growth, imagine a landowner who plans to build a motel. Money is tight at first, so she starts with just a few rooms. As her business grows, however, she adds more units to the basic plan. Plant growth is similar. Shoots become larger by adding units ("modules") consisting of repeated nodes and internodes.

Some plants, such as dandelions, stop growing after they reach their mature size, a pattern called **determinate growth.** Most plants, however, can grow indefinitely by adding module after module. Such **indeterminate growth** can persist as long as environmental conditions allow it.

Modular growth enhances a plant's ability to respond to the environment. For example, plants produce the most root tips in pockets of soil with the richest nutrients. Likewise, a shrub growing in partial shade can add new branches where it receives the most sunlight, while the shaded limbs remain unchanged.

In addition, modular growth means that the loss of a branch or root does not harm a plant as much as, say, the loss of a leg affects a cat. Neighboring branches can add modules to compensate for a broken tree limb, but the cat's body cannot regenerate a leg. Modular growth is one feature that distinguishes plants from animals.

B. Plant Growth Occurs at Meristems

All of a plant's new cells come from **meristems,** regions that undergo active mitotic cell division (see chapter 8). Meristems are patches of "immortality" that allow a plant to grow, replace damaged parts, and respond to environmental change.

Three main types of meristems occur in plants (table 21.2). **Apical meristems** are small patches of actively dividing cells near the tips of roots and shoots. When cells in the apical meristem divide, they give rise to new cells that differentiate into all of the tissue types described in section 21.3. **Primary growth** lengthens the root or shoot tip by adding cells produced by the apical meristems.

Many plants also have **lateral meristems,** which produce cells that thicken a stem or root. A lateral meristem is usually an internal cylinder of cells extending along most of the length of the plant. When the cells divide, they typically produce tissues to both the inside and the outside of the meristem. **Secondary growth** is the resulting increase in girth, a process described in section 21.5.

Table 21.2   Meristem Types: A Summary

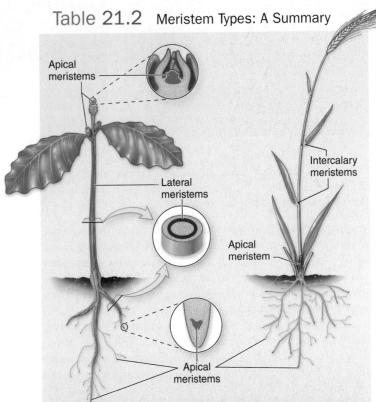

Type	Locations	Function
Apical	Terminal and axillary buds of shoots; root tips	Produces tissues that lengthen the tips of shoots and roots
Lateral	Internal cylinder along length of roots and stems of woody plants	Thickens roots and stems
Intercalary	Between nodes of mature stems in grasses and other monocots	Regrowth of tissue at base of stem or leaf if tip is removed

In some monocots, **intercalary meristems** occur between the nodes of a mature stem, usually at the base of an internode. (*Intercalary* means "inserted between," referring to the position of these meristems). Grasses, for example, tolerate grazing and mowing because they have intercalary meristems whose cells divide to regrow a leaf from its base when the tip is munched off.

21.2 | Mastering Concepts

1. What is the difference between determinate and indeterminate growth?
2. What are the locations and functions of meristems?

21.3 | Plant Cells Build Tissues

A cactus, an elm tree, and a dandelion have distinctly different stems, leaves, roots, and growth patterns. They may seem to have little in common, but a closer look reveals that all consist of the same types of cells and tissues. This section examines these building blocks. (It may be helpful to review the structure of a plant cell in figure 3.9 and of the primary and secondary cell wall in figure 3.25).

A. Plants Have Several Cell Types

Plants consist of several cell types, all arising from meristems. This section lists the most common ones.

Ground Tissue Cells   **Ground tissue** makes up the majority of the primary body of a plant. It consists of three main cell types: parenchyma, collenchyma, and sclerenchyma (figure 21.4).

Parenchyma cells are the most abundant cells in the primary plant body. Parenchyma cells are alive at maturity, and they retain the ability to divide, which enables them to differentiate in response to injury or a changing environment. It is parenchyma cells, for example, that divide to produce the adventitious roots in a house-plant cutting. Although structurally unspecialized, parenchyma cells have vital functions, including respiration, photosynthesis, and storage.

Collenchyma cells are elongated living cells with unevenly thickened primary walls that can stretch as the cells grow. These cells provide elastic support without interfering with the growth of young stems or expanding leaves. Collenchyma strands often form near the angular ridges on the stems of some plants, such as those in the mint family. Collenchyma is perhaps most familiar, however, as the tough, flexible "strings" in celery stalks.

Sclerenchyma cells provide inelastic support to parts of a plant that are no longer growing. These cells, which are dead at maturity, have thick, rigid secondary cell walls that occupy most of the cell's volume. The secondary walls of most sclerenchyma cells contain **lignin,** a tough, complex molecule that adds great strength to the cell walls.

Two types of sclerenchyma are fibers and sclereids. **Fibers** are elongated cells that usually occur in strands. Linen, for

Cell Type	Description	Alive at Maturity	Functions	
Parenchyma	• Most abundant cell type in primary plant body • Thin primary cell walls • Unspecialized • Can divide at maturity	Yes	Make up most nonwoody tissues; carry out photosynthesis, gas exchange, secretion, wound repair, and storage	Starch grain Buttercup root LM 125 µm
Collenchyma	• Elongated cells • Unevenly thickened primary cell walls	Yes	Elastic support for growing stems and leaves	Sunflower plant stem LM 40 µm
Sclerenchyma: Fiber	• Long, slender cells • Thick secondary cell walls high in lignin	No	Inelastic support for nongrowing plant parts	Spruce wood LM 580 µm
Sclerenchyma: Sclereid	• Variable shapes, generally not elongated • Thick secondary cell walls high in lignin	No	Inelastic support for nongrowing plant parts	Pear flesh LM 100 µm

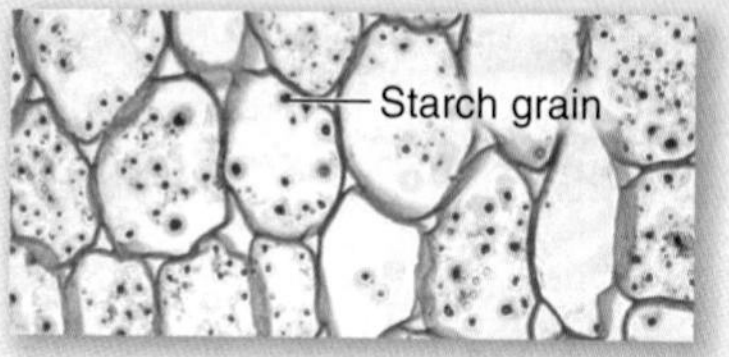

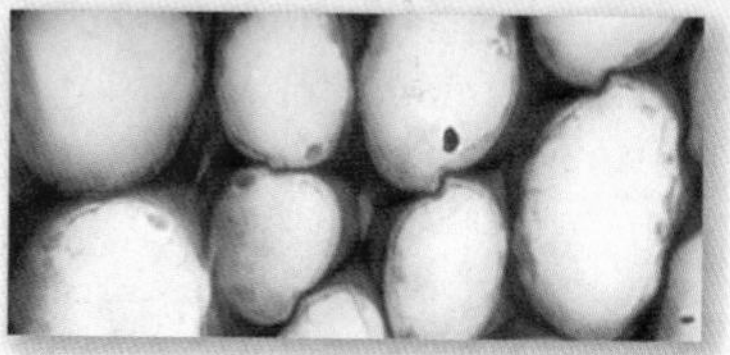

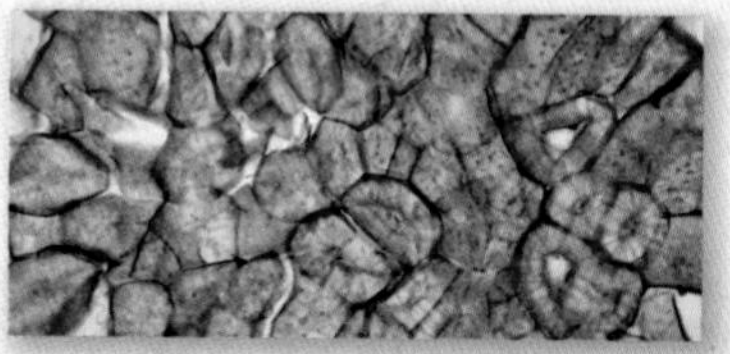

Figure 21.4 Plant Cell Types: A Summary. Parenchyma, collenchyma, and sclerenchyma cells make up the majority of the primary plant body.

example, comes from the soft fibers of the stems of *Linum usitatissimum*, or flax. Sisal, which comes from the leaves of *Agave sisalana*, is a hard fiber used in coarse fabrics and rope. The other sclerenchyma cells, **sclereids,** occur in many shapes, although they are generally shorter than fibers. Small groups of sclereids create a pear's gritty texture. Sclereids also form hard layers in nutshells, apple cores, and the pits of cherries and plums.

Conducting Cells in Xylem and Phloem

Vascular tissues transport water, minerals, carbohydrates, and other dissolved compounds throughout the plant. Two types of vascular tissue are xylem and phloem; figure 21.5 summarizes the most important cells in each.

Xylem transports water and dissolved minerals from the roots to all parts of the plant. The water-conducting cells of xylem are elongated and have thick, lignin-rich secondary walls. They are dead at maturity, which means that no cytoplasm blocks water flow.

The two kinds of water-conducting cells in xylem are tracheids and vessel elements. **Tracheids** are long, narrow cells that overlap at their tapered ends. Water moves from tracheid to tracheid through pits, where the secondary cell walls are absent and the primary cell walls are thin. Water moves slowly in tracheids because of their small diameters and overlapping end walls.

Vessel elements are short, wide, barrel-shaped conducting cells that stack end to end, forming long, continuous tubes. Their side walls have pits, but their end walls either are perforated or disintegrate completely. Water moves much faster in vessels than in tracheids, both because of their greater diameter and because water can pass easily through each vessel element's end wall. All vascular plants have tracheids, but only angiosperms also have vessel elements.

Phloem transports dissolved organic compounds, primarily sugars produced in photosynthesis. Phloem sap also contains proteins, RNA, hormones, ions, and sometimes viruses.

The main phloem conducting cells are **sieve tube elements,** which align end to end to make a single functional unit called a sieve tube. Most sieve tube elements have **sieve plates,** where modified plasmodesmata form large pores at the ends of the cells. Their other walls also have thin areas perforated by many pores. Strands of cytoplasm pass through the openings, allowing carbohydrates to pass from cell to cell. Sieve tube elements are alive at maturity but lack nuclei, many other organelles, and much of their cytoplasm.

Adjacent to each sieve tube element is at least one **companion cell,** a specialized parenchyma cell that retains all of its organelles. Companion cells not only transfer carbohydrates

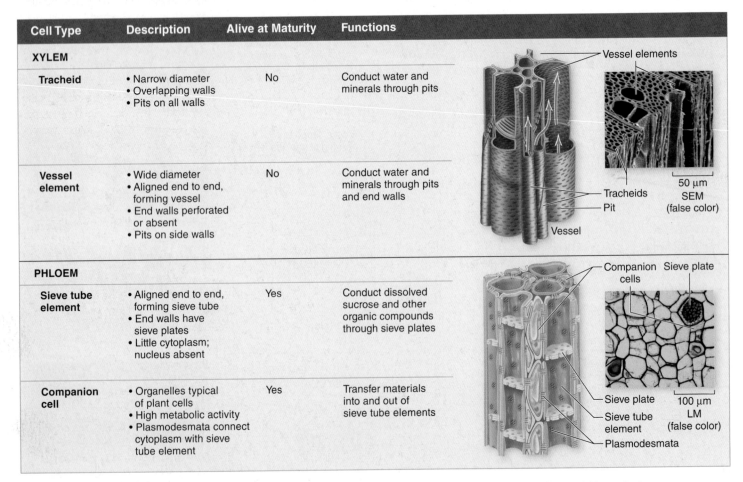

Cell Type	Description	Alive at Maturity	Functions
XYLEM			
Tracheid	• Narrow diameter • Overlapping walls • Pits on all walls	No	Conduct water and minerals through pits
Vessel element	• Wide diameter • Aligned end to end, forming vessel • End walls perforated or absent • Pits on side walls	No	Conduct water and minerals through pits and end walls
PHLOEM			
Sieve tube element	• Aligned end to end, forming sieve tube • End walls have sieve plates • Little cytoplasm; nucleus absent	Yes	Conduct dissolved sucrose and other organic compounds through sieve plates
Companion cell	• Organelles typical of plant cells • High metabolic activity • Plasmodesmata connect cytoplasm with sieve tube element	Yes	Transfer materials into and out of sieve tube elements

Figure 21.5 Transport Cells: A Summary. Xylem and phloem are vascular tissues that transport materials within a plant. This figure summarizes the main cell types in each tissue.

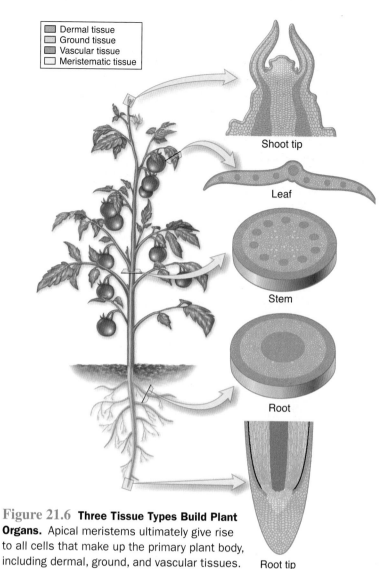

Dermal tissue
Ground tissue
Vascular tissue
Meristematic tissue

Shoot tip

Leaf

Stem

Root

Root tip

Figure 21.6 Three Tissue Types Build Plant Organs. Apical meristems ultimately give rise to all cells that make up the primary plant body, including dermal, ground, and vascular tissues.

into and out of the sieve tube elements but also provide energy and proteins to the conducting cells.

B. Plant Cells Form Three Main Tissue Systems

The cells that make up a plant form three main tissue systems: ground tissue, dermal tissue, and vascular tissue (**figure 21.6** and **table 21.3**). For comparison, animals have four types of tissues (see chapter 24). Each tissue type derives its properties from a unique combination of specialized cells. Together, these cells carry out all of the plant's functions.

Ground Tissue   Ground tissue often fills the spaces between more specialized cell types inside roots, stems, leaves, fruits, and seeds. For example, the pulp of an apple, the photosynthetic area inside a leaf, and the starch-containing cells of a potato all consist of ground tissue composed mainly of parenchyma cells. Although most ground tissue cells are structurally unspecialized, they are important sites of photosynthesis, respiration, and storage.

Dermal Tissue   **Dermal tissue** covers the plant. In the primary plant body, dermal tissue consists of the **epidermis,** a single layer of tightly packed, flat, transparent parenchyma cells that cover leaves, stems, and roots. In plants with secondary growth, however, lateral meristems produce tissues that replace the epidermis in stems and roots (see section 21.5).

Natural selection has shaped dermal tissue over hundreds of millions of years. When plants first moved onto land, they were at risk for drying out. The threat of desiccation selected for new water-conserving adaptations. For instance, in land plants, epidermal cells secrete a **cuticle,** a waxy layer that coats the epidermis of the aboveground parts of the plant (**figure 21.7**). The cuticle conserves water and protects the plant from predators and fungi.

Table 21.3   Plant Tissue and Cell Types: A Summary

Tissue System	Location	Cell Types	Functions
Ground	Forms bulk of interior of stems, leaves, and roots, including cortex, pith, and leaf mesophyll	Parenchyma	Photosynthesis, storage, respiration
		Collenchyma	Elastic support
		Sclerenchyma (fibers and sclereids)	Inelastic support
Dermal	Surface of roots and stems	Epidermal parenchyma	Protects primary plant body; controls gas exchange in stems and leaves; absorbs water and minerals in roots
		Periderm derived from parenchyma cells	Protects woody roots and stems
Vascular: xylem	Vascular bundles in stems; veins in leaves; vascular bundles or cylinder in roots	Tracheids, vessel elements	Conduct water and minerals in stems, leaves, and roots
		Parenchyma	Storage
		Sclerenchyma (fibers and sclereids)	Support
Vascular: phloem	Vascular bundles in stems; veins in leaves; vascular bundles or cylinder in roots	Sieve tube elements	Transport of organic molecules
		Companion cells (specialized parenchyma)	Transfer carbohydrates to/from sieve tube elements
		Sclerenchyma (fibers and sclereids)	Inelastic support

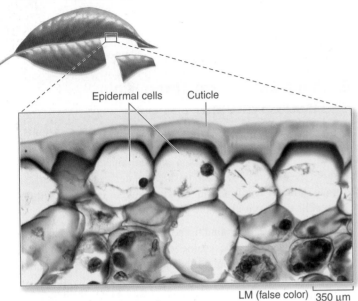

Epidermal cells   Cuticle

LM (false color) 350 μm

Figure 21.7 Cuticle. A waterproof cuticle protects the epidermis of the primary shoot.

The cuticle is impermeable not only to water but also to gases such as O_2 and CO_2. How do these gases pass through the cuticle? The answer is **stomata** (singular: stoma), pores through which leaves and stems exchange gases with the atmosphere (**figure 21.8**). A pair of specialized **guard cells** surrounds each stoma and controls its opening and closing (see section 22.2).

Stomata occupy about 1% to 2% of the leaf surface area, allowing for a balance between gas exchange and water conservation. Open stomata let CO_2 diffuse into a leaf for photosynthesis but also allow water to diffuse out; the pores typically close when conditions are too dry. Moreover, the arrangement of stomata on a leaf's surface can also minimize water loss. In many species, for example, stomata are most abundant on the shaded undersides of horizontal leaves; in pine needles, stomata are recessed in grooves.

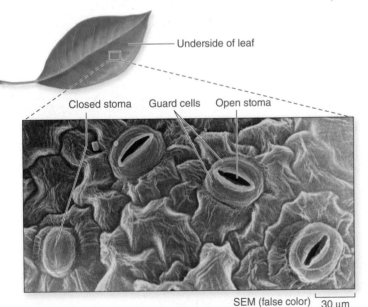

— Underside of leaf

Closed stoma   Guard cells   Open stoma

SEM (false color) 30 μm

Figure 21.8 Stomata. Plants exchange gases with the atmosphere through open stomata. Guard cells control whether each stoma is open or closed.

Vascular Tissue   Vascular tissues—xylem and phloem—form a continuous distribution system embedded in the ground tissue of shoots and roots. In stems and leaves, a **vascular bundle** is a strand of tissue containing xylem and phloem, often together with collenchyma or sclerenchyma fibers. The tough fibers prevent animals from tapping the rich sugar supply in the phloem.

The transition from water to land selected for vascular tissues, which accommodate the division of labor between roots and shoots. Roots absorb water and minerals; shoots produce food. Xylem and phloem form the transportation system that shuttles these materials throughout the plant's body. (Chapter 22 further explores the function of vascular tissue).

Vascular tissue also has another function: support. Lignin strengthens the walls of both xylem and fibers. As described in chapter 18, this additional physical support enables vascular plants to tower over their nonvascular counterparts, an important adaptation in the intense competition for sunlight.

Table 21.3 on page 461 summarizes the locations, cell types, and functions of the tissue systems in plants.

21.3 | Mastering Concepts

1. Describe the structures and functions of the cell types that make up a plant body.
2. Where in the plant does ground tissue occur?
3. What are the functions of vascular tissue?
4. How does the structure of dermal tissue contribute to its functions?

? Burning Question

What's the difference between fruits and vegetables?

Have you ever heard the rumor that tomatoes are really fruits? If so, you may also have wondered about a larger question: the difference between fruits and vegetables.

Botanically speaking, tomatoes are fruits because they contain seeds. Any seed-bearing structure produced by a flowering plant is a fruit. Apples, cherries, oranges, peaches, and raspberries are fruits. So are foods that we think of as vegetables, such as zucchinis, green beans, bell peppers, and pumpkins. But the term *vegetable* also includes plant parts that don't contain seeds, such as spinach leaves or carrot roots or Brussels sprouts.

Unlike fruits, vegetables do not have a botanical definition—just a culinary one. That is, foods that we consider vegetables may come from any part of a plant, but they are typically used in salty or savory dishes, not sweet ones.

Submit your burning question to:
marielle_hoefnagels@mcgraw-hill.com

21.4 | Tissues Build Stems, Leaves, and Roots

The tissues described in section 21.3 make up the stems, leaves, and roots of vascular plants. We now return to these organs to examine their structures more closely.

A. Stems Support Leaves

Stems grow and differentiate at their tips, with new cells originating at apical meristems (**figure 21.9**). The daughter cells eventually give rise to ground tissue, the epidermis, and vascular tissue. The stem elongates as the vacuoles of the new cells absorb water, pushing the apical meristem upward.

Remnants of the apical meristem remain, however, in the axillary buds that form at stem nodes. These buds may either remain dormant or "awaken" to form side branches. When a shoot loses its terminal bud, cells in one or more dormant axillary buds begin to divide. The result is a bushy growth form. Gardeners exploit this phenomenon by pinching off the tips of young tomato or basil stems, a practice that promotes the growth of side branches and therefore greatly increases yields. ▶ apical dominance, p. 500

Ground tissue occupies most of the volume of the stem of a herbaceous plant. The ground tissue, which consists mostly of

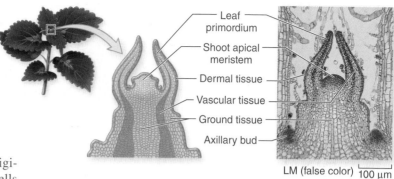

LM (false color) ⌐100 µm⌐

Figure 21.9 Shoot Apical Meristem. The apical meristem at the tip of a growing shoot gives rise to the specialized tissues that make up the mature shoot.

parenchyma cells, stores water and starch. The cells are often loosely packed, allowing for gas exchange between the stem interior and the atmosphere.

Numerous vascular bundles are embedded in ground tissue. The vascular bundles, which typically have phloem to the outside and xylem toward the inside, are arranged differently in monocots and eudicots (**figure 21.10**). In most monocot stems, vascular bundles are scattered throughout the ground tissue. Most eudicot stems, in contrast, have a single ring of vascular bundles.

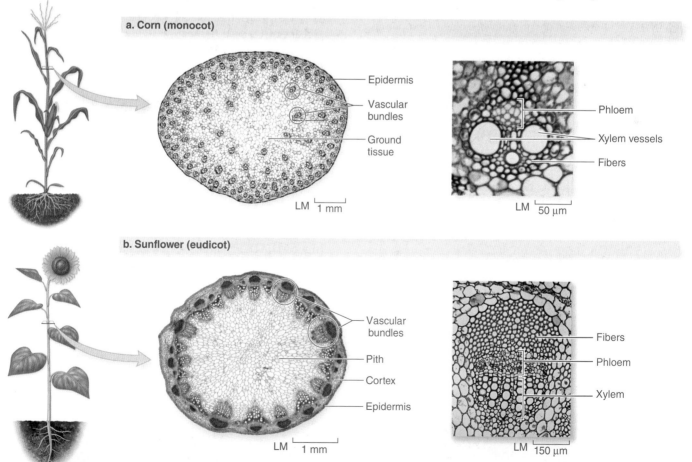

Figure 21.10 Anatomy of Primary Stems. (a) The cross section of a monocot stem (corn) features vascular bundles scattered in ground tissue. (b) A eudicot stem (sunflower) has a ring of vascular bundles surrounding a central pith.

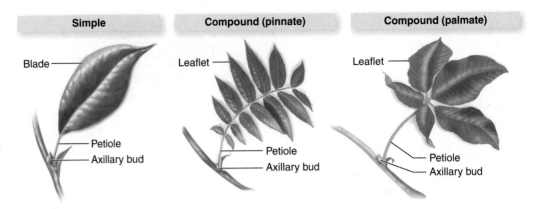

| Simple | Compound (pinnate) | Compound (palmate) |

Figure 21.11 Leaf Forms. A simple leaf has an undivided blade, whereas compound leaves consist of multiple leaflets. An axillary bud defines the base of each leaf.

Ground tissue occupies most of the rest of the eudicot stem: the **cortex** fills the area between the epidermis and vascular tissue, and **pith** occupies the center.

B. Leaves Are the Primary Organs of Photosynthesis

Most leaves have two main parts: the flattened **blade** and the supporting, stalklike **petiole.** The broad, flat blade maximizes the surface area available to capture solar energy. For example, a large maple tree has approximately 100,000 leaves, with a total surface area that would cover six basketball courts (about 2500 square meters).

Biologists categorize leaves according to their basic forms (**figure 21.11**). A **simple leaf** has an undivided blade. **Compound leaves** are divided into leaflets, typically either paired along a central line or all attached to one point at the top of the petiole, like fingers on a hand. How can you tell the difference between a simple leaf and one leaflet of a compound leaf? A leaf has an axillary bud at its base, whereas an individual leaflet does not.

Veins are vascular bundles inside leaves, and they are often a leaf's most prominent external feature. Networks of veins occur in two main patterns (**figure 21.12**). Many monocots have parallel veins, with several major longitudinal veins connected by smaller minor veins. Most eudicots have netted veins, with minor veins branching in all directions from larger, prominent midveins.

In a developing shoot, leaves originate as bumps called "leaf primordia" on the flanks of the apical meristem (see figure 21.9). Each leaf primordium develops into a small, dormant leaf bud at a stem node, where vascular bundles from the stem pass into the leaf's petiole. When it is time for the leaf to expand, meristem cells begin dividing. The new cells eventually absorb water, elongate, and specialize to become the cells of the mature leaf.

The ground tissue inside a leaf is called **mesophyll,** and it is composed mostly of parenchyma cells (**figure 21.13**). Most mesophyll cells have abundant chloroplasts and produce sugars by photosynthesis. The long, column-shaped mesophyll cells along the upper side of a leaf maximize light absorption. Below this layer are irregularly shaped mesophyll cells separated by large air spaces. When the stomata are open, these mesophyll cells can exchange CO_2 and O_2 directly with the atmosphere.

Mesophyll cells also exchange materials with vascular tissue. Xylem at the ends of the tiniest leaf veins delivers water and minerals to nearby mesophyll cells. Meanwhile, sugars produced in photosynthesis move from the mesophyll cells to the phloem's companion cells and then to the sieve tube elements. The sugars, along with other organic compounds, travel within the phloem to the roots and other nonphotosynthetic plant parts.

C. Roots Absorb Water and Minerals, and Anchor the Plant

Roots grow in two main patterns that differ based on the fate of the primary root, the first root to develop after a seed germinates (**figure 21.14**). In a **taproot system,** the primary root enlarges to

Figure 21.12 Leaf Veins. (a) Leaves of many monocots, such as this lily, have prominent parallel veins. (b) Leaves of eudicots, such as this pumpkin plant, have netlike venation.

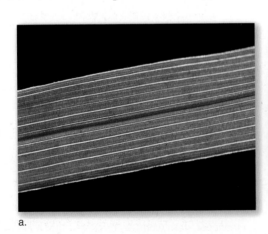

a.

b.

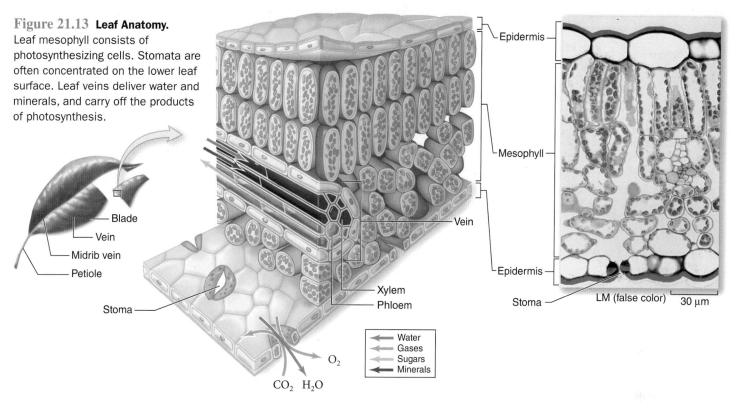

Figure 21.13 Leaf Anatomy. Leaf mesophyll consists of photosynthesizing cells. Stomata are often concentrated on the lower leaf surface. Leaf veins deliver water and minerals, and carry off the products of photosynthesis.

Blade
Vein
Midrib vein
Petiole
Stoma

Epidermis
Mesophyll
Vein
Xylem
Phloem
Epidermis
Stoma
LM (false color)   30 μm

O$_2$
CO$_2$ H$_2$O

| Water |
| Gases |
| Sugars |
| Minerals |

form a thick root that persists throughout the life of the plant. Lateral branches emerge from this main root. Taproots grow fast and deep, maximizing support and enabling a plant to use minerals and water deep in the soil. Most eudicots develop taproot systems. In a **fibrous root system,** on the other hand, slender adventitious roots arise from the base of the stem and replace the short-lived primary root. Grasses and other monocots usually have fibrous root systems. Because they are relatively shallow, these roots rapidly absorb minerals and water near the soil surface and prevent erosion.

In both taproot and fibrous root systems, countless root tips explore the soil for water and nutrients. Just behind the tip of each actively growing root is an apical meristem (**figure 21.15**). Some of the cells produced at the apical meristem differentiate into the **root cap,** which protects the meristem from abrasions.

Taproot system   Fibrous root system

Figure 21.14 Root Systems. A taproot system has a central, thick root with lateral branches. A fibrous root system features numerous slender, adventitious roots.

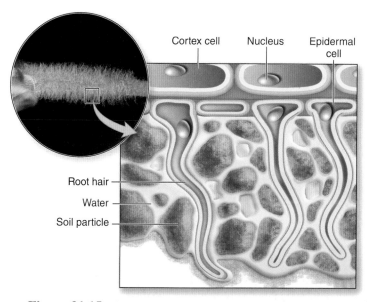

Cortex cell   Nucleus   Epidermal cell
Root hair
Water
Soil particle

Figure 21.15 Root Hairs. These epidermal cell outgrowths greatly increase the surface area of this radish seedling's root.

Root cap cells, which slough off and are continually replaced, secrete a slimy substance that lubricates the root as it grows. The root cap also plays a role in sensing gravity. ▸ gravitropism, p. 505

Other cells produced at the apical meristem are destined to become ground, dermal, and vascular tissues. The newly produced daughter cells elongate by absorbing water into their vacuoles. As the cells become larger, the root grows farther into the soil. Beyond this zone of elongation is a zone of maturation, in which cells complete their differentiation into the functional tissue systems that make up the root.

The epidermis surrounds the entire primary root except the root cap, and its adaptations maximize the absorption of water and minerals from soil. In contrast to the stem and leaf, the root epidermal cells have a very thin cuticle or none at all. In addition, **root hairs** in the zone of maturation provide extensive surface area for absorption (**figure 21.16**).

Just internal to the epidermis is the cortex, which makes up most of the primary root's bulk. It consists of loosely packed, interconnected parenchyma cells that may store starch or other materials. The spaces between the cells allow for both gas exchange and the free movement of water.

The **endodermis** is the innermost cell layer of the cortex. The walls of its tightly packed cells contain a ribbon of waxy, waterproof material. These waxy deposits form the **Casparian strip,** a barrier that blocks the passive diffusion of water and dissolved substances into the xylem (see figure 22.7). Instead, all materials entering the root's vascular tissue must first pass through the selectively permeable membranes of the endodermal

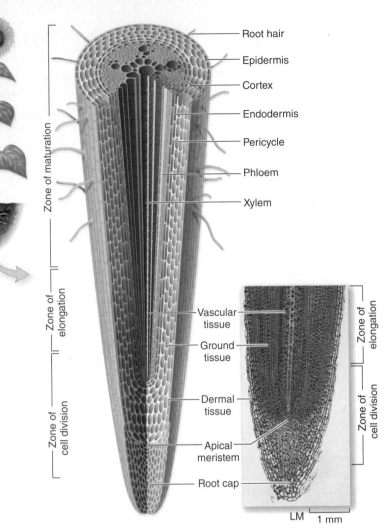

Figure 21.16 The Root Apical Meristem. The apical meristem at the tip of a growing root produces root cap cells, ground tissue, vascular tissue, and the epidermis.

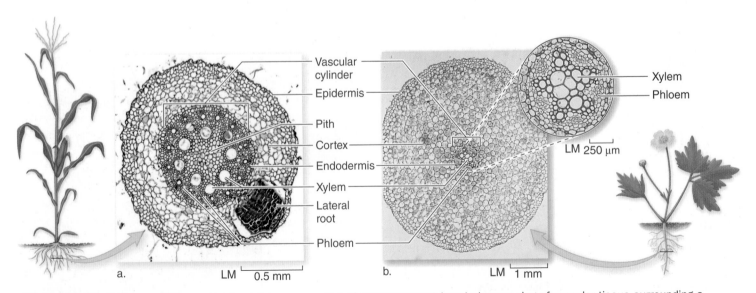

Figure 21.17 Anatomy of Primary Roots. (a) A cross section of a monocot root (corn) shows a ring of vascular tissue surrounding a central pith. (b) A eudicot root (buttercup) has a central cylinder of vascular tissue surrounded by the cortex.

cells. This filtering system enables the plant to exclude toxins and control the concentrations of some minerals.

Internal to the endodermis is the **pericycle,** the outermost layer of the root's vascular cylinder. Cells in the pericycle divide to produce lateral roots. As each branch root develops, it reserves at its tip a small patch of dividing cells that become the apical meristem. The lateral root grows as these cells divide, pushing through the cortex and epidermis before entering the soil.

The vascular cylinders of eudicot and monocot roots have different arrangements (**figure 21.17**). In most monocot roots, a ring of vascular tissue surrounds a central core (pith) of parenchyma cells. In most eudicots, the vascular cylinder consists of a solid core of xylem, with ridges that project to the pericycle. Phloem strands are generally located between the "arms" of the xylem core.

In most plant species, roots do not explore the soil alone. Instead, they form mycorrhizal associations with fungi (see figures 19.10 and 19.18). Fungal filaments inside the root feed on the plant's sugars and extend into the soil, greatly increasing the root's surface area for absorption. Fungi and plants are ancient partners; hundreds of millions of years ago, these two groups of organisms moved onto land together. ▶ mycorrhizae, p. 402

In addition, some plant species, such as peas and beans, form root nodules in association with nitrogen-fixing bacteria. These microbes consume some of the sugar produced by their host plants. In exchange, the bacteria act as built-in fertilizer by making nitrogen available to the plants. ▶ nitrogen fixation, p. 346

The structural differences between monocots and eudicots are summarized in **figure 21.18**.

21.4 | Mastering Concepts

1. Name the cell layers that occur in the stem of a monocot and a eudicot, moving from the epidermis to the innermost tissues.
2. List the parts of a simple and a compound leaf.
3. Describe the internal anatomy of a leaf.
4. Compare and contrast the development of taproot and fibrous root systems.
5. How do the roots of monocots and eudicots differ?

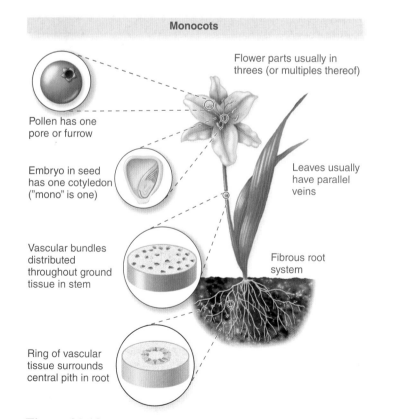

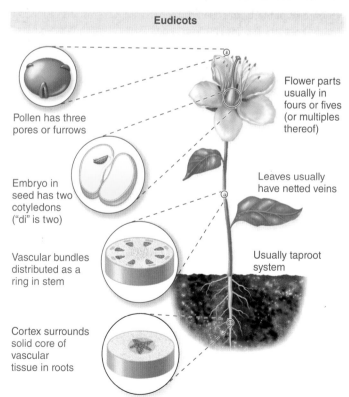

Figure 21.18 Monocots and Eudicots Compared. Although many variations occur, most monocots and eudicots differ from each other in flower and leaf morphology, root and stem anatomy, pollen structure, and seed structure.

21.5 | Lateral Meristems Produce Wood and Bark

In many habitats, plants compete for sunlight. The tallest plants reach the most light, so selection for height has been a powerful force in the evolutionary history of plants. But primary tissue is not strong enough to support a very tall plant. The increasing competition for light therefore selected for additional support, in the form of secondary growth that could increase the girth of stems and roots.

Wood and bark are tissues that arise from secondary growth originating at two types of lateral meristems: vascular cambium and cork cambium. These meristems occur in gymnosperms (conifers and their relatives) and many eudicots but not in monocots.

A. The Vascular Cambium Produces Xylem and Phloem in Woody Plants

The **vascular cambium** is an internal cylinder of meristem tissue that produces most of the diameter of a woody root or stem. This lateral meristem forms a thin layer between the primary xylem and phloem (**figure 21.19**).

When a cell in the vascular cambium divides, it produces two daughter cells, one of which remains a meristem cell. Cells produced to the inside of the cambium become secondary xylem; secondary phloem forms on the outer side. Overall, the vascular cambium produces much more secondary xylem than secondary phloem, so the stem or root acquires most of its girth from growth internal to the cambium. Secondary xylem, more commonly known as **wood**, can accumulate to massive proportions. For example, a giant sequoia tree *(Sequoiadendron giganteum)* in California is 100 meters tall and more than 7 meters in diameter.

The vascular cambium also produces **rays,** bands of parenchyma cells that extend from the center of the stem or root like spokes on a bicycle wheel. Rays transport water and nutrients laterally within secondary xylem and phloem.

Secondary growth destroys the stem or root's primary phloem, cortex, and epidermis. The secondary phloem takes over responsibility for phloem sap transport; the periderm (described in section 21.5B) replaces the epidermis. Secondary phloem forms the live, innermost layer of **bark,** a collective term for all tissues to the outside of the vascular cambium. Periderm makes up the outer portion of bark.

B. The Cork Cambium Produces the Outer Layer of a Woody Stem or Root

Periderm is the protective dermal tissue that covers a woody stem or root. It originates from the **cork cambium,** a lateral meristem that gives rise to cork to the outside and parenchyma to the inside (see figure 21.19). Periderm thus consists of three layers: a living parenchyma layer, the cork cambium, and the nonliving cork.

Cork consists of layers of densely packed, waxy cells on the surfaces of mature stems and roots. The cells are dead at maturity and form waterproof, insulating layers that protect plants. The cork used to stopper wine bottles comes from oak trees that grow in the Mediterranean region. Every 10 years, harvesters remove much of the cork cambium and cork, which grows back. Cork is also important in the history of biology; in 1665, Robert Hooke became the first person to see cells when he used a primitive microscope to gaze at cork.

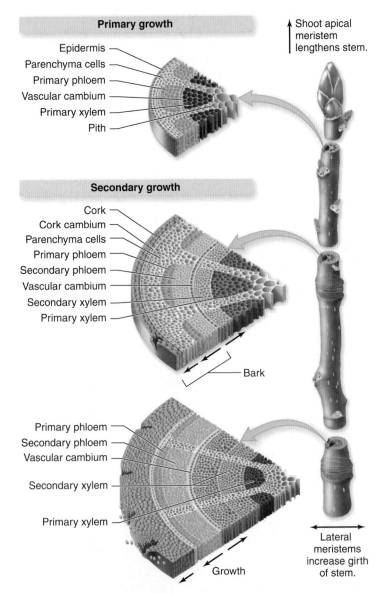

Figure 21.19 Secondary Growth Produces Wood. In a primary root, the microscopic vascular cambium has not yet started producing secondary vascular tissue. The two lower diagrams show the vascular cambium producing secondary xylem toward the inside of the stem and secondary phloem toward the outside. Cork cambium, meanwhile, produces cork cells to the outside and parenchyma to the inside.

C. Wood Is Durable and Useful

Few plant products are as versatile or economically important as wood. Lumber forms the internal frame that supports many buildings. Firewood provides heat and cooking fuel. Most paper comes from wood, as this chapter's Apply It Now box describes. Throughout history, humans have fashioned wood into furniture, pencils, cabinets, boats, baseball bats, serving bowls, roofing shingles, jewelry, picture frames, and countless other items.

Wood varies in hardness. Angiosperms such as oak, maple, and ash are often called hardwood trees, whereas conifers such as pine, spruce, and fir are called softwood trees. Sclerenchyma fibers account for the difference. Angiosperm wood contains fibers in addition to tracheids and vessels, whereas wood from softwood trees consists mainly of tracheids. As a result, eudicot wood is usually stronger and denser than wood from conifers. The soft, light wood from the balsa tree, an angiosperm, is a notable exception.

A quick glance at a cross section of a tree trunk reveals that the innermost wood is darker than the outer portion (figure 21.20a). This color difference arises as the tree ages. As the years pass, the oldest secondary xylem gradually becomes unable to conduct water. This darker colored, nonfunctioning region is called **heartwood.** The lighter colored **sapwood,** located nearest the vascular cambium, transports water and dissolved minerals.

Another feature of a trunk's cross section is tree rings. In temperate climates, cells in the vascular cambium are dormant in winter, but they divide to produce wood during the spring and summer. During the moist days of spring, the vascular cambium produces large, water-conducting cells. During the drier days of summer, new wood has smaller cells. The contrast between the summer wood of one year and the spring wood of the next highlights each annual tree ring (figure 21.20b).

Fortunately, it is not necessary to cut down a tree to see its growth rings; a slender core extracted from the trunk reveals the pattern without harming the tree. By counting the rings, a forester can estimate a tree's age. Growth rings also provide clues about climate and significant events throughout a tree's life. A thick ring indicates plentiful rainfall and good growing conditions. Narrow tree rings may reflect stress from herbivory, disease, or fierce competition for light or water. A fire leaves behind a charred "burn scar." Geologists have even used tree rings to infer the dates of earthquakes that occurred before recorded history. Ground shaking can break stems and change water flow patterns on the land, both of which can leave signals in tree ring data.

Researchers can also combine tree ring data from multiple trees to look into ancient history. For example, the oldest known living tree is a bristlecone pine *(Pinus longaeva)* growing in the White Mountains of California. It is more than 4760 years old. By aligning distinctive patterns in its early tree rings with the same patterns in older, dead trees from the same area, researchers have deduced rainfall data going back 8200 years!

21.5 | Mastering Concepts

1. What are the two lateral meristems in a woody stem or root, and which tissues does each meristem produce?
2. What are the functions of wood and bark?
3. How do softwoods differ from hardwoods?
4. Explain the origin of tree rings.

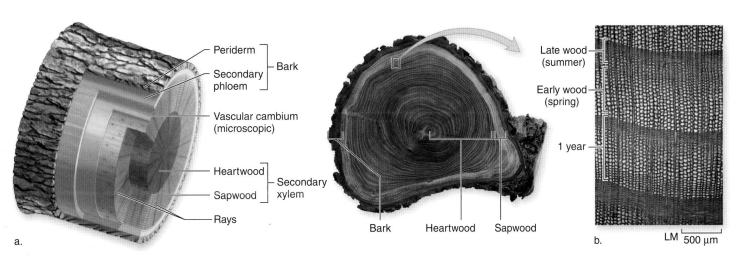

Figure 21.20 Anatomy of a Secondary Stem. (a) Wood is secondary xylem, and bark is all the tissue outside the vascular cambium. At the center of the stem, the darker colored heartwood is nonfunctioning secondary xylem. (b) Differences in available soil moisture when wood forms in spring and summer result in different-sized cells, which are visible as tree rings.

21.6 | Investigating Life: An Army of Tiny Watchdogs

For many people, a dog is a both a friend and a protector, warning of burglars and other intruders. In exchange, we provide our canine companions with shelter and food. Likewise, some plants welcome ants and other invertebrates with open arms—or, more precisely, with open stems.

In the tropics, hundreds of tree species provide ants with room and board. Some of the trees' stems have unique anatomical features such as hollow areas, called domatia, where ants live. Moreover, their young leaves secrete a sugary nectar or other food that their guests eat.

Supplying nectar to ants costs energy, which a plant could otherwise put toward its own reproduction. How can natural selection allow plants to extend such hospitality to ants? Biologist Laurence Gaume, an expert on plant-insect interactions at France's University of Montpellier, wanted to know the answer to that question. Working with colleagues from the Indian Institute of Science and the University of Montpellier, Gaume conducted a series of field experiments in India to find out.

The research team chose a tree called *Humboldtia brunonis* to learn more about ant-plant interactions (**figure 21.21**). In *H. brunonis,* the domatia are swollen, hollowed-out stem inter-

nodes up to 10 centimeters long and 1 centimeter wide. A hole near the node allows ants and other small invertebrates to enter and leave each stem cavity.

In most tree species, either all of the individuals produce domatia or none does. But *H. brunonis* is unusual: within the same population, some individuals have the stem cavities and others do not. This unusual property made *H. brunonis* ideal for an experiment comparing trees with and without domatia.

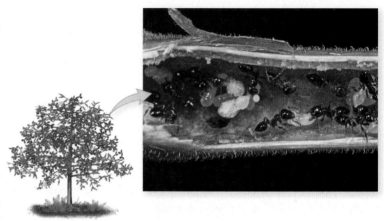

Figure 21.21 Home Sweet Home. These ants make their homes in hollow stems called domatia.

Apply It Now

From Wood to Paper to the Recycling Bin

Along with food, paper is probably the most prevalent plant product in our lives. The average person in the United States uses more than 317 kilograms of paper and paperboard every year. We use it to package food, as currency, to communicate, to decorate walls, to wrap gifts, and absorb fluids, among many other applications.

Manufacturing paper on a commercial scale demands a constant supply of wood pulp from millions of trees—mostly softwoods such as pine, spruce, and hemlock but also hardwoods such as birch. To make paper, wood is cooked in water, washed, and sometimes bleached. The slurry may be mixed with additives such as dyes, then spread on a fine mesh. As the water drains, the cell walls entangle, adjacent cellulose molecules bond, and paper forms. The material is compressed into a sheet, then dried.

Paper recycling programs keep over a third of discarded newspapers, magazines, catalogs, office paper, and junk mail out of landfills. At paper mills, this "postconsumer" waste paper is shredded and placed into huge tanks. Solvents and detergents are added to remove inks, and paper clips and other unwanted objects are removed. The material is then reassembled into new paper. Using recycled paper requires 60% to 70% less energy and about half as much water as using virgin wood pulp. It also produces less water pollution, because recycled pulp requires less bleaching than does wood pulp.

Unfortunately, recycling cannot supply all of the world's paper needs. Pulp deteriorates during recycling, limiting the number of possible uses and reuses. De-inked pulp from newsprint, for example, is too degraded for use in high-quality printer paper. Also, toxins and other contaminants restrict the use of recycled fibers in some types of food packaging.

In the first experiment, Gaume's research team counted the number of fruits on 104 *H. brunonis* trees as a measure of reproductive success. They noted the height of each tree and whether or not domatia were present, and they identified the ants exploring each tree. After entering the data into a statistical analysis program, they found that trees with domatia tended to produce the most fruits. The presence of one ant species, called the tramp ant *(Technomyrmex albipes),* was also correlated with fruit production.

A correlation, however, is not the same thing as a causal relationship. One possible interpretation is that the domatia house tramp ants, which somehow promote fruit production. But an alternative explanation is that trees in the best locations have access to the most resources and therefore can "afford" both domatia and high fruit production. In that case, the apparent link between domatia and reproductive success would be a mere coincidence.

Based on studies of other "ant plants," the researchers hypothesized that the tramp ant indirectly promoted the tree's reproductive success by chasing off insects that would otherwise eat the tree's leaves. They therefore predicted that excluding ants from leaves should increase the amount of herbivory. To set up their experiment, the team randomly selected 20 trees, five of which were patrolled exclusively by the tramp ant; other ant species patrolled the other 15 trees. On each tree, the researchers selected two young compound leaves, each consisting of four leaflets. They dabbed glue at the base of one of the two leaves. Ants would be free to chase herbivores off of the control leaves but would get stuck at the base of a glue-treated leaf.

After 10 days, the researchers scored each leaflet as either intact or damaged. As illustrated in figure 21.22, trees patrolled by the tramp ant suffered less herbivory overall than trees harboring other ant species. In addition, leaves from which the tramp ant were excluded had more damaged leaflets.

In a follow-up experiment, the team observed how tramp ants reacted upon discovering hungry caterpillars on the leaves of "their" trees. The ants took an average of 62 seconds to discover a caterpillar, and the most common reaction was to bite and evict the intruder. Clearly, the ants took an active role in deterring herbivores, presumably to reduce competition for the nectar secreted by the tender young leaves.

Overall, Gaume and the rest of the research team concluded that domatia improve the reproductive success of *H. brunonis* indirectly. Tramp ants that occupy the domatia protect the trees from herbivores. The resulting increased leaf area translates into more photosynthesis, which could, in turn, promote reproductive output.

That finding stimulated another question: Why do *H. brunonis* trees sometimes lack domatia? The researchers speculate that the production of domatia is a heritable trait and that the hollow stems may have a cost as well as a benefit. For example, stems with domatia may break easily or attract birds that damage the trees while exploring the domatia in search of ant prey. Perhaps variation in the production of domatia is an example of a balanced polymorphism, such as the one that maintains the cystic fibrosis and sickle cell anemia traits in the human population. ▶ balanced polymorphism, p. 242

This study helped researchers understand the evolutionary forces that have created ant–plant partnerships. Natural selection has shaped the anatomy of the stem and every other part of the plant body. In this case, the unique hollowed-out stems create a home for ants—an army of tiny watchdogs that protect the plant by chasing off intruders.

Gaume, Laurence, Merry Zacharias, Vladimir Grosbois, and Renee M. Borges. 2005. The fitness consequences of bearing domatia and having the right ant partner: experiments with protective and non-protective ants in a semi-myrmecophyte. *Oecologia,* vol. 145, pp. 76–86.

21.6 | Mastering Concepts

1. What is the overall question the researchers were asking in this study? Summarize the data they used to arrive at their conclusion.

2. Propose an explanation for the observation that the trees secrete nectar only on young leaves and not on tougher, more mature leaves.

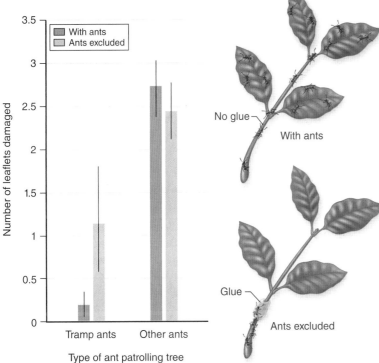

Figure 21.22 Ants on Patrol. When tramp ants were excluded, leaves suffered extra damage from herbivores. Excluding other types of ants did not have an effect. (Error bars represent standard errors).

Chapter Summary

21.1 | Vegetative Plant Parts Include Stems, Leaves, and Roots

- **Anatomy** is the study of an organism's form, and **physiology** is the study of its function. A plant's body consists of **organs** composed of **tissues** with specific functions.
- The **vegetative** plant body includes a **shoot** and **roots** that depend on each other.
- The **stem** is the central axis of a shoot and consists of **nodes,** where **leaves** attach, and **internodes** between leaves, where the stem elongates. An **axillary bud** is located at each node.
- **Herbaceous plants** typically have soft, green stems; **woody plants** have stems and roots strengthened with wood.
- Leaves are the main sites of photosynthesis.
- Roots absorb water and dissolved minerals. **Adventitious roots** arise from stems or leaves.
- Some specialized roots store starch or water, absorb oxygen from air, carry out photosynthesis, or provide support. Stem modifications include stolons, rhizomes, tubers, succulent stems, thorns, and tendrils. Modified leaves include tendrils, storage leaves, cotyledons, spines, and insect traps.

21.2 | Plants Have Flexible Growth Patterns, Thanks to Meristems

A. Plants Grow by Adding New Modules
- Plants with **determinate growth** stop growing when they reach their mature size; plants with **indeterminate growth** grow indefinitely.

B. Plant Growth Occurs at Meristems
- **Meristems** are localized collections of cells that retain the ability to divide throughout the life of the plant. **Apical meristems** at the plant's root and shoot tips provide **primary growth.** **Lateral meristems** are cylinders of cells at the periphery of a stem or root that add girth, or **secondary growth.** In some plants, **intercalary meristems** occur between nodes.

21.3 | Plant Cells Build Tissues

A. Plants Have Several Cell Types
- **Parenchyma** cells are alive at maturity. They are relatively unspecialized and have thin cell walls. They often function in metabolism or storage.
- **Collenchyma** cells are also alive. Their thick primary cell walls provide elastic support to growing shoots.
- **Sclerenchyma** cells, including long **fibers** and shorter **sclereids,** are dead at maturity. Their thick secondary cell walls contain **lignin,** supporting plant parts that are no longer growing.
- Water-conducting cells in **xylem** include long, narrow, less specialized **tracheids** and more specialized, barrel-shaped **vessel elements.** Both cell types have thick walls and are dead when functioning. Water moves through pits in tracheids but through the end walls of vessel elements.
- Sucrose-conducting cells in **phloem** include **sieve tube elements.** Pores cluster at **sieve plates,** allowing nutrient transport between adjacent cells via strands of cytoplasm. **Companion cells** help transfer carbohydrates.

B. Plant Cells Form Three Main Tissue Systems
- Most of the primary plant body consists of **ground tissue,** relatively unspecialized parenchyma cells that fill the space between dermal and vascular tissues.
- **Dermal tissue** includes the **epidermis,** a single cell layer covering the plant. The epidermis secretes a waxy **cuticle** that coats aerial plant parts. Gas and water exchange in the shoot occur through **stomata** bounded by **guard cells.**
- **Vascular tissue** is conducting tissue. Xylem transports water and dissolved minerals from roots upward. Phloem transports dissolved carbohydrates and other substances throughout a plant. Xylem and phloem occur together with other tissues to form **vascular bundles.**

21.4 | Tissues Build Stems, Leaves, and Roots

A. Stems Support Leaves
- Cells in the shoot apical meristem divide to give rise to the dermal tissue, vascular tissue, and ground tissue in stems and leaves.
- Vascular bundles are scattered in the ground tissue of monocot stems but form a ring of bundles in eudicot stems. Between a eudicot stem's epidermis and vascular tissue lies the **cortex,** made of ground tissue. **Pith** is ground tissue in the center of a stem.

B. Leaves Are the Primary Organs of Photosynthesis
- A stalklike **petiole** supports each leaf **blade.** A **simple leaf** has one undivided blade, and a **compound leaf** forms leaflets. **Veins** are vascular bundles in leaves; they may be in either netted or parallel formation.
- Leaves arise at the flanks of the apical meristem. Leaf epidermis is tightly packed, transparent, and mostly nonphotosynthetic. Leaf ground tissue includes **mesophyll** cells that carry out photosynthesis. Stomata enable gas exchange.

C. Roots Absorb Water and Minerals, and Anchor the Plant
- **Taproot systems** have a large, persistent major root, whereas **fibrous root systems** are shallow, branched, and shorter lived.
- The root apical meristem produces cells that differentiate into epidermis, cortex, and vascular tissues. A **root cap** protects the tip of a growing root, and **root hairs** behind the cap provide tremendous surface area. The root cortex consists of storage parenchyma and **endodermis.** The waxy **Casparian strip** ensures that the solution entering the root passes through the cells of the endodermis. The **pericycle** gives rise to branch roots. In monocots, pith fills the center of the root.
- Mycorrhizal fungi help roots take up soil nutrients; nitrogen-fixing bacteria in root nodules provide some plants with nitrogen.

21.5 | Lateral Meristems Produce Wood and Bark

A. The Vascular Cambium Produces Xylem and Phloem in Woody Plants
- The **vascular cambium** increases the girth of the stem or root. It produces **wood** (secondary xylem), secondary phloem, and **rays.**

B. The Cork Cambium Produces the Outer Layer of a Woody Stem or Root
- The **cork cambium** produces the **periderm,** which makes up the majority of a woody plant's **bark.**

C. Wood Is Durable and Useful
- **Heartwood** is the central, nonfunctioning wood in a tree. The light-colored **sapwood** transports water and minerals within a tree.
- Tree rings form in response to seasonal differences in the sizes of conducting cells in wood.

21.6 | Investigating Life: An Army of Tiny Watchdogs

- Many species of tropical plants produce hollowed-out stem internodes called domatia, which often house ants.
- Some types of ants protect their host plants by chasing off herbivores.

Multiple Choice Questions

1. Which of the following is NOT a vegetative organ in a plant?
   a. Stem   c. Flower
   b. Leaf   d. Root

2. The ability of a sunflower plant to become taller is most likely due to its
   a. apical meristem.   c. intercalary meristem.
   b. lateral meristems.   d. Both a and c are correct.

3. The apical meristem of a plant produces
   a. epidermal cells.
   b. mesophyll cells.
   c. xylem cells.
   d. all the cells of the plant.

4. Which of the following is a living cell type that physically supports the growing plant?
   a. Parenchyma
   b. Collenchyma
   c. Sclerenchyma
   d. Both b and c are correct.

5. Sugars and other organic compounds travel through _____, whereas water and dissolved minerals travel through _____.
   a. the cytoplasm of xylem cells; sieve cells
   b. tracheids and vessels; companion cells
   c. the cytoplasm of the phloem sieve tube system; companion cells
   d. the cytoplasm of the phloem sieve tube system; tracheids and vessels

6. If you spray herbicide onto a weed, which barrier might prevent the chemical from entering the leaves?
   a. The stomata
   b. The Casparian strip
   c. The cuticle
   d. The pith

7. A carrot is an example of a(n)
   a. fibrous root.
   b. taproot.
   c. adventitious root.
   d. root nodule.

8. Root hairs are a part of what type of plant tissue?
   a. Ground tissue
   b. Dermal tissue
   c. Vascular tissue
   d. All of the above are correct.

9. What is the function of the Casparian strip?
   a. It allows the endodermis to function as a filter for the plant.
   b. It is made up of meristem cells that produce vascular tissue.
   c. It blocks the movement of water out of the vascular bundle.
   d. It minimizes the loss of water from the leaves of a plant.

10. Which tissue type occupies most of the volume of a woody stem?
    a. Secondary xylem
    b. Secondary phloem
    c. Bark
    d. Vascular cambium

Write It Out

1. List the main vegetative organs of a plant, and explain how each relies on the others.
2. Describe a stem specialization and a leaf specialization that provide protection.
3. How can you tell whether plant growth is determinate or indeterminate?
4. Many biology labs use slides of root tips to demonstrate the stages of mitosis. Why is this a better choice than using a slide of a mature leaf?
5. Compare and contrast tracheids, vessel elements, and sieve elements.
6. Which plant tissues have cells that are dead at maturity? Why is it advantageous to the plant for these cells to die?
7. Corn is a monocot and sunflower is a eudicot. Make a chart that compares the stems, leaves, and roots of these plants.
8. Thorns, spines, and tendrils are so highly modified that it can be difficult to tell whether they derive from leaves or stems. How could a biologist use knowledge of internal plant anatomy to determine the origin of these structures?
9. Compare and contrast the formation of branches in stems and roots.
10. Describe why and how leaves and roots maximize surface area.
11. Where do parenchyma cells occur in stems, leaves, and roots?
12. Fossils show that eudicots appeared about 125 million years ago. What structures would identify a fossilized plant as a eudicot?
13. Mammals exchange gases in the alveoli of the lungs (see figure 30.8). How do the structures and functions of leaf mesophyll compare with those of alveoli?
14. Explain how conditions in the terrestrial environment selected for each of the following adaptations: cuticle, stomata, vascular tissue, roots, stems, and leaves.

15. Predict the anatomical differences you might expect to see among a desert plant, a rain forest plant, and an aquatic plant.
16. Starting at the outside and moving into a eudicot's stem, what tissues do you encounter?
17. Girdling is cutting away or severing the living bark in a ring around a tree's trunk. Which part of a girdled tree do you expect to die first, the roots or the shoot? Why? Would the tree be harmed as much by a vertical gash? Why or why not?
18. A palm tree is a monocot with woody tissues that come from enlarged parenchyma cells in the stem. How is this arrangement different from the secondary growth that occurs in gymnosperms and eudicots?
19. Heartwood often contains secondary metabolites that inhibit microbial growth. How are these secondary metabolites adaptive?
20. Suppose you drive a metal spike from the outermost bark layer to the center of a tree's trunk. Which tissues does your spike encounter as it moves through the stem? What type of meristem produced each type?
21. List a function of each organ, tissue, and cell type described in this chapter, and then list at least one feature that facilitates that function.

Pull It Together

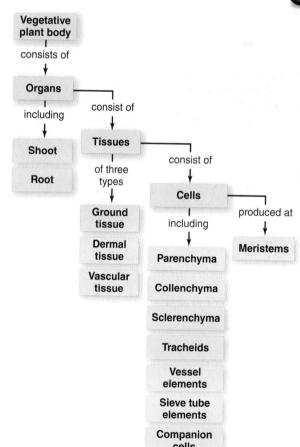

1. What are the three types of meristems, and how do they differ from one another?
2. How do fibers and sclereids fit into this concept map?
3. Add the terms *cortex, pith, mesophyll, endodermis,* and *Casparian strip* to this concept map.
4. Describe a location in which you might find each of the three tissue types.

Carnivorous Plants. The leaves of the Venus flytrap (left) can snap shut to catch prey. A sundew (right) lures insects with sticky globs of nectar.

Carnivorous Plants

LITTLE SHOP OF HORRORS TELLS THE STORY OF AN INNOCENT-LOOKING PLANT THAT GROWS TO ENORMOUS SIZE AND DEVELOPS A TASTE FOR HUMANS. Fortunately, such bloodthirsty plants exist only in science fiction...right?

Although it is true that no plants eat people, some do consume small animals, especially insects and other arthropods. The following examples illustrate just a few of the world's 450 or so species of carnivorous plants:

- The Venus flytrap *(Dionaea muscipula)* has two-sided leaves, each of which has highly sensitive trigger hairs. When an insect wanders onto a leaf and bends the trigger hairs, the two halves snap shut. The leaf's epidermis secretes enzymes that digest the captured animal. These plants, which are native only to North and South Carolina, are endangered in the wild because of habitat loss. Collecting them is illegal. (Most Venus flytraps sold commercially are propagated asexually in greenhouses.)
- The sundew plant (genus *Drosera*) has paddle-shaped leaves with tiny nectar-covered hairs that lure insects. Once the insect alights on the plant and starts dining on its sweet, sticky meal, the surrounding leaf hairs begin to move. Gradually they fold inward, entrapping the helpless visitor and forcing it down toward the leaf's center. Here, powerful digestive enzymes dismantle the insect's body and release its nutrients. Afterward, all that remains of the guest are a few indigestible bits.
- The pitcher plant (genus *Sarracenia*) produces nectar along the rim of a slippery, tube-shaped leaf. Rainwater collects at the bottom of the trap. When a hapless insect falls in and drowns, an army of bacteria, protists, rotifers, mosquito larvae, and other invertebrates goes to work, decomposing the victim. The plant absorbs the nutrients through the thin cuticle lining the pitcher tube. (Section 22.5 examines pitcher plant nutrition in more detail).

It may seem strange that a plant would consume insects. Plants are, after all, photosynthetic, so they can make their own food. The answer to this puzzle is that carnivorous plants use insects as a source of nitrogen or phosphorus, not energy. Carnivorous plants are most abundant in boggy, acidic habitats. Organic matter decomposes slowly in waterlogged soils, and the few nutrients that are released often leach away. The leafy traps of carnivorous plants are adaptations that extract minerals from animals instead of from the soil. As this chapter describes, however, most plants have less exotic ways of acquiring nutrients.

UNIT 5

What's the Point?

Learning Outline

Learn How to Learn
A Quick Once-Over

Unless your instructor requires you to read your textbook in detail before class, try a quick preview. Read the chapter outline to identify the main ideas, then look at the figures and the key terms in the narrative. Previewing a chapter should help you follow the lecture, because you will already know the main ideas. In addition, note-taking will be easier if you recognize new vocabulary words from your quick once-over. Return to your book for an in-depth reading after class to help nail down the details.

22.1 | Soil and Air Provide Water and Nutrients

To stay healthy, a person needs water and the right dietary mix of fat, protein, carbohydrates, vitamins, and minerals (see section 31.4). Plants have similar needs, but they do not acquire these raw materials by eating and drinking. Instead, they are autotrophs that use photosynthesis to assemble organic molecules from elements absorbed from their surroundings.

This section describes the sources of the elements that plants require. The rest of the chapter focuses on how plants absorb and transport water, minerals, and organic substances such as sugar within their tissues.

A. Plants Require 16 Essential Elements

Like every organism, a plant requires certain **essential nutrients,** which are chemicals that are vital for metabolism, growth, and reproduction. Biologists have identified at least 16 elements essential to all plants (**figure 22.1**). Nine are **macronutrients,** meaning that they are needed in fairly large amounts. The macronutrients are carbon (C), oxygen (O), hydrogen (H), nitrogen (N), potassium (K), calcium (Ca), magnesium (Mg), phosphorus (P), and sulfur (S). The others are **micronutrients,** which are required in much smaller amounts.

Among the essential elements, C, H, and O are by far the most abundant, together accounting for about 96% of the dry weight of a plant. The six other macronutrients account for another 3.5%. Of these, N, P, and K are the most common ingredients in commercial fertilizers (see this chapter's Apply It Now box).

Gardeners and farmers use fertilizers to prevent or treat nutrient deficiencies such as those shown in **figure 22.2**. At the other end of the nutritional spectrum are plants that accumulate unusually high concentrations of some elements, such as zinc or nickel. This "hyperaccumulation" of heavy metals is an adaptation that reduces herbivory.

B. Soils Have Distinct Layers

Plants can grow in containers of water with dissolved nutrients, a technique called hydroponics. Most roots, however, extract water and mineral nutrients from **soil,** a complex mixture of rock particles, organic matter, air, and water. Soil is also home to bacteria, archaea, fungi, protists, and animals. The microbes decay organic matter, releasing inorganic nutrients that plant roots absorb.

Most people give little thought to the ground under their feet, but soil is critical to ecosystem function. Plants growing in soil provide food, habitat, and oxygen to humans and other animals. In addition, rain and melting snow seep into soil, which releases the moisture slowly, preventing flooding. Wastewater is also purified as it trickles through soil.

Soil formation begins when rocks disintegrate into small particles. Sandy soils have coarse particles; silty soils have finer ones. Clay particles are the smallest. Soil scientists classify a soil's texture based on the relative amounts of sand, silt, and clay. Soil texture is important to plant growth. The finer the particles, the more surface area per unit of soil volume and the higher the soil's capacity to store water.

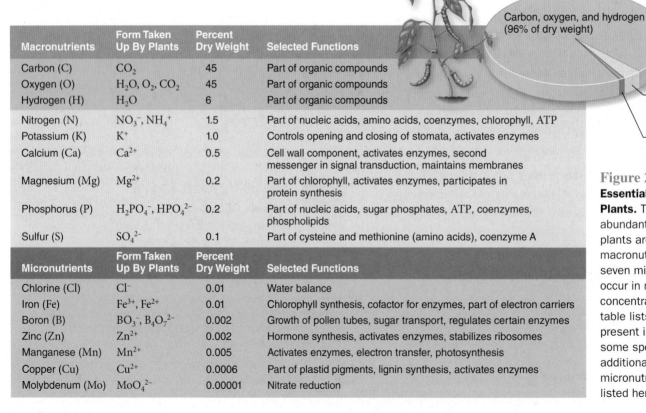

Macronutrients	Form Taken Up By Plants	Percent Dry Weight	Selected Functions
Carbon (C)	CO_2	45	Part of organic compounds
Oxygen (O)	H_2O, O_2, CO_2	45	Part of organic compounds
Hydrogen (H)	H_2O	6	Part of organic compounds
Nitrogen (N)	NO_3^-, NH_4^+	1.5	Part of nucleic acids, amino acids, coenzymes, chlorophyll, ATP
Potassium (K)	K^+	1.0	Controls opening and closing of stomata, activates enzymes
Calcium (Ca)	Ca^{2+}	0.5	Cell wall component, activates enzymes, second messenger in signal transduction, maintains membranes
Magnesium (Mg)	Mg^{2+}	0.2	Part of chlorophyll, activates enzymes, participates in protein synthesis
Phosphorus (P)	$H_2PO_4^-$, HPO_4^{2-}	0.2	Part of nucleic acids, sugar phosphates, ATP, coenzymes, phospholipids
Sulfur (S)	SO_4^{2-}	0.1	Part of cysteine and methionine (amino acids), coenzyme A

Micronutrients	Form Taken Up By Plants	Percent Dry Weight	Selected Functions
Chlorine (Cl)	Cl^-	0.01	Water balance
Iron (Fe)	Fe^{3+}, Fe^{2+}	0.01	Chlorophyll synthesis, cofactor for enzymes, part of electron carriers
Boron (B)	BO_3^-, $B_4O_7^{2-}$	0.002	Growth of pollen tubes, sugar transport, regulates certain enzymes
Zinc (Zn)	Zn^{2+}	0.002	Hormone synthesis, activates enzymes, stabilizes ribosomes
Manganese (Mn)	Mn^{2+}	0.005	Activates enzymes, electron transfer, photosynthesis
Copper (Cu)	Cu^{2+}	0.0006	Part of plastid pigments, lignin synthesis, activates enzymes
Molybdenum (Mo)	MoO_4^{2-}	0.00001	Nitrate reduction

Carbon, oxygen, and hydrogen (96% of dry weight)

Other macronutrients (~3.5%)

Micronutrients (~0.5%)

Figure 22.1

Essential Nutrients for Plants. The nine most abundant elements in plants are macronutrients; the seven micronutrients occur in much lower concentrations. The table lists the nutrients present in all plants; some species require additional micronutrients not listed here.

a.　　　　　　　　b.

Figure 22.2 Nutrient Deficiencies. (a) Phosphorus deficiency causes dark green or purple leaves in seedlings. (b) Iron deficiency causes yellowed leaves, but the veins remain green. A deficiency of manganese produces similar symptoms.

The disintegrating rock particles interact with water, plants, and other organisms. Over time, distinct soil layers develop (figure 22.3). Decomposing leaves and stems, called litter, lie on the soil's surface. Microbes decay the litter and release most of the carbon to the atmosphere as carbon dioxide (CO_2). Some

carbon, however, remains in the upper layer of soil as **humus,** a chemically complex, hard-to-digest, spongy organic substance. Most humus is in the upper soil layer, the **topsoil,** also called the A horizon. Topsoil typically supplies most of a plant's water and nutrients. Below topsoil is the B horizon, which has less organic matter, although roots extend to this depth. Still lower is the C horizon, which consists mostly of partially weathered pieces of rocks and minerals. Below the C horizon is bedrock.

Plant roots normally stabilize the topsoil and prevent **erosion,** the "wearing away" of soil by water and wind. Construction, overgrazing, farming, logging, road-building, and many other human activities remove plants and therefore promote erosion. Bare soil washes away, choking streams and lakes with sediment. Few plants grow in the absence of topsoil, so ecosystem recovery is slow in badly eroded soils.

Soil properties vary around the world; a rich prairie soil, for example, is very different from a desert soil or Arctic permafrost. Why? Part of the answer is that the mineral composition of the underlying bedrock varies. In addition, climate alters how minerals weather. Hot temperatures speed weathering, and rainfall leaches water-soluble nutrients through the soil. Climate also influences the plants and other organisms that shape the soil's properties.

C. Leaves and Roots Absorb Essential Elements

Plants obtain their three most abundant elements (C, H, and O) from water and the atmosphere. Water (H_2O) enters the plant through the roots, as described in section 22.2. Carbon and oxygen atoms come from the atmosphere in the form of CO_2 gas, which diffuses into the leaf or stem through pores called stomata.
▸ diffusion, p. 80

As roots absorb water, they also take up all of the other elements listed in figure 22.1. These nutrients dissolve in the soil's water when rock particles weather or when microbes decompose dead organisms.

Plants often have help from soil organisms in obtaining water and nutrients. Recall from chapter 19 that the roots of most land plants are colonized with mycorrhizal fungi. In exchange for carbohydrates produced in photosynthesis, the fungal filaments explore the soil and absorb water and minerals that a plant's roots could not reach otherwise. In particular, phosphorus is poorly soluble in water and does not move easily to roots. Mycorrhizae therefore especially boost phosphorus absorption.
▸ mycorrhizae, p. 402

The source of nitrogen also deserves special mention. As described in chapter 2, nitrogen atoms occur in proteins, nucleic acids, and chlorophyll. Nitrogen gas (N_2) makes up 78% of the atmosphere, but a strong triple covalent bond holds the two nitrogen atoms together. This bond requires more energy to break than a plant can muster. Instead, roots must take up nitrogen from soil in the form of nitrate (NO_3^-) or ammonium (NH_4^+). These ions can be scarce, so nitrogen availability often limits plant growth.

- Litter
- Topsoil (A horizon)
- Minerals and clay (B horizon)
- Partially weathered rock (C horizon)
- Bedrock

Figure 22.3 Soil Horizons. Soil consists of layers called horizons. Above the rich topsoil (the A horizon) is a litter layer. Water passing through the topsoil deposits clay and minerals in the B horizon, below which is partially weathered rock, the C horizon. Bedrock underlies the C horizon.

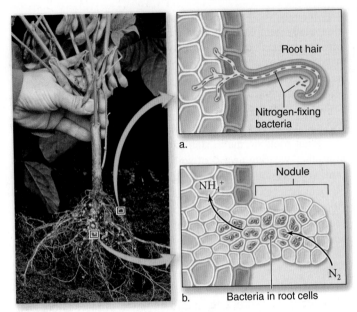

a.

b.

Root hair

Nitrogen-fixing
bacteria

Nodule

NH_4^+

N_2

Bacteria in root cells

Figure 22.4 **Root Nodules.** This soybean plant has abundant root nodules. (a) Nitrogen-fixing bacteria enter through root hairs and trigger the development of nodules. (b) In a mature nodule, the bacteria live symbiotically within the plant's cells. The plant provides the energy the bacteria need to fix nitrogen.

Fortunately for plants (and ultimately animals), several types of bacteria use **nitrogen-fixing** enzymes to convert N_2 to NH_4^+. Many nitrogen-fixing bacteria are free-living, but others live in growths called **nodules** on the roots of some types of plants (figures 22.4 and 16.11). The bacteria consume sugars that the host plant produces by photosynthesis. The plant, in turn, incorporates the nitrogen atoms from NH_4^+ into its own tissues. When the plant dies, decomposers make the nitrogen available to other organisms. The first step—nitrogen fixation—is therefore the key to the entire nitrogen cycle, bringing otherwise inaccessible nitrogen to plants, microbes, and all other life (see figure 37.16 for a detailed look at the nitrogen cycle).

The most famous nitrogen-fixing bacteria, in genus *Rhizobium,* stimulate nodule formation on the roots of plants called legumes (clover, beans, peas, peanuts, soybeans, and alfalfa). A farming technique called crop rotation alternates legume crops with nitrogen-hungry plants such as corn or cotton. The legume replenishes the soil's nitrogen, reducing the need for fertilizer.

A few plants acquire nutrients from more unusual sources. For example, carnivorous plants, described at the beginning of the chapter, often live in waterlogged, acidic soils. They obtain nitrogen and phosphorus from their prey.

22.1 | Mastering Concepts

1. Which macro- and micronutrients do all plants require?
2. How do plants acquire C, H, O, N, and P?
3. Describe the three typical soil horizons.
4. How do bacteria form a critical link in the nitrogen cycle?

Apply It Now

Boost Plant Growth with Fertilizer

Farmers and gardeners often amend soil with commercial synthetic fertilizers or with nutrient-rich organic matter, such as manure or compost. Plant growth surges if the soil amendment provides a nutrient that was previously scarce.

Nitrogen, phosphorus, and potassium are the three elements most commonly deficient in soils. Commercial fertilizer labels prominently display three numbers that indicate the content of these nutrients (figure 22.A). For example, a product marked 6-12-6 contains 6% elemental nitrogen (N), 12% phosphate (P_2O_5), and 6% potash (K_2O) by weight; the rest is filler. The label also lists other macro- and micronutrients in the fertilizer.

Nitrogen promotes leaf growth, whereas potassium and phosphorus encourage flowering and fruit development. A typical lawn fertilizer might therefore have a N-P-K ratio of 29-3-4. A fertilizer formulated for vegetables, on the other hand, would be more balanced at 10-20-20. By comparison, household compost is balanced but not very concentrated, with a N-P-K ratio of about 0.5-0.5-0.5.

Chemically, nutrients from inorganic fertilizer are equivalent to those from organic matter like manure or compost. So why should a gardener bother with organic soil amendments? The answer is that or-

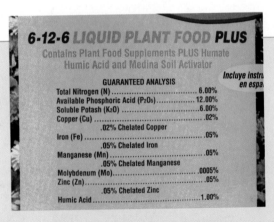

6-12-6 LIQUID PLANT FOOD **PLUS**
Contains Plant Food Supplements PLUS Humate
Humic Acid and Medina Soil Activator

GUARANTEED ANALYSIS	
Total Nitrogen (N)	6.00%
Available Phosphoric Acid (P_2O_5)	12.00%
Soluble Potash (K_2O)	6.00%
Copper (Cu)	.02%
.02% Chelated Copper	
Iron (Fe)	.05%
.05% Chelated Iron	
Manganese (Mn)	.05%
.05% Chelated Manganese	
Molybdenum (Mo)	.0005%
Zinc (Zn)	.05%
.05% Chelated Zinc	
Humic Acid	1.00%

Incluye instru
en espa

Figure 22.A **Read the Label.** Commercial fertilizer labels show which nutrients the product contains and in what quantities.

ganic matter not only improves fertility, but it also aerates the soil, increases the soil's water-holding capacity, and provides food for beneficial microbes and animals. Organic matter also releases nutrients slowly, resulting in less pollution from fertilizer runoff into storm drains, streams, lakes, and other waterways. These environmental benefits explain why organic farmers never amend soil with synthetic fertilizers. ▶ organic food, p. 32

22.2 | Water and Dissolved Minerals Are Pulled Up to Leaves

As you learned in chapter 21, xylem and phloem together form a continuous system of **vascular tissue,** a transportation network that connects the plant's roots, stems, leaves, flowers, and fruits (**figure 22.5**). This section describes how **xylem** transports **xylem sap,** a dilute solution consisting of water and dissolved minerals absorbed from soil. Phloem distributes the products of photosynthesis, as explained in section 22.3.

A. Water Vapor Is Lost from Leaves Through Transpiration

The amount of water that passes through plants each day is staggering. The corn plants occupying an area the size of a football field, for example, might use nearly 20,000 liters in one summer day, and a typical tree might consume 265 liters daily. Throughout a growing season, a plant uses 200 to 1000 liters of water to produce just 1 kilogram of tissue.

Why so much water? A plant's cells are mostly water, which is a medium for most of its metabolic reactions. Water participates in some of those reactions, including hydrolysis and photosynthesis. Furthermore, the watery cytoplasm exerts turgor pressure on the cell wall, which enables plant cells to elongate and helps the entire plant stay upright. Finally, the surfaces of leaf mesophyll cells must remain moist for CO_2 to diffuse inside. These uses, however, add up to only a small fraction of the water that a plant's roots pull in. The rest simply evaporates. ▶▶ hydrolysis, p. 31; turgor pressure, p. 81

The easiest way to visualize how plants acquire and transport water is to begin at the end. Plants lose water through **transpiration,** the evaporation of water from a leaf. Heat from the sun causes water in the cell walls to evaporate into the spaces between the leaf's cells. This evaporation helps cool the leaf, but it also establishes a gradient. That is, the concentration of water molecules inside the leaf is higher than in the air surrounding the leaf. Water vapor therefore diffuses from inside the leaf to the outside air. Most transpiration occurs through pores called stomata, but some water escapes through the cuticle that coats the epidermis.

The bigger the concentration gradient between the leaf interior and the surrounding air, the faster the transpiration rate.

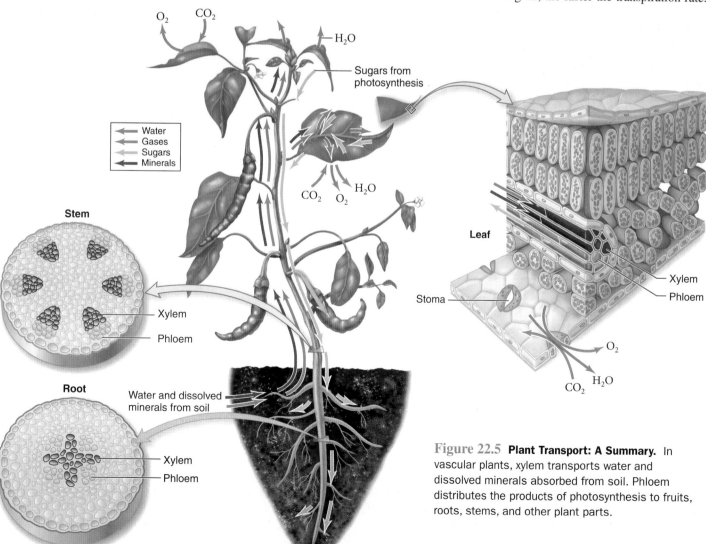

Figure 22.5 Plant Transport: A Summary. In vascular plants, xylem transports water and dissolved minerals absorbed from soil. Phloem distributes the products of photosynthesis to fruits, roots, stems, and other plant parts.

Low humidity, wind, and high temperatures therefore all cause the transpiration rate to soar. As described in section 22.2C, however, the plant's stomata close when air is too hot or too dry. When the stomata close, transpiration slows—but so does photosynthesis.

B. Xylem Transport Relies on Cohesion

The functional cells of the xylem, tracheids and vessel elements, are dead at maturity (see figure 21.5). Metabolic activity in xylem cells therefore cannot drive water transport in a plant. So how does water in the xylem get from roots to the mesophyll in leaves?

The **cohesion–tension theory** explains how xylem sap moves within a plant (**figure 22.6**). As its name implies, the cohesion–tension theory hinges on the cohesive properties of water—the tendency for water molecules to form hydrogen bonds and "cling" together. As water molecules evaporate from the leaf, additional water diffuses out of leaf veins and into the mesophyll. Water

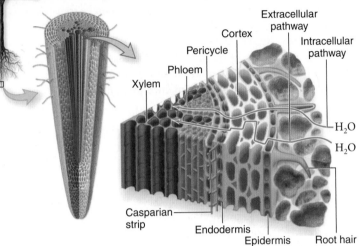

Figure 22.7 Two Routes into Roots. Water and dissolved minerals can travel along two routes through a root's epidermis and cortex: in the spaces between cells or through living cytoplasm. The waxy Casparian strip ensures that all incoming material passes through the living cells of the endodermis to reach the xylem.

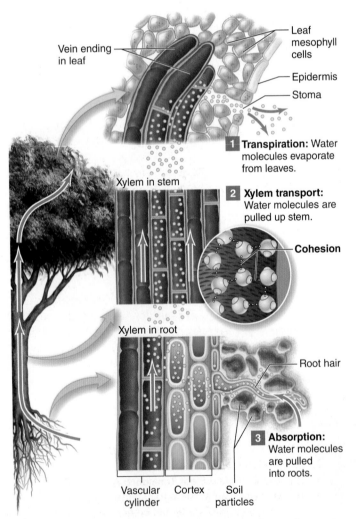

Vein ending in leaf

Leaf mesophyll cells

Epidermis

Stoma

1 Transpiration: Water molecules evaporate from leaves.

Xylem in stem

2 Xylem transport: Water molecules are pulled up stem.

Cohesion

Xylem in root

Root hair

3 Absorption: Water molecules are pulled into roots.

Vascular cylinder   Cortex   Soil particles

Figure 22.6 Xylem Transport. Transpiration of water from leaves pulls water up a plant's stem from the roots. The cohesiveness of water makes xylem transport possible.

molecules leaving the vein attract molecules adjacent to them in the xylem, pulling them under tension (negative pressure) toward the vein ending. Each water molecule tugs on the one behind it. ▸ cohesion, p. 27

As evaporation from leaf surfaces pulls water up the stem, additional water enters roots from the soil. Just behind each growing root tip, the epidermis is fringed with root hairs (see figure 21.15). The plant's millions of root hairs, coupled with filaments of mycorrhizal fungi, add up to an enormous surface area for water and mineral absorption.

Water and dissolved minerals can travel through the root's epidermis and cortex in two ways (**figure 22.7**). In the extracellular pathway, the solution moves in the spaces between cells or along the cell walls. Alternatively, water can move from cell to cell via plasmodesmata in the intracellular pathway. ▸ plasmodesmata, p. 65

The solution flows through the outer portion of the root until it contacts the endodermis, the innermost layer of the cortex. A waxy barrier called the **Casparian strip** forces the materials that had gone around cells to now enter the cells of the endodermis. Water enters by osmosis, because the concentration of solutes in cells is generally higher than in the soil. But not all solutes can enter the endodermal cells, thanks to membrane proteins that admit only certain ions. ▸ osmosis, p. 80

Materials that cross the endodermis continue into the xylem, enter the transpiration stream, and move up the plant. The water eventually returns to the atmosphere through the open stomata in the leaves and stem; the dissolved nutrients are incorporated into the plant's tissues.

As long as sufficient moisture is available in the soil, the cohesion between water molecules is enough to move continuous, narrow columns of xylem sap upward against the force of gravity. Notice that this mechanism exploits the physical properties of water, so the plant does not spend energy hauling xylem sap from soil to the tips of its leaves.

What if the soil is too dry to replace water lost in transpiration? The tension on the xylem sap may become so great that the water column stretches and breaks. The resulting air bubble prevents water movement in the affected tracheid or vessel. The plant's cells quickly lose turgor, and the leaves wilt. Likewise, cut flowers droop when their stems are removed from water, interrupting xylem flow.

Most plants face a trade-off between efficient water movement and the risk of air bubble formation; xylem anatomy reflects this balance. Recall from chapter 21 that tracheids occur in all vascular plants, but only angiosperms have vessels. When water is plentiful, vessels conduct xylem sap more efficiently than do the small-diameter tracheids with their overlapping end walls. But when water is scarce, an air bubble can disable an entire continuous vessel, which may extend for several meters. Tracheids, on the other hand, function individually. An air bubble that forms in a tracheid therefore disrupts the function of just that one cell.

The observation that air bubbles can interrupt xylem function provides important evidence supporting the cohesion–tension theory of water movement. If water were being pushed from below, air bubbles in the xylem stream would not pose a problem. Additional evidence comes from measurements of stem diameters, which decrease when the transpiration rate is highest. Water adheres to xylem walls; water under tension therefore pulls the tubes inward, narrowing their diameter.

C. The Cuticle and Stomata Help Conserve Water

The **cuticle** is an important water-saving adaptation in land plants. This waxy layer covers the epidermis of the leaves and primary stem (see figure 21.7). Impermeable to water and gases, the cuticle prevents the plant's tissues from drying out.

In addition, **stomata** are pores that permeate the cuticle and permit the leaf to exchange gases with the atmosphere. As long as the soil contains enough moisture to replace water lost in transpiration, the plant will flourish. But if the soil dries out, the plant will wilt—unless it closes its stomata and temporarily shuts down the transpiration stream.

A pair of **guard cells** determines whether each stoma is open or closed (figure 22.8). When water is plentiful, membrane proteins in the guard cells import potassium ions (K^+) from adjacent cells in the epidermis. The concentration of K^+ therefore becomes greater inside than outside the guard cells. Water follows by osmosis. The incoming water swells the guard cells, so the stoma opens. During drought stress, on the other hand, a plant hormone called abscisic acid binds to the guard cell membranes. The hormone indirectly triggers the loss of K^+ from the guard cells. As water exits the guard cells by osmosis, the cells lose turgor and collapse against each other. The pore between them closes, and the transpiration stream stops. ▶ abscisic acid, p. 500

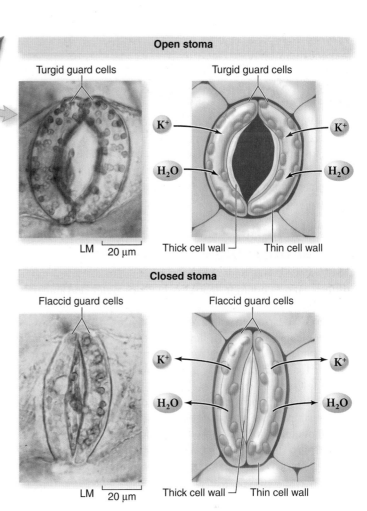

Figure 22.8 Guard Cell Function. When water is abundant, guard cells use energy to import potassium ions (K^+) from adjacent cells. Water follows by osmosis; the guard cells swell, opening the stoma. When water is scarce, hormonal signals stimulate K^+ to leave the guard cells. Water follows, and the stoma closes as the flaccid guard cells collapse.

A plant that closes its stomata, however, also cuts off its supply of CO_2 for use in photosynthesis. Most plants therefore conserve water by closing their stomata after dark, when photosynthesis cannot occur anyway. The C_4 and CAM pathways of photosynthesis also save water, especially in dry habitats. ▶ C_4 and CAM plants, p. 99

22.2 | Mastering Concepts

1. What are the components of xylem sap?
2. How does transpiration occur?
3. Summarize the cohesion–tension theory.
4. How do water and dissolved minerals enter roots and move through the endodermis to the xylem?
5. How do the cuticle and stomata help plants conserve water?

22.3 | Organic Compounds Are Pushed to Nonphotosynthetic Cells

 With sufficient light, water, and nutrients, a photosynthetic cell will produce sugars that can be transported in **phloem** to the plant's nonphotosynthetic cells, which cannot produce food on their own. The major transport structures of phloem are the microscopic sieve tubes (see figure 21.5). Unlike cells of xylem, the cells that make up sieve tubes are alive.

A. Phloem Sap Contains Sugars and Other Organic Compounds

The organic compounds carried in phloem are dissolved in the **phloem sap,** a solution that also includes water and minerals from the xylem. The carbohydrates in phloem sap are mostly dissolved sugars such as sucrose (see figure 2.18). Phloem sap also contains amino acids, hormones, enzymes, and messenger RNA molecules. (Although phloem sap is the most common vehicle for sugar transport, it is not the only one, as this chapter's Burning Question explains.)

Studying phloem composition is difficult. An investigator cannot simply cut a stem and squeeze out the phloem sap because the plant quickly plugs the wound. Biologists receive help, however, from a surprising source: aphids (figure 22.9). These insects feed on phloem sap without triggering the wound response, and they excrete a sticky, sweet material called "honeydew." Scientists found that if they amputated the feeding tube of an aphid

while it was dining on a plant, sugary drops continued to flow from the cut end of the tube. In this way, phloem sap could be harvested and analyzed.

B. The Pressure Flow Theory Explains Phloem Function

The explanation of phloem transport is called the **pressure flow theory.** This theory suggests that phloem sap moves under positive pressure from "sources" to "sinks." A **source** is any plant

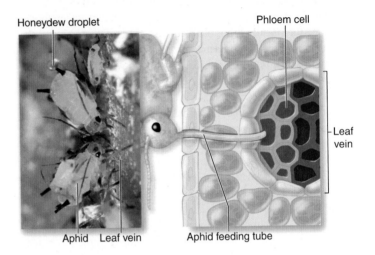

Honeydew droplet · Phloem cell · Aphid · Leaf vein · Aphid feeding tube · Leaf vein

Figure 22.9 Phloem Feeders. Aphids are insects with feeding tubes that penetrate phloem cells. The "honeydew" droplets emitted from their rear ends are derived from phloem sap.

Burning Question

Where does maple syrup come from?

The source of maple syrup is the xylem sap of the sugar maple tree (*Acer saccharum*). During the fall, the tree stores carbohydrates in the living cells of its stem and roots. As the temperature warms in early spring, the stockpiled sugars fuel the production of new leaves.

Most sap flow in sugar maples occurs before noon on warm days that follow a freezing night in winter or spring. This flow was once a bit of a puzzle, especially since the cohesion–tension theory that explains most water movement in the xylem does not apply. (During the sap flow period, sugar maples lack leaves, which are required for transpiration.) Instead, the sap flow apparently results from alternating freezing and thawing of the xylem tissues. During the day, respiring cells in the stem produce CO_2. At night, compressed CO_2 bubbles are trapped in ice that forms in the xylem. When the xylem thaws during the day, the gases expand

once more, pushing the sap up the tree. Thus, sugar maple sap flows under positive, not negative, pressure.

To harvest the sap, collectors drill a hole and insert a spout through the tree's bark and into the xylem. The xylem sap drips off the end of the spout and into a container (figure 22.B). Each tap produces about 40 liters of raw xylem sap each year, which boils down to about 1 liter of finished syrup. The distinctive maple syrup flavor develops as heat alters some of the sap's chemical constituents.

Submit your burning question to:
marielle_hoefnagels@mcgraw-hill.com

Figure 22.B Sap Flow. The sugar maple tree produces sweet xylem sap, the precursor to maple syrup.

part that produces or releases sugars; a **sink** is any plant part that does not photosynthesize. Examples of sinks include flowers, fruits, shoot apical meristems, roots, and storage organs. If these cells do not receive enough sugar to generate the ATP they require, the plant may die or fail to reproduce. ▶ meristems, p. 458

Inside a leaf or other sugar source, companion cells load sucrose into sieve tube elements by active transport. Because sucrose becomes so much more concentrated in the sieve tubes than in the adjacent xylem, water moves by osmosis out of the xylem and into the phloem sap. The resulting increased turgor pressure drives phloem sap through the sieve tubes (**figure 22.10**). ▶ active transport, p. 82

At a root, flower, fruit, or other sink, cells take up the sucrose (and other compounds in the phloem sap) through facilitated diffusion or active transport. As the sucrose is unloaded from the sieve tubes, the concentration of solutes in the phloem sap declines. Water therefore moves by osmosis from the sieve tube to the surrounding tissue (often xylem). Movement of water out of the sieve tube relieves the pressure, so the phloem sap in the sieve tube continues to flow toward the sink.

A given organ may act as either a sink or a source. For example, a developing potato tuber is a sink, storing the plant's sugars in the form of starch. Later, when the plant uses those stored

reserves to fuel the growth of new tissues, that same tuber becomes a source. The starch in the tuber breaks down into simple sugars, which are loaded into phloem sap for transport to other plant parts.

A similar mobilization occurs each spring when a deciduous tree produces new stems and leaves, using carbohydrates stored in roots. Later in the growing season, leaves approach their mature size and produce sugar of their own. The leaves are then sources, and the roots are again sinks.

Evidence for the pressure flow theory includes the observation that phloem sap moves through aphids under positive pressure, a little like toothpaste being squeezed from a tube. Microscopes that produce images of living tissue "in action" have also confirmed some details of phloem function.

22.3 | Mastering Concepts

1. What are the components of phloem sap?
2. Explain the pressure flow theory of phloem transport.
3. What are some examples of sources and sinks in a plant?
4. What is the evidence for the pressure flow theory?

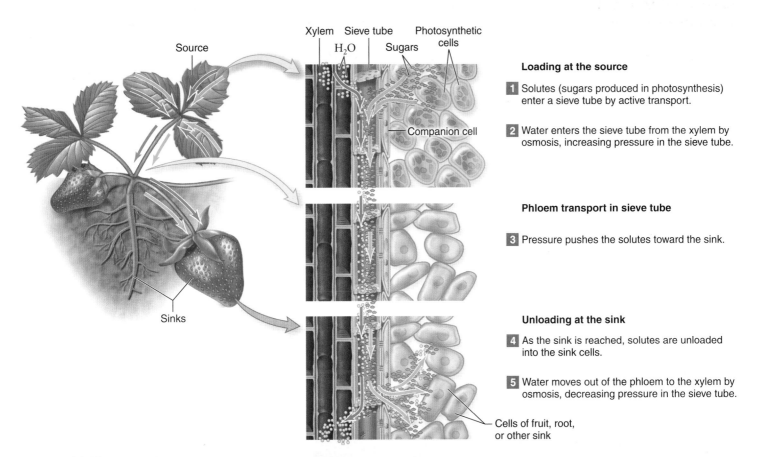

Loading at the source

1 Solutes (sugars produced in photosynthesis) enter a sieve tube by active transport.

2 Water enters the sieve tube from the xylem by osmosis, increasing pressure in the sieve tube.

Phloem transport in sieve tube

3 Pressure pushes the solutes toward the sink.

Unloading at the sink

4 As the sink is reached, solutes are unloaded into the sink cells.

5 Water moves out of the phloem to the xylem by osmosis, decreasing pressure in the sieve tube.

Figure 22.10 Phloem Transport. According to the pressure flow theory, sugar produced in green "source" organs such as leaves moves under positive pressure to roots, fruits, and other nonphotosynthetic sinks.

22.4 | Parasitic Plants Tap into Another Plant's Vascular Tissue

Of the hundreds of thousands of plant species, most are self-sufficient. They produce their own food by photosynthesis, and they absorb their own nutrients and water from soil (often with the help of mycorrhizal fungi). Some plant species, however, are parasites that exploit the hard-won resources of other plants.

Parasitic plants acquire water, minerals, and food by tapping into the vascular tissues of their hosts. The story begins with the parasite's seeds, which are carried to the host by birds or released explosively from seed pods. Either way, a seed germinates, and the seedling secretes an adhesive that sticks the young plant to its host. The seedling's root pushes through the host's epidermis and forms a structure called a haustorium, which connects the parasite's vascular tissues to those of the host.

Dodder is one example of a parasitic plant. Its leafless stems, which lack chlorophyll, drape over host plants like cooked spaghetti. The most common parasites, however, are the many species of mistletoe. These dark green shrubs live in the branches of host trees throughout the United States (**figure 22.11**). Mistletoe produces some of its own food by photosynthesis, but other parasitic plants depend entirely on their hosts.

Most plants infected with mistletoe are weakened but do not die, an observation that makes sense from an evolutionary perspective. The most successful parasites extract enough resources to survive and reproduce—but not so much that the host dies. After all, a dead plant is of no use to a parasite that requires a living host.

22.4 | Mastering Concepts

1. What is a haustorium?
2. Briefly describe how a parasitic plant infects a host.

Figure 22.11 Mistletoe. The branches of this apple tree are heavily infested with parasitic mistletoe plants.

22.5 | Investigating Life: The Hidden Cost of Traps

Carnivorous plants seem to have it made: they live in boggy habitats with plentiful water, their invertebrate prey provide nitrogen and phosphorus, and they can use energy from sunlight to produce their own food.

But if the carnivorous lifestyle is so easy, why don't all plants eat bugs? After all, invertebrates are so abundant that natural selection should favor plants that exploit animal sources of nutrients. Yet rather than taking over the world, most carnivorous plants are limited to sunny, nutrient-poor wetlands.

Biologists hypothesize that these plants cannot compete in most other habitats because carnivory involves an evolutionary trade-off. To catch an animal, a plant must construct a trap. These structures vary from a sticky surface (as in sundews) to a fluid-filled chamber (as in pitcher plants) to a clamshell-shaped snap trap (as in Venus flytraps) to a hollow bladder with a door that closes after an insect enters (as in bladderworts). Although the plant uses the nutrients extracted from the prey, the trap has a cost: reduced photosynthetic efficiency. A plant that invests energy in insect traps therefore has a reduced leaf area available for photosynthesis.

One prediction that emerges from this hypothetical cost-benefit balance is that supplementing a carnivorous plant's habitat with excess nutrients should tip the balance toward leaves that maximize photosynthesis. That is, if a carnivorous plant has sufficient nutrients, then investing in photosynthetic leaf area should benefit the plant more than building new traps.

Aaron Ellison, a biologist at Harvard Forest in Massachusetts, worked with University of Vermont biologist Nicholas Gotelli to test this prediction. They studied the northern pitcher plant, *Sarracenia purpurea* (**figure 22.12**). The prey-capturing pitchers are modified leaves that sprout from the underground rhizomes of this carnivorous plant. The pitcher's tube develops when the edges of a folded leaf fuse together; a flaplike keel reinforces the seam. Once the sealed tube fills with rainwater, invertebrates can fall in, drown, and decay. The pitchers absorb the nutrients through a thin cuticle.

Ellison and Gotelli set up a field experiment in which they manipulated the concentrations of nutrients available to pitcher plants. Ninety adult plants were randomly assigned to one of nine treatments (**table 22.1**). Every two weeks from June through September, the researchers poured 5 milliliters of the assigned solution into every open pitcher of each plant. Afterward, they immediately plugged each pitcher with glass wool, a cottony inorganic substance that prevented prey capture. The nutrient additions continued throughout the growing seasons of three consecutive years.

The researchers knew that leaf shape in pitcher plants follows a continuum, with larger keels generally associated with greater photosynthetic efficiency. A normal pitcher has a narrow keel and a wide pitcher opening, maximizing the potential for

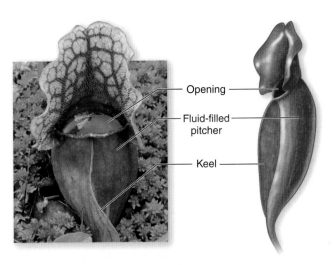

Figure 22.12 Anatomy of a Pitcher. Insects enter the fluid-filled pitcher through the mouthlike opening. The keel is the flap of tissue running the length of the pitcher.

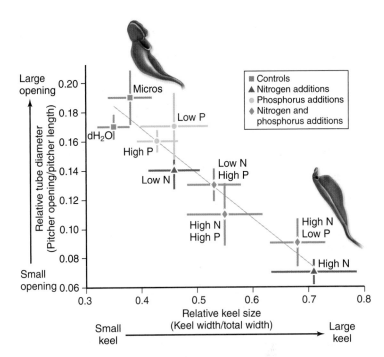

Figure 22.13 Nitrogen Matters. Pitcher plants that received supplemental nitrogen had much larger keels and smaller pitchers than plants that received little nitrogen, which produced normal pitchers. The addition of phosphorus did not have the same effect. In this graph, each point represents a treatment mean; horizontal and vertical bars represent 95% confidence intervals.

trapping prey. On the other hand, a leaf optimized for photosynthesis has a wide keel and a small pitcher opening.

Every September, the researchers measured the shapes of all of the leaves on each study plant; figure 22.13 summarizes the results. The control plants, with low nutrient availability, had prominent pitchers with wide openings and small keels. Supplemental phosphorus did not affect leaf shape, but added nitrogen did. The high-nitrogen treatments were associated with the largest keels and the smallest pitcher openings.

But do leaves with large keels really maximize photosynthesis at the expense of prey capture? To find out, Ellison and Gotelli measured the photosynthetic rate of the largest leaf on each plant. The data supported the prediction: the wider the keel, the greater the rate at which a leaf absorbed CO_2.

Overall, the study showed that a shortage of nitrogen favored prey capture at the expense of photosynthetic leaf area. Nutrient manipulation, coupled with careful leaf measurements, has therefore helped shed light on an interesting twist on plant evolution. Carnivory apparently represents a significant trade-off; in most terrestrial habitats, the traps would simply cost more energy than they are worth.

Ellison, Aaron M., and Nicholas J. Gotelli. 2002. Nitrogen availability alters the expression of carnivory in the northern pitcher plant, *Sarracenia purpurea*. *Proceedings of the National Academy of Sciences*, vol. 99, no. 7, pp. 4409–4412.

22.5 | Mastering Concepts

1. Describe the hypothesis and experimental design in Ellison and Gotelli's study.
2. Explain the purpose of the two control solutions.
3. Predict how the graph in figure 22.13 would look if nitrogen did not affect leaf morphology.

Table 22.1   Treatments in Pitcher Plant Study

Treatment	Nutrient Solution
Control (water only)	Distilled water only
Control (micronutrients)	Micronutrients only
Low N	0.1 mg NH_4-N/liter + micronutrients
High N	1.0 mg NH_4-N/liter + micronutrients
Low P	0.025 mg PO_4-P/liter + micronutrients
High P	0.25 mg PO_4-P/liter + micronutrients
Low N + high P	0.1 mg NH_4-N/liter + 0.25 mg PO_4-P/liter + micronutrients
High N + high P	1.0 mg NH_4-N/liter + 0.25 mg PO_4-P/liter + micronutrients
High N + low P	1.0 mg NH_4-N/liter + 0.025 mg PO_4-P/liter + micronutrients

Chapter Summary

22.1 | Soil and Air Provide Water and Nutrients

A. Plants Require 16 Essential Elements

- Like all organisms, plants require water and **essential nutrients.**
- In all plants, the nine essential **macronutrients** are C, H, O, P, K, N, S, Ca, and Mg. The seven **micronutrients** are Cl, Fe, B, Mn, Zn, Cu, and Mo.

B. Soils Have Distinct Layers

- **Soil** consists of rock and mineral particles mixed with living organisms and decaying organic molecules. Soil layers are called horizons.
- **Topsoil** contains **humus,** a major source of water and nutrients for plants. **Erosion** destroys topsoil and hinders plant growth.

C. Leaves and Roots Absorb Essential Elements

- Plants obtain CO_2 and O_2 from the atmosphere, and they acquire hydrogen and oxygen from H_2O in soil. The other elements also come from soil.
- Mycorrhizal fungi add to the root's surface area for absorbing water and nutrients, especially phosphorus. Several types of bacteria live in root **nodules** and **fix nitrogen,** converting it into forms that plants can use.

22.2 | Water and Dissolved Minerals Are Pulled Up to Leaves

- **Vascular tissue,** which consists of **xylem** and **phloem,** transports materials within plants.

A. Water Vapor Is Lost from Leaves Through Transpiration

- **Xylem sap** consists of water and dissolved minerals.
- Leaves lose water by **transpiration** through open stomata.

B. Xylem Transport Relies on Cohesion

- According to the **cohesion–tension theory,** water molecules evaporating from leaves are replaced by those pulled up from below, a consequence of the cohesive properties of water.
- Water enters roots by osmosis because the solute concentration in the soil is less than that of root cells.
- Water and dissolved minerals move through the root's epidermis and cortex either between cells or through cell interiors. The endodermis, with its impermeable **Casparian strip,** controls which substances enter the nearby xylem.

C. The Cuticle and Stomata Help Conserve Water

- The waxy, waterproof **cuticle** prevents water loss in aerial plant parts.
- Plants can open and close **stomata** in response to water availability. When **guard cells** bordering each stoma absorb water, the pore opens; when the guard cells lose water and collapse, the stoma closes.

22.3 | Organic Compounds Are Pushed to Nonphotosynthetic Cells

A. Phloem Sap Contains Sugars and Other Organic Compounds

- **Phloem sap** includes sugars, hormones, viruses, and other organic molecules, along with water and minerals from xylem.

B. The Pressure Flow Theory Explains Phloem Function

- According to the **pressure flow theory,** phloem sap flows under positive pressure through sieve tubes from a **source** to a **sink.**
- Sources are photosynthetic or sugar-storing parts that load carbohydrates into phloem. Water follows by osmosis. The increase in pressure drives phloem sap to nonphotosynthetic sinks such as roots, flowers, and fruits.

22.4 | Parasitic Plants Tap into Another Plant's Vascular Tissue

- Mistletoe, dodder, and other parasitic plants form a haustorium that absorbs water, minerals, and sugar from a host plant's xylem and phloem.

22.5 | Investigating Life: The Hidden Cost of Traps

- Carnivorous plants produce leaves that specialize in prey capture or photosynthesis. Experiments show that nutrient additions tip the balance in favor of photosynthetic leaves.

Multiple Choice Questions

1. What is the difference between a macronutrient and a micronutrient?
   a. The size of the atoms
   b. The amount required by the plant
   c. The use by multicellular (macroorganisms) versus microorganisms
   d. The source of the nutrient

2. Which soil layer is most important for supporting plant growth?
   a. The A horizon   c. The C horizon
   b. The B horizon   d. Both b and c are correct.

3. How do nitrogen-fixing bacteria contribute to the nitrogen cycle?
   a. They decompose dead plant material, releasing N_2.
   b. They convert NH_4^+ to N_2 for release into the atmosphere.
   c. They reduce N_2 to NH_4^+ that plants can absorb.
   d. They release N_2 from root nodules.

4. What is transpiration?
   a. The evaporation of water from leaves
   b. The movement of water through the xylem
   c. The uptake of water through root hairs
   d. All of the above are correct.

5. According to the cohesion–tension theory, transpiration drives xylem sap movement because of
   a. the presence of the Casparian strip.
   b. hydrogen bond formation between water molecules.
   c. upward pressure from water moving into roots.
   d. the presence of the cuticle.

6. What determines whether a mineral dissolved in the soil becomes part of a plant?
   a. The concentration of the mineral within the soil
   b. The presence of specific transport proteins along the epidermis of the root hair
   c. The presence of specific transport proteins in the endodermis of the root
   d. The osmotic balance between the root and the soil

7. Why do desert plants that live in very hot and dry conditions have stomata?
   a. To allow for transpiration
   b. To allow for gas exchange
   c. To allow for the release of excess heat
   d. Both a and b are correct.

8. Where do the simple sugars in phloem sap ultimately come from?
   a. The roots   c. Photosynthesis
   b. Storage organs such as fruits d. Starch

9. Which of the following is NOT a sink?
   a. A root   c. A flower
   b. A leaf   d. A meristem

10. The haustorium of a parasitic plant
    a. consumes insects as a source of nitrogen.
    b. carries out photosynthesis.
    c. absorbs materials from the host plant.
    d. produces the seeds, which are dispersed by birds.

Write It Out

1. On a planet with an atmosphere similar to Earth's, which elements would have to be present for earthly plants to grow?

2. How do plants obtain carbon and nitrogen?

3. Clover is a legume that requires nitrogen, as does every other plant. How would planting clover in a crop rotation reduce the need for nitrogen fertilizer?

4. How might transgenic technology (see chapter 7) be used to endow plants with the ability to fix nitrogen without the aid of bacteria? In what ways would this new feature change agriculture?

5. Why don't plants grow on a path that has been trampled by repeated foot or vehicle traffic?

6. Use the Internet to learn about soil conservation. How do soil conservation practices work, and why are they important? Research the historical factors and farming practices that led to the Dust Bowl.

7. When Chris mows the grass, she faces a choice between discarding the clippings or leaving them on the lawn. How would each choice influence the nutrient content of the soil? Explain your answer.

8. Some plants "hyperaccumulate" metals such as nickel. Their tissues contain nickel concentrations that would be toxic to most other plants. How might hyperaccumulation be adaptive to the plant? How might people put hyperaccumulators to practical use?

9. Trace the path of water and dissolved minerals from soil, into the root's xylem, and up to the leaves.

10. Explain how cohesion plays a part in xylem transport.

11. Are roots necessary for transpiration to occur? Are leaves? Explain your answers.

12. Suppose you use a rubber band to secure a clear plastic bag around a few leaves on a live plant. What do you think will happen?

13. "Root pressure theory" proposes that water from roots is pushed up a stem. Although root pressure sometimes occurs, it does not account for all xylem flow. What evidence would you look for to determine whether water in xylem moves under tension or pressure?

14. During what time of day is a plant at greatest risk for the formation of air bubbles in the xylem stream?

15. Use osmosis to explain why plants would have a hard time extracting water from extremely salty soil.

16. Review C_3, C_4, and CAM photosynthesis in chapter 5. Explain how plants that use each pathway conserve water.

17. Angiosperms have both tracheids and vessel elements; conifers and ferns have only tracheids. Propose a hypothesis that explains how xylem with vessel elements might be advantageous over xylem that contains only tracheids. In what circumstances might tracheids-only xylem be adaptive?

18. What special water transport problems do plants encounter in habitats subject to periodic flooding or prolonged drought?

19. Distinguish between a source and a sink. How can the same plant part act as both a source and a sink?

20. How does xylem flow influence phloem flow?

21. Peach and nectarine growers remove some flowers and small, immature fruits from trees. Why does this practice yield larger, higher-quality fruit?

22. Suppose that a scientist exposes a leaf to CO_2 labeled with carbon-14, and the radioactive carbon is incorporated into organic compounds in photosynthetic cells. At various times after exposure, the scientist can determine the location of the radioactive carbon in the plant. In what tissues do you expect to find the radioactive material immediately after exposure to the labeled carbon? What about during transport? When transport is complete, will the radioactive material be in plant parts above the leaf, below the leaf, or both?

23. Scientists studying phloem function find that if they use a tiny instrument to damage sieve tubes, the plant begins to plug the wound within a minute. Why is it adaptive for the plant to prevent leakage of phloem sap?

24. Make a chart comparing xylem and phloem transport. Include sap composition, characteristics of the conducting cells, how the sap moves, whether the transport costs energy, and direction of flow within the plant.

25. At which points in the transport of xylem sap and phloem sap does osmosis occur? At which points does active transport occur?

26. Explain two ways that leaf mesophyll cells and root cortical cells interact.

27. Taylor has a tree with mistletoe. He thinks he can solve the problem by knocking the mistletoe off the branches. How would you explain why the only effective treatment is to prune off infected branches?

28. Pitcher plants are threatened in the wild by habitat destruction and collection for plant trade. Researchers are also concerned that air pollution will deposit additional nitrogen on soil. How would additional nitrogen affect pitcher plants?

29. In the early 1600s, Jean Baptista van Helmont investigated how plants acquire new mass as they grow. He weighed and planted a willow shoot into soil that he had also weighed. After 5 years of adding nothing but water to the plant, he found that the soil had lost only a little weight, while the plant had grown from 2 kilograms to about 76 kg. He therefore concluded, incorrectly, that water was the sole source of the added plant material. What other source did he fail to consider in his experiment?

30. Long-term space travel will require the construction of on-board "closed loop" ecosystems that will produce food and oxygen for astronauts. Suppose you are a NASA engineer designing such an ecosystem. Brainstorm the characteristics of the ideal space plant. Which microbes would you need to include? How would the astronauts interact with the ecosystem? What problems can you foresee with growing plants in space?

31. Phytoremediation is the use of plants to treat environmental problems. Search the Internet for applications of phytoremediation. What are the benefits of phytoremediation? If you were trying to discover plants suitable for use in phytoremediation, what qualities would you look for? Can you foresee any problems with phytoremediation?

32. Some architects specialize in building living roofs covered with plants. What are the benefits and risks of living roofs? What raw materials must be supplied to plants living on a roof? What types of plants would be the best choice? How does your answer depend on where you live?

Pull It Together

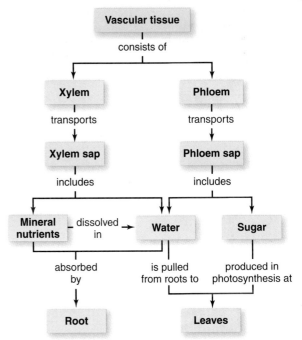

1. Add the terms *soil, source, sink, pressure flow*, and *transpiration* to the concept map.

2. What role do the stomata play in transpiration?

3. What are examples of sources and sinks in phloem transport?

4. Which nutrients do roots absorb from soil, and which do leaves absorb from the atmosphere?

5. Besides sugar, what other organic substances does phloem transport?

Reproduction and Development of Flowering Plants

Partners. This beekeeper is tending to her beehives. The honeybees occupying the hives aid in the reproduction of many types of flowering plants, including crops that humans depend on for food.

BIOLOGY

Enhance your study of
this chapter with practice quizzes,
animations and videos, answer keys,
and downloadable study tools.
www.mhhe.com/hoefnagels

Imperiled Pollinators

NEARLY ALL ANIMALS DEPEND ULTIMATELY ON PLANTS FOR FOOD, HABITAT, AND OXYGEN. Yet many flowering plant species, including many of the crops we raise for food, depend on animals as well. The plants produce nectar, a sweet substance that lures insects, birds, and small mammals. As the animals explore the blooms for food, they unwittingly transfer pollen from one flower to another. As we shall see in this chapter, pollination is a critical link in the life cycle of the flowering plant.

Bees are proficient pollinators. European honeybees *(Apis mellifera),* which were introduced to the United States in 1621, visit many flowering plant species. In fact, honeybees gather pollen and nectar from so many plants in so many habitats that they have displaced native bees.

But bee populations worldwide are plummeting. In 2006, beekeepers began to report a mysterious ailment called colony collapse disorder. The bees in affected hives disappear for no known reason. The insects have weakened defenses and harbor unusually high populations of harmful bacteria and fungi, but so far no one knows exactly why the bees are dying.

Some possible culprits, however, have been ruled out. For example, researchers have concluded that neither radiation from cell phone towers nor insecticide-producing genes from genetically modified plants are responsible for the die-off.

So what is causing colony collapse disorder? The problem is especially difficult to study because a combination of conditions may be to blame. The use of pesticides in agriculture has killed many bees and stressed many more. Moths and mites threaten bees, too. Caterpillars of the greater wax moth *(Galleria mellonella)* tunnel through honeycombs and eat beeswax, pollen, and cocoons. Tracheal mites *(Acarapis woodi)* colonize a bee's respiratory system and consume the host's blood; an external mite *(Varroa)* sucks the blood of young bees and adults. Infectious diseases also take a toll. A bacterial disease called American foulbrood kills bee larvae, and single-celled parasites called microsporidia *(Nosema* spp.) damage the bee's digestive tract and increase susceptibility to other diseases. Researchers hypothesize that the combination of these stresses has made bees vulnerable to a particular virus, but no one knows for sure.

Bee experts, called apiculturists, predict that the falling bee populations will soon lead to declining crops of many edible fruits and seeds. What can we do? Entomologists suggest boosting populations of wild bees, such as bumblebees, to offset the loss of honeybees. Strategies include avoiding the use of pesticides, planting clusters of diverse native plants in yards, providing nesting habitat for bees, and surrounding agricultural fields with native grasses, herbs, and trees.

UNIT 5

What's the Point?

Learning Outline

Learn How to Learn

Know Yourself

Setting aside time to study is one important ingredient for academic success; another is paying attention to your work habits throughout the day and night. Are you most alert in the morning, afternoon, or evening? Block off time to study during periods when you are at your best. Your study time will be much more productive if you are not fighting to stay awake.

23.1 | Angiosperms Reproduce Asexually and Sexually

Flowering plants (angiosperms) dominate many terrestrial landscapes. From grasslands to deciduous forests, and from garden plots to large-scale agriculture, angiosperms have been extremely successful. These plants first evolved only about 130 million years ago, yet their subsequent rapid diversification has been one the most extraordinary examples of adaptive radiation in the history of life. ▶ adaptive radiation, p. 282

Angiosperms owe their success to three adaptations. First, pollen enables sperm to fertilize an egg in the absence of free water. In contrast, the sperm cells of mosses and ferns must swim to the egg, so these plants can reproduce sexually only in moist habitats. Second, the seed protects the embryo during dormancy and nourishes the developing seedling. Third, flowers not only promote pollination but also develop into fruits that help disperse the seeds far from the parent plant.

Most angiosperms reproduce sexually, although some species also reproduce asexually. This section describes asexual reproduction and briefly introduces sexual reproduction, a topic described in detail in section 23.2. (For comparison, chapter 18 describes the life cycles of the mosses, ferns, and conifers).

A. Asexual Reproduction Yields Clones

Many plants reproduce asexually, forming new individuals by mitotic cell division. In **asexual reproduction,** a parent organism produces offspring that are genetically identical to it and to each other—they are clones. Asexual reproduction, also called vegetative reproduction, is advantageous when conditions are stable and plants are well adapted to their surroundings, because the clones will be equally suited to the same environment.

Plants often reproduce asexually by forming new plants from portions of their roots, stems, or leaves (**figure 23.1**). For example, aerial shoots can grow upward from buds on the roots of cherry, pear, apple, and black locust plants. If these shoots, called "suckers," are cut or broken away from the parent plant, they can become new individuals.

Asexual reproduction has many commercial applications. Most houseplants arise from cuttings taken from parent plants. In addition, most fruit and nut trees are produced by grafting a scion (part of a parent tree) to rootstock from a different but closely related plant. The grower selects the scion for the quality of its fruit. Usually the rootstock is either disease- and pest-resistant or especially well adapted to dry or salty soil. Grafting the scion to the rootstock gives the grower the advantages of both.

Scientists can use tissue culture to produce huge numbers of identical plants in a laboratory (**figure 23.2**). The first step is to place pieces of plant tissue in a dish containing nutrients and hormones. After a few days, the plant cells lose their specialized characteristics and form a white lump called a callus. The cells of the callus divide for a few weeks. The lump is then transferred to a dish containing a new mix of hormones, prompting some cells to develop into tiny, genetically identical plantlets with shoots and roots. Each plantlet can then be separated from the others and grown into an individual plant.

Callus growth is unique to plants. The human equivalent would be a cultured skin cell multiplying into a blob of unspecialized tissue and then sprouting tiny humans!

B. Sexual Reproduction Generates Variability

Asexually generated plants have predictable characteristics because they are essentially identical to their parents

a.

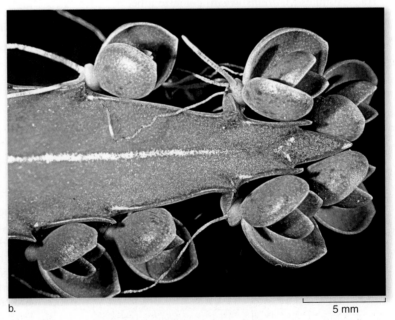

b.

5 mm

Figure 23.1 Asexual Reproduction. (a) Quaking aspen (*Populus tremuloides*) trees are clones connected by a common root system. (b) The leaf of this kalanchoe plant produces genetically identical plantlets, complete with leaves and tiny roots.

Figure 23.2 Asexual Reproduction in a Dish. When plant tissue is cultured with the correct combination of hormones and nutrients, a callus gives rise to genetically identical plantlets.

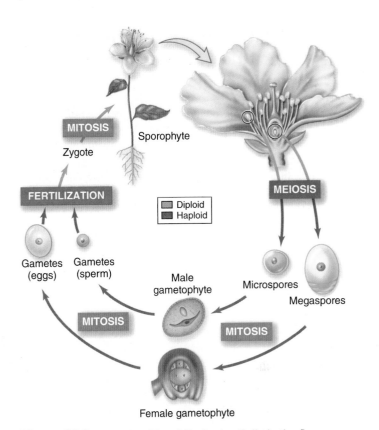

Figure 23.3 Flowering Plant Life Cycle. Cells in the flower undergo meiosis to produce microspores and megaspores, which develop into the gametophytes that produce sperm and egg cells. Fertilization yields the zygote, the first cell of the sporophyte generation.

(except for mutations). In contrast, **sexual reproduction** yields genetically unique offspring with a mix of traits derived from two parents. Sexual reproduction is adaptive in a changing environment. After all, a gene combination that is successful today might not work at all if selective pressures change in the future. Producing variable offspring improves reproductive success in an uncertain world. ▶ why sex?, p. 180

The major groups of multicellular organisms (fungi, animals, and plants) all have the same basic pattern of sexual reproduction. Meiosis produces the cells that begin the haploid generation; fertilization unites the gametes to begin the diploid generation. But organisms vary in the proportion of the life cycle spent as haploid or diploid cells. For example, nearly all of an animal's cells are diploid; sperm and egg cells are usually the only haploid animal cells.

In contrast, the basic plant life cycle has multicellular diploid and haploid stages (see figure 18.5). The **sporophyte,** or diploid generation, produces haploid **spores** by meiosis. A spore divides mitotically to produce a multicellular haploid **gametophyte,** which undergoes mitosis to generate haploid **gametes** (egg cells or sperm). In **fertilization,** these gametes fuse to form a diploid **zygote.** The zygote develops into an embryo as its cells divide mitotically. With additional growth, the embryo becomes a mature sporophyte, and the cycle begins anew.

Figure 23.3 applies this life cycle to angiosperms. Notice that flowering plants produce two types of spores. Microspores give rise to male gametophytes, and the larger megaspores produce female gametophytes. The prefix *mega-* may suggest huge size, but that is only in comparison with the size of microspores.

In flowering plants, the spores, gametophytes, and gametes are all microscopic; the entire haploid gametophyte generation consists of only a few cells.

In fact, one of the many features that characterize angiosperms is the greatly reduced gametophyte. As described in section 18.1, the evolutionary history of plants has seen a gradual reduction in the size of the gametophyte generation, coupled with a corresponding increase in the prominence of the sporophyte. In mosses, the gametophyte is the dominant generation; in angiosperms, the gametophyte is reduced to a few cells that depend on the sporophyte for nutrition.

23.1 | Mastering Concepts

1. What adaptations contribute to the reproductive success of angiosperms?
2. What are some examples of asexual reproduction in plants?
3. When are sexual and asexual reproduction each adaptive?
4. Describe alternation of generations in the plant life cycle.

23.2 | The Angiosperm Life Cycle Includes Flowers, Fruits, and Seeds

When humans reproduce, sexual intercourse brings sperm to an egg cell, and the embryo develops into a fetus. Childbirth separates woman from baby. Clearly, flowering plant sexual reproduction is different from our own. How does the sperm get to the egg in angiosperms? How does the angiosperm embryo develop (along with surrounding tissues) into a seed? How do the seeds separate from the "mother" plant?

This section explains sexual reproduction in flowering plants. **Figure 23.4** summarizes the life cycle; you may find it helpful to refer to this illustration frequently as you study this section. For now, concentrate on the relationship between the two most prominent players: flowers and fruits. **Flowers** are reproductive organs where eggs and sperm unite. Parts of the flower develop into a **fruit,** which protects and disperses the seeds.

A. Flowers Are Reproductive Organs

After a seed germinates (see figure 23.4, step 1), the plant grows and develops to maturity (step 2). Flowers then begin to form. The trigger for flower production is a protein called florigen, which leaf cells load into the phloem. The protein travels in sieve tubes to the shoot tips, where it stimulates the expression of genes that transform a shoot apical meristem into a floral meristem. The meristem cells divide to produce the flower. ▸ meristems, p. 458

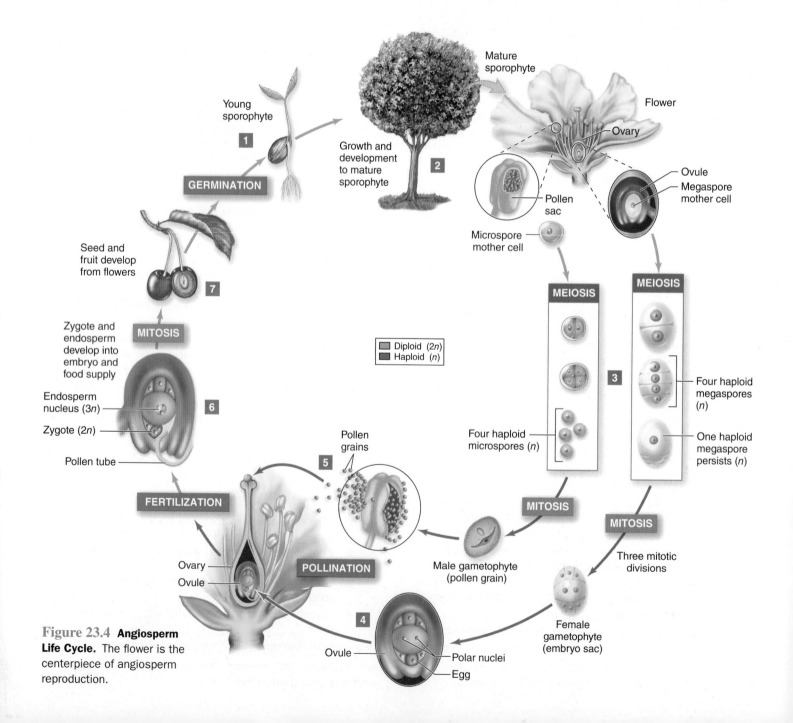

Figure 23.4 Angiosperm Life Cycle. The flower is the centerpiece of angiosperm reproduction.

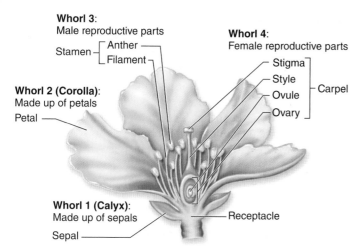

Figure 23.5 **Parts of a Flower.** A complete flower consists of four whorls: sepals, petals, male reproductive parts, and female reproductive parts.

A typical flower has four types of structures, all of which are modified leaves composed mostly of parenchyma cells (see chapter 21). A part of the floral stalk called the **receptacle** is the attachment point for all flower parts (**figure 23.5**). The outermost whorl is called the calyx. It consists of **sepals,** which are leaflike structures that enclose and protect the inner floral parts. Inside the calyx is the corolla, which is a whorl of **petals.** The calyx and corolla do not play a direct role in sexual reproduction, although in many flowers, colorful petals attract pollinators.

The two innermost whorls of a flower are essential for sexual reproduction. The male flower parts consist of **stamens,** which are filaments that bear pollen-producing bodies called **anthers** at their tips. The female part, at the center of a flower, is composed of one or more **carpels,** the structures enclosing the egg-bearing **ovules.** The bases of carpels and their enclosed ovules make up the **ovary.** The upper part of each carpel is a stalklike **style** that bears a structure called a **stigma** at its tip. Stigmas receive pollen.

The flower in figure 23.5 is "complete" because it includes all four whorls, including both male and female parts. In some species, however, the sexes are separate. That is, each flower has either male or female parts but not both. An individual plant may have both types of single-sex flowers, or the plant may produce just one flower type. A holly plant, for example, is either male or female; only the female plants produce the distinctive red berries.

Recall from chapters 18 and 21 that the two largest clades of angiosperms are monocots and eudicots. Flower structure is one feature that distinguishes the two groups. Most monocots, such as lilies and tulips, have petals, stamens, and other flower parts in multiples of three. Most eudicots, on the other hand, have flower parts in multiples of four or five. Buttercups and geraniums are examples of eudicots with five prominent petals on each flower.

B. The Pollen Grain and Embryo Sac Are Gametophytes

Once the flowers have formed, the next step is to produce the microscopic male and female gametophytes (see figure 23.4,

steps 3 through 5). Inside the anther's **pollen sacs,** diploid cells divide by meiosis to produce four haploid **microspores.** Each microspore then divides mitotically and produces a two-celled, thick-walled structure called a **pollen** grain, which is the young male gametophyte. One of the haploid cells inside the pollen grain divides by mitosis to form two sperm nuclei.

Meanwhile, meiosis also occurs in the female flower parts. The ovary may contain one or more ovules, each containing a diploid cell that divides by meiosis to produce four haploid **megaspores.** In many species, three of these cells quickly disintegrate, leaving one large megaspore. The megaspore undergoes three mitotic divisions to form the **embryo sac,** which is a mature female gametophyte. Each female gametophyte therefore consists of eight haploid nuclei but only seven cells. One of these cells is the egg. In addition, a large, central cell contains two **polar nuclei;** as we shall see, both the egg and polar nuclei participate in fertilization.

C. Pollination Brings Pollen to the Stigma

Eventually, the pollen sac opens and releases millions of pollen grains. The next step is **pollination:** the transfer of pollen from an anther to a receptive stigma. Usually, either animals or wind carry the pollen (**figure 23.6**).

Figure 23.6 **Pollination.** Animal pollinators include (a) hummingbirds, (b) butterflies, and (c) bats. (d,e) Some flowers attract insects with distinctive markings visible only in ultraviolet light. (f) The birch tree is wind-pollinated.

Flower color, shape, and odor attract animal pollinators. For example, hummingbirds are attracted to red, tubular flowers. Beetles visit dull-colored flowers with spicy scents, whereas blue or yellow sweet-smelling blooms attract bees. Bee-pollinated flowers often have markings that are visible only at ultraviolet wavelengths of light, which bees can perceive. Butterflies prefer red or purple flowers with wide landing pads. Moths and bats pollinate white or yellow, heavily scented flowers, which are easy to locate at night.

Many animals benefit from their association with plants: they obtain food in the form of pollen or sugary nectar, seek shelter among the petals, or use the flower for a mating ground (see section 6.10). Some plants, however, lure pollinators with a false reward. Skunk cabbage and the "carrion" flowers of South African *Stapelia* plants emit foul odors that attract flies. Flies that land on the flowers gain nothing, but the plant benefits if the insect unwittingly carries pollen grains to another flower of the same species.

Sometimes, the flower and its animal partner have matching parts. For example, some hummingbirds have long, curved bills that fit precisely into the tubular flowers from which they sip nectar. The connection between plant and pollinator may be so strong that the partners directly influence each other's evolution. In **coevolution,** a genetic change in one species selects for subsequent change in the genome of another species. Coevolution is likely to be occurring when a plant has an exclusive relationship with just one pollinator species.

About 10% of angiosperms (and most gymnosperms) use wind, not animals, to carry pollen. Wind-pollinated flowers are small and odorless, and their petals are typically reduced or absent; perfume, nectar, and showy flowers are not necessary for wind to disperse pollen. One advantage of wind pollination is that the plant does not spend energy on nectar or other lures. On the other hand, the wind drops pollen where and when it slackens, whereas an animal delivers pollen directly to another plant. Wind-pollinated plants therefore manufacture abundant pollen. The large quantities of pollen produced by oaks, cottonwoods, ragweed, and grasses provoke allergies in many people. ▸ allergies, p. 686

D. Double Fertilization Yields Zygote and Endosperm

After a pollen grain lands on a stigma of the correct species, a pollen tube emerges (**figure 23.7,** step 1). The pollen grain's two haploid sperm nuclei enter the pollen tube as it grows through the tissue of the style toward the ovary (figure 23.7, step 2). When the pollen tube reaches an ovule, it discharges its two sperm nuclei into the embryo sac.

Then, in **double fertilization,** these sperm nuclei fertilize the egg and the two polar nuclei (figure 23.7, step 3; see also figure 23.4, step 6). That is, one sperm nucleus fuses with the haploid egg nucleus and forms a diploid zygote, which will develop into the embryo. The second sperm nucleus fuses with the two haploid polar nuclei. The resulting triploid nucleus divides to form a tissue called **endosperm,** which is composed of parenchyma cells that store food for the developing embryo.

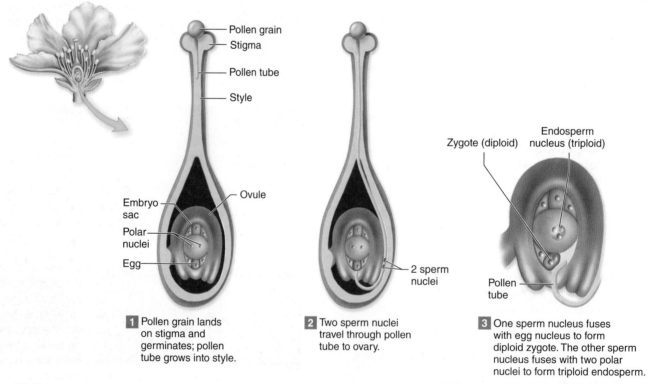

1 Pollen grain lands on stigma and germinates; pollen tube grows into style.

2 Two sperm nuclei travel through pollen tube to ovary.

3 One sperm nucleus fuses with egg nucleus to form diploid zygote. The other sperm nucleus fuses with two polar nuclei to form triploid endosperm.

Figure 23.7 Double Fertilization. (1) Pollen sticks to a stigma on a flower. (2) A pollen tube grows toward the ovule and transports two sperm nuclei. (3) One sperm nucleus fertilizes the egg to form a zygote, and the other fertilizes the polar nuclei to yield the endosperm.

Familiar endosperms are the "milk" and "meat" of a coconut and the starchy part of a rice grain. The starchy endosperm of corn is an important source of not only food but also ethanol, a biofuel. ▶ biofuels, p. 376

Double fertilization reduces the energetic cost of reproduction. In gymnosperms, the female gametophyte stockpiles food for the embryo in advance of fertilization. The investment is wasted if no zygote ever forms. In contrast, double fertilization ensures that an angiosperm devotes energy to endosperm only if sperm fertilizes egg.

E. A Seed Is an Embryo and Its Food Supply Inside a Seed Coat

Immediately after fertilization, the ovule contains an embryo sac with a diploid zygote and a triploid endosperm. The ovule eventually develops into a **seed:** a plant embryo together with its stored food, surrounded by a seed coat (see figure 23.4, step 7). Where do these parts come from?

The zygote divides to form the embryo (**figure 23.8**). Among the first features of the developing embryo are the **cotyledons,** or seed leaves. Soon, the shoot and root apical meristems form at opposite ends of the embryo.

Meanwhile, the endosperm cells divide more rapidly than the zygote and thus form a large multicellular mass. This endosperm supplies nutrients to the developing embryo. In monocots, the single cotyledon transfers stored nutrients from the endosperm to the embryo during germination. The paired cotyledons of many eudicots, on the other hand, become thick and fleshy as they absorb the endosperm.

As the embryo and endosperm develop, the seed coat begins to form. In many species, the **seed coat** is a tough, sclereid-rich outer layer that protects both the embryo and its food supply

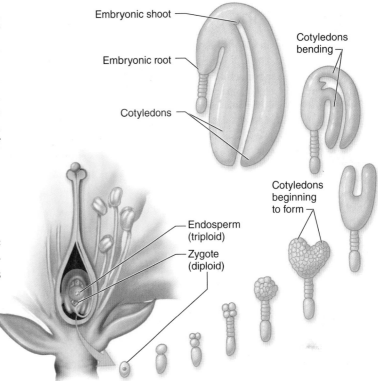

Figure 23.8 **Embryonic Development.** Inside the flower, the zygote divides repeatedly to form the tiny embryonic plant.

from damage and predators. In other plants, including corn, the thin seed coat all but disappears as the seed matures.

At some point in seed development, hormonal signals "tell" cells in the embryo and endosperm to stop dividing, and the seed gradually loses moisture and enters dormancy. The ripe, mature seeds are firm, dry, and ready for dispersal (**figure 23.9**).

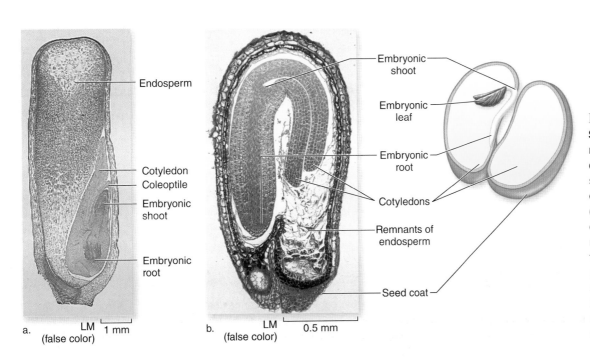

Figure 23.9 **Mature Seeds.** (a) Corn is a monocot. A grain of corn contains a seed, including a starchy endosperm and an embryo with one cotyledon. (b) In a eudicot, the paired cotyledons may absorb much of the endosperm as the seed develops. The photo shows an embryo of shepherd's purse (*Capsella bursa-pastoris);* the diagram shows the corresponding structures in a bean seed.

Burning Question

How can a fruit be seedless?

Given the role of seeds and fruits in plant reproduction, seedless fruits might seem a bit of a puzzle. After all, from a plant's point of view, what's the point of producing a fruit without seeds inside? Natural selection clearly would not favor such a trait in the wild! But the crop plants that people grow do not necessarily live by the same rules as wild plants. Because humans often consider seeds in fruits to be a nuisance, plant breeders have developed many seedless varieties.

Seedless fruits can form in two ways. Often, the fruit develops in the absence of fertilization. Seedless oranges and watermelons are two examples. Alternatively, if fertilization does occur, the embryos die during development. Tiny, immature seeds remain inside the fruit, but they do not interfere with eating. Seedless grapes and bananas illustrate this second path to seedlessness.

Since seedless fruits lack seeds, how do farmers grow more of them? Most are produced asexually by grafting or taking cuttings. But, surprisingly, seedless watermelons do come from seeds! To make "seedless watermelon seeds," plant breeders first cross a regular diploid watermelon plant with a tetraploid plant (with four sets of chromosomes). Farmers then sow the triploid seeds arising from this union. The new triploid plants produce flowers, which the grower pollinates with diploid pollen. Pollination stimulates fruit production, but, like a mule, the triploid plant is sterile. The "seeds" that do develop inside seedless watermelons are actually empty, soft hulls that are easy to chew and swallow. ▶ why mules are sterile, p. 185

Submit your burning question to:
marielle_hoefnagels@mcgraw-hill.com

Dormancy is a crucial adaptation that enables seeds to postpone development if the environment is unfavorable, such as during a drought or frost. Embryo growth resumes only if conditions improve, when young plants are more likely to survive.

F. The Fruit Develops from the Ovary

The rest of the flower also changes as the seeds develop. When a pollen tube begins growing, the stigma produces large amounts of ethylene, a plant hormone discussed in section 23.4. Ethylene stimulates the stamens and petals, which are no longer needed, to wither and fall to the ground. Developing seeds also produce another hormone, auxin, that triggers fruit formation.

In many angiosperms, the ovary grows rapidly to form the fruit, which may contain one or more seeds. (This chapter's Burning Question explores how fruits can develop in the absence of seeds). In some species, additional plant parts also contribute to fruit development. The pulp of an apple, for example, derives from a cup-shaped region of the receptacle. The apple's core is derived from the carpel walls, which enclose the seeds. **Figure 23.10** shows how the parts of an apple flower give rise to the fruit.

Fruits come in many forms (**table 23.1**). A simple fruit develops from a flower with one carpel. The carpel may have one seed, as in a cherry, or many seeds, as in a tomato. An aggregate fruit develops from one flower with many carpels. Strawberries and raspberries are examples of aggregate fruits. A multiple fruit develops from clusters of flowers that fuse into a single fruit as they mature. Pineapples and figs are multiple fruits.

Some fruits are soft and wet; others are tough and dry. Fruits owe their textures to different tissue compositions. Soft, wet fruits such as tomatoes and apples have more parenchyma than

Figure 23.10 Development of a Fruit. After pollination and fertilization, the apple tree's flower begins to develop into a fruit. The fleshy part of an apple develops from the receptacle, which enlarges along with the ovary wall as the fruit develops.

Table 23.1 Types of Fruits: A Summary

Fruit Type		Characteristics	Example(s)
Simple		Derived from one flower with one carpel	Olive, cherry, peach, plum, coconut, grape, tomato, pepper, eggplant, apple, pear
Aggregate		Derived from one flower with many separate carpels	Blackberry, strawberry, raspberry, magnolia
Multiple		Derived from tightly clustered flowers whose ovaries fuse as the fruit develops	Pineapple

G. Fruits Protect and Disperse Seeds

Fruits have two main functions. Besides protecting developing seeds, fruits also promote seed dispersal by animals, wind, and water (**figure 23.11**). The seeds are often deposited far from the parent plant, promoting reproductive success by minimizing competition between parent and offspring.

Many animals disperse fruits and seeds. Colored berries attract birds and other animals that carry the ingested seeds to new locations, only to release them in their feces. Birds and mammals spread seeds when spiny fruits attach to their feathers or fur. Squirrels and other nut-hoarding animals also disperse seeds to their various cache locations. Although the animals consume many of their stored nuts, those that remain uneaten may germinate and establish new plants far from the parent tree.

Wind and water can also distribute seeds. Wind-dispersed fruits, such as those of dandelions and maples, have wings or other structures that catch air currents. Water-dispersed fruits include gourds. These fruits are native to Asia, but they have drifted across the oceans multiple times and established new populations in South America, Australia, and Africa. Likewise, coconuts are water-dispersed fruits that may travel long distances before colonizing distant islands.

23.2 | Mastering Concepts

1. What is the function of each part of a flower?
2. Describe the male and female gametophytes in angiosperms.
3. How does pollen move from one flower to another, and why is this process essential for sexual reproduction?
4. Describe the events of double fertilization.
5. What is the function of the endosperm?
6. What are the components of a seed?
7. Which flower parts develop into a fruit?
8. What are the two main functions of fruits?
9. How do fruits and seeds disperse to new habitats?

any other cell type. But the hard, papery apple core, the outer wall of a dry fruit such as a nut, and the gritty particles in pear flesh are all made of sclereids. The more sclereids, the tougher the fruit. ▶ plant cell types, p. 459

Only shoots give rise to flowers, so it may seem surprising that some fruits develop underground. The yellow flowers of peanut plants, for example, form on the shoot. After fertilization, the petals wither, and the young fruit produces a peg that turns downward and buries itself in the soil. Three to five months later, farmers dig up the plants to harvest the mature fruits. Each fruit consists of a fibrous shell enclosing one to three peanuts—the seeds.

a.

b.

c.

Figure 23.11 Seed Dispersal. (a) This cedar waxwing helps disperse the seeds of winterberry, a type of holly. (b) Hooks on the surface of the burdock fruit easily attach to the fur of a passing animal. (c) Dandelion fruits have fluff that enables them to float on a breeze.

23.3 | Plant Growth Begins with Seed Germination

Germination is the resumption of growth and development after a period of seed dormancy. It usually requires water, oxygen, and a favorable temperature. First, the seed absorbs water. The incoming water swells the seed, rupturing the seed coat and exposing the plant embryo to oxygen. Water also may cause the embryo to release hormones that stimulate the production of starch-digesting enzymes. The stored starch in the seed breaks down to sugars, providing energy for the growing embryo.

Growth and development continue after the embryo bursts out of the seed coat (**figure 23.12**). Rapidly dividing cells in apical meristems add length to both shoots and roots (see section 21.4). The new cells differentiate into the ground tissue, vascular tissue, and dermal tissue that make up the plant body. ▶ plant tissue types, p. 459

The seedling soon begins to take on its mature form. Young roots grow downward in response to gravity, anchoring the plant in the soil and absorbing water and minerals. The shoot produces leaves as it grows upward toward the light. Initially, the energy source for the seedling's growth is stored food inside the seed. By the time the seedling has depleted these reserves, however, the new green leaves are producing food by photosynthesis.

The size of a plant's seeds therefore reflects an evolutionary trade-off. Large, heavy seeds contain ample nutrient reserves to fuel seedling growth but may not travel far. Small seeds, on the other hand, store limited nutrients but tend to disperse far and wide. Interestingly, the crops that humans cultivate typically have larger seeds than do their wild ancestors. We gather and plant the seeds ourselves, removing the selection pressure favoring small seed size.

Depending on the species, a plant may keep growing for weeks, months, years, decades, or even centuries. When the plant reaches reproductive maturity, it too will develop flowers, seeds, and fruits, continuing the life cycle.

23.3 | Mastering Concepts

1. Why must seeds absorb water before germinating?
2. What are the events of early seedling development?
3. How does natural selection influence seed size?

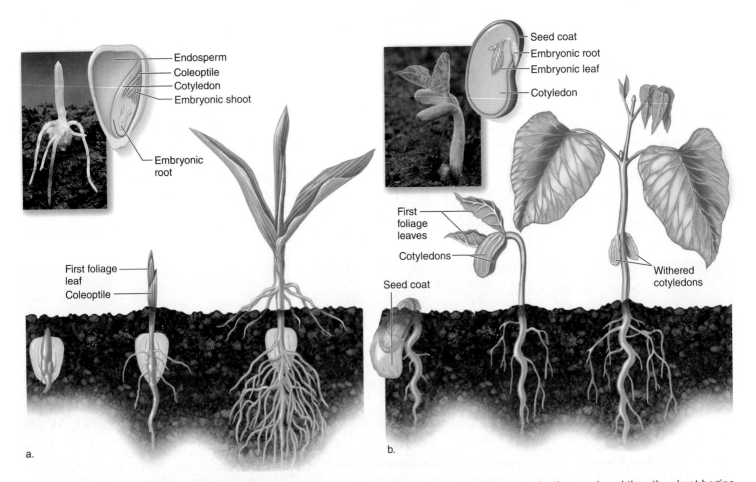

Figure 23.12 Seed Germination and Early Seedling Development. The root emerges first from a germinating seed, and then the shoot begins to elongate. (a) In a monocot such as corn, a sheathlike coleoptile covers the shoot until the first foliage leaves develop. The cotyledon remains belowground. (b) In green beans and some other eudicots, the cotyledons carry out photosynthesis until the first foliage leaves form.

23.4 | Hormones Regulate Plant Growth and Development

Plant responses to environmental stimuli usually seem much more subtle than those of animals. Plants cannot hide, bite, or flee; instead they must adjust their growth and physiology to external conditions. The rest of this chapter explores some of the ways in which a plant responds to the changing environment as it grows and develops, from seed germination through senescence (aging and death).

The environment affects plant growth in many ways. Shoots grow toward light and against gravity; roots grow down. Many plants leaf out in the spring, produce flowers and fruits, then enter dormancy in autumn—all in response to seasonal changes. Other responses are immediate. When the weather is hot, plants reduce transpiration by closing their stomata. A Venus flytrap snaps shut when a fly wanders across a leaf. Plants may even send signals to one another, warning of such dangers as insect infestations.

Chemicals called hormones regulate many aspects of plant growth, flower and fruit development, senescence, and responses to environmental change. Classically, a **hormone** is a biochemical synthesized in small quantities in one part of an organism and transported to another location, where it stimulates or inhibits a response from target cells.

This traditional definition must be modified slightly for plants, however, because some plant hormones act close to or at their site of synthesis. Also, a plant can produce the same hormone at multiple sites throughout the body. Unlike animals, plants do not have glands dedicated to hormone production.

Hormones move within a plant's body either by diffusing from cell to cell or by entering xylem or phloem. When a hormone reaches a target cell, it binds to a receptor protein. This interaction triggers a cascade of chemical reactions that ultimately change the expression of genes in target cells.

The interactions and responses of plant hormones are extremely complex and difficult to study, for several reasons. First, plants produce hormones in extremely low concentrations. Second, each type of hormone exerts a wide variety of effects. Third, the same hormone can either stimulate or inhibit a process, depending on its concentration and the developmental stage of the plant. Fourth, the functions of plant hormones may overlap: at least three different compounds promote stem elongation, for example.

The "classic five" plant hormones are auxins, cytokinins, gibberellins, ethylene, and abscisic acid (table 23.2). A plant must produce auxins and cytokinins if it is to develop at all. Both of these hormones occur in all major organs of all plants at all times, and biologists have never found a mutant plant lacking either one. Other hormones are required for normal development, but plants can complete their life cycles (albeit with altered morphology) without them.

Table 23.2 The "Classic Five" Plant Hormones: A Summary

Class	Synthesis Site(s)	Mode of Transport	Selected Actions
Auxins	Shoot apical meristem, developing leaves and fruits	Diffusion between parenchyma cells associated with vascular tissue	• Stimulate elongation of cells in stem • Control phototropism, gravitropism, thigmotropism • Stimulate growth of adventitious roots from stem cuttings • Suppress growth of lateral buds (apical dominance)
Cytokinins	Root apical meristem	In xylem	• Stimulate cell division in seeds, roots, young leaves, fruits • Delay leaf senescence • Stimulate growth of lateral buds
Gibberellins	Young shoot, developing seeds	In xylem and phloem	• Stimulate cell division and elongation in roots, shoots, young leaves • Break seed dormancy
Ethylene	All parts, especially under stress, aging, or ripening	Diffusion of gas	• Hastens fruit ripening • Stimulates leaf and flower senescence • Stimulates leaf and fruit abscission • Participates in thigmotropism
Abscisic acid	Mature leaves, especially in plants under drought or freezing stress	In xylem and phloem	• Inhibits shoot growth and maintains bud dormancy • Induces and maintains seed dormancy • Stimulates closure of stomata • Promotes leaf, flower, and fruit abscission

A. Auxins and Cytokinins Are Essential for Plant Growth

Auxins (from the Greek meaning "to grow") are hormones that promote cell elongation in stems and fruits but have the opposite effect in roots. Auxins also control plant responses to light and gravity. The most active auxin is indoleacetic acid (IAA), which is produced in shoot tips, embryos, young leaves, flowers, fruits, pollen, and coleoptiles. These hormones act rapidly, spurring noticeable growth in a grass seedling in minutes.

The first plant hormones described were auxins. In the late 1870s, decades before researchers determined the chemical structures of plant hormones, Charles Darwin and his son Francis learned that an "influence" produced at the shoot tip caused plants to grow toward light (**figure 23.13**). This influence was later discovered to be auxin.

Auxins have commercial uses. These hormones stimulate the growth of adventitious roots in cuttings, which is important in the asexual production of plants. A synthetic compound with auxinlike effects, called 2,4-D (2,4-dichlorophenoxyacetic acid), is used extensively as an herbicide. The plant cannot completely break down 2,4-D, which accumulates to lethal levels. For reasons that are not completely understood, 2,4-D kills eudicots ("broadleaf weeds") but not grasses, which are monocots. ▶ weed killers, p. 97

Cytokinins earned their name because they stimulate cytokinesis, or cell division. In flowering plants, most cytokinins affect roots and developing organs such as seeds, fruits, and young leaves. Cytokinins also slow leaf senescence, and they can be used to extend the shelf lives of leafy vegetables.

The actions of cytokinins and auxins compete with each other. Cytokinins are more concentrated in the roots, whereas auxins are more concentrated in shoot tips. Cytokinins move upward within the xylem and stimulate lateral bud sprouting. In a counteracting effect called **apical dominance,** the terminal bud of a plant secretes auxins that move downward and suppress the growth of lateral buds. If the shoot tip is removed, the concentration of auxins in lateral buds decreases, and cytokinins increase. Meristem cells in the buds then begin dividing. Apical dominance explains why gardeners can promote bushier growth by pinching off a plant's tip.

B. Gibberellins, Ethylene, and Abscisic Acid Influence Plant Development in Many Ways

In 1926, Japanese scientists studying "foolish seedling disease" in rice discovered **gibberellins,** another class of plant hormone that causes shoot elongation. A fungus (*Gibberella fujikuroi*) causes affected plants to grow rapidly, becoming so spindly that they fall over and die. The researchers soon discovered that a chemical extract of the fungus produced the same symptoms. In 1934, scientists isolated the active compound and named it gibberellin. We now know of at least 84 naturally occurring gibberellins.

Farmers use gibberellins in agriculture to stimulate stem elongation and fruit growth in seedless grapes (**figure 23.14**). But gibberellins also have other functions. For example, they trigger seed germination by inducing the production of enzymes that digest starch in the seed. After germination, gibberellins also promote cell division.

Ethylene is a gaseous hormone that ripens fruit in many species. Ethylene released from one overripe apple can hasten the ripening, and eventual spoiling, of others nearby, leading to the expression "one bad apple spoils the bushel." Exposure to ethylene also ripens immature fruits after harvest (**figure 23.15**). For example, shipping can damage soft, vine-ripened tomatoes. Farmers therefore pick the fruit while it is still hard and green. Ethylene treatment just before distribution to supermarkets yields ripe-looking (if not good-tasting) tomatoes.

All parts of flowering plants synthesize ethylene, especially the shoot apical meristem, nodes, flowers, and ripening fruits. Like other hormones, ethylene has several effects. In most

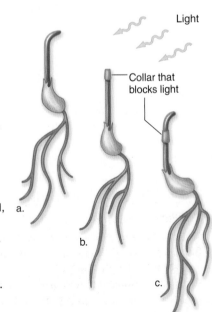

Figure 23.13
Bending Toward Light. (a) The shoot of an oat seedling normally bends toward light. (b) When the coleoptile at the tip of the seedling is blocked, the shoot no longer bends toward the light. (c) The shoot does bend if the collar is placed beneath the tip.

Light

Collar that blocks light

Figure 23.14 Gibberellins and Shoot Elongation. Gibberellins applied to grapes lengthen the stems and increase the size of the fruit. Treated grapes are on the right.

Figure 23.15 **Unripe Fruit.** Exposure to ethylene would turn these green tomatoes red.

species, it causes flowers to fade and wither (see this chapter's Apply It Now box). In addition, a damaged plant part produces ethylene, which hastens aging; the plant then sheds the affected part before the problem spreads. This effect was noticed in Germany in 1864, when ethylene in a mixture of gases in street lamps caused nearby trees to lose their leaves.

A fifth plant hormone, **abscisic acid** (abbreviated ABA), counters the growth-stimulating effects of many other hormones. The name of this hormone comes from its role in promoting the abscission (shedding) of leaves, flowers, and fruits. Stresses such as drought and frost stimulate the production of ABA. One immediate effect is to trigger stomata to close, which helps plants conserve water. ABA also inhibits seed germination, opposing the effects of gibberellins. Commercial growers apply ABA to inhibit the growth of nursery plants so that shipping is less likely to damage them.

C. Biologists Continue to Discover Additional Plant Hormones

Researchers once recognized only five types of plant hormones (see table 23.2). We now know that plants produce several additional hormones. Four examples include:

- **Brassinosteroids:** These steroid hormones occur throughout the plant but are most abundant in pollen and immature seeds. Plants lacking brassinosteroids have dwarfed growth forms, indicating that these hormones are essential for stem elongation.
- **Jasmonic acid:** This stress hormone and its volatile relative, methyl jasmonate, not only induce defenses against insects and pathogens but also signal neighboring plants to beef up their own defenses.
- **Salicylic acid:** This molecule, familiar to most people as aspirin, is best known for its role in plant defenses against viruses and other disease-causing agents. When a plant's cells detect an attack, they release salicylic acid, which induces the surrounding cells to die. This so-called "hyper-

Apply It Now

Cheating Death

In some ways, ethylene is the enemy of agriculture. This plant hormone causes flowers to fade and fruits to spoil, greatly reducing the shelf life of perishable plant products. What's a shipper to do?

One solution is to interfere with the action of ethylene. A chemical called 1-methylcyclopropene, or 1-MCP, does just that. This gas is sold under the brand names EthylBloc and Smart-Fresh. These products preserve cut flowers, potted plants, and fruits by binding to ethylene receptor proteins in the cell membranes of the plants. Ethylene cannot bind to the cells, so spoilage is delayed.

Biologists have also learned to tinker with the ethylene receptor directly. The fresh-looking petunia flowers in figure 23.A, for example, are genetically engineered to have mutant ethylene receptor genes. The effect mimics that of 1-MCP: the modified flowers are unresponsive to ethylene.

Figure 23.A **Ethylene's Effects on Flowers.** All four of these petunia flowers were treated with ethylene gas for 18 hours. The normal flowers on the left have withered, but the fresh-looking flowers on the right were genetically modified to be insensitive to ethylene.

sensitive response" keeps the pathogen from spreading to additional tissues. Moreover, the accumulation of salicylic acid makes the entire plant resistant to a wide variety of pathogens.

- **Florigen:** Leaves produce florigen, the hormone that induces a shoot apical meristem to begin producing flowers. After decades of debate over the chemical identity of florigen, studies have confirmed that the molecule is a protein.

23.4 | Mastering Concepts

1. What is a hormone?
2. How does a plant hormone exert its effects?
3. List the major classes of plant hormones and name some of their functions.
4. Give an example of how plant hormones interact.

23.5 | Light Is a Powerful Influence on Plant Life

Like many other organisms, plants are exquisitely attuned to light. Their lives depend on it, because light is their sole energy source for photosynthesis (see chapter 5). This section illustrates how light influences nearly every aspect of a plant's life.

Angiosperms have several types of **photoreceptors,** molecules that detect the wavelength and intensity of light. All have the same basic structure: a protein bound to a pigment molecule that is "tuned to" certain wavelengths. Light absorption triggers a change in the protein, which then stimulates changes in gene expression that alter the plant's appearance or behavior.

Of course, photoreceptors do not act alone. We have already seen that hormones help regulate fundamental functions in plants. A light-activated photoreceptor may alter the action of any hormone by influencing its production, release, or transport, or by changing the expression of receptors in target cells. These complex photoreceptor-hormone interactions remain an active area of plant biology research.

Figure 23.16
Phototropism. These autumn crocuses *(Colchicum autumnale)* show strong phototropism when the sun is off to the side.

A. Phototropism Is Growth Toward Light

A **tropism** is the orientation of a plant part toward or away from a stimulus such as light, gravity, or touch. All tropisms result from differential growth, in which one side of the responding organ grows faster than the other.

Phototropism, for example, is growth toward or away from light (figure 23.16). Phototropism occurs when cells on the

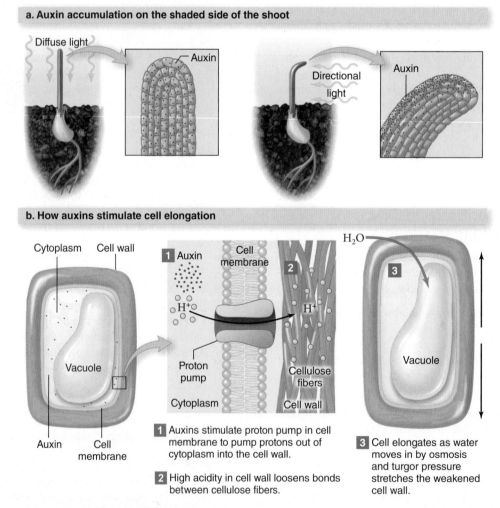

a. Auxin accumulation on the shaded side of the shoot

Diffuse light — Auxin

Directional light — Auxin

b. How auxins stimulate cell elongation

Cytoplasm   Cell wall

1 Auxin   Cell membrane   **2**

H^+ H^+

Proton pump

Cytoplasm   Cellulose fibers

Cell wall

Auxin   Cell membrane   Vacuole

1 Auxins stimulate proton pump in cell membrane to pump protons out of cytoplasm into the cell wall.

2 High acidity in cell wall loosens bonds between cellulose fibers.

H_2O

3

Vacuole

3 Cell elongates as water moves in by osmosis and turgor pressure stretches the weakened cell wall.

Figure 23.17 Auxin and Stem Elongation. (a) Phototropism occurs because auxins move to the shaded side of a shoot. (b) Auxins promote acidification of the cell walls, so the shaded cells can elongate.

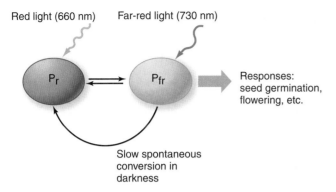

Red light (660 nm) Far-red light (730 nm)

P_r P_{fr}

Responses:
seed germination,
flowering, etc.

Slow spontaneous
conversion in
darkness

Figure 23.18 **Two Forms of Phytochrome.** P_r absorbs red light and rapidly converts into P_{fr}; the reverse occurs when P_{fr} absorbs far-red light. P_{fr} is also converted slowly to P_r in the absence of light. P_{fr} has a variety of biological effects.

shaded side of a stem elongate more than cells on the opposite side. Photoreceptors and auxins participate in the response. Photoreceptors absorb light, which somehow causes auxins to migrate to the shaded side of the stem (**figure 23.17**). The auxin influx causes proteins in the cell membrane to pump protons (H^+) into the cell wall. The protons, in turn, activate enzymes that separate the cellulose fibers, enabling the wall to expand and elongate against turgor pressure. The stem bends toward the light. ▶ cell walls, p. 64

The plant commonly sold as "lucky bamboo" often has a curled stem, illustrating the effects of phototropism. Farmers grow the plants for a year or more in greenhouses, exposing only one side to light. Periodically rotating each plant directs the stem's growth into a twist.

B. Phytochrome Regulates Seed Germination, Daily Rhythms, and Flowering

One type of photoreceptor, **phytochrome,** is a blue pigment molecule that exists in two interconvertible forms (**figure 23.18**). The P_r form absorbs red wavelengths of light (660 nm). The biologically active form, P_{fr}, absorbs in the far-red portion of the electromagnetic spectrum (730 nm). P_r converts to P_{fr} nearly instantaneously in the presence of red light, whereas far-red light prompts the reverse transformation. P_{fr} also converts slowly to P_r in the dark. This back-and-forth transformation of phytochrome influences many important processes.

Figure 23.19 **Growing Up in the Dark.** An overlying log was lifted to reveal these spindly seedlings, which grew in the absence of light. Note the pale, elongated stems and tiny leaves.

Seed Germination Although some seeds will germinate in total darkness, others require light. In seeds of lettuce and many weeds, for example, red light stimulates germination, and far-red light inhibits it. Exposure to red light stimulates the conversion of P_r to P_{fr}, which triggers germination. The presence of P_{fr} apparently "informs" the seed that light is available and induces the transcription of genes required for germination. On the other hand, if seeds are buried too deeply in the soil, then P_r is not converted to P_{fr}, and germination does not occur.

After germination, phytochrome and another photoreceptor, cryptochrome, control early plant growth. Seedlings grown in the dark have elongated stems, small leaves, a pale color, and a spindly appearance (**figure 23.19**). The yellowish white bean sprouts used in Chinese cooking are grown without light. Once exposed to light, normal growth begins: stem elongation slows, root and leaf development accelerate, and chlorophyll synthesis begins.

Circadian Rhythms **Circadian rhythms** are physiological cycles that repeat daily. In most plants, for example, stomata close at night and reopen in the morning. Some plants, such as the four-o'clock and the evening primrose, open their flowers in late afternoon or at nightfall, when their pollinators are most likely to visit. The prayer plant, *Maranta,* is an ornamental houseplant that exhibits nightly "sleep movements." It folds its leaves vertically each night, then moves them to a horizontal position during the day (**figure 23.20**).

Similar rhythms occur in many protists, fungi, and animals. In all species studied, pigments detect light and transmit signals to a "central oscillator," a set of genes and proteins that keep the clock on track. The details of the central oscillator, well understood in animals, remain unclear in plants.

Figure 23.20 **Sleep Movements.** The prayer plant, *Maranta,* exhibits rhythmic sleep movements. When the sun goes down, the plant's leaves move inward in a configuration resembling hands folded in prayer.

Circadian cycles often continue under laboratory conditions of constant light or dark. They are ingrained. Nevertheless, external conditions such as a change in day length can reset (entrain) the plant's internal clock. Phytochromes absorb light and interact with the oscillator, adjusting the clock to match the new conditions.

Flowering　Plants respond in many ways to changes in **photoperiod,** or day length. In the spring, as days grow longer, buds resume growth and rapidly transform a barren deciduous forest into a leafy canopy. Bud dormancy and leaf abscission are responses to the shorter days that accompany the approach of winter. These seasonal changes illustrate the interactions among environmental signals, hormones, other biochemicals, and the plant's genes.

Photoperiod regulates the production of flowers in some plant species. **Long-day plants** flower when days are longer than a critical length, usually 9 to 16 hours. These plants typically bloom in the spring or early summer and include lettuce, spinach, beets, clover, corn, and irises. **Short-day plants** flower when days are shorter than some critical length, usually in late summer or fall. Asters, strawberries, poinsettias, potatoes, soybeans, ragweed, and goldenrods are short-day plants. **Day-neutral plants** flower at maturity, regardless of day length.

Experiments with long- and short-day plants have confirmed that flowering actually requires a defined period of uninterrupted darkness, rather than a certain day length (**figure 23.21**). Thus, short-day plants are really long-night plants, because they flower only if their uninterrupted dark period exceeds a critical length. Similarly, long-day plants are really short-night plants.

Figure It Out

Duckweed is a short-day plant with a critical photoperiod of 14 hours of daylight. Will duckweed flower when there are 15 hours of daylight?

Answer: No.

Phytochrome in the leaves clearly plays a role in flowering in long- and short-day plants (**figure 23.22**). Researchers use flashes of far-red or red light during the dark period to induce the formation of P_r or P_{fr}. The prevailing form of phytochrome, in turn, determines whether flowering will occur. But no one understands exactly how phytochrome interacts with the biological clock to regulate the production or release of florigen, the hormone that induces flowering.

23.5 | **Mastering Concepts**

1. What is a photoreceptor?
2. What is auxin's role in phototropism?
3. How does phytochrome help regulate seed germination, circadian rhythms, and flowering?
4. How do long-day plants differ from short-day plants?

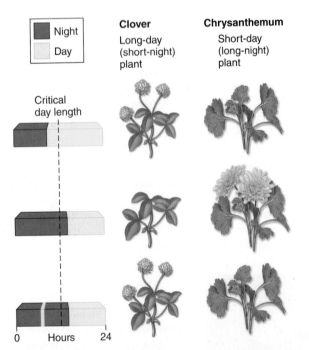

Figure 23.21　Night Length Is Critical. Long-day plants produce flowers only if the dark period is shorter than a critical length. Short-day plants, in contrast, require an uninterrupted dark period longer than a critical period. Interrupting the dark period of a short-day plant inhibits flowering.

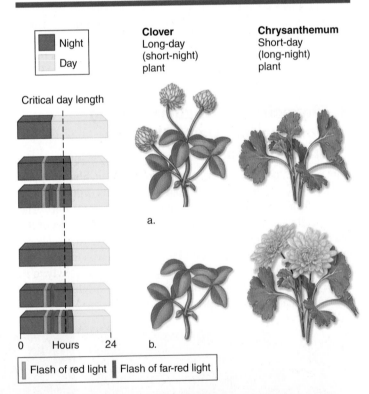

Figure 23.22　The Last Flash Matters. (a) Interrupting night with a flash of red light shortens continuous darkness. A long-day plant flowers, but a short-day plant does not because the time of uninterrupted darkness is too short. (b) A flash of far-red light closely following a flash of red cancels the effect of the red. The last flash determines the prevalent form of phytochrome.

23.6 | Plants Respond to Gravity and Touch

Besides light, a developing plant also responds to countless other environmental cues. For example, the more CO_2 in the atmosphere, the lower the density of stomata on leaves. Likewise, a plant in soil with abundant nitrate (NO_3^-) produces fewer lateral roots than in nutrient-poor soil. Plants can also sense temperature; many require a prolonged cold spell before producing buds or flowering. A warm period in December, before temperatures have really plummeted, does not stimulate apple and cherry trees' buds to "break," but a similar warm-up in late February does induce growth.

Gravity is another important environmental cue. **Gravitropism** is directional growth in response to gravity (**figure 23.23**). As a seed germinates, its shoot points upward toward light, and its root grows downward into the soil. Turn the plant sideways, and the stem and roots bend according to the new direction of gravity.

An asymmetrical distribution of auxins plays a role in gravitropism, as in phototropism. As we have already seen, auxin accumulation on one side of a stem causes cell elongation. In roots, however, auxins have the opposite effect: a high concentration of auxins on the "downward" side of a root inhibits cell elongation.

No one knows exactly how gravitropism works, although it is clear that the root cap must be present for roots to respond to gravity. One hypothesis centers on root cap cells with **statoliths,** starch-containing plastids that function as gravity detectors (see figure 23.23b and c). Statoliths normally sink to the bottoms of the cells, somehow telling the cells which direction is down. Turning a root sideways causes the statoliths to move, redistributing calcium ions and auxins in a way that bends the root. ▶ plastids, p. 60

Besides ever-present gravity, a plant also encounters a changing variety of mechanical stimuli, including contact with wind, rain, and animals. Even roots encounter obstacles as they grow. Over time, repeated touching produces shorter, stockier roots and shoots. The mechanism is unknown, but perhaps plants can detect slight changes in cell shape, which somehow

Figure 23.24 Thigmotropism. A tendril's epidermis is sensitive to touch. This tendril of a passion vine wraps around a blackberry stem.

induces the expression of genes that regulate cell expansion. Because of the touch response, the cells expand radially instead of lengthening.

The coiling tendrils of twining plants exhibit **thigmotropism,** a directional response to touch (**figure 23.24**). Specialized epidermal cells detect contact with an object, which induces differential growth of the tendril. In only 5 to 10 minutes, the tendril completely encircles the object. Auxins and ethylene apparently control thigmotropism.

23.6 | Mastering Concepts

1. How do auxins participate in gravitropism?
2. How does thigmotropism help some plants climb?

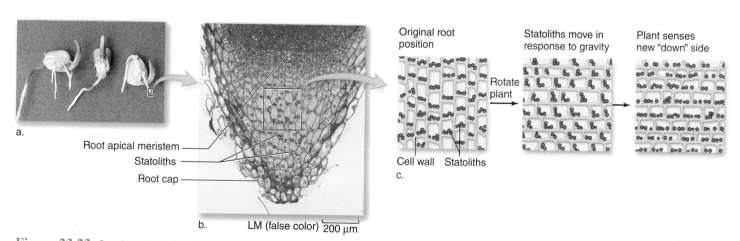

Root apical meristem

Statoliths

Root cap

a.

b.   LM (false color) $\overline{200\ \mu m}$

Original root position

Statoliths move in response to gravity

Plant senses new "down" side

Rotate plant

Cell wall   Statoliths

c.

Figure 23.23 Gravitropism. (a) Shoots grow up and roots grow down, no matter which way a seed is oriented. (b) Starch-filled statoliths in the root cap may help the plant detect gravity. (c) When a root is rotated, statoliths move to occupy the new "down" side, triggering changes that set the root back on its normal downward course.

23.7 | Plant Parts Die or Become Dormant

A normal part of every plant's life is **senescence,** or aging, when metabolism switches from synthesis to breakdown. Annual plants, whose entire lives span just one growing season, senesce and die after they produce seeds. Trees and other perennial plants live for more than a year, sometimes for centuries. But even their tissues senesce, often seasonally.

The most spectacular example of senescence is the changing colors of the leaves of deciduous trees in autumn. As days begin to shorten, enzymes digest proteins, chlorophyll, and other large molecules in the leaves. Yellow, orange, and red carotenoid pigments, previously masked by chlorophyll, become visible. These newly exposed carotenoids, along with the production of purplish anthocyanin pigments, combine to create the spectacular colors of autumn leaves. ▶ why leaves change color, p. 95

Meanwhile, minerals such as nitrogen and phosphorus move out of the leaf for winter storage in the stem and roots. By the time a leaf falls, it is little more than a collection of cell walls and remnants of nutrient-depleted cytoplasm.

The leaf separates from the plant at its **abscission zone,** a specialized layer of cells at the base of the petiole (**figure 23.25**). This cell layer forms early in development, but abscission does not occur until ethylene stimulates the production of digestive enzymes that degrade the cellulose and pectin that bond the cells together.

Why would a deciduous tree spend energy on leaf production each spring, only to discard its investment each autumn? The reason is that leaves would continuously lose moisture to cold, dry winter winds. Roots cannot absorb water from frozen soil, so the tree could not replace the lost water. The seasonal loss of leaves therefore removes a large surface area from which the plant would otherwise lose water. But what about evergreen trees and shrubs? Their leaves also die, but not all at once. Instead, these plants retain leaves for several years, periodically shedding the oldest ones.

As a deciduous tree sheds its leaves before winter, other plant parts often become **dormant,** entering a state of decreased metabolism. Cells synthesize sugars and amino acids, which function as antifreeze that minimizes cold damage. Growth inhibitors accumulate in buds, transforming them into winter buds covered by thick, protective scales (**figure 23.26**).

A variety of cues can end a plant's dormancy. In many desert plants, for example, rainfall alone re-

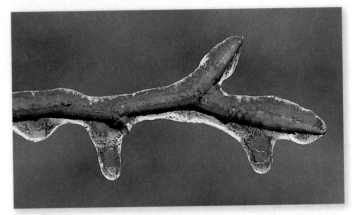

Figure 23.26 Dormancy. Some plants enter a seasonal state of dormancy, which enables them to survive harsh weather. The protective scales give the buds of this species a light brown color.

leases the plant from dormancy. Usually, however, dormancy ends with the arrival of spring. Longer days and warmer temperatures trigger growth in the plant's buds. Initially, stored carbohydrates and minerals in the stem and roots provide the raw materials that build new leaves. Soon, however, the fresh foliage produces its own food by photosynthesis, and carbohydrates begin to flow from the leaves to the rest of the plant.

23.7 | Mastering Concepts

1. What events occur when the leaves of a deciduous tree senesce and fall?
2. How is dormancy different from senescence?

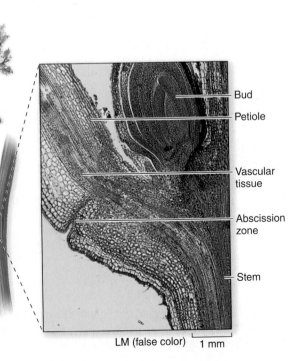

Figure 23.25 Leaf Abscission. The abscission zone is a region of separation that forms near the base of a leaf's petiole, minimizing the risk of infection and nutrient loss when the leaf is shed.

Bud
Petiole
Vascular tissue
Abscission zone
Stem

LM (false color) 1 mm

23.8 | Investigating Life: A Red Hot Chili Pepper Paradox

Chili peppers are famous their hot, spicy taste (**figure 23.27**). Chilies get their kick from a unique family of chemical compounds called capsaicinoids. Glands inside the chili's fruit produce several types of capsaicinoids, including the most famous one: capsaicin. The sensation ranges from the pleasantly mild pimento to the outright painful habanero.

The pungency of chilies is a bit of a paradox. After all, the main function of fruits is to disperse seeds. Many species of flowering plants produce chemicals that make their fruits unpalatable until the seeds are mature. Anyone who has sampled an unripe apple or tomato has experienced this adaptation, which protects the developing seeds from hungry herbivores. Once the seeds mature, however, most fruits lose their defensive chemicals. But chilies retain their pungent arsenal, even after the fruit is mature. Why?

Biologists Joshua Tewksbury, of the University of Washington, and Gary Nabhan, of the University of Arizona, developed a hypothesis to explain this paradox. Tewksbury and Nabhan realized that some fruit-eating animals disperse seeds intact, but others destroy seeds. Perhaps, they suggested, capsaicinoids deter the harmful seed-destroyers without affecting the beneficial dispersers.

The researchers used field observations and laboratory feeding studies to test their hypothesis. Their study population consisted of wild chiltepin chilies *(Capsicum annuum* var. *glabriusculum)* growing in a canyon in southern Arizona. These shrubby plants produce tiny round chilies. First, Tewksbury and Nabhan wanted to learn which animals consumed the fruits. After videotaping chili plants for a total of 146 daylight hours, they found that birds called curve-billed thrashers *(Toxostoma curvirostre)* consumed 72% of fruits that animals removed from the plants.

Thrashers, however, feed during the day. What about small mammals, which are active at night? To learn more about mammalian tastes, the researchers put out fruits of chilies and desert hackberry, the most common fruiting plant in the study area. During the day, chilies and hackberries vanished. At night, however, animals removed only hackberries. This result provided indirect evidence that birds, but not small mammals, eat chili fruits.

A laboratory study settled the question. The team captured five cactus mice *(Peromyscus eremicus),* five pack rats *(Neotoma lepida),* and 10 thrashers from the study site. Each animal was offered three types of fruit: hot chilies from the field sites, nonpungent mutant chilies, and desert hackberries. Although the birds ate all three fruits, the mice and pack rats consumed the hackberries but avoided the chilies (**figure 23.28a**).

Figure 23.27

Chilies. These fruits of the chili plant can be very spicy.

According to Tewksbury and Nabhan's hypothesis, the capsaicin in chilies should repel seed-destroying animals without affecting dispersers. Clearly, chilies deter mammals but not birds. Do mammals harm chili seeds more than the thrashers do? The researchers set up another laboratory test to answer that question. They fed nonpungent chilies to thrashers, mice, and pack rats; control chilies were not offered to any animal. The team then compared the germination rate of the control seeds with germination rates for seeds that had passed through the digestive systems of thrashers or mammals. Figure 23.28b shows the result. The mice and pack rats destroyed the seeds, which presumably were crushed in the mammals' molars. In contrast, the seeds remained viable after consumption by thrashers.

These experiments help explain the evolutionary forces that select for capsaicin production in mature chili peppers. Interestingly, thrashers and other birds seem insensitive to capsaicin, while most mammals avoid it. The real paradox is how humans transformed this innate mammalian aversion into a worldwide love affair with the chili pepper.

Tewksbury, Joshua J., and Gary P. Nabhan. 2001. Directed deterrence by capsaicin in chillies. *Nature,* vol. 412, pages 403–404.

23.8 | Mastering Concepts

1. What did Tewksbury and Nabhan conclude about the function of capsaicin in mature chili fruits?

2. Summarize the evidence that Tewksbury and Nabhan used to arrive at their conclusion.

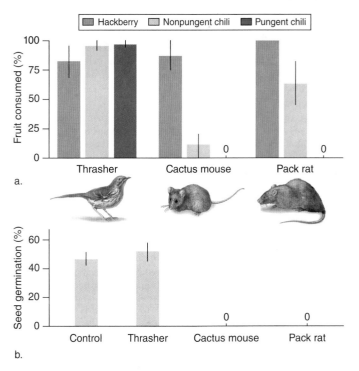

Figure 23.28 Some Like It Hot. (a) In a feeding preference experiment, thrashers readily accepted pungent chilies. Mice and pack rats rejected the spicy fruits. (b) Chili seeds that passed through thrashers remained viable, whereas seeds eaten by mice and pack rats were destroyed.

Chapter Summary

23.1 | Angiosperms Reproduce Asexually and Sexually

A. Asexual Reproduction Yields Clones
- In **asexual reproduction,** clones develop from the roots, stems, or leaves of a parent plant. Asexual reproduction is advantageous in a stable environment where plants are well adapted to their surroundings.
- Grafting is a form of asexual reproduction. Tissue culture techniques help growers produce clones.

B. Sexual Reproduction Generates Variability
- **Sexual reproduction** produces genetically variable offspring, which increases reproductive success in a changing environment. Plant life cycles include alternation of haploid **gametophyte** and diploid **sporophyte** generations.

23.2 | The Angiosperm Life Cycle Includes Flowers, Fruits, and Seeds

A. Flowers Are Reproductive Organs
- **Flowers** are reproductive structures built of whorls of parts attached to a **receptacle.** The calyx, made of **sepals,** and the corolla, made of **petals,** are accessory parts. Inside the corolla, the **stamens** have pollen-containing **anthers** at their tips. **Carpels** occupy the center of the flower. The **ovary** contains one or more **ovules.** The **stigma** tops the **style,** which extends from the ovary.

B. The Pollen Grain and Embryo Sac Are Gametophytes
- Inside the flower, meiosis produces haploid **megaspores** and **microspores,** which develop into haploid female and male gametophytes, respectively. The gametophytes produce haploid **gametes** by mitosis.
- In the anther, specialized cells in **pollen sacs** divide meiotically, each yielding four haploid microspores. The microspores divide mitotically to yield haploid cells, one of which gives rise to two identical sperm nuclei. The **pollen** grain is the immature male gametophyte.
- In ovules within the ovary, specialized cells divide meiotically to yield four haploid cells, one of which persists as a haploid megaspore that divides mitotically three times. The resulting female gametophyte, or **embryo sac,** contains seven cells. One is the egg, and another has two **polar nuclei.**

C. Pollination Brings Pollen to the Stigma
- Animals or wind transfer pollen from anthers to a stigma. Flower structures and odors are usually adapted to either animal or wind **pollination.** In **coevolution,** animal pollinators and flowers select for changes in one another.

D. Double Fertilization Yields Zygote and Endosperm
- Once on a stigma, a pollen grain grows a pollen tube, and its two sperm nuclei move through the tube toward the ovary.
- In the embryo sac, one sperm nucleus **fertilizes** the egg to form the **zygote,** and the other sperm nucleus fertilizes the polar nuclei to form the **endosperm.** This phenomenon is termed **double fertilization.**

E. A Seed Is an Embryo and Its Food Supply Inside a Seed Coat
- A **seed** is an embryo, endosperm, and **seed coat.** The endosperm nourishes the developing embryo as cells in apical meristems divide to produce the embryonic shoot and root. As the embryo grows, **cotyledons** develop.
- Seeds enter a dormancy period in which the embryo postpones development.

F. The Fruit Develops from the Ovary
- After fertilization, nonessential floral parts fall off, and hormones influence the ovary and sometimes other plant parts to develop into a **fruit.** The fruit protects the seeds and aids in dispersal.

G. Fruits Protect and Disperse Seeds
- Animals, wind, and water disperse seeds to new habitats, reducing competition between parent plants and their offspring.

23.3 | Plant Growth Begins with Seed Germination
- Seed **germination** requires oxygen, water, and a favorable temperature. When the embryo bursts from the seed coat, the plant's primary growth begins.

23.4 | Hormones Regulate Plant Growth and Development
- Plants respond to the environment with changes in growth and movement, mediated by the action of **hormones.** Hormone interactions are complex.

A. Auxins and Cytokinins Are Essential for Plant Growth
- **Auxins** stimulate cell elongation in shoot tips, embryos, young leaves, flowers, fruits, and pollen. Auxins are most concentrated at the main shoot tip, which blocks growth of lateral buds (**apical dominance**).
- **Cytokinins** stimulate cell division in actively developing plant parts, including lateral buds.

B. Gibberellins, Ethylene, and Abscisic Acid Influence Plant Development in Many Ways
- **Gibberellins** stimulate cell division and elongation, and they help break seed dormancy.
- **Ethylene** is a gas that speeds ripening, senescence, and leaf abscission.
- **Abscisic acid** counters the growth-inducing effects of other hormones by inducing dormancy and inhibiting shoot growth.

C. Biologists Continue to Discover Additional Plant Hormones
- Brassinosteroids are required for stem elongation. Jasmonic acid and salicylic acid stimulate plant defenses, and florigen triggers flowering.

23.5 | Light Is a Powerful Influence on Plant Life
- **Photoreceptors** absorb light energy of specific wavelengths and influence a plant's growth, development, or response to the environment.

A. Phototropism Is Growth Toward Light
- A **tropism** is a growth response toward or away from an environmental stimulus. Typically, different parts of an organ or structure grow at different rates. In **phototropism,** light sends auxin to the shaded portion of the plant, stimulating growth toward the light.

B. Phytochrome Regulates Seed Germination, Daily Rhythms, and Flowering
- **Phytochrome** controls many plant responses to light. Exposure to red light converts phytochrome to a form that stimulates a response, such as seed germination or flowering.
- Seed germination and normal seedling growth often require exposure to red light wavelengths.
- Internal biological clocks control daily responses, or **circadian rhythms.** Changes in day length can alter, or entrain, these clocks.
- **Photoperiod** is the length of a day. Some plants are **day-neutral,** but others use photoperiod as a cue to produce flowers. **Short-day plants** flower only when the duration of uninterrupted darkness is greater than a critical length. **Long-day plants** require a dark period shorter than a critical length.

23.6 | Plants Respond to Gravity and Touch
- **Gravitropism** is growth toward or away from the direction of the force of gravity. The upward growth of shoots and downward growth of roots are gravitropic responses. The positions of starch-rich **statoliths** in cells apparently help plants detect gravity.
- **Thigmotropism** is growth directed toward or away from a mechanical stimulus such as wind or touch.

23.7 | Plant Parts Die or Become Dormant
- **Senescence** is an active and passive cessation of growth. Senescent leaves detach at an **abscission zone.**
- Growth becomes **dormant** during cold or dry times and resumes when environmental conditions are more favorable.

23.8 | Investigating Life: A Red Hot Chili Pepper Paradox

- The capsaicin in chili pepper fruits deters mammals, which would otherwise destroy the chili's seeds. Capsaicin does not affect birds, which eat chili pepper fruits and disperse the seeds intact.

Multiple Choice Questions

1. The new gene combinations associated with sexual reproduction in plants are the result of
   a. mitosis.
   b. meiosis.
   c. grafting.
   d. Both b and c are correct.

2. Which of the following is NOT a flower structure associated with reproduction?
   a. Anther
   b. Stigma
   c. Sepal
   d. Carpel

3. Where would you find a male gametophyte?
   a. Inside a ripe pollen sac
   b. Inside an ovule
   c. Inside the embryo sac
   d. Both b and c are correct.

4. What are the products of double fertilization?
   a. Two diploid zygotes
   b. A diploid zygote and a triploid endosperm
   c. A haploid sperm and a diploid zygote
   d. A triploid zygote

5. What is the function of a cotyledon?
   a. To support photosynthesis for the embryo
   b. To protect the dormant embryo
   c. To transfer nutrients from the endosperm
   d. To produce more cells for the growth of the embryo

6. Predict what would happen if a plant produced an excess of auxin.
   a. The plant would be very bushy.
   b. The plant would be very tall.
   c. The plant would have very long roots.
   d. The leaves of the plant would not die.

7. How does auxin cause bending in a plant?
   a. It causes an increase in cell division on one side of the plant.
   b. It suppresses cell division on one side of the plant.
   c. It triggers elongation of cells at the apical meristems of the plant.
   d. It triggers elongation of cells on one side of the plant.

8. Spinach plants are typically compact and leafy, but a change in photoperiod can cause bolting, in which the plant's stem elongates and produces flowers. Which plant hormone is most associated with the rapidly lengthening stalk?
   a. Gibberellins
   b. Abscisic acid
   c. Ethylene
   d. Jasmonic acid

9. Why does far-red light inhibit seed germination?
   a. It increases the levels of the P_{fr} form of phytochrome.
   b. It increases the levels of the P_r form of phytochrome.
   c. It prevents the synthesis of gibberellin.
   d. Both b and c are correct.

10. Statoliths play a role in a plant's response to
    a. light.
    b. disease.
    c. touch.
    d. gravity.

Write It Out

1. Give an example of asexual reproduction in a plant.
2. Explain how flowers, fruits, and seeds contribute to the reproductive success of angiosperms.
3. Sketch a flower, indicating the location and function of the following parts: sepal, petal, carpel, stamen, stigma, corolla, calyx, style, anther.
4. Name tissues or cells in an angiosperm that are haploid, diploid, and triploid.

5. What happens to the two sperm cells that form inside a pollen tube?
6. In what ways do plants "manipulate" animals to help them carry out their life cycles?
7. How does an exclusive relationship between a plant and its pollinator benefit each partner? What are the risks of exclusivity?
8. In the study described in section 21.6, why did the scientists count the fruits on the trees as a measure of evolutionary fitness? What might be a more accurate (but also more time-consuming) measure of fitness?
9. If fruit production is a measure of fitness, why wouldn't a plant spend all of its energy producing fruits instead of roots and leaves? Why do you think some annual plants die back as they produce fruits?
10. An oak tree may produce thousands of acorns, which squirrels bury or eat. Why does the tree make so many acorns? Why might a plant whose seeds disperse far from the parent have better reproductive success than one whose seeds fall at the base of the parent plant?
11. List the major plant hormones and describe some of their actions.
12. How does pruning stimulate the growth of new branches on a shrub?
13. Explain why one rotten orange promotes decay of all of the others.
14. What is the function of photoreceptors?
15. Describe four effects of light on plant growth and development.
16. Describe three tropisms. Which hormone is common to all three?
17. How does photoperiod entrain the circadian clock and control flowering?
18. Develop a hypothesis that explains why it might be adaptive for a plant to flower in response to photoperiod rather than temperature.
19. How is gravitropism adaptive?
20. How does seed dormancy promote reproductive success? How does seasonal dormancy in deciduous trees promote reproductive success?
21. How does senescence occur?
22. What are some conditions that can release a plant from dormancy?

Pull It Together

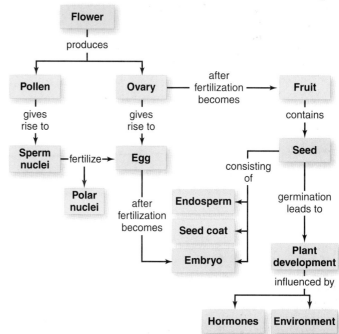

1. Add the following terms to the concept map: *stamen, anther, carpel, ovule, stigma*.
2. What event leads to the formation of the endosperm?
3. What roles do animals play in plant life cycles?
4. What hormones are involved in plant development?
5. What environmental conditions influence plant development?

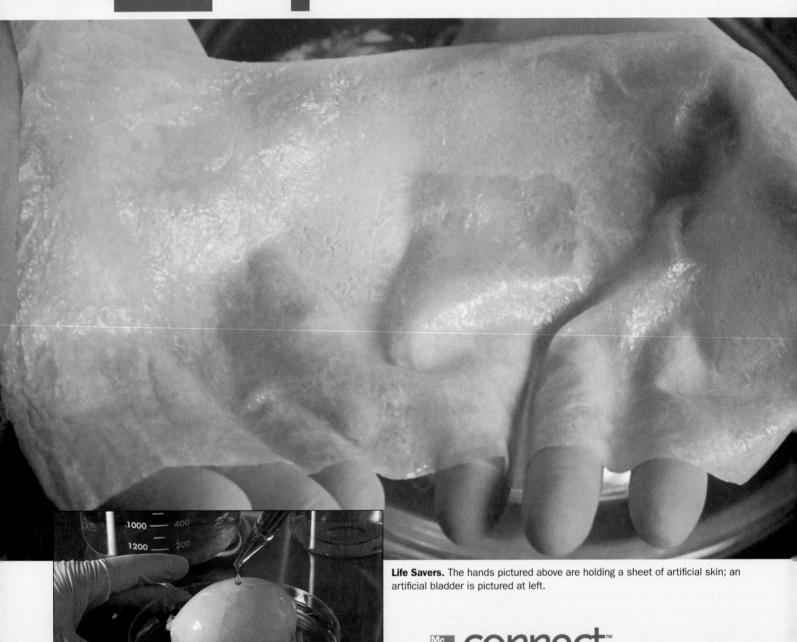

Life Savers. The hands pictured above are holding a sheet of artificial skin; an artificial bladder is pictured at left.

Artificial Tissues and Organs

HEALTHY ORGANS ARE A PRECIOUS RESOURCE. In the United States alone, thousands of people with diseased or damaged kidneys, hearts, lungs, or livers endure an agonizing wait for someone else to die so that they might live. For many, the wait is fruitless. Every day, about 16 people in the United States die awaiting transplants.

The shortage of healthy, compatible, transplantable organs creates a tremendous demand for substitutes. Thanks to a discipline called tissue engineering, however, the production of replacement organs may one day become routine.

One approach to tissue engineering is to grow tissues outside the body. Burn victims, for example, often lose large amounts of skin, increasing their risk of dehydration and infection. Physicians can sometimes graft skin from another part of the body to the burn site, but that is impossible when burns are extensive. Artificial skin helps solve this problem. One method of producing artificial skin begins with cells called fibroblasts that are extracted from donated foreskins removed during circumcision of male babies. These cells are grown on a biodegradable scaffold, a temporary physical structure to which cells cling as they divide. After a few weeks, the cells form a layer of artificial skin, and the scaffold has dissolved. Best of all, the new tissue is free of bacteria and viruses, can be stored frozen, and does not provoke an immune response.

Scaffolds are also being used to grow three-dimensional organs outside the body. The first step in producing a replacement urinary bladder, for example, is to harvest healthy bladder cells and allow them to divide in culture. Biologists then "sow" the cells onto a biodegradable polymer. This scaffold disintegrates after surgeons implant the replacement bladder into a patient. A similar technique might one day produce engineered blood vessels, kidney tubules, livers, or cartilage.

Tissue engineering may also include embedded devices that induce the body's cells to regenerate themselves. For example, physicians may someday be able to treat spinal cord injuries by creating porous "bridges" that would secrete signal molecules to promote nerve cell division. The channels in the implant would provide the architecture needed to guide and support the new nervous tissue.

All of these potentially life-saving advances are possible because of basic research into the biology of cells, tissues, and organs. This chapter introduces these levels of organization in the animal body; subsequent chapters in this unit explore each organ system in detail.

Learning Outline

Learn How to Learn
Flashcard Excellence

While making flashcards, you may be tempted to focus on definitions. For example, after reading this chapter, you might make a flashcard with "simple squamous" on one side and "single layer of flattened cells" on the other. This description is correct, but it won't help you understand the bigger picture. Instead, your flashcards should include realistic questions that cover both the big picture and the small details. Try making flashcards that pose a question, such as "What are the four tissue types in an animal?" or "How do cells, tissues, organs, and organ systems relate to one another?" Write the full answer on the other side, then practice writing the answers on scratch paper until you are sure you know them all.

24.1 | Specialized Cells Build Animal Bodies

 Everywhere we look, form and function are entwined (**figure 24.1**). The broad, flat surface of a plant's leaf maximizes its exposure to light. A neuron's many branches permit the cell-to-cell connections essential to communication in the nervous system. In birds, fluffy down feathers trap pockets of air and conserve warmth.

Anatomy, the study of an organism's structure, describes the parts that compose the body. **Physiology** is a related discipline that considers how those parts work. Unit 5 described the anatomy and physiology of plants; unit 6 turns to animals.

Biologists describe the animal body in terms of an organizational hierarchy (**figure 24.2**). Most animals have specialized cells organized into **tissues,** which are groups of cells that interact and provide a specific function. Blood is a tissue, as are bone and muscle. An **organ** consists of two or more tissues that interact and function as a unit. The human eye, for example, is an organ that consists of light-sensitive nerve cells, blood, muscle, and other types of tissue. (Organ donation is the topic of this chapter's Burning Question.) Still farther up the organizational hierarchy are **organ systems,** which consist of two or more organs that are physically or functionally joined. The human nervous system,

Figure 24.1 **Form and Function.** A penguin is a bird that spends much of its life in cold water. Waterproof feathers prevent moisture from reaching the penguin's warm body. Flipperlike wings and a streamlined body enable the bird to "fly" under water.

? Burning Question

Which types of organs can be transplanted in humans?

Medical technology can help replace body parts damaged by disease or injury. Transplantable organs include corneas, pancreases, kidneys, skin, livers, lungs, bone marrow, parts of the digestive tract, and hearts. Surgeons have even transplanted entire faces.

One of the challenges in transplanting an organ is to prevent the recipient's immune system from rejecting the foreign tissues. Transplant surgeons minimize rejection by matching organ donors with recipients based on cell surface compatibility. In addition, physicians can prescribe drugs that suppress the immune system, but this strategy increases the risk of deadly infection.

Unfortunately, the demand for transplantable organs far exceeds the supply. The chapter opening essay described tissue engineering as one approach to solving this problem. Transplanting organs from other species into humans is another possible solution (**figure 24.A**). Heart valves from pigs, for example, can replace malfunctioning ones in humans. But cross-species transplants may transmit viruses from the donor animal to the recipient. We do not know what effect pig viruses can have on a human body, and because many viral infections take years to cause symptoms, a new infectious disease in the future could be the trade-off for using nonhuman animal parts today.

Submit your burning question to:
marielle_hoefnagels@mcgraw-hill.com

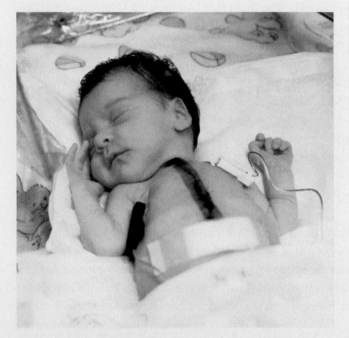

Figure 24.A **Cross-Species Transplant.** In 1984, a California newborn, Baby Fae, lived 20 days with a baboon heart. The left half of her own heart was underdeveloped.

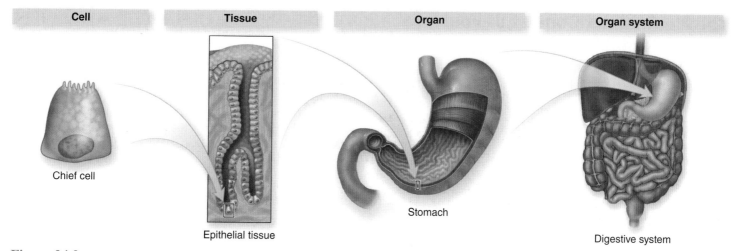

Cell	Tissue	Organ	Organ system

Chief cell

Epithelial tissue

Stomach

Digestive system

Figure 24.2 Organizational Hierarchy. Cells make up tissues, which build the organs that form organ systems. In this example, a chief cell, which secretes gastric juice, is one cell type in the epithelial tissue that lines the stomach. This organ is one of many that make up the digestive system.

for example, is composed of many organs, including the eyes, ears, spinal cord, and brain.

The human body contains dozens of organs composed of trillions of specialized cells. To understand the origin of this complexity, turn back the clock to the beginning of your life. Your father's sperm fertilized your mother's egg. Hours later, that zygote divided and became two cells. Both cells divided again, and the embryo consisted of four cells, and so on. The cells formed a ball, and then the ball hollowed out to form a blastula. (In embryos created in the laboratory, embryonic stem cells may be harvested at this stage; see section 8.7).

Up until this point in development, your cells were unspecialized and appeared pretty much alike. Then changes occurred. As some cells started expressing new combinations of genes, the blastula developed into a gastrula. A few cells collected on the inner face of the ball and spread out to form sheets. The sheets then folded, eventually forming the three primary germ layers that gave rise to all of your body's tissues and organs (see figure 20.4). Ectoderm, for example, developed into your skin and nervous system. Endoderm differentiated into your digestive tract, liver, and lungs. Mesoderm gave rise to your muscles, bones, reproductive organs, and several other structures. ▶ gastrulation, p. 412

The development of the human body, from embryo to fetus to child to adult, is familiar. Of course, other animal bodies have wildly different shapes, from the flattened tapeworm to the squishy squid to the armored lobster to the scaly snake. All of these animals have organs that carry out the same basic functions as our own—they sense their environment, acquire food and oxygen, eliminate wastes, protect themselves from injury and disease, and reproduce. Although this unit describes some notable adaptations in other animals, the focus is mainly on humans.

This chapter introduces the basic parts that build animal bodies. Chapters 25 through 34 then consider the organ systems one at a time. Throughout this unit, anatomical terms that describe the position of structures in relation to the rest of the body are

common (figure 24.3). For example, *dorsal* is a general term that means "toward the back or spine." The opposite of dorsal is *ventral*, which means "toward the belly." Learning these and the other terms in figure 24.3 should make it easier to remember the arrangement of some of the organs in the animal body.

24.1 | Mastering Concepts

1. What is the difference between anatomy and physiology?
2. What is the relationship among cells, tissues, organs, and organ systems?
3. Trace the body's early development from one fertilized egg to a many-celled organism.

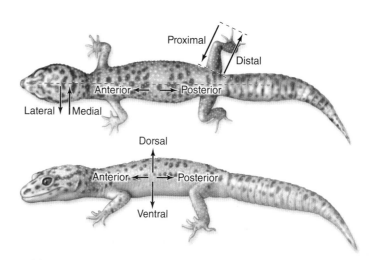

Proximal

Distal

Anterior ← → Posterior

Lateral ↕ Medial

Dorsal

Anterior ← → Posterior

Ventral

Figure 24.3 Anatomical Terms. The names of many structures described in this unit contain these terms, which describe positions relative to the rest of the body.

24.2 | Animals Consist of Four Tissue Types

Animal bodies contain a spectacular diversity of specialized cells (see figure 3.8 to review the structure of a basic animal cell). The vertebrate body, including that of humans, has at least 260 cell types.

When people first began examining the microscopic structure of animal bodies, they noticed patterns: some cell types tended to occur together and share common functions. Eventually, biologists realized that animal tissues fall into four broad categories: epithelial, connective, muscle, and nervous. Table 24.1 summarizes their characteristics, functions, and locations; this chapter's Apply It Now box, on page 521, describes how plastic surgeons reposition these tissues to improve a person's appearance or repair damage.

All animal tissues have a feature in common: the cells are embedded in an **extracellular matrix** consisting of a ground substance and fibers. The ground substance, in turn, is a mixture of water, proteins, carbohydrates, and lipids; in bone, the ground substance also includes minerals. Depending on the ingredients, the ground substance may be solid (as in bone), liquid (as in blood), or rubbery (as in cartilage). In most tissues, the extracellular matrix also contains fibers. The most abundant fiber is a strong, flexible protein called collagen. Elastin protein fibers add resiliency.

The extracellular matrix may take up very little space in a tissue, or it may dominate the tissue's volume and properties. In epithelial and muscle tissues, for example, cells are tightly packed, and the matrix is minimal. In connective tissue, in contrast, the matrix often occupies more volume than the cells and defines the properties of the tissue. Nervous tissue has a watery, indistinct extracellular matrix.

Interestingly, most normal body cells cannot survive or replicate when removed from the extracellular matrix. Proteins in the cell membrane bind to the matrix and to the cytoskeleton inside the cell. A normal cell fails to divide if not anchored in this way. Somehow, cancer cells escape this "anchorage dependence," breaking away from the extracellular matrix yet retaining the ability to divide. These abnormal cells also often secrete enzymes that destroy the fibers of extracellular matrix, clearing the way for the cancer to invade adjacent tissues. ▶ cancer, p. 162

A. Epithelial Tissue Covers Surfaces

Epithelial tissues coat the body's internal and external surfaces with one or more layers of tightly packed cells (figure 24.4). They cover organs and line the inside of hollow organs and body cavities. The diverse functions of epithelial tissues include protection, nutrient absorption along the intestinal tract, and gas diffusion in the lungs. These tissues also form **glands**, organs that secrete substances into ducts or into the bloodstream. Glands release breast milk, sweat, saliva, tears, mucus, hormones, enzymes, and many other important secretions.

Epithelial tissues always have a "free" surface that is exposed either to the outside or to a space within the body. On the opposite side, epithelium is anchored to underlying tissues by a layer of extracellular matrix called the basement membrane. (Despite the name, this structure is not the same as the membrane that surrounds all cells).

Connections called tight junctions often join adjacent epithelial cells into leak-proof sheets. The only way materials can pass through these cell layers is by passing through the cell membranes. The tightly knit structure of epithelial tissue is closely tied to its function as a border between the body's tissues and an open space. ▶ animal cell junctions, p. 65

Table 24.1 Animal Tissue Types: A Summary

Tissue Type	Description	Functions	Selected Locations	Embryonic Origin
Epithelial	Single or multiple layer of flattened, cube-shaped, or columnar cells	Cover interior and exterior surfaces of organs; protection; secretion; absorption	Epidermis, lining of digestive tract	Endoderm, ectoderm, or mesoderm
Connective	Cells scattered in prominent extracellular matrix	Support, adhesion, insulation, attachment, and transportation	Tendons, ligaments, cartilage, bone, blood, fat deposits	Mesoderm
Muscle	Elongated cells that contract when stimulated	Movement	Skeletal muscles, heart, arteries, digestive tract	Mesoderm
Nervous	Cells that transmit electrochemical impulses	Rapid communication among cells	Brain, spinal cord, nerves	Ectoderm

Epithelial tissues are classified partly by the shapes of their cells: squamous (flattened), cuboidal (cube-shaped), or columnar (tall and thin). The number of cell layers is also important. Simple epithelial tissues consist of a single layer of cells, whereas stratified epithelial tissues are made of multiple cell layers. Pseudostratified epithelium contains a single layer of columnar cells, but it appears stratified because of the staggered arrangement of nuclei.

About 90% of human cancers arise in epithelial tissues. Such a cancer is called a carcinoma. The most common carcinomas include cancers of the skin, breast, lung, prostate, and colon.

B. Most Connective Tissues Bind Other Tissues Together

The most widespread tissue type in a vertebrate's body is **connective tissue,** which consists of cells that are scattered within the extracellular matrix rather than being attached to one another. Connective tissues fill spaces, attach epithelium to other tissues, protect and cushion organs, and provide both flexible and firm structural support. Unlike epithelial tissues, connective tissues never coat any body surface.

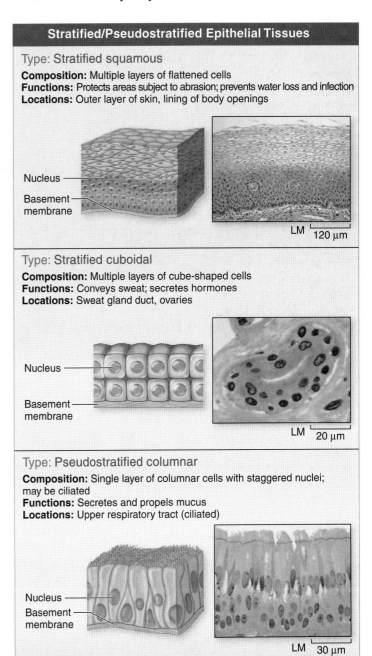

Simple Epithelial Tissues

Type: Simple squamous
Composition: Single layer of flattened cells
Functions: Allows substances to pass by diffusion and osmosis
Locations: Lining of blood vessels, alveoli of lungs, glomeruli of kidneys

Nucleus
Basement membrane
LM 30 µm

Type: Simple cuboidal
Composition: Single layer of cube-shaped cells
Functions: Secretes and absorbs substances
Locations: Glands, lining of kidney tubules

Nucleus
Basement membrane
LM 20 µm

Type: Simple columnar
Composition: Single layer of column-shaped cells; may be ciliated
Functions: Secretes and absorbs substances; sweeps egg/embryo along uterine tube
Locations: Lining of digestive tract, bronchi (ciliated), uterine tubes (ciliated)

Nucleus
Basement membrane
LM 30 µm

Stratified/Pseudostratified Epithelial Tissues

Type: Stratified squamous
Composition: Multiple layers of flattened cells
Functions: Protects areas subject to abrasion; prevents water loss and infection
Locations: Outer layer of skin, lining of body openings

Nucleus
Basement membrane
LM 120 µm

Type: Stratified cuboidal
Composition: Multiple layers of cube-shaped cells
Functions: Conveys sweat; secretes hormones
Locations: Sweat gland duct, ovaries

Nucleus
Basement membrane
LM 20 µm

Type: Pseudostratified columnar
Composition: Single layer of columnar cells with staggered nuclei; may be ciliated
Functions: Secretes and propels mucus
Locations: Upper respiratory tract (ciliated)

Nucleus
Basement membrane
LM 30 µm

Figure 24.4 Epithelial Tissues. Epithelial tissues are composed of tightly packed cells in single or multiple layers that rest atop a basement membrane. These tissues cover body surfaces and line hollow organs and body cavities.

Connective tissues are extremely variable in both structure and function (**figure 24.5**). **Loose connective tissue** binds other tissues together and fills the space between organs, **dense connective tissue** builds ligaments and tendons, and **adipose tissue** stores energy as fat. **Blood** is a connective tissue, as are the **cartilage** and **bone** that make up the vertebrate skeleton. A close look at figure 24.5 reveals that in all connective tissues except adipose tissue, the extracellular matrix occupies more volume than do the cells.

Many cell types occur in connective tissues. In loose connective tissue, dense connective tissue, and adipose tissue, cells called fibroblasts produce and secrete the extracellular matrix. Several types of immune cells patrol these tissues, triggering inflammation and stimulating other defenses against infection and injury (see chapter 33). Adipocytes (fat cells), the cells that dominate adipose tissue, also occur singly or in small clusters in loose and dense connective tissue.

Cartilage, bone, and blood contain still other cell types. Chondrocytes and osteocytes, for example, are cells that secrete the extracellular matrix in cartilage and bone. Blood contains red blood cells, white blood cells, and cell fragments called platelets, all traveling in a liquid ground substance called plasma. The fluid extracellular matrix of blood is also unique in that it lacks fibers.

C. Muscle Tissue Provides Movement

Muscle tissue consists of cells that contract (become shorter) when electrically stimulated. Contraction occurs when protein filaments composed of actin and myosin slide past one another. Abundant mitochondria in muscle cells provide the energy for contraction.

The heat generated by muscle contraction is important in body temperature regulation. The most familiar function of

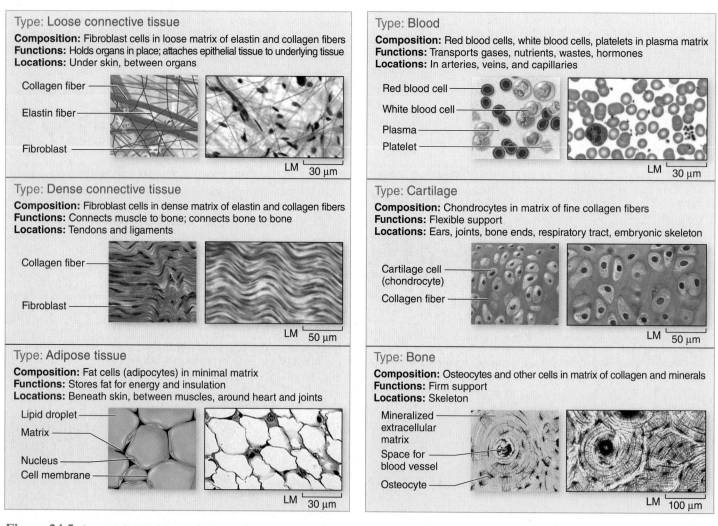

Figure 24.5 Connective Tissues. Connective tissues are highly diverse, but they all consist of cells scattered within an extracellular matrix. In most connective tissues, the matrix occupies more space than the cells themselves.

muscle tissue, however, is to move other tissues and organs. Muscle cells attach to soft tissue or bone; when the cells contract, the body part moves. Digestion, the elimination of wastes, blood circulation, and the motion of the limbs all rely on muscle contraction.

Animal bodies contain three types of muscle tissue (figure 24.6). **Skeletal muscle** tissue consists of long cells called muscle fibers. When viewed with a microscope, this muscle tissue appears striped, or striated, because the protein filaments that fill each fiber align in a repeated pattern. Most skeletal muscle attaches to bone and provides voluntary movements that a person can consciously control. **Cardiac muscle** tissue, which occurs only in the heart, is also striated, but the cells are shorter, and their control is involuntary. Cardiac muscle cells are electrically coupled with one another at connections called intercalated disks. The cells contract simultaneously to produce the heart beat. **Smooth muscle** tissue is not striated, and its contraction is involuntary. This type of muscle pushes food along the intestinal tract, regulates the diameter of blood vessels, and controls the size of the pupil of the eye.

D. Nervous Tissue Forms a Rapid Communication Network

Nervous tissue conveys information rapidly within an animal's body. Sensory cells detect stimuli such as the scent of a rose or a prick of its thorn. Other cells then transmit that information along nerves to the central nervous system (brain and spinal cord), which helps you interpret what you experience.

Two main cell types occur in nervous tissue (figure 24.7). **Neurons** form communication networks that receive, process, and transmit information. The cell may connect to another neuron at a junction called a synapse, or it may stimulate a muscle or gland. **Neuroglia** (also called glial cells) are cells that support neurons and assist in their functioning. Schwann cells, for example, are neuroglia that form insulating sheaths of a lipid called myelin, which speeds nerve impulse conduction along a neuron.

Biologists do not yet understand many of the roles of neuroglia in nervous tissue. Once considered to be merely the "glue" that binds neurons together, neuroglia are now known to guide the early development of the nervous system, regulate the concentrations of important chemicals surrounding the neurons, and participate in the formation of synapses.

24.2 | Mastering Concepts

1. List the four main tissue types in animal bodies.
2. Where do epithelial tissues occur, and how are they named?
3. List and describe six types of connective tissue.
4. Explain the similarities and differences among the three types of muscle tissue.
5. What are the two main cell types in nervous tissue?

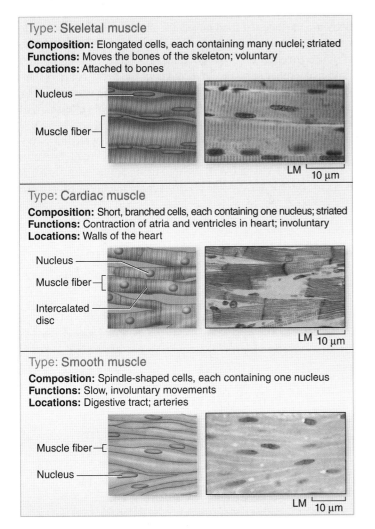

Figure 24.6 Muscle Tissues. Muscle tissue attaches to other body parts; when electrical stimulation causes the muscle cells to contract, the body part moves.

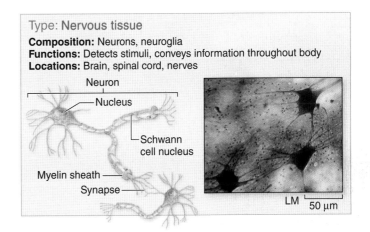

Figure 24.7 Nervous Tissue. Neurons and several types of neuroglia make up nervous tissue, which specializes in rapid communication. Neuroglia called Schwann cells make up the myelin sheath coating parts of many neurons.

24.3 | Organ Systems Are Interconnected

Tissues build organs, which form organ systems. The following sections provide a brief overview of human organ systems, organized by their contributions to the body's function (**figure 24.8**). Each organ system may seem distinct, but the function of each one relies on extensive interactions with the others.

A. The Nervous and Endocrine Systems Coordinate Communication

The human **nervous system** is a vast network composed of trillions of neurons and neuroglia that specialize in rapid communication. Some neurons are sensory receptors that detect stimuli in the environment or inside the body; others relay or interpret the sensory input. Still other neurons carry impulses from the brain or spinal cord to muscles or glands, which contract or secrete products in response.

The **endocrine system** includes glands that secrete hormones, which are communication biochemicals that affect development, reproduction, mental health, metabolism, and many other functions. Hormones travel within the circulatory system and stimulate a characteristic response in target organs. For example, the pituitary gland in the brain produces and secretes the hormone prolactin into the bloodstream. Prolactin binds to target cells in the breasts of a new mother and promotes milk secretion. Hormones act relatively slowly, but their effects last longer than nerve impulses.

B. The Skeletal and Muscular Systems Support and Move the Body

The **skeletal system** consists of bones, ligaments, and cartilage. Bones provide frameworks and protective shields for soft tissues and serve as attachment points for muscles. The marrow within some bones produces the components of blood; bones also store minerals such as calcium.

Individual skeletal muscles are the organs that make up the **muscular system.** When a skeletal muscle contracts, it moves another body part or helps support a person's posture. The heat released by contracting skeletal muscles also helps maintain body temperature.

C. The Digestive, Circulatory, and Respiratory Systems Work Together to Acquire Energy

The organs of the **digestive system** dismantle food, absorb the small molecules, and eliminate indigestible wastes. All cells of the body use the digested food molecules, either to generate energy in cellular respiration or as raw materials in maintenance and growth.

The **circulatory system** transports these food molecules (and many other substances) throughout the body. Nutrients absorbed by the digestive system enter blood at the intestines. The heart pumps the nutrient-laden blood through blood vessels that extend to all of the body's cells.

The **respiratory system** exchanges gases with the atmosphere. Cellular respiration requires not only food but also oxygen gas (O_2), which diffuses into blood at the lungs. The circulatory system delivers the O_2 throughout the body. Blood also

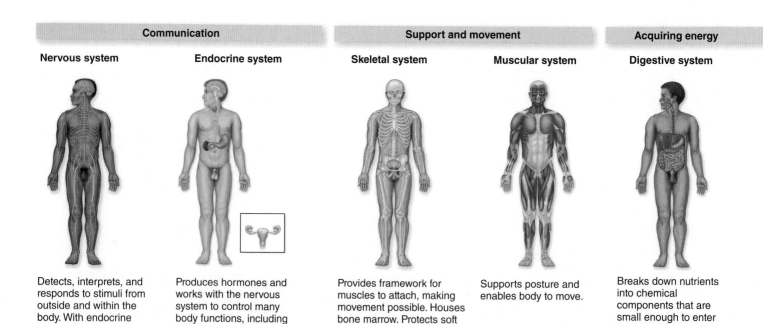

Communication		Support and movement		Acquiring energy
Nervous system	Endocrine system	Skeletal system	Muscular system	Digestive system
Detects, interprets, and responds to stimuli from outside and within the body. With endocrine system, coordinates all organ functions.	Produces hormones and works with the nervous system to control many body functions, including reproduction, response to stress, and metabolism.	Provides framework for muscles to attach, making movement possible. Houses bone marrow. Protects soft organs. Stores minerals.	Supports posture and enables body to move.	Breaks down nutrients into chemical components that are small enough to enter the circulation. Eliminates undigested food.

Figure 24.8 **Human Organ Systems.**

carries carbon dioxide gas (CO_2), a waste product of cellular respiration, to the lungs to be exhaled.

D. The Urinary, Integumentary, Immune, and Lymphatic Systems Protect the Body

Cell metabolism generates many waste products in addition to CO_2. These wastes enter the blood, which circulates through the kidneys. These organs are part of the **urinary system,** the organs that remove the water-soluble wastes and other toxins from blood and eliminate them in urine. The kidneys also have other protective functions; they adjust the concentrations of many ions, balance the blood's pH, and regulate blood pressure.

One line of physical protection is the **integumentary system,** which consists of skin, associated glands, hair, and nails. Skin is a waterproof barrier that helps keep the underlying tissues from drying out, blocks the entry of many disease-causing organisms (pathogens), and helps maintain body temperature.

The body also fights infection, injury, and cancer. The **immune system** is a huge army of specialized cells, organs, and transport vessels. This complex system launches an attack against cancer cells, viruses, microbes, and other foreign substances. Moreover, the immune system quickly neutralizes harmful molecules that it "remembers" from previous infections. Vaccines build upon this immunological memory by "teaching" the immune system about disease-causing agents the body has never actually encountered.

The **lymphatic system** is a bridge between the immune system and the circulatory system. Lymph originates as fluid that leaks out of blood capillaries and fills the spaces around the body's cells. Lymphatic capillaries absorb the excess fluid and pass it through the lymph nodes, where immune system cells destroy foreign substances. The cleansed fluid then returns to the circulatory system.

E. The Reproductive System Produces the Next Generation

The **reproductive system** consists of organs that produce and transport sperm and egg cells (gametes). The female body also can nurture developing offspring. Moreover, hormones from reproductive organs promote the development of secondary sex characteristics, such as facial hair in men and breasts in women.

The reproductive system illustrates how the organ systems are, in a sense, not separate at all. Consider the uterus, the pear-shaped sac that houses the embryo and fetus. The majority of the uterus is muscle. It also contains nervous tissue, which is why a woman feels cramps when it contracts. Hormones from the endocrine system stimulate these contractions. The entire system is richly supplied with the circulatory system's blood vessels, which also deliver cells and biochemicals of the immune system.

24.3 | Mastering Concepts

1. Which organ systems contribute to each of the five general functions of life?
2. What are some examples of interactions between organ systems?

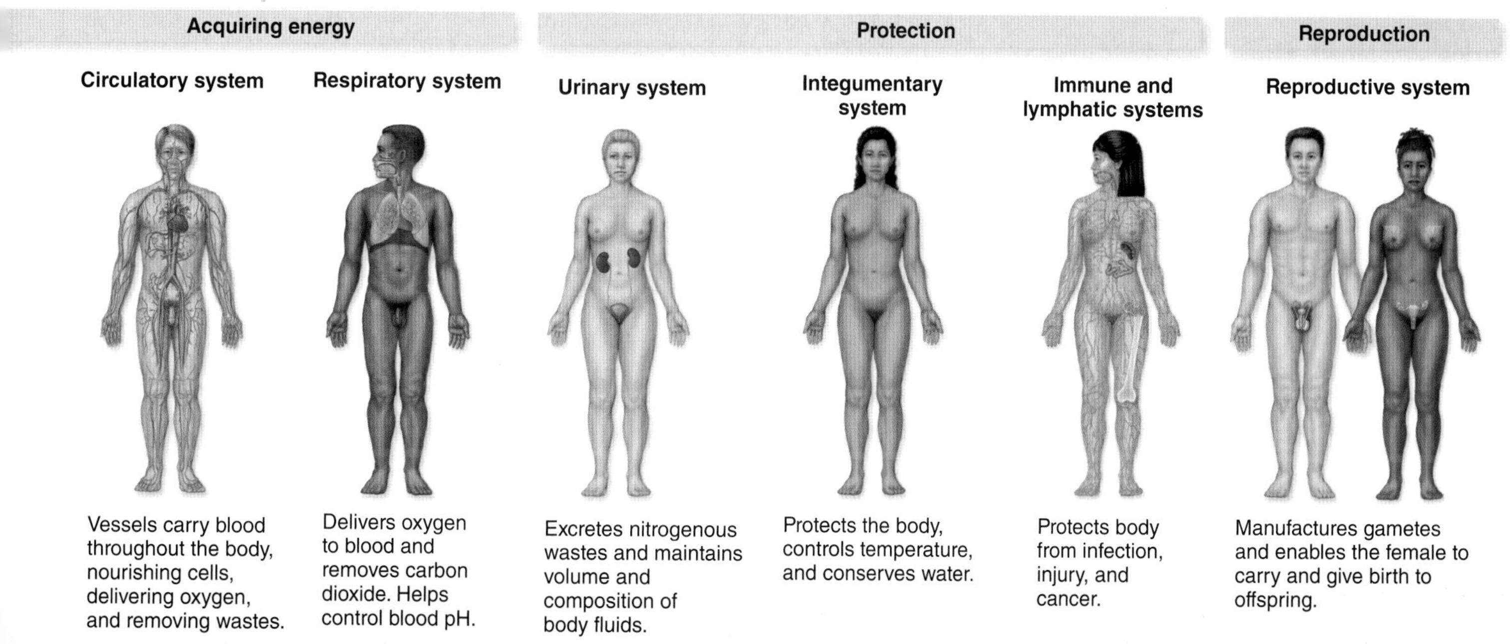

Acquiring energy			Protection		Reproduction
Circulatory system	**Respiratory system**	**Urinary system**	**Integumentary system**	**Immune and lymphatic systems**	**Reproductive system**
Vessels carry blood throughout the body, nourishing cells, delivering oxygen, and removing wastes.	Delivers oxygen to blood and removes carbon dioxide. Helps control blood pH.	Excretes nitrogenous wastes and maintains volume and composition of body fluids.	Protects the body, controls temperature, and conserves water.	Protects body from infection, injury, and cancer.	Manufactures gametes and enables the female to carry and give birth to offspring.

24.4 | Organic System Interactions Promote Homeostasis

So far, this chapter has emphasized cells, tissues, organs, and organ systems. An animal's body, however, consists mostly of water. Some of this moisture makes up the cytoplasmic "soup" that fills every cell. The rest of it forms blood plasma and the **interstitial fluid** that bathes the body's cells. Because interstitial fluid is inside the body but outside the cells, biologists consider it part of the "internal environment." Many organ systems interact to help maintain the correct concentrations of nutrients, salts, hydrogen ions, and dissolved gases in body fluids (figure 24.9).

Yet the external environment, which surrounds the body, changes constantly. Temperatures rise and fall; food may be abundant or scarce; water comes and goes. In the midst of this great variability, an animal's body must maintain its internal tempera-ture, its blood pressure, and the chemical composition of its fluids within certain limits. **Homeostasis** is this state of internal constancy.

If a body system cannot maintain homeostasis, it may stop functioning, and the organism may die. As just one example, consider what happens if the lungs fill with water. The body can no longer acquire O_2 or dispose of CO_2, yet cells continue to respire. Soon, all available O_2 is consumed, and CO_2 accumulates to toxic levels. Cells begin to die, and the person will drown unless rescuers arrive quickly.

A common way to maintain homeostasis is by **negative feedback,** which is an action that counters an existing condition. Figure 24.10 illustrates negative feedback in a familiar situation: maintaining room temperature. When the room gets too warm, the heater turns off. When the temperature is low, the thermostat signals the heater to switch back on.

In all negative feedback systems, **sensors** monitor changes in some parameter. If the value is too high or too low, the sen-

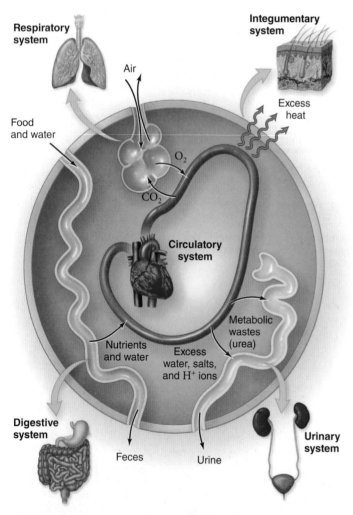

Figure 24.9 Organ System Interactions. This schematic diagram illustrates how organ systems work together to maintain the body's temperature and concentrations of oxygen, carbon dioxide, nutrients, water, and urea (a metabolic waste).

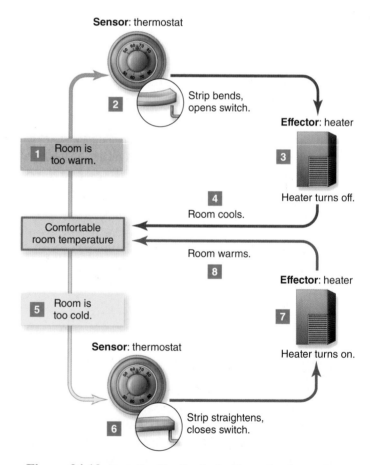

Figure 24.10 Negative Feedback: An Example. A negative feedback system maintains room temperature within comfortable limits. The thermostat is the sensor that controls the effector (the heater). (*1*) If the room is too warm, (*2*) the thermostat sends a signal to (*3*) the heater. (*4*) The heater shuts off, and the room cools. (*5*) If the room becomes too cold, (*6*) the thermostat sends a different signal to (*7*) the heater, which switches on. (*8*) This action warms the room.

sor activates one or more effectors. The **effector** responds by counteracting the original change. In figure 24.10, the thermostat is the sensor, and the heater is the effector.

Likewise, the body uses sensors and effectors to maintain homeostasis. The "supervisor" that coordinates much of the action is an almond-sized part of the brain called the hypothalamus. If blood pressure rises too high, for example, receptors in the walls of blood vessels signal the hypothalamus to slow the contraction of the heart. The pressure drops. If blood pressure falls too low, the hypothalamus signals the heart to speed up, sending out more blood.

The hypothalamus participates in many negative feedback loops. If the concentration of glucose in the blood is too high, the hypothalamus indirectly signals the pancreas to secrete a hormone called insulin. The insulin, in turn, stimulates body cells to absorb more sugar from blood. If the concentration of salts is too high, hormones from the hypothalamus signal the kidney to release more salt into urine. If oxygen is scarce, the hypothalamus stimulates faster breathing, and so on.

Only a few biological functions demonstrate **positive feedback,** in which the body reacts to a change by amplifying it. Blood clotting and milk secretion are examples of positive feedback—once started, they perpetuate their activity. Positive feedback therefore does not maintain homeostasis. Ultimately, however, other controls take over and restore equilibrium.

24.4 | Mastering Concepts

1. Which fluids are part of the body's internal environment?
2. What is homeostasis?
3. What happens to an organism that fails to maintain homeostasis?
4. Distinguish between negative and positive feedback.

24.5 | The Integumentary System Regulates Temperature and Conserves Moisture

 The integumentary system consists mostly of **skin,** the organ that forms the body's outer surface. Hair, nails, and several types of glands complete the system.

This organ system beautifully illustrates the main themes of this chapter. Not only does skin consist of multiple interacting tissue types, but the integumentary system also helps the body maintain homeostasis in several ways.

The most obvious way that skin helps maintain homeostasis is body temperature regulation. Specialized nerve endings in the skin sense the temperature and convey the information to the hypothalamus. If the temperature is too high, the hypothalamus stimulates blood vessels in the skin to dilate, releasing heat from deeper tissues to the surroundings. Meanwhile, perspiration pours out of sweat glands. When the temperature is too cold, the skin's blood vessels constrict, keeping warm blood away from the body's surface.

The integument contributes to homeostasis in other ways as well. For example, skin conserves water. People who suffer extensive burns lose large amounts of skin and may die of dehydration. Moreover, burn patients are vulnerable to infection because intact skin is the first defense against disease-causing microorganisms. Skin even plays a role in nutrition, because the initial step in vitamin D synthesis occurs when ultraviolet light strikes the skin. Vitamin D helps regulate the concentrations of calcium and phosphate in the blood, which in turn influences bone structure. Inadequate exposure to sunlight or a diet low in vitamin D causes a serious bone disease called rickets.

Skin is a surprisingly extensive organ. It is only 1 to 2 mm thick on average, although skin on the soles of the feet and several other places is thicker. Nevertheless, it makes up about 15% by weight of an adult's body.

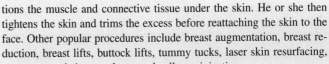
Apply It Now

Two Faces of Plastic Surgery

The "plastic" in "plastic surgery" has nothing to do with the substance that makes up water bottles and margarine tubs. Instead, the word derives from the Greek word *plastikos,* which means "to mold." Plastic surgeons "mold" a person's appearance by moving and reshaping tissues such as fat, bone, and cartilage.

Cosmetic surgery is one field of plastic surgery, and its scope extends far beyond aging movie stars struggling to maintain the illusion of youth. In fact, millions of people undergo cosmetic surgery every year in the United States. Among the most common procedures is a rhinoplasty, commonly called a "nose job," in which a surgeon removes or repositions some of the cartilage and bone of the nose to create a new shape. Liposuction is another popular procedure (**figure 24.B**). The surgeon makes small incisions and uses a surgical vacuum to remove excess fat from the thighs, buttocks, arms, neck, or stomach. Face lifts are also common. In a typical face lift, a surgeon makes a long incision at the hairline, lifts the skin of the face, and reposi-

tions the muscle and connective tissue under the skin. He or she then tightens the skin and trims the excess before reattaching the skin to the face. Other popular procedures include breast augmentation, breast reduction, breast lifts, buttock lifts, tummy tucks, laser skin resurfacing, hair transplants, and collagen injections.

In contrast to cosmetic surgery's attempts to enhance a healthy person's appearance, reconstructive plastic surgery has another goal: to restore the function of damaged body parts. For example, some children are born with a cleft palate (a gap in the bones between the nose and mouth). To repair a cleft palate, a reconstructive plastic surgeon moves tissue on either side of the gap to close the opening. Reconstructive surgery may also include skin grafts (for burn patients) or breast reconstruction (for women who have lost one or both breasts to cancer). Far from frivolous, these procedures can make a tremendous, life-changing difference in a patient's ability to function.

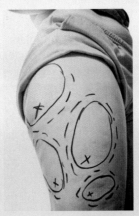

Figure 24.B Liposuction Markings.

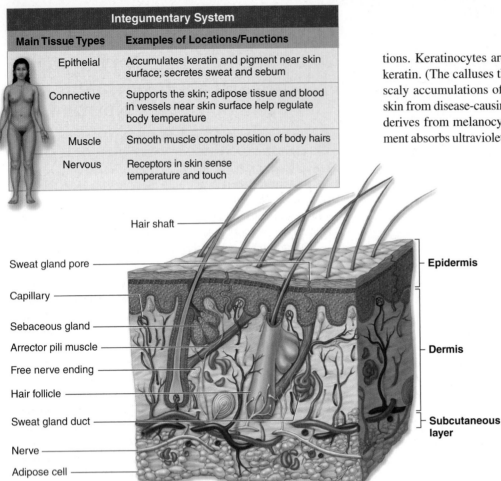

Integumentary System	
Main Tissue Types	**Examples of Locations/Functions**
Epithelial	Accumulates keratin and pigment near skin surface; secretes sweat and sebum
Connective	Supports the skin; adipose tissue and blood in vessels near skin surface help regulate body temperature
Muscle	Smooth muscle controls position of body hairs
Nervous	Receptors in skin sense temperature and touch

Hair shaft

Sweat gland pore

Capillary

Sebaceous gland

Arrector pili muscle

Free nerve ending

Hair follicle

Sweat gland duct

Nerve

Adipose cell

Epidermis

Dermis

Subcutaneous layer

Figure 24.11 The Integumentary System. Skin consists of the epidermis and dermis, with a subcutaneous layer beneath the dermis. Other integumentary structures include hair follicles, sweat glands, and sebaceous glands.

Human skin has two major layers (**figure 24.11**): the epidermis and the dermis. The **epidermis**, or outermost layer, consists mostly of stratified squamous epithelium. Active cell division continuously replaces epidermal cells lost to abrasion at the surface. Below the epidermis is the **dermis,** which is composed mostly of collagen but also contains elastin fibers, nerve endings, smooth muscle, blood vessels, and glands. A thin basement membrane anchors the epidermis to the underlying dermis. Beneath the dermis lies a layer of loose connective tissue and adipose tissue that is not technically part of the skin.

A tattoo looks like it is on the surface of the skin (**figure 24.12**). If it were, however, it would quickly disappear as epidermal cells slough off. Instead, a tattoo needle deposits the ink directly into the dermis. The ink remains in place permanently, just beneath the boundary with the epidermis.

Cells within the epidermis have specialized func-

Figure 24.12 Tattoo. To create a tattoo, a needle is used to deposit ink into the dermis of the skin.

tions. Keratinocytes are abundant cells that produce the protein keratin. (The calluses that form on the soles of the feet are thick, scaly accumulations of keratin.) The tough, dry keratin protects skin from disease-causing organisms and abrasion. The skin's color derives from melanocytes, cells that produce melanin. This pigment absorbs ultraviolet radiation that can damage DNA and therefore protects against some types of skin cancer. Other epidermal cells detect touch, and yet others help fight infection if the skin is injured. ▶ skin cancer, p. 166

The epidermis also produces hair (and fur in other mammals). A hair grows from a group of cells called a hair follicle, which is anchored in the dermis. Cells at the base of a hair follicle divide, pushing daughter cells up. These cells stiffen with keratin and die to form the hair. In humans, hair on the chest and face is a secondary sex characteristic that visibly differentiates men from women.

Hair and fur have many functions. The color of a mammal's fur can provide camouflage, as the white fur of an Arctic fox illustrates. Conversely, the prominent striped fur of a skunk is a warning to other animals. In addition, smooth muscle tissue surrounding each hair follicle can contract and make the hair "stand on end" when skin senses cold—this is the basis of goose bumps. This reaction has no apparent function in humans, but in many other mammals, erect fur provides extra insulation that helps retain body heat.

The tissues that make up the dermis also have multiple functions. Collagen fibers in the dermis support the skin. In addition, humans can generate a wide array of facial expressions, thanks to muscles attached to fibers in the dermis. Nerve endings in the dermis convey information from sensory receptors to the brain.

Several types of glands originate in the dermis as well. Sweat glands produce perspiration. Mammary glands, which produce milk in female mammals, are derived from sweat glands. Many mammals, including humans, have sebaceous glands, which secrete an oily substance that softens and helps protect the skin and hair.

Injuries that damage the dermis, such as a cut from broken glass, can leave a scar. As the injury heals, the replacement skin contains more collagen fibers than the surrounding undamaged tissue. Neither hair follicles nor sweat glands develop in the damaged area. The resulting scar consists of new skin that does not look or work exactly like the old.

24.5 | Mastering Concepts

1. How does the integumentary system help the body maintain homeostasis?
2. What tissue types occur in each layer of human skin?
3. What are some examples of specialized cells of the epidermis and dermis?

24.6 | Investigating Life: Vitamins and the Evolution of Human Skin Pigmentation

Human skin color ranges from the very pale to the deeply pigmented (see figure 10.25). How did this variation arise?

Scientists have long understood that the darker the skin, the higher the concentration of melanin. Moreover, indigenous people from the tropics (near the equator) tend to have darker skin than their counterparts from higher latitudes (toward the poles). Tropical areas receive more ultraviolet (UV) radiation than do locations farther north and south. High melanin production is therefore correlated with exposure to UV radiation.

A question arises from these observations: Have differences in UV radiation selected for variation in pigmentation? More specifically, is there a relationship among UV radiation, melanin production, and reproductive success in the human population?

The risk for skin cancer offers one possible answer. The more melanin in the skin, the less UV radiation penetrates below the epidermis, and the lower the risk for cancer. But skin cancer is rarely fatal, and most cases develop after a person's reproductive years. Since skin cancer usually does not interfere with reproductive success, it is probably not a strong selective force in the evolution of skin color.

Oddly enough, nutrition offers a more convincing explanation for the worldwide distribution of skin pigmentation. Ultraviolet radiation interacts with two vitamins, folic acid (vitamin B_9) and vitamin D, both of which are critical to overall health and reproduction. If a woman's diet is too low in folic acid, she faces a high risk of giving birth to a child with serious birth defects. In men, insufficient folic acid may cause infertility. A vitamin D deficiency can cause death or pelvic deformities that make childbirth difficult.

Exposure to UV radiation affects the concentrations of both vitamins, but in opposite ways. Ultraviolet light destroys folic acid in the skin, whereas the body produces vitamin D when sunlight strikes the skin.

One hypothesis that explains the distribution of skin pigmentation around the world, then, reflects a nutritional trade-off. In the tropics, abundant UV radiation selects for a high concentration of melanin, which blocks light penetration and therefore minimizes the sun-induced loss of folic acid. In higher latitudes, where people are exposed to less UV radiation than in the tropics, lighter skin pigmentation maximizes the potential to synthesize vitamin D.

Pennsylvania State University anthropologists Nina Jablonski and George Chaplin tested the strength of the association between vitamin D nutrition and UV radiation. They began with satellite measurements of UV radiation striking Earth's surface. The pair also searched the scientific literature for measurements of skin pigmentation in indigenous peoples anywhere in the world. They also looked for papers documenting how much UV radiation is needed to synthesize sufficient vitamin D.

Jablonski and Chaplin mapped the average annual exposure to UV radiation around the world. They then superimposed three zones on their map, corresponding to areas where UV exposure is insufficient to produce vitamin D in lightly, moderately, and highly pigmented skin (figure 24.13). Sure enough, the observed patterns of skin pigmentation corresponded to the zones, supporting the hypothesis that vitamin D deficiency becomes an increasingly important selective force at higher latitudes.

This study not only provides insight into an interesting question about human evolution, but it also illustrates two important features of scientific investigation. First, technological advances can offer new ways to test old hypotheses. In this case, satellites enabled Jablonski and Chaplin to peek with unprecedented detail at ultraviolet radiation patterns across the globe. Second, scientists often borrow heavily from the past; dozens of previous studies on skin pigmentation and vitamin D synthesis gave the research team the information they needed to reach their conclusion.

Jablonski, Nina G., and George Chaplin. 2000. The evolution of human skin coloration. *Journal of Human Evolution,* vol. 39, pages 57–106.

24.6 | Mastering Concepts

1. Describe how folic acid and vitamin D nutrition might explain the variation in human skin pigmentation.
2. Why do you think people who migrate from the tropics to urban areas in northern latitudes are at high risk for vitamin D deficiency? How might they decrease the risk?

Annual exposure to UV radiation		Average observed skin reflectance		
High ▓▓▓▓ Low		⧄ 68.9 (light)	▨ 55.0 (moderate)	⠿ 37.2 (dark)

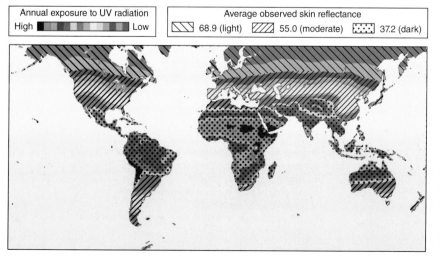

Figure 24.13 UV Radiation and Vitamin D. The colors on this map indicate average annual exposure to UV radiation. Stippling indicates areas where UV radiation is insufficient for vitamin D production in people with lightly pigmented skin, moderately pigmented skin, and highly pigmented skin. Average skin reflectance among indigenous people from each zone is also shown. The lower the skin reflectance value, the darker the pigmentation.

Chapter Summary

24.1 | Specialized Cells Build Animal Bodies

- **Anatomy** and **physiology** are interacting studies of the structure and function of organisms.
- Specialized cells express different genes. These cells aggregate and function together to form **tissues.** Tissues build **organs,** and interacting organs form **organ systems.**

24.2 | Animals Consist of Four Tissue Types

- Animal tissues consist of cells within an **extracellular matrix** consisting of ground substance and (usually) protein fibers. The matrix may be solid, semisolid, or liquid.

A. Epithelial Tissue Covers Surfaces

- **Epithelial tissue** lines organs and forms **glands.** This tissue protects, senses, and secretes.
- Epithelium may be simple (one layer) or stratified (more than one layer), and the cells may be squamous (flat), cuboidal (cube-shaped), or columnar (tall and thin).

B. Most Connective Tissues Bind Other Tissues Together

- **Connective tissues** have diverse structures and functions. Most consist of scattered cells and a prominent extracellular matrix.
- The six major types of connective tissues are **loose connective tissue, dense connective tissue, adipose tissue, cartilage, bone,** and **blood.**

C. Muscle Tissue Provides Movement

- **Muscle tissue** consists of cells that contract when protein filaments slide past one another.
- Three types of muscle tissue are **skeletal, cardiac,** and **smooth muscle.**

D. Nervous Tissue Forms a Rapid Communication Network

- **Neurons** and **neuroglia** make up **nervous tissue.**
- A neuron functions in rapid communication; neuroglia support neurons.

24.3 | Organ Systems Are Interconnected

A. The Nervous and Endocrine Systems Coordinate Communication

- The **nervous system** and **endocrine system** coordinate all other organ systems.
- Neurons form networks of cells that communicate rapidly, whereas hormones produced by the endocrine system act more slowly.

B. The Skeletal and Muscular Systems Support and Move the Body

- The bones of the **skeletal system** protect and support the body, and they act as a reservoir for calcium and other minerals.
- The **muscular system** enables body parts to move and generates body heat.

C. The Digestive, Circulatory, and Respiratory Systems Work Together to Acquire Energy

- The **digestive system** provides nutrients; the **respiratory system** obtains O_2. The **circulatory system** delivers nutrients and O_2 to tissues.
- The body's cells use O_2 to extract energy from food molecules. The circulatory and respiratory systems eliminate the waste CO_2.

D. The Urinary, Integumentary, Immune, and Lymphatic Systems Protect the Body

- The **urinary system** removes metabolic wastes from the blood and reabsorbs useful substances.
- The **integumentary system** provides a physical barrier between the body and its surroundings.
- The **immune system** protects against infection, injury, and cancer.
- The **lymphatic system** connects the circulatory and immune systems, filtering the body's fluids through the lymph nodes.

E. The Reproductive System Produces the Next Generation

- The male and female **reproductive systems** are essential for the production of offspring.

24.4 | Organ System Interactions Promote Homeostasis

- **Homeostasis** is the maintenance of a stable internal environment, including regulation of body temperature and the chemical composition of blood plasma and **interstitial fluid.**
- **Negative feedback** restores the level of a substance or parameter to within a normal range. **Sensors** detect changes in the internal environment and activate **effectors** that counteract the change.
- **Positive feedback** perpetuates an action.

24.5 | The Integumentary System Regulates Temperature and Conserves Moisture

- **Skin** helps regulate body temperature, conserves moisture, and contributes to vitamin D production.
- Skin consists of an **epidermis** over a **dermis,** plus specialized structures such as hairs and sweat glands. A basement membrane joins the epidermis to the dermis, and a layer of connective tissue underlies the dermis.
- In the epidermis, keratinocytes accumulate keratin, and melanocytes provide pigment.

24.6 | Investigating Life: Vitamins and the Evolution of Human Skin Pigmentation

- Variation in exposure to ultraviolet radiation may select for a range of skin pigmentation, reflecting a balance between folic acid and vitamin D nutrition.

Multiple Choice Questions

1. Which of the following represents the correct order of organization of an animal's body?
   a. Cells; organs; organ systems; tissues
   b. Cells; tissues; organ systems; organs
   c. Tissues; cells; organs; organ systems
   d. Cells; tissues; organs; organ systems

2. The nervous system develops from the same embryonic tissue as the
   a. muscles.   c. skin.
   b. bones.   d. digestive tract.

3. The cells of epithelial tissue must _____ to function properly.
   a. form a single layer
   b. attach to one another by tight junctions
   c. secrete substances
   d. transport chloride ions

4. A common property of all types of connective tissue is the
   a. formation of solid or semisolid arrangements of cells.
   b. production of collagen fibers.
   c. arrangement of cells in an extracellular matrix.
   d. arrangement of cells into multiple layers.

5. Blood is an example of what type of tissue?
   a. Epithelial   c. Connective
   b. Nervous   d. Muscle

6. Smooth muscle is *different* from skeletal muscle because
   a. smooth muscle contraction is involuntary.
   b. skeletal muscle is striated.
   c. smooth muscle contains actin and myosin.
   d. Both a and b are correct.

7. Ovaries produce gametes and hormones; these organs therefore belong to the _____ systems.
   a. immune and integumentary
   b. endocrine and reproductive
   c. circulatory and nervous
   d. urinary and lymphatic

8. Which of the following scenarios does NOT illustrate negative feedback?
   a. In childbirth, contractions stimulate the release of oxytocin, which induces more contractions.
   b. Body temperature climbs so high that a person begins to sweat, which cools the body.
   c. The salt concentration in blood is too high, so the kidneys eliminate salt in urine.
   d. Eating a meal causes a rise in blood sugar, which stimulates the pancreas to release insulin.

9. The inner layer of skin is composed of _____, whereas the outer layer is _____ .
   a. epithelial tissue; connective tissue
   b. dermis; epidermis
   c. epidermis; epithelial tissue
   d. epithelial tissue; epidermis

10. How does the integumentary system influence homeostasis?
    a. By preventing water loss
    b. By sensing external temperatures
    c. By preventing infection
    d. All of the above are correct.

Write It Out

1. Distinguish between:
   a. organs and organ systems
   b. simple squamous and stratified squamous epithelial tissue
   c. loose and dense connective tissue
   d. skeletal and cardiac muscle tissue
   e. neurons and neuroglia
   f. negative and positive feedback

2. Marfan syndrome (see chapter 10) and osteogenesis imperfecta are two examples of heritable disorders of connective tissue. Use the Internet to learn about these two diseases. Why do people with connective tissue disorders have many interrelated symptoms?

3. What is homeostasis, and how is it important?

4. When a person gets cold, he or she may begin to shiver. If the weather is too hot, the heart rate increases and blood vessels dilate, sending more blood to the skin. How does each scenario illustrate homeostasis?

5. Observe what happens to the size of your eye's pupil when you leave a dark room and enter the sunshine. What happens in the opposite situation, when you enter a dark room? How do the opposing reactions of your eye illustrate negative feedback?

6. Which tissues make up skin, the largest organ of the body? How do these tissues interact to provide each of the functions of skin? Describe at least one interaction between skin and each of the 10 other organ systems.

7. How would you design an experiment to determine whether a new brand of artificial skin is safe for use in humans?

8. Make a chart that compares and contrasts the organization of the animal body with that of a plant (see chapter 21).

Pull It Together

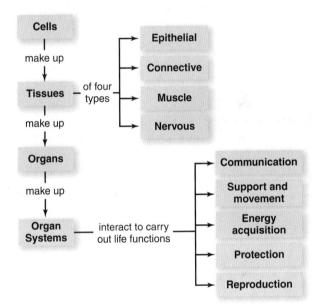

1. What features distinguish the four types of tissue?
2. Add the specific types of epithelial, connective, muscle, and nervous tissue to this concept map.
3. Add the 11 organ systems to this concept map.
4. Describe examples of organ system interactions that maintain homeostasis.

Narcolepsy. Stanford University sleep researcher Dr. William Dement holds a dog, Tucker, before and during an attack of narcolepsy.

BIOLOGY

Enhance your study of
this chapter with practice quizzes,
animations and videos, answer keys,
and downloadable study tools.
www.mhhe.com/hoefnagels

The Narcolepsy Gene

A SLEEPING DOG LOOKS INACTIVE, BUT APPEARANCES CAN BE DECEIVING. A whirlwind of chemical activity is occurring inside the brain of a dozing animal. Although scientists do not fully understand the function of sleep, it is clearly essential to good health. Among other physiological effects, a lack of sleep apparently increases the production of stress hormones and causes high blood pressure. Some studies link sleep deprivation to obesity, heart disease, and other serious illnesses.

Several sleep disorders plague humans. The most familiar is insomnia, the inability to sleep. Causes include everything from worrying to consuming too much caffeine. Another common disorder is sleep apnea, in which a person briefly stops breathing several times a night. In some cases, the brain's "breathing center" briefly quits working and then resumes. Usually, however, an obstructed airway causes sleep apnea.

Of all the sleep disorders, narcolepsy may be the strangest. People with narcolepsy are excessively sleepy during the day, even if they slept enough the night before. Some patients also experience vivid hallucinations or paralysis during or after sleep. The most dramatic symptom, however, is cataplexy, a state in which the person abruptly loses control of his or her muscles, often in response to intense emotion such as joy or surprise. The effects vary widely, from a slight drooping of the head to a total collapse. Cataplexy usually only lasts for a few minutes, but it can be very dangerous. For example, a person who loses muscle control, even for a moment, while driving a car may cause a deadly accident.

Sleep research took a step forward in 1999, when scientists reported the discovery of a defective gene associated with narcolepsy in Doberman pinschers and Labrador retrievers. In healthy dogs, the gene encodes a protein expressed on cell surfaces in a part of the brain called the hypothalamus. The protein is a receptor that receives neurotransmitters, which are chemical messages from other cells. Narcoleptic dogs have a mutated version of the gene, so the receptor protein is absent or fails to bind to the neurotransmitter. Apparently, affected brain cells do not transmit the "stay awake" message, especially when the narcoleptic dogs become excited.

Is this marriage of genetic and neurological research important to anyone other than dog fanciers? Yes, because humans have the gene encoding the receptor, too. The discovery of this gene may someday yield not only better narcolepsy treatments but also, on the opposite end of the sleep disorder spectrum, better sleeping pills.

UNIT 6

What's the Point?

Learning Outline

25.1 The Nervous System Forms a Rapid Communication Network
- A. Invertebrates Have Nerve Nets, Nerve Ladders, or Nerve Cords
- B. Vertebrate Nervous Systems Are Highly Centralized

25.2 Neurons Are Functional Units of a Nervous System
- A. A Typical Neuron Consists of a Cell Body, Dendrites, and an Axon
- B. The Nervous System Includes Three Classes of Neurons

25.3 Action Potentials Convey Messages
- A. A Neuron at Rest Has a Negative Charge
- B. A Neuron Transmitting an Impulse Undergoes a Wave of Depolarization
- C. The Myelin Sheath Speeds Impulse Conduction

25.4 Neurotransmitters Pass the Message from Cell to Cell
- A. Neurons Communicate at Synapses
- B. The Postsynaptic Cell Integrates Signals from Multiple Synapses

25.5 The Peripheral Nervous System Consists of Nerve Cells Outside the Central Nervous System

25.6 The Central Nervous System Consists of the Spinal Cord and Brain
- A. The Spinal Cord Transmits Information Between Body and Brain
- B. The Human Brain Is Divided into Several Regions
- C. Many Brain Regions Participate in Memory Formation
- D. Damage to the Central Nervous System Can Be Devastating

25.7 Investigating Life: The Nerve of Those Clams!

Learn How to Learn

How to Use a Study Guide

Some professors hand out study guides to help students prepare for exams. Used wisely, a study guide can be a valuable tool. One good way to use a study guide is AFTER you have studied the material, when you can go through the guide and make sure you haven't overlooked any important topics. Alternatively, you can cross off topics as you encounter them while you study. No matter which technique you choose, don't just memorize isolated facts. Instead, try to understand how the items on the study guide relate to one another. If you're unclear on the relationships, be sure to ask your instructor.

25.1 | The Nervous System Forms a Rapid Communication Network

Love, happiness, tranquility, sadness, jealousy, rage, fear, and excitement—all of these emotions spring from the cells of the nervous system (**figure 25.1**). So do the ability to understand language, the sensation of warmth, memories of your childhood, and your perception of pain, color, sound, smell, and taste. The muscles that move when you scratch a mosquito bite, chew, blink, or breathe all are under the control of the nervous system, as is the unseen motion that propels food along your digestive tract.

The most critical functions of the nervous system are the "behind the scenes" activities that maintain a state of internal constancy, or homeostasis. The feedback systems that maintain homeostasis require communication between the sensors (the cells that detect internal and external conditions) and the effectors (the muscles and glands that make adjustments). The **nervous system** and the **endocrine system** provide this essential communication. A major difference between these two organ systems is the speed with which they act. The nervous system's electrical impulses travel so rapidly that their effects are essentially instantaneous. The moment you decide to pick up a pencil or type a letter, for example, you can carry out your plan. The endocrine system, the subject of chapter 27, acts much more slowly. Endocrine glands secrete chemical messages called hormones that can take hours (or longer) to exert their effects. ▶ homeostasis, p. 520

Nervous tissue includes two basic cell types: interconnected neurons and their associated neuroglial cells. The **neurons** are the cells that communicate with one another. The more numerous **neuroglia** (also called glial cells) provide physical support, help

Figure 25.1 Great Fun. Thanks to the nervous system, these roller-coaster passengers can experience a range of sensations and emotions.

maintain homeostasis in the fluid surrounding the neurons, guide neuron growth, and play many other roles that researchers are just beginning to discover. ▶ nervous tissue, p. 517

The neurons and neuroglia that make up nervous tissue control mood, appetite, blood pressure, coordination, and the perception of pain and pleasure. Neurons enable animals to sense the environment, screen out unimportant stimuli, move, learn, and remember. Yet despite their diverse functions, all neurons communicate in a similar manner. This chapter explores nervous system diversity, describes how neurons function, and then considers the human nervous system in more detail.

A. Invertebrates Have Nerve Nets, Nerve Ladders, or Nerve Cords

Nervous systems vary greatly in complexity and organization. Depending on the species, a nervous system might be a loose network of relatively few cells or a structure as intricate as the human brain. The more complex the nervous system, the greater the animal's ability to detect multiple stimuli simultaneously and to coordinate responses. Moreover, highly developed nervous systems are associated with complex behaviors and an increased capacity for learning, memory, problem-solving, and language.

Figure 25.2 illustrates some examples of animal nervous systems. The simplest are **nerve nets,** which are diffuse networks of neurons in the body walls of hydras, jellyfishes, sea anemones, and other cnidarians. The nerve cells physically touch one another, so that a stimulus at any point on the body spreads over the entire body surface. Nerve nets stimulate muscle cells near the body surface, enabling the animals to move their tentacles or swim.

Cnidarians are relatively simple, radially symmetrical animals. The evolution of bilateral symmetry, however, meant that one end of the body (the head) took in the most sensory information from the environment. Nervous tissue began to accumulate into a brain and sensory structures at the head end, and the nervous system became increasingly centralized.

A flatworm is a simple bilaterally symmetrical animal. Its brain consists of two **ganglia,** or clusters of neurons, at its head end. Two nerve cords emerge from the ganglia and extend along the length of the body. Transverse nerves connect the nerve cords to each other, forming a **nerve ladder.** Motor structures and the neurons that control them are paired, which allows the worm to move in a coordinated forward motion. In addition, paired eye spots and chemical receptors enable the animal to determine where a stimulus originates; the receptor with the strongest response is closest to the stimulus.

Segmented worms such as leeches and earthworms have a more complex nervous system with a larger brain. Peripheral nerves branch from a **ventral nerve cord** to each segment, so the animal can coordinate its movements. Cells that are sensitive to light, chemicals, and touch occur all over the body surface.

The nervous system of an arthropod (including insects, lobsters, spiders, and scorpions) also consists of a brain and a ventral nerve cord. Arthropods have well-developed sensory organs that detect light, touch, chemicals, sound, and balance. Many of these animals exhibit complex behaviors, such as the famous "waggle dance" by which honeybees communicate the location of new food sources.

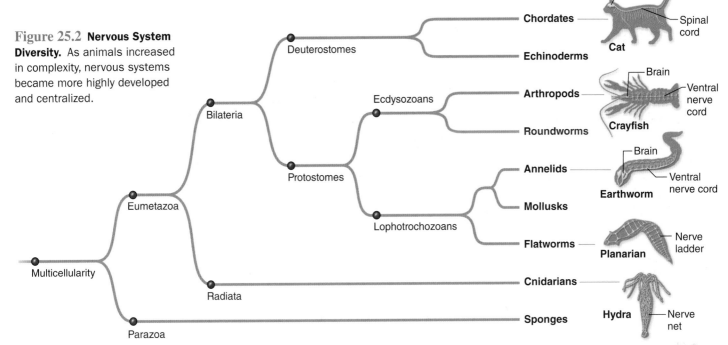

Figure 25.2 Nervous System Diversity. As animals increased in complexity, nervous systems became more highly developed and centralized.

Among invertebrates, the octopuses and squids have the most sophisticated nervous systems. These intelligent animals have large brains, keen eyes, and sensitive tentacles. They can even master complex visual tasks, such as opening food jars!

B. Vertebrate Nervous Systems Are Highly Centralized

The trend toward nervous tissue accumulation in the head is most evident in vertebrates. The vertebrate nervous system has two

Sensory input

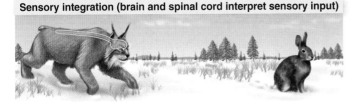

Sensory integration (brain and spinal cord interpret sensory input)

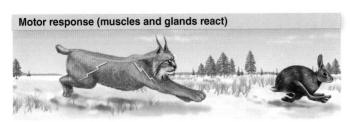

Motor response (muscles and glands react)

Figure 25.3 Roles of the Nervous System. Sensory organs such as eyes, ears, and the nose receive sensory input. The central nervous system integrates the information and sends signals that initiate appropriate motor responses.

main divisions: the central and peripheral nervous systems. The **central nervous system** consists of the **brain** (inside the skull) and the dorsal, tubular **spinal cord.** The brain may be small and relatively simple, as in fishes and amphibians, or it may be large and complex, as in humans and other mammals. The main function of the central nervous system is to integrate sensory information and coordinate the body's response.

All neurons other than those in the central nervous system make up the **peripheral nervous system,** which carries information between the central nervous system and the rest of the body. The sensory pathways of the peripheral nervous system carry information to the spinal cord and brain; the motor pathways transmit nerve impulses from the central nervous system to muscle or gland cells.

To understand how the nervous system regulates virtually all other organ systems, imagine a lynx hunting a hare (**figure 25.3**). Sensory neurons in the peripheral nervous system enable the cat to hear, see, and smell its prey. The lynx's central nervous system interprets this sensory input and decides how to act, and then motor neurons coordinate the skeletal muscles that move the lynx into position to catch the hare. Meanwhile, the cat's heart pumps blood, and its lungs inhale and exhale—all under the control of the central and peripheral nervous systems.

25.1 | Mastering Concepts

1. How is the nervous system's role in maintaining homeostasis different from that of the endocrine system?
2. What are the roles of neurons and neuroglia in the nervous system?
3. Describe how nervous systems changed as animal bodies became more complex.
4. Distinguish between the central and peripheral nervous systems in vertebrates.

25.2 | Neurons Are Functional Units of a Nervous System

The nervous system's function is rapid communication by electrical and chemical signals. Neurons are the cells that do the communicating, both with one another and with muscles and glands. To understand how neurons carry out their function, it helps to first learn about their structure.

A. A Typical Neuron Consists of a Cell Body, Dendrites, and an Axon

All neurons have the same basic parts (figure 25.4). The enlarged, rounded **cell body** contains the nucleus, mitochondria that supply ATP, ribosomes that manufacture proteins, and other organelles. **Dendrites** are short, branched extensions that transmit information toward the cell body. The number of dendrites may range from one to thousands, and each can receive input from many other neurons.

The **axon,** also called the nerve fiber, is typically a single long extension of the cell body. The axon is finely branched at its tip, and each tiny terminal extension communicates with another cell at a junction called a synapse. Usually, the axon conducts nerve impulses from the cell body to a muscle, gland, or other neuron. The axon that permits you to wiggle a big toe, for example, extends about a meter from the base of the spinal cord to the toe.

In many neurons, a **myelin sheath** composed of fatty material coats sections of the axon, speeding nerve impulse conduction (see section 25.3). In the peripheral nervous system, neuroglia called **Schwann cells** form the myelin sheath; other types of

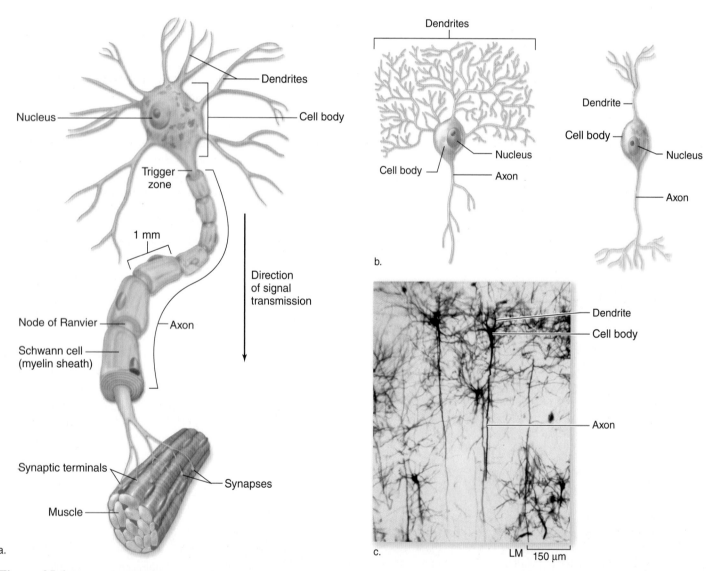

Figure 25.4 Parts of a Neuron. (a) A neuron consists of a rounded cell body, one or more dendrites, and an axon. In many neurons, the axon is encased in a myelin sheath. (b) Neuron shapes vary considerably. A motor neuron with multiple dendrites is on the left, and a sensory neuron with one dendrite is on the right. (c) Several neurons are visible in this micrograph of the brain's outermost layer.

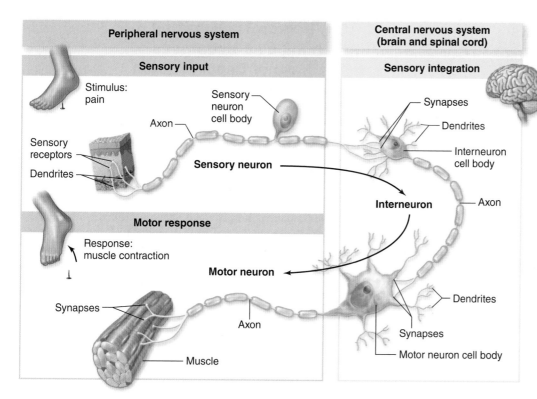

Figure 25.5 Categories of Neurons. Sensory neurons transmit information from sensory receptors to the central nervous system. Interneurons connect sensory neurons to motor neurons, which send information from the central nervous system to muscles or glands.

neuroglia, **oligodendrocytes,** make up the myelin sheaths in the central nervous system. **Nodes of Ranvier** are short regions of exposed axon between sections of the myelin sheath.

To picture the relative sizes of a typical neuron's parts, imagine its cell body is the size of a tennis ball. The axon might then be up to 1.5 kilometers long but only a few centimeters thick. The mass of dendrites extending from the cell body would fill an average-size living room.

B. The Nervous System Includes Three Classes of Neurons

Biologists divide neurons into three categories, based on general function (figure 25.5).

- A **sensory neuron** brings information from the body's organs toward the central nervous system. Sensory neurons respond to light, pressure from sound waves, heat, touch, pain, and chemicals detected as odors or taste. The dendrites, cell body, and most of the axon of each sensory neuron lie in the peripheral nervous system, whereas the axon's endings reside in the central nervous system.

- About 90% of all neurons are **interneurons,** which connect one neuron to another within the spinal cord and brain. Large, complex networks of interneurons receive information from sensory neurons, process this information, and generate the messages that the motor neurons carry to effector organs. Interneurons may also receive signals from other interneurons.

- A **motor neuron** conducts its message from the central nervous system toward an effector: a muscle or gland cell. A motor neuron's cell body and dendrites reside in the central nervous system, but its axon extends into the peripheral nervous system. Thus, motor neurons stimulate muscle cells to contract and glands to secrete. (They are called motor neurons because most lead to muscle cells.)

Figure 25.5 shows a simplified example of how the three types of neurons work together to coordinate the body's reaction to a painful stimulus. The process begins when a person steps on a tack. Sensory neurons whose dendrites are in the skin of the foot convey the information to the spinal cord. There, the sensory neuron synapses with an interneuron, which relays the information to an area of the brain that interprets the sensation as pain. Another interneuron connects the brain's sensory area with the motor area. The motor area, in turn, relays a command to yet another interneuron, which activates the motor neurons that stimulate muscle contraction in the leg and foot. The signals move so quickly that the foot withdraws at about the same time as the brain perceives the pain.

25.2 | Mastering Concepts

1. Describe the parts of a typical neuron.
2. Where is the myelin sheath located?
3. What is the usual direction in which a message moves within a neuron?
4. What are the functions of each of the three classes of neurons?

25.3 | Action Potentials Convey Messages

A sensory neuron, interneuron, or motor neuron sends messages by conveying a neural impulse. These electro-chemical signals result from the movement of charged particles (ions) across the cell membrane. This section describes the distribution of ions in neurons, both when the cell is "at rest" and when it is transmitting a neural impulse. ▶ ions, p. 21

A. A Neuron at Rest Has a Negative Charge

To understand how ions move in a neural impulse, it helps to be familiar with the **resting potential,** which is the difference in electrical charge between the inside and outside of a neuron that is not conducting a message. At rest, a neuron's membrane is polarized (**figure 25.6,** step 1). That is, the inside carries a slightly negative electrical charge relative to the outside. This separation of charge creates an electrical "potential" that measures around -70 millivolts (mV). (A volt measures the difference in electrical charge between two points.)

A neuron has a resting potential because it maintains an unequal distribution of ions, notably potassium (K^+), across its membrane. One membrane protein that helps maintain this gradient is the **sodium–potassium pump,** which pumps three sodium ions (Na^+) out of the cell for every two K^+ that enter, at a cost of one ATP molecule per cycle. The sodium–potassium pump operates continuously. In a neuron at rest, the concentration of K^+ is therefore much higher inside the cell than outside, while the reverse is true for Na^+. ▶ active transport, p. 82

The neuron's resting potential reflects a balance between two opposing forces on K^+. On one hand, the sodium–potassium pump concentrates K^+ inside the cell. Channel proteins in the membrane allow some of this K^+ to diffuse back out of the cell along its concentration gradient. On the other hand, positively charged Na^+ ions outside the cell repel K^+, while large, negatively charged proteins (and other negative ions) inside the cell attract K^+. ▶ diffusion, p. 80

When the two opposing forces—the concentration gradient and charge interactions—are equal, the membrane has a net positive charge on the outside and a net negative charge on the inside. This difference in charge is the resting potential.

The term *resting potential* is a bit misleading because the neuron consumes a tremendous amount of energy while "at rest." In fact, the nervous system devotes about three quarters of its total energy budget to maintaining the resting potential of its neurons. The resulting state of readiness allows the neuron to respond more quickly than it could if it had to generate a potential difference across the membrane each time it received a stimulus. The neuron's resting potential is therefore analogous to holding back the string on a bow to be constantly ready to shoot an arrow.

B. A Neuron Transmitting an Impulse Undergoes a Wave of Depolarization

A neuron's resting potential keeps it primed to convey messages at any moment. If a stimulus does arrive, action potentials may occur along the neuron's axon. An **action potential** is a brief depolarization that propagates like a wave along the membrane of the nerve fiber. A neural impulse is the spread of action potentials along an axon.

The stimulus that triggers a neuron to "fire" may be a change in pH, a touch, or a signal from another neuron. Whatever the stimulus, some sodium channel gates in a neuron's membrane open and then immediately close, usually at the dendrites or cell body. A small amount of Na^+ leaks into the cell through the open channels, causing the interior to become less negative. This local flow of electrical current is called a **graded potential,** both because it weakens with distance from the source of the stimulus and because the magnitude of depolarization depends on the signal's strength.

If the stimulus is very weak, the graded potential is small, and no action potential will occur in the neuron. But if the stimulus is strong enough, the depolarization may spread from the dendrite or cell body to a "trigger zone" at the origin of the axon. Unlike the dendrites and cell body, the trigger zone (along with the nodes of Ranvier on the axon) has densely packed sodium channels that can sustain an action potential.

An action potential occurs, however, only if the depolarization exceeds the cell's **threshold potential.** When enough Na^+ enters to depolarize the membrane to the threshold potential (about -50 mV), additional sodium channels open (figure 25.6, step 2). Driven by both the electrical gradient and the concentration gradient, Na^+ pours into the cell.

The interior of the axon now has a positive charge, but only for an instant. Near the peak of the action potential, membrane permeability changes again (figure 25.6, step 3). Sodium channel gates close, again preventing Na^+ from entering the cell. But permeability to K^+ suddenly increases as delayed potassium gates open. Repelled by the positively charged ions inside the cell, K^+ diffuses out.

The loss of K^+ repolarizes the cell membrane (figure 25.6, step 4). The sodium–potassium pump, which operates continuously, returns Na^+ to the membrane's exterior. For a short time, an unusually large number of K^+ channels are open, so the axon briefly hyperpolarizes to a membrane potential less than -70 mV. The resting potential is quickly restored, however, once potassium permeability returns to normal.

The entire process, from the initial influx of Na^+ to the restoration of the resting potential, takes only 1 to 5 milliseconds to complete. (This chapter's Burning Question, on page 535, describes what happens when venom from a scorpion's sting disrupts the ion channels in a neuron's membrane).

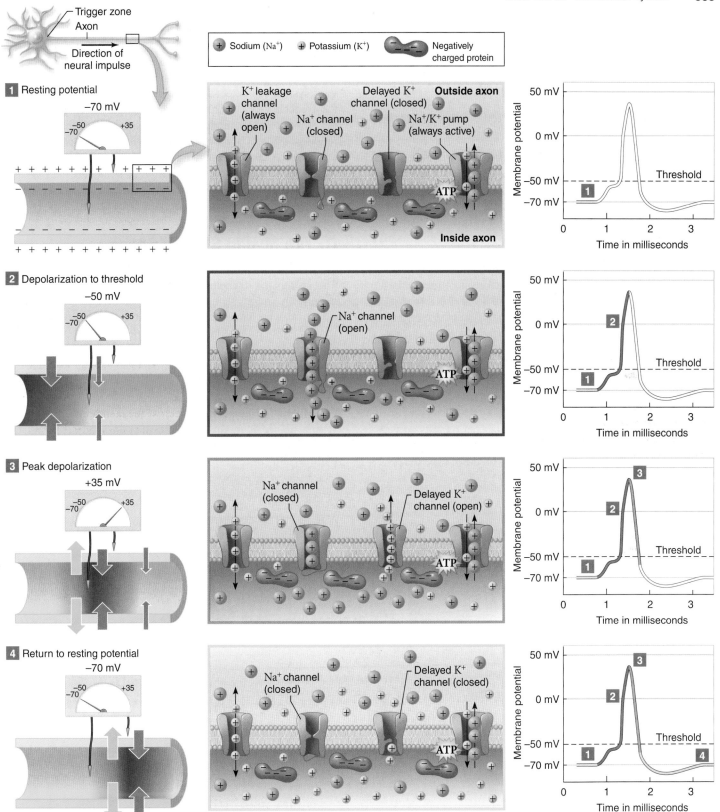

Figure 25.6 The Action Potential. (*1*) At rest, the neuron's membrane potential is about −70 mV. Na⁺ channels are closed, but the Na⁺/K⁺ pump is active and the K⁺ leakage channel is open (as always). (*2*) A stimulus causes some Na⁺ channels to open briefly. Na⁺ diffuses into the cell. This local depolarization of the membrane causes additional voltage-sensitive Na⁺ channels to open. If enough Na⁺ channels open, the cell reaches its threshold potential. (*3*) An all-or-none action potential occurs, and the membrane reverses its polarity. After a split second, Na⁺ channels close, and K⁺ leaves the axon. The membrane begins to repolarize. (*4*) The membrane briefly hyperpolarizes as the resting potential is reestablished.

Figure It Out

If negatively charged chlorine atoms (Cl⁻) move into a neuron, does an action potential become more likely or less likely? Why?

Answer: Less likely; the membrane becomes more polarized.

Figure 25.6 shows how an action potential occurs at one small patch of a neuron's membrane. To transmit a neural impulse, however, the electrical signal must move from the trigger zone to the other end of the axon. How does this occur? During an action potential, some of the Na⁺ ions that rush into the cell diffuse along the interior of the cell. As a result of this local depolarization, the neighboring patch of the axon reaches its threshold potential as well, triggering a new influx of Na⁺. The resulting chain reaction carries the impulse forward. A neural impulse is therefore similar to people "doing the wave" in a football stadium, when successive groups of spectators stand and then quickly sit. The participants do not change their locations, yet the wave appears to travel around the stadium.

A neural impulse usually begins at the trigger zone and moves down the axon toward the synaptic terminal. It does not spread "backward" because of a refractory period during which the membrane reestablishes its resting potential and cannot generate another action potential. The refractory period lasts only a couple of milliseconds, but by the time it is over for one patch of membrane, the neural impulse has moved on down the axon.

Unlike a graded potential, an action potential is an "all-or-none" event; that is, either it proceeds to completion or it doesn't occur at all. Although all action potentials are identical, the nervous system can detect the strength of a stimulus by measuring the frequency of action potentials. A light touch to the skin, for example, might produce 10 impulses per second. A hard hit might generate 100 impulses—along with a more intense sensation. Neurons also distinguish the type of stimulus. For example, we can tell light from sound because light stimulates neurons that transmit impulses to one part of the brain, whereas sound-generated impulses go to another part (see section 25.6).

C. The Myelin Sheath Speeds Impulse Conduction

The greater the diameter of an axon, the faster it conducts an impulse. A squid's "giant axons" are up to 1 mm in diameter. (Much of what biologists know about action potentials comes from studies on these large-diameter nerve fibers.) Axons from vertebrates are a hundredth to a thousandth the diameter of the squid's. Yet even thin vertebrate nerve fibers can conduct impulses very rapidly when they are coated with a myelin sheath (**figure 25.7a**).

Myelin prevents ion flow across the membrane. As a result, at first glance it might seem that myelin should prevent the spread of action potentials. But the entire axon is not coated with myelin. Instead, ions can move across the membrane at

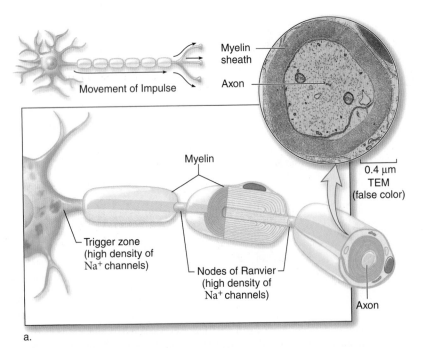

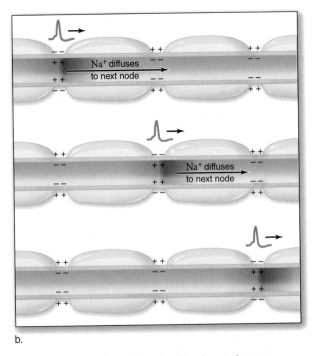

Figure 25.7 The Role of the Myelin Sheath. (a) The myelin sheath on an axon is interrupted by nodes of Ranvier. The inset shows a cross section of a myelinated axon. (b) In myelinated axons, Na⁺ channels occur only at the nodes of Ranvier. Action potentials appear to "jump" from one node to the next, which speeds impulse transmission along the axon.

Burning Question

Why does a scorpion's sting hurt?

The distinctive, venomous stinger on a scorpion's tail is among the most recognizable arthropod structures. Despite their fearsome reputation, however, only one scorpion species in the United States has venom potent enough to threaten human life. Still, it hurts to be stung by any scorpion. Why is it so painful?

Scorpion venom is a complex cocktail of toxins, enzymes, salts, and many other substances. Proteins called neurotoxins (poisons that act on the nervous system) cause most of the sting's worst effects. Some scorpion neurotoxins cause an axon's sodium channels to become stuck in the "open" position, whereas others block potassium channels. The resulting membrane depolarization triggers a continuous barrage of action potentials, accompanied by simultaneous sensations of pain, heat, cold, and touch. The venom also depolarizes the motor neurons that control the body's muscles, sometimes causing convulsions. In the most severe cases, the victim may die of cardiac or respiratory failure because the body can no longer control the muscles required for heartbeat and breathing.

Submit your burning question to:

marielle_hoefnagels@mcgraw-hill.com

---

the nodes of Ranvier. These gaps in the myelin sheath house high densities of sodium channels. When an action potential happens at one node, Na⁺ entering the axon diffuses to the next node. The incoming Na⁺ depolarizes the membrane and stimulates the sodium channels to open at the second node, triggering an action potential there. In this way, when a neural impulse travels along the axon, it appears to "jump" from node to node (figure 25.7b).

The neural impulse moves up to 100 times faster when it leaps between nodes than when it spreads along an unmyelinated axon. Not surprisingly, myelinated fibers occur in neural pathways where speed is essential, such as those that transmit motor commands to skeletal muscles. Thanks to myelin, a sensory message travels from the toe to the spinal cord in less than 1/100 of a second (about one-third the speed of sound). Unmyelinated fibers occur in pathways where speed is less important, such as in the neurons that trigger the secretion of stomach acid.

Illness can result if an axon has too much or not enough myelin. In Tay-Sachs disease, for example, cell membranes accumulate excess lipid and wrap around nerve cells, burying them in fat so that they cannot transmit messages to each other and to muscle cells. An affected child gradually loses the ability to see,

hear, and move. In multiple sclerosis, the reverse happens. The immune system destroys oligodendrocytes in the central nervous system, leading to a loss of myelin. Neural messages are therefore weakened or lost. Symptoms of multiple sclerosis include impaired vision and movement, numbness, and tremors.

25.3 | Mastering Concepts

1. Describe the forces that maintain the distribution of K⁺ and Na⁺ across the cell membrane in a neuron at rest.
2. In what way is the term *resting potential* misleading?
3. Differentiate among a graded potential, the threshold potential, and an action potential.
4. How does an axon generate and transmit a neural impulse?
5. What prevents action potentials from spreading in both directions along an axon?
6. How do action potentials indicate stimulus intensity and type?
7. How do myelin and the nodes of Ranvier speed neural impulse transmission along an axon?

25.4 | Neurotransmitters Pass the Message from Cell to Cell

To form a communication network, a neuron conducting action potentials must convey the impulse to another cell. Most neurons do not touch each other, so the electrical impulse cannot travel directly from cell to cell. Instead, an electrical impulse that reaches the tip of an axon causes the release of a **neurotransmitter,** a chemical signal that travels from a "sending" cell to a "receiving" cell across a tiny space.

A. Neurons Communicate at Synapses

A **synapse** is a specialized junction at which the axon of a neuron communicates with another cell—another neuron, a muscle cell, or a gland cell. The axon usually synapses onto the dendrites or cell body of the cell receiving the signal, but a synapse can also form directly on another axon.

Each synapse has three components. The **presynaptic cell** is the neuron sending the message, and the **postsynaptic cell** receives the message. The **synaptic cleft** is the space between the two cells.

A close look at a synapse reveals that the tip of the axon of a presynaptic cell branches into many **synaptic terminals,** tiny knobs that enlarge at the tips. These knobs contain many small sacs, or vesicles, that hold neurotransmitter molecules. On the postsynaptic cell's membrane, immediately opposite a synaptic terminal, are receptor proteins that can bind to a neurotransmitter.

Figure 25.8 shows how a synapse works. Action potentials travel along the membrane of the presynaptic neuron until they reach the membrane of a synaptic terminal. When the membrane of the synaptic terminal depolarizes, ion channels open, and calcium ions enter the cell. The calcium influx triggers loaded vesicles to dump their neurotransmitter contents into the synaptic cleft by exocytosis. ▶ exocytosis, p. 84

The neurotransmitter molecules released by the presynaptic cell diffuse across the synaptic cleft and attach to receptors on the membrane of the postsynaptic cell. When the neurotransmitter contacts the receptor, the shape of the protein changes. Ion channels then open in the postsynaptic membrane, changing the probability that an action potential will occur.

The interaction between a neurotransmitter and the receptor may be excitatory—that is, the membrane of the postsynaptic cell may become depolarized, increasing the probability of an action potential. In figure 25.8, the Na^+ ions entering the postsynaptic cell are having this effect. Conversely, the interaction may be inhibitory, making the interior of the cell more negative and reducing the likelihood of an action potential.

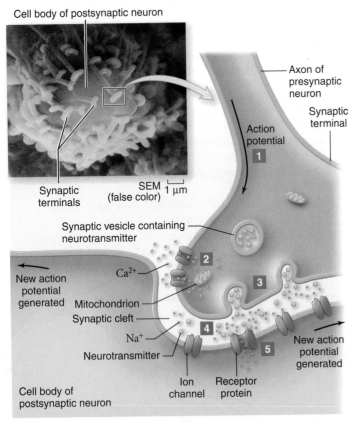

Cell body of postsynaptic neuron

Synaptic terminals

SEM 1 μm (false color)

Axon of presynaptic neuron

Synaptic terminal

Action potential 1

Synaptic vesicle containing neurotransmitter

New action potential generated

Ca²⁺

Mitochondrion

Synaptic cleft

Na⁺

Neurotransmitter

Ion channel

Receptor protein

New action potential generated

Cell body of postsynaptic neuron

a.

1 Action potential arrives at synaptic terminal.

2 Calcium ions enter synaptic terminal.

3 Vesicle loaded with neurotransmitter fuses with presynaptic cell membrane.

4 Neurotransmitters are released into synaptic cleft.

5 Neurotransmitters bind to receptor proteins in postsynaptic cell membrane, stimulating ion channels to open.

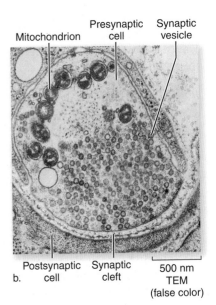

Mitochondrion

Presynaptic cell

Synaptic vesicle

Postsynaptic cell

Synaptic cleft

500 nm TEM (false color)

b.

Figure 25.8 The Synapse. (a) An action potential triggers the release of neurotransmitters from a synaptic terminal. The neurotransmitters diffuse across the synaptic cleft and bind with receptors in the postsynaptic cell membrane. Ion channels then open, changing the likelihood of an action potential in the postsynaptic cell. The inset shows synaptic terminals from many neurons converging on the cell body of a postsynaptic neuron. (b) In this photo, synaptic vesicles are poised to release neurotransmitters into a synaptic cleft.

The human brain uses at least 100 neurotransmitters. The two most common are the amino acids glutamate and GABA (gamma-aminobutyric acid). Other neurotransmitters occur at fewer synapses but still are vital. Serotonin, dopamine, epinephrine, norepinephrine, and acetylcholine are examples.

The same neurotransmitter may excite some neurons but inhibit others. Some receptors, however, are nearly always either excitatory or inhibitory. Most glutamate receptors, for example, are excitatory. Likewise, acetylcholine released from a motor neuron is excitatory, stimulating a muscle cell to contract. In contrast, most GABA receptors are inhibitory.

What happens to the neurotransmitter after it has done its job? If the chemical stayed in the synapse indefinitely, its effect on the receiving cell would be continuous. The nervous system would be bombarded with stimuli. Instead, the neurotransmitter can diffuse away from the synaptic cleft, be destroyed by an enzyme, or be taken back into the presynaptic axon soon after its release, an event called reuptake.

Notice that a synapse is asymmetrical; that is, nerve impulses travel from presynaptic cell to postsynaptic cell and not in the opposite direction. This one-way traffic of information stands in contrast to the cnidarian nerve net, in which nerve impulses travel in all directions (see section 25.1). The unidirectional flow of information was a key adaptation that permitted the evolution of dedicated circuits in which one set of neurons communicated with a limited set of postsynaptic cells. Over time, some circuits became associated with specific functions, controlling complex behaviors and forming the specialized sense organs typical of many animals (including vertebrates).

B. The Postsynaptic Cell Integrates Signals from Multiple Synapses

So far, we have considered the events that occur at one synapse. But the human brain consists of billions of neurons, each of which has synaptic connections to a thousand other neurons (see figure 25.8a). Some of the synapses are inhibitory, whereas others are excitatory. With so many synapses sending potentially conflicting signals, how does a neuron determine whether to pass a neural impulse to the next cell in a pathway?

The cell uses a process called **synaptic integration** to evaluate the incoming messages. If the majority of signals reaching the neuron's trigger zone are excitatory, the membrane of the postsynaptic cell is stimulated to generate action potentials of its own. If, on the other hand, inhibitory signals predominate, the postsynaptic cell will not generate action potentials. Synaptic integration, then, is analogous to a voting system.

Clearly, neurotransmission is very much a matter of balance. Too much or too little of a neurotransmitter can cause serious illness; table 25.1 lists a few examples of disorders associated with neurotransmitter imbalances. Moreover, some drugs can alter the functioning of the nervous system by either halting or enhancing the activity of a neurotransmitter. A drug may bind to a receptor on a postsynaptic neuron, blocking a neurotransmitter from binding there. Alternatively, a drug may activate the receptor and trigger an action potential. The Apply It Now box on page 545 illustrates how drugs such as nicotine, cocaine, and heroin tamper with neurotransmission.

25.4 | Mastering Concepts

1. Describe the structure of a synapse.
2. What event stimulates a presynaptic neuron to release neurotransmitters?
3. What happens to a neurotransmitter after its release?
4. How does synaptic integration determine whether a neuron transmits action potentials?

Table 25.1   Disorders Associated with Neurotransmitter Imbalances

Condition	Imbalance of Neurotransmitter in Brain	Symptoms
Alzheimer disease	Deficient acetylcholine (caused by death of acetylcholine-producing cells)	Memory loss, depression, disorientation, dementia, hallucinations, death
Epilepsy	Excess GABA leads to excess norepinephrine and dopamine	Seizures, loss of consciousness
Huntington disease	Deficient GABA	Uncontrollable movements, dementia, behavioral and personality changes, death
Hypersomnia	Excess serotonin	Excessive sleeping
Insomnia	Deficient serotonin	Inability to sleep
Myasthenia gravis	Deficient receptors for acetylcholine at synapse between motor neuron and muscle cell	Progressive muscle weakness
Parkinson disease	Deficient dopamine	Tremors of hands, slowed movements, muscle rigidity
Schizophrenia	Deficient GABA leads to excess dopamine	Inappropriate emotional responses, hallucinations

25.5 | The Peripheral Nervous System Consists of Nerve Cells Outside the Central Nervous System

The neurons of the brain and spinal cord interact constantly with those of the peripheral nervous system—the nerve cells outside the central nervous system (**figure 25.9**). The peripheral nervous system consists mainly of **nerves,** which are bundles of axons encased in connective tissue. The nerves, in turn, are classified based on where they originate. Twelve pairs of cranial nerves emerge directly from the brain; examples include the optic and olfactory nerves, which transmit information about sights and smells to the brain. The remaining 31 pairs of nerves in the peripheral nervous system are spinal nerves, which emerge

from the spinal cord and control many functions from the neck down.

The peripheral nervous system is functionally divided into sensory and motor divisions (**figure 25.10**). Sensory pathways carry signals to the central nervous system from sensory receptors in the skin, skeleton, muscles, and other organs. Motor pathways, on the other hand, convey information from the central nervous system to muscles and glands. In most nerves, the sensory and motor nerve fibers form a single cable.

The motor pathways of the peripheral nervous system include the somatic (voluntary) nervous system and the autonomic (involuntary) nervous system. The **somatic nervous system** carries signals to voluntary skeletal muscles, such as those that enable you to ride a bicycle, shake hands, or talk. The **autonomic nervous system** transmits impulses to smooth muscle, cardiac muscle, and glands, enabling internal organs to function without conscious awareness.

The autonomic nervous system is further subdivided into the sympathetic and parasympathetic nervous systems (**figure 25.11**). The **sympathetic nervous system** dominates under stress, including emergencies. It accelerates heart rate and breathing rate; shunts blood away from the digestive system and to the heart, brain, and skeletal muscles necessary for "fight or flight"; and dilates airways, easing gas exchange. During more relaxed times ("rest and repose"), the **parasympathetic nervous system** returns body systems to normal; heart rate and respiration slow, and digestion resumes.

Despite the "fight or flight" and "rest and repose" nicknames, the autonomic nervous system is always active, no matter what a person is doing. The parasympathetic and sympathetic divisions continuously work together to maintain homeostasis by having opposite effects on the same organs. The body's moment-to-moment adjustments in blood pressure, for example, occur whether you are napping, standing up to stretch, rushing to class, or sitting down to dinner. Whatever you are doing, the autonomic nervous system regulates your body's vital functions without conscious thought.

Some illnesses interfere with the function of the peripheral nervous system. In Guillain–Barré syndrome, for example, the immune system attacks and destroys the nerves of the peripheral nervous system. The disease can be life-threatening if it causes paralysis, breathing difficulty, and heart problems. In a person with Bell's palsy, another peripheral nervous system disorder, the cranial nerve that controls the muscles on one side of the face is damaged. The facial paralysis typically strikes suddenly and may either resolve on its own or be permanent. The cause is unknown.

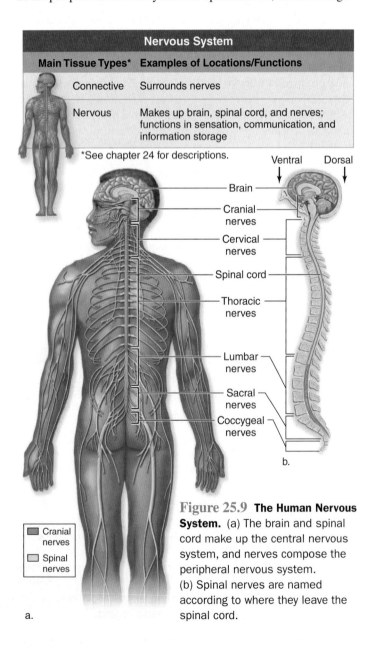

Nervous System	
Main Tissue Types*	**Examples of Locations/Functions**
Connective	Surrounds nerves
Nervous	Makes up brain, spinal cord, and nerves; functions in sensation, communication, and information storage

*See chapter 24 for descriptions.

Ventral   Dorsal

- Brain
- Cranial nerves
- Cervical nerves
- Spinal cord
- Thoracic nerves
- Lumbar nerves
- Sacral nerves
- Coccygeal nerves

b.

☐ Cranial nerves
☐ Spinal nerves

a.

Figure 25.9 The Human Nervous System. (a) The brain and spinal cord make up the central nervous system, and nerves compose the peripheral nervous system. (b) Spinal nerves are named according to where they leave the spinal cord.

25.5 | Mastering Concepts

1. What is the difference between cranial and spinal nerves?
2. How do the sensory and motor pathways of the peripheral nervous system differ?
3. Describe the relationships among the motor, somatic, autonomic, sympathetic, and parasympathetic nervous systems.
4. How do the sympathetic and parasympathetic nervous systems maintain homeostasis?

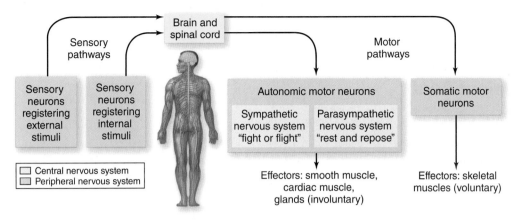

Figure 25.10 **Subdivisions of the Nervous System.** Sensory pathways of the peripheral nervous system provide input to the central nervous system. The brain and spinal cord, in turn, regulate the motor pathways of the peripheral nervous system.

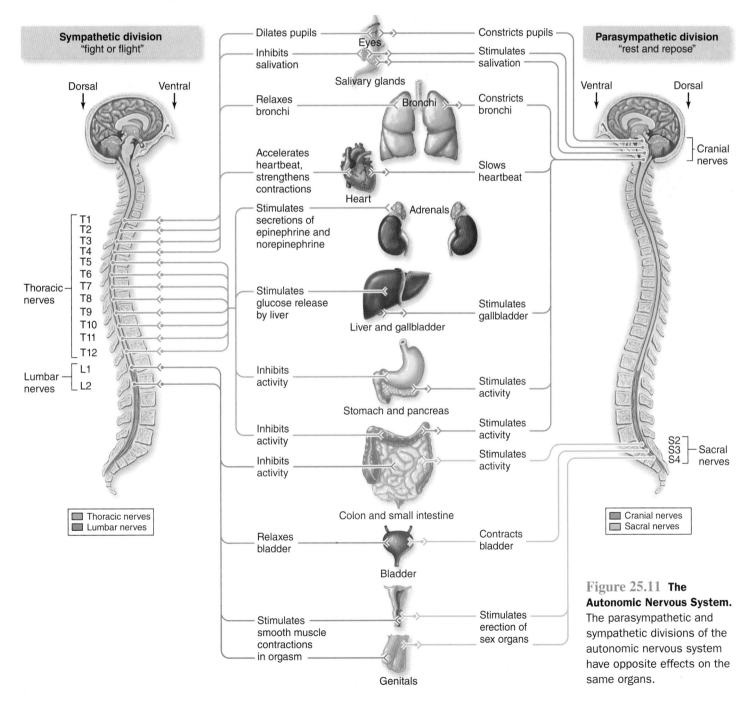

Figure 25.11 **The Autonomic Nervous System.** The parasympathetic and sympathetic divisions of the autonomic nervous system have opposite effects on the same organs.

25.6 | The Central Nervous System Consists of the Spinal Cord and Brain

The nerves of the peripheral nervous system spread across the body, but the brain and spinal cord form the largest part of the nervous system. Table 25.2 summarizes the parts and functions of the central nervous system.

Two types of nervous tissue occur in the central nervous system (figure 25.12). **Gray matter** consists of neuron cell bodies and dendrites, along with the synapses by which they communicate with other cells. The outer surface of the brain, and a few inner structures, are composed of gray matter, as is the central core of the spinal cord. Information processing occurs in the gray matter. **White matter** consists of myelinated axons transmitting information throughout the central nervous system. The periphery of the spinal cord and most inner structures of the brain consist of white matter.

Table 25.2   Major Parts of the Central Nervous System: A Summary

Structure	Selected Functions
Spinal cord	Conducts information to and from the brain; centralizes rapid involuntary responses such as reflexes
Brain	
Brainstem	Connects cerebrum to spinal cord
Medulla oblongata (part of hindbrain)	Regulates essential physiological processes such as blood pressure, heartbeat, and breathing
Pons (part of hindbrain)	Connects forebrain with medulla and cerebellum
Midbrain	Relays information about voluntary movements from forebrain to spinal cord
Cerebellum (part of hindbrain)	Controls posture and balance; coordinates subconscious muscular movements
Forebrain	
Thalamus	Processes information and relays it to the cerebrum
Hypothalamus	Homeostatic control of most organs
Cerebrum	
White matter	Transmits information within brain
Gray matter (cerebral cortex)	Sensory, motor, and association areas

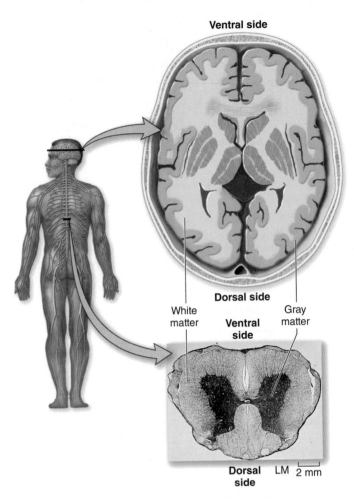

Figure 25.12 Gray Matter and White Matter. Gray matter makes up the exterior of the brain and some internal structures. It also makes up the central core of the spinal cord. Myelin-rich white matter is at the periphery of the spinal cord and forms most of the brain's interior.

A. The Spinal Cord Transmits Information Between Body and Brain

The spinal cord is a tube of neural tissue that emerges from the base of the brain and extends down the dorsal side of the body. This critical component of the central nervous system is encased in the bony armor of the vertebral column, or backbone. The backbone protects the delicate nervous tissue and provides points of attachment for muscles.

A cross section of the spinal cord shows a central H-shaped core of gray matter surrounded by white matter (see figure 25.12). On the dorsal side, the white matter consists of ascending bundles of axons that carry sensory information to the brain. On the ventral side, descending axons transmit motor information from the brain to muscles and glands.

The spinal cord handles reflexes without interacting with the brain (although impulses must be relayed to the brain for awareness to occur). A reflex is a rapid, involuntary response to a stimulus that may come from within or outside the body.

For example, a series of linked neurons causes you to pull your hand away from a painful stimulus such as a thorn (**figure 25.13**). These neurons make up a **reflex arc,** a neural pathway that links a sensory receptor and an effector such as a muscle. The arc begins with a sensory neuron whose dendrites reside in your fingertip. When you touch the sharp thorn, an action potential is generated along the neuron's axon. Inside the spinal cord, the axon synapses with an interneuron, which in turn synapses on a motor neuron. The motor neuron's axon exits the spinal cord, and its activation stimulates a skeletal muscle cell to contract. When enough muscle fibers contract, you pull your hand away from the thorn. The interneuron in the spinal cord also sends action potentials to the brain, perhaps prompting you to yell in pain.

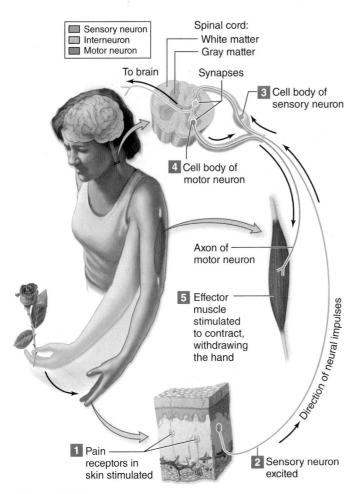

Figure 25.13 **A Reflex Arc.** A reflex arc links a sensory receptor to an effector. (*1*) In response to a painful stimulus, (*2*) the dendrite of a sensory neuron relays an action potential to (*3*) its cell body. In the spinal cord's gray matter, the sensory neuron synapses on multiple neurons, including (*4*) the cell body of a motor neuron. (*5*) The motor neuron, in turn, stimulates a muscle cell to contract. The hand withdraws. The brain perceives the stimulus but does not participate in the reflex.

B. The Human Brain Is Divided into Several Regions

The human brain weighs, on average, about 1.4 to 1.6 kilograms; it looks and feels like grayish pudding. The brain requires a large and constant energy supply to oversee organ systems and to provide the qualities of "mind"—learning, reasoning, and memory. At any time, brain activity consumes 20% of the body's oxygen and 15% of its blood glucose. Permanent brain damage occurs after just 5 minutes of oxygen deprivation.

Three subdivisions of the brain appear early in embryonic development: the hindbrain, the midbrain, and the forebrain (**figure 25.14**). The **hindbrain** is located toward the lower back of the skull. The **midbrain** is a narrow region that connects the hindbrain with the forebrain. The **forebrain** is the front of the brain. As development proceeds, the forebrain grows rapidly, obscuring the midbrain and much of the hindbrain.

The midbrain and parts of the hindbrain make up the **brainstem,** the stalklike lower portion of the brain. The brainstem regulates essential survival functions such as breathing and heartbeat. In addition, 10 of the 12 pairs of cranial nerves emerge from

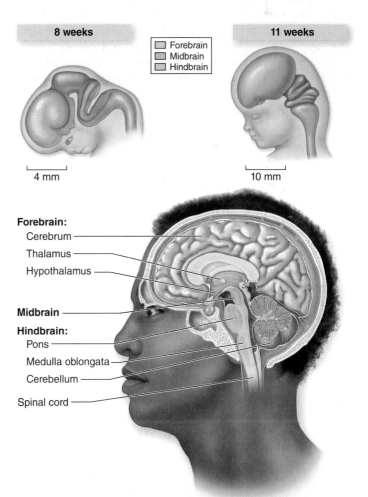

Figure 25.14 **The Human Brain.** The three major areas of the vertebrate brain (the hindbrain, the midbrain, and the forebrain) become apparent early in the development of the human embryo.

the brainstem. Among other functions, these nerves control movements of the eyes, face, neck, and mouth along with the senses of taste and hearing.

The brainstem includes two parts of the hindbrain: the medulla oblongata and the pons. The **medulla oblongata** is a continuation of the spinal cord; this region not only regulates breathing, blood pressure, and heart rate, but it also contains reflex centers for vomiting, coughing, sneezing, defecating, swallowing, and hiccupping. The **pons,** which means "bridge," is the area above the medulla. White matter in this oval mass connects the forebrain to the medulla and to another part of the hindbrain, the cerebellum.

The midbrain is also part of the brainstem. Portions of the midbrain help control consciousness and participate in hearing and eye reflexes. In addition, nerve fibers that control voluntary motor function pass from the forebrain through the brainstem; the death of certain neurons in the midbrain results in the uncontrollable movements of Parkinson disease.

Behind the brainstem is the cerebellum, the largest part of the hindbrain. The neurons of the **cerebellum** ("little brain") refine motor messages and coordinate muscle movements subconsciously. Many routine physical skills, including tying your shoes or brushing your teeth, are difficult to learn at first. But with practice, the cerebellum takes over. Once it does, you can complete those activities without thinking about how to do them each time.

By far the largest part of the human brain is the forebrain, which contains structures that participate in complex functions such as learning, memory, language, motivation, and emotion. Three major parts of the forebrain are the thalamus, hypothalamus, and cerebrum.

The **thalamus** is a mass of gray matter that acts as a relay station for sensory input, processing incoming information and sending it to the appropriate part of the cerebrum. The almond-sized **hypothalamus,** which lies below the thalamus, occupies less than 1% of the brain volume, but it plays a vital role in maintaining homeostasis by linking the nervous and endocrine systems. Cells in the hypothalamus are sensitive not only to neural input arriving via the brainstem but also to hormones circulating in the bloodstream. The autonomic nervous system, in turn, relays neural signals from the hypothalamus to involuntary muscles and glands. Moreover, hormones produced in the hypothalamus coordinate the production and release of many other hormones from the pituitary gland (see chapter 27). All together, neural and hormonal signals from the hypothalamus regulate body temperature, heartbeat, water balance, and blood pressure, along with hunger, thirst, sleep, and sexual arousal.

The other major region of the forebrain is the **cerebrum,** which controls the qualities of what we consider the "mind"— that is, personality, intelligence, learning, perception, and emotion. In humans, the cerebrum is large (occupying 83% of the brain's volume) and highly developed. It is divided into two **hemispheres** that gather and process information simultaneously. The cerebral hemispheres work together, interconnected by a thick band of nerve fibers called the corpus callosum.

Each hemisphere controls the opposite side of the body, so that damage to the left side of the brain affects the right side of the body (and vice versa). In addition, although each side of the brain participates in most brain functions, some specialization does occur. In most people, for example, parts of the left hemisphere are associated with speech, language skills, mathematical ability, and reasoning, whereas the right hemisphere specializes in spatial, intuitive, musical, and artistic abilities.

The cerebrum consists mostly of white matter—myelinated axons that transmit information within the cerebrum and between the cerebrum and other parts of the brain or spinal cord. The outer layer of the cerebrum, the **cerebral cortex,** consists of gray matter where neural integration occurs. The human cerebral cortex is only a few millimeters thick, but it boasts about 10 billion neurons forming some 60 trillion synapses (**figure 25.15**). In humans and other large mammals, deep folds enhance the surface area of the cerebral cortex.

Four main subdivisions are visible at the surface of the cerebrum (**figure 25.16**): the frontal, parietal, temporal, and occipital lobes. The functions of the cerebral cortex overlap across these lobes (**table 25.3**). Sensory areas receive and interpret messages from sense organs (relayed through the thalamus). The primary somatosensory cortex, for example, receives sensory input from the muscles, joints, bones, and skin. Motor areas in the frontal lobes send impulses to skeletal muscles, which produce voluntary movements. (Damage to the motor regions of the brain can cause cerebral palsy, a disorder of posture or muscle movement.) Association areas analyze, integrate, and interpret information from many brain areas. These are the seats of judgment, problem-solving, learning, abstract thought, language, and creativity.

The cerebrum also houses the **limbic system,** a loosely defined collection of structures surrounding the corpus callosum. The limbic system is sometimes called the emotional center of the brain. The thalamus and hypothalamus are part of the limbic system, as are two areas at the edges of the temporal lobes: the hippocampus and the amygdala. The **hippocampus** is essential

LM   20 μm

Figure 25.15 Technicolor Brain. Special labelling techniques reveal the individual neurons that make up the intricate circuits of the cerebral cortex.

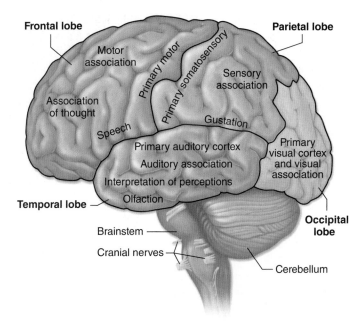

Figure 25.16 Major Subdivisions of the Cerebrum. The surface of the cerebrum is divided into four lobes: the frontal lobe at the front of the head, the temporal lobe above each ear, the occipital lobe at the rear of the head, and the parietal lobe across the top of the head. Each lobe is divided into two hemispheres. Experiments have revealed which areas are associated with sensory, motor or association functions.

for the formation of long-term memories. The **amygdala** is a center for emotions such as pleasure or fear. Information from the amygdala travels to the hypothalamus, which activates the autonomic nervous system and coordinates the physical sensations associated with strong emotions.

Table 25.3 Functional Divisions of the Forebrain

Division	Function(s)	Brain Region(s)
Sensory	Senses of sight, hearing, smell (olfaction), taste (gustation), and touch	Parietal, occipital, and temporal lobes (primary visual cortex, primary auditory cortex, olfactory bulbs, gustatory cortex, primary somatosensory cortex)
Motor	Voluntary movements	Frontal lobe (primary motor cortex)
Association	Judgment, analysis, learning, creativity	Frontal lobe and parts of the parietal, occipital, and temporal lobes
Limbic	Emotion, motivation, memory	Amygdala, hippocampus, hypothalamus

With this perspective on brain anatomy, consider the often-repeated statement that humans use only 10% of their brains. This common notion is a myth. After all, damage to even a tiny part of the brain can have devastating consequences. Moreover, the brain demands a huge amount of energy and oxygen; it does not make sense that the nervous system would waste valuable resources that the body could use in productive ways.

C. Many Brain Regions Participate in Memory Formation

Why is it that you can't remember the name of someone you met a few minutes ago, but you can easily picture your first-grade teacher or your best friend from childhood (figure 25.17)?

The simple answer to that question relates to the difference between short-term and long-term memories. Your brain apparently stored the new acquaintance's name in **short-term memory** only, where it remained available for a few moments before being forgotten. You remember the grade school teacher, however, because you interacted with that person every day for months at a time. This repeated reinforcement meant your brain stored the information in **long-term memory,** which can last up to a lifetime.

Much of what scientists know about memory formation comes from research on people with illnesses or trauma that has destroyed specific parts of the brain. One famous example is a man called Henry Molaison, known in the medical literature by the initials H.M. until his death in 2008. Surgeons removed portions of his temporal lobes and hippocampus in 1953 in an effort to alleviate his severe epilepsy. The surgery accomplished that goal but also had an unintended consequence: H.M. was unable to form new memories. Although he could recall events that occurred before the surgery, he could not remember what he ate for breakfast. Clues from H.M. and other brain-damaged patients suggest that the hippocampus is essential in the formation of long-term memories, but memories are not actually stored there.

No one knows exactly what happens to the brain's neurons and synapses when a new memory forms. For short-term memories, it is possible that neurons in the frontal and parietal lobes connect in a

Figure 25.17 Memories. A glance at old photos can bring back memories from times past.

temporary circuit in which the last cell in the series restimulates the first. As long as the pattern of stimulation continues, you remember the thought. When the reverberation ceases, however, so does the memory. On the other hand, long-term memory probably requires stable, permanent changes in neuron structure or function.

Researchers are trying to learn more about memory formation. Practical applications could include drugs that enhance memory in patients with neurological disorders that cause memory loss, including Alzheimer disease. Conversely, pharmaceuticals that selectively erase memories could help people who are struggling in the aftermath of traumatic experiences.

D. Damage to the Central Nervous System Can Be Devastating

The central nervous system is well protected. The bones of the skull and vertebral column shield nervous tissue from bumps and blows. **Meninges** are layered membranes that jacket the central nervous system. The **blood–brain barrier,** formed by specialized brain capillaries, helps protect the brain from extreme chemical fluctuations. The tilelike epithelial cells that form these capillaries fit so tightly together that only some chemicals can cross into the **cerebrospinal fluid** that bathes and cushions the brain and spinal cord. This fluid, made by cells that line ventricles in the brain, further insulates the central nervous system from injury. ▶ tight junctions, p. 65

Accidents can damage the spinal cord and prevent motor impulses from descending from the brain, causing full or partial paralysis (**figure 25.18**). Actor Christopher Reeve, who broke his neck in a horseback riding accident in 1995, was among the most famous spinal cord injury patients. He died in 2004 because of complications from being bedridden, a common problem in paralyzed people.

Brain damage can result from head trauma or a more subtle killer: stroke. In a stroke, a burst or blocked blood vessel can interrupt the flow of blood to part of the brain. Deprived of oxygen, some brain cells die, often so many that the stroke is fatal. In other cases, the patient may suffer from temporary or permanent muscle weakness, paralysis, loss of speech, blindness, or other impairment. Often, only one side of the body is affected.

Infectious diseases can also affect the central nervous system. Several viruses and bacteria can cause meningitis, or inflammation of the meninges. Bacterial meningitis is rare but can be fatal. Prions, the infectious proteins discussed in chapter 15, also damage the brain. Fungal infections of the central nervous system often afflict people with compromised immune systems. ▶ prions, p. 332

Many serious illnesses of the central nervous system remain a mystery. Parkinson disease is a slowly progressing, degenerative disease in which the death of brain cells causes muscle tremors, weakness, slow movement, and loss of coordination (**figure 25.19**). In most cases, such as that of actor Michael J. Fox, no one knows what triggers the cell death. The death of brain cells also causes Alzheimer disease. The patient, who is usually elderly, suffers from memory loss, confusion, and personality changes. Autopsy reveals tangled neurons and clusters

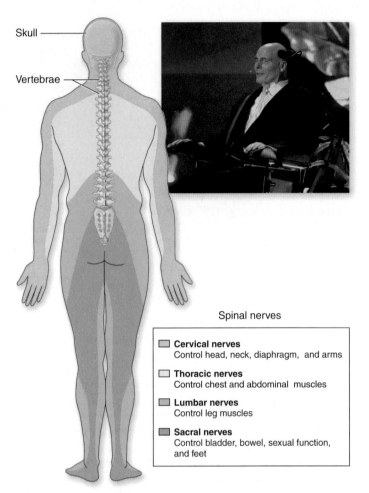

Skull

Vertebrae

Spinal nerves

☐ **Cervical nerves**
Control head, neck, diaphragm, and arms

☐ **Thoracic nerves**
Control chest and abdominal muscles

☐ **Lumbar nerves**
Control leg muscles

☐ **Sacral nerves**
Control bladder, bowel, sexual function, and feet

Figure 25.18   Spinal Cord Injuries. A person with a damaged spinal cord may have full or partial paralysis, depending on the site of the injury. Actor Christopher Reeve's paralysis resulted from a cervical spinal cord injury.

of degenerating nerve endings; neurotransmitter production is also reduced (see table 25.1). Unlike in Parkinson disease, motor function usually remains unaffected.

Another mysterious nervous system disease is amyotrophic lateral sclerosis, also called ALS or Lou Gehrig disease. The first

Figure 25.19   Testimony. Boxer Muhammad Ali (left) and actor Michael J. Fox have Parkinson disease. This photo shows them joking around before asking the U.S. Senate to fund research on the illness.

clues to this disease's onset are often subtle; a person may drop things or suffer from persistent muscle twitches. Eventually, impaired muscle function leaves the patient unable to speak, breathe, swallow, or move, yet the senses and mental function remain unaffected. This devastating disease has no cure.

Whatever its cause, part of the difficulty in reversing nervous system damage is that mature neurons do not divide, so the nervous system cannot simply heal itself by producing new cells as your skin does after a minor cut. The neurons that survive the damage can, however, form some new connections that compensate for the loss. This plasticity explains why rehabilitative therapy can restore some function to injured tissues. Current research on stem cells, growth factors (chemicals that stimulate cell division), and gene therapy may one day improve the outlook for patients with neurological damage or disease. ▸ stem cells, p. 167

25.6 | Mastering Concepts

1. What are the functions of the spinal cord?
2. What is the relationship between gray matter and white matter in the spinal cord?
3. Describe the functions of the neurons that form a reflex arc.
4. What are the major structures in the hindbrain, midbrain, and forebrain? What are their functions?
5. What are the main subdivisions of the cerebral cortex?
6. How do short- and long-term memories differ?
7. List some structures that protect the central nervous system.
8. What are some examples of diseases that affect the central nervous system?
9. To what extent can the nervous system regenerate?

Apply It Now

Drugs and Neurotransmitters

Understanding how neurotransmitters work helps explain the action of some mind-altering illicit and pharmaceutical drugs. The following are some examples, organized by the neurotransmitter affected.

Norepinephrine

Amphetamine drugs are chemically similar to norepinephrine; they bind to norepinephrine receptors and trigger the same changes in the postsynaptic membrane. The resulting enhanced norepinephrine activity heightens alertness and mood. Cocaine, which is chemically related to amphetamine, produces a short-lived feeling of euphoria, in part by blocking reuptake of norepinephrine.

Acetylcholine

Nicotine crosses the blood–brain barrier and reaches the brain within seconds of inhaling from a cigarette. An acetylcholine mimic, nicotine binds to acetylcholine receptor proteins in neuron cell membranes. The nicotine-stimulated neurons signal other brain cells to release dopamine, which provides the pleasurable feelings associated with smoking. Nicotine addiction stems from two sources: seeking the dopamine release and avoiding painful withdrawal symptoms.

On the other hand, excess acetylcholine accounts for the deadly effects of poisonous nerve gases and some insecticides. These toxic chemicals prevent acetylcholine from breaking down in the synaptic cleft. The resulting excess acetylcholine activity overstimulates skeletal muscles, causing them to contract continuously. The twitching legs of a cockroach sprayed with insecticide demonstrate the effects.

Serotonin

Norepinephrine and serotonin are associated with some forms of depression, a debilitating feeling of sadness. Drugs called selective serotonin reuptake inhibitors (SSRIs) block reuptake of serotonin. The neurotransmitter accumulates in the synapse, offsetting a deficit that presumably causes the symptoms (figure 25.A).

Endorphins

Humans produce several types of endorphins, molecules that influence mood and perception of pain. Opiate drugs such as morphine, heroin, codeine, and opium are potent painkillers that bind endorphin receptors in the brain. In doing so, they elevate mood and make the pain easier to tolerate.

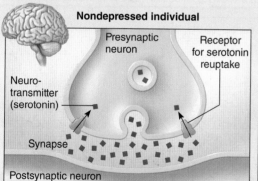
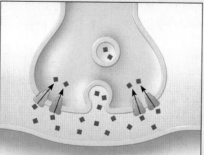
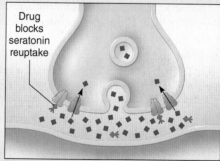
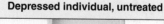

Nondepressed individual

Presynaptic neuron
Receptor for serotonin reuptake
Neurotransmitter (serotonin)
Synapse
Postsynaptic neuron

Depressed individual, untreated

Depressed individual, treated with SSRI

Drug blocks serotonin reuptake

Figure 25.A Anatomy of an Antidepressant. Selective serotonin reuptake inhibitors (SSRIs) block the reuptake of serotonin, making more of the neurotransmitter available in the synaptic cleft. The precise mechanism by which SSRIs relieve the symptoms of depression is not well understood.

25.7 | Investigating Life: The Nerve of Those Clams!

No population is exempt from natural selection—not even clams. These mollusks might seem to have uneventful lives, buried in the mud along coastal waterways. But clams are genetically variable, and not all coastlines are equal. It makes sense that natural selection should produce unique populations of clams that are adapted to their own surroundings. ▶ natural selection, p. 233

Biologists understand how natural selection results in adaptations, from antibiotic resistance in bacteria to the shape of a Galapagos finch's beak. Linking natural selection to adaptive mutations in specific genes, however, is difficult. After all, each species has thousands or tens of thousands of genes, and scientists know little about how most of them function.

Some proteins, however, are exceptionally well-studied. One example is the sodium channels that propagate action potentials in neurons. Nerve function is essential to animal survival, so sodium channels represent a potential target for natural selection. Researchers have now linked a mutation in a gene encoding a sodium channel to an adaptation that promotes the survival of clams.

The adaptation relates to a common enemy of both clams and humans: harmful algal blooms. Abundant light and nutrients stimulate algae called dinoflagellates to "bloom" in coastal waters. Some of these algae release potent neurotoxins that accumulate in filter-feeders such as clams and other shellfish. When humans consume the contaminated clams, the result can be an illness called paralytic shellfish poisoning. ▶ dinoflagellates, p. 356

A multinational research team studied the effects of algal toxins on clam populations. The leader of the team was Monica Bricelj, a Canadian shellfish researcher at the National Research Council Institute for Marine Biosciences in Nova Scotia. She worked with colleagues from the University of Maine, the University of Washington, and the National Oceanic and Atmospheric Administration Northwest Fisheries Science Center in Seattle, Washington.

The researchers collected young, toxin-free softshell clams (*Mya arenaria;* figure 25.20) from two sites along the east coast of North America. Harmful algal blooms occur each summer at one site (Bay of Fundy), but blooms have never been known to occur at the nearby Lawrencetown Estuary site.

First, the researchers wanted to learn whether clams from the two sites are equally susceptible to algal toxins. They set up several tanks containing clams and sediment. Each tank then received one of three treatments: cells of toxic *Alexandrium* algae, cells of nontoxic *Alexandrium* algae, or cells of nontoxic algae of a different species (*Isochrysis galbana*). After 24 hours of exposure, neither of the nontoxic algae affected the clams. But the toxic *Alexandrium* algae paralyzed the muscles of most of the clams from Lawrencetown Estuary, as indicated by their inability to burrow into the sediment (figure 25.21). By contrast, algae did not cause paralysis in the Bay of Fundy clams.

Figure 25.20 **Softshell Clam.** This softshell clam (*Mya arenaria*) has extended one of its siphons. Water enters the tube, and the clam consumes food particles suspended in the water.

The next step was to determine how much of an algal toxin, saxitoxin, it would take to block action potentials in nerves isolated from clams of both populations. The critical saxitoxin concentration was 33 micromoles per liter (μM) for sensitive clams from Lawrencetown Estuary. In contrast, a tenfold higher concentration, 334 μM, failed to block action potentials in resistant clams from the Bay of Fundy.

From the results of the behavioral and physiological studies, Bricelj and her colleagues concluded that annual exposure to harmful algae blooms has selected for toxin-resistant nervous systems in clams from the Bay of Fundy. The team turned to molecular biology to answer the next question: how do the sodium channels in the nerves from the two clam populations differ? DNA sequences revealed that the sodium channel proteins of sensitive and resistant clams differed by just one amino acid. In the susceptible clams from Lawrencetown (and in every other animal studied to date), glutamic acid occurred at a critical position in the protein. In the resistant Bay of Fundy clams, on the other hand, the amino acid at that location was aspartic acid.

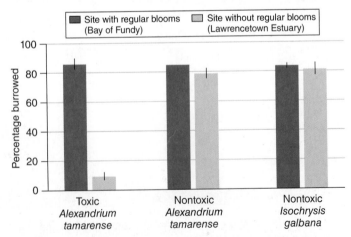

Figure 25.21 **Paralysis.** Toxins caused paralysis only in clams originating from an estuary without regular exposure to harmful algal blooms. (Error bars reflect standard errors.)

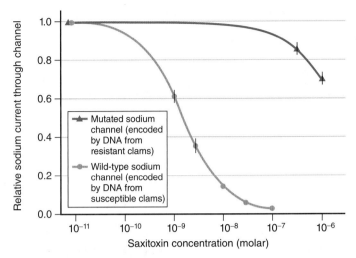

Figure 25.22 Saxitoxin Blocks Susceptible Sodium Channels. Mutant channels from resistant clams supported sodium flow at much higher toxin concentrations than did wild-type channels from susceptible clams. The higher the sodium current, the greater the ability to sustain action potentials. (Error bars reflect standard deviations.)

But did that one difference explain the resistance of the Bay of Fundy clams? To find out, the researchers introduced DNA encoding both versions of the sodium channel protein into cells growing in culture. Once the cells produced the channels on their membranes, the researchers exposed them to varying concentrations of saxitoxin. If sodium continued to flow, the channel was functioning normally. But if the toxin stopped the flow of sodium, then action potentials would also be blocked, so nerve function would be impaired. Figure 25.22 shows the result: the resistant channel continued to function, even at toxin concentrations that blocked sodium flow through the susceptible wild-type channels. A single change in one gene therefore accounted for the difference between sensitive and resistant clams.

This study tells an unusually complete story of evolution. Bricelj and her team have traced natural selection for toxin resistance all the way down to its molecular explanation: a single mutation that causes a tiny difference in the shape of sodium channels in neurons. As so often happens, a new piece of the evolutionary story has emerged from an unexpected place—in this case, from the nerve of a clam.

Bricelj, V. Monica, and six coauthors. 2005. Sodium channel mutation leading to saxitoxin resistance in clams increases risk of PSP. *Nature,* vol. 434, pages 763–767.

25.7 | Mastering Concepts

1. What is the evidence that the presence of algal toxins is a selective force on softshell clam populations?
2. How did Bricelj and her colleagues demonstrate that sodium channel structure explains toxin resistance in some clam populations?

Chapter Summary

25.1 | The Nervous System Forms a Rapid Communication Network

- The **nervous system** and **endocrine system** work together to coordinate the feedback systems that maintain homeostasis. The nervous system's electrical signals produce much more rapid effects than does the endocrine system.
- Nervous tissue consists of **neurons** and **neuroglia.**

A. Invertebrates Have Nerve Nets, Nerve Ladders, or Nerve Cords

- Invertebrate nervous systems vary widely in complexity. The simplest systems, in cnidarians, are **nerve nets.** A flatworm has collections of neurons called **ganglia** at its head end, with a **nerve ladder** running the length of the body. Annelid and arthropod nervous systems are more centralized, with **ventral nerve cords** and an anterior brain.

B. Vertebrate Nervous Systems Are Highly Centralized

- The vertebrate **central nervous system** consists of the **brain** and **spinal cord.** The **peripheral nervous system** conveys information between the central nervous system and the rest of the body.
- Overall, the nervous system receives sensory information, integrates it, and coordinates a response.

25.2 | Neurons Are Functional Units of a Nervous System

A. A Typical Neuron Consists of a Cell Body, Dendrites, and an Axon

- A neuron has a **cell body, dendrites** that receive impulses and transmit them toward the cell body, and an **axon** that conducts impulses away from the cell body.
- Fatty neuroglia called **Schwann cells** or **oligodendrocytes** wrap around portions of some axons to form the **myelin sheath.** The gaps between these insulating cells are **nodes of Ranvier.**

B. The Nervous System Includes Three Classes of Neurons

- A **sensory neuron** carries information toward the central nervous system; an **interneuron** conducts information between two neurons and coordinates responses; a **motor neuron** carries information away from the central nervous system and stimulates an effector (a muscle or gland).

25.3 | Action Potentials Convey Messages

A. A Neuron at Rest Has a Negative Charge

- In a neuron at rest, the K^+ concentration is much greater inside the cell than outside, whereas the Na^+ concentration is greater outside than inside. The **sodium–potassium pump** uses ATP to maintain this chemical gradient.
- The concentration gradient combined with negatively charged proteins within the cell give the interior a negative charge, called the **resting potential.**

B. A Neuron Transmitting an Impulse Undergoes a Wave of Depolarization

- In a **graded potential,** a stimulus causes some Na^+ to enter the cell, depolarizing the membrane in proportion to the strength of the stimulus. If enough Na^+ comes in, the membrane may further depolarize to its **threshold potential.**
- When the membrane reaches its threshold potential, an electrical change called an **action potential** begins. Na^+ and K^+ quickly redistribute across a small patch of the axon's membrane, creating a series of electrochemical changes that propagate like a wave along the nerve fiber.

C. The Myelin Sheath Speeds Impulse Conduction

- Myelination increases the speed of neural impulse transmission. In a myelinated fiber, the neural impulse rapidly "jumps" from one node of Ranvier to the next.

25.4 | Neurotransmitters Pass the Message from Cell to Cell

A. Neurons Communicate at Synapses

- A **synapse** is a junction between a neuron and another cell.
- An action potential reaching the end of an axon causes vesicles in the **synaptic terminals** of the **presynaptic cell** to release **neurotransmitters** into the **synaptic cleft.** These chemicals diffuse across the cleft and bind to receptors on the membrane of the **postsynaptic cell.**
- Used neurotransmitter molecules diffuse away, are enzymatically destroyed, or are reabsorbed into the presynaptic cell.

B. The Postsynaptic Cell Integrates Signals from Multiple Synapses

- An excitatory receptor makes an action potential more probable in the postsynaptic cell; an inhibitory neurotransmitter has the opposite effect. **Synaptic integration** sums excitatory and inhibitory messages, providing fine control of neuron activity.

25.5 | The Peripheral Nervous System Consists of Nerve Cells Outside the Central Nervous System

- **Nerves** are bundles of axons that convey information in the peripheral nervous system.
- The peripheral nervous system is divided into the sensory and motor pathways, and it includes the cranial and spinal nerves that transmit sensations from sensory receptors and stimulate muscles and glands.
- The motor pathways of the peripheral nervous system consist of the **somatic** (voluntary) division and the **autonomic** (involuntary) division. The autonomic nervous system receives sensory information and conveys impulses to smooth muscle, cardiac muscle, and glands.
- Within the autonomic nervous system, the **sympathetic nervous system** controls physical responses to stressful events, and the **parasympathetic nervous system** restores a restful state. Both systems are always active in maintaining homeostasis.

25.6 | The Central Nervous System Consists of the Spinal Cord and Brain

A. The Spinal Cord Transmits Information Between Body and Brain

- **White matter** on the periphery of the spinal cord conducts impulses to and from the brain; the central **gray matter** processes information.
- The spinal cord is a reflex center. A reflex is a quick, automatic, protective response that travels through a **reflex arc.**

B. The Human Brain Is Divided into Several Regions

- The **brainstem** consists of the midbrain and portions of the hindbrain.
- The **hindbrain** includes three main subdivisions: the **medulla oblongata,** which controls many vital functions; the **cerebellum,** which coordinates unconscious movements; and the **pons,** which bridges the medulla and higher brain regions and connects the cerebellum to the cerebrum.
- The **midbrain** conducts information between the hindbrain and the forebrain.
- The major parts of the **forebrain** are the **thalamus,** a relay station between lower and higher brain regions; the **hypothalamus,** which regulates vital physiological processes and levels of some pituitary hormones; and the **cerebrum.** The **limbic system,** which includes the **amygdala** and **hippocampus,** also resides in the gray matter of the forebrain.
- The outer layer of the cerebrum is the **cerebral cortex,** where information is processed and integrated. The cerebrum's two **hemispheres** each receive sensory input from and direct motor responses to the opposite side of the body.

C. Many Brain Regions Participate in Memory Formation

- Biologists have much to learn about memory, but it appears that the brain stores **short-term memories** and **long-term memories** in different ways.
- The formation of long-term memories requires an intact hippocampus, but the memories are stored in multiple lobes of the cerebral cortex.

D. Damage to the Central Nervous System Can Be Devastating

- Bones of the skull and vertebrae, **cerebrospinal fluid,** the **blood–brain barrier,** and **meninges** protect the central nervous system.
- Trauma, infectious agents, strokes, and degenerative diseases all can damage the nervous system.

25.7 | Investigating Life: The Nerve of Those Clams!

- Regular exposure to harmful algal blooms has selected for clams resistant to paralytic shellfish poisoning toxins.
- Biologists have traced toxin resistance to a single difference in the amino acid sequence of a sodium channel protein in the nerves of the clams.

Multiple Choice Questions

1. Some cells of the central nervous system are located in the
   a. spinal cord.
   b. muscles.
   c. glands.
   d. Both a and c are correct.

2. What is the function of an axon?
   a. Metabolic support for the neuron
   b. Insulation to speed impulse conduction
   c. Conduction of an impulse away from a cell body
   d. Input of signals to the nerve cell

3. Which class(es) of neuron would you expect to find in the peripheral nervous system?
   a. Interneurons
   b. Sensory neurons
   c. Motor neurons
   d. Both b and c are correct.

4. The function of the sodium–potassium pump is to move
   a. sodium in and potassium out of the neuron.
   b. potassium into the neuron and sodium out.
   c. both potassium and sodium into the neuron.
   d. both potassium and sodium out of the neuron.

5. What event triggers an action potential?
   a. Opening of sodium channels
   b. Opening of delayed potassium channels
   c. High concentration of negative ions outside the cell
   d. Activation of the sodium–potassium pump

6. What is the likely effect of a loss of myelin along an axon?
   a. It causes the action potential to speed up because more of the membrane is exposed.
   b. It causes the action potential to slow down.
   c. It speeds up the transport of sodium and potassium across the membrane.
   d. It increases the size of the action potential.

7. Which of the following examples of synaptic integration is most likely to lead to an action potential in a postsynaptic cell?
   a. All excitatory and no inhibitory synaptic inputs
   b. An equal mix of excitatory and inhibitory inputs
   c. A majority of inhibitory inputs with only a few excitatory synaptic inputs
   d. Both a and b are correct.

8. Which division of the nervous system would be responsible for a rapid heartbeat?
   a. Autonomic
   b. Sympathetic
   c. Parasympathetic
   d. Both a and b are correct.

9. The part of the human brain involved in coordinating muscle movements is the
   a. cerebrum.
   b. medulla oblongata.
   c. cerebellum.
   d. hypothalamus.

10. Which of the following is NOT among the structures that protects the central nervous system?
    a. Meninges
    b. Vertebrae
    c. Pons
    d. Cerebrospinal fluid

Write It Out

1. How do the nervous and endocrine systems differ in how they communicate?
2. Describe some invertebrate nervous systems. Why do animals with simple nervous systems still exist, even after the more complex vertebrate nervous system evolved?
3. Sketch two neurons, with one synapsing on the other. In your sketch, label the dendrites, axons, cell bodies, myelin sheath, and synapse.
4. How do the locations and functions of sensory neurons, motor neurons, and interneurons differ?
5. Describe the distribution of charges in the membrane of a resting neuron.
6. What causes the wave of depolarization and repolarization constituting an action potential?
7. Make a chart showing whether potassium channels, sodium channels, and the sodium–potassium pump in a neuron are active or inactive during each of the four phases in figure 25.6.
8. Why is an action potential described as an "all-or-none" process?
9. Cyanide is a poison that disables the sodium–potassium pump. Explain how cyanide prevents nerve transmission and causes death.
10. In what ways does an action potential resemble "the wave" in a football stadium? In what ways does a graded potential resemble a cheerleader's attempts to get "the wave" started?
11. How does myelin alter conduction of a neural impulse along a nerve fiber? What would happen to neural impulse transmission in a myelinated axon without nodes of Ranvier? Explain.
12. Sketch a synapse; label the axon and synaptic terminal of the presynaptic cell, the postsynaptic cell, the synaptic cleft, the synaptic vesicles, and the receptor proteins.
13. A synapse is asymmetrical. What structures do the presynaptic and postsynaptic cells have in common? What structures differ between the two cells?
14. Describe how a neuron uses neurotransmitters to communicate with other cells.
15. Suppose that a synapse is like a football game's line of scrimmage and the neurotransmitter is the football. Which structure in the synapse is analogous to the quarterback? To the receiver? What event might be analogous to an interception or a fumble?
16. Why might an overdose of an SSRI drug result in serotonin toxicity?
17. How does a neuron use information from other neurons to determine whether or not to generate a neural impulse?
18. A scientist discovers a way to stop production of a protein required for recycling of synaptic vesicles. What will happen to the amount of neurotransmitter in the synaptic cleft?
19. List the main subdivisions of the human nervous system, along with their functions.
20. What is the relationship between a neuron and a nerve?
21. In carpal tunnel syndrome, a nerve in the wrist becomes compressed, causing numbness or pain in the forearm and hand. Is this a disease of the peripheral or central nervous system? Explain your answer.
22. Why can the loss of reflexes be a possible indication of damage to the central nervous system?
23. Traumatic brain injury can occur when a person receives a strong blow to the head or when an object enters the brain through the skull. Symptoms can include anything from nausea to loss of sight or hearing to memory loss and personality changes. Why do symptoms depend strongly on the location and severity of the injury?
24. Summarize what researchers know and have yet to learn about memory formation and storage.
25. Brain imaging techniques include CT scans, PET scans, MEG, MRI, and fMRI. Search the Internet for information about each of these techniques. How does each technique work? What kinds of images does each one produce? What are the advantages and disadvantages of each?
26. What is a stroke? Use the Internet to learn the symptoms of stroke. What treatments are available for stroke victims? Why is prompt treatment so important to a patient's survival and recovery? Why is it common for only one side of the body to be affected? What are the best ways to prevent stroke?
27. How would you test the hypothesis that a nonhuman animal feels pain or thinks? Which animals would you choose to investigate this question? Why?
28. Albert Einstein's brain was of normal size, but a part of his parietal lobe was about 15% wider than normal. This area controls mathematical reasoning, imagery, and the ability to visualize objects in space. A particular groove in the area appeared much reduced, leading researchers to speculate that this might have allowed more synaptic connections to form than normal. What additional information would help to determine whether Einstein's unique brain features could have accounted for his genius?
29. Neuroglia outnumber neurons in the nervous system by about 10 to 1. In addition, neuroglia retain the ability to divide, unlike neurons. How do these two observations relate to the fact that most brain cancers begin in glial cells?
30. Dentists apply local anesthetics to deaden the pain associated with filling a cavity. These drugs block sodium channels in neurons surrounding the affected tooth. Major surgery requires general anesthetics that act on the brain, causing the patient to become unconscious and unaware of his or her surroundings. Use what you have learned about the nervous system to explain how local and general anesthetics temporarily eliminate pain.
31. Scientists know little about many common illnesses, including migraines and Alzheimer disease. What ethical considerations make research on these diseases difficult? What are the limitations of using animals as models to study the nervous system?

Pull It Together

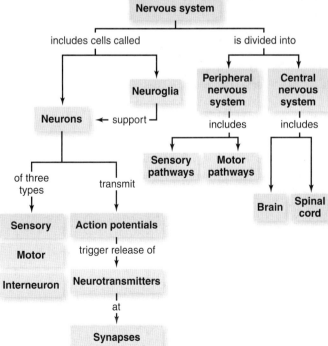

1. How do sensory neurons, motor neurons, and interneurons interact?
2. What are the main parts of a neuron?
3. In what direction do action potentials travel within a neuron?
4. Add myelin and nodes of Ranvier to this concept map.
5. What structures are included in the peripheral nervous system?
6. What are the names, locations, and functions of the main parts of the human brain?
7. Add the somatic, autonomic, sympathetic, and parasympathetic nervous systems to this concept map.

Chapter 26

The Senses

Pectines

Eclectic Senses. A platypus detects electrical fields with its bill; a snake's forked tongue samples odors; and the comblike pectines on a scorpion's belly sense textures and chemicals on the ground.

McGraw Hill **connect** |BIOLOGY

Enhance your study of this chapter with practice quizzes, animations and videos, answer keys, and downloadable study tools.
www.mhhe.com/hoefnagels

Different Views on the Same World

MOST PEOPLE KNOW THAT DOGS HAVE EXTREMELY SENSITIVE NOSES AND THAT CATS SEE WELL IN THE DARK. Here are three lesser-known adaptations that other animals use to sense the world.

Platypus Electroreception

A platypus eats half its body weight daily as it navigates along the river bottoms of Australia, nudging rocks aside with its bill to find crayfish, worms, insect larvae, frogs, and fish. But the bill doesn't sense its targets by touch alone; it can also detect the weak electrical fields emitted by the muscles of their prey. Australian scientists discovered that the sensors in the platypus bill are specialized neurons called electroreceptors. The researchers placed hungry platypuses in pools with live and dead batteries and observed that the animals explored only the live batteries, suggesting an attraction to electricity. Stimulating the platypuses' bills electrically provoked a response in the animals' brain waves. A close look at the bills revealed threadlike ends of nerve fibers deep within the pits that form mucous glands. These were the electroreceptors. The tiniest movement of a shrimp's tail can stimulate the electroreceptors to trigger 20 to 50 nerve impulses per second!

The Snake's Forked Tongue

A snake's tongue is in constant motion. Each flick of the forked tongue represents a sample of the odor molecules left by predators, prey, and potential mates. Inside the mouth, the tips of the forked tongue pass through openings in the palate and contact sensory cells in the snout. The receptors transmit action potentials to the brain, which compares the strength of the signals arriving from each tip. The snake integrates this information, along with the nature of the chemical stimulus, to identify the source. Just as our brains determine the direction of a sound by comparing the stimuli reaching our two ears, the forked tongue allows the snake to sense odors at two points simultaneously. Snakes therefore smell "in stereo."

Scorpion Pectines

A scorpion searches for food and mates by tasting the ground surface with a pair of comb-shaped structures called pectines, located just behind the last pair of legs on the animal's "chest." Each pecten's flexible spine carries a row of "teeth" adorned with thousands of tiny pegs. A close look at each peg reveals an elongated pore. As the animal walks, it taps its pectines on the ground. Chemicals left by other animals enter the pegs' pores and bind to sensory receptors inside, signaling the nervous system that a meal or mate is nearby. Pectines may also detect ground surface texture, which helps males decide where to deposit sperm packets during mating rituals.

The sensory worlds of the platypus, snake, and scorpion may seem strange, but that is only because human sense organs respond to different signals. This chapter focuses on our own senses.

UNIT 6

What's the Point?

Learning Outline

Learn How to Learn
Don't Throw That Exam Away!

Whether or not you were satisfied with your last exam, take the time to learn from your mistakes. Mark the questions that you missed and the ones that you got right but were unsure about. Then figure out what went wrong for each question. For example, did you neglect to study the information, thinking it wouldn't be on the test? Did you memorize a term's definition without understanding how it fits with other material? Did you misread the question? After you have finished your analysis, look for patterns and think about what you could have done differently. Then revise your study plan so that you can avoid making the same mistakes in the future.

26.1 | Diverse Senses Operate by the Same Principles

The human senses paint a complex portrait of our surroundings. Consider, for example, the woman in **figure 26.1.** The tips of her fingers feel the banjo strings, her eyes see the instrument, and her ears hear the music. Her skin senses the warmth of the sun. She also maintains her balance, thanks to both her ability to feel the position of her limbs and her inner ear's sense of equilibrium. When she pauses for a snack, she will be able to smell and taste her food.

As rich as our own senses are, other animals can detect stimuli that are imperceptible to us. The chapter opening essay described the unusual sensory abilities of the platypus, snake, and scorpion. Mammals have their own keen senses as well. Dogs, for example, have an extremely well-developed sense of smell, which explains why these animals are so useful in sniffing out drugs and other contraband. Bats have an entirely different ability, called echolocation. As a bat flies, it emits high-frequency pulses of sound. The animal's large ears pick up the sound waves that bounce off of prey and other objects, and its brain analyzes these echoes to "picture" the surroundings.

An animal's senses are an integral part of its nervous system. As described in chapter 25, the vertebrate nervous system has two main subdivisions: the central and peripheral nervous systems. The central nervous system is composed of the brain and spinal cord. The peripheral nervous system consists of the cranial and spinal nerves that convey information between the central nervous system and the rest of the body.

This chapter turns to the sense organs, which detect stimuli from an animal's own body and its environment. The sensory pathways of the peripheral nervous system send that information to the brain. A **sensation** is the raw input that arrives at the central nervous system. For example, your eyes and hands may inform your brain that a particular object is small, round, red, and smooth. The brain integrates all of this sensory input and consults memories to form a **perception,** or interpretation of the sensations—in this case, of a tomato.

The senses help animals maintain homeostasis, a state of internal constancy that relies on feedback loops (see figure 24.10). Many of the feedback loops operate without our awareness; for example, we can't directly "feel" our blood pH or thyroid hormone concentration. But we are aware of sights, sounds, smells, and many other stimuli. The central nervous system responds to many types of sensory input by coordinating the actions of muscles and glands, which make adjustments as necessary to maintain homeostasis.

A. Sensory Receptors Respond to Stimuli by Generating Action Potentials

All sense organs ultimately derive their information from **sensory receptor** cells that detect stimuli. The human body includes several types of sensory receptors. **Mechanoreceptors** are sensory receptors that respond to physical deflection. **Thermoreceptors** respond to temperature. **Pain receptors** detect tissue damage, extreme heat and cold, and chemicals released from damaged cells. **Proprioceptors** detect the positions of the limbs, head, and other body parts. **Photoreceptors** respond to light, and **chemoreceptors** detect chemicals.

Each of these cell types "translates" sensory information into the language of the nervous system. **Transduction** is the process by which a sensory receptor converts energy from a stimulus into action potentials. Generally, a stimulus alters the shape of a protein embedded in a sensory receptor's cell membrane, causing the membrane's permeability to ions to change. The resulting movement of ions across the membrane triggers a **receptor potential,** which is a graded potential that occurs in a sensory receptor (**figure 26.2**). The green lines in figure 26.2 depict receptor potentials that are below the cell's threshold and therefore do not trigger

Figure 26.1 Sensory Experience. This woman is experiencing many senses at once.

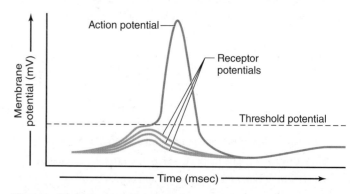

Figure 26.2 Receptor Potentials. Green lines in this figure show receptor potentials that do not exceed the threshold potential and therefore do not trigger action potentials. A larger receptor potential, however, can exceed the threshold potential, stimulating an action potential (red line). The central nervous system detects only stimuli that provoke action potentials.

action potentials; the stimulus remains undetected. If the receptor potential does exceed the threshold potential, however, an action potential occurs in the sensory receptor (red line in the figure). The frequency of action potentials arriving at the brain from specific groups of receptors conveys information about the type and intensity of the stimulus. ▸ action potential, p. 532

B. Continuous Stimulation May Cause Sensory Adaptation

You may have noticed that your perceptions of some stimuli can change over time. Your first thought when you roll out of bed may be "I smell coffee." But by the time you stand up, pull your clothes on, and wander to the kitchen, you hardly notice the coffee odor anymore. Likewise, the steaming water in a bathtub may seem too hot at first, but it soon becomes tolerable, even pleasant.

These examples illustrate **sensory adaptation,** a phenomenon in which sensations become less noticeable with prolonged exposure to the stimulus. The explanation is that sensory receptors generate fewer action potentials under constant stimulation. Generally, the response returns only if the intensity of the stimulus changes.

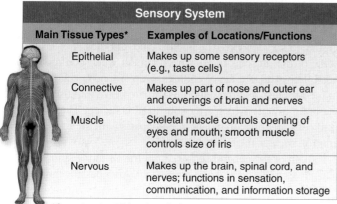

Sensory System	
Main Tissue Types*	**Examples of Locations/Functions**
Epithelial	Makes up some sensory receptors (e.g., taste cells)
Connective	Makes up part of nose and outer ear and coverings of brain and nerves
Muscle	Skeletal muscle controls opening of eyes and mouth; smooth muscle controls size of iris
Nervous	Makes up the brain, spinal cord, and nerves; functions in sensation, communication, and information storage

*See chapter 24 for descriptions.

Special senses
Hearing and equilibrium
Vision
Olfaction
Taste

General senses: touch, temperature, and pain

Figure 26.3 Overview of the Human Senses. The general senses include touch, pain, and other senses with receptors located throughout the body. Receptors for the special senses, such as vision and hearing, are limited to the head.

Without sensory adaptation, we would be distinctly aware of the touch of clothing, along with every sight, sound, and odor. Concentrating on a single stimulus, such as a person speaking, would be difficult. Many receptors therefore adapt quickly. Pain receptors, however, are very slow to adapt. The constant awareness of pain is uncomfortable, but it also alerts us to tissue damage and prompts us to address the source of the pain.

The remainder of the chapter explores the senses in more detail. It first describes the general senses, such as touch and pain, whose receptors are located throughout the body. The chapter then turns to the "special" senses, those that are limited to the head: smell, taste, vision, hearing, and equilibrium. **Figure 26.3** and **table 26.1** summarize the senses.

26.1 | Mastering Concepts

1. What role do the senses play in maintaining homeostasis?
2. Distinguish between sensation and perception.
3. What are the major types of sensory receptors?
4. What is a receptor potential?
5. What is sensory adaptation, and how is it beneficial?

Table 26.1 Sense Organs and Receptors: A Summary

Sense	Sense Organ(s)	Stimulus	Type of Receptor
General senses			
Touch	Skin	Pressure, vibration	Mechanoreceptor
Temperature	Skin	Heat, cold	Thermoreceptor
Pain	Everywhere except the brain	Damage to body tissues	Pain receptor
Position of body parts	Joints, muscles, ligaments	Stretching of muscles and ligaments	Proprioceptor
Special senses			
Smell	Nasal cavity	Airborne molecules	Chemoreceptor
Taste	Mouth and tongue	Dissolved molecules	Chemoreceptor
Vision	Eyes	Light	Photoreceptor
Hearing	Ears	Air pressure waves	Mechanoreceptor
Equilibrium	Ears	Motion of fluid in inner ear	Mechanoreceptor

26.2 | The General Senses Detect Touch, Temperature, Pain, and Position

The general senses allow you to feel sensations such as touch, temperature, or pain with any part of your skin. The receptors are sensory neurons, most of which have their dendrites wrapped (encapsulated) in neuroglia or connective tissue. Some of the neurons, however, have free (unencapsulated) nerve endings.

The sense of touch comes from several types of mechanoreceptors in the skin (figure 26.4). Pacinian and Meissner corpuscles are touch receptors that each consist of a single encapsulated dendrite. A touch pushes the flexible sides of the corpuscle inward, generating an action potential in the nerve fiber. Unencapsulated touch receptors include the dendrites that wrap around each hair follicle and sense when the hair bends. The density of touch receptors varies across the body. As a result, the fingertips and tongue are much more sensitive to touch than, say, the skin of the lower back.

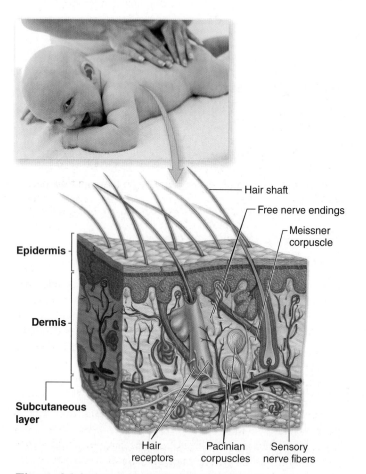

Labels: Hair shaft, Free nerve endings, Meissner corpuscle, Epidermis, Dermis, Subcutaneous layer, Hair receptors, Pacinian corpuscles, Sensory nerve fibers

Figure 26.4 The Skin Senses Many Stimuli. Free nerve endings in skin respond to touch, temperature, and pain. Encapsulated dendrites, including Meissner and Pacinian corpuscles, respond to touch and vibration.

Free nerve endings that act as thermoreceptors in the skin respond to temperature. The brain integrates input from many cold and heat receptors to determine whether a stimulus is cool, hot, or somewhere in between.

Like thermoreceptors, pain receptors are also free nerve endings, but they detect tissue damage. These neurons respond to the mechanical damage that follows a sharp blow, a cut, or a scrape. Pain receptors also detect extreme heat, extreme cold, and chemicals released from damaged cells.

Nearly every part of the body has pain receptors, except the brain. The brain's inability to feel pain in its own tissues has an important practical application. A patient undergoing brain surgery can remain awake and responsive throughout the procedure, which helps the surgeon avoid damage to critical areas of the brain.

Pain is an unpleasant but important response; people who are unable to perceive pain can unknowingly injure themselves. Nevertheless, temporarily suspending the body's pain response with drugs called anesthetics can make some medical treatments tolerable. These drugs work in multiple ways. Local anesthetics such as a dentist's procaine (Novocain) stop pain-sensitive neurons from transmitting action potentials in a limited area of the body. General anesthetics cause a loss of consciousness that prevents the brain from perceiving any pain.

The body also has proprioceptors, which detect the positions of the joints, tendons, ligaments, and muscles. For example, some of the muscles in your neck stretch when you move your head. Encapsulated nerve endings wrap around specialized cells in the muscles, and the dendrites initiate nerve impulses that tell the brain exactly which way your head is facing. These specialized cells are most abundant in body parts with the finest muscle control, such as the hands.

In all of the general senses, input travels along spinal or cranial nerves to the central nervous system. After passing through the thalamus, the information arrives at the primary somatosensory cortex. Here, input from multiple receptors is mapped to each body part. As a result, we can tell where on the body a sensation is originating and identify its characteristics. For example, we can feel that our right hand is touching the hot, smooth hood of a car.

The brain's role in interpreting sensations can lead to strange errors in perception. Consider the "phantom limb" phenomenon, in which a person who has lost an arm, leg, or other body part can feel pain and other sensations arising from the limb. Researchers once attributed the pain to damaged nerve endings in the limb's stump, but they now believe that the problem is in the somatosensory cortex. Areas that once processed input from the missing limb become partially reassigned to other body parts. The brain evidently misinterprets some signals that arrive at the "rewired" part of the cortex as coming from the missing limb.

26.2 | Mastering Concepts

1. Which structures provide the senses of touch, temperature, pain, and position?
2. What is the role of the somatosensory cortex?

26.3 | The Senses of Smell and Taste Detect Chemicals

Chemoreception is probably the most ancient sense. Bacteria and protists use chemical cues to approach food or move away from danger, so the ability to detect external chemicals must have arisen long before animals evolved.

The senses of smell (also called **olfaction**) and taste (**gustation**) both depend on the body's ability to detect chemicals (figure 26.5). Not surprisingly, these two senses have properties in common. In each case, the stimulus molecule must dissolve in a watery solution, such as saliva or the moist lining of a nasal passage. In addition, the molecule must interact with a chemoreceptor on a sensory cell's membrane.

A. Chemoreceptors in the Nose Detect Odor Molecules

The sense of smell begins at the **nose,** which forms the entrance to the nasal cavity inside the head. Specialized olfactory receptor neurons are located in a patch of epithelium high in the nasal cavity (figure 26.6). The human olfactory epithelium houses about 20 million receptor cells. A bloodhound has 10 times as many, and its olfactory epithelium has dozens of times more surface area than ours. The dog's sense of smell is therefore much more acute than our own.

Each olfactory neuron expresses one receptor protein on its cell membrane; each receptor protein, in turn, can bind to a limited set of odorants. An odor molecule that enters the nose in inhaled air binds to a receptor protein. The cell transduces this chemical signal into receptor potentials.

Part of each olfactory receptor cell passes through the skull and synapses with neurons in the brain's olfactory bulb. Unlike with the other senses, information about odors does not first pass through the

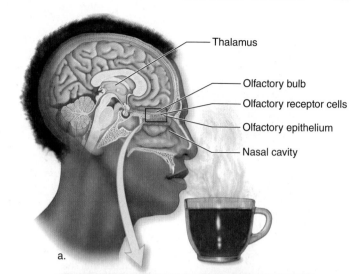
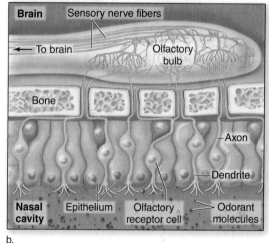

Figure 26.6 The Sense of Smell. (a) Olfaction derives from receptor cells in the nasal cavity. (b) An olfactory receptor cell binds an odorant molecule (such as from coffee) and transmits neural impulses to cells in the olfactory bulb. The axons of these neurons pass the information to the brain, which identifies the scent as coffee.

thalamus. Instead, sensory neurons relay the message directly from the olfactory bulb to the brain's olfactory cortex, which interprets the information from multiple receptors and identifies the odor.

People often associate distinctive odors with vivid memories. A whiff of the perfume Grandma used to wear, for example, may elicit a flood of childhood recollections. The explanation is that the olfactory cortex is embedded in the limbic system, the brain center of memory and emotion. ▶ limbic system, p. 542

Many insects use chemicals in communication. **Pheromones** are volatile chemical substances that elicit specific responses in other members of the same species. For example, female silk moths *(Bombyx mori)* release pheromones that attract males up to several kilometers away. The male's antennae are extremely sensitive to the signal; a single pheromone molecule can trigger an action potential in a sensory receptor. The role of chemical communication in our own species remains an open question, as described in this chapter's Burning Question.

Figure 26.5 Chemical Senses. (a) Chemoreceptors in the nose detect the odor of incense, whereas (b) chemoreceptors in the mouth detect the flavor of spaghetti.

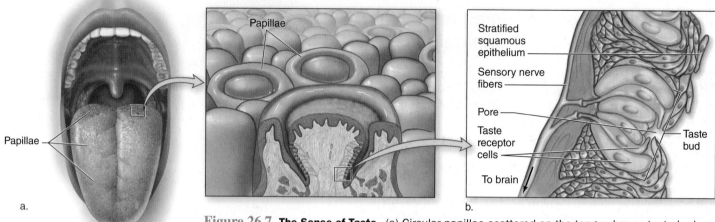

Figure 26.7 **The Sense of Taste.** (a) Circular papillae scattered on the tongue house taste buds, which contain the taste receptor cells. (b) The receptor cells that make up a taste bud synapse on sensory neurons, which convey the information to the brain.

B. Chemoreceptors in the Mouth Detect Taste

The nose detects low concentrations of volatile odor molecules that enter the nose with inhaled air. We can therefore perceive scents originating from near or distant objects. Chemoreceptors in the mouth, on the other hand, can taste items only at very close range.

The human mouth contains about 10,000 **taste buds,** the organs associated with the sense of taste (figure 26.7). The taste buds are most concentrated on the **tongue,** a muscular organ arising from the floor of the mouth. Raised bumps called papillae house numerous taste buds, each containing 50 to 150 chemoreceptor cells that generate action potentials when dissolved food molecules bind to them. The receptors are epithelial cells that synapse at the base of the taste bud onto sensory neurons, which relay information to cranial nerves leading to the medulla and thalamus. From there, input arrives at the gustatory cortex of the brain.

The human tasting experience includes five primary sensations of sweet, sour, salty, bitter, and umami. The umami taste derives from receptors for the amino acid glutamate, which imparts a savory flavor to meat; the food additive monosodium glutamate (MSG) activates these receptors. (Researchers have also identified receptors for fatty acids in taste buds, leading some to argue for recognition of a sixth taste corresponding to fat.) Most taste buds sense all of the primary tastes, but each responds most strongly to one or two. Our sense of taste depends on the pattern and intensity of activity across all taste neurons.

Food's texture, temperature, and aroma also contribute to its flavor. The intimate relationship between olfaction and taste is especially important. Even flavorful foods taste bland when a stuffy or plugged nose blocks a person's sense of smell.

26.3 | Mastering Concepts

1. How does the brain distinguish one odor from another?
2. What are pheromones?
3. How does a taste bud function?
4. What are the five taste sensations that impart a food's flavor?

 ## Burning Question

Do humans have pheromones?

Advertisements for "human pheromone" colognes appeal to the desire to attract the opposite sex. Dab some on, they say, and watch your love life blossom. But are there really human pheromones?

Pheromones are relatively easy to study in insects, whose behavioral repertoire is limited. But mammals, especially humans, have much more complex behaviors, so it is difficult to find chemicals that elicit predictable responses. Nevertheless, at least some mammals do produce pheromones. A male hamster smeared with vaginal secretions from a female will stimulate sexual advances from another male—but only if the responding male has an intact vomeronasal organ, a tiny offshoot of the olfactory system. This structure is apparently the pheromone detector.

In 1998, the journal *Nature* published a study demonstrating that pheromones from human females influence the menstrual cycles of other women. However, researchers still know little about how humans detect pheromones. We do have a vomeronasal organ, but no one has ever shown that it is functional. Researchers are working to discover the genes that encode the pheromone-binding receptors in a rodent's vomeronasal organ. Comparison with the human genome should provide clues about how human pheromones work and whether the vomeronasal organ plays a role in human life or is just a vestige of our evolutionary history.

Submit your burning question to:
marielle_hoefnagels@mcgraw-hill.com

Figure 26.8 **The Sense of Sight.** This boy is using his eyes to explore the world.

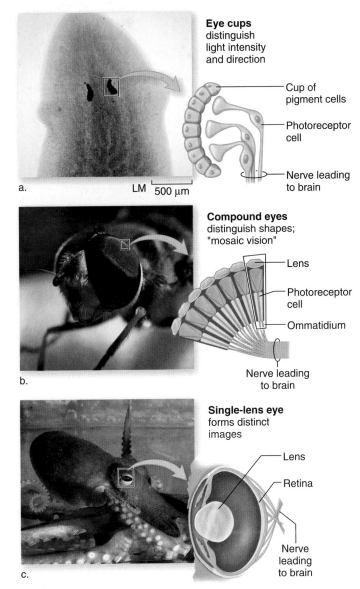

Eye cups
distinguish light intensity and direction

— Cup of pigment cells

— Photoreceptor cell

— Nerve leading to brain

a. LM 500 μm

Compound eyes
distinguish shapes; "mosaic vision"

— Lens

— Photoreceptor cell

— Ommatidium

Nerve leading to brain

b.

Single-lens eye
forms distinct images

— Lens

— Retina

Nerve leading to brain

c.

Figure 26.9 **Invertebrate Eyes.** Invertebrates have several types of eyes, three of which are shown here. (a) Planarian worms have eye cups. (b) Bees and other insects have compound eyes composed of ommatidia. (c) The octopus, a cephalopod, has a single-lens eye.

26.4 | Vision Depends on Light-Sensitive Cells

An **eye** is an organ that produces the sense of sight (figure 26.8). Animal eyes contain dense concentrations of photoreceptors, the sensory cells that respond to light. A photoreceptor contains a pigment molecule associated with a membrane. **Rhodopsin** is a common light-sensitive pigment. When rhodopsin absorbs light, its shape changes, altering the charge across the membrane and possibly generating an action potential.

A. Invertebrate Eyes Take Many Forms

Invertebrates have several types of eyes (figure 26.9). A planarian flatworm's photoreceptor cells, for example, are gathered into two cup-shaped eyes. These simple structures enable the flatworm to detect shadows, which is sufficient for the animal to orient itself in its environment. ▶ flatworms, p. 418

Most adult insects have paired compound eyes that consist of up to 28,000 closely packed photosensitive units called ommatidia. Each ommatidium contains a lens that transmits light to its own or nearby photoreceptor cells, generating a tiny view of the world. The animal's nervous system then integrates the input from many ommatidia to form a clear image. ▶ insects, p. 429

Cephalopods (octopuses and their close relatives) have a single-lens eye that is much like our own. An opening in the eye, the pupil, admits light, which a lens focuses onto photoreceptors at the back of the eye. Action potentials travel along nerves to the brain, where the visual information is interpreted. ▶ cephalopods, p. 421

B. In the Vertebrate Eye, Light Is Focused on the Retina

Figure 26.10 depicts the vertebrate eye, which is composed of several layers. The **sclera** is the white, outermost layer that

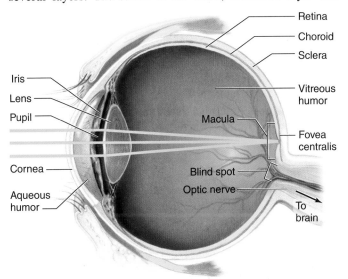

— Retina
— Choroid
— Sclera

Iris —
Lens —
Pupil —

— Vitreous humor

Macula —
— Fovea centralis

Cornea —

Aqueous humor —

Blind spot —
Optic nerve —

To brain

Figure 26.10 **The Vertebrate Eye.** Light passes through the cornea, aqueous humor, pupil, lens, and vitreous humor before striking the retina. Sensory cells in the retina transmit light information to the optic nerve.

protects the inner structures of the eye. Toward the front of the eye, the sclera is modified into the **cornea,** a transparent curved window that bends incoming light rays.

The **choroid** is the layer internal to the sclera. Behind the cornea, the choroid becomes the **iris,** which is the colored part of the eye. The iris regulates the size of the **pupil,** the hole in the middle of the iris. In bright light, the pupil is tiny, shielding the eye from excess stimulation. The pupil grows larger as light becomes dimmer.

A portion of the choroid also thickens into a structure that holds the flexible **lens,** which further bends the incoming light. In a process called visual accommodation, muscles regulate the curvature of the lens to focus on objects at any distance. When a person gazes at a faraway object, the lens is flattened and relaxed. To examine an article closely, however, muscles must pull the lens into a more curved shape.

Blood vessels in the choroid supply nutrients and oxygen to the **retina,** a sheet of photoreceptors that forms the innermost layer of the eye. Reflection of bright light from the choroid's blood vessels produces the "red eye" effect in photographs. A related phenomenon, called eye shine, occurs in cats and other nocturnal vertebrates. If you aim a flashlight at a cat's eyes in darkness, the light bounces off of a reflective layer behind the retina, called the tapetum. The iridescent tapetum gives photoreceptors another chance to transduce light and helps cats see in one-sixth the amount of light required for humans.

The **optic nerve** is a cranial nerve that connects each retina to the brain. The point where the optic nerve exits the retina is called the blind spot because it lacks photoreceptors and therefore cannot sense light.

Each eyeball also contains fluid that helps bend light rays and focus them on the retina. The watery aqueous humor lies between the cornea and the lens. This fluid cleanses and nourishes the cornea and the lens and maintains the shape of the eyeball. Behind the lens is the vitreous humor, which fills most of the eyeball's volume. This jellylike substance presses the retina against the choroid.

Light rays pass through the cornea, lens, and humors of the eye and are focused on the retina. This chapter's Apply It Now box explains how glasses and surgery can improve poor eyesight by redirecting the light that enters the eye. Some vision problems, however, cannot be treated with lenses or surgery. One example is an aging-related disorder called macular degeneration. The macula, a small area near the center of the retina, produces the sharp, central vision required for reading. An indentation called the fovea centralis is especially rich in photoreceptors. In macular degeneration, progressive loss of photoreceptors in the macula gradually causes a loss of central vision. In severe cases, the center of a person's field of view appears as nothing more than a dark spot. Peripheral vision, however, is unaffected.

C. Signals Travel from the Retina to the Optic Nerve and Brain

Oddly, light has to pass through several layers of cells before reaching the photoreceptors at the back of the retina (**figure 26.11**). The photoreceptors are neurons called rods and cones. **Rod cells,**

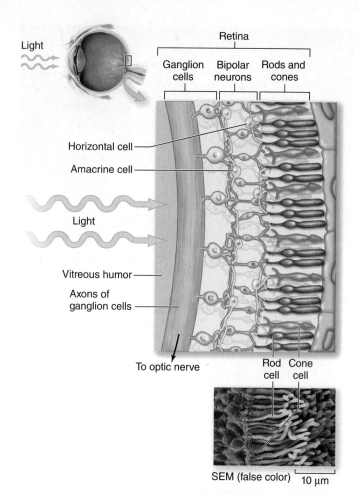

Figure 26.11   The Visual Pathway. Light passes through multiple layers of cells before striking rods and cones, which transduce light energy into action potentials. The photoreceptors transmit the information to bipolar cells, which stimulate the ganglion cells that form the optic nerve. Horizontal and amacrine cells (not shown) process the signals along the way.

which are concentrated around the edges of the retina, provide black-and-white vision in dim light and enable us to see at night. **Cone cells** detect color; they are concentrated toward the center of the retina. The human eye contains about 125 million rod cells and 7 million cone cells. (Section 26.6 explores the evolutionary origin of these cells.)

Both rods and cones consist of many interconnected discs of membranes that are studded with pigment molecules. In a rod cell, the pigment is rhodopsin; cone cells contain rhodopsin-like pigments that absorb light of different wavelengths. Humans have three cone types: "blue" cones absorb shorter wavelengths of light, "green" cones absorb medium wavelengths, and "red" cones absorb long wavelengths. People who lack a cone type entirely, due to a genetic mutation, are color-blind. The most common form of color blindness is a deficiency of pigments sensitive to red and green wavelengths. Because the genes encoding these pigments are on the X chromosome, red–green color blindness is more common in males than females. ▸ X-linked disorders, p. 215

When a rod or cone cell absorbs light energy, the pigment molecule changes shape and triggers receptor potentials that stimulate the retina's bipolar neurons. The bipolar neurons, in turn, transmit the message to the **ganglion cells,** interneurons that make up the innermost layer of the retina. (Two other types of neurons in

Apply It Now

Correcting Vision

The eyeball must have a certain shape for the cornea and lens to focus light rays precisely on the retina. For those of us whose eyeballs are not perfectly formed, corrective lenses (eyeglasses and contact lenses) can alter the path of light (figure 26.A).

A more recent technology for correcting vision problems is laser eye surgery, which vaporizes tiny parts of the cornea, changing the path of light to the retina.

Sometimes, the cornea becomes clouded or misshapen. Surgeons can replace the defective cornea with one taken from a cadaver. Corneal transplant surgery carries a low risk of immune system rejection because, unlike other transplantable organs, the cornea lacks blood vessels. Another common eye disorder is a cataract, in which the lens of the eye becomes opaque. Cataract surgery is a simple procedure that replaces the clouded lens with a plastic implant.

Even people with perfectly shaped eyeballs and corneas usually need reading glasses after the age of about 40. To focus on a very close object, a muscle inside the eye must curve the lens so that it can bend incoming light rays at sharper angles. As we age, the lens becomes less flexible. It therefore becomes difficult for the muscles in the eye to bend the lens enough to clearly focus on nearby objects or printed words. Laser surgery cannot correct this age-related decline in eyesight.

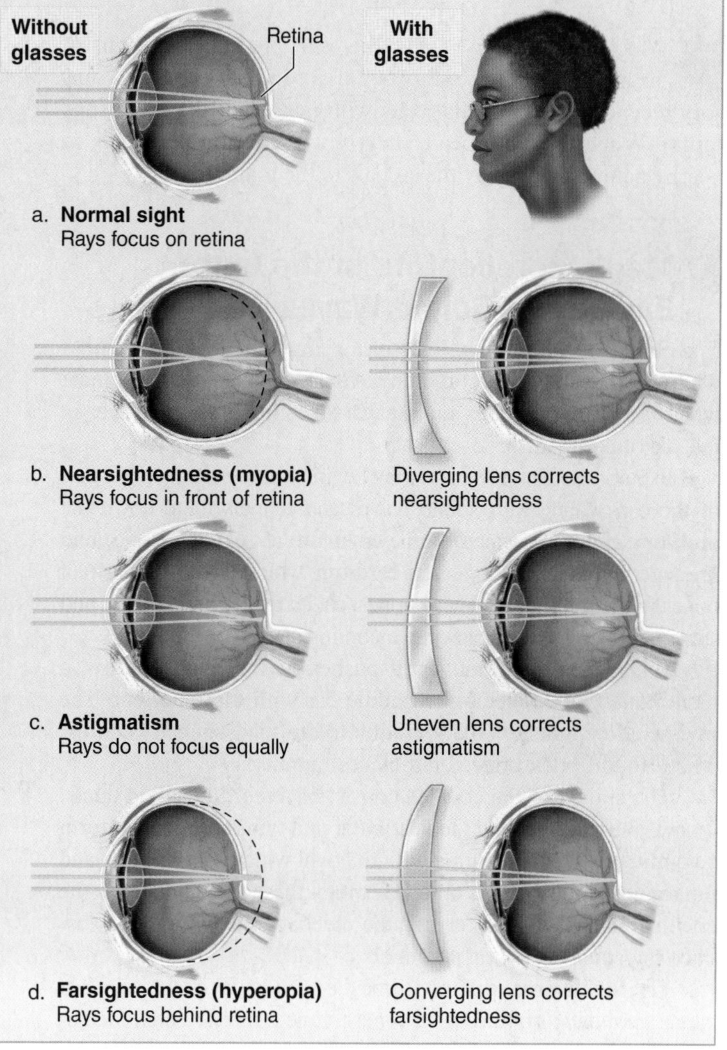

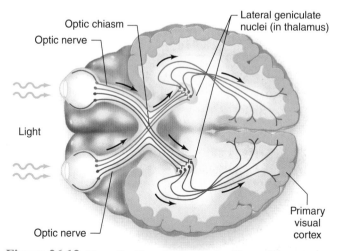

a. Normal sight
Rays focus on retina

b. Nearsightedness (myopia)
Rays focus in front of retina

Diverging lens corrects nearsightedness

c. Astigmatism
Rays do not focus equally

Uneven lens corrects astigmatism

d. Farsightedness (hyperopia)
Rays focus behind retina

Converging lens corrects farsightedness

Figure 26.A **Correcting Vision.** Eyeglasses can correct many common vision problems. (a) A normally shaped eyeball focuses light rays on the retina. (b) In an elongated eyeball, light rays converge in front of the retina, impairing the ability to see distant objects (nearsightedness). (c) In astigmatism, the lens or cornea is misshapen, and incoming light rays do not focus evenly. (d) If the eyeball is short, light focuses beyond the retina, making it difficult to see close objects (farsightedness).

the retina, called horizontal and amacrine cells, form connections that modify the information sent along the visual pathway.) Ganglion cells are the only neurons in the retina that generate action potentials; all others produce only graded potentials.

The axons of the ganglion cells make up the two optic nerves (figure 26.12). Some of these axons criss-cross behind the eyes at a structure called the optic chiasm. The stimulus then passes to the thalamus. From two regions in the thalamus called the lateral geniculate nuclei, the signals go to the primary visual cortex at the rear of the brain for processing and interpretation.

26.4 | Mastering Concepts

1. Describe three types of invertebrate eyes.
2. What are the parts of the vertebrate eye?
3. What are the roles of rod cells, cone cells, and light-sensitive pigments in human vision?
4. Trace the pathway of information from the retina to the visual cortex of the brain.

Figure 26.12 **From the Eyes to the Brain.** The optic nerve collects information from the retina and passes it to the lateral geniculate nuclei of the thalamus. The signal then passes to the primary visual cortex, which processes and integrates the information.

26.5 | The Senses of Hearing and Equilibrium Begin in the Ears

Mechanoreceptors inside the **ear** provide two senses: hearing and balance (**figure 26.13**). In both cases, the sensory receptors are epithelial cells with many hairlike extensions (cilia). When the hairs bend, they provoke action potentials in nearby neurons that relay the signals to the brain.

A. Mechanoreceptors in the Inner Ear Detect Sound Waves

The clatter of a train, the notes of a symphony, a child's wail—what do they have in common? All are sounds that originate when something vibrates and creates repeating pressure waves in the surrounding air.

In humans, the sense of hearing begins with the fleshy outer part of the ear, which traps sound waves and funnels them down the **auditory canal** (ear canal) to the **eardrum** (**figure 26.14**). Sound pressure waves in air vibrate the eardrum, which moves three small bones in the middle ear. These bones, called the hammer, anvil, and stirrup, transmit and amplify the incoming sound.

When the stirrup moves, it pushes on the **oval window,** a membrane that connects the middle ear with the inner ear. The oval window transfers the vibration to the snail-shaped **cochlea,** where sound is transduced into neural impulses.

The spirals of the cochlea consist of three fluid-filled ducts. Two of the ducts, called the vestibular and tympanic canals, form a continuous U-shaped tube with the oval window at one end and the round window at the other. Between these two canals lies the cochlear canal, which contains the mechanoreceptors that transduce the sound to action potentials.

The **basilar membrane** forms the lower wall of the cochlear canal. Embedded in the basilar membrane are **hair cells,** the cochlea's mechanoreceptors. These cells initiate the transduction of mechanical energy to receptor potentials.

Figure 26.13 It's All in the Ears. Sensory receptors in the inner ear confer the ability to (a) listen to music and (b) keep one's balance.

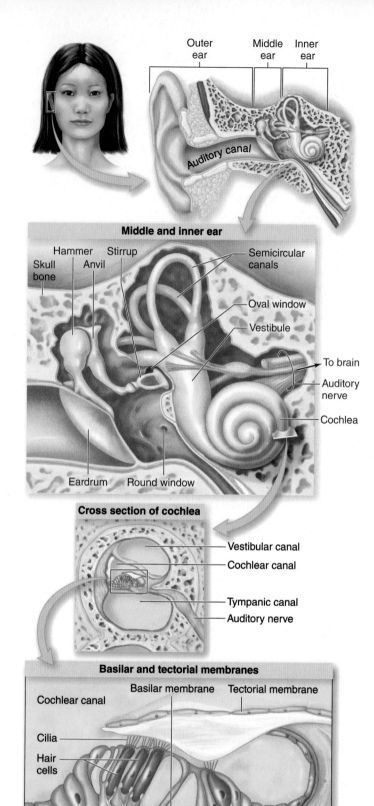

Figure 26.14 The Sense of Hearing. Sound enters the outer ear and vibrates the three bones of the middle ear. The stirrup moves the oval window, setting up vibrations in the fluid of the pea-sized cochlea. Movement of the basilar membrane causes the hair cells to move relative to the tectorial membrane. As the hair cells bend, they transduce the sound waves into action potentials, which travel along the auditory nerve to the brain.

When the oval window vibrates, the fluid inside the cochlea moves, causing a region of the basilar membrane to vibrate as well. When it does so, the hair cells move relative to the **tectorial membrane,** which rests on the hair cells' cilia. As the cilia bend, the hair cells depolarize and initiate action potentials in fibers of the **auditory nerve,** a cranial nerve that carries the impulses to the thalamus. From there, the information passes to the brain's auditory cortex for interpretation.

Each sound stimulates a different region of the cochlea. The high-pitched tinkle of a bell vibrates the narrow, rigid region of the basilar membrane at the base of the cochlea; the low-pitched tones of a tugboat whistle stimulate the wide, flexible end near the cochlea's tip. The brain interprets the input from different regions of the cochlea as sounds of different pitches.

Sound intensity is important as well. Loud sounds cause the basilar membrane to vibrate more than softer sounds. This increased movement stimulates additional hair cells, each of which fires rapid bursts of action potentials. The brain interprets the resulting increase in the rate and number of neurons firing as an increase in loudness.

The sense of hearing requires the interaction of many parts of the ear and nervous system. Deafness can occur if any of those components fails to function correctly. The Apply It Now box on page 563 explores the causes and treatments of hearing loss.

B. The Inner Ear Also Provides the Sense of Equilibrium

The sense of equilibrium includes balance and coordination, both when the body is stationary and when it is in motion. A part of the inner ear called the **vestibular apparatus** contains the receptors for equilibrium. The vestibular apparatus consists of two pouches (the utricle and saccule) and three semicircular canals (figure 26.15).

Hair cells in the utricle and saccule detect whether the velocity of the head is changing. Both chambers contain a dense, jellylike fluid and granules of calcium carbonate. When the body moves forward, the movement of the heavy granules lags slightly behind that of the head. As the fluid moves into place, it bends the cilia on the hair cells lining each chamber. The membrane potential in the hair cells therefore changes. Sensory neurons synapsing on the hair cells send electrical signals along the auditory nerve to the brain, which interprets information from the vestibule of both ears as a change in velocity.

Motion sickness in a car results from contradictory signals from the vestibule and the eyes. The inner ear signals the brain that the person is accelerating. At the same time, the eyes focus on stationary objects in the car and indicate that the person is not moving. The result of these mixed signals is nausea.

The **semicircular canals** are three interconnected, fluid-filled loops that detect whether the head is rotating or tilting. Their perpendicular orientation ensures that movement in any plane will stimulate one or more of the canals. The enlarged bases of the canals, which open into the utricle, are lined with clusters of hair cells covered by a caplike structure called a cupula. As the head tilts, fluid in a canal shifts. The resulting pressure on the cupula

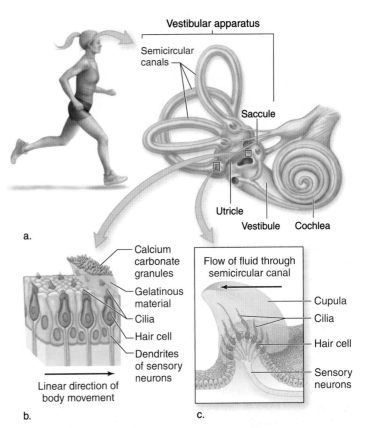

Figure 26.15 The Sense of Equilibrium. (a) The vestibular apparatus in the inner ear provides the sense of equilibrium. (b) When the head is accelerating, calcium carbonate granules move, bending the cilia on the hair cells of the utricle and saccule. This motion provokes action potentials. (c) Rotating the head in any direction causes fluid to move in one or more semicircular canals, deflecting the cilia on hair cells at the base of each loop. Action potentials in either part of the vestibular apparatus are conveyed to the auditory nerve.

bends the cilia on the hair cells. The stimulated hair cells trigger action potentials in the auditory nerve. The brain interprets the impulses from all three canals in each ear to form a perception of head movements in three dimensions.

Like motion sickness, dizziness also traces its origin to contradictory signals. When you spin rapidly, you set the fluid in your semicircular canals into motion. But that motion does not stop as soon as the spinning ceases. As long as the fluid continues to move, the brain senses motion, even if the eyes can see that the spinning has stopped. This sensation of dizziness continues until the fluid eventually stops moving as well.

26.5 | Mastering Concepts

1. What is the role of mechanoreceptors in the senses of hearing and equilibrium?
2. What are the parts of the ear, and how do they transmit sound?
3. How does the vestibular apparatus provide the sense of equilibrium?

26.6 | Investigating Life: Unraveling the Mystery of the Origin of the Eye

In Greek mythology, the Cyclops was a grotesque, one-eyed monster. Of course, the Cyclops never really existed. Modern archaeologists and paleontologists now suggest that the myth originated when ancient Greeks stumbled on the remains of a mastodon fossil. The vertebrae, giant limb bones, and profile of the skull could have resembled a human with a monstrous face and a single eye.

Today, biologists are looking at the origin of the eye itself. In his book *On the Origin of Species*, Charles Darwin calls the notion that the eye formed through natural selection "absurd in the highest possible degree." He follows this statement with great insight, however, when he states, "Yet reason tells me that if numerous gradations from a perfect and complex eye to one very imperfect and simple, each grade being useful to its possessor, can be shown to exist," then this paradox can be overcome. Could a structure as complex as the eye really be the product of evolution?

Animal eyes vary greatly, from the simple eye cups of flatworms to the complex, fluid-filled eyes of cephalopods and vertebrates (see section 26.4A). Nevertheless, all organisms with eyes have a gene (*pax6*) that regulates eye formation. All eyes also have light-sensitive pigments (derived from vitamin A) attached to proteins called opsins in the membranes of the photoreceptor cells. When light strikes the eye and causes the pigment to change shape, the opsin protein triggers the transduction of light to action potentials.

Two major types of cells, called rhabdomeric and ciliary photoreceptors, store opsin pigments. Rhabdomeric photoreceptors express one form of opsin, called r-opsin. These photoreceptors predominantly occur in the eyes of invertebrates. Vertebrate rods and cones, as well as cells in the brain's light-sensitive pineal gland, are ciliary photoreceptors that express a different form of opsin (c-opsin). ▶ pineal gland, p. 577

This distinction between invertebrate and vertebrate photoreceptors raises a question: If our invertebrate ancestors had rhabdomeric photoreceptors, where did the ciliary photoreceptors in the vertebrate eye come from?

Part of the answer has come from studies of *Platynereis dumerilii*, a small, segmented marine worm commonly called a ragworm (figure 26.16). *Platynereis dumerilii* is considered a "living fossil" because it differs little from its 600-million-year-old ancestors. This animal has been the subject of intense molecular and evolutionary research.

Ragworm eyes contain rhabdomeric photoreceptors, as expected for an invertebrate. Embedded in the worm's brain, however, are cells that resemble ciliary photoreceptor cells. In the worm, these cells do not contribute to vision. Nevertheless, a team of researchers led by Detlev Arendt and Joachim Wittbrodt at the European Molecular Biology Laboratory noticed that the

LM   0.5 mm

Figure 26.16   **The marine ragworm, *Platynereis dumerilii*.** Two pairs of eyes are visible as dark spots on the animal's head.

worm's ciliary photoreceptor cells strongly resemble rods and cones in the human eye.

The scientists hypothesized that human rods and cones are homologous to the worm's ciliary photoreceptor cells. To test their hypothesis, the team used a tool called "molecular fingerprinting." In this procedure, researchers analyze the proteins inside living cells and attempt to identify a "fingerprint" (combination of proteins) that is unique to a cell type. Cells with matching fingerprints probably have a common evolutionary origin.

Upon screening the ragworm's cells for opsin proteins, the researchers discovered a new type of opsin in the ciliary cells of the worm's brain. The rhabdomeric photoreceptor cells of the worm's eyes did not contain this new opsin. Moreover, the amino acid sequence of the newly discovered protein grouped with the c-opsins of vertebrates, not with the invertebrate r-opsins (figure 26.17). The team therefore concluded that the ragworm's ciliary cells do indeed share a common evolutionary origin with vertebrate rods and cones. In vertebrates, the ciliary cells apparently took on a new role—vision—that they never had in their invertebrate ancestors.

Arendt and his colleagues propose that early animals had one type of photoreceptor cell containing one type of opsin. Long ago, this ancestral protein diverged into r-opsin and c-opsin, paving the way for photoreceptor cells to specialize into rhabdomeric and ciliary subtypes. In ancient invertebrates, the rhabdomeric cells developed into simple eyes, and the ciliary cells remained in the animals' brains.

Because vertebrates descended from invertebrates, a logical prediction is that vertebrates should also have both ciliary and rhabdomeric photoreceptors. Rods and cones account for the ciliary photoreceptors. But what about the predicted rhabdomeric cells? As it turns out, a subsequent molecular analysis revealed that the ganglion cells of the vertebrate retina contain a type of r-opsin (melanopsin). This homology provides strong evidence that the ganglion cells are the "missing" rhabdomeric cells in vertebrates.

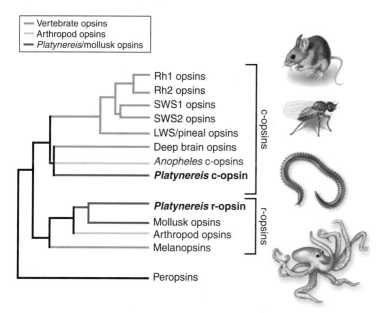

- Vertebrate opsins
- Arthropod opsins
- *Platynereis*/mollusk opsins

Rh1 opsins
Rh2 opsins
SWS1 opsins
SWS2 opsins
LWS/pineal opsins
Deep brain opsins
Anopheles c-opsins
***Platynereis* c-opsin**

c-opsins

***Platynereis* r-opsin**
Mollusk opsins
Arthropod opsins
Melanopsins

r-opsins

Peropsins

Figure 26.17 **Two Groups of Opsins.** Comparison of amino acid sequences shows that the opsin from *Platynereis* rhabdomeric photoreceptors is similar to r-opsins from other invertebrates. The opsin from *Platynereis* ciliary cells, however, groups with vertebrate c-opsins.

For all its marvelous complexity, the basic components of the vertebrate eye were already present hundreds of millions of years ago. Our ever-expanding understanding of the evolution of the eye reinforces and extends the revolutionary ideas of Charles Darwin. Like the myth of the Cyclops, many of our old questions about the evolution of the eye have been put to rest.

Arendt, Detlev, Kristin Tessmar-Raible, Heidi Snyman, Adriaan W. Dorresteijn, and Joachim Wittbrodt. 2004. Ciliary photoreceptors with a vertebrate-type opsin in an invertebrate brain. *Science,* vol. 306, pages 869–871.

26.6 | Mastering Concepts

1. What is the significance of the discovery that ragworms have both rhabdomeric and ciliary photoreceptors?
2. How might you test the researchers' hypothesis that an ancestor of *P. dumerilii* had one photoreceptor and both types of opsins?

Apply It Now

Deafness

The sense of hearing is so complex that it is not surprising to know that deafness can take multiple forms. In conductive deafness, the middle ear fails to transmit sounds to the inner ear. In sensorineural deafness, the inner ear or auditory nerve does not function, or the brain does not respond to input from the nerve.

What causes hearing loss? Some babies are born deaf because of a genetic mutation, chromosomal abnormality, or prenatal exposure to disease. Other people lose their hearing later because of disease, exposure to loud noise, or injury. Earwax or an ear infection can cause short-term deafness. And nearly everyone suffers some hearing loss later in life as the ear becomes less sensitive to higher frequencies.

Hearing aids can treat some forms of hearing loss. By amplifying sounds, the hearing aid moves the eardrum more than normal, helping the person hear more clearly. If the middle ear cannot transmit sound, however, a conventional hearing aid is useless. Bone-conduction aids solve this problem by transmitting sound waves directly to the bones of the skull. The vibrations stimulate the cochlea directly, bypassing the middle ear.

A cochlear implant may restore some hearing to a person who is profoundly deaf (**figure 26.B**). A surgeon places the device under the skin behind the ear. A microphone in the implant picks up sound, then a sound processor decomposes it into separate frequency components. Electrodes placed directly in the cochlea stimulate the parts of the auditory nerve corresponding to each frequency. By sending signals directly to the nervous system, cochlear implants compensate for nonfunctioning parts of the middle and inner ear.

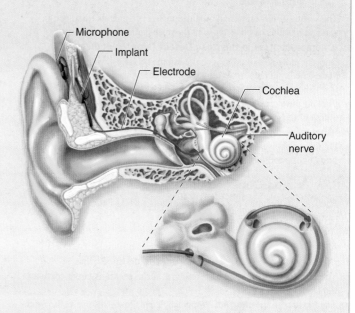

Figure 26.B **Cochlear Implant.** A cochlear implant stimulates the auditory nerve, helping overcome deafness that originates in the middle or inner ear.

Chapter Summary

26.1 | Diverse Senses Operate by the Same Principles

- Sense organs send information about internal and external stimuli to the central nervous system. A **sensation** is the raw input received by the central nervous system; a **perception** is the brain's interpretation of the sensation.
- The senses help the animal body maintain homeostasis.

A. Sensory Receptors Respond to Stimuli by Generating Action Potentials

- **Sensory receptors** are sensory neurons or specialized epithelial cells that detect stimuli. Types of sensory receptors include **mechanoreceptors, thermoreceptors, pain receptors, proprioceptors, photoreceptors,** and **chemoreceptors.**
- A sensory receptor selectively responds to a single form of energy and **transduces** it to **receptor potentials,** which change membrane potential in proportion to stimulus strength. If a receptor potential exceeds the cell's threshold, the cell generates action potentials.

B. Continuous Stimulation May Cause Sensory Adaptation

- In **sensory adaptation,** sensory receptors cease to respond to a constant stimulus.

26.2 | The General Senses Detect Touch, Temperature, Pain, and Position

- The skin's mechanoreceptors, such as Pacinian corpuscles, Meissner corpuscles, and free nerve endings, respond to mechanical deflection. Free nerve endings also include thermoreceptors and pain receptors.
- Sensory neurons in muscles help detect the positions of body parts.

26.3 | The Senses of Smell and Taste Detect Chemicals

- The senses of **olfaction** and **gustation** detect chemicals dissolved in watery solutions, such as those in the **nose** and mouth.

A. Chemoreceptors in the Nose Detect Odor Molecules

- Odorant molecules bind to receptors in the olfactory epithelium of the **nose.** The brain perceives a smell by evaluating the pattern of signals from olfactory receptor cells that bind odorant molecules.
- **Pheromones** are chemicals that many animals use to communicate with others of the same species.

B. Chemoreceptors in the Mouth Detect Taste

- Humans perceive taste when chemicals stimulate receptors within **taste buds** on the **tongue.**

26.4 | Vision Depends on Light-Sensitive Cells

- Photoreceptors in the **eye** contain pigments such as **rhodopsin** associated with membranes.

A. Invertebrate Eyes Take Many Forms

- Invertebrates have eye cups, compound eyes, and single-lens eyes.

B. In the Vertebrate Eye, Light Is Focused on the Retina

- The human eye's outer layer, the **sclera,** forms the transparent **cornea** in the front of the eyeball.
- The next layer, the **choroid,** supplies nutrients to the retina. At the front of the eye, the choroid holds the muscle that controls the shape of the **lens,** which focuses light on the photoreceptors. The **iris** adjusts the amount of light entering the eye by constricting or dilating the **pupil.**
- The innermost eye layer is the **retina,** and the **optic nerve** connects the retina with the brain.

C. Signals Travel from the Retina to the Optic Nerve and Brain

- The retina's photoreceptors are **rod cells,** which provide black-and-white vision in dim light, and **cone cells,** which provide color vision in brighter light.

- Light stimulation alters the pigments embedded in the membranes of rod and cone cells. The resulting change in the charge across the membrane may generate an action potential.
- Photoreceptor cells synapse with bipolar cells that, in turn, synapse with **ganglion cells.** Axons of ganglion cells leave the retina as the optic nerve, which carries information to the thalamus and visual cortex.

26.5 | The Senses of Hearing and Equilibrium Begin in the Ears

- Mechanoreceptors in the **ear** bend in response to sound waves or the motion of fluids.

A. Mechanoreceptors in the Inner Ear Detect Sound Waves

- Sound enters the **auditory canal,** vibrating the **eardrum.** Three bones in the middle ear amplify these vibrations. The movements of these bones are transmitted through the **oval window,** changing the pressure in fluid within the **cochlea.** At the base of the cochlea, vibration of the **basilar membrane** pushes **hair cells** against the **tectorial membrane.** The **auditory nerve** transmits the impulses to the brain.
- The brain perceives the pitch of the sound through the location of the moving hair cells in the cochlea. Louder sounds generate more action potentials than softer ones.

B. The Inner Ear Also Provides the Sense of Equilibrium

- In the inner ear, the **vestibular apparatus** includes the utricle, saccule, and **semicircular canals.**
- When the head accelerates, tilts, or rotates, fluid movement within the vestibular apparatus stimulates sensory hair cells. The brain interprets this information, providing a sense of equilibrium.

26.6 | Investigating Life: Unraveling the Mystery of the Origin of the Eye

- Molecular fingerprinting studies of a marine segmented worm, *Platynereis dumerilii,* show that the rods and cones of the vertebrate eye originated from light-sensitive cells in the invertebrate brain.

Multiple Choice Questions

1. What is the relationship between homeostasis and the sensory system?
   a. Sensory information may be used to help maintain internal conditions of an organism.
   b. Homeostasis controls which sensory inputs are perceived by an organism.
   c. Information from the sensory system may be used to alter the behavior of an organism.
   d. Both a and c are correct.

2. What is a receptor potential?
   a. The potential for a sensory cell to receive a specific type of information
   b. The graded potential generated when a sensory cell receives a stimulus
   c. The change in membrane permeability in a sensory cell
   d. The threshold potential of a receptor

3. When you snuggle into bed, at first you feel the weight of the blankets on your body. Soon, however, you become unaware of the covers. What has happened?
   a. Your skin's touch receptors became unable to receive information about new stimuli.
   b. Your skin's touch receptors adapted to the feeling of the blankets.
   c. All of your body's sensory receptors became unable to receive information about new stimuli.
   d. All of your body's sensory receptors adapted to the feeling of the blankets.

4. Which type of sensory receptor enables you to feel the position of your legs, even if a table hides your legs from sight?
   a. thermoreceptor
   c. chemoreceptor
   b. photoreceptor
   d. proprioceptor

5. Which of the following is the most important limit to a human's perception of smell?
   a. The size of the olfactory epithelium
   b. The number and type of receptor proteins
   c. The number of olfactory receptor cells
   d. All of the above are important.

6. Which of the following is NOT a primary sensation associated with taste?
   a. Spicy
   c. Bitter
   b. Sweet
   d. Sour

7. The structures that enable bees to see flowers are
   a. eye cups.
   c. ommatidia.
   b. single-lens eyes.
   d. maculas.

8. What is the correct order of structures through which light enters the vertebrate eye?
   a. Cornea, lens, pupil, retina
   c. Cornea, pupil, lens, retina
   b. Retina, lens, pupil, cornea
   d. Retina, pupil, lens, cornea

9. What is the function of hair cells in the cochlea?
   a. Transduce sound waves into neural impulses
   b. Interpret and identify sounds
   c. Funnel sounds to the inner ear
   d. Prevent debris from entering the delicate inner ear

10. A region of the ear associated with the sense of equilibrium is the
    a. semicircular canals.
    c. basilar membrane.
    b. cochlear canal.
    d. auditory canal.

Write It Out

1. How does the peripheral nervous system interact with the central nervous system to produce perceptions of stimuli?

2. Distinguish between a sense organ and a sensory receptor.

3. What is the role of transduction in the sensory system? How does transduction occur for each of the senses described in this chapter?

4. Describe an example of sensory adaptation other than the ones listed in this chapter.

5. Why is the least painful place to receive an injection the buttock or upper arm, rather than, say, the inside of the hand?

6. People with Hansen disease (formerly called leprosy) suffer nerve damage that leaves them unable to sense pain in their extremities. How might this situation be dangerous?

7. How does the nervous system differentiate among odors?

8. Suppose that some male moths lack a functional pheromone receptor protein in their antennae. Using what you know about natural selection, would you expect this trait to become more common in the moth population over multiple generations? Explain.

9. Design an experiment to test the hypothesis that pheromones play a role in human sexual attraction. Is your experiment both practical and ethical? Explain.

10. Explain why some people hold their nose when consuming bad-tasting food or medicine.

11. In what ways are the senses of smell and taste similar? In what ways are they different?

12. List the structures of the human eye and their functions.

13. In a disorder called macular degeneration, photoreceptors at the center of the retina die. How does macular degeneration impair vision? Why is peripheral vision unaffected?

14. When you enter a darkened room, the rod cells in your eyes boost their production of rhodopsin, increasing their sensitivity to light over a period of 10 to 30 minutes. How do you think your eyes adjust when you emerge from a movie theater on a sunny day?

15. In what ways do the cochlea and vestibular apparatus function similarly?

16. As we age, our senses of sight, hearing, smell, and taste decline. Brainstorm some of the ways that people compensate for each of these deficits.

Pull It Together

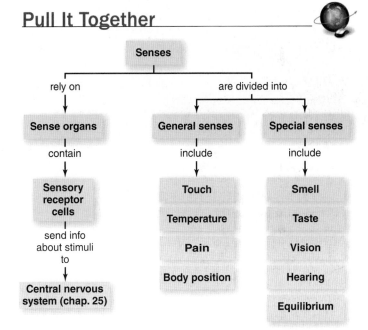

1. Which sense organs are required for each of the general and special senses?

2. Make a chart that lists the types of sensory receptors and the sense organs that use each type.

3. What is the role of rods and cones in the sense of vision?

4. Describe one way that each sense listed in this concept map can help the body maintain homeostasis.

Chapter 27

The Endocrine System

A mother feeds her baby girl from a plastic bottle.

Are Plastics Dangerous?

THE SCENE LOOKS HARMLESS ENOUGH: A MOTHER CRA-DLES HER INFANT WHILE THE CHILD DRINKS FROM A BOTTLE. But the clear plastic bottle may contain more than just breast milk or infant formula. Many health professionals are concerned that plastics could release harmful chemicals that may alter the baby's development.

Much of the controversy over plastics in baby bottles centers on a chemical called bisphenol A (BPA). Manufacturers use BPA to make shatterproof polycarbonate bottles, the linings of food cans, the sealants used in dentistry, and many other items. BPA is so common that everyone on Earth has this chemical in his or her tissues. Not only does BPA accumulate over a person's lifetime, but it can also pass from mother to fetus.

Why the concern over BPA? Research shows that at low doses, BPA acts as an endocrine-disrupting chemical. An endocrine disruptor is any substance that alters hormonal signaling, often by mimicking a natural hormone. BPA, for example, replicates the effect of the sex hormone estrogen. Other endocrine disruptors block the action of natural hormones, and still others stimulate or inhibit the activity of the glands that produce the hormones in the first place.

Low doses of BPA are associated with reproductive problems and developmental abnormalities in laboratory animals. But do these results apply to people? Possibly, but it's hard to say for sure. Testing for long-term effects of endocrine disruptors in humans is extremely difficult. Besides the ethical issues surrounding human experimentation, other complicating factors include the impossibility of finding BPA-free control subjects; the many stages of development at which endocrine disruptors can act; developmental differences between the sexes; and potential interactions between BPA and other endocrine disruptors.

If the problem were limited to BPA, the simple solution would be to ban this chemical and move on. But BPA is just one straw in a massive haystack. Every day, humans release thousands of pesticides, cosmetics, medications, and other products into the air, soil, and water. The environment therefore teems with chemicals that are known or potential endocrine disruptors. They are in our food and in the fatty tissues of our bodies. Determining which are harmful, in which quantities, and at which stages of life is an enormous scientific challenge. Evidence is accumulating, however, that endocrine disruptors have altered the development and reproduction of wild animals including snails, fishes, frogs, alligators, and polar bears.

Hormones are powerful forces throughout an animal's life, and disturbances in hormone levels can have serious consequences. This chapter describes how hormones participate in the complex system of internal signals called the endocrine system.

Learning Outline

Learn How to Learn

Use Those Office Hours

Most instructors maintain office hours. Do not be afraid to use this valuable resource! Besides getting help with course materials, using office hours gives you an opportunity to know your professors personally. After all, at some point you may need a letter of recommendation; a letter from a professor who knows you well can carry a lot of weight. If you do decide to visit during office hours, be prepared with specific questions. And if you request a separate appointment, it is polite to confirm that you intend to come at the time you have arranged.

27.1 | The Endocrine System Uses Hormones to Communicate

Animals communicate with one another in many ways, including color displays, sounds, body language, and communication molecules called pheromones. Likewise, the cells that make up a multicellular organism's body send and receive signals; these intercellular messages coordinate the actions of the body's organ systems. ▶ pheromones, p. 555

In flowering plants, hormones orchestrate growth, development, and responses to the environment (see chapter 23). Animal bodies, in contrast, have two main communication systems. The **nervous system,** described in chapter 25, is a network of cells that specialize in sending speedy signals that vanish as quickly as they arrive. The **endocrine system** is the other main communication system within the animal body. As this chapter explains, the endocrine system does not act with the speed of neural impulses, but its chemical messages have something else: staying power.

A. Endocrine Glands Secrete Hormones That Interact with Target Cells

The endocrine system has two main components: glands and hormones. An **endocrine gland** consists of cells that produce and secrete hormones into the bloodstream, which carries the secretions throughout the body. A **hormone** is a biochemical that travels in the bloodstream and alters the metabolism of one or more cells.

The endocrine system would be ineffective if every hormone acted on every cell in the body. Instead, each hormone has a limited selection of **target cells** that actually respond to the hormone. Inside or on the surface of each target cell is a receptor protein, which binds to the hormone and initiates the cell's response.

Hormones are analogous to the radio signals that multiple stations simultaneously broadcast into the atmosphere. The receptors on the target cells, then, are like individual radios. Even when dozens of signals are present, each radio is tuned to one frequency and therefore picks up the signal of just one station.

Figure 27.1 Some Human Endocrine Glands. The endocrine system includes several glands containing specialized cells that secrete hormones. Additional hormone-secreting cells are scattered among the other organ systems. The hormones circulate throughout the body in blood vessels, which are not shown in this figure.

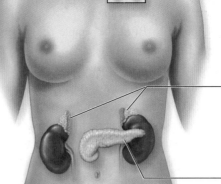

Hypothalamus
Produces hormones that stimulate or inhibit the release of hormones from the pituitary gland

Thyroid gland
Releases thyroid hormones, which regulate metabolism

Pineal gland
Produces melatonin

Pituitary gland
Produces numerous hormones that affect target tissues directly or stimulate other endocrine glands

Parathyroid glands
(behind thyroid)
Secrete parathyroid hormone, which helps regulate blood calcium

Adrenal glands
Produce hormones that regulate kidney function and contribute to the body's stress response

Pancreas
Releases hormones that regulate blood glucose levels

Ovaries (in female)
Produce estrogen and progesterone, which mediate monthly changes in the uterine lining and promote secondary sex characteristics

Endocrine System	
Main Tissue Types*	**Examples of Locations/Functions**
Epithelial	Makes up the bulk of most glands and secretes many types of hormones
Connective	Blood circulates hormones throughout the body
Nervous	Parts of the brain secrete some hormones and control release of others; some neurons secrete hormones

*See chapter 24 for descriptions.

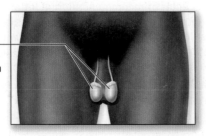

Testes (in male)
Produce testosterone, which promotes sperm maturation and secondary sex characteristics

Figure 27.2 Facial Hair. The growth of a beard and moustache is a sign of puberty in males.

Likewise, each receptor binds to one of the many hormones that may be circulating in the bloodstream. Moreover, just as one house may contain many radios, each tuned to a different station, one cell may also have receptors for many hormones, each of which initiates a unique response.

The main endocrine organs in vertebrates are the hypothalamus, pituitary gland, pineal gland, thyroid gland, parathyroid glands, adrenal glands, pancreas, ovaries, and testes (**figure 27.1**). But not all hormones originate in cells within a gland. The heart, kidneys, liver, stomach, small intestine, and placenta also contain scattered hormone-secreting cells. Together, these organs release dozens of hormones that simultaneously regulate every aspect of our lives, from conception through death.

To illustrate the power of the endocrine system, consider one stage of life that famously involves hormones: puberty. During this period, hormones transform a child's body into that of an adult. Females develop enlarged breasts and wider hips, males acquire a deeper voice and more muscular physique, and new body hair sprouts in both sexes (**figure 27.2**). The same hormones also affect mood, emotions, and feelings of sexual attraction. The endocrine system's effects are not always so extreme, but they are nonetheless present throughout our lives.

Endocrine systems also figure prominently into the lives of other animals. For example, a tadpole undergoes a dramatic metamorphosis as it develops into an adult frog (**figure 27.3**). Steadily increasing levels of hormones secreted by the animal's thyroid gland initiate these changes. Another amazing hormone-orchestrated event is a caterpillar's metamorphosis into a butterfly, as described in this chapter's Burning Question on page 570.

Because hormones are so powerful, an animal's body must strictly regulate the levels of these molecules in the bloodstream.

This tight control often occurs by negative feedback interactions. Recall from chapter 24 that in **negative feedback,** an action counters an existing condition. For example, eating a candy bar can cause blood sugar to rise, which prompts the pancreas to release a hormone called insulin. In response, the body's cells take up sugar from the blood. As blood sugar drops, insulin production slows. Similar feedback loops ensure that endocrine glands adjust the secretion of all hormones as required to maintain homeostasis. ▶ homeostasis, p. 520

B. The Nervous and Endocrine Systems Work Together

Although the nervous and endocrine systems both specialize in communication, they differ in many ways. First, neurons use action potentials and neurotransmitters to send messages, whereas the endocrine system employs hormones. Second, each neuron influences only a few cells at a time, whereas hormones circulate throughout the body in the blood and may affect many different cells. Third, the endocrine system communicates much more slowly than the nervous system. A nervous impulse is virtually instantaneous, and its effects disappear as soon as the stimulus vanishes. Hormones take minutes, hours, or even days to exert their effects, which are generally more prolonged.

Despite these differences, the nervous and endocrine systems are tightly integrated—so much so that some biologists refer to them together as the "neuroendocrine system." The most obvious connection is a physical one: the hypothalamus. This region of the brain is clearly part of the central nervous system. Yet the hypothalamus contains neurons called neurosecretory cells that release hormones directly into the bloodstream. Moreover, the hypothalamus directly or indirectly controls the action of many endocrine glands.

The nervous and endocrine systems also share chemical links. For example, some chemicals, such as epinephrine and norepinephrine, can act as neurotransmitters (if released from a neuron) and hormones (if released from an endocrine gland). In addition, neurotransmitters and hormones share some target cells.

Animal physiologists are still learning how hormones and neurons interact to oversee growth and development, influence appetite, regulate the concentrations of vital nutrients in the blood, and ready the body to confront stress. This chapter therefore cannot paint a complete picture of endocrine action. Instead, the objective is to explain some of the best understood hormonal effects—beginning with section 27.2, which describes how target cells respond to hormones.

27.1 | Mastering Concepts

1. What is the overall function of the endocrine system?
2. Describe the relationship among endocrine glands, hormones, and target cells.
3. What is the role of negative feedback in the endocrine system?
4. How do the nervous and endocrine systems differ?
5. Describe how the nervous and endocrine systems interact.

Figure 27.3 Frog Metamorphosis. Hormones from the thyroid gland initiate the transformation of a tadpole into an adult frog.

27.2 | Hormones Stimulate Responses in Target Cells

Just as a key fits a lock, each hormone affects only target cells bearing specific receptor molecules. The term *target cells* is a little misleading, because it implies that hormones somehow travel straight from their source to a limited set of cells. In reality, the blood circulating throughout the body contains many hormones at once. Each hormone's target cells are simply those with the corresponding receptors.

This section describes how the interaction between a hormone and its receptor initiates the target cell's response. The receptors may occur on the target cell's surface or inside the cytoplasm. In general, receptors for water-soluble hormones are on the surface of the target cell. In contrast, lipid-soluble hormones typically interact with internal receptors.

A. Water-Soluble Hormones Trigger Second Messenger Systems

Peptide hormones, which are chains of a few to several hundred amino acids, are water-soluble. Insulin, the hormone involved in sugar metabolism, is an example. Some hormones that are derived from a single amino acid are also water-soluble, including epinephrine, norepinephrine, and dopamine. ▸ amino acids, p. 36

Water-soluble hormones, which cannot pass readily through the cell membrane's phospholipid bilayer, bind to receptors on the surface of target cells (**figure 27.4a**). This hormone-receptor interaction triggers a chain reaction. The receptor protein contacts a neighboring protein (the "G protein"). The G protein, in turn, stimulates an enzyme called adenyl cyclase to convert ATP to another molecule, cyclic AMP (cAMP). This product is called a second messenger, and it lies at the crux of the entire process. Approximately 1 second after the stimulus arrives, cAMP provokes the cell's response, typically by activating enzymes that produce the hormone's effects. This entire cascade of reactions has the effect of converting the external "message"—the arrival of the peptide hormone—into an internal signal. ▸▸ cell membrane, p. 54; ATP, p. 76

In general, peptide hormones act rapidly, within minutes of their release. Target cells respond quickly because all of the participating biochemicals are already in place when the hormone binds to the receptor.

? Burning Question

How does a caterpillar "remodel" itself into a butterfly?

The changes that sculpt a butterfly from a caterpillar, a fly from a maggot, or a beetle from a grub are among the most dramatic in the animal kingdom. This metamorphosis from larva to adult is a complex, highly coordinated process orchestrated by changing levels of two hormones: juvenile hormone and molting hormone (**figure 27.A**).

A caterpillar is a streamlined eating machine that grows at a spectacular rate. Juvenile hormone predominates during this time. To accommodate rapid growth, the young insect must regularly molt, or shed its exoskeleton. Periodic spikes in levels of molting hormone trigger each molt. Eventually, the cells that produce juvenile hormone shut off, and molting hormone predominates. The insect then secretes a cocoon and becomes a pupa.

The body remodels itself completely inside the cocoon. Many larval cells die. But other cells, set aside in tiny disklike packets early in the larva's life, obtain energy from the degenerating larval cells. The reawakened cells divide, aggregating and differentiating to form adult body parts such as legs, antennae, and wings. Eventually the pupa case splits, and an adult insect emerges—wet and compacted, with its wings plastered against its abdomen. After a few hours, the wings dry out. Fluid flows into the veins of the wings, which expand as they unfurl. Soon the insect is free to take on an entirely new lifestyle. Rather than living to eat, the adult insect lives to mate.

Submit your burning question to:
marielle_hoefnagels@mcgraw-hill.com

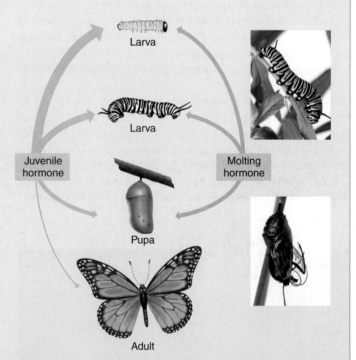

Figure 27.A Insect Metamorphosis. Two hormones control insect metamorphosis. The thickness of the line leading from each hormone denotes the magnitude of its effect at each life stage.

B. Lipid-Soluble Hormones Directly Alter Gene Expression

The body synthesizes lipid-soluble **steroid hormones** from cholesterol, which is one reason humans need at least some cholesterol in their diets. Testosterone, estrogen, and progesterone are steroid hormones. Two other lipid-soluble hormones, the thyroid hormones, are derived from a single hydrophobic amino acid (tyrosine). ▶ lipids, p. 34

Unlike peptide hormones, lipid-soluble hormones easily cross the cell membrane (**figure 27.4b**). Once inside the cell, the hormone may enter the nucleus and bind to a receptor associated with DNA, which triggers the production of proteins that carry out the target cell's response. Alternatively, the hormone may bind to a receptor in the cytoplasm, and the two molecules may travel together to the nucleus. Either way,

response time is much slower than for peptide hormones, because the cell must produce new proteins before the hormone takes effect. ▶ protein synthesis, p. 126

27.2 | Mastering Concepts

1. How does a hormone affect some cells but not others?
2. Describe the locations of the receptors that bind to water- and lipid-soluble hormones.
3. What is the role of second messengers in hormone action?
4. Explain the observation that peptide hormones usually act faster than steroid hormones.

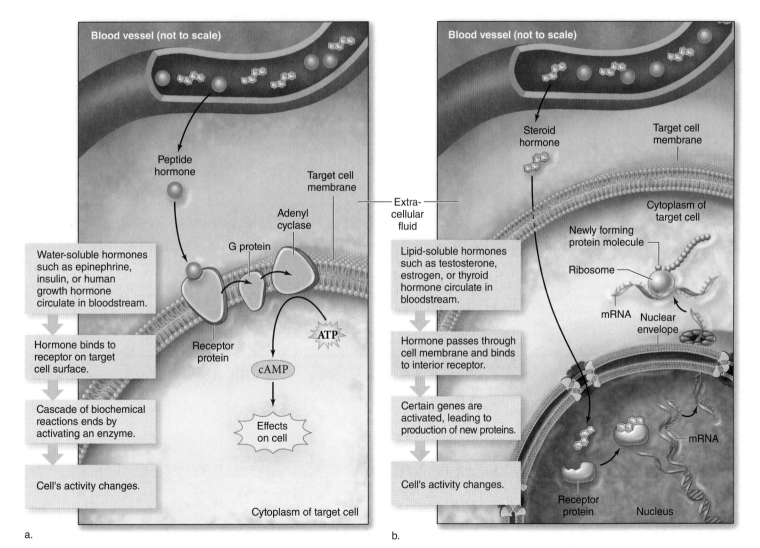

Figure 27.4 Target Cell Responses to Hormones. (a) Peptide hormones and other water-soluble hormones bind to receptors on the surface of target cells. A series of chemical reactions initiates the target cell's response. (b) Steroid hormones are lipid-soluble. They pass through cell membranes and bind to receptors in the cytoplasm or nucleus. The target cell responds by altering the expression of one or more genes.

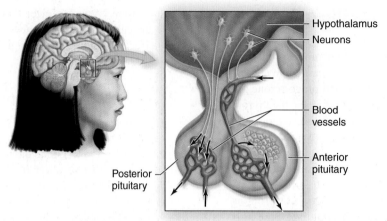

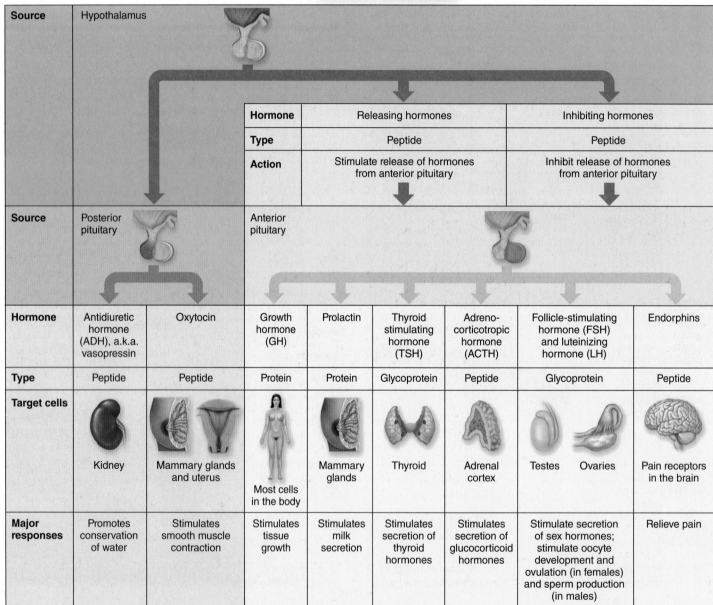

Source	Hypothalamus							

Hormone	Releasing hormones	Inhibiting hormones
Type	Peptide	Peptide
Action	Stimulate release of hormones from anterior pituitary	Inhibit release of hormones from anterior pituitary

Source	Posterior pituitary		Anterior pituitary					

Hormone	Antidiuretic hormone (ADH), a.k.a. vasopressin	Oxytocin	Growth hormone (GH)	Prolactin	Thyroid stimulating hormone (TSH)	Adreno-corticotropic hormone (ACTH)	Follicle-stimulating hormone (FSH) and luteinizing hormone (LH)	Endorphins
Type	Peptide	Peptide	Protein	Protein	Glycoprotein	Peptide	Glycoprotein	Peptide
Target cells	Kidney	Mammary glands and uterus	Most cells in the body	Mammary glands	Thyroid	Adrenal cortex	Testes Ovaries	Pain receptors in the brain
Major responses	Promotes conservation of water	Stimulates smooth muscle contraction	Stimulates tissue growth	Stimulates milk secretion	Stimulates secretion of thyroid hormones	Stimulates secretion of glucocorticoid hormones	Stimulate secretion of sex hormones; stimulate oocyte development and ovulation (in females) and sperm production (in males)	Relieve pain

Figure 27.5 **Hormones of the Hypothalamus and Pituitary: A Summary.** The hypothalamus and the pituitary gland secrete hormones that help coordinate the actions of the other endocrine glands.

27.3 | The Hypothalamus and Pituitary Gland Oversee Endocrine Control

Two structures, the hypothalamus and the pituitary gland, coordinate the vertebrate endocrine system. The almond-sized **hypothalamus** is a part of the forebrain, and the **pituitary gland** is a pea-sized structure attached to a stalk extending from the hypothalamus. The hypothalamus links the nervous and endocrine systems by controlling pituitary secretions (**figure 27.5**).

The pituitary is really two glands in one: the larger **anterior pituitary** (toward the front) and the smaller **posterior pituitary** (toward the back). Anatomically, the posterior pituitary is a continuation of the hypothalamus, whereas the anterior pituitary consists of endocrine cells.

Consequently, the posterior pituitary does not synthesize hormones, but it does store and release two hormones that the hypothalamus produces. The hypothalamus controls the anterior lobe of the pituitary in a different way—by secreting hormones that reach the anterior pituitary through a specialized system of blood vessels.

A. The Posterior Pituitary Stores and Releases Two Hormones

One of the two hormones produced by the hypothalamus and released by the posterior pituitary is **antidiuretic hormone (ADH),** also called vasopressin. This hormone signals target cells in the kidneys to return water to the bloodstream (rather than eliminating the water in urine). If cells in the hypothalamus detect that the body's fluids are too concentrated, the posterior pituitary releases more ADH. Additional water returns to the blood; when balance is restored, the production of ADH slows. Chapter 32 explains kidney function in more detail.

Oxytocin is the other posterior pituitary hormone. When a baby suckles, sensory neurons in the mother's nipple relay the information to the brain, which stimulates the release of oxytocin. The hormone causes cells in the breast to contract, squeezing the milk through ducts leading to the nipple. Oxytocin also triggers muscle contraction in the uterus, which pushes a baby out during labor. For this reason, physicians use synthetic oxytocin to induce labor or accelerate contractions in a woman who is giving birth. Oxytocin also acts on the brain, playing a role in bonding, affection, and social recognition in both sexes.

B. The Anterior Pituitary Produces and Secretes Six Hormones

One of the six hormones that the anterior pituitary gland produces is **growth hormone (GH),** also called somatotropin. This hormone promotes growth and development in all tissues by increasing protein synthesis and cell division rates. Levels of GH peak in the preteen years and, together with rising levels of sex

Figure 27.6 Growth Hormone Abnormality. A pituitary giant poses with his father and young brother.

hormones, cause adolescent growth spurts. A severe deficiency of GH during childhood leads to pituitary dwarfism, which produces extremely short stature. At the other extreme, a child with too much GH becomes a pituitary giant (**figure 27.6**). In an adult, GH does not affect height because the long bones of the body are no longer growing. However, excess GH can cause acromegaly, a thickening of the bones in the hands and face.

Prolactin is an anterior pituitary hormone that stimulates milk production in a woman's breasts after she gives birth. In males and in women who are not breast feeding, a hormone from the hypothalamus suppresses prolactin synthesis. In nursing mothers, however, a suckling infant triggers nerve impulses that overcome this inhibition.

Four other anterior pituitary hormones all influence hormone secretion by other endocrine glands. **Thyroid-stimulating hormone (TSH)** prompts the thyroid gland to release hormones, whereas **adrenocorticotropic hormone (ACTH)** stimulates hormone release from the adrenal glands. The remaining two anterior pituitary hormones stimulate hormone release from the ovaries and testes: **follicle-stimulating hormone (FSH)** and **luteinizing hormone (LH).** Sections 27.4 and 27.5 describe the hormones of the thyroid, adrenal glands, ovaries, and testes in more detail.

The anterior pituitary also produces **endorphins,** which are natural painkillers that bind to receptors on target cells in the brain. Usually, however, endorphins are not detectable in the blood, so their status as hormones is questionable.

27.3 | Mastering Concepts

1. How does the hypothalamus interact with the posterior and anterior pituitary glands?
2. List the names and functions of the hormones released by the posterior and anterior pituitary glands.

27.4 | Hormones from Many Glands Regulate Metabolism

The thyroid gland, parathyroid glands, adrenal glands, and pancreas secrete hormones that influence metabolism (figure 27.7). Hormones from the anterior pituitary control many, but not all, of the activities of these glands (see figure 27.5).

A. The Thyroid Gland Sets the Metabolic Pace

The **thyroid gland** is a two-lobed structure in the neck. The lobes secrete two thyroid hormones, **thyroxine** and **triiodothyronine,** that increase the rate of metabolism in target cells. Under thyroid stimulation, the lungs exchange gases faster, the small intestine absorbs nutrients more readily, and fat levels in cells and in blood plasma decline.

The thyroid hormones illustrate how the hypothalamus and pituitary interact in negative feedback loops (figure 27.8). When blood levels of thyroid hormones are low, the hypothalamus secretes thyrotropin-releasing hormone (TRH), which stimulates the anterior pituitary to increase production of thyroid-stimulating hormone (TSH). In response, epithelial cells in the thyroid secrete thyroxine and triiodothyronine. In the opposite situation, TRH secretion slows, so the thyroid glands reduce their production of hormones.

One disorder that affects the thyroid gland is hypothyroidism, a condition in which the thyroid does not release enough hormones. The metabolic rate slows, the body burns fewer calories,

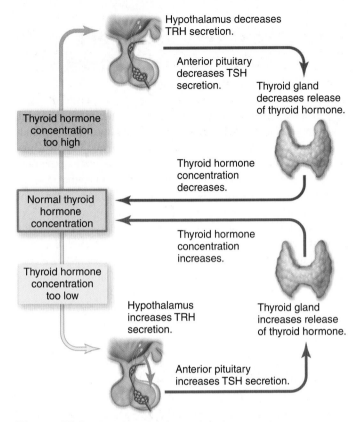

Figure 27.8 Thyroid Hormone Regulation. A negative feedback loop maintains the proper concentration of thyroid hormones in blood.

Source	Thyroid		Parathyroid	Adrenal medulla	Adrenal cortex		Pancreas		Pineal gland
Hormone	Thyroid hormones (thyroxine, triiodothyronine)	Calcitonin	Parathyroid hormone (PTH)	Epinephrine, norepinephrine	Mineralo-corticoids	Gluco-corticoids	Insulin	Glucagon	Melatonin
Type	Amine	Peptide	Peptide	Amine	Steroid	Steroid	Peptide	Protein	Amine
Target cells	All tissues	Bone	Bone, digestive organs, kidneys	Blood vessels	Kidney	All tissues	All tissues	Liver, adipose tissue	Other endocrine glands
Major responses	Increase metabolic rate	Increases rate of calcium deposition	Releases calcium from bone, increases calcium absorption in digestive organs and kidneys	Raise blood pressure, constrict blood vessels, slow digestion	Maintain blood volume and electrolyte balance	Increase glucose levels in blood and brain	Increases uptake of glucose	Stimulates breakdown of glycogen into glucose and of fats into fatty acids	Regulates effects of light–dark cycles

Figure 27.7 Hormones That Regulate Metabolism: A Summary. Hormones from several endocrine glands simultaneously influence many metabolic processes.

and weight increases. Synthetic hormones can treat many cases of hypothyroidism. In the past, the most common cause of hypothyroidism was iodine deficiency. Both thyroid hormones contain iodine; a deficiency of this essential element causes a goiter, or swollen thyroid gland. Today, iodine-deficient goiter is rare in nations where iodine is added to table salt.

An overactive thyroid causes hyperthyroidism. This disorder is associated with hyperactivity, an elevated heart rate, a high metabolic rate, and rapid weight loss. Graves disease is the most common type of hyperthyroidism. Both former President George H. W. Bush and his wife, Barbara, have this disorder.

Scattered cells throughout the thyroid gland produce a third hormone, **calcitonin,** which decreases blood calcium level by increasing the deposition of calcium in bone. Levels of calcitonin greatly increase during pregnancy and milk production, when a woman's body is under calcium stress. The overall physiological importance of calcitonin in adult humans, however, is usually minimal.

B. The Parathyroid Glands Control Calcium Level

The **parathyroid glands** are four small groups of cells embedded in the back of the thyroid gland. **Parathyroid hormone (PTH)** increases calcium levels in blood and tissue fluid by releasing calcium from bones and by enhancing calcium absorption at the digestive tract and kidneys (see figure 28.9). PTH action therefore opposes that of calcitonin.

Calcium is vital to muscle contraction, neurotransmitter release, blood clotting, bone formation, and the activities of many enzymes. Underactivity of the parathyroids can therefore be fatal. Excess PTH can also be harmful if calcium leaves bones faster than it accumulates. This condition, called osteoporosis, is most common in women who have reached menopause (cessation of menstrual periods). The estrogen decrease that accompanies menopause makes bone-forming cells more sensitive to PTH, which depletes bone mass. ▶ osteoporosis, p. 588

C. The Adrenal Glands Coordinate the Body's Stress Responses

The paired, walnut-sized **adrenal glands** sit on top of the kidneys (*ad-* means near, *renal* means kidney). The **adrenal medulla** is the inner portion of each gland, and the **adrenal cortex** is the outer portion. Each region secretes different hormones, mostly in response to stress (**figure 27.9**).

The adrenal medulla's hormones, **epinephrine** (adrenaline) and **norepinephrine** (noradrenaline), help the body respond to exercise, trauma, and other stresses. Signals from the sympathetic nervous system trigger release of these hormones, which cause heart rate and blood pressure to climb. In addition, the airway increases in diameter, making breathing easier. The metabolic rate increases, while digestion and other "nonessential" processes slow.

Epinephrine can save the lives of people with severe allergic reactions to bee stings or specific foods. Moments after contacting the allergen, a massive immune system reaction causes the airway to constrict. People with known allergies may therefore carry a self-injectable dose of epinephrine with them. The epinephrine temporarily reverses the allergic reaction, allowing the person to survive long enough to seek medical help. ▶ allergies, p. 686

Unlike the adrenal medulla, the adrenal cortex secretes steroid hormones, including mineralocorticoids, glucocorticoids, and even a small amount of testosterone. The **mineralocorticoids** maintain blood volume and salt balance. One example, aldosterone, stimulates the kidneys to return sodium ions and water to the blood while excreting potassium ions. This action conserves water and increases blood pressure, which is especially important in compensating for fluid loss from severe bleeding. ▶ aldosterone, p. 667

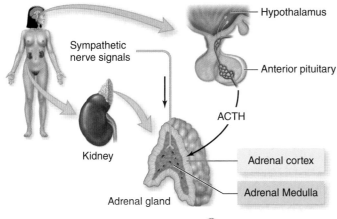

Hypothalamus

Sympathetic nerve signals

Anterior pituitary

ACTH

Kidney

Adrenal cortex

Adrenal Medulla

Adrenal gland

Source	Adrenal medulla		Adrenal cortex	
	Short-term stress		Long-term stress	
Hormone	Epinephrine, norepinephrine		Mineralocorticoids	Glucocorticoids
Major responses	• Increase heart rate and blood pressure • Dilate airways, so breathing rate increases • Increase metabolic rate • Slow digestion		• Maintain blood volume	• Increase glucose synthesis • Constrict blood vessels, raising blood pressure • Suppress immune system

Figure 27.9 Hormones of the Adrenal Glands: A Summary. The adrenal medulla secretes epinephrine and norepinephrine, which help the body respond to short-term stresses. Mineralocorticoids and glucocorticoids from the adrenal cortex enable the body to survive prolonged stress. The adrenal cortex also secretes small amounts of sex hormones (not shown).

Glucocorticoids are essential in the body's response to prolonged stress. Cortisol is the most important glucocorticoid. This hormone mobilizes energy reserves by stimulating the production of glucose from amino acids. Glucocorticoids also indirectly constrict blood vessels, which slows blood loss and prevents tissue inflammation after an injury. These same effects, however, account for the unhealthy effects of chronic stress. Narrowed blood vessels can lead to heart attacks, and the suppressed immune system leaves a person vulnerable to illness.

Prednisone, like other synthetic glucocorticoids, is an anti-inflammatory drug that mimics the effects of cortisol. These drugs can treat arthritis, allergic reactions, and asthma, but they also suppress the immune system. In addition, with long-term use of the drug, the adrenal cortex may stop producing its own glucocorticoid hormones. Abruptly stopping treatment may therefore cause a potentially dangerous "steroid withdrawal" condition, with symptoms including fatigue, low blood pressure, and nausea. In severe cases, the patient may go into shock, which can be fatal.

Figure It Out

Which hormone acts more slowly: cortisol or epinephrine? Why?

Answer: Cortisol; it is a steroid hormone.

D. The Pancreas Regulates Nutrient Use

The **pancreas** is an elongated gland, about the size of a hand, located beneath the stomach and attached to the small intestine. Clusters of cells called **pancreatic islets** (also called islets of Langerhans) secrete insulin and glucagon, two polypeptide hormones that regulate the body's use of nutrients (**figure 27.10**).

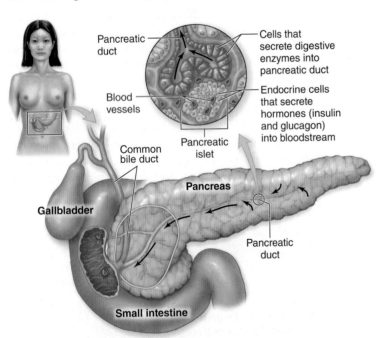

Figure 27.10 Anatomy of the Pancreas. This organ produces hormones and digestive enzymes. The inset shows the hormone-secreting pancreatic islets next to enzyme-secreting cells.

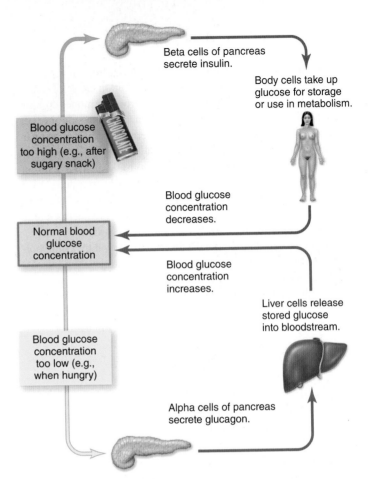

Figure 27.11 Blood Glucose Regulation. When blood sugar is high, insulin prompts cells to absorb glucose from blood. Glucagon has the opposite effect, increasing blood glucose levels by stimulating the release of sugar.

Insulin and glucagon oppose each other in regulating blood glucose levels (**figure 27.11**). After a meal rich in carbohydrates, glucose enters the circulation at the small intestine. The resulting rise in blood sugar triggers beta cells in the pancreas to secrete **insulin,** which stimulates target cells throughout the body to absorb glucose from the bloodstream. The target cells may then store the glucose as glycogen, consume it in cellular respiration to generate energy, or use it as a reactant in other metabolic reactions. As cells take up sugar, the blood glucose concentration declines, and insulin secretion slows. If blood sugar dips too low, however, alpha cells in the pancreas secrete **glucagon,** which stimulates target cells in the liver to release stored glucose into the bloodstream.

Too Much Glucose in Blood: Diabetes   Failure to regulate blood sugar can be deadly. In diabetes mellitus, glucose accumulates to dangerously high levels in the bloodstream. (Although other forms of diabetes exist, the term **diabetes** typically refers to diabetes mellitus.) Centuries ago, before lab tests for blood sugar were available, physicians diagnosed diabetes from the sweet taste of a patient's urine.

Diabetes is a paradox: sugar pours out of the body in urine, yet the body's cells starve for lack of glucose. Symptoms include frequent urination, excessive thirst, extreme hunger, blurred vision,

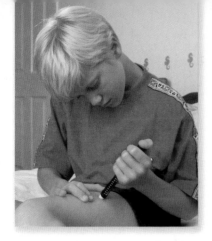

Figure 27.12 Type I Diabetes. This diabetic boy is injecting himself with insulin.

weakness, fatigue, irritability, nausea, and weight loss. If the diabetes remains untreated, complications may occur. For example, elevated blood glucose can eventually cause kidney failure, and damage to the peripheral nervous system may cause blindness or a loss of sensation in the hands and feet. The nerve damage, in turn, can contribute to poor healing of wounds, as undetected cuts and scrapes become infected with bacteria and fungi. Severe diabetes can eventually result in coma and death.

The accumulation of blood sugar that characterizes diabetes can occur for two reasons. In type 1 diabetes, the pancreas fails to produce insulin, so the body's cells never receive the signal to "open the door" and admit glucose. In type 2 diabetes, the body's cells fail to absorb glucose even when insulin is present; this condition is called insulin resistance. In type 2 diabetes, then, insulin "rings the doorbell," but the cell never opens the door.

Fifteen percent of affected individuals have type 1 diabetes, which usually begins in childhood or early adulthood. Typically, the underlying cause is an autoimmune disorder in which immune system cells attack the beta cells of the pancreas, which therefore cannot produce insulin. Type 1 diabetes is sometimes also called insulin-dependent diabetes because insulin injections can replace the missing hormone (figure 27.12). ▶ autoimmune disorders, p. 685

Type 2 diabetes is much more common, and it usually begins in adulthood. The incidence of type 2 diabetes in developed countries (including the United States) is increasing. The disease is strongly associated with obesity, although the cause–effect relationship is unclear (figure 27.13). Nearly all type 2 diabetes patients are overweight.

Medicines can help lower blood glucose levels, but the best way to prevent and treat type 2 diabetes is to maintain a healthy body weight by being physically active, reducing calorie intake, and choosing healthy foods (see section 31.4). Even though type 2 diabetes usually strikes in midlife, diabetes prevention should begin much earlier. After all, losing weight can be especially difficult when it requires changing dietary and exercise habits acquired over many decades.

Not Enough Glucose in Blood: Hypoglycemia The opposite of diabetes is **hypoglycemia,** in which excess insulin production or insufficient carbohydrate intake causes low blood sugar. A person with this condition feels weak, sweaty, anxious, and shaky; in severe cases, hypoglycemia can cause seizures or loss of consciousness. A healthy person might temporarily experience hypoglycemia after strenuous exercise. Frequent, small meals low in carbohydrates and high in protein can prevent insulin surges and help relieve symptoms of hypoglycemia.

E. The Pineal Gland Secretes Melatonin

The **pineal gland,** a small brain structure near the hypothalamus, produces the hormone **melatonin.** Darkness stimulates melatonin synthesis in the pineal gland; exposing the retina of the eye to light, on the other hand, inhibits melatonin production. The amount of melatonin in the bloodstream therefore communicates the amount of light to other cells of the body, setting the stage for the regulation of sleep–wake cycles and other circadian rhythms.

A form of depression called seasonal affective disorder (SAD) may be linked to abnormal melatonin secretion. Exposure to additional daylight (or full-spectrum light bulbs) can elevate mood. Because melatonin levels decrease with age, some people believe that taking extra melatonin might prevent age-related diseases.

27.4 | Mastering Concepts

1. What are the three hormones produced in the thyroid, and what are their functions?
2. What is the function of parathyroid hormone (PTH)?
3. How do the functions of hormones secreted by the adrenal cortex and adrenal medulla differ?
4. Describe the opposing roles of insulin and glucagon.
5. How do darkness and light affect melatonin secretion?

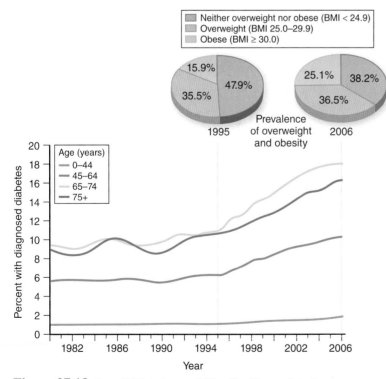

Neither overweight nor obese (BMI < 24.9)
Overweight (BMI 25.0–29.9)
Obese (BMI ≥ 30.0)

1995: 15.9%, 35.5%, 47.9%
2006: 25.1%, 38.2%, 36.5%

Prevalence of overweight and obesity

Figure 27.13 Type 2 Diabetes and Obesity. These data from the Centers for Disease Control show that the prevalence of type 2 diabetes has increased along with obesity since 1995.

27.5 | Hormones from the Ovaries and Testes Control Reproduction

The reproductive organs include the **ovaries** in females and the **testes** in males. These egg- and sperm-producing organs secrete the steroid hormones that not only enable gametes to mature but also promote the development of secondary sexual characteristics (**figure 27.14**). This section briefly introduces the sex hormones; chapter 34 explains their role in reproduction in more detail.

In a woman of reproductive age, the levels of several sex hormones cycle approximately every 28 days. The hypothalamus produces gonadotropin-releasing hormone (GnRH), which stimulates the anterior pituitary to release FSH and LH into the bloodstream. At target cells in the ovary, these two hormones stimulate the events that lead to ovulation. Meanwhile, the cells surrounding the egg release the sex hormones **estrogen** and **progesterone,** which exert negative feedback control on the hypothalamus and pituitary. Estrogen also promotes development of the female secondary sex characteristics, such as breasts and wider hips, whereas progesterone helps prepare the uterus for pregnancy.

In males, FSH stimulates the early stages of sperm formation in the testes. Sperm production is completed under the influence of LH, which also prompts cells in the testes to release the sex hormone **testosterone.** This hormone stimulates the formation of male structures in the embryo and promotes later development of male secondary sexual characteristics, including facial hair, deepening of the voice, and increased muscle growth (see this chapter's Apply It Now box).

Source	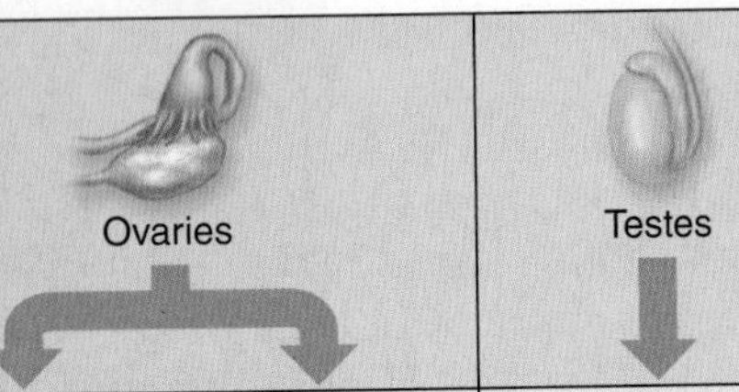 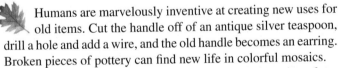 Ovaries		Testes
Hormone	Progesterone	Estrogen	Testosterone
Type	Steroid	Steroid	Steroid
Target cells	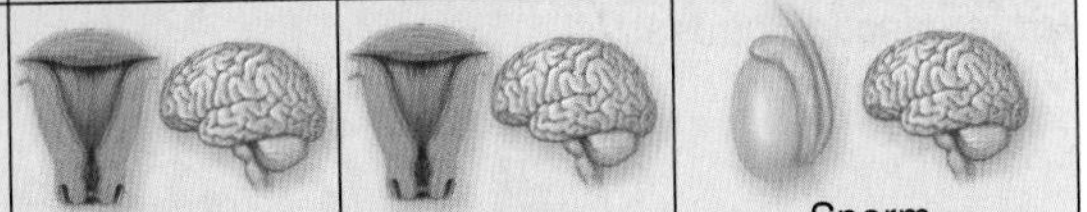 Uterine lining, hypothalamus, pituitary, other tissues	Uterine lining, hypothalamus, pituitary, other tissues	Sperm-producing cells, hypothalamus, pituitary, other tissues
Major responses	Regulates menstrual cycle, prepares body for pregnancy	Regulates menstrual cycle, maintains secondary sex characteristics in females	Promotes sperm development, maintains secondary sex characteristics in males

Figure 27.14 Hormones of the Ovaries and Testes: A Summary. Hormones produced in the ovaries and testes coordinate reproduction and the development of secondary sex characteristics.

27.5 | Mastering Concepts

1. Which organs contain target cells for FSH and LH?
2. What are the functions of estrogen, progesterone, and testosterone?

Apply It Now

Anabolic Steroids in Sports

The anabolic steroids that regularly make headlines in the sporting news are synthetic forms of testosterone. Despite their notoriety, these drugs have a legitimate place in medicine. A physician might prescribe steroids for a person who produces too little testosterone, for example, or for someone with an illness that causes muscles to waste away.

Although anabolic steroids are legal only by prescription, some athletes abuse these drugs as a shortcut to greater muscle mass. Steroid users may improve strength and performance in the short term, but the drugs are harmful in the long run. In males, the body mistakes synthetic steroids for the natural hormone and lowers its own production of testosterone, causing infertility once use of the drug stops. Other possible side effects are impotence, shrunken testicles, and the growth of breast tissue. Females who abuse steroids may develop a masculine physique, a deeper voice, and facial or body hair. In adolescents, steroids hasten adulthood, stunting height and causing early hair loss. Finally, research suggests that high doses of steroids may cause psychological side effects such as aggression, mood swings, and irritability. For all of these reasons, health professionals strongly advise against the use of illegal steroids.

27.6 | Investigating Life: Something's Fishy in Evolution—The Origin of the Parathyroid Gland

Humans are marvelously inventive at creating new uses for old items. Cut the handle off of an antique silver teaspoon, drill a hole and add a wire, and the old handle becomes an earring. Broken pieces of pottery can find new life in colorful mosaics.

Likewise, natural selection can yield new adaptations from existing parts. A bat's wing, for example, contains the same bones as the forelimbs of other mammals. Despite the modifications that enable flight, it is easy to see that a bat's wings are homologous to our own arms; that is, the structures share a common evolutionary history. ▶ homologous structures, p. 262

To discover the origins of some structures, however, researchers need to dig deeper into our ancestry. The parathyroid glands provide one example. Scientists have used a model organism called the zebrafish (*Danio rerio*) to trace the evolutionary roots of these tiny endocrine organs—findings that are all the more surprising because fishes don't even have parathyroid glands!

To understand this research, it helps to know more about the functions of gills. Fishes use gills to acquire O_2 and release CO_2. Besides exchanging gases, gills also absorb calcium dissolved in

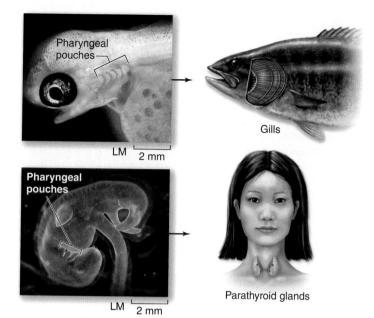

Figure 27.15 Pharyngeal Pouches. The same embryonic structures give rise to the gills of fishes and the parathyroid glands of terrestrial vertebrates.

the water. In air-breathing vertebrates, the situation is different. The lungs take care of air exchange and, as described in section 27.4B, the parathyroid glands help boost calcium levels.

If fishes lack parathyroid glands, where did these new calcium-regulating organs come from in their air-breathing descendants? The answer seems to be that nature found a new use for an ancient structure. That is, some of the genetic instructions for building gills in a fish have been "re-purposed" to build the parathyroid glands of a terrestrial vertebrate.

Researchers Masataka Okabe and Anthony Graham of Kings College London used multiple lines of evidence to make this argument. First, they pointed to an anatomical similarity: the location of the parathyroid glands in the neck is somewhat comparable to the behind-the-head location of a fish's gills. Second, they noted that gills and the parathyroid glands both arise from embryonic structures called pharyngeal pouches (figure 27.15).

Third, Okabe and Graham already knew that a gene called *Gcm-2* is involved in the development of the parathyroid gland in mice. They used molecular biology to see whether the same gene was active in the embryos of zebrafish and chickens (like humans, chickens are air-breathing vertebrates). As described in chapter 7, a gene is a sequence of DNA. Gene expression begins when a cell uses the information in DNA to produce a molecule called messenger RNA (mRNA). The researchers created nucleic acid

"probes" that specifically stuck to the mRNA corresponding to the *Gcm-2* gene. By treating fish and chicken embryos with the probe and then checking where the probe accumulated, researchers could see precisely which cells in each embryo express *Gcm-2*.

Early in the development of both species, the *Gcm-2* gene was expressed in the cells of the pharyngeal pouches (figure 27.16). In the fish, the gene was later expressed in the gill buds that develop from the pharyngeal pouches. In the chicken, the developing parathyroid glands expressed the gene. These results showed that the *Gcm-2* gene somehow participates in the development of gills and parathyroid glands—evidence of a close relationship between the two organs.

Fish may express *Gcm-2*, but does gill formation actually *require* this gene? The final step of the project was to check whether fish would develop gills if *Gcm-2* was turned off. Genes do not come with on–off switches, but it is possible to stop a gene's expression by applying synthetic "antisense" nucleic acids that inactivate the mRNA transcribed from a gene. When the team knocked out *Gcm-2* in zebrafish embryos, the fish failed to develop normal gill buds. *Gcm-2* is therefore required for gill development in fishes, just as it is involved in the development of the parathyroid gland in mice.

This finding adds *Gcm-2* to the long list of genes that have been repurposed as life has diversified over evolutionary time. Moreover, the evidence described in this study demonstrates the intimate relationship between pharyngeal pouches, gills, and the parathyroid glands. Just as the forelimbs of ancient mammals gave rise to arms and bat wings, the pharyngeal pouches have proved to be very versatile actors in vertebrate development.

Okabe, Masataka, and Anthony Graham. December 21, 2004. The origin of the parathyroid gland. *Proceedings of the National Academy of Sciences,* vol. 101, no. 51, pages 17716–17719.

Additional reference: Graham, Anthony, Masataka Okabe, and Robyn Quinlan. November 2005. The role of the endoderm in the development and evolution of the pharyngeal arches. *Journal of Anatomy,* vol. 207, no. 5, pages 479–487.

27.6 | Mastering Concepts

1. Summarize the evidence for an evolutionary relationship between fish gills and a chicken's parathyroid glands.

2. How do you think the embryos in figure 27.16 would look if the probe indicated expression of a gene required for cellular respiration?

a. Zebrafish

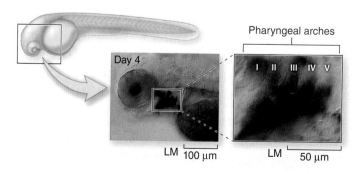

b. Chicken

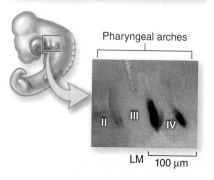

Figure 27.16 Different Animals, Same Gene. The dark blue areas reveal where the gene *Gcm-2* is expressed in 4-day-old embryos of (a) a zebrafish and (b) a chicken. Roman numerals indicate the arches that separate the pharyngeal pouches.

Chapter Summary

27.1 | The Endocrine System Uses Hormones to Communicate

- The **nervous system** and **endocrine system** specialize in the intercellular communication needed to maintain homeostasis in an animal body.

A. Endocrine Glands Secrete Hormones That Interact with Target Cells

- The endocrine system includes several **endocrine glands** and scattered cells, plus the **hormones** they secrete into the bloodstream. Hormones interact with **target cells** to exert their effects.
- **Negative feedback** loops ensure that the levels of a hormone in the bloodstream are not too high or too low.

B. The Nervous and Endocrine Systems Work Together

- The nervous system acts faster and more locally than the endocrine system.
- The hypothalamus physically connects the nervous and endocrine systems. In addition, the two communication systems share many messenger molecules and target cells.

27.2 | Hormones Stimulate Responses in Target Cells

A. Water-Soluble Hormones Trigger Second Messenger Systems

- **Peptide hormones** are water-soluble and bind to the surface receptors of target cells. A second messenger triggers the hormone's metabolic effect.

B. Lipid-Soluble Hormones Directly Alter Gene Expression

- Most lipid-soluble **steroid hormones** cross cell membranes and bind to receptors in the cytoplasm or nucleus. The receptors activate genes, which direct the synthesis of proteins that provide the cell's response.

27.3 | The Hypothalamus and Pituitary Gland Oversee Endocrine Control

- Neurons from the **hypothalamus** influence the release of hormones from the **posterior pituitary** and **anterior pituitary** gland. These hormones influence many other processes in the body.

A. The Posterior Pituitary Stores and Releases Two Hormones

- The hypothalamus manufactures two hormones that are stored in and released from the posterior pituitary. **Antidiuretic hormone (ADH)** regulates body fluid composition, and **oxytocin** contracts muscles in the uterus and milk ducts.

B. The Anterior Pituitary Produces and Secretes Six Hormones

- Hormones from the hypothalamus regulate the production and release of hormones from the anterior pituitary.
- **Growth hormone (GH)** stimulates cell division, protein synthesis, and growth throughout the body.
- **Prolactin** stimulates milk production.
- **Thyroid-stimulating hormone (TSH)** prompts the thyroid gland to release hormones.
- **Adrenocorticotropic hormone (ACTH)** stimulates the adrenal cortex to release hormones.
- **Follicle-stimulating hormone (FSH)** and **luteinizing hormone (LH)** stimulate hormone release from the ovaries and testes.
- **Endorphins** are painkillers with target cells in the brain.

27.4 | Hormones from Many Glands Regulate Metabolism

A. The Thyroid Gland Sets the Metabolic Pace

- **Thyroxine** and **triiodothyronine** from the **thyroid gland** speed metabolism. The thyroid also releases **calcitonin,** which lowers the level of calcium in the blood.

B. The Parathyroid Glands Control Calcium Level

- The **parathyroid glands** secrete **parathyroid hormone (PTH),** which increases blood calcium level by releasing calcium from bone and increasing its absorption in the gastrointestinal tract and kidneys.

C. The Adrenal Glands Coordinate the Body's Stress Responses

- The **adrenal gland** has an inner portion, the **adrenal medulla,** which secretes **epinephrine** and **norepinephrine.** These hormones ready the body to cope with a short-term emergency. The outer **adrenal cortex** secretes **mineralocorticoids** and **glucocorticoids,** which mobilize energy reserves during stress and maintain blood volume and blood composition.

D. The Pancreas Regulates Nutrient Use

- The **pancreatic islets** of the **pancreas** secrete **insulin,** which stimulates cells to take up glucose. **Glucagon** increases blood glucose levels.
- In **diabetes** mellitus, blood sugar concentrations rise to dangerous levels. Type 1 diabetes occurs when the pancreas fails to produce insulin; in type 2 diabetes, the body's cells do not respond to insulin. Type 2 diabetes is associated with obesity.
- **Hypoglycemia** is low blood sugar, which is caused by excess insulin or insufficient carbohydrate intake.

E. The Pineal Gland Secretes Melatonin

- The **pineal gland** may regulate the responses of other glands to light–dark cycles through its hormone, **melatonin.**

27.5 | Hormones from the Ovaries and Testes Control Reproduction

- FSH and LH stimulate the **ovaries** to secrete **estrogen** and **progesterone,** hormones that stimulate development of female sexual characteristics and control the menstrual cycle.
- In males, FSH and LH stimulate the **testes** to secrete **testosterone,** which stimulates sperm cell production and the development of secondary sexual characteristics.

27.6 | Investigating Life: Something's Fishy in Evolution—The Origin of the Parathyroid Gland

- Multiple lines of evidence, including comparative anatomy and studies of gene expression, indicate that the parathyroid gland of vertebrates is derived from the same structure as the gills of fishes.

Multiple Choice Questions

1. Which of the following statements correctly describes a difference between the endocrine and nervous systems?
   a. The nervous system uses hormones; the endocrine system uses neurotransmitters.
   b. Only the endocrine system uses chemicals to send signals between cells.
   c. Nervous signals typically act more rapidly than endocrine signals.
   d. An endocrine signal typically affects fewer cells than a nervous signal.

2. What is a second messenger?
   a. A hormone produced only in adults
   b. A hormone that binds to the cell's exterior
   c. A molecule that initiates a hormone's effects
   d. A molecule participating in a negative feedback loop

3. The effect of a peptide hormone such as insulin is generally quicker than that of a steroid hormone such as estrogen because
   a. peptide hormones exert changes using molecules already present in the target cell.
   b. steroid hormones trigger the synthesis of new proteins.
   c. steroid hormones cannot pass through the cell's plasma membrane.
   d. Both a and b are correct.

4. Which of the following is NOT a hormone produced in the anterior pituitary?
   a. Epinephrine
   b. Thyroid-stimulating hormone
   c. Adrenocorticotropic hormone
   d. Growth hormone

5. Treatment for high blood pressure often involves the use of medication that alters blood volume. Which of the following treatments would decrease a person's blood volume?
   a. Increase in release of thyroid-stimulating hormone
   b. Inhibition of follicle-stimulating hormone
   c. Activation of prolactin synthesis
   d. Inhibition of antidiuretic hormone

6. Besides thyroxine, the thyroid gland also produces
   a. parathyroid hormone.
   b. thyrotropin-releasing hormone.
   c. calcitonin.
   d. juvenile hormone.

7. Glucocorticoids are secreted by
   a. the thyroid.
   b. the pancreas.
   c. the adrenal glands.
   d. the pineal gland.

8. Would an insulin injection help a person with type 2 diabetes?
   a. Yes, because their beta cells are not producing insulin.
   b. No, because their target cells do not respond to insulin.
   c. Yes, because the extra insulin will help with glucose uptake.
   d. No, because the extra insulin will trigger excess release of glucagon.

9. Secretion of melatonin is regulated by
   a. light.
   b. temperature.
   c. stress.
   d. glucose.

10. If researchers develop a new drug that blocks testosterone receptors, a likely effect is a(n)
    a. increase in body hair.
    b. reduction in testosterone production.
    c. reduction in sperm production.
    d. Both a and c are correct.

Write It Out

1. How does the endocrine system interact with the nervous system? The circulatory system?
2. A queen honeybee secretes a substance from a gland in her mouthparts that inhibits the development of ovaries in worker bees. Is this substance most likely a hormone, a neurotransmitter, or a pheromone? Cite a reason for your answer.
3. What prevents a hormone from affecting all body cells equally?
4. Many dairy operators inject their cows with bovine growth hormone to stimulate milk production. Cite two reasons that bovine growth hormone might not stimulate growth in people drinking the milk.
5. How do hormones regulate their own levels?
6. Compare and contrast the endocrine and nervous systems.
7. Sketch the mechanisms of peptide and steroid hormone function.
8. List the hormones released from the posterior pituitary and the anterior pituitary.
9. How are the pituitary and adrenal glands each really two glands in one?
10. Which hormone(s) match each of the following descriptions?
    a. Produced by a woman who is breast feeding
    b. Causes fatigue if too little is present
    c. Causes increase in blood calcium level
    d. Causes decrease in blood glucose level
    e. Synthetic steroids mimic the muscle-building effects of this hormone.
11. Describe how thyroxine, triiodothyronine, TSH, and TRH interact.
12. Ancient Egyptian doctors treated goiters with seaweed, not realizing that it was the iodine in the seaweed that was reversing the condition. Why would iodine help cure a goiter?
13. Why would a physician counsel a patient with high blood pressure to try to reduce the amount of stress in his or her life?
14. Sleep deprivation increases cortisol concentrations in the blood. Why would sleep deprivation be associated with increased risk of illness?
15. Search the Internet for a diabetes risk test. What actions can you take to reduce your risk of type 2 diabetes?

16. How might insulin-producing stem cells transplanted to the pancreas help people with type 1 diabetes? Would the same treatment help people with type 2 diabetes?
17. In healthy adults, the concentration of glucose in blood is 80 to 110 milligrams per deciliter (mg/dl). After a carbohydrate-rich meal, however, the concentration may spike to 140 mg/dl.
    a. Describe the hormonal action that returns blood glucose to normal.
    b. What is the name of the condition in which the glucose concentration drops below 70 mg/dl?
    c. What is the name for the condition in which the glucose concentration before a meal is 130 mg/dl or higher?
18. Identify the target cells and effects of FSH, LH, estrogen, progesterone, and testosterone.
19. Imagine you are a researcher who has just found a small glandlike structure in a human. How would you determine whether it is an endocrine gland? If it is, how might you identify the gland's role in the endocrine system?
20. Consult chapter 25 and this chapter to create a chart that compares hormones with neurotransmitters. Include the distance over which each is active, the connection between the cell releasing the signals and the receiving cell, the response speed, and the duration of the response.
21. An endocrine disruptor is a molecule that either mimics or blocks the activity of a hormone. Propose a way to test the hypothesis that:
    a. a pesticide such as atrazine is an endocrine disruptor.
    b. endocrine disruptors have caused declines in human sperm counts.
    c. microwaving foods in plastic containers releases chemicals that act as endocrine disruptors.
22. Some people advocate treating severely disabled children with hormones that prevent growth, since a small person is much easier to clothe and bathe than a full-grown adult. What factors would you consider in deciding whether to pursue such a treatment for a disabled child?

Pull It Together

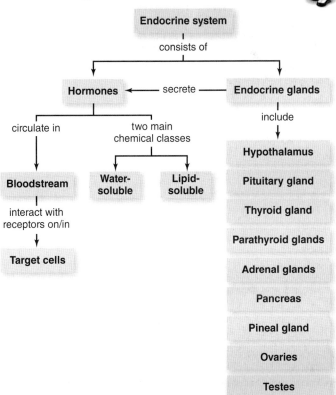

1. Describe a negative feedback loop that controls hormone secretion.
2. What brain structure connects the endocrine and nervous systems?
3. Describe the relationship among the hypothalamus, pituitary, and other endocrine glands.

28

The Skeletal and Muscular Systems

Artificial knee. This woman is recovering from knee replacement surgery.

|BIOLOGY

Enhance your study of
this chapter with practice quizzes,
animations and videos, answer keys,
and downloadable study tools.
www.mhhe.com/hoefnagels

Spare Parts Replace Worn-Out Joints

MANY INJURIES REQUIRE A VISIT WITH AN ORTHOPEDIC SURGEON, A PHYSICIAN WHO REPAIRS DAMAGE TO BONES AND MUSCLES. *Orthopedic* literally means "straight child." This odd-sounding phrase dates to the mid-eighteenth century, when orthopedic surgeons focused on correcting birth defects such as a club foot. Since then, the profession has expanded to include many adult surgeries, including hip and knee replacement and the repair of bone fractures and torn tendons and ligaments.

Joint replacements are among the most dramatic of orthopedic surgeries because they substitute synthetic materials for a patient's original bone. Finding the right material is crucial to the success of the operation. The new part must not trigger rejection by the immune system, and it must be strong and flexible enough to take over the function of the original joint. It must also resist wear and tear—after all, many joints are in almost constant motion.

The most commonly replaced joint is the hip, where the thighbone's spherical head swivels in a cup-shaped socket in the pelvis. Arthritis, or inflammation of the joints, commonly causes deformed, swollen hips that make movement stiff and painful as a person ages. In a hip replacement surgery, a surgeon replaces the top part of the thighbone with a ball-shaped substitute and inserts a cup that fits over the ball into the worn-out area of the pelvis. Early hip replacement surgeries in the nineteenth century used ivory in place of the patient's original bones, but surgeons now use ceramic, plastic, and metals such as titanium.

Arthritis also damages the knees, making walking painful. In a knee replacement operation, a surgeon detaches the quadriceps (thigh) muscle from the knee cap. After moving the knee cap to one side, the surgeon replaces the ends of the thighbone and shinbone with metal parts. A piece of plastic between the two metal parts provides cushioning. The surgeon then puts the knee cap back into place, reattaches the quadriceps muscle, and reconstructs the tendons and ligaments that allow the bones to move.

Complete recovery from hip or knee replacement surgery takes several months. Unfortunately, the metal and plastic parts do not last forever. After 15 to 20 years, depending on how active the patient is, the replacement parts may themselves need replacing.

It is easy to take bones and muscles for granted, at least until they stop working correctly. This chapter describes the structure and function of both components of the interrelated skeletal and muscular systems.

UNIT 6

What's the Point?

Learning Outline

Learn How to Learn
Make Appointments with Yourself

If you prefer to study alone but often find yourself putting off your solo study sessions, try making recurring "appointments" with yourself. That is, use your calendar to block off time that you dedicate to studying each day. You can reread chapters after class, quiz yourself on course materials, make concept maps, or work on homework assignments. Keep those appointments throughout the semester, so you never get behind.

28.1 | Skeletons Take Many Forms

Ask a child what sets animals apart from other organisms, and he or she will likely answer "movement." This response is technically wrong—many bacteria, archaea, protists, fungi, and plants have swimming, creeping, or gliding cells. Yet the child correctly recognizes that animal movements are unmatched in their drama, versatility, and power.

The ability to hop, dig, fly, slither, scuttle, or swim comes from two closely allied, interacting organ systems: the **muscular system** and the **skeletal system,** which together function under the direction of the nervous system. In most animals, organs called **muscles** provide motion; the **skeleton** adds a firm supporting structure that muscles pull against. The skeleton also gives shape to an animal's body and protects internal organs.

The simplest type of skeleton is a **hydrostatic skeleton** (*hydro-* means water), which consists of fluid constrained within a layer of flexible tissue. Many of the invertebrate animals described in chapter 20 have hydrostatic skeletons. The bell of a jellyfish, for example, consists mostly of a gelatinous substance constrained between two tissue layers (**figure 28.1**). To swim, the animal rhythmically contracts the muscles acting on this hydrostatic skeleton, forcibly ejecting water from its body. Other animals with hydrostatic skeletons include earthworms. By alternately contracting and relaxing muscles surrounding the fluid-filled coelom, the worm can burrow through soil. Snails, squids, flatworms, and nematodes also use hydrostatic skeletons in locomotion. ▶ coelom, p. 413

The most common type of skeleton is an **exoskeleton** (*exo-* means outside), which acts as a "suit of armor" that protects the animal from the outside (**figure 28.2**). Internal muscles pull against the exoskeleton, enabling the animal to move. The chemical composition of the exoskeleton varies, depending on the species. Mollusks such as clams and snails produce hard, calcium-containing shells. In lobsters, crabs, insects, and other arthropods, the jointed exoskeleton consists of the complex carbohydrate chitin. ▶ chitin, p. 32

Exoskeletons have advantages and disadvantages. The hard covering protects soft internal organs and provides excellent leverage for muscles. On the other hand, a growing arthropod must periodically molt; until its new skeleton has hardened, the

Figure 28.2
Exoskeleton. Arthropods such as crabs have tough, jointed exoskeletons that consist of chitin.

animal is vulnerable to predators. Moreover, a large animal would be unable to support the weight of an exoskeleton; for this reason, exoskeletons occur only in relatively small animals.

An **endoskeleton** (*endo-* means inner) is an internal support structure. Sea stars and other echinoderms, for example, produce rigid, calcium-rich spines and internal plates. Vertebrate animals have endoskeletons made of cartilage or bone (**figure 28.3**). Fishes such as sharks and rays, for example, produce cartilage skeletons. Other fishes, and all land vertebrates, have skeletons composed primarily of bone.

Like exoskeletons, endoskeletons represent an evolutionary trade-off. On the plus side, an internal skeleton grows with the animal, eliminating molting. Also, an endoskeleton consumes less of an organism's total body mass, so it can support the weight of animals as large as a whale. One disadvantage, however, is that the endoskeleton does not protect soft tissues at the body surface.

The vertebrate skeleton's capacity to change over evolutionary time is striking. Each vertebrate species has a distinctive skeleton, yet all are composed of the same types of cells and have similar arrangements. The characteristics of each species' muscles and skeleton reflect common ancestry and the selective forces in its environment. ▶▶ natural selection, p. 233; homologous structures, p. 262

28.1 | Mastering Concepts

1. How do the skeletal and muscular systems interact?
2. Describe similarities and differences among the three main types of skeletons.
3. How do vertebrate skeletons reveal common ancestry?

Figure 28.1 Hydrostatic Skeleton. When a jellyfish contracts muscles surrounding its bell, water shoots out of its body. The animal moves forward.

Figure 28.3 Vertebrate Skeleton Diversity. A bird, a seal, and a fish have different body forms, yet their skeletons are composed of similar bones.

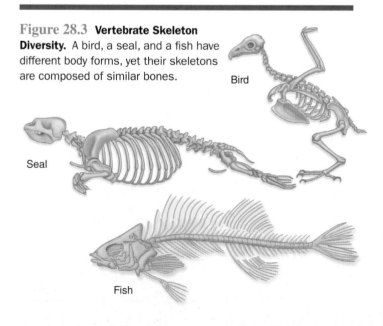

Bird

Seal

Fish

28.2 | The Vertebrate Skeleton Features a Central Backbone

Bones, the organs that compose the vertebrate skeleton, are grouped into two categories (**figure 28.4**). The **axial skeleton,** so named because it is located in the longitudinal central axis of the body, consists of the bones of the head, vertebral column, and rib cage. The **appendicular skeleton** consists of the appendages (limbs) and the bones that support them.

The axial skeleton shields soft body parts. The skull, which protects the brain and many of the sense organs, consists of hard, dense bones that fit together like puzzle pieces. All of the head bones are attached with immovable joints, except for the lower jaw and the middle ear. These movable jaw and ear bones enable us to chew food, speak, and hear.

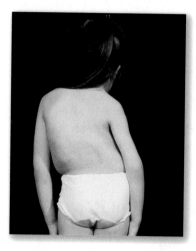

Figure 28.5 Scoliosis. This young girl's spine curves to the side. Girls are more often affected with scoliosis than boys.

The **vertebral column** supports and protects the spinal cord. A human vertebral column consists of 33 vertebrae, separated by cartilage disks that cushion shocks and enhance flexibility. A "slipped," or herniated, disk occurs when these pads tear or rupture, causing a bulge that presses painfully on a nearby nerve. Scoliosis, in which the vertebral column curves to the side, is also a disorder of the axial skeleton (**figure 28.5**).

Attached to the human vertebral column are 12 pairs of ribs, which protect the heart and lungs. Flexible cartilage between the ribs and other bones allows muscles to elevate the ribs, a movement important in breathing.

In the appendicular skeleton, the **pectoral girdle** connects the forelimbs to the axial skeleton; it includes the clavicles (collarbones) and scapulae (shoulder blades). Likewise, the **pelvic girdle** attaches the hind limb bones to the axial skeleton. The hipbones join the sacrum in the rear and meet each other in front, creating a bowl-like pelvic cavity. (The term *pelvis* is Latin for "basin.") The bony pelvis protects the lower digestive organs, the bladder, and some reproductive structures, especially in the female.

This chapter's Apply It Now box illustrates how skeletal features reveal clues that are useful to people in several professions.

28.2 | Mastering Concepts

1. What are the two subdivisions of the human skeleton?
2. What are the locations of the pectoral and pelvic girdles?

Figure 28.4 The Human Skeleton. The axial skeleton in humans includes the bones of the head, vertebral column, and rib cage. The bones that compose and support the limbs constitute the appendicular skeleton.

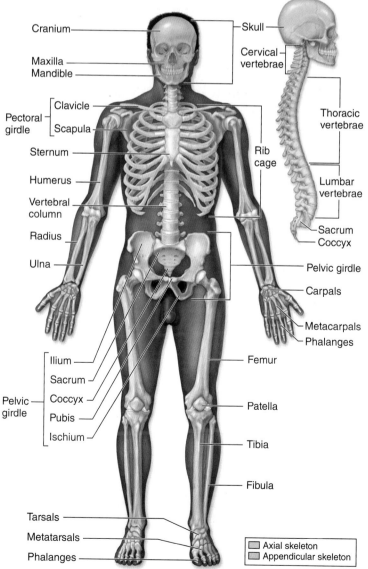

Cranium
Maxilla
Mandible
Clavicle
Pectoral girdle — Scapula
Sternum
Humerus
Vertebral column
Radius
Ulna
Ilium
Sacrum
Pelvic girdle — Coccyx
Pubis
Ischium
Tarsals
Metatarsals
Phalanges

Skull
Cervical vertebrae
Thoracic vertebrae
Rib cage
Lumbar vertebrae
Sacrum
Coccyx
Pelvic girdle
Carpals
Metacarpals
Phalanges
Femur
Patella
Tibia
Fibula

☐ Axial skeleton
☐ Appendicular skeleton

Skeletal System		
Main Tissue Types*	**Examples of Locations/Functions**
Connective	Makes up bone, cartilage, tendons, ligaments, marrow of vertebrate skeleton
Muscle	Skeletal muscle connects to movable bones, enabling voluntary movements
Nervous	Senses body position and controls muscles

*See chapter 24 for descriptions.

28.3 | Bones Provide Support, Protect Internal Organs, and Supply Calcium

 The skeleton not only supports and protects the body, but it also has several other functions that may at first glance seem unrelated (table 28.1). Bones connected to muscles provide movement, and bone minerals supply calcium and phosphorus to the rest of the body. Blood cells also form at the marrow inside bones.

A. Bones Consists Mostly of Bone Tissue and Cartilage

A glance back at figure 28.4 reveals that bones take many shapes. Long bones make up the arms and legs, whereas the wrists and ankles consist mainly of short bones. Flat bones include the ribs and skull. Vertebrae are irregularly shaped.

No matter what their shape, bones are lightweight and strong because they are porous, not solid (figure 28.6). The weight of long bones is further reduced by the **marrow cavity,** a space in the shaft that contains the soft, spongy, red or yellow marrow. **Red bone marrow** is a nursery for blood cells and platelets; in adults, **yellow bone marrow** replaces the red marrow in the limb bones. The fatty yellow marrow does not produce blood, but if the body faces a severe shortage of blood cells, yellow marrow can revert to red marrow.

Besides marrow, bones also contain nerves and blood vessels. But the majority of the vertebrate skeleton consists of two types of connective tissue: bone and cartilage. Figure 28.6 offers a closer look at both of these tissues. ▸ connective tissue, p. 515

Table 28.1 Functions of the Vertebrate Endoskeleton: A Summary

Function	Explanation
Support	The skeleton is a framework that supports an animal's body against gravity. It largely determines the body's shape.
Movement	The vertebrate skeleton is a system of muscle-operated levers. Typically, the two ends of a skeletal (voluntary) muscle attach to different bones that connect in a structure called a joint. When the muscle contracts, one bone is pulled toward the other.
Protection of internal structures	The backbone surrounds and shields the spinal cord, the skull protects the brain, and ribs protect the heart and lungs.
Production of blood cells	Many bones, such as the long bones of the human arm and leg, contain and protect red marrow, a tissue that produces red blood cells, white blood cells, and platelets.
Mineral storage	The skeleton stores calcium and phosphorus.

Bone tissue consists of cells suspended in a hard extracellular matrix. **Osteoblasts** are the bone-forming cells that secrete the matrix; an **osteocyte** is a former osteoblast embedded in the matrix it has produced. Each osteocyte inhabits a small space joined to others by narrow passageways. The matrix itself consists mainly of collagen and minerals. Collagen is a protein that

Apply It Now

Bony Evidence of Murder, Illness, and Evolution

Skeletons sometimes provide useful clues to past events. Hard, mineral-rich bones and teeth remain intact long after a corpse's soft body parts decay. These durable remains can help solve crimes, lend insight into human history, and shed light on evolution.

Detectives can use bones to identify the sex of a decomposed murder victim. This technique relies on the differences between male and female skeletons. Most obviously, the average male is larger than the average female. In addition, the front of the female pelvis is broader and larger than the male's, and it has a wider bottom opening that accommodates the birth of a baby. These same features allow anthropologists to determine the sex of ancient human fossils.

Bones can also reveal events and illnesses unique to each person's life. Healed breaks may

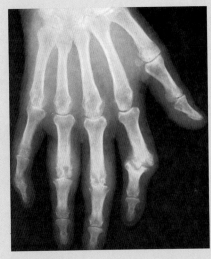

indicate accidents or abuse. Egypt's King Tut, for example, suffered a severe leg break shortly before he died. Crooked joints, such as those in the hand shown in the photo, may be evidence of arthritis, and patterns of bone thickenings tell whether a person spent his or her life in hard physical labor.

The shapes and sizes of fossilized bones also reveal some of the details of human evolution. Section 14.4 explains how the skeletons and teeth of primate fossils provide clues to brain size, diet, and posture in our ancestors. Animal skeletons also tell the larger story of vertebrate evolution. For example, paleontologists can examine skeletal features to determine whether an extinct animal was terrestrial or aquatic. Air is much less supportive than water, so land-dwellers tend to have sturdier skeletons than their aquatic relatives.

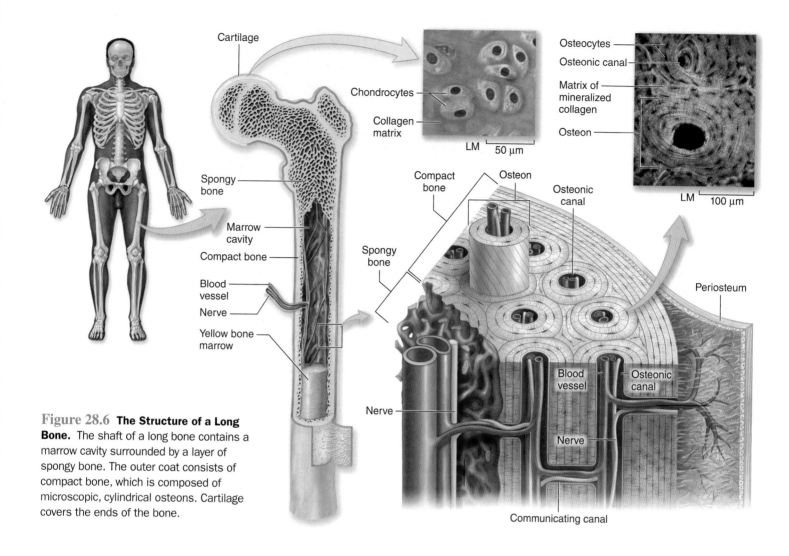

Figure 28.6 The Structure of a Long Bone. The shaft of a long bone contains a marrow cavity surrounded by a layer of spongy bone. The outer coat consists of compact bone, which is composed of microscopic, cylindrical osteons. Cartilage covers the ends of the bone.

gives the bone flexibility, elasticity, and strength. The hardness and rigidity of bone comes from the minerals, primarily calcium and phosphate, that coat the collagen fibers.

A third cell type associated with bones is the **osteoclast,** a cell that degrades the matrix at the bone surface, releasing calcium and phosphorus into the blood. Osteoclasts and osteoblasts therefore have opposing actions.

In **compact bone,** which is hard and dense, osteocytes and the surrounding matrix are organized into functional units called osteons (see figure 28.6). Each **osteon** consists of a set of concentric rings of osteocytes surrounding a central osteonic canal, which contains nervous tissue and a blood supply. Nutrients and oxygen pass from the blood, and from osteocyte to osteocyte, through a "bucket brigade" of up to 15 cells. Other canals connect the entire labyrinth to the outer surface of the bone and to the marrow cavity.

Spongy bone is much lighter than compact bone, thanks to a web of hard, bony struts that enclose large, marrow-filled spaces. Unlike compact bone, spongy bone tissue does not consist of osteons. Instead, the osteocytes secrete layers of matrix as rods

and plates that develop along a bone's lines of stress. Each cell acquires nutrients and oxygen directly from the nearby bone marrow.

Bones include both compact and spongy bone tissue. In most flat bones, spongy bone is sandwiched between two layers of compact bone. Long bones consist mostly of compact bone, although their bulbous tips also contain spongy bone inside. Likewise, in short bones, compact bone covers the spongy bone.

Cartilage is the other main connective tissue in the skeleton (see figure 28.6). This rubbery material covers the ends of bones. Cells called chondrocytes secrete the extracellular matrix of collagen and another protein, elastin. Strong networks of collagen fibers resist breakage and stretching, even when bearing great weight. Elastin provides flexibility.

The protein network in cartilage holds a great deal of water, making cartilage an excellent shock absorber. But cartilage lacks a blood supply. As the body moves, water within cartilage cleanses the tissue and bathes it with dissolved nutrients from nearby blood vessels. Nevertheless, the absence of a dedicated blood supply means that injured cartilage is slow to heal.

B. Bones Are Constantly Built and Degraded

The bones of a developing embryo originate as cartilage "models" (figure 28.7). As the fetus grows, each model's matrix hardens with calcium salts, cutting off diffusion to the cartilage cells. The cartilage matrix therefore degenerates, and capillaries penetrate the degenerating areas. Osteoblasts migrate in via these new blood vessels and secrete bone matrix. These osteoblasts eventually become mature osteocytes.

After birth, bone growth becomes concentrated near the ends of the long bones in thin disks of cartilage ("growth plates"). The bones continue to elongate until the late teens, when bone tissue begins to replace the cartilage growth plates. By the early twenties, bone growth is complete.

Even after a person stops growing, bone is continually being remodeled. Bones become thicker and stronger with strenuous exercise such as weight lifting. On the other hand, less-used bones lose mass, thanks to osteoclasts that slowly dissolve the minerals. For example, astronauts lose bone density if they are in a prolonged weightless environment because their bodies don't have to work as hard as they do against Earth's gravity.

Moreover, broken bones can repair themselves (figure 28.8). Osteoblasts near the site of the fracture produce new bone, so that after several weeks, the injury is all but healed.

C. Bones Help Regulate Calcium Homeostasis

Throughout life, bones are a reservoir for calcium. This mineral is vital for muscle contraction, the transmission of neural impulses, blood clotting, the activity of some enzymes, cell adhesion, and cell membrane permeability. The body constantly shuttles calcium between blood and bone. Hormones from the parathyroid and thyroid glands control this exchange in a negative feedback loop (figure 28.9). ▶▶ negative feedback loop, p. 520; parathyroid glands, p. 575

Bones sometimes lose more calcium than they add, leading to **osteoporosis,** a condition in which bones become less dense

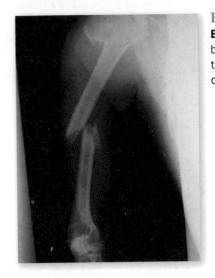

Figure 28.8 **Broken Bone.** A badly broken arm bone can heal, thanks to the bone-building action of osteoblasts.

(figure 28.10). An astronaut's "disuse osteoporosis," mentioned in section 28.3B, is one example. Much more familiar, however, is the age-related osteoporosis that causes shrinking stature, back pain, and frequent fractures in the elderly.

Both men and women can suffer from osteoporosis, but the disorder is most common in females. The bone mass of the average woman is about 30% less than a man's to begin with, and women live longer than men. In addition, bone cells have receptors for estrogen; a postmenopausal decline in the level of this hormone somehow makes a woman more prone to osteoporosis. To prevent bone loss, doctors therefore advise all women to take 1000 to 1500 milligrams of calcium daily and to exercise regularly. Several drugs can increase bone density or slow its loss.

D. Bone Meets Bone at a Joint

A **joint** is an area where two bones meet. The most common and familiar joints are the freely movable synovial joints, such as those of the knees, hips, elbows, fingers, and toes (figure 28.11). A **synovial joint** consists of movable bones joined by a fluid-filled capsule of fibrous connective tissue. The synovial membrane lining

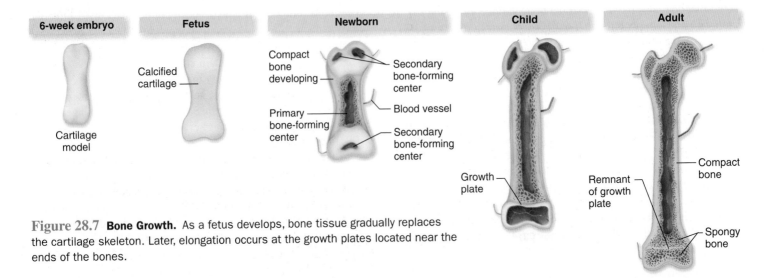

Figure 28.7 **Bone Growth.** As a fetus develops, bone tissue gradually replaces the cartilage skeleton. Later, elongation occurs at the growth plates located near the ends of the bones.

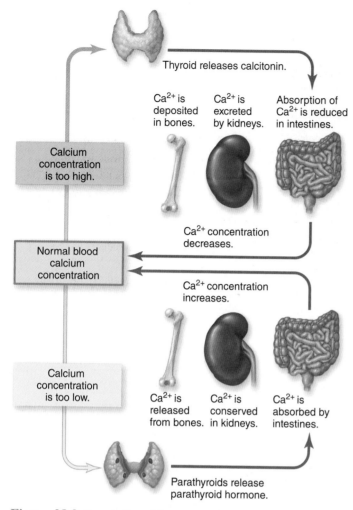

Thyroid releases calcitonin.

Ca^{2+} is deposited in bones.

Ca^{2+} is excreted by kidneys.

Absorption of Ca^{2+} is reduced in intestines.

Calcium concentration is too high.

Ca^{2+} concentration decreases.

Normal blood calcium concentration

Ca^{2+} concentration increases.

Calcium concentration is too low.

Ca^{2+} is released from bones.

Ca^{2+} is conserved in kidneys.

Ca^{2+} is absorbed by intestines.

Parathyroids release parathyroid hormone.

Figure 28.9 **Regulation of Blood Calcium Level.** Bones play a critical role in maintaining calcium homeostasis, absorbing excess calcium if the blood concentration is too high and releasing calcium into blood when the concentration is too low.

the capsule secretes a lubricating synovial fluid inside the capsule. Together, the synovial fluid and slippery cartilage allow bones to move against each other in a nearly friction-free environment.

Tendons and ligaments help stabilize synovial joints. **Tendons** are tough bands of connective tissue that attach bone to

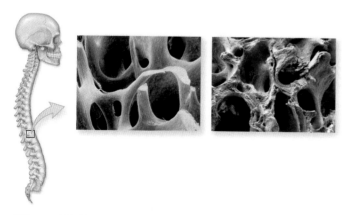

Figure 28.10 **Osteoporosis.** In osteoporosis, calcium loss weakens bones. The bone on the left is normal; osteoporosis has leached away part of the bone on the right.

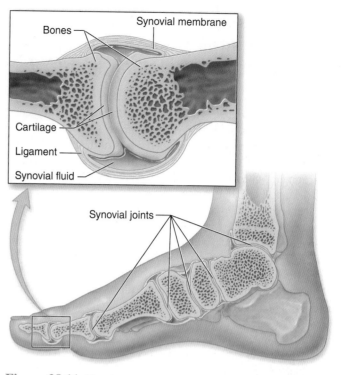

Synovial membrane

Bones

Cartilage

Ligament

Synovial fluid

Synovial joints

Figure 28.11 **The Synovial Joint: Where Bone Meets Bone.** A fluid-filled capsule of fibrous connective tissue surrounds a synovial joint. This illustration shows the synovial joints in the foot and ankle. The enlargement shows bones of the big toe.

muscle; **ligaments** are similar structures that attach bone to bone. A strain is an injury to a muscle or tendon, whereas a sprain is a stretched or torn ligament. A torn anterior cruciate ligament (ACL) is a common type of knee sprain, especially in sports such as basketball and volleyball. The ACL is one of two ligaments that criss-cross at the knee, connecting the thighbone to the shinbone. Surgical reconstruction of the ACL enables many injured athletes to return to their sports.

Arthritis is a common disorder of joints. A very severe form, rheumatoid arthritis, is an inflammation of the synovial membrane, usually at the small joints of the hands and feet. In the more common osteoarthritis, joint cartilage wears away. As the bone is exposed, small bumps of new bone begin to form. Osteoarthritis usually reveals itself as stiffness and soreness after age 40.

28.3 | Mastering Concepts

1. Describe the organization and functions of bone tissue and cartilage.
2. List the names and functions of each of the three cell types in bone tissue.
3. What are osteons, and in what type of bone tissue are they most concentrated?
4. How do osteoclasts and osteoblasts remodel and repair bones throughout life?
5. How do bones participate in calcium homeostasis?
6. What structures form a synovial joint?

28.4 | Muscle Movement Requires Contractile Proteins, Calcium, and ATP

The human muscular system includes more than 600 **skeletal muscles,** which generate voluntary movements. (This number does not include smooth muscle and cardiac muscle, which are not typically considered part of the muscular system). **Figure 28.12** identifies a few of the major skeletal muscles in a human, and **table 28.2** lists some functions of skeletal muscles.

Pairs of muscles often work together to generate body movements. **Figure 28.13,** for example, shows the biceps and triceps muscles. Tendons attach both of these muscles to the bones of the shoulder and lower arm. Each contracting muscle can pull a bone in one direction but cannot push the bone the opposite way. The elbow can bend and straighten because the biceps and triceps operate in opposite directions. When you contract the biceps (the bulge that appears when you "make a muscle"), the arm bends at the elbow joint. Contraction of the triceps muscle extends the arm. Many other skeletal muscles occur in similar antagonistic pairs that permit back-and-forth movements.

A. Actin and Myosin Filaments Fill Muscle Cells

Picture a softball player swinging a bat, an action that requires the contraction of many skeletal muscles. Each muscle moves a different body part, yet all are organized in essentially the same way. **Figure 28.14** illustrates the several levels of muscle anatomy, zooming from the whole organ to the microscopic scale.

The left side of figure 28.14 shows a whole muscle, an organ that consists of multiple tissue types. Connective tissue, for example, makes up tendons and the sheath that wraps around each muscle. Blood vessels within the muscle deliver nutrients and oxygen while removing wastes. Nerves transmit information to and from the central nervous system (see chapter 25).

The bulk of the organ, however, consists of skeletal muscle tissue. The cross sections in the center of figure 28.14 show that

Figure 28.12 The Human Muscular System. The human body has more than 600 skeletal muscles, a few of which are identified here.

Muscular System		
Main Tissue Types*	**Examples of Locations/Functions**	
	Connective	Makes up bone, cartilage, marrow of vertebrate skeleton, and tendons that attach muscles to bone; surrounds muscle cells, bundles, and whole muscles
	Muscle	Connects to bones and soft tissue, enabling movement of body parts
	Nervous	Senses body position and controls muscles

*See chapter 24 for descriptions.

Table 28.2 Functions of Skeletal Muscles: A Summary

Function	Explanation
Voluntary movement	Muscles attached to bones build lever systems under voluntary control.
Control of body openings	Skeletal muscles provide voluntary control of the eyelids, mouth, and anus.
Maintain posture	Muscles attached to bones keep the body upright.
Communication	Skeletal muscle movements enable facial expressions, speech, writing, and gesturing.
Maintain body temperature	Metabolic activity in skeletal muscle generates abundant heat.

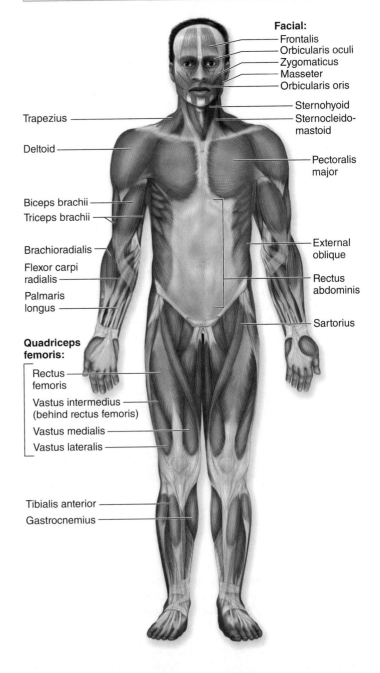

Facial:
- Frontalis
- Orbicularis oculi
- Zygomaticus
- Masseter
- Orbicularis oris
- Sternohyoid
- Sternocleidomastoid

Trapezius

Deltoid

Pectoralis major

Biceps brachii
Triceps brachii

Brachioradialis

Flexor carpi radialis

Palmaris longus

External oblique

Rectus abdominis

Sartorius

Quadriceps femoris:
- Rectus femoris
- Vastus intermedius (behind rectus femoris)
- Vastus medialis
- Vastus lateralis

Tibialis anterior
Gastrocnemius

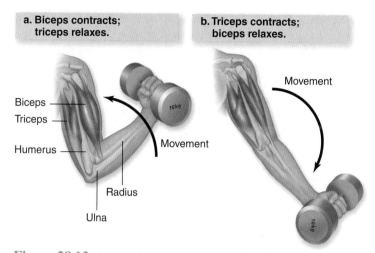

a. Biceps contracts; triceps relaxes.

Biceps
Triceps
Humerus
Radius
Ulna
Movement

b. Triceps contracts; biceps relaxes.

Movement

Figure 28.13 **Antagonistic Muscle Pair.** The biceps and triceps muscles work together to move the lower arm in opposite directions. (a) Contracting the biceps bends the elbow. (b) When the triceps contracts, the arm straightens.

muscle tissue is composed of parallel bundles of **muscle fibers,** which are individual muscle cells. The right half of the figure focuses on one muscle cell. Each cell contains cytoplasm, multiple nuclei, and other organelles. But most of the muscle fiber's volume is occupied by hundreds of thousands of cylindrical **myofibrils,** bundles of parallel protein filaments that run the length of the cell.

The inset at the top of figure 28.14 depicts the proteins that compose each myofibril. A **thick filament** is composed of a protein called **myosin.** A **thin filament** is composed primarily of two entwined strands of another protein, **actin,** along with two regulatory proteins called troponin and tropomyosin. In a resting muscle, troponin holds tropomyosin in a groove between the actin strands. This arrangement keeps the muscle relaxed until the nervous system delivers the signal to contract.

B. Sliding Filaments Are the Basis of Muscle Fiber Contraction

Skeletal muscle tissue appears striped, or striated, because of the alternating arrangement of thick and thin filaments (figure 28.15a). These striations divide each myofibril into many functional units, called **sarcomeres.**

A closer look at each sarcomere reveals characteristic lines and bands (figure 28.15b). Z lines define the boundaries of each sarcomere; they are membranes to which thin (actin) filaments attach. A bands are areas where myosin is present; I bands contain only actin; and H zones contain only myosin.

According to the **sliding filament model,** a muscle fiber contracts when the thin filaments slide between the thick ones

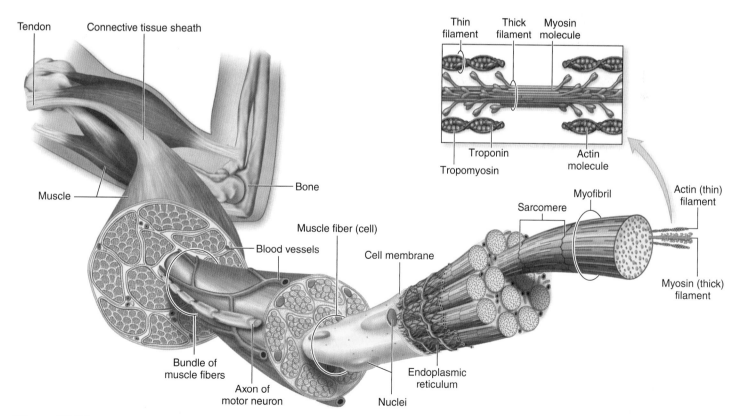

Tendon
Connective tissue sheath
Muscle
Bone
Blood vessels
Bundle of muscle fibers
Axon of motor neuron
Muscle fiber (cell)
Cell membrane
Nuclei
Endoplasmic reticulum
Sarcomere
Myofibril
Actin (thin) filament
Myosin (thick) filament

Thin filament
Thick filament
Myosin molecule
Troponin
Tropomyosin
Actin molecule

Figure 28.14 **Skeletal Muscle Organization.** A muscle is an organ enclosed in connective tissue, fed by blood vessels, and supplied with nerves. Tendons attach skeletal muscles to bones. Bundles of muscle fibers make up most of the muscle's volume; each muscle fiber is a single cell with many nuclei. Most of the cell's volume is occupied by myofibrils, which are, in turn, composed of filaments of the proteins actin and myosin.

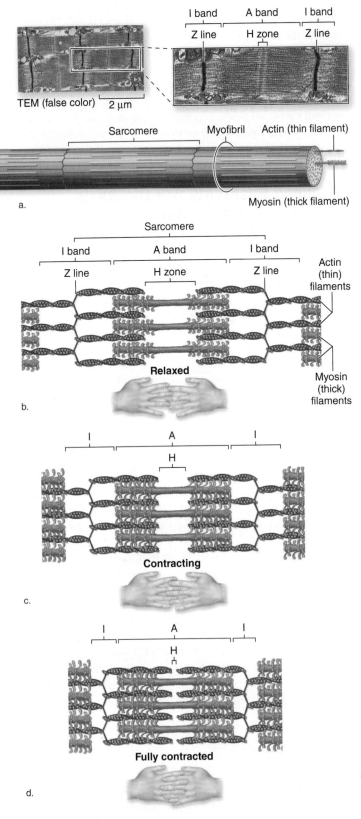

I band   A band   I band
Z line   H zone   Z line

TEM (false color)   2 μm

Sarcomere   Myofibril   Actin (thin filament)

Myosin (thick filament)

a.

Sarcomere

I band   A band   I band

Z line   H zone   Z line

Actin (thin) filaments

Myosin (thick) filaments

Relaxed

b.

I   A   I

H

Contracting

c.

I   A   I

H

Fully contracted

d.

Figure 28.15 The Sliding Filament Model of Muscle Contraction. (a) Myofibrils are divided into functional units called sarcomeres. (b–d) During muscle cell contraction, thick and thin filaments slide past one another, decreasing the length of each sarcomere.

(figure 28.15c and d). This motion shortens the sarcomere; that is, it brings the Z lines closer together while shortening both the I bands and H zones. The length of A bands remains unchanged. The overall effect is a little like fitting your fingers together to shorten the distance between your hands.

For thick and thin filaments to move past each other and contract a muscle fiber, actin and myosin must touch. The physical connection between the two types of filaments is the pivoting, club-shaped "head" portion of each myosin molecule (see figure 28.14). A myosin head forms a **cross bridge** when it swings out to contact an actin molecule. As described in section 28.4C, this action requires two additional ingredients: calcium and ATP.

C. Motor Neurons Stimulate Muscle Fiber Contraction

Muscles do not contract at random; rather, they wait for electrical stimulation from the nervous system. The nervous system delivers the signal to contract at a **neuromuscular junction,** a synapse between a motor neuron and a muscle cell (**figure 28.16**).

The moment a softball player decides to swing a bat, action potentials are conveyed along the motor neuron's axon. These signals stimulate the release of a **neurotransmitter,** acetylcholine, at the neuromuscular junction. When acetylcholine binds to receptors on the cell surface of the muscle fiber, an electrical wave races along the cell membrane. ▶▶ synapse, p. 536; action potential, p. 532

To understand how this electrical signal prompts contraction, look more closely inside the muscle cell in figure 28.16. The membrane of the muscle cell forms tunnel-like structures, called T tubules, that extend deep into the cell's interior. The T tubules touch the muscle cell's endoplasmic reticulum, which surrounds the myofibrils.

Electrical excitation of the muscle fiber's membrane spreads through the T tubules and causes the endoplasmic reticulum to release calcium ions (Ca^{2+}) into the cytoplasm. The Ca^{2+}, in turn, binds to the troponin proteins attached to the actin filaments (**figure 28.17**, step 1). This binding causes troponin to change its shape. As a result, the tropomyosin proteins move aside, allowing myosin cross bridges to bind to actin. The muscle is now free to contract.

Besides calcium ions, the sliding interaction between actin and myosin also requires energy in the form of ATP. A myosin head attaches to an exposed actin monomer on a thin filament (figure 28.17, step 2). The cross bridge bends, which pulls on actin and causes it to slide past myosin in the same way that an oar's motion moves a boat (step 3). The myosin head then binds a molecule of ATP, causing the cross bridge to release the actin (step 4). ATP splits into ADP and a phosphate group (step 5). The myosin head swivels back to its original position, ready to contact an actin monomer farther down the thin filament. ▶ ATP, p. 76

This sliding repeats about a hundred times per second on each of the hundreds of myosin molecules of a thick filament. Although each individual movement is minuscule, a skeletal muscle contracts quickly and forcefully due to the efforts of many thousands of "rowers."

Figure 28.16 **The Neuromuscular Junction.** A synapse between a motor neuron's axon and a muscle fiber is called a neuromuscular junction. Action potentials in the axon stimulate the release of acetylcholine, a neurotransmitter that triggers electrical changes in the muscle cell membrane. T tubules in the cell membrane conduct impulses to the interior of the muscle cell, leading to the release of calcium ions from the muscle fiber's endoplasmic reticulum. Figure 28.17 illustrates the role of calcium ions in muscle contraction.

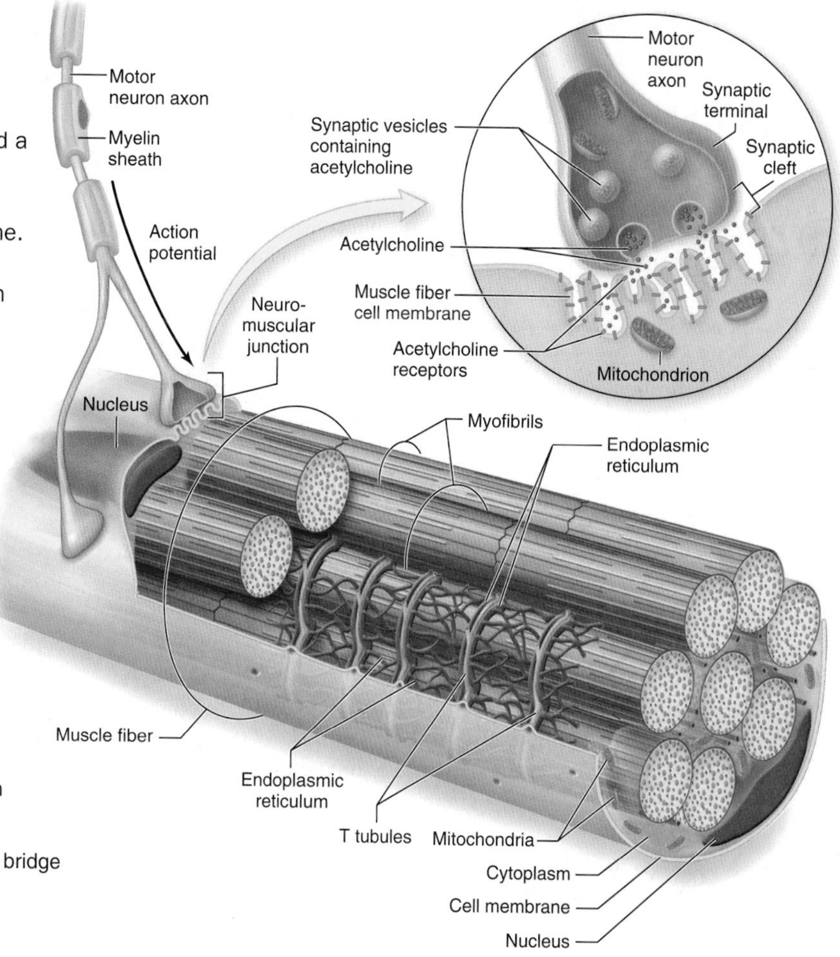

Motor neuron axon
Myelin sheath
Action potential
Neuro-muscular junction
Nucleus
Muscle fiber
Endoplasmic reticulum
T tubules
Mitochondria

Motor neuron axon
Synaptic terminal
Synaptic cleft
Synaptic vesicles containing acetylcholine
Acetylcholine
Muscle fiber cell membrane
Acetylcholine receptors
Mitochondrion
Myofibrils
Endoplasmic reticulum
Cytoplasm
Cell membrane
Nucleus

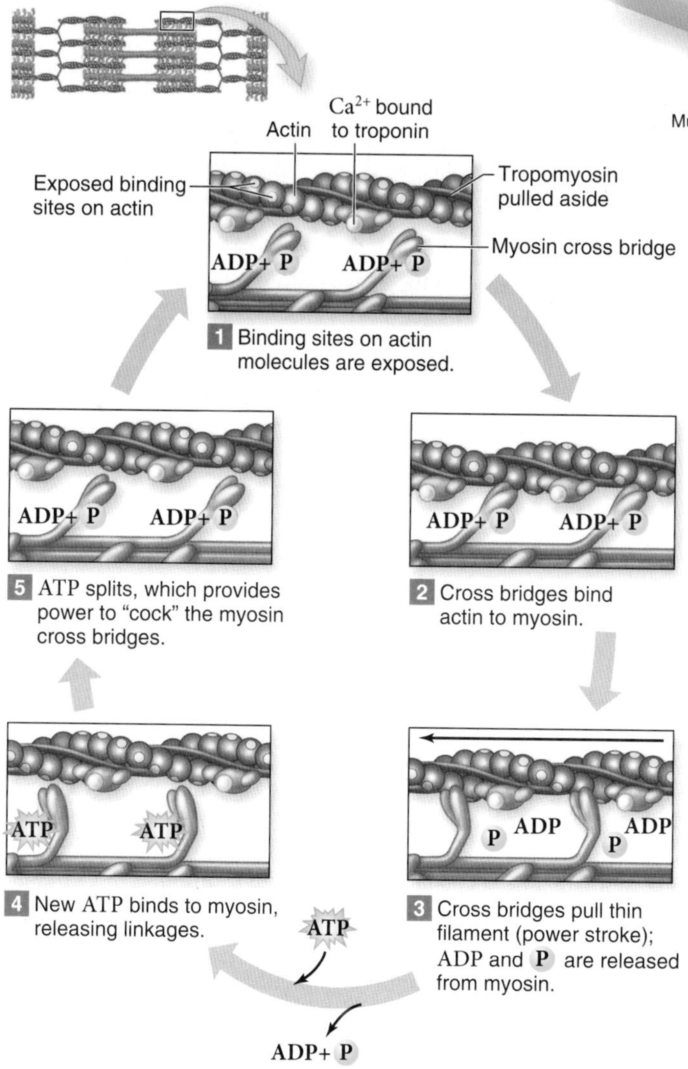

Ca²⁺ bound to troponin
Actin
Exposed binding sites on actin
Tropomyosin pulled aside
Myosin cross bridge
ADP+ P   ADP+ P

1 Binding sites on actin molecules are exposed.

ADP+ P   ADP+ P

5 ATP splits, which provides power to "cock" the myosin cross bridges.

ADP+ P   ADP+ P

2 Cross bridges bind actin to myosin.

ATP   ATP

4 New ATP binds to myosin, releasing linkages.

ATP

P ADP   P ADP

3 Cross bridges pull thin filament (power stroke); ADP and P are released from myosin.

ADP+ P

Figure 28.17 **The Roles of Calcium and ATP in Muscle Contraction.** Calcium ions change the interaction between troponin and tropomyosin in a way that allows myosin to bind to actin. ATP provides the energy required for myosin filaments to "ratchet" past actin filaments as the muscle contracts.

How does a muscle *stop* contracting? Shortly after the release of Ca²⁺, the ions are actively transported back to the endoplasmic reticulum. Troponin and tropomyosin return to their original positions. Even if ATP remains available, actin can no longer interact with myosin. The sarcomere relaxes until the next neural impulse brings a fresh influx of Ca²⁺. ▶ active transport, p. 82

Some diseases interfere with the neural signals that stimulate muscle contraction. The virus that causes polio, for example, causes paralysis by destroying the cell bodies of motor neurons. Another disease-causing organism is the bacterium *Clostridium botulinum,* which produces botulinum toxin (commonly known as Botox). This poison blocks the release of acetylcholine from motor neurons. Affected muscles never receive the signal to contract. Ingesting botulinum in tainted foods can cause paralysis, which can be fatal. Injecting tiny amounts of Botox into the face, however, temporarily helps reduce the appearance of wrinkles by paralyzing selected muscles.

28.4 | Mastering Concepts

1. What is an antagonistic pair of muscles?
2. Describe the levels of organization of a muscle.
3. Describe how sliding filaments shorten a sarcomere.
4. How do motor neurons, calcium ions, and ATP participate in muscle contraction?

28.5 | Muscle Fibers Generate ATP in Many Ways

Skeletal muscle contraction requires huge amounts of ATP, both to power the return of Ca^{2+} to the endoplasmic reticulum and to break the connection between actin and myosin, allowing new cross bridges to form.

Muscle cells have several ways to produce ATP (figure 28.18). Chapter 6 described aerobic respiration, which generates ATP in resting muscle cells. When muscle activity begins, a molecule called **creatine phosphate** rapidly replenishes ATP by donating a high-energy phosphate to ADP. (This chapter's Burning Question discusses the value of creatine phosphate as a dietary supplement.)

Less than a minute after intense exercise starts, the supply of creatine phosphate is depleted. At that point, aerobic respiration can continue to produce ATP for as long as the lungs and blood deliver sufficient oxygen. Otherwise, muscle cells switch to fermentation, a less efficient metabolic route that does not require O_2 but that rapidly leads to muscle fatigue. Lactic acid buildup from fermentation pathways causes muscle pain.

Intense exercise may lead to a period of **oxygen debt,** during which the body requires extra O_2, both to restore resting levels of

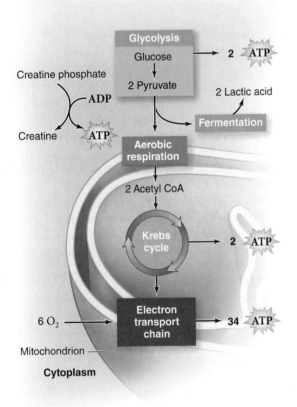

Figure 28.18 Energy Sources. An active muscle cell can use creatine phosphate to replenish its ATP supply for a short time. The cell can also generate ATP in aerobic respiration as long as O_2 is available. If the demand for O_2 exceeds the supply, the cell switches to fermentation, which is far less efficient but does not require O_2.

 ## Burning Question

Is creatine a useful dietary supplement?

You may have seen jars of creatine powder on nutrition store shelves, marketed as a muscle-building aid. But does this supplement really work?

Recall that creatine phosphate donates its P to ADP, quickly regenerating ATP soon after muscle activity starts. The idea behind creatine supplementation is that an increase in creatine levels should help skeletal muscle cells generate ATP, providing an energy boost during brief, intense bouts of exercise.

Although the idea seems logical, people differ in their response to creatine supplementation. For some, taking creatine powder does increase the amount of creatine phosphate inside skeletal muscle cells. (Vegetarians are most likely to see an effect, because an all-plant diet lacks creatine-rich meat and fish.) But not everyone sees improved athletic performance. Results vary from no effect to small but measurable gains in sprints and other short-term intense exercises.

As a note of caution, long-term creatine supplementation may be harmful. Much of the extra creatine ends up in urine, indicating stress on the kidneys. The possibility of kidney toxicity requires further study.

Submit your burning question to:
marielle_hoefnagels@mcgraw-hill.com

ATP and creatine phosphate and to recharge the proteins that carry oxygen in blood and muscle. Heavy breathing for several minutes after intense muscle activity is a sign of oxygen debt.

Shortly after death, muscles can no longer generate any ATP at all. One consequence is rigor mortis, the stiffening of muscles that occurs shortly after death. Without ATP, the cross bridges cannot release from actin. The muscles remain in a stiff position for the next couple of days, until the protein filaments begin to decay.

28.5 | Mastering Concepts

1. Describe the role of creatine phosphate in muscle metabolism.
2. What happens when a muscle cell cannot generate ATP by aerobic respiration?

28.6 | Many Muscle Fibers Combine to Form One Muscle

This chapter has so far described individual muscle fibers, each of which is either contracted or relaxed. But as figure 28.14 illustrates, muscle fibers form bundles that build whole muscles. The muscle, then, is an organ that may contract a lot, a little, or not at all. This section describes how the activities of individual muscle fibers contribute to the behavior of whole muscles.

A. Each Muscle May Contract with Variable Force

External electrical stimulation of an isolated muscle fiber produces a **twitch,** a single rapid cycle of contraction and relaxation. Each twitch yields one jerky movement. Yet in the body, the nervous system interacts with muscles to produce smooth, coordinated movements. Two integrated mechanisms maintain this fine control of whole-muscle movement.

The first mechanism operates at the level of the individual muscle cell. A muscle cell contracts whenever a motor neuron stimulates it, but Ca^{2+} is quickly cleared from the cytoplasm of the muscle fiber after each action potential. For a muscle cell to contract to its full range, repeated action potentials must sustain Ca^{2+} availability. **Tetanus** is the maximal contraction caused by continuous stimulation (this term is not to be confused with the tetanus infection caused by *Clostridium tetani*). With a high enough rate of action potentials, tetanus ensures smooth, prolonged contractions of individual muscle cells (**figure 28.19**).

The second mechanism that fine-tunes muscle movements involves groups of muscle fibers. Each motor neuron's axon branches extensively at its tip, with each branch leading to a different muscle fiber. A **motor unit** consists of one motor neuron and all the muscle fibers it contacts. When a neural impulse arrives, all of these fibers contract at the same time. A motor neuron that controls only a few muscle fibers produces fine, small-scale responses, such as the eye movements required for reading. A motor unit consisting of hundreds of muscle cells produces large, coarse movements, such as those required for throwing a ball.

Within one muscle, motor units vary in size from tens to hundreds of cells per motor neuron. The more motor units activated, the stronger the force of contraction. In this way, by recruiting different combinations of motor units, the same hand can both grip a hammer and pick up a tiny nail.

B. Muscles Contain Slow- and Fast-Twitch Fibers

Most skeletal muscles contain fibers of two main twitch types, distinguished by how quickly they contract and tire. **Slow-twitch fibers** are relatively small and produce twitches of relatively long

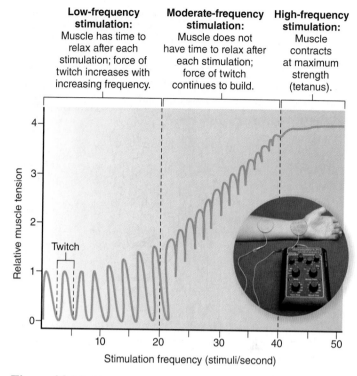

Low-frequency stimulation: Muscle has time to relax after each stimulation; force of twitch increases with increasing frequency.

Moderate-frequency stimulation: Muscle does not have time to relax after each stimulation; force of twitch continues to build.

High-frequency stimulation: Muscle contracts at maximum strength (tetanus).

Figure 28.19 Muscle Twitch and Tetanus. When a muscle is electrically stimulated at low frequency, it contracts and relaxes following each stimulus. Each such cycle is called a twitch. Under higher frequency stimulation, the muscle does not have time to relax between stimuli, so it contracts further. Continuous stimulation leads the muscle to sustain its maximum contraction, a condition called tetanus.

duration (up to 100 milliseconds long). Abundant capillaries bring in oxygen-rich blood; the O_2, in turn, supports the aerobic respiration that regenerates ATP in the fibers' plentiful mitochondria. High-endurance, slow-twitch muscles predominate in body parts that are active for extended periods, such as the flight muscles ("dark meat") of ducks and geese or the back muscles that maintain our upright posture.

Fast-twitch fibers, in contrast, are larger cells that split ATP quickly in short-duration twitches (as short as 7.5 milliseconds). Anaerobic pathways fuel the short bouts of rapid, powerful contraction that are characteristic of fast-twitch fibers. Muscles dominated by these fibers appear white because they have few capillaries and a lower content of myoglobin (the red pigment that stores oxygen inside muscle cells). The white breast muscle of a domesticated chicken, for example, can power barnyard flapping for a short time but cannot support sustained, long-distance flight.

The proportion of fast-twitch to slow-twitch fibers affects athletic performance. People with a high proportion of slow-twitch fibers excel at endurance sports, such as long-distance biking, running, and swimming. Athletes who have a higher proportion of fast-twitch fibers perform best at short, fast events, such as weight lifting, hurling the shot put, and sprinting (**figure 28.20**).

Figure 28.20
Fast-Twitch Fibers.
Sprinters and other athletes that can produce short bursts of strength tend to have muscles with a high proportion of fast-twitch fibers.

C. Exercise Strengthens Muscles

Regular exercise strengthens the muscular system. During the few months after a runner begins training, leg muscles noticeably enlarge. This increase in muscle mass, called hypertrophy, comes from the growth of individual skeletal muscle cells rather than from an increase in their number. Exercise-induced muscle growth is even more pronounced in a weight lifter, because the resistance of the weights greatly stresses the muscles. Anabolic steroids boost muscle growth by activating the genes encoding muscle proteins, but the potential side effects of illicit steroid use are serious. ▶ anabolic steroids, p. 578

A trained athlete's muscle fibers also use energy more efficiently than those of an inactive person. The athlete's cells contain more active and more numerous enzymes, and more mitochondria, so his or her muscles can withstand far more exertion before anaerobic metabolism begins. The athlete's muscles also receive more blood and store more glycogen than those of an untrained person.

Like bones, muscles can degenerate from lack of use, a condition called atrophy. After just two days of inactivity, mitochondrial enzyme activity drops in skeletal muscle cells. After a week without exercise, aerobic respiration efficiency falls by 50%. The number of small blood vessels surrounding muscle fibers declines, lowering the body's ability to deliver O_2 to the muscle. Lactic acid metabolism becomes less efficient, and glycogen reserves fall. After a few months of inactivity, the benefits of regular exercise all but disappear.

28.6 | Mastering Concepts

1. How can the same muscle generate both small and large movements?
2. How do slow- and fast-twitch muscle fibers differ?
3. How does exercise strengthen muscles?

28.7 | Investigating Life: Did a Myosin Gene Mutation Make Humans Brainier?

The old admonition not to "bite off more than you can chew" seems to apply especially well to humans. Our chewing muscles are considerably smaller than those of most primates, including chimpanzees and gorillas. We favor soft foods such as bread or cheese, and we would have a hard time chewing the bark, stems, and seeds that some of our primate relatives savor. Fossils show that our chewing muscles are even more delicate than those of the ancestral hominines *Australopithecus* and *Paranthropus*. These differences have spurred researchers to investigate the evolution of the muscles that connect the lower jaw to the other bones of the skull. ▶ human evolution, p. 312

Part of the evidence that has helped scientists understand the evolution of jaw muscles came from an unexpected source. It all began with the Human Genome Project, which has allowed researchers to comb through human DNA in search of particular genes. Hansell H. Stedman and associates from the University of Pennsylvania and the Children's Hospital of Philadelphia hoped to catalog every myosin gene in the human genome, in an effort to better understand diseases such as muscular dystrophy. During their search of chromosome 7, they stumbled on an inactive gene that encoded a nonfunctional myosin protein. ▶ Human Genome Project, p. 141

Section 28.4 explains myosin's role in muscle contraction. But myosin is not just one protein; it is a family of proteins encoded by at least 40 closely related genes. Myosins participate in cell division, the transport of organelles within a cell, and other cell processes that require movement. Mutations in many of the myosin-encoding genes can lead to loss of muscle function and other serious disorders.

At first, Stedman's team thought the inactive gene they found represented a quirk in the sequences they were searching. They therefore searched the genomes of six widely dispersed human populations originating in locations ranging from Africa to Iceland. All of the human groups had the mutated gene. The team also compared the gene to a homologous DNA sequence in seven species of nonhuman primates, including gorillas and chimpanzees. The results were clear: the human myosin gene contains a mutation that is not present in nonhuman primates (figure 28.21).

Next, the researchers wanted to know which cells in primates express the gene. Recall from chapter 7 that the first step in gene expression is transcription, in which a DNA sequence is transcribed to mRNA. A search for mRNA showed that, in both humans and macaque monkeys, only two muscles express the gene, and both participate in the up-and-down jaw movements required for chewing. Because the protein is nonfunctional in humans but functional in macaques, this finding suggested that an ancient genetic mutation made our chewing muscles small and weak, at least compared with our close relatives.

That result prompted Stedman's research team to learn more about the chewing apparatus in fossil primates. Most fossils consist

Non-human	10	20	30	40	50	60	70	80
Woolly monkey	GAGCAGCTGAACAAGCTGATGACCACCCTCCACAGCACTGTACCCCATTTTGTCCGCTGTATTGTGCCCAATGAGTTTAAGCAGTCAG
Pigtail macaque	GAGCAGCTGAACAAGCTGATGACCACCCTCCACAGCACTGTACCCCATTTTGTCCGCTGTATTGTGCCCAATGAGTTTAAGCAGTCAG
Rhesus	GAGCAGCTGAACAAGCTGATGACCACCCTCCACAGCACTGTACCCCATTTTGTCCGCTGTATTGTGCCCAATGAGTTTAAGCAGTCAG
Orangutan	GAGCAGCTGAACAAGCTGATGACCACCCTCCACAGCACTGTACCCCATTTTGTCCGCTGTATTGTGCCCAATGAGTTTAAGCAGTCAG
Gorilla	GAGCAGCTGAACAAGCTGATGACCACCCTCCACAGCACTGTACCCCATTTTGTCCGCTGTATTGTGCCCAATGAGTTTAAGCAGTCAG
Bonobo	GAGCAGCTGAACAAGCTGATGACCACCCTCCACAGCACTGTACCCCATTTTGTCCGCTGTATTGTGCCCAATGAGTTTAAGCAGTCAG
Chimpanzee	GAGCAGCTGAACAAGCTGATGACCACCCTCCACAGCACTGTACCCCATTTTGTCCGCTGTATTGTGCCCAATGAGTTTAAGCAGTCAG

Human								
Africa (pygmy)	GAGCAGCTGAACAAGCTGATGACCACCCTCCATAGC--CGCACCCCATTTTGTCCGCTGTATTATCCCCAATGAGTTTAAGCAATCGG
Spain (Basque)	GAGCAGCTGAACAAGCTGATGACCACCCTCCATAGC--CGCACCCCATTTTGTCCGCTGTATTATCCCCAATGAGTTTAAGCAATCGG
Iceland	GAGCAGCTGAACAAGCTGATGACCACCCTCCATAGC--CGCACCCCATTTTGTCCGCTGTATTATCCCCAATGAGTTTAAGCAATCGG
Japan	GAGCAGCTGAACAAGCTGATGACCACCCTCCATAGC--CGCACCCCATTTTGTCCGCTGTATTATCCCCAATGAGTTTAAGCAATCGG
Russia	GAGCAGCTGAACAAGCTGATGACCACCCTCCATAGC--CGCACCCCATTTTGTCCGCTGTATTATCCCCAATGAGTTTAAGCAATCGG
South America	GAGCAGCTGAACAAGCTGATGACCACCCTCCATAGC--CGCACCCCATTTTGTCCGCTGTATTATCCCCAATGAGTTTAAGCAATCGG

Figure 28.21 Myosin Mutation. The DNA sequences for a small portion of the myosin gene are shown for nonhuman primates (upper seven sequences) and humans (lower six sequences). Note the two-base deletion just after position 36 in the human gene. This mutation causes the encoded protein to be truncated (shortened) and nonfunctional.

of bones, which bear marks indicating where muscles were once attached. Until about 2 million years ago (MYA), primates had large skull bones and robust chewing muscles. These large complexes occurred in *Australopithecus* and *Paranthropus,* as well as in contemporary primates such as macaques and gorillas (figure 28.22). But a more delicate chewing apparatus appeared in early humans (*Homo erectus/ergaster*) between 1.8 and 2.0 MYA.

The researchers used a "molecular clock" technique to estimate that the mutation in the myosin gene occurred approximately 2.4 MYA—before the migration of *Homo* out of Africa. The timing coincides with a significant trend in human evolution: increasing brain size. Stedman and his colleagues argue that the myosin mutation may have changed the muscles in a way that released a

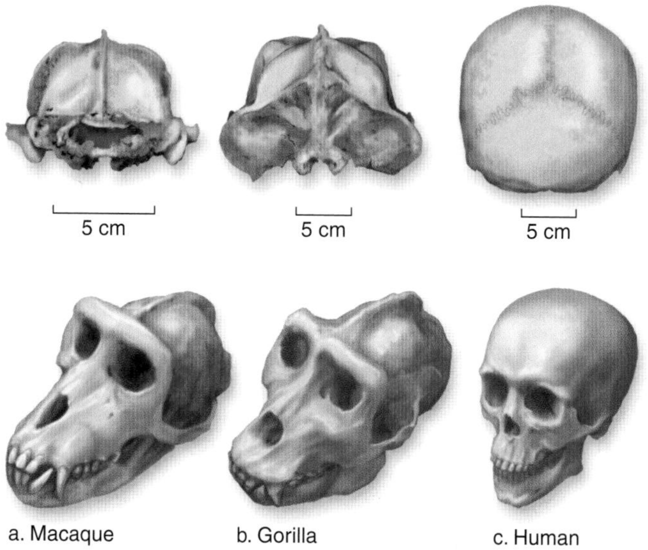

constraint on the size of the brain. Enhanced brain power may have eventually led to the spread of culture, profoundly changing the course of human evolution. ▶ molecular clock, p. 266

Of course, one mutation in a myosin gene is, by itself, not likely to have set in motion the entire course of human history. Other changes in the skeletal and nervous systems must also have occurred to spur the evolution of the brain. Researchers also continue to debate how the myosin gene mutation became "fixed" in the human population. Did it become more common as humans began eating softer foods, or did humans only begin eating softer foods once the mutation occurred?

The story of this myosin gene illustrates how a seemingly routine study can have thoroughly unexpected results. After all, what began as a simple search through the human genome ended up having much wider implications for the study of human evolution. This study also points to the connections among different areas of biology. To tell their story, Stedman and his colleagues needed to understand both the muscular and the skeletal systems. In the process, they combined detailed observations of ancient fossils with modern studies of gene mutation and protein function. Their results give us something to chew on as we contemplate the evolutionary history of our own species.

Stedman, Hansell, Benjamin W. Kozyak, Anthony Nelson, and 7 coauthors. 2004. Myosin gene mutation correlates with anatomical changes in the human lineage. *Nature,* vol. 428, pages 415–418.

5 cm   5 cm   5 cm

a. Macaque   b. Gorilla   c. Human

Figure 28.22 Primate Comparison. The temporalis muscle is one of the four main chewing muscles. (a) In macaques and (b) in gorillas, the temporalis attachment area *(red)* occupies most of the side of the skull. (c) In humans, the corresponding area is much smaller, resulting in a much weaker chewing apparatus.

28.7 | Mastering Concepts

1. Summarize the hypothesized relationship between the myosin gene mutation and brain size in humans and other primates.

2. Use the genetic code in chapter 7 to predict the amino acid sequence corresponding to the two gene fragments in figure 28.21. How does the deletion result in a truncated protein?

Chapter Summary

28.1 | Skeletons Take Many Forms

- The **muscular system** and the **skeletal system** together enable an animal to move.
- An animal's **skeleton** supports its body and protects soft tissues. **Muscles** act on the skeleton to provide motion.
- Animals have three main types of skeletons. A **hydrostatic skeleton** consists of tissue containing constrained fluid; an **exoskeleton** is on the organism's exterior; and an **endoskeleton** forms inside the body.

28.2 | The Vertebrate Skeleton Features a Central Backbone

- The **axial skeleton** consists of the bones of the head, **vertebral column,** and rib cage.
- The **appendicular skeleton** includes the limbs and the limb girdles (**pectoral** and **pelvic**) that attach them to the axial skeleton.

28.3 | Bones Provide Support, Protect Internal Organs, and Supply Calcium

- **Bones** are strong and lightweight because they are porous. Long bones have a **marrow cavity** that contains **red bone marrow** or **yellow bone marrow.**

A. Bones Consists Mostly of Bone Tissue and Cartilage

- Bone tissue derives its strength from collagen; bone hardness comes from minerals. Bone cells include **osteoblasts, osteoclasts,** and **osteocytes.** In **compact bone,** osteocytes are arranged in concentric rings that form **osteons.** **Spongy bone** has few osteons but many spaces separated by hard struts.
- Cartilage consists of chondrocytes embedded in a matrix of collagen and elastin. This tissue entraps a great deal of water, which makes it an excellent shock absorber.

B. Bones Are Constantly Built and Degraded

- Bone continually degenerates and builds up, based on the opposing activities of osteoclasts and osteoblasts.
- Weight-bearing exercise strengthens bones; conversely, bones weaken with disuse.

C. Bones Help Regulate Calcium Homeostasis

- Hormones control the exchange of calcium between blood and bones, maintaining homeostasis. **Osteoporosis** results when a person's bones lose more calcium than they replace.

D. Bone Meets Bone at a Joint

- **Joints** attach bones to each other. Some joints, such as those holding the skull bones in place, are immovable.
- Freely moving **synovial joints** consist of cartilage and a connective tissue capsule that contains lubricating fluid. **Ligaments** and **tendons** stabilize the joint.

28.4 | Muscle Movement Requires Contractile Proteins, Calcium, and ATP

- Many **skeletal muscles** form antagonistic pairs, which enable bones to move in two directions.

A. Actin and Myosin Filaments Fill Muscle Cells

- Skeletal muscle fibers are elongated, multinucleate, striated, and voluntary.
- Each skeletal **muscle fiber** is a long, cylindrical cell that contains **myofibrils** composed of two types of protein filaments. The **thick filaments** are **myosin,** and the **thin filaments** are composed primarily of **actin.**

B. Sliding Filaments Are the Basis of Muscle Fiber Contraction

- A myofibril is a chain of contractile units called **sarcomeres.** Z lines define the borders of each sarcomere. The orderly arrangement of thick and thin filaments within a sarcomere gives skeletal muscle tissue its striated appearance.
- According to the **sliding filament model,** muscle contraction occurs when thick and thin filaments move past one another. Myosin **cross bridges** provide "rowing" movements as they briefly contact actin filaments.

C. Motor Neurons Stimulate Muscle Fiber Contraction

- When a **motor neuron** stimulates a muscle fiber, the **neurotransmitter** acetylcholine is released at a **neuromuscular junction.** Electrical waves spread along the T tubules of the muscle cell membrane, causing the endoplasmic reticulum to release Ca^{2+} ions. Calcium binds to the regulatory protein troponin, which changes shape in a way that moves tropomyosin and allows actin to bind to myosin.
- Muscle contraction also requires ATP. When a myosin cross bridge touches actin, ATP attached to the myosin head splits. The head moves, causing the actin filament to slide past the myosin filament. A new ATP then binds to the myosin head, and the cross bridge separates from actin. The myosin head returns to its original position, and a new cross bridge forms farther along the filament.
- After the electrical impulse, Ca^{2+} is actively pumped back into the endoplasmic reticulum, and tropomyosin once again prevents actin and myosin interactions. The muscle relaxes.

28.5 | Muscle Fibers Generate ATP in Many Ways

- The energy that powers muscle contraction comes first from stored ATP, then from **creatine phosphate** stored in muscle cells, and then from aerobic respiration or fermentation, depending on O_2 availability.
- **Oxygen debt** is a temporary deficiency of O_2 after intense exercise.

28.6 | Many Muscle Fibers Combine to Form One Muscle

A. Each Muscle May Contract with Variable Force

- When a muscle cell is stimulated once, it contracts and relaxes (a **twitch**). At a high rate of stimulation, muscle cells reach a sustained state of maximal contraction, **tetanus.**
- A nerve cell and the muscle fibers it touches form a **motor unit.** The more motor units stimulated, the greater the contraction of the muscle.

B. Muscles Contain Slow- and Fast-Twitch Fibers

- **Slow-twitch fibers** use ATP slowly and regenerate it by aerobic respiration. **Fast-twitch fibers** use ATP quickly and use mostly anaerobic pathways to replenish it.
- People vary in their proportion of fast- and slow-twitch muscle fibers.

C. Exercise Strengthens Muscles

- A muscle exercised regularly increases in size (hypertrophy) because each muscle cell thickens. An unused muscle shrinks (atrophy). Regular exercise causes changes in muscle cells that enable them to use energy more efficiently.

28.7 | Investigating Life: Did a Myosin Gene Mutation Make Humans Brainier?

- Researchers have discovered in humans a mutated myosin gene that is expressed only in the muscles required for chewing. Nonhuman primates lack the mutation.
- The evolution of weaker, smaller chewing muscles may have paved the way for increased brain capacity in humans.

Multiple Choice Questions

1. A hydrostatic skeleton
   a. occurs in multiple types of invertebrates.
   b. is filled with fluid.
   c. changes shape due to muscle contraction.
   d. All of the above are correct.

2. Exoskeletons differ from endoskeletons in
   a. their chemical composition.
   b. their ability to grow along with an organism.
   c. their function as a framework for muscle attachment.
   d. Both a and b are correct.

3. The bones of your hand are part of the _____, whereas your backbone is part of the _____.
   a. axial skeleton; pectoral girdle
   b. appendicular skeleton; axial skeleton
   c. axial skeleton; appendicular skeleton
   d. axial skeleton; vertebral column

4. Which cell type is most likely to cause the bone loss associated with osteoporosis?
   a. Osteoblasts
   b. Osteocytes
   c. Osteoclasts
   d. Both a and b are correct.

5. Which type of joint connects movable bones?
   a. Fibrous joint
   b. Synovial joint
   c. Cartilaginous joint
   d. Both b and c are correct.

6. Muscle tissue is made up of what type of cells?
   a. Sarcomeres
   b. Myofibrils
   c. Actin and myosin filaments
   d. Muscle fibers

7. Which of the following is NOT part of the sliding filament model?
   a. Sarcomeres become shorter as muscle contracts.
   b. Z lines move closer together as muscle contracts.
   c. Actin and myosin filaments slide past one another as muscle contracts.
   d. Calcium ions are released from T tubules as muscle contracts.

8. What is the source of energy for muscle contraction?
   a. ATP
   b. Acetylcholine
   c. Creatine phosphate
   d. Calcium

9. Slow-twitch muscle cells have a high concentration of myoglobin, which
   a. provides oxygen to the mitochondria to promote ATP synthesis.
   b. helps promote the growth of the muscle cells in response to exertion.
   c. prevents tetanus.
   d. slows the rate of ATP hydrolysis by myosin.

10. Which of the following is NOT a value of exercise?
    a. Increased bone density
    b. Enhanced muscle cell metabolism
    c. Increase in the number of muscle cells
    d. Increase in the size of the muscle cells

Write It Out

1. Distinguish among a hydrostatic skeleton, an exoskeleton, and an endoskeleton. What are the advantages and disadvantages of each type of skeleton? Give an example of an animal with each type.

2. Explain the observation that animals with exoskeletons and endoskeletons are better represented in the fossil record than are animals with hydrostatic skeletons. How might this difference alter scientific interpretations of the fossil record?

3. What are the components of the axial and appendicular skeletons?

4. What role does cartilage play in the vertebrate skeletal system?

5. What are the differences between spongy bone and compact bone?

6. What are the two major components of bone matrix? How do they work together to give bone its characteristics?

7. Describe the events of bone development from embryo through adulthood.

8. Would an overactive parathyroid prevent osteoporosis or contribute to it? Explain your answer.

9. How do antagonistic muscle pairs move bones? Give an example of such a pair.

10. Describe four muscle proteins and their functions.

11. What is the role of tetanus when you lift a heavy box with your arms?

12. How do the effects of exercise (or lack thereof) illustrate homeostasis in bones and muscles?

13. How does the muscular system interact with the nervous system? The skeletal system? The respiratory system? The circulatory system?

14. What roles does fluid play in hydrostatic skeletons, cartilage, and synovial joints?

15. Search the Internet for disorders of the skeletal or muscular system. Choose one such illness to research in more detail. Describe how the disorder interferes with bone or muscle function. What causes the disorder? Is a treatment or cure available?

16. What is the role of calcium in bones? In muscle contraction?

17. The following table shows recent men's world-record times for various running events. Graph the distance traveled against the average running speed, in meters per second. How does the use of ATP by muscles over time explain the graph?

Distance (m)	Time	Average m/sec
100	9.78 sec	10.22
200	19.32 sec	10.35
400	43.18 sec	9.26
800	1 min, 41.11 sec	7.91
1500	3 min, 26.00 sec	7.28
5000	12 min, 37.37 sec	6.60

Pull It Together

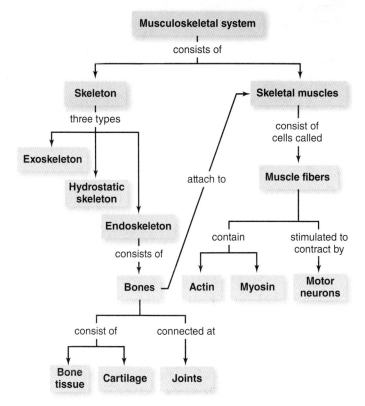

1. How do bones help maintain blood calcium concentrations?

2. How do actin and myosin interact in the sliding filament model of muscle contraction?

3. Add acetylcholine, calcium ions, ATP, and creatine phosphate to this concept map.

4. How do fast-twitch muscle fibers differ from slow-twitch fibers?

Chapter 29

The Circulatory System

Raw Material for Artificial Blood? A technician holds a bag of hemoglobin purified from human blood. The hemoglobin is being tested for use in a blood substitute.

connect |BIOLOGY

Enhance your study of
this chapter with practice quizzes,
animations and videos, answer keys,
and downloadable study tools.
www.mhhe.com/hoefnagels

Using Stem Cells to Replace Hearts and Blood

THE HEART IS THE CIRCULATORY SYSTEM'S ONLY PUMP; IF IT FAILS TO DO ITS JOB, DEATH CAN COME QUICKLY. Treatment options for heart failure currently include surgery, drugs, and lifestyle changes, but in the future, physicians may have a new tool: stem cells.

Scientists may one day be able to coax stem cells to divide and produce daughter cells that differentiate into exactly what is needed to heal damaged tissue. The burgeoning field of tissue engineering, described in chapter 24, has already yielded artificial bladders. No one has yet created a human heart using the same technology, but research looks promising. In one study, scientists "seeded" a mixture of cells onto a scaffold made from a rat's heart. The cells divided and formed a heart-shaped organ that could beat and pump blood.

A heart is useless without blood to transport. Blood transfusions have been common since the nineteenth century, but donors may transmit diseases to the recipients of the blood. Artificial blood, or a blood substitute, would eliminate this risk.

Blood substitutes used (unsuccessfully) in the past include wine, milk, urine, plant resins, and even opium. More recently, scientists have tried to find fluids that mimic the oxygen-carrying capacity of red blood cells, at least temporarily. Chemicals called perfluorocarbons, for example, carry dissolved O_2. In the 1960s, a famous photo showed a mouse swimming in the stuff, literally breathing in the oxygen-rich chemical.

Other blood substitutes are based on hemoglobin, the oxygen-toting protein in red blood cells. Hemoglobin is extracted, chemically stabilized, and mixed with a sterile salt solution to create artificial blood. A cow hemoglobin preparation saved the life of a young woman whose immune system was attacking her own blood, maintaining her circulation for several days until the illness subsided.

Red blood cell replacements may also come from stem cells. Bone marrow, for example, contains stem cells that give rise to all blood cells. A bone marrow transplant replaces a patient's own marrow with that of a healthy person. Such transplants have already saved the lives of people with sickle cell disease, an illness in which red blood cells are misshapen. The transplant procedure entails many risks, but if all goes well, stem cells in the donor's marrow churn out healthy cells in the recipient's body.

Someday, stem cells reared in the laboratory may give us a safe, limitless supply of human blood. But stem cell research is a young science, and researchers still have much to learn before the medical applications become widespread. Success will require a basic understanding of how the body works, including the circulatory system—the subject of this chapter.

UNIT 6

What's the Point?

Learning Outline

Learn How to Learn

Studying in Groups

Study groups can offer a great way to learn from other students, but they can also dissolve into social events that accomplish little real work. Of course, your choice of study partners makes a huge difference; try to pick people who are at least as serious as you are about learning. To stay focused, plan activities that are well-suited for groups. For example, you can agree on a list of vocabulary words and take turns adding them to a group concept map. You can also write exam questions for your study partners to answer, or you can simply explain the material to each other in your own words. Focus on what you need to learn, and your study sessions should be productive.

29.1 | Circulatory Systems Deliver Nutrients and Remove Wastes

Watch a crime drama on TV, and it won't be long until a gunshot wound leaves someone lying in a pool of blood. Unless help arrives immediately, life quickly fades—a vivid reminder of blood's importance (figure 29.1).

Blood is the fluid that the **circulatory system** transports throughout the body. This liquid carries many substances, among them the glucose and oxygen gas (O_2) required in aerobic cellular respiration. Without these resources, the body's cells could not generate the ATP required for life. Blood not only delivers these and other raw materials, but it also carries off wastes such as carbon dioxide (CO_2). Besides blood, the circulatory system includes a system of vessels that contain the blood (or a comparable fluid). The **heart** is the central pump that keeps the blood moving through these vessels.

The circulatory system has extensive connections with organ systems that exchange materials with the environment. For example, blood vessels acquire O_2 and unload CO_2 at gills, lungs, or other organs of the respiratory system (see chapter 30). Nutrients enter the circulatory system at blood vessels near the intestines, which form part of the digestive system (see chapter 31). Blood also circulates through the kidneys, which eliminate many water-soluble metabolic wastes (see chapter 32).

A. Circulatory Systems Are Open or Closed

Some types of animals lack a dedicated circulatory system. Flatworms, for example, use their incomplete digestive tracts not only to absorb nutrients but also for gas exchange. ▶ flatworms, p. 418

Most animals, however, do have a circulatory system, which may be open or closed (figure 29.2). In an **open circulatory system,** a heart pumps fluid through short, open-ended vessels. These vessels lead to open spaces in the body cavity, where the fluid can exchange materials with the body's cells. The fluid then enters other vessels leading back to the heart. Animals with open circulatory systems include most mollusks and arthropods. ▶▶ mollusks, p. 420; arthropods, p. 426

In a **closed circulatory system,** blood remains within vessels that exchange materials with the fluid surrounding the body's tissues. Examples of animals with closed circulatory systems include vertebrates, annelids, and cephalopod mollusks such as squids and octopuses. ▶ annelids, p. 422

Both types of circulatory systems have advantages. Open circulatory systems require fewer vessels, and the blood moves under low pressure. The energetic costs of circulation are therefore relatively low. On the other hand, blood flows at higher pressure in a closed circulatory system, so nutrient delivery and waste removal can occur more rapidly. Moreover, the vessels of

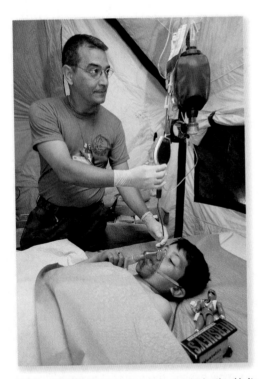

Figure 29.1 Vital Fluid. Millions of people in the United States receive life-saving blood transfusions each year as a result of injury, surgery, or cancer treatment.

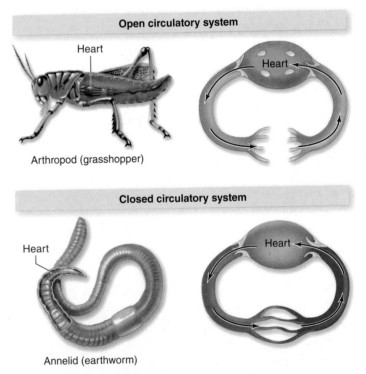

Figure 29.2 Open and Closed Circulatory Systems. In an open circulatory system, fluid leaves vessels and bathes cells directly. In a closed circulatory system, blood is confined within vessels.

a closed system can direct blood flow toward or away from specific areas of the body, depending on metabolic demands. Closed circulatory systems therefore tend to be more efficient than open systems.

B. Vertebrate Circulatory Systems Have Become Increasingly Complex

As vertebrates adapted to life on land, their circulatory systems evolved. One prominent trend was that blood flowing to the organs of gas exchange (gills or lungs) became increasingly separated from blood flowing to the rest of the body (figure 29.3).

Among the vertebrates, fishes and tadpoles have the simplest circulatory systems (see figure 29.3a). A fish's heart has just two chambers: an **atrium** where blood enters, and a **ventricle** from which blood exits. The heart pumps blood through the gills to pick up O_2 and unload CO_2. The blood then circulates to the rest of the body before returning to the heart.

Other vertebrates divide the circulatory system into two interrelated circuits. In the **pulmonary circulation,** blood exchanges gases at the lungs and returns to the heart; in the **systemic circulation,** blood circulates throughout the rest of the body and back to the heart.

In these animals, the heart may have three or four chambers. The three-chambered heart of a frog, for example, has one undivided ventricle and two atria (see figure 29.3b). The left atrium receives

O_2-rich blood from the lungs, and the right atrium receives O_2-depleted blood from the rest of the body. Blood from the two atria mixes in the ventricle, which pumps the blood throughout the body. In turtles, snakes, and lizards, the ventricle is partially divided, an adaptation that increases the separation of oxygenated and deoxygenated blood.

Birds and mammals independently evolved four-chambered hearts with two atria and two ventricles, completing the separation of O_2-rich from O_2-depleted blood (see figure 29.3c). Birds and mammals are endotherms, meaning they spend enormous amounts of energy maintaining a constant body temperature (see chapter 32). Their cells therefore consume much higher quantities of nutrients and O_2 than do those of ectotherms, which have variable body temperatures. The four-chambered heart supports these high metabolic rates by maximizing the amount of O_2 reaching tissues. (Interestingly, the ectothermic crocodiles also have four-chambered hearts, reflecting their shared ancestry with birds.)

29.1 | Mastering Concepts

1. What are the components of a circulatory system?
2. Distinguish between open and closed circulatory systems.
3. Describe the circulatory systems of the major groups of vertebrates.

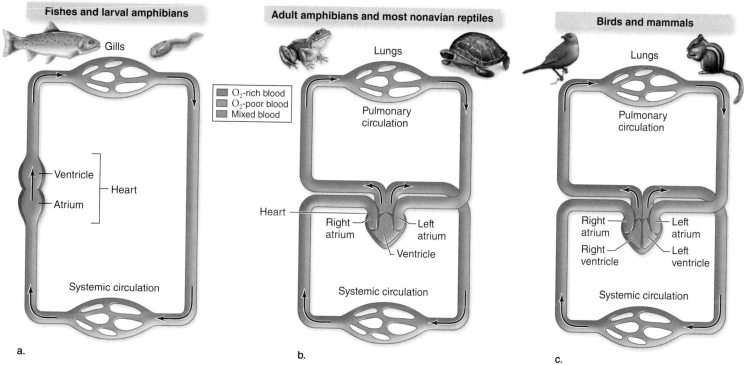

Figure 29.3 Vertebrate Circulatory Systems. (a) A fish's two-chambered heart pumps blood in a single circuit around the body. (b) The three-chambered heart of a frog or turtle has two atria and one ventricle. Blood from the pulmonary and systemic circuits may mix in the ventricle. (c) A bird or mammal has a four-chambered heart, maximizing the separation of the pulmonary and systemic circuits.

29.2 | Blood Is a Complex Mixture

The human circulatory system transports blood, a complex liquid tissue with many functions (table 29.1). Blood not only carries gases, nutrients, hormones, and wastes, but it also participates in immune reactions and helps maintain homeostasis in several ways.

Blood is a connective tissue that consists of several types of cells and cell fragments, all suspended in a liquid extracellular matrix called plasma (table 29.2 and figure 29.4). The cell types are diverse: a cubic millimeter of blood normally contains about 5 million red blood cells, 7000 white blood cells, and 250,000 cell fragments called platelets. This section describes the functions of each component of blood; this chapter's Burning Question box (on page 606) explains how donations of whole blood and plasma can save lives. ▶ connective tissue, p. 515

A. Plasma Carries Many Dissolved Substances

Plasma is the liquid matrix of blood. This fluid, which makes up more than half of the blood's volume, is 90% to 92% water.

Besides water, more than 70 types of dissolved proteins make up the largest component of plasma. These proteins have many functions. For example, antibodies participate in the body's immune response; high- and low-density lipoproteins transport cholesterol; and clotting factors such as fibrinogen help stop bleeding following an injury (see section 29.2D). ▶▶ antibodies, p. 678; cholesterol, p. 35

Table 29.1   Functions of Blood: A Summary

Function	Explanation
Gas exchange	Carries O_2 from lungs to tissues; carries CO_2 to the lungs to be exhaled
Nutrient transport	Carries nutrients absorbed by the digestive system throughout the body
Waste transport	Carries urea (a waste product of protein metabolism) to the kidneys for excretion in urine
Hormone transport	Carries hormones secreted by endocrine glands
Creation of interstitial fluid	Interstitial fluid that surrounds cells originates as blood plasma.
Maintain homeostasis	Regulates blood pH; regulates cells' water content; generates pressure gradient that keeps plasma in capillaries; absorbs heat and dissipates it at the body's surface
Protection	Blood clots plug damaged vessels; white blood cells destroy foreign particles and participate in inflammation.

Table 29.2   Components of Blood: A Summary

Component	Function(s)
Plasma	Liquid component of blood; exchanges water and many dissolved substances with interstitial fluid
Red blood cells (erythrocytes)	Carry O_2
White blood cells (leukocytes)	Destroy foreign substances, initiate inflammation
Platelets	Initiate clotting

About 1% of plasma consists of dissolved salts, hormones, metabolic wastes, CO_2, nutrients, and vitamins. The concentrations of these dissolved molecules are low, but they are critical. For example, blood usually contains about 0.1% glucose; if the concentration falls to 0.06%, the body begins to convulse.

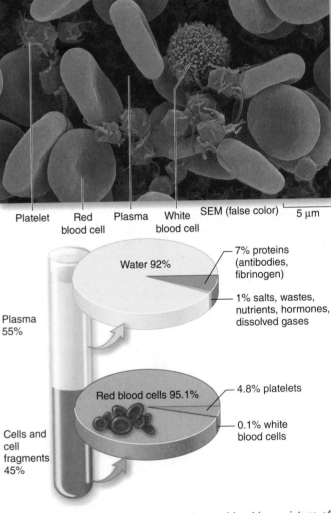

Platelet   Red blood cell   Plasma   White blood cell   SEM (false color)   5 μm

Plasma 55%

Water 92%

7% proteins (antibodies, fibrinogen)

1% salts, wastes, nutrients, hormones, dissolved gases

Cells and cell fragments 45%

Red blood cells 95.1%

4.8% platelets

0.1% white blood cells

Figure 29.4 Blood Composition. Human blood is a mixture of platelets, red blood cells, and white blood cells suspended in a liquid extracellular matrix called plasma.

B. Red Blood Cells Transport Oxygen

Mature **red blood cells,** or erythrocytes, are saucer-shaped disks packed with the pigment **hemoglobin,** a protein that carries O_2. Tucked into each hemoglobin molecule are four iron atoms, each of which can combine with one O_2 molecule picked up in the lungs. Oxygen-saturated hemoglobin is bright red; without bound O_2, hemoglobin has a deeper red color.

Red blood cells originate from stem cells in red bone marrow at a rate of 2 million to 3 million per second. As they fill with hemoglobin, the red blood cells of humans and some other vertebrates lose their nuclei, ribosomes, and mitochondria. This adaptation maximizes the space available for hemoglobin but also means that the mature cells cannot divide or repair damage. They generate ATP by lactic acid fermentation, a pathway that does not require mitochondria. ▶▶ fermentation, p. 115; bone marrow, p. 586

Mature red blood cells leave the bone marrow and enter the circulation. For about 120 days, each red blood cell pounds against artery walls and squeezes through tiny capillaries. Eventually, the spleen destroys the cell and recycles most of its components.

A person's blood type derives from various carbohydrates and other molecules embedded in the outer membranes of red blood cells. The genes dictating the structures of these molecules may have multiple alleles, each corresponding to a different blood type. These molecules are important in blood transfusions because of the immune system's reaction to "foreign" molecules that are not already present in the blood. Antibodies produced against incompatible blood types cause **agglutination,** a reaction in which the cells clump together. Agglutination following a transfusion of incompatible blood can be fatal.

Biologists have identified at least 26 human blood group systems, two of which are familiar to most people: ABO and Rh. ABO blood groups derive from carbohydrate "markers," designated A and B, on red blood cells (**figure 29.5**). A person's cells may express marker A (type A blood), marker B (type B), both A and B (type AB), or neither A nor B (type O). A person with type A blood cannot receive a transfusion of type B or type AB blood, because his or her plasma contains antibodies that will react against any blood carrying marker B. For people with blood type AB, however, neither marker A nor B is "foreign." Type O blood reacts against all blood types except O. Figure 10.15 explains the genetic basis of ABO blood types.

Rhesus (Rh) typing yields the "+" and "−" designations appended to ABO blood groups, such as "A positive" or "O negative." People who express the Rh marker are Rh⁺, whereas those who lack it are Rh⁻. Agglutination caused by Rh incompatibility is most important when an Rh⁻ woman becomes pregnant with an Rh⁺ fetus. Figure 33.17 depicts this immune reaction.

C. White Blood Cells Fight Infection

Blood also contains five types of **white blood cells,** or leukocytes. These immune system cells are larger than red blood cells, retain their nuclei, and lack hemoglobin.

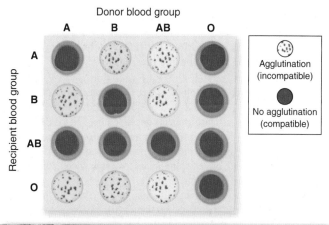

Donor blood group

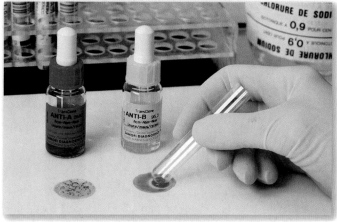

Figure 29.5 ABO Blood Groups. In this chart, the agglutination reactions reveal which blood types are incompatible with one another. The photo shows an agglutination (clumping) reaction on the left and non-agglutination on the right.

White blood cells originate from stem cells in red bone marrow. Although some enter the bloodstream, most either wander in body tissues or settle in the lymphatic system. These cells participate in many immune responses. Some secrete signaling molecules that provoke inflammation, whereas others destroy microbes or produce antibodies. Chapter 33 explains the interactions of white blood cells in much more detail.

White blood cell numbers that are too high or too low can indicate illness. For example, **leukemias** are cancers in which the bone marrow overproduces white blood cells. The abnormal white cells form at the expense of red blood cells, so when the patient's "white cell count" rises, the "red cell count" falls. Thus, leukemia also causes anemia. Having too few white blood cells, on the other hand, leaves the body vulnerable to infection. HIV destroys T cells (a type of white blood cell), causing AIDS. Likewise, exposure to radiation or toxic chemicals can severely damage bone marrow, killing many white blood cells. Death occurs rapidly from rampant infection. ▶▶ cancer, p. 162; HIV, p. 328

Burning Question

What is the difference between donating whole blood and donating plasma?

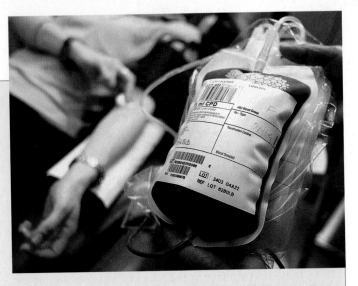

Each time a person "gives blood," he or she donates 450 to 500 milliliters (about a pint) of blood to a nonprofit blood bank. After being screened for disease-causing agents, the blood may go to patients who need transfusions following trauma or surgery. More commonly, however, the blood is separated into its components, such as red blood cells, platelets, or clotting proteins. In this way, a single blood donation can help several different patients.

A person may also donate plasma. In this process, whole blood is removed from a donor's body, then a machine separates out the plasma. The red blood cells and other components are returned to the donor. The plasma center sells the fluid to pharmaceutical companies, which use it to manufacture treatments for hemophilia, hepatitis, and other diseases. A person can donate plasma more frequently than whole blood, because it takes only about a day or two to replenish the fluid lost in plasma donation. Whole blood takes longer to replace.

In the United States, it is illegal to pay a donor for whole blood. This law promotes a safe blood supply, because donors have no finan-

cial incentive to lie about illnesses that might disqualify them from donating. Plasma donors, however, can receive money. The companies that process the plasma purify each fraction separately, removing viruses and other harmful components.

Submit your burning question to:
marielle_hoefnagels@mcgraw-hill.com

D. Blood Clotting Requires Platelets and Plasma Proteins

Platelets are small, colorless cell fragments that initiate blood clotting. A platelet originates as part of a huge cell containing rows of vesicles that divide the cytoplasm into distinct regions, like a sheet of stamps. The vesicles enlarge and join together, "shedding" fragments that become platelets.

In a healthy circulatory system, platelets travel freely within the vessels. Sometimes, however, a wound nicks a blood vessel, or the blood vessel's inner lining may become obstructed. Platelets then "catch" on the obstacle and shatter, releasing biochemicals that combine with plasma proteins called clotting factors. The resulting complex series of reactions ends with the formation of a **blood clot**—a plug of solidified blood (**figure 29.6**).

Blood that clots too slowly can lead to severe blood loss. Hemophilias, for example, are inherited bleeding disorders

caused by absent or abnormal clotting factors. Deficiencies of vitamins C or K can also slow clotting and wound healing. Blood that clots too readily is also extremely dangerous. In atherosclerosis, platelets may snag on rough spots in blood vessel linings. The resulting clot may stay in place or travel in the bloodstream to another location; either way, the obstruction may cut off circulation and sometimes even cause death.

29.2 | Mastering Concepts

1. What are the components of blood?
2. What is the location and function of hemoglobin?
3. Where do red and white blood cells originate?
4. What is the basis for the ABO and Rh blood groups?
5. Describe the process of blood clotting.

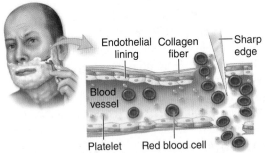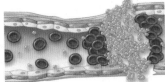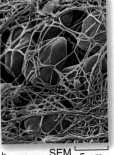

1 Break in vessel wall allows blood to escape; vessel constricts.

2 Platelets adhere to each other, to end of broken vessel, and to exposed collagen. Platelet plug helps control blood loss.

3 Exposure of blood to surrounding tissue activates clotting factors. Trapped red blood cells form a clot.

SEM 5 μm (false color)

Figure 29.6 Blood Clotting. (a) A cut blood vessel immediately constricts. Platelets aggregate at the injured site. Proteins called clotting factors participate in a cascade of reactions, producing a meshwork of protein threads. (b) Red blood cells trapped by protein threads in a clot.

29.3 | Blood Circulates Through the Heart and Blood Vessels

The plasma, cells, and platelets that make up blood circulate throughout the body in an elaborate system of blood vessels, thanks to the relentless pumping of the heart (**figure 29.7**). This entire transportation network is called the **cardiovascular system** (*cardio-* refers to the heart, *vascular* to the vessels).

Figure 29.7 shows only the largest of the body's blood vessels, which are classified by size and the direction of blood flow. **Arteries** are large vessels that conduct blood away from the heart; the left half of the figure lists some of the body's major arteries. These branch into **arterioles**, smaller vessels that then diverge into a network of **capillaries**, the body's tiniest blood vessels.

As described in section 29.5, water and dissolved substances diffuse between each capillary and the **interstitial fluid**, the liquid that bathes the body's cells. The interstitial fluid, in turn, exchanges materials with the tissue cells.

To complete the circuit, capillaries empty into slightly larger vessels, called **venules**, which unite to form the **veins** that carry blood back to the heart. The right half of figure 29.7 lists some major veins.

Notice that in figure 29.7, arteries are red, and veins are blue. This convention, coupled with the bluish appearance of blood vessels under lightly pigmented skin, has led to the misconception that the blood in veins is actually blue. In fact, blood is always red, whether hemoglobin is oxygenated or not. Veins only appear blue because of the way that light of various wavelengths interacts with the skin. As for arteries, these blood vessels tend to be located in deeper tissues, far from the skin's surface. If arteries were visible through skin, they would appear blue, too.

29.3 | Mastering Concepts

1. What is the cardiovascular system?
2. Describe the relationship among arteries, veins, arterioles, venules, and capillaries.

Figure 29.7 Human Circulatory System.

Cardiovascular System	
Main Tissue Types*	**Examples of Locations/Functions**
Epithelial	Forms inner lining of heart wall; lines veins and arteries; makes up capillary walls
Connective	Surrounds heart (part of pericardium); forms outer layers of veins and arteries; blood and lymph are connective tissues
Nervous	Regulates heart rate and blood pressure
Muscle	Heart wall is mostly cardiac muscle; smooth muscle forms middle layer of arteries and veins; skeletal muscle propels blood in veins and lymph in lymph vessels

*See chapter 24 for descriptions.

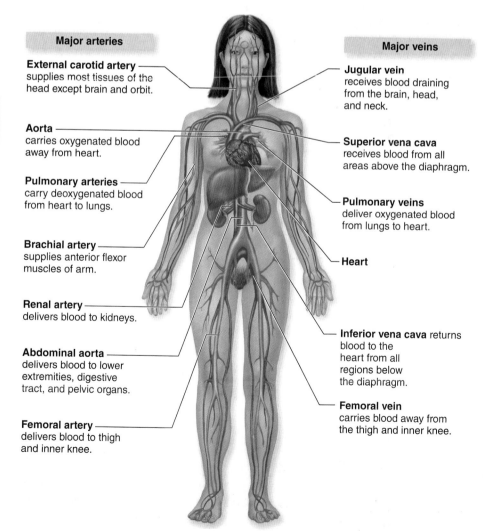

Major arteries

External carotid artery supplies most tissues of the head except brain and orbit.

Aorta carries oxygenated blood away from heart.

Pulmonary arteries carry deoxygenated blood from heart to lungs.

Brachial artery supplies anterior flexor muscles of arm.

Renal artery delivers blood to kidneys.

Abdominal aorta delivers blood to lower extremities, digestive tract, and pelvic organs.

Femoral artery delivers blood to thigh and inner knee.

Major veins

Jugular vein receives blood draining from the brain, head, and neck.

Superior vena cava receives blood from all areas above the diaphragm.

Pulmonary veins deliver oxygenated blood from lungs to heart.

Heart

Inferior vena cava returns blood to the heart from all regions below the diaphragm.

Femoral vein carries blood away from the thigh and inner knee.

29.4 | The Human Heart Is a Muscular Pump

Each day, the human heart sends a volume equal to more than 7000 liters of blood through the body, and it contracts more than 2.5 billion times in a lifetime. This section explores the structure and function of the heart.

A. The Heart Has Four Chambers

Figure 29.8 illustrates the fist-sized human heart. The **pericardium** (meaning "around the heart") is a tough connective tissue sac that encloses the heart and anchors it to surrounding tissues. This protective structure consists of two tissue layers; figure 29.8 shows only the inner layer. Thanks to lubricating fluid between the two layers of the pericardium, the heart is free to move, even during vigorous beating.

The wall of the heart itself consists mostly of the **myocardium,** a thick layer of muscle. Contraction of the **cardiac muscle** that makes up the myocardium provides the force that propels blood. The innermost lining of the heart (and of all blood vessels) consists of **endothelium,** a one-cell-thick layer of simple squamous epithelium. ▶▶ epithelial tissue, p. 514; cardiac muscle, p. 517

The human heart has four chambers: two upper atria and two lower ventricles. The atria are "primer pumps" that send blood to the ventricles, which pump the blood to the lungs or the rest of the body. The **atrioventricular valves (AV valves)** are thin flaps of tissue that prevent blood from moving back into the atrium when a ventricle contracts. The other two heart valves, the **semilunar valves,** prevent backflow into the ventricles from the arteries leaving the heart.

B. The Right and Left Halves of the Heart Deliver Blood Along Different Paths

A schematic view of the circulatory system shows the pathway of blood as it travels to and from the heart (figure 29.9). The two largest veins in the body, the superior vena cava and the inferior vena cava, deliver blood from the systemic circulation to the right atrium. From there, blood passes into the right ventricle and through the **pulmonary arteries** to the lungs, where blood picks up O_2 and unloads CO_2. The **pulmonary veins** carry oxygen-rich blood from the lungs to the left atrium of the heart, completing the pulmonary circuit.

The blood then flows from the left atrium into the left ventricle, the most powerful heart chamber. The massive force of contraction of the left ventricle sends blood into the **aorta,** the largest artery in the body. The blood then circulates throughout the body before returning to the veins that deliver blood to the right side of the heart. The systemic circuit is complete.

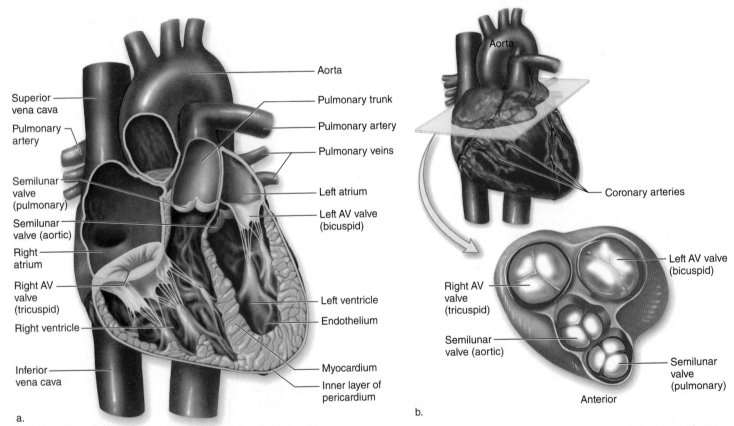

a.

b.

Figure 29.8 A Human Heart. (a) This illustration depicts the four chambers, the valves, and the major blood vessels of the human heart. (b) A cross-sectional view shows the four major heart valves, which prevent the backflow of blood.

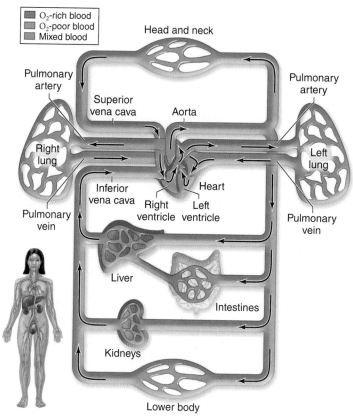

◼ O₂-rich blood
◼ O₂-poor blood
◼ Mixed blood

Head and neck

Pulmonary artery

Superior vena cava

Aorta

Pulmonary artery

Right lung

Left lung

Inferior vena cava

Heart

Pulmonary vein

Right ventricle

Left ventricle

Pulmonary vein

Liver

Intestines

Kidneys

Lower body

Figure 29.9 Blood's Journey in the Circulatory System. Oxygen-depleted blood leaving the right side of the heart goes to the lungs to pick up O₂. The oxygenated blood enters the left side of the heart, which pumps the blood throughout the body.

In some locations, circulating blood detours from the usual route of capillary to venule to vein. In a **portal system,** blood passes from capillaries into a vein that drains into a second set of capillaries before returning to the heart. For example, in the hepatic portal system, veins leaving the intestines diverge into a capillary bed in the liver, which removes toxins. (The hepatic portal vein is the blood vessel "bridge" connecting the intestines and liver in figure 29.9.) From there, veins return the blood to the heart. Another portal system delivers hormones from the hypothalamus to target cells in the anterior pituitary (see figure 27.5).

How does the heart muscle itself receive its blood supply? Blood does not seep from the heart's chambers directly to the myocardium. Instead, vessels that branch off of the aorta, the **coronary arteries,** supply blood to the heart muscle (see figure 29.8b). A smaller vein that enters the right atrium returns blood that has been circulating within the walls of the heart. Blockage of a coronary artery is the most common cause of heart attacks.

C. Cardiac Muscle Cells Produce the Heartbeat

A **cardiac cycle,** or a single beat of the heart, consists of the events that occur with each contraction and relaxation of the heart muscle.

Each heartbeat requires the forceful contraction of cardiac muscle in the myocardium. The sliding filament model of muscle contraction described in chapter 28 applies to cardiac muscle, just as it does to skeletal muscle. Unlike skeletal muscle, however, cardiac muscle does not require stimulation from motor neurons to contract.

Instead, cardiac muscle is "self-excitable"; many cardiac muscle cells contract in unison without input from the central nervous system. Cardiac muscle cells are interconnected, forming an almost netlike pattern (see figure 24.6). Wherever individual cardiac muscle cells meet, gap junctions spread synchronized waves of action potentials from cell to cell.

Figure 29.10 shows how electrical activity in the heart triggers cardiac muscle contraction. The signal begins at the **sinoatrial (SA) node,** a region of specialized cardiac muscle cells in the upper wall of the right atrium. The SA node is also called the **pacemaker** because it sets the tempo of the beat (normally about 75 beats per minute). Each time the cells of the SA node fire, they stimulate the cardiac cells of the atria to contract.

The electrical impulses then race across the atrial wall to the **atrioventricular (AV) node** in the wall of the lower right atrium. After a brief delay, which gives the ventricles time to fill, the AV node conducts electrical stimulation throughout the ventricle walls. The cardiac cells of the ventricles contract in unison. A device called an electrocardiograph records these electrical changes, producing a chart called an electrocardiogram (ECG).
▶ gap junctions, p. 65

The familiar "lub-dup" sound of the beating heart comes from the two sets of heart valves closing, preventing blood from flowing backward during each contraction. A heart murmur is a variation on the normal "lub-dup" sound, and it often reflects abnormally functioning valves.

D. Exercise Strengthens the Heart

After you circle the bases in a softball game, you may notice that your heart is beating faster than normal. The explanation for your elevated heart rate relates to the activity of your skeletal muscles, which require lots of ATP. Regenerating that ATP by aerobic respiration requires O₂; as we have already seen, one function of blood is to deliver this essential gas to the body's cells.

As you exercise, your heart meets the increased demand for O₂ by increasing its **cardiac output,** a measure of the volume of blood that the heart pumps each minute. Cardiac output is a function of the heart rate and the volume of blood pumped per stroke. Elevating the heart rate quickly boosts cardiac output during an exercise session.

With regular exercise, however, the stroke volume will increase. An active person can therefore pump the same amount of blood at 50 beats per minute as the sedentary person's heart pumps at 75 beats per minute. Cyclist Lance Armstrong provides an extreme example of cardiovascular fitness; at the peak of his career, he had a resting heart rate of about 32 beats per minute.

Exercise provides several other cardiovascular benefits as well. The number of red blood cells increases in response to regular exercise, and these cells are packed with more hemoglobin,

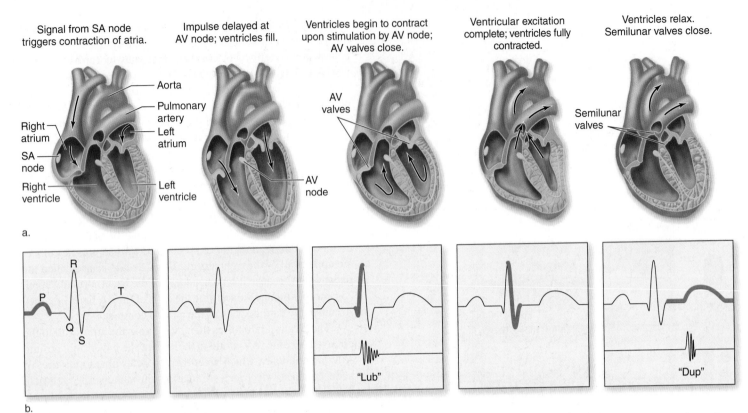

Signal from SA node triggers contraction of atria.

Impulse delayed at AV node; ventricles fill.

Ventricles begin to contract upon stimulation by AV node; AV valves close.

Ventricular excitation complete; ventricles fully contracted.

Ventricles relax. Semilunar valves close.

Aorta
Pulmonary artery
Left atrium
Right atrium
SA node
Right ventricle
Left ventricle

AV valves

AV node

Semilunar valves

a.

R

P   T

Q
S

"Lub"

"Dup"

b.

Figure 29.10 Heartbeat. (a) Electrical changes start in the sinoatrial (SA) node and travel through the atrial wall to the AV node and then to the ventricle walls. (b) An electrocardiogram (ECG) is a chart of electrical activities of the heart. The P wave corresponds to the depolarization of the atria, the QRS complex corresponds to the depolarization of the ventricles, and the T wave corresponds to the repolarization of the ventricles. The closing of the atrioventricular (AV) valves during ventricle contraction produces the "lub" sound of the heartbeat; the closing of the semilunar valves during ventricle relaxation causes the "dup" sound.

delivering more O_2 to tissues. Exercise can also lower blood pressure and reduce the amount of cholesterol in blood. Moreover, regular activity spurs the development of extra blood vessels within the walls of the heart, which may help prevent a heart attack by providing alternative pathways for blood to flow to the heart muscle.

To achieve the most benefit from exercise, the heart rate must be elevated to 70% to 85% of its "theoretical maximum" for at least half an hour three times a week. One way to calculate your theoretical maximum is to subtract your age from 220 (table 29.3). If you are 18 years old, your theoretical maximum is 202 beats per minute; 70% to 85% of this value is 141 to 172 beats per minute. Tennis, skating, skiing, handball, vigorous dancing, hockey, basketball, biking, or brisk walking can elevate heart rate to this level.

Table 29.3 Target Heart Rates by Age

Age	Theoretical Maximum Heart Rate (beats per minute)	Target Heart Rate During Exercise (beats per minute)
20	200	140–170
25	195	137–166
30	190	133–162
40	180	126–153
50	170	119–145
60	160	112–136
70	150	105–128

29.4 | Mastering Concepts

1. Describe the locations of the pericardium, myocardium, and endothelium.
2. Describe the path of blood through the heart's chambers and valves, and through the pulmonary and systemic circulations.
3. What is the function of heart valves?
4. How does heartbeat originate and spread?
5. How does exercise affect the circulatory system?
6. Explain why exercise makes the heart beat faster.

29.5 | Blood Vessels Form the Circulation Pathway

As the heart's ventricles contract, they push blood to the lungs and the rest of the body. This section describes the system of vessels through which blood travels as it delivers supplies and removes wastes.

A. Arteries, Capillaries, and Veins Have Different Structures

As we have already seen, arteries carry blood away from the heart, whereas veins return blood to the heart. Despite these opposite functions, the walls of arteries and veins share some similarities (figure 29.11). The outermost layer is a sheath of connective tissue. The middle layer is made mostly of **smooth muscle,** and endothelium forms the innermost layer. ▶▶ epithelial tissue, p. 514; smooth muscle, p. 517

One feature that characterizes arteries is the thick layer of smooth muscle. The muscular walls of major arteries can withstand the high-pressure surges of blood leaving the heart. As arteries branch farther from the heart, however, their walls become thinner, and the outermost layer of connective tissue may taper away. Arterioles do retain a layer of smooth muscle

that helps regulate blood pressure; section 29.5B describes how this occurs.

Arterioles branch into **capillary beds,** networks of tiny blood vessels that connect an arteriole and a venule (**figure 29.12**). Capillaries are tiny but very numerous, providing extensive surface area where materials are exchanged with the interstitial fluid. Because their walls consist of a single layer of endothelial cells, nutrients and gases easily diffuse into and out of capillaries. ▶ diffusion, p. 80

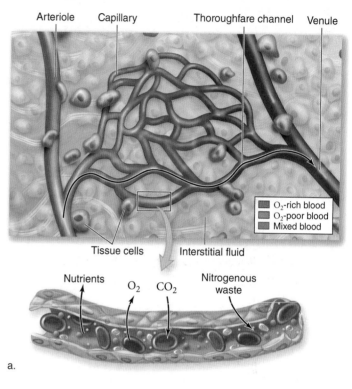

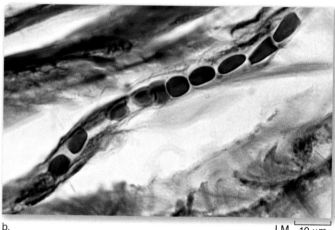

Figure 29.12 Capillaries. (a) A capillary bed is a network of tiny vessels that lies between an arteriole and a venule. At capillaries of the systemic circulation, O_2 and nutrients leave the bloodstream, and CO_2 and other wastes enter the circulation for disposal at the lungs and kidneys. (b) Capillaries are so small that red blood cells move through them in single file.

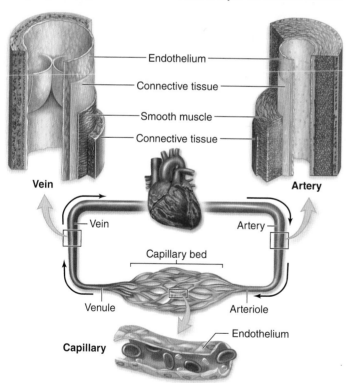

Figure 29.11 Types of Blood Vessels. The walls of arteries and veins consist of connective tissue, smooth muscle, and endothelium. Arteries, which are subject to high blood pressure, are much more muscular than veins. In some veins, valves keep blood moving toward the heart. The capillary wall consists only of endothelium.

From the capillary beds, blood flows into venules, which converge into veins. These vessels receive blood at low pressure. The smooth muscle layer in their walls is much reduced or even absent (see figure 29.11); in fact, unlike an artery, a vein collapses when empty.

If pressure in veins is so low, what propels blood back to the heart, against the force of gravity? In many veins, flaps called venous valves keep blood flowing in one direction (figure 29.13). As skeletal muscles in the leg contract, they squeeze veins and propel blood through the open valves in the only direction it can move: toward the heart. Varicose veins result in part from faulty venous valves that cause blood to pool in the veins of the lower legs. The walls of these distended blood vessels form prominent bulges under the skin.

B. Blood Pressure and Velocity Differ Among Vessel Types

A routine doctor's office visit always includes a blood pressure reading, a good indication of overall cardiovascular health. **Blood pressure** is the force that blood exerts on artery walls. As the heart drives blood through the vessels, you can feel your blood pressure as a "pulse."

A device called a sphygmomanometer measures the changes in blood pressure that occur throughout the cardiac cycle (figure 29.14). The **systolic pressure,** or upper number in a blood pressure reading, reflects the contraction of the ventricles. The **diastolic pressure,** or low point, occurs when the ventricles relax.

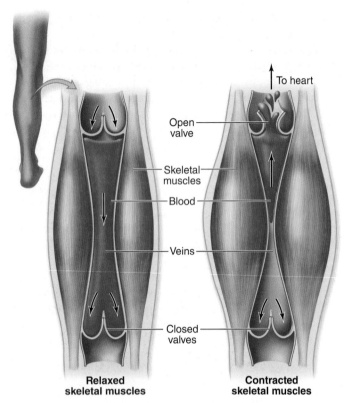

Figure 29.13 Venous Valves. When skeletal muscles are relaxed, valves prevent blood from flowing backward in veins. Contracted skeletal muscles squeeze the veins, propelling blood through the open valves. In this illustration, the veins appear much larger than they are relative to real muscles.

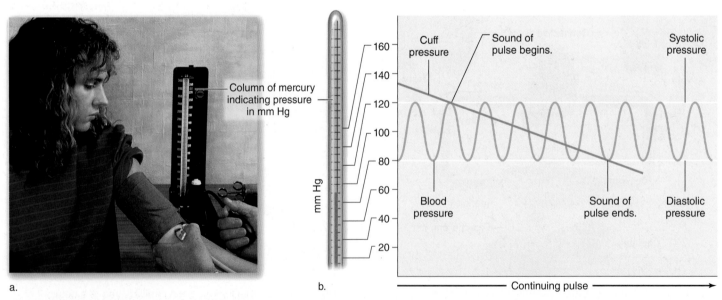

Figure 29.14 Blood Pressure. (a) A sphygmomanometer measures blood pressure. The cuff is wrapped around the upper arm and inflated until no pulse is felt in the inner elbow, which signifies that circulation to the lower arm has been temporarily cut off. A stethoscope placed on the arm just below the cuff detects the sound of returning blood flow when the cuff slowly deflates. (b) The listener notes the pressure on the gauge when a thumping is first audible; this sound is the blood rushing through the arteries past the deflating cuff. The value on the gauge when the sound begins is the systolic blood pressure. The pressure reading when the sound ends is the diastolic blood pressure.

Blood pressure readings are in units of "millimeters of mercury," abbreviated "mm Hg"; this terminology derives from older sphygmomanometers, which measured the distance over which blood pressure could push a column of mercury. A typical blood pressure reading for a young adult is 110 mm Hg for the systolic pressure and 70 mm Hg for the diastolic pressure, expressed as "110 over 70" (written 110/70). "Normal" blood pressure, however, varies with age, sex, race, and other factors.

Blood pressure decreases with distance from the heart; that is, blood in arteries has the highest pressure, followed by capillaries and then veins (**figure 29.15**). Blood velocity, however, is lowest in the capillaries. The reason is that the total cross-sectional area of capillaries is much greater than that of the arteries or veins. Just as the velocity of a river slows as the water spreads out over a delta, so does the flow of blood slow as it is divided among countless tiny capillaries. This leisurely flow of blood through the capillaries enhances nutrient and waste exchange.

Past the capillaries, venules converge into veins. The total cross-sectional area of these blood vessels is again smaller than that of the capillaries. The resulting reduction in cross-sectional area helps speed blood flow back to the heart. To understand why, picture water flowing out of a hose. If you put your thumb over the nozzle, you reduce the area of the opening. What happens? The velocity of water through the nozzle increases.

Overall, a person's blood pressure reflects many factors, including blood vessel diameter, heart rate, and blood volume.

The body regulates blood pressure over the long term by raising or lowering the volume of blood. Chapter 32 describes how the kidneys adjust the blood's volume and salt concentration by controlling the amount of fluid excreted in urine.

In the short term, blood vessel diameter and heart rate are under constant regulation by negative feedback (**figure 29.16**). Pressure receptors within the walls of major arteries detect blood pressure and pass that information to the medulla, in the brainstem. The medulla, via the autonomic nervous system, adjusts both heart rate and the diameter of arterioles to maintain homeostasis. ▶▶ negative feedback, p. 520; autonomic nervous system, p. 538

The role of the arterioles deserves special mention. **Vasoconstriction** is the narrowing of blood vessels that results from the contraction of smooth muscle in arteriole walls. When arteriole diameter decreases, blood pressure rises. The opposite effect, **vasodilation,** is the widening of blood vessels that occurs when the same muscles relax. Altering arteriole diameter allows the body to increase blood delivery to regions that need it most. During physical activity, for example, skeletal muscles receive additional blood at the expense of organs not in immediate use, such as those in the digestive tract.

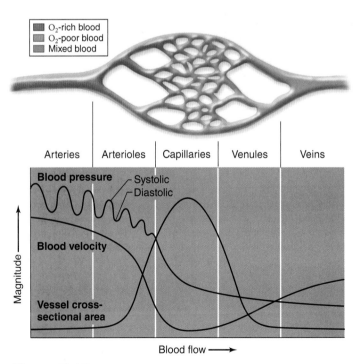

Figure 29.15 Blood Pressure and Velocity. Blood pressure drops with increasing distance from the heart. Blood moves rapidly in the aorta, but its velocity declines steadily as it moves through arteries, arterioles, and capillaries. Blood regains some of its velocity as venules converge into veins.

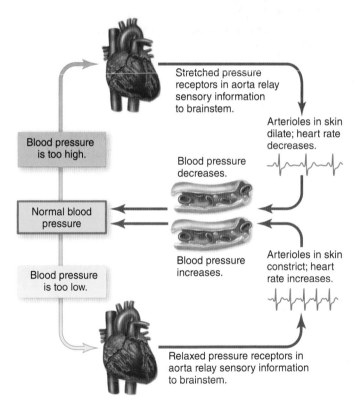

Figure 29.16 Regulation of Blood Pressure. A negative feedback loop regulates blood pressure in the short term. Stretch-sensitive neurons detect pressure in major arteries. If blood pressure climbs too high, signals from the central nervous system cause the heart rate to decrease and blood vessels to dilate. If blood pressure is too low, vessels constrict and the heart beats faster.

Blood pressure that is too low or too high can cause health problems. Hypotension, which is blood pressure that is significantly lower than normal, may cause fainting. At the opposite end of the spectrum, consistently elevated blood pressure, or hypertension, may severely damage the circulatory system and other organs. High blood pressure affects 15% to 20% of adults residing in industrialized nations.

High blood pressure is just one example of illnesses that can affect the human circulatory system; this chapter's Apply It Now box considers several others.

29.5 | Mastering Concepts

1. Compare and contrast the structures of arteries, capillaries, and veins.
2. Trace the path of a red blood cell from the heart to a capillary bed in the foot and back to the heart.
3. Across the human circulatory system, how are blood pressure, blood velocity, and vessel diameter related?
4. How is the regulation of blood pressure an example of negative feedback?

Apply It Now

The Unhealthy Circulatory System

Anemias

Anemias are a collection of more than 400 disorders resulting from a decrease in the oxygen-carrying capacity of blood. One symptom is fatigue, reflecting a shortage of O_2 at the body's cells. Some forms of anemia are inherited; a mutation in the gene encoding hemoglobin, for example, can cause an inherited form of anemia called sickle cell disease. Other types of anemia are related to diet, especially an iron deficiency, which can prevent cells from producing hemoglobin. Iron-deficiency anemia is most common in women because of blood loss in menstruation. In still other forms of anemia, red blood cells may be too small, be manufactured too slowly, or die too quickly. ▶ sickle cell disease, p. 138

Atherosclerosis

Fatty deposits inside the walls of coronary arteries reduce blood flow to the heart muscle (**figure 29.A**). A diet high in fat and cholesterol is associated with this "hardening of the arteries," also called atherosclerosis (*athero-* is from the Greek word for "paste," and *sclerosis* means "hardness"). Atherosclerosis can cause several ailments, including angina pectoris (chest pain), heart attack, arrhythmia, and aneurysm. ▶ bad and good cholesterol, p. 35

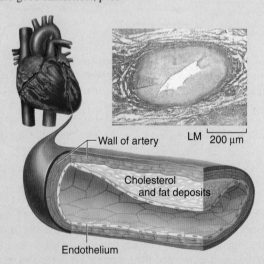

Wall of artery — LM ⌐ 200 μm

Cholesterol and fat deposits

Endothelium

Figure 29.A **Atherosclerosis.** Deposits of cholesterol and other fatty materials beneath the inner lining of the coronary arteries cause atherosclerosis, or "hardening of the arteries."

Heart Attack

Blocked blood flow in a coronary artery kills part of the myocardium, the heart muscle. This is a heart attack (myocardial infarction), and it may come on suddenly. A common treatment for a blocked coronary artery is a bypass operation. A surgeon creates a bridge around the blockage by sewing pieces of blood vessel taken from the patient's chest or leg onto the blocked artery. The procedure's name often includes the number of arteries repaired. A quadruple bypass operation, for example, bridges four obstructed arteries.

Arrhythmia

An arrhythmia is an abnormal heartbeat. Some arrhythmias originate in the atria, causing transient flutters or racing that lasts only a few seconds. An electronic pacemaker implanted under the skin is a common treatment. In ventricular fibrillation, the ventricles twitch but cannot produce a normal contraction, so they pump only a small volume of blood out of the heart. The resulting sudden cardiac arrest can cause death within minutes. First aid for cardiac arrest includes cardiopulmonary resuscitation (CPR); in addition, external defibrillators that shock an erratically beating heart back to normal are available in many public places. People with chronic arrhythmia may have a defibrillator implanted under the skin of the chest.

Aneurysm

Atherosclerosis can weaken the wall of an artery so much that a region of the vessel forms a pulsating, enlarging sac called an aneurysm. Aneurysms may also result from a congenitally weakened area of an arterial wall; trauma; infection; persistently high blood pressure; or an inherited disorder such as Marfan syndrome. The aneurysm may rupture without warning. Depending on the location and volume of bleeding, a burst aneurysm can be fatal.

The Effects of Smoking on Cardiovascular Health

Smoking is the most common preventable cause of death. Most famously, cigarette smoke damages the lungs, impairing their ability to deliver O_2 to blood (and, of course, increasing the chance of lung cancer). Tobacco's effects on the circulatory system are less familiar to most people. Nicotine stimulates the secretion of epinephrine and norepinephrine, increasing both heart rate and blood pressure. Nicotine also damages blood vessels and stimulates the formation of blood clots, increasing the risk of stroke. Cigarette smoking is clearly unhealthy, yet many people find it difficult or impossible to quit, thanks to nicotine's addictive qualities.

29.6 | The Lymphatic System Maintains Circulation and Protects Against Infection

Capillary walls are typically rather porous. Red blood cells remain confined to capillaries, but many other components of blood can pass between the endothelial cells that make up capillary walls. How does the fluid that leaks out of these blood vessels return to the circulatory system? The answer is that the **lymphatic system** collects the fluid, removes bacteria, debris, and cancer cells, and returns the liquid to the blood (**figure 29.17**). The lymphatic system is therefore a bridge between the circulatory and immune systems.

Lymph, the colorless fluid of the lymphatic system, originates in **lymph capillaries**—tiny, dead-end vessels that absorb fluid from the spaces between cells. The chemical composition of lymph is therefore similar to that of blood plasma, minus the proteins that are too large to leave blood capillaries.

The cells of a lymph capillary, however, are not joined as tightly together as those of a blood capillary. Lymph capillaries therefore also admit bacteria, viruses, cancer cells, and other large particles in body tissues. These capillaries then converge into larger lymph vessels that eventually empty into veins in the chest, where the fluid returns to the blood.

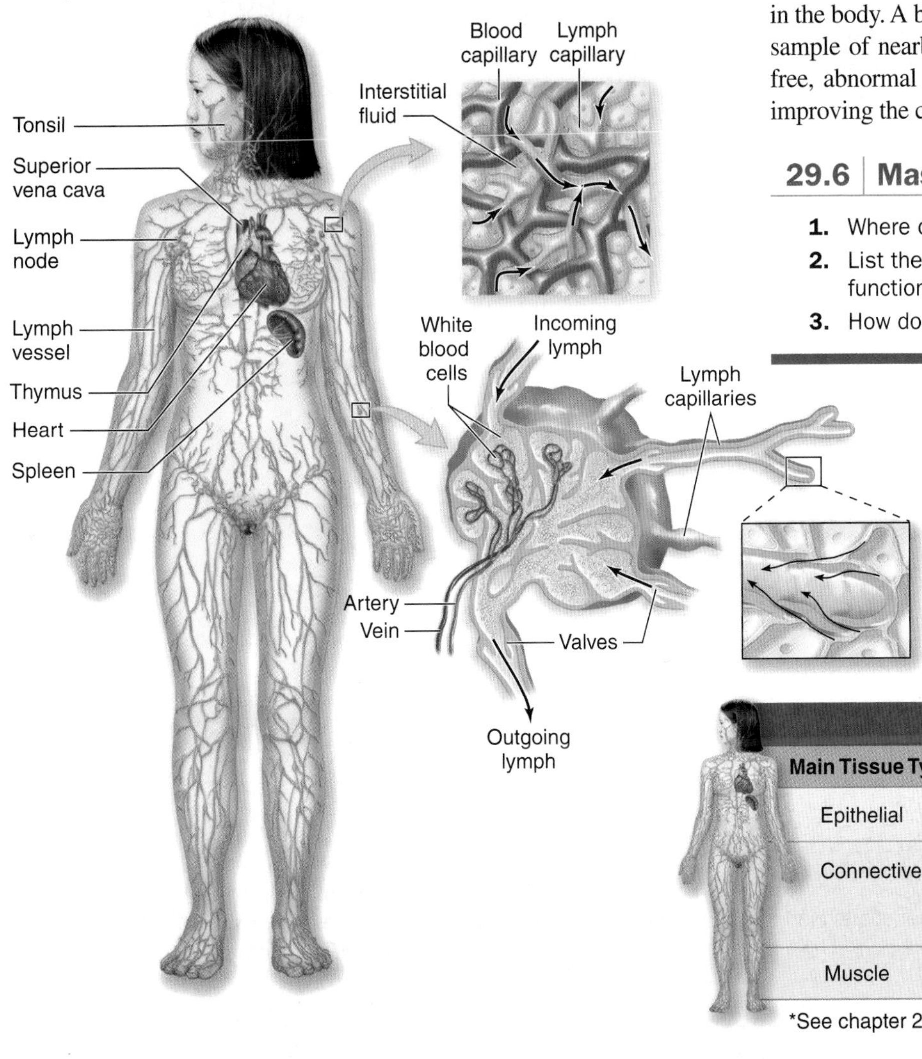

Along the way, lymph passes through **lymph nodes,** kidney-shaped organs that contain millions of white blood cells. These infection-fighting cells intercept and destroy cellular debris, cancer cells, and bacteria in the lymph flow. White blood cells can also migrate from lymph nodes into the lymph, which transports them to the blood.

White blood cells also occur in other organs of the lymphatic system. For example, one of the spleen's many functions is to produce, store, and release white blood cells. The thymus is an organ in which specialized white blood cells called T cells learn to distinguish body cells from foreign cells.

Lymph flow is sluggish, because the lymphatic system has no pump. Instead, contractions of surrounding skeletal muscles and valves in lymph vessels help move the fluid. The lymphatic system normally returns less than 30 milliliters of fluid per minute to the veins, in contrast to the 5 or 6 liters pumped through the circulatory system in the same amount of time.

When lymph flow stops, excess fluid accumulates in the body's tissues and causes swelling (edema). This condition commonly occurs in immobile people; when skeletal muscles do not contract, lymph fails to circulate. Parasites can also cause extreme edema. In an infectious disease called elephantiasis, nematode worms block the flow of lymph and cause grotesquely swollen tissues (see figure 20.22).

Along with the bloodstream, the lymphatic system can carry cancer cells that break off of tumors, seeding new tumors elsewhere in the body. A biopsy of cancerous tissue therefore often includes a sample of nearby lymph nodes. If these lymph nodes are cancer-free, abnormal cells may not have begun to invade other tissues, improving the chance of successful treatment. ▶ cancer, p. 162

29.6 | Mastering Concepts

1. Where does lymph come from?
2. List the components of the lymphatic system and their functions.
3. How does lymph travel within lymph capillaries?

Figure 29.17 Human Lymphatic System. This network of vessels and lymphoid organs collects excess fluid that leaks from the blood capillaries, removes debris and foreign cells, and returns the fluid to the bloodstream.

Lymphatic System	
Main Tissue Types*	**Examples of Locations/Functions**
Epithelial	Makes up lymph capillary walls
Connective	Forms supportive framework for lymph nodes, spleen, thymus, and bone marrow; lymph is a connective tissue
Muscle	Skeletal muscle propels fluid in lymph vessels

*See chapter 24 for descriptions.

29.7 | Investigating Life: In (Extremely) Cold Blood

Sometimes, an obscure discovery can turn into a big story. So it was with a report that appeared in the journal *Nature* in 1954, when a researcher named J. T. Ruud confirmed the existence of a fish with colorless blood. The story was an interesting curiosity, because animal physiologists at the time believed that vertebrate life requires hemoglobin, the oxygen-toting pigment that gives blood its red color. But no one at the time could have predicted that Ruud's fish would end up at the center of genetic studies exploring the evolution of hemoglobin.

The fish in Ruud's report, *Chaenocephalus aceratus,* is an icefish that lives deep in the ocean waters surrounding Antarctica. The blood of this fish is a ghostly white. Microscopic examination of the blood revealed white blood cells, but few if any red blood cells and never any hemoglobin. Apparently, the plasma of an icefish's blood carries enough O_2 to meet the animal's metabolic needs.

Since Ruud's time, researchers have described 16 species of icefishes, none of which has hemoglobin (**figure 29.18**). These animals are unusual in another way as well. The Antarctic water in which they live is extremely cold, with temperatures hovering around 0°C. The cells of icefishes produce antifreeze proteins that protect them from freezing to death. In fact, these animals are so cold-adapted that they die of heat shock if their temperature climbs to 4°C—about as cold as your kitchen refrigerator.

Interestingly, icefishes have red-blooded relatives that share their frigid habitat. The "family tree" of these fishes is well understood, based on studies of ribosomal RNA, mitochondrial DNA, and many morphological traits. Could these relationships help researchers answer questions about the genetic origin of the icefishes' colorless blood?

Biologist Thomas J. Near, of Yale University, collaborated with Northeastern University's Sandra K. Parker and H. William Detrich III in an effort to learn more about icefish blood. They knew that decades of previous research had already established some basic facts about hemoglobin. For example, this protein consists of four polypeptides called globins; two of the chains are designated alpha, and the other two are called beta (see figure 30.11). Also, the genes encoding both alpha and beta globins had already been well-studied in many vertebrates.

Near, Parker, and Detrich collected 44 specimens representing 33 fish species around the Antarctic. All 16 species of icefishes were included in the study, plus three species representing their red-blooded relatives. The researchers extracted DNA from the fishes' cells and used the polymerase chain reaction to amplify the stretch of DNA where the globin genes are normally located. ▶ polymerase chain reaction, p. 171

After sequencing this stretch of DNA for all of the specimens, the team discovered that the fishes fell into three categories (**figure 29.19**). One group consisted of the red-blooded fishes with functioning genes encoding both alpha and beta globins. The second category included 15 of the 16 icefish species. All of these fishes had just a small fragment of the alpha globin gene; the beta globin gene was missing entirely. The third category included just one icefish species, *Neopagetopsis ionah,* which had a unique gene arrangement. In this animal's DNA, both the alpha and beta genes were mutated in a way that leaves the fish unable to produce hemoglobin. ▶ DNA sequencing, p. 170

The genetic analysis clearly explained why icefishes have colorless blood: unlike their red-blooded relatives, icefishes simply cannot produce hemoglobin.

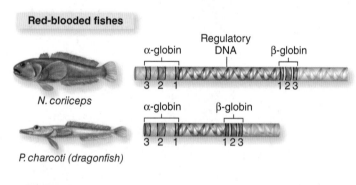

Red-blooded fishes

N. coriiceps

P. charcoti (dragonfish)

Icefishes

Typical icefishes

N. ionah

Figure 29.19 Shattered Genes. Analysis of DNA sequences revealed that red-blooded fishes have intact alpha and beta globin genes. Most icefishes, however, have just a fragment of the alpha gene and lack the beta gene entirely. The icefish *N. ionah* has nonfunctional versions of both genes.

Figure 29.18 Icefish. A scuba diver approaches an icefish in the Southern Ocean near Antarctica.

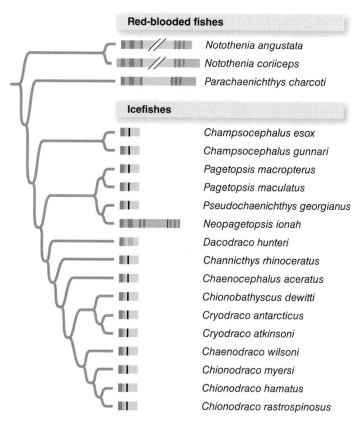

Figure 29.20 Icefish Family Tree. The icefish with the unique globin gene cluster, *N. ionah,* is not ancestral to all other icefishes. Researchers are still studying how *N. ionah* got its strange genes.

But a surprising result emerged when the researchers mapped the gene configurations onto their previously existing phylogenetic tree (figure 29.20). The icefish with the unique globin genes, *N. ionah,* was not a "missing link" that gave rise to the other icefishes, as one might expect from its globin gene configuration. Instead, *N. ionah* was right in the middle of the family tree.

So why doesn't *N. ionah* have the same alpha globin gene remnant as every other icefish? The researchers proposed two possible explanations. Identical deletion mutations in the alpha and beta globin genes could have occurred four separate times in the 8 million years since icefishes diverged from their red-blooded relatives. The alternative explanation, which is perhaps more likely, is that *N. ionah*'s ancestors acquired the unusual hybrid hemoglobin gene by interbreeding with red-blooded relatives.

The story of the Antarctic icefishes reminds us that scientific inquiry is a journey. What began as a report of an unusual fish in 1954 has blossomed into an in-depth study of the evolution of hemoglobin. And, although we now know exactly why icefishes have colorless blood, a new question has emerged: What accounts for *N. ionah*'s unusual globin genes? The answer may lie with future biologists—people whose desire to understand evolution is "in their blood."

Near, Thomas J., Sandra K. Parker, and H. William Detrich III. 2006. A genomic fossil reveals key steps in hemoglobin loss by the Antarctic icefishes. *Molecular Biology and Evolution,* vol. 23, no. 11, pages 2008–2016.

29.7 | Mastering Concepts

1. How did researchers use DNA evidence to determine why icefishes have colorless blood?
2. In figure 29.20, mark the nodes at which independent, identical mutations could have occurred to produce the pattern of gene deletions revealed in this study.

Chapter Summary

29.1 | Circulatory Systems Deliver Nutrients and Remove Wastes

- A **circulatory system** consists of **blood** or a similar fluid, a network of vessels, and a **heart.** The fluid delivers nutrients and O_2, removes metabolic wastes, and transports other substances.

A. Circulatory Systems Are Open or Closed

- In many animals, a heart pumps the fluid to the body cells. In an **open circulatory system,** the fluid bathes tissues directly in open spaces before returning to the heart.
- In a **closed circulatory system,** such as that of vertebrates, the heart pumps the fluid to and from cells through a system of vessels.

B. Vertebrate Circulatory Systems Have Become Increasingly Complex

- A fish has a two-chambered heart, with an **atrium** that receives blood and a **ventricle** that pumps blood out.
- In other vertebrates, the **pulmonary circulation** delivers oxygen-depleted blood to the lungs, and the **systemic circulation** brings freshly oxygenated blood to the rest of the body.
- Most land vertebrates have a three- or four-chambered heart.

29.2 | Blood Is a Complex Mixture

- Human blood is a mixture of water, proteins and other dissolved substances, cells, and cell fragments.

A. Plasma Carries Many Dissolved Substances

- **Plasma** is the fluid component of blood; it transports all other blood components.

B. Red Blood Cells Transport Oxygen

- **Red blood cells** contain abundant **hemoglobin,** a pigment that binds O_2 molecules. Like other blood cells, these cells originate in red bone marrow.
- Surface markers on red blood cells react with antibodies in **agglutination** reactions that reveal a person's blood type.

C. White Blood Cells Fight Infection

- The five types of **white blood cells** provoke inflammation, destroy infectious organisms, and secrete antibodies. **Leukemia** is a type of cancer in which red bone marrow produces too many white blood cells.

D. Blood Clotting Requires Platelets and Plasma Proteins

- **Platelets** are cell fragments that collect near a wound. Damaged tissue activates plasma proteins that trigger the formation of a network of fibers, trapping additional platelets and perpetuating **blood clot** formation.

29.3 | Blood Circulates Through the Heart and Blood Vessels

- The heart is the muscular pump that drives blood through the vessels of the human **cardiovascular system.**
- **Arteries** are blood vessels that carry blood away from the heart. Arteries branch into smaller **arterioles,** which lead to tiny capillaries.
- **Capillaries** exchange materials with the **interstitial fluid** surrounding the body's cells.
- Capillaries empty into **venules,** which converge into the **veins** that return blood to the heart.

29.4 | The Human Heart Is a Muscular Pump

A. The Heart Contains Four Chambers

- A sac of connective tissue, the **pericardium,** surrounds the heart. **Cardiac muscle** makes up most of the **myocardium,** the thickest portion of the heart wall. **Endothelium** lines the inside of the heart and all of the body's blood vessels.
- The heart has two atria that receive blood and two ventricles that propel blood throughout the body. The heart's **atrioventricular (AV) valves** and **semilunar valves** ensure one-way blood flow.

B. The Right and Left Halves of the Heart Deliver Blood Along Different Paths

- **Pulmonary arteries** and **pulmonary veins** transport blood between the right side of the heart and the lungs.
- Blood exits the left side of the heart at the **aorta,** the artery that carries blood toward the rest of the body.
- Blood typically flows from artery to arteriole, capillary bed, venule, vein, and back to the heart. In a **portal system,** however, blood passes from a capillary bed to a vein to a second set of capillaries before returning to the heart.
- **Coronary arteries** supply blood to the heart muscle itself.

C. Cardiac Muscle Cells Produce the Heartbeat

- A **cardiac cycle** consists of a single contraction and relaxation of the heart muscle.
- The **pacemaker** or **sinoatrial (SA) node,** a collection of specialized cardiac muscle cells in the wall of the right atrium, sets the heart rate. From there, heartbeat spreads to the **atrioventricular (AV) node** and then along specialized fibers through the ventricles.

D. Exercise Strengthens the Heart

- Exercise increases the heart's **cardiac output** and lowers blood pressure.

29.5 | Blood Vessels Form the Circulation Pathway

A. Arteries, Capillaries, and Veins Have Different Structures

- The walls of arteries and veins consist of an inner layer of endothelium, a middle layer of **smooth muscle,** and an outer layer of connective tissue. Arteries have thicker, more elastic walls than veins.
- Nutrient and waste exchange occur at the **capillary beds,** where blood vessels consist of a single layer of endothelium.

B. Blood Pressure and Velocity Differ Among Vessel Types

- The pumping of the heart and constriction of blood vessels produces **blood pressure. Systolic pressure** reflects the force exerted on blood vessel walls when the ventricles contract. The low point, **diastolic pressure,** occurs when the ventricles relax.
- Blood pressure is highest in the arteries and lowest in the veins. Because of high total cross-sectional area, blood velocity is lowest in the capillaries.

- The autonomic nervous system speeds or slows heart rate. **Vasoconstriction** and **vasodilation** adjust blood pressure.

29.6 | The Lymphatic System Maintains Circulation and Protects Against Infection

- The **lymphatic system** includes a network of **lymph capillaries** that collect fluid from the body's tissues.
- **Lymph nodes** cleanse the resulting **lymph,** which subsequently returns to the bloodstream.

29.7 | Investigating Life: In (Extremely) Cold Blood

- Antarctic icefishes are unique among vertebrates in that they lack hemoglobin. Researchers have traced this unusual characteristic to major deletions in the alpha and beta hemoglobin genes.

Multiple Choice Questions

1. A property of an open circulatory system includes
   a. the absence of a heart.
   b. movement of fluid into spaces in the tissues of an organism.
   c. the absence of vessels.
   d. All of the above are correct.

2. What is the advantage of a four-chambered heart?
   a. It uses the least energy of any type of heart.
   b. It maximizes the amount of O_2 reaching tissues.
   c. It enhances the mixing of blood from the pulmonary and systemic circulation.
   d. Both a and c are correct.

3. Why are people with type O blood considered to be universal donors?
   a. Because type O cells do not carry marker A or B.
   b. Because there are more people with type O than any other blood type.
   c. Because everyone has the O marker.
   d. Because type O cells lack an Rh group.

4. Which component of blood is responsible for clot formation?
   a. Red blood cells   c. Plasma
   b. White blood cells   d. Platelets

5. The function of the pericardium is to
   a. prevent the flow of blood back into an atrium.
   b. contract, causing the blood to move.
   c. allow blood to flow into a ventricle from an atrium.
   d. protect the heart.

6. Which chamber of the human heart collects the oxygenated blood from the lungs?
   a. Left atrium   c. Right atrium
   b. Left ventricle   d. Right ventricle

7. How would the sinoatrial node of an athlete at rest differ from that of a sedentary person?
   a. It would establish a higher rate of contraction.
   b. It would not be any different.
   c. It would establish a lower rate of contraction.
   d. It would trigger a stronger contraction.

8. Which type of blood vessel uses smooth muscle contractions to help control blood flow?
   a. A capillary   c. A vein
   b. A venule   d. An arteriole

9. If a man suffers from high blood pressure, it is possible that his
   a. blood vessels are abnormally dilated.
   b. blood vessels are abnormally constricted.
   c. kidneys are removing too much fluid.
   d. medulla is underactive.

10. How is lymph related to blood plasma?
    a. Both contain white blood cells.
    b. Both are fluids that are propelled by the heart.
    c. Lymph is plasma that has leaked out of blood vessels.
    d. Both a and c are correct.

Write it Out

1. How are open and closed circulatory systems similar? How are they different?

2. Describe the circulatory systems of fishes, lizards, and mammals. What is the advantage of separating the pulmonary and systemic circulatory pathways?

3. Maintaining the proper proportions of cells and platelets in the blood is essential for health. What can happen when the blood contains too few or too many red blood cells? Too few or too many white blood cells? Too few or too many platelets?

4. Why can a person with type A blood not receive a transfusion of type B blood? If a person has type O blood, what blood type(s) can he or she receive in a transfusion?

5. Why is blood clotting that happens too quickly or too slowly dangerous?

6. One effect of aspirin is to prevent platelets from sticking together. Why do some people take low doses of aspirin to help prevent a heart attack?

7. Trace the pathway of an O_2 molecule from the lungs to a respiring cell at the tip of your finger.

8. Describe the relationship between blood and interstitial fluid.

9. Why is the heart sometimes called "two hearts that beat in unison"?

10. Describe the events that occur during one cardiac cycle.

11. Athletes usually are slim and strong, have low blood pressure, do not smoke, and alleviate stress through exercise. How might these characteristics complicate a study designed to assess the effects of exercise on the circulatory system?

12. Make a chart that compares systemic arteries, capillaries, and systemic veins in terms of the following properties: overall structure; amount of smooth muscle; presence of valves; cross-sectional area; blood pressure; blood velocity; direction of blood flow relative to the heart; O_2 content of blood.

13. Endothelial cells lining the heart and blood vessels are the only cells to receive nutrients and O_2 directly from blood. How do the rest of the body's cells receive these resources and dispose of wastes?

14. Explain why blood pressure is highest in the arteries and lowest in the veins.

15. What types of changes in blood vessels would raise blood pressure?

16. The carotid artery extends from the heart to the head. Some of the body's blood pressure receptors are located in the carotid sinus, where the carotid artery passes through the neck. If you press lightly on the carotid sinus, what do you predict should happen to your heart rate? What if you press lightly on a spot just *below* the carotid sinus? *Hint:* Figure 29.16 may help you answer this question.

17. Describe a process discussed in the chapter that illustrates negative feedback.

18. Where does lymph originate? What propels lymph through the lymphatic vessels? How is the lymphatic system connected with the circulatory system? The immune system?

19. A common treatment for breast cancer is to surgically remove the cancerous tissue along with associated lymph nodes. Women who undergo this surgery sometimes develop swollen arms. Explain this phenomenon.

20. What are the roles of cardiac, smooth, and skeletal muscle in the circulatory system?

21. Describe the interactions between the circulatory system and the respiratory, immune, digestive, and endocrine systems.

22. Name three ways that the circulatory system helps maintain homeostasis.

23. For reasons that scientists do not fully understand, diabetes (see chapter 27) impairs the circulatory system. Explain how damage to the blood vessels and peripheral nerves of the kidneys, eyes, and extremities causes many of the complications of diabetes, including blindness, kidney failure, and poor wound healing.

24. How is the human cardiovascular system similar to and different from the vascular tissue in a plant?

25. Section 8.9 describes the rationale behind cancer treatments called angiogenesis inhibitors, drugs that inhibit the growth of blood vessels toward a tumor. Why would tumors without a blood supply quit growing (or even shrink)?

26. Use the Internet to learn more about disorders of the cardiovascular or lymphatic system. Choose one to investigate in more detail. What causes the disease you chose? Who is affected, and what are the consequences? Are there ways to prevent, treat, or cure the disease?

Pull It Together

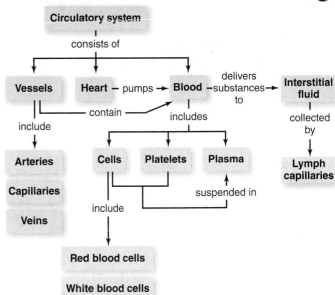

1. How do pulmonary and systemic circulation fit into this concept map?
2. What other substances are in blood besides those listed in this concept map?
3. What are the functions of red and white blood cells?
4. Where is interstitial fluid located?
5. Add arteriole, venule, lymph, lymph node, and lymphatic system to this concept map.

The Respiratory System

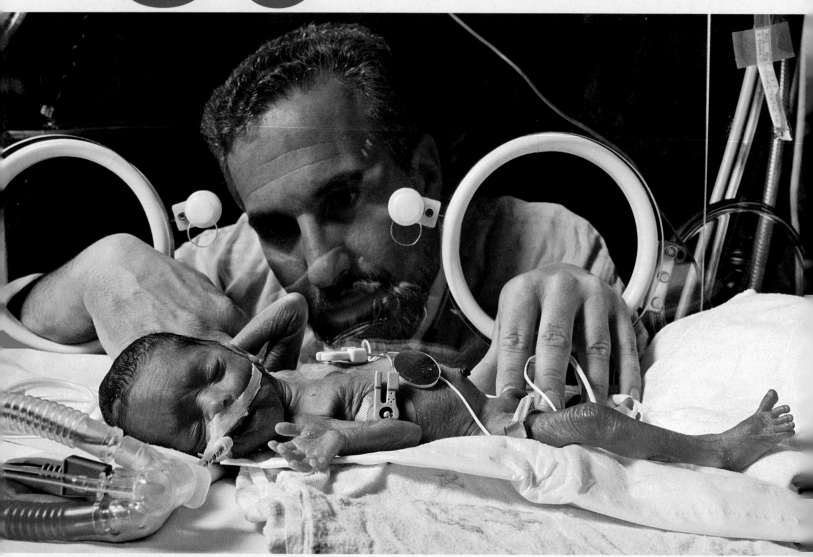

Early Arrival. A premature baby is tiny and fragile and requires around-the-clock medical care.

Premature Babies Have a Tough Time Breathing

WHILE IN THE WOMB, A FETUS GETS ALL THE OXYGEN IT NEEDS THROUGH THE PLACENTA. A newborn baby, however, must acquire all of his or her own oxygen. This transition occurs moments after birth.

The first breath of a newborn requires 15 to 20 times the strength needed for subsequent breaths. The infant must force air into millions of partially inflated air sacs, called alveoli, for the first time. In fully developed lungs, these tiny, moist air sacs would collapse if not for a chemical called lung surfactant that counters the cohesive force of water molecules. This surfactant is especially important at birth. Once the first breath expands the lungs, the surfactant that has accumulated for the preceding weeks keeps the air sacs open, easing subsequent breaths.

Breathing is especially difficult for premature babies. Normally, a pregnancy lasts about 9 months, or from 38 to 42 weeks. A baby is considered premature if it is born at 37 weeks or earlier. The shorter the gestation, the more health problems the infant will have. Usually a baby born after about 25 weeks of gestation can survive with intensive care; a fetus younger than that is usually not mature enough to survive.

One problem that premature babies face is that their lungs are not yet ready to breathe air. They have not yet produced enough surfactant to prevent the alveoli from collapsing after each breath, which causes infantile respiratory distress syndrome. The premature newborn must fight as hard for every breath as a full-term infant does for the first one. For this reason, a pregnant woman who knows she will deliver a premature baby may receive hormones that speed the production of surfactant in the fetus. After birth, synthetic surfactant dripped into the respiratory tract can help these tiny patients breathe.

In addition, a mechanical ventilator or fine breathing tubes may help keep a premature infant's lungs inflated. A monitor constantly measures the amount of oxygen in the blood—too much can be as harmful as too little. Even if the baby can breathe without help, its underdeveloped brain may not control breathing as it should. Neonatal intensive care wards therefore continuously monitor babies for apnea, the cessation of breathing.

The challenges that premature babies face illustrate the importance of the respiratory system in supplying the oxygen that all of the body's tissues need to function properly. This chapter describes the structure and function of this vital organ system.

Learning Outline

Learn How to Learn
How to Use a Tutor

Your school may provide tutoring sessions for your class, or perhaps you have hired a private tutor. How can you make the most of this resource? First, meet regularly with your tutor for an hour or two each week; don't wait until just before an exam. Second, if possible, tell your tutor what you want to work on before each session, so he or she can prepare. Third, bring your textbook, class notes, and questions to your tutoring session. Fourth, be realistic. Your tutor can discuss difficult concepts and help you practice with the material, but don't expect him or her to simply give you the answers to your homework.

30.1 | Gases Diffuse Across Respiratory Surfaces

Each of us breathes some 20,000 times a day. Most of the time, you inhale and exhale without thinking—unless you happen to be using your breath to fog a mirror, spin a pinwheel, play the trumpet, or make soap bubbles (**figure 30.1**). Fun aside, breathing is obviously a vital function; if a person is deprived of air, death can occur within minutes.

Breathing is so automatic that it is easy to forget why we do it. Cells use ATP to power protein synthesis, movement, DNA replication, cell division, growth, reproduction, and countless other activities that require energy. As described in chapter 6, animal cells generate ATP in **aerobic respiration,** which consumes oxygen gas (O_2) and generates carbon dioxide gas (CO_2) as a waste product. In most animals, the **respiratory system** is the organ system that does the breathing; it works with the circulatory system (the topic of chapter 29) to deliver O_2 to cells and to eliminate the CO_2 waste. Without gas exchange, cells die.
▶ ATP, p. 76

The term *respiration* also has two additional meanings, besides its use in aerobic cellular respiration. One is breathing (ventilation), the physical movement of air into and out of the body. Respiration can also mean the act of exchanging gases. External respiration is gas exchange between an animal's body and its environment; internal respiration is gas exchange between tissue cells and the bloodstream.

An animal's **respiratory surface** is the area of its body where external respiration occurs. In humans and other terrestrial vertebrates, the respiratory surface is inside the lungs, but other animals use other surfaces. Regardless of their form, however, all respiratory surfaces share three characteristics.

First, their surface area must be relatively large. Not only must the respiratory surface meet the animal's demand for O_2, but it must also eliminate CO_2 fast enough to keep this waste gas from accumulating to toxic levels.

Second, an animal's respiratory surface must come into contact with either air or water; both of these substances can be a source of O_2 and a "dumping ground" for CO_2. In general, air offers two advantages over water. Air has a higher concentration

![Blowing bubbles photo]

Figure 30.1 Blowing Bubbles. This girl is using air exhaled from her lungs to make a soap bubble.

Figure 30.2 Respiratory Surfaces. Gas exchange may occur across the body surface or via specialized respiratory organs such as tracheae, gills, or lungs.

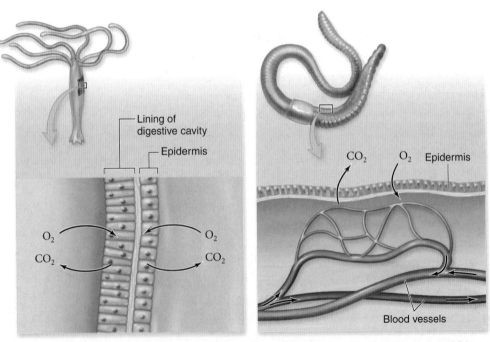

Lining of digestive cavity
Epidermis
O_2
CO_2
O_2
CO_2

CO_2 O_2 Epidermis
Blood vessels

Body surface: Gases diffuse through the body's surface. Found in many invertebrates and amphibians.

of O_2 than does water; in addition, air is lighter than water. Less energy is therefore required to move air across a respiratory surface than to move an equal volume of water.

Third, respiratory surfaces consist of moist membranes across which O_2 and CO_2 diffuse. This requirement for moisture puts air-breathing organisms at a disadvantage: a respiratory surface exposed to air may dry out, rendering it useless. ▶ diffusion, p. 80

This section describes a variety of respiratory surfaces, ranging from the relatively unspecialized to the complex (**figure 30.2**). It may be helpful to refer back to figure 30.2 as you read through the rest of this section.

A. Some Invertebrates Exchange Gases Across the Body Wall or in Internal Tubules

In animals that are very small or very flat (or both), all cells are close to the external environment. Gas exchange occurs across the moist body wall, as gases simply diffuse into and out of each cell. For example, cnidarians such as *Hydra* and sea anemones have thin, extended body parts with a large surface area for gas exchange. Likewise, flatworms are thin enough for gas exchange and distribution to occur without specialized respiratory or circulatory systems. ▶▶ cnidarians, p. 416; flatworms, p. 418

In larger animals, the body's external surface area is insufficient to exchange gases with cells in deeper tissues. These animals often have a circulatory system that transports gases between interior cells and the skin. In an earthworm, for example, blood vessels connect the body's surface with deeper tissue layers. O_2 from the surrounding air diffuses across the skin and into the blood vessels; CO_2 moves in the opposite direction, out of the blood and into the atmosphere. ▶ annelids, p. 422

In insects and many other arthropods, the circulatory system plays only a minor role in the transportation of gases. These animals have **tracheae,** which are internal, air-filled tubes that connect to the atmosphere through openings, called spiracles, along each side of the abdomen. Inside the animal, tracheae branch into tiny tubules that extend around individual cells. Tracheae therefore bring the outside environment close to every cell of the body. ▶ insects, p. 429

Chitin, the material that makes up the exoskeleton, strengthens the walls of the tracheae, but these tubules could not withstand the crushing weight of a very large arthropod. Tracheal breathing may therefore be one factor that constrains insect size; the largest insects are only about 15 centimeters long, and most are much smaller.

B. Gills Exchange Gases with Water

Most aquatic organisms have **gills,** highly folded structures containing blood vessels that exchange gases with water across a thin layer of epithelium. Gills occur in some invertebrates, such as mollusks, and in vertebrates such as fishes and amphibians.

A bony fish's gills work by a mechanism called **countercurrent exchange,** in which two adjacent currents flow in opposite directions and exchange materials with each other (chapter 32

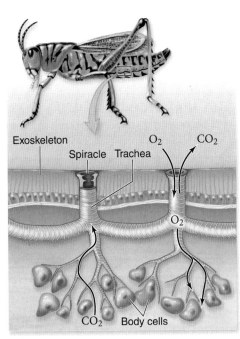

Tracheae: Gases enter tracheae and diffuse directly into the cells; they do not enter capillaries first as in most other respiratory systems. Found in many arthropods, including insects.

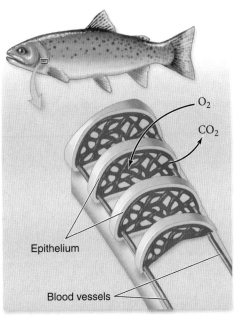

Gills: Gases are exchanged at blood vessels covered by a thin layer of epithelium. Found in fishes, amphibian larvae, and many invertebrates.

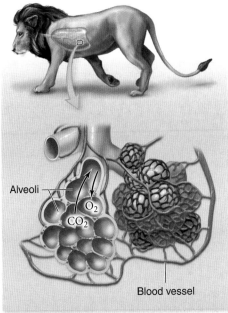

Lungs: Two-way airflow. Found in terrestrial vertebrates.

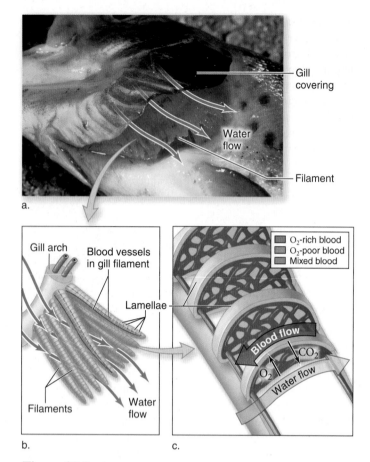

a.

b.   c.

O₂-rich blood
O₂-poor blood
Mixed blood

Gill arch
Blood vessels
in gill filament

Lamellae

Filaments

Water
flow

Blood flow

CO₂

O₂

Water flow

Gill
covering

Water
flow

Filament

Figure 30.3 Fish Gills. (a) The feathery gills of a fish lie beneath a protective cover. (b) Each gill arch consists of many filaments. (c) A filament is divided into many platelike lamellae that house blood vessels. The direction of water flow across the lamellae opposes that of blood flow in the capillaries. This "countercurrent" relationship maximizes gas exchange between water and the blood.

describes other examples of countercurrent exchange). Examine the gills illustrated in **figure 30.3**. Each gill arch is composed of many filaments. Each filament includes platelike lamellae that house a dense network of **capillaries,** the tiniest of blood vessels. In figure 30.3c, water is flowing from left to right across the gill membrane. Blood, however, is flowing through the capillaries in the opposite direction. This arrangement maximizes gas exchange because the concentration gradient favors O_2 diffusion from water to blood along the entire length of the capillary bed. Meanwhile, CO_2 diffuses from blood to water. ▶ fishes, p. 438

Bony fishes are not the only vertebrates with gills; tadpoles and other amphibian larvae have external gills that close off once the lungs begin working after metamorphosis. Some salamanders, however, retain external gills through adulthood. ▶ amphibians, p. 440

C. Terrestrial Vertebrates Exchange Gases in Lungs

As vertebrates ventured onto the land hundreds of millions of years ago, gills became useless; these delicate, feathery structures simply dry up when exposed to air. Instead, terrestrial habitats selected for **lungs,** saclike organs that exchange gases. The location of lungs inside the body helps keep the respiratory surfaces moist.

Where did lungs come from in the first place? Anatomical evidence suggests that lungs are homologous to the gas bladders of bony fishes. In lungfishes, these sacs allow fishes to gulp air, supplementing gas exchange through gills during periods of drought. In the ancestors of terrestrial vertebrates, however, these gas bladders developed into air-breathing lungs. ▶▶ homologous structures, p. 262; lungfishes, p. 439

A look at the lungs of modern vertebrate species gives an idea of how these organs have increased in complexity (**figure 30.4**). The simplest lungs, such as those in lungfishes and

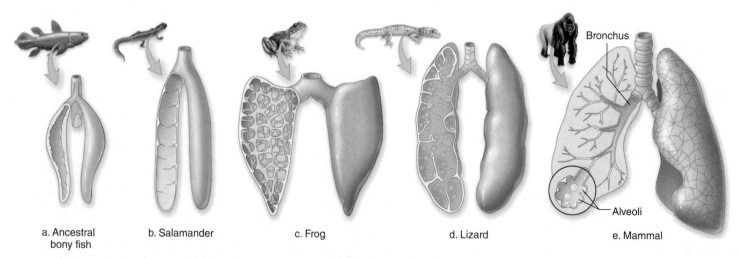

a. Ancestral
bony fish

b. Salamander

c. Frog

d. Lizard

e. Mammal

Bronchus

Alveoli

Figure 30.4 Lung Evolution. (a) The first lungs were little more than paired sacs with capillary networks in their smooth walls. (b) Lungs in salamanders have a few partitions. (c) Lungs become more extensively subdivided in frogs and toads. (d) The lizard lung is even more clearly compartmentalized. (e) The mammalian lung takes maximization of surface area to an extreme, with millions of microscopic air sacs.

salamanders, are little more than air-filled pouches. The lungs of frogs and toads, however, are somewhat partitioned, an adaptation that increases the surface area available for external respiration. In addition, an amphibian supplements gas exchange in the lungs by obtaining oxygen through its moist skin and the lining of its mouth.

A lizard's lungs are much more extensively subdivided. The mammalian lung, however, is truly spectacular in its maximization of surface area; millions of tiny air sacs sit within baskets of capillaries. Mammals have a high metabolic rate, both because they maintain a constant body temperature and because they tend to be very active animals. The complex mammalian lung provides the enormous respiratory surface necessary to support this high metabolic rate. ▶▶ reptiles, p. 442; mammals, p. 445

Lungs did, however, introduce a new challenge: moving air in and out to renew the O_2 supply and expel CO_2. Vertebrates accomplish this task in a variety of ways (**figure 30.5**). For example, frogs force air into their lungs under positive pressure and then expel it by contracting muscles in the body wall. Birds, on the other hand, have an extensive respiratory system through which air flows in one direction. O_2-rich air therefore never mixes with O_2-depleted air; moreover, O_2 remains in the respiratory system even during exhalation. Thanks to this arrangement, the bird's respiratory system is extremely efficient, delivering the O_2 required to support the enormous metabolic demands of flight. ▶ birds, p. 443

The lungs of mammals, including humans, move air in yet another way: a muscular diaphragm expands the chest area, so air flows passively into the lungs under negative pressure. Section 30.3 describes this topic in more detail.

30.1 | Mastering Concepts

1. What is the main function of the respiratory system?
2. What is the relationship between the circulatory and respiratory systems?
3. Differentiate between aerobic cellular respiration, internal respiration, and external respiration.
4. What characteristics do all respiratory surfaces share?
5. List four examples of respiratory surfaces and describe how each participates in gas exchange.
6. Describe how lungs differ among the main groups of terrestrial vertebrates.

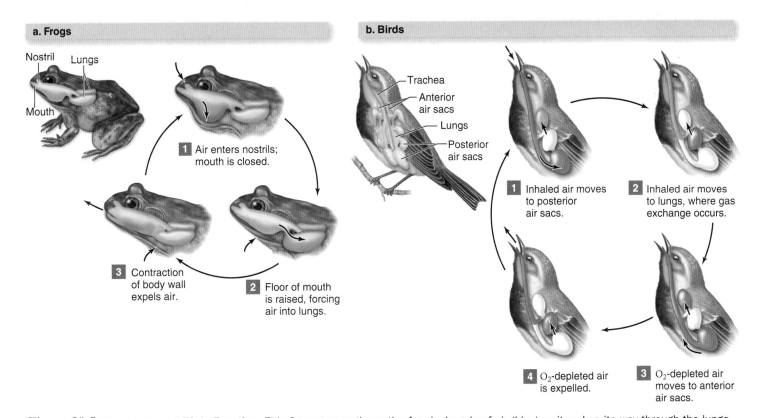

a. Frogs

Nostril Lungs

Mouth

1 Air enters nostrils; mouth is closed.

3 Contraction of body wall expels air.

2 Floor of mouth is raised, forcing air into lungs.

b. Birds

Trachea
Anterior air sacs
Lungs
Posterior air sacs

1 Inhaled air moves to posterior air sacs.

2 Inhaled air moves to lungs, where gas exchange occurs.

4 O_2-depleted air is expelled.

3 O_2-depleted air moves to anterior air sacs.

Figure 30.5 How Frogs and Birds Breathe. This figure traces the path of a single gulp of air *(blue)* as it makes its way through the lungs of two types of vertebrates. (a) A frog draws air into its nostrils with its mouth shut and forces the air into the lungs. The air is expelled when the body wall contracts and the lungs recoil. (b) In a bird, air moves in one direction through the lungs. Inhaled air goes first to the posterior air sacs, then passes through the lungs and enters the anterior air sacs before being released. Compare these animals to the human in figure 30.9.

30.2 | The Human Respiratory System Delivers Air to the Lungs

The human respiratory system is a continuous network of tubules that delivers O_2 to the circulatory system and unloads CO_2 into the lungs. **Figure 30.6** presents an overview of the system, and **table 30.1** summarizes its functions.

A. The Nose, Pharynx, and Larynx Form the Upper Respiratory Tract

The **nose,** which forms the external entrance to the nasal cavity, functions in breathing, immunity, and the sense of smell. Stiff hairs at the entrance of each nostril keep dust and other large particles out. If a large particle is inhaled, a sensory cell in the nose may signal the brain to orchestrate a sneeze, which forcefully ejects the object. ▸ sense of smell, p. 555

Epithelial tissue in the nose secretes a sticky mucus that catches most airborne bacteria and dust particles that manage to pass the hairs. Enzymes in the mucus destroy some of the would-be invaders, and immune system cells under the epithelial layer await any disease-causing organisms that penetrate the mucus.

The nasal cavity also adjusts the temperature and humidity of incoming air. Blood vessels lining the nasal cavity release heat, and mucus contributes moisture to the air. This function ensures that the respiratory surface of the lungs remains moist.

The back of the nose and mouth leads into the **pharynx,** or throat. Both swallowed food and inhaled air pass through the pharynx. Just below and in front of the pharynx is the **larynx,** or

Adam's apple, a boxlike structure that produces the voice. Stretched over the larynx are the **vocal cords,** two elastic bands of tissue that vibrate as air from the lungs passes through a slit-like opening called the **glottis.** Vibrations of the vocal cords produce the sounds of speech. A male's voice becomes deeper during puberty because the vocal cords grow longer and thicker. The cords therefore vibrate more slowly during speech, producing lower-frequency sounds that we perceive as a deeper voice.

Another function of the larynx is to direct ingested food and drink away from the respiratory system. During swallowing, a cartilage flap called the **epiglottis** covers the glottis so that food enters the digestive tract instead of the lungs.

Table 30.1   Functions of the Human Respiratory System: A Summary

Function	Explanation
Gas exchange	Lungs exchange oxygen (O_2) and carbon dioxide (CO_2) with blood.
Olfaction (sense of smell)	Breathing moves air across the nose's olfactory epithelium, which detects odors.
Production of sounds, including speech	Movement of air across the vocal cords in the larynx produces sounds.
Maintaining blood pH homeostasis	Breathing more slowly or rapidly adjusts the concentration of CO_2 in blood, which affects blood pH.

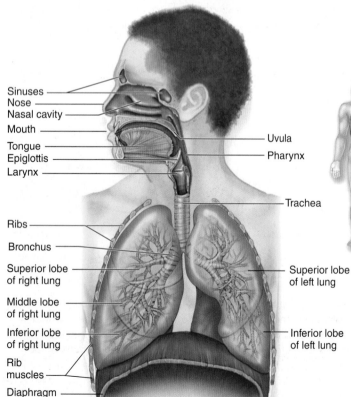

Sinuses
Nose
Nasal cavity
Mouth
Tongue
Epiglottis
Larynx
Uvula
Pharynx
Ribs
Bronchus
Trachea
Superior lobe of right lung
Superior lobe of left lung
Middle lobe of right lung
Inferior lobe of right lung
Inferior lobe of left lung
Rib muscles
Diaphragm

	Respiratory System	
Main Tissue Types*	**Examples of Locations/Functions**	
Epithelial	Enables diffusion across walls of alveoli and capillaries; secretes mucus along respiratory tract.	
Connective	Blood (a connective tissue) exchanges gases with lungs; cartilage makes up part of the nose, trachea, bronchi, and larynx.	
Nervous	Autonomic nervous system controls smooth muscle in bronchi.	
Muscle	Smooth muscle in lungs regulates airflow to alveoli; skeletal muscle in diaphragm expands lungs.	

*See chapter 24 for descriptions.

Figure 30.6 The Human Respiratory System. Inhaled air passes through the mouth and trachea and then into increasingly narrow tubes until it arrives in the alveoli, where gas exchange occurs.

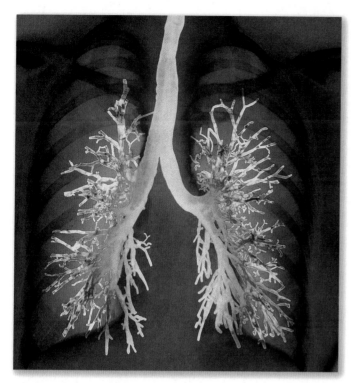

Figure 30.7 The Bronchial Tree. The respiratory passages of the lungs form a complex branching pattern. This is an X-ray image called a bronchogram.

Figure 30.8 Alveoli. The human lung contains some 300 million alveoli, which make the lung's structure similar to that of foam rubber. Gas exchange occurs at the lush capillary network that surrounds each cluster of alveoli.

The entire upper respiratory tract is lined with epithelium that secretes mucus. Dust and other inhaled particles trapped in the mucus are swept out by waving cilia. Coughing brings the mucus up, to be either spit out or swallowed.

B. The Lower Respiratory Tract Consists of the Trachea and Lungs

The **trachea,** or windpipe, is the tube just beneath the larynx. C-shaped rings of cartilage hold the trachea open and accommodate the expansion of the esophagus—the tube leading from the mouth to the stomach—during swallowing. You can feel these rings in the lower portion of your throat. Cilia and mucus coat the trachea's inside surface, trapping debris and moistening the incoming air.

The trachea branches into two **bronchi,** one leading to each lung. The bronchi branch repeatedly, each branch decreasing in diameter and wall thickness (**figure 30.7**). **Bronchioles** ("little bronchi") are the finest branches. The bronchioles have no cartilage, but their walls contain smooth muscle. The autonomic nervous system controls contraction of these muscles, adjusting airflow in response to metabolic demands. ▸ autonomic nervous system, p. 538

Each bronchiole narrows into several alveolar ducts, and each duct opens into a grapelike cluster of alveoli, where gas exchange occurs (**figure 30.8**). Each **alveolus** is a tiny sac with a wall of epithelial tissue that is one cell layer thick. A vast network of capillaries surrounds each cluster of alveoli. Oxygen and CO_2 diffuse through the thin walls of the alveoli and the neighboring capillaries. The interface between the alveoli and the capillaries is the respiratory surface in humans, and it is enormous: the total surface area of the alveoli in one pair of lungs is about 50 times the area of a person's skin.

30.2 | Mastering Concepts

1. List the components of the upper and lower respiratory tracts.
2. How does the nose function in the respiratory system?
3. Which structures prevent swallowed food from entering the respiratory system?
4. Describe the relationships among the trachea, bronchi, bronchioles, and alveoli.

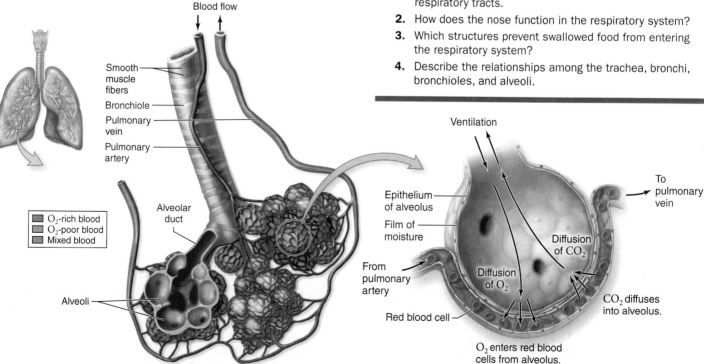

Blood flow

Smooth muscle fibers
Bronchiole
Pulmonary vein
Pulmonary artery

O$_2$-rich blood
O$_2$-poor blood
Mixed blood

Alveolar duct

Alveoli

Ventilation

Epithelium of alveolus
Film of moisture

To pulmonary vein

Diffusion of CO_2

From pulmonary artery

Diffusion of O_2

Red blood cell

CO_2 diffuses into alveolus.

O_2 enters red blood cells from alveolus.

30.3 | Breathing Requires Pressure Changes in the Lungs

Pay attention to your own breathing for a moment. Each **respiratory cycle** consists of one inhalation and one exhalation (**figure 30.9**). Each time you **inhale,** air moves into the lungs; when you **exhale,** air flows out of the lungs.

What drives this back-and-forth motion? The answer is that air flows from areas of high pressure to areas of low pressure. Therefore, air enters the body when the pressure inside the lungs is lower than the pressure outside the body. Conversely, air moves out when the pressure in the lungs is greater than the atmospheric pressure.

The body generates these pressure changes by altering the volume of the chest cavity. As a person inhales, skeletal muscles of the rib cage and diaphragm contract, expanding and elongating the chest cavity. The resulting increase in volume lowers the air pressure within the space between the lungs and the outer wall of the chest. The lungs therefore expand, lowering pressure in the alveoli. Air rushes in. Inhalation requires energy because muscle contraction uses ATP. ▸ muscle contraction, p. 590

The muscles of the rib cage and the diaphragm then relax. As a result, the rib cage falls to its former position, the diaphragm rests up in the chest cavity again, and the elastic tissues of the lung recoil. The pressure in the lungs now exceeds atmospheric pressure, so air flows out. At rest, exhalation is passive—that is, it requires only muscle relaxation, not contraction—and therefore does not require ATP.

Hiccups briefly interrupt the respiratory cycle. In a hiccup, the diaphragm contracts unexpectedly, causing a sharp intake of air; the "hic" sound occurs as the epiglottis closes. Chapter 12's Apply It Now box explores the evolutionary origin of hiccups, which have no known function.

Illness or injury may prevent contraction of the diaphragm and rib muscles, so breathing stops. Old-fashioned "iron lungs" and modern mechanical ventilators compensate for this loss of function (**figure 30.10**). An iron lung is an airtight chamber in which the patient lies with his or her head sticking out. Every few seconds, the air pressure inside the machine drops, drawing air into the lungs through the nose or mouth. Conversely, raising the pressure causes exhalation.

Medical professionals can test lung function by having a patient blow into a device called a spirometer. One measure of lung capacity is the **tidal volume,** or the amount of air inhaled or exhaled during a quiet breath taken at rest. In a young adult male, the tidal volume is about 500 milliliters. **Vital capacity,** on the other hand, is the total amount of air that a person can exhale after taking the deepest possible breath, about 4700 mL in a young man.

As we age, vital capacity declines. Illnesses can also interfere with a person's lung function by reducing the elasticity of the lungs, obstructing the airways, or weakening the muscles of the chest. This chapter's Apply It Now box describes some breathing disorders.

30.3 | Mastering Concepts

1. What is the relationship between the volume of the chest cavity and the air pressure in the lungs?
2. Describe the events of one respiratory cycle.
3. Define tidal volume and vital capacity.

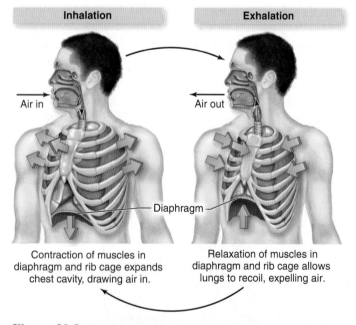

Inhalation | **Exhalation**

Air in | Air out

Diaphragm

Contraction of muscles in diaphragm and rib cage expands chest cavity, drawing air in. | Relaxation of muscles in diaphragm and rib cage allows lungs to recoil, expelling air.

Figure 30.9 How We Breathe. Inhalation requires contraction of the muscles of the diaphragm and rib cage. The expanding chest cavity has lower air pressure than the atmosphere, so air moves into the lungs. When we exhale, the diaphragm relaxes and the rib cage lowers, reversing the process and pushing air out of the lungs.

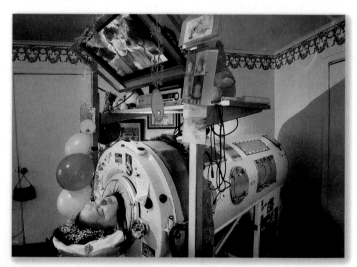

Figure 30.10 Breathing Machine. When this photo was taken, this 60-year-old woman had been in an iron lung for 57 years. She was a small child when she contracted polio, a viral disease that caused her muscles to deteriorate. Because her diaphragm is paralyzed, she cannot inhale on her own.

Apply It Now

The Unhealthy Respiratory System

Asthma

During an asthma attack, spasms occur in the smooth muscle lining the lung's bronchi, slowing airflow and causing wheezing. An allergy to pollen, dog or cat dander (skin particles), or dust mites triggers most asthma attacks. Inhalant drugs that treat asthma usually relax the bronchial muscles.

Lung Cancer

Eighty-five percent of lung cancer cases occur in smokers. Cigarette smoke contains chemicals that mutate DNA in lung cells; these altered cells may divide and form tumors. Patients with lung cancer experience chest pain, chronic coughing, and shortness of breath. Large tumors may obstruct the airway. Moreover, cancerous cells may break away from the tumor and spread throughout the body. Besides smoking, risk factors for lung cancer include exposure to asbestos or radioactivity. ▶ cancer, p. 162

Apnea

Apnea is the cessation of breathing. It can be voluntary for a short time, as when you hold your breath. Premature infants may stop breathing because the part of the brain that regulates breathing is not fully developed (see the chapter opening essay). Overweight adults are especially susceptible to sleep apnea. The fleshy folds of the throat sag into the airway, causing breathing to stop for a short time. The person reflexively clears the throat, and breathing resumes. Because sleep is interrupted many times a night, people with sleep apnea may be extremely tired during the day.

Pneumonia

Pneumonia is an inflammation of the alveoli, usually resulting from an infection. Many bacteria and viruses cause pneumonia. Symptoms include green or yellow mucus, fever, chest pain, coughing, and shortness of breath.

Common Cold

Viruses that cause colds infect cells in the upper respiratory tract (see chapter 15). A day or so after infection, the immune system's efforts to get rid of the invading viruses cause the typical symptoms of a cold: coughs, runny nose, and sneezes. There is no cure. Fortunately, colds are usually not a serious health problem, although clogged sinuses can be vulnerable to bacterial infections.

Tuberculosis

Infection with the bacterium *Mycobacterium tuberculosis* causes tuberculosis (also called TB). Symptoms of advanced TB include painful coughs, bloody phlegm, fever, and weight loss. When a patient with an active TB infection coughs or sneezes, he or she expels bacteria-laden droplets that nearby people can inhale. Once in the lungs, the bacteria replicate inside immune system cells. Other immune system cells clump around the infected cells. This response keeps the bacteria from spreading, but these clusters also provide a place for the bacteria to become dormant. The resulting latent infection may later reemerge as full-blown (active) TB. Tuberculosis is especially important today because of the emergence of *Mycobacterium* strains that resist antibiotic treatment.

COPD (Emphysema and Chronic Bronchitis)

Emphysema, a term derived from the Greek word for "inflate," is an abnormal accumulation of air in the lungs. Long-term exposure to cigarette smoke and other irritants causes a loss of elasticity in lung tissues. The walls of the alveoli tear, impeding airflow and reducing the surface area for gas exchange. The patient experiences shortness of breath, an expanded chest, and hyperventilation. Emphysema is often accompanied by chronic bronchitis—that is, inflammation of the bronchi. Together, these two illnesses are called chronic obstructive pulmonary disease (COPD).

Cystic Fibrosis

Cystic fibrosis is among the most common inherited diseases. A faulty transport protein in the membranes of epithelial cells causes sticky mucus to accumulate in the lungs. The thick mucus prevents cilia from beating. Bacteria also thrive in the warm, moist mucus, causing persistent infections. (The same faulty protein may confer resistance to cholera when expressed in intestinal cells, as described in section 4.6.)

The Effects of Smoking on the Respiratory System

Besides increasing the risk of heart attack and stroke (see chapter 29), tobacco use also disrupts the respiratory system (**figure 30.A**). The very first inhalation of cigarette smoke slows the beating of cilia. Over time, the cilia become paralyzed, and they eventually vanish. Without cilia to remove mucus, coughing alone must clear particles from the airways. Smoking also causes excess mucus production, which favors the reproduction of disease-causing microorganisms. Smokers are therefore especially susceptible to respiratory infections.

The smoker's cough leads to emphysema and chronic bronchitis. As the linings of the bronchioles thicken and become less elastic, they can no longer absorb the pressure changes that accompany coughing. A cough can therefore rupture the delicate walls of alveoli, causing a worsening cough, fatigue, wheezing, and impaired breathing. Meanwhile, cancers of the lungs, mouth, larynx, esophagus, and jaw can develop and spread throughout the body.

The addictive properties of nicotine make quitting difficult, but it pays to stop smoking. Cilia may reappear, and the thickening of alveolar walls can reverse, although ruptured alveoli are gone forever.

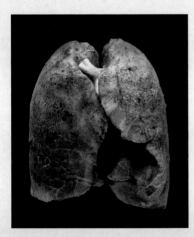

Figure 30.A Smoker's Lungs. These lungs have turned black after years of exposure to cigarette smoke.

30.4 | Blood Delivers Oxygen and Removes Carbon Dioxide

As we breathe, O_2 from the atmosphere enters the capillaries surrounding the alveoli; CO_2 moves in the opposite direction, from the capillaries to the alveoli. As a result, the air we exhale has a different gas composition than inhaled air (table 30.2). This section describes how blood transports O_2 and CO_2, and how the brain uses blood gas concentrations to set the breathing rate.

A. Blood Carries Gases in Several Forms

Blood is a connective tissue consisting of cells and platelets suspended in **plasma,** the liquid component of blood (see figure 29.4). Cells and plasma interact in different ways to carry O_2 and CO_2.

Only about 1% of O_2 in blood is dissolved in plasma; **red blood cells** transport the rest. These cells are packed with **hemoglobin,** an iron-rich pigment protein that binds with O_2 (figure 30.11). Each hemoglobin molecule can carry up to four O_2 molecules. The hemoglobin in a red blood cell becomes almost completely saturated with O_2 after spending only a second or two in the alveolar capillaries.

Not surprisingly, illness or death results when hemoglobin cannot bind O_2. Carbon monoxide (CO), for example, is a colorless, odorless gas in cigarette smoke and in exhaust from car engines, kerosene heaters, wood stoves, and home furnaces. CO binds to hemoglobin more readily than O_2 does, and it is less likely to leave the hemoglobin molecule. When 30% of the hemoglobin molecules carry CO instead of O_2, a person loses consciousness and may go into a coma or even die.

Blood carries CO_2 absorbed at tissue cells in three ways. About 5% to 10% is dissolved in plasma, and hemoglobin transports about 23%. (Hemoglobin can carry both O_2 and CO_2 at the same time because the gases attach to different sites on the molecule.) But most CO_2, about 70%, is transported as bicarbonate ions (HCO_3^-). Bicarbonate ions are the product of the following sequence of chemical reactions (the two double arrows indicate that the reactions are reversible):

$$CO_2 + H_2O \longleftrightarrow H_2CO_3 \text{ (carbonic acid)}$$

$$H_2CO_3 \longleftrightarrow H^+ \text{ (hydrogen ions)} + HCO_3^- \text{ (bicarbonate)}$$

Table 30.2 Comparison of Inhaled and Exhaled Air

Gas	Proportion by Volume (%)	
	Inhaled Air (at sea level)	Exhaled Air
O_2	21	16.5
CO_2	0.03	4
Nitrogen (N_2)	78	79.5
Other gases	≈ 1	≈ 1

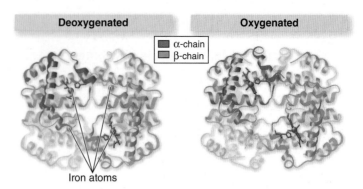

Figure 30.11 Hemoglobin. This protein consists of four chains (two alpha globin and two beta globin). When iron atoms bind to O_2, the shape of the protein changes slightly. Deoxygenated hemoglobin appears deep red; the oxygenated form is bright red.

An enzyme in red blood cells speeds the second reaction, but most bicarbonate ions subsequently diffuse into plasma. At the lungs, these reactions reverse, liberating CO_2 into the air to be exhaled. ▸ enzymes, p. 78

Carbonic acid and bicarbonate are an important part of blood's ability to buffer pH changes and maintain homeostasis. Ordinarily, the respiratory and excretory systems work together to maintain the pH of blood at about 7.4. Some metabolic processes, however, release H^+ into the blood; an example is the production of lactic acid in strenuously exercising muscles. If blood pH declines, excess H^+ reacts with HCO_3^- and forms H_2CO_3, raising the pH. ▸ pH, p. 30

Gas exchange in the alveoli and at the body's other tissues relies on simple diffusion (figure 30.12). In external respiration, which occurs at the lungs, O_2 diffuses from the alveoli into the bloodstream, where the concentration of O_2 is lower. At the same time, CO_2 diffuses from the blood to the air spaces of the lungs. The heart then pumps the freshly oxygenated blood to the rest of the body. In internal respiration, O_2 diffuses from the bloodstream to the tissue fluid and then into the body's respiring cells, which have the lowest O_2 concentration. CO_2 diffuses in the opposite direction. ▸ diffusion, p. 80

B. Blood Gas Levels Help Regulate the Breathing Rate

The heart has an internal pacemaker, so it can beat without stimulation from the nervous system. In contrast, breathing involves coordinated contractions of multiple skeletal muscles (see section 30.3). The brain's breathing control centers in the pons and medulla send out the neural signals that control these muscles. ▸ brainstem, p. 541

The regulation of the depth and rate of breathing helps maintain homeostasis in blood gas concentrations (figure 30.13). Chemoreceptors in the medulla and in the body's largest arteries monitor blood CO_2 levels. Because CO_2 is a byproduct of aerobic respiration, monitoring blood CO_2 levels is a good way to determine how quickly cells use oxygen. When the blood CO_2 level increases, the resulting decline in pH stimulates chemoreceptors

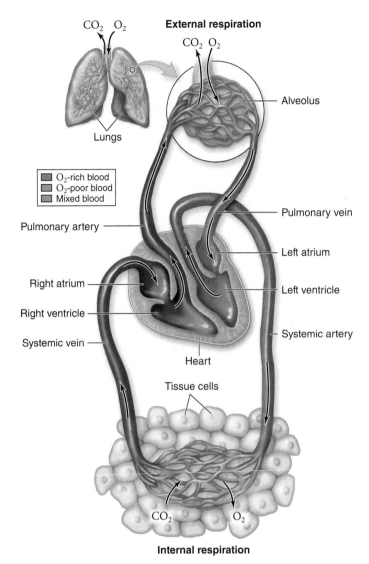

Figure 30.12 Gas Exchange at the Lungs and Tissues. CO_2 diffuses out of red blood cells and plasma and into the alveoli of the lungs. O_2 moves in the opposite direction. In the rest of the body, O_2 diffuses from blood and to tissues, while CO_2 moves into the bloodstream.

to send messages to the medulla. Neurons in the medulla subsequently trigger an increase in the breathing rate. Small changes in the CO_2 concentration of the air we breathe can trigger a tremendous increase in breathing rate. For example, an increase to 0.5% CO_2 in the air (from the normal 0.03%) makes us breathe 10 times faster. ▸ negative feedback, p. 520

Oxygen level is less important than CO_2 level in regulating breathing. In fact, the concentration of O_2 in blood affects breathing rate only if it falls dangerously low. (This chapter's Burning Question discusses mountaintop breathing.) Normally, however, the large amount of O_2 bound to hemoglobin provides an ample margin of safety. Nevertheless, sudden infant death syndrome, in which a baby dies while asleep, may occur when chemoreceptors fail to detect low oxygen levels in arterial blood.

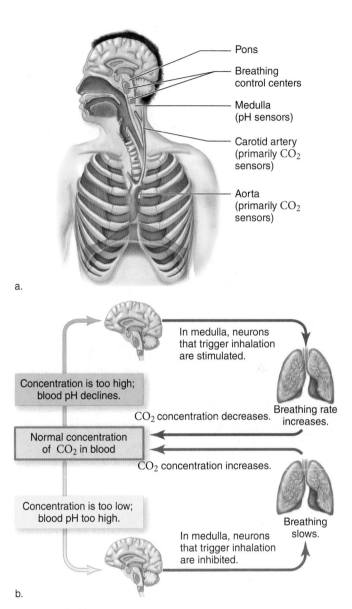

Figure 30.13 Breathing Control. (a) Multiple chemoreceptors send information about blood pH and CO_2 concentrations to the brain. Neurons in the medulla integrate this information and regulate the contraction of rib and diaphragm muscles. (b) The control of breathing is an example of a negative feedback loop.

30.4 | Mastering Concepts

1. What protein in a red blood cell delivers oxygen to the body's tissues?
2. In what forms does blood transport O_2 and CO_2?
3. How does carbon monoxide reduce blood's ability to carry O_2?
4. Describe the diffusion gradients for O_2 and CO_2 in the lungs and in the rest of the body.
5. Describe how the brain's breathing control centers use blood CO_2 concentration to alter the breathing rate.

30.5 | Investigating Life: Why Do Bugs Hold Their Breath?

 A child having a tantrum may scream, "I'll hold my breath until I turn blue!" in an effort to get his way. Luckily, the parents have nothing to fear; even if the child does begin to carry out his plan, his brain's breathing control centers will eventually override his effort to hold his breath. In insects, the situation is different. An insect can hold its breath much longer than we can, and its parents might actually approve: evidence suggests that an insect that periodically holds its breath may live a long, healthy life.

An insect's respiratory system consists of air-filled tracheae extending inward from valvelike openings called spiracles (see figure 30.2). Gases diffuse significantly more rapidly in air than in water or blood, and calculations show that an insect can easily exchange all the O_2 and CO_2 it needs by keeping its spiracles open all the time. Yet some insects close their spiracles for extended periods; in effect, the animal periodically holds its breath. In one breathing cycle, the spiracles may be open for 15 minutes, closed for 20 minutes, and then "flutter" open and closed for over an hour before opening again for 15 minutes.

Scientists have speculated for years about how this strange pattern benefits an insect. Some thought that discontinuous breathing reduces water loss, much as plants conserve water by closing pores (stomata) in their leaves during hot or windy weather. Yet subsequent studies did not support this hypothesis. For example, grasshoppers close their spiracles at night, when water loss is minimal, but not during the drier daytime. Another idea was that discontinuous breathing helps underground insects such as ants cope with low O_2 and high CO_2 conditions. That explanation, however, leaves out aboveground insects that breathe discontinuously.

Biologists Stefan Hetz (from Humboldt University in Berlin) and Timothy Bradley (from the University of California, Irvine) suggested a third explanation. They proposed that discontinuous breathing guards against *too much* O_2. Oxygen, of course, is necessary for aerobic respiration; without it, an animal dies. But O_2 is also toxic. The higher the concentration of O_2, the more harmful free radicals produced in the mitochondria. The more free radicals, the more damage to cells; in fact, studies have shown that insects experimentally exposed to the highest concentrations of O_2 live the shortest time.

? Burning Question

How does the body respond to high elevations?

The human respiratory system functions best near sea level. At high elevations, air density and oxygen availability fall gradually, so that at about 3000 meters above sea level, an individual inhales a third less O_2 with each breath. Each year, 100,000 mountain climbers and high-altitude exercisers experience altitude sickness (table 30.A). More than 180 climbers have died attempting to reach Mount Everest's 8850-meter summit, often from the effects of low oxygen. At high altitudes, the body's effort to get more O_2, by increasing breathing and

heart rate, cannot keep pace with declining O_2 concentration. Altitude sickness can be cured by descending slowly at the first appearance of symptoms.

The body also reacts to the low O_2 supply at high altitudes by increasing red blood cell production. When cells in the kidneys do not receive enough O_2, they trigger production of the hormone erythropoietin (also called EPO), which stimulates red blood cell production. Synthetic erythropoietin is a treatment for anemia (a reduction in the O_2-carrying capacity of blood), but some athletes inject themselves with EPO to gain a performance edge. This illicit practice is a form of blood doping, which improves an athlete's stamina but is difficult for sports authorities to detect. Doping with EPO can be dangerous because it thickens the blood and therefore strains the heart.

Submit your burning question to:
marielle_hoefnagels@mcgraw-hill.com

Table 30.A The Effects of High Altitude

Altitude (meters)	Condition	Symptoms
1800	Acute mountain sickness	Headache, weakness, nausea, poor sleep, shortness of breath
2700	High-altitude pulmonary edema (fluid accumulation in lungs)	Severe shortness of breath, cough, gurgle in chest, stupor, weakness; person can drown in accumulated fluid in lungs
4000	High-altitude cerebral edema (fluid accumulation in brain)	Severe headache, vomiting, altered mental status, loss of coordination, hallucinations, coma, and death

Figure 30.14 The Atlas Moth, *Attacus atlas.*

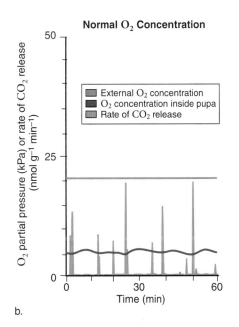

Adult

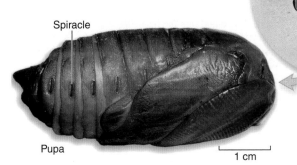

Spiracle

Pupa

1 cm

Hetz and Bradley thought that discontinuous breathing might help the insect's body maintain a constant, safe level of O_2. To test this idea experimentally, they studied pupae of the Atlas moth (*Attacus atlas*; **figure 30.14**). The pupa is the stage of the life cycle during which the animal transforms from a caterpillar into an adult moth. Hetz and Bradley poked tiny oxygen sensors into spiracles of the Atlas moth pupa and measured the concentration of O_2 inside the tracheae. At the same time, they changed the external concentration of O_2. The results of the experiment supported the "oxygen-guarding" hypothesis: the internal oxygen concentration stayed the same regardless of how much O_2 the insects were exposed to (**figure 30.15**).

The oxygen-guarding hypothesis explains one puzzling aspect of discontinuous breathing: it happens only in resting insects. During periods of high metabolic activity, the O_2 concentration inside the tracheae remains low because cells consume O_2 as quickly as it arrives. An active insect therefore has no reason to close its spiracles. At rest, however, the insect respiratory system has excess capacity; this situation is somewhat comparable to a race car's engine, which is much too powerful for everyday driving.

An inactive insect compensates for its supercharged respiratory system by opening its spiracles only when necessary to prevent toxic buildup of CO_2.

A toddler who refuses to breathe does nothing but annoy his parents. In contrast, a logical conclusion of Hetz and Bradley's work is that discontinuous breathing helps insects live longer by guarding against excess O_2 exposure. Animals that live longer may have more mating opportunities, which may mean greater reproductive success—and an explanation for how natural selection maintains the behavior in insect populations.

Hetz, Stefan K. and Timothy J. Bradley. 2005. Insects breathe discontinuously to avoid oxygen toxicity. *Nature*, vol. 433, pages 516–519.

30.5 | Mastering Concepts

1. What does this study suggest is the function of discontinuous breathing in the Atlas moth pupa?
2. Cells respire more rapidly at higher temperatures, increasing the demand for O_2. Do you predict that Atlas moth pupae would close their spiracles more frequently or less frequently as the temperature increases? Design an experiment to test your hypothesis.

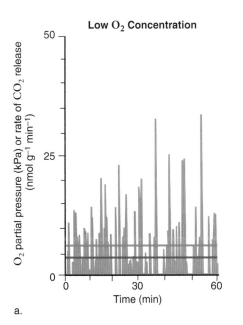

Low O_2 Concentration

O_2 partial pressure (kPa) or rate of CO_2 release (nmol g^{-1} min^{-1})

a.

Time (min)

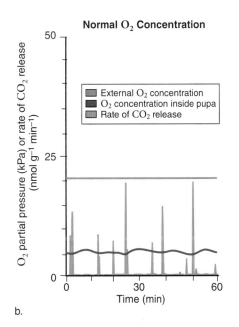

Normal O_2 Concentration

O_2 partial pressure (kPa) or rate of CO_2 release (nmol g^{-1} min^{-1})

☐ External O_2 concentration
■ O_2 concentration inside pupa
▨ Rate of CO_2 release

b.

Time (min)

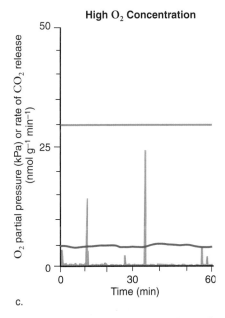

High O_2 Concentration

O_2 partial pressure (kPa) or rate of CO_2 release (nmol g^{-1} min^{-1})

c.

Time (min)

Figure 30.15 The Benefit of Discontinuous Breathing. The concentration of O_2 inside an Atlas moth pupa stays constant, regardless of the external concentration of O_2. The researchers made these measurements during the flutter phase, in which spiracles rapidly open and close; CO_2 release indicates open spiracles. Notice that spiracles open most frequently when O_2 is in extremely short supply.

Chapter Summary

30.1 | Gases Diffuse Across Respiratory Surfaces

- Cells use oxygen gas (O_2) in **aerobic respiration** to release the energy in food and store the energy in ATP. Carbon dioxide (CO_2) forms as a byproduct of respiration and must be eliminated from the body.
- **Respiratory systems** exchange O_2 and CO_2 with air or water, often in conjunction with a circulatory system that transports gases within the body.
- O_2 and CO_2 are exchanged by diffusion across a moist **respiratory surface.** Body size, metabolic requirements, and habitat have affected the form and function of respiratory surfaces.

A. Some Invertebrates Exchange Gases Across the Body Wall or in Internal Tubules

- Cnidarians, earthworms, and other invertebrates exchange O_2 and CO_2 directly across the body surface.
- Most terrestrial arthropods bring the atmosphere into contact with almost every cell through a highly branched system of **tracheae.**

B. Gills Exchange Gases with Water

- Many aquatic animals exchange gases across **gill** membranes enclosing networks of **capillaries.** In bony fishes, water flows over the gills in the direction opposite that of blood flow, an arrangement called **countercurrent exchange.**

C. Terrestrial Vertebrates Exchange Gases in Lungs

- Vertebrate **lungs** contain a moist internal respiratory surface. Amphibians have the simplest lungs; birds and mammals have the most complex.

30.2 | The Human Respiratory System Delivers Air to the Lungs

A. The Nose, Pharynx, and Larynx Form the Upper Respiratory Tract

- The **nose** purifies, warms, and moisturizes inhaled air. The air then flows through the **pharynx** and **larynx.**
- **Vocal cords** stretched over the larynx produce the voice as air passes through the **glottis.** The **epiglottis** prevents food from entering the trachea through the glottis.

B. The Lower Respiratory Tract Consists of the Trachea and Lungs

- Cartilage rings hold open the **trachea,** which branches into **bronchi** that deliver air to the lungs. The bronchi branch extensively and form tinier air tubules, **bronchioles,** which end in clusters of thin-walled, saclike **alveoli.**
- Many capillaries surround each alveolus. O_2 diffuses into the blood from the alveolar air, while CO_2 diffuses from the blood into the alveoli.

30.3 | Breathing Requires Pressure Changes in the Lungs

- A **respiratory cycle** consists of one **inhalation** and one **exhalation.**
- When the diaphragm and rib cage muscles contract, the chest cavity expands. This reduces air pressure in the lungs, drawing air in. When these muscles relax and the chest cavity shrinks, the pressure in the lungs increases and pushes air out.
- Measurements of lung function include **tidal volume** and **vital capacity.**

30.4 | Blood Delivers Oxygen and Removes Carbon Dioxide

A. Blood Carries Gases in Several Forms

- **Blood** consists of cells suspended in a matrix of **plasma.**
- Almost all oxygen transported to cells is bound to **hemoglobin** in **red blood cells.** Carbon monoxide (CO) poisoning occurs when CO prevents O_2 from binding hemoglobin.

- Some CO_2 in the blood is bound to hemoglobin or dissolved in plasma. Most CO_2, however, is transported as bicarbonate ion, generated from carbonic acid that forms when CO_2 reacts with water.
- In the lungs, hemoglobin binds O_2 and releases CO_2 for exhalation. At the body's other tissues, the concentration of O_2 is low, so O_2 diffuses out of the blood and into the respiring cells. CO_2 moves in the opposite direction along its concentration gradient.

B. Blood Gas Levels Help Regulate the Breathing Rate

- The brain's breathing control centers adjust the breathing rate. Chemoreceptors in the medulla and major arteries sense a rise in CO_2 concentration and blood acidity. The result: faster breathing.
- Only critically low O_2 levels trigger a change in the breathing rate.

30.5 | Investigating Life: Why Do Bugs Hold Their Breath?

- At rest, some insects breathe discontinuously by alternately opening and closing the spiracles that connect their tracheae to the atmosphere.
- Experimental evidence suggests that discontinuous breathing helps the insects maintain a safe, low level of O_2 inside their bodies.

Multiple Choice Questions

1. Why do humans breathe?
   a. To eliminate CO_2
   b. To support aerobic respiration in mitochondria
   c. To keep the respiratory surface dry
   d. Both a and b are correct.

2. What do tracheae and gills have in common?
   a. Countercurrent exchange
   b. High surface area
   c. Diffusion across the body surface
   d. Lungs

3. Which of the following does NOT have lungs?
   a. Earthworm   c. Lizard
   b. Frog   d. Chicken

4. The voice is produced at the
   a. olfactory epithelium.   c. larynx.
   b. bronchioles.   d. trachea.

5. Gas exchange in a human lung occurs at the
   a. bronchioles.   c. bronchi.
   b. alveoli.   d. trachea.

6. Air flows from areas of _____ to areas of _____.
   a. high O_2 concentration; low O_2 concentration
   b. low O_2 concentration; high O_2 concentration
   c. high pressure; low pressure
   d. low pressure; high pressure

7. In humans,
   a. vital capacity is typically less than tidal volume.
   b. a respiratory cycle lasts at least 1 minute.
   c. contraction of the diaphragm causes exhalation.
   d. inhaling requires more energy than exhaling.

8. What happens to CO in the blood?
   a. It is converted to carbonic acid and bicarbonate.
   b. It competes with O_2 for binding to hemoglobin.
   c. It is used instead of O_2 to generate ATP.
   d. Both a and c are correct.

9. Which reaction would most likely occur near the alveoli of a lung?
   a. CO_2 and water are formed from carbonic acid.
   b. Carbonic acid is converted to bicarbonate.
   c. Hemoglobin splits into alpha and beta globins.
   d. Hemoglobin releases O_2.

10. What information does the brain normally use to regulate the breathing rate?
    a. Level of O_2
    b. Level of CO_2
    c. Hemoglobin content of blood
    d. Demand for ATP

Write It Out

1. What is the function of breathing?
2. What is the connection between breathing and aerobic cellular respiration?
3. Make a chart that lists four types of respiratory surfaces (body surface, tracheae, gills, lungs). In your chart, describe each surface, list which animals have each, and say whether the surface can exchange gases with air or water or both.
4. A single-celled organism can exchange gases by simple diffusion through its cell membrane, which provides enough surface area for oxygen to diffuse into all parts of the cell. Animals, however, have many cells. Why do most animals have a respiratory system that performs a function that one-celled organisms can do alone?
5. How do an animal's size, activity level, and environment influence the structure and function of its respiratory surface?
6. An earthworm prefers moist soil. Why does this animal die if it dries out on a sidewalk?
7. How is air cleaned, warmed, and humidified before it reaches the lungs?
8. It is well below freezing outside, but the dedicated runner bundles up and hits the roads anyway. "You're crazy," shouts a neighbor. "Your lungs will freeze." How is the well-meaning neighbor wrong?
9. How does breathing through the mouth instead of the nose dry out the throat?
10. The trachea conducts air to the lungs, whereas the esophagus is a tube that conducts food to the stomach. Explain the observation that the trachea has cartilage, but the esophagus does not.
11. Trace the path of an O_2 molecule from a person's nose to a red blood cell at an alveolar capillary.
12. A person can choke if a hard candy or other small object obstructs the airway. In drowning, a person's lungs fill with water. Explain how each of these events can cause death. In most circumstances, how does the body react to prevent either of these disasters from occurring?
13. Describe the events that happen during inhalation and exhalation.
14. What is the difference between tidal volume and vital capacity? What does each measurement indicate about lung function?
15. How does blood transport O_2 to cells?
16. How does blood transport most CO_2? In what other ways is CO_2 transported?
17. One recommended treatment for anxiety-related hyperventilation (overbreathing) is to breathe into a paper bag for several minutes. After several breaths, how does the composition of air inside the bag compare with that in the atmosphere? How would breathing air from the bag help relieve hyperventilation? How might it be dangerous to breathe into a paper bag for more than a few minutes?

18. A human fetus produces hemoglobin with a higher affinity for O_2 than adult hemoglobin. As fetal blood mixes with the mother's blood in the placenta, in what direction does O_2 move? What would happen to a fetus with adult hemoglobin? What would happen if fetal hemoglobin had a lower affinity for O_2 than adult hemoglobin?
19. Describe an example of homeostasis mentioned in this chapter.
20. How does the brain establish the breathing rhythm?
21. Why can't a person commit suicide by holding his or her breath?
22. The concentration of O_2 in the atmosphere declines with increasing elevation. Why do you think the times of endurance events at the 1968 Olympics, held in Mexico City (elevation: 2200 m), were relatively slow?
23. How are the lungs similar to the stomata of plants, and how are they different?
24. Search the Internet to learn more about diseases that affect the respiratory system. Choose one to study in more detail. How does the disease affect respiratory function? What are the symptoms of the disease? What causes the disease, and how is it transmitted? Who is most likely to be affected? How is the disease diagnosed? Is a vaccine, treatment, or cure available?

Pull It Together

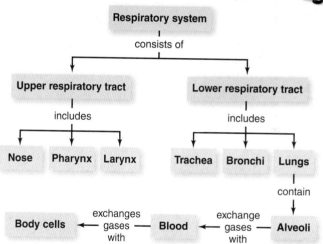

1. Add the following terms to this concept map: *diaphragm, brain, heart, capillaries, hemoglobin, plasma, O_2, CO_2.*
2. Describe the relationships among the parts of the upper and lower respiratory tracts.
3. Explain the diffusion gradients for O_2 and CO_2, both at the alveoli and at the rest of the body's cells.

Chapter 31

Digestion and Nutrition

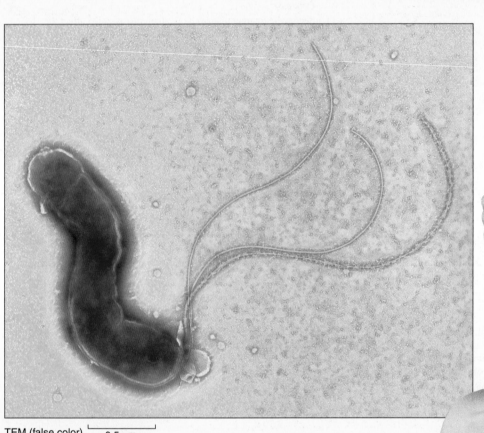

TEM (false color) |⎯⎯⎯⎯| 0.5 µm

Take a Pill. Many patients with ulcers now take antibiotics plus antacids to combat infection with *Helicobacter pylori* (inset).

McGraw Hill **connect** |BIOLOGY

Enhance your study of
this chapter with practice quizzes,
animations and videos, answer keys,
and downloadable study tools.
www.mhhe.com/hoefnagels

Bacteria Can Cause Gastric Ulcers

A GASTRIC ULCER IS A COMMON, PAINFUL CONDITION IN WHICH THE STOMACH WALL BECOMES IRRITATED AND ERODED. From 1910 until the mid-1990s, physicians thought ulcers were the direct result of excess stomach acid, which they blamed on stress and eating spicy foods—in short, "hurry, worry, and curry." In the 1970s, studies showed that drugs that block acid production could temporarily relieve ulcers. The standard treatment for ulcers was therefore a bland diet, stress reduction, acid-blocking drugs, or surgery.

In the 1980s, however, a bacterium called *Helicobacter pylori* was identified in the laboratory of J. Robin Warren at Royal Perth Hospital in Western Australia. Warren had found *H. pylori* in stomach tissue samples from people suffering from gastritis (an inflamed stomach). Were the bacteria attracted to inflamed tissue, or did they cause the inflammation? Warren's assistant, medical resident Barry Marshall, helped choose between the hypotheses. Marshall knew he had a healthy stomach and had never had gastritis or an ulcer. If the bacteria caused the irritation, they would do so in him. So on a hot July day in 1984, Marshall drank a brew containing about a billion of the microbes.

Marshall suffered for two weeks with mild gastritis, which cleared up without treatment. A second volunteer, however, was ill for several months. Although bismuth subcitrate (Pepto-Bismol) gave temporary relief, the pain returned. Only when he took two antibiotic drugs for a few weeks did the gastritis vanish completely.

These results were suggestive, but they hardly constituted scientific proof. Soon, however, other researchers conducted controlled experiments that confirmed the role of *H. pylori* in gastritis and ulcers. In the meantime, Warren, Marshall, and others continued to evaluate treatments that would eradicate the infection and prevent the recurrence of ulcers and gastritis. Finally, in 1994, the National Institutes of Health in the United States advised doctors that a 2-week course of two antibiotic drugs plus an antacid should become the standard treatment for gastritis and ulcers in patients with an infection. In recognition of their discovery, Warren and Marshall won the Nobel Prize in Physiology or Medicine in 2005.

H. pylori inhabits the stomach, part of an extensive internal network of food-processing tubes and compartments. This collection of organs—the digestive system—is the subject of this chapter.

Learning Outline

Learn How to Learn
Avoid Distractions

Despite your best intentions, constant distractions may take you away from your studies. Friends, music, TV, phone calls, text messages, video games, and the Internet all offer attractive diversions. How can you stay focused? One answer is to find your own place to study where no one can find you. Turn your phone off for a few hours; the world will get along without you while you study. And if you must use your computer, create a separate user account with settings that prevent you from visiting favorite websites during study time.

31.1 | Digestive Systems Derive Nutrients from Food

Is it true that "you are what you eat"? In some ways, the answer is yes. After all, the atoms and molecules that make up your body came from food that you ate (or that your mother ate before you were born). But in other ways, the answer is no. The woman in **figure 31.1** may enjoy eating fruit, yet she looks and acts nothing like a mango. Clearly, food is not incorporated whole into her body, even though atoms and molecules derived from food compose her and every other animal.

The resolution of this paradox lies in the **digestive system,** the organs that ingest food, break it down, absorb the small molecules, and eliminate undigested wastes. As this section describes, some of the molecules absorbed from food do become part of the animal body, but others are used to generate the energy needed for life.

A. Animals Eat to Obtain Energy and Building Blocks

All animals, along with fungi and many other microbes, are heterotrophs. A **heterotroph** is an organism that must consume food—organic matter—to obtain carbon and energy. The opposite of a heterotroph is an **autotroph,** such as a plant or alga, which uses inorganic raw materials and an energy source such as sunlight to build its own organic molecules.

An animal's food is its source of **nutrients,** which are substances that the organism uses for metabolism, growth, maintenance, and repair of its tissues. The six main types of nutrients in food are carbohydrates, proteins, lipids, water, vitamins, and minerals.

These nutrients contains two important resources. One is the potential energy stored in the chemical bonds of carbohydrates, proteins, and lipids. As described in chapter 6, cells can use each

of these fuels in respiration to generate ATP, the energy-rich molecule that powers most cellular activities. ▶ ATP, p. 76

The second resource that food contains is the chemical building blocks that make up the animal's body. Simple sugars, fatty acids, amino acids, nucleotides, water, vitamins, and minerals are the raw materials that build, repair, and maintain all parts of the body, from blood to bone. To the extent that your body incorporates these building blocks, you really are what you eat.

B. How Much Food Does an Animal Need?

An animal's metabolic rate largely determines its need for energy and nutrients. Endotherms such as birds and mammals use a lot of energy to maintain a constant body temperature; these animals have relatively high metabolic rates. In contrast, an ectotherm's body temperature fluctuates with the environment. Examples of ectotherms include invertebrate animals, fishes, amphibians, and nonavian reptiles such as lizards, snakes, and crocodiles. Thanks to this difference in metabolic rate, a sparrow or mouse weighing 25 grams requires much more food than does a 25-gram lizard. ▶ endotherms and ectotherms, p. 658

Body size also influences the need for food. In general, the larger the animal, the more it needs to eat. When corrected for body size, however, the smallest animals typically have the highest metabolic rates. A hummingbird, for example, has a much higher surface area relative to its body mass than does an elephant. Because the tiny bird loses much more heat to the environment, it must consume much more food to maintain a constant body temperature. A hummingbird therefore eats its own weight in food every day; an elephant takes three months to do the same.

Another factor that affects metabolic rate is an animal's physiological state. Growth and reproduction require more energy and nutrients than simply maintaining the adult body. Baby animals therefore often have ravenous appetites, and a new parent may spend much of its time finding food to fuel its offsprings' rapid growth (**figure 31.2**).

C. Animals Process Food in Four Stages

The overall process by which animals use food has four major steps. First, **ingestion** is the assimilation of food into the digestive tract. The second stage, **digestion,** is the physical and chemical breakdown of food. In mammals, this process begins with chewing, which tears food into small pieces mixed with saliva. Chewing therefore both softens food and increases the surface area exposed to digestive enzymes. In chemical digestion, enzymes split large nutrient molecules into their smaller components. In the third stage, **absorption,** the nutrients enter the cells lining the digestive tract and move into the bloodstream to be transported throughout the body (see chapter 29). Fourth, in

Figure 31.1
Delicious.
Eating provides raw materials and energy for life.

Figure 31.2 **Ravenous.** These chicks are begging for the food that will fuel their rapid growth.

elimination, the animal's body expels (egests) undigested food. **Feces** are the solid wastes that leave the digestive tract. ▶ enzymes, p. 78

One potential source of confusion is the distinction between feces and urine, both of which are animal waste products. Feces are composed partly of undigested food that never enters the body's cells. Urine, on the other hand, is a watery fluid containing dissolved nitrogen and other metabolic wastes produced by the body's cells. Chapter 32 describes how the kidneys produce urine.

D. Animal Diets and Feeding Strategies Vary Greatly

Biologists divide animals into categories based on what they eat and how they eat it (**figure 31.3**). **Herbivores,** such as cows and rabbits, eat only plants. Eagles, cats, wolves, and other **carnivores** hunt other animals for food. **Detritivores** consume decomposing organic matter; dung beetles and earthworms illustrate this diet. Finally, **omnivores** eat a broad variety of foods, including plants and animals. Humans are omnivores, as are raccoons and chickens.

Many animals rely on one or a few kinds of food. Insectivores such as anteaters, bats, flycatchers, praying mantises, and most spiders eat only insects. A piscivore eats fish, and a frugivore eats fruits. The giant panda is a folivore (leaf-eater). Because it eats only bamboo, the giant panda can survive only where that plant thrives. Animals with flexible diets, such as raccoons, can live in a broader range of habitats.

Animals also differ in how they acquire food. Humans and most other animals are **bulk feeders,** ingesting large chunks of food. **Fluid feeders,** in contrast, drink their food: a mosquito takes a "blood meal" after piercing human skin with its mouthparts. **Suspension feeders** are animals that strain particles from water. Sponges are suspension feeders that pass water through their porous bodies, filtering out the suspended organic matter.

Figure 31.3 **Essential Food.** (a) A giant panda munches on bamboo. (b) This leopard is eating its kill. (c) A mosquito takes a blood meal. (d) The basking shark filters food from water.

Corals, clams, and mussels, on the other hand, have feeding appendages that extend into the water column. Suspension feeding is adaptive in animals that do not move about much, because water carries the food to them. Nevertheless, some suspension feeders, such a baleen whale or a basking shark, are active swimmers.

Still other animals, called **substrate feeders,** live in their food and eat it from the inside. A female parasitic fly, for example, may lay eggs inside a live grasshopper. When the eggs hatch, the larvae feed on the host's tissues. A **deposit feeder** such as an earthworm is also a type of substrate feeder. These animals eat partially decayed organic matter in soil or other sediments.

Regardless of whether an animal eats plants, animals, or decomposing organic matter, note that all food ultimately comes from autotrophs. Chapter 37 explores this ecological principle in more detail.

31.1 | Mastering Concepts

1. What are two reasons that animals must eat?
2. Explain three factors that affect an animal's metabolic rate.
3. What four processes does food undergo when an animal eats?
4. Define the terms *herbivore, carnivore, detritivore,* and *omnivore.*
5. List four ways that animals obtain food.

31.2 | Animal Digestive Tracts Take Many Forms

Because heterotrophs and their food are composed of the same types of chemicals, digestive enzymes could just as easily attack an animal's body as its food. Digestion therefore occurs within specialized compartments that are protected from enzyme action.

These compartments may be located inside or outside of the body's cells (figure 31.4). One-celled organisms, including protists such as *Paramecium*, take in nutrients by phagocytosis and enclose the food in a food vacuole. A loaded food vacuole fuses with another sac containing digestive enzymes that break down nutrient molecules. Digestion occurs entirely inside the cell, but the process is physically separated from other functions. Sponges are the only animals that rely solely on intracellular digestion. ▶▶ phagocytosis, p. 83; sponges, p. 415

More complex animals use extracellular digestion, producing hydrolytic enzymes in a digestive cavity connected with the outside world. The enzymes dismantle large food particles, then cells lining the cavity absorb the products of digestion. In this system, food remains outside the body's cells until it is digested and absorbed. Extracellular digestion eases waste removal because indigestible components of food never enter the body's cells. Instead, the digestive tract simply ejects the waste.

The cavity in which extracellular digestion occurs may have one or two openings (figure 31.5). An **incomplete digestive tract** has only one opening: a mouth that both ingests food and ejects wastes. The animal must digest food and eliminate the residue before the next meal can begin. This two-way traffic limits the potential for specialized compartments that might store, digest, or absorb nutrients. Cnidarians such as jellyfish and *Hydra* have incomplete digestive tracts, as do flatworms. In these organisms, the digestive tract is also called a **gastrovascular cavity** because it doubles as a circulatory system that distributes nutrients to the body cells. ▶▶ cnidarians, p. 416; flatworms, p. 418

Most animals have a **complete digestive tract** with two openings; the mouth is the entrance, and the **anus** is the exit. This tubelike digestive cavity is called the **alimentary canal** or **gastrointestinal (GI) tract.** Notice that food passes through in one direction. One advantage of the two-opening digestive system is that regions of the tube can develop specialized areas that break food into smaller particles, digest it, absorb the nutrients into the bloodstream, and eliminate wastes. A complete digestive tract therefore extracts nutrients from food more efficiently than does an incomplete digestive tract.

Diet and lifestyle differences select for digestive system adaptations in all animals, including mammals. Figure 31.6 shows the digestive tracts of two herbivores and a carnivore. An herbivore's diet is rich in hard-to-digest cellulose from the cell walls of plants. The long digestive tract allows extra time for digestion. The diet of a carnivore, on the other hand, consists mostly or entirely of highly digestible meat. The overall length of the digestive tract is therefore short.

Ruminants such as cows, sheep, deer, and goats are herbivores with a complex, four-chambered organ that specializes in the digestion of grass. Saliva mixed with chewed grass enters the

Figure 31.4

Intracellular and Extracellular Digestion.

(a) Protists such as *Paramecium* use intracellular digestion, in which a vacuole containing food fuses with a sac containing digestive enzymes. (b) Most multicellular animals, including *Hydra*, digest food extracellularly. Cells lining the digestive cavity absorb the nutrients.

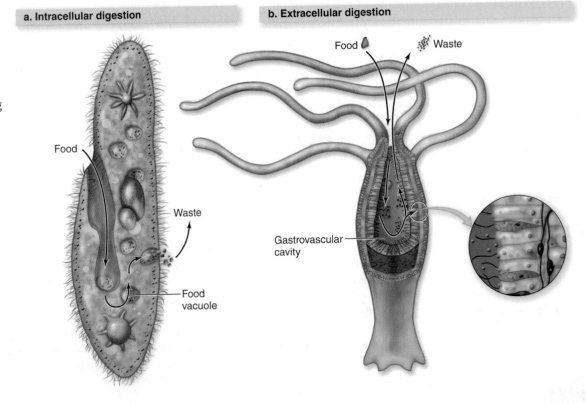

a. Intracellular digestion

Food

Waste

Food vacuole

b. Extracellular digestion

Food

Waste

Gastrovascular cavity

a. One opening: gastrovascular cavity

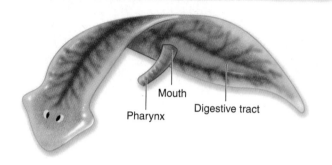

Mouth

Pharynx   Digestive tract

b. Two openings: alimentary canal

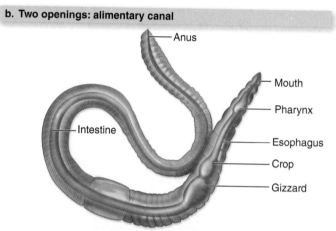

Anus

Mouth

Pharynx

Esophagus

Crop

Gizzard

Intestine

Figure 31.5 One-Opening and Two-Opening Digestive Systems. (a) In an animal with a gastrovascular cavity such as this flatworm, food enters and wastes exit through the mouth. (b) An earthworm's alimentary canal has two openings: the mouth and anus.

first and largest chamber, the rumen, where fermenting microorganisms break the plant matter down into balls of cud. The animal regurgitates the cud into its mouth; chewing the cud breaks the food down further. When the animal swallows again, the food bypasses the rumen, continuing digestion in the remaining chambers. ▶ fermentation, p. 115

Figure 31.6 also depicts another structure whose size varies with diet: the pouchlike **cecum,** which forms the entrance to the large intestine. In herbivores, the cecum is large, and it houses bacteria that ferment plant matter. Carnivores have a small or absent cecum. The cecum is medium-sized in omnivores, reflecting a diet based partly on plants.

31.2 | Mastering Concepts

1. Distinguish between intracellular and extracellular digestion.
2. How is an incomplete digestive tract different from a complete digestive tract?
3. Compare and contrast the digestive systems of a deer, a chipmunk, and a wolf.

a. Ruminant herbivore

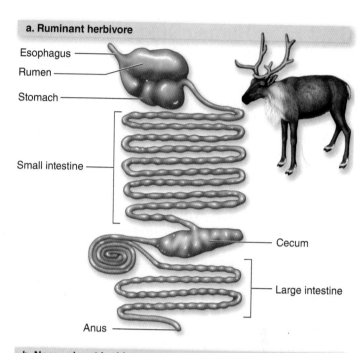

Esophagus
Rumen
Stomach

Small intestine

Cecum

Large intestine

Anus

b. Nonruminant herbivore

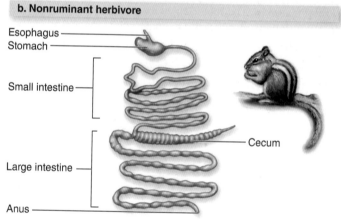

Esophagus
Stomach

Small intestine

Cecum

Large intestine

Anus

c. Carnivore

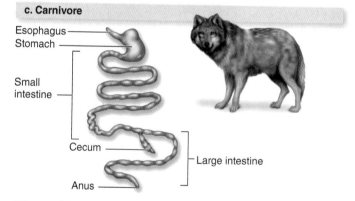

Esophagus
Stomach

Small intestine

Cecum

Large intestine

Anus

Figure 31.6 Digestive System Adaptations. (a) Ruminant herbivores have a rumen and an extensive gastrointestinal tract. Microorganisms break down the cellulose in their fibrous diet. (b) The intestines of nonruminant herbivores such as rabbits and rodents also harbor anaerobic microbes that break down cellulose in plants. (c) The protein-rich diet of carnivores is easy to digest, so the digestive system is much shorter and features a reduced cecum.

31.3 | The Human Digestive System Consists of Several Organs

The human digestive system consists of the gastrointestinal tract and accessory structures (**figure 31.7**). Some of the accessory structures, such as the salivary glands and pancreas, produce digestive enzymes. Two others, the liver and gallbladder, produce and store bile, which assists in fat digestion. The teeth and tongue are also accessory organs.

Muscles control the movement of food along the alimentary canal. Layers of smooth muscle that underlie the entire digestive tract undergo **peristalsis,** or rhythmic waves of contraction (**figure 31.8**). These contractions not only propel food in one direction but also churn the food, mixing it with enzymes to form a liquid. Unlike skeletal muscle, smooth muscle contraction is involuntary and does not require input from motor neurons. Instead, the autonomic nervous system stimulates smooth muscle to contract.
▶▶ smooth muscle tissue, p. 517; autonomic nervous system, p. 538

Muscles also control the openings between digestive organs. **Sphincters** are muscular rings that can contract to block the passage of materials. The sphincters at the mouth and anus are composed of skeletal muscle and are under voluntary control, so we can decide when to open our mouth or eliminate feces. The

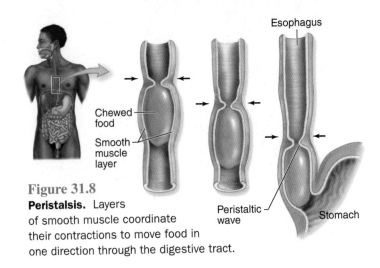

Figure 31.8
Peristalsis. Layers of smooth muscle coordinate their contractions to move food in one direction through the digestive tract.

remaining sphincters within the digestive tract, however, are composed of involuntary smooth muscle.

A. Digestion Begins in the Mouth and Esophagus

A tour of the digestive system begins at the mouth. The taste of food triggers salivary glands in the mouth to secrete saliva. This fluid contains an enzyme, salivary amylase, that starts to break down starch into maltose. Meanwhile, the **teeth**—

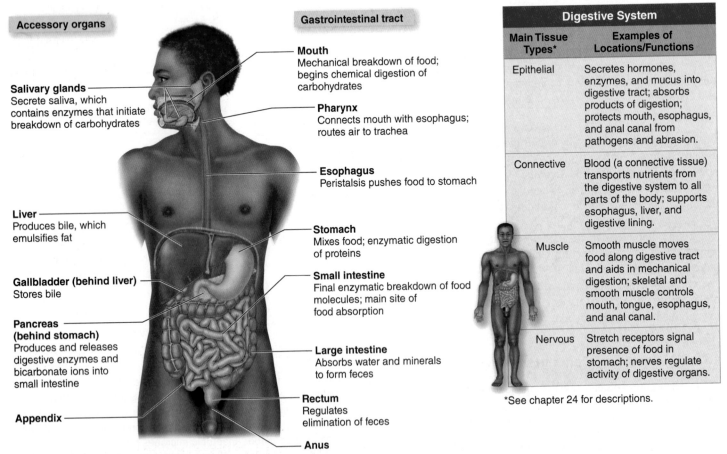

Accessory organs

Salivary glands
Secrete saliva, which contains enzymes that initiate breakdown of carbohydrates

Liver
Produces bile, which emulsifies fat

Gallbladder (behind liver)
Stores bile

Pancreas (behind stomach)
Produces and releases digestive enzymes and bicarbonate ions into small intestine

Appendix

Gastrointestinal tract

Mouth
Mechanical breakdown of food; begins chemical digestion of carbohydrates

Pharynx
Connects mouth with esophagus; routes air to trachea

Esophagus
Peristalsis pushes food to stomach

Stomach
Mixes food; enzymatic digestion of proteins

Small intestine
Final enzymatic breakdown of food molecules; main site of food absorption

Large intestine
Absorbs water and minerals to form feces

Rectum
Regulates elimination of feces

Anus

Digestive System	
Main Tissue Types*	**Examples of Locations/Functions**
Epithelial	Secretes hormones, enzymes, and mucus into digestive tract; absorbs products of digestion; protects mouth, esophagus, and anal canal from pathogens and abrasion.
Connective	Blood (a connective tissue) transports nutrients from the digestive system to all parts of the body; supports esophagus, liver, and digestive lining.
Muscle	Smooth muscle moves food along digestive tract and aids in mechanical digestion; skeletal and smooth muscle controls mouth, tongue, esophagus, and anal canal.
Nervous	Stretch receptors signal presence of food in stomach; nerves regulate activity of digestive organs.

*See chapter 24 for descriptions.

Figure 31.7 The Human Digestive System. Food breaks down as it moves through the chambers of the digestive tract. Accessory organs aid in digestion.

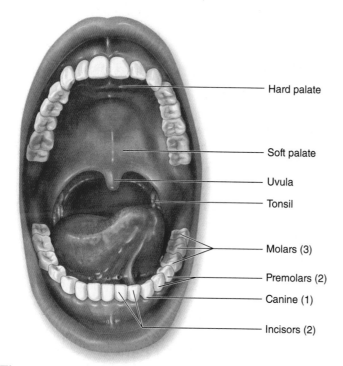

Figure 31.9 Human Mouth. Mechanical digestion begins with the teeth, which grasp, cut, tear, shred, crush, and grind food into pieces small enough to be swallowed.

mineral-hardened structures embedded in the jaws—grasp and chew the food (figure 31.9). Chewing is a form of mechanical digestion: water and mucus in saliva aid the teeth as they tear food into small pieces, increasing the surface area available for chemical digestion. The muscular **tongue** at the floor of the mouth mixes the food with saliva and pushes it to the back of the mouth to be swallowed.

The chewed mass of food passes first through the **pharynx,** or throat, the tube that also conducts air to the trachea. During swallowing, the **epiglottis** temporarily covers the opening to the trachea so that food enters the digestive tract instead of the lungs. From the pharynx, swallowed food and liquids pass to the **esophagus,** a muscular tube leading to the stomach. Food does not merely slide down the esophagus under the influence of gravity; instead, contracting muscles push it along in a wave of peristalsis.

B. The Stomach Stores, Digests, and Pushes Food

The **stomach** is a J-shaped muscular bag that receives food from the esophagus (figure 31.10). The stomach is about the size of a large sausage when empty, but when very full, it can expand to hold as much as 3 or 4 liters of food. Ridges in the stomach's lining can unfold like the pleats of an accordion to accommodate a large meal.

The stomach absorbs very few nutrients, but it can absorb some water and salts (electrolytes), a few drugs (e.g., aspirin), and, like the rest of the digestive tract, alcohol. As a result, we feel alcohol's intoxicating effects quickly.

The stomach's main function, however, is to continue the mechanical and chemical digestion of food. Swallowed chunks of food break into smaller pieces as waves of peristalsis churn the stomach's contents. At the same time, the stomach lining produces **gastric juice,** a mixture of water, mucus, salts, hydrochloric acid, and enzymes.

This gastric juice comes from mucous, chief, and parietal cells housed in gastric pits in the epithelium of the stomach. Mucous cells near the entrance to each pit produce mucus.

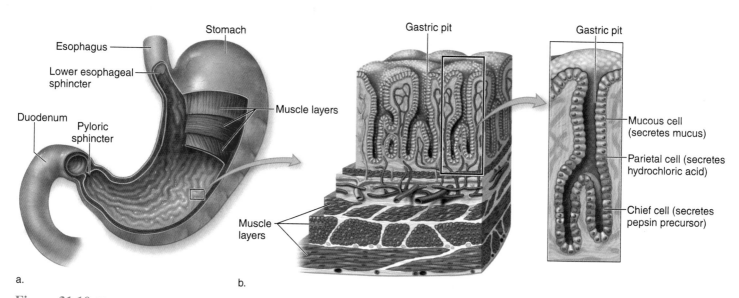

Figure 31.10 The Stomach. (a) The stomach receives food from the esophagus, mixes it with gastric juice, and passes it to the duodenum. Layers of muscle enable the organ to move food and break it into smaller pieces. Two sphincters control the entrance and exit of food. (b) The stomach lining contains cells that secrete mucus and gastric juice.

Chief cells secrete a protein that becomes **pepsin,** an enzyme that digests proteins. Parietal cells release hydrochloric acid upon stimulation by a hormone called gastrin. Thanks to this acid, the pH of the gastric juice is low, about 1.5 or 2. The acidity denatures the proteins in food, kills most disease-causing organisms, and activates pepsin so that protein digestion can begin. ▶▶ epithelial tissue, p. 514; pH, p. 30

If gastric juice breaks down protein in food, how does the stomach keep from digesting itself? First, the stomach produces little gastric juice until food is present. Second, mucus coats and protects the stomach lining. Tight junctions between cells in the stomach lining also prevent gastric juice from seeping through to the tissues below.

Chyme is the semifluid mixture of food and gastric juice in the stomach. Small amounts of chyme squirt through the pyloric sphincter that links the stomach and the upper part of the small intestine (the duodenum). A negative feedback loop controls the opening and closing of the pyloric sphincter (**figure 31.11**). ▶ negative feedback, p. 520

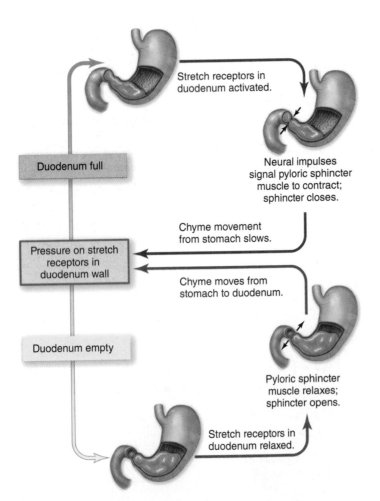

Figure 31.11 Chyme Movement to the Small Intestine.
Stretch receptors signal the pyloric sphincter to open or close, depending on the quantity of food in the duodenum.

C. The Small Intestine Digests and Absorbs Nutrients

The **small intestine** is a tubular organ that completes digestion and absorbs nutrients. Although narrow in comparison with the large intestine, the small intestine's 3- to 7-meter length makes it the longest organ in the digestive system.

Anatomy of the Small Intestine   The small intestine contains three main regions. The duodenum makes up the first 25 centimeters. Glands in the wall of the duodenum secrete mucus that protects and lubricates the small intestine. In addition, ducts from the pancreas and liver open into the duodenum. The next two regions, the jejunum and the ileum, form the majority of the small intestine. The 2.5-meter-long jejunum absorbs most carbohydrates and proteins, and the 3.5-meter-long ileum absorbs most water, fats, vitamins, and minerals.

Close examination of the hills and valleys along the small intestine's lining reveals millions of **villi,** tiny fingerlike projections that absorb nutrients (**figure 31.12**). The epithelial cells on the surface of each villus bristle with **microvilli,** extensions of the cell membrane. Each villus cell and its 500 microvilli increase the surface area of the small intestine at least 600 times, allowing for the efficient extraction of nutrients from food. The capillaries that snake throughout each villus take up the newly absorbed nutrients and water, then empty into veins that carry the nutrient-laden blood to the liver. Also inside each villus is a lymph capillary that receives digested fats.

The Role of the Pancreas, Liver, and Gallbladder
The small intestine absorbs water, minerals, free amino acids, cholesterol, and vitamins without further digestion. Most molecules in food, however, require additional processing.

Digestive enzymes released by cells lining the small intestine act on short polysaccharides and disaccharides to release simple sugars, which the small intestine immediately absorbs. These enzymes are called carbohydrases. People who lack one such enzyme, lactase, cannot digest milk sugar; this chapter's Burning Question discusses this condition.

Most of the digestive enzymes in the small intestine, however, come from the **pancreas,** an accessory organ (see figure 31.7) that sends about a liter of pancreatic "juice" to the duodenum each day. This fluid contains many enzymes. Trypsin and chymotrypsin break polypeptides into amino acids; pancreatic amylase digests starch; pancreatic lipase breaks down fats; and nucleases split nucleic acids such as DNA into nucleotides. In addition to these enzymes, pancreatic juice also contains alkaline sodium bicarbonate, which neutralizes the acid from the stomach.

Fats present an interesting challenge to the digestive system. Lipase enzymes are water soluble, but fats are not. Therefore, lipase can only act at the surface of a fat droplet, where it contacts water. **Bile** is a greenish-yellow biochemical that helps solve this problem by dispersing the fat into tiny globules suspended in

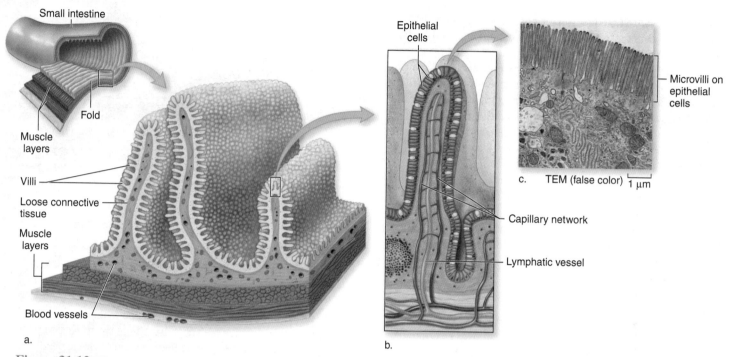

Figure 31.12 **The Lining of the Small Intestine.** (a) Each ridge of the small intestine's lining is folded into villi. (b) Within each villus, blood-filled capillaries absorb digested carbohydrates and proteins, and lymph capillaries absorb digested fats. Epithelial cells coat each villus. (c) Microvilli extending from the epithelial cells add tremendous surface area for absorption.

water. The resulting mixture, called an emulsion, increases the surface area exposed to lipase.

Bile comes from the **liver,** a large accessory organ with more than 200 functions. The liver's only *direct* contribution to digestion is the production of bile. The **gallbladder** is an accessory organ that stores this bile until chyme triggers its release into the small intestine. The cholesterol in bile can crystallize, forming gallstones that partially or completely block the duct to the small intestine. Gallstones are very painful and may require

removal of the gallbladder. This chapter's Apply It Now box describes some other examples of illnesses affecting the gastrointestinal tract. ▸ cholesterol, p. 35

The liver's other functions include detoxifying alcohol and other harmful substances in the blood, storing glycogen and fat-soluble vitamins, and synthesizing blood-clotting proteins. The liver receives nutrient-laden blood from the intestines via the hepatic portal vein (see figure 29.9). The blood passes through the liver's extensive capillary beds, which remove bacteria and toxins.

 Burning Question

What's lactose intolerance?

Lactose, or milk sugar, is a disaccharide in milk. In infants, the lactase enzyme breaks down lactose in the small intestine. Most people stop producing lactase after infancy, when milk is no longer part of the typical mammalian diet. If a person without lactase consumes milk, bacteria in the large intestine ferment the undigested sugar, producing abdominal pain, gas, diarrhea, bloating, and cramps.

A person with lactose intolerance can avoid these symptoms by consuming fermented dairy products such as yogurt, buttermilk, and cheese instead of fresh milk;

bacteria have already broken down the lactose in those foods. Taking lactase tablets can also prevent symptoms of lactose intolerance.

Interestingly, people with roots in northern Europe and a few other locations continue to produce lactase as adults, thanks to a long-ago genetic mutation that was adaptive in dairy-herding regions of the world.

Submit your burning question to:
marielle_hoefnagels@mcgraw-hill.com

The liver also gets "first dibs" on the nutrients in the blood before it is pumped to the rest of the body. In a condition called cirrhosis, scar tissue blocks this vital blood flow through the liver.

Like the stomach, the small intestine protects itself against self-digestion by producing digestive biochemicals only when food is present. In addition, mucus protects the intestinal wall from digestive juices and neutralizes stomach acid. Nevertheless, many cells of the intestinal lining die in the caustic soup. Rapid division of the small intestine's epithelial cells compensates for the loss, replacing the lining every 36 hours.

Figure 31.13 summarizes the locations of the major digestive enzymes, along with their products. Ultimately, as we have seen, the small intestine absorbs these nutrients and passes them to the bloodstream. Cells throughout the body then use the nutrients to generate energy and to build new proteins, carbohydrates, lipids, and nucleic acids.

D. The Large Intestine Completes Nutrient and Water Absorption

The material remaining in the small intestine moves next into the **large intestine,** which extends from the ileum to the anus while forming a "frame" around the small intestine (**figure 31.14**). At 1.5 meters long, the large intestine is much shorter than the small intestine, but its diameter is greater (about 6.5 centimeters). Its main functions are to absorb water and salts and to eliminate the remaining wastes as feces.

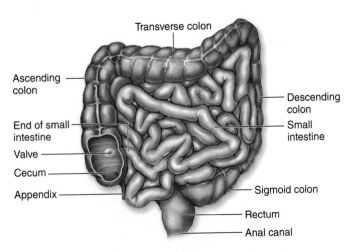

Figure 31.14 The Large Intestine. The large intestine receives chyme from the small intestine, absorbs water, salts, and minerals, and eliminates the remainder as feces.

The start of the large intestine is the pouchlike cecum. Dangling from the cecum is the appendix, a thin, worm-shaped tube. Trapped bacteria or undigested food can cause the appendix to become irritated, inflamed, and infected, producing severe pain. A burst appendix can spill its contents into the abdominal cavity and spread the infection.

The colon forms the majority of the large intestine. Here, the large intestine absorbs most of the water, electrolytes, and

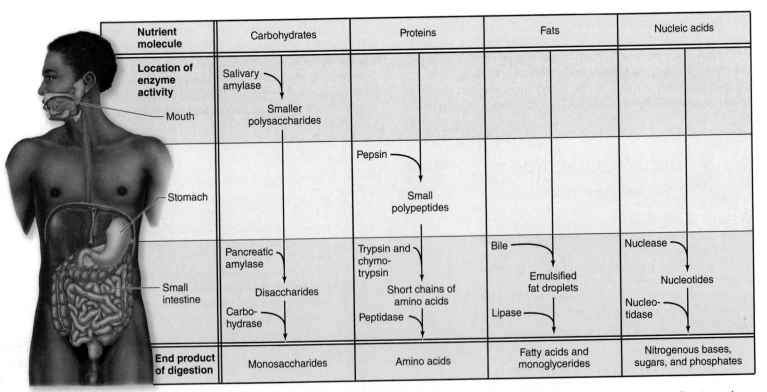

Figure 31.13 Overview of Chemical Digestion. Although digestion begins in the mouth and stomach, the small intestine digests and absorbs most molecules.

Apply It Now

The Unhealthy Digestive System

The entire length of the human digestive tract is subject to numerous disorders. Some are a nuisance or easily treated, whereas others can be deadly. A few are listed here:

- **Tooth decay:** Bacteria living in the mouth secrete acids that eat through a tooth's surface, causing cavities. This decayed area can extend to the interior of the tooth, eventually killing the tooth's nerve and blood supply.

- **Acid reflux:** Gastric juice, normally confined to the stomach, sometimes emerges through the esophageal sphincter and burns the esophagus. This painful condition is commonly known as "heartburn." Antacids can neutralize the acidity.

- **Vomiting:** When a person vomits, the medulla (in the brainstem) coordinates several simultaneous events. The muscles of the diaphragm and abdomen contract, while the sphincter at the entrance of the stomach relaxes. These actions force chyme out of the stomach and up the esophagus. Alcohol, bacterial toxins from spoiled foods, and excessive eating can trigger queasiness and vomiting.

- **Hepatitis:** Hepatitis literally means inflammation of the liver. Many conditions can cause hepatitis, including alcohol abuse, some pharmaceutical drugs, and eating poisonous mushrooms. Some forms of hepatitis are contagious; five viruses (hepatitis A–E) cause most cases of infectious hepatitis.

- **Appendicitis:** The appendix can become inflamed or infected. If it ruptures, it can release bacteria into the abdominal cavity, causing serious infection and sometimes even death when left untreated.

- **Diarrhea:** If the intestines fail to absorb as much water as they should, the feces become loose and watery. The risk of dehydration is high. Many food- and waterborne viruses, bacteria, and protists cause diarrhea, as can treatment with some antibiotics. Diarrhea is a major cause of death in underdeveloped countries. Section 4.6 describes the faulty protein implicated in cystic fibrosis, which may protect against some causes of diarrhea.

- **Constipation:** A constipated person eliminates feces less frequently than three times a week, and the feces are hard, dry, and difficult to eliminate. Causes of constipation include an obstructed large intestine, loss of peristalsis, dehydration, starvation, and anxiety.

- **Colon (colorectal) cancer:** Cancerous tumors may arise in the rectum, colon, or appendix. This is among the most common cancer types, and it is a leading cause of death worldwide. A diet high in fiber helps prevent colon cancer.

- **Hemorrhoids:** These swollen, distended veins protrude into the rectum or anus, causing painful defecation and bleeding.

---

minerals from chyme. Veins carry blood from vessels surrounding the large intestine to the liver.

The remnants of digestion consist mostly of bacteria, undigested fiber, and intestinal cells. These materials, plus smaller amounts of other substances, collect in the rectum as solid or semisolid feces. When the rectum is full, receptor cells trigger a reflex that eliminates the feces through the anus.

Our partnership with our intestinal microbes deserves special mention. Trillions of bacteria, representing about 500 different species, are normal inhabitants of the large intestine. These microscopic residents produce the characteristic foul-smelling odors of intestinal gas and feces, but they also provide many benefits. Most notably, they help prevent infection by other microorganisms that are harmful—an interesting application of ecology's competitive exclusion principle. They also decompose cellulose and some other nutrients, produce B vitamins and vitamin K, and break down bile and some drugs. Antibiotic drugs often kill these normal bacteria and allow other microorganisms to grow. This change in the intestinal ecosystem causes the diarrhea that sometimes accompanies treatment with antibiotics. ▶ competitive exclusion, p. 763

Interestingly, a baby is born with a microbe-free digestive tract. The infant begins acquiring its microbiota with its first meal of milk or formula. Bacteria on the mother's skin and in the environment gradually enter the baby and colonize the intestines, establishing populations that will persist throughout the person's life.

31.3 | Mastering Concepts

1. Explain the action and importance of peristalsis and sphincters in digestion.
2. Describe the functions of saliva, teeth, and the tongue in digestion.
3. How does food move from the mouth to the stomach?
4. Describe the mechanical and chemical digestion that occurs in the stomach.
5. What is the function of the small intestine?
6. How does the small intestine maximize surface area?
7. How do secretions of the pancreas, liver, and gallbladder aid digestion?
8. Describe the events that occur as food passes through the large intestine.
9. How does undigested food leave the body?

31.4 | A Healthy Diet Includes Essential Nutrients and the Right Number of Calories

Healthy eating has two main components. First, the diet must include all of the nutrients necessary to sustain life. A second consideration is the calorie content of food, which must balance a person's activity level.

A. A Varied Diet Is Essential to Good Health

Nutrients fall into two categories. **Macronutrients** are required in large amounts. Water is a macronutrient; all living cells require water as a solvent and as a participant in many reactions. Organisms use three other macronutrients—carbohydrates, proteins, and lipids—to build cells and to generate ATP. Despite the many nonfat foods on grocery store

Table **31.1** Minerals in the Human Diet

Mineral	Food Sources	Function(s)	Selected Deficiency Symptoms
Bulk Minerals			
Calcium	Milk products, green leafy vegetables	Electrolyte, bone and tooth structure, blood clotting, hormone release, nerve transmission, muscle contraction	Muscle cramps and twitches; weakened bones, heart malfunction
Chlorine	Table salt, meat, fish, eggs, poultry, milk	Electrolyte, part of stomach acid	Muscle cramps, nausea, weakness
Magnesium	Green leafy vegetables, beans, fruits, peanuts, whole grains	Muscle contraction, nucleic acid synthesis, enzyme activity	Tremors, muscle spasms, loss of appetite, nausea
Phosphorus	Meat, fish, eggs, poultry, whole grains	Bone and tooth structure; part of DNA, ATP, and cell membranes	Weakness, mineral loss from bones
Potassium	Fruits, potatoes, meat, fish, eggs, poultry, milk	Electrolyte, nerve transmission, muscle contraction, nucleic acid synthesis	Weakness, loss of appetite, muscle cramps, confusion, heart arrhythmia
Sodium	Table salt, meat, fish, eggs, poultry, milk	Electrolyte, nerve transmission, muscle contraction	Muscle cramps, nausea, weakness
Sulfur	Meat, fish, eggs, poultry	Hair, skin, and nail structure, blood clotting, energy transfer	Brittle hair and nails, stunted growth
Trace Minerals			
Chromium	Yeast, pork kidneys	Regulates glucose use	High blood sugar
Cobalt	Meat, eggs, dairy products	Part of vitamin B_{12}	Weakness, fatigue, nerve degeneration
Copper	Organ meats, nuts, shellfish, beans	Part of many enzymes, iron metabolism in red blood cells	Anemia, low white blood cell counts
Fluorine	Water (in some areas)	Mineralization of bones and teeth	Tooth decay
Iodine	Seafood, iodized salt	Part of thyroid hormone	Goiter (enlarged thyroid gland)
Iron	Meat, liver, fish, shellfish, egg yolk, peas, beans, dried fruit, whole grains	Part of hemoglobin and myoglobin, part of some enzymes	Anemia, learning deficits in children
Manganese	Bran, coffee, tea, nuts, peas, beans	Part of some enzymes, bone and tendon structure	Impaired growth, bone abnormalities
Selenium	Meat, milk, grains, onions	Part of some enzymes, heart function	Weakened heart, degeneration of cartilage
Zinc	Meat, fish, egg yolk, milk, nuts, some whole grains	Part of some proteins, regulates gene expression	Stunted growth, diarrhea, impaired immune function

shelves, the diet must include *all* of these nutrients, including moderate amounts of fat.

Unlike macronutrients, **micronutrients** are required in very small amounts. Two examples are minerals and vitamins, neither of which are used as fuel. Instead, these micronutrients participate in many aspects of cell metabolism. Tables 31.1 and 31.2 provide details on the sources and functions of

minerals and vitamins. Many processed foods are fortified with vitamins and minerals, so that deficiencies in developed countries are rare.

The best way to acquire the required macro- and micronutrients is to eat a varied diet. The U.S. government's food pyramid emphasizes whole grains, fresh vegetables, and low-fat dairy products, along with a variety of fruits and limited amounts of meat and

Table 31.2 Vitamins in the Human Diet

Vitamin	Food Sources	Function(s)	Selected Deficiency Symptoms
Water-Soluble Vitamins			
B complex vitamins			
Thiamine (vitamin B_1)	Pork, beans, peas, nuts, whole grains	Growth, fertility, digestion, nerve cell function, milk production	Beriberi (confusion, loss of appetite, muscle weakness, heart failure)
Riboflavin (vitamin B_2)	Liver, leafy vegetables, dairy products, whole grains	Energy use	Cracked skin at corner of mouth
Pantothenic acid*	Liver, eggs, peas, potatoes, peanuts	Growth, cell maintenance, energy use	Headache, fatigue, poor muscle control, nausea, cramps
Niacin	Liver, meat, peas, beans, whole grains, fish	Growth, energy use	Pellagra (diarrhea, dementia, dermatitis)
Pyridoxine (vitamin B_6)*	Red meat, liver, corn, potatoes, whole grains, green vegetables	Protein metabolism, production of neurotransmitters	Mouth sores, dizziness, nausea, weight loss, neurological disorders
Folic acid (folate)	Liver, navy beans, dark green vegetables	Manufacture of red blood cells, metabolism	Weakness, fatigue, diarrhea, neural tube defects in fetus
Biotin*	Meat, milk, eggs	Metabolism	Skin disorders, muscle pain, insomnia, depression
Cobalamin (vitamin B_{12})	Meat, organ meats, fish, shellfish, milk	Manufacture of red blood cells, maintenance of myelin sheath	Weakness, fatigue, nerve degeneration
Ascorbic acid (vitamin C)	Citrus fruits, tomatoes, peppers, strawberries, cabbage	Antioxidant, production of connective tissue and neurotransmitters	Scurvy (weakness, gum bleeding, weight loss)
Fat-Soluble Vitamins			
Retinol (vitamin A)	Liver, dairy products, egg yolk, vegetables, fruit	Night vision, new cell growth	Blindness, impaired immune function
Cholecalciferol (vitamin D)	Fish liver oil, milk, egg yolk	Bone formation	Skeletal deformation (rickets)
Tocopherol (vitamin E)*	Vegetable oil, nuts, beans	Antioxidant	Anemia in premature infants
Vitamin K*	Liver, egg yolk, green vegetables	Blood clotting	Internal bleeding

*Deficiencies in these vitamins are rare in humans, but they have been observed in experimental animals.

Figure 31.15 One Healthful Diet. The U.S. Department of Agriculture's food pyramid emphasizes whole grains, fruits, vegetables, and low-fat dairy products, along with exercise.

fat (figure 31.15). The Harvard School of Public Health suggests a somewhat different diet that minimizes dairy products, red meat, and starchy processed grains. Whole grains and vegetable oils, along with abundant vegetables, make up the base of this pyramid.

The indigestible components of food help maintain good health, too. Dietary fiber, for example, is composed of cellulose from plant cell walls. Humans do not produce cellulose-digesting enzymes, so fiber contributes only bulk—not nutrients—to food. This increased mass eases movement of the food through the digestive tract, so that harmful ingredients in food contact the walls of the intestines for a shorter period. People who consume abundant fiber in their food therefore have a lower incidence of colorectal cancer. A high-fiber diet also reduces blood cholesterol and helps regulate blood sugar. ▸ cellulose, p. 32

A balanced diet delivers many long-term health benefits, including a reduced risk of type 2 diabetes, cancer, osteoporosis, high blood pressure, and heart disease. Fortunately, packaged foods have labels that depict the nutrient content in

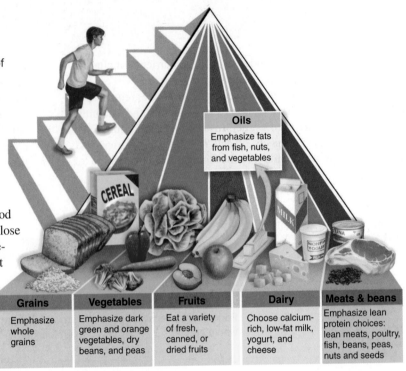

Oils
Emphasize fats from fish, nuts, and vegetables

Grains	Vegetables	Fruits	Dairy	Meats & beans
Emphasize whole grains	Emphasize dark green and orange vegetables, dry beans, and peas	Eat a variety of fresh, canned, or dried fruits	Choose calcium-rich, low-fat milk, yogurt, and cheese	Emphasize lean protein choices: lean meats, poultry, fish, beans, peas, nuts and seeds

Source: U.S. Dept. of Agriculture

each serving of food (figure 31.16). Some labels may indicate additional information; for example, chapter 2's Burning Question explains the distinction between "natural" and "organic."

B. Body Weight Reflects Food Intake and Activity Level

Nutritional labels also list a food's caloric content. This information is determined by burning food in a bomb calorimeter, a chamber immersed in water and designed to measure heat output. Energy released from the food raises the water temperature: 1 **kilocalorie** (1 food **Calorie;** kcal) is the energy needed to raise 1 kilogram of water from 14.5°C to 15.5°C under controlled conditions.

Bomb calorimetry studies have shown that 1 g of carbohydrate yields 4 kcal, 1 g of protein yields 4 kcal, and 1 g of fat yields 9 kcal. Although the body cannot extract all of the potential energy in food, these values help explain the link between a fatty diet and weight gain; fats supply more energy than most people can use. When we take in more kilocalories than we expend, weight increases; those who consume fewer kilocalories than they expend lose weight and may even starve. Most young adults require 2000 to 2400 kcal per day, depending on gender and level of physical activity.

Figure It Out

Consider the following nutritional facts. Bacon cheeseburger: 23 g fat, 25 g protein, 2 g carbohydrates; large fries: 25 g fat, 6 g protein, 63 g carbohydrates; large soda: 86 g carbohydrates. How many kcal are in this meal?

Answer: 1160 kcal

What constitutes a "healthful" weight? The most common measure is the body mass index, or BMI (figure 31.17). To calculate BMI, divide a person's weight (in kilograms) by his or her

Nutrition Fact

Serving Size 1/2 cup (103g)
Servings Per Container 4

Amount Per Serving

Calories 310   Calories from Fat 18

	% Daily Value*
Total Fat 20g	31%
Saturated Fat 12g	60%
Trans Fat 0g	
Cholesterol 95mg	32%
Sodium 125mg	5%
Total Carbohydrate 29g	10%
Dietary Fiber 0g	0%
Sugars 24g	
Protein 4g	

| Vitamin A 10% | • | Vitamin C 0% |
| Calcium 10% | • | Iron 2% |

Figure 31.16 Nutritional Information. The packaging of processed foods includes a standard nutritional label that indicates the calorie and nutrient content in each serving.

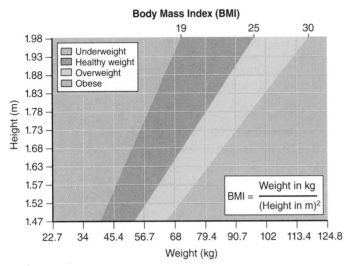

Source: U.S. Department of Agriculture: Dietary Guidelines for Americans

Figure 31.17 Body Mass Index. This chart is a quick substitute for a BMI calculation; the intersection of a person's height (in meters) and weight (in kilograms) indicates the BMI range.

squared height (in meters): BMI = weight/(height)². Alternatively, multiply weight (in pounds) by 704.5, and divide by the square of the height (in inches).

Many health professionals consider a person whose BMI is less than 19 underweight. A BMI between 19 and 25 is healthy; an overweight person has a BMI greater than 25; and a BMI greater than 30 denotes **obesity.** Morbid obesity is defined as a BMI greater than 40. One limitation of BMI is that it cannot account for many of the details that affect health. An extremely muscular person, for example, will have a high BMI because muscle is denser than fat, yet he or she would not be considered overweight.

Another useful measure is the ratio of waist diameter to hip diameter. People whose fat accumulates at the waistline ("apples") are more susceptible to health problems such as insulin resistance than are "pears," who are bigger around the hips.

C. Starvation: Too Few Calories to Meet the Body's Needs

A healthy human can survive for 50 to 70 days without food—much longer than without air or water. In some areas of the world, famine is a constant condition, and millions of people starve to death every year. Hunger strikes, inhumane treatment of prisoners, and eating disorders can also cause starvation (figure 31.18).

The starving human body essentially digests itself. After only a day without food, reserves of sugar and glycogen are gone, and the body begins extracting energy from stored fat and muscle protein. By the third day, hunger eases as the body uses energy from fat reserves. Gradually, metabolism slows, blood pressure drops, the pulse slows, and chill sets in. Skin becomes dry and hair falls out as the proteins that form these structures are digested. When the body dismantles the immune system's antibody proteins, protection against infection declines. Mouth sores and anemia

Figure 31.18 Two Forms of Starvation. The malnourished girl on the left is a famine victim. Her thin legs and distended belly are signs of a severe protein deficiency. The girl on the right suffers from anorexia nervosa, or self-imposed starvation.

develop, the heart beats irregularly, and bones begin to degenerate. Near the end, the starving human is blind, deaf, and emaciated.

Anorexia nervosa, or self-imposed starvation, is refusal to maintain normal body weight. The condition affects about 1 in 250 adolescents, more than 90% of whom are female. The sufferer perceives herself as overweight and eats barely enough to survive, losing as much as 25% of her original body weight. She may further lose weight by vomiting, taking laxatives and diuretics, or exercising intensely. Intravenous feedings, psychotherapy, and nutritional counseling may help, but 15% to 21% of people with anorexia die from the disease.

Bulimia is another eating disorder that mainly affects females. Rather than avoiding food, a person with **bulimia** eats large quantities and then intentionally vomits or uses laxatives shortly afterward, a pattern called "binge and purge." A person with bulimia may or may not be underweight.

D. Obesity: More Calories Than the Body Needs

Obesity is increasingly common in the United States (see figure 27.13), and the health consequences can be serious. People who accumulate fat around their waists are susceptible to type 2 diabetes, high blood pressure, and atherosclerosis. High body weight is also correlated with congestive heart failure, acid reflux disease, urinary incontinence, low back pain, stroke, sleep disorders, and many other health problems. In addition, obese people also face higher risk of cancers of the colon, breast, and uterus. ▶ diabetes, p. 576

Excess body weight accumulates when a person consumes more calories than he or she expends. For most people, the main culprits are a diet loaded with sugar and fat, coupled with an inactive lifestyle. Nevertheless, many genes contribute to appetite, digestion, and metabolic rate, so the combination of alleles that a person inherits plays at least a small part in his or her risk for obesity.

Some of these genes encode hormones, which interact in complex and poorly understood ways to maintain the balance

Figure 31.19 Hormone Deficiency. In the normal mouse on the right, leptin secreted by fat cells affects target cells in the hypothalamus, helping the brain regulate appetite and metabolism. The obese mouse on the left cannot produce leptin and therefore has a ravenous appetite.

between food intake and energy expenditure. Two weight-related hormones are leptin and ghrelin. Stored fat (adipose tissue) releases leptin into the bloodstream. At the hypothalamus, leptin interacts with receptors and triggers a signal cascade that inhibits food intake and increases metabolic activity. This action explains why leptin-deficient mice become extremely obese (figure 31.19). Leptin deficiency, however, is very rare in humans; in fact, in most overweight people, leptin is abundant, but the hypothalamus is somehow leptin-resistant.

Whereas leptin is an indicator of relatively long-term fat storage, ghrelin helps to stimulate hunger in the short term. Cells in the stomach release ghrelin into the bloodstream before a meal, and ghrelin production slows after eating. This hormone interacts with receptors in the hypothalamus and elsewhere.

Research into leptin, ghrelin, and other appetite-related hormones may someday yield new treatments for obesity. In the meantime, the concern over expanding waistlines has fueled demand for low-calorie artificial sweeteners and fats (see chapter 2). Although fad diets remain popular, the most healthful way to lose weight and reverse cardiovascular disease and diabetes is to exercise and reduce calorie intake while maintaining a balanced diet. For people who have difficulty losing weight in this way, stomach-reduction surgery and drugs that either reduce appetite or block fat absorption offer other options.

31.4 | Mastering Concepts

1. Which nutrients are macronutrients and which are micronutrients?
2. How does indigestible fiber contribute to a healthy diet?
3. Describe the relationship of body weight to calorie intake and energy expenditure.
4. What is body mass index?
5. Describe the events of starvation.
6. What are some of the causes and effects of obesity?

31.5 | Investigating Life: The Ultimate Sacrifice

Of all of the surprising behaviors in the animal kingdom, among the most puzzling is sexual cannibalism. A cannibal, of course, is an organism that eats other members of its species. Sexual cannibalism is a special case in which a female eats a male before, during, or after copulation.

Sexual cannibalism occurs in arthropods, including praying mantises, scorpions, and some groups of spiders. The Australian redback spider, *Latrodectus hasselti* (figure 31.20), however, offers a particularly interesting example. This arachnid, which is a relative of the notorious North American black widow, has a bizarre mating ritual in which the male volunteers his body for the female to consume.

The male redback spider has two mating structures, called palps, on his head. The palps resemble boxing gloves with coils at their tips. The male fills each palp with sperm that he has deposited on a web. He courts a female; if she accepts him, he inserts a palp into one of her two sperm storage organs and begins delivering sperm. A few seconds later, with the palp still inserted, he does something that no other animal is known to do: he offers up his abdomen for her to eat. Using the palp as a sort of pivot, he somersaults his body onto her fangs. Observations of spiders in the field suggest that about 65% of the time, the female eats the male.

The most unusual feature of the redback spider's behavior is that the male appears to cooperate in his own demise. The female does not position the male's body into her jaws; rather, he moves his own body toward her mouth. And unlike other arthropods, which often try to avoid being eaten during mating, the male redback spider does not struggle or try to escape.

This behavior is puzzling. After all, by allowing his mate to kill him, he forfeits all possible opportunities to mate again in the

Figure 31.20 Australian Redback Spiders. The female of this species is much larger than the male resting on her abdomen.

future. Natural selection should weed out behaviors that decrease an organism's reproductive success. Yet the male redback spider's behavior is consistent and predictable.

Scientists have proposed many hypotheses to explain sexual cannibalism, not only in the redback spider but also in other species. One possibility is that the female uses the nutrients from her meal to help support the growth of her eggs and offspring, increasing both her and her mate's reproductive success. Or perhaps she simply mistakes the male for prey.

These explanations may apply to other species, but not to the Australian redback spider. After all, the female is much larger than the male: the average female is about 1 centimeter long and weighs some 256 milligrams; the male's body, in contrast, is only 3 to 4 millimeters long and weighs just over 4 milligrams. Moreover, laboratory studies suggest that consuming one tiny male neither boosts the mass of a female's egg sac nor increases the number of eggs. Researchers have therefore ruled out a nutritional explanation for sexual cannibalism in this species.

It is also unlikely that the female accidentally mistakes the male for prey. Spiders have elaborate courtship rituals involving visual displays and distinctive vibrations. These signals all but eliminate the chance for a female spider to confuse a male suitor for a prey species.

How, then, does the male redback spider's strange behavior increase his fitness? Evolutionary biologist Maydianne Andrade investigated this question while she was a graduate student at the University of Toronto at Mississauga. Andrade set up a series of two consecutive matings between virgin females and either normal or irradiated (sterile) virgin males. The order in which the matings occurred—normal male first or sterile male first—was randomized. Andrade carefully observed how long each pair copulated and noted whether or not each male was cannibalized. After the second mating, Andrade counted the number of hatched and unhatched eggs in each female's egg sac. Andrade assumed that all of the unhatched eggs were "fertilized" by sterile males. These counts therefore enabled Andrade to estimate the proportion of the offspring that normal males father on the first or second copulation.

Overall, copulation lasted from 6 to 31 minutes, but Andrade noticed an interesting pattern. A male that was cannibalized spent about 25 minutes in copulation, more than twice as long as one that survived the mating (about 11 minutes). While the female occupied herself with her meal, the dying male evidently put the extra time to good use: longer copulation predicted a big boost in paternity, from 45% for a noncannibalized male to 92% for a cannibalized male (figure 31.21).

A cannibalized male also received an indirect benefit. If the female ate her first mate, she rejected a subsequent male 67% of the time. In contrast, if the first male survived, the female rejected a second male just 4% of the time. A male who was eaten therefore had a higher likelihood of being the sole father to the female's offspring than one who survived the mating.

Andrade's study showed that sexual cannibalism boosts a male spider's fitness in two ways: longer copulation time, plus

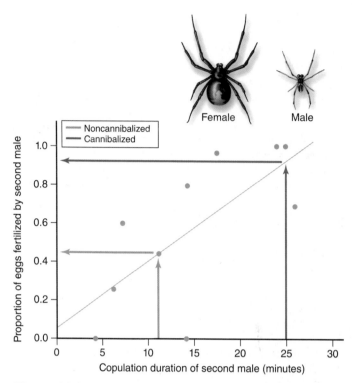

Figure 31.21 **Time Matters.** The longer a male copulates, the more offspring he fathers. Since cannibalized males copulate longer, they should leave more offspring than noncannibalized males.

fewer mating opportunities for subsequent males. But wouldn't the male be even better off by escaping and pursuing other females? In this species, the answer appears to be no. Males that venture off of a web in search of females are vulnerable to ants and other predators. And even when a wandering male does find a female's web, he still has to compete with as many as five other males for the chance to mate. As a result, fewer than 20% of males ever get to mate at all. In Australian redback spiders, mating opportunities for males are exceedingly rare; apparently, a copulating male can make the best of his good fortune by offering up his own body—all of it—as the ultimate sacrifice to the next generation.

Andrade, Maydianne C. B. 1996. Sexual selection for male sacrifice in the Australian redback spider. *Science,* vol. 271, pages 70–72.

31.5 | Mastering Concepts

1. Explain why sexual cannibalism represents an evolutionary paradox.

2. Use figure 31.21 to predict the percentage of a female's offspring that a male redback spider will sire if he copulates for 20 minutes.

Chapter Summary

31.1 | Digestive Systems Derive Nutrients from Food

- The **digestive system** acquires and breaks down food.

A. Animals Eat to Obtain Energy and Building Blocks

- Plants and algae produce their own food and are therefore **autotrophs,** whereas animals are **heterotrophs** that consume food. **Nutrients** in food provide energy and raw materials needed for growth and maintenance.

B. How Much Food Does an Animal Need?

- An animal's metabolic rate determines its need for food. Metabolic rate, in turn, reflects the animal's body temperature regulation, body size, and physiological state.

C. Animals Process Food in Four Stages

- Food is **ingested, digested,** and **absorbed** into the bloodstream; indigestible wastes are **eliminated** as **feces.**

D. Animal Diets and Feeding Strategies Vary Greatly

- **Herbivores** eat plants, **carnivores** eat meat, **detritivores** consume decaying organic matter, and **omnivores** have a varied diet.
- Animals obtain food in diverse ways. They may be **bulk feeders, fluid feeders, suspension feeders,** or **substrate feeders.** A **deposit feeder** is a type of substrate feeder.

31.2 | Animal Digestive Tracts Take Many Forms

- Protists and sponges have intracellular digestion. Their cells engulf food and digest it in food vacuoles.
- Animals that are more complex have extracellular digestion in a cavity within the body. An **incomplete digestive tract** (also called a **gastrovascular cavity**) has one opening. A **complete digestive tract,** or **alimentary canal (gastrointestinal tract),** has two openings. Food enters through the mouth and is digested and absorbed; undigested material leaves through the **anus.**
- The length of the digestive tract, number of stomach chambers, and size of the **cecum** are adaptations to specific diets. In **ruminants,** bacteria inhabiting the rumen help break down hard-to-digest plant matter.

31.3 | The Human Digestive System Consists of Several Organs

- Waves of contraction called **peristalsis** move food along the digestive tract. Muscular **sphincters** control movement from one compartment to another.

A. Digestion Begins in the Mouth and Esophagus

- In the mouth, **teeth** break food into small pieces. Salivary glands produce saliva, which moistens food and begins starch digestion.
- With the help of the **tongue,** swallowed food moves past the **pharynx** and through the **esophagus** to the stomach. The **epiglottis** prevents food from entering the trachea.

B. The Stomach Stores, Digests, and Pushes Food

- The **stomach** stores food, mixes it with **gastric juice,** and churns it into liquefied **chyme.** Hydrochloric acid in the gastric juice kills microorganisms and denatures proteins. The protein-splitting enzyme **pepsin** begins protein digestion.

C. The Small Intestine Digests and Absorbs Nutrients

- The **small intestine** is the main site of digestion and nutrient absorption. **Villi** absorb the products of digestion; **microvilli** on each villus provide tremendous surface area.
- The **pancreas** supplies pancreatic amylase, trypsin, chymotrypsin, lipases, and nucleases to the small intestine. These enzymes break down carbohydrates, polypeptides, lipids, and nucleic acids.
- The **liver** produces **bile,** which emulsifies fat; the **gallbladder** stores the bile and releases it to the small intestine.

D. The Large Intestine Completes Nutrient and Water Absorption

- Material remaining after absorption in the small intestine passes to the **large intestine,** which absorbs water, minerals, and salts, leaving feces. Bacteria digest the remaining nutrients and produce useful vitamins that are then absorbed. Feces exit the body through the anus.

31.4 | A Healthy Diet Includes Essential Nutrients and the Right Number of Calories

A. A Varied Diet Is Essential to Good Health

- Metabolism, growth, maintenance, and repair of body tissues all require nutrients from food. **Macronutrients** include carbohydrates, proteins, fats, and water, whereas **micronutrients** include vitamins and minerals.

B. Body Weight Reflects Food Intake and Activity Level

- **Kilocalories** measure the energy stored in food. Fat has more **Calories** per gram than either carbohydrate or protein.

C. Starvation: Too Few Calories to Meet the Body's Needs

- If a person does not eat enough over a long period, the body uses reserves of fat and protein to fuel essential processes. **Anorexia nervosa** and **bulimia** are eating disorders that may reduce calorie intake to dangerously low levels.

D. Obesity: More Calories Than the Body Needs

- A person who eats more calories than he or she expends will gain weight. **Obesity** is associated with many health problems.

31.5 | Investigating Life: The Ultimate Sacrifice

- During copulation, a male Australian redback spider offers his body for the female to eat.
- Males that are cannibalized copulate longer and therefore father more offspring than those that survive mating.

Multiple Choice Questions

1. The protein you eat is mostly
   a. incorporated into your body without modification.
   b. eliminated in feces.
   c. converted to fat before digestion.
   d. dismantled into individual amino acids.

2. At which stage does an organism's cells gain nutrients?
   a. Ingestion   c. Absorption
   b. Digestion   d. Elimination

3. A human is a
   a. suspension feeder.   c. fluid feeder.
   b. deposit feeder.   d. bulk feeder.

4. A flatworm has a(n)
   a. incomplete digestive tract.
   b. mouth and anus.
   c. alimentary canal.
   d. Both b and c are correct.

5. If a person is hanging upside down, food can still move along the esophagus to the stomach, thanks to
   a. microvilli.   c. the epiglottis.
   b. chyme.   d. peristalsis.

6. Making the pH of the stomach closer to neutral (pH 7) would prevent the
   a. digestion of proteins.
   b. movement of chyme to the duodenum.
   c. absorption of nutrients.
   d. Both a and c are correct.

7. What is the function of bile?
   a. To break proteins into amino acids
   b. To digest fat molecules
   c. To break fat into small droplets
   d. To digest nucleic acids

8. What is an important function of the normal intestinal microbiota?
   a. Prevention of infection by disease-causing bacteria
   b. Production of digestive enzymes that degrade protein
   c. Decomposition of dietary fats
   d. Absorption of excess vitamins

9. What is the value of dietary fiber?
   a. It is a carbohydrate source.
   b. It helps material move along the digestive tract.
   c. It helps trigger the production of gastric juice.
   d. All of the above are correct.

10. A person's body mass index is calculated based on
    a. caloric intake.   c. weight.
    b. height.   d. Both b and c are correct.

Write It Out

1. What are the two main functions of food in an animal's body?
2. Name an organism that has each of the following:
   a. Extracellular digestion   c. An alimentary canal
   b. A gastrovascular cavity   d. Suspension feeding
3. Explain why a horse has a lower metabolic rate than a squirrel, relative to its body size.
4. Would an alligator require more, less, or the same amount of food as a horse of the same size? Explain.
5. What are the four stages of food use in animals, and where in the human digestive tract does each of these stages occur?
6. List four ways that animals obtain food.
7. Compare and contrast the digestive systems of a whale and *Paramecium*.
8. Biologists estimate that carnivores assimilate 90% of the mass in their diet; in contrast, most herbivores assimilate only 30% to 60% of their food. How do these differences in assimilation efficiency relate to the morphology of their digestive systems?
9. How does mechanical breakdown of food speed chemical digestion?
10. Identify the part of the digestive system that includes the following: duodenum; cecum, appendix, rectum, and anus; villi and microvilli; pyloric sphincter.
11. How does the structure of the small intestine maximize surface area?
12. What are the digestive products of carbohydrates, proteins, and fats?
13. Compare and contrast the alveoli of the lungs (see chapter 30) with the villi of the small intestine.
14. List the main parts of the human digestive system, then create a chart that compares the locations, anatomy, and functions of each part.
15. Trace the movement of food in the digestive tract from mouth to anus.
16. What is the difference between a macronutrient and a micronutrient? Give examples of each.
17. What determines whether a person will gain, lose, or maintain weight?
18. Orlistat (Alli) is a weight-loss drug that inhibits the activity of lipases in the small intestine. Why would this be more effective than a drug that blocks absorption of proteins or carbohydrates? Given that four essential vitamins are fat-soluble, what might be a side effect of blocking fat absorption?
19. Calculate your body mass index using the formula in the text. How could you change your BMI?
20. Nutritional scoring systems rate foods according to their nutritional value. The table at the top of the next column lists sample values from the NuVal system, which rates foods on a scale of 1 (least healthy) to 100 (healthiest). Search the Internet to learn what information researchers use in calculating the score for each food. Browse the scores for your favorite foods. Which foods that you eat have the highest scores? The lowest? How might you use these scores to adjust your shopping and eating habits?

Food	Nutritional Score
Chocolate granola bar	9
Ground beef	26
Egg	33
Canned corn	50
Whole wheat bread	81
Frozen broccoli	100

21. Fructose and glucose are both monosaccharides, but the body metabolizes these sugars differently. For example, glucose stimulates insulin release from the pancreas (see chapter 27); fructose does not. Moreover, insulin stimulates leptin release. Use this information to propose an explanation for the correlation between the skyrocketing consumption of high fructose corn syrup since 1970 and the rise in obesity during the same period.
22. How do the circulatory and muscular systems interact with the digestive system?
23. How does it benefit an organism to have a digestive system with an extensive surface area?
24. Design an experiment to test the hypothesis that intestinal bacteria are essential to nutrient absorption in mice.
25. Many children believe that a piece of swallowed chewing gum will remain in the body for 7 years. Chewing gum is made of an indigestible polymer that does not dissolve in water. Since the gum cannot be digested, what happens to it after it is swallowed? Given your answer, does the 7-year timescale make sense? Propose an alternative prediction for how long it might take instead, and explain your reasoning.
26. The chapter opening essay describes how Barry Marshall induced gastritis by swallowing *H. pylori* bacteria. Why didn't this experiment provide sufficient support for the hypothesis that *H. pylori* causes gastritis?

Pull It Together

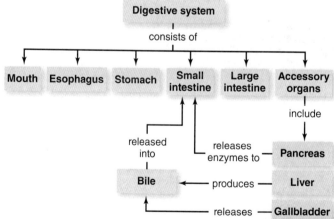

Digestive system *consists of*: Mouth, Esophagus, Stomach, Small intestine, Large intestine, Accessory organs. Small intestine — released into — Bile; releases enzymes to — Pancreas. Accessory organs — *include* — Pancreas. Liver — produces — Bile. Gallbladder — releases — Bile.

1. Add the terms *ingestion, digestion, absorption, elimination, chyme, bacteria,* and *peristalsis* to this concept map.
2. Other than enzymes, what does the pancreas release into the small intestine?
3. In which organ(s) are carbohydrates, proteins, and fats digested?

Chapter

32

Regulation of Temperature and Body Fluids

Sea Lizards. Glands on a marine iguana's head eliminate excess salt ingested with food. The salt accumulates as a crust on the lizard's forehead.

connect™
|BIOLOGY

Enhance your study of
this chapter with practice quizzes,
animations and videos, answer keys,
and downloadable study tools.
www.mhhe.com/hoefnagels

A Day in the Life of a Marine Iguana

MANY ENVIRONMENTS ARE HOSTILE TO CELL SURVIVAL—THEY ARE TOO HOT, TOO COLD, TOO SALTY, OR NOT SALTY ENOUGH. Humans cope with some of these extremes in familiar ways; for example, we sweat when we are hot and shiver when we are cold. We can eat salty pretzels, but only to the extent that our kidneys can dispose of the excess salt. We cannot drink seawater. Other animals, however, have radically different behavioral and physiological adaptations to both temperature and salt.

Consider the marine iguana lizard. These reptiles spend much of their time regulating their body temperatures. Iguanas and other nonavian reptiles are sometimes called "cold-blooded" (as opposed to "warm-blooded" birds and mammals). But this term is misleading. An iguana's body temperature is not necessarily low; rather, its temperature fluctuates as the lizard derives heat from the environment.

Marine iguanas on the Galápagos Islands begin their days basking in the rising sun, draped on boulders and hardened lava, sunning their backs and sides. After about an hour, they turn, raise their bodies, and aim their undersides at the sun.

By midmorning, the air temperature is rising rapidly. Iguanas cannot sweat as humans do; instead, they must escape the blazing sun. They lift their bodies by extending their short legs, which removes their bellies from the hot rocks and allows breezes to fan them. By noon, these push-ups are insufficient to stay cool. The iguanas retreat to the shade of rock crevices.

The animals are hungry by midday. The iguanas hang off rocks to reach seaweed by the shore, or they dive into the ocean and eat green algae on the ocean floor. The water is so cold that the lizards' body temperatures rapidly drop. Arteries near their body surfaces constrict, helping conserve heat. Even so, the water is too cold for them to stay in for more than a few minutes.

After feeding, the iguanas stretch out on the rocks again, warming sufficiently to digest their meal. They continue basking as the day ends, absorbing enough heat to sustain them through the cooler night temperatures until a new day begins. Meanwhile, the iguana's body must cope with the high salt content of its diet. A gland in the lizard's head rids the animal's body of extra salt. Cells lining the interior of the salt gland secrete a fluid that travels in branching tubules, which empty through the nostrils. This fluid carries the salts with it.

Marine iguanas, and every other animal, maintain homeostasis in many ways. This chapter describes how animals regulate their temperatures and control the composition of their body fluids.

Learning Outline

Learn How to Learn
Pay Attention in Class

It happens to everyone occasionally: your mind begins to wander while you are sitting in class, so you doodle or doze off. Before you know it, class is over, and you got nothing out of it. How can you keep from wasting your class time this way? One strategy is to get plenty of sleep and eat well, so your mind stays active instead of drifting off. Another is to prepare for class in advance, since getting lost can be an excuse for drifting off. When you get to class, sit near the front, listen carefully, and take good notes. Finally, a friendly reminder can't hurt; make a small PAY ATTENTION sign to put on your desk, where you can always see it.

32.1 | Animals Regulate Their Internal Temperature

Animals live nearly everywhere on Earth (**figure 32.1**). The extremely dry, scorching home of a camel contrasts sharply with a frigid Arctic habitat or with the perpetual humidity of a tropical rain forest. These habitats select for very different ways of regulating body temperature, conserving water, and disposing of wastes. In hot, dry areas, an animal must conserve enough water for cells to function and, at the same time, produce enough sweat to tolerate the heat. At another extreme, frigid water surrounds an animal living in an Arctic freshwater lake. An animal in this habitat must continually pump excess water out of its body and conserve salts, all while surviving the cold.

Most animals live in more moderate environments, but each species has adaptations that enable it to maintain homeostasis. This chapter begins by describing how animals regulate body temperature and the concentrations of salt and water in their tissues. It moves next to the types of nitrogenous wastes that animals produce, then concludes with the structure and function of the human urinary system.

A. Heat Gains and Losses Determine an Animal's Body Temperature

Whether an animal lives in the Arctic circle or the Amazonian rain forest, its body temperature must remain within certain limits. Part of the reason is that extreme temperatures alter biological molecules. Excessive heat can alter a protein's three-dimensional shape and disrupt its function. Extreme cold solidifies lipids and therefore inhibits the function of biological membranes. In addition, if the temperature inside a cell deviates from an animal's customary body temperature, enzymes function less efficiently and vital biochemical reactions proceed too slowly to sustain life. ▶▶▶ protein shape, p. 37; lipids, p. 34; enzymes, p. 78

Thermoregulation is the control of body temperature, and it requires the ability to balance heat gained from and lost to the environment. An animal can also maintain homeostasis by controlling how much heat it produces. When cells generate ATP in aerobic respiration (the topic of chapter 6), they also produce metabolic heat. The more active the animal, the higher its metabolic rate and the more heat it produces.

The main source of an animal's body heat may be external or internal (**figure 32.2**). An **ectotherm** lacks an internal temperature-regulating mechanism. It thermoregulates by moving to areas where it can gain or lose heat, so its temperature varies with external conditions. The vast majority of animals are ectotherms, including all invertebrates, plus fishes, amphibians, and nonavian reptiles such as the iguana described in this chapter's opening essay.

An **endotherm** regulates its body temperature internally. Most endotherms maintain a relatively constant body temperature by balancing heat generated in metabolism (especially in the

Figure 32.1 Different Habitats. Penguins thrive on Antarctic ice, whereas parrots inhabit the warm, humid tropics. The adaptations of each bird reflect its diet, the temperature of its habitat, and many other selective forces.

Figure 32.2 Ectotherm and Endotherm. An ectotherm such as a snake alters its behavior to manage the exchange of heat between its body and the environment. In contrast, the mouse is an endotherm; its metabolism generates most of its body heat.

muscles) with heat lost to the environment. Mammals and birds are endotherms; insulation in the form of fat, feathers, or fur helps retain their body heat.

Some animals combine features of both ecto- and endothermy. In some of these "heterotherms," body temperature fluctuates with the animal's activity level. A bat, for example, maintains a high temperature while flying, but its temperature drops while the animal roosts. In other heterotherms, an animal's limbs are much colder than the rest of its body. A penguin's feet are nearly as cold as ice, while the bird's core body temperature is much higher.

Both ectothermy and endothermy have advantages and disadvantages. The ectotherm uses much less energy, and therefore requires less food and O_2, than an endotherm. However, an ectothermic animal must be able to obtain or escape environmental heat. An injured iguana that could not squeeze into a crevice to avoid the broiling noonday sun would cook to death. Ectotherms also become sluggish when the temperature is low, which can make it hard for them to escape from predators.

On the other hand, an endotherm maintains its body temperature even in cold weather or in the middle of the night. But this internal constancy comes at a cost. The metabolic rate of an endotherm is generally five times that of an ectotherm of similar size and body temperature. Endotherms therefore require much more food than do ectotherms.

Figure It Out

As the sun sets and the external temperature gets colder, will a frog's body temperature go down, stay the same, or go up?

Answer: It will go down.

B. Several Adaptations Help an Animal to Adjust Its Temperature

The hypothalamus detects blood temperature, receives information from thermoreceptors in the skin and other organs, and controls many of the responses that maintain homeostasis (**figure 32.3**). These responses may be physiological or behavioral. Sweating is a physiological reaction to heat, but a person who is too hot might also move into the shade—a behavioral response. Thermoregulation therefore depends on the interactions of the musculoskeletal, nervous, endocrine, respiratory, integumentary, and circulatory systems. ▸ hypothalamus, p. 542

One adaptation to extreme cold is a **countercurrent exchange** system, in which two adjacent currents flow in opposite directions and exchange heat with each other (**figure 32.4**). Recall from chapter 29 that arteries carry blood away from the heart, whereas veins transport blood in the opposite direction.

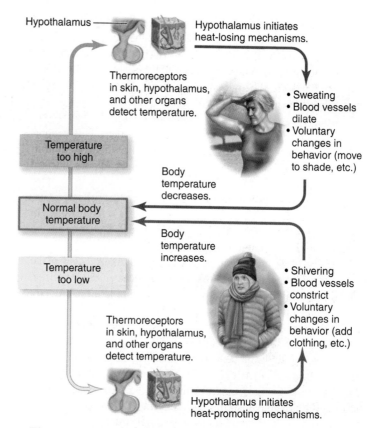

Figure 32.3 Thermoregulation in Humans. Thermoreceptors signal the hypothalamus to trigger the responses that keep body temperature within a certain range.

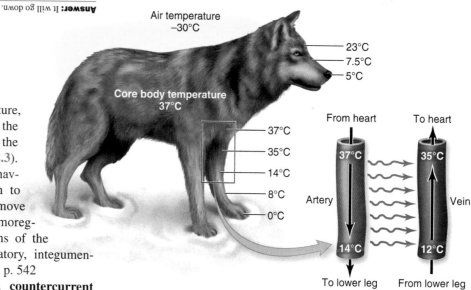

Figure 32.4 Countercurrent Heat Exchange. In the wolf's leg, arteries carrying warm blood lie near veins carrying cooler blood in the opposite direction. Heat moves along its gradient from artery to vein, rather than being lost at the extremities.

The heart is located in the warm core of the body, so arterial blood is warmer than blood in veins near the body's surface. In the limbs of the wolf in figure 32.4, arteries and veins are adjacent to each other. Because the blood flows in opposite directions in the two vessels, the gradient favors heat transfer from artery to vein along the entire length of the limb.

Countercurrent exchange therefore conserves heat rather than allowing it to escape through the extremities. This adaptation not only enables Arctic mammals such as wolves to hunt in extreme cold, but it also allows penguins to spend hours in frigid water. (Note that animals use countercurrent exchange in other ways as well; figure 30.3, for example, depicts the gills of a fish, which use countercurrent exchange to extract O_2 from water. In addition, section 32.5 explains countercurrent exchange in the context of kidney function). ▶ gills, p. 623

Many birds and mammals, including humans, also make other physiological changes that conserve heat. In cold weather, blood vessels in the extremities constrict, retaining more blood in the warmer core of the body. The animal may also shiver; the contraction of skeletal muscle generates heat. At the same time, muscles in the skin cause feathers or fur to stand erect, trapping an insulating air layer next to the skin. (In humans, this hair-raising response is useless because we have so little body hair. Nevertheless, goose bumps form when the hair muscles contract.) Animals may also migrate, hibernate, or huddle together to conserve heat (**figure 32.5a**). ▶ vestigial structures, p. 262

Ectotherms also have a repertoire of behaviors that retain heat, such as seeking sunlight, sprawling on warm rocks or roadways, building insulated burrows, and tucking wings and other extremities near their bodies.

So far, the focus has been on conserving heat, but endotherms also must maintain homeostasis when the environment is too hot (figure 32.5b). Evaporative cooling from the skin or respiratory surfaces is one way to lower body temperature. For example, humans sweat to cool off, whereas a panting dog allows water to evaporate from the moist lining of its mouth. Likewise, an owl flutters loose skin under its throat to move air over moist surfaces in the mouth.

In addition, when the environment is warm, blood vessels in the extremities dilate and allow more blood to approach the relatively cool body surface. Tiny veins in the face and scalp also reroute blood cooled near the body's surface toward the brain. This adaptation explains why vigorous exercise causes the face of a light-skinned person to turn red.

Behavioral strategies can also help both ectotherms and endotherms to cool off. Many animals escape the sun's heat by swimming, covering themselves with cool mud, or retreating to the shade. Some burrow underground and emerge only at night; others extend their wings to promote cooling. Humans shed extra layers of clothing, and we consume cold food and drinks. We also swim or fan ourselves to increase heat loss by convection.

32.1 | Mastering Concepts

1. Differentiate between ectotherms and endotherms.
2. Name examples of animals that are ectotherms and those that are endotherms.
3. What are the advantages and disadvantages of ectothermy and endothermy?
4. What are some physiological and behavioral adaptations to extreme cold and extreme heat?

a.

b.

Figure 32.5 Too Cold or Too Hot. (a) These snow monkeys (Japanese macaques) are adapted to cold winters. Their thick fur retains body heat, as does their huddling behavior. Some snow monkeys soak in warm springs to escape winter weather. (b) A dog loses excess heat through his mouth as he pants, while the man is displaying a behavioral adaptation to the hot weather: he has shed his excess clothing.

32.2 | Animals Regulate Water and Ions in Body Fluids

Besides regulating its temperature, an animal must also maintain homeostasis in the composition of its body fluids. The balance of solutes and water between a cell's interior and its surroundings is critical to life. Cells must retain water even when conditions are dry, yet too much water is damaging. Sodium, chlorine, hydrogen, and other ions are vital to life, but not in excess. ▶▶ ions, p. 21; water, p. 27

In most habitats, therefore, organisms must **osmoregulate;** that is, they control the concentration of ions in their body fluids as the environment changes. Osmoregulation requires managing the gain and loss of water, ions, or both.

A brief review of how water and ions move across membranes will help explain how osmoregulation works. Osmosis is the diffusion of water across a semipermeable membrane; the net direction of water movement is toward the side with the highest concentration of dissolved solutes. If a cell's environment is saltier than the cell itself, water moves out of the cell. In the opposite situation, water moves into the cell. ▶ membrane transport, p. 80

Unlike water, most ions cross membranes via protein channels. Osmoregulation often requires cells to move ions against their concentration gradient. This process, called active transport, requires energy in the form of ATP. Water may follow the ions by osmosis; as described in section 32.5, our kidneys exploit this mechanism to conserve water during the production of urine.

Bony fishes that live in the ocean and in fresh water face opposite challenges in osmoregulation (figure 32.6). Ocean water is much saltier than a fish's cells, so the animal loses water by osmosis, mostly at the gills. The fish therefore drinks seawater, produces little urine, and uses active transport at the gills to get rid of excess salts. In contrast, fresh water is much more dilute than the cells of a fish. A freshwater fish therefore constantly takes in water at its gills and through its skin by osmosis, while losing salts to its surroundings. Its kidneys shed excess water in dilute urine, and the gills take up ions from water by active transport.

Land animals use a combination of strategies to obtain and conserve water (figure 32.7). Whereas humans ingest most of their water in food and drink, desert kangaroo rats derive most of their water as a byproduct of metabolism, especially cellular respiration. Animals lose water through evaporation from lungs and the skin surface, in feces, and in urine.

32.2 | Mastering Concepts

1. What is osmoregulation, and why is it important?
2. Describe osmoregulation in marine and freshwater fishes.
3. How do land animals gain and lose water?

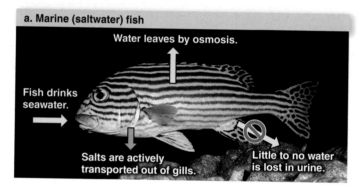

a. Marine (saltwater) fish

Water leaves by osmosis.

Fish drinks seawater.

Salts are actively transported out of gills.

Little to no water is lost in urine.

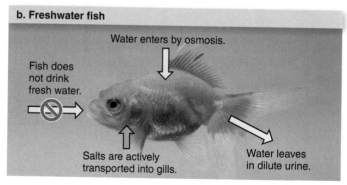

b. Freshwater fish

Water enters by osmosis.

Fish does not drink fresh water.

Salts are actively transported into gills.

Water leaves in dilute urine.

Figure 32.6 Osmoregulation in Fishes. Marine and freshwater fishes face opposite challenges in managing ions and water. (a) Seawater contains a higher solute concentration than the cells of a marine fish. The animal therefore constantly loses water by osmosis and pumps ions out by active transport. (b) Fresh water contains few dissolved solutes; a fish in that habitat therefore constantly gains water by osmosis and must use active transport to acquire essential ions.

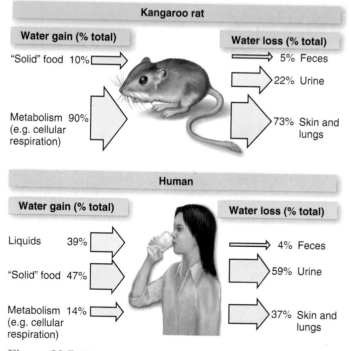

Kangaroo rat

Water gain (% total)

"Solid" food 10%

Metabolism 90% (e.g. cellular respiration)

Water loss (% total)

5% Feces
22% Urine
73% Skin and lungs

Human

Water gain (% total)

Liquids 39%

"Solid" food 47%

Metabolism 14% (e.g. cellular respiration)

Water loss (% total)

4% Feces
59% Urine
37% Skin and lungs

Figure 32.7 Water Gain and Loss. The desert-dwelling kangaroo rat gets most of its water from its metabolism and loses little through feces and urine. In contrast, a human gets most water from food and drink, and loses most of it through urine.

32.3 | Nitrogenous Wastes Include Ammonia, Urea, and Uric Acid

Animals produce two main categories of waste. One type is feces, which contains undigested food that passes through the digestive tract without ever entering the body's cells (see chapter 31). The other category is waste produced by the body's cells; **excretion** is the elimination of these metabolic wastes. For example, as described in chapter 30, the respiratory system excretes CO_2, a byproduct of aerobic cellular respiration.

An animal's body also must excrete nitrogen-containing (nitrogenous) wastes, which cells produce mainly during the breakdown of proteins (**figure 32.8**). Proteins are composed of amino acids, which can enter the energy-generating pathways described in chapter 6. During this process, amino groups (—NH_2) are stripped from amino acids. Each amino group then picks up a hydrogen ion and becomes **ammonia,** NH_3.

Excreting ammonia consumes little energy, but this waste product is very toxic. Animals must therefore excrete ammonia in a dilute solution. Not surprisingly, animals that excrete nitrogen in this form—aquatic invertebrates, most bony fishes, tadpoles, and salamanders—live in habitats with abundant water. In these animals, ammonia enters the water by simply diffusing out of the blood that flows through the gills.

In contrast, land animals expend energy to change ammonia to urea or uric acid. Both of these substances store higher concentrations of nitrogen than ammonia. Moreover, because urea and uric acid are less toxic than ammonia, they require less dilution in water. Excreting either of these nitrogenous wastes therefore helps conserve water.

Mammals, adult frogs and toads, turtles, and cartilaginous fishes (sharks and rays) form **urea,** which the liver produces as cells break down proteins. In most of these animals, urea moves to the bloodstream and is eliminated with water in urine. Sharks, however, retain the urea, which helps with osmoregulation in their salty habitat.

Insects, land tortoises, lizards, and birds change most of their nitrogenous wastes to **uric acid,** which is relatively insoluble in water. These animals therefore excrete uric acid in an almost solid form, losing the least water of all. In birds, uric acid mixes with undigested food to form the familiar "bird dropping."

Insects and spiders also eliminate uric acid with undigested food. The excretory system of these animals consists of structures called Malpighian tubules that empty into the gut (**figure 32.9**). The cells lining the Malpighian tubules use active transport to move uric acid and ions, especially potassium (K^+), into the tubules. Water follows by osmosis. Fluid in the Malpighian tubules enters the intestine, where cells lining the rectum reabsorb most of the K^+ and water. The uric acid mixes in the intestine with undigested food, which the insect eliminates through its anus.

32.3 | Mastering Concepts

1. What are three nitrogenous wastes that animals produce?
2. Describe the function of Malpighian tubules in insects.

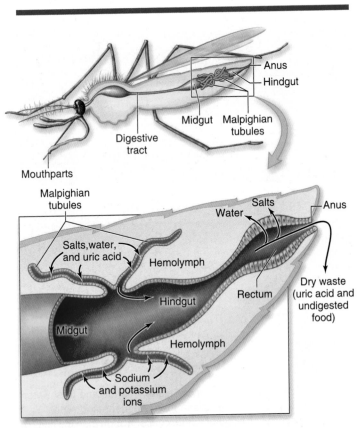

Figure 32.9 Insect Excretion. Insect excretory organs, called Malpighian tubules, collect salts, water, and uric acid from the animal's cells. The tubules empty into the digestive tract, where the contents mix with undigested food. After most of the water and salts are reabsorbed, the combined wastes are eliminated.

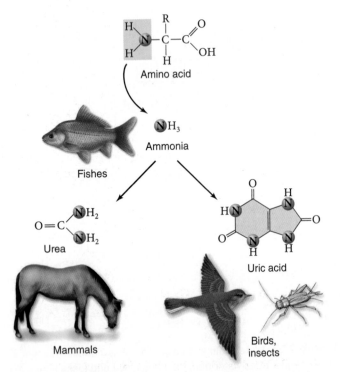

Figure 32.8 Three Nitrogenous Wastes. Ammonia is the simplest byproduct of protein breakdown, but it is toxic. Urea and uric acid are more costly to produce but much less toxic.

32.4 | The Urinary System Produces, Stores, and Eliminates Urine

The human **urinary system** filters blood, eliminates nitrogenous wastes, and helps maintain the ion concentration of body fluids. The paired **kidneys** are the major excretory organs in the urinary system (**figure 32.10**). Located near the rear wall of the abdomen, each kidney is about the size of an adult fist and weighs about 230 grams.

As the kidneys cleanse blood, a liquid waste called **urine** forms; section 32.5 describes this process in detail. Besides eliminating urea and other toxic substances, kidneys have other functions as well. These organs conserve water, salts, glucose, amino acids, and other valuable nutrients. They also help regulate blood pH.

One other function of the kidneys is to regulate the volume of blood. Look back at the water balance in figure 32.7. What happens if we drink too many fluids, or if we lose too much moisture in sweat and breath? Rather than swelling like a balloon or drying up like a leaf, the body adjusts: the kidneys maintain the volume of blood by controlling the amount of water lost in urine. This function is important because blood volume is one factor that influences blood pressure. ▶ blood pressure, p. 612

The urine from each kidney drains into a **ureter,** a muscular tube about 28 centimeters long. Waves of smooth muscle contraction, called peristalsis, squeeze the fluid along the two ureters and squirt it into the **urinary bladder,** a saclike muscular organ that collects urine.

The **urethra** is the tube that connects the bladder with the outside of the body. In females, the urethra opens between the clitoris and vagina. In males, the urethra extends the length of the penis. The urethra also carries semen in males (see chapter 34). The term *urogenital tract* reflects the intimate connection between the urinary and reproductive systems.

Two rings of muscle (sphincters) guard the exit from the bladder. Both sphincters must relax before urine can leave the bladder. A spinal reflex involuntarily controls the innermost sphincter, which is made of smooth muscle. The outer sphincter consists of skeletal muscle. Its relaxation is under voluntary control in most people over 2 years old.

The adult bladder can hold about 600 milliliters of urine; however, as little as 300 mL of accumulating urine stimulates stretch receptors in the bladder. The receptors send impulses to the spinal cord, which stimulates sensory neurons that contract the bladder muscles, generating a strong urge to urinate. We can suppress the urge to urinate for a short time by contracting the external sphincter. Eventually, the cerebral cortex directs the sphincters to relax, and bladder muscle contractions force urine out of the body.

Weakened sphincter muscles, overactive bladder muscles, and nerve damage are among the underlying causes of urinary incontinence—the loss of bladder control. Urinary tract infections, pregnancy, prostate enlargement, spinal cord injuries, and many other conditions are associated with incontinence. Treatments in-

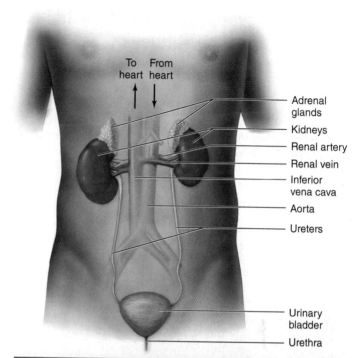

Urinary System	
Main Tissue Types*	**Examples of Locations/Functions**
Epithelial	Enables diffusion across glomerular capsules and kidney tubules; also lines ureters and bladder.
Connective	Blood (which kidneys filter) is a connective tissue.
Muscle	Smooth muscle controls flow of blood to and from nephrons; smooth and skeletal muscle sphincters control urine release.
Nervous	Sensory cells in hypothalamus coordinate negative feedback loops that maintain homeostasis.

*See chapter 24 for descriptions.

Figure 32.10 Human Urinary System. The human urinary system includes the kidneys, ureters, urinary bladder, and urethra. This generalized depiction omits the differences between male and female organs. Figures 34.4 and 34.7 show the sex differences in more detail.

clude everything from behavioral changes to drugs to the surgical implantation of an artificial sphincter.

32.4 | Mastering Concepts

1. List the organs that make up the human urinary system.
2. What are the functions of the kidneys?

32.5 | The Nephron Is the Functional Unit of the Kidney

The body's entire blood supply courses through the kidney's blood vessels every 5 minutes. At that rate, the equivalent of 1600 to 2000 liters of blood passes through the kidneys each day. Most of the fluid that the nephrons process, however, is reabsorbed into the blood and not released in urine. As a result, a person produces only about 1.5 liters of urine daily.

A. Nephrons Interact Closely with Blood Vessels

Each kidney contains 1.3 million tubular **nephrons**—the functional units of the kidney. As illustrated in figure 32.11, each nephron is entwined with a network of capillaries. Our tour of the kidney begins by tracing the flow of blood into and out of these vessels.

Each kidney receives blood via a renal artery, which branches into multiple arteries and arterioles. An arteriole delivers blood to a **glomerulus,** a tuft of capillaries where blood is filtered into the nephron. The capillaries of the glomerulus then converge into another arteriole, which leads to the **peritubular capillaries** that snake around part of each nephron. These blood vessels empty into a venule, which joins the renal vein carrying cleansed blood out of the kidney and (ultimately) to the heart.

Now examine the anatomy of the nephrons in figures 32.11b and c. Each nephron consists of two main parts: a glomerular capsule and a renal tubule. In the kidney's outer portion, or cortex, the glomerular (Bowman's) capsule receives fluid from the glomerulus. From there, the solution travels along the **renal tubule,** a winding passageway consisting of three functional regions: the proximal convoluted tubule, the nephron loop, and the distal convoluted tubule. The **proximal convoluted tubule** leads from the glomerular capsule to the hairpin-shaped **nephron loop** (also called the loop of Henle). The descending limb of the nephron loop dips into the renal medulla toward the kidney's center, and the ascending limb returns to the **distal convoluted tubule** in the cortex. The entire nephron, when stretched out, is about 3 to 4 centimeters long.

A **collecting duct** receives the fluid from several nephrons. Urine from many collecting ducts accumulates in the funnel-like renal pelvis before entering the ureter and urinary bladder and moving out of the body through the urethra.

B. Urine Formation Includes Filtration, Reabsorption, and Secretion

The chemical composition of urine reflects three processes (figure 32.12):

1. **Filtration:** Water and dissolved substances are filtered out of the blood at the glomerular capsule.

2. **Reabsorption:** Useful materials such as salts, water, glucose, and amino acids return from the nephron to the blood.

3. **Secretion:** Toxic substances, drug residues, hydrogen ions (H^+), and surplus ions are secreted into the nephron to be eliminated in urine.

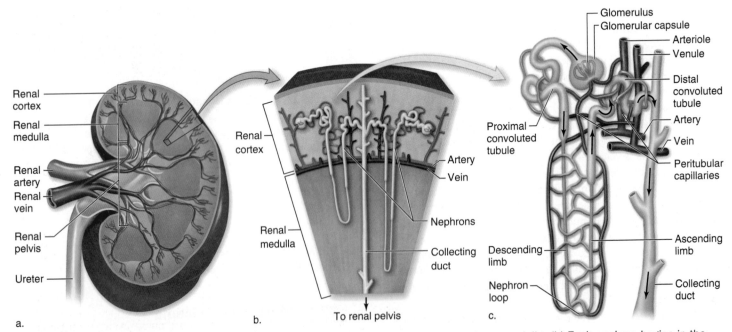

Figure 32.11 Anatomy of a Kidney. (a) The kidney is divided into an outer cortex and an inner medulla. (b) Each nephron begins in the cortex, descends into the medulla, and returns to the cortex, where it empties into a collecting duct. (c) Nephrons are intimately associated with blood vessels.

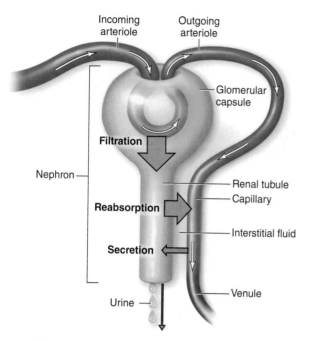

Figure 32.12 Overview of Urine Formation. Urine forms from three processes: filtration, reabsorption, and secretion. Arrows represent the flow of water and dissolved substances; each arrow's size reflects the magnitude of the flow. This stylized, schematic view does not reflect nephron anatomy.

When the nephrons fail to do their job, nitrogenous wastes and other toxins accumulate in the blood to harmful levels; in addition, a person may lose too much water and become dehydrated. Without treatment, kidney failure can be fatal (see this chapter's Apply It Now box on page 667).

The remainder of this section describes the events that occur at the nephron. Figure 32.13 summarizes the exchange of materials between blood vessels and a nephron; refer to this figure frequently as you go along.

C. The Glomerular Capsule Filters Blood

Like a baseball mitt catching a ball, the glomerular capsule surrounds the glomerulus. The function of the glomerulus, however, is more like that of a coffee filter: its pores allow water, urea, glucose, salts, amino acids, and creatinine (a byproduct of muscle contraction) to pass into the glomerular capsule, but large structures such as plasma proteins, blood cells, and platelets remain in the bloodstream. ▶ blood, p. 604

Blood pressure drives substances out of the glomerulus. The incoming arteriole has a larger diameter than the arteriole that leaves the glomerulus. This change in diameter raises the pressure in the glomerular capillaries. The increased pressure, in turn, forces fluid and small dissolved molecules across the capillary walls.

The liquid entering the glomerular capsule, called the filtrate, is the product of the first step in urine manufacture. The 1600 to 2000 liters of blood per day that pass through the kidneys produce approximately 180 liters of glomerular filtrate.

D. Reabsorption and Secretion Occur in the Renal Tubule

The Proximal Convoluted Tubule   Most selective reabsorption occurs at the proximal convoluted tubule. All along the tubule, specialized cells transport sodium ions (Na^+), glucose, amino acids, and other useful solutes into the interstitial fluid surrounding the tubule. Water follows by osmosis. Overall, blood vessels along the proximal convoluted tubule reabsorb almost two thirds of the salts and water that were present in the original filtrate, along with nearly all of the glucose and amino acids.

Secretion also occurs in the proximal convoluted tubule. The antibiotic penicillin, for example, enters the filtrate by secretion. Patients must take penicillin several times a day to compensate for this loss.

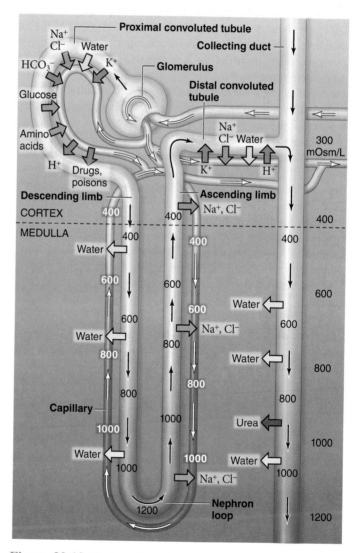

Figure 32.13 Nephron Function. This schematic view shows the movement of materials into and out of the nephron. The countercurrent flow of fluids in the tubules and adjacent blood vessels improves the exchange of substances. (Numbers indicate milliosmoles per liter, abbreviated mOsm/L, a measure of ion concentration.)

Reabsorption and secretion work together to maintain the pH of blood between 7 and 8; any variance from this range is deadly. Acids produced in metabolism lower blood pH. To prevent the pH from dipping too low, H^+ is secreted into the proximal convoluted tubule, and alkaline bicarbonate ions (HCO_3^-) are reabsorbed into the blood. ▶ pH, p. 30

The Nephron Loop   Next, the filtrate moves into the nephron loop, which makes a hairpin turn into the medulla. Notice in figure 32.13 that the medulla's solute concentration becomes progressively higher with depth; as you will see, this gradient is critical to the nephron's ability to conserve water and produce concentrated urine. Also, note that the filtrate in the nephron loop flows in the opposite direction of the blood in the surrounding capillaries—another example of countercurrent exchange.

As the descending limb of the nephron loop dips into the medulla, the solute concentration in the surrounding fluid is higher than that of the filtrate inside. Water therefore leaves by osmosis along the entire length of the descending limb. The cells of the descending limb, however, are impermeable to ions and urea. As water leaves the descending limb, the solute concentration inside the tubule gradually rises until it reaches its maximum at the bottom of the loop.

The filtrate then flows around the bend and into the ascending limb. Here, the filtrate is more concentrated than the surrounding capillaries, but the wall of the ascending limb is impermeable to urea and water. Therefore, water cannot reenter the filtrate. Ions, however, can leave. Along most of the ascending limb, Na^+ and Cl^- diffuse from the filtrate into the capillaries, which have a lower salt concentration. As the filtrate approaches the end of the nephron loop, Na^+ and other ions move out of the filtrate by active transport. As a result, the filtrate again becomes less concentrated than the tissues and blood in the surrounding cortex.

The Distal Convoluted Tubule   At the distal convoluted tubule, Na^+ and Cl^- ions continue to move out of the filtrate and into the blood by active transport. The walls of the distal convoluted tubule are again permeable to water, which moves by osmosis into the capillaries. At the same time, excess K^+ and H^+ in the blood may be secreted into the distal convoluted tubule and collecting duct. As described below, this fine-tuning of urine composition is under delicate hormonal control.

Once the filtrate has passed through the distal convoluted tubule, 97% of the water that was in the original glomerular filtrate has been reabsorbed, and little salt remains.

E. The Collecting Duct Conserves More Water

The collecting duct descends back into the salty medulla, as did the nephron loop. Much of the remaining water in the filtrate therefore leaves the collecting duct by osmosis. Some urea also diffuses out of the collecting duct and contributes to the high solute concentration surrounding the nephron loop deep in the medulla.

After reabsorption and secretion, the filtrate is urine. Urine contains water, urea, a small amount of uric acid, creatinine, and several ions; nearly all of the glucose and amino acids present in the original glomerular filtrate have been returned to the blood. The exact chemical composition of urine can vary, however, as described in the Burning Question on page 668.

F. Hormones Regulate Kidney Function

Kidney function adjusts continuously to maintain homeostasis. For example, if water is scarce, we produce more concentrated urine. When we drink too much water, our urine is abundant and dilute.

Hormones help make these adjustments (**figure 32.14**). When we are dehydrated, osmoreceptor cells in the hypothalamus send impulses to the posterior pituitary gland, which secretes a peptide hormone called **antidiuretic hormone (ADH),** or **vasopressin.** ADH triggers the formation of additional water channels in the walls of the distal convoluted tubule and collecting

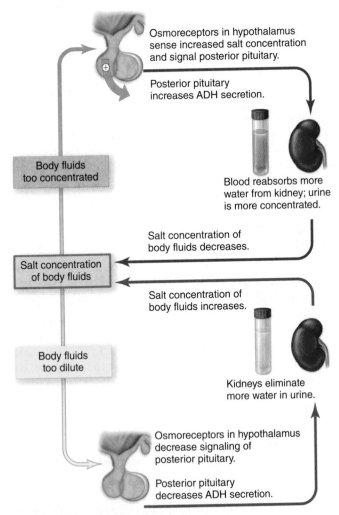

Osmoreceptors in hypothalamus sense increased salt concentration and signal posterior pituitary.

Posterior pituitary increases ADH secretion.

Body fluids too concentrated

Blood reabsorbs more water from kidney; urine is more concentrated.

Salt concentration of body fluids decreases.

Salt concentration of body fluids

Salt concentration of body fluids increases.

Body fluids too dilute

Kidneys eliminate more water in urine.

Osmoreceptors in hypothalamus decrease signaling of posterior pituitary.

Posterior pituitary decreases ADH secretion.

Figure 32.14 Osmoregulation in Humans. In response to the solute concentration of body fluids, this feedback loop regulates the amount of water reabsorbed from the kidneys into the bloodstream.

duct. As a result, the blood reabsorbs more water, and the urine becomes very concentrated. Conversely, if blood plasma is too dilute, ADH production stops, and more water is eliminated in the urine. ▶ posterior pituitary, p. 573

A diuretic is a substance that increases the volume of urine. The ethyl alcohol in alcoholic beverages is one example. Alcohol stimulates urine production partly by reducing ADH secretion, thereby decreasing the permeability of the tubules to water. By increasing water loss to urine, alcoholic beverages actually intensify thirst; dehydration also causes the discomfort of a hangover.

Hormones also act on the kidneys to regulate blood pressure. For example, when blood pressure and blood volume dip too low, the adrenal cortex releases the steroid hormone **aldosterone.** This mineralocorticoid stimulates the production of sodium

channels in the distal convoluted tubule. Water follows Na^+ ions by osmosis, so blood pressure rises. ▶ adrenal glands, p. 575

32.5 | Mastering Concepts

1. Trace the path of blood as it moves through a kidney.
2. What are the parts of a nephron?
3. What three processes occur in urine formation?
4. Describe the events that occur at the glomerular capsule, proximal convoluted tubule, nephron loop, and distal convoluted tubule.
5. What is the function of the collecting duct?
6. Describe the roles of antidiuretic hormone and aldosterone in regulating kidney function.

Apply It Now

Kidney Failure, Dialysis, and Transplants

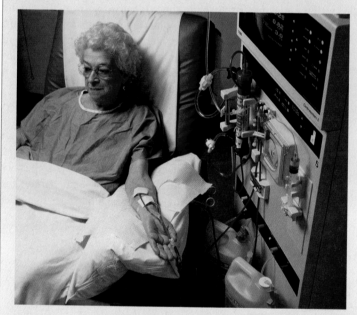

Figure 32.A **Kidney Dialysis.** A dialysis machine can take over some, but not all, of a healthy kidney's functions.

Kidneys can fail for many reasons. The most common causes are diabetes, high blood pressure, and chronic inflammation that causes progressive loss of nephrons. Other causes are polycystic kidney disease (the accumulation of cysts in the kidneys), scarring from childhood kidney infections, and obstructed urine flow.

One common treatment option for kidney patients is dialysis, in which a machine takes over the kidney's function (figure 32.A). The dialysis machine pumps blood out of the patient's body and past a semipermeable membrane. Wastes and toxins, along with water, diffuse across the membrane to a waste fluid called "dialysate," but blood cells

do not. The cleaned blood then circulates back to the patient's body. The procedure requires hours a day, several times a week.

Dialysis membranes cannot replace all of the kidney's functions. For example, nephrons selectively recycle useful components such as glucose and salts to the blood. The dialysis machine cannot do this, although a technician can adjust the concentrations of these dissolved compounds in the dialysate to promote or inhibit diffusion from the blood.

Transplantation is a second option. Kidneys are among the most commonly transplanted organs, and the success rate is very high. The transplanted kidney may come from a cadaver, or a living person may donate one healthy kidney to a recipient. (The remaining kidney has enough blood-cleansing power for the donor to live.) Surgeons connect the new kidney to the recipient's blood supply and attach its ureter to the bladder, usually leaving the old kidneys in place (figure 32.B).

Figure 32.B **Three Is Not a Crowd.** Surgeons usually insert a new kidney and ureter in a cavity low in the abdomen, leaving the patient's own kidneys in place.

Burning Question

What can urine reveal about health and diet?

Urinalysis is a routine part of many medical examinations. Laboratory tests of the chemical components of urine can reveal many health problems:

- More than a trace of glucose may be a sign of diabetes, a high-carbohydrate diet, or stress. Stress causes the adrenal glands to release excess epinephrine, which stimulates the liver to break down more glycogen into glucose. ▶ diabetes, p. 576

- Albumin may be a sign of damaged nephrons, since this plasma protein does not normally fit through the pores of intact glomerular capsules.

- Pus and an absence of glucose indicate a urinary tract infection. The pus consists of infection-fighting white blood cells along with bacteria, which consume the glucose.

- Traces of marijuana, cocaine, and other substances may appear in the urine. Athletes and employees of some organizations undergo routine drug testing.

Although urine is usually odorless, some foods impart a distinctive aroma. Asparagus, for example, contains a sulfur-rich molecule called mercaptan. Many people produce an enzyme that breaks down mercaptan, and the reaction products have a strong odor.

What about urine's color—why is it normally pale yellow? The pigment that lends its color to urine is a byproduct of the liver's breakdown of dead blood cells. Colorless urine usually indicates excessive water intake or the ingestion of diuretics such as coffee or beer. A reddish tinge may suggest anything from bleeding in the urinary tract to beet consumption to mercury poisoning. Either vitamin C or carrot consumption can color urine orange. It is always best to check with a physician when urine changes color.

Submit your burning question to:
marielle_hoefnagels@mcgraw-hill.com

32.6 | Investigating Life: Sniffing Out the Origin of Fur and Feathers

While visiting the zoo and appreciating the beauty of peacocks, parrots, and flamingos, have you wondered about the origin of their beautiful plumage (figure 32.15)? Have you ever admired the fur coats of a giraffe, jaguar, or red fox and asked yourself how hair evolved? Evolutionary biologists have been working to answer the same questions.

Fossil and DNA evidence clearly indicate that mammals and birds arose from two different lineages of reptiles. Existing non-avian reptiles are ectotherms, whereas both birds and mammals are endotherms. Because the ancestors of birds were not closely related to mammalian ancestors, researchers surmise that endothermy evolved independently in these two animal lineages.

It makes sense that feathers and fur evolved at the same time as the elevated metabolic rates associated with birds and mammals. After all, these integumentary adaptations provide insulation that helps the animals maintain their body temperature. To test this hypothesis of simultaneous evolution, scientists need to trace animal ancestry back in time. Ideally, they would examine fossils of animals from several points along the mammal and bird lineages, before and after the evolution of endothermy, fur, and feathers. Such an analysis would not prove a direct cause-and-effect relationship between endothermy and insulation, but it would show whether the adaptations arose at the same time.

This task is easier said than done. Both fur and feathers occasionally show up in fossils, but determining the metabolic style of an extinct animal is difficult. How can we know whether an animal that no longer exists was an endotherm or an ectotherm?

John A. Ruben and his colleagues at Oregon State University believe they have hit on the ideal indicator of endothermy: the inside of the nose. Endothermy requires a high metabolic rate,

Figure 32.15 Pretty Bird. Feathers have many functions, including insulation, flight, and communication. Scientists are working to determine how these structures evolved.

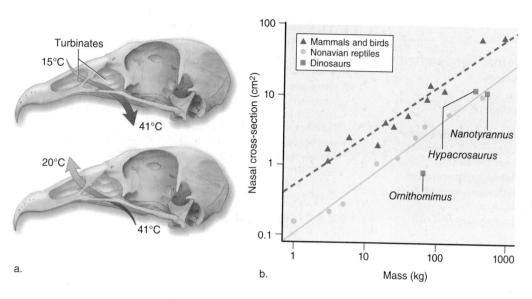

a.

b.

Figure 32.16 The Nose Knows.
(a) Turbinates direct air flow within the nasal cavity of endotherms, conserving both heat and water. (b) The nasal cavities of existing endotherms (ranging in size from herons to African cape buffaloes) have a higher cross-sectional area than those of ectotherms of equal size. Dinosaurs had nasal cavities similar in size to those of existing ectotherms.

which in turn means a huge demand for O_2 to fuel respiration. Mammals and birds therefore have high breathing rates, compared with their ectothermic counterparts. Along with this greater volume of air moving in and out of the lungs, however, comes a higher potential loss of both heat and water from an animal's body. Endotherms minimize this loss with the help of structures called turbinates that direct the flow of air within the nasal cavity (figure 32.16a).

Turbinates are internal ridges made of bone or cartilage covered with epithelium. These structures warm and humidify inhaled air before it enters the lungs; they also cool and remove water vapor from air as it is exhaled. The nasal cavities of more than 99% of all existing birds and mammals contain turbinates; no known ectotherms have them. Ruben and his colleagues reasoned that turbinates are therefore the next best thing to direct evidence of endothermy in extinct animals. They might be useful for approximating when endothermy first appeared in the ancestors of mammals and birds.

The researchers began by looking for evidence of bony turbinates in numerous fossils of extinct reptiles thought to be ancestors of mammals, including therapsids. They found turbinates in therapsids that lived during the Permian period, some 250 million years ago. Yet the first evidence of insulating fur appears only in fossils of true mammals, which did not arise until millions of years later, during the Triassic period. The researchers therefore concluded that the reptilian ancestors of mammals developed endothermy long before the origin of fur. ▶ geologic time scale, p. 254

But birds posed a problem. The earliest known bird, *Archaeopteryx*, lived some 150 MYA and clearly had feathers. Did this animal have turbinates? It is hard to know, because turbinates in birds are made of cartilage, a soft tissue that does not leave fossils. Nevertheless, Ruben's team suggested an indirect way to detect whether turbinates were present. The researchers predicted that, in general, endotherms should have broader nasal cavities than ectotherms, both to accommodate the presence of turbinates and to allow for a high breathing rate.

To make sure this was the case, Ruben's research team measured the cross-sectional areas of the nasal cavities of 21 living species of birds, mammals, and nonavian reptiles. In every case, the cross-sectional area of the nasal cavity was larger for an endotherm than for an ectotherm of equal size (figure 32.16b). The consistent, strong relationship suggested that measuring the nasal cavity in a fossil should be a good way to learn if an extinct animal was an endotherm or an ectotherm.

The researchers used modern imaging technologies to measure nasal cavities inside the fossilized skulls of three dinosaur species that lived about 70 MYA and are closely related to modern birds. The results indicate that the reptilian ancestors to birds probably were ectotherms. Feathers therefore preceded endothermy in birds by tens of millions of years.

Fur and feathers are adaptations that help mammals and birds stay warm, so it is easy to assume they evolved hand-in-hand with endothermy. Ruben's team paired old-fashioned comparative anatomy with modern technology to turn this assumption on its head. Endothermy was evidently present in mammalian ancestors before there was fur, and feathers apparently existed long before birds became endothermic. Thanks to the efforts of this research team, we now have part of the answer to the puzzle—and it is right in front of our nose.

Ruben, John A., Willem J. Hillenius, Nicholas R. Geist, et al. 1996. The metabolic status of some late Cretaceous dinosaurs. *Science,* vol. 273, pages 1204–1207.

Ruben, John A. and Terry D. Jones. 2000. Selective factors associated with the origin of fur and feathers. *American Zoologist,* vol. 40, pages 585–596.

32.6 | Mastering Concepts

1. How did Ruben and his colleagues use turbinates and nasal cavities to draw their conclusions?

2. Suppose researchers discover a fossil of a 100-kilogram dinosaur with a nasal cavity greater than 10 square centimeters. According to figure 32.16, how would that finding affect Ruben's conclusions?

Chapter Summary

32.1 | Animals Regulate Their Internal Temperature

A. Heat Gains and Losses Determine an Animal's Body Temperature

- Animals regulate their body temperatures (**thermoregulate**) with physiological and behavioral adaptations.
- **Ectotherms** regulate body temperature by seeking an environment with the appropriate temperature. **Endotherms** use internal metabolism to generate heat. Some animals, heterotherms, combine ecto- and endothermy.

B. Several Adaptations Help an Animal to Adjust Its Temperature

- Adaptations to cold include **countercurrent exchange** systems of blood vessels; fat, feathers, and fur; and hibernation.
- Evaporative cooling, circulatory specializations, and behaviors that help cool the body are adaptations to heat.

32.2 | Animals Regulate Water and Ions in Body Fluids

- **Osmoregulation** is the control of ion concentrations in body fluids. Depending on its habitat, an animal may need to conserve or eliminate water and ions.

32.3 | Nitrogenous Wastes Include Ammonia, Urea, and Uric Acid

- Animals must **excrete** metabolic wastes. Most nitrogenous wastes come from protein breakdown and include **ammonia, urea,** and **uric acid.**
- The excretion of ammonia requires little energy, but ammonia is highly toxic and therefore eliminated with copious amounts of water. Urea and uric acid require more energy to produce, but they are less toxic and therefore can be excreted in less water.

32.4 | The Urinary System Produces, Stores, and Eliminates Urine

- The human **urinary system** excretes nitrogenous wastes (mostly urea) and regulates water and electrolyte levels.
- The **kidneys,** each of which drains into a **ureter,** produce **urine.** Ureters drain urine into the **urinary bladder** for temporary storage. Urine leaves the body through the **urethra.**

32.5 | The Nephron Is the Functional Unit of the Kidney

A. Nephrons Interact Closely with Blood Vessels

- The **nephron** receives filtered blood from the **glomerulus** and exchanges materials with the **peritubular capillaries.**
- The two main regions of the nephron are the **glomerular capsule** and the **renal tubule.** The renal tubule, in turn, includes the **proximal convoluted tubule, nephron loop,** and **distal convoluted tubule.**
- Fluid moves from the nephron into a **collecting duct.**

B. Urine Formation Includes Filtration, Reabsorption, and Secretion

- The nephron **filters** blood and **secretes** wastes and other substances into the filtrate, while blood capillaries **reabsorb** useful components.

C. The Glomerular Capsule Filters Blood

- Blood pressure forces some components of blood from the glomerulus into the glomerular capsule.

D. Reabsorption and Secretion Occur in the Renal Tubule

- The proximal convoluted tubule returns glucose, amino acids, water, ions, and other solutes to the blood; H^+ and HCO_3^- adjustments in this region help maintain blood pH.
- The nephron loop dips into the medulla of the kidney, then returns to the cortex. A concentration gradient between the nephron loop and surrounding fluid drives water out of the filtrate in the descending limb. In the ascending limb, salt ions leave the filtrate by diffusion or active transport.
- At the distal convoluted tubule, additional water and salts are reabsorbed into the blood. K^+ and H^+ may be secreted into the filtrate.

E. The Collecting Duct Conserves More Water

- The fluid entering the collecting duct is called urine.
- Collecting ducts deliver urine to the funnel-like renal pelvis, which drains the fluid into the ureter.

F. Hormones Regulate Kidney Function

- **Antidiuretic hormone (ADH),** secreted by the posterior pituitary gland, regulates water reabsorption from the distal convoluted tubule. ADH increases permeability of the distal convoluted tubule and the collecting duct so more water is reabsorbed and urine is more concentrated.
- The adrenal cortex releases **aldosterone** in response to either low Na^+ concentration in the plasma or low blood pressure. Aldosterone causes additional Na^+ to be reabsorbed into the blood; water follows by osmosis, raising blood pressure.

32.6 | Investigating Life: Sniffing Out the Origin of Fur and Feathers

- Researchers have tested the hypothesis that fur and feathers evolved at the same time as endothermy. Their findings suggest that whereas endothermy preceded fur in the ancestors of mammals, feathers came first in birds.

Multiple Choice Questions

1. Why do endotherms require more energy than ectotherms?
   a. Because they need more energy to move between hot and cold environments
   b. Because they use metabolic energy to maintain an internal temperature
   c. Because they are more active
   d. Both b and c are correct.

2. Countercurrent heat exchange in birds or mammals functions to
   a. warm the blood at the surface of the organism.
   b. cool the blood at the surface of the organism.
   c. equalize the temperature of the blood throughout the organism.
   d. warm the blood as it returns to the core of the organism.

3. What would happen to a deep-sea fish if it were placed in fresh water?
   a. Water would move into the organism.
   b. Ions would move into the organism.
   c. Water would move out of the organism.
   d. There would be no effect.

4. What is the primary form of nitrogenous waste you would expect to find produced by an insect?
   a. Amino acids   c. Urea
   b. Ammonia   d. Uric acid

5. Urine passes through the _____ as it exits the body.
   a. ureter   c. urethra
   b. bladder   d. kidney

6. Which of the following is the correct order of the structures through which filtrate moves through the nephron?
   a. distal convoluted tubule ⟶ glomerular capsule ⟶ nephron loop ⟶ proximal convoluted tubule
   b. distal convoluted tubule ⟶ proximal convoluted tubule ⟶ nephron loop ⟶ glomerular capsule
   c. glomerular capsule ⟶ nephron loop ⟶ proximal convoluted tubule ⟶ distal convoluted tubule
   d. glomerular capsule ⟶ proximal convoluted tubule ⟶ nephron loop ⟶ distal convoluted tubule

7. The proximal convoluted tubule is responsible for
   a. filtration.   c. secretion.
   b. reabsorption.   d. Both b and c are correct.

8. Water moves out of the filtrate of the descending limb of the nephron loop because
   a. there is a high concentration of solutes in the surrounding capillaries.
   b. the cells of the descending limb are permeable to ions.

c. the difference in the diameter of the arterioles creates a pressure gradient.

d. there is a high concentration of solutes in the descending limb.

9. Which of the following components of glomerular filtrate is NOT normally found in urine?

   a. amino acids   c. urea

   b. water   d. Na$^+$

10. A common treatment for high blood pressure is the use of a *diuretic* medication. How does this help control blood pressure?

   a. By increasing water permeability of the distal convoluted tubule and collecting duct

   b. By decreasing water permeability of the distal convoluted tubule and collecting duct

   c. By enhancing the production of aldosterone

   d. By increasing Na$^+$ reabsorption from the distal convoluted tubule and collecting duct

Write It Out

1. How do humans and iguanas differ in body temperature regulation?

2. Birds and insects frequently collect nectar from plants. Birds are endothermic, and insects are ectothermic. Do you think that a given amount of nectar can support a greater mass of insects or of birds?

3. On a per-kilogram basis, why does a small mammal such as a shrew require so much more energy than does an elephant?

4. Cite three adaptations that help keep an animal's feet warm.

5. Whales and seals are mammals that swim in ice-cold water. Explain how they might use countercurrent exchange to keep the blood in their flippers from freezing.

6. Woolly mammoths are extinct relatives of modern-day elephants. The mammoths were heavier and shaggier than elephants, their ears were smaller, and they had a thick fat layer under their skin. Explain each of these differences in light of the fact that today's elephants originate in Asia and Africa, whereas mammoths lived on the tundra.

7. Imagine you are adrift at sea. If you drink seawater, you will dehydrate much faster than if you have access to fresh water. Explain.

8. What are the three types of nitrogenous wastes, and under which conditions is each most likely to be used?

9. How are Malpighian tubules similar to and different from a kidney?

10. List the organs that make up the human urinary system. What is the function of each?

11. Kidney stones are calcium-rich crystals that form inside the kidney. What symptoms would you expect if the stones lodge in a ureter?

12. Urinary tract infections frequently accompany sexually transmitted diseases, especially in women. Why?

13. Draw a nephron and label the parts.

14. Shortly after you drink a large glass of water, you will feel the urge to urinate. Explain this observation. Begin by tracing the path of the water, starting at the glomerulus and ending with the arrival of urine in the bladder.

15. How does the kidney reduce the volume of urine to a small fraction of the volume of filtrate that enters the glomerular capsule?

16. Why is protein in the urine a sign of kidney damage? What structures in the kidney are probably affected?

17. How could very low blood pressure impair kidney function?

18. The fluids deep in the medulla of the kidney have a much higher solute concentration than those in the cortex. Where do the solutes in the medulla come from? What is the role of that high solute concentration in the functioning of the nephron loop?

19. An amphibian's kidneys lack nephron loops. Why can't a human also live without nephron loops?

20. Many pharmaceutical drugs leave the body in urine. As we age, the number of nephrons in the kidneys declines. Do you predict that an older person would need a higher or lower dose of a drug to compensate for the amount lost in urine? Explain your answer.

21. How is the descending limb of the nephron loop different from the ascending limb? Considering their roles in transport of sodium ions, which cells do you think have the most mitochondria: the cells lining the descending limb or the cells lining the ascending limb? Explain.

22. Which of the substances in the table are excreted, and which are reabsorbed into the bloodstream?

	Concentrations (mg/100 mL)		
Substance	Plasma	Glomerular Filtrate	Urine
Glucose	100	100	0
Urea	26	26	1820
Uric acid	4	4	53
Creatinine	1	1	196

23. Review the action of steroid and peptide hormones in chapter 27. Which hormone should act faster, ADH or aldosterone? Why?

24. In a disease called diabetes insipidus, ADH activity is insufficient. Would a person with this disease produce more or less urine than normal? Explain.

25. Use the Internet to find a list of diseases of the kidneys or other organs of the urinary system. Select one to research further. What causes the illness you chose, and what are the symptoms? How does the disease interfere with the function of the urinary system? Is there a treatment or cure? Who is most at risk of the disease?

Pull It Together

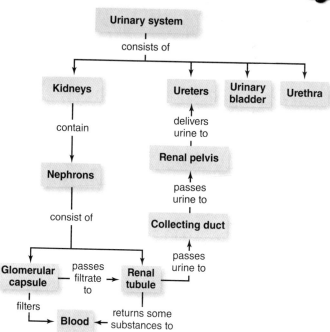

1. Add the terms *glomerulus, proximal convoluted tubule, nephron loop,* and *distal convoluted tubule* to this concept map.

2. Where do filtration, secretion, and reabsorption occur in the nephron?

3. What is the relationship between the kidneys, ureters, bladder, and urethra?

4. How do aldosterone and antidiuretic hormone influence kidney function?

Chapter 33

The Immune System

Waiting for Shots. A crowd waits in line for flu vaccinations.

Enhance your study of
this chapter with practice quizzes,
animations and videos, answer keys,
and downloadable study tools.
www.mhhe.com/hoefnagels

Vaccine Myths and Realities

A CHILD ENTERING DAY CARE OR KINDERGARTEN MUST ALREADY BE VACCINATED AGAINST A HOST OF DISEASES. Targeted illnesses include diphtheria, tetanus, whooping cough (pertussis), hepatitis A and B, *Haemophilus influenzae* type B, measles, mumps, rubella, polio, rotavirus, and chickenpox. Each vaccine relies on the immune system's amazing ability to recognize and destroy foreign particles, including viruses and bacteria. A vaccine containing weakened measles viruses, for example, "trains" the immune system to recognize measles. Then, if the person later encounters the full-strength virus, the immune system is primed to pounce.

Vaccines have saved millions of lives and prevented enormous suffering since Edward Jenner developed the first smallpox vaccine in 1796. For example, the Centers for Disease Control estimates that smallpox, diphtheria, measles, polio, and rubella once killed nearly 650,000 people a year in the United States. Many more people suffered from lingering harm such as paralysis, deafness, blindness, and mental retardation. Ever since routine childhood vaccinations started, however, the overall incidence of these five diseases has declined by 99% in the United States, and mortality has plunged below 100 per year.

Nevertheless, some people refuse to vaccinate their children. They cite a variety of arguments. One is that vaccines can have harmful side effects. It is true that no vaccine is 100% safe, and some children do have medical conditions that preclude vaccines. But for a healthy child, the risk of a deadly reaction to a vaccine is much smaller than the risk of death from infectious disease.

A second objection stems from the suspicion that vaccines cause autism, sudden infant death syndrome, diabetes, asthma, and other illnesses. This argument is based primarily on anecdotal evidence from parents who noticed signs of illness shortly after a child received a vaccination. It is important to realize, however, that many vaccines are administered at the same age as these childhood diseases typically appear. The events may be correlated, but that does not mean that one causes the other. In fact, numerous epidemiological studies have found no connection between vaccines and autism.

A third argument against vaccination is that the shots are unnecessary. It is true that many infectious diseases are now very rare in the United States and in other countries with high vaccination rates. People in many parts of the world, however, are not so fortunate. Global air travel means that diseases easily spread from one part of the world to another. Outbreaks can and do occur. When people refuse to vaccinate their children, they increase the chance that many now-rare diseases will flare up again.

Vaccines represent one tangible benefit of scientific research on the immune system. This complex set of cells, tissues, and organs is the subject of this chapter.

UNIT 6

What's the Point?

Learning Outline

Learn How to Learn
Find a Good Listener

For many complex topics, it is not easy to know how well you really understand what is going on. One tip is to try explaining what you think you know to somebody else. Choose a subject that takes a few minutes to explain. As you describe the topic in your own words, your partner should ask follow-up questions and note where your explanation is vague. Those insights should help draw your attention to important details that you overlooked.

33.1 | Many Cells, Tissues, and Organs Defend the Body

Disease-causing agents—**pathogens**—are nearly everywhere. Viruses, bacteria, protists, fungi, and worms are in water, food, soil, and air. These pathogens can enter our bodies whenever we eat, drink, breathe, or interact with people and other animals. Yet we are not constantly sick. The explanation is that the **immune system** enables the body to recognize its own cells and to defend against infections, cancer, and foreign substances. The vertebrate immune system consists of a network of cells, defensive chemicals, and fluids that permeate the body (figure 33.1).

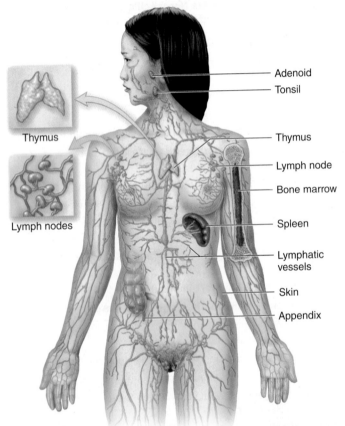

Adenoid

Tonsil

Thymus

Thymus

Lymph node

Bone marrow

Lymph nodes

Spleen

Lymphatic vessels

Skin

Appendix

Immune System	
Main Tissue Types*	**Examples of Locations/Functions**
Epithelial	Thymus, spleen, and tonsils consist partly of epithelial tissue; lines lymphatic and blood vessels; lymphoid tissue lies beneath epithelial tissues of the digestive, respiratory, and urinary tracts, guarding potential points of entry for pathogens.
Connective	Lymphocytes, antibodies, and other immune system chemicals circulate in blood and lymph, which are connective tissues; bone marrow is connective tissue that produces lymphocytes; thymus, spleen, and lymph nodes consist partly of connective tissue.

*See chapter 24 for descriptions.

Figure 33.1 The Human Immune System.

A. White Blood Cells Play Major Roles in the Immune System

Blood is critical to immune function. Plasma, the liquid matrix of blood, carries defensive proteins called antibodies. In addition, infection-fighting **white blood cells** are suspended in blood plasma and occupy the interstitial fluid between cells. Stem cells in **red bone marrow,** the spongy tissue inside bones, give rise to white blood cells. ▶ composition of blood, p. 604

White blood cells fall into five categories and play many roles in the body's defenses (figure 33.2). **Neutrophils** and **eosinophils** are white blood cells that function primarily as **phagocytes,** which are scavenger cells that travel in the bloodstream or wander through body tissues, engulfing bacteria and debris.

Basophils release chemical signals that trigger inflammation and allergies. Their close relatives, **mast cells,** share similar functions. Like basophils, mast cells originate in red bone marrow. But mast cells do not circulate in blood. Rather, they settle in tissues, especially those near the skin, digestive tract, and respiratory system.

Lymphocytes include several types of white blood cells. **B cells** mature in red bone marrow, then migrate to lymphoid tissues and into the blood. **T cells** also originate in red bone marrow, but they mature in the **thymus** ("T" is for thymus). From there, T cells migrate throughout the body. Together, B cells and T cells coordinate the body's responses to specific pathogens, as

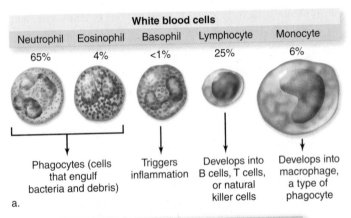

White blood cells				
Neutrophil	Eosinophil	Basophil	Lymphocyte	Monocyte
65%	4%	<1%	25%	6%

Phagocytes (cells that engulf bacteria and debris)

Triggers inflammation

Develops into B cells, T cells, or natural killer cells

Develops into macrophage, a type of phagocyte

a.

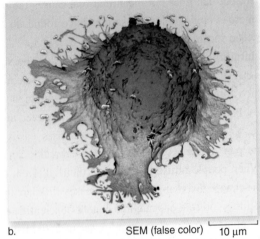

b.   SEM (false color)   10 μm

Figure 33.2 White Blood Cells. (a) Human blood contains five types of white blood cells. (b) A phagocyte engulfing bacteria.

described in section 33.3. Another type of lymphocyte, the **natural killer cell,** attacks cancerous or virus-infected body cells.

White blood cells called **monocytes** give rise to **macrophages,** which also function as phagocytes. Some types of macrophages wander throughout the body; others remain in just one tissue. As described in sections 33.2 and 33.3, macrophages that consume foreign particles play versatile roles in triggering the body's defenses. ▶▶ bone marrow, p. 586; phagocytosis, p. 83

B. The Lymphatic System Consists of Several Tissues and Organs

We have already seen that the **lymphatic system** helps regulate the volume of tissue fluid (see chapter 29) and absorbs fat from the small intestine (see chapter 31). A third role, described in this chapter, is to help defend the body against pathogens.

The lymphatic system transports **lymph,** a colorless fluid that forms when plasma seeps out of blood vessels and into the interstitial fluid. Lymph capillaries throughout the body absorb this fluid, along with bacteria, viruses, and white blood cells. The lymphatic system cleanses the lymph and returns it to the bloodstream.

Other components of the lymphatic system include the lymphoid organs that produce, accumulate, or aid in the circulation of lymphocytes. Red bone marrow and the thymus are two examples of lymphoid organs. In addition, the **spleen** is a large lymphoid organ containing masses of lymphocytes and macrophages, which destroy pathogens in the blood.

A lymph node is one of the many small, bean-shaped lymphoid organs located along the lymph vessels (see figure 29.17). Inside each lymph node, millions of lymphocytes engulf dead cells

and pathogens circulating in lymph. Lymph nodes also release B and T cells to lymph, which carries them to the blood. When you have an infection, you can sometimes feel your enlarged lymph nodes as "swollen glands" in the neck, armpits, or groin.

Scattered concentrations of lymphoid tissues also guard the mucous membranes where pathogens may enter the body. Examples include the tonsils (near the throat), appendix, and patches of lymphoid tissue in the small intestine.

C. The Immune System Has Two Main Subdivisions

Biologists divide the human immune system into two parts: innate defenses and adaptive immunity (figure 33.3). Together, innate defenses and adaptive immunity interact in highly coordinated ways to defend the body against pathogens.

Innate defenses provide a broad defense against any infectious agent. *Innate* refers to the fact that these defenses are always present and ready to function, as described in section 33.2. In **adaptive immunity,** the body's immune cells not only recognize specific parts of a pathogen, but they also "remember" previous encounters. Section 33.3 describes adaptive immunity in detail.

33.1 | Mastering Concepts

1. List the cell types that participate in the body's defenses, along with some of their functions.
2. What is the relationship between lymph and blood?
3. What are the functions of the main lymphoid organs?
4. What are the two subdivisions of the immune system?

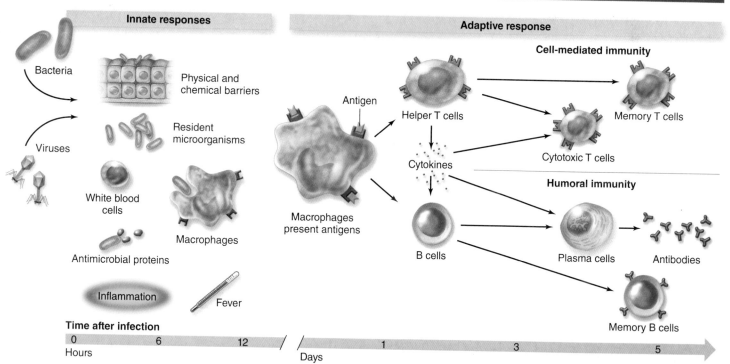

Figure 33.3 Overview of Body Defenses. Innate defenses prevent bacteria, viruses, and other pathogens from entering the body, or they attack those that do breach the physical and chemical barriers. In the adaptive immune response, macrophages displaying antigens (molecules stripped from pathogens) stimulate helper T cells, which trigger both cell-mediated and humoral immunity.

33.2 | Innate Defenses Are Nonspecific and Act Early

The body's innate defenses include many components, as summarized in figure 33.3. Each defense acts against any type of invader.

A. Barriers Form the First Line of Defense

Physical barriers block pathogens and foreign substances from entering the body. Unpunctured skin is the most extensive and obvious wall. Other physical barriers include mucus that traps inhaled dust particles in the nose; wax in the ears; tears that wash irritants from the eyes and contain an antimicrobial substance called lysozyme; and cilia that sweep bacteria out of the respiratory system. In addition, a bath of strong acid awaits microbes that manage to reach the stomach. ▶ ▶ integumentary system, p. 521; stomach, p. 643

An often underappreciated component of this first line of defense is the body's normal microbiota. Resident microbes on the skin, in the gut, and elsewhere help prevent colonization by pathogens. Moreover, studies of lab animals reared in special microbe-free environments show that the immune system does not develop correctly without stimulation from our microscopic companions. ▶ normal microbiota, p. 347

B. White Blood Cells and Macrophages Destroy Invaders

White blood cells play many roles in the body's innate defenses. Neutrophils and eosinophils destroy bacteria by phagocytosis or by secreting substances that destroy pathogens. Basophils provoke inflammation, attracting additional white blood cells. Natural killer cells destroy cancerous or virus-infected cells. Meanwhile, macrophages consume pathogens and promote fever. And, as described in section 33.3, macrophages also play a critical role in activating the body's adaptive immune response.

C. Redness and Swelling Indicate Inflammation

Inflammation is an immediate, localized reaction to an injury or to any pathogen that breaches the body's barriers. The body's familiar response to a minor cut illustrates the effects of inflammation: the area immediately surrounding the wound becomes red, warm, swollen, and painful. This nonspecific defense recruits immune system cells, helps clear debris, and creates an environment hostile to microorganisms.

Figure 33.4 illustrates the events of inflammation in response to an injury. Tissue damage provokes basophils and mast cells to release **histamine,** a biochemical that dilates (widens) blood vessels and causes them to become "leakier"—that is, more permeable to fluids and white blood cells.

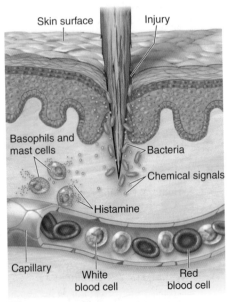

1 After a splinter penetrates the skin, damaged cells trigger mast cells and basophils to release histamine and other chemical signals.

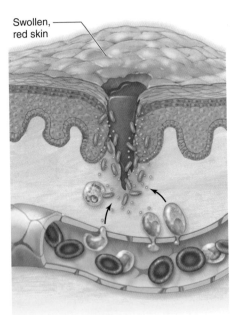

2 Histamine causes blood vessels to dilate and become more permeable, causing warmth, redness, and swelling. White blood cells squeeze between the cells in the capillary walls and move into the damaged area.

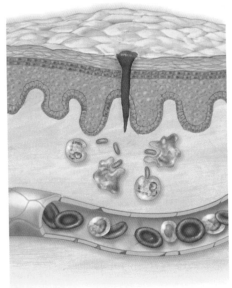

3 White blood cells engulf and destroy bacteria and damaged cells.

Figure 33.4 Inflammation. Immediately after a splinter pierces the skin, chemicals released from damaged cells trigger the inflammatory response.

As capillaries near the injury become dilated, additional blood arrives, turning the area warm and red. Blood plasma, which carries antimicrobial substances, leaks out of the blood vessels. This fluid dilutes the toxins secreted by bacteria and causes localized swelling. Pressure on the swollen tissues, coupled with chemical signals released from the injured cells, stimulates pain receptors in the skin.

Meanwhile, macrophages and neutrophils squeeze through the blood vessel walls and move into the area, engulfing and destroying bacteria and damaged cells. Pus may accumulate; this whitish fluid contains white blood cells, bacteria, and debris from dead cells.

In medical terminology, the suffix -itis indicates inflammation. For example, dermatitis (a rash) signifies inflamed skin, often resulting from direct contact with an irritant such as poison ivy. Appendicitis is inflammation of the appendix, usually caused by bacterial infection. Aspirin and ibuprofen reduce pain and swelling after an injury by blocking the enzymes required for inflammation to occur.

Inflammation may be acute or chronic. Acute inflammation is an adaptation that prevents infection after an injury. The effects usually last only a few days or less, as illustrated by the short-term discomfort of a minor burn or "paper cut." Chronic inflammation, on the other hand, is a prolonged response that may last for months or years. The persistent presence of pathogens or toxins can cause any tissue in the body to become chronically inflamed; genetic mutations may also play a role. Medical problems associated with chronic inflammation include rheumatoid arthritis, atherosclerosis, gum disease, heart disease, diabetes, Alzheimer disease, depression, cancer, and many other serious illnesses.

D. Complement Proteins and Cytokines Are Chemical Defenses

As inflammation begins, white blood cells produce and release many antimicrobial biochemicals. For example, the 25 or so **complement** proteins all help to destroy pathogens in the body. When activated, some trigger a chain reaction that punctures bacterial cell membranes (**figure 33.5**). Others cause mast cells to release histamine, and still others attract phagocytes.

Other chemical defenses include **cytokines,** messenger proteins that bind to immune cells and promote cell division, activate defenses, or otherwise alter their activity. For example, cells infected by viruses release **interferons,** which are cytokines that alert other components of the immune system to the infection. White blood cells release **interleukins,** the largest group of cytokines. Their name comes from their role in communicating (*inter-*) between leukocytes, or white blood cells (*-leukins*). One type of interleukin, produced by macrophages, activates B and T cells, provokes inflammation, and causes fever.

E. Fever Helps Fight Infection

Cytokines travel throughout the body in the bloodstream. At the hypothalamus, they can trigger a temporary increase in the set point of the body's thermostat. **Fever,** a rise in the body's temperature, is therefore a common reaction to infection. Although the shivering and chills that accompany fever feel uncomfortable, a mild fever can help fight infection in several ways. A higher body temperature kills some bacteria and viruses. Fever also counters microbial growth indirectly, because higher body temperature reduces the iron level in the blood. Bacteria and fungi require more iron as the temperature rises, so a fever stops the replication of these pathogens. Phagocytes also attack more vigorously when the temperature climbs. ▸ hypothalamus, p. 542

33.2 | Mastering Concepts

1. List five categories of innate defenses.
2. What are some barriers to infection in the human body?
3. How do white blood cells and macrophages destroy invaders?
4. Describe the events of inflammation.
5. What are some examples of antimicrobial biochemicals?
6. How is fever protective?

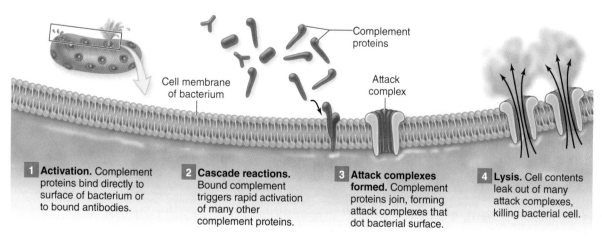

Cell membrane of bacterium

Complement proteins

Attack complex

1 Activation. Complement proteins bind directly to surface of bacterium or to bound antibodies.

2 Cascade reactions. Bound complement triggers rapid activation of many other complement proteins.

3 Attack complexes formed. Complement proteins join, forming attack complexes that dot bacterial surface.

4 Lysis. Cell contents leak out of many attack complexes, killing bacterial cell.

Figure 33.5 How Complement Kills. Complement proteins combine to riddle a bacterium's cell membrane with holes. The bacterial cell quickly dies.

33.3 | Adaptive Immunity Defends Against Specific Pathogens

The innate defenses described in section 33.2 are broad-spectrum weapons. Adaptive immune responses, on the other hand, act against individual targets. Two classes of lymphocytes, B cells and T cells, provide the ammunition in these precision defenses.

The target in an adaptive immune response is an **antigen,** which is any molecule that stimulates an immune reaction by B and T cells. Most antigens are carbohydrates or proteins. Examples include parts of a bacterial cell wall or a virus, proteins on the surface of a mold spore or pollen grain, and unique molecules on the surface of a cancer cell.

The word *antigen* (short for *anti*body-*gen*erating) reflects a crucial part of adaptive immunity: the production of **antibodies,** which are Y-shaped proteins that recognize specific antigens. As described later in this section, each B and T cell is genetically programmed to recognize and bind to only one target antigen. But because every foreign particle contains dozens of molecules that can act as antigens, many sets of lymphocytes respond to invasion by one pathogen.

A. Macrophages Trigger Both Cell-Mediated and Humoral Immunity

One of the first cell types to respond to infection is the macrophage, which both participates in innate defenses and triggers adaptive immunity. The macrophage engulfs a pathogen, dismantles it, and links each antigen to a self protein on the macrophage surface (**figure 33.6**, steps 1–3).

The **major histocompatibility complex (MHC)** is the cluster of genes that encode the body's self proteins. The MHC encodes two classes of cell surface proteins. One class marks all cells of an individual. The other class marks only macrophages and lymphocytes, which participate in the adaptive immune response.

A macrophage displaying an antigen on its surface travels in lymph to a lymph node, where the cell encounters collections of T and B cells (figure 33.6, step 4). **Helper T cells** are "master cells" of the immune system because they initiate and coordinate the adaptive immune response. When an antigen-presenting macrophage meets a helper T cell with receptors specific to the antigen it is displaying, the two cells bind (figure 33.6, step 5). This event simultaneously initiates the two arms of the adaptive immune response: cell-mediated and humoral immunity (figure 33.6, step 6).

In **cell-mediated immunity,** defensive cells attack and kill invaders by direct cell-to-cell contact. The activated helper T cell secretes cytokines that activate **cytotoxic T cells,** which destroy virus-infected, cancerous, or damaged cells. The cytokines also activate B cells specific to the same antigen and trigger

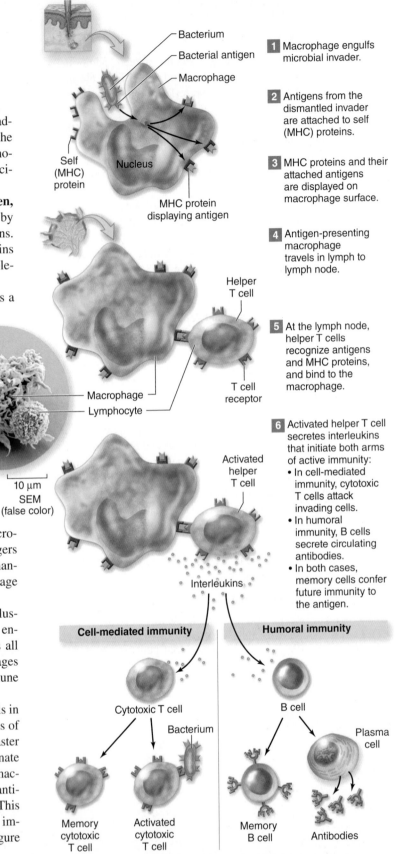

1 Macrophage engulfs microbial invader.

2 Antigens from the dismantled invader are attached to self (MHC) proteins.

3 MHC proteins and their attached antigens are displayed on macrophage surface.

4 Antigen-presenting macrophage travels in lymph to lymph node.

5 At the lymph node, helper T cells recognize antigens and MHC proteins, and bind to the macrophage.

6 Activated helper T cell secretes interleukins that initiate both arms of active immunity:
• In cell-mediated immunity, cytotoxic T cells attack invading cells.
• In humoral immunity, B cells secrete circulating antibodies.
• In both cases, memory cells confer future immunity to the antigen.

Figure 33.6 Adaptive Immunity: A Summary. A macrophage engulfs a bacterium, then displays bacterial antigens on its surface. At a lymph node, a helper T cell binds to the macrophage. The activated helper T cell sets into motion the cell-mediated and humoral immune responses.

Apply It Now

HIV Tests

Viral infections can be hard to diagnose. Unlike most bacteria, which can be identified based on their growth in Petri dishes, viruses cannot reproduce in pure culture. Instead, diagnostic tests may rely on the detection of virus-specific antibodies in a person's saliva, blood, or urine.

The oral swab test is one tool that determines whether a person is producing antibodies against HIV, the virus that causes AIDS. A technician collects saliva by soaking a cotton swab between a person's cheek and gums, then puts the fluid into the wells of a test plate containing a special solution. The mixture contains antigens from HIV, plus a substance that changes color when these antigens bind to antibodies from the saliva (**figure 33.A**). The results are available in about 20 minutes. If the test is positive for antibodies, then other tools can confirm the presence of HIV proteins.

For most people, 2 to 3 months elapse between HIV infection and the production of detectable antibodies. Although antibody tests for HIV are both accurate and sensitive, people who have recently been

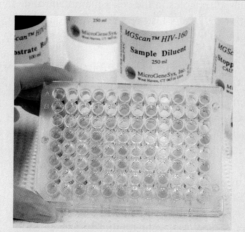

Figure 33.A HIV Test. A person who has been exposed to HIV produces antibodies to the virus. When the person's body fluid is applied to a specially treated plate, a color change reveals the presence of these antibodies.

exposed to HIV may have "false-negative" test results. To be safe, a person who suspects recent exposure to HIV should be tested both immediately and a few months later.

humoral immunity, which relies primarily on secreted antibodies (the term *humoral* refers to substances that circulate in body fluids). On stimulation by T cells, B cells proliferate explosively and differentiate into plasma cells and memory B cells. The **plasma cells** immediately secrete huge numbers of antibodies. The **memory cells,** on the other hand, remain in the body long after the initial infection subsides, launching a quick immune response upon subsequent exposure to the antigen. In effect,

memory cells "remember" antigens the immune system has already encountered.

B. T Cells Coordinate Cell-Mediated Immunity

Cytotoxic T cells provide cell-mediated immunity (**figure 33.7**). Receptors on the surface of a cytotoxic T cell bind to antigens on

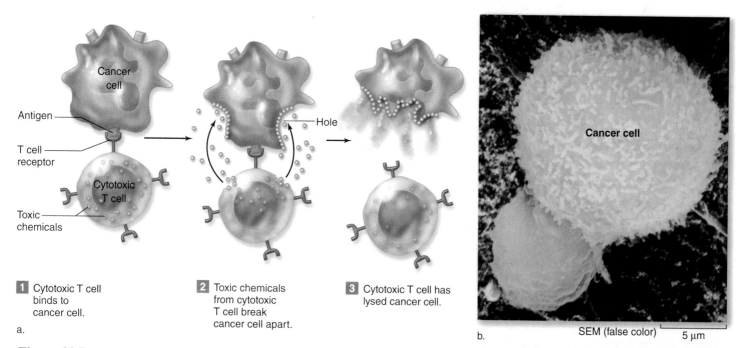

1 Cytotoxic T cell binds to cancer cell.

2 Toxic chemicals from cytotoxic T cell break cancer cell apart.

3 Cytotoxic T cell has lysed cancer cell.

a.

b. SEM (false color) 5 µm

Figure 33.7 Cytotoxic T Cells. (a) An activated cytotoxic T cell binds to a cancer cell (step 1) and injects a protein that pokes holes in the cancer cell's membrane (step 2). The cancer cell dies (step 3), leaving behind debris that macrophages clear away. (b) A small cytotoxic T cell attacks a large cancer cell above it.

the surface of an invader or a cancer cell. The T cell then releases a protein, perforin, that pokes holes in the cell's membrane and kills the invader. Cytotoxic T cells also recognize and destroy cells infected with viruses. By destroying the cell before the virus can replicate, the infection is stopped.

After the attack, memory T cells help the immune system retain a long-term "memory" of a pathogen. If the body encounters the same pathogen again, memory T cells differentiate immediately into cytotoxic T cells.

Cell-mediated immunity is one factor that complicates the transplantation of organs from one individual to another. The body perceives any foreign object, including a donated kidney, heart, or skin graft, as something to be destroyed. Cytotoxic T cells bind to and destroy the transplanted cells, provoking tissue rejection. A regimen of drugs can suppress this immune response, but the cost is increased vulnerability to cancer and infectious disease (see section 33.5).

C. B Cells Direct the Humoral Immune Response

The humoral immune response includes millions of different B cells, each producing a unique antibody.

Antibodies Are Defensive Proteins Antibodies (also called immunoglobulins) are the main weapons of humoral immunity (figure 33.8). These large proteins circulate freely in blood plasma, lymph, and interstitial fluid. Some antibodies also act as antigen receptors on the membranes of B cells.

The simplest antibody molecule consists of four polypeptides: two identical light chains and two identical heavy chains. Together, the four chains form a shape like the letter Y. Each chain has both constant and variable regions. The **constant regions** have amino acid sequences that are very similar in all antibody molecules, but the **variable regions** differ a great deal among antibodies. These variable regions determine the specific target antigen to which an antibody binds.

As illustrated in figure 33.8, biologists divide antibodies into five classes, designated IgA, IgD, IgE, IgG, and IgM, based on the structures of their heavy chains (the "Ig" stands for immunoglobulin). Each class has unique chemical properties and predominates in different circumstances. For example, IgE antibodies are responsible for allergies (see section 33.5), whereas IgG is the predominant class of antibody circulating in blood after exposure to a pathogen.

Antibodies are potent weapons that attack pathogens in many ways. The binding of an antibody to an antigen can in-

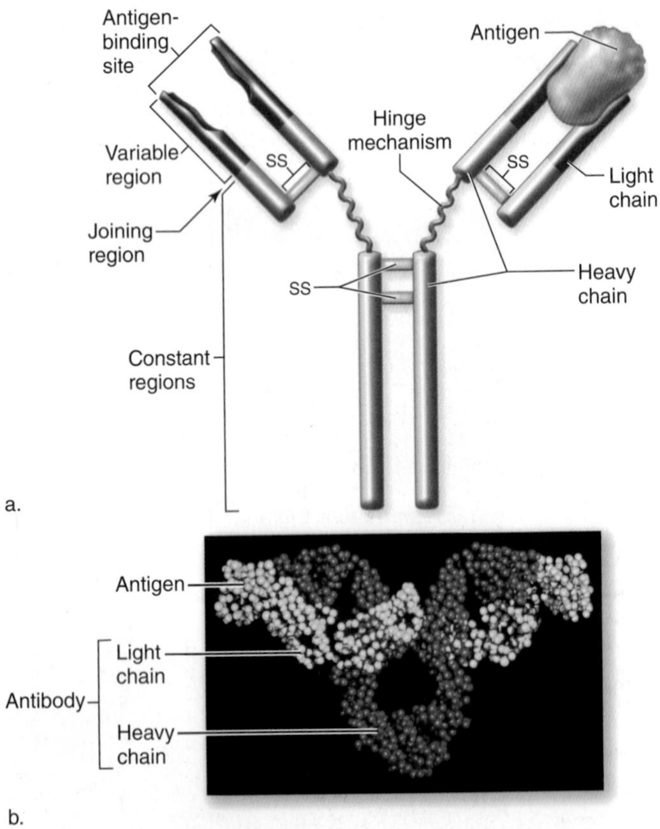

a.

b.

Antibody Class	Location	Sample Functions
IgA	Lining of digestive, respiratory, and urogenital tracts; also in milk, saliva, and tears	Protects points of entry into the body
IgD	B cell membranes	Antigen receptor; regulates B cell activation
IgE	Tonsils, skin, mucous membranes	Stimulates mast cells and basophils to release histamine in inflammatory and allergic responses
IgG	Circulate in blood plasma; can cross placenta to confer immunity on fetus	Predominant antibody secreted in secondary immune response
IgM	B cell membranes; circulate in blood plasma	Antigen receptor; predominant antibody secreted in primary immune response

c.

Figure 33.8 Antibody Structure. (a) The simplest antibody molecule consists of four polypeptide chains, each of which has constant and variable regions. The variable portions form the antigen-binding sites. (b) A computer-generated view of an antibody molecule bound to an antigen. (c) The five types of antibodies have distinct locations and functions.

activate a microbe or neutralize its toxins. Antibodies can cause pathogens to clump, making them more apparent to macrophages. They can coat viruses, preventing them from contacting target cells. Antibodies also activate complement proteins, which destroy microorganisms.

Genetic Recombination Yields a Huge Variety of Antibodies and Antigen Receptors

Of the human genome's 25,000 or so genes, fewer than 250 encode proteins that specifically bind to antigens. How can one person's lymphocytes produce enough unique antibody proteins and antigen receptor proteins to defeat millions of potential pathogens? As it turns out, generating these diverse molecules is a little like using the limited number of words in a language to compose an infinite variety of stories.

The genes that encode antibodies contain hundreds of small DNA segments that are rearranged in developing lymphocytes (**figure 33.9**). The result: countless lineages of cells

that each produce a unique antigen receptor and antibody. Most lymphocytes will never encounter a pathogen with the corresponding antigen, but a few will. Thanks to genetic recombination, the immune system can respond to even newly emerging pathogens.

Some of these antigen receptors, however, will no doubt correspond to the body's own molecules. In a process called **clonal deletion,** lymphocytes that recognize the body's own cell surfaces and molecules are "weeded out" by apoptosis, or programmed cell death (**figure 33.10**). This process begins before birth. The developing immune system somehow also learns not to attack antigens in food. ▶ apoptosis, p. 167

Figure It Out

If DNA can recombine to encode 320 unique light chains and 10,530 unique heavy chains, how many unique antibodies can be produced?

Answer: $320 \times 10,530 = 3,368,600$.

Clonal Selection Explains the Surge of Identical Antibodies

On the surface of each B cell is a version of the antibody that the cell will produce. Until a B cell encounters the antigen it is genetically programmed to recognize, it remains dormant. But when an antigen binds to this surface antibody, the B cell begins to activate. Cytokines from the corresponding helper T cell complete the activation.

In a process called **clonal selection,** a stimulated B cell divides rapidly, generating an army of memory cells and plasma

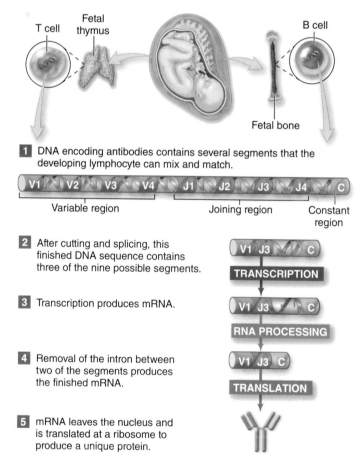

1 DNA encoding antibodies contains several segments that the developing lymphocyte can mix and match.

| V1 | V2 | V3 | V4 | J1 | J2 | J3 | J4 | C |

Variable region | Joining region | Constant region

2 After cutting and splicing, this finished DNA sequence contains three of the nine possible segments.

V1 J3 C

TRANSCRIPTION

3 Transcription produces mRNA.

V1 J3 C

RNA PROCESSING

4 Removal of the intron between two of the segments produces the finished mRNA.

V1 J3 C

TRANSLATION

5 mRNA leaves the nucleus and is translated at a ribosome to produce a unique protein.

Figure 33.9 Gene Shuffling. In this simplified illustration, a developing lymphocyte mixes and matches gene segments to generate a unique antibody or antigen receptor protein. Refer to chapter 7 for the details of protein synthesis.

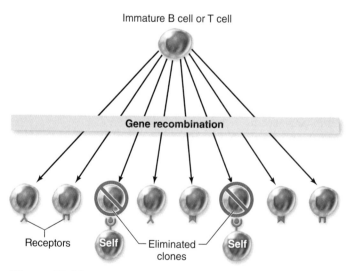

Figure 33.10 Clonal Deletion. As lymphocytes develop, they are tested against proteins and polysaccharides already present on the body's own cell surfaces. Clones that match self antigens are eliminated by programmed cell death (apoptosis).

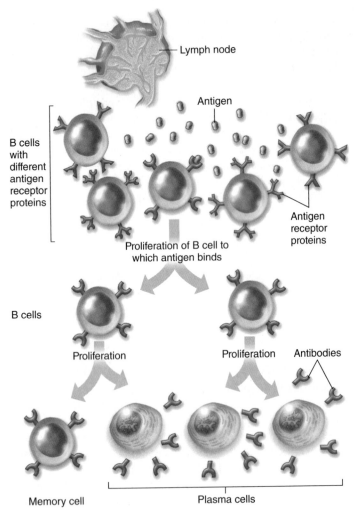

Lymph node

Antigen

B cells with different antigen receptor proteins

Proliferation of B cell to which antigen binds

Antigen receptor proteins

B cells

Proliferation

Proliferation

Antibodies

Memory cell

Plasma cells

Figure 33.11 Clonal Selection. According to clonal selection theory, only the lymphocyte that binds an antigen proliferates; its descendants develop into memory cells or plasma cells.

cells that are clones of the original B cell (figure 33.11). The plasma cells are efficient protein factories, making thousands of antibody molecules each second. At first exposure to the pathogen, B cells produce mostly IgM antibodies. Later, some daughter cells of the activated B cell switch to IgG or another class of antibodies specific to the same antigen.

The body's immune response is complicated by the fact that each invading virus, bacterium, or other cell has many different antigens on its surface (figure 33.12). The immune response therefore involves multiple antibodies, each made by a specific B cell and its clones.

Humoral Immunity Is Active or Passive   The humoral immune response may be passive or active (table 33.1). In **passive immunity,** a person receives intact antibodies from another individual. For example, an infant acquires antibodies from its mother in breast milk. The administration of antivenom to a victim of a snake bite also illustrates passive immunity.

Active immunity results from the body's own production of antibodies after exposure to antigens in the environment. A person who is bitten by a tick, for example, may begin producing antibodies against the bacteria that cause Lyme disease.

D. The Immune Response Turns Off Once the Threat Is Gone

Turning off an immune response once an infection has been halted is as important as turning it on because powerful immune biochemicals can also attack the body's healthy tissues.

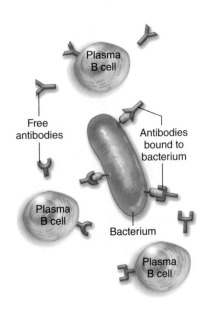

Plasma B cell

Free antibodies

Antibodies bound to bacterium

Plasma B cell

Bacterium

Plasma B cell

Figure 33.12 One Cell, Many Antigens. A bacterial cell surface includes many antigens, each of which stimulates the production of a different antibody.

Table 33.1 Ways to Acquire Immunity: A Summary

Type	Description	Examples
Passive immunity	One individual acquires antibodies from another individual.	• Fetus acquires antibodies from mother via placenta or milk. • Dog bite victim receives injections of antibodies to rabies virus. • Snake bite victim receives intravenous antivenom (antibodies to snake venom).
Active immunity	Individual produces antibodies to an antigen.	• Having chickenpox confers future immunity to that disease ("natural" active immunity). • Influenza vaccine triggers production of memory cells specific to antigens in the vaccine ("artificial" active immunity).

Immunologists continue to learn more about the precise cytokine combinations that signal the immune system to "back down" after a threat is removed.

Figure 33.13 shows a simplified version of the negative feedback that regulates lymphocyte and antibody concentrations in blood. Once a threat is removed, regulatory T cells somehow reduce the number of dividing B and T cells, and the army of plasma cells shrinks until only the memory cells remain.

E. The Secondary Immune Response Is Stronger Than the Primary Response

The **primary immune response** is the adaptive immune system's first reaction to a nonself antigen. Because the clonal selection process takes time, days or even weeks may elapse before antibody concentrations reach their peak. During this time, the pathogen can cause severe damage or death. If a person survives, however, the memory B cells and memory T cells leave a lasting impression—that is, immunological memory.

As a result, the **secondary immune response**—the immune system's reaction the next time it detects the same foreign antigen—is much stronger than the primary response (figure 33.14). Memory B cells transform into rapidly dividing plasma cells. Within hours, billions of antigen-specific IgG antibodies are circulating throughout the host body, destroying the pathogen before it takes hold. Usually, there is no hint

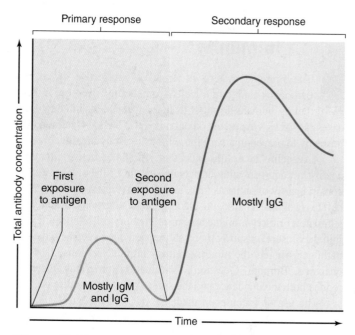

Figure 33.14 Primary and Secondary Immune Responses. The primary immune response leaves memory cells that stimulate a strong secondary immune response following subsequent exposure to the foreign antigen.

that a second infection ever occurred. As described in section 33.4, vaccines create this immunological memory without risking an initial infection.

Figure 33.14 illustrates another difference between the primary and secondary immune responses: the predominant classes of antibodies. During the initial stages of a primary response, IgM predominates, but some B cells quickly switch to producing IgG antibodies. In the secondary response, IgG antibodies dominate. The levels of these antibodies in blood may therefore be useful in diagnosing illness. High concentrations of IgM typically indicate a new infection, while high IgG is consistent with chronic infection.

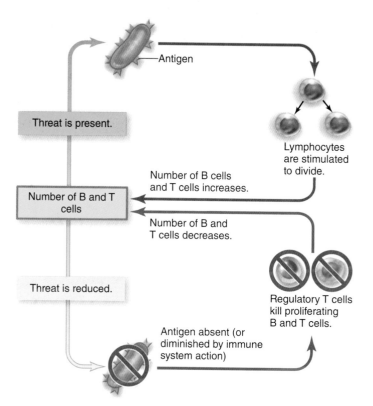

Figure 33.13 Homeostasis in the Immune System. This simplified negative feedback loop summarizes how the immune system regulates itself so that the immune attack stops when an antigen's concentration declines.

33.3 | Mastering Concepts

1. What is the relationship between antigens and antibodies?
2. What is the role of macrophages in adaptive immunity?
3. What are the two subdivisions of adaptive immunity, and which cell types participate in each?
4. What do cytotoxic T cells do?
5. Describe the structure and function of an antibody.
6. How can a relatively small number of genes encode billions of antibodies and antigen receptors?
7. What happens after a B cell is stimulated?
8. What happens if an immune reaction persists after a pathogen is eliminated?
9. Explain the difference between the primary and secondary immune response.

33.4 | Vaccines Jump-Start Immunity

The immune system is remarkably effective at keeping pathogens and tumor cells from taking over our bodies. Nevertheless, humans do suffer from many incurable infectious diseases caused by viruses and other pathogens. Besides sanitation, the best way to prevent many of these illnesses is vaccination.

A **vaccine** is a substance that stimulates active immunity against a pathogen without actually causing illness. The vaccine consists partly of antigens that "teach" the recipient's immune system to recognize a pathogen. Once taken into the body, the antigens simulate a primary immune response. Memory cells linger after this initial exposure, ensuring that a subsequent encounter with the real pathogen triggers the rapid secondary immune response (see this chapter's Burning Question). Vaccination programs therefore reduce both the incidence and the spread of infectious disease.

The use of vaccines in medicine dates to the late eighteenth century, when an incurable viral disease called smallpox was ravaging the human population. The fatality rate among infected people was about 25%, and about two thirds of the survivors were left horribly scarred and sometimes blind. But in 1796, a British country physician named Edward Jenner invented a vaccine against smallpox. The preparation contained vaccinia viruses, which cause a mild infection called cowpox. The two viruses—smallpox and vaccinia—are related to each other, and they share some antigens. Exposure to the vaccinia virus therefore confers immunity to smallpox.

This vaccine became the centerpiece of a worldwide smallpox eradication campaign, which began in 1967. By 1980, the World Health Organization declared that "smallpox is dead." The declaration heralded a milestone in medicine—the eradication of a disease. Today, all known stocks of smallpox virus reside in two labs, one in the United States and the other in Russia. Only the threat of bioterrorism maintains the demand for the smallpox vaccine among some military personnel and "first responders."

Scientists have developed vaccines against many other illnesses as well. The antigens in the vaccines take several different forms (table 33.2). Measles and mumps vaccines, for example, are weakened viruses. Others, such as the vaccine against polio, contain inactivated viruses that cannot cause an infection. Diphtheria and tetanus vaccines incorporate a toxin that bacterial pathogens produce. Still others, such as the hepatitis A and hepatitis B vaccines, incorporate only a part of the pathogen's surface. The "cervical cancer" vaccine, approved by the U.S. Food and Drug Administration in 2006, contains proteins identical to those on the human papillomavirus, a pathogen that is strongly correlated with cervical cancer.

Although vaccines have saved countless lives since Jenner's time, they cannot prevent all infectious diseases. For example, researchers have been unable to develop a vaccine against HIV, the rapidly evolving virus that causes AIDS. Influenza viruses also mutate rapidly, so each vaccine is effective for only one flu season. And it has so far proved impossible to develop one vaccine that will prevent infection by the many viruses that cause the common cold. ▶ influenza, p. 323

33.4 | Mastering Concepts

1. What is a vaccine?
2. List the main types of vaccine formulations.
3. Why haven't scientists been able to develop vaccines against HIV and the common cold?

Burning Question

Why do we need multiple doses of the same vaccine?

Vaccines stimulate the immune system to produce memory cells, which "remember" their exposure to the antigens in the vaccine. These memory cells can last for decades, triggering the production of antibodies when the real pathogen comes along. Yet most childhood vaccines require a series of shots, spaced out over months or years. And the tetanus and diphtheria vaccines require booster shots at least every 10 years. Why isn't one dose enough?

The answer has to do with the number of memory cells that the body produces after exposure to the vaccine. After just one dose, the number of memory cells may be too small to launch an effective immune response against a pathogen. Repeated doses ensure a stronger immune response because they trigger the formation of more memory cells.

The vaccine's formulation therefore helps determine the need for booster shots. A vaccine that consists of active viruses, such as the one Jenner used against smallpox, triggers a mild infection in the body. During this period, which lasts for about a week, new viruses are produced. This lengthy exposure to viral antigens stimulates a robust immune response, so booster shots are not usually necessary. On the other hand, when a vaccine contains toxins or inactivated viruses, no infection occurs. The body's exposure to the antigens is therefore relatively brief. In that case, repeated shots help boost the number of memory cells over time.

Submit your burning question to:
marielle_hoefnagels@mcgraw-hill.com

Table 33.2 Types of Vaccines

Vaccine Formulation	Examples
Live, weakened (attenuated) pathogen	Polio (oral vaccine), measles, mumps, rubella, chickenpox
Inactivated pathogen	Polio (injectable vaccine), influenza, hepatitis A
Inactivated toxins	Tetanus, diphtheria
Subunits of pathogens	Cholera, whooping cough (pertussis)
Recombinant vaccines; component vaccines	Lyme disease, hepatitis B, human papillomavirus

33.5 | Several Disorders Affect the Immune System

The immune system may turn against the body's own cells, or it may fail to respond to disease-causing organisms. In addition, harmless substances sometimes trigger an immune response.

A. Autoimmune Disorders Are Devastating and Mysterious

Ideally, the immune system does not attack the body's own cells; as a person develops, lymphocytes corresponding to molecules already present in the body should be eliminated by clonal deletion. In an **autoimmune disorder,** however, the immune system attacks the body's "self antigens." The resulting damage to tissues and organs may be severe (table 33.3).

Immunologists have much to learn about autoimmunity. For example, the body develops self-tolerance against some antigens by deleting lymphocytes in the red bone marrow and thymus. Other mechanisms of self-tolerance act on mature lymphocytes. And the immune system simply ignores some self antigens for unknown reasons. Which self antigens induce which type of tolerance and how? The answers will lead to a better understanding of, and perhaps better treatments for, autoimmunity.

B. Immunodeficiencies Lead to Opportunistic Infections

An **immunodeficiency** is a condition in which the immune system lacks one or more essential components. A weakened immune system leaves a person vulnerable to **opportunistic pathogens** and cancers that do not normally affect people with healthy immune systems (figure 33.15). Viruses such as HIV can weaken the immune system, as can some inherited diseases and pharmaceutical drugs.

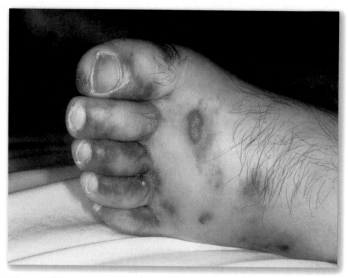

Figure 33.15 Opportunistic Illness. Kaposi sarcoma is a form of cancer that is prevalent in people with reduced immune system function (including AIDS).

Human immunodeficiency virus (HIV) causes acquired immune deficiency syndrome (AIDS). A person can acquire HIV by sexual contact or by using contaminated needles when injecting drugs. (Routine screening of the blood supply has virtually eliminated blood transfusions as an exposure route.) A mother can also transmit HIV to her baby, either during delivery or in breast milk.

When HIV enters the body, it initially targets helper T cells (see figure 15.4). Viral RNA enters the cell, where the enzyme reverse transcriptase transcribes the RNA to DNA. The freshly synthesized DNA inserts itself into the cell's chromosome. HIV's genes encode the raw materials needed to manufacture new viruses. Some types of infected cells, including helper T cells, die as they assemble and release new viruses. These viral particles infect additional helper T cells, spreading the infection within the body.

For months to a decade or more, however, no AIDS symptoms appear, because the body can produce enough new T cells to

Table 33.3 Examples of Autoimmune Disorders

Disorder	Symptoms	Target(s) of Antibody Attack
Glomerulonephritis	Lower back pain, kidney damage	Kidney cell antigens that resemble *Streptococcus* antigens
Graves disease	Restlessness, weight loss, irritability, increased heart rate and blood pressure	Thyroid gland
Type I diabetes	Thirst, hunger, weakness, emaciation	Pancreatic beta cells
Myasthenia gravis	Muscle weakness, difficulty speaking and swallowing	Nerve message receptors on skeletal muscle cells
Rheumatic heart disease	Weakness, shortness of breath	Heart valve cell antigens that resemble *Streptococcus* antigens
Rheumatoid arthritis	Joint pain and deformity	Cells lining joints
Scleroderma	Thick, hard, pigmented skin patches	Connective tissue cells
Systemic lupus erythematosus	Red rash on face, prolonged fever, weakness, kidney damage	DNA, neurons, blood cells

compensate for the loss. During this latent period, B cells manufacture antibodies to the virus; rapid tests for HIV exposure detect these proteins (see the Apply It Now box in section 33.3). Unfortunately, the antibodies do not prevent new viruses from forming.

HIV-positive people track their disease progress with blood tests that measure the number of helper T cells. As helper T cell counts decline, the ability of cytotoxic T cells to destroy infected cells also weakens. Eventually, the immune system fails entirely, and the infections and cancers of AIDS begin. Kaposi sarcoma and pneumocystis pneumonia are two common AIDS-associated illnesses in North America, but AIDS patients are susceptible to many other infectious diseases as well.

AIDS is a consequence of viral infection, but immune deficiency can also be inherited. Each year, a few children are born defenseless against infection due to **severe combined immunodeficiency (SCID),** a disorder in which neither T nor B cells function. David Vetter was the most famous SCID patient. Born in Texas in 1971, David had no thymus gland and spent the 12 years of his life in a vinyl bubble, awaiting a treatment that never came. Today, most children born with SCID receive bone marrow transplants before they are 3 months old. In addition, gene therapy has been used to replace faulty genes in some SCID patients. ▸ gene therapy, p. 144

Immunodeficiency is also common in organ transplant recipients. To avoid rejection of a donated organ, transplant recipients must take immune-suppressing drugs such as cyclosporine for the rest of their lives. Like other people with immunodeficiencies, these patients are vulnerable to opportunistic infections.

C. Allergies Misdirect the Immune Response

In an **allergy,** the immune system is overly sensitive, launching an exaggerated attack on a harmless substance (**figure 33.16**). Common **allergens,** or antigens that trigger an allergy, include foods, dust mites, pollen, fur, and oils in the leaves of plants such as poison ivy. The allergens activate B cells to produce IgE antibodies. A first exposure to the allergen initiates a step called sensitization, in which IgE antibodies bind to mast cells. On subsequent exposure, the allergens bind to the molecules attached to the mast cells, causing them to explosively release histamine and other allergy mediators.

The symptoms of an allergic response depend on where in the body the mast cells release mediators. Many mast cells are in the skin, respiratory passages, and digestive tract, so allergies often affect these organs. The result: hives, runny nose and eyes, asthma, nausea, vomiting, and diarrhea. Antihistamine drugs relieve these symptoms by preventing the release of histamine or blocking its binding to target cells.

Some individuals react to allergens with **anaphylactic shock,** a rapid, widespread, and potentially life-threatening reaction in which mast cells release allergy mediators throughout the body. The person may at first feel an inexplicable apprehension. Then, suddenly, the entire body itches and erupts in hives. Histamine causes blood vessels to dilate, lowering blood pressure. As blood rushes to the skin, not enough of it reaches the brain, and

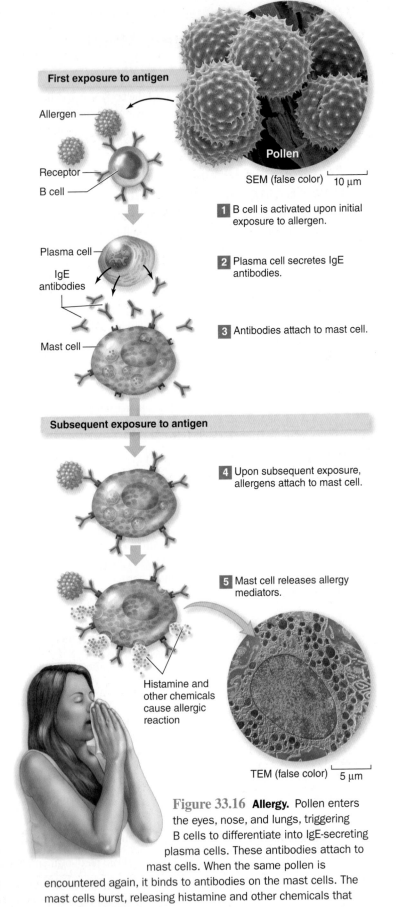

First exposure to antigen

Allergen

Receptor

B cell

Pollen

SEM (false color)   10 μm

1 B cell is activated upon initial exposure to allergen.

Plasma cell

IgE antibodies

2 Plasma cell secretes IgE antibodies.

3 Antibodies attach to mast cell.

Mast cell

Subsequent exposure to antigen

4 Upon subsequent exposure, allergens attach to mast cell.

5 Mast cell releases allergy mediators.

Histamine and other chemicals cause allergic reaction

TEM (false color)   5 μm

Figure 33.16 Allergy. Pollen enters the eyes, nose, and lungs, triggering B cells to differentiate into IgE-secreting plasma cells. These antibodies attach to mast cells. When the same pollen is encountered again, it binds to antibodies on the mast cells. The mast cells burst, releasing histamine and other chemicals that cause the allergic reaction.

the person becomes dizzy and may lose consciousness. Breathing becomes difficult as the bronchioles in the lungs become constricted. Meanwhile, the face, tongue, and larynx begin to swell. Unless the person receives an injection of epinephrine and sometimes an incision into the trachea to restore breathing, death can come within minutes.

Anaphylactic shock most often results from an allergy to penicillin, insect stings, or foods. Peanut allergy, for example, affects 6% of the U.S. population and is on the rise. The allergens are seed storage proteins that enter the bloodstream undigested. The fact that the initial allergic reaction to peanuts occurs at an average age of 14 months, typically after eating peanut butter, suggests that sensitization occurs even earlier, during breast feeding or before birth.

Early exposure to microorganisms and viruses may be crucial to the development of the immune system. A growing body of evidence has led to the "hygiene hypothesis," which suggests that excessive cleanliness has contributed to recent increases in the prevalence of asthma and some allergies (see section 33.6). Apparently, ultraclean surroundings decrease stimulation of the immune system early in life.

D. A Pregnant Woman's Immune System May Attack Her Fetus

Since the immune system should reject nonself cells, it may seem surprising that a woman's body does not destroy her fetus. After all, the developing child is not genetically identical to its mother. In general, the female immune response dampens during pregnancy so that it doesn't reject the embryo and fetus. Full immune function returns after the woman gives birth.

One possible source of problems, however, traces to an antigen called the Rhesus (Rh) factor. Some people produce this protein on the surfaces of their red blood cells. A person can be Rh-positive (Rh$^+$) or Rh-negative (Rh$^-$). If your blood type is positive (such as "A-positive"), the Rh antigen is present; if your blood is Rh-negative (such as "O negative"), your cells lack the Rh antigen. ▸ blood types, p. 605

Suppose an Rh$^-$ woman becomes pregnant with an Rh$^+$ baby (figure 33.17). Some of the fetus's cells enter the mother's bloodstream, so her immune system produces antibodies to the Rh antigen. In all subsequent Rh$^+$ pregnancies, these antibodies can cross the placenta and destroy the fetus's blood.

A transfusion of Rh$^-$ blood at birth can save the newborn's life, but this is rarely necessary. Instead, blood tests identify couples who face this potential problem. During and after each pregnancy, the woman receives an injection of a drug, Rho(D) immune globulin, which prevents the immune response.

33.5 | Mastering Concepts

1. What events might lead to autoimmunity?
2. Which immune system cells does HIV attack, and what is the consequence?
3. How does SCID differ from HIV infection?
4. Which cells and biochemicals participate in an allergic reaction?
5. How is the Rh-factor important in determining whether a pregnant woman's immune system attacks her fetus?

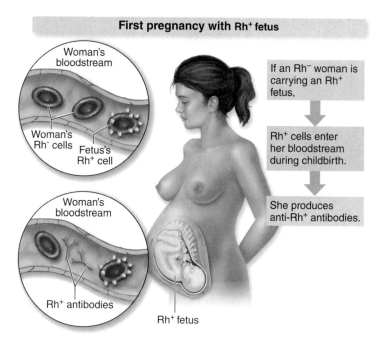

First pregnancy with Rh$^+$ fetus

Woman's bloodstream

Woman's Rh$^-$ cells Fetus's Rh$^+$ cell

Woman's bloodstream

Rh$^+$ antibodies

Rh$^+$ fetus

If an Rh$^-$ woman is carrying an Rh$^+$ fetus,

Rh$^+$ cells enter her bloodstream during childbirth.

She produces anti-Rh$^+$ antibodies.

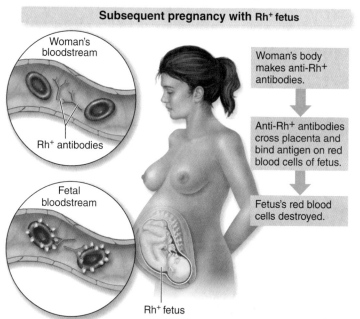

Subsequent pregnancy with Rh$^+$ fetus

Woman's bloodstream

Rh$^+$ antibodies

Fetal bloodstream

Rh$^+$ fetus

Woman's body makes anti-Rh$^+$ antibodies.

Anti-Rh$^+$ antibodies cross placenta and bind antigen on red blood cells of fetus.

Fetus's red blood cells destroyed.

Figure 33.17 Rh Incompatibility. A woman who is Rh$^-$ may become pregnant with an Rh$^+$ fetus. During the first such pregnancy, fetal cells entering the woman's bloodstream can stimulate her immune system to make anti-Rh antibodies. In a subsequent pregnancy, these antibodies can attack the baby's blood.

33.6 | Investigating Life: The Hidden Cost of Hygiene

Most people do not consider tapeworms, flukes, and other parasitic worms to be friends of the human species. But our long shared history with these animals has shaped the evolution of our immune systems in ways that we are just now beginning to understand.

Healthy immune function requires a delicate balance. On the one hand, the immune system must be strong enough to protect the body from dangerous pathogens. But if the immune response is too strong, we run the risk of overreacting to harmless substances—as in allergies—or launching an autoimmune attack against our own cells and tissues.

This fine balance reflects our evolutionary history. For millions of years, the human immune system has coevolved with countless bacteria, viruses, parasitic worms, and other infectious agents. Many of these hidden residents produce substances that suppress our immune systems, an adaptation that allows them to "fly under the radar" and maintain long-term, chronic infections.

In the past, nearly all humans were exposed to a wide range of infectious agents from birth. In developed countries, however, the situation has changed. Thanks to improved sanitation and easy access to vaccines and antibiotics, children in developed nations suffer far less from infectious disease than their counterparts in developing countries.

Many biologists believe that this reduced exposure to pathogens has changed the immune equation. According to the hygiene hypothesis, by parting ways with many of our long-term parasites, people in developed countries have removed a critical check on our immune system's activities.

One compelling line of evidence for the hygiene hypothesis is the observation that people in developed countries have a relatively high incidence of allergies, asthma, and autoimmune disorders such as type 1 diabetes, inflammatory bowel disease, and multiple sclerosis. But correlation does not imply causation, and developed countries differ from less-industrialized nations in many ways other than sanitation. To further test the hygiene hypothesis, we therefore need to know whether chronic infections really do rein in the immune system.

A research team led by Maria Yazdanbakhsh at Leiden University in the Netherlands investigated this question. Along with colleagues from the Netherlands, the United Kingdom, Germany, and the west African nation of Gabon, Yazdanbakhsh studied the incidence of allergies and chronic worm infection among schoolchildren in Gabon. If the hygiene hypothesis is correct, children infected with worms should have a lower incidence of allergies than their uninfected schoolmates. One objective of this study was to test this prediction.

The study population consisted of 513 children, aged 5 to 14 years, attending school in Gabon. The researchers tested each child for allergies to several substances, including extracts of the European house dust mite (*Dermatophagoides pteronyssinus*).

The team scratched each potential allergen into the skin of each child's forearm. If the child was allergic to a substance, the scratched area turned red and swollen within about 10 minutes (figure 33.18).

The team also checked for infection with several types of parasitic worms, including a flatworm called *Schistosoma haematobium*. This worm, a type of fluke, has a complex life cycle and uses both humans and aquatic snails as hosts (see figure 20.16 for a sample schistosome life cycle). In humans, the infection cycle begins when larvae in contaminated water burrow through a person's skin and hitch a ride in the bloodstream. The worms settle in the liver, grow to adulthood, and migrate to tissues near the urinary bladder. After mating, female flukes release fertilized eggs into the bladder. Diagnosis of the infection is based on the detection of these eggs in urine. ▶ flatworms, p. 418

Of the 513 schoolchildren, 57 tested positive for dust-mite allergies. Only 11 of these allergic children were infected with *S. haematobium* worms; the other 46 were worm-free (figure 33.19). Statistical analysis of the data indicated that harboring worms significantly lowers the risk for allergies—more evidence supporting the hygiene hypothesis.

Yazdanbakhsh and her colleagues then went a step further. Could they connect worms and allergies with specific immunological events? To find out, the researchers collected blood samples from 130 of the children. They grew cultures of white blood cells extracted from the samples. The scientists then exposed the cells either to antigens from *S. haematobium* or to a control growth medium lacking the worm antigens. After a few days, the researchers measured the production of four interleukins in both groups of cells.

One of these interleukins turned out to be especially important: interleukin-10 (abbreviated IL-10). Cells extracted from children infected with worms produced far more of this substance than did cells from the worm-free children. At the same time, the

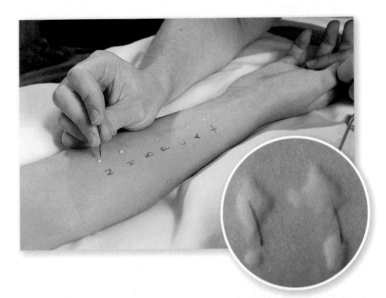

Figure 33.18 Skin Test. Raised welts indicate an allergic reaction to a substance scratched into the skin.

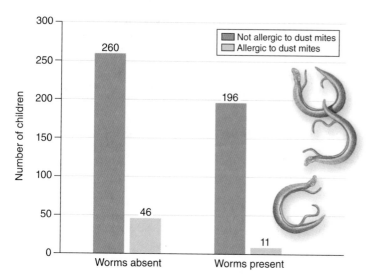

Figure 33.19 Protective Worms? Of the 57 schoolchildren with dust mite allergies, only 11 (about 20%) were infected with *S. haematobium* worms.

team found that children with high IL-10 concentrations were the least likely to have dust-mite allergies.

This result made sense in light of IL-10's function. Previous studies have shown that IL-10 acts as a sort of "brake" on the immune system. The researchers surmise that worms induce immune cells to release IL-10, dampening the overall immune response and making the host's body more hospitable for the worms. But the reined-in immune system also has a secondary consequence: the suppression of allergies.

This finding suggests the intriguing possibility that parasitic worms might have a role to play in the fight against allergies. Of course, many allergy sufferers would be reluctant to intentionally infect themselves with worms. Instead, biologists hope to identify the chemical substances that stimulate the release of IL-10 from immune cells. The long-term outcome of this research may be new drugs to treat allergies and autoimmune diseases.

In considering these results, it is important not to conclude that hygiene is harmful. After all, we cannot ignore the suffering and loss of life caused by infectious disease. No one is suggesting that we return to the days before sanitation and clean water or that we deny these resources to people who lack them now. At the same time, it is interesting to know that allergies and autoimmune diseases may be the price we pay to live a cleaner life.

Anita H. J. van den Biggelaar and six coauthors, including Maria Yazdanbakhsh. 2000. Decreased atopy in children infected with *Schistosoma haematobium*: a role for parasite-induced interleukin-10. *Lancet,* vol. 356, pages 1723–1727.

33.6 | Mastering Concepts

1. In this study, how did researchers document the inverse relationship between worm infection and allergies?
2. Describe the role of interleukin-10 in explaining the worm–allergy relationship.

Chapter Summary

33.1 | Many Cells, Tissues, and Organs Defend the Body

- The **immune system** protects the body against **pathogens** and cancer cells.

A. White Blood Cells Play Major Roles in the Immune System

- Five types of **white blood cells—neutrophils, eosinophils, monocytes, lymphocytes,** and **basophils**—participate in immune responses, as do several types of noncirculating cells, such as **mast cells.**
- Neutrophils, eosinophils, and **macrophages** (which arise from monocytes) are all **phagocytes,** cells that engulf and destroy bacteria and debris.
- **B cells** and **T cells** are lymphocytes that mature in the **red bone marrow** and the **thymus,** respectively. **Natural killer cells** are also lymphocytes.

B. The Lymphatic System Consists of Several Tissues and Organs

- The **lymphatic system** plays a crucial role in the immune response. The vessels of the lymphatic system collect and distribute a fluid called **lymph.**
- Besides the thymus, other lymph organs include the **spleen** and **lymph nodes.** Immune cells are also concentrated in the tonsils, appendix, and digestive tract.

C. The Immune System Has Two Main Subdivisions

- **Innate defenses** provide broad protection against all pathogens, whereas **adaptive immunity** is directed against specific pathogens. Only adaptive immunity produces immunological memory.

33.2 | Innate Defenses Are Nonspecific and Act Early

A. Barriers Form the First Line of Defense

- Intact skin, mucous membranes, tears, earwax, and cilia block pathogens.

B. White Blood Cells and Macrophages Destroy Invaders

- White blood cells destroy bacteria and promote inflammation; natural killer cells destroy cancerous or virus-infected cells; macrophages consume pathogens, promote fever, and activate the immune response.

C. Redness and Swelling Indicate Inflammation

- Basophils and mast cells trigger **inflammation,** which is an immediate reaction to injury. These cells release **histamine,** a biochemical that causes blood vessels to dilate.
- Redness, warmth, swelling, and pain are associated with inflammation.

D. Complement Proteins and Cytokines Are Chemical Defenses

- **Complement** proteins interact in a cascade that ends with the destruction of bacterial cells.
- **Cytokines** are antimicrobial molecules that communicate with immune system cells and trigger fever. **Interferons** and **interleukins** are examples of cytokines.

E. Fever Helps Fight Infection

- The elevated body temperature of a mild **fever** helps discourage microbial replication.

33.3 | Adaptive Immunity Defends Against Specific Pathogens

- Adaptive immunity is directed against specific **antigens.**

A. Macrophages Trigger Both Cell-Mediated and Humoral Immunity

- Macrophages display "self proteins" encoded by genes of the **major histocompatibility complex (MHC).** A macrophage that engulfs a pathogen links antigens from the microbe to the MHC markers.
- A **helper T cell** binding to the antigen-presenting cell triggers the **cell-mediated** and **humoral** components of the adaptive immune response. Activated helper T cells activate other T cells and B cells.

B. T Cells Coordinate Cell-Mediated Immunity

- **Cytotoxic T cells** release biochemicals that kill bacteria and cells infected with viruses. Memory T cells help provide long-term immunity.

C. B Cells Direct the Humoral Immune Response

- An **antibody** is a Y-shaped protein composed of two heavy and two light polypeptide chains. Each chain has a **constant region** and a **variable region.** Antibodies bind antigens and form complexes that attract other immune system components.
- Antibodies and antigen receptors are incredibly diverse because DNA segments shuffle during early lymphocyte development. **Clonal deletion** subsequently eliminates lymphocytes corresponding to self antigens.
- In **clonal selection,** an activated B cell multiplies rapidly, generating an army of identical **plasma cells** that all churn out the same antibody. Some also differentiate into **memory cells.**
- In **passive immunity,** a person receives antibodies from someone else. In **active immunity,** a person makes his or her own antibodies.

D. The Immune Response Turns Off Once the Threat Is Gone

- Tissue damage can occur if the immune response does not dampen after eliminating a pathogen.

E. The Secondary Immune Response Is Stronger Than the Primary Response

- The first encounter with an antigen provokes the **primary immune response,** which is relatively slow. Its legacy is memory cells that greatly speed the **secondary immune response** on subsequent exposure to the same antigen.

33.4 | Vaccines Jump-Start Immunity

- A **vaccine** "teaches" the immune system to recognize specific components of a pathogen, bypassing the primary immune response.

33.5 | Several Disorders Affect the Immune System

A. Autoimmune Disorders Are Devastating and Mysterious

- An **autoimmune disorder** occurs when the immune system produces antibodies that attack the body's own tissues.

B. Immunodeficiencies Lead to Opportunistic Infections

- An **immunodeficiency** is the absence of one or more essential elements of the immune system. These disorders leave patients vulnerable to cancer and **opportunistic pathogens.**
- The **human immunodeficiency virus (HIV)** kills helper T cells, causing AIDS.
- **Severe combined immune deficiency (SCID)** is an inherited disease. Immune deficiency can also be induced by drugs that prevent organ transplant rejection.

C. Allergies Misdirect the Immune Response

- An **allergy** is an immune reaction to a harmless substance. An **allergen** triggers production of antibodies, which bind mast cells. On subsequent exposure, these mast cells release allergy mediators such as histamine.
- **Anaphylactic shock** is a life-threatening allergic reaction.

D. A Pregnant Woman's Immune System May Attack Her Fetus

- The immune system of a woman whose cells lack the Rh factor can reject an Rh-positive fetus. A drug can prevent this problem.

33.6 | Investigating Life: The Hidden Cost of Hygiene

- According to the hygiene hypothesis, reduced exposure to parasitic worms and other pathogens is associated with an increase in allergies.
- Researchers investigating the immunological basis of the hygiene hypothesis found that in Gabonese schoolchildren, infection with parasitic worms induced immune cells to release interleukin-10. When levels of this cytokine were high, allergies were suppressed.

Multiple Choice Questions

1. Which types of white blood cells play important roles in both inflammation and allergy?
   a. Neutrophils and eosinophils
   b. Memory cells and plasma cells
   c. Basophils and mast cells
   d. B cells and T cells

2. Histamine acts on the _____ , causing redness and swelling.
   a. white blood cells
   b. cells lining blood vessels
   c. smooth muscle cells
   d. red blood cells

3. How do complement proteins contribute to the innate immune response?
   a. They cause bacteria to burst.
   b. They attract phagocytes to an injury.
   c. They cause mast cells to release histamine.
   d. All of the above are correct.

4. The innate immune response is characterized by its
   a. rapid response to invading pathogens.
   b. ability to "remember" pathogens it has already encountered.
   c. ability to produce antibodies.
   d. Both b and c are correct.

5. What is the function of a cytotoxic T cell?
   a. It displays antigens to a helper T cell.
   b. It secretes proteins that destroy foreign cells.
   c. It secretes cytokines that stimulate antibody production.
   d. It triggers clonal selection of B cells.

6. Antibody function requires that the shape of the _____ matches the shape of the antigen.
   a. constant region   c. variable region
   b. light chain   d. heavy chain

7. In what process is clonal selection important?
   a. Complement function   c. Inflammation
   b. B cell activation   d. Phagocytosis

8. Why is the secondary immune response so much stronger than the primary response?
   a. Because high concentrations of antibodies are already present
   b. Because the phagocytes present the antigens more rapidly
   c. Because memory cells can rapidly convert to plasma cells
   d. Because protein synthesis occurs more quickly in memory cells

9. HIV causes immunodeficiency by attacking
   a. memory B cells.   c. plasma cells.
   b. helper T cells.   d. cytotoxic T cells.

10. How do vaccines prevent infectious disease?
    a. By killing bacteria and viruses
    b. By boosting overall immune function
    c. By triggering a primary immune response
    d. By passive immunity

Write It Out

1. List and describe the components of the lymphatic system.
2. Explain the observation that lymphoid tissues are scattered in the skin, lungs, stomach, and intestines.
3. How does the immune system interact with the circulatory system?
4. List five types of innate defenses.
5. Dead phagocytes are one component of pus. Why is pus a sure sign of infection?

6. How can inflammation be both helpful and harmful?

7. If you take an antiinflammatory drug after spraining your ankle, what symptoms should be reduced?

8. State the functions of antibodies, cytokines, and complement proteins.

9. What do a plasma cell and a memory cell descended from the same B cell have in common, and how do they differ?

10. In your own words, write a paragraph describing the events of adaptive immunity, beginning with a macrophage engulfing a bacterial cell and ending with the production of memory cells.

11. How do the innate defenses and adaptive immunity cooperate to eradicate an infection?

12. Which do you think would be more dangerous, a deficiency of T cells or a deficiency of B cells? Explain your reasoning.

13. One benefit of sexual reproduction is a genetically variable population. According to the red queen hypothesis, genetic variability helps a population stay "one step ahead" of pathogen populations. Describe how genetic variability can enhance immunity.

14. What is a vaccine, and how is a vaccine different from an antibiotic?

15. Explain why it might be dangerous for a person with a weakened immune system to receive a vaccine consisting of a live, weakened pathogen.

16. Influenza viruses mutate rapidly, whereas the chickenpox virus does not. Why are people encouraged to receive vaccinations against influenza every year, whereas immunity to chickenpox lasts for decades?

17. If worm infections suppress the immune system as suggested in section 33.6, do you think vaccines should be more or less effective in children infected with worms? Explain your answer.

18. What is an opportunistic infection? Explain the statement that opportunistic infections, not HIV alone, cause death in an end-stage AIDS patient.

19. How does each of the following illnesses affect immunity?
    a. SCID   c. Hay fever
    b. AIDS

20. What part do antibodies play in allergic reactions and in autoimmune disorders?

21. How might a drug advertised as a "histamine blocker" relieve allergy symptoms?

22. Explain the difference between: clonal deletion and clonal selection; a natural killer cell and a cytotoxic T cell; antibodies and antigens; cell-mediated and humoral immunity; an autoimmune disorder and an immunodeficiency.

23. Search the Internet for evidence for and against the hypothesis that autism is associated with the thimerosal preservative that was once added to childhood vaccines (keep in mind that not all websites are equally credible). Do you find the evidence for or against the hypothesis most compelling? What sort of evidence would it take to change your position to the opposite side of the issue?

24. Search the Internet for information about immune system disorders. Choose one illness to study in more detail. What are the characteristics of the disorder? Who is primarily affected? What causes the illness, and is a treatment or cure available?

25. Search the Internet for ads for commercial products that claim to "boost the immune system." Choose a product to investigate in detail. What specific claims do ads for the product make? What scientific evidence does the manufacturer offer in support of its claims? Based on what you know about the immune system, does the scientific evidence seem convincing?

Pull It Together

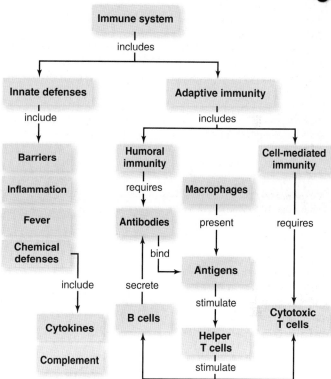

1. How do innate defenses and adaptive immunity differ?
2. What are examples of barriers that contribute to innate defenses?
3. How do cytokines interact with the adaptive immune system?
4. How do lymph and lymph nodes fit into this concept map?
5. What are the roles of memory B cells and plasma B cells?

Chapter 34

Animal Reproduction and Development

Different Roles for Males. A female desert-grassland whiptail lizard with her eggs (left); a male sea horse giving birth (right).

Enhance your study of this chapter with practice quizzes, animations and videos, answer keys, and downloadable study tools.

www.mhhe.com/hoefnagels

What Are Males For?

WHEN IT COMES TO REPRODUCTION, THE QUESTION OF MALE FUNCTION SEEMS EASY TO ANSWER. Everyone knows that in humans, males provide sperm and (often) parental care. The same goes for many other species. But a few types of animals test the limits of the male's role in reproduction.

At one extreme is the desert-grassland whiptail lizard in genus *Cnemidophorus*. In several species in this genus, including *C. uniparens,* all of the individuals are female—there are no males at all! Instead of producing eggs to be fertilized by sperm, as occurs in most animals, these lizards produce genetically identical eggs by a process called parthenogenesis. The eggs, which do not require fertilization, hatch into identical females that again reproduce by parthenogenesis.

Even though males are obsolete among these whiptail lizards, some of the species retain courtship behaviors that are reminiscent of a more conventional lifestyle. Two *C. uniparens* females, for example, will engage in a behavior called "pseudocopulation." One female imitates a male lizard's mating behavior, mounting another female and initiating contact between their cloacas (reproductive openings). The behavior resembles courtship and copulation in sexually reproducing lizards that are closely related to *C. uniparens.* But pseudocopulation plays no direct role in reproduction; after all, both participants are female. No one knows exactly why the lizards do it, but studies indicate that individuals that pseudocopulate lay more eggs than those that do not.

At the opposite end of the male-function spectrum lie sea horses, bony fishes in genus *Hippocampus*. These animals live in coastal areas, drifting among coral reefs or sea grasses. They hardly resemble fish, with their elongated horselike snouts, upright swimming posture, and prehensile tails that curl around grasses and other stationary objects for stability.

Beyond their looks, reproduction in sea horses is also unique: the males become pregnant! Their leisurely courtship ritual lasts several days, during which the male and female swim and "dance" together and hold one another's tails. The female then produces eggs, which she deposits into a brood pouch on the male's abdomen. He then releases sperm, fertilizing the eggs.

The young sea horses develop in the wall of the male's brood pouch for 2 to 4 weeks. In the meantime, the female does not pursue other mates, and she visits the male each day. At the end of the pregnancy, he gives birth to dozens or hundreds of miniature sea horses, which receive no further parental care. He can then accept more eggs, beginning the process anew.

Why the role reversal? Researchers speculate that this arrangement speeds reproduction, since the female can prepare her next batch of eggs while the male is pregnant. One thing is sure: the animal kingdom is full of interesting reproductive strategies. This chapter explores both reproduction and animal development, with a focus on our own species.

UNIT 6

What's the Point?

Learning Outline

Learn How to Learn
Don't Waste Old Exams

If you are lucky, your instructor may make old exams available to your class. If so, it is usually a bad idea to simply look up and memorize the answer to each question. Instead, use the old exam as a chance to test yourself before it really counts. Put away your notes and textbook, and set up a mock exam. Answer each question without "cheating," then check how many you got right. Use the questions you got wrong—or that you guessed right—as a guide to what you should study more.

34.1 | Animal Development Begins with Reproduction

A monarch butterfly emerges from its chrysalis; a baby bird hatches from an egg; a kitten becomes a full-grown cat. All of these familiar examples illustrate growth and development. Together, reproduction and development are shared features of all multicellular life. Chapter 23 described how flowering plants reproduce and grow; this chapter picks up the topic for animals.

A. Reproduction Is Asexual or Sexual

Like plants, animals may reproduce asexually or sexually (or both). In **asexual reproduction,** the offspring contain genetic information from only one parent. Aside from mutations that occur during replication, all offspring are identical to the parent and to one another. Animals that undergo asexual reproduction include sponges, corals, aphids, and some types of lizards (see the chapter opening essay). In general, asexual reproduction is advantageous in environments that do not change much over time.

Sexual reproduction requires two parents, each of which contributes half the DNA in each offspring. In many species, sexual reproduction entails high energy costs for attracting mates, copulating, and defending against rivals (see section 34.7). Nevertheless, the benefits of genetic diversity apparently outweigh these costs, especially in a changing environment. Sexual reproduction is therefore extremely common among animals. ▶ why sex? p. 180

In organisms that reproduce sexually, haploid **gametes** are the sex cells that carry the genetic information from each parent. The gametes—sperm cells from males and egg cells from females—are the products of meiosis, a specialized type of cell division. In meiosis, a diploid cell containing two sets of chromosomes divides into four haploid cells, each containing just one chromosome set. **Fertilization** is the union of two gametes; the product of fertilization is the **zygote,** the diploid first cell of the new offspring.

Sperm and egg come together in a variety of ways. In **external fertilization,** males and females release gametes into the same environment, and fertilization occurs outside the body (**figure 34.1a**). Salmon, for example, spawn in streams. Females lay eggs in gravelly nests, and then males shed sperm over them. Other animals with external fertilization include sponges, corals, and some amphibians. Unique "recognition" proteins on the surfaces of the gametes ensure that sperm cells fertilize eggs of the correct species.

In **internal fertilization,** a male deposits sperm inside a female's body, where fertilization occurs (**figure 34.1b**). Birds, nonavian reptiles, and mammals use internal fertilization. After copulation, the female may lay hard-shelled, fertilized eggs that provide both nutrition and a protected environment in which young develop. A chicken egg is a familiar example. Alternatively, the female may bear live young, as humans and other mammals do.

a.

b.

Figure 34.1 External and Internal Fertilization. (a) A tropical sea urchin releases sperm cells into the water. Meanwhile, females release eggs, and fertilization is external. (b) A male black-necked stilt mates with a female. The offspring will develop inside the female's body until she lays three or four hard-shelled eggs in a nest.

B. Gene Expression Dictates Animal Development

No matter what the reproductive strategy, development of sexually reproducing animals begins with the zygote. That first cell begins to divide soon after fertilization is complete. As the embryonic animal grows, cells divide and die in coordinated ways to shape the body's distinctive form and function. Developmental biologists study the stages of the animal's growth as cells specialize and interact to form tissues, organs, and organ systems (**figure 34.2**). The similarities and differences among developing animals can yield important clues to evolution, as illustrated in figure 12.15.

One key to animal development is **differentiation,** the process by which cells in organs such as the skin, brain, eyes, and heart activate unique combinations of genes to acquire their specialized functions. Another essential process is **pattern formation,**

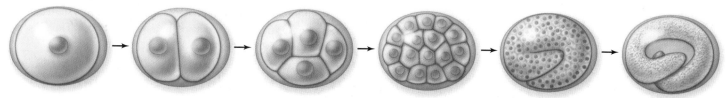

Figure 34.2 Tiny Worm. The life of the nematode *Caenorhabditis elegans* begins as a single fertilized egg cell. Researchers understand nematode development in detail, describing and mapping the fates of all cells produced from the original zygote.

in which genes determine the overall shape and structure of the animal's body, such as the number of segments or the placement of limbs. ▶ control of gene expression, p. 134

Differentiation and pattern formation involve complex interactions between the DNA inside cells and external signals such as hormones. Many of the molecules that influence animal development in general, and human development in particular, remain unknown. But researchers have found important clues in animals as small as the fruit fly, *Drosophila*. For example, female flies pack a protein called bicoid into one side of each developing egg cell. Bicoid instructs the embryo where to develop a head end. Normally, a high concentration of bicoid protein in the front end of the embryo causes the tissues of the head to differentiate. At the rear of the animal, higher concentrations of another protein, nanos, specify tail formation. When the gene encoding bicoid is mutated in the mother, the embryo develops two rear ends!

These protein gradients are signals that stimulate cells to produce yet other proteins, which ultimately regulate the formation of each structure. Section 7.7 described the importance of homeotic genes in establishing the correct placement of a developing animal's parts. Scientists first discovered homeotic genes by studying mutant flies with legs growing out of their heads. Since that time, additional studies have verified that homeotic genes orchestrate development in humans and all other animal species.

C. Development Is Indirect or Direct

Although many details of animal development remain undiscovered, clear patterns do emerge on a broader scale. For example, biologists distinguish between indirect and direct development. An animal that undergoes **indirect development** spends the early part of its life as a **larva,** an immature stage that looks different from the adult. A caterpillar, for example, is a larva that looks nothing like its butterfly parents; likewise, a tadpole resembles a fish, not the adult frog or salamander that it will grow up to become. Caterpillars, tadpoles, and other larvae often spend most of their time eating and growing. Then, during a process called **metamorphosis,** the larva matures into an adult (see figures 20.8 and 27.A).

Humans and many other familiar animals undergo **direct development:** our species has no larval stage, so at birth, an infant resembles a smaller version of its parents. The hatching

turtle in **figure 34.3**, for example, is a miniature version of the adult.

This chapter combines reproduction and development, starting with the reproductive anatomy of human males and females. Before you begin, you may find it helpful to review mitosis (chapter 8), meiosis (chapter 9), and the basics of hormone function (chapter 27). The second half of the chapter describes where babies come from—that is, how a fertilized egg grows into a fully formed infant.

34.1 | Mastering Concepts

1. What is the difference between asexual and sexual reproduction?
2. How is internal fertilization different from external fertilization?
3. How do genes participate in differentiation and pattern formation?
4. Differentiate between indirect and direct development.

Figure 34.3 Tiny Tortoise. A baby tortoise emerges from its egg. Like other animals that undergo direct development, the newly hatched tortoise is a miniature version of the adult.

34.2 | Males Produce Sperm Cells

Both the male and female **reproductive systems** consist of the organs that produce and transport gametes. Each system includes primary sex organs: the paired **gonads** (testes or ovaries), which contain the **germ cells** that give rise to gametes. Secondary sex organs include the network of tubes that transport the gametes.

In both sexes, hormones control reproduction and promote the development of **secondary sex characteristics,** features that distinguish the sexes but do not participate directly in reproduction. Examples include enlarged breasts and menstruation in adult females and facial hair and deep voices in adult males.

Although the male and female reproductive systems share some similarities, there are also obvious differences. This section details the features and processes that are unique to males.

A. Male Reproductive Organs Are Inside and Outside the Body

Figure 34.4 illustrates the human male reproductive system. The paired **testes** (singular: testis) are the male gonads. The testes lie in a sac called the **scrotum.** Their location outside of the abdominal cavity allows the testes to maintain a temperature about 3°C cooler than the rest of the body, which is necessary for sperm to develop properly. Muscles surrounding each testis can bring the scrotum closer to the body, conserving warmth when the

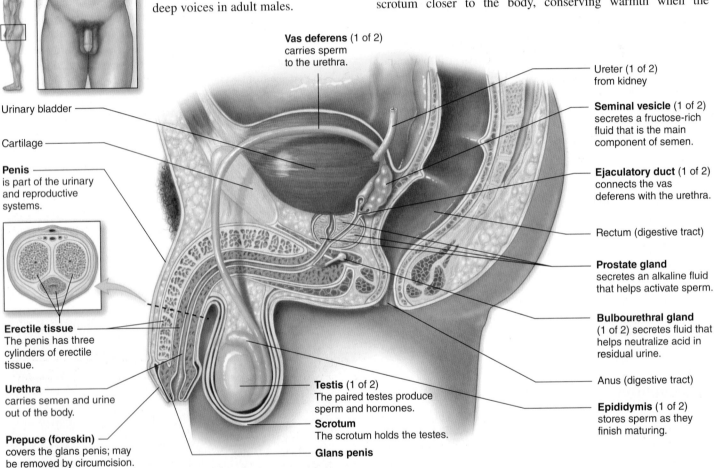

Frontal view

Urinary bladder

Cartilage

Penis is part of the urinary and reproductive systems.

Erectile tissue The penis has three cylinders of erectile tissue.

Urethra carries semen and urine out of the body.

Prepuce (foreskin) covers the glans penis; may be removed by circumcision.

Vas deferens (1 of 2) carries sperm to the urethra.

Ureter (1 of 2) from kidney

Seminal vesicle (1 of 2) secretes a fructose-rich fluid that is the main component of semen.

Ejaculatory duct (1 of 2) connects the vas deferens with the urethra.

Rectum (digestive tract)

Prostate gland secretes an alkaline fluid that helps activate sperm.

Bulbourethral gland (1 of 2) secretes fluid that helps neutralize acid in residual urine.

Anus (digestive tract)

Epididymis (1 of 2) stores sperm as they finish maturing.

Testis (1 of 2) The paired testes produce sperm and hormones.

Scrotum The scrotum holds the testes.

Glans penis

Figure 34.4 The Human Male Reproductive System. The paired testes manufacture sperm cells, which travel through a series of ducts before exiting the body via the urethra in the penis. Various glands add secretions to the semen.

Reproductive System (Male)	
Main Tissue Types*	**Examples of Locations/Functions**
Epithelial	Lines ducts of reproductive tract; produces sperm cells in seminiferous tubules; produces secretions in accessory glands.
Connective	Makes up walls of testes; makes up erectile cylinders in penis.
Nervous	Penis contains sensory nerve fibers and nerve endings; hypothalamus secretes hormones that affect the anterior pituitary.
Muscle	Smooth muscle surrounds ducts of reproductive tract, propelling sperm out of the body.

*See chapter 24 for descriptions.

temperature is too cold. Conversely, if conditions are too warm, the scrotum descends away from the body. Moreover, a network of tiny veins surrounding each testis helps regulate its temperature by exchanging heat with nearby arteries in a countercurrent arrangement. ▶ countercurrent exchange, p. 659

A maze of small ducts carries developing sperm to the left or right **epididymis,** a tightly coiled tube that receives and stores sperm from one testis. From each epididymis, sperm cells move into a **vas deferens,** a duct that travels upward out of the scrotum, bends behind the bladder, and connects with the left or right **ejaculatory duct.** The two ejaculatory ducts empty into the **urethra,** the tube that extends the length of the cylindrical **penis** and carries both urine and semen out of the body. Although these two fluids share the urethra, a healthy male cannot urinate when sexually aroused because a ring of smooth muscle temporarily seals the exit from the urinary bladder.

Semen, the fluid that carries sperm cells, includes secretions from several accessory glands. The two **seminal vesicles,** one of which opens into each vas deferens, secrete most of the fluid in semen. The secretions include fructose (a sugar that supplies energy) and prostaglandins. Prostaglandins are hormone-like lipids that may stimulate contractions in the female reproductive tract, helping to propel sperm. In addition, the single, walnut-sized **prostate gland** wraps around part of the urethra and contributes a thin, milky, alkaline fluid that activates the sperm to swim. The two **bulbourethral glands,** each about the size of a pea, attach to the urethra where it passes through the body wall. These glands secrete alkaline mucus, which coats the urethra before sperm are released.

During sexual arousal, the penis becomes erect, enabling it to penetrate the vagina and deposit semen in the female reproductive tract. (Erectile dysfunction is the inability to maintain an erection; chapter 2's opening essay describes the drugs that treat this condition.) At the peak of sexual stimulation, a pleasurable sensation called **orgasm** occurs, accompanied by rhythmic muscular contractions that eject the semen through the urethra and out the penis. **Ejaculation** is the discharge of semen from the penis. One human ejaculation typically delivers more than 100 million sperm cells.

Two cancers of the male reproductive system originate in the prostate gland and testes. Prostate cancer is the second most common type of cancer in men (behind lung cancer). The resulting prostate enlargement constricts the urethra and may interfere with urination and ejaculation. Prostate cancer usually strikes men older than 50; noncancerous (benign) prostate enlargement affects many older men as well. ▶ cancer, p. 162

In testicular cancer, which usually occurs in men younger than 40, mutated germ cells in the testes divide out of control, forming lumps that may be detected in a self-examination. Testicular cancer has a very high cure rate, if detected before it spreads to other parts of the body.

B. Spermatogenesis Yields Sperm Cells

Spermatogenesis, the production of sperm, is a continuous process that begins when a male reaches puberty and continues throughout life.

Figure 34.5 illustrates the internal anatomy of a testis. Each testis contains about 200 tightly coiled, 50-centimeter-long **seminiferous tubules,** which produce the sperm cells. Large **sustentacular cells** extend the entire thickness of a seminiferous tubule; these cells surround, support, and nourish developing sperm cells. **Interstitial cells** fill the spaces between the seminiferous tubules; they are endocrine cells that secrete male sex hormones.

Sperm production begins with germ cells called **spermatogonia** that reside within the wall of a seminiferous tubule (see figure 34.5b). The spermatogonia are diploid, so their nuclei contain 46 chromosomes. When a spermatogonium divides

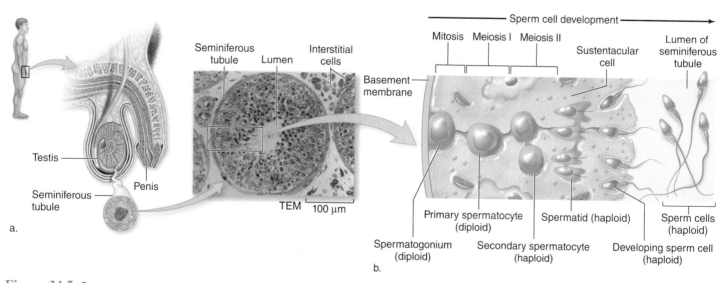

Figure 34.5 Spermatogenesis. (a) Anatomy of a testis. (b) In the first stage of sperm production, diploid spermatogonia divide mitotically in the linings of the seminiferous tubules. Some of the daughter cells undergo meiosis. The products of meiosis are four haploid secondary spermatocytes that differentiate into spermatids, which mature into sperm cells.

mitotically, one daughter cell remains in the tubule wall and acts as a stem cell, continually giving rise to cells that specialize into sperm. The other cell becomes a diploid **primary spermatocyte** that accumulates cytoplasm and moves closer to the tubule's lumen (central cavity).

In the wall of the seminiferous tubule, the primary spermatocyte undergoes the first division of meiosis, yielding two haploid **secondary spermatocytes.** These cells undergo meiosis II, forming four round, haploid cells called **spermatids** that each contain 23 chromosomes. (It may be helpful to refer back to figure 9.16, which summarizes how meiosis I and II apply to spermatogenesis.)

As the spermatids move into the lumen of the seminiferous tubule, they complete their differentiation into **spermatozoa,** or mature sperm cells. They separate into individual cells and develop flagella, although they will not be able to swim until they reach the epididymis. They also lose much of their cytoplasm, acquire a streamlined shape, and package their DNA into a distinct head (**figure 34.6**). Mitochondria just below the head region generate the ATP the sperm needs to move toward an egg cell. The caplike **acrosome** covers the head and releases enzymes that will help the sperm penetrate the egg cell. The entire process, from spermatogonium to spermatozoon, takes 74 days in humans. ▶ ATP, p. 76

Figure It Out

A single ejaculate may contain 240 million sperm cells. How many primary spermatocytes must enter meiosis to produce this many sperm cells, and how many secondary spermatocytes are produced along the way?

Answer: 60 million primary spermatocytes; 120 million secondary spermatocytes.

C. Hormones Influence Male Reproductive Function

Hormones play a critical role in male reproduction. In the brain, the hypothalamus secretes **gonadotropin-releasing hormone (GnRH).** This peptide hormone travels in the bloodstream to the anterior pituitary, where it stimulates the release of two other peptide hormones: **follicle-stimulating hormone (FSH)** and **luteinizing hormone (LH).** Blood carries FSH and LH throughout the body. ▶ peptide hormones, p. 570

LH signals interstitial cells in the testes to release the steroid hormone **testosterone** and other male sex hormones (androgens). In the presence of FSH, testosterone affects the body in multiple ways. In adolescents, the hormone stimulates the development of secondary sex characteristics. The testes and penis begin to enlarge at puberty, and hair grows on the face, in the armpits, and at the groin. Testosterone also stimulates the secretion of growth hormone, causing a growth spurt that increases height, boosts muscle mass, and deepens the voice. In adults, testosterone stimulates sperm production, sustains the libido, and controls the activity of the prostate gland. ▶ steroid hormones, p. 571

Negative feedback loops maintain homeostasis in the concentrations of these hormones. When sperm production is too high, sustentacular cells release a hormone called inhibin, which blocks the release of additional FSH from the pituitary. On the other hand, low sperm production suppresses inhibin release, so FSH levels rise. A negative feedback loop also helps explain one well-known consequence of abusing anabolic steroids: low sperm counts (see chapter 27's Apply It Now box). The body mistakes the synthetic steroids for testosterone, causing the testes to produce less of the real sex hormone. Without testosterone, sperm do not form. ▶ negative feedback, p. 520

34.2 | Mastering Concepts

1. What is the relationship among gonads, germ cells, gametes, and the zygote?
2. Describe the role of each part of the male reproductive system.
3. Where in the testes do sperm develop?
4. What are the stages of spermatogenesis?
5. What are the parts of a mature sperm cell?
6. How do hormones regulate sperm production?

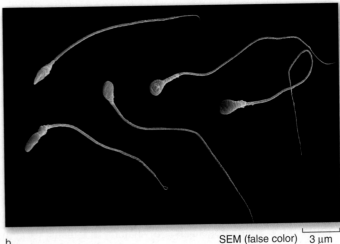

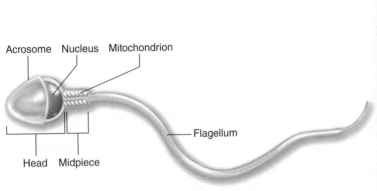

Acrosome   Nucleus   Mitochondrion

Flagellum

Head   Midpiece

a.

b.

SEM (false color)   3 μm

Figure 34.6 Human Sperm. (a) A sperm has distinct structures that assist in delivering DNA to an egg cell. (b) Scanning electron micrograph of human sperm cells.

34.3 | Females Produce Egg Cells

Egg cell production in females is somewhat more complicated than is sperm formation in males, for at least two reasons. First, in females, meiosis begins before birth, pauses, and resumes at sexual maturity. Meiosis does not complete until after a sperm cell fertilizes the egg cell. Second, egg cell production is cyclical, under the control of several interacting hormones whose levels fluctuate monthly during a woman's reproductive years. Keep these differences in mind while reading this section.

A. Female Reproductive Organs Are Inside the Body

Female sex cells develop within the **ovaries,** which are paired gonads in the abdomen (**figure 34.7**). Ovaries produce both egg cells and sex hormones. They do not contain ducts comparable to the seminiferous tubules of the male's testes. Instead, within each ovary of a newborn female are about a million oocytes, the cells that give rise to mature egg cells. Nourishing **follicle cells** surround each oocyte.

Approximately once a month, beginning at puberty, one ovary releases the single most mature oocyte. Beating cilia sweep the mature oocyte into the fingerlike projections of one of the two **uterine tubes** (also called Fallopian tubes or oviducts). If sperm are present, fertilization occurs in a uterine tube. The tube carries the oocyte or zygote into a muscular saclike organ, the **uterus.** During pregnancy, the fetus develops inside the uterus, also called the womb. The **endometrium,** or inner lining of the uterus, has a rich blood supply that is important in both menstruation and pregnancy.

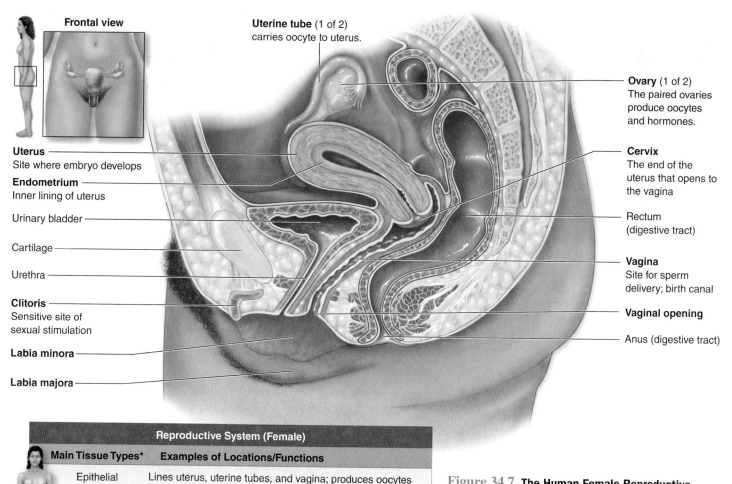

Frontal view

Uterine tube (1 of 2) carries oocyte to uterus.

Ovary (1 of 2) The paired ovaries produce oocytes and hormones.

Uterus Site where embryo develops

Endometrium Inner lining of uterus

Urinary bladder

Cartilage

Urethra

Clitoris Sensitive site of sexual stimulation

Labia minora

Labia majora

Cervix The end of the uterus that opens to the vagina

Rectum (digestive tract)

Vagina Site for sperm delivery; birth canal

Vaginal opening

Anus (digestive tract)

Reproductive System (Female)		
Main Tissue Types*	**Examples of Locations/Functions**	
Epithelial	Lines uterus, uterine tubes, and vagina; produces oocytes in ovaries; forms external surface of umbilical cord.	
Connective	Makes up walls of ovaries, uterus, and vagina.	
Nervous	Clitoris contains sensory nerve fibers and nerve endings; hypothalamus secretes hormones that affect the anterior pituitary.	
Muscle	Smooth muscle surrounds uterine tubes, uterus, and vagina.	

*See chapter 24 for descriptions.

Figure 34.7 The Human Female Reproductive System. One of the two ovaries releases an oocyte each month. This egg cell is drawn into a nearby uterine tube. If a sperm cell fertilizes the oocyte, the offspring develops in the uterus and is delivered through the vagina. If fertilization does not occur, the oocyte is expelled with the uterine lining as menstrual flow through the cervix and vagina.

The **cervix** is the necklike narrowing at the lower end of the uterus. The cervix opens into the **vagina,** the tube that leads outside the body. The vagina receives the penis during intercourse, and it is also the birth canal. Like many other areas of the body, the vagina harbors a community of resident microorganisms. These bacteria lower the pH of the vagina, which helps prevent colonization by harmful bacteria and the yeast *Candida albicans.* Taking antibiotics can disrupt this microbial community and create an opportunity for *Candida* to overgrow, causing a vaginal yeast infection. ▸ normal microbiota, p. 347

Two pairs of fleshy folds protect the vaginal opening on the outside: the labia majora (major lips) and the thinner, underlying flaps of tissue they protect, called the labia minora (minor lips). The **clitoris** is a 2-centimeter-long structure at the upper junction of both pairs of labia. Rubbing the clitoris stimulates females to experience orgasm. Together, the labia, clitoris, and vaginal opening constitute the **vulva,** or external female genitalia.

The female secondary sex characteristics include the wider and shallower shape of the pelvis, the accumulation of fat around the hips, a higher pitched voice than that of the male, and the breasts. Each **breast** contains fatty tissue, collagen, and milk ducts. The nipple delivers milk to the nursing infant.

Cancers of the female reproductive system often develop in the breasts, cervix, or ovaries. Breast cancer is the most common cancer type in women. The abnormally dividing cells may originate in the milk-forming tissues or in the milk ducts of the breast. Some, but not all, forms of breast cancer have a strong genetic component. A family history is also the leading risk factor for ovarian cancer, which usually originates in the outer lining of the ovary. In contrast, nearly all cases of cervical cancer are associated with the human papillomavirus (see section 34.4). Pap tests detect the abnormal cells of cervical cancer. In 2006, the U.S. Food and Drug Administration approved for girls and young women a "cervical cancer vaccine" that contains proteins of the human papillomavirus. ▸ vaccines, p. 684

B. Oogenesis Yields Egg Cells

The making of an egg cell—**oogenesis**—begins with an **oogonium,** a diploid germ cell containing 46 chromosomes (figure 34.8). Each oogonium grows, accumulates cytoplasm, replicates its DNA, and divides mitotically, becoming two **primary oocytes**. The ensuing meiotic divisions partition the cytoplasm unequally, so that oogenesis (unlike spermatogenesis) produces cells of different sizes. By the end of meiosis I, the primary oocyte has divided into a small haploid **polar body** and a larger haploid **secondary oocyte,** each containing 23 chromosomes.

Meiosis II halts at metaphase II and does not resume unless fertilization occurs. In that case, the secondary oocyte again divides unequally to produce a small additional polar body and the mature haploid egg cell (or ovum), which contains 23 chromosomes and a large amount of cytoplasm. The polar body produced in meiosis I may divide into two additional polar bodies, or it may decompose.

The egg cell, in receiving most of the cytoplasm, contains all of the biochemicals and organelles that the zygote will use until its own DNA begins to function. The polar bodies normally play no further role in development. Rarely, however, sperm can fertilize polar bodies, and a mass of tissue that does not resemble an embryo grows until the woman's body rejects it. A fertilized polar body accounts for about 1 in 100 miscarriages.

Between puberty and menopause (the cessation of menstruation), monthly hormonal cues prompt an ovary to release one secondary oocyte into a uterine tube. If a sperm penetrates the oocyte membrane, meiosis in the oocyte completes, and the two nuclei combine to form the diploid zygote. If the secondary oocyte is

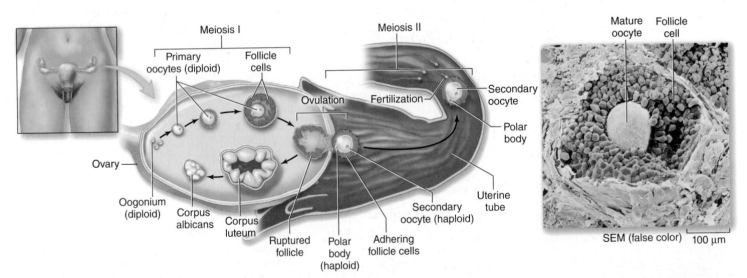

Figure 34.8 Oogenesis. Diploid oogonia in each ovary divide mitotically to produce primary oocytes, which develop in protective follicles. The products of meiosis I are a secondary oocyte and a tiny polar body. Every month between puberty and menopause, the most mature follicle ruptures and the secondary oocyte bursts out, an event called ovulation. The secondary oocyte completes meiosis II only if fertilized by a sperm cell.

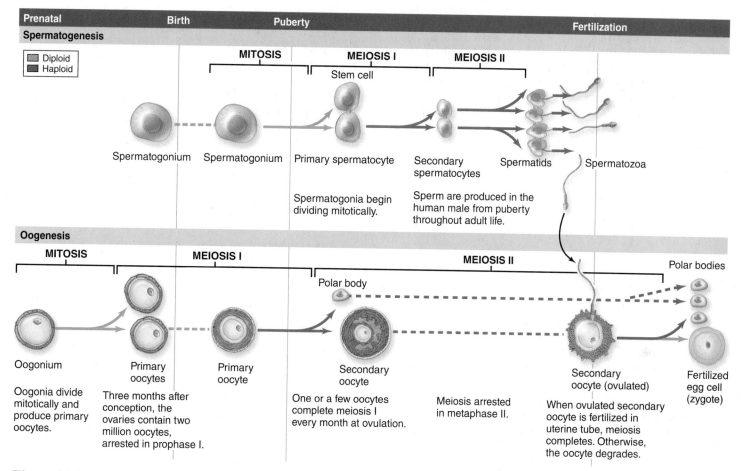

Figure 34.9 Gamete Formation in Males and Females. The steps of sperm and oocyte formation are similar, and the products of each process are haploid. But the male and female gametes look different, and they form according to different timetables.

not fertilized, it leaves the body with the endometrium in the menstrual flow.

In some ways, oogenesis is similar to spermatogenesis (figure 34.9 and table 34.1). Each process starts with a diploid germ cell (spermatogonium or oogonium) that eventually gives rise to the haploid gametes. Also, both testes and ovaries contain gametes in various stages of development. Of course, the two processes also differ. For example, spermatogenesis gives rise to four equal-sized sperm cells, whereas one oogonium yields one functional egg cell and three smaller polar bodies.

Also, the timetable for oogenesis differs greatly from that of spermatogenesis. A male takes about 74 days to produce a sperm cell. In contrast, oogenesis stretches from before birth until after puberty. The ovaries of a 3-month-old female fetus contain 2 million or more primary oocytes. From then on, the number declines. At birth, a million primary oocytes are present, their development arrested in prophase I. Only about 400,000 remain by the time of puberty, after which one or a few oocytes complete meiosis I each month. These secondary oocytes stop meiosis again, this time at metaphase II. Meiosis is completed only if fertilization occurs.

Table 34.1   Spermatogenesis and Oogenesis Compared

Diploid Starting Cell	Product of Mitosis (Diploid)	Products of Meiosis I (Haploid)	Products of Meiosis II (Haploid)
Spermatogonium in seminiferous tubule	Primary spermatocyte	Two secondary spermatocytes	Four equal-sized spermatids
Oogonium in ovary	Primary oocyte	One large secondary oocyte + one small polar body	One large egg cell + three small polar bodies

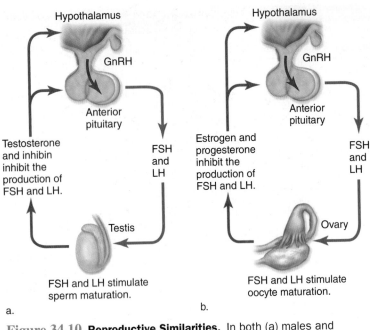

a.   b.

Figure 34.10 **Reproductive Similarities.** In both (a) males and (b) females, GnRH from the hypothalamus stimulates the release of FSH and LH from the anterior pituitary. These hormones, which promote the maturation of gametes, are under negative feedback control.

C. Hormones Influence Female Reproductive Function

The male and female reproductive systems rely on many of the same hormones, but in different quantities and on a different schedule (figure 34.10 and table 34.2).

In females, hormonal fluctuations produce two interrelated cycles. The **ovarian cycle** controls the timing of oocyte maturation in the ovaries, and the **menstrual cycle** prepares the uterus for pregnancy. Figure 34.11 tracks changes in the ovarian follicle, the uterine lining, and the levels of four hormones during the ovarian and menstrual cycles.

On the first day of the menstrual cycle, when menstruation begins, low blood levels of estrogen and progesterone signal the

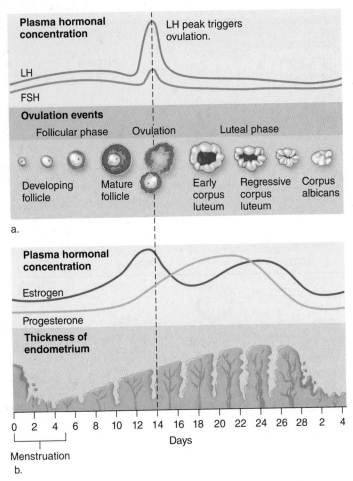

Figure 34.11 **The Ovarian and Menstrual Cycles.** (a) In the ovarian cycle, elevated concentrations of FSH stimulate follicle maturation. A burst of LH triggers ovulation and converts the follicle into a corpus luteum, which releases estrogen and progesterone. These hormones inhibit LH and FSH production. (b) The menstrual cycle begins with release of the endometrial lining. As estrogen rises and ovulation occurs, the endometrium thickens; meanwhile, progesterone prepares the lining for pregnancy. If fertilization does not occur, estrogen and progesterone secretion from the corpus luteum ceases, and the cycle begins anew.

Table 34.2   The Roles of GnRH, LH, and FSH in Human Males and Females

Hormone	Source	Target Cells	Function(s) in Males	Function(s) in Females
GnRH (gonadotropin-releasing hormone)	Hypothalamus	Anterior pituitary	Secreted at constant frequency; stimulates release of LH and FSH	Amount secreted depends on stage of menstrual cycle; stimulates release of LH and FSH
LH (luteinizing hormone)	Anterior pituitary	Testes, ovaries	Stimulates production of testosterone	Stimulates production of estrogens; LH surge midway through menstrual cycle triggers ovulation; LH stimulates development of corpus luteum, which secretes progesterone
FSH (follicle-stimulating hormone)	Anterior pituitary	Testes, ovaries	Enhances production of testosterone-binding protein in sustentacular cells; required for spermatogenesis	Stimulates follicle maturation

hypothalamus to secrete GnRH. This hormone prompts the anterior pituitary to release FSH and LH into the bloodstream. In the ovaries, receptors on the surfaces of follicle cells bind to FSH, stimulating follicles to mature and to release estrogen.

At low concentrations, estrogen inhibits the release of both LH and FSH. But at around day 14 of the cycle, estrogen accumulation triggers the release of additional LH and FSH from the anterior pituitary. This LH surge in the bloodstream triggers **ovulation,** the release of an oocyte from an ovarian follicle into the nearby uterine tube. LH also transforms the now-ruptured follicle into a gland called a **corpus luteum.**

The corpus luteum, in turn, secretes progesterone and estrogen, which have multiple effects. These two hormones act together to promote the thickening of the endometrium, preparing the uterus for possible pregnancy. Progesterone and estrogen also feed back to the hypothalamus, inhibiting the production of GnRH, LH, and FSH. Over the next several days, the corpus luteum degenerates into the corpus albicans, an inactive scar made of collagen.

If pregnancy does not occur, levels of progesterone and estrogen gradually decline (section 34.5 describes what happens if pregnancy does occur). The reduced levels of these hormones no longer maintain the endometrium, which then exits the body through the cervix and vagina as menstrual flow. Lowered progesterone and estrogen levels also release their inhibition of GnRH, LH, and FSH in the brain, and the cycle begins anew.

D. Hormonal Fluctuations Can Cause Discomfort

Fluctuating concentrations of hormones trigger a variety of conditions that are unique to women (figure 34.12). One example is premenstrual syndrome (PMS), a collection of symptoms that appears in the second half of the menstrual cycle. In the days before her menstrual period begins, a woman may experience headache, breast tenderness, cramping, depression, irritability, or dozens of other signs of PMS. The exact cause of each symptom is unknown, but the correlation with hormonal changes is clear. Over-the-counter drugs provide relief from many symptoms.

Figure 34.12 Shifting Hormones. Hormonal fluctuations cause premenstrual syndrome, cramps, endometriosis, and hot flashes.

Menstrual cramps may occur with or without PMS. This abdominal pain arises when prostaglandins induce smooth muscle contractions in the uterus. These contractions help to squeeze the menstrual fluid out of the body; heavy bleeding is correlated with severe cramps.

Cramps may also be related to endometriosis or another underlying problem. In endometriosis, cells from the endometrium travel within the body and become established somewhere other than the uterus, usually within the pelvic cavity. The cells may migrate during "retrograde menstruation," in which menstrual fluid enters the uterine tubes rather than leaving the body through the vagina. The displaced cells, which still respond to hormonal signals, trigger inflammation and pain, especially during menstruation. Surgery to remove the tissue is the standard treatment; for reasons that are poorly understood, untreated endometriosis can lead to infertility.

As a woman nears the end of her reproductive years, her hormonal fluctuations cease. Ironically, the estrogen withdrawal that accompanies menopause can cause symptoms as well. For example, hot flashes affect most women going through menopause. A hot flash, which typically lasts a few minutes, is a temporary sensation of heat in the face and upper body, coupled with flushed skin and sweating. The exact cause is unknown, but researchers speculate that the withdrawal of estrogen somehow affects the temperature set point of the hypothalamus.

E. Contraceptives Prevent Pregnancy

Pregnancy requires the union of sperm and egg; **contraception** is the use of devices or practices that work "against conception"; that is, they prevent pregnancy. Table 34.3, on the following page, summarizes some of the most common methods. Each has advantages and disadvantages.

Besides abstinence, the most effective contraceptives are surgical; next are those that adjust hormone concentrations in the woman's body. Birth control pills, patches, vaginal rings, injections, and implants all contain a synthetic form of progesterone, which mimics the hormonal effects of pregnancy. If used correctly, each of these methods prevents ovulation and therefore precludes fertilization.

Other birth control methods kill sperm, block the meeting of sperm and oocyte, or prevent a developing embryo from implanting in the lining of the uterus. Only latex condoms, however, simultaneously prevent pregnancy and protect against sexually transmitted diseases—the subject of section 34.4.

34.3 | Mastering Concepts

1. What is the role of each part of the female reproductive system?
2. Where do egg cells develop?
3. What are the stages of oogenesis?
4. What is the role of polar bodies in oogenesis?
5. How do hormones regulate the ovarian and menstrual cycles?
6. List examples of hormone-induced female problems.
7. Describe some of the most common contraceptives.

Table 34.3 Birth Control Methods

Method	Mechanism	Advantage(s)	Disadvantage(s)	Likelihood of Success (%)
Barriers and Spermicides				
Condom used with spermicide	Worn over penis or inserted into vagina; keeps sperm out of vagina and kills sperm that escape	Protects against sexually transmitted diseases	Disrupts spontaneity, reduces sensation	95–98
Diaphragm used with spermicide	Kills sperm and blocks cervix	Inexpensive	Disrupts spontaneity, must be fitted	83–97
Cervical cap used with spermicide	Kills sperm and blocks cervix	Inexpensive, can be kept in for 24 hours	May slip out of place, must be fitted	80–95
Spermicidal foam or jelly	Kills sperm and blocks cervix	Inexpensive	Messy	78–95
Spermicidal suppository	Kills sperm and blocks cervix	Easy to use and carry	Irritation in 25% of users	85–97
Hormonal				
Combination birth control pill or patch (estrogen and progesterone)	Prevents ovulation and implantation, thickens cervical mucus	Does not interrupt spontaneity, lowers cancer risk, lightens menstrual flow	Raises risk of heart disease in some women, causes weight gain and breast tenderness	90–100
Vaginal ring (estrogen and progesterone)	Prevents ovulation and implantation, thickens cervical mucus	Continuous hormone release; lower estrogen exposure than with pill or patch	May cause vaginitis or be accidentally dislodged	92–99.7
Minipill (progesterone only)	Blocks implantation, deactivates sperm, thickens cervical mucus	Fewer side effects than other birth control pills	Weight gain, irregular periods	87–100
Depo-Provera (progesterone only)	Prevents ovulation, alters uterine lining	Easy to use, lasts 3 months	Menstrual changes, weight gain, requires injection	99
Subdermal implant (progesterone only)	Prevents ovulation, thickens cervical mucus	Easy to use, lasts 3 years	Menstrual changes, doctor must implant under skin	99.8
Emergency contraception ("morning-after pill")	Prevents ovulation, thickens cervical mucus, prevents implantation	Used up to 5 days after intercourse	Nausea; effectiveness declines with time elapsed after intercourse	75–89
Behavioral				
Abstinence	No intercourse	No cost	Difficult to do	100
Rhythm method	No intercourse during fertile times, sometimes inferred from body temperature and other clues	No cost	Difficult to do, hard to predict timing	79–87
Withdrawal	Removal of penis from vagina before ejaculation	No cost	Difficult to do; sperm may leak from penis before it is withdrawn, even without ejaculation	75–91
Surgical				
Vasectomy	Cuts vas deferentia, so sperm cells never reach urethra	Permanent, does not interrupt spontaneity	Requires minor surgery, difficult to reverse	99.85
Tubal ligation	Cuts uterine tubes, so oocytes never reach uterus	Permanent, does not interrupt spontaneity	Requires surgery, risk of infection, difficult to reverse	99.6
Other				
Intrauterine device, with or without progesterone	Prevents implantation of preembryo	Does not interrupt spontaneity	Severe menstrual cramps, risk of infection	95–99

34.4 | Sexual Activity May Transmit Disease

Disease-causing viruses, bacteria, protists, and fungi lurk everywhere: in food, air, water, and, of course, in and on the human body. Some of these microbes cause **sexually transmitted diseases (STDs),** which spread to new hosts during sexual contact. Interestingly, humans are not the only ones to suffer from STDs. Other animal species have STDs of their own; so do plants, which can pass viroids in pollen. ▶ viroids, p. 332

Vaginal intercourse, oral sex, and anal sex all can provide direct, person-to-person transmission for a wide variety of disease-causing agents. Table 34.4 lists a sampling of some of the most common STDs.

In the United States, the most common STD is genital warts, which can be caused by more than 40 strains of the human papillomavirus (HPV). These viruses infect millions of new hosts every year. Many infections remain symptomless, but others trigger the growth of visible bumps on the male or female genitals. Some strains of HPV that do not cause warts can create a much bigger problem: cervical cancer. Nearly all cervical cancer tumors test positive for DNA from HPV.

Another STD that infects millions of people each year is trichomoniasis, caused by a protist called *Trichomonas vaginalis* (see figure 17.15b). Males can transmit this organism, usually without visible symptoms. Infected women may notice unusual vaginal discharge along with pain, irritation, and itching.

Chlamydia, a third common STD, is caused by the bacterium *Chlamydia trachomatis.* Typical symptoms include discharge from the penis or vagina. An infected person may also experience pain or a burning sensation when urinating. Chlamydia infections, however, are sometimes called a "silent disease" because some 75% of infected women and 50% of infected men remain symptomless.

Sexually transmitted diseases are especially relevant to young adults, who account for 50% of new infections but only 25% of the sexually active population. Females age 15 to 19, for example, lead the nation in infection rates for both chlamydia and gonorrhea.

Infections with HPV, trichomoniasis, chlamydia, and other agents listed in table 34.4 often remain invisible, but that does not mean they are harmless. We have already seen that HPV infection is associated with a high risk of cervical cancer. In addition, untreated chlamydia infections in women can spread to the uterine tubes. Pelvic inflammatory disease, one possible complication, can cause pain and infertility. Moreover, syphilis, gonorrhea, and chlamydia can pass to infants during childbirth.

Most STDs are also associated with an elevated risk for acquiring and transmitting HIV. One explanation is that many STDs cause sores through which viral particles can enter or leave the body. A second reason is that any infection can cause inflammation and other defensive reactions. As described in chapter 33, these responses attract the types of white blood cells that HIV infects. Efforts to prevent STDs in general can therefore also help reduce infections by HIV in particular.

The best way to prevent sexually transmitted diseases is to abstain from any type of sexual activity. The second best way is to develop a long-term, monogamous relationship with a partner who has been recently tested for STDs and is therefore known to be disease-free. The third best way is to properly use a latex condom throughout a sexual encounter. In addition, vaccines can protect against some sexually transmitted viruses, including HPV and hepatitis B.

Table 34.4 Examples of Sexually Transmitted Diseases

Disease	Agent	Treatment
Viruses		
HIV/AIDS	Human immunodeficiency virus (HIV)	Combination of drugs that reduce viral replication
Genital warts	Human papillomavirus (HPV)	Removal of warts
Genital herpes	Herpes simplex virus	Medications that reduce outbreak frequency and duration
Hepatitis B	Hepatitis B virus	None
Bacteria		
Chlamydia	*Chlamydia trachomatis*	Antibiotics
Gonorrhea	*Neisseria gonorrhoeae*	Antibiotics
Syphilis	*Treponema pallidum*	Antibiotics
Protists		
Trichomoniasis	*Trichomonas vaginalis*	Antiprotozoan drugs
Fungi		
Candidiasis	*Candida albicans*	Antifungal drugs

34.4 | Mastering Concepts

1. List and describe three common STDs.
2. Describe two reasons that a symptomless infection with an STD can be harmful.

34.5 | The Human Infant Begins Life as a Zygote

So far, this chapter has described gamete production in the human male and female reproductive systems. This section now turns to the development of a human baby. As you shall see, fertilization produces the zygote; this single cell divides many times as it develops into a preembryo, embryo, fetus, and newborn baby (table 34.5).

A. Fertilization Joins Genetic Packages and Initiates Pregnancy

After intercourse, sperm cells swim toward the oocyte. Of the 100 million sperm that begin the journey, only about 200 approach the egg cell in a uterine tube.

Those that make it must penetrate two layers to contact the ovum. An outer layer of follicle cells surrounds a thin, clear layer of proteins and carbohydrates that encases the oocyte (figure 34.13). On contact with the follicle cells surrounding the oocyte, each sperm's acrosome bursts, spilling enzymes that digest both outer layers.

Fertilization begins when the outer membranes of one sperm cell and the secondary oocyte touch. At that time, a wave of electricity spreads physical and chemical changes across the oocyte surface, preventing other sperm from fertilizing the same egg cell. As the sperm's head enters the secondary oocyte, its nuclear membrane disperses, liberating the sperm's DNA. Meanwhile, the female cell completes meiosis and becomes a fertilized egg cell, or zygote. The nuclear membrane of the ovum then disappears, and the two sets of chromosomes mingle. A new nuclear membrane soon encloses the combined DNA. This diploid cell has 23 pairs of chromosomes; one chromosome of each pair comes from each parent. ▶ homologous chromosomes, p. 181

Occasionally a woman ovulates two or more egg cells at once, and a different sperm fertilizes each one. If all the zygotes complete development, the result is twins, triplets, quadruplets, or even higher-order multiple births. Because each zygote results

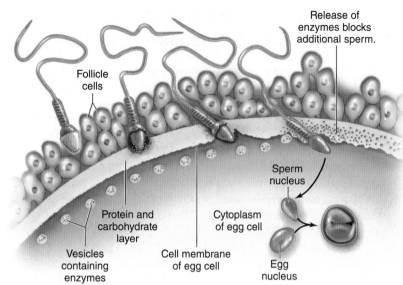

Figure 34.13 Fertilization. Before fertilization occurs, the sperm's acrosome bursts, spilling enzymes that help the nucleus enter the oocyte. A series of chemical reactions helps ensure that only one sperm fertilizes the egg cell.

from a separate sperm and egg cell, the siblings will not be genetically identical, and they may be of different sexes. At the opposite end of the spectrum are couples that have trouble conceiving any child at all; the Apply It Now box on page 712 explains how technology can help. ▶ multiple births, p. 195

B. Preembryonic Events Include Cleavage, Implantation, and Gastrulation

The first 2 weeks of prenatal development have a variety of names, but we will call this period the **preembryonic stage.**

About 3 hours after fertilization, the zygote divides for the first time, beginning a period of rapid mitotic cell division called **cleavage** (figure 34.14). The result is a **morula,** a solid ball of

Table 34.5   Stages in Development from Fertilization to Birth: A Summary

Name	Duration	Description
Zygote	About 24 hours after ovulation	Fertilized egg cell
Preembryonic stage		
Morula	3–4 days after fertilization	Solid ball of 16 or more cells
Blastocyst	4 days to 2 weeks after fertilization	Hollow sphere consisting of outer layer (trophoblast) and inner cell mass; implants into endometrium
Gastrula	2 weeks after fertilization	Three germ layers form: endoderm, mesoderm, ectoderm
Embryonic stage	From implantation until about 8 weeks after fertilization	Germ layers differentiate into organ systems
Fetal stage	From end of eighth week until birth (end of first trimester, plus second and third trimesters)	Organ systems become functional

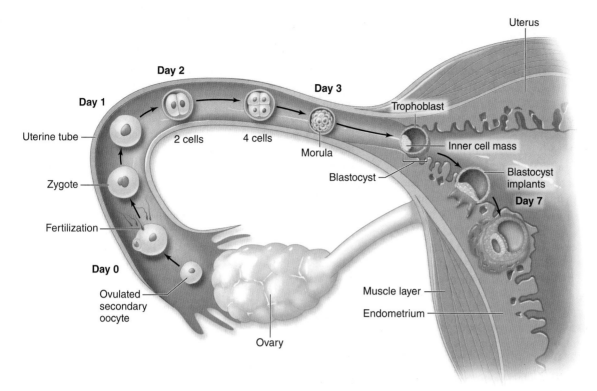

Figure 34.14 From Ovulation to Implantation. Fertilization occurs in the uterine tube, producing the zygote. The first mitotic divisions occur while the preembryo moves toward the uterus. By day 7, the preembryo begins to implant in the uterine lining.

16 or more cells. *Morula* is Latin for "mulberry," which the preembryo resembles. Three days after fertilization, the morula is still within the uterine tube but moving toward the uterus.

At first, the morula is about the same size as the zygote. The reason is that the initial cleavage divisions split the zygote—an unusually large cell—into smaller and smaller cells. Once the morula consists of cells that are "normal"-sized, however, cell size stabilizes, and the morula begins to enlarge as additional cells accumulate.

The mass of cells reaches the uterus 3 to 6 days after fertilization. The preembryo then hollows out, its center filling with fluid that seeps in from the uterus. This fluid-filled ball of cells is called a **blastocyst.** The blastocyst's outer layer, the trophoblast, will eventually form the fetal portion of the placenta. The cells inside the blastocyst form the **inner cell mass,** the cells that will develop into the embryo itself. (The inner cell mass is also the source of embryonic stem cells.) ▶ stem cells, p. 167

In a process called **implantation,** the blastocyst becomes embedded in the lining of the uterus (figure 34.15). Implantation occurs within a week of fertilization. The trophoblast not only secretes digestive enzymes that eat through the outer layer of the uterine lining, but it also sends

Figure 34.15 Implantation and Gastrulation. About a week after fertilization, the inner cell mass settles against the uterine lining. The trophoblast sends extensions into the endometrium. Meanwhile, the enlarging preembryo begins to fold into the three primary germ layers. Implantation is complete about 2 weeks after fertilization.

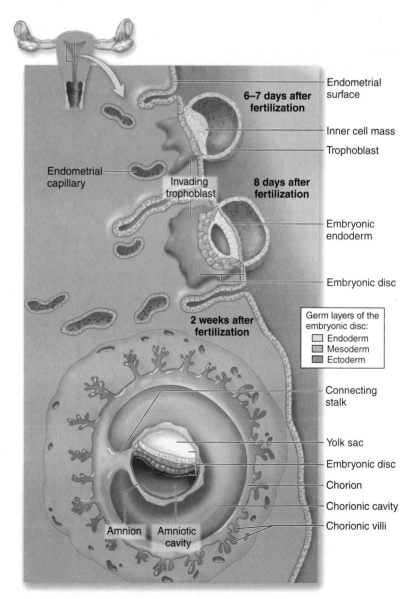

projections into the uterine lining. Until the placenta forms, the preembryo obtains nutrients from these digested endometrial cells.

The trophoblast cells now secrete **human chorionic gonadotropin (hCG),** the hormone that is the basis of pregnancy tests. For a while, hCG keeps the cells of the corpus luteum producing progesterone, which prevents menstruation and further ovulation. In this way, the blastocyst helps to ensure its own survival; if the uterine lining were shed, the blastocyst would leave the woman's body, too. The trophoblast cells produce hCG for about 10 weeks.

During the second week of development, the blastocyst completes implantation. A space called the amniotic cavity forms within a sac called the amnion, which lies between the inner cell mass and the trophoblast (see figure 34.15). The inner cell mass flattens and forms the **embryonic disc,** which will develop into the embryo.

As the preembryo continues to develop, one layer of the embryonic disc becomes **ectoderm,** and an inner layer becomes **endoderm.** Soon, a middle **mesoderm** layer will form from the ectoderm. The **gastrula** is the resulting three-layered structure.

The formation of the gastrula is called gastrulation, and its start marks the end of the preembryonic stage at about 2 weeks post-fertilization (see figure 34.15). Although the woman has not yet missed her menstrual period, she might notice effects of her shifting hormones, such as swollen, tender breasts and fatigue. By now, her urine contains enough hCG for an at-home pregnancy test to detect. ▶ germ layers, p. 412

The preembryo may split during the first 2 weeks of development, forming identical twins. Depending on when the split occurs, the twins may or may not share the same amnion and placenta; the later the split, the more structures the twins will share. If a preembryo splits after day 12 of pregnancy, the twins are unlikely to separate completely, and they may be conjoined.

C. Organs Take Shape During the Embryonic Stage

The **embryonic stage** lasts from the end of the second week until the end of the eighth week. During this stage of development, cells of the three layers continue to divide and differentiate,

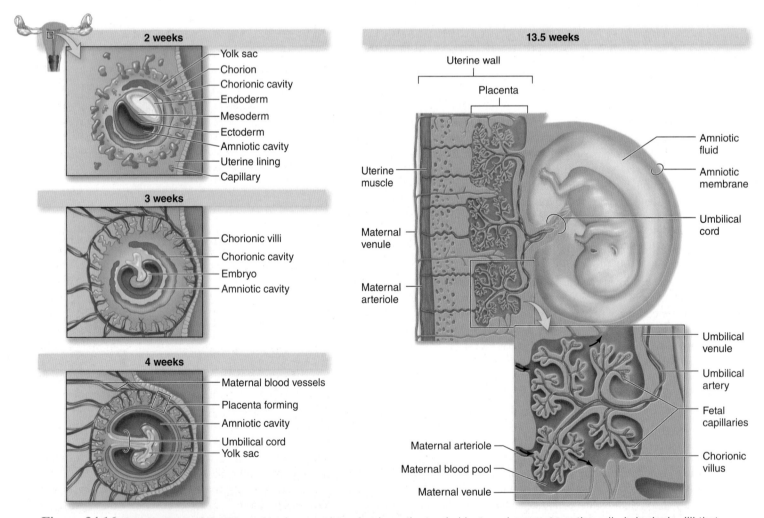

Figure 34.16 Development of the Placenta. As an embryo develops, the trophoblast produces outgrowths called chorionic villi that extend into the uterine lining. Eventually, blood vessels from the embryo grow into the chorionic villi and exchange materials with pools of maternal blood across a thin membrane. Note the locations of the yolk sac, amniotic membrane, and chorion; the allantois is not visible.

forming all of the body's organs. Structures that support the embryo also develop during this period.

Four Membranes, the Placenta, and the Umbilical Cord Support the Embryo

Four extraembryonic membranes support, protect, and nourish the embryo. Their presence is an important clue to our evolutionary history because each occurs not only in humans but also in other mammals, in birds, and in nonavian reptiles. One membrane, the **yolk sac,** manufactures blood cells until about the sixth week; at that point, the liver takes over, and the yolk sac starts to shrink. Parts of the yolk sac, however, eventually develop into the intestines and germ cells. By the third week, an outpouching of the yolk sac forms the **allantois,** another extraembryonic membrane. It, too, manufactures blood cells, and it gives rise to blood vessels in the umbilical cord. The **amnion** is the transparent sac that contains the amniotic fluid. This fluid cushions the embryo, maintains a constant temperature and pressure, and protects the embryo if the woman falls. ▶ amnion, p. 434

Some cells of the trophoblast develop into the **chorion,** the outermost extraembryonic membrane. **Chorionic villi** are finger-like projections from the chorion that extend into the uterine lining. These structures establish the beginnings of the **placenta,** a structure that will connect the developing embryo with its mother's uterus (**figure 34.16**). One side of the placenta comes from the embryo; the other side consists of endometrial tissue and blood from the pregnant woman's circulation. Cells collected from amniotic fluid, or sampled chorionic villi cells, are used in many prenatal medical tests.

The placenta begins to develop in the embryonic stage, but it is not completely developed until 10 weeks after fertilization. At that point, substances can diffuse across the chorionic villi cells between the woman's circulatory system and the fetus. Arteries and veins in the **umbilical cord** connect the embryo or fetus to the placenta. Umbilical cord blood is a rich source of stem cells used to treat an ever-expanding list of disorders.

The placenta blocks most large proteins and red blood cells, but many other substances can cross. Carbon dioxide, urea, and other wastes from the fetus's circulation pass to the mother's blood. Glucose, amino acids, oxygen, ions, and some antibodies from the mother travel in the opposite direction, nourishing and protecting the embryo. At the same time, harmful substances can also cross the placenta; examples include some disease-causing agents and drugs such as cocaine, alcohol, and nicotine.

Organ Formation Begins in the Third Week of Development

As the extraembryonic membranes, placenta, and umbilical cord develop, so does the embryo itself. Beginning during the third week of prenatal development, distinct organs form. Over several months, ectoderm cells develop into the nervous system, sense organs, outer skin layers, hair, nails, and skin glands. Splits in the mesoderm form the coelom, which becomes the chest and abdominal cavities. Mesoderm cells develop into bone, muscle, blood, the inner skin layer, and the reproductive organs. Endoderm cells form the organs of the digestive and respiratory systems. ▶ coelom, p. 413

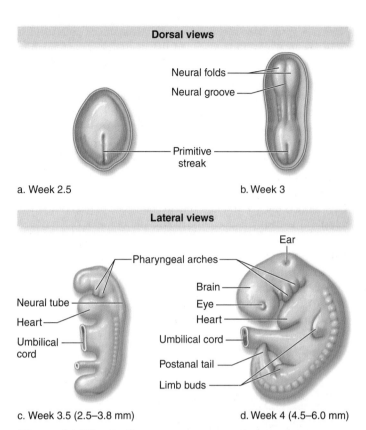

Figure 34.17 **Early Embryos.** (a) The primitive streak establishes the embryo's longitudinal axis. (b) At a signal from the notochord, ectoderm folds around the neural groove to form the neural tube. (c) The heart becomes prominent during the middle of week 3. (d) By week 4, the embryo is beginning to look like a typical vertebrate, complete with pharyngeal pouches and a postanal tail.

During the third week of development, a furrow called the **primitive streak** appears along the back of the embryonic disc, forming a longitudinal axis around which other structures organize as they develop (**figure 34.17**). For example, the primitive streak gives rise to the **notochord,** a flexible rodlike structure that forms the basic framework of the vertebral column. The notochord also induces overlying ectoderm to differentiate into the central nervous system. The neural groove in figure 34.17 is the first visible sign of this process. The neural groove folds into a hollow **neural tube,** which eventually develops into the brain and spinal cord. Then the central nervous system begins to expand and become more complex. At about the same time, a bulge containing the heart appears. It begins to beat around day 22. ▶▶ notochord, p. 432; central nervous system, p. 540

The fourth week of the embryonic period is a time of rapid growth and differentiation. Blood cells begin to form and to fill developing blood vessels. Immature lungs and kidneys appear. Small buds appear that will develop into arms and legs. If the neural tube does not close normally at about day 28, a neural tube defect such as spina bifida results. (In a neural tube defect, nervous tissue protrudes from an open area of the spine, causing paralysis from that site downward.) Meanwhile, cells detach from an area of ectoderm called the neural crest and migrate throughout the body, forming the foundation of the peripheral nervous system and many other organs.

The 4-week embryo has a distinct head and jaw and early evidence of eyes, ears, and nose. The digestive system appears as a long, hollow tube that will develop into the intestines. A woman carrying this embryo, which is now only about 6 millimeters long, may suspect that she is pregnant because her menstrual period is about 2 weeks late.

By the fifth week, the embryo's head appears disproportionately large. Limbs extending from the body end in platelike structures. Tiny ridges run down the plates, and by week 6, the ridges deepen as certain cells die, molding fingers and toes. The eyes open, but they do not yet have eyelids or irises. Cells in the brain are rapidly differentiating. The embryo is now about 1.3 centimeters from head to rump.

All early embryos have unspecialized reproductive structures. At week 7, however, a gene on the Y chromosome, called *SRY* (for "sex-determining region of the Y"), is activated in male embryos. Hormones then begin to stimulate development of male

reproductive organs and glands. If there is no *SRY* gene (or if it never switches on), female reproductive structures develop (**figure 34.18**). Overall, many enzymes and membrane proteins participate in sex determination. This chapter's Burning Question describes how abnormalities in these molecules cause a condition called intersex.

During the seventh and eighth weeks, a cartilage skeleton appears. The placenta is now almost fully formed and functional, secreting estrogen and progesterone that maintain the blood-rich uterine lining. The embryo is about the size and weight of a paper clip. By the end of the eighth week, all organ systems are in place, and the embryonic stage of prenatal development is complete. The offspring is now a fetus.

Many pregnancies end during the embryonic stage; the Apply It Now box on page 711 explains why.

 ## Burning Questions

Do human hermaphrodites exist?

Many people believe that there are just two kinds of people: males and females. But the truth is considerably more interesting. The development of genitalia actually depends on a complex interplay of genes, hormones, and embryonic tissues.

In a child conceived as a boy, genes on the Y chromosome signal the embryonic testes to produce testosterone and other masculinizing hormones (androgens). If cells in the early embryo have a membrane receptor that binds to the androgens, the fetus develops normal male genitalia. Sometimes, however, a child inherits a mutated version of the gene encoding the receptor. The testes produce androgens, but the target tissues cannot bind to the hormones because the receptor is misshapen or absent. This condition is called androgen insensitivity syndrome. The child (conceived as a male) may develop external genitalia that appear female, ambiguous, or male.

Androgen insensitivity syndrome is just one of many so-called *intersex* conditions in which a person's anatomy does not match the that of a "standard" male or female. Others include 5-alpha reductase deficiency, Klinefelter syndrome, and Turner syndrome. Intersex conditions affect people who are genetically male (XY) or female (XX) or who have sex chromosome abnormalities (such as XO or XXY). ▶ sex chromosome abnormalities, p. 191

Although it is tempting to confuse the terms *intersex* and *hermaphrodite,* they are not the same. A true hermaphrodite, such as a flatworm or earthworm, is an organism with both ovaries and testes. In humans, such a condition is exceedingly uncommon; when it does occur, the gonads rarely if ever function.

Submit your burning question to:
marielle_hoefnagels@mcgraw-hill.com

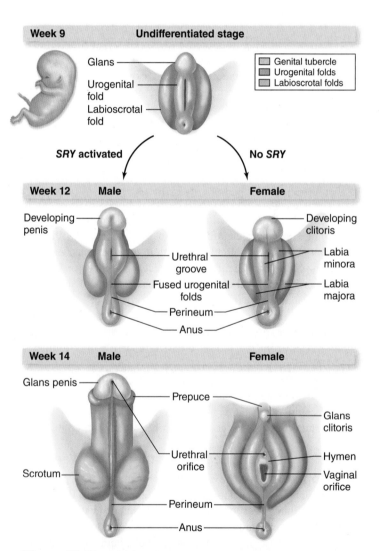

Figure 34.18 Development of Male and Female Genitalia. Until about week 9 of development, embryos have undifferentiated genitalia. If the *SRY* gene on the Y chromosome is activated, development continues as a male. Otherwise, female genitalia develop.

Apply It Now

When a Pregnancy Ends

Many conditions can cause miscarriage (also called a spontaneous abortion), in which an undeveloped fetus separates from the uterus. The woman expels the fetus and placenta (if present) from her body, ending the pregnancy. A miscarriage may occur any time before 7 months of development, when a fetus can survive outside the womb. But most occur early in a pregnancy, often reflecting a chromosomal abnormality in the embryo or fetus. If a miscarriage occurs later, during the second trimester, the problem usually originates with the woman, not the fetus. Diabetes, high blood pressure, hormonal abnormalities, infectious diseases, uterine problems, or drug abuse may be to blame, but often the cause of a miscarriage is unclear. ▶ nondisjunction, p. 189

Just as a miscarriage is the loss of a pregnancy before the fetus is viable, a stillbirth occurs if a baby dies in the uterus after about the twenty-fourth week. The same conditions that contribute to miscarriage are also associated with stillbirths.

An induced abortion is the deliberate termination of a pregnancy. During the first 15 weeks of pregnancy, the most common method of abortion is for a medical professional to use a manual syringe or electric pump to remove the embryo or fetus from the uterus. If the woman terminates the pregnancy between the fifteenth and eighteenth weeks, the cervix must first be dilated before removing the fetus. In the third trimester, a variety of chemical substances can be administered to induce premature delivery, or the fetus can be surgically removed from the uterus.

D. Organ Systems Become Functional in the Fetal Stage

The third stage of prenatal development is the **fetal stage,** which lasts from the beginning of the ninth week through the full 38 weeks of development. The fetal stage therefore begins near the end of the first trimester (3-month period) of pregnancy and ends 9 months after fertilization. During this time, the fetus grows considerably (**figure 34.19**). Organs begin to function and coordinate, forming organ systems.

As the first trimester progresses, the body proportions of the fetus begin to appear more like those of a newborn. Bone begins to form and will eventually replace most of the cartilage, which is softer. Soon, as the nerves and muscles coordinate their actions, the fetus will move its arms and legs. Physical differences between the sexes are usually detectable by ultrasound after the twelfth week. From this time, the fetus sucks its thumb, kicks, and makes fists and faces, and baby teeth begin to form in the gums. More than half of all pregnant women experience the nausea and vomiting of morning sickness in their first trimester.

During the second trimester, body proportions become even more like those of a newborn. By the fourth month, the fetus has hair, eyebrows, lashes, nipples, and nails. Bone continues to replace the cartilage skeleton. The fetus's muscle movements become stronger, and the woman may begin to feel a slight fluttering in her abdomen. By the end of the fifth month, the fetus curls into the classic head-to-knees "fetal position." As the second trimester ends, the woman feels distinct kicks and jabs and may even detect a fetal hiccup. The fetus is now about 30 centimeters long.

In the final trimester, fetal brain cells rapidly connect into networks, and organs differentiate further and grow. A layer of fat develops beneath the skin. The digestive and respiratory systems mature last, which is why infants born prematurely often have difficulty digesting milk and breathing. The fetus may move vigorously, causing back pain and frequent urination in the woman as it presses against her bladder. About 266 days (38 weeks) after a single sperm burrowed into an oocyte, a baby is ready to be born.

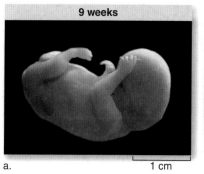

9 weeks

a.   1 cm

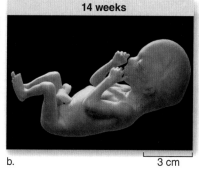

14 weeks

b.   3 cm

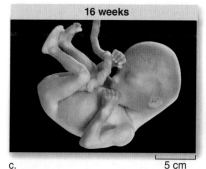

16 weeks

c.   5 cm

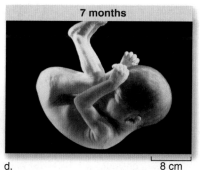

7 months

d.   8 cm

Figure 34.19 The Fetal Stage. These photographs show the development of a fetus from (a) 9 weeks to (b)14 weeks to (c) 16 weeks to (d) 7 months postfertilization.

Apply It Now

Assisted Reproductive Technologies

What is something that you desperately want but cannot have? Money? A new car? Better health? Less stress? One more chance to talk to a loved one who has passed away? For millions of couples, the answer is "a baby." About 12% of women of child-bearing age in the United States have trouble conceiving a child.

To make a baby, sperm and egg must meet and merge; this process ordinarily occurs in the woman's reproductive tract. Abnormal gametes or blockages that impede this meeting of cells can result in infertility (the inability to conceive). Assisted reproductive technologies such as those described below can sometimes help.

Artificial Insemination: Donated Sperm

In artificial insemination, a doctor places donated sperm in a woman's reproductive tract. A woman might seek artificial insemination if her partner is infertile or carries an allele for an inherited illness. Alternatively, she may want to be a single parent or to raise a child with a partner other than the biological father.

Artificial insemination is simple and has a low risk of side effects. The woman can select sperm from a catalog that lists the potential donors' personal characteristics (not all of which are actually heritable). Alternatively, she may know the sperm donor personally. Either way, a doctor first confirms that the woman is ovulating, then uses a catheter to place donated sperm directly into her uterus.

Surrogate Motherhood: A "Leased" Uterus

If a man produces healthy sperm but his partner's uterus is absent or cannot maintain a pregnancy, a surrogate mother may be able to help. She is artificially inseminated with the man's sperm and carries the pregnancy, but the couple raises the child. Typically, the surrogate is both the genetic and the gestational mother. Another type of surrogate mother, however, lends only her uterus: she receives a fertilized egg from a woman who has healthy ovaries but lacks a functional uterus.

In Vitro Fertilization: Test Tube Babies

In *in vitro* fertilization (IVF), which means "fertilization in glass," sperm meets oocyte outside the woman's body (hence the common name of "test tube babies" to refer to infants conceived by this method). A woman might undergo IVF if her ovaries and uterus work but her uterine tubes are blocked, or if she is using donated eggs instead of her own.

To begin, a woman is injected with hormones that hasten the maturity of several oocytes. When the oocytes are ready, a physician removes a few of the largest ones from an ovary and transfers them to a dish. After applying chemicals that mimic those in the female reproductive tract, the physician adds sperm. The fertilized eggs divide two or three times and are introduced into the oocyte donor's (or a surrogate's) uterus. If all goes well, a pregnancy begins—often with twins or triplets.

If a man's sperm cannot penetrate an oocyte *in vitro*, a technician use a tiny syringe to inject one sperm cell directly into an egg. This variant of IVF is called intracytoplasmic sperm injection (ICSI). It can help men with very low sperm counts, abnormal sperm, or injuries or illnesses that prevent ejaculation. Viable sperm are first isolated from the testes and then injected into oocytes. A day or so later, a physician transfers some of the resulting balls of 8 or 16 cells to the woman's uterus.

Gamete or Zygote Intrafallopian Transfer

A procedure called gamete intrafallopian transfer (GIFT) returns fertilization to the woman's body. A woman takes a superovulation drug for a week and then has several of her largest oocytes removed. A man donates a sperm sample. The collected oocytes and sperm are deposited together in the woman's uterine tube at a site past any obstruction, so that implantation can occur.

A related procedure combines elements of IVF and GIFT. In zygote intrafallopian transfer (ZIFT), a physician places an *in vitro* fertilized ovum in a woman's uterine tube. This technique is unlike IVF because the zygote is introduced to the uterine tube, not the uterus. And it is unlike GIFT because fertilization occurs in the laboratory.

Oocyte Donation

College newspapers often run ads recruiting healthy young women to become egg donors. In exchange for a payment, the donor receives several daily injections of a superovulation drug, has her blood checked for hormones, and then undergoes minor surgery to collect the oocytes. The prospective father's sperm and the donor's oocytes are then placed in the recipient's uterus.

Assisted reproductive technologies have helped many couples to conceive, but many people eventually give up. The expense and emotional roller coaster of infertility treatment simply become overwhelming.

E. Muscle Contractions in the Uterus Drive Labor and Childbirth

Labor refers to the strenuous work a pregnant woman performs in the hours before giving birth. Labor may begin with an abrupt leaking of amniotic fluid as the fetus presses down and ruptures the sac ("water breaking"). Labor may also begin with a discharge of blood and mucus from the vagina, or a woman may feel mild contractions in her lower abdomen about every 20 minutes.

As labor proceeds, hormones prompt the smooth muscle that makes up the wall of the uterus to contract with increasing frequency and intensity. During the first stage of labor, the cervix dilates (opens) a little more each time the baby's head presses against it. By the end of the first stage of labor, the cervix stretches open to about 10 centimeters. The second stage of labor is delivery (**figure 34.20**), during which the baby typically descends head-first through the cervix and vagina. In the third and last stage of labor, the uterus expels the placenta.

The events of childbirth illustrate **positive feedback,** in which a process reinforces itself (as opposed to negative feedback, in which a process counteracts an existing condition). At the onset of labor, the cervix begins to stretch. Sensory receptors at the cervix relay the message to the hypothalamus, which triggers release of the hormone oxytocin from the posterior pituitary. Oxytocin travels in the bloodstream and stimulates muscles in the uterus to contract, pushing the baby out (for this reason, physicians sometimes use synthetic oxytocin to induce labor). As the baby emerges, the cervix stretches farther and stimulates more oxytocin production, which intensifies the uterine contractions, and so on. When the baby is born, other hormonal changes stop the cycle.

Not all births go according to plan. For example, about 3% of babies have a "breech presentation," in which the baby's feet, knees, or buttocks appear first instead of the head. Breech deliveries are more difficult than head-first deliveries and account for some cesarean sections ("C-sections"). In this procedure, a surgeon removes the child from the uterus through an incision in the abdomen. About one fourth of all babies in the United States are delivered by cesarean section, either by choice or because the life of the mother or the baby is at risk.

Sometimes, a pregnancy ends with a premature birth. An infant is premature if it is born before completing 35 weeks (about 8 months) of gestation. A fetus born before 22 weeks of gestation is not viable, but babies born after that time may survive. Premature infants are at risk for many health problems, so they typically spend their first weeks or months in incubators at a neonatal intensive care unit. The incubators control the temperature and protect the infants from infection. ▶ premature babies, p. 621

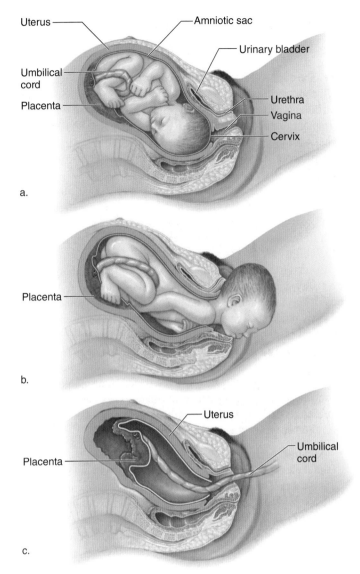

Figure 34.20 Labor and Birth. (a) About 2 weeks before birth, the fetus "drops" in the woman's pelvis, and the cervix may begin to dilate. (b) At the onset of labor, the amniotic sac may break as the baby begins to emerge. (c) The baby is pushed out of the birth canal, followed by the placenta and extraembryonic membranes.

34.5 | Mastering Concepts

1. What is the role of the acrosome in fertilization?
2. What is the relationship among the zygote, morula, blastocyst, inner cell mass, and gastrula?
3. What is implantation, and when does it occur?
4. Which supportive structures develop during the embryonic period? What are their functions?
5. When do sex differences appear, and what triggers them?
6. What are the events of the second and third trimesters?
7. Which events make up the three stages of labor?

34.6 | Birth Defects Have Many Causes

The birth of a live, healthy baby seems against the odds, considering the complexity of human development. Of every 100 egg cells exposed to sperm, 84 are fertilized. Of these, 69 implant in the uterus, 42 survive a week or longer, 37 survive 6 weeks or longer, and only 31 are born alive. Of those that do not survive, about half have chromosomal abnormalities too severe to maintain life.

About 97% of newborns are apparently normal. In the remaining cases, genetic abnormalities, vitamin deficiencies, toxins, or viruses disrupt prenatal development and cause a **birth defect**—any abnormality that causes death or disability in the child.

Some birth defects result from a chromosomal abnormality or a faulty gene that acts during prenatal development. An extra copy of chromosome 21, for example, is the most common cause of Down syndrome. Faulty genes cause Tay-Sachs disease, phenylketonuria, and other metabolic disorders. ▶ trisomy 21, p. 190

Teratogens are substances that cause birth defects. People encounter teratogens everywhere, including the workplace. Women who work with textile dyes, lead, some photographic chemicals, semiconductor materials, mercury, and cadmium face increased risk of miscarriage and birth defects in their children. Several other types of teratogens include:

- **Alcohol:** A pregnant woman who consumes alcohol risks fetal alcohol syndrome. A child with fetal alcohol syndrome has a small head, misshapen eyes, and a flat face and nose. The child has impaired intellect, ranging from minor learning disabilities to mental retardation. Because everyone metabolizes alcohol slightly differently, physicians advise all pregnant women to avoid alcohol.

- **Cigarettes:** Smoking during pregnancy increases the risk of miscarriage, stillbirth, and prematurity. Carbon monoxide crosses the placenta and robs rapidly growing fetal tissues of oxygen. Other chemicals in cigarette smoke prevent nutrients from reaching the fetus. The placentas of women who smoke lack important growth factors, thus lowering birth weight.

- **Illicit drugs:** A pregnant woman's use of cocaine, heroin, methamphetamine, and other illicit drugs is associated with poor fetal growth and low birth weight. In addition, cocaine increases the risk of urinary tract defects. However, drug-addicted women also tend to have poor nutrition and poor health overall, so it is often difficult to isolate the effect of each drug on fetal health.

- **Prescription drugs:** During the late 1950s, a drug called thalidomide was prescribed as a treatment for morning sickness in pregnant women. The drug caused severe limb shortening in the children of these women (**figure 34.21**). From the 1940s through the 1960s, a drug prescribed to pregnant women with a history of miscarriages was

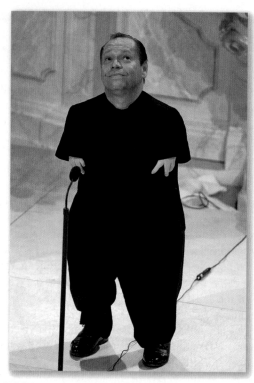

Figure 34.21 Teratogenic Effects of Thalidomide. Grammy award-winning baritone Thomas Quasthoff receives the European Culture Award in 2006. Quasthoff's mother took thalidomide to combat morning sickness early in pregnancy. The limb bones in his arms and legs are unusually short, and his hands resemble flippers.

diethylstilbestrol (DES). Their children—and perhaps their children's children—face an elevated risk of cancer and reproductive tract abnormalities.

- **Excess or insufficient vitamins:** The acne medicine isotretinoin (Accutane), derived from vitamin A, causes miscarriages and defects of the heart, nervous system, and face. Vitamin deficiencies can also cause birth defects. Neural tube defects such as spina bifida, for example, are associated with insufficient folic acid in the mother's diet. ▶ vitamins, p. 649

- **Starvation:** Inadequate nutrition during pregnancy increases the incidence of miscarriage and damages the placenta, causing low birth weight, short stature, tooth decay, delayed sexual development, learning disabilities, and possibly mental retardation. Starvation greatly raises the risk of a child developing obesity, type 2 diabetes, and other conditions later in life. ▶ starvation, p. 651

- **Viral infection:** HIV, the virus that causes AIDS, infects 15% to 30% of infants born to HIV-positive women. This risk drops sharply if an infected woman takes antiviral drugs while pregnant. Fetuses infected with HIV are at risk

When structures develop

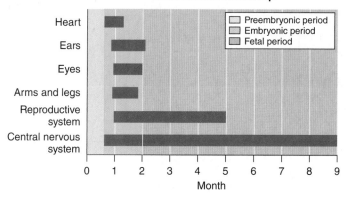

Sensitivity to teratogens during pregnancy

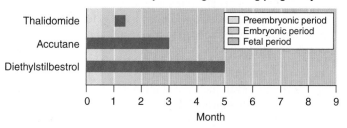

Figure 34.22 Critical Periods of Development. Body parts differ in their critical periods. The type of birth defect resulting from each drug depends on which structures are developing at the time of the exposure.

for low birth weight, prematurity, and stillbirth. In addition, the virus that causes rubella (German measles) causes birth defects such as deafness, cataracts, and heart disease; herpes simplex viruses can infect the fetal nervous system and cause neurological disabilities such as mental retardation or cerebral palsy.

The "critical period" of a structure is the time during which its development is most susceptible to damage by a teratogen (figure 34.22). About two thirds of birth defects stem from a disruption during the embryonic period, because developing organs are especially sensitive to damage. The brain is vulnerable throughout prenatal development, as well as during the first 2 years of life. Because of the brain's long critical period, many birth defect syndromes include mental retardation. The continuing sensitivity of the brain after birth explains why toddlers who accidentally ingest lead-based paint suffer impaired learning.

34.6 | Mastering Concepts

1. What is a birth defect?
2. What are some examples of genetic conditions and teratogens that cause birth defects?
3. What is the critical period in the development of a structure?

34.7 | Investigating Life: The "Cross-Dressers" of the Reef

The scene on TV is familiar: Two bighorn sheep rear up on their hind legs and bash their heads together with bone-crunching force. Only one will win the prize—the exclusive "right" to mate with a female. Throughout the animal kingdom, however, fighting is not the only form of competition that determines a male's reproductive success. If a female mates with many males in a short period, the competition shifts to a much smaller field: her reproductive tract. Only one sperm will fertilize each egg cell. Which will it be?

Sometimes, numbers decide the winner. Males that produce the most sperm are most likely to fertilize an egg cell, just as the owner of several raffle tickets has a better chance of winning a prize than someone who buys only one. But behavior also often plays a role. In some species, for example, a male may try to block other males from approaching a female he has already mated with; in effect, the male prevents his rivals from buying raffle tickets. Males also do the equivalent of destroying raffle tickets that others have purchased. That is, they remove other males' sperm before depositing their own.

Sperm competition has selected for unusual behavioral adaptations in a squidlike mollusk called the Australian giant cuttlefish *(Sepia apama)*. These animals, which grow to nearly a meter long, usually live alone. During the winter mating season, however, they congregate by the hundreds of thousands on the reefs of southern Australia. ▶ mollusks, p. 420

The two cuttlefish sexes look different from each other (figure 34.23). A female has shorter arms than a male, and her skin has dark patches on a white background. The male's arms are longer and whiter, and he displays moving patterns of zebralike stripes on his skin during courtship rituals. Specialized pigment sacs called chromatophores create the ever-changing

Figure 34.23 The Australian Giant Cuttlefish. A large male guards a female, preventing her from mating with other males.

Figure 34.24 **Hanging Eggs.** A female Australian giant cuttlefish has deposited these eggs on the roof of her den.

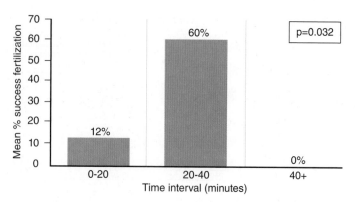

Figure 34.25 **Mate Guarding Works.** Researchers timed the mate-guarding activities of 22 male cuttlefish and used DNA to determine the paternity of the resulting fertilized eggs. On average, the 10 males who guarded their mates from 20 to 40 minutes had greater reproductive success than the eight who guarded for less than 20 minutes or the four who did so for more than 40 minutes.

color palette. Muscles surround each chromatophore, which holds yellow, reddish orange, or brownish black pigment. Rapid expansion and contraction of different-colored chromatophores creates the appearance of shifting patterns.

When a female accepts a male's mating attempt, the two animals align head-to-head, and he inserts a sperm packet into a pouch near her mouth. He also tries to flush out the sperm packets that other males have deposited. After mating, she retreats to her den and removes eggs, one by one, from her mantle cavity. She passes the eggs through her sperm pouch, then hangs them from the roof of her den (figure 34.24).

With a sex ratio of at least four males to every female, the competition to father offspring is fierce. The largest males mate with females and guard them afterward, fighting off rivals to prevent subsequent insemination of their mates. A study led by Marié-Jose Naud of Australia's Flinders University and Roger Hanlon of the Marine Biological Laboratory at Woods Hole, Massachusetts, showed that males who guarded a female for up to 40 minutes after mating fertilized significantly more eggs than those who guarded for less than 20 minutes (figure 34.25).

Nevertheless, smaller males do get to mate, and they sometimes use surprising strategies to gain access to females. A team of Australian biologists, led by Mark Norman, noticed that some males are female impersonators. A small male can convincingly disguise himself as a female by hiding the arms that reveal his sex and manipulating his chromatophores to change his skin color. (The ploy is so realistic that other males, including other female impersonators, often try to mate with the mimic.) Norman's team observed more than 20 examples of sexual mimicry, but they were unable to tell how often mimics slipped past a guard, approached a female, mated, and successfully fertilized eggs.

Hanlon, Naud, and their colleagues began to answer these questions. They videotaped sexual encounters on the reef, but they also used DNA analysis to determine which male fathered each female's next laid egg. They found that impersonators deceived a guarding male and approached a female about 30 times out of 62 attempts. Five of the 30 mimics tried to mate with the female. The guard male interrupted one attempt, and the female rejected another, but three were successful. DNA analysis showed that two of these three female impersonators fathered the first egg laid after mating. The third did not. Because females

store sperm, however, the team could not rule out the possibility that sperm from the third mimic fertilized a subsequent egg.

Sexual mimicry illustrates an important point: in many species, there is more than one path to reproductive success. In polygamous species such as the cuttlefish, natural selection favors the male with the most successful sperm—and he may not be the largest or the strongest individual. True, the brawniest males can fight off rivals and preserve their own access to a female, but the fierce competition and constant distractions often leave females with additional mating opportunities. A small male has little chance of winning a head-to-head competition with a larger rival, but he does have another tool—deception. Thanks to his changeable skin, even a small cuttlefish has an excellent shot at winning in the great raffle of life.

Hanlon, Roger T., Marié-Jose Naud, Paul W. Shaw, and Jonathan N. Havenhand. 2005. Transient sexual mimicry leads to fertilization. *Nature*, vol. 433, page 212.

Naud, Marié-Jose, Roger T. Hanlon, Karina C. Hall, et al. 2004. Behavioural and genetic assessment of reproductive success in a spawning aggregation of the Australian giant cuttlefish, *Sepia apama*. *Animal Behaviour*, vol. 67, pages 1043–1050.

Norman, Mark D., Julian Finn, and Tom Tregenza. 1999. Female impersonation as an alternative reproductive strategy in giant cuttlefish. *Proceedings of the Royal Society of London B.*, vol. 266, pages 1347–1349.

34.7 | Mastering Concepts

1. List two questions that Hanlon and Naud investigated.
2. How would you test the hypothesis that the offspring of a mate-guarding male is more likely to guard his mate than the offspring of a "sneaker" male? What does this hypothesis assume about the inheritance of mate guarding?

Chapter Summary

34.1 | Animal Development Begins with Reproduction

A. Reproduction Is Asexual and Sexual
- **Asexual reproduction** does not require a partner and yields identical offspring.
- In **sexual reproduction,** two haploid **gametes** unite and form a **zygote,** the first cell of a new offspring. **Fertilization** may occur in the environment (**external fertilization**) or inside an animal's body (**internal fertilization**).

B. Gene Expression Dictates Animal Development
- Development requires **differentiation,** the formation of specialized cells. In **pattern formation,** the animal takes on its overall shape and structure.

C. Development Is Indirect or Direct
- Animals that undergo **indirect development** have a **larva** stage that does not resemble the adult. **Metamorphosis** transforms the larva into an adult.
- In **direct development,** young animals resemble miniature adults.

34.2 | Males Produce Sperm Cells
- The **reproductive systems** of both males and females include **gonads,** which house the **germ cells** that give rise to gametes. Other sex organs deliver or nurture the gametes. Males and females have different **secondary sex characteristics,** traits that do not directly participate in reproduction.

A. Male Reproductive Organs Are Inside and Outside the Body
- Developing sperm originate in **seminiferous tubules** within the paired **testes.** Also inside the testes are **sustentacular cells** that surround the seminiferous tubules and **interstitial cells** that secrete hormones. A pouch called the **scrotum** contains the testes.
- Sperm travel through the **epididymis, vas deferens,** and **ejaculatory duct,** and they exit the body with **semen** through the **urethra** (within the **penis**) during **orgasm** and **ejaculation.**
- The **prostate gland, seminal vesicles,** and **bulbourethral glands** add secretions to semen.

B. Spermatogenesis Yields Sperm Cells
- **Spermatogenesis** begins with diploid **spermatogonia,** which divide mitotically to yield a stem cell and a **primary spermatocyte.** The first meiotic division produces two haploid **secondary spermatocytes.** In meiosis II, the secondary spermatocytes divide, yielding four **spermatids.** The spermatids develop into **spermatozoa,** or mature sperm cells.
- A mature sperm cell has a flagellum and a head covered with a caplike **acrosome.**

C. Hormones Influence Male Reproductive Function
- **Gonadotropin-releasing hormone (GnRH)** from the hypothalamus stimulates the anterior pituitary gland to release **follicle-stimulating hormone (FSH)** and **luteinizing hormone (LH).** In males, these hormones affect the testes, triggering the release of **testosterone** necessary for sperm formation and the development of secondary sex characteristics.

34.3 | Females Produce Egg Cells

A. Female Reproductive Organs Are Inside the Body
- Egg cells (and their nourishing **follicle cells**) originate in the **ovaries.** Each month after puberty, one ovary releases an egg cell into a **uterine tube,** which leads to the **uterus.** A blood-rich **endometrium** lines the inside of the uterus. The **cervix** leads to the **vagina,** which connects the uterus with the outside of the body.
- The **vulva,** or external genitalia, consists of the labia, **clitoris,** and vaginal opening. **Breasts** deliver milk to infants.

B. Oogenesis Yields Egg Cells
- In **oogenesis, oogonia** divide mitotically to form two **primary oocytes.** In meiosis I, the cytoplasm of the primary oocyte divides unevenly as it splits into one large, haploid **secondary oocyte** and a much smaller **polar body.** In meiosis II, the secondary oocyte again divides unequally, yielding the large ovum and another small polar body.
- Meiosis in the female begins before birth and completes at fertilization.

C. Hormones Influence Female Reproductive Function
- The ovaries secrete **estrogen** and **progesterone,** hormones that stimulate development of female sexual characteristics. GnRH, FSH, LH, estrogen, and progesterone control the **ovarian cycle** and the **menstrual cycle.**
- After **ovulation,** the ruptured follicle develops into the **corpus luteum,** which secretes hormones that prepare the body for pregnancy. If pregnancy does not occur, the endometrium is shed in menstrual flow.

D. Hormonal Fluctuations Can Cause Discomfort
- Premenstrual syndrome, menstrual cramps, endometriosis, and hot flashes are associated with changing hormone concentrations.

E. Contraceptives Prevent Pregnancy
- **Contraception** is the use of behaviors, barriers, hormones, spermicidal chemicals, or other devices that prevent pregnancy.

34.4 | Sexual Activity May Transmit Disease
- A **sexually transmitted disease (STD)** spreads via sexual contact. Viruses and bacteria cause most STDs, which may cause infertility.

34.5 | The Human Infant Begins Life as a Zygote

A. Fertilization Joins Genetic Packages and Initiates Pregnancy
- In fertilization, a sperm cell burrows through the two layers surrounding a secondary oocyte. The two united cells constitute the diploid zygote.

B. Preembryonic Events Include Cleavage, Implantation, and Gastrulation
- The **preembryonic stage** lasts from fertilization until the end of the second week of development.
- After fertilization, **cleavage** divisions produce the **morula.** Between days 3 and 6, the morula arrives at the uterus and hollows out, forming a **blastocyst.** An outer layer of cells (the trophoblast) and an **inner cell mass** form. Cells of the trophoblast secrete **human chorionic gonadotropin (hCG),** which prevents menstruation. **Implantation** occurs between days 6 and 14.
- During the second week, the amniotic cavity forms as the inner cell mass flattens, forming the **embryonic disc.** Ectoderm and endoderm form, and then **mesoderm** appears, establishing the three primary germ layers of the **gastrula.**

C. Organs Take Shape During the Embryonic Stage
- The **embryonic stage** lasts from the second week through the eighth week of development.
- **Chorionic villi** extending from the **chorion** start to develop into the **placenta.** The **yolk sac, allantois,** and **umbilical cord** form as the **amnion** swells with fluid.
- Organs form throughout the embryonic period. The **primitive streak** forms the longitudinal axis around which new structures appear, including the **notochord, neural tube,** arm and leg buds, heart, facial structures, skin specializations, sex organs, and skeleton.

D. Organ Systems Become Functional in the Fetal Stage
- Structures continue to elaborate during the **fetal stage,** which lasts from the ninth week of development until the baby is ready for birth.

E. Muscle Contractions in the Uterus Drive Labor and Childbirth
- Labor begins as the fetus presses against the cervix. In a **positive feedback** loop, cervical stretching triggers the release of oxytocin, which stimulates uterine contractions that push the baby farther out, causing even more stretching. Eventually the mother expels baby and placenta.

34.6 | Birth Defects Have Many Causes

- Genetic abnormalities, dietary deficiency, or exposure to **teratogens** such as chemicals or viruses can cause **birth defects.**
- Each body structure has a different critical period, which is the time in development during which it is especially vulnerable to teratogens.

34.7 | Investigating Life: The "Cross-Dressers" of the Reef

- In many animal species, females store sperm from multiple males. Natural selection favors the males that produce the most sperm or that prevent other males from fertilizing the female's eggs.
- In Australian giant cuttlefish *(Sepia apama)*, some males disguise themselves as females, increasing their opportunities to mate.

Multiple Choice Questions

1. What is a zygote?
   a. The male gamete
   b. The product of fertilization
   c. The female germ cell
   d. The female gamete

2. What is the relationship between a primary spermatocyte and a spermatogonium?
   a. They are genetically identical.
   b. The spermatocyte is haploid and the spermatogonium is diploid.
   c. The spermatocyte has a flagellum; the spermatogonium does not.
   d. Both b and c are correct.

3. In both males and females, the main function of GnRH is to
   a. inhibit the production of androgens.
   b. stimulate the release of LH and FSH.
   c. inhibit the development of the corpus luteum.
   d. stimulate the development of the trophoblast.

4. Why can't a fertilized polar body develop into a fetus?
   a. Because it has too few chromosomes
   b. Because it has too many chromosomes
   c. Because it lacks cytoplasm and organelles
   d. Because it carries excess flagella

5. Which of the following STDs can be treated with an antiviral drug?
   a. Genital herpes   c. Chlamydia
   b. Trichomoniasis   d. All of the above are correct.

6. The acrosome reaction allows a sperm cell to penetrate the
   a. proteins and carbohydrates encasing the oocyte.
   b. follicle cells surrounding the oocyte.
   c. plasma membrane of the secondary oocyte.
   d. Both a and b are correct.

7. How does the size of the blastocyst compare to that of the morula and zygote?
   a. Larger than the morula but smaller than the zygote
   b. Larger than both the morula and the zygote
   c. Smaller than the morula but larger than the zygote
   d. Smaller than both the morula and the zygote

8. The umbilical cord's function is to:
   a. transport nutrients to the fetal digestive system.
   b. transfer nutrients and oxygen from maternal blood to fetal blood.
   c. carry the blood of the mother to the developing fetus.
   d. transfer nutrients from the yolk to the developing fetus.

9. The hormones that determine the development of male or female reproductive structures first become active during
   a. fertilization.   c. the embryonic stage.
   b. the preembryonic stage.   d. the fetal stage.

10. A chemical that can cause a birth defect is a
    a. toxin.   c. mutagen.
    b. trisomy.   d. teratogen.

Write It Out

1. What are the advantages and disadvantages of asexual and sexual reproduction? Of internal and external fertilization?

2. How do protein gradients cause cells to differentiate?

3. How are the human male and female reproductive tracts similar, and how are they different? How are the structures of the testis and ovary similar and different?

4. Make a chart that lists the major hormones involved in human reproduction. Complete the chart by listing the organ that releases each hormone; the location of target cells; the function of each; and whether each is most important in males, females, or both.

5. How are the timetables different for oogenesis and spermatogenesis in humans?

6. Point mutations usually occur during interphase of mitosis, but most chromosomal abnormalities arise during meiosis. Given the differences between gamete production in males and females, why is it reasonable to predict that more point mutations occur during sperm production and more chromosomal abnormalities appear in egg cells?

7. Is each of the following cell types haploid or diploid? How does each cell type relate to the others?
   a. An oogonium
   b. A primary spermatocyte
   c. A spermatid
   d. A secondary oocyte
   e. A polar body derived from a primary oocyte

8. How do the structures of the male and female human gametes aid them in performing their functions?

9. A vasectomy is a surgical procedure in which a physician cuts the vas deferens leading away from each testis. Does a vasectomy stop sperm production entirely, block testosterone production, block sperm delivery to the ejaculate, or prevent ejaculation? Explain your answer.

10. Use the Internet to learn more about sexually transmitted diseases. Choose one to study in detail. What type of infectious agent causes the disease? What are the symptoms and long-term consequences of infection? Is a treatment available? Who is most affected?

11. What would happen if two sperm fertilized the same egg cell? If two sperm fertilized two egg cells?

12. How does menstruation stop when a woman becomes pregnant?

13. This chapter used the term *villus* in describing part of the chorion; the same term appeared in chapter 31's description of the small intestine. How are the chorionic and intestinal villi similar and different in structure and function?

14. Arrange these structures in order from youngest to oldest: morula, gastrula, zygote, fetus, blastocyst.

15. Fetal red blood cells contain a version of hemoglobin (an oxygen-binding protein) that is slightly different from that in adult red blood cells. Given the relationship between fetus and mother, which hemoglobin do you predict should have a higher affinity for oxygen? Explain your answer.

16. Consult a website that describes and illustrates fetal development. What technology do you think would be necessary to enable a fetus born in the fourth month to survive in a laboratory setting?

17. What are the events of childbirth?

18. List some of the causes of birth defects. What is the significance of the critical period in determining the type and severity of a birth defect?

19. What kinds of studies and information would be necessary to determine whether exposure to a potential teratogen during a war can cause birth defects a year later? How would such an analysis differ if it were a man or a woman who was exposed?

20. Chapter 27's opening essay describes examples of endocrine disruptors. What are endocrine disruptors? Would these chemicals be considered teratogens? Why or why not?

21. One risk factor for developing breast cancer or ovarian cancer is a family history of either disease. Researchers have identified some alleles associated with inherited breast and ovarian cancers, making genetic tests possible. If a female close to you had a family history of either cancer, what considerations would help her decide whether to have her DNA tested for these alleles?

22. Use this textbook or the Internet to learn more about human cloning and stem cell therapies. How does each topic relate to human reproduction and development? Why are these techniques controversial?

Pull It Together

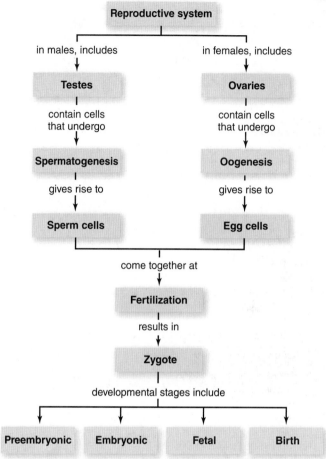

1. How are spermatogenesis and oogenesis similar, and how are they different?
2. Which hormones influence reproductive function in males and females?
3. What are the events of fertilization?
4. What events occur during each of the main developmental stages?
5. Add the terms *placenta, follicle, polar body, gonad, gamete, meiosis,* and *mitosis* to this concept map.

Chapter 35

Animal Behavior

A Stotting Thomson's Gazelle.

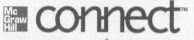

Enhance your study of
this chapter with practice quizzes,
animations and videos, answer keys,
and downloadable study tools.
www.mhhe.com/hoefnagels

Risky Business

IT IS A HOT MORNING IN THE AFRICAN BUSH, AND A HERD OF THOMSON'S GAZELLES GRAZES ON SHOOTS OF NEW GRASS. The gazelles are alert as they feed. Their ears swivel and their tails twitch away flies. Every now and then, one pauses, looks up, and scans the horizon. Death crouches in the tall grasses in the form of a hungry cheetah. The wind shifts, and one gazelle jerks up its head. Then it does something that looks awfully odd. Instead of running away, the gazelle leaps into the air. Its nose points to the ground as it bounces up and down on stiff legs. In a flash it races away, but the odd bouncing, called stotting, seems to be contagious. As they race away, other gazelles also stot, sometimes in midstride. The cheetah charges, but it is a half-hearted attempt. The spotted cat stops short and watches the herd disappear.

Mobbing is another risky animal behavior, and you can see it closer to home. Terns (a type of gull) will mob any animal that intrudes into their nesting colony, from cats to foxes to humans. When a peregrine falcon sails over a colony of terns, for example, the birds fly up from their nests and join the attack. Screaming, they dive-bomb the falcon. This truly is risky business. The peregrine is such an agile flier that it is perfectly capable of rolling over on its back in midair and striking at a mobber or grabbing it in lethal talons.

Other birds also mob predators. If a crow spies a great horned owl or red-tailed hawk, it will send a loud alarm call. The sound draws other crows and often smaller birds that gather where the predator is perched. Sometimes a huge mixed flock is recruited. Giving their different alarm calls, the birds settle on tree branches near the predator. Some fly straight at it, veering away only at the last second. Eventually, the predator flies away, out of the nesting territories of the birds.

Stotting and mobbing are spectacular behaviors, but if you watch any animal long enough, something interesting is bound to happen. It might spin a web, curl into a ball, start barking, groom itself, hide, jump into the air, yawn, or swish its tail. Even the most mundane behavior can prompt interesting questions: Why does the animal do that? How does it do that? Was it born knowing how to perform the behavior, or did it learn by imitating other animals?

In pondering such questions, the mind may wander to some of the strange things we do, like kissing, playing basketball, or whistling. How did those behaviors ever start, and why do they persist? The answers to many such questions remain elusive, but this chapter explores some of what we do know about a wide spectrum of animal behaviors, both simple and complex.

UNIT 7

What's the Point?

Learning Outline

Learn How to Learn
What's Your Learning Style?

Students differ in how they prefer to receive information: some love to hear lectures, others like reading, and still others thrive on hands-on activities in lab. You may find it helpful to use an online learning styles inventory to discover more about your own preferences. At the very least, the advice on the website may alert you to study techniques that you have not tried before.

35.1 | Animal Behaviors Have Proximate and Ultimate Causes

Animal behavior is an ideal topic to bridge the animal physiology and ecology units of this textbook. An animal's behavior represents the integration of its muscular, nervous, endocrine, reproductive, and other physiological systems. At the same time, the animal interacts with other members of its own species and with the plants and other organisms in its habitat. These interactions form part of the subject of ecology.

The study of animal behavior has evolved throughout human history. To early humans, the behaviors of animals must have been familiar. After all, a hunter's knowledge of an animal's habits might mean the difference between survival and starvation, and dangerous animals might lurk around the bend of any trail. Nowadays, few people have firsthand experience with wild animals, although many own pets (see this chapter's Apply It Now box).

As popular knowledge of animal behavior has moved to the periphery, the science of animal behavior has developed and blossomed into several related disciplines, including behavioral ecology and evolutionary psychology. In the 1930s, naturalists Niko Tinbergen, Karl von Frisch, and Konrad Lorenz introduced the term *ethology* for the scientific study of animal behavior, especially in the natural environment. An ethologist must understand what animals do, beginning with firsthand observations that do not interfere with the animals. The emphasis in ethological studies, however, is on experiments. Karl von Frisch, for example, used experiments to help decode the meaning of the honeybees' famous "waggle dance." Tinbergen, von Frisch, and Lorenz received the Nobel Prize in Physiology or Medicine in 1973 for their research.

Animal behavior encompasses many fields of research. If an observer of the stotting gazelle in the chapter opening photo were to ask several biologists what the animal is doing and why, he or she might receive very different answers. A functional morphologist might point to the muscles, tendons, and ligaments involved in the springing motor pattern that enables stotting to occur. A physiologist might explain how the liver mobilizes glucose to provide a burst of energy. A neurobiologist might consider the actions of the motor neurons and the role of the central nervous system. All of these explanations are directed at **proximate** causes, or the "how" mechanisms that bring about the behavior.

On the other hand, an evolutionary biologist or behavioral ecologist would explain how the behavior contributes to an animal's fitness and how it might have evolved by natural selection. These latter explanations are directed at **ultimate** causes, which explore the evolutionary basis ("why") of a behavior; that is, how the behavior promotes the animal's survival or reproductive success. Biologists have proposed at least 11 hypotheses to explain stotting; for example, perhaps the stotter is signaling fellow members of the herd that danger is present or indicating to the predator that the stotter is healthy and can outrun the predator. Section 35.3 addresses these hypotheses further. ▶ fitness, p. 239

Studying animal behavior is complicated by the fact that animals do not necessarily share our perception of the world. If you have ever seen dogs respond to a dog whistle, it becomes clear that some animals live in a different sensory world from ours; dogs can detect much higher sound frequencies than we can. Other animals have special senses as well. Rattlesnakes, for example, have infrared heat detectors that can detect prey in total darkness. Sharks detect the electromagnetic fields emitted by fish; bees see ultraviolet light; and bats use ultrasonic "sonar" to find food. Researchers must take these sensory differences into account when trying to understand how and why each animal behaves as it does.

This chapter begins with approaches at the proximate level of causation—descriptions of innate and learned behavior patterns—followed by the interplay of genes and environment in behavior. The chapter then leads into ultimate causation, focusing on the role of behavior in survival and reproduction. Although we treat these topics separately, a recent trend in animal behavior research is to combine both levels of causation in the same study. Teams including geneticists, physiologists, ecologists, and evolutionary biologists can work together to arrive at a much more complete understanding of the causes of a behavior, from its control at the cellular and molecular level to its evolutionary history.

35.1 | Mastering Concepts

1. What is ethology?
2. Distinguish between the proximate and ultimate causes of an animal's behavior.

Apply It Now

Puppy Love

If you own—and love—a pet, you may wonder how the animal's experience of your relationship compares with your own. The cat may jump in your lap, curl up, and begin to purr. The dog may greet you exuberantly when you come home. Do these behaviors mean that your pet loves you?

However tempting it is to project love onto our animal companions, it is unscientific to do so. Scientists have no way to measure objectively the love that humans feel, let alone the emotions that other animals may have. Your cat may seem to enjoy her time in your lap, but is she simply seeking a soft, warm, comfortable place to rest? Your dog may seem happy to see you, but perhaps he is simply behaving in ways that he has learned will earn attention or treats.

This question of animal emotions illustrates the sometimes frustrating reality that science cannot answer every question that interests us. Our inability to peer into the animal mind adds mystery to our relationships with our nonhuman companions. Perhaps this mystery helps make pet ownership one of the world's most popular hobbies.

35.2 | Animal Behaviors Combine Innate and Learned Components

Biologists traditionally classify animal behaviors based on the way they are acquired and how changeable they are. All animals are born with the ability to perform some behaviors; other must be learned.

A. Innate Behaviors Do Not Require Experience

The ethological approach to animal behavior has stressed the importance of **innate,** or instinctive, behavior patterns. An animal can perform innate behaviors at birth; learning is not involved. These behaviors tend to be stereotypic; that is, they are performed in a predictable way each time.

Two simple examples of innate behaviors are reflexes and taxes (singular: taxis). A **reflex** is an instantaneous, automatic response to a stimulus. If you touch a hot pan on the stove, you quickly withdraw your hand without even thinking about it. In fact, the reflex circuitry is located in the spinal cord and does not involve the brain at all (see figure 25.13); only the sensation of pain, which comes later, requires the brain. Another innate behavior is a **taxis,** a movement toward or away from a stimulus. An earthworm, for example, tends to crawl toward the high humidity of moist soil, and a cockroach scurries away from light. ▶ reflex arc, p. 541

A more complex behavior sequence is a **fixed action pattern (FAP),** a motor response that is initiated by an environmental stimulus and that continues to completion, even if the stimulus is withdrawn. For example, Tinbergen studied fish called three-spined sticklebacks. In this species, breeding males have bright red undersides and defend territories in which females lay eggs. The sight of an intruding male elicits an aggressive response consisting of lunges and bites.

Tinbergen focused on the specific stimuli that triggered the aggressive threat display (figure 35.1). By presenting male sticklebacks with a series of models, Tinbergen determined that many shapes would elicit the threat display, as long as the object was red on the bottom. Accurate clay models of sticklebacks without the color red, however, were not effective. In male sticklebacks, red serves as the sign stimulus that releases the threatening FAP. Apparently, a male stickleback is more or less

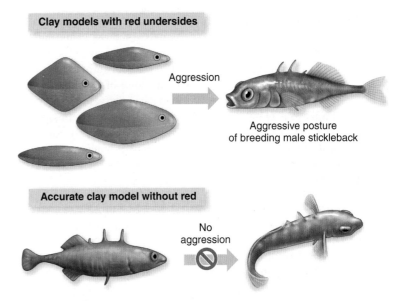

Figure 35.1 Color Matters. Male sticklebacks will act aggressively toward clay models that look nothing like fish, as long as the undersides are red. They will not, however, display aggression toward a realistic model of a fish lacking a red belly.

"programmed" to threaten any intruding breeding male. This behavior not only helps ensure that he is the one that fertilizes eggs laid in his territory but also helps protect the eggs from cannibalism.

Behavioral geneticists have delved into the proximate causes of some FAPs. A favorite model species is the fruit fly, *Drosophila melanogaster.* One fruit fly gene with powerful behavioral effects is named fruitless (*fru,* for short). Normal *fru* function is required for proper development of the motor neurons that innervate muscles involved in courtship. Mutations in *fru* disrupt the complex, six-part FAP that leads to copulation (figure 35.2). One mutant form of *fru,* for example, causes males to court other males, moving in conga lines several flies long.

B. Learning Requires Experience

In **learning,** an animal alters its behaviors as an outcome of its experiences. Psychologists use laboratory animals to conduct much of the fundamental research in this area. This section describes several types of learning, emphasizing the roles each might play in survival.

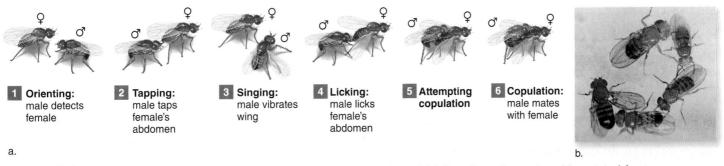

a.   b.

1 Orienting: male detects female
2 Tapping: male taps female's abdomen
3 Singing: male vibrates wing
4 Licking: male licks female's abdomen
5 Attempting copulation
6 Copulation: male mates with female

Figure 35.2 Fruit Fly Courtship. (a) Sequence of normal courtship behaviors. (b) A line of courting males with mutated *fru* genes.

Habituation

Habituation occurs when an animal learns *not* to respond to a stimulus. Have you ever noticed how you tune out the familiar sounds, sights, and smells of your own home? A behaviorist would say that you are habituated to your surroundings. A city apartment dweller hardly hears the noise of traffic in the streets, but he may not be able to sleep in the country: the sounds of nocturnal insects will keep him awake. Likewise, pigeons or squirrels in city parks are habituated to people, especially compared with their counterparts in habitats where humans rarely venture. Habituation allows animals to ignore normal stimuli while remaining alert to the out-of-the-ordinary.

Associative Learning

In **associative learning,** an animal learns the relationship between two events. Two types of associative learning are classical conditioning and operant conditioning.

Classical conditioning occurs when a behavior is modified ("conditioned") by the pairing of two stimuli. In the famous early experiments of Ivan Pavlov, a dog presented with meat powder would invariably salivate. If Pavlov repeatedly accompanied the meat stimulus with the sound of a bell, the dog would soon salivate upon hearing the bell, even with no meat present. The dog formed an association between the two stimuli. Classical conditioning plays an important role in the real world as well. For example, an animal may become sick shortly after eating a noxious food. If the animal associates the two events, it may avoid that food in the future.

Another type of associative learning is **operant conditioning,** in which an animal learns to associate a behavior with its consequences. For example, a bird may learn that pressing a bar or pecking a key results in a reward or punishment (called a reinforcer). This type of learning is involved in most animal training. For example, a dog learns to associate a command ("Sit!") with the behavioral response that earns a food reward. Operant conditioning also undoubtedly plays a role in the lives of animals in the wild.

Imprinting

Imprinting is a kind of rapid learning that occurs during a restricted time early in an animal's life without obvious reinforcement. The animal retains the memory throughout life. In the 1930s, Lorenz found that during a "sensitive period" of their lives, goslings would imprint on the first moving object that they encountered. In one experiment, goslings raised in an incubator accepted Lorenz as "mother" and followed him wherever he went (**figure 35.3a**). Animals that have not been properly imprinted, however, may have trouble recognizing their own species when it comes time to mate. Biologists who raise endangered species in captivity therefore may use hand puppets resembling parent birds to feed the young (**figure 35.3b**).

Observational Learning

Some animals can learn by **observational learning,** in which they watch what others do and then imitate the behavior. Observational learning saves the time, energy, and risk involved in trial-and-error learning. Sometimes a learned behavior pattern spreads through a group and persists as a tradition, as when one young female Japanese snow monkey began washing sweet potatoes in the ocean. This practice removed sand from the food and perhaps added salt flavoring. Although older

a.

b.

Figure 35.3 Imprinting. (a) Konrad Lorenz and his flock of greylag geese. (b) A hand puppet modeled on an adult condor is used to feed an endangered condor chick.

monkeys were slow to adopt the washing behavior, the younger animals apparently learned by observing their companions. Over several years, the practice spread throughout the entire group.

Animal Cognition

Some animals, especially many mammals and birds, have cognitive abilities that extend to reasoning, problem solving, tool use, and symbolic communication. Much of the research in this area suggests that animals use a combination of operant conditioning, observational learning, and insight to develop these abilities.

Although tool use was once thought to be a hallmark of humans, nonhuman animals also learn to use objects as aids in obtaining food in the wild. Tool-using animals include woodpecker finches and crows that use sticks to probe for insects, sea otters that use rocks to open shellfish, and chimpanzees that modify objects to collect termites. In early lab experiments, chimpanzees even figured out how to stack boxes on top of one another to reach bananas suspended high above.

A few types of animals can learn to use words or symbols to communicate. The great apes cannot make the sounds of human speech, but they can communicate using symbols. One of the early students was a chimp named Washoe that learned about 250 words in American Sign Language. Washoe could string words together into phrases and even invented new combinations of signs. Other chimps that were trained on computer keyboards learned to communicate with one another about such matters as the location of hidden food.

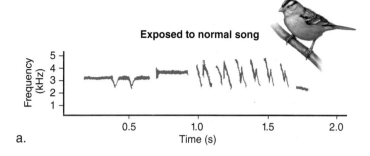

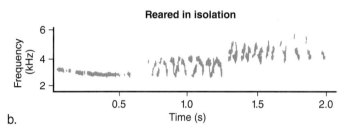

b.

Figure 35.5 Song Learning. (a) If a male white-crowned sparrow hears its own species's song while young, the bird performs the song correctly. (b) If the bird is reared in isolation without ever hearing its species's song, its own song is abnormal.

Figure 35.4 Talkative Bird. Scientist Irene Pepperberg used Alex, an African grey parrot, in studies of cognition. Alex could name more than 100 objects, recognized the concepts of "same" and "different," and could count. Alex was more than 30 years old when he died in 2007.

The use of words is not limited to primates; an African grey parrot named Alex learned to communicate by speech. Alex had an English vocabulary of more than 150 words, could count, and could answer questions about an object's shape, color, and material (figure 35.4). It's not clear that animals actually use such abilities in the wild, however, and researchers disagree about whether nonhuman animals can learn true language.

C. Genes and Environment Interact to Determine Behavior

Biologists once placed innate and learned behavior patterns at opposite ends of a spectrum (formerly called "nature versus nurture"). Genes were assumed to control innate behavior, whereas experience in the environment governed learning. We now know that reality is considerably more complicated: innate behavior patterns do not occur without interaction with the environment, and learning has a strong genetic component. ▶ gay gene? p. 137

Animal behavior therefore does not follow a strict "innate or learned" dichotomy. Every animal begins life with a range of possible behavior patterns, some of which have more genetic input than others. The social, biological, and physical context, plus the experiences that an animal acquires as it matures, help determine how it actually behaves at any time.

The interplay of genes and environment is well illustrated by studies of the development of bird song. White-crowned sparrows are widespread across North America and have been a favorite subject of animal behaviorists. Under natural conditions, a young male hears the songs of his father and other neighbors during his first year. A species-typical song is crucial to a male's ability to defend a territory and attract a mate. During the first few months of life, the young male instinctively begins to sing. At first, he produces a "subsong" consisting of variable notes that sound little like the typical full song. During the year, he adds elements to the song that are more like those used by adults, so that by the first breeding season, his song has crystallized to sound much like his father's.

Unlike a wild bird, a white-crowned sparrow reared in isolation in the laboratory develops an abnormal song (figure 35.5). If it hears taped or live adult male songs between 10 and 50 days after hatching, however, it develops a normal song at breeding time, demonstrating a sensitive period for song learning. When hearing the song of a different species instead, such as that of a song sparrow, the young birds end up no better than those that heard nothing. Therefore, some sort of genetic template seems to guide the bird to learn the correct song, but only if exposed to it at a certain time in development. Moreover, birds that were deafened after hearing normal song during the sensitive period sang abnormally as adults, indicating that they need to hear themselves sing to produce normal song.

Research on another bird species, the zebra finch, has revealed some of the neural substrates of song learning. In these birds, exposure to song activates a gene called *zenk,* but *only* if it is zebra finch song. Several areas in the brain serve as song centers. One such region adds new neurons as a song is being learned. Thus, as we discover more about how a behavior pattern actually emerges, the traditional dichotomy of "innate versus learned" seems like a gross oversimplification.

35.2 | Mastering Concepts

1. What is the traditional distinction between innate and learned behaviors?
2. Describe three examples of innate animal behaviors.
3. Explain the connection between the *fru* gene and courtship behavior in *Drosophila*.
4. Name four types of learning.
5. What are some examples of animal cognition?
6. How does song learning in the white-crowned sparrows illustrate the relationship between genes and experience in determining an animal's behavior?

35.3 | Many Behaviors Improve Survival

Behavioral ecology is a relatively recent approach to animal behavior that explores the survival value and evolution of behavior patterns. Most behavior can be linked to adaptations that increase survival, reproduction, or both (see this chapter's Burning Question). This section explores survival: how animals find their way around their habitats, how they locate food, and how they avoid becoming food for another animal. Section 35.4 then takes up behaviors that promote reproductive success.

A. Some Animals Can Find Specific Locations

In many animal species, members of one sex disperse from the place of birth and move to a new habitat. Once there, they need to be able to find their way to resources such as food and mates. Many animals also return regularly to a home refuge.

Consider the desert ant, *Cataglyphis fortis,* that feeds on dead arthropods in the barren salt pans of the Sahara desert in Tunisia. For their size, these ants travel some of the longest distances of any animal. They follow a meandering path as they search hundreds of meters from their nest, but once they find a prey item, they take a direct route back home (**figure 35.6**). Likewise, many species of birds migrate thousands of miles each year, returning to the same nesting location where they bred the previous year. How do animals perform such feats? Researchers have identified several types of cues.

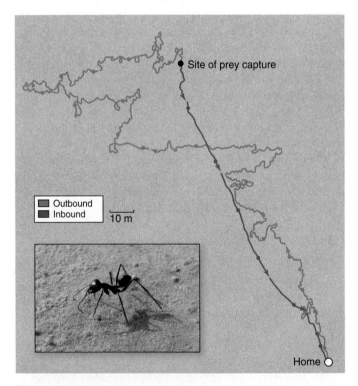

Figure 35.6 Shortcut. A foraging desert ant takes a meandering path as it leaves its nest in search of food but takes a much more direct route upon returning home.

Piloting
In a process called piloting, an animal uses distant objects as landmarks when finding its way, just as you might use a distant mountain as a landmark when on a hike. Landmarks are also useful for locating objects in familiar areas close to home, as when a squirrel retrieves a food cache. Ethologist Niko Tinbergen showed experimentally that beewolf wasps use nearby landmarks to find their nest hole when returning from a foraging trip. Homing pigeons also use landmarks when in the vicinity of their loft. Piloting requires a sort of "mental map" of the terrain; one limitation of this approach is that it does not work in unfamiliar terrain.

Sun, Star, or Magnetic Orientation
Some animals can orient themselves to the landscape by maintaining a constant angle to a celestial object, such as the sun. This strategy may also require an internal clock if the object's angle changes during the day. Thus, the sun's position is one of several clues that desert ants can use to take a straight path home. A desert ant that is leaving its nest also keeps track of its turns, accelerations, decelerations, and number of steps taken in each direction. The ant integrates this information as an independent means of monitoring its whereabouts. Once the animal is near its home, it uses visual and olfactory landmarks to find the exact nest location.

Night migrants use the moon or the star patterns in the night sky to maintain their heading during their nocturnal flights. Indigo buntings judge their location by referring to the position of the north star, which always appears stationary in the sky. A young bunting does not develop this ability, however, unless it is exposed to the northern sky before beginning its migratory flight in autumn.

Many animals can detect magnetic fields; experimentally shifting the field produces corresponding shifts in homing. For example, African mole rats *(Cryptomys hottentotus)* are rodents that live in underground colonies and are nearly blind. They dig the longest tunnels of any mammal, more than 200 meters, and they typically orient their tunnels in a southerly direction. Researchers confirmed that the mole rats detect magnetic cues by allowing the animals to build nests in a light-proof arena surrounded by electric coils that shifted the magnetic field. The mole rats altered the orientation of their nests in proportion to the shift in the magnetic field (**figure 35.7**).

True Navigation
Navigation requires that an animal in unfamiliar terrain can understand its current position and make its way to a destination *without* using information that it collected on its way to its current position or relying on any stimulus that comes from the destination. The animal must therefore use other information to determine its latitude and longitude. Only a few animal species, including the homing pigeon, sea turtle, and spiny lobster, are known to use true navigation; researchers dispute the mechanisms these animals use.

B. Animals Balance the Energy Content and Costs of Acquiring Food

Animals spend much of their time **foraging,** or searching for and collecting food. Their foraging techniques vary greatly. Some animals, such as the desert ant, wander in search of dead insects. Spiders spin webs that ensnare prey. Packs of African wild dogs work

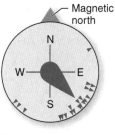  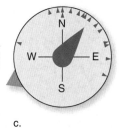

Positions of the nests in the local magnetic field

Positions of the nests with magnetic north turned to geographic S

Positions of the nests with magnetic north turned to geographic WSW

Magnetic north

a.   b.   c.

Figure 35.7 Mole Rats Detect Magnetic Fields. In each of these compasses, the green triangle indicates magnetic north, and the blue "pointer" represents the average nest position. (a) Positions of nests in the local (unmanipulated) magnetic field. (b) Nest positions when researchers reversed the magnetic field and (c) shifted magnetic north toward the southwest.

Lure

Figure 35.8 Deception. The angler fish's lure is a modified spine that resembles a tiny fish.

cooperatively to hunt large, dangerous prey such as zebras. Some species of fish have modified fin spines that they use to entice prey; the angler fish attracts passing animals by waving a lure that closely resembles a small fish (figure 35.8). And, as we have already seen, a variety of bird and mammal species use tools to collect food.

For many animals, learning plays an important role in finding food. They may form a **search image,** which reflects an improvement in a predator's ability to detect inconspicuous prey. In laboratory studies, blue jays were trained to peck a key for a reward when shown a slide of a moth hidden on a tree trunk; they received no reward if they pecked when no moth was present. The blue jays

Burning Question

Do lemmings really commit mass suicide?

The 1958 Disney documentary *White Wilderness* purports to show the "mass suicide" behavior that makes lemmings famous. Masses of lemmings scurry toward the sea, while the narrator describes the scene. "They've become victims of an obsession, a one-track thought: Move on, move on, keep moving on," he says. Then, as the rodents approach the cliff's edge, the narrator intones, "This is the last chance to turn back. Yet over they go, casting themselves bodily into space." Dozens of lemmings hurtle over the cliff, splash into the water, swim a short distance, and drown.

This evocative imagery solidified the popular notion that lemmings commit mass suicide when their populations grow too large. In reality, however, the mass suicide of lemmings is a myth. These rodents have populations that fluctuate wildly, peaking every 3 to 4 years. As the population reaches its height, lemmings begin to disperse to new habitats. Sometimes, an obstacle such as a lake or river blocks their path. Lemmings can swim, but if the distance is too great, they will drown before reaching the shore.

From an evolutionary viewpoint, mass suicide makes no sense. How can a population inherit the desire to kill itself? Even if such a trait were to arise in some lemmings, the ones lacking the self-destructive desire would presumably have the greatest reproductive success. Over many generations, a genetic tendency toward mass suicide should gradually disappear.

And what of the *White Wilderness* footage? Researchers who investigated the history of the film discovered that the entire lemming sequence was faked. The filmmakers purchased the lemmings from schoolchildren in Manitoba and transported the animals to landlocked Alberta, Canada. The crew then forced the lemmings into the water to create footage of the supposed mass suicide.

Submit your burning question to:
marielle_hoefnagels@mcgraw-hill.com

greatly improved their performance on this task over time, but not if more than one species of moth was included in the trials. This result is consistent with the formation of a search image.

Food contains the energy that an animal needs to live, but finding, acquiring, and handling food costs energy. Over the long term, an animal cannot spend more energy collecting food than it gains by consuming it. Behavioral ecologists have developed models to predict which foraging decisions maximize an animal's fitness.

Optimal foraging theory predicts that an animal's food-finding strategy should maximize the amount of energy (measured in calories) collected per unit time. For example, if given the choice between a plate of celery and an equal-sized plate of chocolate cake, a person seeking to maximize energy gain would choose the cake. But if the cake were broken into tiny crumbs and scattered over a wide area, the additional foraging costs for the cake might make celery the better deal.

In predicting an animal's optimal foraging strategy, it is important to keep in mind that each animal has a limited ability to remember a previous visit to a site or how profitable it was. A bumblebee, for example, moves from flower to flower as it forages for nectar and pollen. How does it decide which flower to visit next? If the bee had a perfect memory, it might plan an efficient path that considered the potential gain from each flower in a field. Instead, models that account for the bee's limited brain power predict that the insect should follow a simple rule: visit the closest flower other than the one that was most recently visited. That is pretty much what bees actually do.

Once an animal secures its food, it may incur considerable handling costs before it can actually eat. Consider, for example, the northwestern crows that search along the shore at low tide for snails called whelks. When they spot a whelk, they pick it up, fly over a rocky area, swoop upward, and drop the snail on the rocks below. If the shell breaks open, the crow eats the contents. If not, the bird retrieves the whelk and repeats the dropping process until the shell finally breaks. To forage optimally, the crows should choose the snails that are easiest to break open and yet yield the most food.

In an effort to find out how close these crows get to their optimal behavior, researchers dropped whelks of different sizes from platforms of known height. They found that smaller whelks needed to be dropped from much greater heights than large whelks. So it was no surprise that crows prefer larger whelks, which also contain more food than smaller ones. But how high should the crow fly to get the most food for the least effort? The platform tests revealed that drops from 5 meters minimized the total effort (figure 35.9). Drops from higher than 5 m weren't much more likely to break the shell but were more costly to the crow.

When the researchers analyzed their observations of the crows, they found that the average height from which the birds actually dropped whelks was 5.23 m, close to the 5-m height predicted from the platform tests. Thus, in this instance, crows seem to be foraging optimally: they were maximizing food intake while minimizing costs.

C. Avoiding Predation Is Another Key to Survival

Animals have many morphological and behavioral adaptations that reduce their chances of becoming someone else's meal. Body markings often come into play. Many animals have camouflage, which helps animals become less visible by matching their background. Others have shapes and markings that allow them to masquerade as something inedible, such as a leaf or a twig (figure 35.10). Still others frighten would-be predators by looking and acting like predators themselves (figure 35.11). At least one type of caterpillar, for example, resembles a snake head. Particularly common are fake eyespots, such as those on the underwings of moths; when displayed suddenly, the eyespots trigger innate fright responses in predators.

Some animals protect themselves with poisons or other weapons, such as the quills of a porcupine or the noxious spray of a skunk. The bombardier beetle has an especially unusual weapon (figure 35.12): its plumbing system resembles a liquid-fueled rocket. Separate reservoirs inside the beetle's body hold quinones and hydrogen peroxide. If a threat arises, these two liquids enter

Height of Drop (m)	Average Number of Drops Required to Break Shell	Total Flight Height (Number of Drops × Height per Drop)
2	55	110
3	13	39
5	6	30
7	5	35
15	4	60

Figure 35.9 Optimal Foraging in Crows. To determine optimal behavior, researchers repeatedly dropped whelks from a platform until the snail's shell broke. The height that required the least total effort was 5 meters. Crows actually dropped whelks from an average of 5.23 m, close to the most efficient height.

Figure 35.10 Camouflage. The wings of the brimstone butterfly look like leaves.

b.

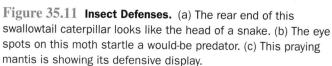

a.

c.

Figure 35.11 Insect Defenses. (a) The rear end of this swallowtail caterpillar looks like the head of a snake. (b) The eye spots on this moth startle a would-be predator. (c) This praying mantis is showing its defensive display.

an outer compartment where, in the presence of enzymes, the mixture heats to the boiling point and is sprayed on the attacker.

Instead of hiding, many noxious species have distinctive warning colors or patterns that make them conspicuous to potential predators. Examples include the black-and-white stripes on a skunk and the bright orange-and-black patterns on the wings of a monarch butterfly.

In addition to weapons, behaviors can also deter predators. Individuals may use **distraction displays** that divert a predator's attention away from a nest or den. One example is the broken-wing display of killdeers, which lures the predator away from young.

Figure 35.12 Bombardier Beetle. When disturbed, this beetle mixes, heats, and sprays a mix of noxious chemicals.

Figure 35.13
Meerkat Sentries. These meerkats are on the lookout for predators. When a sentry barks the alarm call, the entire community dashes underground to safety.

A conspicuous display in gazelles and their relatives is stotting, the stiff-legged jumping display depicted in the chapter opening photo. Researchers have found that cheetahs are more likely to abandon hunts when gazelles stot than when they don't. Gazelles also stot when alone, indicating that stotting is probably not used to warn other gazelles or to divert the attack toward the stotter. Rather, stotting seems to signal: "I see you, I'm in good shape, and I can outrun you." Other functions for stotting are also possible, however, such as distracting the predator from a fawn or simply attempting to get a better view of the predator in tall vegetation.

Group living can also offer a defense. An advantage of living in groups is the "many eyes" effect; that is, many individuals are more likely to spot a predator than one individual searching alone (figure 35.13). Groups can also protect vulnerable young, as when musk oxen form a circle around their offspring when wolves or polar bears threaten. Groups may also mob a predator, as when ground squirrels surround a rattlesnake, make an alarm call, and kick sand in the snake's face.

Perhaps the simplest advantage of being in a group is the dilution effect: by getting into a large group, one's chances of being picked off by a predator are reduced. For example, many species of fish quickly form schools in the presence of a predator: the greater the threat, the tighter the school. Such groups have been called **selfish herds:** individuals behaving selfishly try to position themselves so that as many of their companions as possible are between them and the predator. In a school of fish, the result is an ever-tightening group.

Competition for the relative safety at the center of a group also plays a part in the distribution of nesting territories of colonially breeding sunfish. The larger, older males get the central territories; their nests of eggs are subject to less predation than are those at the periphery.

35.3 | Mastering Concepts

1. What are three mechanisms by which animals find their way to a specific location?
2. Describe how optimal foraging theory explains an animal's foraging decisions.
3. What are some behaviors that help animals avoid predation?

35.4 | Many Behaviors Promote Reproductive Success

Section 35.3 demonstrated that behavioral adaptations are important for an animal's survival. This section turns to reproductive behaviors. Most reproductive activities fall into the category of **social behaviors,** which are interactions among members of the same species. Besides courtship and mating, other social behaviors include interactions among animals that live in groups (the subject of section 35.5). A field of study called **sociobiology** attempts to understand social behavior in the context of an animal's fitness.

A. Courtship Sets the Stage for Mating

Many species have elaborate courtship behavior patterns, often involving a long chain of actions leading to copulation; recall the fruit flies illustrated in figure 35.2. If you have seen the movie *March of the Penguins,* you might recall some of the calls, postures, promenades, and bows that occur when pair bonds are formed or when mates reunite after a long absence.

Courtship rituals have at least three functions. First, these displays are generally species-specific, so they help to ensure that an animal is directing its mating attempts at the correct species. Matings between closely related species often yield no offspring or offspring of reduced fitness. ▶ reproductive barriers, p. 276

A second function of courtship is to coordinate the physiological and hormonal states of the participants. One of the best-studied examples is the ring dove (**figure 35.14**). If nesting materials are available, the male courts the female using a "bow-coo" display; this behavior stimulates the female's pituitary gland to release follicle stimulating hormone (FSH). FSH stimulates follicle development in her ovaries. The follicles secrete estrogen, and within a day or two, both birds begin nest construction. The doves soon copulate. The presence of a nest stimulates progesterone synthesis, followed by egg laying. Both sexes incubate the eggs, a process that stimulates the production of prolactin. This hormone inhibits secretion of FSH and LH, and it stimulates crop development and the production of crop milk for the hatched young. The actual interactions between behavior and hormones are more complicated than described here; researchers continue to explore the details more than 50 years after the ritual was first described. ▶ reproductive hormones, p. 578

In many mammals, courtship is necessary to trigger ovulation. If you have ever observed cats mating, you may have wondered about all the yowling. The male not-so-gently bites the back of the female's neck prior to copulation. These stimuli lead to a surge of hormones that trigger ovulation.

A third function of courtship is to provide each participant with information about the suitability of the other as a mate. A female bird, for example, may evaluate the courtship rituals of many males before selecting a mate.

B. Sexual Selection Leads to Differences Between the Sexes

Females of many species seem especially picky when it comes to mate choice. Recall from chapter 34 that females produce relatively small numbers of large gametes. A female's egg cells contain yolk and other substances used for the growing embryo. Males, on the other hand, produce many tiny sperm that each contain little more than a haploid set of chromosomes. This disparity in gamete size is thought to set the stage for the tendency of females to be the more choosy sex, given their large investment in a relatively small number of eggs. Males, with their small investment in a large number of relatively cheap sperm, tend to be less discriminating and also to compete more strongly than females for access to mates.

Charles Darwin noticed this tendency, and he argued that traits that improved an individual's chances of getting a mate could be selected for independently from selection for survival traits. He coined the term **sexual selection** for the form of natural selection that results from variation in the ability to obtain mates (see section 11.5). Sexual selection can lead to extreme **sexual dimorphism,** a situation in which the two sexes look very different. Many sexually selected traits seem to be of no value to survival; they may even be harmful.

Some sexually selected traits result from intrasexual selection, which occurs when males compete with one another for access to females. Northern elephant seals offer a spectacular example (**figure 35.15**). Males are several times larger than females, have an elongated snout that they use to make loud roaring sounds, and have tough skin around their neck that protects them from the blows of other males. Winners of the competition gain access to the large numbers of females hauled out on the beach. Reproductive success therefore varies tremendously among adult males. Fewer

1 Courtship   2 Nest building and copulation   3 Egg laying   4 Incubation   5 Feeding young squabs

Figure 35.14 Ring Dove Courtship and Mating. Given a cage with nesting materials and a nesting bowl, a male–female pair of ring doves carries out a predictable sequence of courtship, mating, and rearing young. The sequence involves complex interactions between behaviors and hormones.

Figure 35.15 Northern Elephant Seals. These two large males are fighting to establish dominance. The winner mates with the females in the territory.

than one third get to copulate at all in a given season; those that do mate may sire more than 50 pups.

Antlers are also subject to sexual selection. Mammals of the deer family grow antlers before each breeding season. These structures, which typically occur only on males, are made entirely of bone. They are therefore costly to produce. Males use their antlers in fights that establish dominance in bachelor herds. As with the elephant seals, winners gain access to females for breeding; the antlers are discarded at the end of the breeding season.

Intersexual selection results in traits that are attractive to members of the opposite sex. One spectacular example is the peacock; in contrast to the normal-sized, drab tail feathers of the female, those of the male are hugely elongated and have iridescent spots (**figure 35.16**). The male spreads his tail in the

Figure 35.16 Spectacular Tail. The remarkable tail of the peacock is a sexually selected trait.

presence of a female, who may or may not cooperate with his mating attempts. Tails with more spots attract more females, possibly explaining this example of "runaway selection."

Sexually selected traits can be very costly, to the point where they compromise survival. Why should a female prefer a mate who may look (or sound) attractive if he is actually less likely to survive? Part of the answer is that some traits are so costly that they may be reliable indicators of fitness. For instance, the brightness of a rooster's comb turns out to be an honest indicator of the male's parasite load: males with the brightest combs have the fewest parasites, and females prefer them as mates. It is also possible that an extravagant trait such as a peacock's tail serves as a handicap: a male that can afford such a tail must be fit indeed!

C. Animals Differ in Mating Systems and Degrees of Parental Care

Animals display almost every imaginable mating system. Some species, such as corals, simply shed eggs and sperm into the environment. At the opposite extreme are some birds that form pairs for life.

Species that are strongly sexually dimorphic tend to have polygamous mating systems. **Polygamy** is a general term that encompasses all forms of multiple mating, in which either males or females have multiple sexual partners. Polygyny is a form of polygamy in which one male has exclusive sexual access to more than one female. In polygynous animals, the male is not likely to help care for the young. The dominant male elephant seal, for example, mates with many females and provides no care for his offspring. Males are even known to trample pups from the previous year's matings as they attempt to copulate with females.

On the other hand, species in which males and females resemble each other tend to have monogamous mating systems. **Monogamy** means that both male and female have one sexual partner. In monogamous species, both parents typically provide care for offspring. Most bird species are monogamous, with both parents needed to feed and defend the vulnerable nestlings. Some small mammals are also monogamous, as described in section 35.6.

In birds and mammals, males rarely care for their young alone. Indeed, extensive paternal care is relatively rare in mammals; after all, males cannot produce milk. Nevertheless, males can protect the group and may even supply food to the mother and young. In contrast, among many species of amphibians and fish, the male parent is the one that typically tends the nest and eggs. (A male seahorse even becomes pregnant, as described in the opening essay of chapter 34.)

Why should birds and mammals behave differently from fishes and amphibians? The answer likely involves confidence of paternity, which relates, in turn, to the events of fertilization. Most species of fish and amphibians have external fertilization. In frogs, for instance, the male grasps the female from behind and literally squeezes the eggs out of her, fertilizing them as they are shed (**figure 35.17**). Because the male's confidence of paternity is high, he enhances his reproductive success by investing his time and energy protecting the egg mass.

Figure 35.17 **High Confidence of Paternity.** Sperm cells from a male frog cover a female's eggs as she releases them into the water; the male is on top. The chance that another frog fertilizes her eggs is small.

Birds and mammals, on the other hand, have internal fertilization. The female lays eggs or gives birth some time after copulation, so a male's confidence of paternity is relatively low. Animals with internal fertilization, including some species of insects, spiders, fishes, and reptiles, increase their confidence of paternity in many ways. Some males guard the female before and after copulation (see section 34.7). Others insert a plug after insemination, reducing the chance that another male will follow. Some spiders enhance their probability of paternity by offering their own bodies as a "nuptial gift" (see section 31.5). And some dragonflies even have a penis with a special "sperm scoop" that can remove a previous male's sperm!

D. Human Reproductive Choices May Reflect Natural Selection

How do the reproductive behaviors of nonhuman animals compare to those of humans? The range of human sexual behavior

obviously varies greatly from person to person, but it is possible to use what we know about other animals to make some general predictions. Evolutionary psychologists and other researchers can then test these predictions by gathering evidence about human behavior.

First, consider the disparity in reproductive investment between human females and males. As in other mammals, human females make a larger investment in gamete production than do males. A pregnant female also undergoes 9 months of gestation, followed by milk production for up to several years. We might therefore expect females to be the choosier sex, while males might be less discriminating when it comes to choice in a mate.

What do the data reveal? One survey of college undergraduates asked how many sexual partners they would like to have across different time spans. The result: men wanted many more partners than did women (**figure 35.18a**). These subjects were also asked how likely they would be to have sexual intercourse with a person after having known him or her for periods ranging from 1 hour to 5 years. The sexes were similar at 5 years, but men were much more inclined than women to have sex at the shorter time spans (figure 35.18b). Other studies suggest that men value most highly physical attractiveness in a potential mate. Women most highly value a male's resources (wealth), although his looks are also important.

These types of surveys are subject to the criticism that the preferences simply reflect cultural norms and do not necessarily stem from underlying selection pressures. Nevertheless, the results do seem to hold up in a wide range of cultures. In addition, other types of studies quantify human behavior and fitness. In a number of cultures, both a woman's and a man's reproductive success are positively correlated with the husband's socioeconomic status. This finding is consistent with the value women place on wealth in choosing a mate.

A second prediction arises from the degree of sexual dimorphism in humans. Although sexual dimorphism in our species is nothing like that of elephant seals, it is still substantial. Males of most races are taller, heavier, and more muscular than females,

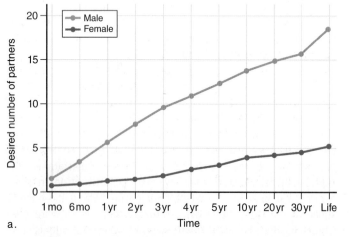

a.

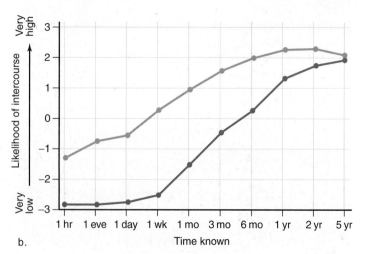

b.

Figure 35.18 **Partner Preferences.** (a) On average, men would prefer to have more sexual partners over a lifetime than would females. (b) On average, males are more likely than females to agree to sexual intercourse after having known a potential partner for shorter time.

suggesting the importance of male–male competition in the past. Given the association between mating system and degree of dimorphism, we might expect a certain degree of polygyny in humans. In fact, anthropologists have reported that some 85% of known societies worldwide have either occasional or frequent polygyny, typically where men of high status have more than one wife.

Third, like all other mammals, humans have internal fertilization. Therefore the female, but not the male, has high confidence of parentage. Researchers working in the United Kingdom have found that the percentage of women copulating with males other than their primary partner increased the more time the primary partner spent away from the woman, thus lowering his confidence of paternity of any resulting offspring (figure 35.19). Thus we might expect mate guarding or other tactics so that males can increase their certainty of paternity prior to investing in offspring. Indeed, evidence of mate guarding by men comes in many forms. In medieval times, chastity belts prevented women from having intercourse while their husbands were away. And in many cultures, adultery by a woman is considered a punishable offense against her husband.

35.4 | Mastering Concepts

1. List three functions of animal courtship rituals.
2. Describe examples of sexual selection between males and between males and females.
3. What is the relationship between sexual dimorphism and mating systems?
4. Explain how confidence of paternity influences the degree of parental care by a male.
5. How do observations of reproductive behavior in nonhuman animals enable researchers to make predictions about humans?

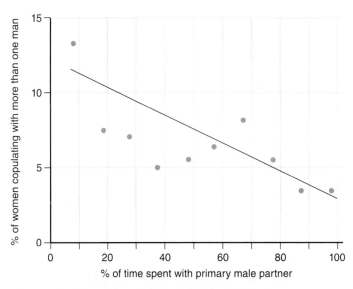

Figure 35.19 Mate Guarding in Humans? In one study, the more time a female spent with her primary male partner, the less likely she was to have copulated with other males.

35.5 | Social Behaviors Often Occur in Groups

Lots of animals live in flocks, herds, prides, schools, or other types of groups. But not all groups are considered societies. A society is a group of individuals of the same species that is organized in a cooperative manner, extending beyond sexual and parental behavior. This section describes some of the social behaviors that occur in groups.

A. Group Living Has Costs and Benefits

Animals profit in several ways by living in groups. One advantage is in finding food, as we have already seen. For example, groups can share information about the location of food. One striking example is the dance language of honeybees, in which foragers return to the hive and communicate the location of food sources (figure 35.20). In addition, groups can cooperate in the capture of prey not available to a lone individual.

Another advantage is protection. Animals can huddle together to conserve body heat, providing protection from cold, wind, or other environmental conditions. (Anonymous aggregations, however, might not fit the biological definition of a society.) Similarly, being part of a large group helps protect against predators, as with schools of fish. Groups of individuals are also more likely to spot predators, allowing others to spend more time foraging or in other activities.

Living in groups, however, does have costs. Predators may be more likely to detect groups of animals than dispersed individuals. Also, members of the same species are likely to compete for the same resources, leading to an increase in aggressive behavior. This aggression may interfere with the rearing of young, as when a new cadre of males takes over a lion pride and kills the cubs sired by the previous males.

Additionally, group living may promote the spread of diseases and parasites. For example, in colonies of cliff swallows, the number of blood-sucking bugs on each nestling increases as a function of colony size. Infested chicks grow much more slowly than do uninfested chicks. Likewise, dense colonies of black-tailed prairie dogs are more vulnerable to deadly outbreaks of bubonic plague than are the less-social white-tailed prairie dogs.

B. Dominance Hierarchies and Territoriality Reduce Competition

One way group members reduce the cost of competing with one another is by establishing dominance hierarchies. A **dominance hierarchy** is a social order with dominant and submissive members. Individuals learn their place in the "pecking order" from previous interactions, reducing the time, energy, and risk of fighting. In general, higher-ranking individuals have higher fitness than do lower-ranking individuals. Although members at the bottom of the hierarchy may suffer reduced fitness, they presumably fare better than by striking out on their own, and their status may improve with time.

For example, rhesus monkeys live in relatively closed social groups of 20 to 50 individuals. Adult females have a linear

Figure 35.20 **Waggle Dance.** Bees observing a waggle dance interpret the dancer's movements relative to the angle of the sun. If the food is toward the sun (1), the straight run of the dance points directly up. If the food source lies at a 45-degree angle relative to the sun (2), the straight run of the dance is 45 degrees relative to vertical. If the food source is away from the sun (3), the straight run points directly down.

dominance order that is family-based: offspring rank just below their mother in the hierarchy and above the next adult female and her family. Once established, the hierarchy is maintained by threat and submission displays that rarely lead to costly fights (figure 35.21). Males have a separate linear order that is based on seniority in the group. Males leave the natal group at puberty (about 4 years) and eventually join another group, where they start

Submissive grin	Open-mouthed threat

Figure 35.21 **Maintaining Dominance.** Rhesus monkeys display (a) submission to monkeys that occupy higher positions and (b) threats to those that are lower in the hierarchy.

out at the bottom and work their way up as more senior males die or leave. Rank is generally positively correlated with reproductive success in both sexes.

Another way to reduce competition between or within a group is to occupy a **territory,** which is space that an animal defends against intruders. The key to territory formation is economic defendability, such that the benefits (exclusive access to resources) outweigh the costs (energy expenditure, risk of injury, and loss of feeding or mating opportunities). Note that territoriality is not exclusively associated with group living; solitary animals such as bobcats establish territories as well.

Olfactory or vocal signals often establish the boundaries of a territory. For example, hyena clans mark their borders with latrine areas, whereas male gibbons signal theirs with loud whoop-gobble calls. Within breeding colonies of gulls, each pair defends a small area around the nest from marauding neighbors.

Territorial animals spend much time and energy patrolling their boundaries, especially in newly established territories (figure 35.22). Eventually, neighbors become conditioned and need only occasional reminders to keep out. If the owner should leave or die, however, neighbors or immigrants will quickly move in.

C. Kin Selection and Reciprocal Altruism Explain Some Acts of Cooperation

Group living incurs both costs and benefits, and it is easy to see how social behavior might be maintained in a population if individuals in groups have higher fitness than those living alone. But social living often involves cooperation with group mates, and cooperation can be costly. Occasionally, a social behavior even seems to be **altruistic;** that is, the behavior seems to lower the animal's fitness for the good of others.

The definition of *altruism* does not include parental behavior, in which parents risk their own lives for the sake of their offspring. After all, parents increase their own fitness by helping their offspring survive. Instead, altruism *reduces* an individual's fitness. Behavioral ecologists have therefore reasoned that altruistic individuals should be selected against in favor of selfish individuals. But if that is the case, how can altruistic behaviors be maintained?

Part of the answer may lie in the distinction between direct fitness and indirect fitness. An individual's own offspring account for its direct fitness. But what about other relatives that are not direct descendants? It's true that we directly pass genes on to

Figure 35.22 **Keep Out.** Two great crested grebes fight over territory on a lake in Greece.

only our own offspring, but we share genes through common descent with all our close relatives. In fact, we share half our genes with our full siblings and with our own offspring. We share one quarter of our genes with our full nieces and nephews, and one eighth of our genes with first cousins. Another way to look at it is that on average, we share as many genes with one of our own offspring as we do with two of our nieces and nephews.

An individual's **inclusive fitness** is the sum of direct and indirect fitness. Individuals, then, can be expected to behave so as to maximize their inclusive fitness. The most familiar way to maximize fitness is to help one's own offspring to survive and reproduce, but another possibility is to maximize indirect fitness.

In **kin selection,** an individual reduces its own direct fitness but assists the survival and reproduction of nondescendant relatives. Kin selection seems to explain the helping behavior seen in a variety of animals. For instance, Belding's ground squirrels put themselves at risk by giving alarm calls when they spot a predator such as a weasel or a coyote (**figure 35.23**). Females with relatives in the vicinity are most likely to call. Males, which disperse each year, are unlikely to have relatives in the group and rarely sound alarm calls.

Researchers have rigorously tested the role of kin selection in the Florida scrub jay, where some birds forego breeding and help their parents raise siblings. Why? Part of the answer may lie in the observation that this species occupies a limited habitat in which all available territories may be filled. Stay-at-home helpers may increase their direct fitness by inheriting their natal territory should the parents leave or die. In the meantime, they increase their

a.

b.

Figure 35.23 Sound the Alarm. (a) A female ground squirrel gives an alarm call. (b) The bars on the left show the distribution of alarm calls expected if ground squirrels called in proportion to their abundance in the population. Instead, the bars on the right show that adult females and young females made the majority of the calls.

indirect fitness because their parents raise more offspring than they would without help.

Not all altruistic behavior, however, occurs among close relatives. In **reciprocal altruism,** individuals might help others at a cost to themselves if it is likely that they will be paid back at a later time. For reciprocal altruism to work, the individuals must have a high likelihood of encountering one another for possible repayment. For example, vampire bats feeding at night on blood from cattle and horses return to a community roost during the day. At the roost, mothers often regurgitate blood to their offspring, but adults also regurgitate to unrelated individuals who have not fed the night before. The latter, however, only get fed if they have associated with the donor frequently in the past and are thus likely to return the favor at a later time.

You may have noticed that neither kin selection nor reciprocal altruism represent "pure" altruism. In the former, the helper increases its inclusive fitness by helping relatives; in the latter, the helper gets repaid, possibly with interest and possibly when it is needed most. It should not be surprising that pure altruism is rare in animals; such traits will lower fitness and be selected out of the population.

Humans may be uniquely prone to altruistic behavior. In laboratory tests, infants at 18 months of age immediately assist an unrelated adult whose hands are full and who needs help picking up a dropped object. Even at the age of 12 months, infants will point toward objects that an adult appears to have lost. Starting at age 3, human children become more selective in who they assist, being more likely to help a child who has previously helped them. Because altruism appears so early in development, across many cultures, and without reward, some researchers think it is innate.

D. Eusocial Animals Have Highly Developed Societies

The epitome of organization is **eusociality,** a social structure that includes extensive division of labor, especially in reproduction. The characteristics of eusociality are cooperative care of young by nonparents; the presence of nonreproductive adults (often called sterile castes) that care for the breeders; and overlap of generations such that offspring assist parents in raising siblings.

In a honey bee colony, the queen does nothing but lay eggs. Her daughters are workers who will never reproduce; instead, they forage and tend the queen and her offspring. Males (drones) contribute little to colony life, flying off and trying to mate with queens that are forming new colonies. Darwin had difficulty explaining how sterile castes could evolve and decided that a bee colony should be considered a single "superorganism." Thought of this way, the sterile workers are no different than organs in the body, such as the liver. The liver helps maintain the entire organism, but it can't exist on its own or reproduce.

Eusociality is relatively common in insects of the order Hymenoptera, including ants, bees, and wasps. One possible explanation is their unusual mechanism of sex determination. Whereas sex chromosomes such as XX and XY determine the sex of many animals, hymenopterans use a method called haplodiploidy. Males in this order develop from unfertilized eggs and are therefore haploid; females develop from fertilized eggs and are diploid.

Since males are already haploid, they produce genetically identical sperm by mitotic cell division. If only one male mates with the queen in a colony and there is only one queen, all of the female worker offspring get the same genes from their father and are therefore related by three quarters instead of the usual one half. As a result, the workers are more closely related to one another than they would be to their own offspring. This unusual condition may explain why the workers give up reproducing for the good of the colony. ▶ sex chromosomes, p. 181

There are, however, animals that have the usual method of sex determination and have evolved eusociality anyway; termites and naked mole rats are two examples (**figure 35.24**). Thus, haplodiploidy is not necessary for eusociality to evolve.

35.5 | Mastering Concepts

1. What are some costs and benefits of group living?
2. Explain the roles of dominance hierarchies and territoriality in animal societies.
3. Define *altruism* and explain why evolution should select against it.
4. How do kin selection and reciprocal altruism explain some examples of cooperation among animals?
5. What is eusociality?

a.

b.

Figure 35.24 Eusocial Animals. (a) A queen termite is surrounded by nonreproductive workers. (b) Pups suckle from a naked mole rat queen.

35.6 | Investigating Life: Addicted to Affection

The sexual behavior of animals fascinates many people, perhaps because of the insight it can lend into human relationships. In our own species, sexual attraction and feelings of love are often intertwined. Love is difficult to study in humans (and impossible to study in other animals), but scientists can examine patterns of sexual behavior in many species other than our own.

Some animals are faithful to their sexual partners for life, whereas others are much more promiscuous. One of the rarest and least understood social behaviors is monogamy. To qualify as monogamous, an animal must mate exclusively with one partner, live with its mate, help with care of the young, and defend the family against intruders. Only a tiny fraction (about 3%) of mammal species are monogamous.

Two closely related species of snub-nosed rodents have given scientists the opportunity to investigate the biological basis of monogamy. Whereas prairie voles (*Microtus ochrogaster*) are monogamous and highly social (**figure 35.25**), montane voles (*Microtus montanus*) are promiscuous and solitary. What accounts for the difference in lifestyle? A chemical messenger called antidiuretic hormone (abbreviated ADH and also called vasopressin) apparently plays a role. In the 1990s, researchers learned that if male prairie voles were given ADH, they rapidly formed social attachments, even with females they had not mat-

Figure 35.25 Prairie Vole. These small rodents live in underground colonies. They are highly social and typically monogamous, so biologists study them to learn more about the genetic basis of sexual fidelity.

ed with. When ADH's effects were blocked, pair bonds did not form. ▸ antidiuretic hormone, p. 573

One logical explanation for the difference in sexual behavior between the two species is that the monogamous prairie voles have naturally higher levels of ADH than do the promiscuous montane voles. Yet this is not the case; both vole species have similar levels of the pair-bonding hormone. The location of receptors for ADH in the brain, however, does differ between the two species. In monogamous prairie voles, ADH receptors occur in the same brain region where addictive drugs act, and males seem to derive feelings of reward from being with their mates and young. In contrast, ADH receptors in promiscuous montane voles are located in brain regions associated with aggression.

A team of researchers, led by Lauren Pitkow and Larry Young from Emory University in Atlanta, wondered what would happen to prairie voles with extra ADH receptors in the brain areas associated with pleasure and rewards. Would the animals form social attachments even more readily? To find out, the researchers inserted the gene encoding the ADH receptor into a virus. They injected the modified virus into the reward areas of the brains of prairie voles, effectively increasing the number of ADH receptors. Two groups of control voles also received injections. One control group received a different gene (one not related to ADH) into the reward area of the brain. Voles in the second control group received the ADH receptor gene but in a different brain area.

Afterward, each male spent 17 hours in a cage with an adult female that was not sexually receptive. The voles then underwent "partner-preference" tests. The researchers placed each male into a choice chamber, where he was free to spend as much time as he wanted with his female "partner" or with a different female who was a stranger to him. Male prairie voles do not ordinarily pair-bond with females after less than 24 hours together, unless they have mated. Nevertheless, the voles with the extra ADH receptors in the reward region of the brain spent much more time in contact with their partners than did either group of control voles (figure 35.26). The extra ADH receptors evidently made the male prairie voles especially likely to form pair bonds.

Of course, the evolution of complex mating behaviors required more than just one change in the brain chemistry of the vole. Nevertheless, this study is interesting because it explicitly links genes, brain chemistry, and social behavior. This research also raises the startling possibility that a genetically modified virus can transmit genes that increase social attachment, at least in prairie voles. Could such a "love bug" infect humans, too? Researchers already know that the location of ADH receptors in the brain may play a role in human social behaviors; examples include not only sexual fidelity but also autism, a disorder in which individuals have difficulty forming social attachments. Researchers now are investigating primate ADH receptors to learn more about the biochemistry of attachment in our closest relatives.

1

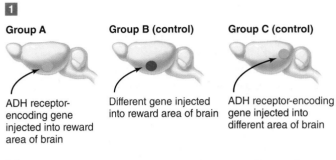

Group A   **Group B (control)**   **Group C (control)**

ADH receptor-encoding gene injected into reward area of brain

Different gene injected into reward area of brain

ADH receptor-encoding gene injected into different area of brain

2 Each male spends 17 hours in cage with non-sexually receptive female

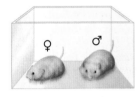

3 In a 3-hour partner-preference test, the male can choose to spend time with his "partner" or a stranger

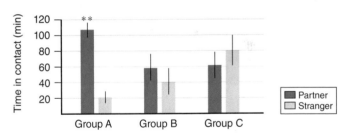

Figure 35.26 More Receptors, More Bonding. Prairie voles injected with ADH receptor genes in the reward center of the brain spent significantly more time with a female partner than with a stranger. Control voles were less likely to bond with their partners. (Asterisks indicate a statistically significant difference.)

Pitkow, Lauren, and five co-authors (including Larry Young). 2001. Facilitation of affiliation and pair-bond formation by vasopressin receptor gene transfer into the ventral forebrain of a monogamous vole. *Journal of Neuroscience,* vol. 21, no. 18, pages 7392–7396.

35.6 | Mastering Concepts

1. How did the researchers test the hypothesis that the number of ADH receptors in the brain's reward area influences pair-bonding?

2. Could the researchers have drawn the same conclusions if they had omitted one of the control groups? Why or why not?

Chapter Summary

35.1 | Animal Behaviors Have Proximate and Ultimate Causes

- The science of animal behavior has its roots in observations by people whose everyday environment included wild animals.
- Ethology emphasizes experiments into the causes of animal behavior in natural environments. The focus may be on the **proximate** causes, such as the hormones that contribute to a behavior. Questions about **ultimate** causes reflect a behavior's evolution or adaptive significance.

35.2 | Animal Behaviors Combine Innate and Learned Components

A. Innate Behaviors Do Not Require Experience
- **Innate** behaviors are instinctive; the animal performs innate behaviors from hatching or birth.
- Examples of innate behaviors include **reflexes,** involuntary responses that do not vary between individuals of a species; **taxes,** automatic movements toward or away from an environmental stimulus; and **fixed action patterns,** sequences of behaviors that are triggered by a specific stimulus.
- Mutations in the *fru* gene cause male fruit flies to court other males.

B. Learning Requires Experience
- **Learned** behaviors result from an animal's experiences.
- In **habituation,** an animal learns *not* to respond to a stimulus. An animal pairs two stimuli in **associative learning,** which includes both **classical conditioning** and **operant conditioning.**
- **Imprinting** is a type of learning that occurs only during a restricted time in an animal's life.
- In **observational learning,** an animal learns by watching what others do.
- Animal cognition reflects multiple types of learning.

C. Genes and Environment Interact to Determine Behavior
- All learned behaviors have a genetic component, and all innate behaviors are triggered by interactions with the environment.
- Song-learning in white-crowned sparrows includes both innate and learned components.

35.3 | Many Behaviors Improve Survival

- **Behavioral ecology** is the study of how behaviors contribute to an animal's fitness.

A. Some Animals Can Find Specific Locations
- Animals use piloting, celestial bodies, and navigation to find their way home or to another specific location.

B. Animals Balance the Energy Content and Costs of Acquiring Food
- **Foraging** behaviors are directed at finding and handling food. An animal that hunts visually may develop a **search image** of its prey.
- According to **optimal foraging theory,** an animal's behavior reflects a balance between calories gained from food and energy spent in foraging.

C. Avoiding Predation Is Another Key to Survival
- Adaptations that help animals avoid predation include deceptive markings, noxious weapons, **distraction displays,** and living in groups.
- A **selfish herd** is a group of individuals that look out only for their own self-interest.

35.4 | Many Behaviors Promote Reproductive Success

- Reproductive behaviors are included among **social behaviors,** which occur between members of the same species. **Sociobiology** is the study of the evolutionary significance of social behaviors.

A. Courtship Sets the Stage for Mating
- Courtship behaviors help animals recognize others of the same species, trigger the release of hormones essential to reproduction, and help potential partners evaluate each other's suitability.

B. Sexual Selection Leads to Differences Between the Sexes
- In animals that are **sexually dimorphic,** males look different from females. Sexual dimorphism reflects **sexual selection.**

C. Animals Differ in Mating Systems and Degrees of Parental Care
- Sexually dimorphic species are often **polygamous;** species in which male and female look alike are often **monogamous.**
- Male parental care typically reflects confidence of paternity: a male will commit more resources to rearing offspring if he has high confidence of paternity.

D. Human Reproductive Choices May Reflect Natural Selection
- Observations of animal reproductive behaviors have enabled researchers to form and test hypotheses about human sexual behaviors.

35.5 | Social Behaviors Often Occur in Groups

A. Group Living Has Costs and Benefits
- An animal that lives in a group may be more conspicuous to predators but may also receive information about food and threats to the community.

B. Dominance Hierarchies and Territoriality Reduce Competition
- **Dominance hierarchies** reinforce social status within a group, whereas **territorial** behaviors help an animal retain exclusive rights to a subset of resources.

C. Kin Selection and Reciprocal Altruism Explain Some Acts of Cooperation
- **Altruism** occurs when an animal reduces its own fitness while improving the fitness of an unrelated individual.
- An animal's **inclusive fitness** includes not only its offspring but also its close relatives. Behavioral ecologists propose that **kin selection** explains why animals are most likely to sacrifice themselves for close relatives.
- In **reciprocal altruism,** an animal that shares resources with an unrelated individual receives future benefits.

D. Eusocial Animals Have Highly Developed Societies
- **Eusocial** animals have the most complex societies, often including workers that support the reproduction of only a few members of the society.

35.6 | Investigating Life: Addicted to Affection

- Monogamy is unusual among animals, but the prairie vole forms a monogamous bond with its mate. Researchers have traced pair-bonding behavior to a receptor that binds antidiuretic hormone in the pleasure-seeking area of the vole's brain.
- Increasing the number of ADH receptors made male voles more likely to bond with a female partner.

Multiple Choice Questions

1. Which of the following studies involves an ultimate cause of a behavior?
   a. A study of how a squid alters its color to blend in with its environment
   b. A study of the aerodynamics involved in a hummingbird's ability to hover
   c. A study of the effect of alarm calls on prairie dog fitness
   d. A study of the cues monarch butterflies use during migration

2. A fixed action pattern is an example of
   a. a learned behavior.   c. a reflex.
   b. an innate behavior.   d. a taxis.

3. Although the trains that rumble through your town once bothered you, now you barely even hear them. This example illustrates
   a. habituation.   c. a reflex.
   b. associative learning.   d. imprinting.

4. Why is it difficult to classify a human behavior as being either innate or learned?
   a. Because it is impossible to fully know the history of a person
   b. Because behavior is a blend of genetics and experience

c. Because it is unethical to test hypotheses about human behavior
d. Both a and b are correct.

5. Niko Tinbergen investigated how female digger wasps find their nests. After a wasp emerged, Tinbergen moved the pine cones surrounding the nest to a nearby location. Upon its return, the wasp flew to the pine cones, not the nest. What system was the wasp using to find its nest?
   a. Navigation
   b. Compass orientation
   c. Imprinting
   d. Piloting

6. Optimal foraging theory considers the
   a. energy content of food.
   b. handling cost of food.
   c. distribution of food in the habitat.
   d. All of the above are correct.

7. Male stag beetles have mandibles (jaws) that are greatly enlarged compared to those of the females. This difference is likely to result from
   a. distraction displays.
   b. monogamy.
   c. sexual selection.
   d. external fertilization.

8. In most mammal species, a male is _____ to provide parental care because his confidence of paternity is _____.
   a. Likely; low
   b. Unlikely; low
   c. Likely; high
   d. Unlikely; high

9. What is the advantage of kin selection?
   a. It helps the individual to attract mates.
   b. It maximizes sexual dimorphism in a population.
   c. It helps the individual maximize its inclusive fitness.
   d. It increases the incidence of mating between related individuals.

10. Why is parental care NOT considered to be altruistic behavior?
    a. Time spent rearing offspring increases the fitness of the individual.
    b. Time spent rearing offspring does not represent a cost to the individual.
    c. Rearing offspring involves indirect fitness, not direct fitness.
    d. The offspring will reciprocate upon reaching adulthood.

Write It Out

1. Differentiate between the proximate and ultimate causes of a particular behavior.
2. People tend to see what they expect to see. How can ethologists prevent themselves from falling into this trap when observing an animal's behavior?
3. How is a taxis different from a reflex? How are they similar?
4. In what ways is a taxis similar to and different from a tropism in a plant (see chapter 23)?
5. Give an example of a fixed action pattern.
6. Explain how biologists identify genes controlling behavior in a fruit fly. Why might it be difficult to identify similar genes in humans?
7. How is observational learning different from imprinting?
8. Once accustomed to captivity, praying mantids tend to stop making the startle response when their caregivers come near. Given what you now know about animal behavior, construct a hypothesis to explain this change. How would you test your hypothesis?
9. Male thynnine wasps approach and attempt to mate with female wasps that release pheromones. Some orchid flowers, however, emit a chemical that mimics the female wasp's pheromone so convincingly that male wasps will attempt to copulate with the flower. Once a male makes that mistake, however, he avoids the flower in the future. How does this example illustrate the interaction between innate and learned behaviors? What type of learning has occurred in the male?
10. How is it incorrect to refer to a gene "for" alcoholism?
11. Suppose an American couple moves to Australia and raises all of their children in their new country. The children all speak with an Australian accent. What does this mean about the influence of environment on speech? Does this mean that speech does not have a genetic component?

12. How is the idea of "nature versus nurture" outdated? How might an investigator use studies of identical twins raised apart to sort out the contribution of genes and environment to a particular behavior?
13. For three weeks in October 2002, two snipers terrorized people in the Washington, D.C., area. Bystanders observed that pedestrians waiting to cross the street would stay under cover until a group formed in the crosswalk. They would then dash out and join the group to cross. Explain this observation with respect to the selfish herd hypothesis.
14. European cuckoos and American cowbirds are "brood parasites" that lay their eggs in the nests of other birds. The chicks are generally raised by "adoptive parents" whose own chicks receive less nourishment and often are killed by the larger intruders. What kinds of behaviors does natural selection favor in the cuckoo and cowbird chicks? In the adoptive parents?
15. Think of an example of a sexually dimorphic species other than those mentioned in this chapter. How might sexual selection have led to this sexual dimorphism?
16. Use the Internet to investigate the mating behaviors of corals, sea horses, chimpanzees, and salmon. Arrange these species in order from lowest to highest confidence of paternity. Does the degree of paternal care follow the same pattern? Explain your answer.
17. Read section 31.5 to learn about a male spider that allows his mate to eat his body during copulation. How does this unusual behavior improve the male's confidence of paternity?
18. What are the costs and benefits of associating with a group? Can you think of costs and benefits not mentioned in the chapter?
19. Are humans solitary or social animals? Explain your answer.
20. At home, you are usually a sound sleeper, but when you travel, you toss and turn and get little sleep. Give a behavioral explanation for this observation.
21. How is it advantageous for a species to have a fixed action pattern that controls courtship or mating behavior?
22. What is kin selection, and how can it explain some types of altruistic behavior? Can you think of examples of true altruism in humans or nonhuman animals?
23. In the 1930s, the population geneticist J. B. S. Haldane famously said, "I would lay down my life for two brothers or eight cousins." What was he talking about?

Pull It Together

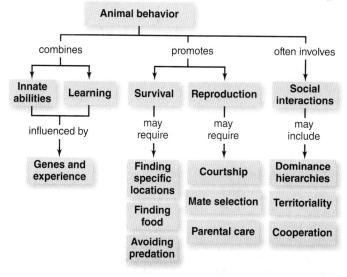

1. Compare and contrast innate and learned behaviors.
2. Add *foraging, habituation, kin selection, fixed action pattern, eusocial, inclusive fitness, imprinting,* and *reciprocal altruism* to this concept map.
3. Describe some of the behaviors that help animals avoid predation.

The Brown Tree Snake, *Boiga irregularis*, decimated much of the wildlife on Guam after being accidentally released onto the island.

A Population Out of Control

WHEN A NEW PREDATOR COLONIZES AN ISLAND, POP-ULATIONS OF NATIVE SPECIES CAN PLUMMET. This fate befell the island of Guam, thanks to the brown tree snake, *Boiga irregularis*. The brown tree snake likely arrived on Guam during World War II, hitching a ride on cargo shipped to the island from its native East Asia. Once there, the population grew explosively. Because this predator was new to the island, nature had not selected for defensive adaptations that might have enabled more birds, lizards, and other reptiles to escape. The snakes feasted. As a result, more than a dozen species of birds and small reptiles have disappeared from Guam.

It wasn't until the 1960s that ecologists began to notice that something was wrong on Guam. Birds had vanished, first in the south, then farther north. At first, biologists thought a pesticide or disease was at fault. Further study, however, revealed several pieces of evidence implicating the brown tree snake:

- The snake is extremely common, and it stalks and kills a variety of prey.
- Areas where snakes were common matched areas where birds disappeared.
- Birds and lizards were not disappearing on nearby snake-free islands.
- No known insecticide or disease was killing the animals.

Once researchers began surveying the forests of Guam, they found an enormous snake population. Whereas most snakes live in populations of 1 to 10 animals per hectare, on Guam, 100 brown tree snakes occupied each hectare! The birds and lizards of Guam were hopelessly outnumbered.

Many efforts to control the snake population on Guam focus on increasing the death rate. The most common approach is to capture and kill the snakes; researchers are developing pheromone-based attractants to make the traps more alluring. Another tactic is leaving dead rats laced with acetaminophen, a commonly used painkiller that destroys the snakes' livers. Since a population's growth rate reflects both birth and death rates, however, researchers are also investigating ways to either prevent reproduction or stop juveniles from reaching reproductive age. To keep brown tree snakes from leaving Guam and causing disasters elsewhere, specially trained Jack Russell terriers inspect outgoing cargo for snakes.

The story of the brown tree snake illustrates the importance of studying how populations grow, stabilize, and decline. This chapter introduces the factors that affect populations, including that of our own species.

Learning Outline

Learn How to Learn
Use Your Head

Many students make the mistake of memorizing information without really thinking about what they are learning. This strategy not only makes schoolwork boring but also limits your ability to answer questions. Instead of repeating someone else's words as you study, try elaborating on what you have learned. Here are some ideas: explain how the topic relates to other course material or to your own life; draw a picture of important processes; discuss the topic with someone else; explain it to yourself out loud or in writing; or think of additional examples or analogies.

36.1 | A Population Consists of Individuals of One Species

Take a moment to think of all the ways you have interacted with the world today. You have certainly breathed air, and you have probably sipped a beverage, eaten one or more meals, put on some clothes, and greeted other people. You may have stepped on the grass or driven a car or played with a pet. Such interactions are part of the science of **ecology,** the study of the relationships among organisms and the environment.

Ecologists classify these relationships at several levels (see figure 1.2). The brown tree snakes on Guam, described in the chapter opening essay, represent a **population:** a group of interbreeding organisms of one species occupying a location at the same time. A **community** includes all of the populations, representing multiple species, that interact in a given area. An **ecosystem** is a community plus its nonliving environment, including air, water, minerals, and fire. This chapter describes population ecology, whereas chapter 37 describes community- and ecosystem-level processes.

The study of population ecology is relevant to disease prediction, land management, the protection of endangered species, and many other fields. Typical questions that population ecologists might ask include: "Which weather conditions favor reproduction among rodents that transmit human diseases?" "How large a patch of old-growth forest does a breeding pair of spotted owls require?" "How many deer should hunters cull to keep the herd healthy?" "How many humans can Earth support?" To answer such questions, ecologists begin by describing the population.

A. Density and Distribution Patterns Are Static Measures of a Population

The **habitat** is the physical location where the members of a population normally live. The ocean, desert, or rain forest are typical examples, but an organism's habitat might even be another organism. Your body, for example, is home to billions of microbes. ▶ normal microbiota, p. 347

Within a habitat, the boundaries of a population can be difficult to define. Natural features sometimes provide important clues. For example, a lake may be home to a population of catfish, or an island may house all known members of a unique orchid species. Sometimes, however, ecologists use arbitrary lines to define a population's boundaries, such as the borders of a park, state, or country.

Population density is the number of individuals of a species per unit area or unit volume of habitat. Density varies greatly among species. Billions of bacteria can occupy a gram of soil, whereas the largest trees may live at densities of one trunk per 10 or more square meters. Not surprisingly, the general trend is for the smallest organisms to live at the highest densities; one redwood seedling, for example, requires far fewer resources than does a mature tree. This chapter's Apply It Now box describes some of the ways that biologists estimate population density.

Population distribution patterns describe how individuals are scattered through the habitat space (**figure 36.1**). Organisms may occur in a random pattern if individuals neither strongly attract nor repel one another and the environment is relatively uniform (for example, on an exposed rock surface or in the deep shade under a dense forest canopy). More uniform spacing may occur if individuals repel one another, as in the case of strongly territorial animals. Most often, however, a population's distribution is somewhat clumped, with individuals being attracted to one another or to the most favorable patches of habitat.

Population density and distribution measurements provide static "snapshots" of a population at one time. By repeatedly using the same method to measure a population's size over multiple generations, ecologists can determine whether the population is growing, shrinking, or stable. Section 36.2 explains the factors that determine the population's fate.

B. Isolated Subpopulations May Evolve into New Species

The overall area occupied by a population is not necessarily fixed. Habitat destruction can eliminate some or all of a population,

a. Random

b. Uniform

c. Clumped

Figure 36.1 Patterns of Population Distribution. (a) These lichens illustrate random spacing. Resources are evenly distributed within the habitat, and the organisms neither attract nor repel one another. (b) These penguins defend the space immediately surrounding their nests, leading to a uniform distribution pattern. (c) Clumps of palms growing in Tunisian sand dunes reflect unevenly distributed resources—especially water.

as explained in chapter 39. Conversely, individuals in a population can disperse to new habitats. Such range expansion is most obvious for invasive species that spread from their point of introduction. Examples include brown tree snakes, fire ants, zebra mussels, gypsy moths, European starlings, and kudzu. But dispersal also occurs as new habitats become available. For example, glaciers gradually retreat after an ice age, and new volcanic islands occasionally emerge from the sea. Plants, animals, and microbes from nearby areas typically colonize the new habitat.

As organisms disperse to new habitats, they may form local subpopulations. Individuals belonging to a subpopulation are more likely to breed with one another than with members outside the group. But that does not necessarily mean that each subpopulation is isolated from the others. For example, butterflies may flit from meadow to meadow, or the wind may carry a plant's seeds from one subpopulation to another.

Sometimes, however, a subpopulation becomes physically separated from its parent population, limiting the further exchange of individuals and setting the stage for the evolution of a new species. How might this occur? The isolated subpopulation often encounters a novel combination of selective forces, such as differences in sunlight, moisture, or nutrient availability. Previously rare variants may therefore have the greatest reproductive success in the new habitat. Over many generations, natural selection is likely to favor a unique combination of adaptations, shaping a genetically distinct subpopulation. If these genetic differences lead to new reproductive barriers, the subpopulation will be unable to interbreed with the parent population. The result: an entirely new species. Chapter 13 describes this process, which is called speciation.

36.1 | Mastering Concepts

1. What are the relationships among ecosystems, communities, and populations?
2. What is population density?
3. Distinguish among the types of distribution in a habitat.
4. Explain how dispersal creates subpopulations that may evolve into new species.

Apply It Now

Counting Crows (or Other Organisms)

Collecting long-term data on plant and animal populations can help scientists learn which species are healthy and which are threatened with extinction. Population data can also help ecologists measure the effects of catastrophes such as fires or oil spills. Hunting, trapping, and fishing regulations are based on population estimates, as are decisions on where to build (or not to build) houses, dams, bridges, and pipelines. But how do researchers know how many individuals of each species inhabit an area?

Unless a species is extremely rare or restricted to a limited range, it is usually not possible to count each individual. Instead, most population estimates rely on sampling techniques. For example, one common way to estimate the size of a plant population is to count the number of stems in randomly selected locations, such as within a 1-meter square or along a 50-meter line.

For animal populations, simple counts are sometimes possible. For example, aerial photos can reveal the number of seals on an island or caribou in a herd. But because animals tend to move around or hide, population estimates often rely on sampling techniques. For example, a stream ecologist might count the invertebrates swept into a net placed in the stream's current, or a soil ecologist might study the insects or spiders that stumble into a buried pitfall trap.

One widely used technique to estimate animal populations is called mark–recapture. Suppose researchers want to know how many squirrels inhabit a park. They might place baited nest boxes in trees. The researchers periodically collect the boxes and record the weight, sex, age, and health status of the squirrels inside. After receiving a unique tattoo (the "mark"), each squirrel is released. The proportion of tattooed squirrels that are subsequently recaptured can help the biologists estimate the population size.

For example, assume the park researchers mark and release 50 squirrels (M = number marked = 50) in a month. During the next month,

they capture 25 squirrels, of which five bear the tattoo (f = fraction of marked animals in "recapture" sample = 5/25 = 0.20). The number of marked squirrels divided by this fraction estimates the total squirrel population size. For the area sampled, the estimated total squirrel population is M/f = 50/0.20 = 250.

The potential for sampling bias is one weakness of the mark–recapture technique. Baited traps may repeatedly lure some individuals in search of a free meal. On the other hand, if an animal finds the capture or marking technique unpleasant, it may learn to avoid the traps.

Another way to estimate the size of an animal population is to count the number of droppings within a defined area. The more excrement piles, the higher the population. Alternatively, researchers sometimes set up animal-triggered camera traps. Bait is draped in a tree, along with a heat detector attached to a camera. When a bear or other animal saunters over to collect the treat, the detector picks up the animal's body heat, and the camera snaps its picture. By setting up multiple camera traps, ecologists can compare the population densities in multiple locations.

36.2 | Births and Deaths Help Determine Population Size

Some populations grow, whereas others remain stable. Still others decline, sometimes to extinction. **Population dynamics** is the study of the factors that influence these changes in a population's size (table 36.1).

Births and deaths are the most obvious ways to add to and subtract from a population. If a population adds more individuals than are subtracted, the population grows. If the opposite happens, the population shrinks (and may become extinct).

Table 36.1 Factors Affecting Population Growth: A Summary

Factor	Affected by ...
Additions	
Births	• Number of reproductive episodes per lifetime • Number of offspring per reproductive episode • Age at first reproduction • Population age structure (proportion at reproductive age)
Immigration	• Availability of dispersal mechanism • Availability of suitable habitat
Subtractions	
Deaths	• Nutrient availability • Predation • Accidents • Genetic/infectious disease
Emigration	• Availability of dispersal mechanism

The population size remains unchanged if additions exactly balance subtractions.

Regionally, dispersal to new habitats can also increase or decrease a population; **immigration** is the movement of individuals into a population, and **emigration** occurs when individuals leave. For example, 200 or so years of immigration has tremendously increased the human population in the United States. Most Americans are either immigrants or descended from immigrants, and today migration accounts for over 30% of population growth in this country. Likewise, nonhuman species also disperse to new habitats. They may actively swim, fly, or walk; alternatively, wind or water currents may move individuals into or out of a population.

A. Births Add Individuals to a Population

A population's **birth rate** is the number of new individuals produced per individual in a defined time period. For example, the human birth rate worldwide is about 20 per 1000 people per year.

The number of offspring an individual produces over its lifetime depends on many variables. The number of times it reproduces in its lifetime and the number of offspring per reproductive episode are important, as is the age at first reproduction. All other things being equal, the earlier reproduction begins, the faster the population will grow.

A population's **age structure,** or distribution of age classes, helps determine whether a population is growing, stable, or declining (figure 36.2). A population with a large fraction of prereproductive individuals will grow. As these individuals enter their reproductive years and produce offspring, the prereproductive age classes swell further, building a foundation that ensures future growth. Conversely, a population that consists mainly of

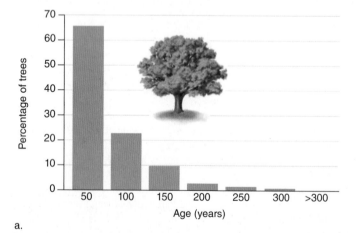

a.

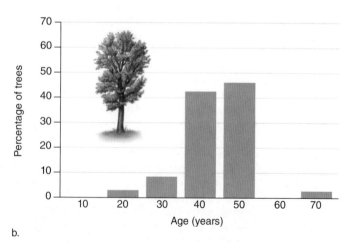

b.

Figure 36.2 Age Structures. (a) The white oak population is dominated by younger individuals, indicating high potential for future growth. (b) This population of cottonwoods has few individuals in the youngest age classes. Lacking young trees, the population is probably doomed.

older individuals will be stable or may even decline. This situation can doom a population of endangered plants if, for example, habitat destruction makes it impossible for seedlings to establish themselves. With few individuals of reproductive age, the population may go extinct.

B. Survivorship Curves Show the Probability of Dying at a Given Age

A population's **death rate** is the number of deaths per unit time, scaled by the population size. The causes of death may include accidents, disease, predation, and competition for scarce resources. In the human population, for example, the overall death rate is approximately 8 per 1000 people per year; section 36.5 describes human mortality in more detail.

Each individual in a population will eventually die; the only question is when. But species vary tremendously in their patterns of survivorship. In species that produce many offspring but invest little energy or care in each one, the probability of dying before reaching reproductive age is very high. In other species, heavy parental investment in a small number of offspring means that most individuals survive long enough to reproduce.

To help interpret which pattern might apply to a particular species, population biologists developed the **life table,** a chart that shows the probability of surviving to any given age. (Life insurance companies use life tables to compute premiums for clients of different ages.) Table 36.2, for example, shows a life table for yellow-eyed penguins. The values in the life table account for predation, disease, food availability, and all other factors that prevent an individual from reaching its theoretical life span.

The values in a life table are often plotted onto a **survivorship curve,** a graph of the proportion of surviving individuals at each age. Figure 36.3, for example, shows a graph of the values from the penguin life table. Note that the y-axis data are plotted on a logarithmic scale, not a linear one. The log scale makes it easier to see trends along the entire range of values, from 0 to 1000.

The survivorship curve reveals important details about the lives of yellow-eyed penguins. For example, only about 32% of the chicks survive their first year, although the death rate slows thereafter. Moreover, combining life table data with information about fertility at each age can help fill in additional details. For example, reproduction typically begins in year 3 of a yellow-eyed penguin's life. According to the survivorship curve, only about 25% of the penguins that hatch in any given year ever reach reproductive age. Compare this figure to our own species, in which about 99% of offspring survive long enough to reproduce.

Table 36.2 Life Table for the Yellow-Eyed Penguin

Age (Years)	Number of Survivors (Out of 1000)	Age (Years)	Number of Survivors (Out of 1000)
0	1000	9	122
1	324	10	110
2	278	11	98
3	242	12	87
4	217	13	75
5	193	14	71
6	172	15	63
7	154	16	57
8	138		

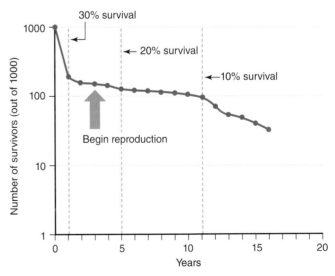

Figure 36.3 Penguin Survivorship. In this population of yellow-eyed penguins, most offspring die before age 1, after which mortality levels off.

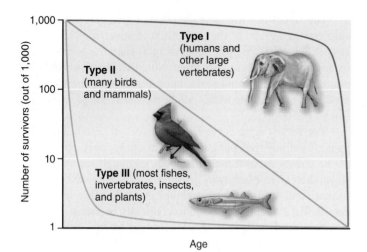

Figure 36.4 Three Survivorship Curves. This graph depicts the number of survivors out of a group of 1000 individuals as age increases. The scale of the *y*-axis is logarithmic, which means that straight-line portions of the curves reflect a constant survivorship rate.

The survivorship curves of many species follow one of three general patterns (**figure 36.4**). Type I species, such as humans and elephants, invest a great deal of energy and time into each offspring. The mortality rate is highest as individuals approach the maximum life span. Type II species, including many birds and mammals, have an equal probability of dying at any age. Type III species, such as most marine fishes and invertebrates, most insects, and many plants, may produce many offspring but invest little in each one. Most offspring of type III species therefore die at a very young age. Section 36.4 looks more closely at the evolutionary trade-offs that these survivorship curves reflect.

Of course, these generalized examples do not describe all populations. Many species have survivorship curves that fall between two patterns. The yellow-eyed penguin, for example, has a survivorship curve that combines features of both type I and type II curves.

36.2 | Mastering Concepts

1. Under what conditions will a population grow?
2. What factors determine the birth and death rates in a population?
3. What is the relationship between a life table and a survivorship curve?
4. Describe the three patterns of survivorship curves.

36.3 | Population Growth May Be Exponential or Logistic

Any population will grow if the number of individuals added exceeds the number removed. The **per capita rate of increase,** *r,* is the difference between the birth rate and the death rate. (Note that this simplified formula for *r* ignores the effects of immigration and emigration.) For example, a population with 35 births and 10 deaths per 1000 individuals per year is growing at a rate (*r*) of 25 people per 1000 (0.025), or 2.5% per year. Any population with a positive value of *r* is growing. On the other hand, a negative *r* means the population is shrinking. If *r* equals zero, the population is stable.

A. Growth Is Exponential When Resources Are Unlimited

In the simplest model of population growth, the number of individuals added during any time interval (*G*) depends only on the per capita rate of increase (*r*) and the initial size of the population (*N*), as expressed by this equation:

$$G = rN$$

This pattern describes **exponential growth,** in which the number of new individuals is proportional to the size of the population. For example, suppose that 100 aquatic animals called rotifers are placed in a tank under ideal growth conditions. If *r* is 0.22 per day, then how many rotifers are added each day? The answer depends on the size of the population: the more rotifers in the tank, the larger the capacity to add offspring. **Figure 36.5** shows how to calculate the size of each generation and plots the running total on a graph.

As figure 36.5 shows, a **J-shaped curve** emerges when exponential growth is plotted over time. Growth resulting from repeated doubling (1, 2, 4, 8, 16, 32 . . .), such as in bacteria, is exponential. Species introduced to an area where they are not native may also proliferate exponentially for a time, since they often have no natural population controls. **Figure 36.6** shows another example of exponential growth: the production of grey seal pups on Sable Island, Nova Scotia.

Figure It Out

Suppose the last data point in figure 36.6 is 25,000. If the grey seal pups have an *r* of 0.12 per year, how many additional pups would you expect to be born in the following year?

Answer: $G = rN = 0.12 \times 25,000 = 3000.$

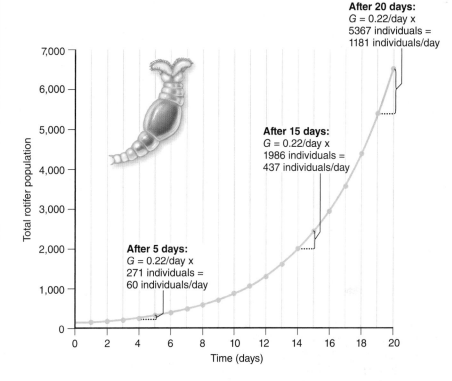

Time (days)	r	N	G = rN	Total population at end of time interval
0	0.22	100	22	122
1	0.22	122	27	149
2	0.22	149	33	182
3	0.22	182	40	222
4	0.22	222	49	271
5	0.22	271	60	331
6	0.22	331	73	404
7	0.22	404	89	493
8	0.22	493	109	602
9	0.22	602	132	734
10	0.22	734	162	896
11	0.22	896	197	1093
12	0.22	1093	241	1334
13	0.22	1334	294	1628
14	0.22	1628	358	1986
15	0.22	1986	437	2423
16	0.22	2423	533	2956
17	0.22	2956	650	3606
18	0.22	3606	793	4399
19	0.22	4399	968	5367
20	0.22	5367	1181	6548

r = Per capita rate of increase
N = Number of individuals at start of time interval
G = Growth rate (number of individuals added per unit time)

After 20 days:
G = 0.22/day x 5367 individuals = 1181 individuals/day

After 15 days:
G = 0.22/day x 1986 individuals = 437 individuals/day

After 5 days:
G = 0.22/day x 271 individuals = 60 individuals/day

Figure 36.5 Exponential Population Growth. For a population of rotifers with unlimited resources, the number of individuals added each day (G) is the product of the per capita rate of increase (r) and the number of individuals at the beginning of the interval (N). With each generation, G increases, even though r remains constant.

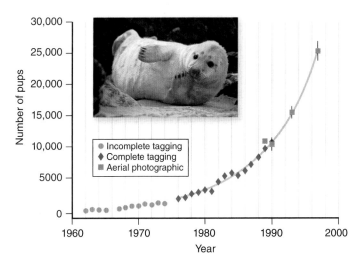

Figure 36.6 Seal Pup Population. Forty years of tagging and aerial surveys of the grey seal pups on Sable Island, Nova Scotia, reveal a consistent pattern of exponential population growth.

B. Population Growth Eventually Slows

Exponential growth may continue for a short time, but it cannot continue indefinitely because some resource is eventually depleted. **Environmental resistance** is the combination of external factors that keep a population from reaching its maximum growth rate. Environmental resistance includes competition, predation, and anything else that reduces birth rates or increases mortality. (Section 36.3C describes these factors in more detail.)

Thanks to environmental resistance, every habitat has a **carrying capacity,** which is the maximum number of individuals that the habitat can support indefinitely. This carrying capacity imposes an upper limit on a population's size. How does this limit affect the population's growth rate? According to the **logistic growth** model, the early growth of a population may be exponential, but growth slows as the population approaches the habitat's carrying capacity (**figure 36.7**). The resulting **S-shaped**

Time (days)	r	N	K	$\left(\frac{K-N}{K}\right)$	$G = rN\left(\frac{K-N}{K}\right)$	Total population at end of time interval
0	0.22	100	2000	0.95	21	121
5	0.22	253	2000	0.87	49	302
10	0.22	578	2000	0.71	90	668
15	0.22	1088	2000	0.46	109	1197
20	0.22	1578	2000	0.21	73	1651
25	0.22	1850	2000	0.07	31	1881
30	0.22	1954	2000	0.02	10	1964
35	0.22	1987	2000	0.01	3	1990
40	0.22	1996	2000	0.00	1	1997

r = Per capita rate of increase
N = Number of individuals at start of time interval
K = Carrying capacity
G = Growth rate (number of individuals added per unit time)

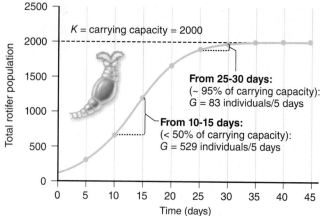

Figure 36.7 Logistic Population Growth. When resources become limited, the number of individuals added each day declines as the carrying capacity approaches. Note that r and the initial value of N are the same as in figure 36.5.

curve depicts the leveling off of a population in response to environmental resistance.

In logistic growth, the number of new individuals added after each time interval follows this simple equation:

$$G = rN\left(\frac{K-N}{K}\right)$$

G equals the number of individuals added per unit time; r is the per capita rate of increase; N is the number of individuals in the population at a given time; and K is the carrying capacity. On the right side of the equation, the term rN is the growth rate that would occur if resources were not limiting. The term (K−N)/K, however, accounts for the increasing environmental resistance as the population approaches the carrying capacity. When N is very small relative to K, this expression yields a numerical value near 1, so the population growth rate is high. When N is near K, the value of the expression drops to near 0, and the growth rate is low.

Figure It Out

Suppose that the carrying capacity for the grey seal pup population on Sable Island is 60,000. If r = 0.12 per year and the current pup population is 25,000, how many new pups will be added next year?

Answer: (K−N)/K = (60,000−25,000)/60,000 = 0.58. Therefore, G = 0.12 × 25,000 × 0.58 = 1740.

Figure 36.8 shows a logistic curve that describes the spread of a rose disease over several years. The S-shaped curve reflects the simple fact that the disease cannot continue to spread after all available hosts are infected. In this case, the carrying capacity is determined largely by host availability.

In most situations, however, the carrying capacity of a habitat is not fixed. A drought that lasts for a decade may be followed by a year of exceptionally heavy rainfall, causing a flush of new plant growth and a sudden increase in food availability.

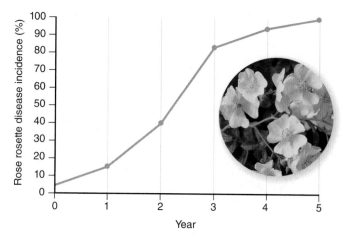

Figure 36.8 Rose Disease. The incidence of rose rosette disease showed a logistic growth pattern over a 5-year period.

Alternatively, the food on which a species relies may disappear, or a catastrophic flood can drastically reduce the carrying capacity as habitat is destroyed.

In studying population growth, it is important to understand that some species do not fit neatly into either the exponential or logistic models. In collared lemmings, for example, the population fluctuates on a 4-year cycle (figure 36.9). The origin of such boom-and-bust cycles remains mysterious, but predation by stoats (a type of weasel) appears to be one of the main factors regulating the ups and downs of the lemming population.

C. Many Conditions Limit Population Size

A combination of factors regulates the size of most populations. Consider the population of songbirds in your town. Some lose their lives to cold weather or food shortages, whereas others succumb to infectious disease or the jaws of a cat. These and other limits on the growth of the bird population fall into two general categories: density-dependent and density-independent (figure 36.10).

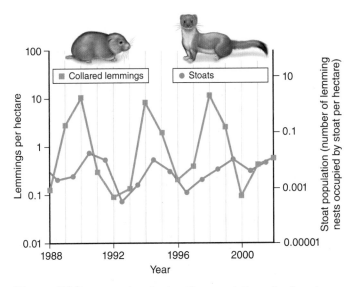

Figure 36.9 Population Cycle. The population of collared lemmings fluctuates regularly over a 4-year period. Research suggests that a major influence on the lemming population is the number of stoats, a type of weasel that preys on the lemmings.

a.

b.

Figure 36.10 Factors Limiting Populations. (a) Density-dependent factors become more important as the habitat becomes crowded. These lesser flamingoes are feeding on a mud flat. (b) Density-independent factors such as forest fires limit populations at all densities.

Density-dependent factors are conditions whose growth-limiting effects increase as a population grows. Most density-dependent limits are **biotic,** meaning they result from interactions with living organisms. Within a population, competition for space, nutrients, sunlight, food, mates, and breeding sites is density-dependent. When many individuals share limited resources, few may be able to reproduce, and population growth slows or even crashes.

For example, competition for food may finally slow the growth of the invasive brown tree snake on Guam (see this chapter's opening essay). After depleting the island of small prey, the snake population is finally showing signs of stress and overcrowding, and reproduction is apparently slowing down.

Besides intraspecific competition, other species can also exert density-dependent limits on a population's growth. Many species often share the same resources in a habitat, so interspecific competition for nutrients and space can be fierce (see section 37.1). Infectious disease also takes a toll. Many viruses, bacteria, fungi, and protists spread by direct contact between infected individuals and new hosts. The higher the host population density, the more opportunity for these disease-causing organisms to find new hosts. Likewise, a higher population density may lead to a higher probability of death by predation, as depicted in figure 36.9.

Density-independent factors exert effects that are unrelated to population density. Most density-independent limits are **abiotic,** or nonliving. Natural disasters such as earthquakes, floods, volcanic eruptions, and severe weather are typical density-independent factors. The high winds of a hurricane, for example, destroy many birds' nests, regardless of whether the bird population is high or low. Likewise, a lava flow kills everything in its path. The effects of oil spills and other industrial accidents are density-independent, too. In addition, as described in chapter 39, habitat destruction related to human activities is a density-independent factor that is pushing many species to the brink of extinction.

No matter what combination of factors controls a population's size, it is worth remembering the importance of these limits in natural selection and evolution. Many individuals do not survive long enough to reproduce. Those that do manage to breed have the adaptations—and the luck—that allow them to escape the biotic and abiotic challenges that claim the lives of many of their counterparts. As these individuals pass their genes on, the next generation contains a higher proportion of offspring with those same adaptations.

36.3 | Mastering Concepts

1. What is the per capita rate of increase?
2. What conditions support exponential population growth?
3. What is environmental resistance, and what is its relationship to the carrying capacity?
4. How does logistic growth differ from exponential growth?
5. Distinguish between density-dependent and density-independent factors that limit population size, and give three examples of each.

36.4 | Natural Selection Influences Life Histories

As we have already seen, a population's size depends in part on birth and death rates. Species differ widely, however, in the timing of these events. Population ecologists therefore find it useful to document a species' **life history,** which includes all events of an organism's life from conception through death. The main focus of life history analysis, however, is the adaptations that influence reproductive success.

Population growth curves and life tables, described in section 36.3, document two elements of an organism's life history. Other aspects include the life span, developmental rate, reproductive timing, social behaviors, mate selection, number and size of the offspring, the number of reproductive events, and the amount of parental care. This section explores how natural selection shapes some of these life history traits. ▸ social behaviors, p. 730

A. Organisms Balance Reproduction Against Other Requirements

A species' life history reflects a series of evolutionary trade-offs; after all, supplies of time, energy, and resources are always limited. As a juvenile, an organism divides its efforts among growth, maintenance, and survival. After reaching maturity, another competing demand—reproduction—joins the list.

Reproduction is extremely costly. Many animals, for example, devote time and energy to attracting mates, building and defending nests, and incubating or gestating offspring. Once the young are hatched (or born), the parents may feed and protect them. All of these activities limit a parent's ability to feed itself, defend itself, or reproduce again. In plants, the reproductive investment is also substantial. Flowers, fruits, and defensive chemicals cost energy to produce and take away from a plant's photosynthetic area, reducing the ability to capture sunlight.

Besides the total effort allocated to reproduction, the timing is also critical. An organism that delays reproduction for too long may die before producing any offspring at all. On the other hand, reproducing too early diverts energy away from the growth and maintenance that may be crucial to survival.

The solutions to these trade-offs vary tremendously among species. A bacterial cell, for example, may grow for just 20 minutes before dividing asexually. A female winter moth mates once and lays hundreds of fertilized eggs just before she dies. A pigweed plant sheds 100,000 seeds in the one summer of its life. These and many other organisms mature early, produce many offspring in a single reproductive burst, and die. Humans, elephants, and century plants, on the other hand, mature late and produce only a few offspring throughout their long lives.

B. *r*- and *K*-Selected Species Differ in the Trade-Off Between Quantity and Quality

Although each species is unique, ecologists have discovered that life histories fall into patterns shaped by natural selection. One prominent trade-off is the balance between offspring quality and quantity.

At one extreme are ***r*-selected species,** in which individuals tend to be short-lived, reproduce at an early age, and have many offspring that receive little care (figure 36.11a). The population's growth rate can be very high; the *r* in "*r*-selected" reflects a large per capita rate of increase. Each offspring, however, has a very low probability of surviving to reproduce; this pattern is typical of species with type III survivorship curves (see figure 36.4).

Weeds, crop pests such as insects, and many other invertebrates are *r*-selected. Their populations can skyrocket when conditions are favorable, but density-independent factors such as frost or drought soon limit their growth.

At the other extreme are ***K*-selected species,** in which individuals tend to be long-lived, to be late-maturing, and to produce a small number of offspring that receive extended parental care (figure 36.11b). High parental investment in each offspring means that most live long enough to reproduce. The *K* in "*K*-selected" stands for the carrying capacity; density-dependent factors such as competition keep these populations close to the carrying capacity. Many birds and large mammals (organisms with survivorship curves that approximate type I or type II) live in *K*-selected populations.

Figure 36.11 *r*- and *K*-Selection. (a) Grasses such as rice plants produce many small offspring and invest little in each one. These species are *r*-selected. (b) Coconut trees produce large fruits that each represent a relatively large investment of energy. These trees are therefore *K*-selected.

a.   b.

The concept of *r*- and *K*-selection dates to the 1960s, and ecologists have since found that strict adherence to *r*- or *K*-selection is the exception, not the rule. For example, even populations of large animals with *K*-selected life histories fluctuate greatly in response to changes in their environments. Nevertheless, these two strategies illustrate the important point that natural selection shapes a species' life history characteristics. Each species' life history must balance the competing demands of reproduction and survival.

C. Guppies Illustrate the Importance of Natural Selection

Experiments show that natural selection directly influences the details of a life history, including age at maturity and the number of offspring per reproductive event. For example, section 11.7 described how changes in fishing regulations can alter the life histories of commercially fished species.

Caribbean fish called guppies *(Poecilia reticulata)* offer another well-studied example. These fish live in rivers and streams on the South American mainland and on the islands of Trinidad and Tobago. In streams with high predation intensity, predators called cichlids eat adult guppies. In a second set of streams, with moderate predation intensity, the predators are omnivorous fish called *Rivulus,* 10% of whose diet consists of juvenile guppies. In a third set of streams, the guppies do not face significant predation.

The researchers hypothesized that predation would be a selective force influencing the guppies' life histories. For example, they predicted that guppies from streams with cichlids would reach reproductive maturity at a younger age and devote more energy to reproduction than guppies from other streams. To test their hypothesis, they collected guppies from 16 sites in Trinidad: seven with high predation intensity, five with moderate predation intensity, and four with low predation intensity. They dissected each female fish and weighed any offspring she carried inside her, and they measured the mature males and females from each stream to estimate the age at maturity.

Guppies from the cichlid (high-predation) stream matured earlier and had smaller offspring than guppies in the other streams (figure 36.12). They also had more offspring and reproduced more frequently. Other studies confirmed that these differences were genetic and therefore were subject to natural selection. The investigators also used field and laboratory experiments to confirm that the predators (not some other condition in the streams) were responsible for the life history differences.

36.4 | Mastering Concepts

1. Distinguish between *r*-selected and *K*-selected species.
2. How have studies with guppies shown that natural selection shapes life histories?

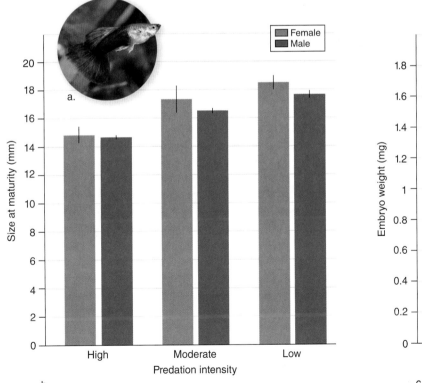

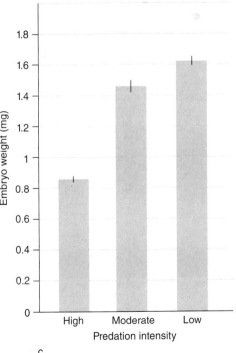

Figure 36.12 Guppy Life History. (a) Guppies *(Poecilia reticulata)* are brightly colored fish that are popular subjects for studies of population ecology. (b) High predation intensity prompted early maturity in male and female guppies. (c) Predation intensity was also correlated with smaller offspring.

36.5 | The Human Population Continues to Grow

The human population has grown exponentially in the last 2000 years (**figure 36.13**). So far, humans have found ways to escape many of the forces that limit the growth of other animal species, but exponential growth cannot continue indefinitely. This section describes how the principles of population growth apply to the human population.

A. Population Dynamics Reflect the Demographic Transition

By early 2010, the world's human population was approaching 6.8 billion. Since Earth's land area is about 150 million square kilometers, the average population density is about 45 people per square kilometer of land area, but the distribution is far from random. For the most part, the highest population densities worldwide occur along the coastlines and in the valleys of major rivers; very few people live on high mountains, in the middle of the world's major deserts, or on Antarctica. Two countries—China and India—alone account for one third of all humans.

Overall, the human population growth rate is about 1.2% per year and declining. Demographers project that zero population growth may happen during the twenty-second century, but no one is certain. Nor do we know how many people will inhabit Earth when that occurs (see this chapter's Burning Question).

Clearly, however, less-developed countries are growing at much faster rates than are more-developed countries (**figure 36.14**).

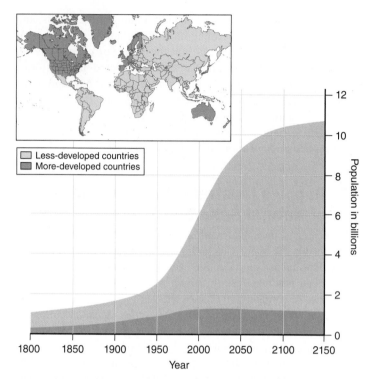

Figure 36.14 Projected Human Population Growth. Future population growth will continue to be concentrated in less-developed countries.

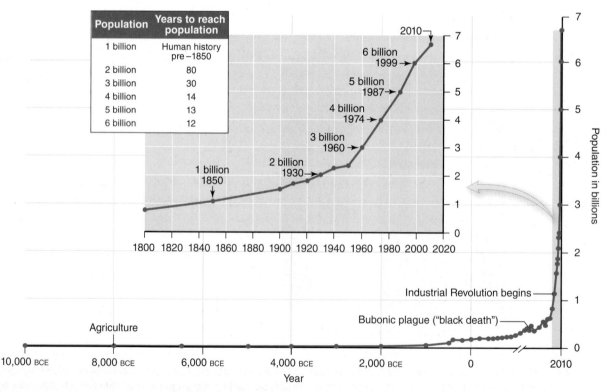

Figure 36.13 Historical Growth of the Human Population. The J-shaped curve indicates exponential growth, with the most rapid population growth occurring in the past 200 years (*inset*). (Data from U.S. Census Bureau.)

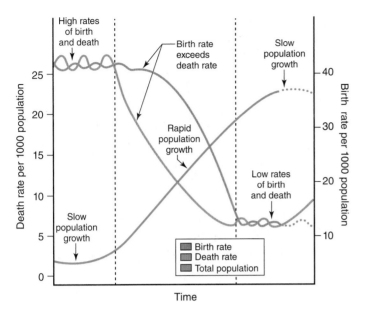

Figure 36.15 The Demographic Transition. During the early phase of the demographic transition, both birth and death rates are high, so the population remains low. Population growth is rapid during the second stage, when birth rates exceed death rates. In the third phase, birth and death rates are both low, and the population stabilizes.

Why the difference? The answer is that each country's economic development influences its progress along the **demographic transition,** during which birth and death rates shift from high to low (figure 36.15). In the first stage of the demographic transition, population growth is minimal because both birth and death rates are high. Then, during the second stage, improved living conditions and disease control lower the death rate, but birth rates remain high. This transitional period therefore sees the rapid population growth typical of the world's less-developed countries.

During the third stage of the demographic transition, birth rates fall; the difference between birth and death rates is once again small. The world's more-developed countries have entered this stage, and a few even have declining populations because death rates exceed birth rates.

Factors Affecting Birth Rates As described in section 36.2, a population's age structure helps predict its future birth rate. Figure 36.16 shows the age structures for the world's three most populous countries. In India and many other less-developed countries, a large fraction of the population is entering its reproductive years, suggesting a high potential for future growth. In the United States, as in many other developed countries, the population consists mainly of older individuals. Such populations are stable.

As recently as 1990, China's age structure resembled that of present-day India. But China's population now shows a decline in the youngest age classes, so its future growth rate should decline. The Chinese government has controlled runaway population growth with drastic measures, rewarding one-child families and revoking the first child's benefits if a second child is born. Although China remains the world's most populous nation, biologists expect India to take the lead by 2025.

A close look at China's age structure diagram also reveals another consequence of that country's childbirth policy. Although the numbers of older males and females are approximately equal, the diagram reveals a bias toward male children in younger age classes. According to cultural tradition in China, parents prefer male children. Couples who are allowed just one child may end pregnancies with female fetuses in hopes of conceiving a male later. Population biologists are closely studying the social and demographic consequences of this gender shift.

Overall, why do birth rates tend to decline as the demographic transition progresses? One explanation for this trend is the availability of family planning programs, which are relatively inexpensive

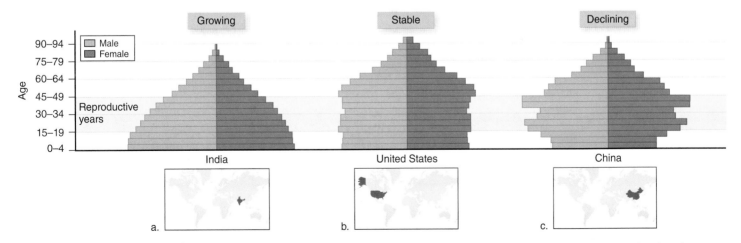

Figure 36.16 Age Structures for Three Human Populations. In age structure diagrams, the width of each bar is proportional to the percent of individuals in that age class. As time passes, younger age classes shift into the reproductive years and beyond. (a) India's population is likely to continue to grow because a high proportion of individuals are in prereproductive age classes. (b) The population of the United States is stable, with roughly equal numbers of people of each age group. (c) China's growth rate should decline because relatively few of its members are in prereproductive age classes. (Data from U.S. Census Bureau, International Data Base.)

and have immediate results. Social and economic factors play an important role as well. Educated women are most likely to learn about and use family planning services, have more opportunities outside the home, and may delay marriage and childbearing until after they enter the work force. Delayed childbearing often means fewer children and therefore slows population growth.

Factors Affecting Mortality   Besides birth rate, mortality is the other major factor influencing the demographic transition. Overall, average life expectancy has increased steadily throughout history, from 22 during the Roman Empire to 33 in the Middle Ages. Today, average life expectancy is about 75 in the most-developed countries and about 50 in the least-developed countries, reflecting a substantial difference in mortality rates.

Table 36.3 shows the top 10 causes of death in developed and developing countries. Heart disease, stroke, lung cancer, and other lung problems top the list in the developed world. Infectious diseases rank relatively low, thanks in large part to sanitation, antibiotics, and vaccines (figure 36.17).

In contrast, deadly infectious diseases such as diarrhea, HIV/AIDS, tuberculosis, and malaria are more prominent in developing countries. Crowded conditions facilitate the spread of tuberculosis and waterborne diseases, especially in areas with limited access to clean drinking water. AIDS has taken an especially high toll in sub-Saharan Africa, where the epidemic has significantly increased mortality rates, and life expectancy has actually begun to decline. ▶ HIV, p. 328

B. The Ecological Footprint Is an Estimate of Resource Use

The human population cannot continue to grow exponentially because living space and other resources are finite. Worldwide, increasing numbers of people will mean greater pressure on land,

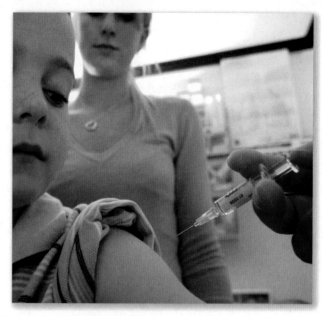

Figure 36.17 Lifesaver. Disease prevention, including widespread vaccination, is one factor that has allowed the human population to continue to grow.

water, air, and fossil fuels as people demand more resources and generate more waste.

Ecologists summarize each country's demands on the planet by calculating an ecological footprint. Just as an actual footprint shows the area that a shoe occupies with each step, an **ecological footprint** measures the amount of land area needed to support a country's overall lifestyle. The calculation includes, among other measures, energy consumption and the land area used to grow crops for food and fiber, produce timber, and raise cattle and other animals. The area occupied by streets, buildings, and other infrastructure is also part of the ecological footprint, as is land used for waste disposal. Not surprisingly, the world's wealthiest and most populous countries have the largest ecological footprints (figure 36.18).

Table 36.3   Top 10 Causes of Death in High- and Low-Income Countries

Rank	High-Income Countries	Low-Income Countries
1	Heart disease (coronary artery blockage)	Lower respiratory infection (pneumonia, acute bronchitis)
2	Stroke	Heart disease (coronary artery blockage)
3	Lung cancer	Diarrhea (e.g., cholera, rotavirus)
4	Lower respiratory infection (pneumonia, acute bronchitis)	HIV/AIDS
5	Chronic obstructive pulmonary disease (chronic bronchitis, emphysema)	Stroke
6	Dementia (including Alzheimer disease)	Chronic obstructive pulmonary disease (chronic bronchitis, emphysema)
7	Colorectal cancer	Tuberculosis
8	Diabetes mellitus	Neonatal infections
9	Breast cancer	Malaria
10	Stomach cancer	Prematurity and low birth weight

Source: World Health Organization Fact Sheet, "The 10 Leading Causes of Death by Broad Income Group."

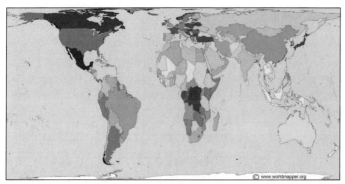

a.

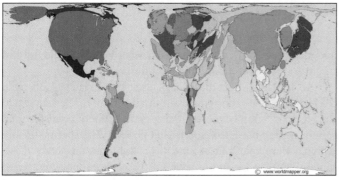

b.

Figure 36.18 Ecological Footprint. (a) In a traditional world map, each country's size is proportional to its land area. (b) In this map, each country's size is proportional to its ecological footprint.

Energy consumption accounts for about half of the ecological footprint. The wealthiest countries make up less than 20% of the world's population yet consume more than half of the energy. Less-developed countries, however, will take a larger share of energy supplies as their populations grow and their economies become more industrialized. Since the vast majority of the energy comes from burning oil, coal, and natural gas, the result will be increased air pollution, acid rain, and carbon dioxide buildup in the atmosphere. The accumulation of CO_2 and other greenhouse gases is implicated in global climate change.

Food production is another significant element of the ecological footprint. Overall, agricultural productivity rises annually. However, increased demand for food, coupled with competition for corn and soybeans from the biofuels and animal feed markets, has pushed food prices beyond the reach of many of the world's poor. To boost food production, people often expand their farms into forests, destroying habitat and threatening biodiversity.
▶ biofuels, p. 376

Because the ecological footprint focuses on land area, it does not include water consumption. Nevertheless, the availability of fresh water has declined worldwide as people have demanded more water for agriculture, industry, and household use. In many poor countries, less than half the population has access to safe water for drinking and cooking. As a result, waterborne diseases such as cholera periodically surge through the dense populations of India and many African and South American countries.

Burning Question

What will happen to the human population?

No one knows exactly how growth rates will change in the future, so it is impossible to predict when or at what level the human population will stabilize. In 2006, the United Nations issued three projections for the world's population, assuming high, medium, and low growth rates. The highest projection says the Earth's population will be around 10.8 billion (and still growing) in 2050. The medium estimate shows the population leveling off at around 9 billion in 2050. The low estimate predicts that the population will peak at about 7.8 billion in 2040, then it will decline to just over 7.7 billion by 2050.

These projections are only as good as their assumptions. Will birth rates in developing countries continue to decline? If so, by how much, and how fast will it happen? Will birth rates decline by the same amount in all countries? Will more-developed countries be willing and able to provide family planning services even as the rural populations of less-developed countries continue to grow? The answers to these questions will determine the future of Earth's human population.

Submit your burning question to:
marielle_hoefnagels@mcgraw-hill.com

Chapter 39 further explores deforestation, species extinctions, increased fuel consumption, global climate change, and other environmental problems related to the expanding human population.

36.5 | Mastering Concepts

1. Which parts of the world have the highest and lowest rates of human population growth?
2. How does the demographic transition reflect changes in mortality and birth rates?
3. What factors affect birth and mortality rates worldwide?
4. What are some of the environmental consequences of human population growth?

36.6 | Investigating Life: Let Your Love Light Shine

In sexually reproducing organisms, the production of new individuals—the foundation of population growth—requires that sperm find their way to egg cells. The life histories of many species include elaborate games of seduction in which the spoils of mating go to those who play their cards just right. But why does the game exist in the first place? What prompts a ram to butt heads with a rival, a frog to croak, or a peacock to flourish its magnificent tail? The answer is that these actions draw the attention of choosy mates (usually females) who often decide which males will get to pass their genes to the next generation.

In some species, the signals of the mating game are easy to interpret. The strongest ram gets the ewe by driving out rivals with jaw-jarring furor. His display of brute strength suggests that his offspring also will be able to fend for themselves. In other species, however, winning strategies are more difficult to sort out. Why do ornamental tail feathers turn the drab female peahen's eye instead of, say, a large beak? And what is it about a male frog's croak that is so irresistible to the female of the species? Couldn't she just as easily make a mating decision based on the color of his belly or the length of his toes?

The mystery of sexual selection has long intrigued evolutionary biologists. Chapters 11 and 35 described sexual selection as a version of natural selection that results from variation in the ability to get mates. Not surprisingly, individuals with the most mating opportunities usually leave the most offspring. In sexual selection, the female's choice is critical. Her reproductive success—that is, her fitness—depends in part on a mate that offers the best provisions for the next generation. Clearly, there is strong selective pressure on females to choose mates based on honest signals. Yet every male conveys many possible signals, including his size, weight, strength, color, calls, and movements. Sorting out which features are key to the female's choice is a challenge.

Tufts University biologists Christopher Cratsley and Sara Lewis sought to unravel the mystery of the sexual dialogue. They studied flashes of light from the firefly *Photinus ignitus* (figure 36.19). Though the light charms us on a warm summer's evening, the real function of the bioluminescent flashes is to deliver precise mating information between the sexes.

The consummate act in firefly mating is the transfer of a protein-rich sperm packet, called a spermatophore, from the male to the female's reproductive tract. Because fireflies (which are actually beetles) do not feed as adults, the spermatophore is a nuptial gift complete with genetic information and nutritious amino acids for the developing eggs.

Cratsley and Lewis figured that a fitness advantage went to females that choose males with larger spermatophores. The larger the nuptial gift, the more protein would be available for the offspring and the greater the female's reproductive success. But how would a female know which male packed the largest

Figure 36.19
Firefly. A male firefly lets his love light shine.

spermatophore? The researchers hypothesized that male fireflies might code this information in their flashes.

To test their hypothesis, the researchers watched 36 male fireflies copulate with females in the laboratory. They dissected the females immediately after spermatophore transfer and removed each packet of sperm and protein. The spermatophore weights were analyzed against male flash duration, body mass, lantern width, and lantern area. Of these, flash duration carried the most reliable information: the longer the flash, the heavier the spermatophore (figure 36.20a).

Next, the team let female fireflies tell them what they liked about the males. Would they prefer the long flashes that signal a large nuptial gift? One way to answer this question would have been to let the females choose between males with different flash durations. However, because males differ in other ways as well, it would have been difficult to conclude that the female preference was for flash duration and not some other variable that the researchers never thought to measure.

The researchers therefore created artificial males. In a series of clever experiments, they flashed a yellow light-emitting diode, and the females responded to the artificial males by giving a return flash if they liked what they saw. The researchers tested females against a range of flash durations: from very fast, 55-millisecond flashes to ones much longer than those the males of the species can produce (130 milliseconds). They found that females preferred long flashes, but only up to a limit (figure 36.20b). Apparently, very long flashes resemble those of predatory fireflies. Furthermore, wasting time and energy responding to flashes from the wrong species would reduce the female's reproductive success.

Cratsley and Lewis' experiments support the notion that firefly flashes are much more than the sweet nothings of a romantic rendezvous. Males advertise the size of their spermatophore gifts, and females respond when the size and flash duration are right. In some ways, these advertisements are similar to commercial messages on TV. Manufacturers of everything from deodorant to dishwashing detergents clamor for your attention, each claiming that their products are best. If you are dissatisfied with your purchase, the worst that can happen is that you buy another

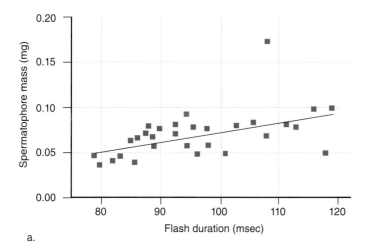

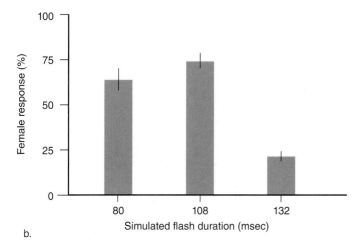

Figure 36.20 The Firefly Flash: An Honest Signal. (a) Flash duration is positively correlated with the mass of a male firefly's spermatophore, which contains both sperm and nutrition for the future offspring. (b) Females responded the most to flashes that were long—but not too long.

brand on your next trip to the store. In the mating game, however, the consequence of a "wrong" decision may make or break an individual's contribution to the next generation.

Cratsley, Christopher K., and Sara M. Lewis. 2003. Female preferences for male courtship flashes in *Photinus ignitus* fireflies. *Behavioral Ecology,* vol. 14, pages 135–140.

36.6 | Mastering Concepts

1. How did researchers test the hypothesis that a firefly's flash duration is correlated with spermatophore mass?

2. A male firefly with a small spermatophore could "lie" by using a long flash. What might be the consequence in the short term? In the long term, what would happen to the reliability of the flash duration signal if many males "cheated" in this way?

Chapter Summary

36.1 | A Population Consists of Individuals of One Species

- **Ecology** considers interrelationships between organisms and their environments. It includes interactions at the **population, community,** and **ecosystem** levels.

A. Density and Distribution Patterns Are Static Measures of a Population

- A **habitat** is the location where an individual normally lives.
- **Population density** is a measure of the number of individuals per unit area, and **population distribution** describes how individuals are distributed within the habitat.

B. Isolated Subpopulations May Evolve into New Species

- Dispersal to new habitats may produce subpopulations that exchange individuals on a limited basis. An isolated subpopulation may evolve into a new species if genetic isolation leads to one or more reproductive barriers.

36.2 | Births and Deaths Help Determine Population Size

A. Births Add Individuals to a Population

- The factors that determine a population's size over time are part of the study of **population dynamics.**
- A population grows when more individuals are added through birth or **immigration** than leave due to death or **emigration.**
- The **birth rate** is the number of new individuals produced per capita in a defined time period. A population's birth rate depends on many factors, including the **age structure.**

B. Survivorship Curves Show the Probability of Dying at a Given Age

- The **death rate** reflects the number of deaths per unit time.
- A **life table** shows the number of survivors remaining in a cohort at each age. **Survivorship curves** fall into three general patterns, reflecting the balance between the number of offspring and the amount of parental investment in each.

36.3 | Population Growth May Be Exponential or Logistic

- The difference between the birth rate and the death rate is *r,* the **per capita rate of increase.**

A. Growth Is Exponential When Resources Are Unlimited

- Population growth that is proportional to the size of the population is **exponential** and produces a **J-shaped curve.**

B. Population Growth Eventually Slows

- In response to **environmental resistance,** the population may stabilize indefinitely at the habitat's **carrying capacity.** A plot of the resulting **logistic growth** produces a characteristic **S-shaped curve.**

C. Many Conditions Limit Population Size

- **Density-dependent factors** such as infectious disease, predation, and competition have the greatest effect on crowded populations. Most such factors are **biotic,** or living.
- **Density-independent factors** such as natural disasters kill the same fraction of the population regardless of the population's density. Most such factors are **abiotic** (nonliving).

36.4 | Natural Selection Influences Life Histories

- The **life history** of a species includes all events from birth to death but typically emphasizes the factors that affect reproduction.

A. Organisms Balance Reproduction Against Other Requirements

- Organisms must allocate limited time, energy, and resources among growth, maintenance, survival, and reproduction.

B. r- and K-Selected Species Differ in the Trade-Off Between Quantity and Quality

- *r*-selected species produce many offspring but expend little energy on each, whereas *K*-selected species invest heavily in rearing relatively few young.

C. Guppies Illustrate the Importance of Natural Selection

- Predation experiments on guppies show that natural selection influences a population's life history characteristics.

36.5 | The Human Population Continues to Grow

A. Population Dynamics Reflect the Demographic Transition

- Fertility, mortality, and age structure differ worldwide, reflecting different stages in the **demographic transition.** In less-developed countries, birth rates are high and death rates are low, producing rapid population growth. As economic development increases, birth rates decline and population growth slows.
- Education, access to contraceptives, and government policies affect birth rates.
- The top causes of death vary around the world. Although sanitation and medical advances have increased average life expectancy overall, HIV/ AIDS is reducing life expectancy in some parts of the world.

B. The Ecological Footprint Is an Estimate of Resource Use

- Countries differ in their **ecological footprint.** Sustained population growth will continue to strain supplies of natural resources such as fossil fuels, farmland, and clean water.

36.6 | Investigating Life: Let Your Love Light Shine

- Males use many signals to display their qualities to females. Understanding the relationships among reproductive success, a suitor's signal, and the female's detection system is the key to understanding sexual selection.
- Female fireflies prefer males that generate long flashes of light, a signal that is positively correlated with the size of the male's spermatophore. The larger the spermatophore, the more nutrition available to the firefly pair's offspring.

Multiple Choice Questions

1. Which of the following is correctly arranged from least inclusive to most inclusive?
   a. Community < ecosystem < population < subpopulation
   b. Subpopulation < population < community < ecosystem
   c. Ecosystem < community < subpopulation < population
   d. Population < subpopulation < ecosystem < community

2. A declining growth rate can be attributed to a large percentage of a population in its
   a. postreproductive years.
   b. reproductive years.
   c. prereproductive years.
   d. Both b and c are correct.

3. Using table 36.2, predict the age that about 10% of the penguin cohort should reach.
   a. 1.5 years   c. 11 years
   b. 3 years   d. 20 years

4. A population that has a consistent survivorship rate is a _____ species.
   a. Type I   c. Type III
   b. Type II   d. Type IV

5. Subtracting the death rate from the birth rate yields
   a. *G,* the number of individuals added in a generation.
   b. *K,* the carrying capacity.
   c. *N,* the number of individuals in the population.
   d. *r,* the per capita rate of increase.

6. In a population with an *r* of 1.5 per day and a starting population of 20, about how many individuals will there be after 2 days of exponential growth?
   a. 125   c. 45
   b. 60   d. 313

7. What happens to a population as *N* approaches *K*?
   a. The death rate exceeds the birth rate.
   b. The value of *r* increases.
   c. The value of *K* increases.
   d. *G* becomes smaller.

8. As a habitat becomes more crowded, which of the following is likely to increase?
   a. Predation
   b. Disease
   c. Competition for food
   d. All of the above are correct.

9. An *r*-selected species emphasizes
   a. large offspring size.
   b. high offspring quantity.
   c. late reproduction.
   d. type I survivorship.

10. The demographic transition describes how
    a. life histories shift as predation intensity increases.
    b. a population grows exponentially.
    c. birth and death rates change with economic development.
    d. populations shift between *r*- and *K*-selection.

Write It Out

1. List some of the ways you have interacted with your surroundings today. Categorize each item on your list as a population-, community-, or ecosystem-level interaction.

2. Describe the difference between population density and distribution. Why aren't organisms always distributed evenly throughout their habitat?

3. Biologists often use rectangular sampling plots called quadrats in studies of plant population ecology. Suppose you want to determine which type of habitat is most favorable for the growth of a particular weed species. How could you use a quadrat of known area to measure the weed's population density in different habitats? Since plants are rarely distributed uniformly throughout a habitat, how might you ensure random sampling?

4. According to the *CIA World Factbook,* in 2009, the birth rate in the United States was 13.82 births per 1000 population, and the death rate was 8.38 deaths per 1000. Immigration added 4.31 migrants per 1000 population. Calculate the overall growth rate of the U.S. population in 2009.

5. Rats, mustangs, koalas, and deer are examples of mammals that are overpopulated in at least some parts of the world. Use the Internet to learn about some proposed strategies for addressing each overpopulation problem. Is each strategy aimed at increasing death rates, reducing birth rates, or changing distribution patterns?

6. Population biologists often tag animals with radio collars equipped with Geographic Positioning System units. What type of useful information might ecologists gain by learning more about animal movements?

7. Decades of overfishing led to the collapse of the cod fishing industry off the coast of North America in the 1980s and 1990s. Given the effect of birth rates, death rates, and age structure on a population, propose an explanation for the decline of the cod population. What sorts of policies might protect the cod from extinction?

8. How does a population's age structure help predict the population's future?

9. Define the following terms: *per capita rate of increase, environmental resistance,* and *carrying capacity.*

10. The rat *(Rattus norvegicus)* has a per capita rate of increase *(r)* of 5.48 per year. Assuming exponential growth and an initial population of 50 rats, what will be the population size after 1 year? List three factors that prevent rats from achieving this reproductive potential.

11. In the 1890s, a group of Shakespeare enthusiasts released 100 European starlings in New York City. Descendants of the birds have spread across North America, and they now number in the hundreds of millions. Propose an explanation for the exponential growth of the starling population. How might you test your hypothesis?

12. What are the differences between the conditions that result in exponential versus logistic population growth?

13. Give three examples of how a habitat's carrying capacity might change over time.

14. Describe a method for estimating the number of elk in an area. What clues might help you understand whether the elk were below or above their habitat's carrying capacity?

15. Bird feeders and bird baths attract dense crowds of songbirds, which, in turn, attract hawks and other predatory birds. Droppings from the birds can spread disease. How do these conditions illustrate density-dependent limits on songbird populations? What are some examples of density-independent factors that might also limit songbird populations?

16. What prevents predator populations from eliminating prey populations?

17. Cite three recent environmental upheavals that may have had a density-independent influence on wildlife populations.

18. Research suggests that salmon farming threatens wild salmon in multiple ways. For example, farmed salmon may spread parasitic lice; interbreeding with farmed salmon can change the genetic structure of wild populations; and contaminants released from the salmon farms may harm wild salmon. Identify each of these problems as density-dependent or density-independent.

19. The grey seal population on Sable Island cannot continue its exponential growth forever. What are three examples of density-dependent factors that might eventually limit population growth for these marine mammals?

20. Some animal behaviors seem at odds with survival and reproduction. For example, when food is scarce, a female scorpion may consume her offspring. In addition, section 31.5 describes a male spider that offers his own body for his mate to consume. Explain each of these behaviors in terms of the trade-offs described in section 36.4A.

21. Distinguish between *r*-selection and *K*-selection, and give an example of an organism with each type of life history.

22. Review section 11.7, which describes how fishing regulations can affect fish populations. How does commercial fishing act as a selective force that alters the life histories of harvested fish?

23. According to the *CIA World Factbook,* in Pakistan, 37% of the population is younger than age 15, and 4% is older than 65. In the United Kingdom, 17% of the population is younger than 15, and 16% is older than 65. Which population will increase faster in the future? Explain.

24. Consult the websites for the U.S. Census and *CIA World Factbook.* What sorts of demographic information do these two sites compile? Why might it be important to understand human population trends within the United States and in the rest of the world?

25. Because of historical, cultural preferences for boys, an unintended result of China's "one child" policy has been a skewing of the sex ratio toward males. In China, about 110 boys are born for every 100 girls. What effects might this have on future population growth in China? Can you foresee any social consequences of having so many females "missing" from the population?

26. How does the human population illustrate the concept of subpopulations? How have improvements in global transportation influenced these subpopulations?

27. What is the relationship between the human population and the ecological footprint? Why does the ecological footprint vary around the world? Overall, humans are using many resources at a faster rate than they can be replenished. What actions would be necessary to reduce humanity's ecological footprint to a sustainable level?

28. The rock pocket mouse *(Chaetodipus intermedius)* is a small mammal that lives in the southwestern United States. Its fur comes in two colors. Light variants typically live on sandy-colored rocks, and the dark mice live on black basalt. Propose a hypothesis that explains this observation, then design one or more experiments to test your hypothesis.

29. Each winter, the Audubon Society encourages birdwatchers to participate in an annual Christmas bird count. Use the Internet to find information about this project. What methods are used to count birds? Why is it useful to repeat counts annually?

Pull It Together

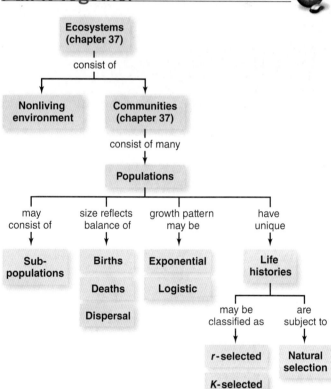

1. What are the additions to and subtractions from a population?

2. Add the following terms to this concept map: *density-dependent, density-independent, carrying capacity,* and *age structure.*

3. How do exponential and logistic growth differ?

4. How do *r*- and *K*-selection reflect life history trade-offs?

5. Describe the trends for human birth and death rates in more- and less-developed countries.

Sea otters are a keystone species in the Pacific Northwest.

Mc Graw Hill **connect**™

|BIOLOGY

Enhance your study of
this chapter with practice quizzes,
animations and videos, answer keys,
and downloadable study tools.
www.mhhe.com/hoefnagels

Mysterious Otter Deaths in the Pacific Northwest

SEA OTTERS, *ENHYDRA LUTRIS*, ARE AMONG THE SMALLEST AND MOST CHARISMATIC OF THE MARINE MAMMALS. But otters are disappearing in ocean and coastline ecosystems near the Aleutian Islands in western Alaska. Between the 1950s and 2000, the population of otters declined by a factor of 10 over an 800-kilometer expanse of coastline. What is going on?

A population of sea otters—or any other species—goes down when birth rates decrease, deaths increase, or individuals move out of the population (see chapter 36). However, studies suggested that otter births were not changing, and researchers detected no redistribution of radio-collared otters from one island to another.

The only remaining explanation was an increase in mortality. Disease and pollution did not seem to account for the change in death rates, so could a new predator have been responsible? One piece of evidence supporting this idea was that no dead otters were washing up on beaches. Another clue came from an observant researcher who saw an orca (killer whale) kill an otter. This event was unusual because the two species had lived together for decades. Then other researchers noticed similar attacks.

Circumstantial evidence that orcas could be responsible came from observations at Adak Island, which had two lagoons with otters: one where killer whales could enter and one where they could not. Where orcas had access, the otter population plunged by two thirds, but where they didn't, only 12% of the otter population perished.

Why would orcas suddenly start eating otters? Marine ecologist James Estes and his coworkers developed a hypothesis to explain the shifting food web. According to their hypothesis, the problem started with a decrease in populations of oily, nutrient-rich fishes such as herring and ocean perch. When Steller sea lions and harbor seals ate a less nutrient-packed fish called pollock instead of their usual fare, they began to die from malnutrition. The orcas that normally eat sea lions and seals sought another food source, and they turned to the coastal ecosystem and its abundant otters.

The orca hypothesis remains controversial. But whatever the cause of the otters' demise, the consequences are dire. Sea otters play a critical role in the vast underwater kelp forests in the ocean waters along the Pacific Northwest coast. Otters eat sea urchins, which devour kelp. In the absence of otters, sea urchin populations explode. The loss of kelp, in turn, eliminates the habitat for many species of marine shrimps, fishes, sea stars, snails, and corals. The otter deaths could therefore topple an entire marine ecosystem.

Killer whales, otters, sea urchins, and kelp form just a few strands in a complex web of interactions; similar webs sustain every ecosystem on Earth. This chapter describes the ties that bind species together.

UNIT 7

What's the Point?

Learning Outline

Learn How to Learn

Studying for Exams

Last-minute cramming for exams may be a classic college ritual, but it is not usually the best strategy. If you try to memorize everything right before an exam, you may become overwhelmed and find yourself distracted by worries that you'll never learn it all. Instead, work on learning the material as the course goes along. Then, on the night before the exam, get plenty of rest, and don't forget to eat on exam day. If you are too tired or too hungry to think, you won't be able to give the exam your best shot.

37.1 | Multiple Species Interact in Communities

Chapter 36 described the ecology of **populations,** which consist of members of the same species that inhabit the same area. Douglas fir trees, for example, form a population in the Pacific Northwest. But no population lives in isolation; this chapter therefore extends the study of ecology to communities and ecosystems.

A **community** is a group of interacting populations. For example, the forest pictured in **figure 37.1** houses a bustling **biotic,** or living, community of trees, mosses, wood decay fungi, worms, beetles, and bacteria. Moreover, each community lives within the context of its physical and chemical surroundings. The **ecosystem** includes the community plus the **abiotic,** or nonliving, environment within a defined area.

Each species in a community has a characteristic home and way of life. Recall from chapter 36 that a **habitat** is the physical place where members of a population typically live, such as a rain forest canopy or the bottom of a river. The habitat is one part of the **niche,** which is the total of all the resources a species requires for its survival, growth, and reproduction. In addition to the physical habitat, the niche also includes the salinity, temperature, light, water availability, and other abiotic conditions where the species lives. Biotic interactions, such as an organism's place in the food chain, are part of the niche as well.

From oceans to mountaintops, many species share each habitat. As described in section 38.3, however, communities vary greatly in diversity. How do ecologists describe these differences? One way is to measure **species richness,** which is simply the total number of species occupying a habitat. A patch of prairie, for instance, may contain about 100 plant species, whereas an equal-sized area of desert might house only six types of plants. In this example, the prairie is more diverse than the desert. But two communities with the same species richness may not be equally diverse. **Relative abundance** describes the proportion of the community that each species occupies. In our patch of prairie, for example, suppose that one type of plant accounts for 90% of the individuals in the community, with 99 species making up the remaining 10%. Because one species has such high relative abundance, that community is less diverse than one in which, say, each of the 100 species makes up 1% of the community.

The first half of this chapter describes community-level interactions, whereas sections 37.3 and 37.4 consider ecosystem-level processes. In studying these topics, keep in mind that population-, community-, and ecosystem-level interactions are the selective forces that shape the evolution of each species.

Figure 37.1 The Forest Floor. The community on the forest floor includes diverse plants and fungi, plus many unseen animals and microbes.

Plants, animals, and every other organism must be able to defend themselves and to acquire resources for growth, maintenance, and reproduction. The adaptations that characterize each species—the ability to produce thorns or live in salt water or catch a gazelle—trace their origins to genetic mutations. But these features persist over multiple generations because they have enhanced reproductive success in a dangerous and competitive world.

A. Populations Interact in Many Ways

Communities usually consist of enormous numbers of species. Some are easily visible, whereas others are microscopic. One on one, their interactions may seem simple—an orca eats an otter, or a wasp kills a caterpillar. But an attempt to map all interactions within a community quickly becomes complicated. Individuals of different species compete for limited resources, live in or on one another, eat one another, and try to avoid being eaten (**table 37.1**).

Table 37.1   Types of Species Interactions: A Summary

Type	Definition and Fitness Effects	Example(s)
Competition	Two or more species vie for the same limited resource (–/–)	Two species of barnacles compete for space in the intertidal zone
Mutualism	Symbiosis in which both partners benefit (+/+)	Algae in coral animals; mycorrhizal fungi in plant roots; animal pollinators of flowering plants
Commensalism	Symbiosis in which one partner benefits with no effect on the other (+/0)	Moss plants on tree bark
Parasitism	Symbiosis in which one partner benefits and the other is harmed (+/–)	Tick on a deer; tapeworm in a human (see chapters 16, 17, 19, and 20 for many examples of disease-causing organisms)
Herbivory	Animal consumes a plant or other photosynthetic organism (+/–)	Cattle grazing on grassland; caterpillar eating tree leaves
Predation	Animal consumes another animal (+/–)	Cat eating birds; bat eating insects

Competition   **Competition** occurs when two or more organisms vie for the same limited resource, such as food, shelter, nutrients, water, or light. Assuming that neither participant obtains all of the resource, the effects of competition are negative for both.

As described in section 36.3, individuals of the same species compete for food, mates, breeding sites, and other resources. Organisms also compete with other species, especially those that occupy a similar niche.

According to the **competitive exclusion principle,** two species cannot coexist indefinitely in the same niche. The two species will compete for the limited resources that they both require, such as food, nesting sites, or soil nutrients. The species that acquires more of the resources will eventually replace the other, less successful species. (Competitive exclusion explains how the microbes that normally occupy the human intestinal tract provide such good protection against harmful invaders; see chapter 31.)

Figure 37.2 illustrates a classic example of competitive exclusion, which restricts two species of barnacles to limited regions within the intertidal zone along Scotland's shoreline. By itself, either type of barnacle can survive throughout the intertidal zone. When both species are present, the faster-growing species *(Balanus)* crowds out its competitor *(Chthamalus),* but only in the lower, moister region of the intertidal zone. The slower growing *Chthamalus* better tolerates dehydration while the tide is out. It can therefore more efficiently use the resources of the upper region of the intertidal zone.

Introduced species sometimes displace native species by competitive exclusion. Zebra mussels *(Dreissena polymorpha),* for example, are native to the Caspian Sea in Asia. These mollusks were accidentally introduced to the Great Lakes in the 1980s and have since spread to many waterways in the United States and Canada. The tiny filter feeders reproduce rapidly and have crowded out native mussel species, with which they compete for food and oxygen. The effects of the zebra mussel invasion have rippled through the rest of the lake community as well. Zebra mussels have greatly increased water clarity, which has changed aquatic plant communities. In turn, the altered plant species composition has triggered changes in the community of fishes. ▸ invasive species, p. 812

Competitive exclusion, however, is not inevitable; coexistence in overlapping niches is also possible. After all, if interspecific competition reduces fitness, then natural selection should favor organisms that avoid competition. Therefore, another possible outcome of competition is **resource partitioning,** in which multiple species use the same resource in a slightly different way or at a different time. For example, 29 warbler species diverged from a common ancestor some 5 million years ago. Five of the species coexist in the same trees in New England forests. The birds occupy similar niches: they all appear similar, and they all eat small insects. Careful observations of the birds, however, revealed that the five species feed in different ways and in different parts of the tree **(figure 37.3).** Their feeding locations and behaviors reduce competition and therefore improve the reproductive success of all five species.

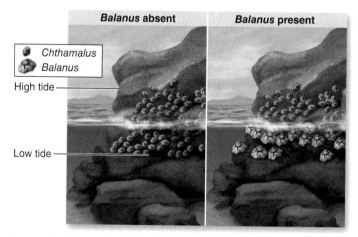

Figure 37.2 Competitive Exclusion. When the barnacle *Balanus* is absent, *Chthamalus* adults occupy the entire intertidal zone. *Balanus* grows faster than *Chthamalus,* however, and when *Balanus* is present, competition for space limits *Chthamalus* to the upper intertidal zone.

Cape May warbler (tips of branches, high tree tops)

Blackburnian warbler (tips of branches, tree tops)

Black-throated green warbler (tips of branches, mid- to upper tree)

Bay-breasted warbler (midstory branches)

Myrtle warbler (treetops, tree trunks, lower foliage, ground)

Figure 37.3 Resource Partitioning. These five species of warblers eat the same foods in conifer forests, but each has a different primary feeding zone.

Symbiotic Relationships

In a **symbiosis** (literally, "living together"), one species lives in or on another. The relationship between symbiotic species may take several forms, defined by the effect on each participant.

Mutualistic relationships improve the fitness of both partners. Flowering plants and their reward-seeking pollinators are a classic example, as are mycorrhizal fungi. In the latter case, the fungus acquires nutrients and water for its host plant; the plant feeds sugars to its live-in partner. Many of the bacteria in our intestines are also mutualistic; we share our food with these microbes, which produce vitamins and defend our bodies against disease. ▶▶ mycorrhizal fungi, p. 402; normal microbiota, p. 347

Commensalism is a type of symbiosis in which one species benefits, but the other is not significantly affected. Most humans, for example, never notice the tiny mites that live, eat, and breed in our hair follicles. Similarly, the reproductive success of a tree is neither helped nor harmed by the moss plants that grow on its trunk and branches. ▶ tiny companions, p. 428

In **parasitism,** one species benefits at the expense of another. The most familiar parasites are disease-causing bacteria, protists, fungi, and worms. Plants may also be parasites. Mistletoe, for example, is a parasitic plant that taps into the water- and nutrient-conducting "pipes" of a host plant. ▶ parasitic plants, p. 484

Herbivory and Predation

All animals must obtain energy and nutrients by eating other organisms, living or dead. An **herbivore** is an animal that consumes plants; a **predator** is an animal that eats other animals, also called **prey.** As in the case of parasitism, the fitness of the herbivore or predator increases at the expense of organism being consumed. In some cases, predator–prey interactions are directly responsible for fluctuations in an animal's population size (see figure 36.9).

Herbivores may eat leaves, roots, stems, flowers, fruits, or seeds. The loss of leaf and root tissue reduces a plant's ability to carry out photosynthesis; consumption of flowers or immature fruits and seeds compromises the plant's reproductive success. Natural selection therefore favors plant defenses against herbivory. Some plant species produce thorns, a milky sap, or distasteful or poisonous chemicals that deter herbivores. The spicy hot chemicals in chili peppers, for example, discourage attack by both fungi and small mammals (see section 23.8). At the same time, many herbivores have adaptations that correspond to the plant's defenses. The caterpillars of monarch butterflies, for example, tolerate the noxious chemicals in milkweed plants.

Likewise, predation exerts strong selective pressure on prey animals, which often have adaptations that help them avoid being eaten (**figure 37.4**). Camouflage and warning coloration are two examples. An interesting variation on the theme of warning coloration is mimicry, in which different species develop similar appearances. For example, a harmless species of fly may have predator-deterring yellow and black stripes similar to those of a stinging bee.

Prey species also commonly have weapons and structural defenses. Many animals have hard shells, pincers, stingers, or other

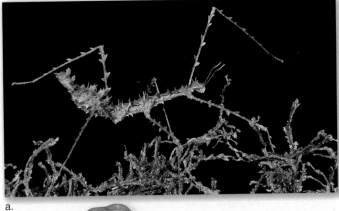
a.

b.

c.

Figure 37.4 Prey Defenses. (a) Camouflage helps prey species hide. This insect from Madagascar resembles the leaves in its habitat. (b) Warning coloration advertises a poison dart frog's defenses. (c) These animals look like ants, but they are actually ant-mimicking jumping spiders. The spiders have none of the weaponry of the ants they resemble.

defensive adaptations (some of which are also useful in capturing their own prey). And as described in chapter 35, prey animals also have a repertoire of defensive behaviors, including stotting, fleeing, fighting, releasing noxious chemicals, or forming a tight group.

Only those predators that can defeat prey defenses will live long enough to reproduce and care for their young. Acute senses, agility, sharp teeth, and claws are common among predators. Camouflage is adaptive in predators as well as prey. Tigers and other big cats, for example, have markings that hide their shape against their surroundings, which helps them sneak up on their prey. Hunting in groups is a behavioral adaptation that helps predators capture large prey efficiently (**figure 37.5**).

Figure 37.5 **Predator Cooperation.** By working together, a group of lions can bring down prey animals that are much larger than the lions themselves.

B. A Keystone Species Has a Pivotal Role in the Community

Sometimes, many species in a community depend on one type of organism, such as the sea otter described in the chapter opening essay. Because otters keep kelp-eating sea urchins in check, their presence is critical to the entire kelp forest community.

The sea otter is a **keystone species:** it makes up a small portion of the community by weight, yet its influence on community diversity is large. Note that *keystone* does not simply mean *essential*. For example, the grasses that underlie the prairie community are not considered keystone species because they make up the bulk of the community.

How do ecologists identify keystone species? One strategy is to measure what happens to a community after artificially removing each species, one at a time. Using this method, researchers discovered that sea stars are keystone predators that maintain species diversity in tide pools. The sea stars normally prey on diverse invertebrates. Removing sea stars, however, allowed mussels and barnacles to take over, crowding out algae and other invertebrates. Without the keystone predator, the tide pools lost seven out of 15 species, and the community collapsed.

Many keystone species, including otters and sea stars, are versatile predators. But mutualists may also be keystone species. For example, mycorrhizal fungi help coniferous trees acquire nutrients from soil, and they produce underground fruiting bodies that small rodents eat. Owls and other predators hunt these small mammals. The fungi are considered keystone species because their small biomass is disproportionate to their enormous influence on community structure.

C. Closely Interacting Species May Coevolve

Some connections between species are so strong that the species directly influence one another's evolution. In **coevolution,** a genetic change in one species selects for subsequent change in the genome of another species. Of course, all interacting species in one community have the potential to influence one another, and they are all "evolving together." These genetic changes are considered coevolution only if scientists can demonstrate that adaptations specifically result from the interactions among the species.

Consider, for example, garter snakes that eat poisonous newts (a type of amphibian). The newts produce a toxin that binds to sodium channels in the snake's muscles, causing paralysis and death. Some of the newts produce exceedingly high concentrations of the toxin. At the same time, some garter snakes have modified sodium channels that are resistant to the poison. Snakes with resistant sodium channels have greater reproductive success than those with typical channels; on the other hand, resistant snakes select for newts that produce increasingly potent toxins.

Coevolution also links flowering plants and insects. As described in chapter 23, a plant may rely exclusively on one insect species for pollination, and the insect may eat nothing but nectar from that plant. The plant–insect relationship may also select for specialized structures and behaviors. Some species of ants and acacia trees, for example, have adaptations not found among related species that lack the relationship. The acacia produces swollen, hollow thorns in which the ants live, along with nectar and nutritious nodules that the ants consume (figure 37.6). These adaptations have, in turn, selected for ant behaviors that defend "their" tree from other herbivores. Section 21.6 describes a similar mutualism, in which ants occupy specialized compartments in branches and attack caterpillars that might otherwise eat the tree's leaves.

37.1 | Mastering Concepts

1. Distinguish among communities, ecosystems, and populations.
2. Name some abiotic and biotic components of your environment.
3. Distinguish between habitat and niche.
4. How do ecologists measure species diversity in a community?
5. What is the competitive exclusion principle?
6. List three examples of symbiotic relationships.
7. Describe some adaptations that protect against herbivory and predation.
8. How are keystone species important in communities?
9. Define *coevolution* and describe an example.

Figure 37.6 **Coevolution of Ants and Plants.** Acacias have adaptations, including these nutritious nodules, that suggest they coevolved with ants.

37.2 | Communities Change Over Time

Communities may appear stable, but that is only because we usually only observe them over a relatively short period. **Succession** is a gradual change in a community's species composition. Succession occurs as competing organisms, especially plants, respond to and modify the physical environment. When pine trees invade a site, for example, they simultaneously shade out lower-growing plants while attracting species that grow or feed on pines.

Ecologists define two major types of succession: primary and secondary. **Primary succession** occurs in an area where no community previously existed. When a volcano erupts, for example, lava may obliterate existing life, a little like suddenly replacing an intricate painting with a blank canvas. Road cuts and glaciers that scour the landscape also expose virtually lifeless areas on which new communities eventually arise.

A patch of bare rock also provides a clear view of primary succession (figure 37.7). **Pioneer species** are first to colonize the area. These hardy organisms, such as lichens and mosses, can grow on smooth rock. The lichens produce organic acids that erode the rock, so that sand and dust accumulates in the crevices. Decomposing lichens add organic material, eventually forming a thin covering of soil. Then rooted plants such as herbs and grasses invade. Soil continues to form, and larger plants, such as shrubs, appear. Larger animals move into the area. Next come aspens and conifers, such as jack pines or black spruces. Finally, hundreds of years after lichens first colonized bare rock, the soil becomes rich enough to support other deciduous trees, and a stable oak-hickory forest community may develop. ▶ lichens, p. 403

The succession that occurs among plants is easiest to see, but the microbial and animal communities change as well. As the first thin soil begins to form, microorganisms, worms, and insects colonize the area. Later, as soil becomes richer and larger plants take root, a changing variety of birds, mammals, and other vertebrates joins the community as well.

In contrast to primary succession, **secondary succession** occurs where a community is disturbed but not destroyed. Because some soil and life remain, secondary succession occurs faster than primary succession. Fires, hurricanes, and agriculture commonly trigger secondary succession (figure 37.8).

An abandoned farm in the eastern United States illustrates secondary succession. This "old field" succession begins when the original deciduous forest is cut down for farmland. As long as crops are cultivated, succession stops. Once farming ends, however, fast-growing pioneer species such as dandelion move in. Slower-growing goldenrod and perennial grasses follow. In a few years, trees such as pin cherries and aspens arrive; pine and oak eventually replace them. A century or so later, the forest community of beech and maple may again be well developed.

Primary and secondary succession share a common set of processes. First, pioneer species colonize a bare or disturbed site. Recall from chapter 36 that the hardy pioneer species are usually *r*-selected, with rapid reproduction and efficient dispersal. These early colonists often alter the physical conditions in ways that

| Bare rock | Lichens | Mosses | Herbs, weeds | Grasses | Shrubs | Pines, hickories, immature oaks | Oaks, hickories, black walnuts, maples, tulip poplars, beeches |

⟵ Time (hundreds of years) ⟶

Figure 37.7 Primary Succession. It takes centuries for a mature forest community to develop on a patch of bare rock. The changing plant species attract an ever-changing variety of animals. The example shown here includes plant species typical of New England.

Figure 37.8 Secondary Succession. A forest fire can devastate an existing community. But it does not take long for seedlings to sprout and take advantage of the nutrients in the tree's ashes. Soon, fresh green foliage will obscure the burned stumps.

enable other species to become established. These new arrivals continue to change the environment. Some early colonists do not survive the new challenges, further altering the community. Later in succession, the dominant species are usually long-lived, late-maturing, *K*-selected species that are strong competitors in a stable environment.

A century ago, ecologists hypothesized that primary and secondary succession eventually lead to a so-called **climax community,** which is a community that remains fairly constant. More recently, however, it has become clear that few (if any) communities ever reach true climax conditions. In the Pacific Northwest, for example, old-growth forests are 500 to 1000 years old, yet they are still changing in their structure and composition. Major disturbances such as fire, disease, and severe storms can leave a mark that lasts for centuries. On a smaller scale, pockets of local disturbance, such as the area affected when a large tree blows over, create a patchy distribution of successional stages across a landscape.

37.2 | Mastering Concepts

1. What is succession?
2. Distinguish between primary and secondary succession.
3. What processes and events contribute to primary and secondary succession?
4. How do disturbances prevent true climax communities from developing?

37.3 | Ecosystems Require Continuous Energy Input

 Sections 37.1 and 37.2 described the biotic interactions among members of a community. We now turn to ecosystems, which represent a still larger scale of ecology. Ecosystems include everything from polar ice to the open ocean to the tropical rain forest, each with its own unique community and set of abiotic conditions. (This chapter's Burning Question considers the difficulties of building artificial ecosystems.)

? Burning Question

Could human life be supported in space or on Mars?

A human mission to Mars would take as long as 10 months just to get there. During that time, the space capsule would have to be a fully self-contained ecosystem. It would have to have a constant energy source, probably the sun, to fuel the growth of primary producers, probably plants.

Without a way to make "pit stops" to acquire new resources, the capsule's occupants would have to be fanatical about recycling all essential elements. Water would be purified and reused again and again. Human wastes would likely be composted into a form that onboard plants could use. Luckily, the byproduct of photosynthesis is O_2. But imagine the catastrophe if an unwanted stowaway, such as a plant pathogen, killed all the onboard plants!

The Biosphere II project in Oracle, Arizona, was an attempt to create an artificial ecosystem that space travelers could eventually replicate on Mars or the moon (figure 37.A). Eight crew members lived in Biosphere II for 2 years beginning in September 1991. They recycled all water and wastes to grow crops, which supplied about 80% of their food (the rest came from an initial 3-month supply grown in Biosphere II before the crew's residence began). Still, the system did not remain self-contained. Soil bacteria produced more CO_2 than anticipated, so Biosphere II's operators had to pump in O_2 from outside. The Biosphere II story illustrates just how hard it will be to design a new human habitat on another planet.

Submit your burning question to:
marielle_hoefnagels@mcgraw-hill.com

Figure 37.A   Biosphere II in Oracle, Arizona.

To understand ecosystem-level interactions, it is useful to recall from unit 1 that all organisms consist of both matter and energy. One way to remember this is to picture a candy bar's nutrition label, which lists the nutrient and calorie (energy) content of the snack. The food label mirrors the contents of a living cell. Inside each cell are organic molecules such as fats, sugars, and proteins. These molecules consist of carbon, hydrogen, oxygen, nitrogen, and other elements. Moreover, the covalent bonds of organic molecules store potential energy that cells can use to do work.
▶ organic molecules, p. 31

Both energy and nutrients are critical to the two properties shared by all ecosystems on Earth. First, energy flows through ecosystems in one direction only. All ecosystems therefore rely on a continuous supply of energy from some outside source, usually the sun. Second, biogeochemical cycles constantly recycle the atoms that make up every object in an ecosystem. This section and the next describe these two properties and their consequences for ecosystem function.

A. Food Webs Depict the Transfer of Energy and Atoms

Many energy and nutrient transfers occur in the context of food chains and food webs. A **food chain** is a series of organisms that successively eat one another (**figure 37.9**). Each organism's **trophic level** describes its position in the food chain.

Trophic levels are defined relative to the ecosystem's energy source. The first trophic level in any food chain is a primary producer. A **primary producer,** or **autotroph** ("self-feeder"), is any organism that can use energy, CO_2, H_2O, and other inorganic substances to produce all the organic material it requires. For most ecosystems, the energy source is sunlight. Plants, algae, cyanobacteria, and some other microorganisms are primary producers that use photosynthesis to trap solar energy in the bonds of organic chemicals such as glucose (see chapter 5).

Nevertheless, a few ecosystems rely on energy sources other than sunlight. For example, some bacteria and archaea are primary producers that can extract energy from inorganic chemicals such as iron or manganese. Countless numbers of these microbes form the bases of complex hydrothermal vent communities that never see the sun. The opening essay for chapter 38 describes these unique ecosystems in more detail.

All of the other trophic levels consist of **consumers,** or **heterotrophs** ("other eaters"), which obtain energy either from producers or from other consumers. For example, the primary consumers in figure 37.9 are herbivores, which eat the primary producers. Secondary consumers are carnivores (meat-eaters) that eat primary consumers, and tertiary consumers eat secondary consumers.

Decomposers, such as many fungi, bacteria, insects, and worms, are consumers that break down **detritus** (dead organisms

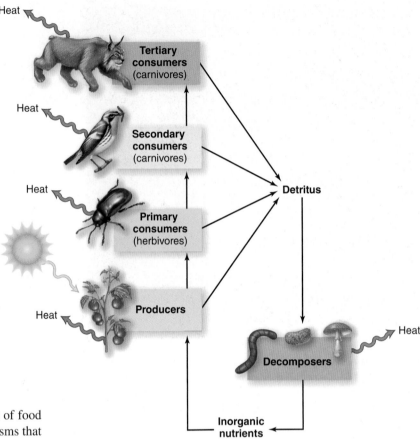

Figure 37.9 Trophic Levels. The cat eats the bird, which ate the beetle, which ate the tomato plant. All organisms contribute organic wastes and dead bodies (detritus) to the ecosystem. Decomposers recycle the detritus back into inorganic nutrients that producers can use.

and feces). This chapter's Apply It Now box describes how we employ decomposers in community wastewater treatment facilities. Without decomposers, dead bodies and organic wastes would tie up all useful nutrients, and ecosystems would grind to a halt.

Autotrophs and decomposers have opposite roles in ecosystems. Whereas autotrophs absorb inorganic nutrients and produce organic molecules, decomposers return the elements in those organic molecules to their inorganic form. Both roles are critical to ecosystem function.

Of course, feeding relationships in an ecosystem are more complex than a simple food chain might suggest. A **food web** is a network of interconnected food chains, such as the Antarctic web in **figure 37.10**. Keep in mind that this diagram is highly simplified. Not shown, for example, are the worms, crabs, hagfishes, sharks, and microbes that feed for months or years on the carcasses of dead whales that sink to the seafloor.

Careful examination of a food web diagram reveals that even the fiercest top predator, such as a killer whale, relies on other organisms, many of them microscopic. The same principle

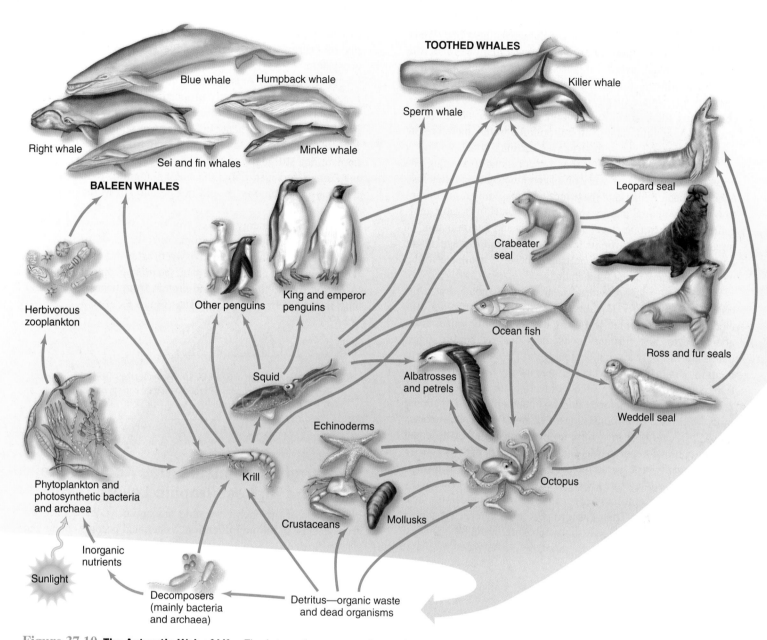

Figure 37.10 The Antarctic Web of Life. The interactions among Antarctic residents form a very complex network. Note that producers, consumers, and scavengers are all present, and that decomposers release inorganic nutrients for producers to use. For simplicity, losses to heat are not illustrated.

applies to the food web in which humans participate. Think of everything you have eaten today: in all likelihood, your meals and snacks have included both plant and animal products. Humans have developed an extremely complex global food chain that includes organisms harvested from land and water all over the world. Some of the connections are surprising. For example, pigs and chickens raised on commercial farms around the world eat millions of tons of fish harvested near South America each year.

B. Every Trophic Level Loses Energy

The total amount of energy that is trapped, or "fixed," by all autotrophs in an ecosystem is called gross primary production. Autotrophs use much of this energy to generate ATP for their own growth, maintenance, and reproduction. As they do so, they lose heat energy (the second law of thermodynamics explains this loss). The remaining energy in the producer level is called **net primary production;** it is the amount of energy available for consumers to eat. ▶ laws of thermodynamics, p. 72

Primary productivity varies widely across the globe. Water availability and the prevailing temperature often determine the potential for plant growth on land. The warm, wet tropical rain forests therefore have among the highest rates of net primary productivity per square meter; deserts have the lowest. In aquatic ecosystems, the availability of inorganic nutrients such as phosphorus is more important. Nutrient-polluted waters therefore often become overgrown with algae (see section 37.4).

Consumers are no better than producers at conserving energy; they cannot digest every bit of food they eat, and they lose energy to heat. Because of these inefficiencies, only a small fraction of the potential energy stored in the bonds of organic molecules at one trophic level fuels growth and reproduction of organisms at the next trophic level. As an overall average, about one tenth of the energy at one trophic level is available to the next highest rank in the food chain.

The "10% rule" provides a convenient estimate, but it ignores the fact that food quality varies widely. The transfer efficiency from one trophic level to the next actually ranges from about 2% to 30%. Primary consumers that eat hard-to-digest plants convert only a small percentage of the energy available to them into animal tissue, whereas meat is easy to digest. In addition, ectothermic animals such as invertebrates and reptiles use energy much more efficiently than do endothermic mammals and birds. A trophic level consisting of lizards therefore consumes much less energy than a trophic level consisting of an equal weight of birds. ▶ endotherms and ectotherms, p. 658

Eventually, as organic molecules pass from trophic level to trophic level, all the stored energy is lost as heat. The heat energy leaves the ecosystem forever, because no organism can use heat as its energy source. Thus, energy flows through an ecosystem in one direction: from source (usually the sun), through organisms, to heat. For the ecosystem to persist, it must have a continual supply of energy. If the energy source goes away, so does the ecosystem.

A **pyramid of energy** represents each trophic level as a block whose size is directly proportional to the energy stored in biomass per unit time (**figure 37.11a**). Because every organism loses energy to heat, the energy pyramid explains why food chains rarely extend beyond four trophic levels. An organism in a still higher trophic level would have to expend tremendous effort just to find the small amount of food available, and that small amount would not be enough to make all that effort pay off.

The loss of energy at each trophic level suggests a way to maximize the benefit we get from crops we grow for food. The most energy available in an ecosystem is at the producer level. The lower we eat on the food chain, the more people we can feed. A person can do this by getting protein from beans, grains, and nuts instead of from meat and dairy (figure 37.11b).

Figure It Out

Consult the food web in figure 37.10. Assuming the "10% rule" is correct, about how many kilograms of krill would it take to support one 40-kilogram emperor penguin?

Answer: 4000 kg

C. Harmful Chemicals May Accumulate in the Highest Trophic Levels

The shape of the energy pyramid has another consequence for ecosystems. In **biomagnification,** a chemical becomes more and

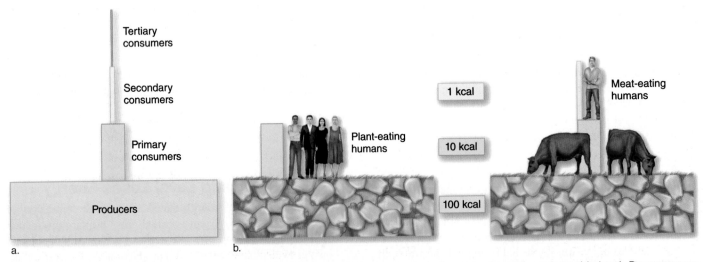

Figure 37.11 **Pyramid of Energy.** (a) Each level in this pyramid depicts the amount of energy stored in each trophic level. Decomposers are omitted in this figure. (b) Humans who derive energy by eating meat are getting only a small fraction of the energy originally present in grain. This example assumes that an average of 10% of the energy in any trophic level is available to the next.

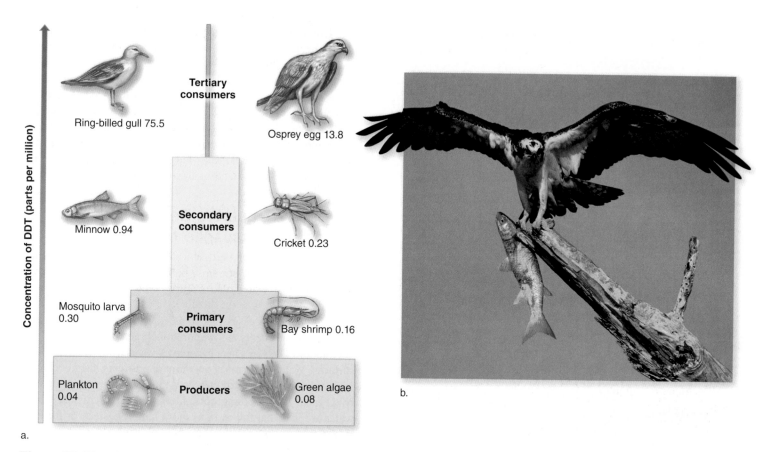

Figure 37.12 Biomagnification. (a) The concentration of DDT in organisms' bodies increases with each successive trophic level. Numbers represent concentrations of DDT per unit of tissue, measured in parts per million (ppm). (b) An osprey with its prey.

more concentrated in organisms at successively higher trophic levels. Biomagnification happens for pollutants and other chemicals that share two characteristics. First, they dissolve in fat. Animals eliminate water-soluble chemicals in their urine but retain fat-soluble chemicals in fatty tissues. Second, chemicals that biomagnify are not readily degraded. A highly degradable chemical would not persist long enough in the environment to ascend food chains.

DDT is a persistent chemical that illustrates biomagnification (figure 37.12). This insecticide was once widely used to kill pests such as mosquitoes and body lice. Researchers soon found that it also harmed other organisms indirectly. Imagine DDT being sprayed over a waterway for mosquito control. Its concentration in the water is initially low. But DDT is fat-soluble, and once it enters an organism's body, it is not eliminated in urine or other watery wastes. As one animal eats another, all of the DDT stored in the prey ends up in the predator. Each predator eats many prey, so the DDT concentrates in the predator's tissues. In organisms at the fourth level of the food chain, DDT concentrations may be 2000 times greater than in organisms at the base of the food web.

The United States banned use of DDT in 1971, after much evidence showed it to become biomagnified in birds of prey such as bald eagles, peregrine falcons, and osprey. High concentrations of DDT caused the birds to produce weak eggshells, which broke when adult birds attempted to incubate the eggs. Few young birds hatched, and populations plummeted until the ban on DDT took effect. Nevertheless, DDT is still used in other countries, especially to control the mosquitoes that transmit malaria.

37.3 | Mastering Concepts

1. Identify the trophic levels of a food chain.
2. What sources of energy sustain ecosystems?
3. What roles do primary producers and decomposers play in ecosystems?
4. How efficient is energy transfer between trophic levels in food webs?
5. Draw an energy pyramid for an ecosystem with three levels of consumers.
6. Explain how biomagnification disproportionately affects organisms at the top of a food chain.

37.4 | Chemicals Cycle Within Ecosystems

Energy flows in one direction, but all life must use the elements that were present when Earth formed. In **biogeochemical cycles,** interactions of organisms and their environments continuously recycle these elements. If not for this worldwide recycling program, supplies of essential elements would have been depleted as they became bound in the bodies of organisms that lived eons ago.

Whatever the element, all biogeochemical cycles have features in common (figure 37.13). Each element is distributed among four major storage reservoirs: organisms, the atmosphere, water, and rocks and soil. Depending on the reservoir, the element may combine with other elements and form a solid, liquid, or gas.

These "pools" are not isolated; instead, transfers among the pools form the basis of each biogeochemical cycle. Along the way, the elements undergo chemical changes. For example, recall from figure 37.9 that autotrophs such as plants take up the inorganic forms of elements and incorporate them into organic molecules such as lipids, carbohydrates, proteins, and nucleic acids. If an animal eats the plant, the elements may become part of animal tissue. If another animal eats the herbivore, the elements may be incorporated into the predator's body. Eventually, decomposers consume the dead bodies and organic wastes, releasing inorganic forms of the elements into the environment.

These transfers and transformations are the root of the term *biogeochemical cycle*: *bio-* refers to the role of organisms; *geo-* stands for the inorganic components, such as those found in the earth; *-chemical* relates to the chemical transformations; and *cycle* is for the endless transfers from reservoir to reservoir.

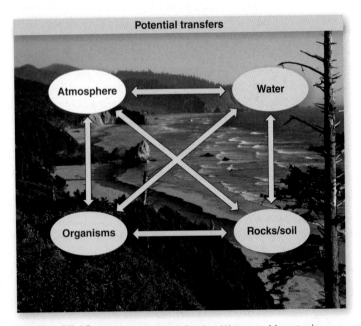

Figure 37.13 Biogeochemical Cycle. Water and inorganic nutrients cycle among four basic storage reservoirs. The photo in the background shows all four.

This section describes the water, carbon, nitrogen, and phosphorus cycles. All of these substances are essential to life and abundant in cells. As you will see, however, the water cycle is somewhat different from the other three cycles. Most processes in the water cycle are physical, not biological. Its status as a biogeochemical cycle is therefore tenuous. Nevertheless, carbon, nitrogen, and phosphorus compounds can dissolve in water, and water movement is important in transporting these elements among the storage pools. Knowledge of the water cycle is therefore essential to understanding the three other nutrient cycles.

A. Water Circulates Between the Land and the Atmosphere

Water covers much of Earth's surface, primarily as oceans but also as lakes, rivers, streams, ponds, swamps, snow, and ice (figure 37.14). Water also occurs below the land surface as groundwater.

The main processes that transfer water among these major storage compartments are evaporation, precipitation, runoff, and percolation. The sun's heat evaporates water from land and water surfaces. Water vapor rises on warm air currents, then cools and forms clouds. If air currents carry this moisture higher or over cold water, more cooling occurs, and the vapor condenses into water droplets that fall as rain, snow, hail, fog, sleet, or freezing rain. Some of this precipitation falls on land, where it may run along the surface. Streams unite into rivers that lead back to the ocean, where the sun's energy again heats the surface, evaporating the water and continuing the cycle.

Rain and melted snow may also soak (percolate) into the ground, restoring soil moisture and groundwater. This underground water feeds the springs that support many species. Spring water evaporates or flows into streams, linking groundwater to the overall water cycle.

Although most processes in the water cycle are physical, organisms do participate; after all, water is essential to life. Animals drink water and consume it with their food, returning it to the environment through evaporation and urination. In addition, plant roots absorb water from soil and release much of it from their leaves in transpiration. The lush plant life of the tropical rain forests draws huge amounts of water from soil and returns it to the atmosphere. ▶ transpiration, p. 479

Because trees play such a critical role in the global water cycle, the wholesale destruction of forests may have important consequences. Deforestation reduces the potential for transpiration, an important avenue by which water returns to the atmosphere. This decrease in transpiration may trigger changes in the global water cycle and, by extension, in the global climate. In addition, in an intact forest, rain and melting snow seep slowly through the forest floor and into the soil, helping to recharge groundwater. Forest destruction removes the spongy duff on the forest floor, so rainwater runs off into streams rather than soaking into the soil. The excess runoff hurts both forests and streams. As the forest soil erodes away, waterways become choked with nutrients and sediments.

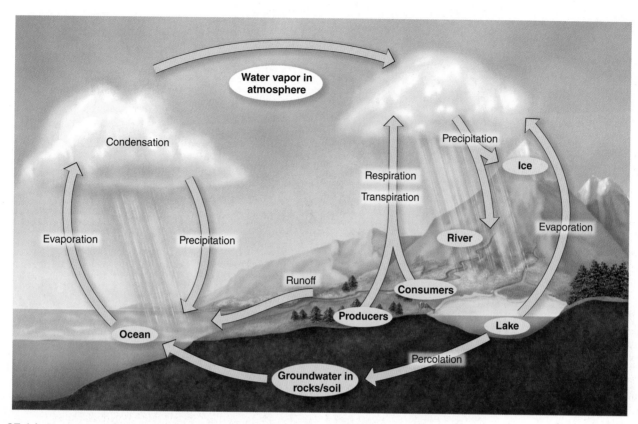

Figure 37.14 The Water Cycle. Water falls to Earth as precipitation. Organisms use some water, and the remainder evaporates, runs off into streams and the ocean, or enters the ground. Animals return water to the environment by respiring and urinating, and plants do so by transpiring.

Apply It Now

What Happens After You Flush

Most people probably give little thought to what happens to whatever they flush down the toilet. But the engineers and technicians who run community sewage treatment facilities think about it all the time.

Decades ago, many towns simply released untreated sewage into rivers and oceans. Because human waste harbors disease-causing organisms, this practice posed an obvious threat to public health. It was also harmful for another reason: the raw sewage killed fish and other aquatic animals. The problem was not that human diseases spread to aquatic organisms. Instead, aquatic microbes decayed the feces and other organic wastes in the sewage. As the microbes respired, they used all the available O_2 dissolved in the water. Huge numbers of fishes and other aquatic organisms suffocated and died.

Federal law now mandates that communities treat wastewater before releasing it. Treatment plants harness the power of microorganisms to consume the organic matter in sewage before it enters waterways. In trickling filters, for example, sewage-eating bacteria and archaea are given "dream homes"—all the organic matter they can eat, along with plenty of moisture and O_2 (see figure 16.13c). After the microbes have done their job, the treated water contains a very low concentration of organic matter.

The presence of these microscopic workers explains why communities prohibit dumping used motor oil or organic solvents such as paint thinner down the drain. Toxic chemicals can poison the bacteria and archaea that degrade sewage, making water treatment impossible.

Thanks in part to water quality laws, U.S. river ecosystems have largely recovered from the past onslaught of untreated sewage. The most commonly used forms of sewage treatment do not, however, remove all chemicals from the waste stream. In particular, a wide array of household chemicals and pharmaceutical drugs often remain in the water discharged from a sewage treatment plant. The antibiotics in soaps and hand sanitizers are especially common, as are hormones excreted in the urine of women taking birth control pills. Ecologists are still studying the effects of these chemicals on wildlife. In the meantime, experts recommend against flushing medications—or anything other than human waste—down the toilet.

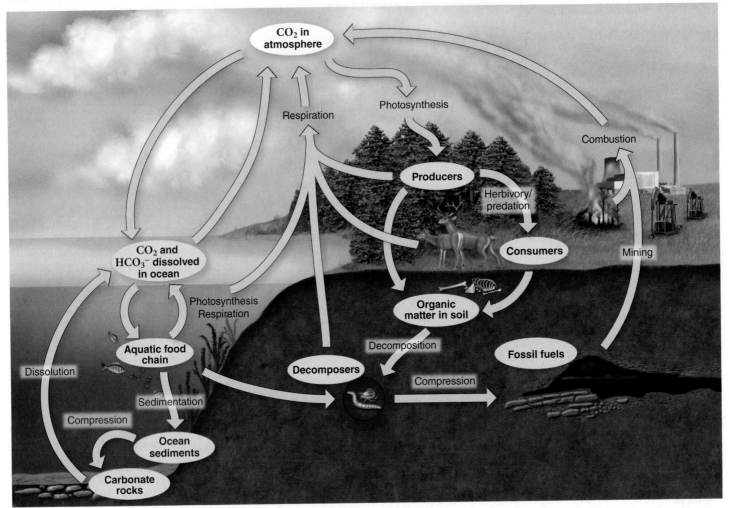

Figure 37.15 **The Carbon Cycle.** Carbon dioxide in the air and water enters ecosystems through photosynthesis and then passes along food chains. Respiration and combustion return carbon to the abiotic environment. Carbon can be retained in geological formations and fossil fuels for long periods. The archaea that produce methane (CH_4) are omitted from the cycle for simplicity.

B. Autotrophs Obtain Carbon as CO_2

Carbon is a part of all organic molecules. Organisms continually exchange carbon with the atmosphere, where the concentration of CO_2 is about 387 parts per million. Autotrophs absorb atmospheric CO_2 and synthesize organic compounds that they incorporate into their tissues (figure 37.15). Cellular respiration releases carbon back to the atmosphere as CO_2. Dead organisms and wastes deposit organic carbon in soil or water. Invertebrates, bacteria, and fungi decompose these organic compounds and release CO_2 to the soil, air, and water.

Some types of archaea also participate in the carbon cycle by metabolizing organic compounds and releasing methane (CH_4) into the atmosphere. Most of these microbes live in anaerobic habitats such as wetlands, marine sediments, and the intestines of humans, cattle, and other animals.

The exchange of carbon between organisms and the environment is relatively rapid, but more stable pools of carbon also exist in soils, carbonate rocks (including limestone), fossil fuels, and ocean sediments. A substantial fraction of soil carbon consists of persistent, decay-resistant organic matter called humus. Limestone consists mostly of the calcium carbonate exoskeletons and shells of ancient sea inhabitants. Fossil fuels, such as coal and oil, form from the remains of long-dead organisms. When these fuels burn, carbon returns to the atmosphere as CO_2. Decades of accumulation of CO_2 and other so-called greenhouse gases in the atmosphere are likely responsible for Earth's gradually warming climate. Chapter 39 describes this topic in more detail.

One of the largest reservoirs of carbon is the ocean. CO_2 from the atmosphere dissolves in ocean water. Most of the dissolved gas reacts with the water to form carbonic acid (H_2CO_3), which dissociates into hydrogen ions (H^+), bicarbonate ions (HCO_3^-), and carbonate ions (CO_3^{2-}). Some of the carbonate ions react with calcium to form calcium carbonate, which precipitates into sediments on the ocean floor. These sediments are one of the major stable repositories of carbon on the planet.

The reaction between CO_2 and water, coupled with the accumulation of CO_2 in the atmosphere, means that the H^+ concentration of the ocean is gradually increasing. The ocean is therefore becoming more acidic, joining global climate change as another side effect of greenhouse gas buildup. Ocean acidification harms coral reefs by dissolving the calcium carbonate skeletons of coral animals. ▶ pH scale, p. 30

C. The Nitrogen Cycle Relies on Bacteria

Nitrogen is an essential component of proteins, nucleic acids, and other biochemicals in living cells. Although the atmosphere is about 78% nitrogen gas (N_2), most organisms cannot use this form of nitrogen.

The nitrogen cycle, shown in **figure 37.16**, therefore depends on **nitrogen fixation,** the process by which some bacteria and archaea convert N_2 into ammonia (present in soil as ammonium ion, NH_4^+). Plants and other autotrophs can absorb ammonium and incorporate it into organic molecules.

Examples of nitrogen-fixing bacteria include *Rhizobium,* which lives in nodules on the roots of legume plants such as beans, peas, and clover (see figure 16.11). Many farmers alternate nonlegume crops, such as corn, with legumes to enrich the soil with biologically fixed nitrogen. In addition, many farmers boost plant growth by applying nitrogen fertilizers to their fields. These synthetic fertilizers rely on an industrial-scale form of nitrogen fixation in which N_2 reacts with hydrogen gas. The

product, ammonia, can be further processed to generate ammonium nitrate or urea fertilizer.

Whatever the source, plants absorb some of the available ammonium. In addition, some microorganisms can generate energy by using the ammonium as a substrate in additional chemical reactions. In a process called **nitrification,** bacteria and archaea convert ammonia to nitrites (NO_2^-) and eventually to nitrates (NO_3^-), another form that plants can use. This process occurs both in soil and in the ocean.

The combustion of fossil fuels such as coal, oil, and natural gas also releases NO_2 gas and other nitrogen oxides into the atmosphere. These compounds return to Earth in precipitation. Excess nitrogen deposition from the atmosphere may be altering the ecology of some low-nitrogen ecosystems, including the bogs where carnivorous plants thrive. ▶ carnivorous plants, p. 475

Figure 37.16 shows that plants and other producers incorporate nitrogen into the organic molecules that make up their own bodies. Consumers then acquire the nitrogen by eating the producers,

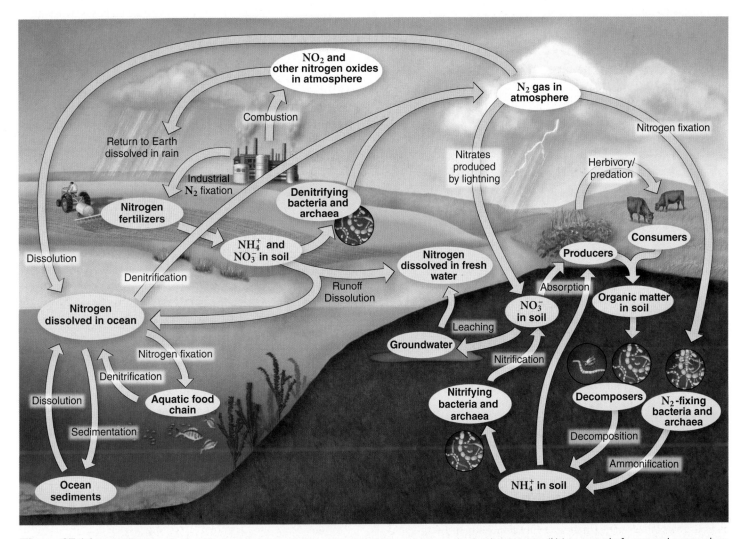

Figure 37.16 **The Nitrogen Cycle.** Nitrogen-fixing bacteria and archaea convert atmospheric nitrogen gas (N_2) to organic forms and ammonium ions, which plants can absorb. Nitrogen returns to the abiotic environment in urine and during the decomposition of organic matter. Bacteria and archaea convert ammonium ions to nitrate (another form plants can use); microbes also convert nitrate to nitrogen gas, completing the cycle.

and so on up the food chain. Decomposers release some ammonia when they decay the dead bodies and wastes.

Yet another group of bacteria completes the cycle. In **denitrification,** bacteria return nitrogen to the atmosphere as they convert nitrites and nitrates to N_2. This process, which is a form of anaerobic respiration, occurs where O_2 is scarce. Wetlands, water-saturated soils, groundwater, and ocean sediments are therefore prime sites for denitrification. ▶ anaerobic respiration, p. 114

Nitrification and denitrification occur in a fish aquarium, a miniature ecosystem with a nitrogen cycle of its own. Fish release toxic ammonia as a waste product, which begins to accumulate soon after fish are placed into a new aquarium. Given enough time, however, populations of nitrifying bacteria grow on the gravel and in the aquarium's filter. These microbes convert the ammonia to nitrites and nitrates. Other microbes carry out denitrification, eliminating the nitrogen as N_2 gas. When starting a new aquarium, one way to jump-start this process is to borrow a used filter or some gravel from an established aquarium. The microbes in the transplanted biofilms will give the new tank a head start in establishing a functioning nitrogen cycle.

D. The Phosphorus Cycle Begins with the Weathering of Rocks

Phosphorus occurs in nucleic acids, ATP, and membrane phospholipids; in vertebrates, this element is also a major component of bones and teeth.

Unlike in the carbon and nitrogen cycles, the atmosphere plays little role in the phosphorus cycle (**figure 37.17**). Phosphorus therefore circulates on a more local scale than do water, carbon, and nitrogen.

The main storage reservoirs for phosphorus are marine sediments and rocks. As phosphate-rich rocks erode, they gradually release phosphate ions (PO_4^{-3}). Autotrophs absorb some of the available phosphorus, and consumers move the element throughout the food web. Decomposers eventually return inorganic phosphates to soil and water. Much of the phosphate, however, joins the sediments raining down onto the ocean floor. After many millions of years, geological uplift returns some of this underwater sedimentary rock to the land.

Because phosphorus is one of the three elements that most commonly limit plant growth (along with nitrogen and potassium), humans mine phosphate rocks to produce plant fertilizers. Animal waste also contains abundant phosphorus. Humans therefore sometimes harvest the guano (droppings) of birds and bats for use as fertilizer.

Excess phosphorus in water, however, can pose environmental problems. Fertilizer from croplands and wastes from chicken and hog farms sometimes enter waterways. In addition, wastewater from sewage treatment plants may contain phosphate detergents that were flushed down household drains. In a process called **eutrophication,** these excess nutrients enter a stream or lake and trigger the explosive growth of algae populations (see figure 38.19). When the algae die, microbes consume their bio-

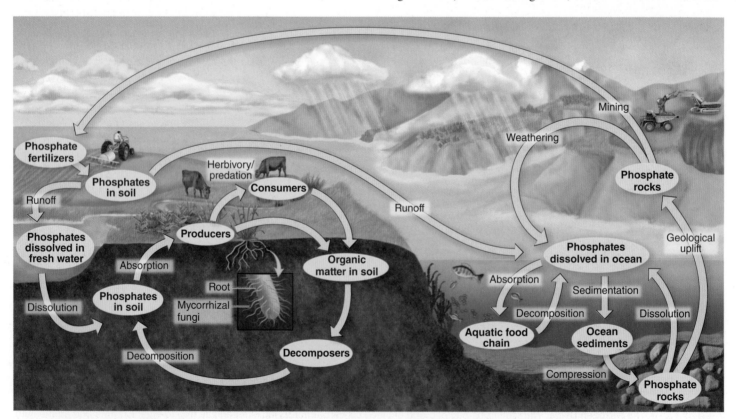

Figure 37.17 **The Phosphorus Cycle.** As phosphate-rich rocks erode, they release phosphorus that plants can absorb and pass to the rest of the food chain. Decomposers return phosphorus to the abiotic environment. Fertilizers have increased phosphorus availability to both terrestrial and aquatic organisms.

。

mass; respiration by these decomposers depletes the water of O_2. The resulting anaerobic conditions kill many aquatic organisms, including fish.

E. Terrestrial and Aquatic Ecosystems Are Linked in Surprising Ways

Sometimes, nutrients in an ecosystem can come from unexpected sources. Consider, for example, a mountain stream in Alaska. The eggs of Alaska salmon hatch in the streambed, and over the next year or so, the hatchlings develop into juveniles as they make their way to the mouths of their home rivers. As they transform into adults, their bodies adjust to seawater, and the fish swim into the Pacific Ocean. They spend as many as 4 years in the ocean, eating crustaceans and small fish. During this portion of their life cycle, the salmon (and their prey) ultimately rely on nutrients that well up from the ocean bottom.

Eventually, the fully grown salmon readjust to fresh water and swim upstream, back to the same waters where they hatched. As they make their way to their home steams to spawn, their bodies carry nutrients from the ocean to the heart of Alaska. Bears and eagles feast on the salmon (figure 37.18), and decomposers consume the remains. The nutrients they release support the growth of algae, which form the foundation of the stream's food chain. The adult salmon therefore form a link between the biogeochemical cycles of the ocean and the land.

37.4 | Mastering Concepts

1. What features do biogeochemical cycles share?
2. Describe the steps of the water, carbon, nitrogen, and phosphorus cycles.
3. Describe how a terrestrial ecosystem can interact with a faraway aquatic ecosystem.

Figure 37.18 Fish Feast. This brown bear is carrying a sockeye salmon, which contains nutrients from the sea.

37.5 | Investigating Life: Two Kingdoms and a Virus Team Up to Beat the Heat

In his 1733 poem entitled *On Poetry, a Rhapsody*, Jonathan Swift wrote:

So, naturalists observe, a flea
Has smaller fleas that on him prey;
And these have smaller still to bite 'em;
And so proceed ad infinitum.

Swift's observation, though not literally true, showed amazing insight into the relationships between the merely small and the truly microscopic. We now know that symbiotic arrangements can cross all phylogenetic lines. Symbioses exist between animals and plants (ants and acacias), plants and fungi (mycorrhizae), protists and animals (protozoa in termite guts), and bacteria or protists and fungi (lichens). Mutualism is a particularly interesting form of symbiosis, in which both species benefit from the deal. It is not that each species is advocating the success of the other's genome. Rather, if an alliance gives both parties a selective advantage, then *vive la différence*!

Researchers at the Noble Foundation in Ardmore, Oklahoma, have discovered an unusual three-way symbiosis and documented its survival benefit to the partners. Biologists Luis Márquez and Marilyn Roossinck knew of a type of panic grass (*Dichanthelium lanuginosum*) that endured the scorching geothermal soils of Yellowstone National Park. Previous studies had shown that the grass benefited from a fungus growing in its roots. The fungus (*Curvularia protuberate*) is an endophyte, an organism that lives between a plant's cells without causing disease. By itself, neither plant nor fungus could grow at temperatures above 38°C (100°F). The grass–fungus relationship somehow allows the plant to survive in soil temperatures up to 65°C (about 150°F). As long as the grass survives, so does the fungus, but each is doomed without the other.

By itself, a plant–endophyte relationship is not unusual; section 19.8 described endophytes that help cacao plants fight off other fungi. But Márquez and his team dug a bit deeper. They knew that many fungi are themselves infected with viruses. For example, a mycovirus (a virus that infects fungi) helps shape the biology of the fungus responsible for a disease of chestnut trees. The team wondered whether the fungus within the grass might also have a partnership with a virus.

The first thing to do was to see if a mycovirus was present. Mycoviruses usually have genomes made of double-stranded RNA, which is distinctive from the fungus's own double-stranded DNA. The researchers extracted nucleic acids from the fungal cells and found telltale fragments of double-stranded RNA within the fungal tissue. ▶ nucleic acids, p. 40

Next, the team wanted to see if the presence of the mycovirus within the fungal endophytes helped plants survive in hot soils. They found a colony of the fungus that contained very low amounts of viral RNA. By repeatedly drying, freezing, and thawing this fungus in the laboratory, they "cured" the fungus of its virus. The researchers then set up an experiment with three treatment groups: grass with the wild-type (virus-infected) fungus, grass without any fungus at all, and grass inoculated with virus-free fungus. They submitted all three groups of plants to a tough regimen of 65°C soil temperatures for 10 hours a day, followed by 14 hours at 37°C. After 14 days, they checked to see how the plants were doing. The results were clear: only the plants with the virally infected fungus survived the ordeal (**figure 37.19**).

The results seemed promising, but the team had to be certain it was the mycovirus that did the trick. After all, some unseen difference between the two fungi (something other than the virus) may have accounted for the heat tolerance. They therefore inserted a special genetic marker in one of their virus-free fungus samples and grew this marked fungus on a plate next to a fungus that contained the mycovirus. They let the two fungi grow across each other and took samples from the entwined hyphae. They moved these samples to new plates, and in one of the 35 subcultures they found that the marked fungus had picked up the virus. They inoculated grass with this newly infected fungus and turned up the heat. Sure enough, these plants survived, too (**figure 37.20**).

Finally, the scientists decided to see if the mycovirus–fungus partnership could work its magic in tomato plants, a distant relative of grass. They repeated the treatments and

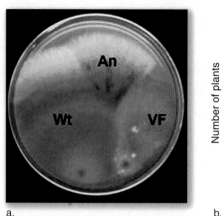
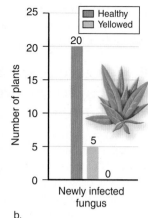

a.
b.

Figure 37.20 Reinfecting a "Cured" Fungus. (a) The researchers grew virus-infected (Wt, or wild-type) and virus-free (VF) fungi on agar in a petri dish. Wt and VF fungi are growing together in the zone marked "An," where previously virus-free fungi reacquired the mycovirus. (b) When the virus was reintroduced into fungi that were previously virus-free, the plants were heat-tolerant once more.

found that the plant–fungus–virus trio survived the heat, but the others did not.

Nearly 300 years after Jonathan Swift penned his whimsical poem, biologists are coming to realize that symbiotic relationships are the rule rather than the exception. Over hundreds of millions of years of shared evolutionary history, the struggle for existence has generated some unusual working relationships. The grass–fungus–virus partnership apparently allows all of the partners to colonize habitats that were previously unavailable. No one knows how the mycovirus-infected endophytes help their hosts beat the heat; perhaps they scavenge harmful oxygen free radicals produced by the heat-stressed plants. Whatever the mechanism, however, the relationship appears to be deeply "rooted" in evolutionary history.

Luis M. Márquez, Regina S. Redman, Russell J. Rodriguez, and Marilyn J. Roossinck. 2007. A virus in a fungus in a plant: Three-way symbiosis required for thermal tolerance. *Science,* vol. 315, pages 513–515.

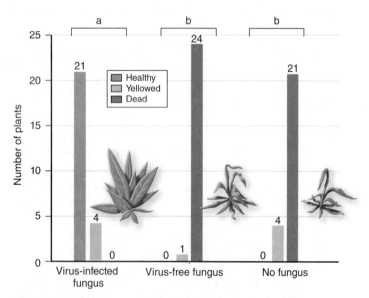

Figure 37.19 The Virus Matters. Plants that were inoculated with the normal virus-infected fungus survived the heat. Not so for plants infected with virus-free fungi or for nonsymbiotic plants without fungal endophytes.

37.5 | Mastering Concepts

1. Describe the relationship between panic grass, the fungus, and the virus.

2. The fungus–virus partnership helped young grass plants survive at very high temperatures. What would be the benefits of inoculating all of our food crop plants with the fungus–virus team? What else would you need to know before recommending that strategy? Can you think of any possible drawbacks?

Chapter Summary

37.1 | Multiple Species Interact in Communities

- **Communities** are composed of coexisting **populations** of multiple species.
- An **ecosystem** consists of a **biotic** community plus its **abiotic** environment.
- Each species in a community has a place where it normally lives (**habitat**) and a set of resources necessary for its life activities (**niche**).
- Ecologists describe the diversity of a community by measuring **species richness** (the number of species) and the **relative abundance** of each species.

A. Populations Interact in Many Ways

- Populations that share a habitat often **compete** for limited resources. Competition reduces the fitness of both species.
- According to the **competitive exclusion principle,** two species cannot indefinitely occupy the exact same niche.
- If multiple species with similar niches share a habitat, competition may restrict each species to only some of the resources available. This process is called **resource partitioning.**
- **Symbiotic** relationships include **mutualism** (both species benefit), **commensalism** (one species benefits, whereas the other is unaffected), and **parasitism** (one species benefits, and the other is harmed).
- **Herbivory** is an interaction in which a consumer eats a plant; a **predator** is an animal that eats another animal (its **prey**).
- Plants and prey animals have defenses against herbivores and predators; in addition, animals have adaptations that help them capture food.

B. A Keystone Species Has a Pivotal Role in the Community

- A **keystone species** makes up a small proportion of a community's biomass but has a large influence on the community's composition.

C. Closely Interacting Species May Coevolve

- In **coevolution,** the interaction between species is so strong that genetic changes in one population select for genetic changes in the other.

37.2 | Communities Change Over Time

- As species interact with one another and their physical habitats, they change the composition of the community. This process is called **succession.**
- **Primary succession** occurs in a previously unoccupied area, beginning with **pioneer species** that allow soil to develop, paving the way for additional organisms to thrive.
- **Secondary succession** is more rapid than primary succession because soil does not have to build anew.
- Succession may lead toward a stable **climax community,** but true long-term stability is rare. Pockets of local disturbance mean that most communities are a patchwork of successional stages.

37.3 | Ecosystems Require Continuous Energy Input

A. Food Webs Depict the Transfer of Energy and Atoms

- Within a community, each species occupies one or more **trophic levels.** An organism's trophic level depends on the number of steps between it and the ultimate source of energy in the ecosystem.
- At the base of each **food chain** are **primary producers (autotrophs)** that harness energy from the sun or inorganic chemicals.

- **Consumers (heterotrophs)** make up the next trophic levels. Primary consumers (herbivores) eat the primary producers. A secondary consumer may eat the primary consumer, and a tertiary consumer may eat the secondary consumer. **Decomposers** break down **detritus** (nonliving organic material) into inorganic nutrients.
- Interconnected food chains form **food webs.**

B. Every Trophic Level Loses Energy

- The total amount of energy converted to chemical energy is gross primary production. After subtracting energy devoted to growth, maintenance, and reproduction, the energy remaining in the producer trophic level is **net primary production.**
- Food chains rarely extend beyond four trophic levels because only a small percentage of the energy in one trophic level transfers to the next level.
- A **pyramid of energy** is a diagram that depicts the amount of energy at each trophic level in a food chain.

C. Harmful Chemicals May Accumulate in the Highest Trophic Levels

- **Biomagnification** concentrates stable, fat-soluble chemicals in the highest trophic levels because the chemical passes to the next consumer rather than being metabolized or excreted.

37.4 | Chemicals Cycle Within Ecosystems

- **Biogeochemical cycles** are geological and chemical processes that recycle chemicals essential to life.
- Each cycle is characterized by transfers among four main storage reservoirs: organisms, the atmosphere, water, and rocks and soil.

A. Water Circulates Between the Land and the Atmosphere

- Water is transferred from the atmosphere to the land or water as precipitation. Organisms release water in transpiration, evaporation, or urination.

B. Autotrophs Acquire Carbon as CO_2

- Autotrophs use carbon in CO_2 to produce carbohydrates. Cellular respiration and burning fossil fuels release CO_2. Decomposers release carbon from once-living material.

C. The Nitrogen Cycle Relies on Bacteria

- **Nitrogen-fixing** microbes convert atmospheric nitrogen to ammonia, which plants can incorporate into their tissues. Decomposers convert the nitrogen in dead organisms back to ammonia. **Nitrification** converts the ammonia to nitrites and nitrates, whereas **denitrification** converts these compounds to nitrogen gas.

D. The Phosphorus Cycle Begins with the Weathering of Rocks

- Rocks release phosphorus as they weather. Autotrophs absorb the phosphates and incorporate them into organic molecules; decomposers return inorganic phosphates to the environment.

E. Terrestrial and Aquatic Ecosystems Are Linked in Surprising Ways

- During the salmon life cycle, nutrients are transferred from the ocean to inland ecosystems.

37.5 | Investigating Life: Two Kingdoms and a Virus Team Up to Beat the Heat

- Researchers have discovered a three-way symbiosis between a grass plant, a fungus, and a virus that enables plants to survive extremely high temperatures.

Multiple Choice Questions

1. What is the difference between a population and a community?
   a. A population consists of a single species; a community includes multiple species.
   b. A population consists of all members of a species; a community includes only those individuals that share a habitat.
   c. A population includes only plants; a community includes both plants and animals.
   d. A population includes a single species; a community includes populations plus abiotic features.

2. Picocyanobacteria are photosynthetic bacteria that live in the ocean. Two species, isolated from the same habitat, use different wavelengths of light in photosynthesis. This is an example of
   a. competitive exclusion.
   b. commensalism.
   c. resource partitioning.
   d. biomagnification.

3. Monarch butterflies contain a substance that causes predators to vomit. Viceroy butterflies look like monarchs but lack the substance that sickens predators. The viceroys are an example of
   a. warning coloration.
   b. camouflage.
   c. mimicry.
   d. structural defense.

4. A moth that eats plants is a(n) _____, whereas the bat that eats the moth is a _____.
   a. heterotroph; decomposer
   b. secondary consumer; decomposer
   c. primary consumer; secondary consumer
   d. autotroph; secondary consumer

5. Which of the following examples illustrates mutualism?
   a. Barnacles attached to a whale's skin neither harm nor help the whale.
   b. Sparrows and chickadees fight for access to the seeds in a homeowner's bird feeder.
   c. A mosquito consumes a person's blood.
   d. Fig wasps help pollinate fig trees and reproduce inside the tree's fruit.

6. Primary succession would occur
   a. on a newly formed volcanic island.
   b. following a forest fire.
   c. where bare rock is exposed following the retreat of a glacier.
   d. Both a and c are correct.

7. Primary producers are at the first trophic level because they are the
   a. organisms that directly use the ecosystem's ultimate energy source.
   b. least complex organisms in an ecosystem.
   c. smallest organisms in an ecosystem.
   d. first organisms to appear within an ecosystem.

8. In a prairie ecosystem, which trophic level should have the highest biomass?
   a. Predatory birds
   b. Grasses
   c. Seed-eating rodents
   d. Soil fungi that decompose organic matter

9. Which statement best explains why persistent, fat-soluble chemicals such as DDT accumulate in the highest trophic levels?
   a. Animals at the highest trophic levels eat the lowest-quality food.
   b. The amount of biomass increases in each trophic level, and large organisms accumulate the most DDT.
   c. Large organisms are often herbivores, so they would consume DDT directly.
   d. DDT does not leave an animal's body; each predator therefore consumes the DDT in many prey.

10. Which cycle relies the least on decomposers?
    a. Carbon cycle
    b. Water cycle
    c. Phosphorus cycle
    d. Nitrogen cycle

Write It Out

1. How does a community differ from an ecosystem?
2. How does a habitat differ from a niche? Describe your own habitat and niche.
3. How can two species share the same habitat without one driving the other to extinction?
4. Describe and give examples of three types of symbiotic relationships.
5. List examples of adaptations that enable an organism to compete with other species, live inside another species, find food, and avoid herbivory or predation. How does each adaptation contribute to the organism's reproductive fitness?
6. In one type of mimicry, a harmless species such as a jumping spider physically resembles a noxious species such as an aggressive type of ant. Explain why this type of mimicry can exist only if the spiders are less abundant than the ants.
7. What is a keystone species? How do researchers design experiments that help them identify keystone species?
8. How is natural selection apparent in ecological succession?
9. Why are true climax communities rare?
10. After fires destroyed much of Yellowstone National Park in 1988, forest managers suggested humans could help the areas recover by feeding deer, bringing in plants, and planting trees. What are some advantages and disadvantages of intervening in recovery from a disaster?
11. When Mount St. Helens erupted in 1980, areas nearest the volcano were totally devastated. Farther away from the blast site, however, the disturbance was more mild. Predict how the rate of forest succession would change with distance from the volcano, and explain your prediction.
12. Can you think of an example of an organism that is both a producer and a consumer?
13. Identify a consumer in an ecosystem not mentioned in the text.
14. Krill are tiny, shrimplike plankton that are abundant in Antarctic waters. Some people have suggested that we "farm" the krill in Antarctic waters and use it to feed people who are starving. Predict the effects of krill farming on the Antarctic ecosystem.
15. In a long-term study of a tropical rain forest in Puerto Rico, researchers studied the aftermath of hurricanes Hugo and Georges. Tree debris supported the growth of detritivores. Beetles and flies ate the detritivores. Propose a food chain and trace how the booming insect populations might eventually benefit top predators such as birds and snakes in the forest. What do you predict happened to fruit- and nectar-eating animals immediately after the storm?
16. As the human population in the United States has increased, we have reduced or eliminated the populations of top predators such as wolves and coyotes. How might this history relate to the recent explosion of deer populations? Using the ecological principles you have learned in this chapter and chapter 36, what are some possible ways to reduce deer populations? What are the pros and cons of each approach?
17. Pigs and chickens in commercial farms may eat food made from herrings, anchovies, and other fish. What are some possible consequences to marine food webs?
18. In the United States alone, tens of millions of domesticated cats consume hundreds of millions of songbirds. Use the Internet to learn more about songbird diets, then predict how predation by cats might affect the rest of the food web.

19. Some farmers and ranchers kill the prairie dogs on their land. Use the Internet to research the ecology of prairie dogs and some of the techniques used to control prairie dog populations. How does each technique affect the prairie ecosystem?

20. Use the second law of thermodynamics and a pyramid of energy to explain why most food chains have four or fewer levels.

21. Some forms of mercury biomagnify in aquatic food chains. High concentrations of mercury in fish are prompting health advisories that warn pregnant women not to eat tuna and some other fish species. From this information, do you think tuna is an herbivore or a carnivore? Explain your reasoning. What properties do you predict that mercury shares with DDT and other pollutants that biomagnify as they ascend food chains?

22. How do organisms return water, carbon, nitrogen, and phosphorus to the abiotic environment?

23. Use the carbon cycle to trace a hypothetical path of a carbon atom from Abraham Lincoln's body into your own.

24. Suppose your friend says "I hate germs! I wish we could kill all the bacteria in the world!" What would happen to your friend if she didn't harbor bacteria in her own body? What would happen to nutrient cycles without bacteria?

25. Several sites on the Internet offer water- or carbon-footprint calculators. Choose one and compute your footprints. What are the assumptions that go into these calculations? What strategies could you use to reduce either footprint?

26. Use the Internet to look up the components of a so-called "low-carbon" diet. How does each element of a low-carbon diet help reduce CO_2 emissions into the atmosphere?

27. Review the structures of organic molecules in chapter 2. How do the molecules in living cells create interconnections between the carbon, nitrogen, and phosphorus cycles?

Pull It Together

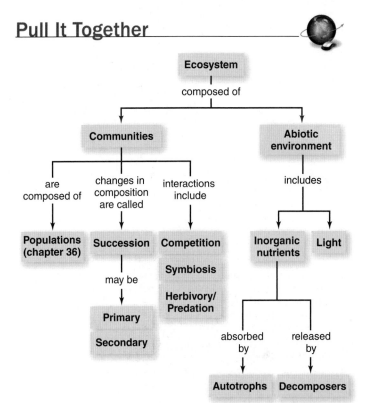

1. Where do mutualism, commensalism, and parasitism fit into this concept map?
2. How are decomposers and autotrophs essential to ecosystem function?
3. Distinguish between primary and secondary succession.
4. List some additional components of your abiotic environment.
5. Where do primary, secondary, and tertiary consumers fit into this concept map?

The submersible *Alvin* completes a research mission in the Atlantic Ocean. The inset shows deep ocean fish and giant tube worms near a hydrothermal vent.

Enhance your study of
this chapter with practice quizzes,
animations and videos, answer keys,
and downloadable study tools.
www.mhhe.com/hoefnagels

Life in Deep-Sea Hydrothermal Vents

IN 1977, THE SUBMERSIBLE *ALVIN* TOOK A GROUP OF GEOLOGISTS TO THE SEA BOTTOM NEAR THE GALÁPAGOS ISLANDS, DIRECTLY ABOVE CRACKS IN EARTH'S CRUST CALLED DEEP-SEA HYDROTHERMAL VENTS. In these areas of intense pressure and temperature extremes where Earth's crust is born, the researchers were astounded to see abundant life. Tubelike worms waved, anemones clung, crabs crawled, and shrimp grazed. Most of these animals were unknown to marine biologists.

Since that time, many more vent ecosystems have been discovered along the ridge of undersea volcanoes that dot the floor of the Pacific Ocean. In the early 1980s, researchers discovered chimneylike structures extending from some hydrothermal vents. Black, intensely hot, mineral-laden water shot through the chimneys, which formed as molten minerals from Earth's mantle crystallized in the cold water.

The environment at a hydrothermal vent could hardly be more different from our own familiar surroundings. The pressure is 300 times greater than at Earth's surface, and the water temperature can exceed 400°C. No animals can survive in the hottest water, but tube worms can live at 100°C. Moreover, the pH can be as low as 2.8, about as acidic as orange juice. There is no oxygen, and the hydrogen sulfide that spews from the Earth would be toxic to most organisms.

Too far beneath the surface to use sunlight, the producers in these vibrant communities are bacteria and archaea that tap energy from inorganic chemicals, predominantly hydrogen sulfide. Thick mats of these microorganisms cover the rock surfaces. Consumers include tube worms and giant clams that harbor mutualistic bacteria inside their tissues. Some of these animals have no digestive tracts; instead, they absorb nutrients directly from their live-in microbes. Other organisms prey on the tube worms, and crabs are scavengers and decomposers.

Deep-sea hydrothermal vents may hold clues to conditions at the time when life began. The discovery of these ecosystems has also helped to dispel the long-held idea that life exists only on the thin layer between bedrock and Earth's atmosphere. Nevertheless, life on and near Earth's surface is by far more familiar. This chapter describes the forests, grasslands, waters, and other ecosystems that are critically important to our own survival.

▸ origin of life, p. 298

Learning Outline

Learn How to Learn

Use Those Campus Resources

Many first-time college students struggle to keep up with reading and writing assignments. Time management can be part of the problem, but poor reading and writing skills may also share the blame. Most campuses offer free resources that can help. For example, look for workshops aimed at improving reading comprehension. Writing centers are also common; the staff should include consultants trained to help you be a more proficient writer.

38.1 | The Physical Environment Determines Where Life Exists

Picture the following landscapes in your mind: a prairie, the seashore, the Sahara desert, a jungle, and the top of Mount Everest. These and all other locations on Earth form a patchwork of unique habitats, each with its own set of conditions. Fire regularly ravages the prairie but not the beach; water is scarce in the desert but not in the jungle. The species that are native to each location have adaptations that correspond to these conditions. The same basic evolutionary process—natural selection—has produced unique populations and communities of organisms in nearly every possible habitat (**figure 38.1**).

Indeed, life abounds almost everywhere on Earth, even in places once thought to be much too harsh to support it (see chapter 16's Burning Question on page 345). Scientists have discovered life in Arctic ice, salt flats, hot springs, mines that plunge miles below Earth's surface, and hydrothermal vents (see the chapter opening essay). All of these areas are part of the **biosphere,** the portion of Earth where life exists.

The biosphere is one huge **ecosystem,** an interconnected community of organisms and their physical environment. Recall from chapter 37 that community ecologists study the **biotic** interactions among species, such as competition, predation, and mutualism. Ecosystem ecologists incorporate the community's interaction with its **abiotic,** or nonliving, environment. As we have already seen, both biotic and abiotic interactions shape the adaptations that contribute to the survival and reproductive success of each species.

Ecologists divide the biosphere into **biomes,** which are the major types of ecosystems. Forests, deserts, and grasslands are examples of terrestrial biomes. Lakes, streams, and oceans are water-based ecosystems. Each biome is characterized by a particular climate and a distinctive group of species. Although it is convenient to classify each ecosystem as belonging to one biome or another, keep in mind that no ecosystem exists in isolation. Water, air, sediments, and organisms travel freely from one part of the biosphere to another.

Gardeners know that every species is adapted to a limited set of conditions; no "superorganism" can exist everywhere. A plant adapted to bright sunlight will not thrive on the shady side of the house, nor will a cactus survive in a pond. Fish need water, and earthworms cannot live on a sandy beach. This chapter begins by describing some of the "rules" that determine how distinctive communities develop in each biome, beginning with abiotic features.

Many abiotic factors determine the limits of each species' distribution. The ultimate abiotic factor is an energy source, since no ecosystem can exist without one (see section 37.3). **Primary producers,** also called autotrophs, are organisms that can harvest this energy source as they build organic molecules. These primary producers provide the energy, nutrients, and habitats that support animals, fungi, and the rest of the food web.

Sunlight is the energy source for most ecosystems. On land, plants are the dominant primary producers. In water, however, most photosynthesis occurs courtesy of **phytoplankton:** microscopic, free-floating, photosynthetic organisms such as cyanobacteria and algae. A few ecosystems, such as deep-sea hydrothermal vents, are based on chemical energy, not sunlight. At these vents, the producers are microbes that extract energy from hydrogen sulfide and other inorganic chemicals.

Figure 38.1 Habitat Variety. (a) A kelp bed along the coast of southern California is home to a damselfish. (b) This young manzanita tree lives in the dry, rocky Huachuca mountains of southeastern Arizona. (c) A saguaro cactus is another Arizona resident. Its stem expands like an accordion, helping this desert plant to store water during rare rainstorms. (d) The broad, flat leaf of a lily pad soaks up sunshine in a habitat that is always wet.

a.

b.

c.

d.

Besides sunlight, the major abiotic factors that determine the numbers and types of plants on land are temperature and moisture. All organisms are adapted to a limited temperature range; trees, for example, cannot live where the temperature is too low (see this chapter's Burning Question). In addition, all life requires water. The plants that grow where water is abundant, such as the tropic rain forest, have very different adaptations from the vegetation that characterizes a desert ecosystem.

Nutrient availability is another crucial abiotic factor that often determines an ecosystem's productivity. On land, soil is the source of essential nutrients such as nitrogen and phosphorus. In aquatic ecosystems, both nutrients and sunlight are often scarce, especially with increasing depth and distance from the shoreline.

Other abiotic factors may also be important in some ecosystems. For example, the amount of dissolved oxygen influences the types of animals that can live in water. Aquatic plants and phytoplankton release oxygen into the water column as they carry out photosynthesis; oxygen also dissolves into water at the interface with the atmosphere. But respiration by aquatic organisms can deplete water of oxygen. Some microbes can tolerate water with little or no oxygen, but most fishes and other animals will suffocate and die.

Another important abiotic factor in water and on land is salinity. Many organisms are adapted to seawater or salty soils, but others are not. The problem of life in a salty habitat lies mostly in the difficulty of extracting water from a saline solution. This is why human castaways adrift on the ocean can die of thirst, even while surrounded by water. Likewise, saline soils prevent agriculture in many parts of the world, because most crop plants are not salt-tolerant.

Fire is an essential abiotic condition in some terrestrial biomes. In grasslands, for example, periodic fires kill trees that might otherwise take over. In coniferous forests, many adult trees die in fires, but their cones open only after prolonged exposure to heat. The seeds germinate after the fire, and the young trees thrive with little competition for sunlight or nutrients.

The rest of this chapter explains the distribution of the main biomes on land and in water. As you read this material, remember that the biomes we see today have not been in place forever. Over hundreds of millions of years, the continents have moved, and sea levels have risen and fallen. The central United States, for example, was once under the sea, which explains why fossils of marine animals are abundant in landlocked states such as Oklahoma. Likewise, 375 million years ago, the land mass that now includes the islands of Arctic Canada was once very near the equator. Long-buried fossils tell the tales of these tremendous ecosystem shifts. ▶ continental drift, p. 260

38.1 | Mastering Concepts

1. What are the relationships among ecosystems, communities, biomes, and the biosphere?
2. What abiotic conditions influence the distribution of species in the biosphere?
3. How do the main factors affecting primary production differ between land and water?

Burning Question

Why is there a "tree line" above which trees won't grow?

The tree line, or timberline, is an edge beyond which trees will not grow. The Arctic and Antarctic tree lines are the farthest points north and south, respectively, that trees can grow; the alpine tree line is the highest elevation at which they can grow (see the photo at right). In each case, the tree line usually defines the point at which the environment simply becomes too cold to support trees.

Most species near the tree line are evergreen conifers such as pine, spruce, larch, and fir. Their needles have a waxy coating and an arrangement of stomata that minimizes water loss in the thin, dry air. Eventually, however, chilly temperatures and biting winds get the best of even these hardy trees. At the timberline, they shorten to low, stunted bushes, and beyond the tree line, it is simply too cold year-round for seeds to germinate. ▶ stomata, p. 462

Wind, salt, and a dry climate can also produce different types of tree lines. For example, along coasts, a tree line can result from high winds and salt spray that make life impossible for trees. Beyond the desert tree line, rainfall is insufficient to support trees.

Submit your burning question to:
marielle_hoefnagels@mcgraw-hill.com

38.2 | Earth Has Diverse Climates

Earth has a wide variety of climates, from the year-round warmth and moisture of the tropics to the perpetually chilly poles. Why does each part of the planet have a different climate? The answer relates to the curvature of the planet and to the tilt of its axis.

Earth is a sphere that rotates at an angle as it orbits the sun (**figure 38.2**). Regions near the equator receive the most intense sunlight, and the temperature does not change much over the course of a year. Thanks to Earth's curvature, however, the sun's rays hit other parts of the surface at oblique angles. The average temperature falls with distance from the equator because the same amount of sunlight is distributed over a larger area.

The tilt of Earth's axis accounts for seasonal temperature changes in nonequatorial regions. From March through September, the northern hemisphere tilts toward the sun and experiences the warm temperatures of spring and summer. During the other half of the year, cooler temperatures prevail as the northern hemisphere tilts away from the sun. The seasons are the opposite in the southern hemisphere.

Equatorial regions get not only the most solar radiation but also the most precipitation. When intense sunlight heats the air over the equator, the air rises, expands, and cools. Because cool air cannot hold as much moisture as warm air, the excess water vapor condenses, forming the clouds that pour rain over the tropics.

Air that rises near the equator also travels north and south (**figure 38.3**). As the air cools at higher latitudes and elevations, its density increases, and it sinks back down to Earth at about

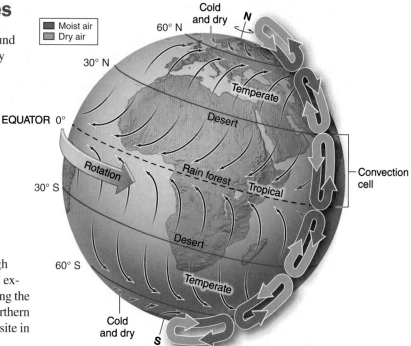

Figure 38.3 Patterns of Air Circulation and Moisture. Earth's air circulation pattern includes six convection cells. In each one, air cools as it rises and releases its moisture as rain. Conversely, descending air masses pick up warmth and absorb moisture from the land. These shifting air masses produce winds that appear to deflect toward the east and west because Earth rotates beneath the convection cells.

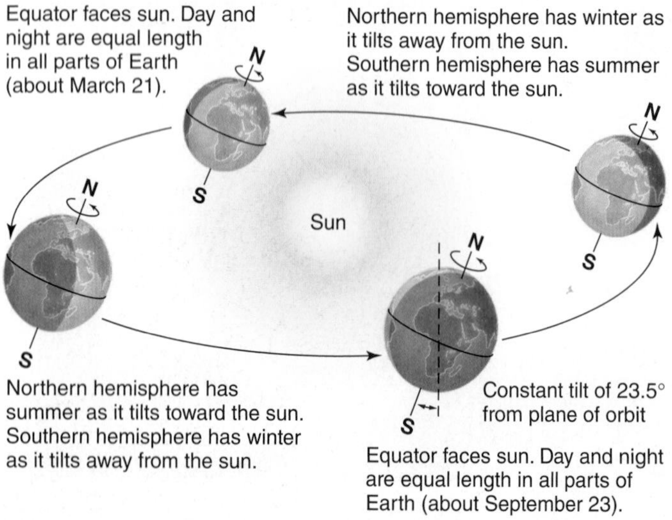

Equator faces sun. Day and night are equal length in all parts of Earth (about March 21).

Northern hemisphere has winter as it tilts away from the sun. Southern hemisphere has summer as it tilts toward the sun.

Sun

Northern hemisphere has summer as it tilts toward the sun. Southern hemisphere has winter as it tilts away from the sun.

Constant tilt of 23.5° from plane of orbit

Equator faces sun. Day and night are equal length in all parts of Earth (about September 23).

Figure 38.2 Earth's Seasons. The tilt of Earth's axis results in distinct seasons in the northern and southern hemispheres as Earth travels around the sun.

30° North and South latitude. Here the warming air absorbs moisture from the land, creating the vast deserts of Asia, Africa, the Americas, and Australia. Some of the air continues toward the poles, rising and cooling at about 60° North and South latitude. The resulting rains support temperate (midlatitude) forests in these areas. Some of the rising air spreads toward the poles, where precipitation is quite low. Some returns to the equator, where the air heats up again, and the cycle begins anew.

A cycle of heating and cooling, rising and falling air is called a convection cell. The planet has three such convection cells north of the equator and three south. As figure 38.3 illustrates, Earth's major winds correspond to these convection cells. (From Earth, the winds appear deflected toward the east and west because the planet rotates beneath them.) Together, these winds power major ocean currents (**figure 38.4**); an example is the Gulf Stream that flows along the eastern United States.

Ocean currents, in turn, influence coastal climates. Part of the reason is that ocean currents transport heat around the globe. In addition, large bodies of water have a huge heat capacity, so they heat up and cool down much more slowly than does the

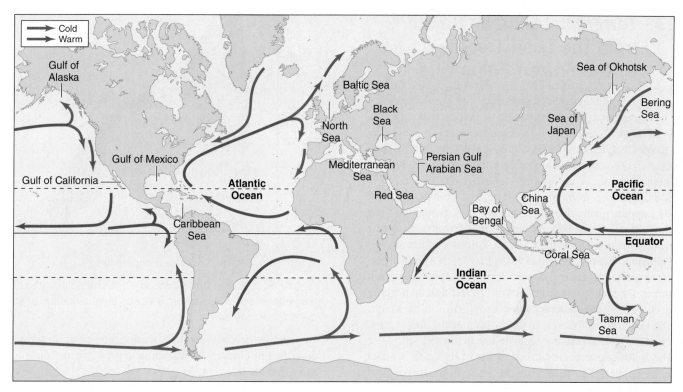

Figure 38.4 Ocean Currents. Earth's prevailing winds produce the major ocean currents, which redistribute water and nutrients throughout the oceans.

land. Coastal regions therefore often have milder climates than do inland areas at the same latitude. On a hot summer day, beach-goers notice this effect as they enjoy cooling breezes from the sea. Conversely, during the winter, the ocean releases stored heat that helps counteract cold inland temperatures. Changes in major ocean currents can therefore trigger ecological upheaval, as described in the Apply It Now box on page 799.

Mountain ranges also add variation to large-scale climatic patterns in two ways. First, the top of a mountain is generally cooler than its base because air cools as it loses pressure at high elevations. This is why many residents of Phoenix, Arizona, escape the broiling summer heat by migrating to nearby mountains. Second, mountains often block wind and moisture-laden clouds on their upwind side. The resulting **rain shadow** on the down-wind side of the mountain has a much drier climate (figure 38.5).

Although Earth's features explain the broad patterns of climate, it is extremely difficult to predict the weather at a given location more than a few days in advance. Computerized climate models must integrate complex events occurring simultaneously in the atmosphere, on land, and in water across the entire planet. Beyond the accuracy of day-to-day weather predictions, the prospect of global climate change poses another challenge to these models. Most scientists agree that human activities are causing Earth's average temperature to rise, but no one knows exactly how this increase will alter regional climates. Section 39.3 describes some of what we do know.

38.2 | Mastering Concepts

1. Explain why sunlight is most intense at the equator.
2. Moving outward from the equator, what are the major climatic regions of the world?
3. How do prevailing winds, ocean currents, and mountain ranges affect climate?

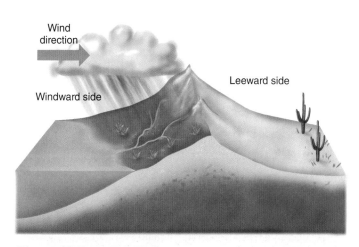

Figure 38.5 Rain Shadow. Precipitation falls on the windward side of the mountain, leaving the leeward side with a dry climate.

38.3 | Terrestrial Biomes Range from the Lush Tropics to the Frozen Poles

Earth's climatic zones give rise to huge bands of characteristic types of vegetation, which correspond to the terrestrial biomes (figure 38.6). Temperature and moisture are the main factors that determine the dominant plants in each location (figure 38.7). The overall pattern of vegetation, in turn, influences which microorganisms and animals can live in a biome.

Soils form the framework of terrestrial biomes, for they directly support plant life. Although soil may seem like "just dirt," it is actually a complex mixture of rock fragments, organic matter, and microbes (see figure 22.3). Climate influences soil development in many ways. Heavy rain may leach soluble materials from surface layers and deposit them in deeper layers, or it may remove them entirely from the soil. In addition, in a warm, moist climate, rapid decomposition may leave little humus (organic material) in the soil. In cold, damp areas, on the other hand, undecomposed peat may accumulate in the soil.

The following sections consider some of the world's major terrestrial biomes. The map in figure 38.6 shows the original range of each biome. It is important to remember, however, that humans have drastically reduced many natural biomes, replacing

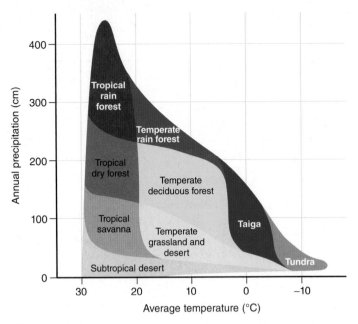

Figure 38.7 Biome Classification. This chart illustrates how temperature and precipitation influence the distribution of some of the major biomes.

them with farmland, suburban housing, and cities. In addition, as described in chapter 39, human activities threaten much of the native habitat that remains.

In studying these biomes, pay attention to the unique communities that occupy each one. Notice, also, that the number of species

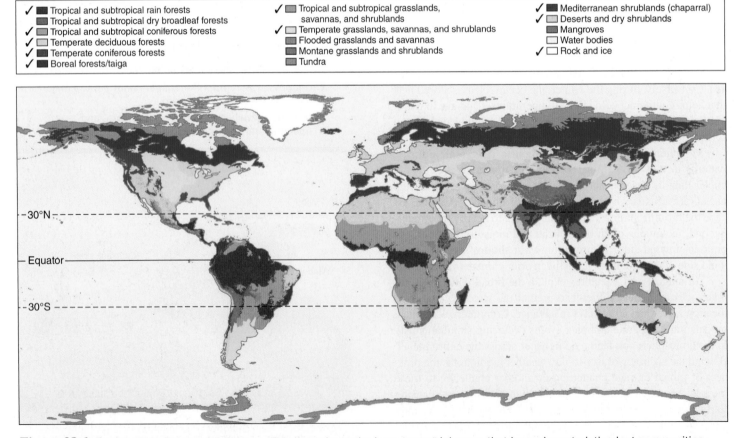

Figure 38.6 Earth's Major Terrestrial Biomes. This map shows the large terrestrial areas that have characteristic plant communities. Biomes marked with a check are described in this chapter.

occupying each biome generally declines with distance from the equator. No one understands the full explanation for this phenomenon. Perhaps the high primary production near the equator provides the most opportunities for specialized consumers to evolve. Perhaps year-round favorable conditions at the equator drive few species to extinction. Or perhaps some other mechanism is at work.

A. Towering Trees Dominate the Forests

Forests supply many of the resources that we use every day: lumber, paper, furniture, and foods such as wild mushrooms and nuts. The trees in the forest also provide wildlife habitat and protect soil from erosion. Moreover, like all plants, trees absorb CO_2 from the atmosphere and use it in photosynthesis, a process that also releases O_2. The wood of a living tree is an especially important long-term "carbon storehouse" that helps offset CO_2 released when humans burn fossil fuels. ▶ carbon cycle, p. 774

Tropical Rain Forests

The **tropical rain forests** encircle the equator in Africa, Southeast Asia, and Central and South America; their location ensures that the climate is almost constantly warm and moist. These forests are home to a stunning diversity of species connected by intricate networks of relationships.

The warm, wet equatorial climate means favorable conditions for plant growth year-round (**figure 38.8**). A 10-square-kilometer area of tropical rain forest is likely to house 750 tree species, including broadleaf evergreen trees with tall, straight trunks that form the forest canopy. Vegetation in the shade beneath the canopy includes climbing vines and epiphytes, which are small plants that grow on the branches, bark, or leaves of another plant. Tree saplings and countless smaller species grow in the deep shade of the forest floor. These understory plants often have large leaves that maximize the capture of scarce light.

Animal life is similarly diverse. That same forest patch might contain 60 species of amphibians, 100 species of lizards and snakes, 125 species of mammals, 400 species of birds, and thousands or even millions of insect species. The herbivores eat leaves, fruits, and other plant parts. Large cats, birds of prey, snakes, and other carnivores consume the herbivores. Many of the animals have similar adaptations to life in the trees: bright colors that maximize visibility, coupled with loud calls that echo throughout the forest canopy.

Despite the lush plant growth, tropical rain forest soils are usually nutrient-poor and low in organic matter. The heat and humidity speed decomposition, but the mycorrhizal roots of giant trees absorb the nutrients efficiently. Animals recycle nutrients, too. Leaf-cutting ants, for example, cultivate gardens of fungi on decomposing leaf fragments.

Worldwide, people are logging and burning tropical rain forests to make room for crops and domesticated animals. As described in chapter 39, tropical rain forest destruction threatens indigenous people and global water and carbon cycles. As plants near extinction, we lose potential medicines and the ancestors of many of our most important domesticated plants. These older varieties are a

Figure 38.8 **The Tropical Rain Forest.** Plants in the tropical rain forest form distinct layers, from the tallest trees emerging from the canopy to the tiniest residents of the shady forest floor. This rain forest is on the island of Borneo, Malaysia.

valuable resource to plant breeders, who are always searching for disease-resistance genes to breed into modern crop plants.

Temperate Deciduous and Coniferous Forests

The world's temperate forests occupy large areas between 30° and 60° North latitude. **Temperate deciduous forests** are dominated by trees that shed their foliage in autumn, whereas **temperate coniferous forests** contain mostly evergreen conifers that lose only a few leaves at a time (**figure 38.9**). These forests once covered parts of Asia, Western Europe, North America, South America, Australia,

a.   b.

Figure 38.9 **Temperate Forests.** (a) Trees that lose their leaves each autumn dominate temperate deciduous forests such as this one in Pennsylvania. (b) Temperate coniferous forests have evergreen trees such as these in Wyoming's Grand Teton National Park.

and New Zealand, but logging, agriculture, and urbanization have decimated most of the world's native temperate forests.
▶ deforestation, p. 806

Deciduous trees occur where summers are warm, winters are cold, and rainfall is approximately constant throughout the year. Usually one or two tree species predominate, as in oak–hickory or beech–maple forests. Shade-tolerant shrubs grow beneath the towering trees. Below them, small flowering plants grow in early spring, when light penetrates the leafless tree canopy. Herbivores include seed- and nut-eating mice and birds, whitetail deer, and gray squirrels. Red foxes and snakes are common carnivores. Raccoons eat a wide variety of foods, including insect grubs, acorns, frogs, and bird eggs. Many of these animals adjust to the seasons by putting on fat in the summer and hibernating in the winter, or they may store caches of seeds and nuts that sustain them when food is scarce. Still others migrate to warmer areas for the winter.

Mild winters, cool summers, and abundant rain and fog favor the temperate coniferous forest (sometimes called the temperate rain forest). Most trees in the temperate coniferous forest are evergreens such as spruce, pine, fir, and hemlock, all of which have waxy, needlelike leaves adapted to year-round photosynthesis. Understory shrubs include alder and hazelnut. Herbivores include birds, deer, and squirrels. Snakes, bobcats, weasels, and coyotes hunt the herbivores, and black bears are omnivores.

The temperate coniferous forests include the world's largest trees, the giant sequoias, which dominate the ancient redwood forests along the west coast of North America. Scientists studying the redwood forest canopy began climbing redwood trees in the 1990s and were astonished to find "forests in the air." Organic matter had accumulated into thick soils high on the branches of the giant redwoods. Berry bushes, ferns, and entire trees grew in the soil. The researchers also found insects, other invertebrates, chipmunks, birds, and even salamanders in the aerial ecosystem.

Taiga (Boreal Forest)   North of the temperate zone in the northern hemisphere lies the cold, snowy **taiga** (figure 38.10). This biome is also called the boreal forest or the northern coniferous forest. The long, harsh winter can last more than 6 months, so the growing season is short. Moisture can be scarce in winter, when water may remain frozen for months.

Soils in the taiga are cold, damp, acidic, and nutrient-poor. The low temperature and acidic pH slow decomposition, and nutrients tend to stay in the leaf litter above the soil, rather than entering the topsoil. Spruce, fir, pine, and tamarack (larch) are the dominant trees. The needles of these evergreens resist water loss and can carry out photosynthesis whenever the weather is warm enough to support it. Mycorrhizal fungi help the plants maximize nutrient uptake from the leaf litter. Seed- or leaf-eating herbivores

Figure 38.10   Taiga. Woodland caribou forage for food in the Canadian boreal forest.

include the woodland caribou, porcupines, red squirrels, chipmunks, snowshoe hares, and moose. As in the deciduous forest, many of these animals hibernate or store food for the winter. Carnivores include lynx, gray wolves, and wolverines. Migratory birds visit during the short growing season.

Caribou, reindeer, berries, and fungi have provided food to humans and other animals for millennia. Expanded logging, hunting, and trapping, however, are rapidly depleting the boreal forests.

B. Grasslands Occur in Tropical and Temperate Regions

Vast seas of grasses sustain huge herds of large, grazing animals such as bison and zebras. But the rich soils that support the grasslands also make these prime areas for agriculture. Grasslands are therefore endangered around the world.

Tropical Savannas   Tropical **savannas** are grasslands with scattered trees or shrubs and bands of woody vegetation along streams (figure 38.11). The weather is warm year-round, with distinct wet and dry seasons. Perennial grasses dominate the savanna along with patches of drought- and fire-resistant trees

Figure 38.11 **Tropical Savanna.** A herd of wildebeest roams the savanna in Kenya's Masia Mara National Reserve.

and shrubs such as palms, acacias, and baobabs. These plants have deep roots, thick bark, and trunks that store water.

The Australian savanna is home to many birds and kangaroos, whereas the grassland in Africa features herds of grazers such as zebra, giraffes, wildebeests, gazelles, and elephants. Lions, cheetahs, wild dogs, birds of prey, and hyenas prey on the herbivores, and vultures and other scavengers eat the leftovers. During the dry season, great herds of animals migrate enormous distances in search of water. In many tropical savannas, termites are major detritivores, and their huge nest mounds dot the landscape. It was from the African savanna that our own ancestors evolved, emerged, and spread. ▶ human evolution, p. 312

Widespread overgrazing by domesticated animals threatens to turn large areas of tropical savanna to desert. Hunting and poaching also endanger some populations of large animals. ▶ desertification, p. 806

Temperate Grasslands The **temperate grasslands** are also known as the prairies of North America, the steppes of Russia, and the pampas of Argentina (figure 38.12). These ecosystems have few if any trees, partly because annual rainfall is often not sufficient to support them. Grazing and fire also suppress tree growth. The tips of a tree's branches, where growth occurs, are easily destroyed by fire or herbivores. In contrast, the perennial buds of grasses lie protected below the soil surface and resprout soon after the fire passes.

Wind-pollinated grasses dominate this biome, although sunflowers, thistles, and other flowering plants are also common. Bison, elk, and pronghorn antelope, whose teeth and digestive systems are adapted to a grassy diet, were originally the large,

grazing herbivores of North America; hunting has caused their numbers to dwindle. Other herbivores include prairie chickens, insects such as grasshoppers, and rodents such as prairie dogs and mice. Some of these small animals burrow into the soil to hide from predators, whereas others have camouflage. Coyotes, bobcats, snakes, and birds of prey feed on the herbivores.

The deep, black prairie soils of the North American grasslands are famously fertile and rich in organic matter. Because these soils are ideal for cultivation, only small patches remain of the vast grasslands that once occupied three continents. Farmland has replaced prairie, with wheat and corn taking the place of diverse grasses. Most prairie remnants are too small to sustain the herds of large herbivores that once roamed the plains.

C. Whether Hot or Cold, All Deserts Are Dry

All **deserts** are dry, receiving less than 20 centimeters of rainfall per year. They ring the globe at 30° North and South latitude, and additional deserts occur in the rain shadows of tall mountains. Sparse desert life means soils are low in organic matter.

Although deserts have a reputation for being hot, the temperature can vary dramatically. In a hot desert such as the Sonoran, which spans parts of Arizona and Mexico, few clouds filter the sun's strongest rays, and the days can be scorchingly hot. In China's cold Gobi desert, in contrast, the average annual temperature is below freezing.

Figure 38.12 **Temperate Grassland.** Bison once dominated the North American prairie, a temperate grassland.

Parts of the Sahara are so dry that they are nearly devoid of life. Other desert habitats, however, are species-rich (figure 38.13). Desert plants often have long taproots, quick life cycles that exploit the brief rainy periods, fleshy stems or leaves that store water, and spines or toxins that guard against thirsty herbivores. Many use CAM photosynthesis, a water-saving variation on the photosynthetic pathway. ▸ CAM photosynthesis, p. 100

Desert herbivores include jack rabbits and kangaroo rats, which eat seeds and leaves. Snakes and cougars hunt the herbivores. Most desert animals burrow or seek shelter during the day, then become active when the sun goes down. Standing water is scarce, so most water comes from the animal's food.

Humans are drawn to the warm climates of the desert southwest of the United States. Urban development, however, changes the desert ecosystem drastically by diverting huge amounts of water from faraway rivers. Even so, deserts are among the few biomes that are expanding worldwide, as unsustainable agriculture eats away at forests and savannas.

Figure 38.13 Desert. Saguaro cacti and many other desert plants have water-conserving adaptations. This is the Sonoran Desert in the southwestern United States.

D. Fire- and Drought-Adapted Plants Dominate Mediterranean Shrublands (Chaparral)

Despite the name, **Mediterranean shrublands** (also called **chaparral**) occur not only around the Mediterranean Sea but also in other small areas along the west coasts of North and South America, Australia, and South Africa. Summers are hot and dry; winters are mild and moist. As the name suggests, the dominant vegetation is shrubby plants (figure 38.14), including poison oak, manzanita, and scrub oak. The plants have thick bark and small, leathery, evergreen leaves with thick cuticles and hairs that slow moisture loss during the dry summers. Herbivores include jackrabbits, mule deer, and birds and rodents that forage for seeds under the shrub canopy; some of their predators include coyotes, foxes, snakes, and hawks.

Mediterranean shrublands are especially susceptible to fires because the vegetation dries out during the summer. The sandy soils retain little water. The fire-adapted plants resprout from underground parts or produce seeds that germinate only after the heat of a fire. People once burned chaparral to make room for grazing livestock. Since then, large human communities have moved onto the shrublands, which nevertheless remain at great risk for fire. Burning the plant cover makes the soil susceptible to erosion, raising the risk of mudslides and further property damage.

Figure 38.14 Mediterranean Shrubland. Dry summers, wet winters, and fire shape this chaparral ecosystem in California.

E. Tundras Occupy High Latitudes and High Elevations

The **tundra** is a low-temperature biome. The largest region of tundra occurs in a band across the northern parts of Asia, Europe, and North America. The Antarctic continent also has a small amount of tundra. In addition, at middle latitudes, alpine tundra occupies high mountaintops between the tree line and areas of permanent snow and ice cover.

Snow covers the Arctic and Antarctic tundra during the bitterly cold and dark winter. Temperatures venture above freezing for a few months each year, and summer sunlight is intense. Because cold temperatures slow decomposition, tundra soils are rich in organic matter. Below the top layer of soil is a zone called **permafrost,** where the ground remains frozen year-round. Permafrost limits rooting depth, which prevents the establishment of large plants (**figure 38.15**). The shallow tundra soil supports reindeer lichens, mosses, dwarf shrubs, and low-growing perennial plants such as sedges, grasses, and broad-leafed herbs.

Penguins, seals, and other vertebrates visit the Antarctic tundra, but the Arctic tundra has much more diverse animal life. Inhabitants include caribou, musk oxen, reindeer, lemmings, hares, foxes, and wolverines, all of which have thick, warm fur. Polar bears sometimes visit coastal areas of the Arctic tundra to den. In the summer, migratory birds raise their young and feed on the insects that flourish in the tundra and its ponds.

The shallow soil, short growing season, and slow decomposition make the tundra a very fragile environment that recovers slowly from disturbance. In recent decades, oil drilling in the Arctic tundra has significantly increased the human presence.

F. The Polar Ice Caps House Cold-Adapted Species

Of all of the world's terrestrial biomes, the polar ice caps are the least explored (**figure 38.16**). Although the North and South Poles seem superficially similar, they differ from each other. The northern ice cap is a relatively thin ice layer that covers the Arctic Ocean. Antarctica, the southernmost continent, is a land mass covered with a thick layer of ice. Both ice caps interact extensively with water and therefore share characteristics of both terrestrial and aquatic biomes.

Life is difficult at both poles. Both Antarctica and the Arctic ice cap are extremely cold, dry, and windy year-round. The primary producers in ice are phytoplankton that eke out a living in brine-filled channels in the ice. The light passing through the ice is dim, even in the summer. Yet these phytoplankton support a unique food web consisting of bacteria, archaea, worms, crustaceans, and ice fishes (see section 29.7). All of these organisms have antifreeze chemicals that prevent ice crystals from forming in their cells.

Larger animals exploit the Arctic and Antarctic food webs as well. Arctic animals include polar bears, seals, whales, and birds that use ice for migration, hunting, and protection. On Antarctica, vertebrates include penguins, seals, and whales. Squid ply the waters surrounding Antarctica as well.

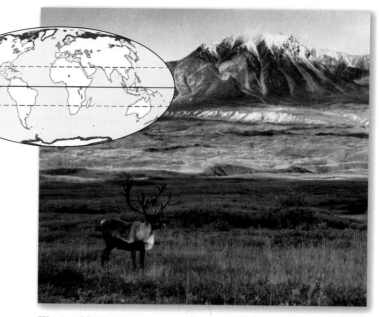

Figure 38.15 Arctic Tundra. In the Alaskan tundra, permafrost limits plant life to shallow-rooted shrubs. Lichens are also abundant.

Scientists are increasingly concerned that human-induced global climate change will continue to cause the polar ice caps to melt. In addition to altering the ocean ecosystem and driving rare species to extinction, the resulting rise in sea level could inundate coastal cities. ▶ global climate change, p. 809

38.3 | Mastering Concepts

1. How do climate and soil composition determine the characteristics of terrestrial biomes?
2. List and describe the climate, soils, and inhabitants of each of the major terrestrial biomes.
3. Which biomes are supported by fire and grazing?

Figure 38.16 Polar Ice. These emperor penguins live in the frozen desert of Antarctica, Earth's coldest continent.

38.4 | Freshwater Biomes Include Lakes, Ponds, and Streams

Although terrestrial biomes are most familiar to us, aquatic ecosystems occupy much more space. In fact, water covers about 71% of Earth's surface. This water moves continuously among the ocean, atmosphere, land surface, and groundwater, providing vital connections among all biomes.
▶ water cycle, p. 773

As illustrated in **figure 38.17**, most of Earth's water is in the ocean. Only about 3% of all water has a low enough salt content to be considered "fresh." Glaciers and the great polar ice sheets of Greenland and Antarctica tie up more than two thirds of this fresh water, and most of the rest is in underground aquifers. Lakes and rivers on Earth's surface therefore contain only about 0.3% of the freshwater supply.

Groundwater, lakes, and rivers make up only a tiny sliver of the global water "pie," but that sliver is vital to humans for drinking water and irrigation. Most other terrestrial species rely on surface fresh water as well.

A. Lakes and Ponds Contain Standing Water

Standing water includes lakes and ponds, which differ from one another mainly in size and depth. A lake is generally larger and deeper than a pond.

Light penetrates the regions of a lake to differing degrees, creating zones with characteristic groups of organisms (**figure 38.18**). The lake's **photic zone,** where light is sufficient

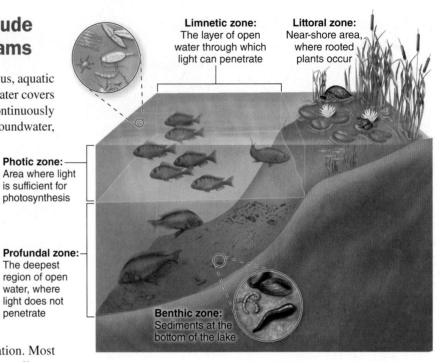

Limnetic zone: The layer of open water through which light can penetrate

Littoral zone: Near-shore area, where rooted plants occur

Photic zone: Area where light is sufficient for photosynthesis

Profundal zone: The deepest region of open water, where light does not penetrate

Benthic zone: Sediments at the bottom of the lake

Figure 38.18 Zones of a Lake. Distance from the shore and the availability of light define the zones of a lake. Each zone houses a unique assortment of organisms.

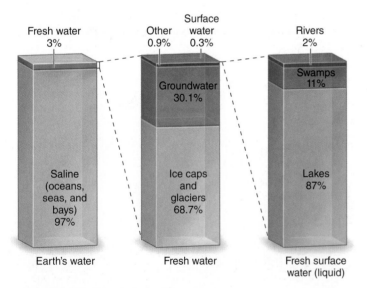

Fresh water 3%

Surface water 0.3%

Other 0.9%

Rivers 2%

Groundwater 30.1%

Swamps 11%

Saline (oceans, seas, and bays) 97%

Ice caps and glaciers 68.7%

Lakes 87%

Earth's water

Fresh water

Fresh surface water (liquid)

Figure 38.17 World Water Resources. Most water on Earth is either in the ocean or frozen. Only a small percentage of fresh water is in lakes or rivers on Earth's surface.

for photosynthesis, is subdivided into littoral and limnetic zones. The **littoral zone** is the shallow shoreline region where rooted plants occur. Productivity is high, thanks to abundant nutrients from the shore. Phytoplankton and plants such as cattails thrive here, providing food and shelter for myriad invertebrates, fishes, amphibians, and other animals. The **limnetic zone** is the layer of open water where light penetrates. Phytoplankton are the dominant producers in the limnetic zone.

In contrast to the photic zone, the **profundal zone** is the deep region of water where light does not penetrate. Scavengers and decomposers, both here and in the **benthic zone** (the sediment at the lake bottom), rely on a gentle rain of organic material from above to supply both energy and nutrients.

Lakes change over time (**figure 38.19**). Younger lakes are often deep, steep-sided, and low in nutrients. These lakes are **oligotrophic,** which means they are nutrient-poor and therefore low in productivity. They are clear and sparkling blue, because phytoplankton aren't abundant enough to cloud the water.

As a lake ages, however, nutrients accumulate from decaying organisms and sediment. Such a lake is **eutrophic,** which means it is nutrient-rich and high in productivity. The rich algal growth turns the water green and murky. Not surprisingly, nutrient-rich urban wastewater and farm runoff carrying phosphate-rich fertilizers can speed this process. The nutrients promote the excessive growth of algae, which sink to the lake bottom after they die. As the dead algae decompose, deep waters are rapidly depleted of oxygen. Many fish and other animals die. In addition, microbes respond by switching to anaerobic metabo-

Rivers are the largest streams. They change as they flow toward the ocean (figure 38.20). At the headwaters, the water is relatively clear, and the stream channel is narrow. Where the current is swift, turbulence mixes air with water, so the water is rich in oxygen. As the river flows toward the ocean, it continues to pick up sediment and nutrients from the channel. The river widens as small streams draining additional land areas contribute more water. As the land flattens, the current slows. The river is now murky, restricting photosynthesis to the banks and water surface. As a result, the oxygen content is low relative to the river upstream.

Rivers and streams depend on runoff from the land for water and nutrients. Dead leaves and other organic material fall into the river and add to the nutrients. On the other hand, rivers also return nutrients to the land. Many rivers flood each year, swelling with meltwater and spring runoff, and spreading nutrient-rich silt onto floodplains. When a river approaches the ocean, its current slows and deposits fine, rich soil that forms new delta lands at the mouth of the river.

38.4 | Mastering Concepts

1. Describe the types of organisms that live in each zone of a lake or pond.
2. What is the difference between an oligotrophic and a eutrophic lake?
3. How does a river change from its headwaters to its mouth?

a.

b.

Figure 38.19 Nutrients Make a Difference. (a) Oregon's Crater Lake is oligotrophic. It formed about 7000 years ago, and few nutrients have made their way into the crystal-clear water. (b) Algae are accumulating in the shallow, nutrient-rich waters of this pond.

lism; the byproducts are gases with unpleasant, "rotten egg" odors. ▶ eutrophication, p. 776

In time, a lake continues to fill with sediments and transforms into a freshwater **wetland,** where the soil is permanently or seasonally saturated with water. These nutrient-rich swamps, bogs, or marshes often host spectacularly diverse assemblages of plants and animals that rely on the interface between land and water. Eventually, the wetland fills in completely and becomes dry land.

B. Streams Carry Running Water

Streams include brooks, creeks, and rivers that carry water and sediment from all portions of the land toward the ocean (or an interior basin such as the Great Salt Lake). Along the way, streams provide moisture and habitat to aquatic and terrestrial organisms.

Figure 38.20 A River Changes Along Its Course. A narrow, swift stream in the mountains becomes a slow-moving river as it accumulates water and sediments along its journey to the ocean.

38.5 | Oceans Make Up Earth's Largest Ecosystem

The oceans, covering 70% of Earth's surface and running 11 kilometers deep in places, form the world's largest biome. Most photosynthesis on Earth occurs in the vast oceans, contributing enormous amounts of oxygen to the atmosphere. Moreover, oceans absorb so much heat from the sun that they help stabilize Earth's climate.

Food webs in the ocean need nutrients and sunlight. Supplies of both of these resources—and therefore primary production—are highest in shallow waters near the coasts. In deep water away from the coasts, however, both energy and nutrients can be scarce.

A. Land Meets Sea at the Coast

For humans, the world's coasts are a vital source of food, transportation, recreation, and waste disposal. Most of the world's largest cities are located along coastlines, and human populations in those cities are expected to increase in the future. As the human population expands, concerns about coastal ecosystems will likely increase. Several types of ecosystems border shorelines.

Estuaries   An **estuary** is an area where the fresh water of a river meets the salty ocean (figure 38.21a). When the tide is out, the water may not be much saltier than water in the river. The returning tide, however, may make the water nearly as salty as the sea. Organisms that can withstand these extremes in salinity receive nutrients from both the river and the tides. Estuaries therefore house some of the world's most productive ecosystems.

In the open water of an estuary, phytoplankton account for most of the productivity. In addition, salt-tolerant grasses and other flowering plants dominate the salt marshes that often occur along the fringes of an estuary. Together, these producers support many animals. For example, more than half of the commercially important fish and shellfish species spend some part of their life cycle in an estuary. Migratory waterfowl feed and nest here as well. Moreover, almost half of an estuary's photosynthetic products go out with the tide and nourish coastal communities.

Intertidal Zone   Along coastlines, the **intertidal zone** is the area between the high tide and low tide marks. This region of constant change, which is also called the littoral zone, is alternately exposed and covered with water as the tide rises and falls.

Species-rich mangrove forests occupy intertidal areas in the tropics (figure 38.21b). An entirely different intertidal zone is a sandy beach, which features long strips of bare sand that make for beautiful vistas and all-day exposure to the sun. Constantly shifting sands mean that few producers can take root on the beach, but ocean water delivers a constant supply of dead organisms and other nutrients. Crabs feed on the organic debris and then burrow into the sand to escape the pounding waves, while shorebirds forage for worms and other small invertebrates.

Some intertidal zones are rocky, not sandy (figure 38.21c). Here, organisms often attach to rocks, preventing wave action from carrying them away. Holdfasts attach large marine algae (seaweeds) to rocks. Threads and suction fasten mussels to rocks. Sea anemones, sea urchins, sea stars, and snails live in pools of water that form between rocks.

Coral Reefs   Colorful and highly productive coral reef ecosystems border tropical coastlines where the water is clear and sediment-free. **Coral reefs** are vast underwater structures of calcium carbonate built by coral animals. The living coral is but a thin layer atop the remains of ancestors. A coral reef, then, is at the same time an immense graveyard and a thriving ecosystem.

The tissues of the coral animals house symbiotic algae that are essential for the coral's—and the ecosystem's—survival. The sun penetrates the clear, shallow water, allowing photosynthesis to occur, and constant wave action brings in additional nutrients.

a.

b.

c.

d.

Figure 38.21 Coastal Ecosystems. (a) Salt and fresh water meet at estuaries such as this one along the coast of St. Croix. (b) Mangrove trees provide habitat for many other organisms. These mangroves are in Papua New Guinea. (c) The rocky intertidal zone along the coast of Washington state is frequently pounded by waves. (d) Coral reefs in the Red Sea house a spectacular diversity of fish.

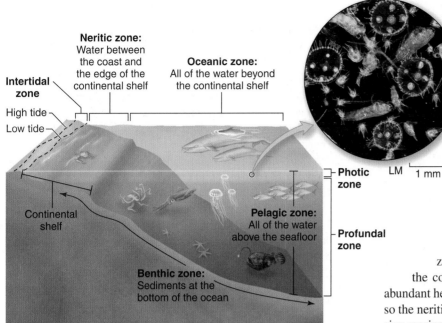

Neritic zone: Water between the coast and the edge of the continental shelf

Oceanic zone: All of the water beyond the continental shelf

Intertidal zone

High tide
Low tide

LM   1 mm

Photic zone

Pelagic zone: All of the water above the seafloor

Profundal zone

Continental shelf

Benthic zone: Sediments at the bottom of the ocean

Figure 38.22 **Zones of the Ocean.** The intertidal, neritic, pelagic, and benthic zones of the ocean each harbor unique types of organisms. The inset shows the sea surface microlayer, which brims with phytoplankton, zooplankton, tiny animals, eggs, larvae, and nutrients.

The nooks and crannies of a coral reef provide food and habitat for a huge variety of organisms (figure 38.21d). The Great Barrier Reef of Australia, for example, is composed of some 400 species of coral and supports more than 1500 species of fishes, 400 of sponges, and 4000 of mollusks. Other residents include algae, snails, sea stars, sea urchins, sea turtles, and countless types of microorganisms. This remarkably rich biodiversity has prompted some people to call coral reefs "the rain forests of the sea."

B. The Open Ocean Remains Mysterious

The oceans cover most of Earth's surface, but we know less about marine life than we do about biodiversity in a single tree in a tropical rain forest. Not only is this ecosystem vast, but it also houses populations that are sometimes small, usually very dispersed, and nearly always difficult for us to observe. Biologists have explored only 5% of the ocean floor and 1% of the huge volume of water above.

Marine biologists divide the ocean's surface into horizontal zones (figure 38.22). The intertidal zone is the shoreline. The **neritic zone** is the area from the coast to the edge of the continental shelf. Sunlight is abundant here, and sediments from the coast contribute nutrients, so the neritic zone supports high primary productivity and extensive marine food webs. The great kelp forests that fringe many cool-water coastal areas are examples of neritic ecosystems. The **oceanic zone** extends beyond the continental shelf.

The ocean can also be subdivided according to depth. The **pelagic zone** consists of all of the water above the seafloor. The upper layer of the pelagic zone, the photic zone, is the only area where photosynthesis can occur. At the very top of the pelagic zone, the sea surface houses phytoplankton and the zooplankton that feed on them (see figure 38.22, inset). Large sea animals such as fishes and whales scoop up vast quantities of krill and other zooplankton.

Below the photic zone is the profundal zone, where light is too dim for photosynthesis. Nevertheless, the bodies and wastes of top-dwelling organisms provide a continual rain of nutrients to the species below, including great numbers and varieties of jellyfishes, fishes, whales, dolphins, mollusks, echinoderms, crustaceans, and organisms yet to be discovered. The unusual communities that occupy hydrothermal vents add biodiversity to the benthic zone (see the chapter opening essay).

The most productive ocean environments arise in zones of **upwelling,** where deep currents force cold, nutrient-rich lower layers of water to move upward. The influx of nutrients causes phytoplankton to "bloom," and with this widening of the food web base, many ocean populations grow. Upwelling generally occurs on the western side of continents, such as along the coasts of southern California, South America, parts of Africa, and the Antarctic. A climatic event called El Niño periodically shifts these ocean currents, greatly disrupting coastal ecosystems (see this chapter's Apply It Now box on page 799).

38.5 | Mastering Concepts

1. Describe some of the adaptations that characterize organisms in estuaries, intertidal zones, and coral reefs.
2. List and define the major zones of the ocean.
3. How is upwelling important to ocean ecosystems?

38.6 | Investigating Life: Some Like It Hot

Coral reefs are among the world's most fascinating and important ecosystems. Countless species of microbes, algae, invertebrates, and fishes live and breed in the reef's cracks and crevices. Reefs also protect coastlines from erosion and attract the fishing boats and tourists that support entire economies.

A coral is a type of animal called a cnidarian, a group that also includes jellyfishes. Corals secrete stony calcium carbonate skeletons that protect their soft bodies from injury and predation. Their stinging tentacles grab small animals and stuff them into a saclike digestive cavity. But corals also feed in another way. The animal's tissues harbor huge populations of symbiotic algae. As long as light is abundant, the algae carry out photosynthesis, providing their hosts with organic carbon in exchange for shelter and a stable home. ▶▶ cnidarians, p. 416; mutualism, p. 764

The relationship between coral animals and their photosynthetic residents is surprisingly vulnerable to disruption. For example, the animals expel their symbionts if the water temperature increases by just a few degrees. The coral reef then "bleaches"; that is, the animals lose their algal partners and die, leaving behind only the white calcium carbonate skeleton (figure 38.23).

The prospect of global climate change therefore raises the possibility of widespread coral bleaching and the eventual collapse of the reef ecosystem. As chapter 39 describes in more detail, human activities are increasing the concentration of CO_2 in the atmosphere. The additional CO_2 is causing the average surface temperature on Earth to rise. As sea surface temperatures go up, coral bleaching may become a more widespread problem.

Or will it? In 1993, a group of scientists proposed the "adaptive bleaching hypothesis," which suggests that bleaching allows coral animals to expel algae that are poorly suited for warm water and replace them with other, better-adapted varieties. This idea makes sense. After all, coral animals appeared in the fossil record more than 500 million years ago; somehow, they must have adjusted to many periods of heating and cooling.

One way to test the adaptive bleaching hypothesis is to study reefs during and after El Niño events, when the sea surface temperature spikes in some coastal areas. Biologist Andrew Baker of

Figure 38.23 Coral Bleaching. This colony of staghorn coral, near Papua New Guinea, has a large bleached area at the center.

the Wildlife Conservation Society, working with colleagues from Columbia University and the University of Miami in Florida, used the 1997–1998 El Niño event to do just that. They knew that *Symbiodinium* is the predominant genus of symbiotic algae in the tissues of coral animals. Previous studies had revealed at least eight genetically different varieties, or genotypes, of *Symbiodinium*.

Baker and his team compared the genotypes of *Symbiodinium* in hundreds of coral samples collected from several locations. Reefs in the Persian Gulf and along the coast of Kenya had bleached during the El Niño event; in contrast, reefs in the Red Sea and off the island of Mauritius had not bleached. The researchers also took samples on a Panamanian reef before, during, and after the El Niño event. Figure 38.24a and b shows that recent bleaching was associated with the presence of one algal genotype (type D) and a shift away from another (type C).

Does this result mean that *Symbiodinium* group D functions better in warm water than does group C? One way to find out is to measure photosynthetic rates at different temperatures in controlled conditions. Rob Rowan of the University of Guam collected corals in genus *Pocillopora* and identified colonies that harbored *Symbiodinium* types C or D. After allowing the animals to become acclimated to 28.3°C water for 14 days, Rowan divided the corals randomly into two groups. Controls remained at 28.3°C. The water temperature of the treatment group, however, climbed to 31.3°C for 4.5 hours every day for 5 days. For 5 days after that, the temperature spiked to 32°C for 4.5 hours.

Rowan monitored O_2 concentrations to estimate the rates of photosynthesis and respiration at each temperature. A healthy coral produces more O_2 in photosynthesis than it consumes in respiration, so the photosynthesis-to-respiration ratio should exceed 1. That is exactly what happened at 28.3°C, regardless of the alga's genotype (figure 38.24c). But when Rowan cranked up the heat, corals colonized with type C algae had a photosynthesis-to-respiration ratio of about 1. Those harboring type D were unaffected. Rowan concluded that type D is "a high-temperature specialist."

Together, these two studies illustrate the importance of genetic diversity within a species. If all of the symbionts were identical, corals might be unable to adjust to temperature shifts in their habitat. Instead, changing conditions select for different genotypes of *Symbiodinium* algae. This adaptability is good news for the coral reefs. But a big question remains: will the corals be able to adjust to the current warming period, which is apparently occurring more rapidly than previous climate shifts? Only time will tell.

Baker, Andrew C., Craig J. Starger, Tim R. McClanahan, and Peter W. Glynn. 2004. Corals' adaptive response to climate change. *Nature,* vol. 430, page 741.

Rowan, Rob. 2004. Thermal adaptation in reef coral symbionts. *Nature,* vol. 430, page 742.

38.6 | Mastering Concepts

1. Explain how researchers tested the adaptive bleaching hypothesis.
2. What are the benefits and limitations of field studies and lab experiments in testing this hypothesis?

a. Corals sampled after 1997-98 El Niño

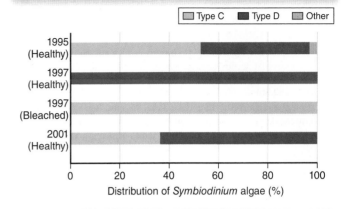

Legend: ■ Type A □ Type C ■ Type D □ Other

- Lamu, Kenya
- Vipingo, Kenya
- Kanamai, Kenya
- Mombasa, Kenya
- Kisite, Kenya
- Persian Gulf — *Recent bleaching*
- Mauritius
- Red Sea — *No bleaching*

Distribution of *Symbiodinium* algae (%)

b. Corals in Panama before, during, and after 1997-98 El Niño

Legend: □ Type C ■ Type D □ Other

- 1995 (Healthy)
- 1997 (Healthy)
- 1997 (Bleached)
- 2001 (Healthy)

Distribution of *Symbiodinium* algae (%)

Figure 38.24 Adjustments to Warm Water. (a, b) A survey of coral reefs suggests that warm waters prompt coral animals to shift from *Symbiodinium* type C to type D. (c) Laboratory experiments show that *Symbiodinium* type D tolerates warm water temperatures better than type C. The asterisk indicates a statistically significant difference between types C and D.

c. Corals hosting *Symbiodinium* algae in a laboratory experiment

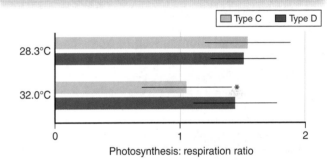

Legend: □ Type C ■ Type D

- 28.3°C
- 32.0°C *

Photosynthesis: respiration ratio

Apply It Now

El Niño Years

An El Niño is a periodic global climate shift in which the easterly winds that normally blow at the equator go slack. As a result, warm water drifts eastward from the western Pacific toward South America (figure 38.A).

The event usually occurs every 2 to 7 years and typically lasts a year or two. *El Niño* is Spanish for "the child," a reference to the observation that the phenomenon typically begins in December (around Christmas). An El Niño is often followed by a reversal of conditions called a La Niña.

Although El Niño results from events in the western Pacific, the effects are widespread. Shifting wind directions can cause powerful storms and raise sea surface temperatures along a wide belt in the Pacific around the equator. Severe or unusual weather may occur worldwide: monsoons in the central Pacific, typhoons in Hawaii, torrential rains in South America, blizzards in the Rockies, warm winters in the Northeast, flooding in southern California, and severe droughts in southern Africa.

The effects on coastal ecosystems are dramatic. As warm water moves east, it prevents cold, nutrient-rich water below from rising to the surface ("upwelling"). Huge numbers of anchovies and sardines that live in the colder waters move down to lower depths. With the fishes gone, the fishing industry collapses. Moreover, the birds, seals, and sea lions that normally scoop these smaller fishes from the water can no longer reach them. Many animals starve, grimly illustrating the connection between nutrients, producers, and consumers.

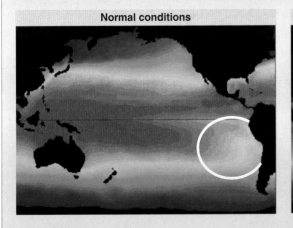

Normal conditions

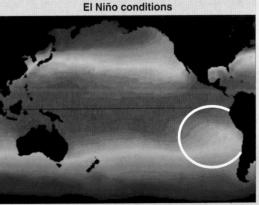

El Niño conditions

Figure 38.A El Niño. An El Niño brings warm water from west to east. In these maps, red areas represent the warmest water. In particular, note the increase in sea surface temperatures along the coast of South America.

Chapter Summary

38.1 | The Physical Environment Determines Where Life Exists

- The **biosphere** is subdivided into **biomes,** which are major types of **ecosystems** that occupy large geographic areas and share a characteristic climate and group of species. Each ecosystem incorporates **biotic** (living) and **abiotic** (nonliving) interactions.
- Autotrophs support each biome. On land, the most important **primary producers** are plants; in water, **phytoplankton** are most important.
- The major abiotic factors that limit a species' distribution include sunlight, temperature, moisture, salinity, and fire.

38.2 | Earth Has Diverse Climates

- Solar radiation is most intense at the equator and least intense at the poles. The resulting uneven heating creates the patterns of precipitation, prevailing winds, and ocean currents that influence climate.
- A mountain range can affect local climate by producing a **rain shadow.**

38.3 | Terrestrial Biomes Range from the Lush Tropics to the Frozen Poles

A. Towering Trees Dominate the Forests
- The **tropical rain forest** is warm and wet, with diverse life. Nutrients cycle rapidly, leaving soils relatively poor.
- In **temperate deciduous forests,** rainfall is evenly distributed throughout the year, summers are warm, and winters are cold. Their soils are fertile. **Temperate coniferous forests** usually have somewhat poorer soils, more rainfall, milder winters, and cooler summers.
- The **taiga** is a very cold northern (boreal) coniferous forest. Conifers conserve moisture during the long winters, when freezing temperatures mean liquid water is scarce.

B. Grasslands Occur in Tropical and Temperate Regions
- Tropical **savannas** have alternating dry and wet seasons and are dominated by grasses, with sparse shrubs and woody vegetation. Migrating herds of herbivores graze on the plants.
- Fire, grazing, and seasonal drought keep **temperate grasslands** free of trees.

C. Whether Hot or Cold, All Deserts Are Dry
- **Deserts** are dry; plants in desert biomes have adaptations that help them obtain and store water. Many desert animals seek shelter during the day and become active at night.

D. Fire- and Drought-Adapted Plants Dominate Mediterranean Shrublands (Chaparral)
- **Mediterranean shrublands** such as California's **chaparral** have dry summers and moist, mild winters. Many of the plants are fire-adapted.

E. Tundras Occupy High Latitudes and High Elevations
- **Tundra** can occur in the Arctic, in the Antarctic, or high on mountaintops in temperate zones.
- Arctic and Antarctic tundras have very cold, long winters. A layer of frozen soil called **permafrost** lies beneath the surface and prevents the growth of trees.

F. The Polar Ice Caps House Cold-Adapted Species
- At the poles, the climate is extremely cold and dry. Phytoplankton living in melted ice support a food web that includes zooplankton and many types of vertebrates.

38.4 | Freshwater Biomes Include Lakes, Ponds, and Streams

A. Lakes and Ponds Contain Standing Water
- In a lake's **photic zone,** light is sufficient for photosynthesis. The **littoral zone** of a lake is the shallow area where rooted plants occur, whereas phytoplankton are the primary producers in the **limnetic zone,** the upper layer of open water.
- The **profundal zone** is the dark deeper layer below the photic zone, and the lake bottom is the **benthic zone.** Nutrients arriving from the upper layers support life in the profundal and benthic zones.
- Young, deep, **oligotrophic** lakes are clear blue, with few nutrients to support algae. Over time, however, nutrients gradually accumulate, and algae tint the water green. The lake becomes a productive, or **eutrophic,** lake. As sediments accumulate, the lake transforms into a **wetland.**

B. Streams Carry Running Water
- In rivers, organisms are adapted to local currents. Near the headwaters, the channel is narrow, and the current is swift. As the river accumulates water and sediments, the current slows, and the channel widens.

38.5 | Oceans Make Up Earth's Largest Ecosystem

A. Land Meets Sea at the Coast
- An **estuary** is a highly productive area where rivers empty into oceans, and life is adapted to fluctuating salinity. Residents of the rocky **intertidal zone** cling tightly to the substrate as the tide ebbs and flows. In tropical regions, **coral reefs** support many thousands of species.

B. The Open Ocean Remains Mysterious
- The region of ocean near the shore is the **neritic zone.** Open water is the **oceanic zone.** Vertical subdivisions of the ocean include the **pelagic zone** (open water above the ocean floor) and the benthic zone (the bottom). The most productive areas are in the neritic zones, where **upwelling** occurs.

38.6 | Investigating Life: Some Like It Hot

- Coral animals maintain populations of symbiotic algae, often of genus *Symbiodinium,* in their tissues. High water temperature causes coral bleaching as the corals expel their symbionts.
- Some genotypes of *Symbiodinium* tolerate rising temperatures better than others, which may help coral reefs survive the current period of climate change.

Multiple Choice Questions

1. Which of the following is a biome?
   a. A rotting log   c. A forest
   b. A freshwater lake   d. Both b and c are correct.

2. Which abiotic factors likely determine the types of organisms found in Utah's Great Salt Lake?
   a. Sunlight   c. Temperature
   b. Salinity   d. All of the above are correct.

3. If Earth's axis were not tilted, there would be no
   a. equator.
   b. seasons.
   c. difference in the intensity of solar radiation between the equator and the poles.
   d. precipitation.

4. A biome composed of trees that seasonally shed their leaves is a
   a. chaparral.   c. temperate deciduous forest.
   b. tropical rain forest.   d. taiga.

5. People who cut down the tropical rain forests for agriculture often abandon the land after a few years. Why?
   a. Because they want to minimize the effect on biodiversity
   b. Because the soil is nutrient poor
   c. Because the climate is unsuitable for plant growth
   d. Because only a few crop species will grow

6. What is a distinguishing feature of the tundra?
   a. Permafrost   c. Wind-pollinated grasses
   b. Infertile soil   d. Year-round ice cover

7. An organism that lives in lake sediment would be in the _____ zone.
   a. limnetic
   b. profundal
   c. benthic
   d. littoral

8. Why can't the phytoplankton of a eutrophic lake maintain deep-water oxygen levels?
   a. Because oxygen production by photosynthesis only occurs where there is light
   b. Because phytoplankton only occur in the littoral zone
   c. Because nutrient levels are too low to support photosynthesis in deep water
   d. Both a and b are correct.

9. Why do estuaries have such high primary productivity?
   a. The changing salinity of the water
   b. High nutrient input from the land
   c. The absence of consumers that eat producers
   d. All of the above are correct.

10. The deepest waters of the ocean occur in
    a. the profundal zone.
    b. the neritic zone.
    c. the photic zone.
    d. Both a and c are correct.

Write It Out

1. How does the fact that Earth is a sphere tilted on its axis influence the distribution of life?

2. Explain why the climate on the west side of Oregon's Cascade Mountains is much wetter than on the east side of the mountain range.

3. List adaptations that characterize organisms in each of the following biomes: tropical rain forest, savanna, temperate grassland, tundra, desert, taiga, the rocky intertidal zone, the bottom of a lake.

4. How can the tropical rain forest support diverse and abundant life with such poor soil?

5. What is permafrost, where does it occur, and how does it affect primary production in that biome?

6. Polar bears live on the ice cap near the North Pole. Their numbers are dwindling, apparently because of both pollution and global climate change. Ice on Canada's Hudson Bay, for example, is melting about 3 weeks earlier in the year than it was some 30 years ago. List some specific ways that this change in habitat might affect polar bear populations.

7. A watershed is an area of land in which all of the precipitation drains into the same body of water. The central United States, for example, is in the Mississippi River's watershed, which itself consists of many smaller watersheds. Why would cities and towns that share a watershed want to cooperate to manage water resources? Why might one part of a watershed face different issues from another part?

8. Poultry farmers apply large amounts of nutrient-rich animal waste onto the land, where it runs off into nearby lakes and streams. What effect might this nutrient input have on the aquatic ecosystems? Lawmakers in some states have debated whether animal waste, a natural substance, should legally qualify as a hazardous waste. Do you think it should?

9. Nuisance aquatic plants such as *Hydrilla* can disrupt the ecology of the littoral zone of a lake. Two of the most common ways to control nuisance aquatic plants are herbicides (chemicals that kill plants) and biological control (introducing fungi or animals that consume the plants). How might each strategy help or harm the lake ecosystem?

10. Describe how and why photosynthetic activity differs in the zones of a lake or ocean.

11. Describe the physical and chemical differences between the water in a mountain stream and the water near the mouth of the Mississippi River.

12. Make a concept map depicting the relationships among the zones of the ocean. What is the main energy source in each zone?

13. The biomes described in this chapter do not include those that humans create, such as cities, villages, croplands, rangelands, and tree farms. How are these biomes similar to and different from the biomes in this chapter?

14. Use the Internet to learn about cave ecosystems. What are some possible energy sources in a cave? How do cave ecosystems interact with other ecosystems? How might the abiotic conditions in a cave select for unique adaptations in cave-dwelling organisms? How do human activities affect caves?

15. Some scientists are currently attempting to catalog all of the world's biodiversity. What are some of the technical problems they may encounter?

16. Suppose you are exploring the chaparral ecosystem in California. You encounter a shrub species that you think may be fire-adapted, and you wonder whether the plant can reproduce in the absence of fire. Design an experiment that would help answer your question.

17. Researchers and citizens in Prairie City, Iowa, are reconstructing the prairie by planting seeds collected from remnants of native grasslands and reintroducing animals. Which other biomes discussed in the chapter might it be possible to reconstruct and which not?

18. Hundreds of millions of years ago, Earth's land masses were joined into one supercontinent, Pangaea. If Pangaea had never broken up, do you think there would be more biodiversity, less biodiversity, or the same amount of diversity as today? Explain your answer.

Pull It Together

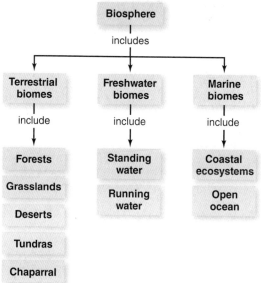

1. What factors determine the location of each biome on Earth?
2. What types of forests occur on Earth, and what combination of conditions favors each type?
3. How do tropical savannas differ from temperate grasslands?
4. List examples of coastal ecosystems.

Chapter 39 Preserving Biodiversity

Herons, crayfish, grasses, trees, and countless other species make up the biodiversity of the watery Everglades.

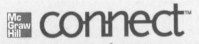

The Endangered Everglades

ALTHOUGH THE WORLD HAS NO SHORTAGE OF DEGRADED ECOSYSTEMS, THE FLORIDA EVERGLADES REGION IS REMARKABLE FOR THE SCOPE OF ITS DESTRUCTION. Before 1900, much of Florida was a continuous waterway. This vast area included estuaries, saw grass plains, mangrove swamps, and tropical hardwood forests. The Everglades was home to marsh grasses, cypress trees, egrets, eagles, herons, panthers, and alligators.

Damage to the Everglades began in the twentieth century. From 1903 until 1917, the "Everglades reclamation project" built four large canals that drained the area to prevent flooding. In 1930, following severe hurricanes in 1926 and 1928, levees were built around Lake Okeechobee, a huge lake that supplies water to the Everglades. Farmers, reassured of their safety from floods, doubled their sugarcane crops.

In 1947 and 1948, hurricanes dumped 274 centimeters of rain, nearly twice the normal rainfall. Many farms and homes were submerged. Congress decided to control the flooding with a new water supply system. Engineers built a system of canals, levees, and pumps that covered nearly 1600 kilometers to direct water to agricultural and urban areas. Meanwhile, the human population in the area skyrocketed.

Then came disaster—nature changed. In the early 1960s, a drought started that would last until 1975. Nearly 300,000 hectares of land burned. The great effort to prevent flooding now backfired, causing a terrible water shortage. The situation worsened with a drought in 1981 and torrential rains in 1982–1983. In 1986, a fifth of Lake Okeechobee fell under an algae bloom, the result of nitrogen and phosphorus from fertilizer and dairy operations washing into the water. Native saw grass could not compete with species that could tolerate the nutrient pollution. Moreover, the influx of fresh water robbed mangroves of salt water, which led to declines in populations of sea grasses and various animals. Wading bird populations plummeted to 10% of their levels a century earlier. Dozens of species became endangered.

Efforts have been underway to undo the damage to the Florida ecosystem. Consider, for example, the Kissimmee River, which drains into Lake Okeechobee. The goal of the Kissimmee River Restoration Project is to backfill canals and remove levees, gradually restoring flow to 69 kilometers of the river's channel and the surrounding floodplain wetlands. Native vegetation, waterfowl, aquatic invertebrates, and fishes have already returned to areas where the project is complete.

The Kissimmee River project is just one part of the much larger Comprehensive Everglades Restoration Plan. With a price tag estimated at $9.5 billion, the CERP is "the world's largest ecosystem restoration effort." Monitoring its progress will require a small army of conservation biologists. This chapter describes some of the many challenges scientists such as these face worldwide.

UNIT 7

What's the Point?

Learning Outline

Learn How to Learn

Make Your Own Review Sheet

If you are facing a big exam, how can you make sense of everything you have learned? One way is to make your own review sheet. The best strategy will depend on what your instructor expects you to know, but here are a few ideas to try: make lists; draw concept maps that link ideas within and between chapters; draw diagrams that illustrate important processes; and write mini-essays that explain the main points in each chapter's learning outline.

39.1 | Earth's Biodiversity Is Dwindling

For more than 3 billion years, evolution has produced an extraordinary diversity of life, both obvious and unseen; unit 4 provided an overview of Earth's inhabitants. Humans simply cannot live without these other species (figure 39.1). We use other organisms for food, shelter, energy, clothing, and drugs. Microbes carry out indispensable tasks, from digesting food in our intestines, to decaying organic matter, to fixing nitrogen, to

Figure 39.1 We Need Other Species. This child and water buffalo are in a rice field. Humans have cultivated rice for thousands of years, using the grain as a source of protein and starch, and feeding the rice straw to livestock. The water buffalo is not only a work animal but also a source of dairy products. The animal's dung fertilizes the rice fields.

producing oxygen (O_2). Plants and microbes together absorb carbon dioxide (CO_2) and purify the air, soil, and water. Wetland plants reduce the severity of floods. Insects pollinate our crops. The remains of species that lived millions of years ago provide the fossil fuels that sustain our economies. The list goes on and on.

Clearly, our existence as a species depends on **biodiversity**—the variety of life on Earth. Biologists measure biodiversity at three levels: genetic, species, and ecosystem. Genetic diversity is the amount of variation that exists within a species. This aspect of biodiversity is essential for populations to adapt to changing conditions. The next level, species diversity, accounts for the number of species that occupy the biosphere. Finally, ecosystem diversity means the variety of ecosystems on Earth, such as deserts, rain forests, grasslands, and mountaintops.
▶ species richness, p. 762

One way to monitor species biodiversity is to count how many known species are at risk of extinction. **Extinction** means that the last individual of a species has perished. Some species, such as the dodo, are extinct altogether; others, such as a bird called the Guam rail, are extinct in the wild but still exist in captivity. A species that is **endangered** has a high risk of extinction in the near future, and a **vulnerable** species is likely to become extinct in the more distant future. The International Union for Conservation of Nature (IUCN) combines endangered and vulnerable species into one umbrella category ("threatened").

The data suggest that Earth is in the midst of a biodiversity crisis (table 39.1, figure 39.2). The current extinction rate of vertebrates is some 100 to 1000 times the "background" species extinction rate, which estimates how quickly species disappeared before human intervention. According to the IUCN, about 30% of amphibian species, 28% of nonavian reptiles, 21% of mammals, and 12% of birds are threatened. Other taxonomic groups are similarly imperiled. Overall, the current biodiversity crisis is

Table 39.1 A Few of the World's Endangered Species

Species	Former Range	Threat(s)
California condor (*Gymnogyps californianus*)	Western United States	Habitat loss, shooting, lead poisoning, toxic substances in environment
Green pitcher plant (*Sarracenia oreophila*)	Alabama, Tennessee, Georgia	Habitat loss, overharvesting for commercial trade
Red-cockaded woodpecker (*Picoides borealis*)	Eastern United States from Florida to New Jersey and Maryland, inland to Texas, Oklahoma, Missouri, Kentucky, and Tennessee	Habitat loss
American burying beetle (*Nicrophorus americanus*)	Arkansas, Kansas, Oklahoma, Nebraska, South Dakota	Competition for food (carrion)
Guam rail (bird) (*Gallirallus owstoni*)	Guam	Invasive species (brown tree snake)
Sumatran rhino (*Dicerorhinus sumatrensis*)	Indonesia, Malaysia	Poaching
Northern bluefin tuna (*Thunnus thynnus*)	Atlantic Ocean, Mediterranean and Black Seas	Overfishing

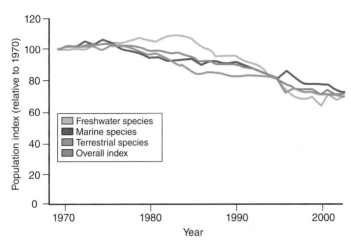

Figure 39.2 **Biodiversity Crisis.** The World Wildlife Fund, an environmental advocacy group, uses scientific reports to monitor 1313 species of freshwater, marine, and terrestrial vertebrates from around the world. The population index measures overall population trends relative to a baseline year of 1970. Overall, the index has declined by about 30% since 1970.

comparable to that of the five major mass extinctions that have occurred in the past 500 million years (see figure 13.13).

Conservation biologists study the preservation of biodiversity at all levels. These scientists try to determine why species disappear, and they develop strategies for maintaining diversity. This final chapter focuses first on the main causes of the loss of biodiversity: habitat destruction and degradation, the introduction of nonnative species, and overexploitation. Each of these threats to biodiversity will only become worse as the human population continues to grow. Nevertheless, the chapter ends on a hopeful note, with some ways people can help counteract the biodiversity crisis.

39.1 | Mastering Concepts

1. What is the value of diversity to humans and to ecosystems as a whole?
2. Describe the three types of biodiversity.
3. Differentiate among extinct, endangered, and vulnerable species.
4. What is conservation biology?

39.2 | Human Activities Destroy Habitats

Habitat destruction is the primary cause of diminishing biodiversity (figure 39.3). Humans have altered nearly 50% of the land, replacing prairies, wetlands, and forests with farms, rangeland, and cities. The link to biodiversity is obvious: destroying

a.

b.

c.

Figure 39.3 **Habitat Destruction.** (a) A road into the rain forest brings new human settlement and agriculture. Poor soils mean the crops will not last long. (b) Farmland has replaced the native prairie in Iowa. (c) Large cities such as San Francisco put pavement and buildings in place of the original ecosystem.

habitat makes it difficult or impossible for its occupants to survive and reproduce.

One form of habitat destruction is **deforestation,** the removal of all tree cover from a forested area. Forests harbor a tremendous diversity of plant, animal, and microbial life, which may become extinct as the forests vanish. Deforestation also promotes soil erosion, which reduces soil fertility and pollutes water. Transpiration by forest plants contributes water vapor to the atmosphere, affecting global climates. When humans burn a forest we not only remove an important component of the global water cycle but also release stored carbon into the atmosphere, adding to the greenhouse effect (see section 39.3C). ▶ ▶ water cycle, p. 773; forests, p. 789

Nearly half of the world's moist tropical forests have already been cleared, mostly to make room for subsistence agriculture. Ironically, the same soils that support the lush tropical rain forest produce poor crop yields. Warm temperatures near the equator promote rapid decomposition of organic matter, and heavy rains deplete soil nutrients. Once native plants give way to crops or grazing animals, the nutrient-poor soils harden into a cementlike crust. Species disappear and food webs topple, threatening biodiversity in the entire region.

The disappearance of the native North American temperate forest has paralleled settlement by Europeans. Since the early eighteenth century, people have cleared the land from east to west to create farmland, obtain fuel, and make room for railroads and towns. Today, although vast areas of managed forests and tree plantations occupy the region, less than 1% of the original temperate forest survives.

Prairies and other temperate grasslands have also disappeared as the human population has expanded. Grassland soils are among the most fertile in the world, and their rolling hills are ideal for cultivation. Fields of corn, soybeans, wheat, and other crops have replaced nearly all of the North American prairie. ▶ temperate grasslands, p. 791

Whereas forests and grasslands are shrinking worldwide, deserts are expanding into surrounding areas in a process called **desertification.** For example, desert is displacing savanna in Africa, where the Sahara is the driest and largest desert in the world. Along the Sahara's southern edge and in countries farther south, drought and overgrazing have destroyed grasses and compacted the soil. New plants cannot establish themselves, so patches of desert enlarge and join. As the desert spreads, farmers cut down more forest to create grazing land. This land is destined to become desert too, if these practices continue.

Freshwater habitats are also vulnerable to destruction. Damming for flood control or power generation, for example, completely alters river ecosystems (**figure 39.4**). Worldwide, the number of large dams (over 15 meters high) is estimated at more than 48,000. How do dams reduce biodiversity? Deep reservoirs replace waterfalls, rapids, and wetlands, where birds and many other species breed. Areas that were once seasonally flooded become dry. Water temperature, oxygen content, and nutrient levels all change, triggering shifts in species diversity and food webs both above and below the dam. Dams also disrupt the migration of fishes and other aquatic animals.

Figure 39.4 Big Dam. The Ataturk Dam on the Euphrates River provides power to Turkey. Dams help control flooding and provide irrigation water, but they also eliminate streamside habitat and disrupt the migration of fishes and other animals.

Another threat to freshwater biodiversity is alterations to a river's path. Along the banks of the Mississippi River, for example, levees built to prevent flooding alter the pattern of sediment deposition. Channelization increases the water's flow rate, eroding sediments and choking out downstream communities. Nutrients that once spread over the floodplain during periodic floods are now confined to the river channel, which carries them to the Gulf of Mexico. There, the nutrients feed algae, causing additional problems described in section 39.3.

Coastlines are also suffering from habitat destruction. Many fishes and invertebrates spend part of their lives in estuaries, and diverse algae and flowering plants support the food web. Yet humans have drained and filled estuaries for urbanization, housing, tourism, dredging, mining, and agriculture. These activities affect life in the oceans, too. The loss of coastal habitats can threaten populations of commercially important species of marine animals such as bluefin tuna, grouper, and cod. ▶ estuaries, p. 796

39.2 | Mastering Concepts

1. Which human activities account for most of the loss of terrestrial habitat?
2. How do dams and channelization alter river ecosystems?
3. Why is damage to estuaries especially devastating?

39.3 | Pollution Degrades Habitats

Pollution is any chemical, physical, or biological change in the environment that harms living organisms. Pollution degrades the quality of air, water, and land, threatening biodiversity worldwide.

A. Water Pollution Threatens Aquatic Life

A diverse array of pollutants affects rivers, lakes, and groundwater (table 39.2). For example, mining operations often release inorganic pollutants such as heavy metals or cyanide into the water, whereas shipping accidents and leaking oil wells add petroleum.

Raw sewage can also be a major pollutant. In addition to carrying disease-causing organisms, sewage also contains organic matter and nutrients such as nitrogen and phosphorus. When released into waterways, organic matter fuels the growth of bacteria, whose respiration depletes the water of oxygen. Fish and other organisms die. At the same time, in a process called **eutrophication,** nutrients from the sewage fertilize phytoplankton in the water (see figure 38.19). The resulting algal blooms are unsightly. Moreover, when the algae die, the microbes that decompose their dead bodies further deplete dissolved oxygen in the water.

Fertilizer and animal wastes that enter waterways also cause eutrophication. For example, section 39.2 explained how nutrients from lands drained by the Mississippi River find their way into the Gulf of Mexico. There, the nutrients fuel algae blooms. On a local scale, the resulting red and brown tides may kill fishes, manatees, and other sea life. A larger-scale problem, however, is the oxygen-depleted zone that forms each summer near the seafloor off the coast of Louisiana (figure 39.5). The seasonal lack of oxygen in this so-called "dead zone" kills

Figure 39.5 Dead Zone. Nutrient-rich runoff enters the five major river systems that drain into the Mississippi River. The combined waters then pour into the Gulf of Mexico. The result is a zone of seasonal oxygen depletion off the coast of Louisiana.

many animals, disrupting not only the Gulf's food web but also its economy: commercially important fish and shrimp cannot live in oxygen-depleted water. ▶ red tide, p. 356

Some pollutants seem deceptively harmless. Sediments, for example, reduce photosynthesis by blocking light penetration into water. Even heat can be a pollutant. Hot water discharged from power plants reduces the ability of a river to carry dissolved oxygen, harming fishes and other aquatic organisms.

Toxic chemicals and trash also pollute the open ocean. For example, ocean currents have concentrated millions of tons of plastic in a huge area dubbed the "Great Pacific Garbage Patch." Floating just below the water surface in the northern Pacific Ocean are tiny pellets, called "nurdles," that form when plastic debris disintegrates. Fishes, sea turtles and sea birds mistake the plastic for their natural food. The pellets can lodge in intestines and kill the animals outright, or chemicals from the plastic may accumulate in their tissues.

The chemicals leached from plastic are a small subset of human-made **persistent organic pollutants,** carbon-containing molecules such as polychlorinated biphenyls (PCBs), polycyclic aromatic hydrocarbons (PAHs), and DDT. These substances contaminate ecosystems over long periods. Some of these compounds cause cancer; others are hormone mimics that disrupt reproduction (see the opening essay for chapter 27). Because they do not biodegrade, these fat-soluble chemicals become more concentrated as they ascend the food chain. This process, called biomagnification, accounts for the high concentrations of toxic chemicals in the fatty tissues of tunas, polar bears, and other top predators. ▶ biomagnification, p. 770

Organisms living in polluted areas are often exposed to several toxins. In the aftermath of Hurricane Katrina in 2005, floodwaters contained raw sewage, garbage, crude oil, gasoline, lead and other heavy metals, pesticides, and countless other

Table 39.2 Examples of Chemical Water Pollutants

Organic	Inorganic
Sewage	Chloride ions
Detergents	Heavy metals (mercury, lead, chromium, zinc, nickel, copper, cadmium)
Pesticides	
Wood-bleaching agents	Nitrogen from fertilizer
Petroleum	Phosphorus from fertilizer and sewage
Humic acids	
Polychlorinated biphenyls (PCBs)	Cyanide
	Selenium
Polycyclic aromatic hydrocarbons (PAHs)	

Figure 39.6 Deadly Waters. These cars were submerged in the aftermath of Hurricane Katrina, which devastated the city of New Orleans in 2005. The floodwaters contained toxic chemicals that will threaten ecosystems for some time.

pollutants that had spilled from damaged refineries and chemical plants (**figure 39.6**). The toxic soup threatened not only human health but also the microbes, producers, and animals in the aquatic food chain. Moreover, when the floodwaters receded, they left behind their poisonous residues in soil. Katrina's deadly legacy may extend far into the future, well beyond the lives and homes lost in the hurricane itself.

B. Air Pollution Causes Many Types of Damage

Smog is a type of air pollution that forms a visible haze in the lower atmosphere (**figure 39.7**). Industrial smog occurs in urban

Figure 39.7 Smog. Air pollution continues to plague many cities. This is Santiago, Chile.

and industrial regions where power plants, factories, and households burn coal and oil. The resulting smoke and sulfur dioxide (SO_2) may form a dark haze. Photochemical smog forms when nitrogen oxides and emissions from vehicle tailpipes undergo chemical reactions in the presence of light, producing ozone (O_3) and other harmful chemicals that injure plants and cause severe respiratory problems in humans. Warm, sunny areas with heavy automobile traffic have the most photochemical smog, but winds may carry the pollutants to sparsely populated areas.
▶ nitrogen cycle, p. 775

Air also carries suspended **particulates,** tiny bits of matter that float in the air. Examples include road dust, volcanic ash, soot from partially burned fossil fuels, mold spores, pollen, and acidic particles. The damage they cause extends beyond the occasional need to dust off bookshelves and window sills. Most harmful are particles that are 2.5 μm in diameter or smaller. Not only do they become trapped deep within the lungs, but the heavy metals and toxic organic compounds in these particles make them especially likely to trigger inflammation, shortness of breath, asthma, or even cancer.

Other forms of air pollution are less visible but perhaps more harmful than smog. One example is **acid deposition:** acidic rain, snow, fog, dew, or dry particles. Because the atmosphere contains CO_2 and water, all rainfall includes some carbonic acid (H_2CO_3) and is therefore slightly acidic, with a pH around 5.6. The combustion of fossil fuels, however, releases sulfur and nitrogen oxides (SO_2 and NO_2) into the atmosphere. These compounds react with water, forming sulfuric acid and nitric acid. The acids return to the Earth as acid deposition. ▶ pH scale, p. 30

Coal-burning power plants release the most sulfur and nitrogen oxides, although emissions of these pollutants have declined since the 1980s. In the United States, winds carry airborne acids hundreds of miles east and northeast of the power plants in the Midwest. As a result, rainfall in the eastern United States has an average pH of about 4.6. Acid deposition also affects the Pacific Northwest, the Rockies, Canada, Europe, East Asia, and the former Soviet Union.

Most lakes have a pH between 6 and 8; acid deposition may lower it to 5 or less. The acid can leach toxic metals such as aluminum or mercury from soils and sediments, causing fish eggs to die or yield deformed offspring. Lake-clogging algae replace aquatic flowering plants. Organisms that feed on the doomed species must seek alternative food sources or starve, which disrupts or topples food webs. Eventually, lake life dwindles to a few species that can tolerate increasingly acidic conditions.

Acid deposition alters forests, too. As soil pH drops, aluminum ions released from soil enter roots and stunt tree growth. Affected trees become less able to resist infection or to survive harsh weather. As a result, acid deposition is thinning high-elevation forests throughout Europe and on the U.S. coast from New England to South Carolina (**figure 39.8**).

Chemicals that destroy ozone can also be extremely harmful air pollutants. Ozone is an atmospheric molecule with two faces. As we have already seen, ozone in photochemical smog at Earth's

Figure 39.8 Acid Deposition. Acid rain has severely damaged this fir forest in the Czech Republic.

surface is harmful. In the upper atmosphere, however, the stratospheric **ozone layer** blocks damaging ultraviolet (UV) radiation from the sun. Located about 10 to 50 kilometers above Earth's surface, the ozone layer forms when UV radiation from the sun reacts with oxygen gas (O_2). UV radiation damages biological molecules such as DNA, causing genetic mutations. The stratospheric ozone layer thus protects organisms from much of the harmful radiation that would otherwise strike Earth.

In the past 40 years, the ozone layer has thinned over parts of Asia, Europe, North America, Australia, and New Zealand, and a "hole" has formed over Antarctica (figure 39.9). As the ozone layer thins, UV radiation is increasing at Earth's surface. In humans, exposure to short-wavelength UV radiation can cause skin cancer or cataracts. Ozone depletion may also indirectly contribute to species extinctions. For example, UV radiation can harm the phytoplankton that support aquatic food webs. Also, increasing UV radiation may also be one of many factors causing amphibian populations to plummet.

What is the source of damage to the ozone layer? The main culprits are chlorine, fluorine, and bromine gases, some of which enter the atmosphere as persistent, human-made chlorofluorocarbon (CFC) compounds. These compounds were once used in refrigerants such as Freon, as propellants in aerosol cans, and to produce foamed plastics. They can persist for decades in the upper atmosphere, catalyzing chemical reactions that break down ozone. An international treaty signed in 1987, the Montreal Protocol, banned the use of CFCs. If all countries comply with the treaty, experts estimate that the ozone layer should recover by 2050.

C. Global Climate Change Alters and Shifts Habitats

We now turn to air pollutants with the potential to do the most harm of all: greenhouse gases. In the past, scientists debated whether human activities could actually change something as complex as Earth's overall climate. Now, the scientific consensus is clear: we can and do.

Greenhouse Gases Warm Earth's Surface CO_2 is a colorless, odorless gas present in the atmosphere at a concentration of about 387 parts per million. Although it is a minor atmospheric constituent, CO_2 is one of several gases that contribute to the **greenhouse effect,** an increase in surface temperature caused by heat-trapping gases in Earth's atmosphere. As illustrated in

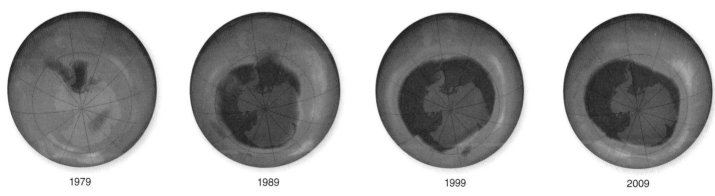

| 1979 | 1989 | 1999 | 2009 |

Figure 39.9 The Antarctic Ozone Hole. Satellite images reveal the ozone hole over Antarctica from 1979 to 2009. Each image was taken on September 15, a time when the ozone hole nears its maximum size. In these images, purple and blue represent thinned areas of the ozone layer.
Source: NASA Ozone Hole Watch.

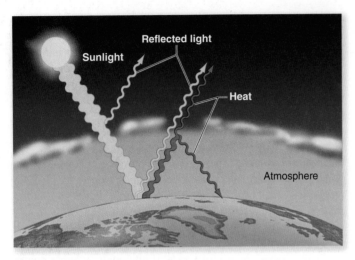

Figure 39.10 The Greenhouse Effect. Solar radiation heats Earth's surface. Some of this heat energy is reradiated to the atmosphere, but some is trapped near the surface by CO_2 and other greenhouse gases.

figure 39.10, sunlight passes through the atmosphere and reaches Earth's surface. Some of the energy is reflected, but some is absorbed and reradiated as heat. So-called "greenhouse gases" block the escape of this heat from the atmosphere, much as do the glass panes of a greenhouse.

Carbon dioxide is a greenhouse gas; others include methane, nitrous oxide, and CFCs (see the Burning Question on page 811).

These other gases actually trap heat much more efficiently than does CO_2, but because they are less abundant, they contribute only half as much to the greenhouse effect.

In a sense, the greenhouse effect supports life, because Earth's average temperature would be much lower without its blanket of greenhouse gases. But CO_2 has been steadily accumulating in the atmosphere since monitoring began in the 1950s (figure 39.11). The increase in atmospheric concentration of CO_2 is largely caused by burning fossil fuels such as coal, oil, and gas. Tropical deforestation and other combustion activities also add a share. All together, human activities release some 6 million metric tons of CO_2 into the atmosphere each year. Photosynthesis temporarily removes some of this carbon from the atmosphere. Overall, however, more CO_2 is added than is removed. ▶ carbon cycle, p. 774

This accumulation of CO_2—along with climbing levels of other greenhouse gases—was accompanied by an increase in average global temperatures in the twentieth century (see figure 39.11). Computer models predict that these trends will continue. Depending on the future concentration of CO_2 in the atmosphere, the Intergovernmental Panel on Climate Change estimates that the global average surface temperature could increase by anywhere from 1.8 to 4.0°C by the end of the twenty-first century, accompanied by a sea level rise of 18 to 59 centimeters.

Because Earth's average temperature is rising, this phenomenon is often called "global warming." In reality, however, some areas will become warmer and others will become cooler.

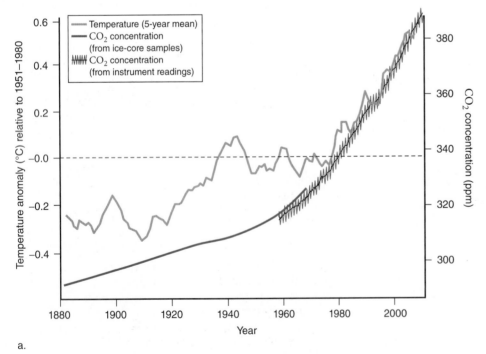

a.

b.

Figure 39.11 CO_2 and Global Average Temperature. (a) Measures of CO_2 concentrations from ice cores and instrument readings at Mauna Loa, Hawaii, show that CO_2 continues to accumulate. The low point of each year's CO_2 concentration reflects peak photosynthesis in the northern hemisphere. As CO_2 accumulates, the average global temperature is also increasing. (b) Coal-fired power plants are among the main culprits in the rising levels of CO_2.
Sources: NOAA Earth System Research Laboratory; NASA Goddard Institute for Space Studies Surface Temperature Analysis.

Global climate change is therefore a more accurate term for past and future alterations in Earth's weather patterns.

Global Climate Change Has Severe Consequences

An increase of 1.8 to 4.0°C may seem too small to make much difference. Yet the warming that occurred in the twentieth century has been associated with many measurable effects, including the shrinking of polar ice sheets and alpine glaciers (figure 39.12). When floating sea ice melts, its effect on sea level is minimal (just as ice melting in a full glass of lemonade does not cause the liquid to overflow). However, the loss of ice from Greenland and Antarctica has contributed to a rise in sea level, causing coastal erosion and flooding in low-lying areas. The warming poles have also seen a decrease in the amount of permafrost.

Weather conditions in tropical and temperate regions are also changing, thanks in part to a rise in sea surface temperature. In the tropics, warmer water means higher winds and more evaporation, which increases the amount of rainfall in a severe storm. In dry areas, on the other hand, global climate change may mean more intense and longer droughts, fewer cold snaps, and more heat waves.

These changes kill some organisms outright, whereas others become stressed and vulnerable to disease. Still others may migrate to higher elevations or higher latitudes. Warming in Yosemite National Park, for example, has driven some small mammals to cooler mountaintop habitats. Elsewhere, the ranges of at least 34 species of butterflies are moving northward. At the southern ends of their ranges, where temperatures are rising, some species have become locally extinct. In addition, scientists are tracking events known to occur at the same time each year. In the United Kingdom, butterflies are emerging and amphibians are mating a few days earlier than usual; in North America, many plants are flowering and birds are migrating earlier.

Continued climate change will affect not only wild organisms but also agriculture and public health. Growing seasons in

Figure 39.12 Shrinking Glaciers. (a) Switzerland's Steigletscher glacier in the summer of 1994. (b) By the summer of 2006, the glacier had retreated substantially.

temperate areas are lengthening, and the southern United States may become too dry to sustain many traditional crops. Water shortages worldwide may affect more than a billion people. Infectious disease patterns may also shift as tropical diseases such as malaria, African sleeping sickness, dengue fever, and river blindness move into more temperate areas.

Ocean life is also vulnerable to global climate change. One problem is that the accumulation of CO_2 causes ocean water to become more acidic. The lower pH causes the calcium carbonate shells of oysters, clams, and other mollusks to dissolve, along with the exoskeletons of coral animals. At the same time, high sea temperatures threaten coral reefs by triggering coral bleaching (see section 38.6).

39.3 | Mastering Concepts

1. How do toxic chemicals, nutrients, sediments, and heat affect aquatic ecosystems?
2. What are major sources of industrial smog, photochemical smog, and acid deposition?
3. What effects do smog, acid deposition, particulates, and the thinning ozone layer have on life?
4. Why is CO_2 accumulating in Earth's atmosphere?
5. Describe how and why Earth's climate changed during the past century.
6. How does global climate change threaten biodiversity?

Burning Question

What does the ozone hole have to do with global climate change?

Many people confuse global climate change and the ozone hole. These two problems are largely separate, but they do share two common threads. First, the chlorofluorocarbon gases that deplete the ozone layer are also greenhouse gases, contributing to a warmer atmosphere. Second, the greenhouse effect may cause the hole in the ozone layer to grow. A thick, heat-trapping "blanket" of greenhouse gases in the troposphere (the lowest part of the atmosphere) means less heat reaches the stratosphere, where the ozone layer is. A cooler stratosphere, in turn, extends the time that stratospheric clouds blanket the polar regions in winter. These clouds of ice and nitric acid speed the chemical reactions that deplete stratospheric ozone.

Submit your burning question to:
marielle_hoefnagels@mcgraw-hill.com

39.4 | Exotic Invaders and Overexploitation Devastate Many Species

In addition to habitat destruction and pollution, two other important threats to biodiversity are invasive species and overexploitation.

A. Invasive Species Displace Native Organisms

An introduced species (also called a nonnative, alien, or exotic species) is one that humans bring to an area where it did not previously occur. When people move from one location to another, we often bring along our pets, crops, livestock, and ornamental plants. We also unintentionally introduce microorganisms, parasites, and stowaways such as rodents and insects on ships, cars, and planes.

This transport may seem harmless at first, and many introduced species die. Even if they survive in their new homes, they may not cause problems. For example, the house sparrow, *Passer domesticus,* was introduced to the United States from Europe in the 1850s. Although it has spread throughout the North American continent, it has not caused obvious ecological problems. In addition, at least 5000 nonnative plant species live in U.S. ecosystems, introduced from agriculture and urbanization. Most have done no apparent harm.

If a nonnative species becomes invasive, however, it can cause immense destruction. To be considered **invasive,** an introduced species must begin breeding in its new location and spread widely from the original point of introduction. In addition, according to some definitions, the species must harm the environment, human health, or the economy. Of every 100 species introduced, only one persists to take over a niche. Nonetheless, the Global Invasive Species Database lists 484 invasive plants, animals, and microorganisms in the United States alone, including fire ants and the fast-growing plant called kudzu. Many more invasive species occur worldwide.

Some invasive species are staggeringly destructive. The opening essay of chapter 36 describes how brown tree snakes have devastated Guam's wildlife, and section 37.1 refers to the invasion of zebra mussels in the Great Lakes. Other examples of invasive species in North America include:

- European starlings *(Sturnus vulgaris)* are birds that were released in New York City's Central Park in 1890. Huge flocks of starlings now reside all across North America, fouling cities with their droppings.

- The marine toad *(Bufo marinus)* is a voracious omnivore that competes with and preys on native amphibians in Florida.

- Hydrilla *(Hydrilla verticillata)* is an aquatic plant that alters nutrient cycles, affects aquatic animals, and reduces recreational use of lakes and rivers.

- Purple loosestrife *(Lythrum salicaria)* is a wetland plant that displaces native plants. The disappearance of the native plants, in turn, threatens the turtles and other animals that would otherwise feed on them.

- Hungry caterpillars of the gypsy moth *(Lymantria dispar)* strip the foliage from hundreds of tree species in North America (**figure 39.13**).

- The larvae of an Asian beetle called the emerald ash borer *(Agrilus planipennis)* destroy the bark of ash trees. Since they arrived in 2002, these beetles have killed millions of ash trees in the midwestern and eastern United States and Canada.

- Fungi have all but eradicated American chestnut and American elm trees.

When an invasion does occur, the harm may be ecological and economic. A nonnative species not only changes the composition of a community but also may carry diseases that spread to

Figure 39.13 Voracious Eater. Gypsy moths were introduced to the United States from Europe or Asia in the late 1860s. Their larvae strip the foliage from hundreds of hardwood tree species. These caterpillars are eating oak leaves.

native species. The economic costs include everything from the purchase of herbicides that kill invasive weeds, to the loss of grain to hungry birds and rodents, to declining tax revenue when invasive aquatic plants interfere with boating and recreation.

B. Overexploitation Can Drive Species to Extinction

Another cause of species extinction is **overexploitation:** harvesting a species faster than it can reproduce (figure 39.14). The market for exotic pets, for example, is harming populations of mammals, birds, snakes, lizards, amphibians, and fishes (see chapter 20's Apply It Now box on page 441).

Many of the most famous examples of species extinctions result from overhunting of terrestrial animals. The dodo, for example, was a flightless bird that once lived on the Indian Ocean island of Mauritius. In the late seventeenth century, humans hunted the dodo for food while introducing other species to the island. The dodo soon went extinct. In the United States, commercial-scale hunting nearly drove the American bison to extinction in the 1800s. The passenger pigeon and Carolina parakeet did go extinct in the 1900s, victims of overhunting and habitat destruction.

The best illustration of widespread overexploitation is the recent collapse of the ocean fisheries (see section 11.7). Since the 1950s, some 90% of the world's large, predatory ocean fishes have disappeared, including tuna, flounder, halibut, swordfish,

Figure 39.15 Bycatch. Nontarget animals sometimes get caught in nets meant for other species. Fishermen often discard these "bycatch" animals, dead or alive.

and cod. Superefficient fishing boats harvest the adults faster than the fishes can reproduce. Moreover, fishing pressure has shifted to other species as predatory fishes have vanished. Fishing equipment scrapes and scours the seafloor, destroying habitat for many other species. In addition, many marine mammals, seabirds, sea turtles, and nontarget fish species are killed accidentally when they are caught up as "bycatch" in the nets set for the target species (figure 39.15). Even farmed seafood may contribute to the problem, because ocean fishes are harvested and used to feed farmed shrimp and salmon, depleting marine food webs.

Improved management practices can help an overharvested population recover its numbers. Consider, for example, the Chesapeake Bay blue crab. The federal government declared the crab fishery a disaster in 2008. The state of Virginia had already stopped issuing new commercial crabbing licenses in the 1990s, but in 2009, the state began buying back existing licenses as well. Combined with other new regulations, the overall goal is to reduce pressure on the dwindling number of crabs, which have been depleted not only by overharvesting but also by pollution and habitat loss. Thanks to these new management practices, the blue crab population has begun to rebound.

Figure 39.14 Unsustainable Harvest. Customs officials in Lhasa, the capital of Tibet, inspect confiscated tiger, leopard, and otter skins. Trade in these skins is illegal, yet poaching remains one of the main threats to mammal populations in Asia.

39.4 | Mastering Concepts

1. What features characterize an invasive species?
2. How do invasive species disrupt ecosystems?
3. List examples of species declines caused by overexploitation.

39.5 | Some Biodiversity May Be Recoverable

As the human population continues to grow, pressure on natural resources will only increase. One key to reversing environmental decline will therefore be to slow the growth of the human population. In addition, although some species are gone forever, humans may have the power to undo some of our past mistakes (see this chapter's Apply It Now box on page 816). For example, thanks to the Endangered Species Act of 1973, some species that once faced extinction, such as the bald eagle, have recovered (figure 39.16).

One important conservation tool is to set aside parks, wildlife refuges, and other natural areas and to protect them from destruction, invasive species, and hunting. Preserving critical habitat is a good conservation tool because it saves not just one endangered species but also the many other species that share its habitat. For example, the red-cockaded woodpecker (*Picoides borealis*) is native to the southeastern United States, where it builds its nest in a cavity that it excavates high on the trunk of a mature pine tree (figure 39.17). Although most of its preferred habitat has been logged, private property owners and the government have cooperated to save some of the remaining habitat. Other animals that use the nesting cavities, including birds, mammals, snakes, amphibians, and insects, also benefit from the woodpecker recovery plan.

Reversing habitat destruction is a second conservation tool. Major restoration projects, such as the Everglades plan described in this chapter's opening essay, show that recovery, although costly and difficult, may yet reverse some species declines. On a smaller scale, we can also help species bypass degraded habitats by supplying wildlife corridors through housing developments or building "fish ladders" over dams.

A third strategy is to breed a species in captivity and return it to its former habitat. The California condor, red wolf, and black-footed ferret are notable examples. But this solution does not work for species whose habitat is gone (submerged after dam construction,

Figure 39.17 At Home in the Pines. The endangered red-cockaded woodpecker builds its nest in a hole that it excavates in the trunk of a longleaf pine.

for example) or still besieged by the same pressures that threatened the species in the first place. Native birds will not recover on Guam as long as the brown tree snake roams the island.

A fourth conservation tool is to manage harvests. For example, it is illegal to collect endangered carnivorous plant species such as Venus flytraps and pitcher plants in the wild. In the ocean, northern and southern right whales were nearly hunted to extinction for their blubber in the 1800s. They remain endangered, but it is now illegal to kill them. Likewise, the catastrophic decline of Atlantic cod in the past few decades prompted the closure of some fisheries off Newfoundland's coast, along with strict quotas for the overall catch. Whether conservation efforts are on time to save the cod fishery remains to be seen.

Predator control programs are a fifth conservation tool. Introduced predators are the greatest threat to some species. For example, rats, weasels, and other predators brought by European settlers endanger the great spotted kiwi (*Apteryx haastii*), a flightless bird native to New Zealand. Eliminating these predators from a nature preserve has made life much easier for the endangered kiwis.

The biotechnology revolution plays a role in a sixth approach to species conservation. In one project, researchers are using DNA to identify bison whose genes are uncontaminated with those of domesticated cattle. The "purest" bison are set aside as the best candidates for reestablishing wild bison herds. In the future, it may be possible to recover extinct species using DNA extracted from preserved specimens. Scientists who are sequencing DNA from a frozen baby mammoth, for example, may one day be able to recreate a live mammoth by using a cloning technique similar to the one used to make Dolly the sheep. One possible method would be to replace the DNA in a fertilized

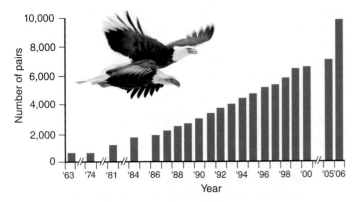

Figure 39.16 Good News for Bald Eagles. After nearing extinction in the 1960s, bald eagles have made a steady recovery. Federal officials removed the bird from the endangered species list in 2007.

elephant egg with mammoth DNA and then implant the resulting embryo into an elephant's uterus. After gestation, a woolly mammoth would be born—some 10,000 years after its species went extinct. ▶ cloning, p. 169

In areas where wildlife poaching is a problem, changing the local economic incentives can be a powerful seventh conservation tool. Poaching is profitable if a dead animal is worth more money than a live one. But ecotourism can turn this economic calculation on its head. By attracting visitors who pay to see endangered animals, conservation may bring in more money than poaching. A black rhino conservation program in Namibia, for example, hires former poachers as armed guards that protect the animals from hunters. Meanwhile, guides lead tourists hoping to spot a rhino in the wild (**figure 39.18**).

Regardless of which tools are used to preserve biodiversity, all conservation efforts require a scientific approach. To get a true measure of Earth's biodiversity, taxonomists must continue to catalog all organisms, not just vertebrates and plants (**figure 39.19**). Evolutionary biologists must continue to analyze the relationships among all species. Preserving biodiversity also requires an understanding of which species need help, whether current conservation efforts are working, and the consequences to ecosystems as species disappear.

But not every important question has a scientific answer. Are the only species worth saving the photogenic ones, such as giant pandas? Or do we also commit to saving the worms, algae, bacteria, and fungi so essential to global ecology? How much money should we spend on conservation? Should developed countries help poor nations with their efforts? How do we balance the need for conservation with the need for

Figure 39.19 **Counting Flies.** This entomologist is placing bait to check the numbers of phorid flies at a research site in Texas. The small, inconspicuous flies may hold the key to controlling fire ant populations in the southern United States.

economic growth? Which of the tangle of threads that tie all life together should we sacrifice to other interests?

Life has had many millions of years to adapt, diversify, and occupy nearly every part of the planet's surface. It would be very difficult to halt life on Earth completely, short of a global catastrophe such as a meteor collision or a nuclear holocaust. Just the presence of life, however, does not guarantee that the surviving species will have the diversity that humans value. It is safest to try to protect the remaining resources for the future, while maintaining a reasonable standard of living for all people. Scientists and politicians, as well as ordinary citizens, share this heavy burden. Part of the solution lies within you and how you choose to live (see the Burning Question on page 818). Do whatever you can to preserve the diversity of life, for in diversity lies resiliency and the future of life on Earth.

39.5 | Mastering Concepts

1. What is the relationship between human population growth and conservation biology?
2. List and describe seven tools that conservation biologists use to preserve biodiversity.
3. How can scientists, governments, and ordinary citizens work together for conservation?

Figure 39.18 **Black Rhino.** Ecotourism may provide some hope for the survival of the endangered black rhinoceros.

Apply It Now

Environmental Legislation

President Richard Nixon created the Environmental Protection Agency by executive order in 1970. Since that time, Congress has passed several laws to combat some of the worst environmental problems in the United States. Below are a few major pieces of environmental legislation:

- The **Endangered Species Act** of 1973 requires that the U.S. Secretary of the Interior identify threatened and endangered species. The overall goals are to prevent extinction and to help endangered species recover their numbers. Since the act was implemented, more than 575 species of vertebrate and invertebrate animals and nearly 750 species of plants and lichens have been classified as threatened or endangered. Only a few dozen species have been removed from the list because they have either recovered or become extinct, or because new information revealed that their populations are larger than had been thought.

- The **Clean Air Act,** passed in 1970 and amended several times since, sets minimum air quality standards for many types of air pollutants. Since 1970, emissions of nitrogen and sulfur oxides, lead, carbon monoxide, particulates, and other pollutants have declined, leading to significant improvements in regional air quality. In the United Kingdom, similar measures have decreased acid precipitation by about half over the past 15 years.

- Among other provisions, the **Clean Water Act** of 1972 required nearly every city to build and maintain a sewage treatment plant, drastically reducing discharge of raw sewage into rivers and lakes. The 1987 **Water Quality Act** followed up on the Clean Water Act, regulating water pollution from industry, agricultural runoff, sewage during storms, and runoff from city streets. Many of the nation's surface waters have recovered from past unregulated discharge of phosphorus, other nutrients, and toxic chemicals.

39.6 | Investigating Life: The Case of the Missing Frogs: Is Climate the Culprit?

With so many different threats to biodiversity, how can we ever know what drives a species to extinction? It's one thing to document that a species that once occupied a habitat has vanished; it's another thing to explain how it happened. Not knowing what has gone wrong in the past makes it difficult to prevent future extinctions.

The decline of amphibians worldwide has been especially difficult to unravel. Amphibians are frogs, salamanders, and caecilians. All of these animals require fresh water to reproduce. Most breathe through their thin skin, which must therefore remain moist. According to some estimates, about one third of the world's amphibians, representing thousands of species, have declined. Hundreds of species are critically endangered, and hundreds more have already vanished. Researchers have pointed at pollution, habitat loss, and overhunting as possible causes. But what are we to make of widespread extinctions in pristine habitats and in species that humans do not hunt for food or medicine? ▸ amphibians, p. 440

A multinational research team, led by J. Alan Pounds of the Monteverde Cloud Forest Preserve and Tropical Science Center in Costa Rica, may have found part of the answer. They studied the extinction of harlequin frogs in the genus *Atelopus* (**figure 39.20**). These animals are a good choice for this type of research because their populations are relatively easy to monitor. Harlequin frogs are brightly colored and active during the day, so they are much easier to observe than amphibians that either blend in with their surroundings or are active at night.

Pounds and his team examined a database that cataloged the last reported sighting of more than 100 species of harlequin frogs in mountainous regions of Central and South America. Many had not been observed since the 1980s or 1990s and were apparently extinct. The pace of the extinctions was puzzling, considering that these species have thrived in the same habitat for millions of years. What caused them all to vanish in such a short time?

Figure 39.20 Harlequin Frog. A Pebas stubfoot toad, *Atelopus spumarius,* is one of many vulnerable amphibian species.

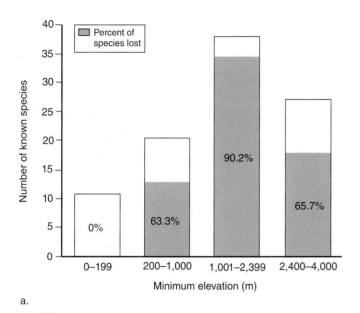

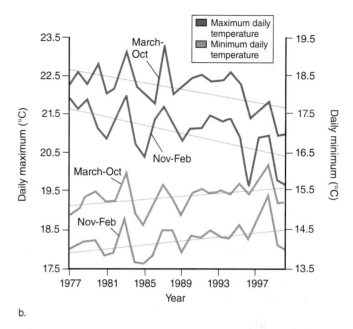

a.   b.

Figure 39.21 **Elevation Matters.** (a) In Central and South America, the harlequin frogs (genus *Atelopus*) at middle elevations are at greatest risk of extinction. (b) Days are getting cooler and nights are getting warmer at the Monteverde Cloud Forest Preserve in Costa Rica.

One clue came from looking closely at where the extinctions were occurring (figure 39.21a). None of the species that had vanished was from lowland areas (elevation less than 200 meters). The most vulnerable species occupied the middle elevations, between 1000 and 2400 meters.

Another clue was the timing of the extinctions, which coincided with increasing temperatures in the tropics. The researchers decided to take a closer look at weather patterns in one study area, Costa Rica's Monteverde Cloud Forest Preserve. This area is perpetually shrouded in fog, which forms when moist air ascends the mountainside. As the air rises, it cools, causing moisture in the air to condense into clouds. At Monteverde, the cloud forest extends from an elevation of about 1500 meters to about 1850 meters, which closely overlaps the habitats of the amphibians that were most likely to go extinct.

Pounds and his team scrutinized weather station records collected at Monteverde between 1977 and 1997. They learned that days had gotten cooler and nights had gotten warmer during those two decades (figure 39.21b). This made sense. Air temperatures and sea surface temperatures were climbing throughout the tropics at that time, causing more evaporation and therefore more clouds. The researchers surmised that the thicker cloud cover blocked the sun during the day but trapped heat at night.

How did these changes affect the frogs? According to Pounds and his team, the new conditions made the amphibians more susceptible to a skin disease caused by a type of fungus called a chytrid (see figure 19.6). This fungus, *Batrachochytrium dendrobatidis,* has an optimal growth temperature that matches the cooler days and warmer nights at Monteverde. Furthermore, the chytrid dies if the temperature climbs above 30°C. The additional cloud cover may have meant fewer sunny spots where the frogs could raise their skin temperatures high enough to kill the fungi. Without this defense,

perhaps the chytrids finally gained the advantage over their amphibian hosts, driving many to extinction.

Overall, the researchers concluded that higher temperatures in the tropics created conditions that favored the spread of skin disease, producing a wave of midelevation amphibian extinctions. According to Pounds, "The disease was the bullet killing the frogs, but climate was pulling the trigger." Nevertheless, one critical link in this chain of events is missing: no one knows whether the frogs that vanished in recent decades actually died from the chytrid. Without evidence of a direct cause-and-effect relationship, the meaning of this study remains open to debate.

This study illustrates the difficulty that scientists will face in assessing the biological effects of global climate change. Clearly, Earth's average temperature is rising, local weather patterns are shifting, and habitats are changing. Will it ever be possible to "connect the dots," linking global climate change directly to species extinctions? Scientists around the globe continue to study these issues, knowing that the first step toward preserving biodiversity is to understand what affects it.

Pounds, J. Alan, Martín R. Bustamante, Luis A. Coloma, et al. 2006. Widespread amphibian extinctions from epidemic disease driven by global warming. *Nature*, vol. 439, pages 161–167.

39.6 | Mastering Concepts

1. Summarize the evidence that climate change is a factor in the extinction of harlequin frogs.

2. Other than climate change, what is an alternative hypothesis that could explain that extinction of harlequin frogs? How would you test your hypothesis?

Burning Question

What can an ordinary person do to help the environment?

Even ordinary citizens can join forces to clean and preserve the environment. The list of small actions that together can make a big difference is endless, but here are a few ideas:

- Use less stuff. Manufacturing, packaging, and transporting consumer goods uses energy and raw materials and produces waste. The less you buy, the less resources you consume and the less waste you discard.

- Choose foods, lumber, and other products that reflect sustainable practices. Buying organic food, for example, reduces the use of pesticides in agriculture. Shade-grown coffee plantations provide habitat for a wide variety of tropical plants and animals. Before purchasing seafood, consult a Seafood Watch guide for your area to learn which fish species are threatened. Lumber and paper certified by the nonprofit Forest Stewardship Council has been harvested and produced in an environmentally responsible way.

- Conserve energy. Replace conventional light bulbs with compact fluorescent bulbs, recycle, carpool, drive a fuel-efficient car, ride a bicycle, or turn down the thermostat. Pouring filtered tap water into a reusable bottle rather than buying bottled water not only saves energy but also reduces landfill waste.

- Check with your electric company to see whether you can select renewable energy sources, such as wind or solar energy.

- Eat less meat. Farm animals and their manure emit copious greenhouse gases—especially methane—into the atmosphere.

- Pay attention to what you discard and pour down the drain. Pharmaceutical drugs, petroleum products, and harsh chemicals can end up in waterways and harm ecosystems.

- Encourage your local governments to set aside land for parks.

- Write to state and federal lawmakers to ask them to support legislation that can help protect the environment. The house.gov and senate.gov websites have contact information for your House and Senate representatives.

- If you garden, avoid invasive species. Instead, choose native plant species that attract wildlife.

- Don't buy rare or exotic species as pets.

- Donate time or money to groups that save critical habitat.

Submit your burning question to:

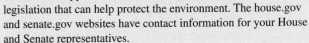

marielle_hoefnagels@mcgraw-hill.com

Chapter Summary

39.1 | Earth's Biodiversity Is Dwindling

- **Biodiversity** means the variety of life on Earth, and it includes diversity at the genetic, species, and ecosystem levels.
- Increasing numbers of species are threatened with **extinction** or are **endangered** or **vulnerable. Conservation biologists** study and attempt to preserve biodiversity.

39.2 | Human Activities Destroy Habitats

- Agriculture, logging, and urbanization contribute to **deforestation** in tropical and temperate regions. Prairies and other grasslands also have been destroyed, especially for agriculture.
- Drought and agriculture promote **desertification,** the expansion of desert into surrounding areas.
- Dams and levees alter the species that live in and near rivers.
- Preserving estuaries is important because they are breeding grounds for many species.

39.3 | Pollution Degrades Habitats

- **Pollution** is any change in the environment that harms living organisms.

A. Water Pollution Threatens Aquatic Life

- Excessive nutrient levels cause **eutrophication.** Sediments and heat also pollute aquatic ecosystems.
- Water and sediments can be contaminated by a mixture of toxic substances such as **persistent organic pollutants,** heavy metals, spilled oil, and plastics.

B. Air Pollution Causes Many Types of Damage

- Air pollutants include heavy metals, **particulates,** and emissions from fossil fuel combustion in automobiles and industries. Some of these pollutants react in light to form photochemical **smog.**
- **Acid deposition** forms when nitrogen and sulfur oxides react with water in the upper atmosphere to form nitric and sulfuric acids. These acids return to Earth as dry particles or in precipitation.
- Particulates are bits of matter suspended in the air. When inhaled, the tiniest particles can cause lung problems.
- Use of chlorofluorocarbon compounds has thinned the stratospheric **ozone layer,** which protects life from damaging ultraviolet radiation.

C. Global Climate Change Alters and Shifts Habitats

- The **greenhouse effect** results from CO_2 and other gases that trap heat near Earth's surface. Agriculture, the combustion of fossil fuels, and the destruction of tropical rain forests generate greenhouse gases that are accumulating in the atmosphere.

- As Earth's average temperature increases, polar ice is melting and sea level is rising.
- Shifting vegetation patterns and species ranges are responses to **global climate change.**

39.4 | Exotic Invaders and Overexploitation Devastate Many Species

A. Invasive Species Displace Native Organisms

- **Invasive** species reduce biodiversity by preying on or outcompeting native organisms.

B. Overexploitation Can Drive Species to Extinction

- **Overexploitation** means individuals are harvested faster than they can reproduce. Extreme fishing pressure means that global fisheries are in danger of collapse.

39.5 | Some Biodiversity May Be Recoverable

- Protected reserves, habitat restoration, captive breeding, harvest management, predator exclusion, biotechnology, and economic incentives are important tools for conservation biology.
- In the future, scientists may be able to clone extinct or endangered species from preserved DNA.
- Every human can choose actions that preserve or deplete biodiversity.

39.6 | Investigating Life: The Case of the Missing Frogs: Is Climate the Culprit?

- Researchers studying the disappearance of amphibians in Central and South America have suggested a direct link to climate change.
- Increasing cloud cover has altered temperatures in a way that favors the spread of a fungus that infects frogs.

Multiple Choice Questions

1. Which of the following is not one of the main causes of today's biodiversity crisis?
   a. Natural disasters, such as earthquakes
   b. Habitat destruction and degradation
   c. Overexploitation
   d. Introduction of nonnative species

2. Which human activity has been the most harmful to biodiversity in tropical forests?
   a. Housing
   b. Transportation
   c. Agriculture
   d. Tourism

3. What is the connection between agriculture in the midwestern United States and the Gulf of Mexico's "dead zone"?
   a. Pesticides from farmlands are killing ocean life.
   b. Nutrient enrichment causes oxygen depletion in the waters of the Gulf.
   c. River sediments block out the light needed for photosynthesis in the Gulf.
   d. Farmlands use up all of the nitrogen in the water, so the Gulf waters are starved for nutrients.

4. The burning of fossil fuels produces
   a. smog.
   b. acid deposition.
   c. particulates.
   d. All of the above are correct.

5. How does destruction of the ozone layer affect life on Earth?
   a. It alters global temperatures.
   b. It leads to DNA damage.
   c. It changes the spectrum of light reaching the surface of the Earth.
   d. It reduces the amount of O_2 in the atmosphere.

6. Based on the data in figure 39.9, what can you conclude about the effectiveness of the 1987 Montreal Protocol banning the use of CFCs?
   a. It has helped to stabilize the ozone layer.
   b. It has accelerated the destruction of the ozone layer.
   c. It has helped to restore the ozone layer.
   d. It has caused the ozone hole to disappear.

7. What is the greenhouse effect?
   a. The filtering of specific wavelengths of light by Earth's atmosphere
   b. The increase in global plant growth due to enhanced photosynthesis
   c. The trapping of heat by gases in the atmosphere
   d. The reduction in the amount of CO_2 in Earth's atmosphere

8. Why is deforestation associated with global climate change?
   a. Because burning forests adds to the CO_2 levels in the atmosphere
   b. Because loss of trees reduces the amount of photosynthesis occurring on the planet
   c. Because the loss of forest makes more land available for agriculture
   d. Both a and b are correct.

9. An *invasive* species is one that
   a. is introduced into a new habitat.
   b. causes the extinction of a native species.
   c. establishes a breeding population in a new habitat.
   d. Both a and c are correct.

10. What might limit the effectiveness of a captive breeding program for the restoration of an extremely rare endangered species?
    a. A very rare species has limited genetic diversity.
    b. Captive breeding programs do not preserve the habitat of an organism.
    c. Not all animals will breed in captivity.
    d. All of the above are correct.

Write It Out

1. List the main threats to biodiversity worldwide.
2. Describe the relationships among the three levels of biodiversity. Why is each level important?
3. In an article in *Nature* magazine, Sean Nee wrote: "Earth's real biodiversity is invisible, whether we like it or not." What does that statement mean?
4. Which human activities promote habitat destruction?
5. How can too much of a nutrient alter an aquatic ecosystem?
6. How does the Gulf of Mexico's "dead zone" demonstrate the connections among the world's ecosystems?
7. Visit the waterfootprint.org website to see how much water is required to produce various foods and beverages. Do you think it is important for an individual to choose products with a low water footprint? Why or why not?
8. How does the combustion of fossil fuels influence acid deposition and global climate change?
9. Particulate air pollution damages animal respiratory tracts. How might dust-covered leaves harm a plant?
10. In what ways is the greenhouse effect both beneficial and detrimental?
11. Cite biological evidence of global climate change.
12. Explain the logic behind planting trees as a way to reduce global climate change.
13. Select a biome from chapter 38, and list three ways that an earlier spring and later fall resulting from global climate change might affect biodiversity in that biome.
14. Why are invasive species harmful?

15. One way to combat invasive species is to kill the invaders. In Hawaii, officials shoot feral cats, goats, and pigs. In Australia, the government fought zebra mussels by adding chlorine and copper to a bay, killing everything living in the water. Do you think that these approaches are reasonable? Suggest alternative strategies.

16. Which of the strategies described in section 39.5 are being used in the Everglades restoration project?

17. Name three ways you can alter your lifestyle in a way that promotes conservation practices.

18. Search the Internet for information on the Convention on Biological Diversity and on the international agreement called CITES. How does each approach tackle the biodiversity crisis on a global scale?

19. Give an example of an environmental problem that can immediately reduce biodiversity and one that has a delayed effect.

20. Several polar bears were discovered that have reproductive organs of both sexes. These bears also have high concentrations of PCBs in their blood, but researchers do not know whether exposure to PCBs is related to the disturbed sexual development. Given that many heavily polluted ecosystems are tainted with several types of pollutants, how could you link exposure to one chemical to a specific biological effect?

21. In the southeastern United States, several species of freshwater mussels are extinct or threatened because of habitat destruction. In the past, they were also harvested for the button trade. How would a population ecologist (see chapter 36) approach the problem of species recovery for these animals?

22. Your friend reads an article about scientists who are collecting DNA from endangered species. She says, "I don't know why we spend so much money to save these species when we can just sequence their DNA." Given what you know about the importance of biodiversity, how would you respond?

23. Use the Internet to search for examples of conflicts involving the Endangered Species Act and the rights of private property owners. If you owned land that housed an endangered species, what would you be willing to sacrifice to save the species? What information would you need before answering this question?

24. The red-cockaded woodpecker is an example of an "umbrella species" because conserving this bird's habitat will also protect many other species. Can you think of other species in your area that might qualify as umbrella species?

25. Use the Internet to learn about so-called "biodiversity hot spots." Choose one to investigate in detail. Why is the area you chose a hot spot? What are the main threats to biodiversity in the area? What efforts are being made to preserve the area?

26. Refer back to section 11.6, which describes the bottleneck effect. With this information in mind, why might recovery be difficult for species, such as cheetahs, that are nearly extinct?

Pull It Together

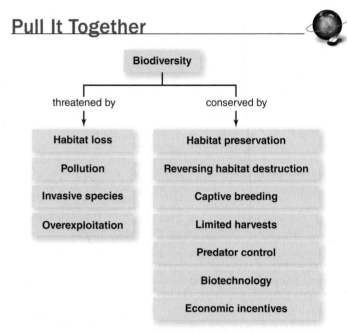

1. What are examples of pollutants in air and in water? Which of these pollutants eventually reach land?

2. How does human population growth contribute to each of the main factors causing species extinctions?

3. Give examples of government actions that threaten biodiversity and examples of government actions that preserve biodiversity.

Appendix A | Answers to Multiple Choice Questions

Chapter 1
1. d 2. b 3. c 4. a 5. d 6. c 7. c 8. d 9. c 10. a

Chapter 2
1. d 2. b 3. c 4. c 5. a 6. b 7. d 8. a 9. d 10. d

Chapter 3
1. b 2. c 3. b 4. d 5. a 6. c 7. b 8. c 9. a 10. b

Chapter 4
1. b 2. b 3. a 4. c 5. a 6. c 7. a 8. d 9. b 10. b

Chapter 5
1. c 2. b 3. a 4. b 5. c 6. d 7. b 8. c 9. b 10. d

Chapter 6
1. d 2. c 3. b 4. d 5. c 6. d 7. c 8. a 9. c 10. c

Chapter 7
1. c 2. a 3. b 4. d 5. c 6. a 7. c 8. d 9. d 10. a

Chapter 8
1. d 2. c 3. d 4. a 5. b 6. c 7. c 8. a 9. b 10. c

Chapter 9
1. c 2. b 3. a 4. d 5. a 6. c 7. c 8. b 9. a 10. b

Chapter 10
1. d 2. b 3. d 4. a 5. b 6. b 7. c 8. a 9. d 10. c
11. c 12. d

Chapter 11
1. b 2. c 3. c 4. d 5. d 6. c 7. b 8. c 9. a 10. b

Chapter 12
1. d 2. c 3. c 4. d 5. c 6. b 7. a 8. c 9. a 10. a

Chapter 13
1. a 2. c 3. a 4. c 5. b 6. d 7. b 8. c 9. b 10. b

Chapter 14
1. c 2. d 3. a 4. c 5. d 6. d 7. b 8. c 9. d 10. d

Chapter 15
1. a 2. d 3. a 4. a 5. b 6. a 7. c 8. d 9. c 10. b

Chapter 16
1. c 2. b 3. a 4. b 5. b 6. c 7. b 8. c 9. d 10. a

Chapter 17
1. a 2. d 3. b 4. d 5. c 6. c 7. c 8. d 9. d 10. c

Chapter 18
1. d 2. a 3. b 4. b 5. d 6. b 7. c 8. d 9. c 10. c

Chapter 19
1. b 2. b 3. c 4. c 5. b 6. a 7. c 8. c 9. b 10. a

Chapter 20
1. b 2. b 3. d 4. a 5. b 6. c 7. c 8. d 9. d 10. b
11. c 12. b 13. a 14. d 15. c

Chapter 21
1. c 2. a 3. d 4. b 5. d 6. c 7. b 8. b 9. a 10. a

Chapter 22
1. b 2. a 3. c 4. a 5. b 6. c 7. d 8. c 9. b 10. c

Chapter 23
1. b 2. c 3. a 4. b 5. c 6. b 7. d 8. a 9. b 10. d

Chapter 24
1. d 2. c 3. b 4. c 5. c 6. d 7. b 8. a 9. b 10. d

Chapter 25
1. a 2. c 3. d 4. b 5. a 6. b 7. a 8. d 9. c 10. c

Chapter 26
1. d 2. b 3. b 4. d 5. d 6. a 7. c 8. c 9. a 10. a

Chapter 27
1. c 2. c 3. d 4. a 5. d 6. c 7. c 8. b 9. a 10. c

Chapter 28
1. d 2. d 3. b 4. c 5. b 6. d 7. d 8. a 9. a 10. c

Chapter 29
1. b 2. b 3. a 4. d 5. d 6. a 7. c 8. d 9. b 10. d

Chapter 30
1. d 2. b 3. a 4. c 5. b 6. c 7. d 8. b 9. a 10. b

Chapter 31
1. d 2. c 3. d 4. a 5. d 6. a 7. c 8. a 9. b 10. d

Chapter 32
1. b 2. d 3. a 4. d 5. c 6. d 7. d 8. a 9. a 10. b

Chapter 33
1. c 2. b 3. d 4. a 5. b 6. c 7. b 8. c 9. b 10. c

Chapter 34
1. b 2. a 3. b 4. c 5. a 6. d 7. b 8. b 9. c 10. d

Chapter 35
1. c 2. b 3. a 4. d 5. d 6. d 7. c 8. b 9. c 10. a

Chapter 36
1. b 2. a 3. c 4. b 5. d 6. a 7. d 8. d 9. b 10. c

Chapter 37
1. a 2. c 3. c 4. c 5. d 6. d 7. a 8. b 9. d 10. b

Chapter 38
1. d 2. d 3. b 4. c 5. b 6. a 7. c 8. a 9. b 10. a

Chapter 39
1. a 2. c 3. b 4. d 5. b 6. a 7. c 8. d 9. d 10. d

Appendix B | A Brief Guide to Statistical Significance

Experiments often yield numerical data, such as the height of a plant or the incidence of illness in vaccinated children (see, for example, figure 1.10). But how are we to know whether an observed difference between two samples is "real"? For example, if we do find that 100 vaccinated children become sick slightly less often than 100 unvaccinated ones, how can we make sure that this outcome does not simply reflect random variation between samples of 100 children?

A statistical analysis can help. The dictionary definition of *statistics* is "the science that deals with the collection, analysis, and interpretation of numerical data, often using probability theory." Note that the analysis is grounded in probability theory, a branch of mathematics that deals with random events. A statistical test is therefore a mathematical tool that assesses variation, with the goal of determining whether any observed differences between treatments could be explained by the variation that random events would produce.

Researchers use many types of statistical tests, depending on the type of data collected and the design of the experiment. A de-

scription of these tests is beyond the scope of this appendix. For now, it is enough to understand that in each statistical test, the researcher computes a value (the "test statistic") that takes into account the sample size and the variability in the data. The researcher then determines the likelihood that the observed test statistic could be explained by chance alone.

An imaginary experiment will help you understand the role of variability in accepting or rejecting a hypothesis. Suppose that you have two friends, Pat and Kris, both of whom play softball. Pat claims to be able to hit a ball farther than Kris, but Kris disagrees. You therefore set up a test of the null hypothesis, which is that Pat and Kris can hit the ball equally far. You ask both of your friends to hit the ball one time, and Pat's ball does go farther. But Kris wants to re-do the test. This time, Kris's ball goes farther. Evidently, two hits apiece is not sufficient for you to settle the matter.

You therefore decide to improve the experiment (figure B.1). This time, each player gets to hit 10 balls, and you use a tape measure to determine how far each ball traveled from home plate. Figure B.2 shows two possible outcomes of the contest. In each scenario,

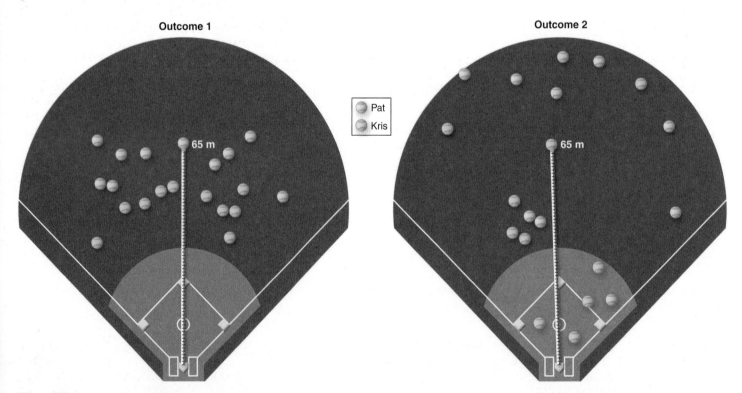

Figure B.1 Hitting Competition. These illustrations show two possible outcomes in a hitting contest between Pat and Kris. Note that the batting distances in Outcome 1 were much less variable than they were in Outcome 2.

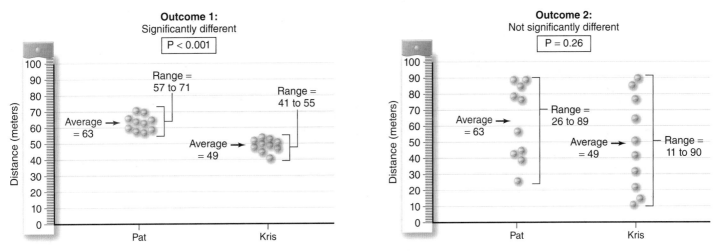

Figure B.2 Statistical Significance. In Outcome 1, the difference between Pat and Kris is considered highly significant at P<0.001; in Outcome 2, however, the variability in hitting distances means that the difference between the two batters is not statistically significant.

Pat's average distance is 63 meters, compared with 49 meters for Kris. Pat therefore appears to be the better hitter.

But look more closely at the data. In Outcome 1, the hitting distances are much less variable than they are in Outcome 2. How does this variability influence our conclusions about Pat and Kris?

The answer lies in statistical tests. The goal of each statistical test is to calculate the probability that an observed outcome would arise *if the null hypothesis were true*. This probability is called the P value. The lower the P value, the greater the chance that the difference between two treatments is "real."

Generally, biologists accept a P value of 0.05 or smaller as being statistically significant. In other words, a difference between two treatments is statistically significant when the probability that we would observe that result by chance alone is 5% or less. Moreover, a P value of 0.01 or smaller is considered highly significant. Note that the meaning of the word *significant* is different from its use in everyday language. Ordinarily, "significant" means "important." Not so in statistics, where "significant" simply means "likely to be true," and "highly significant" means "very likely to be true."

Returning to our experiment, the null hypothesis is that Pat and Kris are equally good hitters. The amount of variability in Outcome 1 was small, and the calculated value of our test statistic suggests that we can reject this null hypothesis with a P value of <0.001. In other words, there is a 99.999% chance that Pat really is a better hitter than Kris and that our results are not simply due to chance (see figure B.2). In Outcome 2, however, the hitting distances were much more variable, and the calculated P value is 0.26. We therefore cannot reject the null hypothesis with confidence. After all, given these data, the chance that Pat really is the better batter is only 74%. In science, a 26% chance of incorrectly rejecting the null hypothesis is unacceptably high.

Scientists often include information about statistical analyses along with their data (figure B.3). The most common technique is to graph the average for each treatment and then add error bars that reflect the amount of variability in the data. The longer the error bar, the greater the amount of variability and the less confident we are in the accuracy of our result.

An error bar usually indicates either the standard error or the 95% confidence interval; you may wish to consult a statistics reference to learn more about each. Because these two measures have slightly different meanings, researchers should always note the type of error bar depicted on a graph.

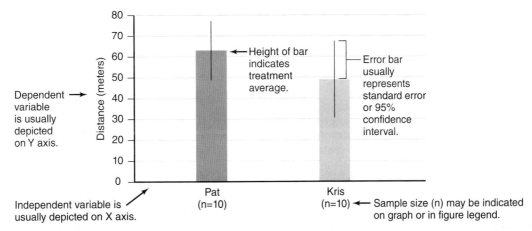

Figure B.3 Representing Statistics on a Graph. The basic parts of a bar graph include the axes, treatment averages, and error bars.

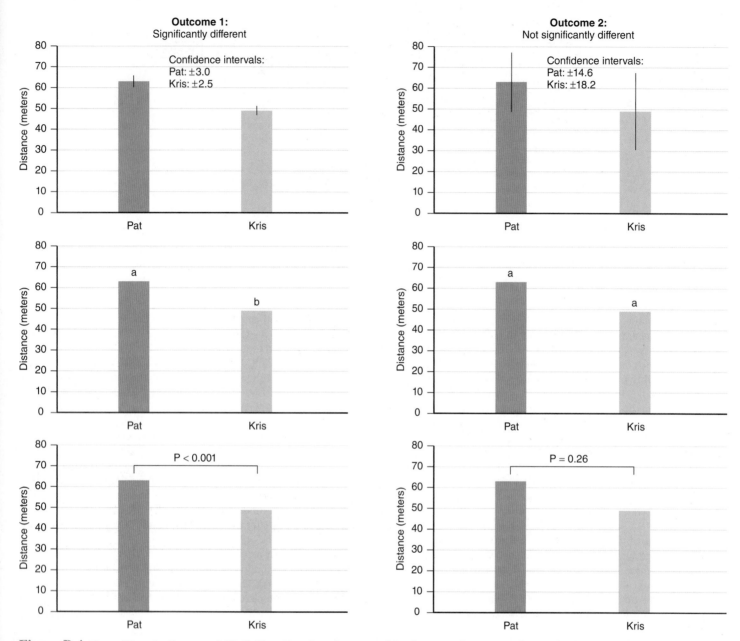

Figure B.4 **Three Ways to Represent Statistics.** Error bars, letters, and P values are all ways to indicate whether the difference between two treatments is statistically significant.

Figure B.4 shows three common ways to indicate whether an observed difference is statistically significant. Examine the error bars on the top pair of graphs in figure B.4. These bars indicate the 95% confidence intervals that were calculated for the two possible outcomes in our hitting competition. Because the data in Outcome 1 are much less variable than for Outcome 2, the error bars for Outcome 1 are smaller than for Outcome 2. The overlapping error bars for Outcome 2 indicate that the difference between Pat and Kris is not statistically significant.

Figure B.4 also shows that researchers sometimes use other methods to illustrate which data points are significantly different from one another. The middle pair of graphs in figure B.4 shows one approach. In Outcome 1 of our batting experiment, Pat was a significantly better hitter than Kris, so the bars depicting their averages are topped with different letters (*a* and *b*). In Outcome 2, however, Pat and Kris were not significantly different, so their bars have the same letter, *a*. A third possibility is to simply report the P values between pairs of treatments; the last pair of graphs in figure B.4 illustrates this strategy.

Appendix C | Metric Units and Conversions

Metric Prefixes

Symbol	Prefix	Increase or Decrease	
G	giga	One billion	1,000,000,000
M	mega	One million	1,000,000
k	kilo	One thousand	1,000
h	hecto	One hundred	100
da	deka (or deca)	Ten	10
d	deci	One-tenth	0.1
c	centi	One-hundredth	0.01
m	milli	One-thousandth	0.001
μ	micro	One-millionth	0.000001
n	nano	One-billionth	0.000000001

Metric Units and Conversions

	Metric Unit	Metric to English Conversion	English to Metric Conversion
Length	1 meter (m)	1 km = 0.62 mile 1 m = 1.09 yards = 39.37 inches 1 cm = 0.394 inch 1 mm = 0.039 inch	1 mile = 1.609 km 1 yard = 0.914 m 1 foot = 0.305 m = 30.5 cm 1 inch = 2.54 cm
Mass	1 gram (g) 1 metric ton (t) = 1,000,000 g = 1,000 kg	1 t = 1.102 tons (U.S.) 1 kg = 2.205 pounds 1 g = 0.0353 ounce	1 ton (U.S.) = 0.907 t 1 pound = 0.4536 kg 1 ounce = 28.35 g
Volume (liquids)	1 liter (l)	1 l = 1.06 quarts 1 ml = 0.034 fluid ounce	1 gallon = 3.79 l 1 quart = 0.95 l 1 pint = 0.47 l 1 fluid ounce = 29.57 ml
Temperature	Degrees Celsius (°C)	°C = (°F − 32)/1.8	°F = (°C × 1.8) + 32
Energy and Power	1 joule (J)	1 J = 0.239 calorie 1 kJ = 0.239 kilocalorie ("food calorie")	1 calorie = 4.186 J 1 kilocalorie ("food calorie") = 4186 J
Time	1 second (sec)		

Appendix D | Periodic Table of Elements

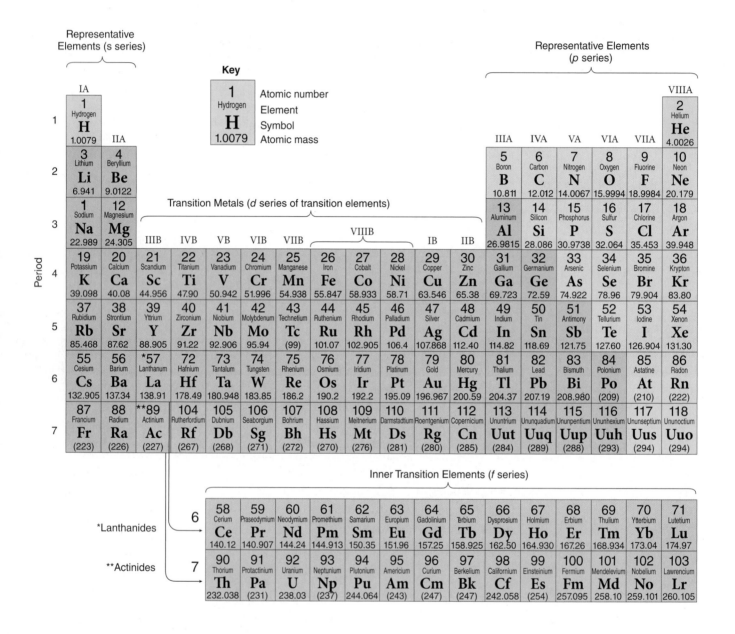

Representative Elements (s series)

Representative Elements (p series)

Key

1
Hydrogen
H
1.0079

1 — Atomic number
Hydrogen — Element
H — Symbol
1.0079 — Atomic mass

Transition Metals (d series of transition elements)

Period

Inner Transition Elements (f series)

*Lanthanides

**Actinides

IA																	VIIIA
1 Hydrogen **H** 1.0079	IIA											IIIA	IVA	VA	VIA	VIIA	2 Helium **He** 4.0026
3 Lithium **Li** 6.941	4 Beryllium **Be** 9.0122											5 Boron **B** 10.811	6 Carbon **C** 12.012	7 Nitrogen **N** 14.0067	8 Oxygen **O** 15.9994	9 Fluorine **F** 18.9984	10 Neon **Ne** 20.179
1 Sodium **Na** 22.989	12 Magnesium **Mg** 24.305	IIIB	IVB	VB	VIB	VIIB		VIIIB		IB	IIB	13 Aluminum **Al** 26.9815	14 Silicon **Si** 28.086	15 Phosphorus **P** 30.9738	16 Sulfur **S** 32.064	17 Chlorine **Cl** 35.453	18 Argon **Ar** 39.948
19 Potassium **K** 39.098	20 Calcium **Ca** 40.08	21 Scandium **Sc** 44.956	22 Titanium **Ti** 47.90	23 Vanadium **V** 50.942	24 Chromium **Cr** 51.996	25 Manganese **Mn** 54.938	26 Iron **Fe** 55.847	27 Cobalt **Co** 58.933	28 Nickel **Ni** 58.71	29 Copper **Cu** 63.546	30 Zinc **Zn** 65.38	31 Gallium **Ga** 69.723	32 Germanium **Ge** 72.59	33 Arsenic **As** 74.922	34 Selenium **Se** 78.96	35 Bromine **Br** 79.904	36 Krypton **Kr** 83.80
37 Rubidium **Rb** 85.468	38 Strontium **Sr** 87.62	39 Yttrium **Y** 88.905	40 Zirconium **Zr** 91.22	41 Niobium **Nb** 92.906	42 Molybdenum **Mo** 95.94	43 Technetium **Tc** (99)	44 Ruthenium **Ru** 101.07	45 Rhodium **Rh** 102.905	46 Palladium **Pd** 106.4	47 Silver **Ag** 107.868	48 Cadmium **Cd** 112.40	49 Indium **In** 114.82	50 Tin **Sn** 118.69	51 Antimony **Sb** 121.75	52 Tellurium **Te** 127.60	53 Iodine **I** 126.904	54 Xenon **Xe** 131.30
55 Cesium **Cs** 132.905	56 Barium **Ba** 137.34	*57 Lanthanum **La** 138.91	72 Hafnium **Hf** 178.49	73 Tantalum **Ta** 180.948	74 Tungsten **W** 183.85	75 Rhenium **Re** 186.2	76 Osmium **Os** 190.2	77 Iridium **Ir** 192.2	78 Platinum **Pt** 195.09	79 Gold **Au** 196.967	80 Mercury **Hg** 200.59	81 Thalium **Tl** 204.37	82 Lead **Pb** 207.19	83 Bismuth **Bi** 208.980	84 Polonium **Po** (209)	85 Astatine **At** (210)	86 Radon **Rn** (222)
87 Francium **Fr** (223)	88 Radium **Ra** (226)	**89 Actinium **Ac** (227)	104 Rutherfordium **Rf** (267)	105 Dubnium **Db** (268)	106 Seaborgium **Sg** (271)	107 Bohrium **Bh** (272)	108 Hassium **Hs** (270)	109 Meitnerium **Mt** (276)	110 Darmstadtium **Ds** (281)	111 Roentgenium **Rg** (280)	112 Copernicium **Cn** (285)	113 Ununtrium **Uut** (284)	114 Ununquadium **Uuq** (289)	115 Ununpentium **Uup** (288)	116 Ununhexium **Uuh** (293)	117 Ununseptium **Uus** (294)	118 Ununoctium **Uuo** (294)

6	58 Cerium **Ce** 140.12	59 Praseodymium **Pr** 140.907	60 Neodymium **Nd** 144.24	61 Promethium **Pm** 144.913	62 Samarium **Sm** 150.35	63 Europium **Eu** 151.96	64 Gadolinium **Gd** 157.25	65 Terbium **Tb** 158.925	66 Dysprosium **Dy** 162.50	67 Holmium **Ho** 164.930	68 Erbium **Er** 167.26	69 Thulium **Tm** 168.934	70 Ytterbium **Yb** 173.04	71 Lutetium **Lu** 174.97
7	90 Thorium **Th** 232.038	91 Protactinium **Pa** (231)	92 Uranium **U** 238.03	93 Neptunium **Np** (237)	94 Plutonium **Pu** 244.064	95 Americium **Am** (243)	96 Curium **Cm** (247)	97 Berkelium **Bk** (247)	98 Californium **Cf** 242.058	99 Einsteinium **Es** (254)	100 Fermium **Fm** 257.095	101 Mendelevium **Md** 258.10	102 Nobelium **No** 259.101	103 Lawrencium **Lr** 260.105

Appendix E | Amino Acid Structures

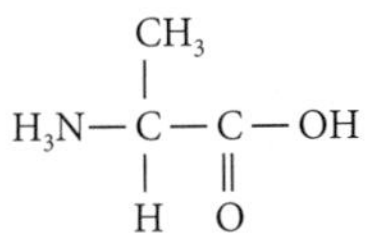

Alanine
(Ala; A)

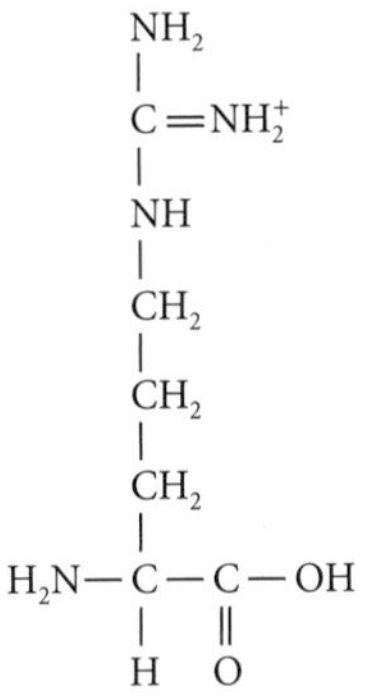

Arginine
(Arg; R)

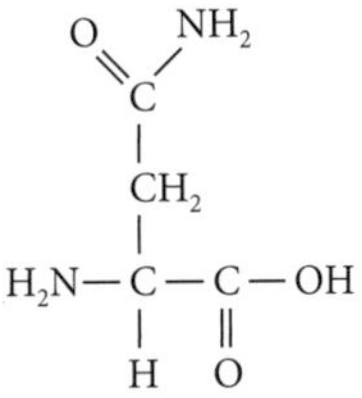

Asparagine
(Asn; N)

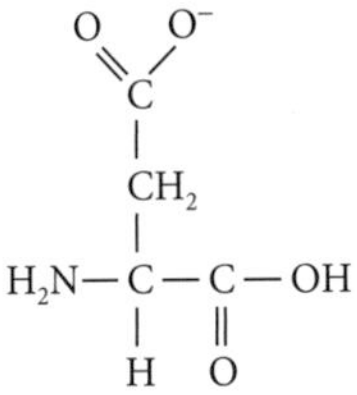

Aspartic
acid (Asp; D)

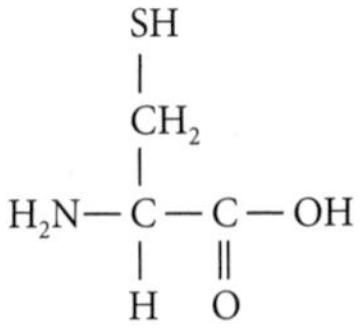

Cysteine
(Cys; C)

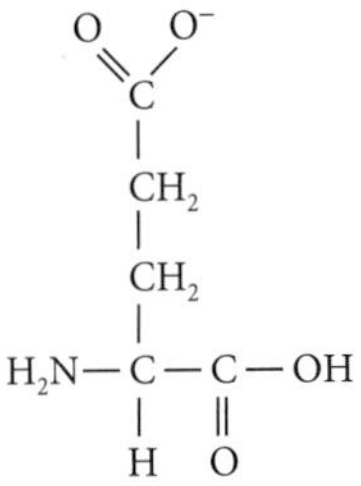

Glutamic
acid (Glu; E)

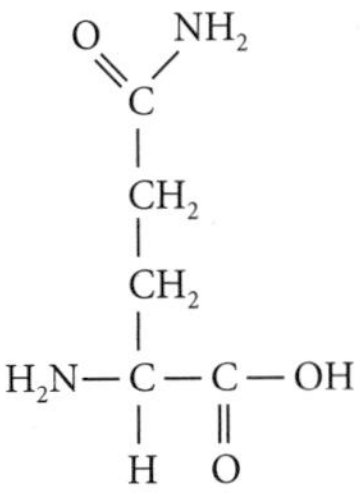

Glutamine
(Gln; Q)

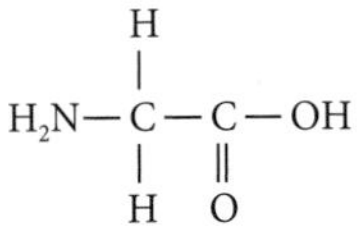

Glycine
(Gly; G)

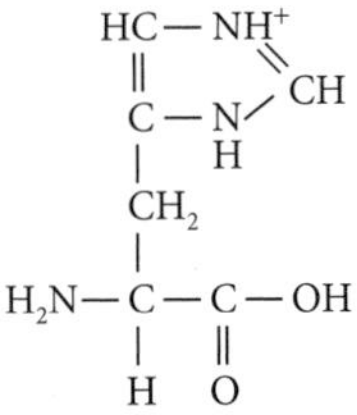

Histidine
(His; H)

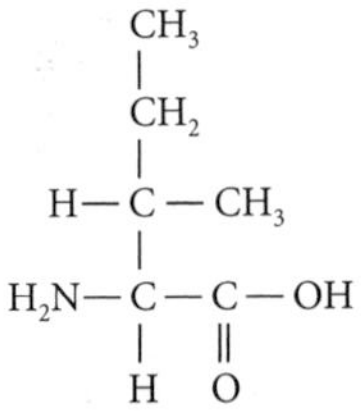

Isoleucine
(Ile; I)

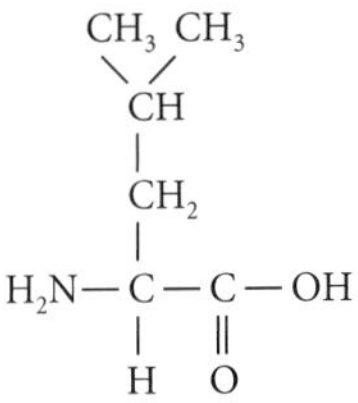

Leucine
(Leu; L)

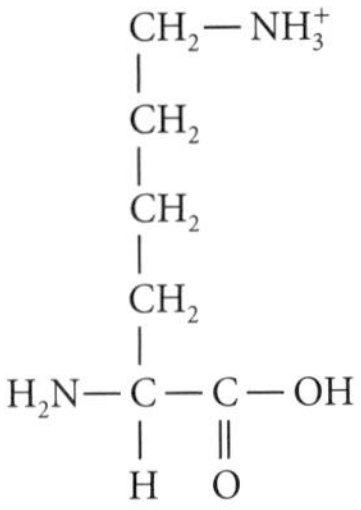

Lysine
(Lys; K)

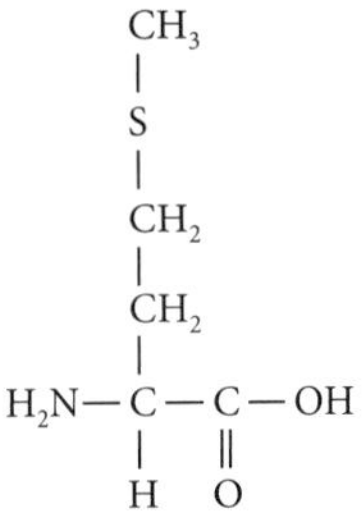

Methionine
(Met; M)

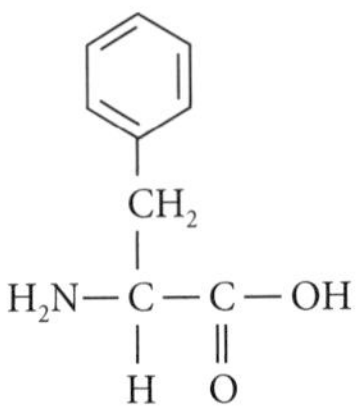

Phenylalanine
(Phe; F)

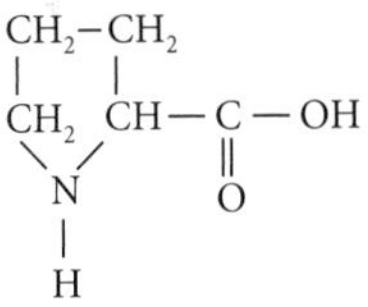

Proline
(Pro; P)

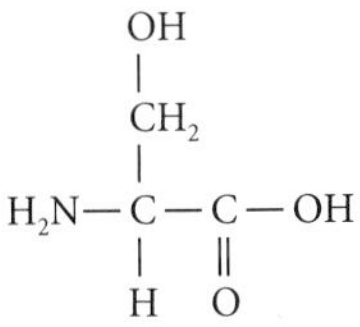

Serine
(Ser; S)

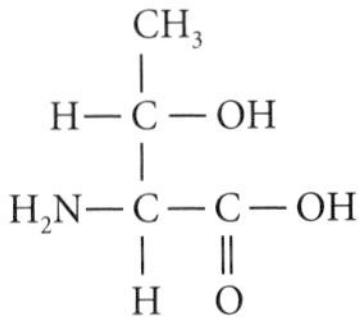

Threonine
(Thr; T)

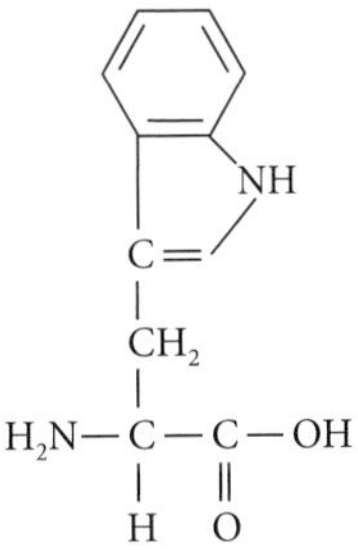

Tryptophan
(Trp; W)

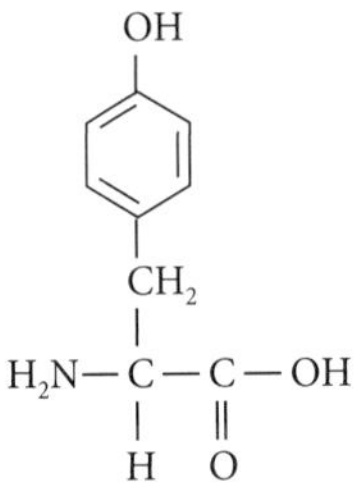

Tyrosine
(Tyr; Y)

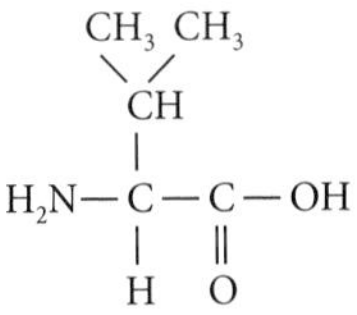

Valine
(Val; V)

Glossary

A

abiotic: nonliving

abscisic acid: plant hormone that inhibits seed germination and plant growth

abscission zone: specialized layer of cells from which leaf petiole detaches from plant

absolute dating: determining the age of a fossil in years

absorption: the process of taking in and incorporating nutrients or energy

accessory pigment: photosynthetic pigment other than chlorophyll *a* that extends the range of light wavelengths useful in photosynthesis

acetyl CoA: molecule that enters the Krebs cycle in cellular respiration; product of partial oxidation of pyruvate

acid: a molecule that releases hydrogen ions into a solution

acid deposition: low pH precipitation or particles that form when air pollutants react with water in the upper atmosphere

acrosome: protrusion covering the head of a sperm cell, containing enzymes that enable the sperm to penetrate layers around the oocyte

actin: protein that forms thin filaments in muscle cells; also part of cytoskeleton

action potential: all-or-none electrochemical change across the cell membrane of a neuron; the basis of a nerve impulse

active immunity: immunity generated by an organism's production of antibodies

active site: the part of an enzyme to which substrates bind

active transport: movement of a substance across a membrane against its concentration gradient, using a carrier protein and energy from ATP

adaptation: inherited trait that permits an organism to survive and reproduce

adaptive immunity: defense system in vertebrates that recognizes and remembers specific antigens

adaptive radiation: divergence of multiple new species from a single ancestral type in a relatively short time

adenosine triphosphate (ATP): a molecule whose high-energy phosphate bonds power many biological processes

adhesion: the tendency of water to hydrogen bond to other compounds

adipose tissue: type of connective tissue consisting of fat cells embedded in a minimal matrix

adrenal cortex: outer portion of an adrenal gland; secretes mineralocorticoids, glucocorticoids, and small amounts of sex hormones

adrenal gland: one of two endocrine glands atop the kidneys

adrenal medulla: inner portion of an adrenal gland; secretes epinephrine and norepinephrine

adrenocorticotropic hormone (ACTH): hormone produced in the anterior pituitary

adult stem cell: cell that can give rise to a limited subset of cells in the body

adventitious root: roots arising from stem or leaf

aerobic respiration: complete oxidation of glucose to CO_2 in the presence of O_2, producing ATP

age structure: distribution of age classes in a population

agglutination: clumping together of cells

alcoholic fermentation: metabolic pathway in which NADH from glycolysis reduces acetaldehyde, producing ethanol and CO_2

aldosterone: mineralocorticoid hormone produced in the adrenal cortex

alga: aquatic, photosynthetic protist

alimentary canal: two-opening digestive tract; also called gastrointestinal (GI) tract

alkaline: having a pH greater than 7

allantois: extraembryonic membrane that forms as an outpouching of the yolk sac

allele: one of two or more alternative forms of a gene

allele frequency: number of copies of one allele, divided by the number of alleles in a population

allergen: antigen that triggers an allergic reaction

allergy: exaggerated immune response to a harmless substance

allopatric speciation: formation of new species after a physical barrier separates a population into groups that cannot interbreed

alternation of generations: the sexual life cycle of plants and many green algae, which alternates between a diploid sporophyte stage and a haploid gametophyte stage

altruism: behavior that lowers an animal's fitness for the good of others

alveolate: protist with flattened sacs, or alveoli, just beneath the cell membrane

alveolus (pl. alveoli): microscopic air sac in the mammalian lungs, where gas exchange occurs; sac beneath cell membrane in alveolate protist

amino acid: an organic molecule consisting of a central carbon atom bonded to a hydrogen atom, an amino group, a carboxyl group, and an R group

amino group: a nitrogen atom single-bonded to two hydrogen atoms

ammonia: nitrogenous waste (NH_3) generated by breakdown of amino acids

amnion: extraembryonic membrane that contains amniotic fluid

amniote: vertebrate in which protective membranes surround the embryo (amnion, chorion, and allantois); reptiles and mammals

amniotic egg: reptile or monotreme egg containing fluid and nutrients within membranes that protect the embryo

amoeboid protozoan: unicellular protist that produces pseudopodia

amphibian: tetrapod vertebrate that can live on land, but requires water to reproduce

amygdala: forebrain structure involved in emotions

anaerobic respiration: cellular respiration using an electron acceptor other than O_2

analogous: similar in function but not in structure because of convergent evolution, not common ancestry

anaphase: stage of mitosis in which the spindle pulls sister chromatids toward opposite poles of the cell

anaphase I: anaphase of meiosis I, when spindle fibers pull homologous chromosomes toward opposite poles of the cell

anaphase II: anaphase of meiosis II, when centromeres split and spindle fibers pull sister chromatids toward opposite poles of the cell

anaphylactic shock: rapid, widespread, severe allergic reaction

anatomy: the study of an organism's structure

ancestral character: characteristic already present in the ancestor of a group being studied

anchoring (or adhering) junction: connection between two adjacent animal cells that anchors intermediate filaments in a single spot on the cell membrane

angiosperm: a seed plant that produces flowers and fruits; includes monocots and eudicots

Animalia: kingdom containing multicellular eukaryotes that are heterotrophs by ingestion

annelid: segmented worm; phylum Annelida

anorexia nervosa: eating disorder characterized by refusal to maintain normal body weight

antenna pigment: photosynthetic pigment that passes photon energy to the reaction center of a photosystem

anterior pituitary: the front part of the pituitary gland

anther: pollen-producing structure at tip of stamen

antibody: protein that binds to an antigen

anticodon: a three-base portion of a tRNA molecule; the anticodon is complementary to one codon

antidiuretic hormone (ADH): hormone released from the posterior pituitary; also called vasopressin

antigen: molecule that elicits an immune reaction by B and T cells

anus: exit from a complete digestive tract

aorta: the largest artery leaving the heart

apical dominance: intact terminal bud of a plant suppresses growth of lateral buds

apical meristem: meristem at tip of root or shoot

apicomplexan: non-motile protist with cell containing an apical complex; obligate animal parasite

apoptosis: programmed cell death that is a normal part of development

appendicular skeleton: the limb bones and the bones that support them in the vertebrate skeleton

arachnid: type of chelicerate arthropod; spiders, ticks, mites, and scorpions

Archaea: one of two domains of prokaryotes

arteriole: small artery

artery: vessel that carries blood away from the heart

arthropod: segmented animal with an exoskeleton and jointed appendages; phylum Arthropoda

artificial selection: selective breeding strategy in which a human allows only organisms with desired traits to reproduce

ascomycete: fungus that produces sexual spores in a sac called an ascus

ascospore: haploid sexual spore produced in an ascus

ascus (pl. asci): in ascomycetes, a saclike structure in which ascospores are produced

asexual reproduction: form of reproduction in which offspring arise from only one parent

associative learning: type of learning in which an animal connects two stimuli or a stimulus and a behavior

atom: a particle of matter; composed of protons, neutrons, and electrons

atomic mass: the average mass of all isotopes of an element

atomic number: the number of protons in an atom's nucleus

ATP: adenosine triphosphate; a molecule whose high-energy phosphate bonds power many biological processes

ATP synthase: enzyme complex that admits protons through a membrane, where they trigger phosphorylation of ADP to ATP

atrioventricular (AV) node: specialized cardiac muscle cells that delay the heartbeat, giving the ventricles time to fill

atrioventricular (AV) valve: flap of heart tissue that prevents the flow of blood from a ventricle to an atrium

atrium: heart chamber that receives blood

auditory canal: ear canal; funnels sounds from the outer ear to the eardrum

auditory nerve: nerve fibers that connect the cochlea and vestibular apparatus to the brain

autoimmune disorder: immune reaction to the body's own cells

autonomic nervous system: in the peripheral nervous system, motor pathways that lead to smooth muscle, cardiac muscle, and glands

autosomal dominant: inheritance pattern of a dominant allele on an autosome

autosomal recessive: inheritance pattern of a recessive allele on an autosome

autosome: a nonsex chromosome

autotroph: organism that produces organic molecules by acquiring carbon from inorganic sources; primary producer

auxin: plant hormone that promotes cell elongation in stems and fruits

axial skeleton: the central axis of a vertebrate skeleton; consists of the bones of the head, vertebral column, and rib cage

axillary bud: undeveloped shoot in the angle between stem and petiole

axon (nerve fiber): extension of a neuron that transmits messages away from the cell body and toward another cell

B

B cell: type of lymphocyte that produces antibodies

bacillus (pl. bacilli): rod-shaped prokaryote

background extinction rate: steady, gradual loss of species through natural competition or loss of genetic diversity

Bacteria: one of two domains of prokaryotes

bacteriophage (phage): a virus that infects bacteria

balanced polymorphism: condition in which multiple alleles persist indefinitely in a population

bark: tissues outside the vascular cambium

base: a molecule that either releases hydroxide ions into a solution or removes hydrogen ions from it

basidiomycete: fungus that produces sexual spores on a basidium

basidiospore: haploid sexual spore produced on a basidium

basidium (pl. basidia): in basidiomycetes, a club-shaped structure that produces basidiospores

basilar membrane: the lower wall of the cochlear canal; vibrates in response to sound

basophil: type of white blood cell that triggers inflammation and allergy

behavioral ecology: study of the survival value and the evolution of behavioral patterns

benign tumor: mass of abnormal cells that does not have the potential to spread

benthic zone: sediment at the bottom of an ocean or lake

bilateral symmetry: body form in which only one plane divides the animal into mirror image halves

bile: digestive biochemical that emulsifies fats

binary fission: type of asexual reproduction in which a prokaryotic cell divides into two identical cells

biodiversity: the variety of life on Earth

biogeochemical cycle: geological and biological processes that recycle elements vital to life

biogeography: the study of the distribution patterns of species across the planet

biological species: a population, or group of populations, whose members can interbreed and produce fertile offspring

biomagnification: increasing concentrations of a chemical in higher trophic levels

biome: one of several major types of ecosystems

biosphere: part of Earth where life can exist

biotic: living

bird: tetrapod vertebrate with feathers, wings, and an amniotic egg

birth defect: abnormality that causes death or disability in a newborn

birth rate: the number of new individuals produced per 1000 individuals per unit time

bivalve: type of mollusk

blade: flattened part of a leaf

blastocyst: preembryonic stage consisting of a fluid-filled ball of cells

blastula: stage of early animal embryonic development; a sphere of cells surrounding a fluid-filled cavity

blood: type of connective tissue consisting of cells and platelets suspended in a liquid matrix

blood clot: plug of solidified blood

blood-brain barrier: close-knit cells that form capillaries in the brain, limiting the substances that can enter

blood pressure: force that blood exerts against artery walls

bone: type of connective tissue consisting of osteocytes and other cells embedded in a mineralized matrix; also, in the vertebrate skeleton, an organ consisting of bone tissue, cartilage, and other tissues

bony fish: jawed fish with a skeleton made of bone; ray-finned fishes and lobe-finned fishes

bottleneck: sudden reduction in the size of a population

brain: a distinct concentration of nervous tissue, often encased within a skull, at the anterior end of an animal

brainstem: continuation of the spinal cord into the vertebrate hindbrain

breast: milk-producing organ in female mammals

bronchiole: small branched airway that connects bronchi to alveoli

bronchus (pl. bronchi): one of two large tubes that branch from the trachea

brown alga: multicellular photosynthetic aquatic protist with swimming spores and brownish accessory pigments

bryophyte: plant that lacks vascular tissue; includes liverworts, hornworts, and mosses

buffer system: weak acid/base pair that resists changes in pH

bulbourethral gland: small gland near the male urethra that secretes mucus

bulimia: eating disorder in which a person eats large quantities and then intentionally vomits or uses laxatives shortly afterward

bulk element: an element that an organism requires in large amounts

bulk feeder: animal that ingests large pieces of food

bundle-sheath cell: thick-walled plant cell surrounding veins; site of Calvin cycle in C_4 plants

C

C_3 pathway: the Calvin cycle

C_4 pathway: a carbon fixation pathway in which CO_2 combines with a three-carbon molecule to form a four-carbon compound

caecilian: type of amphibian

calcitonin: thyroid hormone that decreases blood calcium levels

Calorie: one kilocalorie

calorie: the energy required to raise the temperature of 1 gram of water by 1°C under standard conditions

Calvin cycle: in photosynthesis, metabolic pathway in which CO_2 is fixed and incorporated into a three-carbon carbohydrate

CAM pathway: carbon fixation that occurs at night; CO_2 is later released for use in the Calvin cycle during the day

cancer: class of diseases characterized by uncontrolled division of cells that invade or spread to other tissues

capillary: tiny vessel that connects an arteriole with a venule

capillary bed: network of capillaries

capsid: protein coat of a virus

carbohydrate: compound containing carbon, hydrogen, and oxygen in a ratio 1:2:1

carbon fixation: the initial incorporation of carbon from CO_2 into an organic compound

carbon reactions: the reactions of photosynthesis that use ATP and NADPH to synthesize carbohydrates from carbon dioxide

carboxyl group: a carbon atom double-bonded to an oxygen and single-bonded to a hydroxyl group

cardiac cycle: sequence of contraction and relaxation that makes up the heartbeat

cardiac muscle: involuntary muscle tissue composed of branched, striated, single-nucleated contractile cells

cardiac output: the volume of blood that the heart pumps each minute

cardiovascular system: circulatory system

carnivore: animal that eats animals

carpel: leaflike structure enclosing an angiosperm's ovule

carrying capacity: maximum number of individuals that a habitat can support indefinitely

cartilage: type of connective tissue consisting of cells surrounded by a rubbery collagen matrix

cartilaginous fish: jawed fish with a skeleton made of cartilage; sharks, skates, and rays

Casparian strip: waxy barrier in root endodermis

catastrophism: the idea that brief upheavals such as floods, volcanic eruptions, and earthquakes are responsible for most geological formations

cecum: the entrance to the large intestine

cell: smallest unit of life that can function independently

cell body: enlarged portion of a neuron that contains most of the organelles

cell cycle: sequence of events that occur in an actively dividing cell

cell membrane: the boundary of a cell, consisting of proteins embedded in a phospholipid bilayer

cell plate: in plants, the materials that begin to form the wall that divides two cells

cell theory: the ideas that all living matter consists of cells, cells are the structural and functional units of life, and all cells come from preexisting cells

cell wall: a rigid boundary surrounding cells of many prokaryotes, protists, plants, and fungi

cell-mediated immunity: branch of adaptive immune system in which defensive cells kill invaders by direct cell-cell contact

cellular slime mold: protist in which feeding stage consists of individual cells that come together as a multicellular "slug" when food runs out

centipede: type of mandibulate arthropod

central nervous system (CNS): brain and the spinal cord

centromere: small section of a chromosome where sister chromatids attach to each other

centrosome: part of the cell that organizes microtubules

cephalization: development of sensory structures and a brain at the head end of an animal

cephalopod: type of mollusk

cerebellum: area of the hindbrain that coordinates subconscious muscular responses

cerebral cortex: outer layer of the cerebrum

cerebrospinal fluid: fluid that bathes and cushions the central nervous system

cerebrum: region of the forebrain that controls intelligence, learning, perception, and emotion

cervix: lower, narrow part of the uterus

chaparral: Mediterranean shrubland

charophyte: type of green algae thought to be most closely related to terrestrial plants

chelicerate: arthropod with clawlike mouthparts (chelicerae); horseshoe crabs and arachnids

chemical bond: attractive force that holds atoms together

chemical equilibrium: condition in which a chemical reaction proceeds in both directions at the same rate

chemical reaction: interaction in which bonds break and new bonds form

chemiosmotic phosphorylation: reactions that produce ATP using ATP synthase and the potential energy of a proton gradient

chemoreceptor: sensory receptor that responds to chemicals

chemotroph: organism that derives energy by oxidizing inorganic or organic chemicals

chiton: type of mollusk

chlorophyll *a*: green pigment that plants, algae, and cyanobacteria use to harness the energy in sunlight

chloroplast: organelle housing the reactions of photosynthesis in eukaryotes

chordate: animal that at some time during its development has a notochord, hollow nerve cord, pharyngeal pouches or slits, and a postanal tail; phylum Chordata

chorion: outermost extraembryonic membrane

chorionic villus: fingerlike projection extending from the chorion to the uterine lining

choroid: middle layer of the eyeball, between the sclera and the retina

chromatid: one of two identical DNA molecules that make up a replicated chromosome

chromatin: collective term for all of the DNA and its associated proteins in the nucleus of a eukaryotic cell

chromosome: a continuous molecule of DNA wrapped around protein in the nucleus of a eukaryotic cell; also, the genetic material of a prokaryotic cell

chyme: semifluid mass of food and gastric juice that moves from the stomach to the small intestine

chytridiomycete (chytrid): microscopic fungus that produces motile zoospores

ciliate: protist with cilia-covered cell surface

cilium (pl. cilia): one of many short, movable protein projections extending from a cell

circadian rhythm: physiological cycle that repeats daily

circulatory system: organ system that distributes blood (or a comparable fluid) throughout the body

clade: monophyletic group of organisms consisting of a common ancestor and all of its descendants

cladistics: phylogenetic system that defines groups by distinguishing between ancestral and derived characters

cladogram: treelike diagram built using shared derived characteristics

classical conditioning: type of associative learning in which an animal pairs two stimuli

cleavage: period of rapid cell division following fertilization

cleavage furrow: in animals, the initial indentation between two daughter cells in mitosis

climax community: community that persists indefinitely if left undisturbed

clitoris: small, highly sensitive female sexual organ

clonal deletion: elimination of lymphocytes with receptors for self antigens

clonal selection: rapid division of a stimulated B cell, generating memory B cells and plasma cells that are clones of the original B cell

cloning vector: a self-replicating structure that carries DNA from cell to cell

closed circulatory system: circulatory system in which blood remains confined to vessels

club moss: a type of seedless vascular plant

cnidarian: animal with radial symmetry, two germ layers, a jellylike interior, and cnidocytes; phylum Cnidaria

cnidocyte: cell in cnidarians that can fire a toxic barb for predation or defense

coccus (pl. cocci): spherical prokaryote

cochlea: spiral-shaped part of the inner ear, where vibrations are translated into nerve impulses

codominance: mode of inheritance in which two alleles are fully expressed in the heterozygote

codon: a triplet of mRNA bases that specifies a particular amino acid

coelacanth: type of lobe-finned fish

coelom: fluid-filled animal body cavity that forms completely within mesoderm

coevolution: genetic change in one species selects for subsequent change in another species

cofactor: inorganic or organic substance required for activity of an enzyme

cohesion: the attraction of water molecules to one another

cohesion-tension theory: theory that explains how water moves under tension in xylem

collecting duct: tubule in the kidney into which nephrons drain urine

collenchyma: elongated living plant cells with thick, elastic cell walls

commensalism: type of symbiosis in which one member benefits without affecting the other member

community: group of interacting populations that inhabit the same region

compact bone: solid, hard bone tissue consisting of tightly packed osteons

companion cell: parenchyma cells adjacent to sieve tube elements

competition: struggle between organisms for the same limited resource

competitive exclusion principle: the idea that two or more species cannot indefinitely occupy the same niche

competitive inhibition: change in an enzyme's activity occurring when an inhibitor binds to the active site, competing with the enzyme's normal substrate

complement: group of proteins that help destroy pathogens

complementary: in DNA and RNA, the precise pairing of purines (A and G) to pyrimidines (C, T, and U)

complete digestive tract: digestive tract through which food passes in one direction from mouth to anus

compound: a molecule including different elements

compound leaf: leaf that is divided into leaflets

concentration gradient: difference in solute concentrations between two adjacent regions

cone: a pollen- or ovule-bearing structure in many gymnosperms

cone cell: photoreceptor cell in the retina that detects colors

conidium: asexually produced fungal cell

conifer: a type of gymnosperm

conjugation: a form of horizontal gene transfer in which one cell receives DNA via direct contact with another cell

connective tissue: animal tissue consisting of widely spaced cells in a distinctive extracellular matrix

conservation biology: study of the preservation of biodiversity

constant region: amino acid sequence that is the same for all antibodies

consumer (heterotroph): organism that uses organic sources of energy and carbon

contact inhibition: property of most noncancerous eukaryotic cells; inhibits cell division when cells contact one another

contraception: use of devices or practices that prevent pregnancy

control: untreated group used as a basis for comparison with a treated group in an experiment

convergent evolution: the evolution of similar adaptations in organisms that do not share the same evolutionary lineage

coral reef: underwater deposit of calcium carbonate formed by colonies of coral animals

cork cambium: lateral meristem that produces cork cells and parenchyma in woody plant

cornea: in the eye, a modified portion of the sclera that forms a transparent curved window that admits light

coronary artery: artery that provides blood to the heart muscle

corpus luteum: gland formed from a ruptured ovarian follicle that has recently released an oocyte

cortex: ground tissue between epidermis and vascular tissue in roots and stems

cotyledon: seed leaf in angiosperms

countercurrent exchange: arrangement in which two adjacent currents flow in opposite directions and exchange materials or heat

coupled reactions: two simultaneous chemical reactions, one of which provides the energy that drives the other

covalent bond: type of chemical bond in which two atoms share electrons

craniate: animal with a cranium

cranium: part of the skull that encloses the brain

creatine phosphate: molecule stored in muscle fibers that can donate its high-energy phosphate to ADP, regenerating ATP

crista (pl. cristae): fold of the inner mitochondrial membrane along which many of the reactions of cellular respiration occur

crocodilian: type of reptile

cross bridge: contact between a myosin molecule's head and an actin molecule in a myofibril

crossing over: exchange of genetic material between homologous chromosomes during prophase I of meiosis

crustacean: type of mandibulate arthropod

culture: the knowledge, beliefs, and behaviors that humans transmit from generation to generation

cuticle: waterproof layer covering the aerial epidermis of a plant

cycad: a type of gymnosperm

cytokine: messenger protein synthesized in immune cells that influences the activity of other immune cells

cytokinesis: distribution of cytoplasm into daughter cells in cell division

cytokinin: plant hormone that stimulates cell division

cytoplasm: the watery mixture that occupies much of a cell's volume. In eukaryotic cells, it consists of all materials, including organelles, between the nuclear envelope and the cell membrane

cytoskeleton: framework of protein rods and tubules in eukaryotic cells

cytotoxic T cell: lymphocyte that kills invading cells by binding them and releasing chemicals

D

day-neutral plant: plant that does not rely on photoperiod to stimulate flowering

death rate: the number of deaths per 1000 individuals per unit time

decomposer: organism that consumes wastes and dead organic matter, returning inorganic nutrients to the ecosystem

deforestation: removal of tree cover from a previously forested area

dehydration synthesis: formation of a covalent bond between two molecules by loss of water

deletion: loss of one or more genes from a chromosome

demographic transition: phase of economic development during which birth and death rates shift from high to low

denaturation: modification of a protein's shape so that its function is destroyed

dendrite: thin neuron branch that receives neural messages and transmits information to the cell body

denitrification: conversion of nitrites and nitrates to N_2

dense connective tissue: type of connective tissue with fibroblasts and dense collagen tracts

density-dependent factor: population-limiting condition whose effects increase when populations are large

density-independent factor: population-limiting condition that acts irrespective of population size

dependent variable: response that may be under the influence of an independent variable

deposit feeder: type of substrate feeder that strains partially decayed organic matter from sediment

derived character: characteristic not found in the ancestor of a group being studied

dermal tissue: tissue covering a plant's surface

dermis: layer of connective tissue that lies beneath the epidermis in vertebrate skin

descent with modification: Darwin's term for evolution, describing gradual change from an ancestral type of organism

desert: type of terrestrial biome; very low precipitation

desertification: encroachment of desert into surrounding areas

determinate growth: growth that halts at maturity

detritivore: animal that eats decomposing organic matter

detritus: feces and dead organic matter

deuterostome: clade of bilaterally symmetrical animals in which the first opening in the gastrula develops into the anus

diabetes: disease resulting from an inability to produce or use insulin

diastolic pressure: lower number in a blood pressure reading; reflects relaxation of the ventricles

diatom: photosynthetic aquatic protist with two-part silica wall

differentiation: process by which cells acquire specialized functions

diffusion: movement of a substance from a region where it is highly concentrated to an area where it is less concentrated

digestion: the physical and chemical breakdown of food

digestive system: organ system that dismantles food, absorbs nutrient molecules, and eliminates indigestible wastes

dihybrid cross: mating between two individuals that are heterozygous for two genes

dikaryotic: containing two genetically distinct haploid nuclei

dinoflagellate: unicellular aquatic protist with two flagella of unequal length; many have cellulose plates

diploid cell: cell containing two full sets of chromosomes, one from each parent; also called $2n$

direct development: gradual development of a juvenile animal into an adult, without an intervening larval stage

directional selection: form of natural selection in which one extreme phenotype is fittest, and the environment selects against the others

disaccharide: a simple sugar that consists of two bonded monosaccharides

disruptive selection: form of natural selection in which the two extreme phenotypes are fittest

distal convoluted tubule: region of the renal tubule that connects the nephron loop to the collecting duct

distraction display: behavior that distracts a predator away from a nest or young

DNA (deoxyribonucleic acid): genetic material consisting of a double helix of nucleotides

DNA microarray: collection of short DNA fragments of known sequence placed in defined spots on a small square of glass

DNA polymerase: enzyme that adds new DNA nucleotides and corrects mismatched base pairs in DNA replication

DNA profiling: biotechnology tool that uses DNA to detect genetic differences between individuals

domain: broadest (most inclusive) taxonomic category

dominance hierarchy: social order with dominant and submissive members

dominant allele: allele that is expressed whenever it is present

dormancy: state of decreased metabolism

dorsal, hollow nerve cord: tubular nerve cord that forms dorsal to the notochord; one of the four characteristics of chordates

double-blind: type of experiment in which neither participants nor researchers know which subjects received a placebo and which received the treatment being evaluated

double fertilization: in angiosperms, one sperm nucleus fertilizes the egg and another fertilizes the polar nuclei

duplication: chromosomal abnormality that produces multiple copies of part of a chromosome

E

ear: sense organ of hearing and equilibrium

eardrum: structure that transmits sound from air to the middle ear

echinoderm: unsegmented deuterostome with a five-part body plan, radial symmetry in adults, and a spiny outer covering; phylum Echinodermata

ecological footprint: measure of the land area needed to support one or more humans

ecology: study of relationships among organisms and the environment

ecosystem: a community and its nonliving environment

ectoderm: outermost germ layer in an animal embryo

ectotherm: animal that lacks an internal mechanism that keeps its temperature within a narrow range; invertebrates, fishes, amphibians, and nonavian reptiles

effector: any structure that responds to a stimulus

ejaculation: discharge of semen through the penis

ejaculatory duct: tube that deposits sperm into the urethra

electromagnetic spectrum: range of naturally occurring radiation

electron: a negatively charged particle that orbits the atom's nucleus

electron transport chain: membrane-bound molecular complex that shuttles electrons to slowly extract their energy

electronegativity: an atom's tendency to attract electrons

element: a pure substance consisting of atoms containing a characteristic number of protons

elimination: the expulsion of waste from the body

embryo sac: mature female gametophyte in angiosperms

embryonic disc: in the preembryo, a flattened, two-layered mass of cells that develops into the embryo

embryonic stage: stage of human development lasting from the end of the second week until the end of the eighth week of gestation

embryonic stem cell: stem cell that can give rise to all types of cells in the body

emergent property: quality that results from interactions of a system's components

emigration: movement of individuals out of a population

endangered species: species facing a high risk of extinction in the near future

endergonic reaction: a chemical reaction that requires a net input of energy

endocrine gland: concentration of hormone-producing cells in an animal

endocrine system: organ system consisting of glands and cells that secrete hormones

endocytosis: form of transport in which the cell membrane engulfs extracellular material

endoderm: innermost germ layer in an animal embryo

endodermis: the innermost cell layer of root cortex

endomembrane system: eukaryotic organelles that exchange materials in transport vesicles

endometrium: inner uterine lining

endophyte: fungus that colonizes a plant without triggering disease symptoms

endoplasmic reticulum: interconnected membranous tubules and sacs in a eukaryotic cell

endorphin: pain-killing protein produced in the anterior pituitary

endoskeleton: skeleton on the inside of an animal

endosperm: triploid tissue that stores food for the embryo in an angiosperm seed

endospore: dormant, thick-walled structure that enables some bacteria to survive harsh conditions

endosymbiont theory: the idea that mitochondria and chloroplasts originated as free-living bacteria engulfed by other prokaryotic cells

endothelium: layer of epithelial tissue that lines blood vessels and the heart

endotherm: animal that maintains its body temperature by using heat generated from its own metabolism; birds and mammals

energy: the ability to do work

energy of activation: energy required for a chemical reaction to begin

energy shell: group of electron orbitals that share the same energy level

enhancer: DNA sequence that helps regulate gene expression and lies outside the promoter

entropy: randomness or disorder

envelope: the membrane layer surrounding the protein coat of some viruses

environmental resistance: combination of factors that limit population growth

enzyme: an organic molecule that catalyzes a chemical reaction without being consumed

eosinophil: type of white blood cell that functions as a phagocyte

epidermis: in animals, the outermost layer of skin; in plants, cells covering the leaves, stems, and roots

epididymis: tube that receives and stores sperm from one testis

epiglottis: cartilage that covers the glottis, routing food to the digestive tract during swallowing

epinephrine (adrenaline): hormone secreted by the adrenal medulla; also can act as a neurotransmitter

epistasis: one gene masks another gene's expression

epithelial tissue (epithelium): animal tissue consisting of tightly packed cells that form linings, coverings, and glands

erosion: wearing away of soil by water and wind

esophagus: muscular tube that leads from the pharynx to the stomach

essential nutrient: substance vital for an organism's metabolism, growth, and reproduction

estrogen: steroid hormone produced in ovaries of female vertebrates

estuary: area where fresh water in a river meets salty water of an ocean

ethylene: volatile plant hormone that ripens fruit

eudicot: one of the two main clades of angiosperms

euglenoid: unicellular flagellated protist with elongated cell

Eukarya: domain containing eukaryotes

eukaryote: organism composed of one or more cells containing a nucleus and other membrane-bounded organelles

eumetazoan: animal with true tissues

eusociality: social organization reflecting extensive division of labor in reproduction

eutrophic: nutrient-rich

eutrophication: addition of nutrients to a body of water

evaporation: the conversion of a liquid to a vapor (gas)

evolution: descent with modification; change in allele frequencies in a population over time

excretion: elimination of metabolic wastes

exergonic reaction: an energy-releasing chemical reaction

exhalation: movement of air out of the lungs

exocytosis: form of transport in which vesicles containing cell secretions fuse with the cell membrane

exon: portion of an mRNA that is translated after introns are removed

exoskeleton: skeleton on the outside of an animal

expanding repeat mutation: type of mutation in which the number of copies of a three- or four-nucleotide sequence increases over several generations

experiment: a test of a hypothesis under controlled conditions

exponential growth: population growth pattern in which the number of new individuals is proportional to the size of the population

external fertilization: release of gametes by males and females into the same environment

extinction: disappearance of a species

extracellular matrix: nonliving substances that surround animal cells; includes ground substance and fibers

eye: organ that detects light and produces the sense of sight

F

F₁ (first filial) generation: the offspring of the P generation in a genetic cross

F₂ (second filial) generation: the offspring of the F₁ generation in a genetic cross

facilitated diffusion: form of passive transport in which a substance moves down its concentration gradient with the aid of a transport protein

facultative anaerobe: organism that can live with or without O_2

fast-twitch fiber: large muscle cell that produces twitches of short duration

fatty acid: long-chain hydrocarbon terminating with a carboxyl group

feather: in birds, an epidermal outgrowth composed of keratin

feces: solid waste that leaves the digestive tract

fermentation: metabolic pathway in the cytoplasm in which NADH from glycolysis reduces pyruvate

fertilization: the union of two gametes

fetal stage: stage of human development lasting from the beginning of the ninth week of gestation through birth

fever: rise in the body's temperature

fiber: elongated sclerenchyma cell

fibrous root system: branching root system arising from stem

filtration: removal of water and solutes from the blood, as occurs at the glomerulus

first law of thermodynamics: energy cannot be created or destroyed, just converted from one form to another

fish: vertebrate animal with fins and external gills

fitness: an organism's contribution to the next generation's gene pool

fixed action pattern: type of innate behavior; stereotyped sequence of events that is performed to completion once initiated

flagellated protozoan: unicellular heterotrophic protist with one or more flagella

flagellum (pl. flagella)**:** a long whiplike appendage a cell uses for motility

flatworm: unsegmented worm lacking a coelom; phylum Platyhelminthes

flower: the reproductive structure in angiosperms that produces pollen and eggs

fluid feeder: animal that drinks its food

fluid mosaic: two-dimensional structure of movable phospholipids and proteins that form biological membranes

fluke: type of parasitic flatworm

follicle cell: nourishing cell surrounding an oocyte

follicle stimulating hormone (FSH): reproductive hormone produced in the anterior pituitary

food chain: series of organisms that successively eat each other

food web: network of interconnecting food chains

foot: ventral muscular structure that provides movement in mollusks

foraging: animal behavior related to finding and collecting food

foraminiferan: amoeboid protozoan with a calcium carbonate shell

forebrain: front part of the vertebrate brain

fossil: any evidence of an organism from more than 10,000 years ago

founder effect: genetic drift that occurs when a small, nonrepresentative group of individuals leaves their ancestral population and begins a new settlement

frameshift mutation: type of mutation in which nucleotides are added or deleted by any number other than a multiple of three, altering the reading frame

free-living flatworm: planarian or marine flatworm

frog: type of amphibian

fruit: seed-containing structure in angiosperms

fruiting body: multicellular spore-bearing organ of a fungus

Fungi: kingdom containing mostly multicellular eukaryotes that are heterotrophs by external digestion

G

G$_0$ phase: resting phase of the cell cycle in which the cell continues to function but does not divide

G$_1$ phase: gap stage of interphase in which the cell grows and carries out its basic functions

G$_2$ phase: gap stage of interphase in which the cell synthesizes and stores membrane components and spindle proteins

gallbladder: organ that stores bile from the liver and releases it into the small intestine

gamete: a sex cell; sperm or egg cell

gametophyte: haploid, gamete-producing stage of the plant life cycle

ganglion: cluster of neuron cell bodies

ganglion cell: interneuron in the retina that generates action potentials

gap junction: connection between two adjacent animal cells that allows cytoplasm to flow between them

gastric juice: mixture of water, mucus, salts, hydrochloric acid, and enzymes produced at the stomach lining

gastrointestinal (GI) tract: two-opening digestive tract; also called alimentary canal

gastropod: type of mollusk

gastrovascular cavity: digestive chamber with a single opening

gastrula: stage of early animal embryonic development during which three tissue layers form

gene: sequence of DNA that codes for a specific protein or RNA molecule

gene flow: the movement of alleles between populations

gene pool: all of the genes and their alleles in a population

gene therapy: treatment that replaces a faulty gene in a cell with a functioning version of the gene

genetic code: correspondence between specific nucleotide sequences and amino acids

genetic drift: change in allele frequencies that occurs purely by chance

genome: all the genetic material in an organism

genotype: an individual's combination of alleles for a particular gene

genotype frequency: number of individuals of one genotype, divided by the number of individuals in the population

genus: taxonomic category that groups closely related species

geologic time scale: a division of Earth's history into eons, eras, periods, and epochs defined by major geological or biological events

germ cell: cell that gives rise to gametes in an animal

germination: resumption of growth after seed dormancy is broken

gibberellin: plant hormone that promotes shoot elongation

gill: highly folded respiratory surface containing blood vessels that exchange gases in water

ginkgo: a type of gymnosperm

gland: organ that secretes substances into the bloodstream or into a duct

global climate change: long-term changes in Earth's weather patterns

glomeromycete: fungus lacking sexual spores that form mycorrhizae

glomerular (Bowman's) capsule: cup-shaped end of a nephron; surrounds the glomerulus

glomerulus: ball of capillaries containing blood to be filtered at a nephron

glottis: slitlike opening between the vocal cords

glucagon: pancreatic hormone that raises blood sugar level by stimulating liver cells to break down glycogen into glucose

glucocorticoid: hormone secreted by the adrenal cortex

glycerol: a three-carbon alcohol that forms the backbone of triglycerides and phospholipids

glycocalyx: sticky layer composed of proteins and/or polysaccharides that surrounds some prokaryotic cell walls; slime layer or capsule

glycolysis: a metabolic pathway occurring in the cytoplasm of all cells; one molecule of glucose splits into two molecules of pyruvate

gnetophyte: a type of gymnosperm

golden alga: photosynthetic aquatic protist with two flagella and carotenoid accessory pigments

Golgi apparatus: a system of flat, stacked, membrane-bounded sacs that packages cell products for export

gonad: gland that manufactures hormones and gametes in animals; ovary or testis

gonadotropin-releasing hormone (GnRH): reproductive hormone produced in the hypothalamus

graded potential: a local flow of electrical current that occurs when Na$^+$ leaks into a neuron, weakening with distance from the source of the stimulus

gradualism: theory that proposes that evolutionary change occurs gradually, in a series of small steps

Gram stain: technique for classifying bacteria into two main groups based on cell wall structure

granum (pl. grana)**:** a stack of flattened thylakoid discs in a chloroplast

gravitropism: directional growth response to gravity

gray matter: nervous tissue in the central nervous system; consists mostly of neuron cell bodies, dendrites, and synapses

green alga: photosynthetic protist that has pigments, starch, and cell walls similar to those of land plants

greenhouse effect: increase in surface temperature caused by carbon dioxide and other atmospheric gases

ground tissue: plant tissue that makes up most of the primary plant body; composed mostly of parenchyma

growth factor: a protein that signals a cell to divide

growth hormone (GH): hormone produced in the anterior pituitary

guard cells: pair of cells flanking a stoma

gustation: the sense of taste

gymnosperm: a plant with seeds that are not enclosed in a fruit; includes conifers, *Ginkgo*, gnetophytes, and cycads

H

habitat: physical place where an organism normally lives

habituation: type of learning in which an animal learns not to respond to irrelevant stimuli

hagfish: jawless animal with a cranium but not vertebrae

hair cell: mechanoreceptor that initiates sound transduction in the cochlea

half-life: the time it takes for half the atoms in a sample of a radioactive substance to decay

haploid cell: cell containing one set of chromosomes; also called *n*

Hardy–Weinberg equilibrium: situation in which allele frequencies do not change from one generation to the next

heart: muscular organ that pumps blood (or a comparable fluid) throughout the body

heartwood: dark-colored, nonfunctioning secondary xylem in woody plant

helper T cell: lymphocyte that coordinates activities of other immune system cells

hemisphere: one of two halves of the cerebrum

hemoglobin: pigment that carries oxygen in red blood cells

herbaceous plant: plant with green, soft stem at maturity

herbivore: animal that eats plants

heterotroph: organism that obtains carbon and energy by eating another organism; consumer

heterozygote advantage: condition in which a heterozygote has greater fitness than homozygotes, maintaining balanced polymorphism in a population

heterozygous: possessing two different alleles for a particular gene

hindbrain: lower, posterior portion of the vertebrate brain

hippocampus: forebrain structure involved in memory formation

histamine: biochemical that dilates blood vessels and increases their permeability; involved in inflammation and allergies

homeostasis: the ability of an organism to maintain a stable internal environment despite changes in the external environment

homeotic: describes any gene that, when mutated, leads to organisms with structures in the wrong places

hominid: any of the "great apes" (orangutans, gorillas, chimpanzees, and humans)

hominine: extinct or modern human

hominoid: any ape, including humans

homologous: similar in structure or position because of common ancestry

homologous pair: two chromosomes that look alike and have the same sequence of genes

homozygous: possessing two identical alleles for a particular gene

horizontal gene transfer: transfer of genetic information from one cell to another cell that is not its descendant

hormone: biochemical synthesized in small quantities in one place and transported to another

hornwort: a type of bryophyte

horseshoe crab: type of chelicerate arthropod

horsetail: a type of seedless vascular plant

host range: the kinds of organisms or cells that a virus can infect

human chorionic gonadotropin (hCG): hormone secreted by an embryo; prevents menstruation

human immunodeficiency virus (HIV): virus that causes acquired immune deficiency syndrome (AIDS)

humoral immunity: branch of adaptive immune system in which B cells secrete antibodies in response to a foreign antigen

humus: chemically complex jellylike organic substance in soil

hydrogen bond: weak chemical bond between opposite partial charges on two molecules or within one large molecule

hydrolysis: splitting a molecule by adding water

hydrophilic: attracted to water

hydrophobic: repelled by water

hydrostatic skeleton: skeleton consisting of constrained fluid in a closed body compartment

hypertonic: describes a solution in which the solute concentration is greater than on the other side of a semipermeable membrane

hypha (pl. hyphae): a fungal filament; the basic structural unit of a multicellular fungus

hypoglycemia: low blood sugar caused by excess insulin or insufficient carbohydrate intake

hypothalamus: small forebrain structure beneath the thalamus that controls homeostasis and links the nervous and endocrine systems

hypothesis: a testable, tentative explanation based on prior knowledge

hypotonic: describes a solution in which the solute concentration is less than on the other side of a semipermeable membrane

I

immigration: movement of individuals into a population

immune system: organ system consisting of cells that defends the body against infections, cancer, and foreign substances

immunodeficiency: condition in which the immune system lacks one or more components

impact theory: idea that mass extinctions were caused by impacts of extraterrestrial origin

implantation: embedding of the blastocyst into the uterine lining

imprinting: type of learning that occurs during a sensitive period early in life and occurs without obvious reinforcement

inclusive fitness: the sum of an individuals direct and indirect fitness

incomplete digestive tract: digestive tract with one opening that takes in food and ejects wastes

incomplete dominance: mode of inheritance in which a heterozygote's phenotype is intermediate between the phenotypes of the two homozygotes

independent variable: hypothesized influence on a dependent variable

indeterminate growth: growth that persists indefinitely

indirect development: development of a juvenile animal into an adult while passing through intervening larval stages

inflammation: immediate, localized reaction to an injury or any pathogen that breaches the body's barriers

ingestion: the act of taking food into the digestive tract

inhalation: movement of air into the lungs

inheritance of acquired characteristics: the idea that an organism can inherit the traits that its parent acquired during its lifetime

innate: describing an instinctive behavior that develops independently of experience

innate defense: cell or substance that provides generalized protection against all infectious agents

inner cell mass: cells in the blastocyst that develop into the embryo

insect: type of mandibulate arthropod

insulin: pancreatic hormone that lowers blood sugar level by stimulating body cells to take up glucose from the blood

integumentary system: organ system consisting of skin and its outgrowths

intercalary meristem: meristem between the nodes of a mature stem

interferon: type of cytokine released by a virus-infected cell

interleukin: type of cytokine involved in communication between white blood cells

intermediate filament: component of the cytoskeleton; intermediate in size between a microtubule and a microfilament

intermembrane compartment: the space between a mitochondrion's two membranes

internal fertilization: use of a copulatory organ to deposit sperm inside a female's body

interneuron: neuron that connects one neuron to another in the central nervous system

internode: stem areas between the points of leaf attachment

interphase: stage preceding mitosis or meiosis, when the cell carries out its functions, replicates its DNA, and grows

intersexual selection: choice of mates by one sex from among competing members of the opposite sex

interstitial cell: testosterone-secreting endocrine cell in a testis

interstitial fluid: liquid that bathes cells in a vertebrate's body

intertidal zone: region along a coastline between the high and low tide marks

intrasexual selection: competition between members of the same sex for access to the opposite sex

intron: portion of an mRNA molecule that is removed before translation

invasive species: introduced species that establishes a breeding population in a new location and spreads widely from the original point of introduction

inversion: abnormality in which a portion of a chromosome flips and reinserts itself

invertebrate: animal without a backbone

ion: an atom or group of atoms that has lost or gained electrons, giving it an electrical charge

ionic bond: attraction between oppositely charged ions

iris: colored part of the eye; regulates the size of the pupil

isotonic: condition in which a solute concentration is the same on both sides of a semipermeable membrane

isotope: any of the forms of an element, each having a different number of neutrons in the nucleus

J

jaws: bones that frame the entrance to the mouth

joint: area where two bones meet

J-shaped curve: plot of exponential growth over time

K

karyotype: a size-ordered chart of the chromosomes in a cell

keystone species: species whose effect on community structure is disproportionate to its biomass

kidney: excretory organ in the vertebrate urinary system

kilocalorie (kcal): one thousand calories; one food Calorie.

kin selection: the sacrifice of one individual's fitness for the sake of the genes it shares with related animals

kinetic energy: energy being used to do work; energy of motion

kinetochore: protein that attaches a chromosome to the spindle in cell division

kingdom: taxonomic category below domain

Krebs cycle: stage in cellular respiration that completely oxidizes the products of glycolysis

K-selected species: species consisting of long-lived, late-maturing individuals that have few offspring, with each receiving heavy parental investment

L

***lac* operon:** in *E. coli*, three lactose-degrading genes plus the promoter and operator that control their transcription

lactic acid fermentation: metabolic pathway in which NADH from glycolysis reduces pyruvate, producing lactic acid

lamprey: type of jawless fish

lancelet: type of invertebrate chordate

large intestine: part of the digestive tract that connects the small intestine to the anus

larva: immature stage of animal development; does not resemble the adult of the species

larynx: boxlike structure in front of the pharynx

latent: describes an infection in which viral genetic material in a host cell does not cause symptoms

lateral line: network of canals that extends along the sides of fishes and houses receptor organs that detect vibrations

lateral meristem: meristem whose daughter cells thicken a root or stem

law of independent assortment: Mendel's law stating that during gamete formation, the segregation of the alleles for one gene does not influence the segregation of the alleles for another gene

law of segregation: Mendel's law stating that the two alleles of each gene are packaged into separate gametes

leaf: flattened organ that carries out photosynthesis

learning: alteration of an animal's behavior to reflect experience

leech: type of annelid

lens: structure in the eye that bends incoming light

leukemia: cancer in which bone marrow overproduces white blood cells

lichen: association of a fungus and a green alga or cyanobacterium

life history: the events of an organism's life, especially those that are related to reproduction

life table: chart that shows the probability of surviving to any given age

ligament: band of fibrous connective tissue that connects bone to bone across a joint

ligase: enzyme that catalyzes formation of covalent bonds in the DNA sugar-phosphate backbone

light reactions: photosynthetic reactions that harvest light energy and store it in molecules of ATP or NADPH

lignin: tough, complex molecule that strengthens the walls of some plant cells

limbic system: collection of forebrain structures involved in emotion and memory

limnetic zone: layer of open water in a lake or pond where light penetrates

linkage group: group of genes that tend to be inherited together because they are on the same chromosome

linkage map: diagram of gene order and spacing on a chromosome, based on crossover frequencies

linked genes: genes on the same chromosome

lipid: hydrophobic organic molecule consisting mostly of carbon and hydrogen

littoral zone: shallow region along the shore of a lake or pond where rooted plants occur

liver: organ that produces bile, detoxifies blood, stores glycogen and fat-soluble vitamins, synthesizes blood proteins, and monitors blood glucose level

liverwort: a type of bryophyte

lizard: type of reptile

lobe-finned fish: type of bony fish; coelacanths and lungfishes

locus: physical location of a gene on a chromosome

logistic growth: the leveling-off of a population in response to environmental resistance

long-day plant: plant that flowers when light periods are longer than a critical length

long-term memory: memory that can last from hours to a lifetime

loose connective tissue: type of connective tissue with widely spaced fibroblasts and a loose network of fibers

lung: sac-like structure where gas exchange occurs in air-breathing vertebrates

lungfish: type of lobe-finned fish

luteinizing hormone (LH): reproductive hormone produced in the anterior pituitary

lymph: fluid in lymph vessels

lymph capillary: dead-end vessel that collects lymph

lymph node: lymphatic structure located along lymph capillary; contains white blood cells and fights infection

lymphatic system: organ system consisting of lymphoid organs and lymph vessels that recover excess tissue fluid and aid in immunity

lymphocyte: type of white blood cell; T cell, or B cell, or natural killer cell

lysogenic infection: type of viral infection in which the genetic material of a virus is replicated along with the host cell's chromosome

lysosome: organelle in a eukaryotic cell that buds from the Golgi apparatus and enzymatically dismantles molecules, bacteria, and worn-out cell parts

lytic infection: type of viral infection in which a virus enters a cell, replicates, and causes the host cell to burst (lyse) as it releases the new viruses

M

macroevolution: large-scale evolutionary change

macronutrient: nutrient required in large amounts

macrophage: type of phagocyte

major histocompatibility complex (MHC): cluster of genes that encode "self" proteins on cell surfaces

malignant tumor: mass of abnormal cells that has the potential to invade adjacent tissues and spread throughout the body

mammal: amniote with hair and mammary glands

mammary gland: milk-producing gland in mammals

mandibulate: arthropod with jawlike mouthparts (mandibles); crustaceans, insects, centipedes, and millipedes

mantle: dorsal fold of tissue that secretes a shell in most mollusks

marrow cavity: space in a bone shaft that contains marrow

marsupial: mammal that bears live young after a short gestation

marsupium: pouch in which the immature young of many marsupial mammals nurse and develop

mass extinction: the disappearance of many species over relatively short expanses of time

mass number: the total number of protons and neutrons in an atom's nucleus

mast cell: immune system cell that triggers inflammation and allergy

matrix: the inner compartment of a mitochondrion

matter: substance that takes up space and is made of atoms

mechanoreceptor: sensory receptor sensitive to physical deflection

Mediterranean shrubland: type of terrestrial biome; rainy winters and dry summers (also called chaparral)

medulla oblongata: part of the brainstem nearest the spinal cord

medusa: free-swimming form of a cnidarian

megaspore: in seed plants, spore that gives rise to female gametophyte

meiosis: division of genetic material that halves the chromosome number and yields genetically variable gametes

melatonin: hormone produced in the pineal gland

memory cell: lymphocyte produced in an initial infection; launches a rapid immune response upon subsequent exposure to an antigen

meninges: membranes that cover and protect the central nervous system

menstrual cycle: hormonal cycle that prepares the uterus for pregnancy

meristem: localized region of active cell division in a plant

mesoderm: embryonic germ layer between ectoderm and endoderm in an animal embryo

mesophyll: photosynthetic ground tissue in leaves

messenger RNA (mRNA): a molecule of RNA that encodes a protein

metabolism: the biochemical reactions of a cell

metamorphosis: developmental process in which an animal changes drastically in body form between the juvenile and the adult

metaphase: stage of mitosis in which chromosomes are aligned down the center of a cell

metastasis: spreading of cancer

metaphase I: metaphase of meiosis I, when homologous chromosome pairs align down the center of a cell

metaphase II: metaphase of meiosis II, when replicated chromosomes align down the center of a cell

microevolution: relatively short-term changes in allele frequencies within a population or species

microfilament: component of the cytoskeleton; made of the protein actin

micronutrient: nutrient required in small amounts

microspore: in seed plants, spore that gives rise to male gametophyte

microtubule: component of the cytoskeleton; made of subunits of tubulin protein

microvillus (pl. microvilli): extension of the plasma membrane of epithelial cells of a villus

midbrain: part of the brain between the forebrain and hindbrain

millipede: type of mandibulate arthropod

mineral: essential element other than C, H, O, or N

mineralocorticoid: hormone secreted by the adrenal cortex

mitochondrion: organelle that houses the reactions of cellular respiration in eukaryotes

mitosis: division of genetic material that yields two genetically identical cells

mitotic spindle: a structure of microtubules that aligns and separates chromosomes in mitosis

modern evolutionary synthesis: the idea that genetic mutations create the variation upon which natural selection acts

molecular clock: application of the rate at which DNA mutates to estimate when two types of organisms diverged from a shared ancestor

molecule: two or more atoms joined by chemical bonds

mollusk: unsegmented animal with a soft body, mantle, muscular foot, and visceral mass; phylum Mollusca

monocot: one of the two main clades of angiosperms

monocyte: type of white blood cell that gives rise to macrophages

monogamy: mating system in which males and females have one sexual partner

monohybrid cross: mating between two individuals that are heterozygous for the same gene

monomer: a single unit of a polymeric molecule

monophyletic: describes a group of organisms consisting of a common ancestor and all of its descendants

monosaccharide: a sugar that is one five- or six-carbon unit

monotreme: egg-laying mammal

morula: preembryonic stage consisting of a solid ball of cells

moss: a type of bryophyte

motor neuron: neuron that transmits a message from the central nervous system toward a muscle or gland

motor unit: a motor neuron and all the muscle fibers it contacts

muscle: organ that powers movements in animals by contracting; consists of muscle tissue and other tissue types

muscle fiber: muscle cell

muscle tissue: animal tissue consisting of contractile cells that provide motion

muscular system: organ system consisting of skeletal muscles whose contractions form the basis of movement and posture

mutagen: any external agent that causes a mutation

mutant: a genotype, phenotype, or allele that is not the most common in a population or that has been altered from the "typical" (wild type) condition

mutation: a change in a DNA sequence

mutualism: type of symbiosis that improves the fitness of both partners

mycelium: assemblage of hyphae that forms an individual fungus

mycorrhiza: mutually beneficial association of a fungus and the roots of a plant

myelin sheath: fatty material that insulates some nerve fibers in vertebrates, speeding nerve impulse transmission

myocardium: thick, muscular middle layer of the heart wall

myofibril: cylindrical subunit of a muscle fiber, consisting of parallel protein filaments

myosin: protein that forms thick filaments in muscle cells

N

NADH: coenzyme that carries electrons in glycolysis and respiration

NADPH: coenzyme that carries electrons in photosynthesis

natural killer cell: type of lymphocyte that participates in innate defenses

natural selection: differential reproduction of organisms based on inherited traits

negative feedback: pathway in which the product of a reaction inhibits the enzyme that controls its formation; also, an action that counters an existing condition to maintain homeostasis

nephron: functional unit of the kidney

nephron loop: hairpin-shaped region of the renal tubule that connects the proximal and distal convoluted tubules

neritic zone: region of an ocean from the coast to the edge of the continental shelf

nerve: bundle of nerve fibers (axons) bound together in a sheath of connective tissue

nerve ladder: nervous system consisting of two nerve cords connected by transverse nerves

nerve net: diffuse network of neurons

nervous system: organ system that specializes in rapid communication

nervous tissue: tissue type whose cells (neurons and neuroglia) form a communication network

net primary production: energy available to consumers in a food chain, after cellular respiration and heat loss by producers

neural tube: embryonic precursor of the central nervous system

neuroglia: one of two cell types in nervous tissue

neuromuscular junction: synapse of a neuron onto a muscle cell

neuron: one of two cell types in nervous tissue

neurotransmitter: chemical passed from a neuron to receptors on another neuron or on a muscle or gland cell

neutral: neither acidic nor basic. Also, not electrically charged

neutron: a particle in an atom's nucleus that is electrically neutral

neutrophil: type of white blood cell that functions as a phagocyte

niche: all resources a species uses for survival, growth, and reproduction

nitrification: conversion of ammonia to nitrites and nitrates

nitrogen fixation: conversion of N_2 to NH_4^+, a form plants can use

nitrogenous base: a nitrogen-containing compound that forms part of a nucleotide

node: point at which leaves attach to a stem

node of Ranvier: short region of exposed axon between two sections of myelin sheath

nodule: root growth housing nitrogen-fixing bacteria

noncompetitive inhibition: change in an enzyme's shape occurring when an inhibitor binds to a site other than the active site

nondisjunction: failure of chromosomes to separate at anaphase I or anaphase II of meiosis

nonpolar covalent bond: a covalent bond in which atoms share electrons equally

norepinephrine (noradrenaline): hormone secreted by the adrenal medulla; also can act as a neurotransmitter

nose: organ that forms the entrance to the nasal cavity inside the head; functions in breathing and olfaction

notochord: flexible rod that forms the framework of the vertebral column and induces formation of the neural tube; one of the four characteristics of chordates

nuclear envelope: the two membranes bounding a cell's nucleus

nuclear pore: a hole in the nuclear envelope

nucleic acid: a long polymer of nucleotides; DNA or RNA

nucleoid: the part of a prokaryotic cell where the DNA is located

nucleolus: a structure within the nucleus where components of ribosomes are assembled

nucleosome: the basic unit of chromatin; consists of DNA wrapped around eight histone proteins

nucleotide: building block of a nucleic acid, consisting of a phosphate group, a nitrogenous base, and a five-carbon sugar

nucleus: central part of an atom; also, the membrane-bounded sac that contains DNA in a eukaryotic cell

nutrient: any substance that an organism uses for metabolism, growth, maintenance, and repair of its tissues

O

obesity: unhealthy amount of body fat; body mass index greater than 30

obligate aerobe: organism that requires O_2 for generating ATP

obligate anaerobe: organism that must live in the absence of O_2

observational learning: type of learning in which an animal imitates the behavior of others

oceanic zone: open sea beyond the continental shelf

olfaction: the sense of smell

oligochaete: type of annelid

oligodendrocyte: type of neuroglia that forms a myelin sheath around some axons in the central nervous system

oligosaccharide: intermediate-length carbohydrate consisting of 3 to 100 monosaccharides

oligotrophic: describes water containing few nutrients

omnivore: animal that eats many types of food, including plants and animals

oncogene: gene that normally stimulates cell division but when overexpressed leads to cancer

oogenesis: the production of egg cells

oogonium: diploid germ cell that divides mitotically to yield two primary oocytes

open circulatory system: circulatory system in which blood circulates freely through the body cavity

operant conditioning: type of associative learning in which an animal connects a behavior with its consequences

operator: in an operon, the DNA sequence between the promoter and the protein-encoding regions

operon: group of related bacterial genes plus a promoter and operator that control the transcription of the entire group at once.

opportunistic pathogen: infectious agent that cannot cause disease in a healthy individual

optic nerve: nerve fibers that connect the retina to the brain

optimal foraging theory: theory that predicts an animal should maximize energy collection per unit time

orbital: volume of space where a particular electron is likely to be

organ: two or more tissues that interact and function as an integrated unit

organ system: two or more physically or functionally linked organs

organelle: compartment of a eukaryotic cell that performs a specialized function

organic molecule: compound containing both carbon and hydrogen

organism: a single living individual

orgasm: pleasurable sensation, accompanied by involuntary muscle contractions, associated with sexual activity

osmoregulation: control of an animal's ion concentration

osmosis: simple diffusion of water through a semipermeable membrane

osteoblast: bone cell that secretes matrix

osteoclast: bone cell that degrades matrix

osteocyte: mature bone cell (former osteoblast) surrounded by matrix

osteon: one set of concentric rings of osteocytes

osteoporosis: condition in which bones become less dense

ostracoderm: extinct jawless fish

outgroup: basis for comparison in a cladistics analysis

oval window: membrane between the middle ear and the inner ear

ovarian cycle: hormonal cycle that controls the timing of oocyte maturation in the ovaries

ovary: in angiosperms, base of carpel plus their enclosed ovules; in animals, the female gonad

overexploitation: harvesting a species faster than it can reproduce

ovulation: release of an oocyte from an ovarian follicle

ovule: egg-bearing structure that develops into a seed in gymnosperms and angiosperms

oxidation: the loss of one or more electrons by a participant in a chemical reaction

oxidation-reduction (redox) reaction: chemical reaction in which one reactant is oxidized and another is reduced

oxygen debt: after vigorous exercise, a period in which the body requires extra oxygen to restore ATP and creatine phosphate to muscle and to recharge oxygen-carrying proteins

oxytocin: hormone released from the posterior pituitary

ozone layer: atmospheric zone rich in ozone gas (O_3), which absorbs the sun's ultraviolet radiation

P

P (parental) generation: the first, true-breeding generation in a genetic cross

pacemaker: specialized cardiac cells that set the tempo of the heartbeat

pain receptor: sensory receptor that detects mechanical damage, temperature extremes, or chemicals released from damaged cells

paleontology: the study of fossil remains or other clues to past life

pancreas: gland between the spleen and the small intestine; produces hormones, digestive enzymes, and bicarbonate

pancreatic islet (islet of Langerhans): cluster of cells in the pancreas that secretes insulin and glucagon

parapatric speciation: formation of new species when part of a population enters a habitat bordering the parent species' range, and the two groups become reproductively isolated

paraphyletic: describes a group of organisms that contains a common ancestor and some, but not all, of its descendants

parasitism: type of symbiosis in which one member increases its fitness at the expense of the other

parasympathetic nervous system: part of the autonomic nervous system; dominates during relaxed times and opposes the sympathetic nervous system

parathyroid gland: one of four small groups of cells behind the thyroid gland

parathyroid hormone (PTH): hormone produced in the parathyroid gland

parenchyma: unspecialized plant cells making up majority of ground tissue

parental chromatid: chromatid containing genetic information from only one parent

parsimonious: in cladistics, describes the evolutionary tree that requires the fewest steps to construct from a set of observations

particulate: small piece of matter suspended in air

passive immunity: immunity generated when an organism receives antibodies from another organism

passive transport: movement of a solute across a membrane without the direct expenditure of energy

pathogen: disease-causing agent

pattern formation: developmental process that establishes the body's overall shape and structure

pectoral girdle: bones that connect the forelimbs to the axial skeleton

pedigree: chart showing family relationships and phenotypes

peer review: evaluation of scientific results by experts before publication in a journal

pelagic zone: all water above the ocean floor

pelvic girdle: bones that connect the hind limbs to the axial skeleton

penis: male organ of copulation and urination

pepsin: enzyme that begins the digestion of proteins in the stomach

peptide bond: a covalent bond between adjacent amino acids; results from dehydration synthesis

peptide hormone: water-soluble, amino acid–based hormone that cannot freely diffuse through a cell membrane

peptidoglycan: material in bacterial cell wall

per capita rate of increase (r): difference between the birth rate and the death rate in a population

perception: the brain's interpretation of a sensation

pericardium: connective tissue sac that encloses the heart

pericycle: outermost layer of root vascular cylinder; produces branch roots

periderm: dermal tissue covering woody plant part

periodic table: chart that lists elements according to their properties

peripheral nervous system: neurons that transmit information to and from the central nervous system

peristalsis: waves of muscle contraction that propel food along the digestive tract

peritubular capillaries: blood vessels that surround a nephron

permafrost: permanently frozen ground in tundra

peroxisome: membrane-bounded sac that houses enzymes that break down fatty acids and dispose of toxic chemicals

persistent organic pollutant: carbon-based chemical pollutants that remain in ecosystems for long periods

petal: flower part interior to sepals

petiole: stalk that supports a leaf blade

pH scale: a measurement of how acidic or basic a solution is

phagocyte: cell that engulfs and digests foreign material and cell debris

phagocytosis: form of endocytosis in which the cell engulfs a large particle

pharyngeal pouch (or slit): opening in the pharynx of a chordate embryo; one of the four characteristics of chordates

pharynx: tube just behind the oral and nasal cavities; the throat

phenotype: observable characteristic of an organism

pheromone: volatile chemical an organism releases that elicits a response in another member of the species

phloem: vascular tissue that transports sugars and other dissolved organic substances in plants

phloem sap: solution of water, minerals, sucrose, and other biochemicals in phloem

phospholipid: molecule consisting of two hydrophobic fatty acids and a hydrophilic phosphate group

phospholipid bilayer: double layer of phospholipids that forms in water; forms the majority of a cell's membranes

phosphorylation: the addition of a phosphate to a molecule

photic zone: region in a water body where light is sufficient for photosynthesis

photon: a packet of light or other electromagnetic radiation

photoperiod: day length

photoreceptor: molecule or cell that detects quality and quantity of light

photorespiration: a metabolic pathway in which rubisco reacts with O_2 instead of CO_2, counteracting photosynthesis

photosynthesis: biochemical reactions that enable organisms to harness sunlight energy to manufacture organic molecules

photosystem: cluster of pigment molecules and proteins in a chloroplast's thylakoid membrane

phototroph: organism that derives energy from sunlight

phototropism: directional growth response to unidirectional light

phylogenetics: field of study that attempts to explain the evolutionary relationships among species

phylogeny: graphical depiction of evolutionary relationships among species

physiology: the study of the functions of organisms and their parts

phytochrome: a type of photoreceptor in plants

phytoplankton: microscopic photosynthetic organisms that drift in water

pilus: short projection made of protein on a prokaryotic cell

pineal gland: small gland in the brain that secretes melatonin

pioneer species: the first species to colonize an area devoid of life

pith: ground tissue inside a ring of vascular bundles in roots and stems

pituitary gland: pea-sized endocrine gland attached to the hypothalamus

placebo: inert substance used as an experimental control

placenta: structure that connects the developing fetus to the maternal circulation in placental mammals

placental mammal: mammal in which the developing fetus is nourished by a placenta

placoderm: extinct armored fishes with jaws

Plantae: kingdom consisting of multicellular, eukaryotic autotrophs

plasma: watery, protein-rich fluid that forms the matrix of blood

plasma cell: B cell that secretes large quantities of one antibody

plasmid: small circle of double-stranded DNA separate from a cell's chromosome

plasmodesma (pl. plasmodesmata): connection between plant cells that allows cytoplasm to flow between them

plasmodial slime mold: protist in which feeding stage consists of a multinucleated plasmodium

plate tectonics: theory that Earth's surface consists of several plates that move in response to forces acting deep within the planet

platelet: cell fragment that orchestrates clotting in blood

pleiotropy: multiple phenotypic effects of one genotype

point mutation: type of mutation in which one DNA base substitutes for another

polar body: small cell produced in female meiosis

polar covalent bond: a covalent bond in which electrons are attracted more to one atom's nucleus than to the other

polar nucleus: one of two nuclei fertilized to yield endosperm in angiosperms

pollen: immature male gametophyte in seed plants (gymnosperms and angiosperms)

pollen sac: pollen-producing cavity in anther

pollination: transfer of pollen to female reproductive part

pollution: physical, chemical or biological change in the environment that harms organisms

polychaete: type of annelid

polygamy: mating system in which males or females have multiple sexual partners

polygenic: caused by more than one gene

polymer: a long molecule composed of similar subunits (monomers)

polymerase chain reaction (PCR): biotechnology tool that rapidly produces millions of copies of a DNA sequence of interest

polyp: sessile form of a cnidarian

polypeptide: a long polymer of amino acids

polyphyletic: describes a group of organisms that excludes the most recent common ancestor of all members of the group

polyploid cell: cell with extra chromosome sets

polysaccharide: carbohydrate consisting of hundreds of monosaccharides

pons: oval mass in the brainstem where white matter connects the forebrain to the medulla and cerebellum

population: interbreeding members of the same species occupying the same region

population density: number of individuals of a species per unit area or volume of habitat

population distribution: pattern in which individuals are scattered throughout a habitat

population dynamics: study of the factors that influence changes in a population's size

portal system: arrangement of blood vessels in which a capillary bed drains into a vein that drains into another capillary bed

positive feedback: a process that reinforces an existing condition

postanal tail: muscular tail that extends past the anus; one of the four characteristics of chordates

posterior pituitary: the back part of the pituitary gland

postsynaptic cell: neuron, muscle cell, or gland cell that receives a message at a synapse

postzygotic reproductive isolation: separation of species due to selection against hybrid offspring

potential energy: stored energy available to do work

prebiotic simulation: experiment that attempts to recreate the conditions on early Earth that gave rise to the first cell

predator: animal that eats other animals

preembryonic stage: first two weeks of human development

pressure flow theory: theory that explains how phloem sap moves from source to sink

presynaptic cell: neuron that releases neurotransmitters into a synaptic cleft

prey: animal that a predator eats

prezygotic reproductive isolation: separation of species due to factors that prevent the formation of a zygote

primary growth: growth from apical meristems

primary immune response: immune system's response to its first encounter with a foreign antigen

primary oocyte: in oogenesis, a diploid cell that undergoes the first meiotic division and yields a haploid polar body and a haploid secondary oocyte

primary producer: species forming the base of a food web; autotroph

primary spermatocyte: a diploid cell that undergoes the first meiotic division and yields two haploid secondary spermatocytes

primary structure: the amino acid sequence of a protein

primary succession: appearance of organisms in an area previously devoid of life

primate: mammal with opposable thumbs, eyes in front of the skull, a relatively large brain, and flat nails instead of claws; includes prosimians, simians, and hominoids

primitive streak: furrow that appears along the back of the embryonic disc in the third week of human development

principle of superposition: the idea that lower rock layers are older than those above them

prion: infectious protein particle

producer (autotroph): organism that uses inorganic sources of energy and carbon

product: the result of a chemical reaction

product rule: the chance of two independent events occurring equals the product of the individual chances of each event

profundal zone: deep region of a lake or ocean where light does not penetrate

progenote: collection of nucleic acid, protein, and lipids that was the forerunner to cells

progesterone: steroid hormone produced in ovaries of female vertebrates

prokaryote: a cell that lacks a nucleus and other membrane-bounded organelles; bacteria and archaea

prolactin: hormone produced in the anterior pituitary

prometaphase: stage of mitosis just before metaphase, when the nuclear membrane breaks up and spindle fibers attach to kinetochores

promoter: a control sequence at the start of a gene; attracts RNA polymerase and (in eukaryotes) transcription factors

prophage: DNA of a lysogenic bacteriophage that is inserted into a host cell's chromosome

prophase: stage of mitosis when chromosomes condense and the mitotic spindle begins to form

prophase I: prophase of meiosis I, when chromosomes condense and become visible, and crossing over occurs

prophase II: prophase of meiosis II, when chromosomes condense and become visible

proprioceptor: sensory receptor that detects the position of body parts

prosimian: type of primate; a lemur, aye-aye, loris, tarsier, or bush baby

prostate gland: male structure that produces a milky, alkaline fluid that activates sperm

protein: a polymer consisting of amino acids and folded into its functional three-dimensional shape

protista: eukaryotic organism that is not a plant, fungus, or animal

proton: a particle in an atom's nucleus carrying a positive charge

protostome: clade of bilaterally symmetrical animals in which the first opening in the gastrula develops into the mouth

protozoan: unicellular protist that is heterotrophic and (usually) motile

proximal convoluted tubule: region of the renal tubule between the glomerular capsule and the nephron loop

proximate: describing the mechanistic causes of behavior

pseudocoelom: fluid-filled animal body cavity lined by endoderm and mesoderm

pseudogene: a DNA sequence that is very similar to that of a gene; a pseudogene is transcribed but its mRNA is not translated

pulmonary artery: artery that leads from the right ventricle to the lungs

pulmonary circulation: blood circulation between the heart and lungs

pulmonary vein: vein that leads from the lungs to the left atrium

punctuated equilibrium: theory that life's history has been characterized by bursts of rapid evolution interrupting long periods of little change

Punnett square: diagram that uses the genotypes of the parents to reveal the possible results of a genetic cross

pupil: opening in the iris that admits light into the eye

pyramid of energy: diagram depicting energy stored at each trophic level at a given time

pyruvate: the three-carbon product of glycolysis

Q

quaternary structure: the shape arising from interactions between multiple polypeptide subunits of the same protein

R

R group: an amino acid side chain

radial symmetry: body form in which any plane passing through the body from the mouth to the opposite end divides the body into mirror images

radioactive isotope: atom that emits particles or rays as its nucleus disintegrates

radiolarian: amoeboid protozoan with a silica shell

radiometric dating: a type of absolute dating that uses known rates of radioactive decay to date fossils

radula: a chitinous, tonguelike strap in many mollusks

rain shadow: downwind side of a mountain, with a drier climate than the upwind side

ray: parenchyma cells extending from center of woody stem or root

ray-finned fish: type of bony fish

reabsorption: renal tubule's return of useful substances to the blood

reactant: a starting material in a chemical reaction

reaction center: a molecule of chlorophyll *a* (and associated proteins) that participates in the light reactions of photosynthesis

receptacle: attachment point for flower parts

receptor potential: localized change in membrane potential (a graded potential) in a sensory receptor

recessive allele: allele whose expression is masked if a dominant allele is present

reciprocal altruism: altruistic act performed in anticipation of payback in the future

recombinant chromatid: chromatid containing genetic information from both parents as a result of crossing over

recombinant DNA: genetic material spliced together from multiple sources

red alga: multicellular, photosynthetic, marine protist with red or blue accessory pigments

red bone marrow: marrow that gives rise to blood cells and platelets

red blood cell: disc-shaped blood cell that contains hemoglobin

reduction: the gain of one or more electrons by a participant in a chemical reaction

reflex: type of innate behavior; an instantaneous, automatic response to a stimulus

reflex arc: neural pathway that links a sensory receptor and an effector

relative abundance: measure of biodiversity; considers the relative sizes of the populations that make up a community

relative dating: placing a fossil into a sequence of events without assigning it a specific age

renal tubule: portion of a nephron that adjusts the composition of filtrate

repressor: in an operon, a protein that binds to the operator and prevents transcription

reproductive system: organ system that produces and transports gametes and may nurture developing offspring

reptile: tetrapod vertebrate with an amniote egg and a dry scaly body covering

reservoir: an organism that acts as a source of the virus to infect other species

resource partitioning: use of the same resource in different ways or at different times by multiple species

respiratory cycle: one inhalation followed by one exhalation

respiratory surface: part of an animal's body that exchanges gases with the environment

respiratory system: organ system that acquires oxygen gas and releases carbon dioxide

resting potential: electrical potential inside a neuron not conducting a nerve impulse

restriction enzyme: enzyme that cuts double-stranded DNA at a specific base sequence

retina: sheet of photoreceptors that forms the innermost layer of the eye

reverse transcriptase: enzyme that uses RNA as a template to construct a DNA molecule

rhodopsin: pigment that transduces light into an electrochemical signal in photoreceptors

ribosomal RNA (rRNA): a molecule of RNA that, along with proteins, forms a ribosome

ribosome: a structure built of RNA and protein where mRNA anchors during protein synthesis

ribulose bisphosphate (RuBP): the five-carbon molecule that reacts with CO_2 in the Calvin cycle

RNA (ribonucleic acid): nucleic acid typically consisting of a single strand of nucleotides

RNA polymerase: enzyme that uses a DNA template to produce a molecule of RNA

RNA world: the idea that the first independently replicating life form was RNA

rod cell: photoreceptor in the retina that provides black-and-white vision

root: belowground part of most plants

root cap: cells that protect root apical meristem from abrasion

root hair: epidermal outgrowth that increases root surface area

rough endoplasmic reticulum: ribosome-studded portion of the ER where secreted proteins are synthesized

roundworm: unsegmented worm with a pseudocoelom; phylum Nematoda

***r*-selected species:** species consisting of short-lived, early-maturing individuals that have many offspring, with each receiving little parental investment

rubisco: enzyme that adds CO_2 to ribulose bisphosphate in the carbon reactions of photosynthesis

ruminant: herbivore with a four-chambered organ specialized for grass digestion

S

S phase: the synthesis phase of interphase, when DNA replicates

S-shaped curve: plot of logistic growth over time

salamander: type of amphibian

sample size: number of subjects in each experimental group

sapwood: light-colored, functioning secondary xylem in woody plant

sarcomere: one of many repeated units in a myofibril of a muscle cell

saturated fatty acid: a fatty acid with single bonds between all carbon atoms

savanna: type of terrestrial biome; grassland with scattered trees

Schwann cell: type of neuroglia that forms a myelin sheath around some axons in the peripheral nervous system

scientific method: a systematic approach to understanding the natural world based on evidence and testable hypotheses

sclera: the outermost layer of the eye; the white of the eye

sclereid: relatively short sclerenchyma cell

sclerenchyma: rigid plant cells that support mature plant parts

scrotum: the sac containing the testes

search image: improvement in a predator's ability to detect inconspicuous prey

second law of thermodynamics: every reaction loses some energy to the surroundings as heat; entropy always increases

secondary growth: increase in girth from cell division in lateral meristem

secondary immune response: immune system's response to subsequent encounters with a foreign antigen

secondary oocyte: haploid cell that undergoes the second meiotic division and yields a polar body and a haploid egg cell

secondary sex characteristic: trait that distinguishes the sexes but does not participate directly in reproduction

secondary spermatocyte: haploid cell that undergoes the second meiotic division and yields two haploid spermatids

secondary structure: a "substructure" within a protein, resulting from hydrogen bonds between parts of the peptide backbone

secondary succession: change in a community's species composition following a disturbance

secretion: addition of substances to the fluid in a renal tubule

seed: in gymnosperms and angiosperms, a plant embryo packaged with a food supply inside a tough outer coat

seed coat: protective outer layer of seed

seedless vascular plant: plant with vascular tissue but not seeds; includes true ferns, club mosses, whisk ferns, and horsetails

segmentation: division of an animal body into repeated subunits

selfish herd: group in which each animal tries to move as close to the center as possible

semen: fluid that carries sperm cells out of the body

semicircular canal: one of three perpendicular fluid-filled structures in the inner ear; provides information on the position of the head

semilunar valve: flap of heart tissue that prevents the flow of blood from an artery to a ventricle

seminal vesicle: structure that contributes fluid, fructose, and prostaglandins to semen

seminiferous tubule: tubule within a testis where sperm form and mature

senescence: aging

sensation: information that reaches the central nervous system about a stimulus

sensor: in homeostatic responses, a structure that monitors changes in a parameter

sensory adaptation: lessening of sensation with prolonged exposure to a stimulus

sensory neuron: neuron that transmits information from a stimulated body part to the central nervous system

sensory receptor: cell that detects stimulus information

sepal: part of the outermost whorl of a flower

severe combined immunodeficiency (SCID): inherited immune system disorder in which neither T nor B cells function

sex chromosome: a chromosome that carries genes that determine sex

sex pilus: a bridge of cytoplasm that transfers DNA from one cell to another in conjugation

sex-linked: describes genes or traits on the X or Y chromosome

sexual dimorphism: difference in appearance between males and females

sexual reproduction: the combination of genetic material from two individuals to create a third individual

sexual selection: type of natural selection resulting from variation in the ability to obtain mates

sexually transmitted disease: illness that spreads during sexual contact

shoot: aboveground part of a plant

short-day plant: plant that flowers when light periods are shorter than a critical length

short tandem repeat (STR): short DNA sequences that vary in length among individuals in a population

short-term memory: memory only available for a few moments

sieve plate: porous area at end of sieve tube element

sieve tube element: conducting cell that makes up sieve tube in phloem

simian: type of primate; a monkey

simple diffusion: form of passive transport in which a substance moves down its concentration gradient without the use of a transport protein

simple leaf: leaf with undivided blade

sink: plant part that does not photosynthesize

sinoatrial (SA) node: specialized cardiac muscle cells that set the pace of the heartbeat; the pacemaker

skeletal muscle: voluntary muscle tissue consisting of long, unbranched, striated, multinucleated cells; also, an organ composed of bundles of skeletal muscle cells and other tissue types that generates voluntary movements between pairs of bones

skeletal system: organ system consisting of bones and ligaments that support body structures and attach to muscles

skeleton: structure that supports an animal's body

skin: the outer surface of the body

sliding filament model: sliding of actin and myosin past each other to shorten a muscle cell

slow-twitch fiber: small muscle fiber that produces twitches of long duration

small intestine: part of the digestive tract that connects the stomach with the large intestine; site of most chemical digestion and absorption

smog: type of air pollution that forms a visible haze in the lower atmosphere

smooth endoplasmic reticulum: portion of the ER that produces lipids and detoxifies poisons

smooth muscle: involuntary muscle tissue consisting of nonstriated, spindle-shaped cells

snake: type of reptile

sodium-potassium pump: protein that uses energy from ATP to transport Na^+ out of cells and K^+ into cells

social behavior: interaction between members of the same species

sociobiology: study of social behavior in the context of an animal's fitness

soil: rock and mineral particles mixed with organic matter, air, and water

solute: a chemical that dissolves in a solvent, forming a solution

solution: a mixture of a solute dissolved in a solvent

solvent: a chemical in which other substances dissolve, forming a solution

somatic cell: body cell that does not give rise to gametes

somatic cell nuclear transfer: technique used to clone a mammal from an adult cell

somatic nervous system: in the peripheral nervous system, motor pathways carrying signals to skeletal (voluntary) muscles

source: plant part that produces or releases sugar

speciation: formation of new species

species: a distinct type of organism

species richness: measure of biodiversity; number of species in a community

spermatid: haploid cell produced after meiosis II in spermatogenesis

spermatogenesis: the production of sperm

spermatogonium: diploid germ cell that divides mitotically to yield a stem cell and a primary spermatocyte

spermatozoon (pl. spermatozoa): mature sperm cell

sphincter: muscular ring that contracts to close an opening

spinal cord: tube of nervous tissue that extends through the vertebral column

spirillum (pl. spirilla): spiral-shaped prokaryote

spleen: abdominal organ that produces and stores lymphocytes and destroys worn-out red blood cells

sponge: simple, asymmetrical animal lacking true tissues and gastrulation; phylum Porifera

spongy bone: bone tissue with large spaces between a web of bony struts

spore: reproductive cell of a plant or fungus

sporophyte: diploid, spore-producing stage of the plant life cycle

stabilizing selection: form of natural selection in which extreme phenotypes are less fit than the optimal intermediate phenotype

stamen: male flower part interior to petals

standardized variable: any factor held constant for all subjects in an experiment

statistically significant: unlikely to be attributed to chance

statolith: starch-containing plastid in root cap cell that functions as a gravity detector

stem: part of a plant that supports leaves

stem cell: undifferentiated cell that divides to give rise to additional stem cells and cells that specialize

steroid hormone: a lipid-soluble hormone that can freely diffuse through a cell membrane and bind to receptor inside the cell

sterol: lipid consisting of four interconnected carbon rings

stigma: in angiosperms, pollen-receiving tip of style

stoma (pl. stomata): pore in a plant's epidermis through which gases are exchanged with the atmosphere

stomach: J-shaped compartment in the digestive tract that receives food from the esophagus

stramenopile: protist with two flagella, one smooth and one covered with tubular "hairs"

stroma: the fluid inner region of the chloroplast

style: in angiosperms, the stalklike upper part of a carpel

substrate feeder: animal that lives in its food and eats it from the inside

substrate-level phosphorylation: ATP formation from transferring a phosphate group from a high-energy donor molecule to ADP

succession: change in the species composition of a community over time

survivorship curve: graph of the proportion of individuals that survive to a particular age

suspension feeder: animal that eats by straining particles out of water

sustentacular cell: cell within a seminiferous tubule that supports developing sperm cells

symbiosis: one species living in or on another

sympathetic nervous system: part of the autonomic nervous system; mobilizes the body to respond quickly to environmental stimuli and opposes the parasympathetic nervous system

sympatric speciation: formation of a new species within the boundaries of a parent species

synapse: junction at which a neuron communicates with another cell

synapsid: vertebrate with a single opening behind each eye orbit; mammals and their immediate ancestors

synaptic cleft: space into which neurotransmitters are released between two cells at a synapse

synaptic integration: a neuron's overall response to incoming excitatory and inhibitory neural messages

synaptic terminal: enlarged tip of an axon; contains synaptic vesicles

synovial joint: joint between two freely movable bones connected by a fluid-filled capsule of fibrous connective tissue

systematics: field of study that includes taxonomy and phylogenetics

systemic circulation: blood circulation between the heart and the rest of the body, except the lungs

systolic pressure: upper number in a blood pressure reading; reflects contraction of the ventricles

T

T cell: type of lymphocyte that coordinates adaptive immune response and destroys infected cells

taiga: type of terrestrial biome; the northern coniferous forest (also called boreal forest)

tapeworm: type of parasitic flatworm

taproot: large central root that persists throughout the life of the plant

target cell: cell that expresses receptors for a particular hormone

taste bud: cluster of cells that detect chemicals in food

taxis: directed movement toward or away from a stimulus

taxon: a group of organisms at any rank in the taxonomic hierarchy

taxonomy: the science of describing, naming, and classifying organisms

tectorial membrane: membrane in contact with hair cells of the cochlea

telomerase: enzyme that extends telomeres, enabling cells to divide continuously

telomere: non-coding DNA at the tip of a eukaryotic chromosome

telophase: stage of mitosis in which chromosomes arrive at opposite poles and nuclear envelopes form

telophase I: telophase of meiosis I, when homologs arrive at opposite poles

telophase II: telophase of meiosis II, when chromosomes arrive at opposite poles and nuclear envelopes form

temperate coniferous forest: type of terrestrial biome; coniferous trees dominate

temperate deciduous forest: type of terrestrial biome; deciduous trees dominate

temperate grassland: type of terrestrial biome; grazing, fire, and drought restrict tree growth

template strand: the strand in a DNA double helix that is transcribed

tendon: band of fibrous connective tissue that attaches a muscle to a bone

teratogen: substance that causes birth defects

terminator: sequence in DNA that signals where the gene's coding region ends

territory: a space that an animal defends against intruders

tertiary structure: the overall shape of a polypeptide, resulting mostly from interactions between amino acid R groups and water

test cross: a mating of an individual of unknown genotype to a homozygous recessive individual to reveal the unknown genotype

testis: male gonad

testosterone: steroid hormone produced in the testes of male vertebrates

tetanus: maximal muscle contraction caused by continual stimulation

tetrapod: vertebrate with four limbs

thalamus: forebrain structure that relays sensory input to the cerebrum

theory: well-supported scientific explanation

thermoreceptor: sensory receptor that responds to temperature

thermoregulation: control of an animal's body temperature

thick filament: in muscle cells, filament composed of myosin

thigmotropism: directional growth response to touch

thin filament: in muscle cells, filament composed of actin

threshold potential: potential to which a neuron's membrane must be depolarized to trigger an action potential

thylakoid: disclike structure that makes up the inner membrane of a chloroplast

thylakoid space: the inner compartment of the thylakoid

thymus: lymphoid organ in the upper chest where T cells learn to distinguish foreign antigens from self antigens

thyroid gland: gland in the neck that secretes two thyroid hormones (thyroxine and triiodothyronine) and calcitonin

thyroxine: one of two thyroid hormones that increases the rate of cellular metabolism

tidal volume: volume of air inhaled or exhaled during a normal breath

tight junction: connection between two adjacent animal cells that prevents fluid from flowing past the cells

tissue: group of cells that interact and provide a specific function

tongue: muscular structure on the floor of the mouth

tooth: mineral-hardened structure embedded in the jaw

topsoil: uppermost soil layer

trace element: an element that an organism requires in small amounts

trachea (pl. tracheae): in vertebrates, the respiratory tube just beneath the larynx; the "windpipe." In invertebrates, a branched tubule that brings air in close contact with cells, facilitating gas exchange

tracheid: long, narrow conducting cell in xylem

transcription: production of RNA using DNA as a template

transcription factor: in a eukaryotic cell, a protein that binds a gene's promoter and regulates transcription

transduction: transfer of DNA from one cell to another via a virus; also, conversion of energy from one form to another

transfer RNA (tRNA): a molecule of RNA that binds an amino acid at one site and an mRNA codon at its anticodon site

transformation: type of horizontal gene transfer in which an organism takes up naked DNA without cell-to-cell contact

trans fat: unsaturated fat with straight fatty acid tails

transgenic: containing DNA from multiple species

translation: assembly of an amino acid chain according to the sequence of nucleotides in mRNA

translocation: exchange of genetic material between nonhomologous chromosomes

transpiration: evaporation of water from a leaf

transposable element (transposon): DNA sequence that can move within a genome

triglyceride: lipid consisting of one glycerol bonded to three fatty acids

triiodothyronine: one of two thyroid hormones that increases the rate of cellular metabolism

trilobite: extinct type of arthropod

trophic level: an organism's position along a food chain

tropical rain forest: type of terrestrial biome; year-round high temperatures and precipitation

tropism: orientation toward or away from a stimulus

true breeding: always producing offspring identical to the parent for one or more traits; homozygous

true fern: a type of seedless vascular plant

trypanosome: flagellated protist that causes human diseases transmitted by biting flies

tuatara: type of reptile

tumor: abnormal mass of tissue resulting from cells dividing out of control

tumor suppressor gene: gene that normally prevents cell division but when inactivated or suppressed causes cancer

tundra: type of terrestrial biome; low temperature and short growing season

tunicate: type of invertebrate chordate

turgor pressure: the force of water pressing against the cell wall

turtle: type of reptile

twitch: contraction and relaxation of a muscle cell following a single stimulation

U

ultimate: describing the evolutionary origin or adaptive advantage of an animal's behavior

ultraviolet (UV) radiation: portion of the electromagnetic spectrum with wavelengths shorter than 400 nm

umbilical cord: ropelike structure that connects an embryo or fetus with the placenta

uniformitarianism: the idea that modern geological processes of erosion and sedimentation have also occurred in the past, producing changes in Earth over time

unsaturated fatty acid: a fatty acid with at least one double bond between carbon atoms

upwelling: upward movement of cold, nutrient-rich lower layers of a body of water

urea: nitrogenous waste derived from ammonia

ureter: muscular tube that transports urine from the kidney to the bladder

urethra: tube that transports urine (and semen in males) out of the body

uric acid: nitrogenous waste derived from ammonia

urinary bladder: muscular sac where urine collects

urinary system: organ system that filters blood and helps maintain concentrations of body fluids

urine: liquid waste produced by kidneys

uterine tube: tube that conducts an oocyte from an ovary to the uterus

uterus: muscular, saclike organ where embryo and fetus develop

V

vaccine: substance that initiates a primary immune response so that when an infectious agent is encountered, the secondary immune response can rapidly deactivate it

vacuole: membrane-bounded storage sac in a cell, especially the large central vacuole in a plant cell

vagina: conduit from the uterus to the outside of the body

valence shell: outermost occupied energy shell of an atom

variable: any changeable element in an experiment

variable region: amino acid sequence that is different for every antibody

vas deferens: tube that transports sperm from an epididymis to an ejaculatory duct

vascular bundle: collection of xylem, phloem, parenchyma, and sclerenchyma in plants

vascular cambium: lateral meristem that produces secondary xylem and phloem

vascular tissue: conducting tissue for water, minerals, and organic substances in plants

vasoconstriction: decrease in the diameter of a blood vessel

vasodilation: increase in the diameter of a blood vessel

vegetative plant parts: nonreproductive parts (roots, stems, and leaves)

vein: vascular bundle inside leaf; also, a vessel that returns blood to the heart

ventral nerve cord: nerve cord that runs along the ventral side of an animal

ventricle: heart chamber that pumps blood out of the heart

venule: small vein

vertebra: one unit of the vertebral column composed of bone or cartilage that supports and protects the spinal cord

vertebral column: bone or cartilage that supports and protects the spinal cord

vertebrate: animal with a backbone

vertical gene transfer: passage of DNA from one generation to the next; cell division

vesicle: a membrane-bounded sac that transports materials in a cell

vessel element: short, wide conducting cell in xylem

vestibular apparatus: structure in the inner ear that provides information on the position of the head and changes in velocity

vestigial: having no apparent function in one organism, but homologous to a functional structure in another species

villus (pl. villi): tiny projection on the inner lining of the small intestine

viroid: infectious RNA molecule

virus: infectious agent that consists of genetic information enclosed in a protein coat

visceral mass: part of a mollusk that contains the digestive and reproductive systems

vital capacity: maximal volume of air that can be forced out of the lungs during one breath

vocal cord: elastic tissue band that covers the larynx and vibrates as air passes, producing sound

vulnerable species: species facing a high risk of extinction in the distant future

vulva: external female genitalia

W

water mold: filamentous, heterotrophic protist; also called an oomycete

water vascular system: system of canals in echinoderms; provides locomotion and osmotic balance

wavelength: the distance a photon moves during a complete vibration

wax: lipid consisting of fatty acids connected to alcohol or other molecules

wetland: ecosystem in which soil is saturated with water

whisk fern: a type of seedless vascular plant

white blood cell: one of five types of blood cells that help fight infection

white matter: nervous tissue in the central nervous system; consists of myelinated axons

wild type: the most common phenotype, genotype, or allele

wood: secondary xylem

woody plant: plant with stems and roots made of wood and bark

X

X inactivation: turning off all but one X chromosome in each cell of a mammal (usually female) early in development

X-linked: describes traits controlled by genes on the X chromosome

xylem: vascular tissue that transports water and dissolved minerals in plants

xylem sap: solution of water and dissolved minerals in xylem

Y

yeast: unicellular fungus

yellow bone marrow: fatty marrow that replaces red bone marrow as bones age

Y-linked: describes traits controlled by genes on the Y chromosome

yolk sac: extraembryonic membrane that forms beneath the embryonic disc and manufactures blood cells

Z

zoospore: flagellated spore produced by chytrids

zygomycete: fungus that produces zygospores

zygospore: diploid resting spore produced by fusion of haploid cells in zygomycetes

zygote: the fused egg and sperm that develops into a diploid individual

Credits

Photographs

Chapter 1

Opener: © Science VU/DOE/Visuals Unlimited; 1.1: © SMC Images/The Image Bank/Getty Images; 1.2 (population): © Gregory G. Dimijian, M.D./Photo Researchers; 1.2 (community): © Todd Gustafson/Danita Delimont; 1.2 (ecosystem): © Manoj Shah/The Image Bank/Getty Images; 1.2 (biosphere): © Corbis (RF); 1.5a: © Dennis Kunkel/Phototake; 1.5b: © Brand X Pictures/Getty Images (RF); 1.5c: © Corbis Animals in Action CD; 1.6a–b: © Michael and Patricia Fogden/Animals Animals–Earth Scenes; 1.7a (left): © Dennis Kunkel Microscopy, Inc.; 1.7a (inset): © Ron Occalea/The Medical File/Peter Arnold/Photolibrary; 1.8 (bacteria): © Kwangshin Kim/Photo Researchers; 1.8 (archaea): © Ralph Robinson/Visuals Unlimited/Getty Images; 1.8 (protista): © Melba Photo Agency/PunchStock (RF); 1.8 (animalia): Courtesy of The National Human Genome Research Institute; 1.8 (fungi): © Corbis (RF); 1.8 (plantae): © Photo by Keith Weller/USDA; 1.9: © Patrick Landmann/Photo Researchers; 1.11: © Geoff McIlleron/Firefly Images/photographersdirect.com; p. 13: © Getty Images/Photodisc (RF); p. 14 (sweetener): © Martin Bond/Photo Researchers; p. 14 (rat): © Photodisc Collection/Getty Images; p. 15 (Darwin): © Richard Milner/Handout/epa/Corbis; p. 15 (Wallace): © Hulton Archive/Getty Images; 1.12: © Mitsuhiko Imamori/Minden Pictures; p. 16: © Bettmann/Corbis.

Chapter 2

Opener: © Roy Morsch/Corbis; 2.1: © Getty Images/Digital Vision (RF); 2.7a: © Pixtal/SuperStock (RF); 2.7b: © F. Schussler/PhotoLink (RF)/Getty Images; 2.7c: © Getty Images (RF); 2.7d: © Jonelle Weaver/Getty Images (RF); 2.9: © Corbis (RF); 2.10c: © The McGraw-Hill Companies, Inc./Jacques Cornell photographer; 2.11: © Herman Eisenbeiss/Photo Researchers; 2.12: © Lester V. Bergman/Corbis; 2.15: © Corbis (RF); TA2.5: © Ingram Publishing/Alamy (RF); p. 32: © Masterfile RF; 2.18c (top): © Dennis Kunkel Microscopy, Inc.; 2.18c (middle): © Gary Gaugler/Visuals Unlimited; 2.18c (bottom): © Marshall Sklar/SPL/Photo Researchers; 2.19 (left): © D. Hurst/Alamy (RF); 2.19 (right): © The McGraw-Hill Companies, Inc./Jacques Cornell photographer; p. 35 (top): © The McGraw-Hill Companies, Inc./Jill Braaten photographer; 2.21: © Peter Arnold/Digital Vision/Getty Images (RF); 2.23: © David E. Volk; 2.24a: © Topham/The Image Works; 2.24b: © Jack Hollingsworth/Corbis (RF); 2.24c: © John Cancelosi/Peter Arnold/Photolibrary; 2.27: © Scott Camazine/Alamy.

Chapter 3

Opener: © AP Photo/The Columbian, Janet L. Mathews; 3.1a: © Kathy Talaro/Visuals Unlimited; 3.1b: © Dr. Jeremy Byrgess/SPL/Photo Researchers; p. 47: © Image Source (RF)/Getty Images; 3.3a: © Comstock (RF)/Alamy; 3.3a (inset): © Michael Abbey/Visuals Unlimited; 3.3b: © Inga Spence/Visuals Unlimited; 3.3b (inset): © Dr. Dennis Kunkel/Visuals Unlimited; 3.3c: © Inga Spence/Visuals Unlimited; 3.3c (inset): © Microworks Color/Phototake; 3.3d: © Inga Spence/Visuals Unlimited; 3.3d (inset): © Steve Gschmeissner/SPL/Photo Researchers; 3.4b: © Wim van Egmond/Getty Images; 3.6b: © Dr. Martin Oeggerli/Visuals Unlimited; 3.6c: © David McCarthy/Photo Researchers; 3.6d: © Dennis Kunkel Microscopy, Inc./Visuals Unlimited; 3.7: © Dr. T.J. Beveridge/Visuals Unlimited; 3.A: © Dr. Philippa Uwins, Whister Research Pty/SPL/Photo Researchers; 3.8: © Dr. Gopal Murti/Visuals Unlimited; 3.9: © Dr. George Chapman/Visuals Unlimited; 3.13: © Tim Flach/Getty Images; 3.14c: © David M. Phillips/The Population council/Science Source/Photo Researchers; 3.15: © Prof. J. L. Kemeny/ISM/Phototake; 3.16: © Biophoto Associates/Photo Researchers; 3.17: © Prof. P. Motta & T. Naguro/Science Photo Library/Photo Researchers; 3.18a: © Dr. Donald Fawcett/Visuals Unlimited; 3.18b: Reproduced from *The Journal of Cell Biology*, 1969, 43: 343–353. © 1969 The Rockefeller University Press.; 3.19: © Biophoto Associates/Photo Researchers; 3.20: © Bill Longcore/Photo Researchers; 3.21a: © Innerspace Imaging/Photo Researchers; 3.21b: © Francois Paquet–Durand/Photo Researchers; 3.21c: © Edwin A. Reschke/Peter Arnold/Photolibrary; 3.21d: © Kevin & Betty Collins/Visuals Unlimited; 3.22: © Visuals Unlimited/Corbis; 3.23 (left): © Tomasz Szul/Visuals Unlimited/Corbis; 3.23 (middle): © Albert Tousson/Phototake; 3.23 (right): © Jennifer Waters/Photo Researchers; 3.24a (top): © Dr. David Phillips/Visuals Unlimited; 3.24a (bottom): © Dennis Kunkel Microscopy, Inc./Phototake; 3.24b: © D.W. Fawcett/Photo Researchers; 3.24c: © Dr. Tony Brain/Photo Researchers; 3.25a–b: © BioPhoto Associates/Photo Researchers; 3.27a–b: Courtesy Prof. Jeffery Errington.

Chapter 4

Opener: © Tyler Stableford/Getty Images; 4.3 (both): © Ryan McVay/Getty Images; 4.4 (top): © Blair Seitz/Photo Researchers; 4.4 (bottom): © ImageSource/Corbis (RF); 4.A: © Darwin Dale/Photo Researchers; 4.17a–c: © Dr. David M. Phillips/Visuals Unlimited; 4.18a–b: © Nigel Cattlin/Photo Researchers; 4.21: © Biology Media/Photo Researchers; TA4.1: © Corbis (RF).

Chapter 5

Opener: © Andrew J. Martinez/Photo Researchers; 5.5 (leaves): © Steve Raymer/NGS Image Collection; 5.5 (mesophyll): Courtesy of Dr. Eldon Newcomb, University of Wisconsin, Madison; 5.6: Courtesy Professor So Iwata; p. 95: © Corbis (RF); p. 97: © image100/Corbis (RF); 5.13 (left): © Tony Sweet/Digital Vision (RF)/Getty Images; 5.13 (middle): © Joeseph Sohm–Visions of America/Getty Images (RF): © Digital Vision (RF)/Getty Images; 5.14: Plant Physiology, May 2000, Vol. 123, pp. 29–38, "Solar–Powered Sea Slugs. Mollusc/Algal Chloroplast Symbiosis." Mary E. Rumpho, Elizabeth J. Summer, and James R. Manhart; 5.15: From Rumpho, Mary E. et al November 18, 2008 *PNAS* vol. 105, no. 46: 17867–17871, "Horizontal gene transfer of the algal nuclear gene psbO to the photosynthetic sea slug Elysia chlorotica", used by permission © 2010 National Academy of Sciences, U.S.A.

Chapter 6

Opener: © Gunter Ziesler/Bruce Coleman/Photoshot; 6.1: © Thomas Deerinck, NCMIR/SPL/Photo Researchers; 6.3 (leaves): © Steve Raymer/NGS Image Collection; 6.3 (mesophyll): Courtesy of Dr. Eldon Newcomb, University of Wisconsin, Madison; TA6.1: © Creatas/PunchStock (RF); 6.9: © Digital Vision (RF)/Getty Images; p. 114: © Photodisc/Getty Images (RF); 6.11a: © Adam Woolfitt/Corbis; 6.11b: © Scimat/SPL/Photo Researchers; 6.13 (both): © Marc Gibernau, CNRS, Toulouse, France.

Chapter 7

Opener: © n44/ZUMA Press/Newscom; 7.1 (poster): © Universal Pictures/Photofest; 7.1 (mosquito): © Louie Psihoyos/Science Faction; 7.3: © Oliver Meckes/MPI–Tubingen/Photo Researchers; 7.4a–b: © Science Source/Photo Researchers; 7.4c: © Bettmann/Corbis; 7.5: © Photodisc/Photolibrary (RF); 7.8: © Dr. Gopal Murti/SPL/Photo Researchers; 7.14a: © Tom Pantages/Phototake; 7.15: A. Korostelev, S. Trakhanov, M. Lauerberg and H.F. Noller (2006) "Crystal Structure of a 70S Ribosome–tRNA Complex Reveals Functional Interactions and Rearrangements." Cell 126:1065–1077.; 7.17: © Kiseleva and Donald Fawcett/Visuals Unlimited; p. 133: © Jean-Louis Le Moigne/Peter Arnold/Photolibrary; p. 137: © Getty Images (RF); 7.21a: © Andrew Syred/SPL/Photo Researchers; 7.21b: © Science VU/Dr. F. R. Turner/Visuals Unlimited; 7.22a: © Micro Discovery/Corbis; 7.22b: © Dr. Gopal Murti/SPL/Photo Researchers; 7.24a: © Erich Schlegel/Dallas Morning News/Corbis; 7.24b: © Pallava Bagla/Corbis; 7.24c:

Chapter 17

Chapter 18

Chapter 19

Chapter 20

31.12: Courtesy of David H. Alpers, M.D.; p. 645 (bottom): © Burke/Triolo Productions (RF)/Getty Images; p. 647: © Comstock/Alamy (RF); 31.16: © The McGraw-Hill Companies, Inc./Jill Braaten, photographer; 31.18 (left): © AP Photo/Brennan Linsley; 31.18 (right): © AP Photo/The Grand Island Independent, Barrett Stinson; 31.19: © Science VU/Jackson/Visuals Unlimited; 31.20: © Tim Wimborne/Reuters/Corbis.

Chapter 32

Opener: © Fred Bavendam/Minden Pictures; 32.1 (top): © Johnny Johnson/The Image Bank/Getty Images; 32.1 (bottom): © IT Stock (RF)/PunchStock; 32.2 (left): © IT Stock (RF)/PunchStock; 32.2 (right): © John A.L. Cooke/Animals Animals–Earth Scenes; 32.5a: © Akira Kaede/Digital Vision (RF)/Getty Images; 32.5b: © Dynamic Graphics/PictureQuest (RF); 32.6a–b: © Brand X Pictures/PunchStock (RF); 32A: © Y. Bealieu, Publiphoto Diffusion/SPL/Photo Researchers; p. 668: © Corbis (RF); 32.15: © Pat Pendarvis.

Chapter 33

Opener: © Tracy A. Woodward/The Washington Post/Getty Images; 33.2: © Dr. David M. Phillips/Visuals Unlimited/Getty Images; 33.6: © Dr. Olivier Schwartz, Institute Pasteur/SPL/Photo Researchers; 33.A: © Hank Morgan/Photo Researchers; 33.7: © Dr. Andrejs Liepins/Photo Researchers; 33.8: © Len Lessin/Peter Arnold/Photolibrary; p. 684: © TRBfoto/Getty Images (RF); 33.15: © Dr. Ken Greer/Visuals Unlimited; 33.16 (top): © David Scharf/Peter Arnold/Photolibrary; 33.16 (bottom): © Institut Pasteur/Phototake; 33.18: © Paul Rapson/Photo Researchers; 33.18 (inset): © SIU/Visuals Unlimited/Getty Images; 34.13: © Getty Images (RF).

Chapter 34

Opener (left): © Paul Hobson/Nature Picture Library; (right): © Dr. Paul Zahl/Photo Researchers; 34.1a: © Andrew J. Martinez/Photo Researchers; 34.1b: © Tim Fitzharris/Minden Pictures; 34.3: © Daniel Heuclin/Photo Researchers; 34.5: Larry Johnson, Dept. of Veterinary Anatomy and Public Health; 34.6: © Eye of Science/Photo Researchers; 34.8: © Prof. P.M. Motta, G. Macchiarelli, S.A. Nottola/Photo Researchers; 34.12: © Image Source/PunchStock (RF); 34.19a–d: © Ralph Hutchings/Visuals Unlimited; p. 712 (left): © Digital Vision (RF); p. 712 (right): © Nancy R Cohen/Getty Images (RF); 34.21: © AP Photo/Peer Grimm, Pool; 34.23 (fish): © Georgette Douwma/Photo Researchers; 34.23 (eggs): © Brandon Cole/Peter Arnold/Photolibrary.

Chapter 35

Opener: © Gerald Hinde/ABPL/Animals Animals–Earth Scenes; p. 722: © Getty Images/Digital Vision (RF); 35.02b: Courtesy of Professor Daisuke Yamamoto, Tohoku University/JST–ERATO project; 35.03a: © Nina Leen//Time Life Pictures/Getty Images; 35.03b: © Corbis; 35.04: © Rick Friedman/Corbis; 35.06: © Dr. Matthias Wittlinger; 35.07: © Raymond Mendez/Animals Animals–Earth Scenes; 35.08: © Visuals Unlimited/Corbis; p. 727: © Jim Cartier/Photo Researchers; 35.10: © Stefano Stefani/Getty Images (RF); 35.11a: © Jeff Lepore/Getty Images; 35.11b: © Leroy Simon/Getty Images; 35.11c: © Konrad Wothe/Minden Pictures; 35.12: © Satoshi Kuribayashi/Minden Pictures; 35.13: © Jim Tuten/Animals Animals–Earth Scenes; 35.15: © Roberta Olenick/All Canada Photos/Corbis; 35.16: © Wilfried Krecichwost/Photographer's Choice (RF)/Getty Images; 35.17: © Stealth500; 35.21 a–b: Courtesy of Stephen H. Vessey; 35.22: © Pierre Huguet/Peter Arnold/Photolibrary; 35.23a: © Joe McDonald/Corbis; 35.24a: © Peter Johnson/Corbis; 35.24b: © Neil Bromhall/Nature Picture Library; 35.25: © Gary Meszaros/Visuals Unlimited.

Chapter 36

Opener: © David M. Dennis/Animals Animals–Earth Scenes; 36.1a: © Digital Vision/Getty Images (RF); 36.1b: © John A. Novak/Animals Animals–Earth Scenes; 36.1c: © PhotoAlto/PunchStock (RF); p. 743: © PhotoDisc/Getty Images (RF); 36.3: © Kevin Schafer/Corbis; 36.6: © Ronald Wittek/Getty Images; 36.8: © blickwinkel/Alamy; 36.10 (top): © Digital Vision/PunchStock (RF); 36.10 (bottom): Courtesy of John McColgan, Alaska Fire Service/Bureau of Land Management; 36.11a: © Inga Spence/Visuals Unlimited/Getty Images; 36.11b: © Flat Earth Images (RF); 36.12: © Reinhard/ARCO/Nature Picture Library; 36.17: © Peter Cade/Iconica/Getty Images; p. 755: © Phil Degginger/Alamy (RF); 36.19: © Phil Degginger/Alamy.

Chapter 37

Opener: © Johnny Johnson/Photographer's Choice/Getty Images; 37.1: © IT Stock Free/Alamy (RF); 37.4a: © Kevin Schafer/The Image Bank/Getty Images; 37.4b: © Thomas Marent/Visuals Unlimted/Getty Images; 37.4c: © Simon D. Pollard/Photo Researchers; 37.5: © Beverly Joubert/National Geographic/Getty Images; 37.6: © Mark Moffett/Minden Pictures; 37.8: © Jerry Dodrill/Aurora/Getty Images; 37.A: © Joseph Sohm/Visions of America/Corbis; 37.12–13: © Corbis (RF); p. 773: © C Squared Studios/Getty Images (RF); 37.18: © Digital Vision/Getty Images (RF); 37.20: From "A virus in a fungus in a plant: three-way symbiosis required for thermal tolerance." Luis M. Márquez, Regina S. Redman, Russell J. Rodriguez, and Marilyn J. Roossinck. *Science* 26 January 2007: Vol. 315. no. 5811, pp. 513–515.

Chapter 38

Opener: © Perry Thorsvik/National Geographic Image Collection; (inset): © Dr. Ken MacDonald/SPL/Photo Researchers; 38.1a: © David Hall/Photo Researchers, Inc; 38.1b: © Zandria Muench Beraldo; 38.1c: © Kristy-Anne Glubish/Design Pics (RF)/Corbis; 38.1d: © Terry W. Eggers/Corbis; p. 785: © Corbis (RF); 38.8: © Frans Lanting/Corbis; 38.9a: © Digital Archive Japan/Alamy (RF); 38.9b: © Comstock/PunchStock (RF); 38.10: © Thomas Kitchin/Photo Researchers; 38.11: © Arthur Morris/Corbis; 38.12: © PhotoDisc/Getty Images (RF); 38.13: © Ed Reschke; 38.14: © Andrew Brown; Ecoscene/Corbis; 38.15: © Michael DeYoung/Corbis; 38.16: © Corbis (RF); 38.19a: © ML Sinibaldi/Corbis; 38.19b: © The McGraw-Hill Companies, Inc./Pat Watson, photographer; 38.21a: © Joe/Arrington/Visuals Unlimited; 38.21b: © Mike Parry/Minden Pictures; 38.21c: © D. Brown/Animals Animals–Earth Scenes; 38.21d: © Digital Vision/Getty Images (RF); 38.22: © D. P. Wilson/FLPA/Minden Pictures; 38.23: © Fred Bavendam/Minden Pictures; 38.A: © Science VU/NOAA/Visuals Unlimited.

Chapter 39

Opener (left): © Pat Pendarvis; (right): © PhotoDisc/Getty Images (RF); 39.1: © Keren Su/Stone/Getty Images; 39.3a: © Wayne Lawler; Ecoscene/Corbis; 39.3b: Photo by Lynn Betts, courtesy of USDA Natural Resources Conservation Service; 39.3c: © Corbis (RF); 39.4: © Ed Kashi/Corbis; 39.6: © AP Photo/Robert Galbraith, pool; 39.7: © Reuters/Corbis; 39.8: © Oliver Strewe/Stone/Getty Images; 39.9a–d: NASA; 39.11b: © Corbis (RF); 39.12a-b: © Dr. Juerg Alean/SPL/Photo Researchers; 39.13: © George Grall/National Geographic/Getty Images; 39.14: © China Photos/Getty Images News; 39.15: © CC Lockwood/Animals Animals–Earth Scenes; 39.17: © William Leaman/Alamy (RF); 39.18: © Comstock/PunchStock (RF); 39.19: © Francesco Tomasinelli/Photo Researchers; p. 816: © Brand X Pictures/PunchStock (RF); 39.20: © Thomas Marent,/Visuals Unlimited/Corbis; p. 818: © PhotoDisc/Getty Images (RF).

Index